VIRAL INFECTIONS AND ANTIVIRAL THERAPIES

VIRAL INFECTIONS AND ANTIVIRAL THERAPIES

Edited by

AMAL KUMAR DHARA

Department of Pharmacy, Contai Polytechnic, Purba Medinipur, West Bengal, India

AMIT KUMAR NAYAK

Department of Pharmaceutics, Seemanta Institute of Pharmaceutical Sciences, Jharpokharia, Odisha, India

ACADEMIC PRESS

An imprint of Elsevier

ELSEVIER

Academic Press is an imprint of Elsevier
125 London Wall, London EC2Y 5AS, United Kingdom
525 B Street, Suite 1650, San Diego, CA 92101, United States
50 Hampshire Street, 5th Floor, Cambridge, MA 02139, United States
The Boulevard, Langford Lane, Kidlington, Oxford OX5 1GB, United Kingdom

Notices

Knowledge and best practice in this field are constantly changing. As new research and experience broaden our understanding, changes in research methods, professional practices, or medical treatment may become necessary.

Practitioners and researchers must always rely on their own experience and knowledge in evaluating and using any information, methods, compounds, or experiments described herein. In using such information or methods they should be mindful of their own safety and the safety of others, including parties for whom they have a professional responsibility.

To the fullest extent of the law, neither the Publisher nor the authors, contributors, or editors, assume any liability for any injury and/or damage to persons or property as a matter of products liability, negligence or otherwise, or from any use or operation of any methods, products, instructions, or ideas contained in the material herein.

ISBN: 978-0-323-91814-5

For Information on all Academic Press publications
visit our website at https://www.elsevier.com/books-and-journals

Publisher: Stacy Masucci
Acquisitions Editor: Kattie Washington
Editorial Project Manager: Timothy J. Bennett
Production Project Manager: Selvaraj Raviraj
Cover Designer: Christian J. Bilbow

Typeset by MPS Limited, Chennai, India

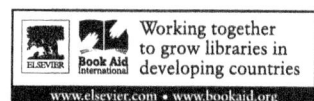

Working together
to grow libraries in
developing countries

www.elsevier.com • www.bookaid.org

Contents

4. Mode of viral infections and transmissions   55

Sora Yasri and Viroj Wiwanitkit

5. Transmission and intervention dynamics of SARS-CoV-2   69

Maame A. Korsah, Caleb Acquah and Michael K. Danquah

6. Sexually transmitted viral infections   85

Aparajita Dasgupta

7. Testing viral infections   99

Shalini Upadhyay

8. Electron microscopic methods for virus diagnosis   121

Nadun H. Madanayake, Ryan Rienzie and Nadeesh M. Adassooriya

13. Antiretroviral therapy   263

Suman Ganguly and Debjit Chakraborty

14. Rotavirus and antirotaviral therapeutics: trends and advances   285

Ujjwal Kumar De, Yashpal Singh Malik, Gollahalli Eregowda Chethan, Babul Rudra Paul, Jitendra Singh Gandhar, Varun Kumar Sarkar, Srishti Soni and Kuldeep Dhama

15. Current therapeutic strategies and novel antiviral compounds for the treatment of nonpolio enteroviruses   303

Angeline Jessika Suresh and Regina Sharmila Dass

16. Antiviral agents against flaviviruses 315

Érica Erlanny S. Rodrigues, Ana Beatriz Souza Flor dos Santos,
Manuele Figueiredo da Silva, João Xavier de Araújo-Júnior and
Edeildo Ferreira da Silva-Júnior

17. Pathophysiology of HIV and strategies to eliminate AIDS as a public health threat 339

Omar Sued and Tomás M. Grosso

18. Herbal drugs to combat viruses 377

Benil P.B., Rajakrishnan Rajagopal, Ahmed Alfarhan and
Jacob Thomas

19. Strategies for delivery of antiviral agents 407

Vuyolwethu Khwaza, Buhle Buyana, Xhamla Nqoro,
Sijongesonke Peter, Zintle Mbese, Zizo Feketshane,
Sibusiso Alven and Blessing A. Aderibigbe

20. Nanovesicles for delivery of antiviral agents 493

Yasmine Radwan, Ali H. Karaly and Ibrahim M. El-Sherbiny

21. Antiviral biomaterials 519

Sandhya Khunger

22. Antiviral biomolecules from marine inhabitants 537

Ishwarya Ayyanar, Subidsha Suyambu Krishnan, Akila Ravindran,
Sunandha Jeeva Bharathi Gunasekaran and
Balasubramanian Vellaisamy

23. Plant polysaccharides as antiviral agents 567

Bulu Mohanta, Amit Kumar Nayak and Amal Kumar Dhara

24. Antiviral peptides against dengue virus 581

Michelle Felicia Lee, Mohd Ishtiaq Anasir and Chit Laa Poh

29. Clinical Trials and Regulatory considerations of Antiviral agents   709

Samir Bhargava, Bhavna, Neeraj Sethiya, Amal Kumar Dhara, Jagannath Sahoo, H. Chitme, Mayuri Gupta, Navraj Upreti and Yusra Ahmad

30. Future perspectives of antiviral therapy   741

Debesh Chandra Bhattacharya

List of contributors

Caleb Acquah Faculty of Health Sciences, University of Ottawa, Ottawa, ON, Canada

Nadeesh M. Adassooriya Department of Chemical & Process Engineering, University of Peradeniya, Peradeniya, Sri Lanka

Blessing A. Aderibigbe Department of Chemistry, University of Fort Hare, Alice, Eastern Cape, South Africa

Yusra Ahmad Faculty of Pharmacy, Uttarakhand Technical University, Dehradun, Uttarakhand, India

Ahmed Alfarhan Department of Botany and Microbiology, College of Science, King Saud University, Riyadh, Saudi Arabia

Sibusiso Alven Department of Chemistry, University of Fort Hare, Alice, Eastern Cape, South Africa

Mohd Ishtiaq Anasir National Institutes of Health (NIH), Ministry of Health Malaysia, Shah Alam, Selangor, Malaysia

Vahid Reza Askari Department of Pharmaceutical Sciences in Persian Medicine, School of Persian and Complementary Medicine, Mashhad University of Medical Sciences, Mashhad, Iran; Applied Biomedical Research Center, Mashhad University of Medical Sciences, Mashhad, Iran; Pharmacological Research Center of Medicinal Plants, Mashhad University of Medical Sciences, Mashhad, Iran

Ishwarya Ayyanar Department of Microbiology, Alagappa University, Science Block, Karaikudi, Tamil Nadu, India

Vafa Baradaran Rahimi Pharmacological Research Center of Medicinal Plants, Mashhad University of Medical Sciences, Mashhad, Iran; Department of Cardiovascular Diseases, Faculty of Medicine, Mashhad University of Medical Sciences, Mashhad, Iran

Samir Bhargava Faculty of Pharmacy, DIT University, Dehradun, Uttarakhand, India

Debesh Chandra Bhattacharya Department of Microbiology, Vidyasagar University, Midnapore, West Bengal, India

Bhavna Faculty of Pharmacy, DIT University, Dehradun, Uttarakhand, India

Buhle Buyana Department of Chemistry, University of Fort Hare, Alice, Eastern Cape, South Africa

Debjit Chakraborty Division of Epidemiology, ICMR-National Institute of Cholera and Enteric Disease, Kolkata, West Bengal, India

Sambuddha Chakraborty Department of Microbiology, Tripura University (A Central University), Suryamaninagar, Tripura West, India

Debprasad Chattopadhyay ICMR-NICED Virus Unit, ID and BG Hospital, Kolkata, West Bengal, India; ICMR-National Institute of Traditional Medicine, Belagavi, Karnataka, India

Ashwini Chauhan Department of Microbiology, Tripura University (A Central University), Suryamaninagar, Tripura West, India

Gollahalli Eregowda Chethan Department of Veterinary Medicine, College of Veterinary Science and Animal Husbandry, Central Agriculture University, Aizawl, Mizoram, India

H. Chitme Faculty of Pharmacy, DIT University, Dehradun, Uttarakhand, India

Manuele Figueiredo da Silva Institute of Pharmaceutical Sciences, Federal University of Alagoas, Maceió, Brazil

Edeildo Ferreira da Silva-Júnior Institute of Chemistry and Biotechnology, Federal University of Alagoas, Maceió, Brazil

Michael K. Danquah Department of Chemical Engineering, University of Tennessee, Chattanooga, TN, United States

Aparajita Dasgupta Department of Preventive and Social Medicine, All India Institute of Hygiene and Public Health, Kolkata, West Bengal, India

Regina Sharmila Dass Fungal Genetics and Mycotoxicology Laboratory, Department of Microbiology, School of Life Sciences, Pondicherry University, Pondicherry, India

Ujjwal Kumar De Division of Medicine, ICAR-Indian Veterinary Research Institute, Izatnagar, Uttar Pradesh, India

João Xavier de Araújo-Júnior Institute of Chemistry and Biotechnology, Federal University of Alagoas, Maceió, Brazil; Institute of Pharmaceutical Sciences, Federal University of Alagoas, Maceió, Brazil

Kuldeep Dhama Division of Pathology, ICAR-Indian Veterinary Research Institute, Izatnagar, Uttar Pradesh, India

Ranjithkumar Dhandapani Chimertech Private Limited, Chennai, Tamil Nadu, India

Amal Kumar Dhara Department of Pharmacy, Contai Polytechnic, Purba Medinipur, West Bengal, India

Ana Beatriz Souza Flor dos Santos Institute of Pharmaceutical Sciences, Federal University of Alagoas, Maceió, Brazil

Ibrahim M. El-Sherbiny Nanomedicine Research Laboratories, Center for Materials Science, Zewail City of Science and Technology, 6 of October City, Giza, Egypt

Zizo Feketshane Department of Chemistry, University of Fort Hare, Alice, Eastern Cape, South Africa

Jitendra Singh Gandhar Division of Medicine, ICAR-Indian Veterinary Research Institute, Izatnagar, Uttar Pradesh, India

Suman Ganguly West Bengal State AIDS Prevention and Control Society, Kolkata, West Bengal, India

Tomás M. Grosso Laboratory of Immunology, National University of Luján, Buenos Aires, Argentina; Clinical Trials Unit, Huésped Foundation, Buenos Aires, Argentina

Sunandha Jeeva Bharathi Gunasekaran Department of Microbiology, Alagappa University, Science Block, Karaikudi, Tamil Nadu, India

Mayuri Gupta Faculty of Pharmacy, DIT University, Dehradun, Uttarakhand, India

Pradip Kumar Jana Virus Research and Diagnostic Laboratory, ICMR-National Institute of Cholera and Enteric Diseases, Kolkata, West Bengal, India

Ali H. Karaly Nanomedicine Research Laboratories, Center for Materials Science, Zewail City of Science and Technology, 6 of October City, Giza, Egypt

Sandhya Khunger Department of Microbiology, Faculty of Allied Health Sciences, Shree Guru Gobind Singh Tricentenary University, Gurugram, Haryana, India

Vuyolwethu Khwaza Department of Chemistry, University of Fort Hare, Alice, Eastern Cape, South Africa

Maame A. Korsah Department of Mathematics, University of Tennessee, Chattanooga, TN, United States

Michelle Felicia Lee Centre for Virus and Vaccine Research, School of Medical and Life Science, Sunway University, Bandar Sunway, Selangor, Malaysia

Nadun H. Madanayake Department of Botany, University of Peradeniya, Peradeniya, Sri Lanka

Agniva Majumdar Virus Research and Diagnostic Laboratory, ICMR-National Institute of Cholera and Enteric Diseases, Kolkata, West Bengal, India

Yashpal Singh Malik College of Animal Biotechnology, Guru Angad Dev Veterinary and Animal Sciences University, Ludhiana, Punjab, India

Keshab C. Mandal Department of Microbiology, Vidyasagar University, Midnapore, West Bengal, India

Zintle Mbese Department of Chemistry, University of Fort Hare, Alice, Eastern Cape, South Africa

Bulu Mohanta Department of Pharmacology, Seemanta Institute of Pharmaceutical Sciences, Jharpokharia, Mayurbhanj, Odisha, India

Joy Mondal ICMR-NICED Virus Unit, ID and BG Hospital, Kolkata, West Bengal, India; Department of Microbiology, Vidyasagar University, Midnapore, West Bengal, India

Igor José dos Santos Nascimento Institute of Chemistry and Biotechnology, Federal University of Alagoas, Maceió, Brazil

Amit Kumar Nayak Department of Pharmaceutics, Seemanta Institute of Pharmaceutical Sciences, Jharpokharia, Odisha, India

Ram Gopal Nitharwal Department of Biotechnology, Central University of Haryana, Mahendergarh, Haryana, India

Xhamla Nqoro Department of Chemistry, University of Fort Hare, Alice, Eastern Cape, South Africa

Benil P.B. Department of Agadatantra, Vaidyaratnam P.S. Varier Ayurveda College, Kottakkal, Kerala, India

Babul Rudra Paul Division of Medicine, ICAR-Indian Veterinary Research Institute, Izatnagar, Uttar Pradesh, India

Sijongesonke Peter Department of Chemistry, University of Fort Hare, Alice, Eastern Cape, South Africa

Chit Laa Poh Centre for Virus and Vaccine Research, School of Medical and Life Science, Sunway University, Bandar Sunway, Selangor, Malaysia

Yasmine Radwan Nanomedicine Research Laboratories, Center for Materials Science, Zewail City of Science and Technology, 6 of October City, Giza, Egypt

Rajakrishnan Rajagopal Department of Botany and Microbiology, College of Science, King Saud University, Riyadh, Saudi Arabia

Saikishore Ramanthan Medical Microbiology Unit, Department of Microbiology, Alagappa University, Karaikudi, Tamil Nadu, India

Akila Ravindran Department of Microbiology, Alagappa University, Science Block, Karaikudi, Tamil Nadu, India

Ryan Rienzie Department of Agricultural Biology, Faculty of Agriculture, University of Peradeniya, Peradeniya, Sri Lanka

Érica Erlanny S. Rodrigues Institute of Chemistry and Biotechnology, Federal University of Alagoas, Maceió, Brazil; Institute of Pharmaceutical Sciences, Federal University of Alagoas, Maceió, Brazil

Vrenda Roy Department of Indian System of Medicine, Government of Kerala, Kerala, India

Jagannath Sahoo Faculty of Pharmacy, DIT University, Dehradun, Uttarakhand, India

Varun Kumar Sarkar Division of Medicine, ICAR-Indian Veterinary Research Institute, Izatnagar, Uttar Pradesh, India

Anamika Sengupta Department of Health Science Education & Pathology, College of Medicine, University of Illinois, Peoria, IL, United States

Neeraj Sethiya Faculty of Pharmacy, DIT University, Dehradun, Uttarakhand, India

Leandro Rocha Silva Institute of Chemistry and Biotechnology, Federal University of Alagoas, Maceió, Brazil

Srishti Soni Division of Medicine, ICAR-Indian Veterinary Research Institute, Izatnagar, Uttar Pradesh, India

Omar Sued Pan American Health Organization, Washington, DC, United States

Angeline Jessika Suresh Fungal Genetics and Mycotoxicology Laboratory, Department of Microbiology, School of Life Sciences, Pondicherry University, Pondicherry, India

Subidsha Suyambu Krishnan Department of Microbiology, Alagappa University, Science Block, Karaikudi, Tamil Nadu, India

Sathiamoorthi Thangavelu Medical Microbiology Unit, Department of Microbiology, Alagappa University, Karaikudi, Tamil Nadu, India

Jacob Thomas Department of Botany and Microbiology, College of Science, King Saud University, Riyadh, Saudi Arabia

Shalini Upadhyay A.S.J.S.A.T.D.S. Medical College, Fatehpur, Uttar Pradesh, India

Navraj Upreti Faculty of Pharmacy, DIT University, Dehradun, Uttarakhand, India

Balasubramanian Vellaisamy Department of Microbiology, Alagappa University, Science Block, Karaikudi, Tamil Nadu, India

Palanivel Velmurugan Center for Materials Engineering and Regenerative Medicine, Bharath Institute of Higher Education and Research, Chennai, Tamil Nadu, India

Viroj Wiwanitkit Dr DY Patil University, Pune, Maharashtra, India

Roghayeh Yahyazadeh Departments of Pharmacodynamics and Toxicology, School of Pharmacy, Mashhad University of Medical Sciences, Mashhad, Iran

Sora Yasri Private Academic Consultant Center, Bangkok, Thailand

Preface

Once virus infection is established, antiviral therapy is the only option to control the infection. In principle, all the steps in the virus life cycle ranging from entry to release can be explored as molecular targets for antiviral therapy. Thus, antiviral therapy is one of the most exciting aspects of virology, since it has successfully employed basic science to generate very effective treatments for serious viral infections. In fact, the vast majority of antiviral drugs developed since the 1980s. These antiviral drugs block diverse steps of the viral life cycle, including entry, reverse transcription, viral protein processing, and integration of various viruses causing influenza, herpes, hepatitis, AIDS, COVID, etc. In this regard, the current book entitled *Viral Infections and Antiviral Therapies* aims to provide a thorough insight into the comprehensive and up-to-date trends of antiviral therapeutics and its mechanisms. This book contains four sections with 30 chapters and presents a systematic discussion on various key topics about viral infections and transmissions, antiviral agents, and therapeutics by some of the leading expert academicians and researchers around the globe. A concise and snappy account on the contents of each chapter has been described to make available a glimpse of the book to the readers.

Section I contains Chapter 1 that presents introductory overviews of viral infections, transmission, and management of viral infections by using various antiviral agents.

Section II contains seven chapters. Chapter 2 describes various important emerging viral diseases, namely ebola virus, dengue virus, chikungunya virus, West Nile virus, zika virus, yellow fever virus, nipah virus, influenza virus, and corona viruses. In addition, the ever changing landscape of infectious diseases, factors affecting emergence of viral diseases, prevention, and control of viral diseases have been addressed in this chapter. Chapter 3 provides a comprehensive understanding of various genetic and ecological changes and their effects on viral evolution through various approaches and different modes of transmission of viruses. Chapter 4 summarizes the modes of viral infections and transmissions. This chapter covers epidemiological triad and viral infection, transmission of viral infections, and various modes of transmission of viral infections. Chapter 5 analyzes SARS-CoV-2, its transmission, impact, varied intervention to mitigate its spread, the potential dynamics of similar pandemics, and recommendations for future studies. Chapter 6 discusses about various common sexually transmitted diseases caused by viral infections are genital warts, genital herpes and human immuno-deficiency virus, human T-cell lymphotropic virus, hepatitis A, B, and C. Additionally, prevention of sexually transmitted viral infections has been addressed in this chapter. Chapter 7 overviews various diagnostic modalities and developments in this field that may augment our decision-making capacity for best possible modality to be used in future. This chapter covers the purpose of laboratory diagnosis of viral infections, sample collection, packaging, and transport of collected samples for testing, and methods in diagnostic virology including virus isolation, direct microscopy, detection of viral antigens and antibody (serology), and molecular techniques. Chapter 8 elaborates on the electron microscopy as diagnostic tool in virology and its structure and functions. In addition, preparation of biological specimens for analyses by electron microscopy has been discussed.

Section III contains 19 chapters. Chapter 9 focuses on recent advancements and clinical trials underway in the past 5 years from 2016 to 2021 explaining various strategies employed by using oncolytic viruses as therapeutic sources to combat the most common forms of cancer. Chapter 10 addresses comprehensive discussions involving the significant development of antiviral drugs against the most challenging viruses that humankind has faced. In addition, aspects associated with the perspective of medical need, technical possibilities, and economic restrictions have been covered. Chapter 11 highlights about the antiviral agents currently available in the market as well as those under clinical trials, emphasizing their antiviral mechanisms. In addition, the chapter gives an outline of compounds that are in the research stage, focusing on phytochemicals that are currently being studied for their efficacy to combat influenza. Chapter 12 not only focuses on currently FDA approved antiherpes virus agents but also reveals the prospects of ethnomedicines, which will be the major lead molecules in the development of new antiherpes virus agents in near future. Chapter 13 deals with a comprehensive discussion on antiretroviral therapies. This chapter covers commonly used ART drugs and formulation of antiretroviral treatment, general principles for antiretroviral therapy initiation,

considerations before initiation of antiretroviral therapy, monitoring on the patient on antiretroviral therapy immune reconstitution inflammatory syndrome (IRIS), antiretroviral failure, drug interaction, antiretroviral drug resistance, pre- and postexposure prophylaxis, and prevention of mother child transmission. Chapter 14 discusses on the trends and advances in the specific therapeutic approaches being targeted against rotavirus This chapter highlights various therapies including fluid therapy, immune-therapies, probiotics, phytoconstituents, antioxidants, etc. Chapter 15 demonstrates researches focusing on novel therapeutic interventions for nonpolio enteroviruses that have emerged as a growing concern in the past decade. Chapter 16 highlights recent progress in medicinal chemistry field focused on relevant compounds, as antiviral agents against flaviviruses. In addition, related molecular modeling, virtual screening, drug repurposing, and biological evaluations have been covered. Chapter 17 addresses the backgrounds of HIV and AIDS, epidemiology, transmission and establishment of HIV infection, HIV life cycle, physiopathogenesis, response to HIV infection, natural history of HIV infection, transmission routes, manifestation of acute HIV infection, laboratory diagnosis, antiretroviral treatment, and various effective strategies to eliminate HIV as a public health threat. Chapter 18 explores important drug targets for curbing viral infections with phytochemicals derived from medicinal plants (herbal drugs). This chapter covers phytochemicals preventing attachment of virus to host cell, phytochemicals preventing penetration and uncoating of viruses, phytochemicals inhibiting replication of viral nucleic acids and phytochemicals preventing assembly and release of virus. Chapter 19 reports the efficacy of hybrid molecules, drug delivery systems incorporated with antiviral agents, and metal-based nanoparticles in the treatment of viral infections, such as HIV, herpes, hepatitis, Ebola, human papillomavirus, viral pneumonia, common cold, COVID-19, and Middle East respiratory syndrome. Chapter 20 expounds on the research and application of nanovesicles for delivery of antiviral agents to attack viral infections such as influenza, HIV, and the most recent COVID-19. Highlights on antiviral monotherapies, currently approved or still undergoing clinical investigations and future perspectives aiming to translate the outstanding research outcome into clinical settings have also been discussed. Chapter 21 presents an overview and better understanding of the current knowledge in the arena of antiviral biomaterials. In addition, the chapter illustrates the multidisciplinary approaches of antiviral biomaterials in terms of applications, recent advancements, and challenges associated with antiviral biomaterials. Chapter 22 mainly focuses on the antiviral benefits of most commonly reported marine polysaccharides such as chitin, chitosan, carrageenan, alginate, etc., and marine-based peptides as antiviral biomaterials. Chapter 23 explores antiviral activities of various plant polysaccharides. The antiviral mechanisms of polysaccharides (of different sources like plant, animal, marine, etc.) and their derivatives have been summarized in the current chapter with the goal of providing a sound platform for future research on the antiviral properties of plant polysaccharides and their chemically modified derivatives. Chapter 24 highlights the current status of development of antiviral peptides targeting dengue virus, strategies that were utilized to design antiviral peptides, interactions that were identified between antiviral peptides and dengue virus host cell receptors

or enzymes, advantages and disadvantages of antiviral peptides, as well as potential ways to overcome their limitations. Chapter 25 summarizes the history of development and delivery of these two mRNA vaccines with emphasis on crucial aspects of design strategy, delivery approach, and their noteworthy roles in inducing immune responses postvaccination. Chapter 26 explains some of the unique existing approaches to prevent and treat viral infectious disease via immunotherapies, such as mAb-based therapies, vaccines, T-cell-based therapies, utilizing cytokine levels, and checkpoint inhibition as well as defensins. Chapter 27 discusses the potential role of nutraceuticals as immunomodulators in combating viral infections. The chapter covers immunity and its classification, virus evasion of the host immune system, mechanism of action, definition, classifications, and immunomodulatory actions of nutraceuticals against viruses.

Section IV contains three chapters. Chapter 28 insights into both in vitro and in vivo approaches that are generally employed to test efficacy of new antiviral drugs. Chapter 29 provides a collected evidence of the undergoing clinical trials on different types of antiviral agents or their combinations. The discussion and recommendations for new molecules approved by the FDA (2015−21) have been included. Chapter 30 discusses the general developments with the antivirals along with the modern perspectives for which these compounds can usher better and holistic therapeutic approaches utilizing state-of-the-art technology, software, and related tools in an omics era.

We, the editors, sincerely convey our thanks to all the distinguished authors for their invaluable contributions of quality chapters in a timely manner. This book could not have been published without cooperation and wholehearted supports of Elsevier, Inc., Andre Gerhard Wolff, Kattie Washington (Acquisitions Editor), Timothy Bennett (Editorial Project Manager), and Selvaraj Raviraj (Production Manager) in the book-editing process. We would like to express our sincere thanks to Dinesh Natarajan (Copyright Coordinator) for their outstanding support in obtaining copyright permissions. All copyright contents and reprinting licenses from different copyright sources have duly been gratefully acknowledged. Finally, we must appreciate our family members, friends, colleagues, and students for their continuous encouragements, inspirations, and moral supports during the book-editing process. We will be extremely happy along with our contributing authors and publishers, if our efforts fulfill the needs of the students, researchers, academicians, and others.

Amal Kumar Dhara
Amit Kumar Nayak

Introduction

Introduction to antiviral therapy

Amal Kumar Dhara[1] *and Amit Kumar Nayak*[2]

[1]Department of Pharmacy, Contai Polytechnic, Purba Medinipur, West Bengal, India
[2]Department of Pharmaceutics, Seemanta Institute of Pharmaceutical Sciences, Jharpokharia, Odisha, India

1.1 Introduction

The current COVID-19 pandemic, caused due to SARS-CoV-2 virus, once again proved that viruses are still a mystery and human beings could be helpless before finding the treatment. Viruses are acellular obligate intracellular parasites that can infect humans, plants, animals, yeast, bacteria, protozoa, archaea, as well as other viruses [1,2]. For multiplication, they require host cells [3]. Viruses are unique as they are alive and can multiply only in living host cells [4]. Viruses are highly complex in producing diseases [5,6]. They consist of nucleic acids deoxyribonucleic acid (DNA) and ribonucleic acid (RNA) covered by a capsid (a protein coat) [7]. The combination of capsid and genome is known as nucleocapsid [8]. Envelope surrounding the nucleocapsid is also found in viruses as an additional component [7,8]. Spike proteins are one of the most important structural features of viruses, which bind to specific receptors of the host cell and cause infections [9]. Some viruses also contain viral enzymes that can cause infection to the host cell. Viruses with all the essential components required to infect the host cell are called virions [7,10]. There are two categories of viruses: DNA viruses and RNA viruses, depending on their replication by using DNA or RNA, respectively [7]. Examples of DNA viruses include adenoviruses, herpes viruses, etc., and RNA viruses include hepatitis C virus (HCV), influenza virus, human immunodeficiency virus (HIV), SARS-CoV-2, etc.

1.2 Virus replication cycle

To replicate, a virus makes use of cell and viral replication cycle that leads to extensive structural and biochemical changes in the infected host cell [11]. Cells infected by virus may die due to lysis or apoptosis and by the release of virions [10,11]. The observed symptoms caused by viral diseases are due to immune response to the virus. There are some variations in replication cycle of viruses; however, in most cases there are five steps in the viral replication cycle [10−12] (Fig. 1.1):

1. Attachment
2. Penetration
3. Synthesis
4. Assembly
5. Release.

Attachment: Without a host cell, viruses are inactive metabolically. The attachment of the virus to a receptor of host cell is highly specific, just like lock and key system [12]. Due to this specificity, a particular virus can infect only a particular cell type.

Penetration or entry: In case of bacteriophage (virus that infects bacteria), nucleic acid enters the host cells and leaves the capsid outside of the cell [13]. In case of eukaryotic viruses, the entire capsid enters the host cell [14,15]. Viruses also enter the cell by endocytosis, when capsid is engulfed by the host cell [16]. Once virus enters the cell, the virus nucleic acid is released due to degradation of viral capsid and finally becomes available for replication as well as transcription [17,18].

Synthesis: This is directed by the viral genome. RNA viruses synthesize mRNAs and viral genomic RNAs usually by using RNA core as a template [19,20]. DNA viruses make mRNAs by using host cell protein and enzymes through direct protein synthesis [21].

Assembly: The capsid is composed of different types of proteins, whose numbers are varied (3−60) [22,23]. Fewer the number of proteins involved, simple is the virus. The different types of proteins are arranged in a specified order.

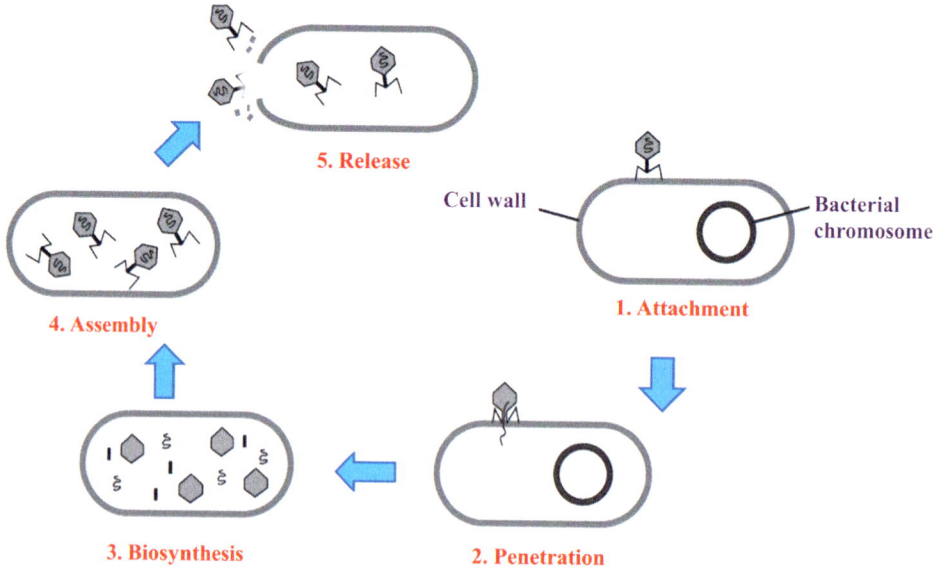

FIGURE 1.1   Virus replication cycle.

Release: At the end of replication, most of the viruses lyse host cell and as a result, new virions are released [18]. Budding is another process, common in enveloped viruses, where one virus is released at a time through the membrane without lysis of the cell [24].

1.3 Virus transmission and types of viral infections

Transmission of viruses takes place in a variety of ways [25]. Some viruses are transmitted through touch, air, saliva, and others are transmitted by insects (e.g., mosquitoes), sexual contact, sharing contaminated needle/blood, contaminated water, food, etc. [26,27]. There are various important modes of viral transmission:

1. Vertical transmission: In this mode, transmission of pathogens occurs from mother to child either before or after childbirth [28–30]. The vertically transmitted viruses include HIV, Zika, etc.
2. Transmission through breast milk: During breast feeding, the pathogens may be transmitted and cause infantile viral infection [31–33]. Example of viruses transmitted via breast milk include HIV, cytomegalovirus (CMV), and human T-cell lymphotropic virus.
3. Transmission via respiratory tract: This is the most common mode of viral transmission that leads to a number of viral infections [34]. Very recent example is COVID-19. The surrounding environment is very important for transmission of viruses via respiratory contact. Droplet transmission (during sneezing, talking, etc.) and airborne transmission are the major causes of transmission [26].
4. Transmission caused by transplantation: This is caused by transplantation of organ obtained from infected donor (living or dead) [35,36]. The viruses transmitted via transplantation include hepatitis B virus (HBV), HCV, and HIV.
5. Transmission via blood and blood transfusion: Viral pathogens may transmit with plasma or cellular components of blood, for example, HIV. Other examples include HBV, HCV, HGV, etc. [37,38].
6. Transmission via vector: Some animals, mainly insects like mosquitoes (from infected host to non-infected host), act as vectors for viral transmission [39–41]. Examples of vector-borne transmission of viral infections are Zika virus (ZIKV), Chikungunya, etc.
7. Transmission via sexual contact: Sexual contact (more specifically sexual intercourse) is also another important cause of transmission of viral infection, such as HIV, HBV, HCV, ZIKV infection, etc. [42,43].
8. Foodborne and waterborne transmission: Food and drinks can also transmit viral infections, which have been the case recently in SARS-CoV-2 [44–46].
9. Orofecal transmission: Transmission of virus through feces due to poor sanitation and improper hand washing may cause viral infections, for example, polio virus and hepatitis virus infection [47–49].
10. Transmission via eye, skin, and mucosal contact: Herpes simplex, human papilloma, and molluscum contagiosum can be transmitted via skin contact [50–52]. Adenovirus and SARS-CoV-2 are also transmitted by hand-to-eye contact [6,43].

Different types of viral infections include:

1. Respiratory viral infections: These are the most common viral infections. The organs affected are nose, throat, lungs, and upper respiratory tract [53,54]. These spread from one person to other by inhaling droplets (containing viruses) [26]. The examples of various respiratory viral infections are [53–58] (a) Respiratory syncytial virus—both upper and lower respiratory tract infection. (b) Rhinovirus—usual symptoms of common cold include sneezing, coughing, sore throat, mild headache, etc. (c) Seasonal influenza—every year certain percentage of people suffer from this illness and the symptoms are usually more severe than common cold, including fatigue, mild fever, aches throughout the body, etc. (d) SARS-CoV-2—this is popularly known as COVID-19, as the infection started to spread during the November 2019 and in 2020, it led to a pandemic. Millions of people across the globe died due to this severe acute respiratory syndrome. Common symptoms of SARS-CoV-2 infection include fever, cough, shortness of breath, and pneumonia [56].

2. Viral skin infection: The skin is affected due to infections such as [50–52,58]: (a) Herpes simplex virus-1 (HSV-1)—this can usually be transmitted through saliva during kissing, sharing of lip balm, or by infected food sharing. (b) *Molluscum contagiosum*—it is another type of skin infection, usually observed in children (age group: 1–10 years) and there is appearance of small white- or pink-colored bumps on the skin anywhere on the body, including neck, face, abdomen, legs, arms, etc. (c) Varicella-zoster virus (VZV)—it is a DNA virus that causes chicken pox. This virus causes fatigue, itchy and oozing blisters along with high fever.

3. Sexually transmitted viral infections: Infections spread through body fluid contact and also through blood. Different sexually transmitted viral infections include [43,59–61]: (a) HIV—these viruses attack the immune system of the body and are transmitted via semen, vaginal fluid, and infected blood. The symptoms of AIDS include fever, weight loss, fatigue, and recurrent infections (due to deficiency of immunity). The incidence of HIV infection can be reduced by avoiding sexual activity with unknown partners, testing blood for HIV before transfusion, and using unused/sterile needle for injection of drugs. (b) Genital herpes—it is caused by herpes simplex virus-2 (HSV-2) and is a very common sexually transmitted infection. The symptoms of HSV-2 include itching, pain, (on the penis or vagina), and small sores. To avoid HSV-2 infection, person should abstain from vaginal, anal, and oral sex (with the partners having HSV-2 infection).

4. Viral hepatitis: Viral hepatitis is caused by different viruses including hepatitis A, B, C, D, and E [62]. In this infection, liver is damaged due to inflammation. Various symptoms of viral hepatitis include fever, loss of appetite, fatigue, abdominal pain, dark urine, nausea, vomiting, etc. [63,64]. Viral hepatitis is transmitted through contaminated water, food, or via body fluids and blood. Hepatitis A and E are transmitted, when a person ingests contaminated food or water (usually contamination comes through fecal matter from infected person). Hepatitis B and C are transmitted through blood and body fluids, which cause acute and chronic hepatitis. Newborn baby can also be infected with hepatitis B and C during childbirth transmitted from mother. Sex with infected person, transfusion of infected blood, equipment sharing like needles, syringes, etc., can also cause hepatitis B and/or C viral infections [62].

5. Viral gastroenteritis: A number of different viruses, including rotaviruses, noroviruses, adenoviruses, sapoviruses, and astroviruses, are responsible for viral gastroenteritis [63]. Vomiting, diarrhea, headache, abdominal pain, fever, etc., are the main symptoms of such infection and is also known as "stomach flu" [65,66]. The above viruses can be transmitted through contaminated water, foods, beverages, and close contact with the infected person and touching contaminated utensils or surfaces. Contamination of food and water is usually caused by sewage, persons (having viral gastroenteritis) engaged in food production and handling, and poor hygienic condition. Rotavirus usually affects young children as well as infants. Children or infants suffer from severe diarrhea, vomiting, and anorexia [66,67]. Norovirus can affect a person of any age and is highly contagious [68,69]. It is transmitted through contaminated water, foods, and also by an infected person. Diarrhea, fever, body aches are the common symptoms. Adenovirus can also affect anyone and is usually transmitted through coughing, sneezing, and by touching contaminated surfaces or objects [70]. Symptoms of adenovirus infection include fever, coughing, bronchitis, sore throat, runny nose, pink eye, etc. Astrovirus usually affects children and observed symptoms include diarrhea, mild dehydration, headache, stomach pain, etc. [71,72].

6. Flavivirus infection: Flaviviruses include ZIKV, West Nile virus (WNV), Dengue virus (DENV), Japanese encephalitis virus (JEV), etc., and belong to Flaviviridae family [73,74]. These viruses usually spread in underdeveloped countries and more commonly among the poor. Different flaviviruses produce varieties of symptoms and represent major threat to human health resulting in high morbidity and mortality rates. The transmission of flaviviruses occurs by mosquitoes, ticks, person-to-person, and also by vertical transmission, that is, mother to child in pregnant women. The symptoms of ZIKV are microcephaly in newborn and malformation of fetus in pregnant mother [75,76]. The other symptoms include congenital abnormalities like cerebral palsy, epilepsy, intellectual disability and also, audio-visual and behavioral disturbances. The symptoms of WNV include headache, fever, myalgia, rash, lymphadenopathy, gastrointestinal disorders, and also meningitis and encephalitis [77,78]. The symptoms associated with DENV include mild fever, severe hemorrhagic fever, and dengue shock syndrome [79,80].

1.4 Antiviral agents

Antiviral agents are drugs used for the control and treatment of viral infections [81]. A schematic presentation of the management of viral infection with antiviral agents is given below (Fig. 1.2) [81−86]:

On the basis of viral infection, antiviral drugs are categorized into following groups [82,85−89]:

1. Antiherpes virus agents
2. Anti-HIV agents
3. Antiviral drugs used for the treatment of hepatitis
4. Anti-influenza agents
5. Antirotavirus agents

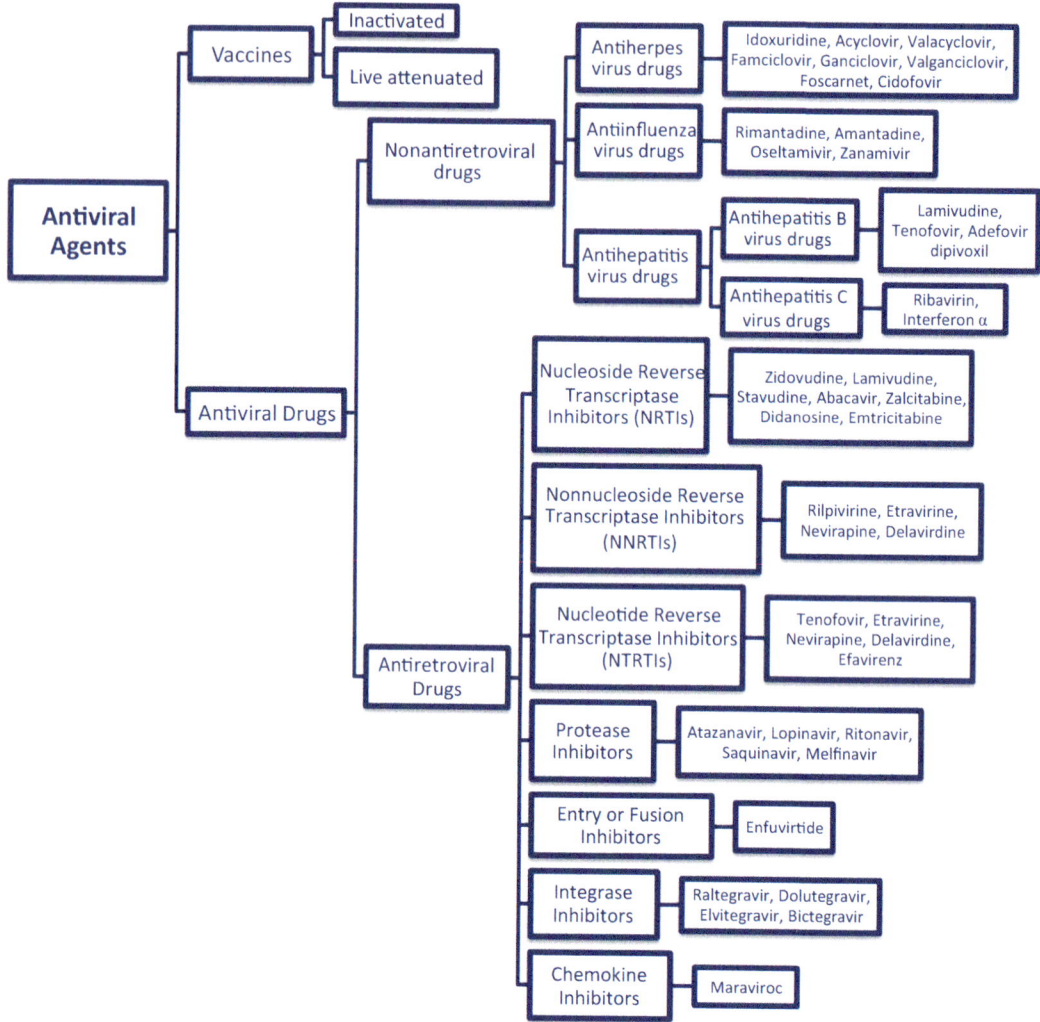

FIGURE 1.2   Antiviral agents for the managements of viral infections.

6. Antiviral agents against enteroviruses
7. Antiviral agents against flaviviruses
8. Antiviral agents used for the treatment of respiratory infection.

1.4.1 Antiherpes virus agents

These drugs are effective against herpes virus (DNA viruses) like HSV-1, HSV-2, Epstein Barr virus (EBV), VZV, CMV, Roseolo virus, Kaposi's sarcoma-associated virus, etc. [90−92]. Like all other viral infections, these can also be managed by minimizing the risk of infection,

TABLE 1.1 Antiherpes virus agents.

Antiviral agent	Effective against	Mechanism of action
Acyclovir	HSV, VZV, CMV	Inhibits viral DNA polymerase competitively, incorporates into the growing viral DNA chain, which results in termination of viral replication
Foscarnet	HSV, CMV, VZV	Non-competitive inhibition of herpes virus DNA polymerase. Inhibits pyrophosphate binding on viral DNA polymerases
Famciclovir	HSV, VZV, Epstein-Barr virus (EBV)	Acts as a substrate for viral DNA polymerase and thereby inhibits DNA synthesis
Ganciclovir	HSV, CMV, EBV	Competitively incorporates during viral DNA synthesis, leading to the DNA chain termination
Valacyclovir	CMV	Inhibits viral replication as a competitive substrate for viral DNA polymerase
Boceprevir	HCV	Reversibly binds with the HCV NS3 to inhibit viral replication
Oseltamivir	Influenza A and B viruses	Neuraminidase inhibitor of Influenza virus
Telaprevir	HCV	Inhibits HCV replication by binding reversibly to NS3 serine protease
Rimantadine	Influenza virus	Inhibits the ion channel function of M_2, thereby, the viral uncoating

vaccination, and suitable antiviral agents. The varieties of anti-herpes virus agents available in the market with varying clinical efficacy are presented in Table 1.1.

1.4.2 Anti-HIV agents

HIV is a typical retrovirus and a single-stranded RNA virus [93–96]. Two major forms of HIV are HIV-1 (widely distributed and found worldwide) and HIV-2 (it is most common in western Africa). Emphasis should be given on prevention of transmission of diseases. Different antiviral drugs are available for the treatment of AIDS, which are shown in Table 1.2.

1.4.3 Antiviral drugs used for the treatment of hepatitis

Hepatitis cause damage to the liver due to inflammation, resulting in impairment of liver function [48]. Inflammatory disorders of liver (hepatitis) may be acute or chronic. There are five types of hepatitis. Hepatitis A caused by hepatitis A virus (HAV), similarly hepatitis B, hepatitis C, hepatitis D, and hepatitis E are caused by HBV, HCV, hepatitis D virus, and hepatitis E virus, respectively [48,62]. Out of these five types, HAV, HBV, and HCV are very common and clinically significant. Several important antiviral drugs as well as vaccines are available in the market for treating hepatitis [97–99].

TABLE 1.2   Antiviral drugs are available for the treatment of AIDS.

Class	Antiviral agent	Mechanism of action
Nucleoside reverse transcriptase inhibitors (NRTIs)	Zidovudine (ZDV, AZT), Abacavir (ABC), Lamivudine (3TC), Emtricitabine (FTC), Stavudine(D4T)	Inhibit reverse transcription by blocking the enzyme causing chain termination after they have been incorporated into viral DNA. The drug is activated by phosphorylating intracellularly. The resultant enzyme inhibition cause development of incomplete DNA and new virus formation is halted
Non-nucleoside reverse transcriptase inhibitors (NNRTIs)	Rilpivirine, Nevirapine (NVP), Etravirine, Delavirdine	Bind (noncompetitively) directly onto reverse transcriptase, RNA to DNA conversion is prevented and stop viral replication
Nucleotide reverse transcriptase inhibitors (NTRTIs)	Tenofovir (TDF), Nevirapine (NVP), Efavirenz (EFV), Etravirine, Rilpivirine, Delavirdine	Nucleotide analog of reverse transcriptase inhibitor works in the similar fashion as nucleosides, but they have a nonpeptidic chemical structure
Protease inhibitors (PIs)	Lopinavir, Atazanavir, Darunavir, Ritonavir, Saquinavir, Nelfinavir, Indinavir, Fosamprenavir, Tipranavir	HIV protease is a viral enzyme, catalyzes proteolysis, breaking down proteins into smaller polypeptides or single amino acids, and spurring the formation of new protein products. It is mainly important for post-translational modification. Protease Inhibitors work at the last stage of the viral replication by preventing HIV from being successfully assembled and released from the infected host cell
Entry or fusion inhibitor	Enfuvirtide (T20)	Disrupts the fusion of HIV-1 with the target cell, uninfected cells from becoming infected is prevented. HIV binds to the host CD4 + T cell receptor via the viral protein gp120, gp41, a viral transmembrane protein, then undergoes a conformational change that assists in the fusion of the viral membrane to the host cell membrane. Enfuvirtide binds to gp41, preventing the creation of an entry pore for the capsid of the virus, keeping it out of the cell
Integrase inhibitors	Raltegravir (RAL), Dolutegravir (DTG), Elvitegravir, Bictegravir	Block the action of the enzyme integrase, a viral enzyme that facilitates insertion of the viral genome into the genome of the host cell
Chemokine inhibitor or CCR5 inhibitor	Maraviroc	Maraviroc binds to CCR5 receptors expressed on the cell of human body, which results in interference with fusion of HIV protein gp120 and human cells like human macrophages and T cells

TABLE 1.3 List of drugs recommended in the treatment of hepatitis C virus.

Class	Drugs
NS3 inhibitors	Grazoprevir, Paritaprevir, Asunaprevir
NS5 A inhibitors	Ombitasvir, Daclatasvir, Ledipasvir
NS5 B inhibitors	Sofosbuvir, Dasabuvir
Pregylated interferons α	PegIFNα-2a and -2b (adult)
Ribavirin	Ribavirin (adult and child)

Hepatitis A: Hepatitis A is caused by HAV. It is not a chronic liver disease, but some complications may occur in the person infected with HAV [100]. Prevention is the main approach of infection control, as there is no specific antiviral drug available [101]. However, for prevention, hepatitis vaccine can be recommended.

Hepatitis B: HBV produces one of the most serious health problems till now and the annual deaths due to HBV infection is over 1 million [102,103]. The therapeutic approach is the suppression of replication of HBV and improvement of quality of life. Lamivudine is widely used, low-cost, and well-tolerated drug, which acts by inhibiting viral polymerase/reverse transcriptase [104,105]. The antiviral drugs used to treat HBV infection include telbivudine, entecavir, adefovir dipivoxil, tenofovir disoproxil, etc. [97,102,105,106]. Interferon-based therapy (recombinant and lymphoblastoid IFN-α) has also been exhibited significant activity in chronic hepatitis B (CHB) infection [107,108]. Inhibition of HBV life cycle is another important strategy in controlling HBV infection. The different HBV life cycle inhibitors include inhibitor of HBV DNA polymerase (e.g., lagociclovir), viral entry inhibitor (e.g., Myrcludex-B), cccDNA synthesis inhibitor, inhibitor of nucleocapsid assembly, etc. [105−109]. Immunomodulators also play a significant role in CHB treatment [110].

Hepatitis C: HCV infection is usually asymptomatic or mildly symptomatic and thus its diagnosis is difficult [111]. Chronic HCV infection causes serious liver problems, including cirrhosis or hepatocellular carcinoma. Chronic HCV must be treated as early as possible [112,113]. List of some important drugs are mentioned in Table 1.3.

1.4.4 Anti-influenza agents

In general, there are three types influenza viruses: Influenza viruses A, B, C, and D are found in diseased pigs [114]. Out of these, influenza viruses A and B infect human beings [115]. In the past, influenza pandemic was observed in 1918 as Spanish flu [116], in 1957 Asian flu [117], in 1968 Honk Kong flu [118], and in 2009 swine flu [119]. Due to these pandemics, millions of lives were lost and severe socioeconomical damage occurred. The anti-influenza viral agent are classified as follows [120−123]:

1. Matrix protein Z ion channel inhibitors, for example, Amantadine, Rimantadine.
2. Neuraminidase inhibitors, for example, Zanamivir, Oseltamivir, Peramivir, Laninamivir.

3. RNA-dependent RNA polymerase (RdRp) inhibitor, for example, Favipiravir.
4. Polymerase acidic protein cap-dependent endonuclease inhibitor, for example, Bloxavir marboxil.
5. Hemagglutinin (HA) inhibitor/Fusion inhibitor (HA is a glycoprotein found in envelope of influenza virus, which plays important role in the process of viral binding, fusion, and entry), for example, Umifenovir or Arbidol.

Some anti-influenza agents, obtained from traditional medicinal plants, have played important role in the management of influenza virus infection [124,125]. Some important examples of such plants include *Cephalotaxus harringtonia* (NA inhibitor) [126], *Glycyrrhiza uralensis* (NA inhibitor) [127], *Echinaceae purpurea* (HA inhibitor) [128], *Sanicula europaea* (RdRp inhibitor) [129], etc. Among the several antiviral drugs, vaccines are the very important agents to prevent influenza virus. Both live attenuated and inactivated vaccines are available in the market.

1.4.5 Antiviral agents against flavivirus

More than 70 viruses belong to genus *Flavivirus*, including DENV, ZIKV, Yellow fever virus, JEV, WNV, etc. [74]. Dengue fever is mosquito-borne disease caused by DENV and spreads very fast [130]. In 2007, ZIKV outbreak was reported in Micronesia [131]. No specific agents are available for the treatment of flavivirus infections. Several research works across the globe are being conducted to obtain a novel antiviral agent(s), which target the different stages in replication cycle.

Much research has already been done as well as is ongoing to find out suitable antiviral agents that can be used for the treatment of dengue [130,132]. Unfortunately, neither directly acting antivirals (DAA) nor host directed antivirals have shown clinical significance. Clinical trials of balapiravir, a DAA, failed to exhibit significant efficacy as compared to placebo [133]. A group of researchers studied narasin, a novel antiviral agent effective against DENV [134]. The investigational results suggested that narasin caused inhibition of DENV replication (in vitro), however, further studies are needed to establish narasin as a potential anti-DENV agent.

1.5 Antiviral agents obtained from plant sources

Like ancient people, people nowadays also show the tendency to use herbal drugs for treatment of a variety of diseases [125–129]. A plethora of valuable phytochemicals is available. During the present pandemic (COVID-19) situation, scientists put in their sincere efforts to find out novel antiviral drug(s) or agents that can be used as an adjuvant therapy or as an immunomodulator to combat viral infections [126,135].

The different phytoconstituents exhibited their activity in different stages of virus replication cycle:

1. Some potential phytochemicals prevent attachment of virus to host cell. Examples include loliolide obtained from *Phyllanthus urinaria* that caused inhibition of attachment

to the host cell of HCV [136]. Ginsenosides from *Panax quinquefolium* prevent attachment of influenza A H1N1 to the host cell receptors [137]. Quercetagetin also exhibited significant activity by inhibiting chikungunya virus attachment to host cell [138].

2. Phytoconstituents prevent penetration and uncoating of viruses. Withanone, an important constituent obtained from *Withania somnifera*, interferes with the entry of SARS-CoV-2 [139]. Epigallocatechin gallate caused inhibition of viral entry and showed activity against ZIKV (Mr766 strain) [140]. Jatrophane esters from *Euphorbia dendroides* latex showed antiviral activity against replication of chikungunya virus [141]. Aqueous extract of bark of *Azadirachta indica* prevented the entry of HSV-1 into the host cell [142].

3. Phytochemicals caused inhibition of viral nucleic acid replication. In a study, C. D. Lee and his coworkers observed that *Phyllanthus amavus* may be used as a potential antiviral agent in the treatment of HBV. The plant caused downregulation of the DNA replication of HBV [143]. Lignan, an important constituent of Chinese plant Radix, caused disruption of replication machinery in respiratory syncytial virus [144].

4. Phytoconstituents caused prevention of assembly and release of virus. Kaempferol from *Ficus benjamina* significantly inhibited HSV-1, HSV-2, and coronavirus by interfering their release via 3a channel inhibition [145]. In a study, it was reported that naringenin showed antiviral activity by impairing the assembly of virion of ZIKV [146].

1.6 Antiviral vaccines

Vaccines play an important role in the effective control and prevention of viral infections. Polio and small pox eradication are ideal examples that guided researchers in the development of vaccines for prevention and control of different types of viral infections [147,148]. Later, many vaccines like measles, mumps, rubella, diphtheria, pertusis, tetanus, BCG., etc., vaccines were developed [149]. The most important objective for developing antiviral vaccines is the development of immunity, which resembles natural immunity for lifelong [150]. Examples of some antiviral vaccines available in the market are mentioned in Table 1.4.

Various types of antiviral vaccines have been developed:

1. Live viral vaccines: These are developed from virus strains but are attenuated [151]. They are able to replicate in the host cell and thus boosting immunity, for example, polio virus vaccine.

2. Inactivated vaccines: Whole particles are inactivated by heat, UV irradiation, and sometimes by treating with special chemicals (formalin and beta propiolactone) [152], for example, polio vaccines.

3. Recombinant viral proteins: By using bacteria, yeast, mammalian cell lines, insect, etc. desired viral proteins are manufactured, which are used as vaccine antigens [153], for example, vaccines for influenza viruses.

TABLE 1.4 Antiviral vaccines available in the market.

Vaccine	Platform	Route of administration
Smallpox	Live attenuated	Percutaneous
Measles, mumps, and rubella	Live attenuated	Subcutaneous
Rotavirus	Live attenuated	Oral
Polio	Inactivated	Intramuscular or subcutaneous
Hepatitis A	Inactivated	Intramuscular
Hepatitis B	Inactivated	Intramuscular
Rabies	Inactivated	Intramuscular

4. Subunit vaccines: Purified preparations are produced instead of whole inactivated vaccines [154]. The purified preparations containing main targets of protective immune responses were developed [150], for example, hepatitis B vaccines.
5. Virus-like particles: From structural proteins of virus, multimeric structure are assembled, which provide immune response against viruses [155,156], for example, hepatitis B vaccines.

Some antiviral vaccines have been developed successfully. However, in some cases, this is not possible due to many factors, such as [150,151,157]: (i) some viruses show wide genetic variations, (ii) some viruses integrate their genomes in the host, (iii) multiple mechanisms involve for evading host-immune detection and response, and (iv) ability for developing latency. In the present pandemic situation (COVID-19), some vaccines have been developed and approved on emergency basis [158]. However, more studies are required regarding safety and efficacy.

1.7 Immunotherapy and role of nutraceuticals in viral infection

The importance of immunotherapy in the management of viral infections is increasing day by day. Immunotherapy regulates the host's adaptive (activated by exposure to pathogen) and innate immune responses against the viral infections [159]. Different viral infections, for example, COVID-19, are significantly controlled by immunotherapy, including T-cell-based therapies, vaccines, monoclonal antibodies (mAbs)-based therapies, etc. [160,161]. mAbs have been successfully employed, produced in the laboratory, and widely used for the treatment of various types of diseases [162–164]. They are safe, have high specificity, have capacity to target specific epitopes, and also have high potency. Use of vaccines is the most popular and cost-effective strategy and also one of the oldest therapies for viral infections. Chimeric antigen receptor T cells are also effective for immunotherapy in case of malignancies as well as viral infections [165]. Importance of nutraceuticals as immunomodulators is increasingly investigated to treat viral infections [166,167]. Nutraceuticals are usually food or food products having nutritional value and

TABLE 1.5 Nutraceuticals effective against different types of viral infection.

Nutraceuticals	Source	Effective against
Glycyrrhizin	*Glycyrrhiza glabra*	Hepatitis C virus [168]
Garlic	*Allium sativum*	Influenza, HIV, herpes simples [169]
Black pepper	*Piper nigrum*	Coxsackie virus [170]
Spirulina	Blue green algae (*Spirulina platensis*)	HIV-1, mumps, measles, and herpes virus [171,172]

provide health and medical benefits and also can be used for the prevention and treatment of diseases. Nutraceuticals having specific health benefits can be called as potential nutraceuticals, for example, probiotic, prebiotic, antioxidants, dietary fibers, polyunsaturated fatty acids, etc. [166]. They are obtained from plant, animal, and microbial sources. Various types of nutraceuticals, their sources, and activities against viral infections are shown in Table 1.5.

1.8 Challenges in the development of antiviral agents

Many antimicrobial agents are being developed on a regular basis. The rate of development of new antiviral agents is less as compared to emerging viruses. Despite the simplicity in the structures of viruses, there are so many challenges like antiviral drug resistance, mutations, etc. that limit the development of new molecules. Other issues include the cost of production of antiviral agents, preference of vaccination over treatment of viral infections with antiviral drugs, and the prolonged and time-consuming process of development of new antiviral agents (there may be a chance of occurrence of new variant due to mutation as observed in SARS-CoV-2 before the new drug is launched in the market) [173].

To overcome the above challenges in the development of effective antiviral agents, many strategies have been designed [174]. Attempts have been made to target host cell factors or the viruses (viral proteins/enzymes) [175,176], interference of the virus attachment with the host cells [177], inhibition of viral entry (fusion inhibitor) [177,178], inhibition of replication and transcription (polymerase inhibitors) [179,180] and inhibition of proteases [181,182] responsible for viral replication/transcription and maturation. Monoclonal antiboldies (mAbs) are also being developed as significant therapeutic agents against viral infections [162,163]. The antibody variable gene sequencing techniques and B-cell isolation in the evaluation of several agents against some viruses, for example, ZIKV, INFV, HIV, SARS-CoV-2, etc., have been identified [164,183].

1.9 Conclusion

With time, the number of diseases due to viruses is increasing. The genetic variation of viruses is wide and most of the viruses have a great tendency for mutation, which caused difficulty in the development of antiviral agents. The viruses are transmitted very fast by various means and lead to death of a large number of people within a very short period. Vaccines are being developed to control many such viral infections. Some antiviral agents have already been developed and successfully employed in the management of viral infections. Many more suitable antiviral agents are yet to be developed for effective control of different types of viral infections.

References

[1] Crawford DH. Viruses: a very short introduction. US: Oxford University Press; 2011, p. 16.

[2] Fields BN, Knipe DM, Howley P, Chanock RM, et al., editors. Fields' virology. 3rd ed. New York: Raven Press; 1995.

[3] Pellett PE, Mitra S, Holland TC. Basics of virology. Handb Clin Neurol 2014;123:45−66.

[4] Koonin EV, Starokadomskyy P. Are viruses alive? The replicator paradigm sheds decisive light on an old but misguided question. Stud Hist Philos Biol Biomed Sci 2016;59:125−34.

[5] Mourya DT, Yadav PD, Ullas PT, Bhardwaj SD, Sahay RR, Chadha MS, et al. Emerging/re-emerging viral diseases & new viruses on the Indian horizon. Indian J Med Res 2019;149(4):447−67.

[6] Herrington CS, Coates PJ, Duprex WP. Viruses and disease: emerging concepts for prevention, diagnosis and treatment. J Pathol 2015;235(2):149−52.

[7] Gelderblom HR. Structure and classification of viruses. In: Baron S, editor. Medical microbiology. 4th ed. Galveston: University of Texas Medical Branch at Galveston; 1996.

[8] Louten J. Virus structure and classification. Essential human virology. 2016, p. 19−29.

[9] Du L, He Y, Zhou Y, Liu S, Zheng BJ, Jiang S. The spike protein of SARS-CoV−a target for vaccine and therapeutic development. Nat Rev Microbiol 2009;7(3):226−36.

[10] Chappell JD, Dermody TS. Biology of viruses and viral diseases. Mandell, Douglas, Bennett's Princ Pract Infect Dis 2015;1681−93 e4.

[11] Villar E. Cell response to viral infection: search for new therapeutric targets. Virus Res 2015;209:1−3.

[12] Nagy PD, Richardson CD. Viral replication-in search of the perfect host. Curr Opin Virol 2012;2(6):663−8.

[13] Kasman LM, Porter LD. Bacteriophages. StatPearls [Internet]. Treasure Island, FL: StatPearls Publishing; 2022.

[14] Cohen FS. How viruses invade cells. Biophys J 2016;110(5):1028−32.

[15] Dolgin E. The secret social lives of viruses. Nature 2019;570(7761):290−2.

[16] Cossart P, Helenius A. Endocytosis of viruses and bacteria. Cold Spring Harb Perspect Biol 2014;6(8): a016972.

[17] Cann AJ. Replication of viruses. Encycl Virol 2008;406−12.

[18] Louten J. Virus replication. Essent Hum Virol 2016;49−70.

[19] Payne S. Introduction to RNA viruses. Viruses 2017;97−105.

[20] Venkataraman S, Prasad BVLS, Selvarajan R. RNA dependent RNA polymerases: insights from structure, function and evolution. Viruses 2018;10(2):76.

[21] Rampersad S, Tennant P. Replication and expression strategies of viruses. Viruses 2018;55−82.

[22] Prasad BV, Schmid MF. Principles of virus structural organization. Adv Exp Med Biol 2012;726:17−47.

[23] Lidmar J, Mirny L, Nelson DR. Virus shapes and buckling *transitions in spherical shells*. Phys Rev E 1910;68 (5 Pt 1):05.

[24] Rheinemann L, Sundquist WI. Virus budding. Encycl Virol 2021;519−28.

[25] Louten J. Virus transmission and epidemiology. Essent Hum Virol 2016;71−92.

[26] Jayaweera M, Perera H, Gunawardana B, Manatunge J. Transmission of COVID-19 virus by droplets and aerosols: a critical review on the unresolved dichotomy. Env Res 2020;188:109819.

[27] Rodríguez-Lázaro D, Cook N, Ruggeri FM, Sellwood J, Nasser A, Nascimento MS, et al. Virus hazards from food, water and other contaminated environments. FEMS Microbiol Rev 2012;36(4):786−814.

[28] Riad A, Hockova B. Vertical transmission. Br Dent J 2020;229(2):71.

[29] Tripathi S, Awasthi S, Singh SN, Kumar M. Vertical transmission of COVID-19. Indian J Pediatr 2021;88 (10):1058.

[30] Martinez-Portilla RJ. Vertical transmission of coronavirus disease 2019. Am J Obstet Gynecol 2021;224 (3):328−9.

[31] Lawrence RM, Lawrence RA. Breast milk and infection. Clin Perinatol 2004;31(3):501−628.

[32] Michie CA, Gilmour J. Breast feeding and the risks of viral transmission. Arch Dis Child 2001;84(5):381−582.

[33] Michie CA, Gilmour JW. Breastfeeding and viral transmission: risks, benefits and treatments. J Trop Pediatr 2000;46(5):256−7.

[34] Kutter JS, Spronken MI, Fraaij PL, Fouchier RA, Herfst S. Transmission routes of respiratory viruses among humans. Curr Opin Virol 2018;28:142−51.

[35] Razonable RR, Eid AJ. Viral infections in transplant recipients. Minerva Med 2009;100(6):479−501.

[36] Eastlund T. Infectious disease transmission through cell, tissue, and organ transplantation: reducing the risk through donor selection. Cell Transpl 1995;4(5):455−77.

[37] Irshad M, Joshi YK, Sharma Y, Dhar I. Transfusion transmitted virus: a review on its molecular characteristics and role in medicine. World J Gastroenterol 2006;12(32):5122−34.

[38] Bihl F, Castelli D, Marincola F, Dodd RY, Brander C. Transfusion-transmitted infections. J Transl Med 2007;5:25.

[39] Rosenberg R, Beard CB. Vector-borne infections. Emerg Infect Dis 2011;17(5):769−70.

[40] Huntington MK, Allison J, Nair D. Emerging vector-borne diseases. Am Fam Physician 2016;94(7):551−7.

[41] Shaw WR, Catteruccia F. Vector biology meets disease control: using basic research to fight vector-borne diseases. Nat Microbiol 2019;4(1):20−34.

[42] Hills SL, Russell K, Hennessey M, Williams C, Oster AM, Fischer M, et al. Transmission of Zika Virus through sexual contact with travelers to areas of ongoing transmission—continental United States, 2016. MMWR Morb Mortal Wkly Rep 2016;65(8):215−16.

[43] Burrell CJ, Howard CR, Murphy FA. Epidemiology of viral infections. Fenner White's Med Virology 2017;185−203.

[44] Olaimat AN, Shahbaz HM, Fatima N, Munir S, Holley RA. Food safety during and after the era of COVID-19 pandemic. Front Microbiol 2020;11:1854.

[45] Godoy MG, Kibenge MJT, Kibenge FSB. SARS-CoV-2 transmission via aquatic food animal species or their products: a review. Aquaculture. 2021;536:736460.

[46] La Rosa G, Bonadonna L, Lucentini L, Kenmoe S, Suffredini E. Coronavirus in water environments: occurrence, persistence and concentration methods—a scoping review. Water Res 2020;179:115899.

[47] Bloomfield SF, Aiello AE, Cookson B, O'Boyle C, Larson EL. The effectiveness of hand hygiene procedures in reducing the risks of infections in home and community settings including handwashing and alcohol-based hand sanitizers. Am J Infect Control 2007;35(10):S27−64.

[48] Robotis JF, Boleti H. Viral hepatitis. xPharm: The Comprehensive Pharmacology Reference. 2007;1−10.

[49] Sathyamala C, Mittal O, Dasgupta R, Priya R. Polio eradication initiative in India: deconstructing the GPEI. Int J Health Serv 2005;35(2):361−83.

[50] Adams BB. New strategies for the diagnosis, treatment, and prevention of herpes simplex in contact sports. Curr Sports Med Rep 2004;3(5):277−83.

[51] Brianti P, De Flammineis E, Mercuri SR. Review of HPV-related diseases and cancers. N Microbiol 2017;40 (2):80−5.

[52] Connell CO, Oranje A, Van Gysel D, Silverberg NB. Congenital molluscum contagiosum: report of four cases and review of the literature. Pediatr Dermatol 2008;25(5):553−6.

[53] Sun CB, Wang YY, Liu GH, Liu Z. Role of the eye in transmitting human coronavirus: what we know and what we do not know. Front Public Health 2020;8:155.

[54] Gern JE. Viral respiratory infection and the link to asthma. Pediatr Infect Dis J 2004;23(1 Suppl.):S78−86.

[55] Ciencewicki J, Jaspers I. Air pollution and respiratory viral infection. Inhal Toxicol 2007;19(14):1135−46.

[56] Woodby B, Arnold MM, Valacchi G. SARS-CoV-2 infection, COVID-19 pathogenesis, and exposure to air pollution: what is the connection? Ann N Y Acad Sci 2021;1486(1):15−38.

[57] Zhao C, Fang X, Feng Y, Fang X, He J, Pan H. Emerging role of air pollution and meteorological parameters in COVID-19. J Evid Based Med 2021;14(2):123−38.

[58] Leung NHL. Transmissibility and transmission of respiratory viruses. Nat Rev Microbiol 2021;19(8):528−45.

[59] German Advisory Committee Blood (Arbeitskreis Blut), Subgroup. Assessment of pathogens transmissible by blood'. Human Immunodeficiency Virus (HIV). Transfus Med Hemother 2016;43(3):203−22.

[60] Esser S, Krotzek J, Dirks H, Scherbaum N, Schadendorf D. Sexual risk behavior, sexually transmitted infections, and HIV transmission risks in HIV-positive men who have sex with men (MSM)—approaches for medical prevention. J Dtsch Dermatol Ges 2017;15(4):421−8.

[61] Johnston C, Corey L. Current concepts for genital herpes simplex virus infection: diagnostics and pathogenesis of genital tract shedding. Clin Microbiol Rev 2016;29(1):149−61.

[62] Valenzuela P. Hepatitis A, B, C, D and E viruses: structure of their genomes and general properties. Gastroenterol Jpn 1990;25(Suppl. 2):62−71.

[63] Lanini S, Ustianowski A, Pisapia R, Zumla A, Ippolito G. Viral hepatitis: etiology, epidemiology, transmission, diagnostics, treatment, and prevention. Infect Dis Clin North Am 2019;33(4):1045−62.

[64] Pisano MB, Giadans CG, Flichman DM, Ré VE, Preciado MV, Valva P. Viral hepatitis update: progress and perspectives. World J Gastroenterol 2021;27(26):4018−44.

[65] Bylund J, Toljander J, Simonsson M. Symtomprofiler—bra verktyg för smittspårning vid magsjukeutbrott [Symptom profiles—good tool for tracing stomach flu outbreaks]. Lakartidningen. 2015;112.

[66] Orenstein R. Gastroenteritis viral. Encycl Gastroenterol 2020;652−7.

[67] Parashar UD, Nelson EA, Kang G. Diagnosis, management, and prevention of rotavirus gastroenteritis in children. BMJ 2013;347:f7204.

[68] Robilotti E, Deresinski S, Pinsky BA. Norovirus. Clin Microbiol Rev 2015;28(1):134−64.

[69] de Graaf M, van Beek J, Koopmans MP. Human norovirus transmission and evolution in a changing world. Nat Rev Microbiol 2016;14(7):421−33.

[70] Rajaiya J, Saha A, Ismail AM, Zhou X, Su T, Chodosh J. Adenovirus and the cornea: more than meets the eye. Viruses 2021;13(2):293.

[71] ohnson C, Hargest V, Cortez V, Meliopoulos VA, Schultz-Cherry S. Astrovirus pathogenesis. Viruses 2017;9 (1):22.

[72] Wohlgemuth N, Honce R, Schultz-Cherry S. Astrovirus evolution and emergence. Infect Genet Evol 2019;69:30−7.

[73] Boldescu V, Behnam MAM, Vasilakis N, Klein CD. Broad-spectrum agents for flaviviral infections: dengue, Zika and beyond. Nat Rev Drug Discov 2017;16(8):565−86.

[74] Zhao R, Wang M, Cao J, Shen J, Zhou X, Wang D, et al. Flavivirus: from structure to therapeutics development. Life (Basel) 2021;11(7):615.

[75] Zorrilla CD, Rivera-Viñas JI, De La Vega-Pujols A, Garcia-Garcia I, Rabionet SE, Mosquera AM. The Zika virus infection in pregnancy: review and implications for research and care of women and infants in affected areas. P R Health Sci J 2018;37(Spec Issue):S66−72.

[76] Zorrilla CD, García García I, García Fragoso L, De La Vega A. Zika virus infection in pregnancy: maternal, fetal, and neonatal considerations. J Infect Dis 2017;216(Suppl. 10):S891−6.

[77] Petersen LR, Brault AC, Nasci RS. West Nile virus: review of the literature. JAMA. 2013;310(3):308−3015.

[78] Ligon BL. Emerging and re-emerging infectious diseases: review of general contributing factors and of West Nile virus. Semin Pediatr Infect Dis 2004;15(3):199−205.

[79] Halstead S. Recent advances in understanding dengue. F1000Res. 2019;8:F1000 Faculty Rev-1279

[80] Guzman MG, Harris E. Dengue lancet 2015;385(9966):453−465.

[81] Kausar S, Said Khan F, Ishaq Mujeeb Ur Rehman M, Akram M, Riaz M, Rasool G, et al. A review: mechanism of action of antiviral drugs. Int J Immunopathol Pharmacol 2021;35 20587384211002621.

[82] Menéndez-Arias L, Gago F. Antiviral agents: structural basis of action and rational design. Subcell Biochem 2013;68:599−630.

[83] Yang PL. Antiviral therapeutics. ACS Infect Dis 2021;7(6):1297.

[84] Richman DD, Nathanson N. Antiviral therapy. Viral Pathog 2016;271−87.

[85] Adamson CS. Antiviral agents: discovery to resistance. Viruses 2020;12(4):406.

[86] Adalja A, Inglesby T. Broad-spectrum antiviral agents: a crucial pandemic tool. Expert Rev Anti Infect Ther 2019;17(7):467−70.

[87] De Clercq E. Antiviral drugs in current clinical use. J Clin Virol 2004;30(2):115−33.

[88] De Clercq E. Antiviral drugs: current state of the art. J Clin Virol 2001;22(1):73−89.

[89] De Clercq E. Emerging antiviral drugs. Expert Opin Emerg Drugs 2008;13(3):393−416.

[90] Poole CL, James SH. Antiviral therapies for herpesviruses: current agents and new directions. Clin Ther 2018;40(8):1282−98.

[91] Field HJ, Vere Hodge RA. Recent developments in anti-herpesvirus drugs. Br Med Bull 2013;106:213−49.

[92] Dong H, Wang Z, Zhao D, Leng X, Zhao Y. Antiviral strategies targeting herpesviruses. J Virus Erad 2021;7 (3):100047.

[93] Arts EJ, Hazuda DJ. HIV-1 antiretroviral drug therapy. Cold Spring Harb Perspect Med 2012;2(4):a007161.

[94] Farooq T, Hameed A, Rehman K, Ibrahim M, Qadir MI, Akash MS. Antiretroviral agents: looking for the best possible chemotherapeutic options to conquer HIV. Crit Rev Eukaryot Gene Expr 2016;26 (4):363−81.

[95] Boone LR, Koszalka GW. Antiretroviral drug development for HIV: challenges for the future. Curr Opin Investig Drugs 2010;11(8):863−7.

[96] Saag MS. HIV infection—screening, diagnosis, and treatment. N Engl J Med 2021;384(22):2131−43.

[97] Buti M, Riveiro-Barciela M, Esteban R. Treatment of chronic hepatitis B virus with oral anti-viral therapy. Clin Liver Dis 2021;25(4):725−40.

[98] Okada M, Enomoto M, Kawada N, Nguyen MH. Effects of antiviral therapy in patients with chronic hepatitis B and cirrhosis. Expert Rev Gastroenterol Hepatol 2017;11(12):1095−104.

[99] Ogholikhan S, Schwarz KB. Hepatitis vaccines. Vaccines (Basel) 2016;4(1):6.

[100] Langan RC, Goodbred AJ. Hepatitis A. Am Fam Physician 2021;104(4):368−74.

[101] Migueres M, Lhomme S, Izopet J, Hepatitis A. Epidemiology, high-risk groups, prevention and research on antiviral treatment. Viruses. 2021;13(10):1900.

[102] Hou J, Liu Z, Gu F. Epidemiology and prevention of hepatitis B virus infection. Int J Med Sci 2005;2(1):50−7.

[103] MacLachlan JH, Cowie BC. Hepatitis B virus epidemiology. Cold Spring Harb Perspect Med 2015;5(5): a021410.

[104] Dienstag JL, Schiff ER, Wright TL, Perrillo RP, Hann HW, Goodman Z, et al. Lamivudine as initial treatment for chronic hepatitis B in the United States. N Engl J Med 1999;341(17):1256−63.

[105] Jarvis B, Faulds D. Lamivudine. A review of its therapeutic potential in chronic hepatitis B. Drugs 1999;58 (1):101−41.

[106] Tang LSY, Covert E, Wilson E, Kottilil S. Chronic hepatitis B infection: a review. JAMA 2018;319 (17):1802−13.

[107] Brunetto MR, Bonino F. Interferon therapy of chronic hepatitis B. Intervirology 2014;57(3−4):163−70.

[108] Ye J, Chen J. Interferon and hepatitis B: current and future perspectives. Front Immunol 2021;12:733364.

[109] Wang XY, Chen HS. Emerging antivirals for the treatment of hepatitis B. World J Gastroenterol 2014;20 (24):7707−17.

[110] Akbar SM, Al-Mahtab M, Khan MS, Raihan R, Shrestha A. Immune therapy for hepatitis B. Ann Transl Med 2016;4(18):335.

[111] Gupta E, Bajpai M, Choudhary A. Hepatitis C virus: screening, diagnosis, and interpretation of laboratory assays. Asian J Transfus Sci 2014;8(1):19−25.

[112] Khullar V, Firpi RJ. Hepatitis C cirrhosis: new perspectives for diagnosis and treatment. World J Hepatol 2015;7(14):1843−5285.

[113] Trinchet JC, Ganne-Carrié N, Nahon P, N'kontchou G, Beaugrand M. Hepatocellular carcinoma in patients with hepatitis C virus-related chronic liver disease. World J Gastroenterol 2007;13(17):2455−60.

[114] Pleschka S. Overview of influenza viruses. Curr Top Microbiol Immunol 2013;370:1−20.

[115] Kalil AC, Thomas PG. Influenza virus-related critical illness: pathophysiology and epidemiology. Crit Care 2019;23(1):258.

[116] Taubenberger JK, Morens DM. Influenza: the mother of all pandemics. Emerg Infect Dis 2006 1918;12:15−22.

[117] Menon IG. The 1957 pandemic of influenza in India. Bull World Health Organ 1959;20(2−3):199−224.

[118] Viboud C, Grais RF, Lafont BA, Miller MA, Simonsen L, Multinational Influenza Seasonal Mortality Study Group. Multinational impact of the 1968 Hong Kong influenza pandemic: evidence for a smoldering pandemic. J Infect Dis 2005;192(2):233−48.

[119] Nellore A, Fishman J. Pandemic Swine flu 2009. Xenotransplantation 2009;16(6):463−5.

[120] Gaitonde DY, Moore FC, Morgan MK. Influenza: diagnosis and treatment. Am Fam Physician 2019;100 (12):751−8.

[121] Świerczyńska M, Mirowska-Guzel DM, Pindelska E. Antiviral drugs in influenza. Int J Env Res Public Health 2022;19(5):3018.

[122] Terrier O, Slama-Schwok A. Anti-influenza drug discovery and development: targeting the virus and its host by all possible means. Adv Exp Med Biol 2021;1322:195–218.

[123] Shie JJ, Fang JM. Development of effective anti-influenza drugs: congeners and conjugates—a review. J Biomed Sci 2019;26(1):84.

[124] Zhang ZJ, Morris-Natschke SL, Cheng YY, Lee KH, Li RT. Development of anti-influenza agents from natural products. Med Res Rev 2020;40(6):2290–338.

[125] Wang X, Jia W, Zhao A, Wang X. Anti-influenza agents from plants and traditional Chinese medicine. Phytother Res 2006;20(5):335–41.

[126] Adhikari B, Marasini BP, Rayamajhee B, Bhattarai BR, Lamichhane G, Khadayat K, et al. Potential roles of medicinal plants for the treatment of viral diseases focusing on COVID-19: a review. Phytother Res 2021;35 (3):1298–312.

[127] Wang L, Yang R, Yuan B, Liu Y, Liu C. The antiviral and antimicrobial activities of licorice, a widely-used Chinese herb. Acta Pharm Sin B 2015;5(4):310–15.

[128] Pleschka S, Stein M, Schoop R, Hudson JB. Anti-viral properties and mode of action of standardized Echinacea purpurea extract against highly pathogenic avian influenza virus (H5N1, H7N7) and swine-origin H1N1 (S-OIV). Virol J 2009;6:197.

[129] Turan K, Nagata K, Kuru A. Antiviral effect of Sanicula europaea L. leaves extract on influenza virus-infected cells. Biochem Biophys Res Commun 1996;225(1):22–36.

[130] Sabir MJ, Al-Saud NBS, Hassan SM. Dengue and human health: a global scenario of its occurrence, diagnosis and therapeutics. Saudi J Biol Sci 2021;28(9):5074–80.

[131] Gubler DJ, Vasilakis N, Musso D. History and emergence of Zika virus. J Infect Dis 2017;216(Suppl. 10): S860–7.

[132] Hosseini S, Muñoz-Soto RB, Oliva-Ramírez J, Vázquez-Villegas P, Aghamohammadi N, Rodriguez-Garcia A, et al. Latest updates in dengue fever therapeutics: natural, marine and synthetic drugs. Curr Med Chem 2020;27(5):719–44.

[133] Nguyen NM, Tran CN, Phung LK, Duong KT, Huynh Hle A, Farrar J, et al. A randomized, double-blind placebo controlled trial of balapiravir, a polymerase inhibitor, in adult dengue patients. J Infect Dis 2013;207 (9):1442–50.

[134] Low JS, Wu KX, Chen KC, Ng MM, Chu JJ. Narasin, a novel antiviral compound that blocks dengue virus protein expression. Antivir Ther 2011;16(8):1203–18.

[135] Silveira D, Prieto-Garcia JM, Boylan F, Estrada O, Fonseca-Bazzo YM, Jamal CM, et al. COVID-19: is there evidence for the use of herbal medicines as adjuvant symptomatic therapy? Front Pharmacol 2020;11:581840.

[136] Chung CY, Liu CH, Burnouf T, Wang GH, Chang SP, Jassey A, et al. Activity-based and fraction-guided analysis of Phyllanthus urinaria identifies loliolide as a potent inhibitor of hepatitis C virus entry. Antivir Res 2016;130:58–68.

[137] Dong W, Farooqui A, Leon AJ, Kelvin DJ. Inhibition of influenza A virus infection by ginsenosides. PLoS One 2017;12(2):e0171936.

[138] Lani R, Hassandarvish P, Shu MH, Phoon WH, Chu JJ, Higgs S, et al. Antiviral activity of selected flavonoids against Chikungunya virus. Antivir Res 2016;133:50–61.

[139] Balkrishna A, Pokhrel S, Singh H, Joshi M, Mulay VP, Haldar S, et al. Withanone from Withania somnifera attenuates SARS-CoV-2 RBD and host ACE2 interactions to rescue spike protein induced pathologies in humanized zebrafish model. Drug Des Devel Ther 2021;15:1111–33.

[140] Imanishi N, Tuji Y, Katada Y, Maruhashi M, Konosu S, Mantani N, et al. Additional inhibitory effect of tea extract on the growth of influenza A and B viruses in MDCK cells. Microbiol Immunol 2002;46 (7):491–4.

[141] Esposito M, Nothias LF, Nedev H, Gallard JF, Leyssen P, Retailleau P, et al. Euphorbia dendroides Latex as a source of jatrophane esters: isolation, structural analysis, conformational study, and anti-CHIKV activity. J Nat Prod 2016;79(11):2873–82.

[142] Tiwari V, Darmani NA, Yue BY, Shukla D. In vitro antiviral activity of neem (Azardirachta indica L.) bark extract against herpes simplex virus type-1 infection. Phytother Res 2010;24(8):1132–40.

[143] Lee CD, Ott M, Thyagarajan SP, Shafritz DA, Burk RD, Gupta S. Phyllanthus amarus down-regulates hepatitis B virus mRNA transcription and replication. Eur J Clin Invest 1996;26(12):1069–76.

[144] Xu H, He L, Chen J, Hou X, Fan F, Wu H, et al. Different types of effective fractions from Radix Isatidis revealed a multiple-target synergy effect against respiratory syncytial virus through RIG-I and MDA5 signaling pathways, a pilot study to testify the theory of superposition of traditional Chinese Medicine efficacy. J Ethnopharmacol 2019;239:111901.

[145] Yarmolinsky L, Huleihel M, Zaccai M, Ben-Shabat S. Potent antiviral flavone glycosides from *Ficus benjamina* leaves. Fitoterapia 2012;83(2):362–7.

[146] Cataneo AHD, Kuczera D, Koishi AC, Zanluca C, Silveira GF, Arruda TB, et al. The citrus flavonoid naringenin impairs the *in vitro* infection of human cells by Zika virus. Sci Rep 2019;9(1):16348.

[147] Chumakov K, Ehrenfeld E, Wimmer E, Agol VI. Vaccination against polio should not be stopped. Nat Rev Microbiol 2007;5(12):952–8.

[148] Belongia EA, Naleway AL. Smallpox vaccine: the good, the bad, and the ugly. Clin Med Res 2003;1(2):87–92.

[149] Greenwood B. The contribution of vaccination to global health: past, present and future. Philos Trans R Soc Lond B Biol Sci 2014;369(1645):20130433.

[150] Ellebedy AH, Ahmed R. Antiviral vaccines: challenges and advances. Vaccine Book 2016;283–310.

[151] Pollard AJ, Bijker EM. A guide to vaccinology: from basic principles to new developments. Nat Rev Immunol 2021;21(2):83–100.

[152] Stauffer F, El-Bacha T, Da, Poian AT. Advances in the development of inactivated virus vaccines. Recent Pat Antiinfect Drug Discov 2006;1(3):291–6.

[153] Tripathi NK, Shrivastava A. Recent developments in bioprocessing of recombinant proteins: expression hosts and process development. Front Bioeng Biotechnol 2019;7:420.

[154] Gomez PL, Robinson JM. Vaccine manufacturing. Plotkin's Vaccines 2018;51–60 e1.

[155] Syomin BV, Ilyin YV. Virus-like particles as an instrument of vaccine production. Mol Biol 2019;53 (3):323–34.

[156] Roldão A, Silva AC, Mellado MCM, Alves PM, Carrondo MJT. Viruses and virus-like particles in biotechnology: fundamentals and applications. Compr Biotechnol 2017;633–56.

[157] Graham BS. Advances in antiviral vaccine development. Immunol Rev 2013;255(1):230–42.

[158] Har-Noy M, Or R. Allo-priming as a universal anti-viral vaccine: protecting elderly from current COVID-19 and any future unknown viral outbreak. J Transl Med 2020;18(1):196.

[159] Hegde NR, Rao PP, Bayry J, Kaveri SV. Immunotherapy of viral infections. Immunotherapy 2009;1 (4):691–711.

[160] Felsenstein S, Herbert JA, McNamara PS, Hedrich CM. COVID-19: immunology and treatment options. Clin Immunol 2020;215:108448.

[161] Jeyanathan M, Afkhami S, Smaill F, Miller MS, Lichty BD, Xing Z. Immunological considerations for COVID-19 vaccine strategies. Nat Rev Immunol 2020;20(10):615–32.

[162] Michaeli D. Vaccines and monoclonal antibodies. Semin Oncol 2005;32(6 Suppl 9):S82–6.

[163] Quinteros DA, Bermúdez JM, Ravetti S, Cid A, Allemandi DA, Palma SD. Therapeutic use of monoclonal antibodies: general aspects and challenges for drug delivery. Nanostruct Drug Deliv 2017;807–33.

[164] Lu RM, Hwang YC, Liu IJ, Lee CC, Tsai HZ, Li HJ, et al. Development of therapeutic antibodies for the treatment of diseases. J Biomed Sci 2020;27(1):1.

[165] Srivastava S, Riddell SR. Chimeric antigen receptor T cell therapy: challenges to bench-to-bedside efficacy. J Immunol 2018;200(2):459–68.

[166] Costagliola G, Nuzzi G, Spada E, Comberiati P, Verduci E, Peroni DG. Nutraceuticals in viral infections: an overview of the immunomodulating properties. Nutrients 2021;13(7):2410.

[167] Parisi GF, Carota G, Castruccio Castracani C, Spampinato M, Manti S, Papale M, et al. Nutraceuticals in the prevention of viral infections, including COVID-19, among the pediatric population: a review of the literature. Int J Mol Sci 2021;22(5):2465.

[168] Ashfaq UA, Masoud MS, Nawaz Z, Riazuddin S. Glycyrrhizin as antiviral agent against Hepatitis C virus. J Transl Med 2011;9:112.

[169] Rouf R, Uddin SJ, Sarker DK, Islam MT, Ali ES, Shilpi JA, et al. Antiviral potential of garlic (*Allium sativum*) and its organosulfur compounds: a systematic update of pre-clinical and clinical data. Trends Food Sci Technol 2020;104:219–34.

[170] Ahmad S, Zahiruddin S, Parveen B, Basist P, Parveen A, Gaurav, et al. Indian Medicinal plants and formulations and their potential against COVID-19-preclinical and clinical research. Front Pharmacol 2021;11:578970.

[171] Ratha SK, Renuka N, Rawat I, Bux F. Prospective options of algae-derived nutraceuticals as supplements to combat COVID-19 and human coronavirus diseases. Nutrition 2021;83:111089.

[172] Karkos PD, Leong SC, Karkos CD, Sivaji N, Assimakopoulos DA. Spirulina in clinical practice: evidence-based human applications. Evid Based Complement Altern Med 2011;2011:531053.

[173] Otto SP, Day T, Arino J, Colijn C, Dushoff J, Li M, et al. The origins and potential future of SARS-CoV-2 variants of concern in the evolving COVID-19 pandemic. Curr Biol 2021;31(14):R918–29.

[174] Cojocaru FD, Botezat D, Gardikiotis I, Uritu CM, Dodi G, Trandafir L, et al. Nanomaterials designed for antiviral drug delivery transport across biological barriers. Pharmaceutics 2020;12(2):171.

[175] Kaur U, Chakrabarti SS, Ojha B, Pathak BK, Singh A, Saso L, et al. Targeting host cell proteases to prevent SARS-CoV-2 invasion. Curr Drug Targets 2021;22(2):192–201.

[176] Asha K, Sharma-Walia N. Targeting host cellular factors as a strategy of therapeutic intervention for herpesvirus infections. Front Cell Infect Microbiol 2021;11:603309.

[177] Escobedo-Bonilla CM. Mini review: virus interference: history, types and occurrence in crustaceans. Front Immunol 2021;12:674216.

[178] Melby T, Westby M. Inhibitors of viral entry. Handb Exp Pharmacol 2009;189:177–202.

[179] Magden J, Kääriäinen L, Ahola T. Inhibitors of virus replication: recent developments and prospects. Appl Microbiol Biotechnol 2005;66(6):612–21.

[180] Tsai CH, Lee PY, Stollar V, Li ML. Antiviral therapy targeting viral polymerase. Curr Pharm Des 2006;12(11):1339–55.

[181] Zhou Y, Vedantham P, Lu K, Agudelo J, Carrion Jr R, Nunneley JW, et al. Protease inhibitors targeting coronavirus and filovirus entry. Antivir Res 2015;116:76–84.

[182] Chang KO, Kim Y, Lovell S, Rathnayake AD, Groutas WC. Antiviral drug discovery: norovirus proteases and development of inhibitors. Viruses 2019;11(2):197.

[183] Salazar G, Zhang N, Fu TM, An Z. Antibody therapies for the prevention and treatment of viral infections. NPJ Vaccines 2017;2:19.

Further reading

Aktaş O, Aydin H, Timurkan MO. A molecular study on the prevalence and coinfections of rotavirus, norovirus, astrovirus and adenovirus in children with gastroenteritis. Minerva Pediatr 2019;71(5):431–7.

Grebely J, Prins M, Hellard M, Cox AL, Osburn WO, Lauer G, et al. International Collaboration of Incident HIV and Hepatitis C in Injecting Cohorts (InC3). Hepatitis C virus clearance, reinfection, and persistence, with insights from studies of injecting drug users: towards a vaccine. Lancet Infect Dis 2012;12(5):408–14.

Kennedy PGE, Gershon AA. Clinical features of varicella-zoster virus infection. Viruses 2018;10(11):609.

Mesters et al., 2006 Mesters JR, Tan J, Hilgenfeld R. Viral enzymes. Curr Opin Struct Biol 2006;16(6):776–86.

Xiao et al., 2021 Xiao T, Cai Y, Chen B. HIV-1 entry and membrane fusion inhibitors. Viruses 2021;13(5):735.

Viral infections and transmission

Emerging viral diseases

Agniva Majumdar and Pradip Kumar Jana

Virus Research and Diagnostic Laboratory, ICMR-National Institute of Cholera and Enteric Diseases, Kolkata, West Bengal, India

The discovery of the first human virus, the Yellow Fever virus, in 1901 by Walter Reed to the COVID-19 pandemic of today caused by SARS-CoV-2 has proved that the emergence and re-emergence of viral diseases undoubtedly play a major role in shaping the human world. Not only can they cause high mortality as they spread but also have a huge impact on the social and economic life in today's interconnected world. Despite widespread advances in medical science, emerging and re-emerging viral diseases continue to have devastating consequences for human populations across the globe. Unfortunately, most of these viral diseases do not yet have any cure.

An emerging infectious disease is caused by a new pathogen that has appeared and affected a population for the first time, or has existed previously but is rapidly increasing, either in terms of the number of new cases within a population, or its spread to new geographical areas [1]. They are responsible for causing major public health problems either locally or globally. It is seldom possible to know if a disease is novel in humans, or whether it has been present but unrecognized by the scientific community. But many emerging diseases are thought to be due to an increase in the rate of close contact between humans and the pathogen reservoirs in nature, which facilitates the successful "jump" of the agent from the animal or arthropod to human crossing the interspecies barrier. A re-emerging infectious disease is caused due to the reappearance and increase in incidence or spread in a geographic range of infections, which are known but had formerly fallen to levels so low that they were no longer considered a public health problem [2]. In addition to these, there are emerging human-generated diseases released intentionally, for example, bioterrorism, or accidentally, such as vaccine-derived polioviruses which emerge due to spontaneous back-mutations of live virus vaccines, as well as a live human-engineered vaccine like vaccinia that escaped to cause a new epizootic disease [3]. Fig. 2.1 illustrates the newly emerging, re-emerging, and "deliberately emerging" infectious disease from 1981 to 2020.

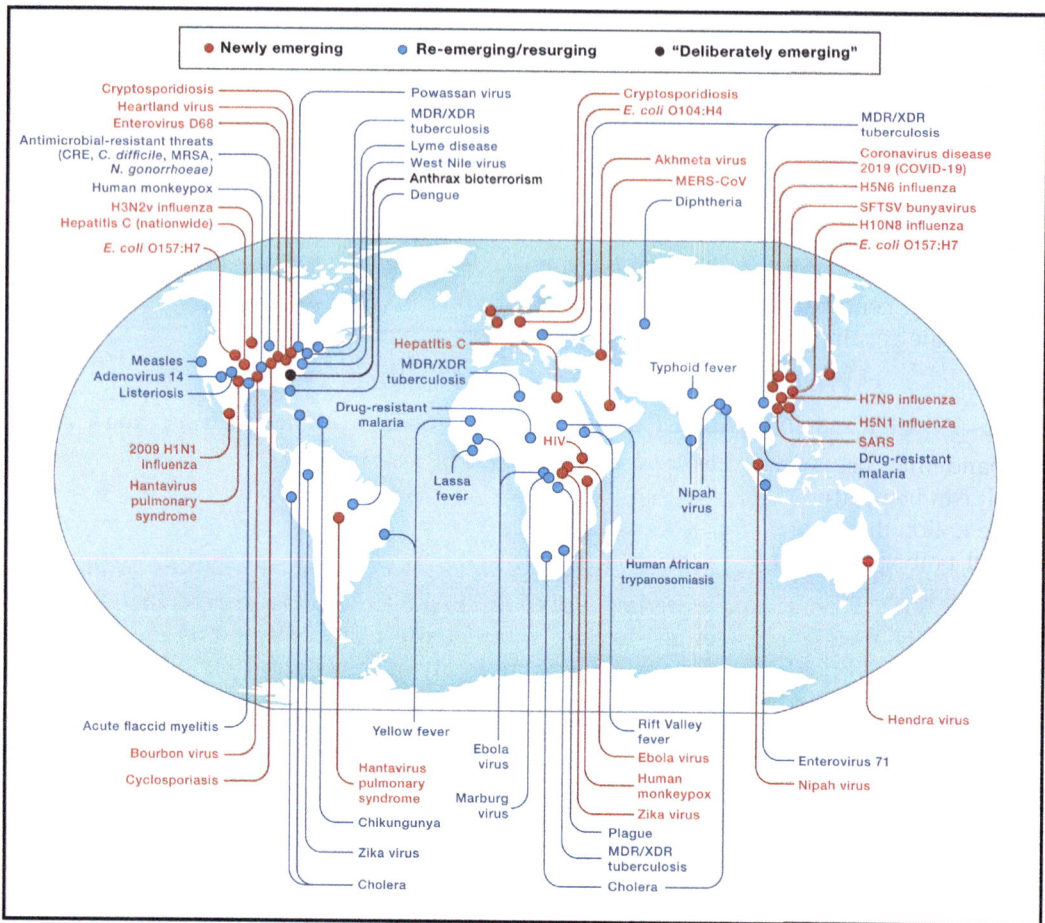

FIGURE 2.1   Emerging infectious diseases (1981 to 2020). Source: Morens DM, Fauci AS. Emerging Pandemic Diseases: How We Got to COVID-19. Cell. 2020 Sep;182(5):1077−92.

The emerging and re-emerging infectious agents are disproportionately viruses. A survey reported 177 species as emerging or re-emerging out of the total 1407 human pathogens, wherein 37% are viruses and prions [4]. RNA viruses form the majority [5]. They are usually zoonotic in origin [1,5], that is, an infectious disease that has jumped from a nonhuman animal to humans. In 2007, WHO warned that infectious diseases are emerging and re-emerging at a rate that has not been seen before [6]. The potential for infectious diseases to spread rapidly results in high morbidity and mortality, causing a potential global public health treat of major concern. In the modern world of globalization and rapid transportation, it is very important that a disease outbreak is quickly identified, and control measures rapidly initiated before it can become an epidemic or pandemic.

2.1 The everchanging landscape of infectious diseases

Emerging and re-emerging infectious diseases have been threatening humans since the agricultural revolution took place some 12,000 years ago when the nomadic hunter-gatherers settled down to cultivate crops and domesticated animals [3]. The history of human civilization is fraught with events like smallpox, measles, bubonic plague, and malaria that killed thousands. The beginning of the last century witnessed the Spanish flu pandemic (1918) that killed more than 50 million people [7]. The HIV pandemic has so far killed 36.3 million since 1981 with an estimated 37.7 million living with the virus globally [8]. The 21st century is witnessing an explosion of events in pandemic proportions: the severe acute respiratory syndrome (SARS) in 2002−03, the H1N1 "swine flu" in 2009, the Middle East respiratory syndrome (MERS) in 2012, chikungunya (2014), Zika (2015), and Ebola fever in various regions of Africa since 2014 [3]. The present COVID-19 pandemic, ongoing since the end of 2019, is the latest example of an unexpected, novel, and devastating pandemic disease with the potential of halting the whole world to a complete standstill. It can be rightly concluded that we are living in a pandemic era [3,9].

The notion that a new disease has emerged does not always reflect the actual situation that a novel pathogen has emerged in circulation. The identification of a novel virus due to the advancement of diagnostic methods often create a perception of an emerging virus. Improved and cost-effective screening tools may also create an impression of increased prevalence of a disease though it may be a matter of fact that the pathogen is already endemic in the community. Notably, human papilloma virus, rotavirus, hepatitis B, and hepatitis C gained widespread recognition due to newer diagnostics [5]. Changes in human behavior is often responsible for changing the epidemiology of an endemic viral disease. Advancements in blood transfusion medicine, dialysis techniques, and parenteral therapy caused outbreaks of hepatitis B which was already endemic. Rigorous screening procedures at blood banks and the use of disposable materials have largely controlled this route of transmission. But there is the simultaneous rise of other blood-borne infections including HIV and hepatitis C due to changing trends in sexual practice and intravenous drug use [10].

2.2 Causes of emergence

Though difficult to predict, it is now clear that the emergence of infectious diseases is caused due to interplay of numerous factors often acting simultaneously causing changes in the dynamics of the host, pathogen, and environment [11]. Some factors increase exposure to pathogens, which are currently obscure; others cause widespread dissemination of localized infections; while others cause changes in characters of virus or host response. Some of the important factors have been listed in Box 2.1 [3,5,12].

Changes in climatic conditions, especially temperature and rainfall, have a profound effect on vector breeding and can result in disproportionate increase in vector-borne diseases [5]. A rise in global temperature causes expansion of the distribution of mosquitoes and in turn the diseases caused by them [13]. Predictive modeling has estimated that global temperature increases of 2°C−3°C would increase the number of people at risk of

BOX 2.1   Factors affecting emergence of viral diseases.

A. Host

1. Cell tropism, alternative, and coreceptors
2. Antibody-dependent enhancement
3. Genetic/inherent susceptibility
4. Nutritional state
5. Herd immunity
6. Immunosuppression, blood transfusions, organ, and tissue transplantation
7. Demographics and behavior - intravenous drug use, changes in sexual activities, or numbers of partners, international air travel

B. Agent

1. Genetic diversity
2. Mutation and selection
3. Genetic evolution - zoonosis or recombination/reassortment

C. Environment

1. Changes in climate and weather
2. Introduction of new medical interventions
3. War, poverty, population growth, overcrowding, inadequate sanitation
4. Flood, drought, famine, natural disasters
5. Deforestation, damming, changes in wetlands
6. Outdoor recreation, explorations into new environments
7. Vector density and exposure
8. Animal exposure, smuggling, global trade in wildlife
9. Occupational exposure
10. Lack of public health infrastructure or effective health programs
11. Lack of political will

malaria by around 3%−5%, which accounts for several hundred millions. Further, the seasonal duration would increase in many currently endemic areas [14].

Changes in land use like urbanization, deforestation, water projects, and agricultural practices can bring new or re-emerging diseases. Rapid deforestation and construction of dams can cause a drastic change in the balance of an ecology resulting in thereby increasing rodent and insect population [15]. Subsequent urbanization brings the human population into direct contact with these potential carriers of the diseases. The huge influx of migrants and laborers to developing cities creates unplanned congested urban slums with poor living conditions which frequently become breeding grounds for diseases. In developing nations poor public health infrastructure makes a large number of people suffer from malnutrition resulting in disease outbreak of diseases such as cholera and diphtheria [2].

Changes in human demographics, like the migration of population from agrarian to crowded urban settings, have brought new diseases like malaria that can be easily transmitted to a susceptible population. In the past, people rarely traveled great distances from their place of residence. With the advent of rapid and mass transportation in modern times especially flights, people can travel from one part of the world to another in a matter of hours thus contributing a major role in the spread of emerging viral diseases. The spread of SARS coronavirus in 2003 from China to more than 17 countries within a week gives an estimate of the power of the spread of virus through air travel [16]. The exponential spread of the emerging virus with each segment of the journey that the index case travels makes containment efforts futile which has again become evident during the current COVID-19 pandemic. Ground transport also plays a vital role in transmission due to the sheer volume of population availing them and the lack effective air filtration methods [5]. Animals and vectors also travel alongside humans contributing to the spread. The introduction of a vector containing the virus from an endemic zone to a location with similar climatic and ecological conditions harboring the same vector can cause rapid proliferation of the virus in the new geography resulting in outbreaks.

Mass gatherings due to religious festivals, sports, and socio-cultural or political events can also play a role in emerging infectious diseases. A large number of people and domestic animals congregating at a site not only puts strain on the resources of the community, city, or nation hosting that particular event but also gives rise to unhealthy living conditions and practices, especially in developing countries. These large-scale gatherings may provide conducive grounds for the exchange of genetic materials emergence of novel pathogens, including viruses [17].

Evidence of zoonoses dates back to prehistoric times. But with urbanization and population explosion following the industrial revolution, humans and animals came in close proximity providing a fertile ground for interspecies transmission of diseases. It is not a coincidence that zoonotic transmission played a pivotal role in the epidemics and pandemics in the last century. Bats, rodents, and domestic livestock like pigs and poultry are frequently involved in not only zoonoses but also frequent reassortments/recombination of wild strains of the virus resulting in new variants [1]. Wild animal populations, companion and captive animals, and even exotic pets play a role [5]. The interspecies transmission highlights the need for interdisciplinary study of the biosphere as a whole to understand human infectious diseases and give due recognition to the central role played by the microbial reservoirs, including those in animals and vectors, and the environment [18].

2.3 Ebola virus

Ebola virus causes life-threatening disease with recurring outbreaks mostly in the African continent. The word "Ebola" originated from the name of a river in Yambuku village in the Democratic Republic of Congo where the first outbreak happened in 1976 with a case fatality rate of 88%. At the same time, another outbreak was reported from Nzara, Sudan where 280 patients died out of 318 infected [19]. In July 2014, the World Health Organization (WHO) declared the disease a Public Health Emergency of International Concern for the first time. Out of the six Ebola virus strains recognized Zaire, Sudan, Taï Forest, and Bundibugyo viruses cause disease in humans, whereas the Reston virus causes disease in nonhuman primates and pigs [20]. Recently Bombali virus has been identified in bats [21]. Ebola virus causes severe viral hemorrhagic fevers affecting humans and bush animals with a high fatality rate (50%−90%) [19]. The potentiality of this disease becoming a pandemic is very low due to the high case-fatality rate. The life cycles of this zoonotic virus are not well known; fruit bats are considered to be the natural hosts. Primarily transmitted by human-to-human direct contact with blood and/or body fluids, the disease starts with a flu-like stage following an incubation period of average of 2−21 days and the most common symptoms are fever, headaches, joint, muscle, and abdominal pain [22]. A symptomatic treatment approach is being used, as there are no approved definitive treatment protocols or vaccines available currently. The only way of managing the Ebola virus outbreak is through a multi-disciplinary approach involving early diagnosis and clinical management of cases, prompt isolation, containment facilities, safe burial practices, and engagement of the community.

2.4 Dengue virus

Dengue fever is also known as "Break-Bone Fever." The virus is transmitted by female mosquitoes mainly of the species *Aedes aegypti* and, to a lesser extent *Aedes albopictus*. Clinically compatible dengue fever was first recorded in a Chinese medical "encyclopaedia of disease symptoms and remedies," in AD 265 to 420 [23]. The clinical dengue-like illness was reported in 1780 from Chennai in India, whereas the first virologically proven epidemic was recorded during 1953–54 in the Philippines [24]. In 1943, dengue virus was isolated first time in Japan; one year later it was isolated in Kolkata from serum samples of the US soldiers [25]. Though mostly prevalent in the tropicals, the disease is now endemic in more than 100 countries in the WHO regions of Africa, the Americas, the Eastern Mediterranean, South-East Asia, and the Western Pacific with an annual incidence of approximately 390 million cases and leading to almost 25,000 deaths per year. As per the WHO recent report the disease has increased over eightfold in the last two decades, from 0.5 million cases in 2000 to over 2.4 million in 2010, and 5.2 million in 2019 [26]. The dengue virus has four closely related serotypes (DENV 1 to 4) and all the serotypes have emerged from sylvatic strains in the forests of South-East Asia [23].

2.5 Chikungunya virus

The name chikungunya means "to become contorted" or "that which bends up" [27]. Chikungunya virus is transmitted mostly by various members of the mosquito genus *Aedes*. The virus was isolated for the first time in East Africa in 1952–53 during an epidemic of nonfatal dengue-like illness that developed along the border between Tanzania and Mozambique [25]. Mutation in the envelope protein increases the efficiency of infecting different mosquito species and this genetic change may be the reason for the high incidence of the disease. Earlier the disease was restricted to countries adjacent to the Indian Ocean; later it circulated in East Africa, many islands of the Indian Ocean, and the Indian subcontinent from the beginning of 2004 [28]. The virus has an incubation period of 4 to 7 days and typically consists of an acute illness with fever, skin rash, fatigue, myalgia, incapacitating arthralgia, and in some rare cases, the mild hemorrhagic syndrome also has been reported. These symptoms resolve automatically in most cases, though some patients develop joint pain that can persist for years leading to a bent or stooping posture.

2.6 West Nile virus

The West Nile virus was first isolated from the West Nile subregion of Uganda in 1937 [29]. The virus was restricted predominantly in the Old World in Africa and the Middle

East until it was introduced into North America. By late July 1999, crows in New York City suddenly started dying in large numbers. A similar incident in birds at Bronx Zoo and reports of encephalitis cases in the city prompted the health officials to investigate the cause which was initially confirmed to be St. Louis encephalitis. Since this virus does not affect the avian population, a reinvestigation was carried out from samples of humans, birds, and mosquitoes by Centers for Disease Control and Prevention and results confirmed the cause as West Nile virus [2]. After that the virus spread rapidly throughout large sections of North America and subsequently into the Caribbean, Central, and South America [30]. Human transmission is mainly through a mosquito bite. There have been reports of cases of transmission associated with blood transfusion, transplantation, breast milk, and transplacental exposure. Only 20%−40% of infected patients suffer from symptoms which include fever, fatigue, headache, malaise, myalgia, lymphadenopathy, vomiting, diarrhea, and skin rash, and the rest 60%−80% remain asymptomatic. However, less than 1% of cases develop severe neurological diseases which manifest as either meningitis or encephalitis.

2.7 Zika virus

The word "Zika" came from a name of a forest in Uganda where the virus was first isolated in 1947 by the Rockefeller Foundation while investigating the ecology of yellow fever [31]. Outside of Africa, the virus was isolated for the first time in Malaysia from mosquitoes in 1966 and human infections were found in 1977 in Central Java, Indonesia. The first known Zika epidemic was reported from the isolated islands of Yap, Federated States of Micronesia in April-May 2007 [32]. The virus is transmitted by *Aedes* (*Stegomyia* subgenus) mosquitoes mainly *Aedes aegypti*. On February 1, 2016, the WHO declared Zika virus disease a Public Health Emergency of international concern due to a large number of outbreaks and related neurological complications across the globe. As of December 2021, the WHO reported 89 countries or territories with current or previous Zika virus transmission [33]. Three genotypes (West African/Nigerian cluster, East African/Mr766 prototype cluster, and Asian cluster) are widely seen [34]. The symptoms are similar to other arbovirus infections and last for 2−7 days. The virus is also known as a cause of Guillain-Barré Syndrome, as well as adverse pregnancy outcomes leading to the most severe form congenital Zika syndrome [34]. There is no specific treatment or vaccine available presently beyond supportive care.

2.8 Yellow fever virus

The term "yellow" in the name signifies jaundice affecting some patients affected by the yellow fever virus [35]. It causes hemorrhagic fever and represents a major threat in low-resource countries. The first yellow fever epidemic was reported in Yucatan in 1648 [36]. It is currently endemic in parts of equatorial Africa and South America, with the previous circulation in parts of North America. The disease is transmitted by *Aedes* and *Haemogogus* mosquitoes. The infection is preceded by an incubation period of 3−6 days and consists of flu-like symptoms in most cases to severe hemorrhage and liver disease. These symptoms resolve automatically, though some patients can become exhibit more toxicity within

24 hours of recovering from initial symptoms. It has a case fatality rate of 20%−50% [27]. No specific treatment exists. Severe cases require aggressive supportive care and hydration. The transmission of yellow fever has declined due to large-scale immunization drives and integrated vector management methods.

2.9 Nipah virus

The zoonotic virus was named Nipah after a river in the village of Sungai Nipah following an outbreak of the disease in pigs and people in Malaysia in April 1999 [37]. The virus can infect new host species because of their very short generation time and faster evolution. The virus has a high fatality rate (40%−75%). Symptoms typically appear after an average incubation period of 4−14 days and generally start with fever, headache, dizziness, cough, sore throat, vomiting, and difficulty breathing [38]. The disease can progress into an encephalitic syndrome, where symptoms like drowsiness, disorientation, and mental confusion can be witnessed, which lead to coma within 24−48 hours. The virus is transmitted through close contact with infected animals or people infected with Nipah virus or their body fluids. Consuming food products of infected animals is another route of transmission. Due to non-specific symptoms of the disease, early diagnosis is quite challenging. Although the virus is included in the WHO research and development blueprint, still now no definitive treatment or vaccine is available. Intensive supportive care for infected people and standard infection control practices are recommended to prevent viral transmission.

2.10 Influenza virus

The influenza virus is notorious for frequently causing unpredictable and recurring pandemics. Based on antigen differences in matrix protein and nucleoprotein, there are four different types of influenza viruses that can infect different vertebrates, that is, influenza A, B, C, and D. Influenza A is the most virulent virus with severe clinical manifestations and often causes pandemic as this group of viruses are more susceptible to antigenic variations [39]. Influenza A pandemics have occurred earlier—Spanish flu in 1918, Asian flu in 1957, Hong Kong flu in 1968, and Russian flu in 1977 [40]. In April 2009, an outbreak of influenza-like illness was reported from Veracruz, Mexico, later identified as a novel strain of influenza A H1N1 (pandemic H1N1/09) infecting healthy young adults and causing deaths in Mexico. Within months it took the shape of a pandemic spreading globally and infecting millions [41]. Prompted by the humanitarian imperative, the international community such as WHO, International Federation of Red Cross and Red Crescent Societies, UN System Influenza Coordination, Office for the Coordination of Humanitarian Affairs, and United Nations Children's Fund came together to work with partnerships like the Global Alliance for Vaccines and Immunisation, Red Cross and Red Crescent Societies, NGOs and civil societies to support governments and communities to reduce the impact from the pandemic H1N1/09. As of 30 May 2010, more than 214 countries and overseas territories had reported laboratory-confirmed cases, including over 18,138 deaths [42]. By the middle of 2010, the curve began to flatten, and the virus was evolving into a pattern similar to that of seasonal influenza prompting the declaration of the official end to the pandemic on August 10, 2010. However, pandemic influenza H1N1/09 virus continues to circulate as a seasonal flu virus, and can lead to illness, hospitalization, and deaths world-wide every year. The WHO Global Influenza Surveillance and Response System (GISRS) currently serves as a global alert mechanism and risk assessment tool for the emergence of novel influenza viruses with pandemic potential through monitoring the evolution of the virus around the world and providing recommendations in areas of laboratory diagnostics, vaccines, and antiviral susceptibility.

2.11 Corona viruses

Coronaviruses belong to a diverse group of RNA viruses infecting different animals, including humans. Endemic coronaviruses, which are the usual causes of the common cold, circulate globally proving the fact that they must have emerged and spread pandemically in the era before their recognition [3]. The rise of two highly pathogenic coronaviruses, the SARS coronavirus in 2002 and the MERS coronavirus in 2012 causing fatal respiratory illness in humans, made emerging coronaviruses a new public health concern of the 21st century. In late 2002, SARS started in Guangdong province of China and within a very short period, it spread in 29 countries among five continents causing 8096 cases and 774 deaths. Intense public health measures contained the spread by July 2003 and there have been no cases since 2004 [16]. In 2012, the outbreak of MERS started in Saudi Arabia from dromedary camels, gradually spreading to 27 countries in the Arabian Peninsula, Europe, Asia, Africa, and North America. It persists though with the inefficient transmission in humans and at the end of February 2022, a total of 2585 laboratory-confirmed cases, including 890 associated deaths (case-fatality ratio of 34.4%) were reported globally [43]. The COVID-19 pandemic is unprecedented in comparison to all other diseases in the past due to the sheer extent of the assault it brought on humanity. The uniqueness of the current pandemic can be attributed to the combination of the characteristics of infecting humans in a sustained manner never done by any pathogen before, the highly efficient human-to-human transmission possibility, and its relatively high rate of morbidity and mortality in the elderly population and individuals with underlying comorbidities. It first broke in Wuhan, China, in December 2019 possibly from a local seafood market and within a few months, almost all countries are affected. As of the end of March 2022, more than 486 million people have been reportedly infected worldwide with 6 million deaths, though the actual figures are far greater [44].

2.12 Prevention and control

Outbreaks of emerging viral diseases vary widely in their presentation due to differences in transmission routes, involvement of vectors, transmissibility, case-fatality rates, and spread of the disease. Therefore, there is no one size fits all control solution. Some can be predicted like the rise of new strains of Influenza regularly, but in most cases, there are no clear indicators.

Early recognition and rapid response are the mainstays of controlling any outbreak. Clinicians need to react quickly to any unusual clinical presentation and should alert the health authorities. The situation needs to be assessed quickly and the virus identity, likely source, transmission routes, and case fatality need to be determined. Access to high standards of diagnostic facilities is essential to characterize the virus. Public health experts need to be mobilized at the onset to study the outbreak. Case definitions need to be formulated to identify, screen, isolate and administer treatment wherever possible. Contact tracing and identification of the source will help to prevent escalation of the cases. Quarantining contacts and imposing travel restrictions where appropriate need to be enforced by the lawmakers. The health infrastructure should be able to handle the increased load of cases, access to antivirals, and carry out vaccination drives. Engagement of the general community in addition to health care staff is essential through mass awareness campaigns to combat the spread [5,45]. Time is of the essence, as any delay can cause an escalation in numbers of cases out of proportion which can quickly overwhelm locally available manpower, capacity, and resources.

Technology can play an important role here improving our ability to respond to viral threats by bringing new biomedical research technologies such as genome sequencing to identify emerging viruses, structure-based vaccine and drug design, global communication networks and rapid diagnostics [46]. Vector-borne diseases can be predicted through meteorological conditions and satellite imagery of changing vegetation patterns in response to rainfall or human activity. The availability of the data in public domains enables researchers to quickly identify novel pathogens and mutants [47]. Development of diagnostic kits, vaccines, and antiviral research is also expedited due to the global sharing of data.

2.13 The global response

Cooperation at a global level is the need of the hour. International bodies like United Nations and its agencies, especially the WHO, the OIE (Office International des Épizooties—World Organisation for Animal Health) must be strengthened; collaborative multinational research to study, predict, and in turn prevent the risk of emergence of newer disease must be supported by the government; proper infrastructure to handle and study high-risk pathogens under appropriate safety and containment conditions must be developed at the regional level, and the world needs to come forward together to act against research and development of bioweapons [3]. The recent pandemic of COVID-19 and the earlier ones like the influenza A (H1N1) pandemic of 2009 and avian influenza have unified the international scientific community showing the importance of effective partnerships to deal with emerging infections. A close liaison between national focal centers and the international bodies is essential for the timely identification of events of public health importance. Alongside that, there is a need to strengthen the existing network of laboratory-based surveillance activities. In 2005, the International Health Regulations were substantially revised by WHO. The countries were required to develop national preparedness capabilities, conduct meaningful surveillance, and report internationally significant events regularly. Several recommendations for national strategies have been made by WHO highlighting the need to strengthen epidemic preparedness and rapid response, public health infrastructure, risk communication, research and its utilization, and advocacy for political commitment and partnership building [48]. The international research community is now involved in creating a unified molecular database of pathogens. As described earlier, the WHO-sponsored GISRS is one such international program continually monitoring isolates from a network of laboratories all over the world serving as an alert mechanism for the emergence of novel influenza virus strains with pandemic potential. As part of the One Health approach, the Global Early Warning System for Major Animal Diseases has been formed in collaboration with the WHO, the Food and Agriculture Organization of the United Nations, and the OIE [49]. This system acts to synergize the alert mechanisms of the three international agencies through data sharing and risk assessment to assist in early warning, prevention, and control of emerging veterinary diseases, including zoonotic ones. [50].

2.14 Conclusions and the way forward

Improved disease surveillance focusing on epidemiology and disease burden is the cornerstone for the fight against emerging diseases. The most important aspects of controlling any new outbreak are by strengthening emergency response which includes early recognition of the outbreak, good access to advanced diagnostic facilities, and proper dissemination and analysis of surveillance data. In addition to it, disease biomics need to be explored, including vector biology and environmental factors influencing the diseases, through cutting-edge research programs. Therefore, this is the right time globally for multisectoral involvement comprising microbiologists, epidemiologists, veterinarians, entomologists, and environmentalists for a "One Health" approach to tackle the problem and improve the health outcomes of all.

References

[1] World Health Organization. Regional office for South-East Asia. A brief guide to emerging infectious diseases and zoonoses. WHO Regional Office South-East Asia; 2014.

[2] Pommerville JC. Infection and disease. Alcamo's fundamentals of microbiology. 10th ed. Jones & Bartlett Learning; 2013.

[3] Morens DM, Fauci AS. Emerging pandemic diseases: how we got to COVID-19. Cell. 2020;182(5):1077−92.

[4] Dikid T, Jain SK, Sharma A, Kumar A, Narain JP. Emerging & re-emerging infections in India: an overview. Indian J Med Res 2013;138(Jul 2013):19−31.

[5] Burrell CJ, Howard CR, Murphy FA. Emerging virus diseases. Fenner White's Med Virology. Elsevier; 2017. p. 217−25.

[6] World Health Organization—WHO. The World Health Report 2007: A safer future global public health security in the 21st century. Glob Public Health. 2007. p. 96.

[7] CDC. Pandemic (H1N1 virus) | Pandemic influenza (flu) | CDC. Centers for Disease Control Control and Prevention. <https://www.cdc.gov/flu/pandemic-resources/1918-pandemic-h1n1.html>; 2018.

[8] WHO. HIV/AIDS factsheet. World Health Organization. <https://www.who.int/news-room/fact-sheets/detail/hiv-aids>; 2021.

[9] The Lancet Planetary Health, A pandemic era. Lancet Planet Heal. 2021;5(1):e1.

[10] World Health Organisation. Global progress report on HIV, viral hepatitis and sexually transmitted infections, 53. WHO; 2021. p. 1689−99.

[11] van Seventer JM, Hochberg NS. Principles of infectious diseases: transmission, diagnosis, prevention, and control. International encyclopedia of public health. Elsevier; 2017. p. 22−39.

[12] Pugliese A, Beltramo T, Torre D. Emerging and re-emerging viral infections in Europe. Cell Biochem Funct 2007;25(1):1−13.

[13] Shope R. Global climate change and infectious diseases. Env Health Perspect 1991;96(1):171.

[14] Patz JA, Githeko AK, McCarty JP, Hussein S, Confalonieri N de W U. Climate change and infectious diseases. In: McMichael AJ, Campbell-Lendrum DH, Corvalán CF, Ebi KL, Githeko AK, Scheraga AW JD, editors. Climate change and human health: risks and responses. Geneva: WHO; 2003. p. 103−32.

[15] Morand S, Lajaunie C. Outbreaks of vector-borne and zoonotic diseases are associated with changes in forest cover and oil palm expansion at global scale. Front Vet Sci 2021;8:230 Mar 24.

[16] Cheng VCC, Lau SKP, Woo PCY, Yuen KY. Severe acute respiratory syndrome coronavirus as an agent of emerging and reemerging infection. Clin Microbiol Rev 2007;20(4):660−94.

[17] Mourya D, Yadav P, Ullas P, et al. Emerging/re-emerging viral diseases & new viruses on the Indian horizon. Indian J Med Res 2019;149(4):447.

[18] Fauci AS, Morens DM. The perpetual challenge of infectious diseases. N Engl J Med 2012;366(5):454−61.

[19] Ghazanfar H, Orooj F, Abdullah MA, Ghazanfar A. Ebola, the killer virus. Infect Dis Poverty 2015;4(1):15.

[20] Yamaoka S, Ebihara H. Pathogenicity and virulence of ebolaviruses with species- and variant-specificity. Virulence, 12. 2021. p. 885−901.

[21] Forbes KM, Webala PW, Jääskeläinen AJ, et al. Bombali virus in Mops condylurus Bat, Kenya. Emerg Infect Dis 2019;25(5):955−7.

[22] Malvy D, McElroy AK, de Clerck H, Günther S, van Griensven J. Ebola virus disease. Lancet 2019;393 (10174):936−48.

[23] Gubler DJ. Dengue/dengue haemorrhagic fever: history and current Status. New treatment strategies for dengue and other flaviviral diseases. John Wiley & Sons, Ltd; 2008. p. 3−22.

[24] Sugawara E, Nikaido H. Properties of AdeABC and AdeIJK efflux systems of acinetobacter baumannii compared with those of the AcrAB-TolC system of *Escherichia coli*. Antimicrob Agents Chemother 2014;58 (12):7250−7.

[25] Ragasudha N, Babu A, Sankari SL, Anitha. Emerging viral diseases in India: a review. Biomed Pharmacol J 2015;8SE:83−9.

[26] World Health Organisation. Dengue and severe dengue. WHO Fact Sheet 2014;117(March):1−4.

[27] Weaver SC, Charlier C, Nikos V, Lecuit M. Viral, other emerging vector-borne. Annu Rev Med 2018;69:395−408.

[28] Weaver SC, Lecuit M. Chikungunya virus and the global spread of a mosquito-borne disease. N Engl J Med 2015;372(13):1231−9.

[29] Schweitzer BK, Chapman NM, Iwen PC. Overview of the flaviviridae with an emphasis on the Japanese encephalitis group viruses. Lab Med 2009;40(8):493−9.

[30] Rossi SL, Ross TM, Evans JD. West Nile virus. Clin Lab Med 2010;30(1):47−65.

[31] Hills SL, Fischer M, Petersen LR. Epidemiology of Zika virus infection. J Infect Dis 2017;216(Suppl. 10): S868−74.

[32] Gubler DJ, Vasilakis N, Musso D. History and emergence of Zika virus. J Infect Dis 2017;216(Suppl. 10): S860−7.

[33] WHO. Zika epidemiology update. WHO; 2019.

[34] Boštíková V, Kuča K, Blažek P, et al. Zika virus—a review. Mil Med Sci Lett 2016;85(3):94−103.

[35] de Oliveira Figueiredo P, Stoffella-Dutra AG, Barbosa Costa G, et al. Re-emergence of yellow fever in Brazil during 2016−2019: challenges, lessons learned, and perspectives. Viruses 2020;12(11):1233.

[36] Staples JE. Yellow fever: 100 years of discovery. JAMA. 2008;300(8):960.

[37] Skowron K, Bauza-Kaszewska J, Grudlewska-Buda K, et al. Nipah virus—another threat from the world of zoonotic viruses. Front Microbiol 2022;12(January):1−13.

[38] Epstein JH, Field HE, Luby S, Pulliam JRC, Daszak P. Nipah virus: impact, origins, and causes of emergence. Curr Infect Dis Rep 2006;8(1):59−65.

[39] Gaitonde DY, Moore FC, Morgan MK. Influenza: diagnosis and treatment. Am Fam Physician 2019;100 (12):751−8.

[40] Saunders-Hastings P, Krewski D. Reviewing the history of pandemic influenza: understanding patterns of emergence and transmission. Pathogens 2016;5(4):66.

[41] World Health Organization. Evolution of a pandemic—A(H1N1) 2009. Health (San Francisco) 2013;48.

[42] WHO. WHO | Pandemic (H1N1) 2009—update 103; 2010. <https://web.archive.org/web/20100609030841/http://www.who.int/csr/don/2010_06_04/en/index.html>

[43] World Health Organization Regional Office for the Eastern Mediterranean. MERS situation update—February 2022. World Health Organization Regional Office for the Eastern Mediterranean; 2022. http://www.emro.who.int.

[44] World Health Organization. WHO Coronavirus (COVID-19) dashboard. WHO Coronavirus (COVID-19) dashboard with vaccination data. WHO; 2021. p. 1–5.

[45] Epidemiology and Control of Viral Diseases. Fenner's veterinary virology. Academic Press; 2017. p. 131–53.

[46] Marston HD, Folkers GK, Morens DM, Fauci AS. Emerging viral diseases: confronting threats with new technologies. Sci Transl Med 2014;6(253):1–6.

[47] Morens DM, Fauci AS. Emerging infectious diseases in 2012: 20 years after the Institute of Medicine Report. MBio. 2012;3:6.

[48] WHO. Asia pacific strategy for emerging diseases and public health emergencies (APSED III): advancing implement international health regulations 2005;2017:1–78.

[49] WHO. A tripartite guide to addressing zoonotic diseases in countries. World organisation for animal health (OIE), 1–. 2019. p. 166.

[50] Chakraborty D, Debnath F, Biswas S, et al. Exploring repurposing potential of existing drugs in the management of COVID-19 epidemic: a critical review. J Clin Med Res 2020;12(8):463–71.

Evolution and transmission of viruses

Shalini Upadhyay

A.S.J.S.A.T.D.S. Medical College, Fatehpur, Uttar Pradesh, India

3.1 Introduction

Viruses have the ability to evolve and adapt to changing environment with short-generation time and high mutation rate. Natural selection allows viruses to evade host immune responses, exploit ecological niches, and evolve new behaviors beneficial for their propagation. As the virus replicates, over time, several mutations accumulate to form a variant of the original virus. A virus variant is better adapted to its environment and this process of changing and selection of successful variants is called virus evolution. Some viruses evolve faster than others, which can be dependent on the virus's internal proof-reading mechanism, which can correct error when they replicate. For example, SARS-CoV-2 (severe acute respiratory syndrome-coronavirus) has shown tremendous evolution and adaptation since the first crossover from animal to human. The reproductive number combined with viral mutation rates and the width of transmission bottleneck determine the evolutionary modifications.

On infection of a new host, the genomes of many viruses undergo rapid adaptive evolution, which may result in escape from host immune responses and increase in viral growth rates. Even though these genetic variations render viruses better competitors within their host, yet they do not essentially increase chances of transmission between hosts. They undergo low key genetic modifications via mutation while there occur major changes through recombination. Mutation happens when an error is integrated in the viral genetic material. Although high rates of mutation help virus to adapt and evolve faster, the small changes in the mutation rate can be determinant of whether virus will be able to evade host immune system. Recombination leads to exchange of genetic material between two infecting viruses, resulting in origin of a novel virus. In spite of a plethora of data generated over the years, very little understanding of the evolutionary mechanisms has been decoded. Furthermore, different studies on virus evolution have different perspectives, either experimental or comparative, and are conducted independently and very often antagonistically. The experimental approaches are mostly focused on short term within

Viral Infections and Antiviral Therapies
DOI: https://doi.org/10.1016/B978-0-323-91814-5.00014-3

the host evolution while comparative approach is based on the phylogenetic analysis of natural virus populations, but is far from the causative evolutionary processes. In this chapter, we will achieve a comprehensive understanding of various genetic and ecological changes and their effects on viral evolution along with the modes through which viral infections can be transmitted.

3.2 Viral Evolution

3.2.1 Genetic basis of virus evolution

The viral genetic information is either encrypted in DNA or RNA of a genome. It is very diverse with constituent nucleic acid being either single stranded (ss) or double stranded (ds) with positive-sense (i.e., the same polarity or nucleotide sequence as the mRNA), negative-sense or ambisense (a mixture of the two), and its structure may vary from linear, circular, or segmented. The size of virus genome may range from approximately 2500 nucleotides to 1.2 million base pairs with bacteriophages being the smallest and least complex and herpesviruses and poxviruses having large ds DNA complex genome. The complexity of the viral genome and lack of proofreading mechanism make it a favorable playground for genetic modifications and remodeling, making itself fit according to the host and environment.

Strains are different lines or isolates of the same virus, which may be isolated from different patients or geographical areas. **Viral types** are various serotypes of the same virus, for example, antibody neutralization phenotypes. A **variant** is a virus whose phenotype varies from the original wild-type strain but the genetic basis for the difference is not known (Fig. 3.1).

I. Several characteristics of viruses like virulence, antigenicity, transmission, severity of infection, etc. are susceptible to genetic modifications. Out of myriad possible modifications in the viral genome, the important genetic changes include mutation, recombination, and interference. The types of changes occur depending on viral genetic makeup, host encounter, and environmental exposure.

3.2.1.1 Mutation

The viral mutation rates may vary from as high as 10^{-3} to 10^{-4} per base pair (e.g., HIV), to as low as 10^{-8} to 10^{-11} (e.g., herpesviruses). The differences are based on the replication mechanism, with chances of error in RNA-dependent RNA polymerases generally being higher than in DNA-dependent DNA polymerases, as most RNA virus polymerases lack proofreading functions. Thus, mutation rates are generally higher in RNA viruses than DNA viruses [1].

Although mutations are very common in viruses, they become apparent only when there are readily observable changes or effect survival of the virus. Mutations may occur spontaneously, by selection pressure, exposure to antivirals, or induced by chemicals (e.g., 5-fluorouracil) or physical agents (UV/Ultraviolet light). Mutations may provide a number of survival advantages to the virus, such as the ability to generate antigenic variants that escape the host immune response and vaccine-induced immunity (e.g., Omicron variant of SARS-CoV-2), but at the same time they may produce defective particles having deleterious consequences.

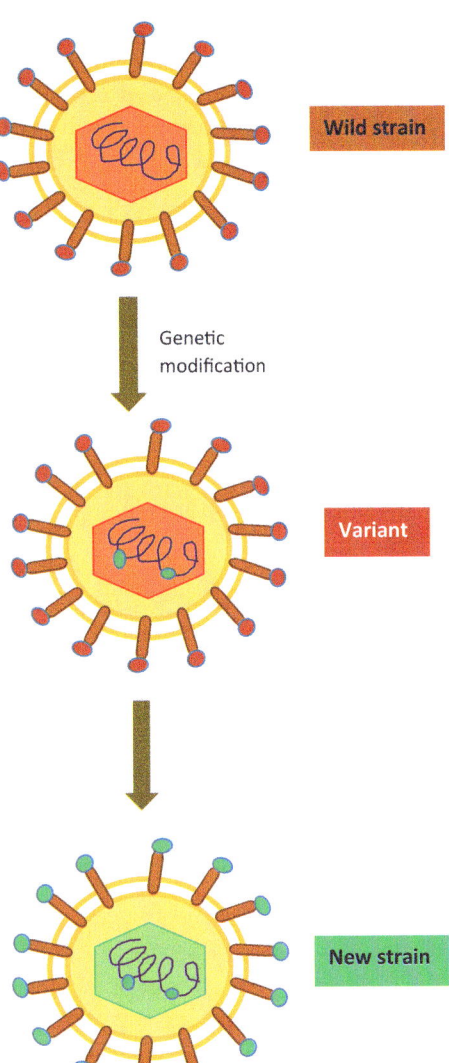

FIGURE 3.1 Evolution of a viral strain. Note: Viruses with genetic modifications (mostly mutations) become variants and when this variant shows divergent physical properties, it develops into a new strain.

The high error rate may lead to production of a mixtures of molecular variants known as quasispecies. Nonetheless, most of these variants being noninfectious are removed out of the replicating population. The precise estimate of virus mutation rate is an important information that can help us follow the evolution of the viruses and control it.

Mutagens can be in vitro mutagens that chemically modify nucleic acids, which do not require replication for their activity. Examples include nitrous acid, alkylating agents (e.g., nitroso-guanidine), hydroxylamine, etc. and some can be in vivo mutagens that need metabolically active nucleic acid for their action. These compounds when get incorporated into freshly replicated nucleic acids induce mutations during subsequent replication cycle. Examples include base analogs resulting in faulty base pairing; UV irradiation leading to formation of pyrimidine dimers, acridine dyes act as intercalating agents. The characteristics of resulting mutant virus depends on the type of mutations and its location. Mutant viruses may revert to their original phenotype by three pathways: back mutation to wild-type, compensatory mutation in the same gene, and suppressor mutation in a different virus gene or a host gene [2].

The mutation rate and selective pressure regulate the rate of evolution and heterogeneity among viruses. Despite having very limited coding capacity, RNA viruses are extremely adaptable, able to adjust to a variety of environments. They accomplish this due to their potential to multiply very rapidly, along with their remarkable degree of genetic heterogeneity. RNA viruses exist as a conglomerate of related variants and this genetic diversity is a vital, feature of their biology. RNA viruses have a variety of mechanisms that act in concert to determine their genetic heterogeneity. These include genomic recombination, polymerase fidelity, and error reduction, and various methods of genome replication. These viral properties help explain how viruses are able to adapt to grow in different tissues within a host, can be passed from one host species to another, and the emergence of novel human pathogens. They also present opportunities for novel antiviral drug and vaccine development strategies.

The RNA viruses are prone to higher mutation rates due to lack of proofreading mechanism and that is why they are more diverse than DNA viruses. The selective pressure is complicated mechanism, which involves various facets of host—pathogen interaction, such as the immune response to viral invasion and its transmission modality. The host immune response acts as positive selective pressure leading to escape mutants and optimization of transmission. Furthermore, replication efficiency provides a negative selective pressure allowing circulation of optimal wild type till there are variations in transmission conditions. Viruses employ various quid pro quos between these two types of selective pressure favoring diversification [3].

3.2.1.2 Genetic recombination

It occurs when two viruses of different parent strains infect a host cell at the same time. They exchange genetic material during replication and generate virus progeny that has hybrid genes from both parents. These hybrids acquire new genes not present in both the parent viruses, and they are genetically stable. Recombination usually takes place between members of the same virus type (e.g., two coronaviruses or two polio viruses). The common mechanisms of recombination observed among viruses are: independent assortment and incomplete linkage. Each of the mechanism is capable of producing new viral serotypes or viruses with altered virulence.

The significance of viral recombination lies in the fact that it creates novel virus progeny which is capable of expressing new antigenic and/or virulence characteristics. For example, the novel progeny viruses may have new surface proteins that permit them to infect previously resistant individuals; they may have altered virulence characteristics; they may have novel combinations of proteins that make them infective to new cells in the original host or to new hosts; or they may carry material of cellular origin that gives them oncogenic potential. Recombination can occur through two major mechanisms as discussed below and depicted in Fig. 3.2.

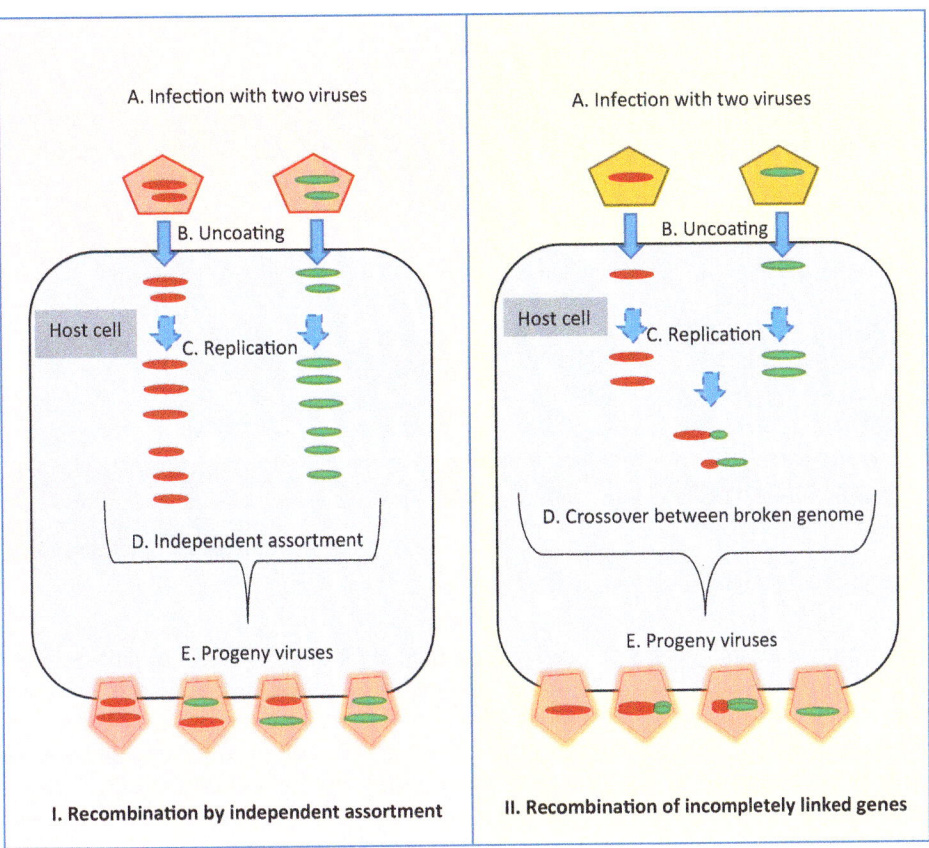

FIGURE 3.2 Evolution through recombination of genetic material.

i. Recombination by independent assortment

It occurs when viruses with segmented genomes exchange genetic material which are unlinked and assorted at random during replication. This method of assortment has been reported, for example, for the orthomyxoviruses (eight segments) like influenza virus, rotavirus, bunyavirus, and reoviruses (10 segments of ds RNA). The frequency of reassortment may vary from 6% to 20% for orthomyxoviruses. History is evident that independent assortment between animal and human influenza virus strain during a hybrid infection can produce an antigenically novel strain. This strain will be able to infect humans, despite carrying animal-strain hemagglutinin (HA) and/or neuraminidase surface molecules. This type of recombination leads to an immediate, major antigenic change known as antigenic shift. Antigenic shifts in viral genome have led to various pandemics of influenza in the history. Such events have happened relatively frequently during recent history (Fig. 3.3). Since the number of HA and neuraminidase serotypes are limited, a given strain reappears from time to time. For example, the H1N1 strain caused influenza pandemics from 1918 to 1919 with death toll of 20 million. This strain caused pandemics in 1934 and in 1947 too, but disappeared after 1958 and reemerged in 1977. The reemergence

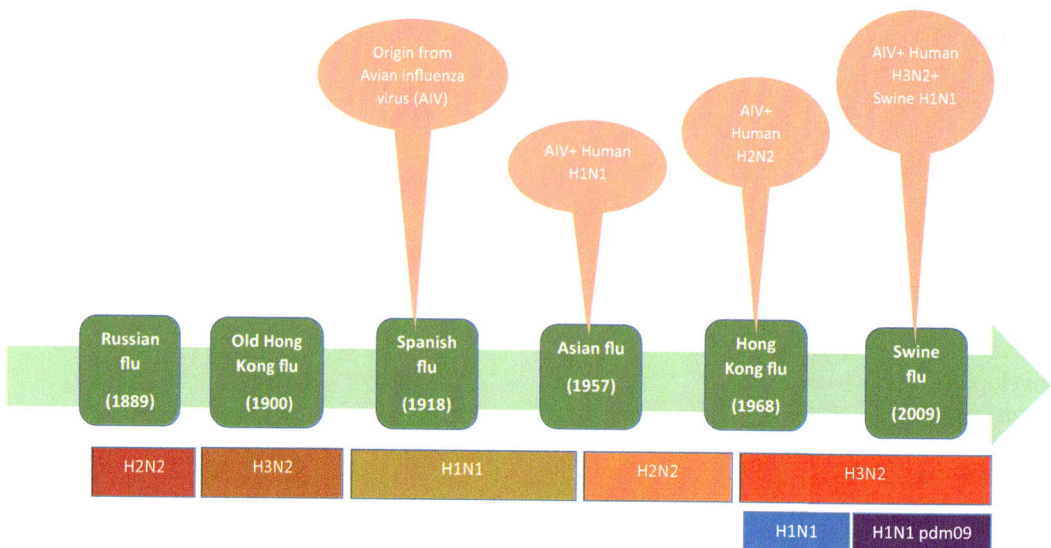

FIGURE 3.3   Timeline of influenza pandemics.

of this strains after an absence of few years can be explained by recombination events involving the independent assortment of genes from two variant viruses.

ii. Recombination of incompletely linked genes

Recombination may also happen between genes present on the same piece of nucleic acid. If recombination occurs between linked genes, the linkage is thought to be incomplete. It occurs in all DNA viruses and in many RNA viruses too. In DNA viruses, recombination happens by break-rejoin involving the actual split of the covalent bonds linking the bases of the two DNA strands. The separated DNA strands then rejoin the DNA strands of a different DNA molecule that has been severed at a similar site. The rate of recombination depends on the distance between a specific pair of genetic loci and may vary from less than 1%−50%. The measure of recombination frequencies for different loci can be helpful in mapping the viral genome. Recombination occurs in several positive-sense ssRNA virus groups like coronaviruses, retroviruses, and picornaviruses. In coronaviruses, recombination takes place at the site of interaction of the viral RNA genomes and not a break-rejoin mechanism [4].

3.2.1.3 *Genetic reactivation*

It can happen in two ways.

a. **Marker rescue:** When a host cell is infected at the same time with an active and an inactive virus, which are different but related. Some part of the inactivated virus genome combines with that of the active virus, so that portion of the inactivated virus is retained with viable progeny viruses. For example, the epidemic influenza strains when combined with inactivated standard laboratory strains, the progeny receives the virulence genes of pathogenic strain and growth behavior of laboratory strain.

b. **Multiplicity reactivation:** When many inactive viruses interact in the same host cell acting as a vessel to produce a stable viable virus. For example, viruses inactivated by UV rays. Since such inactivated viruses may have damages in the genetic material, thus, the healthy genes from the total genetic pool combines to form a viable genome.

3.2.1.4 Viral interference

When two viruses infect a host cell or a cell line, they may either infect and replicate in conjunction or there may occur inhibition of one of the virus. Interference is consequence of the resistance to superinfection in cells infected with another virus. Interference may result from the competitive inhibition, mutations, or by sequestration of virus receptors. Phenotypic combinations can vary from capsid or envelope of one virus enclosing the genome of the other, or the capsid/ envelope of the progeny being an amalgamation of proteins from both viruses. This fusion leads to progeny virus having the phenotypic characters based on the types of proteins incorporated without any genetic modification. However, the following generations of viruses inherit and exhibit the original parental phenotypes. Phenotypic mixing is an important tool that helps in examining biological properties of viruses. For example, vesicular stomatitis virus (VSV) forms pseudotypes with retrovirus with the properties of VSV but the cell tropism of the retrovirus.

The probable mechanisms of viral interference include inhibition of the entry for second virus either by blocking or destroying the host cell receptors, competition for components required for replication, interferons produced by host for one virus may inhibit the second virus. Viral interference can be observed with oral polio vaccine (OPV) having three live attenuated serotypes of poliovirus. It has a mixed effect with OPV serotypes interfering with the spread of wild poliovirus, but at the same time interference between the three OPV serotypes or with enteric viruses may result in vaccine failure.

3.2.2 Ecological basis of virus evolution

3.2.2.1 Reproductive rate and antigenic diversity

Various viruses transmit between different hosts with varying degrees of efficiency. R_0 (basic reproductive number) indicates the average number of cases acquiring infection from a single infected case or a quantitative estimate of the transmission efficiency which varies from ~ 1 to >10. Rapidly evolving viruses that are under a positive selection pressure have low R0, while stable viruses have high R_0 with the exception of antigenically diverse influenza A virus and antigenically stable measles virus. The R_0 value decides the levels of threshold herd immunity, that is, lower R_0 needs lower herd immunity to put an end to outbreak or in other words high R_0 is an important factor in outbreak of a disease. A deviation to this rule is influenza virus, the 2009 pandemic of H1N1 influenza virus had R_0 of 1.3–1.7, similar are the R_0 estimates of other pandemic and seasonal influenza viruses, probably due to its short generation time. New antigenic variants appearing constantly known as "antigenic drift" is a significant issue in influenza vaccine development. Vaccine strains are predicted every 6 months for northern and southern hemispheres depending on extensive surveillance data. Viruses having high R_0 like measles, mumps, rubella, and poliomyelitis contain a limited number of serotypes. These viruses have been causing childhood diseases and due to less number of serotypes, vaccine is highly effective for them [5].

3.2.2.2 *Herd immunity and selective pressure*

Herd immunity or population immunity is the indirect immunity from an infectious disease gained by nonimmune individuals when a significant fraction of the population is immune either through vaccination or immunity derived from previous infection. The indirect protection results from the interruption of transmission chain by immunized population. It can be measured by testing a sample from the population for immune parameter specific to the disease. It is advisable to achieve herd immunity through immunization and not through disease spread to any part of the population, as it would lead to unwanted cases and deaths. The vaccine-induced herd immunity for that vaccine shows geographic variation depending on coverage and efficacy of the vaccine. To terminate an outbreak a threshold fraction of population should be immunized, which depends on the R_0.

This threshold herd immunity can be calculated with the formula $1-1/R_0$, that is, viruses with higher R_0 require a higher herd immunity fraction. Presumably, in a uniform population with evenly distributed contact rate, herd immunity level achieved is high regardless of R_0. However, in a heterogeneous population the threshold herd immunity level may never be attained in reality. A virus with high R_0 will transmit more effectively in a small population with high contact rate, than in a large population with a lower contact rate. Hence, the high herd immunity level needed to terminate an outbreak with a high R0 may not be reached as portions of susceptible hosts will be always available to continue the transmission cycle. Some viruses circulate in many animal species and transmission may occur between two or more species occasionally. This interspecies dissemination may lead to generation of a novel virus into human population, resulting in a pandemic because there is a lack of herd immunity for the new pathogen. Thereof, the course of viral evolution is dependent on the interaction between the virus and human host since there is no back and forth transmission between human and animal species. Examples are influenza virus and more recently SARS-CoV-2 causing years and years of unrest for the human population [6]. There have been some threatening pandemics in history caused by viral infections as shown in Table 3.1 [7].

TABLE 3.1   Pandemics caused due to viral infections.

S. no.	Disease	Years	Pathogen	Source
1	Russian flu	1889−93	H3N8/Influenza A	Avian
2	Spanish flu	1918−19	H1N1/Influenza A	Avian
3	Asian flu	1957−59	H2N2/Influenza A	Avian
4	Hong Kong flu	1968−70	H3N2/Influenza A	Avian
5	Severe acute respiratory syndrome (SARS)	2002−03	SARS-CoV (Coronavirus)	Bats, civets
6	Swine flu	2009−10	H1N1/Influenza A	Pigs
7	Middle Eastern Respiratory Syndrome (MERS)	2015−ongoing	MERS-CoV	Dromedary camels, bats
8	Coronavirus disease-2019 (COVID-19)	2019−ongoing	SARS-CoV-2	Bats, pangolins

3.2.2.3 *The spillover event*

The spillover is the transmission of a virus from one reservoir species, in which it usually resides, to a new host species, in which it can either die or adapt and cause infection with the potential of a pandemic. Examples include SARS-CoV-2, Ebola, HIV (human immuno-deficiency virus), and influenza viruses. This transition requires three distinct phases: contact, infection, and transmission. Wild animals have been implicated in disease spillover to humans for most of zoonotic viruses. Furthermore, the diversity of wildlife on earth provides a rich pool of viruses, a part of which has effectively adapted to infect humans. Human practices have led to increased contact between taxonomically diverse animal hosts and viruses resulting in sharing of zoonotic diseases, for example, arboviruses (Fig. 3.4) [8].

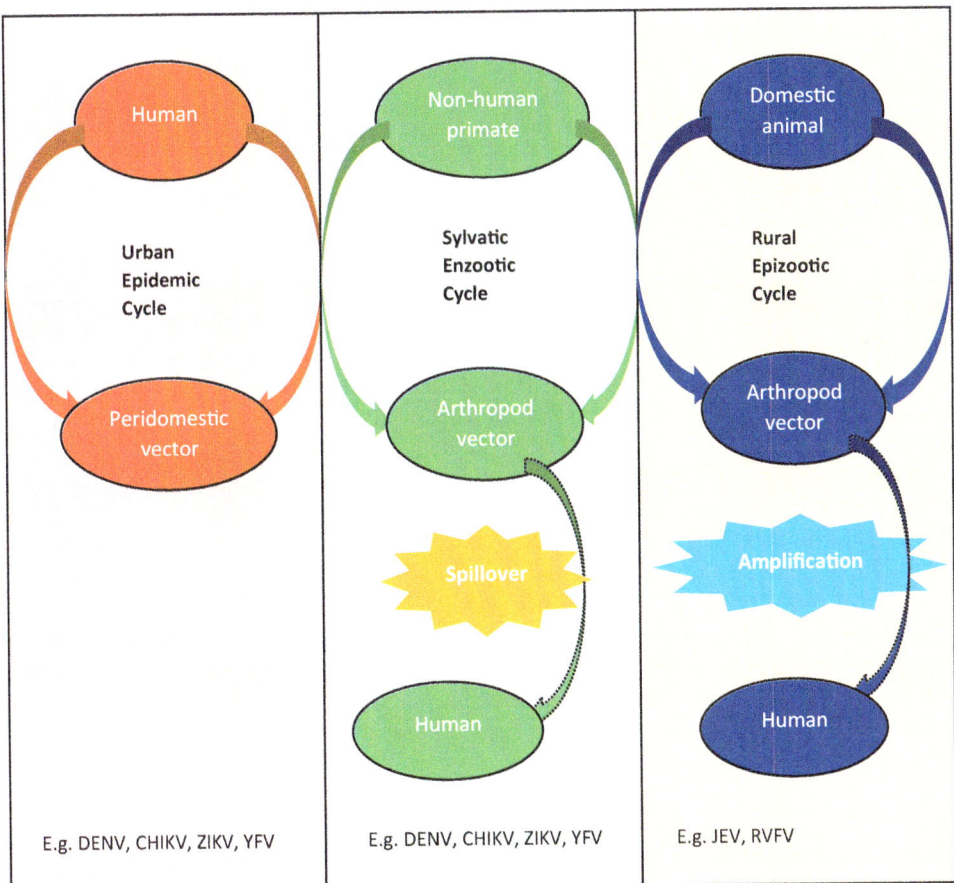

FIGURE 3.4 Emergence mechanisms for arboviruses. Abbreviations: *CHIKV*, chikungunya virus; *DEN V*, dengue virus; *JEV*, Japanese encephalitis virus; *RVFV*, Rift Valley fever virus; *YFV*, yellow fever virus; *ZIKV*, Zika virus. Note: Timeline of influenza pandemics caused by the 1918 H1N1 virus and its descendants produced by reassortment of circulating strains with avian influenza viruses (AIV) and swine H1N1 viruses. The reassortment of genes is shown in parenthesis. The re-emergence of H1N1 virus in 1977 is also shown as it co-circulated with the H3N2 virus before being replaced by the H1N1pdm09. *HA*, hemagglutinin; *M*, matrix proteins; *NA*, neuraminidase; *NP*, nucleoprotein; *NS*, nonstructural proteins; *PA*, polymerase; *PB1*, polymerase; *PB2*, polymerase.

The virus spillover from wildlife has been found to be mostly in and around human dwellings, agricultural fields and occupational exposure to animals (hunters, wildlife management, laboratory workers, veterinarians, and researchers). There are two main barriers for spillover to humans, adapting to a new host environment and effective human-to-human transmission. These evolutionary modifications may occur in an intermediate host species acting as the "mixing vessel." The human-to-human transmission of such zoonotic viruses facilitates continuous spread of infection and tide over the requirement of contact with animal hosts for transmission [9].

The mutant spillover events expedite the evolutionary emergence. When the index case begins with a mixed infection, there is more probability of evolutionary emergence. After the emergence of the highly pathogenic H5N1 influenza virus, it has been a matter of research that this virus would be able to cause sustained transmission in humans. Over the years it has been found that it causes only sporadic spillover infections. Then came highly pathogenic H7N9 virus which has sporadically infected humans. Although actual mortality rate is difficult to ascertain, but it is comprehensible that both these viruses cause quiet high mortality in humans and they could cause serious human epidemics. Such apprehensions have led to use of genomic data to help in pandemic risk assessment. At the subtype level, the polybasic amino acids in the hinge region between the HA1 and HA2 subunits which make up the HA protein facilitates the establishment of a serious systemic infection and consequently acts as a marker of high-virulence. This marker is useful in differentiating potentially low and high-virulence avian influenza viruses. The unresolved mystery is how virulence and transmissibility will change with natural selection, if viruses like H5N1 or H7N9 develop sustained transmission in humans. The phylogenomic analyses reveal a constant set of mutations that differentiate human and avian influenza viruses, even though their effect on host range and virulence is unknown [10].

To understand the different aspects of virus evolution, various phylogenetic studies need to be conducted, most helpful being whole-genome sequencing (WGS). In the COVID-19 pandemic, it has been a major advantage that we could get the WGS done quite early, which led to faster development of vaccines and tracking of subtypes. WGS brings an array of information from evolution to virulence and helps to map the genetic variations accumulating over a period leading to significant change in the genetic makeup of the virus. This facilitates in interpreting the selection pressures acting on virulence mutations and different aspects of virulence evolution.

3.3 Transmission

Viruses may transmit through different routes but for effective transmission, they should have the ability to shed high levels of infectious virions. This needs high replication rate, making the viruses vulnerable to immune response or require to infect large number of cells leading to rapid progressive fatal diseases. The high viral replication rates are not sustainable. So, virus should either replicate at a maximum rate in a short period, or widen the replication time and shedding with a lower replication rate. For long-term persistence, they should be able to evade the host immune reactions which requires an immune escape mechanism. Some of the examples are childhood diseases, which are highly transmissible over a short period; sexually transmitted diseases (STD), with low transmission rate but persistent infection; and influenza virus with high antigenic diversity and low transmission rate.

Viruses get transmitted from host cells either as single virion containing one genome in groups with multiple genomes in same virion or multiple virions. These units of transmission are known as collective infectious units (CIUs), which can comprise larger structures containing multiple virions. These sometimes form free virion clusters after dissemination, through direct contact or through a vector. Different kinds of CIUs have evolved over time, as they exist in various viral families and have diverse forms. CIUs are thought to have important role in viral evolution. The same host cell may get infected by multiple viral genomes simultaneously, allowing interconnections between viruses. These interactions happen even when coinfection are thought to be rare due to strong population bottlenecks or low ratios of infectious viral particles. Such interchange between viral genome can have significant impact on viral diversity, pathogenesis, and evolution of antiviral resistance. Viral social adaptations include cooperation for beneficial changes, but may also facilitate conflict, as in the case of defective interfering genomes, which exploit the cellular machinery of coinfecting viruses [11].

Pathogens require either living organisms or nonliving sites as reservoirs to persist over long periods of time. Nonliving reservoirs include soil, water etc. which help in the survival of virus in the environment. Pathogens may have mechanisms of dormancy or resilience, which helps them survive for long periods in nonliving environments. Although many viruses become nonviable in nonphysiological conditions, but few may persist outside of a living cell for varying amounts of time. For example, in the COVID pandemic we have seen the persistence of SARS-CoV-2 for hours and even days on nonliving surfaces like paper, cloth, plastic, metal etc [6]. Human beings as reservoir may or may not be capable of transmitting the pathogen, depending on the type of infection and the pathogen [12].

An infected/contaminated individual not having any symptoms but capable of transmitting the pathogen is referred to as a carrier. A passive carrier is contaminated with the pathogen and mechanically transmits the infection. For example, a healthcare worker not maintaining proper hand hygiene can transmit an infectious agent. While active carrier is infected and capable of transmitting the infection. Active carriers may transmit the disease during asymptomatic phases of infection like the incubation period or convalescence. Viruses like HIV, hepatitis viruses, and herpes virus are transmitted mostly by asymptomatic carriers. In zoonotic diseases, animals are the reservoirs of infection and transmit the pathogen to humans through direct or indirect contact. The various routes of transmission and viruses transmitted through them are shown in Table 3.2.

3.4 Modes of transmission of viruses

3.4.1 Respiratory tract

The respiratory tract has a very large mucosal surface that allows viral interaction with the host cells. Our breath can act as carrier of aerosolized droplets and virus-containing

TABLE 3.2 Transmission mechanism of various pathogenic viruses.

S. no.	Portal of entry	Route of transmission	Common viruses transmitted
1.	Cutaneous	Close physical contact or touching a contaminated object	HSV1, HSV2
		Vector borne	Dengue virus, West Nile virus, encephalitis virus, Chikungunya virus, Yellow fever virus, Colorado tick fever virus
2.	Oral	Kissing, sharing food or utensils, Feco-oral	EBV, HAV, HEV, Norwalk virus, Rotavirus, Poliovirus
3.	Genital	Unprotected sexual contact	HSV2, HIV, HBV, HCV, HPV
4.	Respiratory tract	Droplet, airborne	Coronavirus, Adenovirus, Measles, Mumps, Rubella, Influenza virus, Rhinovirus, RSV
5.	Systemic	Injection, blood transfusion, during surgical procedure	HIV, HBV, HCV, HDV, Human T-Lymphotrophic Virus-I and -II, Dengue virus, Ebola virus
6.	Eye	Conjunctiva	Adenovirus, Enterovirus 70, Coxsackie virus A-24, HSV, HPV, CMV, Rhinovirus
7.	Placenta	Trans placental	CMV, Variola virus, HSV-1 and -2, Measles virus, Zika virus, Rubella virus
8.	Organ transplant	Transplanted organ	HBV, HCV, HIV, CMV, West Nile virus, Rabies virus, Lymphocytic choriomeningitis virus, HSV, VZV

Notes: *CMV*, Cytomegalovirus; *EBV*, Epstein–Barr virus; *HAV*, Hepatitis A virus; *HBV*, Hepatitis B virus; *HCV*, Hepatitis C virus; *HDV*, Hepatitis D virus; *HEV*, Hepatitis E virus; *HIV*, Human Immunodeficiency virus; *HPV*, Human Papillomavirus; *HSV*, Herpes Simplex virus; *RSV*, Respiratory Eyncytial virus; *VZV*, Varicella Zoster virus.

particles suspended in our immediate environment through a cough or sneeze of an infected person. Viruses carried in larger droplets settle in the upper respiratory tract, while aerosolized particles being smaller reach the lower respiratory tract. There are two protective measures in the upper respiratory tract epithelium, which curtail the spread of viral particles. One is the abundance of goblet cells producing mucus, which traps inhaled particulate matter and second is the ciliary lining that move together to shove the mucus and its contents to the throat. Thus to commence an infection, there must be sufficient viral load and virus must avoid getting trapped in the mucus or elimination by antibodies and macrophages.

Airborne viruses can be either transmitted as droplets or aerosolized particles, which are released when a person speaks, sings, breathes, sneezes, or coughs. Droplets are of about 20 μm size, and they can spread to short distances before they fall off. While aerosols are ≤5 μm, they remain suspended in air for much longer, making it more transmissible in favorable environmental conditions of humidity and temperature. Most of the enveloped respiratory viruses, including SARS-CoV-2, influenza A virus, SARS-CoV, and Middle East respiratory syndrome-CoV, measles virus at lower temperature and humidity are more prone to maintaining airborne virions. On the contrary, nonenveloped viruses like rhinovirus and adenovirus prefer higher humidity environments.

3.4.2 Gastrointestinal tract

The human gastrointestinal tract has adverse conditions for any microorganisms to survive and settle. There is influx of food, water, saliva, and mucus produced by the tract, which act as mechanical barrier to infection. The immune cells like macrophage engulf virions, and antibodies neutralize them to curtail their interaction with host cell receptors. The hostile environment continues with the acidic pH of stomach and the emulsifying qualities of bile produced by the liver. The envelopes of most enveloped viruses are disrupted by bile. If the infecting virus is acid-labile, it will not be able to survive the low pH of the stomach. The acid-resistant viruses have capsid proteins, which cannot be denatured by low pH. For example, in the Picornaviridae family, rhinoviruses are acid-labile while poliovirus is acid-resistant. Some viruses are transmitted by breastmilk to children, for example, HIV and CMV (cytomegalovirus). Most other viruses enter the gastrointestinal tract through the feco-oral route, that is, that virions from the feces of an infected person obtain entry into the oral cavity of another. They may also get aerosolized and inhaled. The neutral pH of human excreta usually keeps the virus viable and organic matter within feces gives protection from the environmental variations. Rotavirus is another virus that is transmitted by the feco-oral route. It is the leading cause of childhood diarrhea worldwide.

3.4.3 Genital tract

Viruses are transmitted via the genital tract predominantly by sexual contact, including vaginal, anal, and oral sex. Out of common STDs, four are viral infections and they are incurable including HBV (Hepatitis B virus), HSV (Herpes Simplex virus), HIV, and human papillomavirus (HPV). Viral infection may spread through either tropism of virus for the epithelium of the cervix or penis (e.g., HPV) or entry via breaks in the genital epithelium or by binding local cell receptors (e.g., HBV or HIV). Viruses infecting via this route should be able to survive the local barriers to infection, such as mucus and acidic pH of the vagina.

The sequelae of viral STDs can be beyond the immediate impact of these infections. Herpes may increase the risk of HIV infection. HPV infection can cause cervical cancer, which is the fourth most common cancer among women. Chronic infection of HBV can lead to cirrhosis and hepatocellular carcinoma.

3.4.4 Skin

Though epidermis supports localized viral infections, entry to the underlying dermis and subcutaneous tissue may lead to dissemination of the virus to the circulatory system and then other parts of the body. Viruses spread through skin via penetrating the epidermis. The subcutaneous tissue can be reached by viruses through needle punctures, animal bites, or infected tattooing or piercing equipment. From here viruses can further access the bloodstream directly or via the draining lymph.

One of the major routes of transmission is through vectors, which can be either mechanical or biological. Mechanical transmission is when an animal transmits a pathogen without being infected itself. For example, a fly which sits on fecal matter and spreads the pathogen from the feces to food; a human eating the food gets infected. Biological transmission occurs when the pathogen reproduces within a biological vector that transmits the pathogen from one host to another. Arthropods are responsible for biological transmission of arboviruses, for example, dengue, chikungunya, and zika virus. Nonarthropod vectors also transmit diseases, including mammals and birds, for example, rabies virus. Chickens and other domestic poultry transmit avian influenza to humans through direct or indirect contact with the birds' saliva, mucous, and feces.

3.4.5 Eyes

Ocular infection by viruses mostly occurs following direct contact with virus, either from infected secretions in the birth canal (HPV, HSV), or airborne particles (rhinovirus), or fomites (adenovirus), or may also happen during viremia (measles virus, CMV). Other mechanisms involve spread from adjoining adnexal disease (HSV), or upper respiratory tract infection via the nasolacrimal duct (rhinovirus) and trans placental infection (rubella virus). In early infection, aqueous tears from the lacrimal glands secrete proinflammatory cytokines, and the conjunctival blood vessels supply both soluble and cellular components of innate immunity. After an established viral infection, aqueous tears bring lacrimal gland-derived monospecific secretory immunoglobulin A [12].

3.4.6 Placenta

Congenital infections are those in which an infection is transmitted from a mother to the fetus before its birth and these are transferred from one generation to the next generation (vertical transmission). The viruses mostly show direct host-to-host transmission, which is known as horizontal transmission. Viruses with direct transmission are dependent upon a high rate of infection to sustain the virus population, in contrast vertical transmission leads to long-term persistence of the virus within the child.

Placenta forms the primary barrier between the mother and the fetus throughout pregnancy. The exact mechanisms by which microbes breach this barrier and infect the fetus are largely unknown. While it is possible that viruses causing congenital infection employ a common mechanism, but it appears more so that they have evolved distinct strategies, which vary with the level of maternal infections or immune responses mounted against them. Some of the viruses that pass through the placenta are zika virus, HIV, HSV-1 & 2, CMV, rubella. The infection may be very mild to severe, leading to miscarriage, abnormal growth pattern, low birth weight, intellectual deficiencies, sensory disability, or even death [13].

3.4.7 Transplants

The transplanted organs and tissues may rarely carry viruses that can be transmitted to the donor. Blood is the most commonly transfused human product and several viruses transmitted through this route include hepatitis A virus, hepatitis C virus, HBV, HIV, etc. The stringent screening methods have significantly lowered the risk of transfusion-transmitted viruses (TTVs). In the early 1980s, before these measures developed and screening was put into force, the risk of TTVs transmission through blood was rampant. There were various such incidents which emphasized the importance of screening techniques. The immunosuppressed condition of the organ recipients make them susceptible to any kind of infection. So, even mild infections and dormant viruses which are usually kept in check by our immune system, may cause serious disease.

3.5 Conclusion

This chapter addresses some of the main aspects of viral evolution, such as the high mutation rates, various genomic modifications, herd immunity, basic reproductive number, and the spillover events. The drivers of viral evolution also include evasion of host defenses, elude from antiviral drugs, and bypassing the vaccine-induced immunity. The spillover from animal to human of new viruses capable of infecting diverse host species can lead to emerging disease events with higher pandemic potential and these amplify by human-to-human transmission. Emerging infectious diseases threaten public health and they are further magnified by global commerce, travel, and disruption of ecological systems.

In the present time, virus evolution can be traced based on computational modeling of experimental data. This type of evolutionary studies assists in comparing related viruses that infect phylogenetically similar hosts and provide insights regarding viral evolution and host responses. Antigenic variations created by the virus help us in deciphering the epidemiologic dynamics and immune escape responses. The ease of sequencing of viral genome has led to better interpretation of viral evolution through laboratory experiments, their adaptation, disease-causing potential, and geographical coverage around the globe. By deciphering the major mechanisms of viral evolution, the globally interconnected research networks are guiding the new strategies that control, treat, and possibly eradicate viral threats.

The knowledge of modes of transmission and the manner in which they affect the trajectory of the infection and their final outcome is of utmost importance. Breaking the chain of transmission is a crucial aspect to interrupt the spread of infection. This is where enhancing the community awareness, infection prevention, and control strategies can be most successful. The understanding of these phenomena and their relationships guides in improving the design of effective and efficient immunization programs aimed to control, eliminate, or eradicate the vaccine-preventable infectious diseases.

References

[1] Sanjuán R, Nebot MR, Chirico N, Mansky LM, Belshaw R. Viral mutation rates. J Virol 2010;84(19):9733–48. Available from: https://doi.org/10.1128/JVI0.00694-10.
[2] Stern A, Andino R. Viral evolution: it is all about mutations. In: Katze M, Korth M, Law GL, Nathanson N, editors. Viral pathogenesis. New York: Academic Press; 2016. p. 233–40.
[3] Sanjuán R, Domingo-Calap P. Mechanisms of viral mutation. Cell Mol Life Sci 2016;73(23):4433–48. Available from: https://doi.org/10.1007/s00018-016-2299-6.

[4] Fleischmann Jr. WR. Viral genetics In: Baron S, editor. Medical microbiology. 4th ed. Galveston, TX: University of Texas Medical Branch at Galveston; 1996Chapter 43. Available from: https://www.ncbi.nlm.nih.gov/books/NBK8439/.

[5] Rodpothong P, Auewarakul P. Viral evolution and transmission effectiveness. World J Virol 2012;1(5):131−4. Available from: https://doi.org/10.5501/wjv.v1.i5.131.

[6] Chakravarti A, Upadhyay S, Bharara T, Broor S. Current understanding, knowledge gaps and a perspective on the future of COVID-19 infections: a systematic review. Indian J Med Microbiol 2020;38(1):1−8. Available from: https://doi.org/10.4103/ijmm.IJMM_20_138.

[7] Piret J, Boivin G. Pandemics throughout history. Front Microbiol 2020;11:631736. Available from: https://doi.org/10.3389/fmicb.2020.631736.

[8] Weaver SC, Charlier C, Vasilakis N, Lecuit MZ. Chikungunya, and other emerging vector-borne viral diseases. Annu Rev Med 2018;69:395−408. Available from: https://doi.org/10.1146/annurev-med-050715-105122.

[9] Johnson CK, Hitchens PL, Evans TS, et al. Spillover and pandemic properties of zoonotic viruses with high host plasticity. Sci Rep 2015;5:14830. Available from: https://doi.org/10.1038/srep14830.

[10] Mourya DT, Yadav PD, Ullas PT, Bhardwaj SD, Sahay RR, Chadha MS, et al. Emerging/re-emerging viral diseases & new viruses on the Indian horizon. Indian J Med Res 2019;149(4):447−67. Available from: https://doi.org/10.4103/ijmr.IJMR_1239_18.

[11] Leeks A, Sanjuán R, West SA. The evolution of collective infectious units in viruses. Virus Res 2019;265:94−101. Available from: https://doi.org/10.1016/j.virusres.2019.03.013.

[12] Louten J. Virus transmission and epidemiology. Essent Hum Virology 2016;71−92. Available from: https://doi.org/10.1016/B978-0-12-800947-5.00005-3.

[13] Arora N, Sadovsky Y, Dermody TS, Coyne CB. Microbial vertical transmission during human pregnancy. Cell Host Microbe 2017;21(5):561−7. Available from: https://doi.org/10.1016/j.chom.2017.04.007.

Further reading

Ferris MT, Heise MT, Baric RS. Chapter 13—Host genetics: it is not just the virus, stupid. In: Katze MG, Korth MJ, Law GL, Nathanson N, editors. Viral pathogenesis. 3rd ed. Boston: Academic Press; 2016. p. 169−79.

Geoghegan JL, Holmes. EC The phylogenomics of evolving virus virulence. Nat Rev Genet 2018;19:756−69. Available from: https://doi.org/10.1038/s41576-018-0055-5.

Emerging viral diseases from a vaccinology perspective: preparing for the next pandemic Barney S. Graham * and Nancy J. Sullivan*

4

Mode of viral infections and transmissions

Sora Yasri[1] and Viroj Wiwanitkit[2]

[1]Private Academic Consultant Center, Bangkok, Thailand [2]Dr DY Patil University, Pune, Maharashtra, India

4.1 Introduction

Many pathogens can cause diseases in human beings. A number of pathogens, such as bacteria, fungi, parasites, and viruses, can cause a medical problem. Virus is a specific group of pathogen that can cause infections. Several pathogenic viruses present global public health threats. Viral infection is still an important global medical problem. For the management of viral disease, basic knowledge on natural history of the infection is necessary [1].

According to the basic epidemiological triad, the viral infection requires interrelationship between a host, virus, and the environment. With an appropriate time and place of contact, one can acquire a viral infection. There are many possible modes of viral infection and different viral diseases might have different disease transmission processes. In this chapter, the authors summarize and discuss the modes of viral infections and transmissions.

4.2 Epidemiological triad and viral infection

In medical epidemiology, the important basic concept is an epidemiological triad. The concept described the three basic requirements for a medical disorder/disease to occur. The three components are host, agent, and environment. The host is the human or animal or living thing that gets attacked by the causative agent. An agent is a causative etiological factor that can lead to disorder/disease. These might be biological, chemical, or physical factors. Environment is the surrounding of host and agent. Conceptually, a disease occurs if there is an appropriate interrelationship between a host, agent, and the environment. This means the three basic components of the epidemiological triad have to exist at the same time and same place and interact altogether. The interaction must occur at an appropriate period and at a specific place that can allow complete fulfillment for disease/disorder development.

This concept is also applicable to viral infections. Regarding the epidemiological triad, a host might be a human or animal, or other living thing. Agent is hereby pathogenic virus and environment is the surrounding of host and virus. The simple diagram showing the epidemiological triad for virus infection is shown in Fig. 4.1.

The concept of epidemiological triad is a basic knowledge for further understanding of pathogenesis of disease. It can be applied to any virus infection. For example, the case of a new emerging virus infection, SARS-CoV-2 infection, will be hereby discussed. First, in the emerging respiratory syndrome caused by SARS CoV2, the host is a human being. At present, human is the main host of the disease in public health focus. Some new reports show that some other animals can also get the infection. Although there are many reports of infection in animals but the clinical significance of animal infection has never been well clarified. For agents, SARS-CoV-2 is hereby a viral pathogen. The SARS-CoV-2 is a new coranavirus, firstly reported from China. In fact, there are many new emerging viral pathogens and many new viruses can cause diseases in human beings. Finally, the environment surrounding is the space where host and agent exist together. When there is an appropriate interrelationship to finally result in infection. Conceptually, these conditions

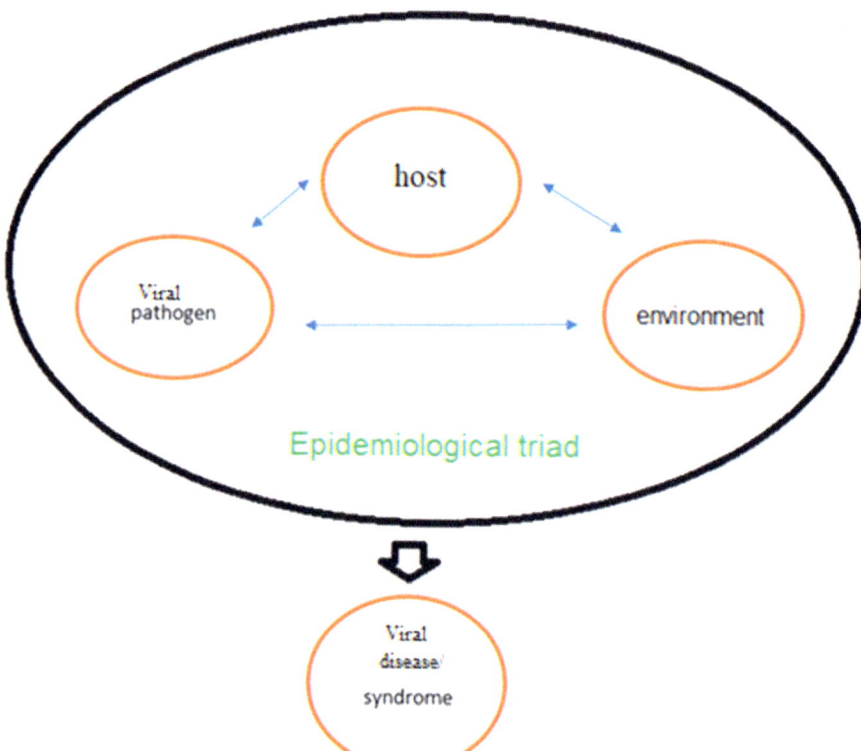

FIGURE 4.1   Epidemiological triad for virus infection.

TABLE 4.1 Examples of interesting reports on epidemiological triad of important viral infections.

Authors	Details
Martínez et al. [2]	Martínez et al. studied the factors of the epidemiological triad that influence the persistence of human papillomavirus infection in women with systemic lupus erythematosus [2]. Martínez et al. found that there was no factor of the epidemiologic triad which was associated with human papillomavirus infection prevalence [2].
Nganwa et al. [3].	Nganwa et al. reported on applying the epidemiologic problem-oriented approach methodology in developing a knowledge base for the modeling of HIV/AIDS [3]. In their study, Nganwa et al. noted that the agent was the causative agent, HIV, and its characteristics, the host was human and the environment included the physical, biological, and socioeconomic environments for both the host and agent [3].
Spicknall et al. [4]	Spicknall et al. discussed informing optimal environmental influenza interventions for influenza [4]. Spicknall et al. noted that influenza could be transmitted through respirable, inspirable, direct-droplet-spray, and contact modes and further discussed on impact of the epidemiogical triad [4]. Spicknall et al. focused on host population (behavior, susceptibility, or shedding profiles), agent (infectivity, survivability, transferability, or shedding profiles), and environment (host density, surface area to volume ratios, or host movement patterns) [4].

must exist: (1) there are both host and pathogen, which are hereby human/animal and SARS-CoV-2, (2) both host and pathogen live at the same place and same time, (3) there must be interrelationship, or contact, which is hereby a respiratory contact. Without completeness of all three requirements, the SARS-CoV-2 infection cannot occur. There are some interesting reports on epidemiological triad of important viral infections and examples are presented in Table 4.1.

For any clinical epidemiological investigation, the three compartments of epidemiological triad have to be assessed. The knowledge on the epidemiological triad of a viral infectious disease is useful for diagnosis, treatment, as well as prevention of those infections.

4.3 Transmission of infection and clinical presentation

Transmission means the passing of a pathogen causing communicable disease from an infected person/animal to another/others. As earlier noted, there are many possible modes of transmission of virus infection. The transmission will occur if there is a fulfillment of the epidemiological triad.

It can conclude that transmission of viral infection requires (1) existence of pathogen, host, and common environment and (2) appropriate interaction at the same common time and place. The transmission of the pathogen from one to one (3) might occur and might or might not further cause disease. The occurrence of disease in pathogen recipient depends on several factors. First, the physiological background of the host is an important determinant for a final clinical problem. When a pathogen invades the body, it will be recognized as a foreign body and there will be a physiological response by nonspecific and specific defense mechanisms. Immunity plays a very important role. However, many conditions can deteriorate immunity. For example, old age can result in impaired immune status. Many medical problems can also cause decreased immune status. Good examples are diabetes mellitus, cancer, posttransplantation, and having an immunosuppressive drug. Also, some specific infections such as HIV can result in immunodeficiency. With strong immunity, the clinical problem might not occur. On the other hand, severe clinical presentation might be observed in an immunocompromised host.

Regarding pathogens, different kinds of pathogens have different invasiveness or virulence. A pathogen with high virulence can cause a severe clinical presentation and the patient might finally die. On the other hand, a disease with low virulence can result in mild symptomatic presentation. The infected person might have a long-term subclinical infection and might be a carrier of the disease. A pathogen might also have its genetic evolution. A mutation might occur and there might be a new sense mutation causing a new problematic variant. Regarding viral infection, many new variants are reported, and become a global public health problem.

Finally, the different environmental backgrounds might have different impacts on disease presentation. Some basic climatic factors, such as rainfall and temperature, might promote some infection and it will be further associated with the incidence of disease. In addition, a different climate can directly affect physiological regulation in body. Based on these facts, transmission of pathogen might occur but the final clinical presentation of the pathogen recipient might be different.

4.4 Modes of transmission of viral infection

As already mentioned, one can have a viral disease only through the pathogen. The viral pathogen can be transmitted from an infected person to another or others. An infection might have one or more mode of transmission (Table 4.2). Some modes of transmission of viral infection are considered common whereas the others are considered atypical modes. The data on mode of transmission of a viral infection is important for diagnostic, therapeutic, and preventive management.

1. Vertical transmission

    Vertical transmission means the direct passage of a pathogen from mother to child during the period immediately before and after birth. Vertical transmission is reported in many viral diseases. HIV and Zika virus infection is well documented for vertical transmission. For a vertical transmission, the host is the child. This is commonly fetus in utero. During pregnancy, the fetus in utero directly connects to the mother via the placenta. The agent is the virus. The environment is a close environment occurring in

TABLE 4.2 Examples of viral infections with single and multiple mode of transmission.

Mode	Examples
Single	Influenza, parainfluenza
Multiple	HIV, dengue, Zika virus

the uterus of the mother. Since it is close contact, it is no doubt that there is a common time and place permitting host and agent to have appropriate interrelationship and further cause clinical problems.

Vertical transmission and congenital infections are reported in many diseases caused by several viruses including Zika virus inf, parvovirus, varicella-zoster virus, Rubella, Cytomegalovirus, HIV, and Herpesvirus [5]. The virus that can cause vertical transmission has to have a small size enough to pass through the natural placental pore barrier. Theoretically, if the pathogen particle is smaller than the barrier of the placenta, the pathogen might cross the placenta to infect the fetus. This is a basic nanophysics principle. Smaller objects can successfully pass through a large passage. Hence, a basic concern for evaluating the possibility of transplancental crossing to cause vertical transmission is size [6−8].

Vertical transmission is the earliest possible mode of transmission of a viral infection in a human lifespan since it occurs at the prenatal stage. Vertical transmission is usually considered important in clinical medicine and obstetrics. When there is a newly emerging viral disease, the common question is whether the disease can be transmitted via vertical transmission or not. There are some interesting reports on some important infections and the mechanism of vertical transmission as presented in Table 4.3.

2. Transmission via breastmilk

Infantile viral infection associated with breastfeeding has already been recognized. Infantile infection due to breastfeeding remains an interesting question. Clinically, the

TABLE 4.3 Examples of some interesting reports on some important infections and mechanism of vertical transmission.

Viral infection	Details
COVID-19	Many studies investigate the possibility of vertical transmission of COVID-19 but there is still no case of vertical transmission [44−46]. This is concordant with the fact that the pathogen is a large virus and should not be able to cross the placenta to cause a vertical transmission [47]. The possibility vertical transmission of COVID-19 is an interesting topic for further research.
Dengue	Vertical transmission is a rare mode of dengue transmission [24]. Basurko et al.suggested that "*if there is a known history of maternal dengue during pregnancy, or if there is fever during the 15 days before term, cord blood and placenta should be sampled after delivery and tested for the virus, and the newborn should be closely monitored during the postpartum period* [48]."
Zika virus	Vertical transmission of Zika virus is widely mentioned and the association with congenital microcephaly is the very important topic in infectious medicine. Brasil et al. noted that "*Zika virus vertical transmission is frequent but laboratory confirmed infection is not necessarily associated with infant abnormalities* [49]."
HIV	Vertical transmission is well studied in HIV infection [50]. Since HIV is very small, it can easily cross placenta and result in a vertical transmission [7]. The present concern in public health is prevention of vertical transmission of HIV. Lynch and Johnson suggested that "*Timely identification, intervention, and treatment are necessary to prevent maternal to child transmission of HIV* [51]" and "*Membrane rupture duration is not associated with higher transmission rates with adequate viral suppression* [51]."

possibility of viral transmission via breastmilk is an important concern in clinical pediatrics. As earlier discussed in the issue of vertical transmission, if the pathogen size is smaller than the breastmilk duct pore, the pathogen might be secreted via breastmilk and might cause a clinical problem.

A transmission via breastmilk and infantile infection is reported in many diseases caused by several viruses. HIV and cytomegalovirus and human T-cell lymphotropic virus type 1are the best examples [9]. Nevertheless, although a virus might not pass through breastmilk, breastfeeding is a closed contact activity. It might promote disease transmission via respiratory contact. This becomes a new concern in the present situation of the COVID-19 pandemic [10].

When there is a newly emerging viral disease, the common question is whether the disease can be transmitted via breastmilk or not. There are some interesting reports on some important infections and the mechanism of vertical transmission as presented in Table 4.4.

3. Transmission via respiratory contact

Transmission via respiratory contact is a common mode of transmission in many diseases. Many viral diseases are known as respiratory viruses and become important

TABLE 4.4 Examples of some interesting reports on some important infections and mechanism of transmission via breastmilk.

Viral infection	Details
COVID-19	Many studies investigate the possibility of transmission of COVID-19 via breastmilk but there is still no confirmation case. This is concordant with the fact that the pathogen is a large virus and should not be able to secrete via breastmilk [10]. Hence, contaminated breastmilk from the mother with COVID-19 is not expected to cause COVID in infants [52], and there should be no opportunity for transmission of COVID-19 via breastmilk. Nevertheless, close contact during breastfeeding, without a good distancing, might result in disease contract COVID-19 in infants from their mothers [36].
Dengue	Transmission via breastmilk is proposed as a possible mode of dengue transmission [24]. A possible referencing case is published by Barthel et al. [53]. In that cse, the virus was detected and detected in sequential blood samples from mother and child as well as in breastmilk, but not in cord blood sample [53].
Zika virus	Breastfeeding transmission of Zika virus is widely studied. The transmission via breastmilk is already confirmed in animal model [54]. Therefore risk and benefit of breastfeeding becomes a widely discussed issue. Mann et al. concluded that *"Because the health advantages of breast feeding are considered greater than the potential risk of transmission, the World Health Organization recommends that mothers with possible or confirmed Zika virus infection or exposure continue to breast feed [55]."*
HIV	Breastfeeding transmission is well studied in HIV infection [56]. Since HIV is very small, it can easily pass breastmilk pore and result in a breastfeeding transmission. John-Stewart et al. suggested that *"Transmission through breast-feeding can take place at any point during lactation, and the cumulative probability of acquisition of infection increases with duration of breast-feeding. HIV-1 has been detected in breastmilk in cell-free and cellular compartments [57]."* The present concern in public health is prevention of breastfeeding transmission of HIV.

public health worldwide. Many pandemics are associated with respiratory viruses. The best examples are Spanish flu and COVID-19.

For transmission via respiratory contact, the host gets an infection in a process that is related to respiration. The agent is the respiratory virus. The environment is the surrounding air of the host. If the space of place is limited, there will be an increasing chance of interrelationship and there might be an increased risk of transmission via the respiratory tract. Hence, the concept for prevention of transmission is good ventilation and no crowding.

Based on infection control principles, the transmission via respiratory contacts can be classified into three subgroups [11].

1. Droplet transmission

Humans can generate fluids and secretions in the airway. The droplets might be produced by several actions such as sneezing, laughing, shouting, coughing, and singing. Large droplets, more than 5 μm, are the main parts of expelled respiratory droplets and those large droplets will fall rapidly on the surrounding land. Hence, environmental contamination is common on land and if a person is in contact with contaminated surface, the transmission might occur.

Hence, facemask will be useful for prevention from expelled contaminated droplets from an infected person and hand washing can help destroy viruses on contaminated hand acquired due to contact with a contaminated surface.

2. Air-borne transmission

Regarding droplets smaller than 5 μm, called droplet nuclei, they can be freely suspended in the air for significant periods of time. Air current might move those small droplets away. This can result in a contamination in surrounding air. If a person stays in that place and directly breathes or inhales the contaminated air, infectious droplet will be passed directly into the airway and might cause disease. This transmission process is called air-borne transmission.

Many diseases are confirmed for air-borne mode of transmission. The most well-known disease is tuberculosis. For viral pathogens, the air-borne mode of transmission is reported in measles and chickenspox. For the new emerging respiratory virus infection, COVID-19, there is a possibility for air-borne transmission [12].

For an air-borne disease, a simple preventive apparatus might not be useful. If the respiratory pathogen is extremely small, the simple face mask might not prevent transmission. The nanostructure of face masks is an important concern in its effectiveness in protection from a viral pathogen. Any kind of face mask including the N95 facemask will be useless if the pathogen is extremely small and can directly penetrate through pores of the facemask [13,14].

4. Blood-borne transmission

Viremia is observable in many viral infections, implying that the virus can be identified in blood and can cause disease transmission. Regarding contamination, a pathogen might exist in plasma or cellular components in blood. However, virus mainly stays in blood cell, hence, blood cell is the main component that can induce blood-borne transmission. At present, the most well-known blood-borne virus is HIsV.

HIV can be contaminated in lymphocyte in blood and if a person contacts and acquires infectious blood in circulation, the transmission of HIV might occur.

Blood banks should implement safety procedures against contamination through blood-borne pathogen. Basic screening for contamination in blood and blood product is necessary.

For a transmission via blood-borne mode, the host gets an infection in a process that is related to blood or blood product contact. The agent is the blood-borne pathogen.

Examples of viral blood-borne pathogens are hepatitis B (HBV), hepatitis C (HCV), hepatitis G (HGV), hepatitis T, and HIV [15,16]. When there is a newly emerging viral disease, the common question is whether the disease can be transmitted via blood-borne mode or not. Regarding the new SARS-CoV-2, viremia exists but therein no confirmed case of blood-borne transmission of COVID-19 is observed. Nevertheless, SARS-CoV-2 is detectable in convalescent patients for a long time [17], hence, using donated convalescent plasma from patients recovering from COVID-19 to treat COVID-19 patients, remains a possible risk [18,19].

5. Transplantation-related transmission

Transplantation-related transmission is a specific mode of transmission, which is considered a complication of medical therapy. Basically, transplantation is indicated for end-stage organ failure. An organ has to be collected from a death or living donor. It Donor might have a viral disease and the pathogen might hide in a donated organ. If there is no good screening, the donated organ product might be contaminated and becomes the source of disease transmission.

It is very important to confirm form safety. The problematic viruses which can cause a transplantation-related transmission are usually blood-borne pathogens. The important ones are HBV, HCV, and HIV [16]. The virus might be present in internal organs, if a contaminated organ is used for transplantation, disease transmission to organ recipient is possible. In general, organ recipient is considered an immunocompromised host due to immunosuppressive therapy during the transplantation process, hence, if a transplantation-related transmission occurs, the disease in the organ recipient is usually serious.

When there is a new emerging viral disease, the common question is whether the disease can be transmitted via blood-borne mode or not. Regarding COVID-19, there are many reports on this topic. Azzi et al. summarized on this topic and concluded that the investigation for COVID-19 was required in transplantation process [20]. Transplantation using SARS-CoV-2 donor is already reported and there is no evidence of transplantation-related transmission [21].

6. Blood transfusion-related transmission

Blood transfusion-related transmission is a specific mode of transmission, which is considered a complication of medical therapy. Basically, a blood transfusion is indicated for anemia or blood loss. However, However, many blood-borne pathogens might be silently hidden in blood and blood product used in the blood transfusion process. If there is no good screening, the blood and blood product might be contaminated and the pathogen might be further transmitted to the blood recipient [22,23]. As an earlier note, blood safety is very important and blood screening is required for any blood bank. The safety process starts with a good selection of blood donors and the introduction of tests identifying virus-infected donors. In general, the main screened viral pathogens include HBV, HCV, and HIV [16]. In fact, other viruses might be transmitted via the blood (such as dengue [24] and cytomegalovirus [25]), however, those viruses might presently be considered of low clinical importance due to their low pathogenicity. Nevertheless, the guideline for screening has to be continuously updated.

7. Needlestick related transmission

Needlestick related transmission is a specific concern in occupational medicine. Needlestick is a common medicine in the clinical practice of medical personnel. Some viruses, especially blood-borne pathogens are transmittable via needle stick injury. Examples of viruses that might be transmitted via this mode are HIV, HBV, HCV, and dengue [16,24].

Regarding needle stick injury, the chance of disease transmission is also related to the severity of the accident [26]. Post-exposure prophylaxis is also important. Safari et al. noted that *"Through adequate management and followup of NSI low transmission rates can be achieved after exposure to blood-borne viruses within the occupational environment [26]"*

8. Vector-borne transmission

Vector is an important carrier of pathogen from an infected host to another noninfected host to cause infection. Concerning on epidemiological triad, vector is considered in the environment. The vector is usually an animal. Many insects act as vector for viral diseases. The mosquito is a good example. A mosquito borne infection is a disease transmitted by mosquito. The mosquito firstly bites a virus infected person and sucked infectious blood. When the mosquito bites another person, the pathogen can be transmitted.

Current lya, there are many well-known vector-borne diseases. The viral diseases that have a mode of vector-borne transmission are dengue, Chikungunya, and Zika virus infection. Concurrent epidemics of many vector-borne diseases are also possible [27]. In a patient, there might be more than one vector-borne viral pathogen infection (such as concurrent dengue and Chikungunya disease [27,28]).

9. Transmission through sexual contact

Sexual activity is a common human behavior that is useful for reproduction. Safe sex is important but non-safe sex is not uncommon and can result in disease. Many pathogens can be transmitted due to sexual intercourse and are considered sexually transmitted diseases. For sexually transmitted disease, the pathogen might exist in semen from a male or vaginal secretion of females. The sexual transmission of the virus is conformed in many viral infections such as HIV, HBV, HCV, and Zika virus infection. The sexually transmittable virus is usually also a blood-borne pathogen.

Sine sexually transmitted virus infection is totally due to sexual activity, sexual intercourse should be safely practiced. Conceptually, if there is no sexual contact, there will be no transmission via sexual contact.

Of interest, some viral diseases might have multiple modes of transmission, which include sexual contact. The best examples are HIV, HBV, HCV, Ebola, [29] and Zika virus [30]. Additionally, sexual intercourse is not a distanced activity and close contact is needed. A sexual contact activity also increases the risk of disease transmission via other modes such as respiratory contact transmission.

10. Zoonosis

Animals can have diseases and many viruses infect animals. The contact with animal might cause a possible transmission of virus disease to human. In general, the virus usually has species-specific pathogenicity. However, many viruses can infect several hosts including humans and animals. This is based on the existence of molecular receptors to pathogens. However, since the virus is genetically stable. A mutation easily occurs and it might result in cross-species infection. A cross-species infection from animal to human is called a zoonosis [31].

Many new emerging diseases, such as bird flu, are considered zoonotic viral diseases. Zoonosis is usually the first mode of transmission bringing a new emerging viral disease. For example, avian flu is believed to be originated in avians and crossed species to human. On the other hand, the reverse zoonosis is also possible. For example, SARS-CoV-2 infection, a new virus infection in human is reported for cross-species infection from a pet owner to animal species, cats [32,33].

11. Orofecal transmission

Regarding orofecal transmission, the first requirement is the existence of pathogen in gastrointestinal tract. The pathogenic virus might exist in the feces of infected patients and contamination of stool is possible. If there is poor toileting sanitation, hand contact with contaminated feces is possible. If there is no good hand washing, the dirty hand might accidentally bring the pathogen into gastrointestinal tract and further cause disease if the pathogen tolerates gastric acidity and gut immunity. The problematic virus in this group is called enteric virus. Poliovirus is a well-known virus that has the orofecal mode of transmission. Another important orofecal pathogen is hepatitis virus [34]. The chance that environmental contamination from virus-containing feces is widely discussed and is an important consideration in sanitation

12. Food-borne and water-borne transmission

These modes of transmission share similar considerations to the possibility of orofecal transmission of COVID-19. However, contamination might or might not be primarily from contaminated feces. Food or drink might be contaminated by infectious particles and becomes infectious food or drink [35]. Hence, contaminated food or drink can act as a vehicle to carry pathogens into a human body. It can further cause disease if the pathogen tolerates gastric acidity and gut immunity.

Another consideration is the possibility of disease transmission from contamination by infectious saliva. The pathogenic virus, especially respiratory pathogen can exist in saliva. The disease transmission might occur through sharing of dish, glasses, spoon, or fork. If the above-mentioned utensil is used for eating by an infected person, it might further spread to others who share that utensil [36].

13. Transmission via skin and muscosa contact

Skin is a barrier for the foreign body. Direct skin contact is not a route for virus disease transmission. However, if there is a skin lesion, such as leakage, it might be the way for the pathogen entrance. In addition, some specific dermatopathogenic viruses such as herpes simplex virus, human papillomavirus, and molluscum contagiosum virus can cause clinical problem via direct skin contact with infectious skin lesion [37].

Similar to skin contact, if there is a mucosal lesion, such as leakage, it might be the way for pathogen entering. This is an important pathological process of sexually transmitted disease via leakage of genital mucosa leakage. Additionally, in case of oral sexual contact, the oral mucosa might be a vulnerable point for getting sexually transmitted virus [38]. In animal model, Zika virus transmission via the oropharyngeal mucosa contact is reported, However, there is still no evidence of human cases [39].

14. Ocular transmission

Ocular transmission of the virus is an interesting issue in clinical ophthalmology. Hand-to-eye contact is the main mode of transmission for adenoviral conjunctivitis [40]. For COVID-19. the potential for ocular transmission of SARS-CoV-2 is mentioned [41]. Barnett et al. noted that *"There is evidence that SARS-CoV-2 may either directly infect cells on the ocular surface, or virus can be carried by tears through the nasolacrimal duct to infect the nasal or gastrointestinal epithelium [41]."*

15. Transmission from a dead body

This possible mode of transmission is widely discussed. In general, many pathogens

can be isolated from the secretions or organs of dead bodies, confirming that it is infectious. However, there is limited evidence of viral transmission from a dead body.

An important consideration occurs during the present COVID-19 pandemic. There are numerous dead bodies and the infectivity of the corpses become a big public health consideration. The first report from Indochina indicated the risk of COVID-19 transmission from dead body contact [42]. At present, safety procedures and personal preventive equipment are required for manipulating death body [43].

4.5 Conclusion

Viral infection is an important group of infectious disease. The epidemiological triad can well explain the viral infection. The virus might be transmitted from an infected case to other(s). The result of viral transmission might be different varying on host background, pathogen characteristics, and environmental conditions. There are several modes of transmission of viruses. An infection might have one or more modes of transmission. Some modes of transmission are considered common whereas the others are considered atypical modes. The data on the mode of transmission of a viral infection is important for diagnostic, therapeutic, and preventive management. New research on the mode of transmission of viral disease are useful and can help improve clinical management of a viral disease.

4.6 Conflict of interest

None

References

[1] Gulis G, Fujino Y. Epidemiology, population health, and health impact assessment. J Epidemiol 2015;25 (3):179—80.

[2] Martínez SM, García-Carrasco M, Jiménez-Herrera EA, Pinto CM, Morales IE, Rubio PWB, et al. Factors of the epidemiological triad that influence the persistence of human papillomavirus infection in women with systemic lupus erythematosus. Lupus 2018;27(9):1542—6.

[3] Nganwa D, Habtemariam T, Tameru B, Gerbi G, Bogale A, Robnett V, et al. Applying the epidemiologic problem oriented approach (EPOA) methodology in developing a knowledge base for the modeling of HIV/ AIDS David. Applying the epidemiologic problem oriented approach (EPOA) methodology in developing a knowledge base for the modeling of HIV/AIDS. Ethn Dis Winter 2010;20(Suppl. 1) S1—173-7.

[4] Spicknall IH, Koopman JS, Nicas M, Pujol JM, Li S, Eisenberg JN. Informing optimal environmental influenza interventions: how the host, agent, and environment alter dominant routes of transmission. PLoS Comput Biol 2010;6(10):e1000969.

[5] Arora N, Sadovsky Y, Dermody TS, Coyne CB. Microbial vertical transmission during human pregnancy. Cell Host Microbe 2017;21(5):561—7.

[6] Wiwanitkit V. New emerging blood-borne hepatitis viral pathogens and the feasibility of passing thorough the placenta: an appraisal. Clin Exp Obstet Gynecol 2006;33(4):213—14.

[7] Wiwanitkit V. Re: HIV transmission from mother to child: an aspect on the placenta barrier at the nano-level. Aust N Z J Obstet Gynaecol 2005;45(6):539—40.

[8] Wiwanitkit V. Can avian bird flu virus pass through the eggshell? An appraisal and implications for infection control. Am J Infect Control 2007;35(1):71.

[9] Prendergast AJ, Goga AE, Waitt C, Gessain A, Taylor GP, Rollins N, et al. Transmission of CMV, HTLV-1, and HIV through breastmilk. Lancet Child Adolesc Health 2019;3(4):264—73.

[10] Joob B, Wiwanitkit V. COVID-19: transmission and breastfeeding. Sri Lanka J Child Health 2020;49:198.

[11] Shiu EYC, Leung NHL, Cowling BJ. Controversy around air-borne vs droplet transmission of respiratory viruses: implication for infection prevention. Curr Opin Infect Dis 2019;32(4):372—9.

[12] Tabatabaeizadeh SA. Air-borne transmission of COVID-19 and the role of face mask to prevent it: a systematic review and *meta*-analysis. Eur J Med Res 2021;26(1):1.

[13] Wiwanitkit V. N-95 face mask for prevention of bird flu virus: an appraisal of nanostructure and implication for infectious control. Lung. 2006;184(6):373—4.

[14] Wiwanitkit V, Sriwijitalai W. N-95 face mask for prevention of wuhan novel coronavirus: it is actually effective? Int J Prev Med 2020;11:81.

[15] Gürtler L. Blood-borne viral infections. Blood Coagul Fibrinolysis 1994;5(Suppl. 3):S5 —10.

[16] Goldmann DA. Blood-borne pathogens and nosocomial infections. J Allergy Clin Immunol 2002;110(2 Suppl.):S21—6.

[17] Carmo A, Pereira-Vaz J, Mota V, Mendes A, Morais C, da Silva AC, et al. Clearance and persistence of SARS-CoV-2 RNA in patients with COVID-19. J Med Virol 2020;92(10):2227—31.

[18] Li Y, Liu S, Zhang S, Ju Q, Zhang S, Yang Y, et al. Current treatment approaches for COVID-19 and the clinical value of transfusion-related technologies. Transfus Apher Sci 2020;59(5):102839.

[19] Joob B, Wiwanitkit V. Convalescent plasma and COVID-19 treatment. Minerva Med 2020;. Available from: https://doi.org/10.23736/S0026-4806.20.06670-7 Online ahead of print.

[20] Azzi Y, Bartash R, Scalea J, Loarte-Campos P, Akalin E. COVID-19 and solid organ transplantation: a review article. Transplantation. 2021;105(1):37—55.

[21] Anurathapan U, Apiwattanakul N, Pakakasama S, Pongphitcha P, Thitithanyanont A, Pasomsub E, et al. Hematopoietic stem cell transplantation from an infected SARS-CoV2 donor sibling. Bone Marrow Transpl 2020;55(12):2359—60.

[22] Goodnough LT, Panigrahi AK. Blood transfusion therapy. Med Clin North Am 2017;101(2):431—47.

[23] Mollison PL, Engelfriet P. Blood transfusion. Semin Hematol 1999;36:48—58 (4 Suppl. 7).

[24] Wiwanitkit V. Unusual mode of transmission of dengue. J Infect Dev Ctries 2009;4(1):51—4.

[25] Ziemann M, Thiele T. Transfusion-transmitted CMV infection—current knowledge and future perspectives. Transfus Med 2017;27(4):238—48.

[26] Safari N, Rabenau HF, Stephan C, Wutzler S, Marzi I, Wicker S. High-risk needlestick injuries and virus transmission: a prospective observational study. Unfallchirurg 2020;123(1):36—42.

[27] Furuya-Kanamori L, Liang S, Milinovich G, Soares Magalhaes RJ, Clements AC, Hu W, et al. Co-distribution and co-infection of chikungunya and dengue viruses. BMC Infect Dis 2016;16:84.

[28] Sookaromdee P, Wiwanitkit V. Dengue fever with chikungunya concurrent infection. Ci Ji Yi Xue Za Zhi 2019;31(1):66.

[29] Rogstad KE, Tunbridge A. Ebola virus as a sexually transmitted infection. Curr Opin Infect Dis 2015;28(1):83—5.

[30] Hastings AK, Fikrig E. Zika virus and sexual transmission: a new route of transmission for mosquito-borne flaviviruses. Yale J Biol Med 2017;90(2):325—30.

[31] McArthur DB. Emerging infectious diseases. Nurs Clin North Am 2019;54(2):297—311.

[32] Newman A, Smith D, Ghai RR, Wallace RM, Torchetti MK, Loiacono C, et al. First reported cases of SARS-CoV-2 infection in companion animals—New York, March—April 2020. MMWR Morb Mortal Wkly Rep 2020;69:710—13.

[33] Ristow LE, Carvalho OV, Gebara RR. COVID-19 in felines, their role in human health and possible implications for their guardians and health surveillance. Epidemiol Serv Saude 2020;29:e2020228.

[34] Pascal JP. Transmission and prevention of viral hepatitis. Rev Prat 1995;45(2):174—9.

[35] Appleton H. Control of food-borne viruses. Br Med Bull 2000;56(1):172—83.

[36] Wiwanitkit V. Atypical modes of COVID-19 transmission: how likely are they? Epidemiol Health 2020;42:e2020059.

[37] Nowicka D, Oleszczuk MB, Maj J. Infectious diseases of the skin in contact sports. Adv Clin Exp Med 2020;29(12):1491—5.

[38] Younai FS. Oral HIV transmission. J Calif Dent Assoc 2001;29(2):142—8.

[39] Newman CM, Dudley DM, Aliota MT, Weiler AM, Barry GL, Mohns MS, et al. Oropharyngeal mucosal transmission of Zika virus in rhesus macaques. Nat Commun 2017;8(1):169.

[40] Jhanji V, Chan TC, Li EY, Agarwal K, Vajpayee RB. Surv adenoviral keratoconjunctivitis. Ophthalmology 2015;60(5):435—43.

[41] Barnett BP, Wahlin K, Krawczyk M, Spencer D, Welsbie D, Afshari N, et al. Potential of ocular transmission of SARS-CoV-2: a review. Vis (Basel) 2020;4(3):40.

[42] Yasri S, Wiwanitkit V. Sharing cigarette smoking and COVID-19 outbreak in a party group. Int J Prev Med 2020;11:50.

[43] Joob B, Wiwanitkit V. COVID-19 and management of the corpse. Pathologica 2020;112:78.

[44] Hijona Elósegui JJ, Carballo García AL, Fernández Risquez AC, Bermúdez Quintana M, Expósito Montes JF. Does the maternal-fetal transmission of SARS-CoV-2 occur during pregnancy? Rev Clin Esp 2020;.

[45] Walker KF, O'Donoghue K, Grace N, Dorling J, Comeau JL, Li W, et al. Maternal transmission of SARS-COV-2 to the neonate, and possible routes for such transmission: a systematic review and critical analysis. BJOG. 2020;127(11):1324—36.

[46] Huntley BJ, Huntley ES, Di Mascio D, Chen T, Berghella V, Chauhan SP. Rates of maternal and perinatal mortality and vertical transmission in pregnancies complicated by severe acute respiratory syndrome coronavirus 2 (SARS-Co-V-2) infection: a systematic review. Obstet Gynecol 2020;136:303—12.

[47] Sriwijitalai W, Wiwanitkit V. Comparative nanostructure consideration on novel coronavirus and possibility of transplacental transmission. Am J Obstet Gynecol 2020;223(6):955.

[48] Basurko C, Matheus S, Hildéral H, Everhard S, Restrepo M, Cuadro-Alvarez E, et al. Estimating the risk of vertical transmission of dengue: a prospective study. Am J Trop Med Hyg 2018;98(6):1826—32.

[49] Brasil P, Vasconcelos Z, Kerin T, Gabaglia CR, Ribeiro IP, Bonaldo MC, et al. Zika virus vertical transmission in children with confirmed antenatal exposure. Nat Commun 2020;11(1):3510.

[50] Ghosn J, Taiwo B, Seedat S, Autran B, Katlama C. HIV. Lancet 2018;392(10148):685—97.

[51] Lynch NG, Johnson AK. Congenital HIV: prevention of maternal to child transmission. Adv Neonatal Care 2018;18(5):330—40.

[52] Yang N, Che S, Zhang J, Wang X, Tang Y, Wang J, et al. Breastfeeding of infants born to mothers with COVID-19: a rapid review. Ann Transl Med 2020;8:618.

[53] Barthel A, Gourinat AC, Cazorla C, Joubert C, Rouzeyrol MD, Descloux E. Breast milk as a possible route of vertical transmission of dengue virus? Anne. Clin Infect Dis 2013;57(3):415—17.

[54] Pang W, Lin YL, Xin R, Chen XX, Lu Y, Zheng CB, et al. Zika virus transmission via breast milk in suckling mice. Clin Microbiol Infect 2021;27(3):469 e1—469.e7.

[55] Mann TZ, Haddad LB, Williams TR, Hills SL, Read JS, Dee DL, et al. Breast milk transmission of flaviviruses in the context of Zika virus: a systematic review. Paediatr Perinat Epidemiol 2018;32(4):358−68.

[56] Black RF. Transmission of HIV-1 in the breast-feeding process. J Am Diet Assoc 1996;96(3):267−74.

[57] John-Stewart G, Mbori-Ngacha D, Ekpini R, Janoff EN, Nkengasong J, Read JS, et al. Working group on HIV in women children. Breast-feeding and transmission of HIV-1. J Acquir Immune Defic Syndr 2004;35 (2):196−202.

Transmission and intervention dynamics of SARS-CoV-2

Maame A. Korsah[1], Caleb Acquah[2] and Michael K. Danquah[3]

[1]Department of Mathematics, University of Tennessee, Chattanooga, TN, United States
[2]Faculty of Health Sciences, University of Ottawa, Ottawa, ON, Canada [3]Department of Chemical Engineering, University of Tennessee, Chattanooga, TN, United States

5.1 Introduction

Pandemics such as the 2019-novel coronavirus disease (COVID-19) are caused by highly infectious pathogens resulting in high rates of morbidity and mortality across continents. Millions of lives have been lost due to the lack of a potent cure(s). A return to the days of prepandemic normalcy, though much awaited, seems rather impossible as normalcy has been redefined. Humans have lived with the knowledge of the existence of infectious diseases for more than thousands of years. Our fight to gain independence from infectious pathogens has led to many technological and sociological advancements. Yet, the fight against infectious diseases is at its peak today as we are faced with the COVID-19 pandemic. In understanding the dynamism of the deadly infectious COVID-19, a discussion on some of the lethal infectious diseases recorded is presented.

The Black Death of the 14th century caused by bubonic, pneumonic, and septicemic plague strains had a rapid worldwide spread [1]. It had a death toll of about 75−200 million people with the death of almost 60% of Europe's population as of then. Like COVID-19, victims of the Black Death experienced symptoms such as severe coughing, which in its violent cases usually resulted in blackouts, emesis, and broken ribs. It was eventually eradicated in 1976 with the emergence of antibiotics [2]. Smallpox of ancient times killed about 300−500 million people, and it was estimated that on average three out of ten infected persons died. It was known as the Variola disease as it was caused by the Variola virus and had flu-like symptoms [3]. It is known to have emerged in the 4th century and was declared eradicated in 1980 by the World Health Organization (WHO) due to the

69

herd immunity gained from vaccinations [3,4]. Finally, cholera of the 19th century was a waterborne disease that first originated in India. It claimed the lives of millions across continents as it rapidly spread to Russia through trade routes before infecting other people in Europe and the rest of the world [5]. Since its rise, seven cholera pandemics have occurred within the past 200 years and the seventh pandemic, which occurred in 1960, started in Indonesia. Cholera is yet to be eradicated and is estimated to affect 3−5 million people with 28,800−130,000 fatalities per annum [5−7].

The 2019 coronavirus disease caused by SARS-CoV-2 has been widespread across countries and continents despite the current state of advancement in technological systems in the world. In less than a year since the first reported case in China, it had disseminated to all six continents resulting in both global health and socioeconomic challenges. Due to its rapid spread and transmission characteristics, the year 2020 saw many lockdown protocols of countries, the shutdown of companies, and schools and work at home policies of several organizations. These intervention programs were created to slow the dissemination of the virus as many healthcare facilities were running on low staff and low equipment capacity. The current pandemic is said to have cost more than the amalgamation of the world's natural disasters within the past 20 years [8] as billions of dollars have been spent in slowing the infection rate, as socioeconomic relief funds, and in the purchase of healthcare equipment. In this chapter, analyses on SARS-CoV-2, its transmission, impact, varied inventions to mitigating its spread, the potential dynamics of similar pandemics, and recommendations for future studies are presented.

5.2 Coronaviruses

Coronaviruses are known to have existed as far back as 55 million years ago [9]. They are generally spherical and may have some variation in the size of their cells and nuclei. Outbreaks of coronaviruses are known to be at their peak during winter as epidemic outbreaks, normally lasting for a few months [10]. According to taxonomists, coronaviruses are classified as members of the kingdom Orthornavirae, phylum Pisuviricota, class Pisoniviricetes, and order Nidovirales. These are viruses with a genome made of RNA and encodes an RNA-dependent RNA polymerase. Coronaviruses also belong to the family Coronaviridae which are enveloped positive-stranded RNA viruses known to infect amphibians, birds, and mammals [9,11,12]. This family consists of two subfamilies:

1. Letovirinae—this subfamily has only one accepted species known to be hosted by the ornate chorus frog. Other species though yet to be accepted are said to exist in the Pacific salmon and Murray River carp [13].
2. Orthocoronavirinae—a subfamily of viruses made up of enveloped positive-sense-single RNA strands known to infect mammals and birds [12]. They are known to have spike projections on their membrane, which are club-shaped. These spikes in electron micrographs produce an image resembling the solar corona, thus its name "coronavirus." The size of the genetic material contained within a copy of a single complete genome is about 26−32 kilobase [14]. The family Orthocoronavirinae is divided into the four genera discussed in Table 5.1.

TABLE 5.1 Classification of common coronavirus species under the four genera of the subfamily orthocoronavirinae.

Genera	Common species	Infection and hosts	References
Alphacoronavirus	Feline coronavirus	Infects the intestines of cats worldwide. It can also cause feline infectious peritonitis	[15]
	Canine coronavirus	Causes a highly infectious intestinal in dogs	[16]
	Porcine epidemic diarrhea virus	Causes diarrhea in pigs	[17]
	Human coronavirus 229E	Bats, humans	[12,18]
	Rhinolophus bat coronavirus HKU2	Bats	[19]
	Miniopterus bat coronavirus 1	Causes severe acute respiratory syndrome in bats	[18]
Betacoronavirus	Murine coronavirus	Infect the liver and causes hepatitis in mice and rats	[20]
	MERS (Middle East Respiratory Syndrome)-related coronavirus	Bats, humans	[18]
	SARS-CoV-1	Bats, humans, and palm civets	[12,18,19]
	SARS-CoV-2	Bats, humans	[12,18,21]
	Bovine coronavirus	Causes enteric and respiratory diseases and transmissible between animals and humans	[22]
	Human coronavirus HKU1	Bats, humans	[12,18]
	Hedgehog coronavirus 1	Infects the lower gastrointestinal tract of hedgehogs	[23]
Gammacoronavirus	Avian coronavirus	Results in respiratory diseases such as bronchitis and causes serious losses in unprotected birds	[24]
	Beluga whale SW1	Causes liver failure in marine mammals	[25]
Deltacoronavirus	Bulbul coronavirus HKU11	Birds	[26]
	Porcine coronavirus HKU15	Causes diarrhea in pigs	[27]

These viruses cause a variety of respiratory illnesses in humans. Seven coronavirus strains have been identified as having infected humans [12]. We discuss the first six strains below and explore the recently discovered SARS-CoV-2 virus in the next section.

- Human Coronavirus OC43 (HCoV-OC43): Originally discovered in the mid-20th century and is known to cause cold, respiratory tract infections, and pneumonia in children, elderly, and immunocompromised persons. It is said to have originated from rodents and passed on to humans through cattle as the intermediate host. It is speculated that the HCoV-OC43 virus is the causative pathogen of the 1889—1890 flu pandemic which resulted in about 1 million deaths worldwide. This speculation is being evidenced by the emergence of COVID-19 as the symptoms of both infections are closely related [28].

- Human Coronavirus HKU1 (HCoV-HKU1): This virus was first discovered in Hong Kong in January 2004. This strain of coronavirus was found in patients with symptoms of acute respiratory distress syndrome and bilateral pneumonia. In the same month, the virus was found in 10 patients with similar symptoms in Australia. It is known to have a worldwide spread through face-to-face interactions and contact with fomites. This species like other human coronavirus species causes upper respiratory disease, symptoms of common cold, and pneumonia bronchiolitis [10,29].
- Human Coronavirus NL63 (HCoV-NL63): The virus is also termed as the New Haven Coronavirus. This worldwide-confirmed virus is known to have originated from bats and palm civets and is known to infect individuals with compromised immune systems, infants and the aged. It is a recombinant virus formed when a CoV-NL63-like virus infects a host cell and combines with a CoV-229E genome already present in the host cell. These viruses are known to circulate in bats and cause upper and lower respiratory tract infections, croup, and bronchiolitis. They spread through direct contact and survive 3 h on dry surfaces and 7 days in an aqueous solution [10,11,29].
- Human Coronavirus 229E (HCoV-229E): The 229E virus was first identified in Chicago in 1965 and was later confirmed to have a worldwide spread as infections were detected in other parts of the world. Like most coronavirus infections, persons infected with the human coronavirus 229E exhibit similar respiratory infection symptoms. The virus is known to infect both bats and humans and is transmitted through the inhalation of infected respiratory droplets and contact with fomites. When introduced in cell cultures, zinc ionophore catalyst and chloroquine were found to impede the viral replication process [11,29].
- Middle East Respiratory Syndrome-related Coronavirus (MERS-CoV): This virus species first emerged in Saudi Arabia in 2012 and is known to infect humans, camels, and bats. Its rapid spread resulted in the 2015 MERS outbreak in the Republic of Korea where infected individuals developed severe respiratory infections with symptoms of fever, cough, and shortness of breath. Statistically, about 4 out of 10 MERS infected patients have died from the infection. Due to this, the WHO has identified the virus as a potential cause of future epidemic outbreaks for advanced research studies on the virus [9,14,30].
- Severe Acute Respiratory Syndrome Coronavirus (SARS-CoV-1): SARS-CoV-1 is the closest to the novel SARS-CoV-2 due to the similarities in viral structure, replication cycle, and infection symptoms. It is known to infect humans, palm civets, and bats with the latter being its natural reservoir. It initially emerged in Shunde, China in November 2002. Like the MERS-CoV and the recently discovered human coronavirus, SARS-CoV-1 resulted in the 2002—2004 SARS worldwide outbreak that caused the death of about 800 people out of the 8,096 confirmed infected cases. Though these cases were reported worldwide in about 29 countries during the outbreak, most of the cases were centered in the East of Asia. Patients infected with the virus developed symptoms such as muscle pain, headache, and pneumonia [10,11,31].

Recently, another human coronavirus known as SARS-CoV-2 has emerged. This zoonotic virus is the causative pathogen of the 2019 coronavirus disease also known as COVID-19. Its emergence and global widespread has resulted in the ongoing COVID-19 pandemic, which remains a major health problem causing over 200 million confirmed cases and 4.8 million deaths at present [32]. The COVID-19 infection first emerged in Wuhan, China, in December 2019 [9]. Though the exact source of the virus remains unknown to date, initial contact tracing of the infection was linked to the Huanan South China Seafood Markets where bats, among other animals, are sold. Thus, many scientists believe SARS-CoV-2 to be a bat coronavirus as bats are a known source of many coronavirus species [18,19].

Each SARS-CoV-2 virion has a diameter of about 50−200 nm [9,33]. The virus has four structural proteins, namely the spike proteins, envelope, virion membrane, and the nucleocapsid proteins, which contain the genetic material of the virus. The spike proteins are glycoproteins that aid in adsorption and fusion stages during the viral replication process. They are divided into two functional subunits S1 and S2 as seen in Fig. 5.1. The S1 subunit is responsible for the attachment of the virion to the host cell surface while the S2 subunits facilitate the fusion process [33].

Since the discovery of the first variant of SARS-CoV-2, different variants of the virus have emerged due to antigenic drift. Virus mutations occur when there is a recombination of the genetic material of the virus in the host cell. Some of the currently identified variants of SARS-CoV-2 by WHO and the globally accepted nomenclature are:

- Alpha variant: The variant is commonly referred to as the British or UK variant by people outside the United Kingdom and the Kent variant by people residing in the UK since the variant was first discovered in the Kent County in the UK in September 2020. It was also referred to as Lineage B.1.1.7 prior to being called the alpha variant on May 31, 2021 by WHO for standardized public communications. It is known to have a rapid dissemination rate than other variants as it has spread to more than 50 countries since its discovery. At present, all approved vaccines are effective against the alpha variant [35,36].

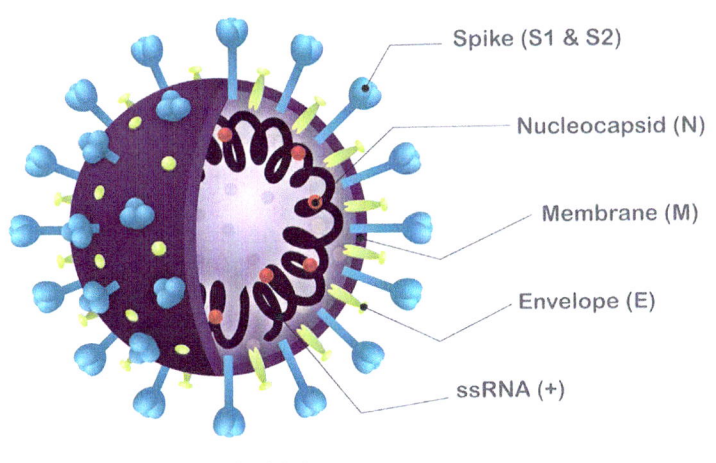

FIGURE 5.1 The schematic structure of SARS-CoV-2 showing the four proteins [34].

SARS-CoV-2

- Beta variant: This variant was first identified in South Africa in May 2020. It has been identified by scientists as the Pango lineages B.1.351, B.1.351.2, and B.1.351.3. It has a more rapid spread than other variants as it has disseminated to more than 20 countries. It was given the beta variant name by WHO on May 31. Like the alpha variant, all authorized vaccines are effective against the beta variant [35,36].
- Gamma variant: The variant was first discovered in Japan and Brazil in November 2020 and identified with the Pango lineages P.1, P.1.1, P.1.2, P.1.4, P.1.6, and P.1.7. Like the alpha and beta variants, all authorized vaccines are known to be effective against the gamma variant [35,37].
- Delta variant: It was first identified in India in December 2020 and has spread to more than 130 countries so far. Currently, the delta strain is the most highly infectious among all the discovered variants in the United States and other parts of the world. The delta variant, also known as the Pango lineages B.1.617.2, AY.1, AY.2, AY.3, and AY.3.1 causes more infections and is one of the variants of concern of most health departments. At present, there have been reports of breakthrough cases with mild symptoms among individuals with double shots of vaccine [37,38].
- Eta variant: It was identified in multiple countries including the UK in December 2020 as the Pango lineage B.1.525 and finally given the eta variant name by WHO on May 31, 2021 [35]. It is currently not a variant of concern as its confirmed cases are easily treated and not highly contagious [35].
- Iota variant: The iota strain of the SARS-CoV-2 virus was first discovered in the United States and identified as the Pango lineage B.1.526 [35].
- Kappa variant: The kappa variant was initially identified as the Pango lineage B.1.617.1. It was discovered in India in October 2020 [35].
- Lambda variant: Also labeled as the Pango lineage C.37, was initially discovered in Peru in December 2020 [37].

The first four variants are variants of concern in the United States and are being studied by health officials to avoid further outbreaks. Other variants include the lineages B.1.621, B.1.621.1, B.1.628, B.1.427, and B.1.429 which are currently of least concern due to their low transmissivity and thus remain unlabeled by WHO [37,38].

Upon entry into the human body through the various transmission pathways, SARS-CoV-2 is transported into the host cell from its surface membrane through a process known as clathrin-mediated endocytosis. In brief, SARS-CoV-2 first undergoes an adsorption stage where it attaches its spike proteins to the receptor angiotensin-converting enzyme 2 (ACE2) on the surface of the host cell membrane. Studies on the SARS-CoV-2 and host cell interactions show that the virus has high affinity toward the human ACE2 receptors and utilizes this affinity as the mechanism for cell entry [39,40]. After the viral attachment, the virion enters the endosomes, fusing the viral and lysosomal membranes, thus separating it from the remaining part of the cell [33]. After the fusion, the virion incorporates its RNA into the host cell's genetic material and induces the replication of the single-stranded RNA. Fig. 5.2 shows the entry into host cells and replication of SARS-CoV-2. The replicated virions are then disseminated to infect other healthy target cells.

Headache, loss of smell and taste, nasal congestion and runny nose, cough, muscle aches and pain, coughing up phlegm, sore throat, fever, diarrhea, and breathing difficulties

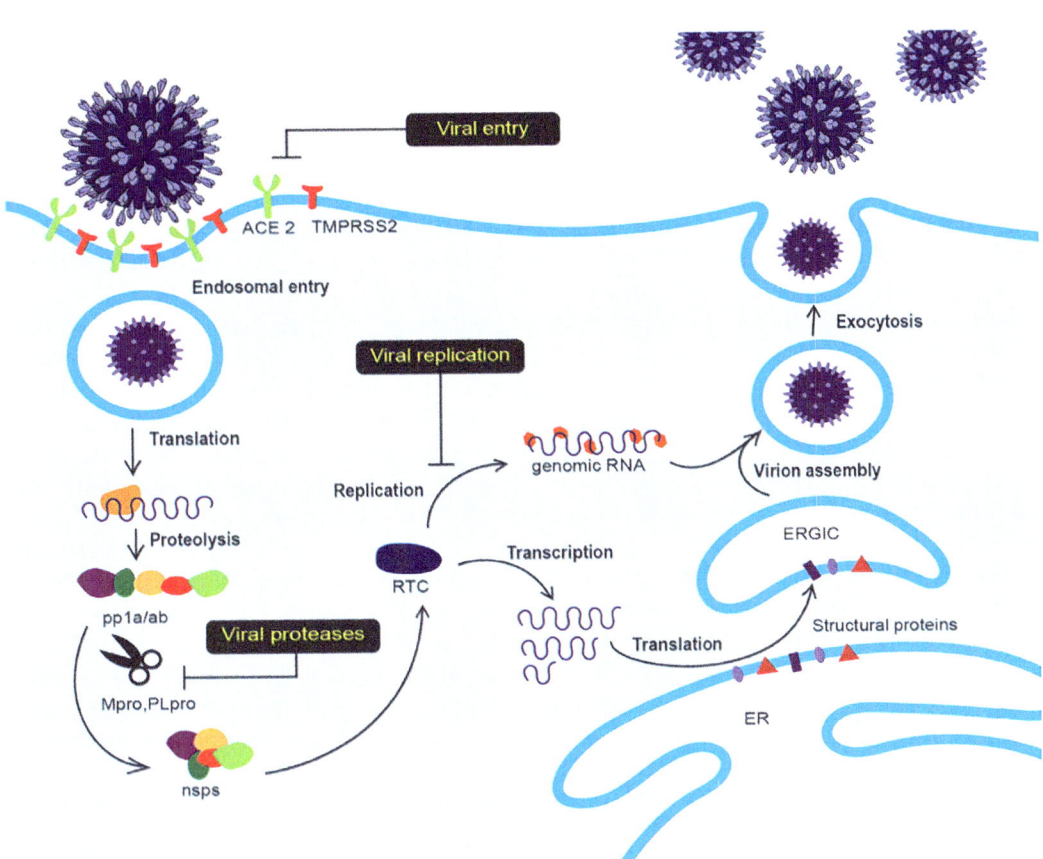

FIGURE 5.2 A pictorial illustration of the replication cycle of SARS-CoV-2 according to Jeong et al. [40]. The invasion of the host cell by the virus ensues through the process known as clathrin-mediated endocytosis.

are some of the usual symptoms of COVID-19 infection. In severe cases, infected patients experience high fever, bluish face or lips, decreased white blood cells, hemoptysis, persistent chest pains, confusion, kidney failure, and sometimes death [40,41]. Though the scientific community has advanced in the quest to stop the spread of COVID-19, there is currently no potent cure for COVID-19 infection yet but there are treatments available for most of the common symptoms experienced by infected patients. Current available treatments for COVID-19 such as Veklury (Remdesivir), which is an all-inclusive antiviral medication and a product of Gilead Sciences, has been authorized for emergency use in severe cases [39,42]. It was first developed to cure hepatitis C and was later investigated as a treatment for other viral diseases. Veklury is also known among scientists as GS-5734 and has the chemical formula $C_{27}H_{35}N_6O_8P$ and a molar mass of 602.585 g/mol. The drug was approved for adults with no prior history of liver or kidney diseases and children above the age of 12 in a COVID-19 emergency. The drug can only be administered by a healthcare professional in a clinical setting to treat any complications that may arise. However, the World Health Organization amended its guidelines in November 2020, advising against the use of Remdesivir in the treatment of COVID-19 [43]. As a result, vaccination is currently highly recommended to avoid the spread of COVID-19.

5.3 Transmission characteristics of SARS-CoV-2

The transmission media of SARS-COV-2 is related to most viral diseases of its kind. Like SARS-CoV-1 and MERS coronaviruses, SARS-CoV-2 is known to be transmissible through physical contact with an infected person or animal. It can also be transmitted via contact with fomites. In addition, SARS-COV-2 also has some peculiar characteristics that make it more difficult to fight or control than other viruses. A unique characteristic of the COVID-19 infection is the high rate of asymptomatic reports among infected individuals relative to previous coronavirus infections. Though infected with SARS-COV-2 and having a high viral load comparable to symptomatic patients', the asymptomatic patients exhibit no sign of infection well beyond the incubation period of the virus [44,45]. Due to the non-exhibition of symptoms, most of the asymptomatic patients can spread the virus as they are not isolated from the general public for treatment. Initially, asymptomatic patients were classified as noncontagious due to the assumption that their immune system has suppressed the virus, rendering it incapable and inactive. Further evidence from recent studies showed that this assumption of no infectivity of asymptomatic patients were false [41,44,46]. Data gathered so far indicates that the asymptomatic patients are contributing to the surge in SARS-COV-2 cases [46,47]. However, asymptomatic patients also have a short duration of viral shedding that may consequently result in less opportunity of infectivity. On this basis, we can say that the potential of viral spread in society by the asymptomatic patients may be comparable to that of the symptomatic and hence must be looked at critically.

Most viral diseases show clear symptoms usually after the period of incubation by which experts identify their possible presence and carry on further tests to confirm the virus as the causative agent for the viral disease. For instance, majority of SARS-CoV-1 or MERS patients had fever or high temperature, which helped in the identification of suspected cases, isolation of infected persons, testing, and the treatment of confirmed infected cases. With SARS-COV-2, however, only 10%−30% have fever, rendering fever screening less potent or effective. This is because reliance on fever screening by hospitals, clinics, and other agencies will at best identify a maximum of 30% of patients leaving about 70% unidentified to increase the spread [41,47].

There are also indications of the fact that SARS-COV-2 can be transmitted from person to person by respiratory droplets aerosolized in the air with an exposure duration above 15 mins in poorly ventilated or overcrowded areas. Both SARS-CoV-1 and MERS had airborne transmission capability [48]. Some notable cases including the SARS outbreak in Amoy Garden and the Prince of Wales hospital in Hong Kong explain the airborne capacity of SARS [49]. A study probing into the airborne transmission of COVID-19 showed that the social distance measure of at least 2 m and hand hygiene were not wholly effective in preventing the spread of SARS-COV-2 in communities [40]. However, with the inclusion of the wearing of protective masks, social distancing practices, handwashing, and surface sanitation being enforced a significant decrease in the spread of SARS-COV-2 infection was observed which indicates that the virus has an airborne transmittal capacity, especially in poorly ventilated or overcrowded areas.

One of the unusual transmission dynamics of some of the coronavirus species is what is described as the super spread effect (SSE). Here a small group of infected individuals can spread the disease pathogen to a far greater number than the average basic reproduction number of the population. Studies have shown that about 10%−20% of infected people are responsible for about 80% of the spread of the virus [50,51]. This percentage consists of asymptomatic or presymptomatic patients who are very active, unaware of their infection status. During the MERS outbreak, a 35-year-old man infected about 82 people within 58 contact hours at a hospital in South Korea [41]. A typical example of a SARS-CoV-2 super spread case was traced in Ghana when a fish factory worker unknowingly infected about 533 of his coworkers [51,52]. The exact factors causing SSEs are currently inconclusive, but it is generally assumed that the presence of asymptomatic or presymptomatic persons in crowded spaces where COVID-19 protocols are not strictly adhered to is the main factor. To curb the spread and surge, wearing masks must be mandatory, as well as all other COVID-19 safety protocols be observed whiles encouraging the ongoing vaccination programs.

In understanding the transmission dynamics of the spread of COVID-19, mathematical models have been effectively used. The SEIR (Susceptible, Exposed, Infectious, and Recovered) deterministic model and its modifications have been used to study the transmission characteristics of the novel viral infection. The SEIR framework is one of the first disease models developed by epidemiologists in efforts to understand and predict the transmission patterns of contagious disease [12,53,54]. The model simply describes the grouping of a population into four compartments: Susceptible, Exposed, Infected, and Recovered. This grouping technique is utilized with the assumption that when a susceptible person is exposed to the virus by means of any of the transmission pathways identified above, the individual moves into the exposed compartment for viral incubation, which is about 14 days for SARS-CoV-2 infection [9]. This period, which is also known as the latency period, is when the individual is infected but is mostly unable to spread the virus, thus the person is infected but not infectious. Individuals at this stage include the asymptomatic and presymptomatic. After the latency period one is moved into the Infected compartment. Here, individuals become infectious and exhibit symptoms of infection. From the infected compartment, one either moves into the Recovered compartment or is removed from the entire population by death. The four subgroups of the model can be represented by four differential equations, which estimate the flow changes of each compartment with respect to time. The SEIR model for COVID-19 can be expressed as;

$$S^{'} = \Lambda - \beta_1 SE - \beta_2 SI - \mu S + \vartheta R,$$
$$E^{'} = \beta_1 SE + \beta_2 SI - (\varepsilon + \mu)E,$$
$$I^{'} = \varepsilon E - (\rho + \mu + \delta)I,$$
$$R^{'} = \rho I - (\vartheta + \mu)R.$$

The parameters used are defined in Table 5.2.

A detailed analysis of the SEIR model together with other SEIR-modified models can be used to predict when the 2019 Coronavirus disease will eventually be eradicated as well as recommend preventive and disease-die-out strategies and programs [12]. It can also

TABLE 5.2   SEIR transmission model parameters and definitions.

Parameters	Definitions
Λ	Population influx
μ	Natural mortality rate
ϑ	Decreasing rate of immunity induced by infection
ρ	Recovery rate
ε^{-1}	Incubation period
β_1	Transmission via exposed individuals
β_2	Transmission via infected individuals
δ	Death rate of infection

help explain some characteristics of the virus/disease. For instance, obtaining the basic reproduction number, R_0 from the model under given conditions can be used to probe the number of susceptible persons each infected individual infects during the stage of infectiousness.

5.4 Intervention, strategies, and impacts

The emergence of COVID-19 and its rapid dissemination has led to the development of strategies and intervention programs intentionally meted out to mitigate the spread of the virus. Most of the early strategies formed were already in practice but were recommended and later enforced based on the conclusions of initial scientific research on the spread of the virus and its mitigation in Wuhan, China. Initial studies confirmed the person-to-person transmission pathway, thus the strategies to decrease the direct transmission were created. Some of these strategies included handwashing and use of alcohol-based sanitizers and disinfectants on surfaces. These practices were adequate in reducing the spread among communities that enforced them. For instance, Turkmenistan is known to have a total of zero cumulative confirmed cases since the emergence of COVID-19 due to its strict implementation of the initial COVID prevention strategies [55,56]. Though the strategies were implemented in most communities, the disregard of COVID-19 intervention strategies by some individuals resulted in the rapid spread among communities, nations, and continents. Due to this rapid spread to many countries in a short time, the World Health Organization on March 11, 2020 declared the COVID-19 infection a pandemic [12]. As a result, many countries closed their borders to prevent infected persons from entering and coming into contact with its residents. This border closure strategy was first implemented by North Korea on January 7, 2020 [57] and the country reportedly had effective results. Thus, when COVID-19 was declared a pandemic most countries such as the United States, Germany, Canada, Russia, Ghana, Australia, Nepal, Benin, Philippines, and New Zealand issued either a full border closure or partial border entry restrictions. Countries such as

China, Iran, South Korea, and Italy with many infected cases had certain parts of their cities and towns on lockdown to prevent the spread of the SARS-CoV-2 within their cities and towns [58]. Citizens from highly infected countries were also barred temporarily from entering other countries. These lockdown procedures, barring of entry of residents of a particular country and border closures, which is currently still in effect, have disturbed international interactions and trading transactions. For instance, the pandemic escalated the civil competition between two of the world's major competing powers and brought about the politics of blaming between the United States and China in 2020 [59].

The spread of the virus has led to a drawback in the socioeconomic well-being of many countries due to pressure on public and private health facilities. There have also been setbacks in many sectors such as education, hospitality, and aviation, and above all the crippling of most economies. Most international organizations have been taking actions to help countries in the fight against the virus. The United Nations issued $15 million from their Central Emergency Response Fund on March 1, 2020 to help developing countries combat the spread of the virus [60]. The WHO has also created a COVID-19 Solidarity Response Fund in collaboration with partners globally to donate to support the fight against COVID-19 [61]. These aids have provided some progress to containing the rampant spread of the virus. They are also assisting low- and middle-income countries in responding to the COVID-19 pandemic and its socioeconomic consequences, as well as in their recovery [60].

Vaccination is one of the major ongoing COVID-19 intervention strategies that has aided in reducing the number of infected cases. Before SARS-CoV-2 was discovered, virologists had advanced in their studies on similar human coronaviruses, such as SARS-CoV-1 and MERS. This knowledge helped in the quick development of vaccines for the COVID-19 pandemic, which were ready for the public in the first quarter of 2021. By August 16, 2021 about 4.8 billion doses of vaccines with reported efficacy as high as 95% in their phase III trials had been administered worldwide [12,62]. There are currently about 22 approved vaccines that are classified under the type of technology used in the creation of the vaccine. Presently, there are five types of technologies being utilized in the manufacture of COVID-19 vaccines:

- mRNA vaccines: These are synthetic messenger RNA that are transfected into cells and causes them to produce immune response against the virus. The two approved mRNA vaccines are Moderna and Pfizer-BioNTech [63,64].
- DNA vaccine: Similar to the mRNA vaccines, immunity is sought by injecting a specific antigen-coding DNA sequence into the body. The only approved DNA vaccine is the ZyCoV-D vaccine [64].
- Adenovirus vector vaccines: These stimulate immune response by introducing virus-expressing pathogen proteins into the body to transport pathogen genes. Oxford-AstraZeneca, Janssen, Sputnik Light, Convidecia, and Sputnik V are COVID-19 viral vector vaccines [65].
- Inactivated virus vaccines: These vaccines are based on virus particles grown in culture and subsequently inactivated before being introduced into the body to reduce infectivity and to gain immunity. Sinopharm (BBIBP), CoronaVac, Covaxin, Sinopharm (WIBP), CoviVac, QazCovid-in, Minhai COVID-19 vaccine, COVIran Barakat vaccine,

and the Chinese Academy of Medical Sciences COVID-19 vaccine are all inactivated virus vaccines [66,67].

- Subunit vaccines: These vaccines use the technology of introducing subunits or antigens (fragments) of the virus particle in the body to produce an immune response. Approved subunit vaccines are ZF2001, Abdala vaccine, Medigen (MVC), Soberana 02, and EpiVacCorona [62,67,68].

5.5 Summary

Since the emergence of the pandemic, efforts have been put in by countries and international organizations to help minimize the spread of the virus. During the first phase of the pandemic, countries had to live through a period of lockdown and restrictions on international travel. Companies worldwide initiated work from home policies, schools incorporated e-learning into their curriculum, and there were mandatory wearing of mask policies for the general public in several communities.

Recently, there has been great improvement in the combat against the COVID-19 pandemic. Vaccines have been developed to minimize the infection potential of the virus. The World Health Organization together with other organizations and research institutes are on the fast track to make sure that vaccines are safe and efficient for deployment across the globe. Also, test kits have been developed and both public and private health institutions have been provided with these kits to aid in the early detection of people infected with the virus. This is to aid the isolation of infected people from the community. Currently, world data show that as of July 19, 2021, approximately 865,422 individuals have received at least one dose of vaccine and 405,971 have received double vaccination worldwide [62].

Organizations such as the United Nations Children's Emergency Fund are actively helping in the sensitization of people across the globe about the virus, its early detection, and vaccination. Hand sanitizing and the mandatory wearing of nose masks still hold in many countries to help mitigate the global spread of the disease.

The world in the past with little technological know-how battled with plagues and other pathogenic contagious diseases. A number of these diseases like the Black Death and smallpox as discussed in Section 5.1 have been eradicated [2–4], the remaining ones like cholera, HIV-AIDS are being managed with vaccinations and better healthcare practices [5]. In this light, one can say that the hope of a COVID-free world is possible. The question then becomes, how long will COVID-19 last to be eradicated for the world to fully recover from the COVID-19 pandemic?

In conclusion, coronaviruses by virtue of being RNA viruses easily undergo genetic mutation during their replication cycle. This has resulted in several SARS-CoV-2 variants of concern. However, the knowledge obtained from previous pandemics and active coronavirus studies has significantly helped in the quick development of a plethora of effective vaccines and the deployment of various effective interventions.

Based on this discussion, the following recommendations are presented;

- Postpandemic studies and surveillance of SARS-CoV-2 and other coronaviruses need to be enhanced to adequately prepare the world for the next coronavirus infection outbreak.

- More funding, resources, and cooperations are needed to build a more vibrant and thorough contact tracing system to trace asymptomatic and presymptomatic patients at a faster rate.
- All mandatory COVID-19 safety protocols must be adhered to full participation of individuals in the ongoing vaccination process as this will help reduce the spread of COVID-19 and aid with a fast recovery from the pandemic.
- Finally, we recommend further studies into the world's recovery from similar pandemics to fast track the recovery process of the world from the COVID-19 pandemic.

References

[1] Horrox R. The black death. Manchester University Press; 2013.

[2] Gottfried RS. Black death. Simon and Schuster; 2010.

[3] Fenner F, Henderson DA, Arita I, Jezek Z, Ladnyi ID. Smallpox and its eradication, Vol. 6. Geneva: World Health Organization; 1988.

[4] Bozzette SA, Boer R, Bhatnagar V, Brower JL, Keeler EB, Morton SC, et al. A model for a smallpox-vaccination policy. N Engl J Med 2003;348(5):416−25.

[5] Ali M, Lopez AL, You Y, Kim YE, Sah B, Maskery B, et al. The global burden of cholera. Bull World Health Organ 2012;90:209−18.

[6] Colwell RR. Global climate and infectious disease: the cholera paradigm. Science 1996;274(5295):2025−31.

[7] World Health Organization. Cholera vaccines: WHO position paper. Wkly Epidemiol. Rec. = Relevé épidémiologique Hebd 2010;85(13):117−28.

[8] Noy I, Doan N. COVID-19 cost more in 2020 than the world's combined natural disasters in any of the past 20 years. The Conversation, 2021 19 April.

[9] Hasöksüz M, Kiliç S, Saraç F. Coronaviruses and sars-cov-2. Turkish J Med Sci 2020;50(SI-1):549−56.

[10] McIntosh K, Peiris J. Coronaviruses no Clinical virology, 3rd ed. American Society of Microbiology; 2009. p. 1155−71.

[11] Monto A, Cowling B, Peiris J. Coronaviruses. Viral infections of humans. Boston, MA: Springer; 2014. p. 199−223.

[12] Korsah M, Assessment of the impact of awareness programs on the transmission dynamics of COVID-19. United States: University of Tennessee at Chattanooga; 2021.

[13] Costa VA, Mifsud JCO, Gilligan D, Williamson JE, Holmes EC, Geoghegan JL. Metagenomic sequencing reveals a lack of virus exchange between native and invasive freshwater fish across the Murray-Darling Basin, Australia. Virus Evol. 2021;7(1):veab034.

[14] Woo PC, Huang Y, Lau SK, Yuen KY. Coronavirus genomics and bioinformatics analysis. Viruses 2010;2(8):1804−20.

[15] Addie DD, Jarrett O. A study of naturally occurring feline coronavirus infections in kittens. Vet Rec 1992;130(7):133−7.

[16] He HJ, Zhang W, Liang J, Lu M, Wang R, Li G, et al. Etiology and genetic evolution of canine coronavirus circulating in five provinces of China, during 2018−2019. Microb Pathog 2020;145:104209.

[17] Wrapp D, McLellan JS. The 3.1-angstrom cryo-electron microscopy structure of the porcine epidemic diarrhea virus spike protein in the prefusion conformation. J Virol 2019;93(23):e00923 19.

[18] Afelt A, Frutos R, Devaux C. ATS, coronaviruses, and deforestation: toward the emergence of novel infectious diseases? Front Microbiol 2018;9:702.

[19] Fan Y, Zhao K, Shi ZL, Zhou P. Bat coronaviruses in China. Viruses 2019;11(3):210.

[20] Bender SJ, Weiss SR. Pathogenesis of murine coronavirus in the central nervous system. J Neuroimmune Pharmacol 2010;5(3):336−54.

[21] Hoseinpour Dehkordi A, Alizadeh M, Derakhshan P, Babazadeh P, Jahandideh A. Understanding epidemic data and statistics: a case study of COVID-19. J Med Virol 2020;92(7):868−82.

[22] Yoshizawa N, Ishihara R, Omiya D, Ishitsuka M, Hirano S, Suzuki T. Application of a photocatalyst as an inactivator of bovine coronavirus. Viruses 2020;12(12):1372.

[23] Lau S, Luk H, Wong A, Fan R, Lam C, Li K, et al. Identification of a novel betacoronavirus (merbecovirus) in amur hedgehogs from China. Viruses 2019;11(11):980.

[24] Laconi A, van Beurden S, Berends A, Krämer-Kühl A, Jansen C, Spekreijse D, et al. Deletion of accessory genes 3a, 3b, 5a or 5b from avian coronavirus infectious bronchitis virus induces an attenuated phenotype both in vitro and in vivo. J Gen Virol. 2018;99(10):1381−90.

[25] Mihindukulasuriya KA, Wu G, Leger St. J, Nordhausen RW, Wang D. Identification of a novel coronavirus from a Beluga Whale by using a panviral microarray. J Virol. 2008;82(10):5084−8.

[26] Chu D, Leung C, Gilbert M, Joyner P, Ng E, Tse T, et al. Avian coronavirus in wild aquatic birds. J Virol. 2011;85(23):12815−20.

[27] Woo P, Lau S, Tsang C, Lau C, Wong P, Chow F, et al. Coronavirus HKU15 in respiratory tract of pigs and first discovery of coronavirus quasispecies in 5′-untranslated region. Emerg Microbes Infect 2017;6:1−7 vol. 1.

[28] Vijgen L, Keyaerts E, Moës E, Thoelen I, Wollants E, Lemey P, et al. Complete genomic sequence of human coronavirus OC43: molecular clock analysis suggests a relatively recent zoonotic coronavirus transmission event. J Virol 2005;79(3):1595−604.

[29] Pyrc K, Berkhout B, Van Der Hoek L. The novel human coronaviruses NL63 and HKU1. J Virol 2007;81 (7):3051−7.

[30] Mackay I, Arden K. MERS coronavirus: diagnostics, epidemiology and transmission. Virol J 2015;12(1):1−21.

[31] Holmes K, Enjuanes L. The SARS coronavirus: a postgenomic era. Science 2003;300(5624):1377−8.

[32] Johns Hopkins University. COVID-19 dashboard by the Center for Systems Science and Engineering (CSSE) at Johns Hopkins University (JHU). 2021.

[33] Shang J, Wan Y, Luo C, Ye G, Geng Q, Auerbach A, et al. Cell entry mechanisms of SARS-CoV-2. Natl Acad Sci 2020;117:11727−34.

[34] Santos Id A, Grosche RV, Bergamini FRG, Sabino-Silva R, Jardim ACG. Antivirals against coronaviruses: candidate drugs for SARS-CoV-2 treatment? Front Microbiology 2020;11(no):1818. Available from: https://doi.org/10.3389/fmicb.2020.01818.

[35] World Health Organization. WHO announces simple, easy-to-say labels for SARS-CoV-2 Variants of Interest and Concern. WHO Departmental News. 2021 31 May.

[36] Wang P, Nair M, Liu L, Iketani S, Luo Y, Guo Y, et al. Ntibody resistance of SARS-CoV-2 variants B. 1.351 and B. 1.1. 7. Nature 2021;593(7857):130−5.

[37] Abdool Karim SS, de Oliveira T. New SARS-CoV-2 variants—clinical, public health, and vaccine implications. N Engl J Med 2021;384(19):1866−8.

[38] Alizon S, Haim-Boukobza S, Foulongne V, Verdurme L, Trombert-Paolantoni S, Lecorche E, et al. Rapid spread of the SARS-CoV-2 Delta variant in some French regions, June 2021. Eurosurveillance 2021;26 (28):2100573.

[39] Zhang Y, Tang LV. Overview of targets and potential drugs of SARS-CoV-2 according to the viral replication. J Proteome Res 2020;20(1):49−59.

[40] Jeong GU, Hanra S, Gun YY, Doyoun K, Young-Chan K. Therapeutic strategies against COVID-19 and structural characterization of SARS-CoV-2: a review. Front Microbiology 2020;11:1723. Available from: https://doi.org/10.3389/fmicb.2020.01723.

[41] Bae S, Lim JS, Kim JY, Jung J, Kim SH. Transmission characteristics of SARS-CoV-2 that hinder effective control. Immune Netw 2021;21(1).

[42] Hsu J. Covid-19: what now for remdesivir? bmj 2020;371.

[43] World Health Organization. 6 July Therapeutics and COVID-19: living guideline. World Health Organization; 2021.

[44] Buitrago-Garcia D, Egli-Gany D, Counotte M, Hossmann S, Imeri H, Ipekci A, et al. Occurrence and transmission potential of asymptomatic and presymptomatic SARS-CoV-2 infections: a living systematic review and meta-analysis. PLoS Med 2020;17:e1003346.

[45] Cao S, Gan Y, Wang C, Bachmann M, Wei S, Gong J, et al. Post-lockdown SARS-CoV-2 nucleic acid screening in nearly ten million residents of Wuhan, China. Nat Commun 2020;11(1):1−7.

[46] Zhao H, Lu X, Deng Y, Tang Y, Lu J. COVID-19: asymptomatic carrier transmission is an underestimated problem. Epidemiol Infect 2020;148.

[47] Oran DP, Topol EJ. Prevalence of asymptomatic SARS-CoV-2 infection. Ann Intern Med 2020;173:362−7.

[48] Centers for Disease Control and Prevention. Scientific brief: SARS-CoV-2 transmission, COVID-19; 2021 7 May.

[49] Yu IT, Li Y, Wong TW, Tam W, Chan AT, Lee JH, et al. Evidence of airborne transmission of the severe acute respiratory syndrome virus. N Engl J Med 2004;350(17):1731−9.

[50] Adam D, Wu P, Wong J, Lau E, Tsang T, Cauchemez S, et al. Clustering and superspreading potential of SARS-CoV-2 infections in Hong Kong. Nat Med 2020;26(11):1714−19.

[51] Mohindra R, Ghai A, Brar R, Khandelwal N, Biswal M, Suri V, et al. Superspreaders: a lurking danger in the community. J Prim Care Commun Health 2021;12:2150132720987432.

[52] Hindustan Times. One person infected 533 with Covid-19 at fish factory in Ghana: President. World News. 2020 11 May.

[53] Ross R. An application of the theory of probabilities to the study of a priori pathometry.—Part I. Proc R Soc Lond Ser A, Contain Pap a Math Phys Character 1916;92(638):204−30.

[54] Kermack WO, McKendrick AG. A Contribution to the Mathematical Theory of Epidemics. Proc R Soc Lond Ser A, Contain Pap a Math Phys Character 1927;115(772):700−21.

[55] Yaylymova A. COVID-19 in Turkmenistan: no data, no health rights. Health Hum Rights 2020;22.2:325.

[56] Dyer O. Covid-19: Turkmenistan becomes first country to make vaccination mandatory for all adults. Br Med J Publ Group 2021;.

[57] Balasa AP. COVID-19 on lockdown, social distancing and flattening the curve—a review. Eur J Bus Manag Res 2020;5(3).

[58] Ghosal S, Bhattacharyya R, Majumder M. Impact of complete lockdown on total infection and death rates: A hierarchical cluster analysis. Diabetes Metab Syndr: Clin Res Rev 2020;14(4):707−11.

[59] Jaworsky B, Qiaoan R. The politics of blaming: the narrative battle between China and the US over COVID-19. J Chin Political Sci 2020;26(2):295−315.

[60] United Nations. The Secretary-General's UN COVID-19 response and recovery fund. United Nations. 2020. Available from: https://unsdg.un.org/resources/secretary-generals-un-covid-19-response-and-recovery-fund.

[61] World Health Organization. COVID-19 Solidarity Response Fund: Help WHO Fight COVID-19. World Health Organization; 2020.

[62] Ritchie H, Mathieu E, Rodés-Guirao L, Appel C, Giattino C, Ortiz-Ospina E et al. Coronavirus pandemic (COVID-19). Our World in Data. 2020. Available from: https://ourworldindata.org/coronavirus.

[63] Kowalski SP, Rudra A, Miao L, Anderson GD. Delivering the messenger: advances in technologies for therapeutic mRNA delivery. Mol Ther 2019;27(4):710−28. Available from: https://doi.org/10.1016/j.ymthe.2019.02.012. PMC 6453548. PMID 3084639.

[64] Park SK, Sun X, Aikins EM, Moon J. Non-viral COVID-19 vaccine delivery systems. Adv Drug Deliv. Rev 2021;169:137−51. Available from: https://doi.org/10.1016/j.addr.2020.12.008. PMC 7744276. PMID 33340620.

[65] U.S. Centers for Disease Control and Prevention (CDC). Understanding viral vector COVID-19 vaccines. U.S. Centers for Disease Control and Prevention (CDC). 2021 13 April. Available from: https://www.cdc.gov/coronavirus/2019-ncov/vaccines/different-vaccines/viralvector.html.

[66] Mulligan MJ. An inactivated virus candidate vaccine to prevent COVID-19. JAMA 2020;324(10):943−5.

[67] Yadav T, Srivastava N, Mishra G, Dhama K, Kumar S, Puri B, et al. Recombinant vaccines for COVID-19. Hum Vaccines Immunother 2020;16(12):2905−12.

[68] Wang N, Shang J, Jiang S, Du L. Subunit vaccines against emerging pathogenic human coronaviruses. Front Microbiol 2020;11:298.

6

Sexually transmitted viral infections

Aparajita Dasgupta

Department of Preventive and Social Medicine, All India Institute of Hygiene and
Public Health, Kolkata, West Bengal, India

6.1 Introduction

Sexually transmitted infections (STIs), as the name suggests, are caused mainly by sexual acts, which may be vaginal, anal, and oral. Beyond the sexual contact, many of these infections may also occur via mother to child which ensues during pregnancy, childbirth, and breastfeeding. Other modes of transmission which may be considered are intravenous drug use and transfusion of contaminated blood and blood products [1]. STIs may be caused by bacteria, virus, protozoa, fungus, and even ectoparasites. STIs caused by virus are disadvantageous due to the fact that they are difficult to manage, are mostly incurable and the patient may carry the virus throughout his/her life with periodic recurrence of active infection [2]. Some very common sexually transmitted diseases caused by viruses are genital warts [human papillomavirus (HPV)], genital herpes [herpes simplex virus (HSV)], AIDS [human immunodeficiency virus (HIV)], human T cell lymphotropic infection (HTLV), and hepatitis A, B, C.

In 2016, worldwide, about 300 million women were suffering from HPV infection, the primary cause of cervical cancer while more than 490 million people were afflicted with genital HSV (herpes) infection [1]. HPV infection causes cervical cancer, which is the fourth most common cancer among women globally, with an estimated 570,000 new cases in 2018 [1]. An estimated 296 million people are living with chronic hepatitis B globally. An estimated 820,000 people died due to cirrhosis and hepatocellular carcinoma, the primary aftermath of hepatitis B [1]. Unlike bacterial STIs, viral STIs are not curable; however, there are some potent vaccines for Hepatitis A, Hepatitis B and HPV infections. Beyond these vaccines, behavioral changes like consistent and correct use of condoms, practice of safe sex, high quality personal hygiene, etc. are some of the important preventive measures.

Vulnerability to STIs is maximum among the most marginalized populations, belonging to the most vulnerable strata of the society—such as commercial sex workers, homosexuals, injectable drug users, prisoners, mobile populations, and adolescents. They are often deprived of adequate healthcare and much needed health education and promotion [1]. Therefore, explicit awareness must be percolated in the general mass and among the above mentioned vulnerable population in terms of prevention and management of STIs.

Viral STIs like HPV, HSV, HTLV, and Hepatitis A, B, C are discussed in this chapter. Other STI-causing viruses include *Haemophilus ducreyi*, *Molluscum contagiosum* virus, Cytomegalovirus, Epstein-Barr virus. HIV infection is a viral STI and an important public health problem with a broad spectrum of signs and symptoms along with its variety of treatment and management, which may be discussed in a separate chapter.

6.2 Human papilloma virus (HPV infection)

HPV infection (HPV infection) is caused by a DNA virus from the *Papilloma viridae* family. It gives rise to genital warts or condylomas, which are small painless growths or lumps around the genital and perianal region often accompanied with itching, discomfort, dyspareunia, and bleeding. The warts, if untreated, coalesce together to give a cauliflower-like appearance [3]. Many HPV infections remain asymptomatic and they even resolve spontaneously [4]. It is very unfortunate that initially patients with HPV infections do not show any signs and symptoms during which there is a very good chance of transmission long before the lesions appear, which may be as long as 6 months after the viral infection [2]. When HPV infections of the genital tract become symptomatic, the infected keratinocytes proliferate in an abnormal fashion and produce genital warts [2]; (Table 6.1; Fig. 6.1). HPV infection is strongly associated with increased risk of cervical cancer, cancers of the vagina, vulva, penis, anal cancer, oral cancer, oropharyngeal cancer and other cancers (Table 6.2).

There are more than 150 subtypes of HPV virus some of which have been detected to be potentially oncogenic as depicted in the table (Table 6.2).

Good immunity status is the most important determinant for the reduction of cancer due to HPV infection. Therefore, all those who have been diagnosed with HPV infection must be counseled for behavioral changes that evoke an ideal lifestyle like regular exercise, healthy and well-balanced diet, which is low in fat and high in fruits and vegetables, and avoidance of tobacco consumption in any form [2]. The latter not only leads to decreased immunity but also is the direct causative factor for cervical cancer and other related cancers following HPV infection [2]. Regular screening for anal and cervical cancer is a must. Testing for HIV should also be suggested. The use of condoms reduces the risk of

TABLE 6.1   Description of four morphological types of genital warts [4].

Warts	Description
1. Condylomata acuminata	Have the shape of a cauliflower
2. Smooth papular warts	Dome-shaped, usually skin epithelial lesions
3. Keratotic genital warts	Have a thick, horny layer
4. Flat warts	Flat to slightly raised papules

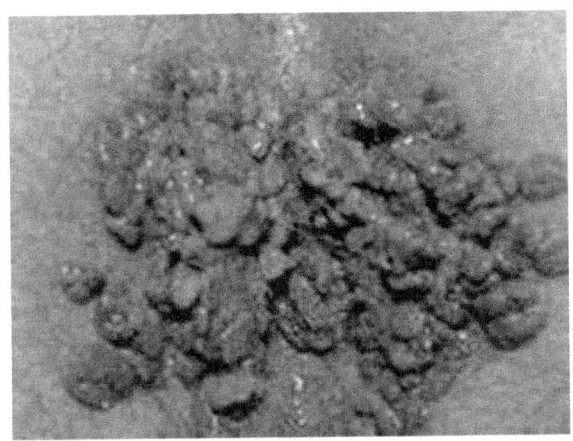

FIGURE 6.1 HPV genital warts (perianal region): shape of cauliflower. *Courtesy: http://dermishealth.com/skin photos of genital warts (no copyright required).*

TABLE 6.2 Potentially oncogenic HPV subtypes [2,4].

HPV	Causes
Types16 and 18	70% of cases of cervical cancer
Types 6 and 11	Laryngeal papillomatosis
Type 16	90% of oropharyngeal cancers in HPV infected
Types 16, 18, 45, and 56	Potential risk of anal, vaginal and vulvar cancer with a relative risk as high as 235 to 296
Types 31, 33, and 35	Oncogenic potential is intermediate for squamous intraepithelial neoplasia (squamous cell carcinoma in situ, Bowenoid papulosis, Bowen's disease of the genitalia).

transmitting HPV, but does not totally prevent it because the virus continues to exist on other skin surfaces [2]. It may be also noted that persons with HPV infection may transmit the infection even after the warts are gone and are asymptomatic [2]. Therefore practice of safe sex is always recommended. Good personal hygiene, especially genital hygiene, for both men and women plays a very important role in the prevention of HPV infection. It has been observed that sexual partners of circumcised men have less chance of contracting HPV infection and also its consequences like cancer of the cervix, anus, etc.

6.3 Effects of human papilloma virus on pregnancy and the neonate [2,4]

- **Mother:** Pregnancy is concordant with reduced immunity resulting in rapid growth of the warts among the pregnant women. Therefore, women who have genital HPV and become pregnant, must be scrutinized more closely since the lesions may grow large, causing vaginal canal obstruction and sometimes profuse bleeding. Treatment during pregnancy is limited to trichloroacetic acid. External genital warts (EGW) may also be treated with surgical excision, loop electrode excision procedure, or cryocautery.

- **Neonates** born to women with HPV are at risk of exposure to the virus during delivery. About 2% to 5% of neonates exposed to HPV types 6 and 11, especially during delivery may develop laryngeal papilloma within the first 5 years of life causing hoarseness, abnormal cry, cough, and stridor, which sometimes may be fatal. So cesarean section is usually indicated in women with these types of HPV.

Treatment: All the EGWs must be removed. However, even after treatment the patient continues to remain infectious and the propensity to have cancer due to HPV infection still remains [2,4]. If left untreated, the warts may grow further both in size and number.

No single treatment has been found to be accurately appropriate for all patients or all warts, and the treatment decision depends on the size, anatomic location, number, and character of the EGWs. Other health conditions such as pregnancy and any other immune deficiency must also be considered. The self-applied therapy (various applicants mentioned below) though is convenient, safe, and effective must be recommended only among those patients empowered with proper HPV-related cognitive affective and psychomotor status. Management of EGW also is influenced by the patient preference, resources, and the expertise of the healthcare provider. Treatment plans, therefore, should be decided after a thorough physical and psychosocial assessment as well as consideration of the available medical resources of the clinician, the patient, and the community.

Most of the treatment options are enlisted in the following table (Table 6.3).

Vaccine: HPV vaccines focus on the prevention of the potentially oncogenic subtypes of HPV like types 16 and 18 [4,5]. There are some vaccines which cover two, four, or nine types of HPV. It is estimated that HPV vaccines may prevent 70% of cervical cancer, 80% of anal cancer, 60% of vaginal cancer, 40% of vulvar cancer, and show more than 90% efficacy in preventing HPV-positive oropharyngeal cancers (Table 6.4).

6.4 Herpes simplex virus type 1 and 2

Herpes simplex type 1 (HSV-1) and Herpes simplex type 2 (HSV-2) are the viral agents of genital herpes, which is a common recurrent viral STI [2]. HSV-1 primarily causes oropharyngeal lesions (cold sores) as well as genital herpes, and transmission is through oral genital sexual act [2]. HSV-2 causes genital herpes through sexual contact. The infected partner may be symptomatic or asymptomatic and the risk of transmission is related to the number

TABLE 6.3 Showing treatments for external HPV warts.

Type	Agent/therapy
Topical (applied locally by self)	Podofilox 0.5% solution Imiquimod 55 cream 5-Fluorouacil cream Cidofovir
Topical (applied locally by provider)	Cryotherapy Podophyllin (10%−25%) Trichloroacetic Acid (80%−90%) Bichloroacdetic acid (80%−90%)
Intralesion injectables systemic office surgery other surgery	Interferon, Electrosurgery Curettage Tangential Scissor/Scalpel excision Laser

TABLE 6.4 Showing Human papillomavirus vaccine recommendations[a].

Age	Dose series
9−14 years	2 doses D1 = 0 months D2 = 6−12 months
15−26 years[b]	3 doses D1 = 0 months D2 = 1−2 Months D3 = 6 months
27−45 years	Shared clinical decision
> 45 years	Not recommended
HIV positive/immune-compromised	3 doses D1 = 0 months D2 = 1−2 months D3 = 6 months

[a]Adapted from CDC immunization schedule.
[b]Only one additional dose after 15 years, if one dose received at 9−14 years, two dose received in a short duration (<5 months) at 9 to 14 years.
Not recommended in pregnancy. However, in case of accidental pregnancy no intervention is required.

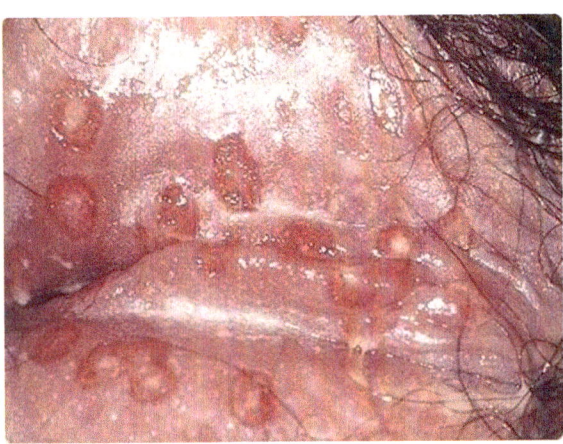

FIGURE 6.2 Genital (vulva) herpes (herpes simplex virus 2). *Courtesy: Blogspot.com, skin disease types (Posted by unknown).*

of lifetime sexual partners [6]. Some remain asymptomatic even after infection while others develop herpes, which include watery blisters in the skin or mucous membranes of the mouth, lips, nose, or genitals. Lesions may heal with a scab. Poor immunity status aggravates the disease. These viruses are neurotropic and neuroinvasive viruses and so they persist in a latent state in the neuronal bodies indefinitely resulting in periodic reactivation [3,6]. Again, increase in incidence of genital herpes in developing countries, particularly in Africa, may be due to increased immunosuppression due to HIV infection [3,6].

Infection with HSV is initiated by contact of virus with mucosa or abraded skin. Cellular destruction of the epidermis occurs followed by inflammation, which results in formation of vesicles on an inflammatory base [3,6]. Microscopically, multinucleated giant cells, focal cellular necrosis, and ballooning of infected cells are noted [6]. Mild flu-like symptoms like fever, headache, malaise, and myalgia are often present [6]. During an initial infection, virus gains access to neurons, allowing establishment of latent infection [3]. Further spread or recurrence occurs according to the patient's immunity status [6]. More women are infected with HSV-2 than men because sexual transmission of HSV is more efficient from men to women than from women to men (**5,6**) (Fig. 6.2).

Moreover, genital herpes HSV can lead to gingivostomatitis, keratoconjunctivitis, cutaneous herpes, genital herpes encephalitis, meningitis, and neonatal herpes [6].

Central nervous system complications of genital herpes include aseptic meningitis, sacral radiculopathy, transverse myelitis, and benign recurrent lymphocytic meningitis (Mollaret's meningitis) [6]. Out of these complications, aseptic meningitis is the most frequent; it is to be noted that about one-third of women and 1 in 10 men with primary infection have meningeal signs, whereas such signs are rare among patients with nonprimary infection.

Neonatal HSV: Neonatal HSV infection is a serious outcome in HSV infected pregnant mothers [6]. The newborn shows signs and symptoms at the age of 2 to 3 weeks. The manifestations are skin vesicles, fever, irritability, seizures, hepatitis, pneumonitis, and disseminated intravascular coagulopathy. Infants born to mothers with primary infection at the time of delivery are at much greater risk than babies of mothers with recurrent genital herpes. Despite advances in the diagnosis and management of neonatal herpes, the occurrence of morbidity and mortality due to this viral infection is high.

Recurrence: The HSV has an affinity for the neuron and is lodged in the dorsal root ganglia where lifelong latency is established [6]. The virus periodically reactivates within the ganglia and travels back down the neuron leading to either asymptomatic or symptomatic infections causing recurrent ulcers. Recurrence due to reactivation of latent virus is more common during pregnancy and among immunocompromised persons [2,6]. Within 12 months after diagnosis with a primary episode of genital HSV-2 infection, 90% of patients have at least 1 recurrence, 38% have 6 or more recurrences, and 20% have 10 or more recurrences. Recurrence is more common in genital HSV-2 infections than genital HSV-1 explaining why most cases of symptomatic HSV-1 genital disease are primary cases.

For herpes, the most effective medications available are oral antivirals that can modulate the course of the disease (reduce the severity and frequency of symptoms), though they cannot cure the disease (Table 6.5).

6.5 Human T-cell lymphotropic virus infection

HTLV-1 (human T-cell lymphotropic virus type 1) is also called human T-cell leukemia type 1. This virus is transmitted sexually or through infected fluids including blood, breastmilk, and semen or vertically from mother to child. HTLV-1 has seven reported subtypes (subtypes A through G) [5]. 4%−5% of infected persons are said to develop cancer

TABLE 6.5   Treatment regime of HSV infection.

Treatment (antivirals)	Doses per day	Treatment duration
Acyclovir 200 mg	5	5−10 days
Acyclovir 400 mg	3	5−10 days
Valaciclovir 500 mg−1 g	2	5−10 days
Famciclovir 250 mg	3	5−10 days

From http://www.drugs.com.

[7,8]. The first human T-lymphotropic virus (HTLV), HTLV-1 was discovered in 1980. Since then, three more subtypes of HTLV have been described (HTLV-2, 3, and 4). The discovery of the HTLV-1 retrovirus took place just prior to the identification of HIV so that one of the early names given to HIV was HTLV-3 [7,8].

Efficiency of sexual transmission is more from men to women than vice versa. In a prospective study of 97 Japanese HTLV-serodiscordant couples were followed for five years, the relative rate of transmission was four times higher from infected males to their uninfected female partners than from females to males.

In reports from infection of HTLV endemic populations it has been observed that the overall rate of MTCT (mother to child transmission) ranges between 4% and 25%. The rates are more among prolonged breast-fed children from infected mothers [7]. This risk of infection in children is closely related to both the viral load in breastmilk, and the duration of breastfeeding [7]. Intrauterine and peripartum transmission of HTLV-1 occurs in less than 5% of children of infected mothers [7,8].

Similar to HIV, HTLV causes chronic infection but unlike those with HIV infection the majority (90%) of HTLV-1-infected individuals remain asymptomatic carriers remaining undiagnosed, thereby contributing to its silent transmission [9]. The other 10% will develop one or more of several diseases causing significant morbidity and mortality. 2%−6% will develop adult T-cell leukemia/lymphoma (ATLL), a highly aggressive T-cell malignancy of four subtypes which are acute, lymphoma, chronic, and smoldering (Shimoyama criteria), while 0.25 to 3%−4% develop a variety of chronic inflammatory syndromes, most notably HTLV-1 associated myelopathy (HAM)/tropical spastic paraparesis [7]. Besides ATLL and HAM, HTLV-1 infection also leads to other diseases, by mechanisms currently unclear, including uveitis, thyroiditis, alveolitis, bronchiectasis, polymyositis, infective dermatitis, and peripheral neuropathies [7].

Since there is no cure, prevention is very vital for HTLV infection. Measures like use of condoms during intercourse and HTLV screening for pregnant women and screening of blood products will go a long way in preventing HTLV infection [7]. Testing for HTLV should also be considered among people with history of unprotected sex with multiple sex partners or intravenous drug use. In addition, testing should occur in those presenting with clinical manifestations such as ATLL and HAM. Greater emphasis must be given on acquiring the knowledge of this infection during the training of healthcare professionals along with dissemination of methods of prevention and control among the general population [9].

6.6 Hepatitis A (HAV infection)

Infection with the hepatitis A virus (HAV) causes hepatitis A. The virus replicates in the liver and is shed in the feces. Fecal-oral route is the commonest mode of transmission for hepatitis A infection, contaminated food or water being the chief sources [2]. Person-to-person transmission is common among people living in crowded living situations with poor personal hygiene [2]. Sexual transmission is less common as a result of which it is often not included in literature and discussions about STIs. Risk for HAV infection is more among persons with certain sexual practices like anal/oral sexual contact, digital-rectal intercourse, anal-receptive intercourse, and group sex. Most acute HAV infections are uncomplicated and do not require hospitalization [2].

Symptoms are nonspecific and usually consist of malaise, anorexia, nausea, vomiting, and fatigue [2]. Fulminant liver failure develops in about 0.1% of cases, and there is an overall mortality rate of 0.3% in persons under age 49. The mortality rate increases to 1.8% in persons over the age of 49 [2].

Treatment consists of supportive therapies related to the symptoms. Persons with acute HAV infection should be advised to avoid medicines that might cause liver damage or that is metabolized in the liver, such as acetaminophen and ethyl alcohol. A well-balanced healthy diet is also recommended [2].

Beyond good personal hygiene and consumption of potable drinking water and uncontaminated food, prevention of this disease is done by using the hepatitis A (inactivated HAV) vaccine through intramuscular (deltoid) injection with an initial dose and a booster dose anytime between 6 and 12 months later [10]. If exposure to HAV has taken place, immunoglobulin may be given intramuscularly and this is about 85% effective in prevention of HAV infection [2].

6.7 Hepatitis B (HBV infection)

Hepatitis B infection can be transmitted parenterally, perinatally, and rarely orally as well as sexually [2]. Up to 6% of those infected with HBV turn into chronic HBV infection, a state where the afflicted person may be asymptomatic but will be able to transmit the virus to others [2]. The outcome of chronic hepatitis B infection may be both morbid and fatal. This infection is implicated for the development of chronic liver disease, cirrhosis, liver cancer, liver failure, and death [2]. People with chronic HBV infection have an 85% 5-year survival rate and a life expectancy of 25 years. HBV infected pregnant mothers may vertically transmit the infection to the newborn, the risk of perinatal transmission being 10%−85% [2]. Unfortunately, around 90% of infected newborns develop chronic HBV infection followed by chronic liver disorders [2]. Children of HBV-infected mothers, who are not themselves infected during the perinatal period, newborns who miss being infected from their HBV-infected mothers are still at risk of chronic HBV infection by person-to-person transmission during the first 5 years of life [2].

Treatment: The mainstay of treatment for acute HBV infection is supportive and symptomatic [2]. Antiviral medications are recommended to ameliorate both the infection and its complications, especially damage to the liver. Beyond this supportive and symptomatic management is done during the acute phase of the disease as per the signs and symptoms of the disease [1].

Prevention: Hepatitis B vaccine is very efficient to prevent the disease in persons who have not been exposed to the virus HBV [2]. Hepatitis B immune globulin is quite effective for prevention of HBV infection among those who have been exposed to hepatitis B, such as family members, close personal contacts, sexual contacts, and infants born to HBV infected mothers. This should be followed by HBV vaccination. Hepatitis B immune globulin must be administered within 14 days of exposure. Screening by detection of hepatitis surface antigen among the above mentioned population and intravenous drug users who are susceptible to the disease must be performed [2] (Table 6.6).

In many countries a dose is given at birth (<24 hours), and three other doses are given as a part of pentavalent vaccine at 6 week, 10 week, and 14 weeks.

TABLE 6.6 Recommended doses of currently licensed formulations of hepatitis B vaccine, by age group and vaccine type[a].

Age group (years)	Dose (µg)	Vol (mL)	Schedule
Recombivax HB			
Infants (<1 year)	5	0.5	3 doses for single antigen vaccine[b] D1 = 0 (at birth <12 h) D2 = 1−2 months D3 = 6−18 months
Children (1−10 years)	5	0.5	3 doses at 0,1,6 months
Adolescent (11−19 years)	5	0.5	3 doses at 0,1,6 months
Adults > −20	10	1	
Patients on haemodialysis and other immune-compromised persons <20 years	5	0.5	
Patients on haemodialysis and other immunocompromised persons ≥ 20 years	40 (Dialysis formulat ion)	1	
Engerix-B			
Infants(<1 year)	10	0.5	3 doses for single antigen vaccine[b] D1 = 0 (at birth <12 h) D2 = 1−2 months D3 = 6−18 months
Children (1−10 years)	10	0.5	3 doses at 0,1−2, 6 months
Adolescent (11−19 years)	20	1	3 doses at 0,1,6 months
Adults > 20	10	0.5	
Patients on haemodialysis and other immune-compromised persons <20 years	20	1	
Patients on haemodialysis and other immune-compromised persons ≥ 20 years	40	2	4 doses at 0,1,2.6 months

[a]*Adapted from CDC Immunization schedule.*
[b]*4 dose schedule can be done for combination vaccine D1 = 0 months (at birth), D2 = 2 months, D3 = 4 months, D4 = 6 months. (As Per CDC).Table 6.6*

6.8 Hepatitis C (HCV infection)

Hepatitis C (HCV) is usually transmitted during blood/blood products transfusion but today this risk has reduced remarkably because of stringent scrutiny of these products. HCV infection is not generally considered a sexually transmitted disease, but for some HCV cases no risk factors can often be identified, thus making sexual transmission a possibility. Sexual and maternal-neonate transmission is small and is usually associated with high levels of circulating HCV. Though majorly acute HCV infection is asymptomatic, its impact is both morbid and fatal since the risk is as high as 90% for the development of chronic hepatitis progressing to chronic liver disease. It may take as long as 10 years for a person to develop symptoms of liver disease resulting from HCV infection. To lower the risk of hepatic complications alpha-interferon is recommended for hepatitis C infection in the acute stage, but 50% of treated patients fail to respond to this therapy [2].

Follow-Up and Education: Patients diagnosed with hepatitis B or C must be constantly and closely checked for the development of chronic liver disease, which should be taken care with timely and appropriate management [2].

Hepatitis B and Hepatitis C are preventable and today they are placed within the ambit of public health problem of the world. As a result, many countries have rolled out efficient national level management programs offering free diagnostics and drugs lifelong to its beneficiaries [9]. The goal of the program is to reduce the complications (cirrhosis and hepato-cellular carcinoma), morbidity, and mortality associated with the infection and finally to eliminate hepatitis B, C by the end of this decade [9].

6.9 Prevention of sexually transmitted viral infections

Prevention of viral STIs remains the mainstay of disease management since there is no or very little scope for curative measures. The dynamics of transmission, rate of spread, and the degree of STIs are determined by some very vital components: (1) status of susceptibility to the infection and to the exposure to the disease; (2) rate of possibility of sexual transmission during such exposures; (3) the duration of infectiousness of individuals who become infected and (4) the viral load in the infected person who transmits the disease. Prevention which is the ultimate and the primary goal focuses largely on the factors that influence these vital components [1].

STIs may be prevented and controlled by the health personnel who must be instrumental in developing robust health educational programs, including patient education literature as well as providing sensitive counseling at both individual and mass level.

6.9.1 Cognizance

Correct, updated, and complete knowledge regarding STI must be disseminated to all status of society irrespective of cast creed or color. Issues on modes of transmission, methods of prevention, approaches of treatment, and all the procedures of control must be an integral part of education program for the benefit of patients and their intimate contacts. There should be no inhibitions regarding depicting a clear picture of the natural history of the disease, the advantages of safe sex with special emphasis on consistent use of condoms and the disadvantages of unsafe sex, risky sex, or even abnormal sex. People should know the basics of signs and symptoms of STIs and the optimum time to visit the doctor for diagnosis and management. Clear and complete knowledge regarding vaccines, correct lifestyle along with all the positive habits and ethos for prevention of the disease must be universally percolated.

6.9.2 Affective

Health education programs must encourage the public to imbibe the positive attitude towards the cure and control of STIs. Special areas of concern are the stigma, humiliation, and disgrace associated with STIs and so the attitude of shaming and blaming those with the disease and reluctance on the part of the victims of STI to share even with the doctor regarding the disease all lead to delayed management and obviously more complications and suffering. STIs should be considered as any other medical condition which must be detected and treated as soon as possible and this very important message should reach the general masses.

MSMs, lesbians, persons practicing abnormal sex, frequent encounters with commercial sex workers are still considered not to be exposed or discussed in the open. Such persons are pushed outside the main stream of the society. They are not only considered to be quirky but also shameful, disgraceful, and even felonious as a result of which they are almost ousted from the society at large. All this affects their health-seeking behavior so that diagnosis, treatment, and management of their disease are often delayed.

Therefore, both sympathy and empathy must be there from the near ones, from the healthcare providers and even from the general mass for those who are afflicted with STI so that they adopt a positive health-seeking behavior with early diagnosis, appropriate treatment with practice of safe sex, and inculcation of high-quality personal hygiene, especially genital hygiene all leading to the sustenance of a healthy body free from STI.

6.9.3 Psychomotor

Sexual behavior: "Abstain," "Be faithful," and "use Condoms" (the ABC interventions) is the dictum of prevention of STI [1]. Decreased transmission of STI including HIV will positively happen if implementation of integrated behavioral prevention interventions is carried out seriously, sincerely, and stringently [1,10]. Safe sex education is particularly important for all starting from the adolescent stage of life [2].

Sex partners of afflicted persons, both symptomatic and asymptomatic, should be counseled about transmission risks and prevention, such as use of condoms with each sexual encounter and abstinence during the known infectious stage of the diseased person and practice of safe sex [2]. Sexual act with multiple partners must be avoided and if not at least use of condom during each sexual act is a must.

When used correctly and consistently for vaginal and anal sex, condoms offer one of the most effective methods of protection against STIs, including HIV, it being the hallmark of protection during unprotected sex not only against STI but also for unintended pregnancy. Although highly effective for prevention of STI at large condoms cannot avoid the occurrence of STI-related extra-genital viral ulcers/eruptions (genital herpes, external genital warts) [1].

Epidemiologic, behavioral, and social determinants including the local contexts must help in setting the guidelines for any prevention program planning for any population. Behavioral elasticity, which signifies the ability to change behaviors for a population is difficult when it is impregnated with poverty, illiteracy, lack of autonomy, and presence of stigma, myths, and misconceptions. For example, there is lack of behavioral elasticity among many women in many developing countries who cannot ask their husbands to use condoms or commercial sex workers who cannot negotiate condom use with clients for fear of losing them [1,10].

Genital Hygiene: Good health prevails in a society that practices high-quality personal hygiene. Genital hygiene plays a very important role in prevention of STIs. While cleaning the genital area with a tissue or even while wiping it with a towel, the area should be patted dry and should be wiped from front to back with a single stroke [1]. This prevents the contamination of the genitals with organisms from the anus.

Undergarments and towels should not be shared with others. Only cotton and loose fitting underwears should be used. Comfort measures for active lesions can include lukewarm sitz baths, nonrestricting clothing, and drying the lesions with a handheld hair dryer set on the low to medium setting [2].

For men the genital area must be washed thoroughly with soap and water. If uncircumcised, the foreskin must be pulled back and cleaned thoroughly. For women genital area should be cleaned with soap and water once every day with a mild and fragrance-free soap. The internal genitalia or vagina does not require too frequent cleaning, especially with soap. Cleaning the genitals just once every day and after sexual intercourse is quite enough. Using a hand-held shower and some cold water is always recommended for cleaning the genital area.

Vaccines: As mentioned there are some very potent, safe, and easily available vaccines for at least three STIs (HPV, Hepatitis A, and Hepatitis B) which should be taken in the right dosage and in the right time. Their contribution to the prevention of certain STIs is indubitable. Presently, the HPV vaccine has been introduced as part of routine immunization programs in 111 countries and it has been established that if high ($>80\%$) vaccination coverage of young women (ages 11–15) is achieved then it will prove to be a big stride in prevention of cervical cancer and all other cancers for which HPV infection has been implicated [1,10].

6.10 Health education

6.10.1 Individual level

Counseling of STI patients and their carers must be an integral part of STI prevention and management and should be provided throughout before during and after the treatment of STI so that knowledge attitude and practice regarding STI is up to the mark. The counselor should be trained to cope with such patients and must possess apt and significant qualities for a successful and effective counseling session. The counselor should be friendly, open-minded, good listener, helpful, knowledgeable, good communicator, energetic, empathetic, patient, sensitive, observant, respectful, and must maintain confidentiality.

6.10.2 Mass level

Health education messages regarding STIs must pervade in all strata of society. STIs can be prevented with little or no medical interventions if people are adequately informed about them and if they are encouraged to take necessary precautions in time. Health education should include not only patients of STI but also the general people the susceptible and the vulnerable segments of the society along with health providers, community leaders, and decision-makers. Health education for the whole community may be given through mass media like television, radio, internet, newspapers, printed material, posters, billboards, and signs. All stigma attached to STIs must be discarded and quashed. All information must be in public instead of putting them under the carpet for fear of embarrassment and mortification. Messages must be clear, comprehensive and complete regarding all STIs while the following must be included in all such messages.

o Modes of transmission
o Signs and symptoms
o Modes of prevention,
o Avoidance of multiple sexual partners,
o Practise of safe sex especially during the period of infectivity and or unknown sexual partner
o High-quality genital hygiene
o Screening and diagnosis
o Health-seeking for healthcare service at the right time and at the right place.

The messages should be in local language, should be easy to understand, should be inspiring and motivating. Short, simple, and easy-to-understand posters should be direct and one that can be understood at a glance. They should be put up in places like bus stops, railway stations, airports, petrol pumps, parks, street corners, hospitals, health centers, etc. where people may be waiting for some time.

6.11 Conclusion

Most sexually transmitted viral infections are incurable and have consequences ranging from mild to severe. Genital warts and HPV infections are oncogenic with special reference to cervical carcinoma. Hepatitis A is a self-limiting disease. However, HBV and HCV infections can cause chronic infection of the liver which culminates in extreme suffering and death. Pregnant mothers with viral STIs transmit the infection to their newborns. Because prevention is the ultimate goal, health personnel at all levels must be knowledgeable about the prevention and treatment options for the STIs so that they can help in developing educational programs and patient education literature at mass level [2]. They also must provide efficient and sensitive counseling about prevention and treatment of various viral STIs at all levels of a healthcare service. It may be noted that simultaneous availability of new and effective preventive and therapeutic interventions like vaccines, effective microbicides, etc. render behavioral disinhibition, lack of adherence, and above all some degree of complacency among the people [1]. All these must be taken care of during the process of control and containment of viral STIs, strongly highlighting the principle of "Prevention is Better than Cure" as propagated by the Dutch philosopher Desiderius Erasmus.

References

[1] WHO. Sexually transmitted infections (STIs) (who.int). 22 November 2021, **(4)**.
[2] Thomas DJ. Sexually transmitted viral infections: epidemiology and treatment. J Obstet Gynecol Neonatal Nurs 2001; **(1)**.
[3] Brugha R, Keersmaekers K, Renton A, Meheus A. Genital herpes infection: a review. Int J Epidemiol 1997;26 (4):698−709 **(6)**.
[4] Beutner KR, Wiley DJ, Douglas JM, et al. Genital warts and their treatment. Clin Infect Dis 1998;28(Suppl. 1) Guidelines for the treatment of sexually transmitted diseases (Jan., 1999).

[5] Siracusano S, Silvestri T, Casotto D. Sexually transmitted diseases: epidemiological and clinical aspects in adults. Urologia 2014;81(4):200−8 **(7)**.

[6] Kimberlin DW, Rouse DJ. Genital herpes. N Engl J Med 2004;350:1970−7 **(8) (5)**.

[7] Caswell RJ, Manavi K. Emerging sexually transmitted viral infections: review of human T-lymphotropic virus-1 disease. Int J STD AIDS 2020; **(2)**.

[8] Verdonck K, González E, Van Dooren S, Vandamme A-M, Vanham G, Gotuzzo E. Human T lymphotropic virus 1: recent knowledge about an ancient infection. Lancet Infect Dis 2007; **(3)**.

[9] Torcia M. Interplay among vaginal microbiome, immune response and sexually transmitted viral infections. Int J Mol Sci. 2019; **(9)**.

[10] Centers for Disease Control and Prevention. Guidelines for treatment of sexually transmitted diseases. Centers for Disease Control and Prevention; 1998. p. 20−6, 88−104. Report No.: 47(RR-1). <https://www.cdc.gov/mmwr/preview/mmwrhtml/00050909.htm> [accessed 15.4.22].

Testing viral infections

Shalini Upadhyay

A.S.J.S.A.T.D.S. Medical College, Fatehpur, Uttar Pradesh, India

7.1 Introduction

Viral infections do not spare any age group and depending on the immune status of the individual, it can lead to the development of acute, chronic, recurrent, or lifelong infection. In the bygone era, the identification of viral pathogen was a tedious process mostly based on virus isolation and culture and so it was restricted to epidemiological purposes or research. Remarkable advances in viral diagnostics in the last few decades have enhanced the relevance of viral diagnosis in patient care. Clinicians now prefer to seek help from viral testing laboratory to make conclusive diagnosis rather than depending on mere clinical judgement for patient management. As a consequence of the confidence of clinicians in laboratories, they have shifted their focus on providing a reliable, rapid, and accurate diagnosis. Molecular and point of care tests with remarkable sensitivity and specificity are sought-after techniques for viral infections. Clinicians working in conjunction with laboratories will go a long way in putting together the findings of a test, its interpretation, and application for specific viruses at different stages of the viral infection. Data collection through viral diagnostic techniques for epidemiological purposes plays a significant part in public health measures. Outbreak investigation, pandemic control, vaccine development, and treatment of emerging viral infections are some of the aspects which are possible only with the advent of revolutionized viral diagnostics.

Viral diagnostics has been considered a challenge since it eludes the simple methods used in bacterial tests like staining and culture as they are too small and do not grow in artificial culture media. But at present, viral infections can be easily diagnosed through commercially available kits based on viral proteins, genetic material, or any other part of the viral structure. The chapter aims to discuss various methods through which viral infections can be diagnosed and recent developments in this field.

7.2 The purpose of laboratory diagnosis of viral infections

1. Better patient care: The antiviral therapy to be initiated depends on the accurate identification of the causative virus. A reliable diagnosis can help in avoiding needless testing, de-escalate antibiotic use, cut the cost of treatment, ensure better prognosis, and shorten the hospital stay of patients thereby decreasing the chances of hospital-acquired infections.
2. Antenatal screening: This screening helps in deciding the modality of delivery. For example, a Cesarean section is advisable in case of primary genital herpes at the time of delivery. In rubella or cytomegalovirus (CMV) infection, education and care should be given to a baby with congenital defects. If rubella is detected in the first trimester, pregnancy termination may be suggested.
3. Occupational hazards: Screening of medical staff who suffered needle stick injuries, contact precautions for particular surgical procedures and testing of donated organs before transplantation depends upon the testing of blood-borne viruses.
4. Specific antiviral therapy: Accurate identification of the causative virus is needed to start the antiviral therapy. For example, rapid and reliable laboratory diagnosis for HIV (human immunodeficiency virus), hepatitis C virus (HCV), and hepatitis B viruse (HBV) is required to decide on the initiation of antiviral therapy. The combinations appropriate for managing individual patient can be tailored based on the resistance profile of the viral strain. Viral load testing is a marker of prognosis and helps monitor the response to the treatment.
5. Public health measures: To establish measures to control public health emergencies. For instance, a novel strain of coronavirus that is SARS-CoV-2 (severe acute respiratory syndrome-coronavirus) led to a pandemic against which vaccine and treatment strategies could be developed only after the accurate identification and knowledge of its genetic makeup. It enabled authorities to promulgate warnings and initiate appropriate respiratory control measures.
6. Monitoring and surveillance: It can help in the natural history, prevalence, and significance of a viral infection in the community. Thus allowing designing of control measures, setting priorities, and monitoring of vaccination campaigns.

7.3 Sample collection, packaging, and transport

An appropriate clinical specimen is one of the important factor influencing the results of a viral diagnostic test. The aspects of a good clinical specimen are; the right site, right time, right packaging, and right temperature. These aspects can be decided based on the global and local epidemiology and pathogenesis of the virus. For viruses causing acute infection, the viremia is relatively short and their detection is possible only in the first ten days from the time of appearance of clinical features, such as coronavirus, rotavirus, arboviruses, etc. The right time for sampling is as early in the infection as possible, since the viral load is maximum when symptoms first develop and it declines further. The right site of the specimen depends on the clinical features, along with an understanding of the pathogenesis of the causative virus. The best site for

TABLE 7.1 Sample collection depending on the site and type of viral infection.

S. no.	Site of infection	Type of infection	Causative virus	Specimens
1.	CNS	Encephalitis, meningitis	Enteroviruses (Coxsackie, echo), mumps, polio, HSV, Arboviruses, rabies	CSF, serum, brain biopsy, stool
2.	Salivary glands	Parotitis	Mumps, parainfluenza, influenza	Nasal/throat/nasopharyngeal swab, serum
3.	Eye	Keratitis, keratoconjunctivitis, conjunctivitis	HSV-1, -2, adenovirus, Measles	Conjunctival/corneal swab
4.	Skin	Rashes	VZV, HSV-1 & 2	Vesicle swab/pus/scraping, biopsy
5.	Upper respiratory tract	URI, flu	Rhinovirus, parainfluenza, influenza, adenovirus, enterovirus, RSV	Nasal/throat/nasopharyngeal swab or aspirate
6.	Lower respiratory tract	Pneumonia, bronchitis, bronchiolitis	Rhinovirus, parainfluenza, influenza, adenovirus, enterovirus, RSV	Nasal/throat/nasopharyngeal swab or aspirate, serum, sputum, BAL, endotracheal secretion, transtracheal aspirate, brochial brush specimen
7.	GI Tract	GI infection	Rotavirus, norovirus, adenovirus, Enterovirus	Stool, urine, serum
8.	Liver	Hepatitis	Hepatitis A, B, C, E, EBV, CMV	Serum, stool, urine
9.	New borns	Congenital infection	CMV, HSV-2, rubella, Parvovirus B19	Nasal/throat/nasopharyngeal swab, serum, stool, pleural fluid, pericardial fluid

BAL, bronchoalveolar lavage, *CMV*, Cytomegalovirus; *CNS*, Central nervous system; *CSF*, cerebrospinal fluid; *EBV*, Epstain Bar virus; *GI*, gastro-intestinal; *HSV*, Herpes simplex virus; *URI*, upper respiratory tract infection; *VZV*, Varicella zoster virus.

collecting specimens is the epithelial surface which acts as the portal of entry and viral replication (Table 7.1).

7.4 Type of specimen

1. *Respiratory samples:* The specimens required to diagnose upper respiratory viruses are nasal, nasopharyngeal, or throat swabs. For lower respiratory infections sputum, endotracheal secretion, broncho-alveolar lavage, trans-tracheal aspirate, and protected brush specimen may be used based on the site which is infected.
2. *Blood/serum/plasma:* The volume required is about two to ten ml depending on the patient's age. The appropriate tube depends on the test to be performed and the blood component needed. EDTA (ethylenediamine tetraacetic acid) tubes are not preferable for molecular testing as anticoagulants might inhibit polymerase chain reaction (PCR).

Clotted blood is collected in serum separation tubes for serological testing as it allows the separation of serum through centrifugation. For virus isolation, whole blood or serum may be used depending on the virus.

3. *Body fluids:* Cerebrospinal fluid which can reveal pathogenic viruses for meningitis or encephalitis is preferred undiluted while other fluids should be placed in a viral transport medium (VTM).

4. *Tissue:* Biopsy or autopsy specimens are usually helpful for virus isolation, or immunofluorescence staining. The tissue is kept in a sterile container with VTM and not formalin.

5. *Swabs:* Dacron or rayon tip swabs transported in VTM are preferred to ensure labile viruses for viral isolation and PCR. Swabs are taken from the eye, ear, genital tract, and various lesions of the skin.

6. *Stool:* At least 4 g of stool is collected and placed in a sterile container. It is useful in the diagnosis of enteric viruses and other viruses causing generalized infections of the gastrointestinal tract.

Storage of specimen: Specimens other than clotted blood must be kept at 4°C (39.2°F) and transported on ice to retain the viability of the viruses and keep nucleic acids intact. If a transit time of more than an hour or so is anticipated the container should be sent refrigerated (but not frozen), with cold packs (4°C) or ice in a thermos flask or styrofoam box. The specimen for molecular testing should reach the laboratory within 72 hours to keep the genetic material intact. For enveloped single-stranded RNA viruses the success rate declines from 48 hours and temperature must be maintained until it reaches the laboratory.

Viral Transport Media: VTM is available commercially and can also be prepared in-house. It requires adding 10 g veal infusion broth and 2 g bovine albumin fraction V to sterile distilled water (to 400 mL). Now add 0.8 mL gentamicin sulfate solution (50 mg/mL) and 3.2 mL amphotericin B (250 μg/mL), then sterilize by filtration.

Packaging, labeling, and documentation for transport: If the testing facilities are not available at the site of sample collection and it has to be sent to a distant laboratory, proper packaging protocols should be followed. The packaging guidelines are given by the UN (United Nation) and are available in International Civil Aviation Organization and International Air Transport Association regulations as Packaging Instructions 602 and 650. The packaging requirements are discussed below.

Triple packaging system: There are three layers as follows

1. *Primary receptacle.* A primary watertight, leak-proof, labeled receptacle containing the specimen. The receptacle is wrapped in enough absorbent material to absorb all fluid in case of breakage.

2. *Secondary receptacle.* A watertight, leak-proof receptacle to enclose and protect the primary receptacle. Several wrapped primary receptacles may be placed in one secondary receptacle. Sufficient absorbent material is added to cushion multiple primary receptacles.

3. *Outer shipping package.* The outer shipping package protects the secondary receptacle from outside influences such as physical damage and water while in transit. The

relevant data, forms, letters, and details of the receiver and sender are attached to the secondary receptacle.

7.5 Labeling/requisition form has the following information

1. The International Infectious Substance Label.
2. An address label with the following information: The receiver's and shipper's name, address, and telephone number, UN shipping name, scientific name of the substance, UN Number and temperature storage requirements (optional).
3. Required shipping documents: The shipper's declaration of dangerous goods, a packing list/proforma invoice which includes the receiver's address, detail of contents, and shipping bill (Table 7.2).

7.6 Methods in diagnostic virology

7.6.1 Virus isolation

It has been considered the "gold standard" in traditional viral diagnostics. Even though enzyme-linked immunosorbent assay (ELISA) and molecular techniques have now become the routine test, virus isolation still remains important for analyzing genotypic and phenotypic changes in virus for vaccine development and epidemiology. It allows monitoring of antigenic modifications, pathogenicity and characteristics of circulating strain to develop vaccines, such as the influenza vaccine, to match circulating strains. Limitations of virus isolation include high cost, labor intensive, risk of personnel hazard, time consuming, false negativity due to improper transport and cell lines losing viability. Specimen quality, the transportation time and conditions are determinants for favorable outcome of viral culture.

7.6.1.1 Tissue culture

In 1913, Steinhardt for the first time grew the vaccinia virus in fragments of rabbit cornea. In the 1930s, yellow fever & small poxviruses were grown in cell culture that aimed for vaccine production. Enders, Weller, and Robbins in 1949 successfully cultured poliovirus in nonneural cells with the production of recognizable cytopathic changes. Although decades passed since mammalian cells were first grown in vitro, it was only after the advent of antibiotics that tissue culture could become widely used.

1. *Organ culture:* It is used for certain viruses that have an affinity to specific organs; for example, tracheal rings for a culture of coronavirus.
2. *Explant culture:* Explants are fragments of minced tissue which are mostly used for research purposes. For example, adenoviruses can be grown in adenoid explants.
3. *Cell line culture:* Cell line isolation is the most used method of viral culture. The discovery of many viruses like adenoviruses and rhinoviruses was directly attributable to cell lines, as was the development of vaccines and advances in molecular-level knowledge of viruses.

TABLE 7.2   Techniques used in diagnostic virology.

S. no.	Viral diagnostic methods
I.	Virus isolation
1.	Tissue culture
2.	Animal inoculation
3.	Egg inoculation
II.	Direct microscopy
1.	Electron microscopy
2.	Fluorescent microscopy
3.	Immunoperoxidase staining
4.	Histology/Cytology staining
III.	Detection of Viral antigen and antibody
1.	Immunochromatography
2.	Enzyme linked immunosorbent assay
3.	Radioimmunoassay
4.	Complement fixation test
5.	Neutralization assay
6.	Latex particle agglutination assay
7.	Haemagglutination inhibition
8.	Western blot
IV.	Molecular assays
1.	Nucleic acid hybridization
2.	Polymerase chain reaction (PCR)
i.	Real time PCR
ii.	Multiplex PCR
iii.	Nested PCR
iv.	PCR for quantitation
v.	Isothermal PCR
3.	Microarray
V.	Gene sequencing
1.	Sanger sequencing
2.	Next generation sequencing
3.	Third generation sequencing
4.	Nanopore sequencing

7.6.1.1.1 Preparation of the cell lines

Cells lines can be grown in vitro either as explants of respiratory, intestinal tissues, or as cell cultures. Explant culture may be used for research purpose or cultivation of viruses, but most of the diagnostic or research activity is carried out through cell cultures including monolayers or suspension cultures. Monolayers is made by cutting tissues into small pieces and placing in a medium containing a proteolytic enzyme such as trypsin or collagenase followed by mechanical shaking to dissociate into individual cells. After the cell suspension, they are washed, counted, diluted in a growth medium, and allowed to settle at the base of glass or plastic container. Cell culture has greatly benefited from media containing most of the nutrients necessary for cell growth.

Eagle's media is an isotonic salt solution with serum, glucose, vitamins and amino acids with a pH of 7.4. It contains antimicrobials to inhibit the growth of other microorganisms. The Madin–Darby canine kidney cell line is grown in serum-free media supplemented with fibronectin or polylysine- or collagen-coated dishes having hormones like insulin, binding proteins like transferrin and albumin. This kind of serum-free media is specially of use for the "hybridoma" cells culture, helpful in the production of monoclonal antibodies. Defined media have advantages of cultivating viruses that may be neutralized by animal serum antibodies except in the case of fetal calf serum. So, fetal calf serum may be incorporated in the media for the initial growth of cells in culture. "Growth medium" is removed after the monolayer is set, then the virus is inoculated, and "maintenance medium" with no serum is added to the culture. Tissue culture bottles are incubated horizontally in a CO_2 environment mostly as a roller drum culture as it provides better aeration useful for the growth of fastidious viruses.

7.6.1.1.2 Types of cell lines

Based on the origin, chromosomal characters and number divisions in vitro before cells die out or propagated indefinitely, there are three main types of tissue culture:

1. *Primary cell lines:* These have diploid karyosome and are derived from animals (often from fetal organs or tissues) containing several cell types which are capable of very limited growth in culture up to 5–10 divisions. This restricts their relevance in routine diagnostics as well as vaccine production due to high cost, difficulty in obtaining fresh tissue each time, and lack of batch-to-batch consistency. Nonetheless, the diverse range of differentiated cell types makes it very sensitive to animal viruses, so it's of great use in veterinary diagnostic virology. For example, Monkey kidney cell line, Human amnion cell line, and Chick embryo cell line.
2. *Secondary or diploid cell lines:* They retain the diploid chromosome number and undergo divisions maximum of up to 10–50 divisions before they undergo senescence depending on the life span of the species. Diploid strains of fibroblasts from humans are widely used in human diagnostic virology like for CMV and vaccine production. Examples are MRC-5 (Medical Research Council cell strain) and WI-38 (Wistar Institute) human embryonic lung cell strain, used for the preparation of various viral vaccines for rabies, chickenpox, hepatitis-A, and MMR (measles, mumps, rubella).
3. *Continuous cell lines:* These are single-type cells capable of indefinite growth in vitro originating from cancers, or by the spontaneous transformation of a diploid cell strain.

They lose resemblance to their cell of origin, as there occur many mutations during maintenance by serial subculturing or indefinite divisions. Since they are easy to maintain in the laboratories by serial subculturing, so they are widely used. These calls are mostly aneuploid in chromosome number, for example, Baby hamster kidney cell line, Vero cell line line (Vervet monkey kidney cell line), KB cell line (subline of the ubiquitous keratin-forming tumor cell line HeLa), Hela cell line (Human carcinoma of cervix cell line), McCoy cell line (Human synovial carcinoma cell line), and HEp-2 cell line (epidermoid carcinoma of the larynx).

An adaption of traditional viral culture is achieved by inoculating the specimen onto a microscope slide and centrifugation which increases the infection rate (Shell vial assays). The enhanced detection rate may result from better contact between specimen cells and the cells in the culture, allowing faster and more substantial infection of the cell lines. The use of fluorescent-labeled monoclonal antibodies to tag the viral antigen can help us visualize the virus through microscopy. Identification of viruses in cell culture is performed by microscopic examination of degenerative morphological changes in the cell monolayer known as cytopathic effect (CPE).

7.6.1.1.3 Cytopathic effect

The rate of CPE appearance is one of the characteristic which can be used to identify viruses. A virus is considered rapid if CPE appears after 1–2 days in cultures while it is slow if CPE appears after 4–5 days. The common types of CPE are:

1. Total destruction: The most severe form of CPE involves rapid shrinkage of all cells in the monolayer. They become dense (pyknosis) and separate from the glass within 72 hours, for example, enteroviruses.
2. Subtotal destruction: Degeneration of some but not all of the cells in the monolayer; for example, togaviruses (alphaviruses), picornaviruses, and paramyxoviruses.
3. Focal degeneration: Localized infection due to direct cell-to-cell transfer of virus rather than diffusion through the extracellular medium. Initially, cells become swollen and refractile, then as the infection spreads concentrically, they separate from the growth surface leaving cleared areas surrounded by round cells. Examples are herpesviruses and poxviruses.
4. Swelling and clumping: Infected cells greatly expand and clump together in "grape-like" clusters before detachment. It is characteristic of adenoviruses.
5. Foamy degeneration: It is due to the production of many large cytoplasmic vacuoles. For example, certain retroviruses, paramyxoviruses, and flaviviruses.
6. Cell fusion: Fusion of plasma membranes of four or more cells to produce an enlarged cell with four or more nuclei (syncytium or polykaryon formation). For example, paramyxoviruses and herpesviruses.
7. Inclusion bodies are areas of altered staining not visualized in live cell cultures. Depending on the causative virus, these inclusions may be characteristics (Table 7.3). Mostly, inclusion bodies indicate areas of the cell where the viral protein or nucleic acid are synthesized or virions are assembled, but in some cases, it can also indicate areas of viral scarring.

TABLE 7.3 Types of Inclusion bodies with examples.

S. no.	Type of inclusion body	Viral infection
I.	*Intranuclear eosinophilic*	
1.	Cowdry type A	
i.	Torres body	Yellow fever
ii.	Lipschultz body	Herpes simples, Varicella zoster
2.	Cowdry type B	Reovirus, Poliovirus, Adenovirus
II.	*Intranuclear basophilic*	
1.	Owl's eye	Hodgkin's lymphoma (Cytomegalovirus)
2.	Cowdry type B	Adenovirus
III.	*Intracytoplasmic eosinophilic*	
1.	Negri bodies	Rabies
2.	Handerson-Patterson bodies	Molluscum contagiosum
3.	Paschen bodies	Variola
4.	Bollinger bodies	Fowl pox
5.	Borrel bodies	Fowl pox
6.	Guarnieri bodies	Poxvirus
IV.	*Intracytoplasmic and Intranuclear*	Measles inclusion-body encephalitis (MIBE)

7.6.1.2 Animal inoculation

Animal inoculation is largely restricted only to research due to ethical issues, high costs, and difficulty of maintenance. It is done to study viral pathogenesis, oncogenesis, vaccine testing, and primary isolation of certain viruses which are difficult to cultivate, such as arboviruses and coxsackie viruses. Rabbit, hamster, guinea pig, suckling mice, and primates are common animals used for the isolation of viruses. Specimens may be inoculated by intracerebral, subcutaneous, or intraperitoneal routes. Animals are observed for signs & symptoms of disease or death. The tissue sections are subjected to histological examination. For example, following intracerebral inoculation into suckling mice, Coxsackie-A produces flaccid paralysis while Coxsackie-B produces spastic paralysis.

7.6.1.3 Egg inoculation

In 1931, Goodpasture used the embryonated hen's egg for the cultivation of the virus for the first time and was later developed by Burnet. For inoculation, eggshell surface is disinfected with iodine and penetrated with a small sterile drill. After inoculation of the virus, it is sealed with gelatin or paraffin and incubated at 36°C for 3–5 days, and the virus is isolated from the tissue of egg. Viral growth and multiplication is indicated by the death of the embryo, cell damage, or by the formation of typical pocks or lesions on the egg membranes. Specimens can be inoculated by four different routes into embryonated 7 to 12 days old eggs.

1. *Yolk sac inoculation*—It is preferred for avian infectious bronchitis virus, where it causes dwarfing of the embryo. It is also used for human viruses and bacteria such as arboviruses (e.g., Japanese B encephalitis virus, West Nile virus), *Rickettsia*, *Chlamydia*, and *Haemophilus ducreyi*. Encephalomyelitis viruses cause the death of the embryo.
2. *Amniotic sac*—Primary isolation of the influenza virus is mainly done in the amniotic sac and its growth is measured by hemagglutination (HA).
3. *Allantoic sac*—Influenza and new castle disease viruses can be grown in the allantoic sacs and they can be detected through haemagglutination. Since this cavity is large. so it can be used for the production of viral vaccines. For example, influenza, yellow fever (17D), and rabies (Flury strain) vaccine.
4. *Chorioallantoic membrane*—It is preferred for HSV (Herpes simplex virus) and Orthopoxviruses. Viruses produce pocks on chorioallantoic membrane (CAM), the pocks can be counted to know the number of viral particles present in the inoculum. The morphology of pocks produced varies for different poxviruses. For example, HSV-2 pocks are larger than HSV-1. Variola pocks are less hemorrhagic and necrotic than pocks of the vaccinia virus.

7.6.2 Assays measuring viral infectivity

The number of infectious virions can be titrated by infecting cell cultures, chick embryos, laboratory animals, or hosts with dilutions of viral suspensions and then detecting the signs of viral replication. There are two types of infectivity assays, quantitative, and quantal.

7.6.2.1 Quantitative assays

1. *Plaque assays*—A serial 10-fold dilution of viral suspension is inoculated onto cultured cell monolayers for virions to adsorb to the cells. An overlay of a medium in an agar or methylcellulose gel is put over infected cells to make sure viral progeny is confined to the immediate vicinity of the originally infected cell. Consequently, each virion gives rise to a localized focus of infected cells with morphological changes which become visible to naked eyes after a few days. They can be stained with neutral red or crystal violet where the living cells take up the stain and the plaques appear as clear areas. A single viable virus particle is enough to form a plaque. The infectivity titer is expressed in terms of plaque-forming units per milliliter.
2. *Transformation assays*—Few viruses especially oncogenic viruses "transform" cells rather than killing them. They show decreased contact inhibition, grow uncontrollably and form a heapedup "microtumor" that proliferates in semisolid agar or methylcellulose media.
3. *Pock assays*—It is an old method, now occasionally used for the poxviruses. Titration of viruses is done on the CAM of the chick embryo. Newly synthesized virus escapes to cells adjacent to infected cells, giving rise to localized lesions, known as pocks. The characteristic morphology of the pock helps in the identification of a group of viruses or even a particular mutant.

7.6.2.2 Quantal assays

This is a quantal assay that detects whether there are any infectious virus particles in the inoculum and not their number. Serial dilutions of the virus are inoculated and virus is allowed to replicate and spread to destroy the whole cell culture. Hence, it is concluded whether or not that particular dilution of the virus could infect the cell lines.

7.6.2.3 Hemagglutination

It was first described by Hirst in 1941, who analyzed HA on influenza virus. Some viruses contain virus-coded proteins capable of binding to erythrocytes in their outer coat. The process of virions bridging together red blood cells to form a lattice is known as HA. Influenza and paramyxoviruses have an enzyme, neuraminidase, which destroys the glycoprotein receptors on the erythrocyte surface allowing the virus to elute making a HA complex. The simplicity of HA makes it a convenient assay, if large amounts of virus are available since its sensitivity is not that good.

Other methods include electron microscopy (EM) or immune staining and visualizing under microscopy.

7.6.3 Direct microscopy

7.6.3.1 Electron microscopy

EM is the oldest method for direct visualization of the virus, and presently it is mostly used for research than diagnostics. In the 1970s, many new noncultivable viruses were discovered through EM. The human coronaviruses, hepatitis A virus, rotavirus, astroviruses, caliciviruses, and adenoviruses were first identified with this technique. EM requires negative staining of the clinical specimen in which the specimen is placed on carbon or formvar-coated grid usually with prior clarification and concentration by ultracentrifugation or antibody-induced clumping. sodium phosphotungstate negatively stains the virions adhered to the surface. There are other variations also through which negative staining can be done.

EM is especially useful for the identification of the fastidious or noncultivable viruses in specimens. The sensitivity of EM depends on viral concentration which should be $\geq 10^6$ viral particles per milliliter of the specimen. The characteristic morphology of most viruses allows identification up to the family level, for further identification other more specific methods are required. Although the major advantage of EM is the speed but the high cost of the instrument and specialized training and expertise needed, coupled with the lack of sensitivity and specificity, do not make this a viable option for routine diagnostics.

7.6.3.2 Fluorescence microscopy

Viral antigens can be detected in fixed, frozen tissue sections, or exfoliated cells. Commonly used indicators are fluorescein, horseradish peroxidase, and alkaline phosphatase. The approach is not only valuable in virus diagnosis but also understanding disease pathogenesis as matching viral material to particular organs or histopathology can be very helpful. The antibody is labeled with a fluorochrome and the antigen—antibody complex emits light of a particular longer wavelength when excited by short-wavelength light.

This can be visualized as fluorescence in an ordinary microscope after the light of all other wavelengths is filtered out. The sensitivity of the method is generally too low to detect complexes of fluorescent antibodies with virions or soluble antigens. The two main variants of this technique are:

1. *Direct immunofluorescence*—A fluorescein-tagged antiviral antibody is layered over a frozen tissue section, monolayer on a coverslip, or an acetone-fixed cell smear. Unbound antibody is washed, and the cells are visualized by light microscopy using a powerful ultraviolet/blue light source. It reveals the apple-green light emitted from the specimen. It is the preferred procedure when large quantities are suspected, for example, RSV (respiratory syncytial virus) in patient sample or viruses growing in cell culture.
2. *Indirect immunofluorescence*—It is a two-step method in which an untagged antiviral antibody binds to an antigen and are recognized by fluorescein-conjugated antiimmunoglobulin. It has proved to be of great value in the detection of infections known to have relatively small quantities of viruses, for example, adult respiratory viruses like paramyxoviruses, orthomyxoviruses, adenoviruses, etc. Direct immunofluorescence is rapid and specific while indirect immunofluorescence is more sensitive.

7.6.3.3 *Immunoperoxidase staining*

Another microscopic method of visualizing viruses in infected cells is to use an antibody directed at viral antigen coupled to horseradish peroxidase; followed by the addition of hydrogen peroxide with a benzidine derivative forms insoluble colored precipitate. These preparations are permanent and a light microscope is required for visualization. Endogenous peroxidase in the leukocytes can interfere with the test and give false-positive results, but this problem can be prevented by right technique and use of controls.

7.6.4 Histology/cytology

Direct microscopy of stained histology or cytology specimens gives the indication of viral infection with cellular changes. Specific cellular changes can be confirmed through staining for particular antigens using antibody or genome sequences through nucleic acid probes, for example, CMV, VZV (varicella-zoster virus), HPV (human papillomavirus), etc.

7.6.5 Detection of viral antigens and antibody (serology)

Apart from the immunological staining and microscopy techniques, there are other less complicated and faster assays for the detection of viral antigens and antibodies in a patient sample. These techniques can target the antigen during the acute phase (first 10 days) of infection or antibody later in infection. IgM antibody is detected within 7–14 days of infection that is a recent infection while IgG antibody is detected after 10–14 days of infection and may remain for life. Paired sera is used to detect the sero-conversion to IgG during the acute and the convalescent phase, 10−14 days apart. A fourfold rise in the titer signifies a new infection. Such immunological diagnosis can be done by the following methods.

7.6.5.1 Immunochromatography

Lateral flow immunoassay or immunochromatography involves the movement of an antigen, or antigen−antibody complexes, through support, for example, nitrocellulose film, filter paper, or agarose. A labeled antibody/antigen reacts with the test antigen/antibody, and then antigen−antibody complex migrates through the solid support by capillary action. A second antibody embedded in the solid support allows detection of the complexes if the antigen is present (Fig. 7.1). For quality control purposes positive and negative controls are included with the tests to ensure the validity of results. In the present time immunochromatography test formats are commercially available for diagnosing most of the viral infections. They may either detect the viral antigens such as SARS-CoV-2, HIV p24, influenza A and B, dengue NS1, RSV, or antibodies (IgM/IgG) such as SARS-CoV-2, dengue, HIV, etc. The significance of this test lies in the fact that it provides rapid results which is essential to guide the early management of patient. Immunochromatography has emerged as a valuable point of care test especially in areas where equipment and trained manpower are limited.

7.6.5.2 Enzyme-linked immunosorbent assay

Before the advent of PCR, ELISA had radically changed diagnostic virology, and it is still widely used. The sensitivity of ELISA is usually high, depending upon the characteristics of the antibody-antigen interaction. The basic principle behind this test is that all the components forming the layers of the reaction are known and defined, except the "unknown" component in the specimen. The "capture" antibody can be attached by

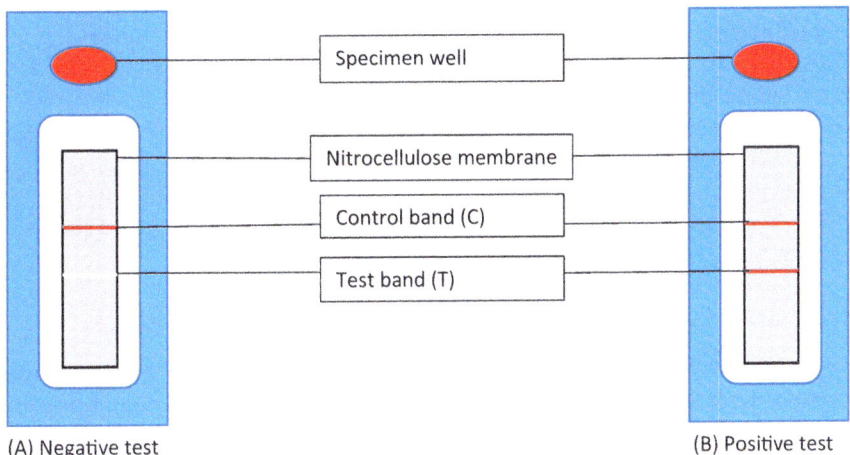

(A) Negative test   (B) Positive test

FIGURE 7.1   Immunochromatography assay.

adsorption or covalent bond to a solid substrate, on the polystyrene or polyvinyl microtiter plate wells, aiding the rinse of the solid surface between applications of reagents.

Virus and soluble viral antigens from the specimen are adsorbed to the captured antibody. After the unbound antigen is washed, an enzyme labeled with a detector antibody (antiviral antibody) is added. Common enzymes linked to antibodies are horseradish peroxidase (HRP), alkaline phosphatase, etc. After the next washing and drying, the appropriate organic substrate for the particular enzyme makes the color change noticeable. The color change can be seen by naked eye or it can be read through a spectrophotometer. Sensitivity can be enhanced by avidin- biotin, high-energy fluorescent substrates (Fluoroimmunoassay), chemiluminescent, or radioimmunoassay (RIA) which are easy to identify in small amounts. RIA and fluorescence immunoassay are highly sensitive and reliable assays that are compatible with automation. Their only disadvantage lies in the cost of the equipment, affordable only for laboratories with a large throughput of specimens.

Monoclonal viral antibodies (mAbs) are commonly used in immunoassays nowadays as detector antibodies. The mAbs represent purified, monospecific, well-characterized, antibodies of a defined class, recognizing only a single epitope, and are free of "natural" and other adventitious antibodies against host antigens or extraneous agents. Since monoclonal antibodies can be manufactured easily, they may serve as reference reagents. It is important to select mAbs of high affinity, but not of such high specificity that some strains of the virus being sought might be missed in the assay. Following are the main four types of ELISA (Fig. 7.2):

1. *Direct ELISA*—The antigen is immobilized on the surface of the plate and detected with its specific antibody which is conjugated to HRP or other detection molecules.
2. *Indirect ELISA*—It uses two-step processes for detection. A primary antibody specific for the antigen binds to the target and a labeled secondary antibody against the host species of the primary antibody binds to the primary antibody for detection.
3. *Sandwich ELISA*—It is the most commonly used format that requires two antibodies specific for different epitopes of the antigen or matched antibody pairs. One of the antibodies is coated and used as a capture antibody to facilitate the immobilization of the antigen while the other is conjugated and facilitates detection.
4. *Competitive ELISA*—It is also known as inhibition ELISA. It measures the concentration of an antigen by detecting signal interference. The antigen in the sample competes with a reference antigen for binding to a specific amount of labeled antibody. The reference antigen is precoated on the plate and the sample preincubated with labeled antibody is added to the wells. Based on the amount of antigen in the sample, free antibodies will be available to bind the reference antigen. The more antigen there is in the sample, less reference antigen will be detected and the weaker will be the signal.

ELISA is routinely used for the diagnosis of HIV, Hepatitis, rubella, measles, mumps, and arboviruses such as dengue or Japanese encephalitis.

7.6.5.3 Neutralization assays

These are highly specific assays which test neutralizing antibodies. They can be used as confirmatory test for other less specific serological assay. It can also be used to evaluate vaccine effectiveness. Antigen and dilutions of antibody are mixed followed by inoculation

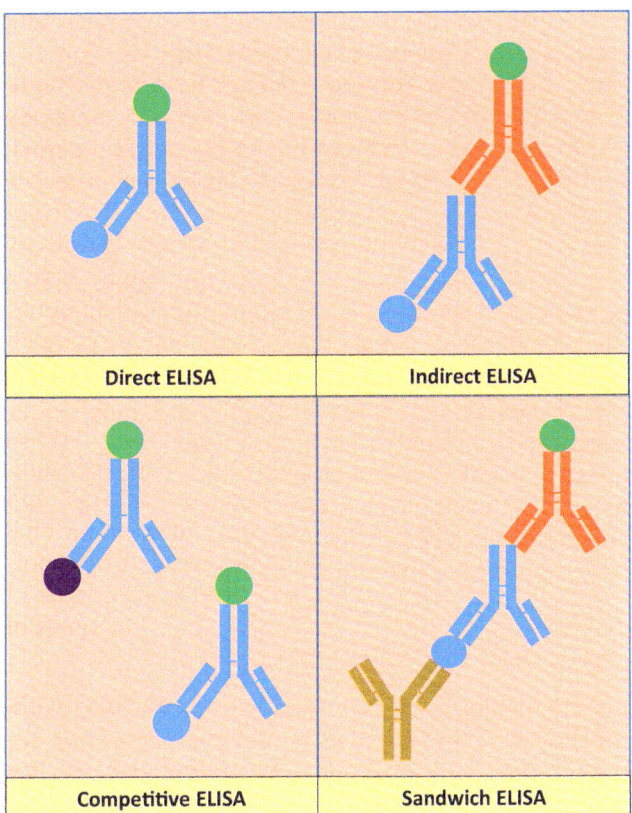

FIGURE 7.2   Different types of Enzyme linked immunosorbent assay (ELISA).

on a tissue monolayer and observation of CPE inhibition or plaque formtion on an overlay of agar for plaque reduction neutralization assays. Micro-neutralization assays can be read by ELISA which reduces the time before infected cells can be detected. Being labor intensive, these tests are not routinely used by diagnostic laboratories.

7.6.5.4 Latex particle agglutination

Perhaps the simplest of all immunoassays is the agglutination by antigen of small latex beads previously coated with antiviral antibodies. This agglutination reaction can be reversed by coating specific antigen on the latex particle, and testing for presence of antibody in the sample; for example, detection of rabies virus antibodies in human and equine sera. The usage of synthetic beads improves the stability, consistency, and uniformity of the reaction. Since it offers rapid results (often within minutes), it has become popular with small laboratories and medical practitioners. Additional advantages are simple design, not need for expensive equipment, easy to perform, flexibility of synthetic beads used, and even small quantities of sample can be tested. However, any contamination during sample preparations may lead to false negatives results. However, these assays suffer from both low sensitivity and low specificity.

These are similarity-searching software that identify homologous DNA sequences and proteins based on the extent of similarity in the sequences. Such similarity between two DNAs or amino acid sequences occurs due to the common lineage-homology. These search tools use a scoring methodology to compare two sequences and provide precise statistical similarities data. BLAST is moainly involved in locating ungapped, locally optimal sequence alignments, while FASTA determines analogy between less similar sequences.

7.6.5.5 *Western blotting*

Western blotting is used to detect antibodies from a mixer of different proteins present in a particular virus preparation. First, the concentrated virus is solubilized and proteins split into discrete bands based on molecular mass by gel electrophoresis. Now, the separated proteins are blotted onto a nitrocellulose paper, the test serum reacts with the viral proteins, and unbound components are washed away. Thereafter, bound antibody can be exhibited using an enzyme-labeled antispecies antibody. Thus, immunoblotting can be used to study the development of antibodies at different phases of infection. It is used as a confirmatory test for samples that are ELISA or ICT positive for anti-HIV antibodies. Western blotting is an important tool for research purposes used to characterize viral antigen mixtures, using the antibody reagent rather than the antigen as the "known" component in the reaction.

7.6.5.5.1 Applications of serology

Serological tests are used for diagnosing acute or chronic viral infection and indicate whether the infection is clinical or subclinical. They can also be used for screening the level of immunity after immunization, and efficacy of vaccination in a population. Testing the risk of the close contact of an individual with a life-threatening infection like the COVID may prevent it's spread to susceptible individuals by contact precautions or immunization and help in monitoring the course of infection.

EIA, RIA, CLIA, and neutralization assays generally display substantially higher sensitivity than immune microscopy techniques and now obsolete serological methods like complement fixation and immunodiffusion. The sensitivity of a test can be improved by the use of purified reagents, updated equipment, and signal amplification.

Signal amplification and use of cloned antigens or synthetic peptides may improve the specificity of the serological test.

7.6.6 Molecular techniques

Routine diagnostics does not require the determination of strain or sequence of the virus causing infection. But, strain identification becomes essential for epidemiological or public health purposes and sometimes for patient management. Influenza virus strains are differentiated by comparing HA-inhibition titers using standard antisera. Complete nucleotide sequencing of a viral genome has become a crucial tool to detect novel viruses, for example identifying the newer mutated versions of coronaviruses. Multiplex molecular methods are gaining momentum in routine diagnostic laboratories. Before the onslaught of the COVID-19 pandemic, most of the molecular and sequencing methods were mainly used for research purposes and epidemiological studies, but now RT-PCR being the backbone of SARS-CoV-2 testing, these methods are now available at most of the diagnostic centers.

7.6.6.1 *Detection of viral nucleic acids*

Nucleic acid detection can be done through PCR assays assisted with standardized automated nucleic acid extraction oligonucleotide synthesis and real-time (RT) observation of PCR products. The advances in genome sequencing methodologies and the availability of standard sequence databases like BLAST (Basic Local Alignment Search Tool), FASTA (fast alignment) etc. have analyzed the obtained sequences easy. These are similarity searching softwares that identify homologous DNA sequences and proteins based on extent of similarity in the sequences. such type of similarity between two DNA or amino acid sequences occur due to the common lineage-homology. these search tools use a scoring methodology to compare two sequences and provide precise statistical similarities data. BLAST is moainly involved in locating ungapped, locally optimal sequence alignments while FASTA detects analogy between less similar sequences.

7.6.6.1.1 Nucleic acid hybridization

A variety of test formats are used for nucleic acid hybridization. Older techniques involved hybridization of immobilized nucleic acid target and a labeled probe, washing unbound probe, and then detection of bound probe. Examples are dot-blot assays where the nucleic acid is immobilized onto filters; Southern blotting where nucleic acids are segregated based on molecular weight by electrophoresis, blotted on a filter, and hybridization is done. Earlier approaches involved the use of radioactive isotopes like 35 S and 32 P to label nucleic acids or oligonucleotides as probes for hybridization assays. Signals were read by a spectrometer or by autoradiograph.

The modern techniques involve nonradioactive labels fluorescein or peroxidase which produce a direct signal and others like biotin or digoxigenin that act indirectly by binding to a labeled ligand that then emitting signal.

7.6.6.1.2 Polymerase chain reaction

In nucleic acid amplification tests, a single copy of a gene sequence is enzymatically amplified into million copies within a few hours and has become a gold standard diagnostic tests. Most of the test platforms have sensitivities and specificities close to 100%. The RT PCR uses florescent labeled probes which help in detection of DNA amplification process as it happens on an instrument that collects the fluorescent data from each PCR cycle. The accumulated data is plotted in the form of a sigmoidal curve is analyzed based on the cycle threshold (Ct) value. Ct value is the point where fluorescence signals from amplicons exceed the background and it is indirectly proportional to the basal concentration of target DNA in the specimen. So, the higher the concentration of viral DNA in the specimen lower will be the Ct value.

Various studies have shown that respiratory virus detection using RT reverse transcriptase PCR (rRT-PCR) is more sensitive than conventional methods. Multiplex rRT-PCR allows simultaneous amplification of several viruses affecting a particular system in a single reaction and assists in the identification of multiple causative agents at once. PCR is done under controlled conditions of temperature, primer concentration, ionic strength, and nucleotide concentration. The three steps of PCR are: melting target DNA at 95°C, cooling at 50°C−60°C for binding of two oligonucleotide primers and extension from oligonucleotide primers catalyzed by DNA polymerase producing complementary copies of target strands. The products from the first cycle are templates for the second cycle and so on for the subsequent cycles increasing the DNA copies exponentially (Fig. 7.3).

Major developments in PCR technology have made it convenient and affordable since it was first described in 1983. These include:

1. Heat-stable DNA polymerase (*Taq*) of *Thermus aquaticus*: The enzyme need not be replenished the in-between cycles, allowing attainment of higher annealing temperatures and thus increasing specificity by decreasing false-positive reactions due to mismatching.
2. Nested PCR: A second amplification with primers internal to the first target increases sensitivity, though with the risk of false-positive reactions.

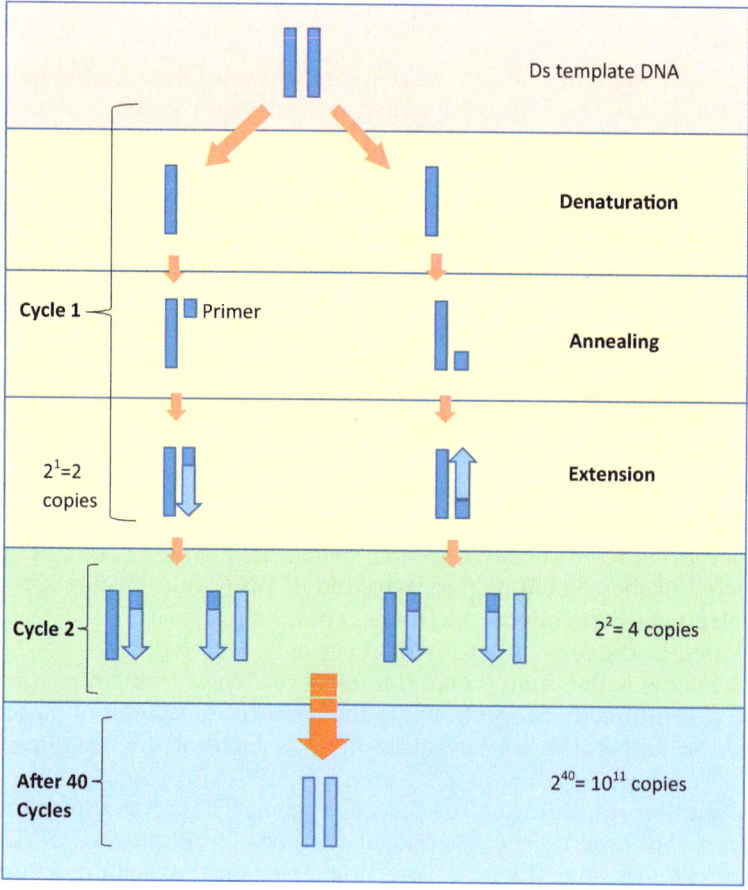

FIGURE 7.3 The steps of PCR shown in the first cycle and then exponential amplification of target DNA after repeated cycling.

3. RT PCR: Detection of PCR products during the reaction requires a thermocycler with an inbuilt fluorimeter. The fluorimeter measures PCR products in the reaction tube and they can be detected by fluorescence using dyes like SYBR Green or reporter probes like TaqMan.
4. Quantitative assays: The generation of standard curves corresponding to the original concentration of the target allows quantitation of viral infection. This aids in clinical interpretation, management, and following the prognosis of a patient. For example, HIV viral load testing helps in detecting the development of drug resistance.
5. Multiplex PCR: Different sets of primers bearing different fluorescent labels for different targets Can be included in the same reaction. This helps in the detection of several different viruses simultaneously.
6. Isothermal amplification: This modification of PCR does not require a high (95°C) temperature cycle to melt the DNA products to allow the binding of a new round of primers and depend on enzymes to displace the strands. Examples are Loop-Mediated Isothermal Amplification, Strand Displacement Amplification, Nicking Enzyme Amplification Reaction, and Helicase-Dependent Amplification. They are fast and do not require thermal cycling equipment.

PCR is particularly important for viruses that cannot be cultured, specimens containing an inactivated virus, due to improper storage or transport, quantitation is required and preventing biohazard category 3 or 4 agent exposure to laboratory staff. Issues associated with PCR include the presence of DNA polymerase inhibitor in leading false negative results, contamination with extraneous DNA and it identifies only agents for which primer is set. Thus, laboratory findings should be correlated to the patient's clinical condition.

7.6.6.2 Microarray technologies

Microarrays or microchips are a solid support matrix on which are "printed" spots, containing thousands of unique oligonucleotides which are representative of conserved sequences from a customized panel of viruses. This technology is based on the principle that these oligonucleotides will capture randomly amplified labeled nucleic acid sequences from clinical specimens. This process is detected by laser scanning followed by software programs that assess the strength of the binding and identifies the virus. For example, it helped in the determination that severe acute respiratory syndrome (SARS) was a coronavirus. Microarray is used as a research tool for the identification of novel viruses, assess host responses to infections, and evaluating the development of antiviral drug resistance.

7.6.6.3 Sequencing

7.6.6.3.1 Sanger sequencing

It is also known as first- generation sequencing or targeted sequencing. This technique uses oligonucleotide primers to target specific DNA regions. The first step involves the denaturation of the double-stranded DNA. The single- strand of DNA is then annealed to oligonucleotide primers and a mixture of deoxynucleotide triphosphates (dNTPs) are used to elongate it. dNTPs provide arginine (A), guanine (G), cytosine (C), and tyrosine (T) nucleotides that are needed to build the new double-stranded structure. A small portion of chain-termination di-dNTPs (ddNTPs) is added to each nucleotide. The sequence continues to elongate using dNTPs until a ddNTP comes into play. Both dNTPs and ddNTPs have an equal probability of attaching to the sequence; every sequence has different length based on site of termination. Each ddNTP has a different fluorescent marker. As a ddNTP attaches to the elongating sequence, the base fluoresces a specific color, which can be detected and nucleotide coding can be accomplished. Sanger sequencing is a vigorous testing methodology, which can detect a point mutation or deletion, duplication etc., with ease. It has been extensively used since many decades for delineating the mutational spectrum of various infections and tumors. It has been a cornerstone technology for finding the viral variants in diagnostic testing. As compared to the newer approaches of sequencing, Sanger sequencing is more costly, time consuming and labor intensive. However, it is still routinely used for sequencing of specific genes or fragment of genes. For example, for viral genotyping or for testing of resistant genes, as these are associated with specific genome regions.

7.6.6.3.2 Next-generation sequencing

Second generation/massively parallel/deep/next-generation sequencing (NGS) is a relatively newer technique of DNA sequencing directly from DNA fragments and it does not require cloning in vectors. It involves four steps; library preparation, cluster generation, sequencing, and alignment & data analysis. Previously, Sanger sequencing was used which had a long turnaround time, and yet the data output was less. NGS is faster and generates an extensive quantity of sequence data at a low cost. This approach has been extremely beneficial for research purposes. Its application to diagnostic virology has come to forefront with the ever-changing and masquerading infection of SARS-CoV-2.

NGS methods are grouped into two major categories, sequencing by synthesis and sequencing by hybridization. Most sequencing by synthesis use technology in which individual DNA molecules are distributed to millions of wells or chambers, or bound to specific locations on a solid substrate. The DNA molecules amplified by PCR undergo DNA synthesis reactions involving labeled nucleotides or chemical reactions based on the incorporation of a specific nucleotide detected by various techniques. For sequencing by hybridization, arrayed DNA oligonucleotides of known sequence are hybridized to labeled fragments of the DNA to be sequenced. Repeated hybridization and washing away the unwanted nonhybridized DNA determine whether the hybridizing labeled fragments match the sequence of the DNA probes on the filter. Analysis through bioinformatics helps to piece together the fragments by mapping the individual sequences to the reference genome. NGS is useful in sequencing entire genomes or specific parts of interest. Some of the platforms of NGS are pyrosequencing, ion torrent sequencing, etc.

7.6.6.3.3 Third-generation sequencing

Large fragment single molecule or third-generation sequencing aims to sequence long DNA or RNA molecules. The current commercial platform is PacBio sequencing or SMRT (Single Molecule Real Time) sequencing which permit very long fragments (up to 30−50 kb) to be sequenced. SMRT involves binding an engineered DNA polymerase with bound DNA to the bottom of a well. An important advantage of SMRT is that the rate of each nucleotide addition can be measured during synthesis.

7.6.6.3.4 Fourth-generation sequencing

Nanopore-based DNA sequencing involves passing long DNA molecules through small diameter "holes" and measuring differing currents as each nucleotide passes by a linked detector. More than a hundred kb of DNA can be passed through the nanopore and hundreds of Gb of sequence could be achieved at a relatively low cost. Because of its small, handheld size, it has the advantage of portability and less space. The high error rate can be overcome by a large number of molecules sequenced.

The applications of sequencing in virology involve discovering new pathogens in undiagnosed illnesses or outbreaks (e.g., SARS-CoV-2), retrospective diagnosis of an undiagnosed illness, screening vaccines for contaminants, analyzing quasispecies sequence composition of the viruses in a clinical sample, identification of mutations in viral genomes and understanding human virome in health and disease.

7.7 Conclusion

The knowledge of viral genetics, stages of infection, pathogenesis, and epidemiology are crucial for making decisions regarding the correct diagnostic modality to be used for a particular disease. There have been advancements in diagnostic platforms, enabling higher specificity and sensitivity of viral diagnostic tests. So, accurate and rapid identification of circulating viruses is more feasible nowadays, making clinical correlation far better. With advances in molecular techniques and DNA sequencing, viral diagnostic has come a long way to provide discrete and authentic results. Progressive and uniform bioinformatic data easily available for analysis, transfer, and storage has been a breakthrough in viral diagnosis. Virus isolation and identification of emerging and re-emerging viruses which was a remote possibility in most parts of the world, has now become a routine test after the incursion of COVID-19. Good quality of specimens, detailed clinical history of a patient, and multidisciplinary committee involving clinicians and laboratory personnel are necessary for reliable diagnosis and improved patient management.

Further reading

Al-Hajjar S. Laboratory diagnosis of viral disease. Textb Clin Pediatrics 2012;923–8. Available from: https://doi.org/10.1007/978-3-642-02202-9_75. PMCID: PMC7123305.

Burrell CJ, Howard CR, Murphy FA. Laboratory diagnosis of virus diseases. Fenner and White's medical virology. 2017. pp. 135–54. Available from: https://doi.org/10.1016/B978-0-12-375156-0.00010-2.

Erica Suchman, Carol Blair. Cytopathic effects of viruses protocols. American Society for Microbiology; 2016. <https://asm.org/ASM/media/Protocol-Images/Cytopathic-Effects-of-Viruses-Protocols.pdf?ext = .pdf> [accessed 11.11.21].

Levin M, Asturias E, Weinberg A. Infections: viral & rickettsial, Vol. 23. New York: McGraw-Hill; 2016.

Louten J. Detection and diagnosis of viral infections. Essent Hum Virol. 2016:111–32. Available from: https://doi.org/10.1016/B978-0-12-800947-5.00007-7. Epub 2016 May 6. PMCID: PMC7150318.

Slatko BE, Gardner AF, Ausubel FM. Overview of next-generation sequencing technologies. Curr Protoc Mol Biol 2018;122(1):e59. Available from: https://doi.org/10.1002/cpmb.59.

Specter S, Hodinka RL, Wiedbrauk DL, Young SA. Diagnosis of viral infections. In: 2nd ed. Richman DD, Whitley RJ, Hayden FG, editors. Clinical virology, Vol. 35. Washington, DC: American Society for Microbiology Press; 2002. p. 243–72.

Storch G. Diagnostic virology. In: 5th ed. Knipe DM, Howley PM, editors. Fields virology, Vol. 5. Philadelphia, PA: Lippincott Williams & Wilkins; 2007. p. 565−604.

Storch GA. Diagnostic virology. Clin Infect Dis 2000;31(3):739−51. Available from: https://doi.org/10.1086/314015.

World Health Organization (WHO). Collecting, preserving and shipping specimens for the diagnosis of avian influenza A(H5N1) virus infection Guide for field operations; 2006. <https://www.who.int/ihr/publications/Annex8.pdf?ua = 1> [accessed 11.11.21].

World Health Organization (WHO). Guidelines for the safe transport of infectious substances and diagnostic specimens. Division of Emerging and Other Communicable Diseases Surveillance and Control; 1997. <https://www.who.int/csr/emc97_3.pdf> [accessed 29.10.21].

Electron microscopic methods for virus diagnosis

Nadun H. Madanayake[1], Ryan Rienzie[2] and Nadeesh M. Adassooriya[3]

[1]Department of Botany, University of Peradeniya, Peradeniya, Sri Lanka [2]Department of Agricultural Biology, Faculty of Agriculture, University of Peradeniya, Peradeniya, Sri Lanka [3]Department of Chemical & Process Engineering, University of Peradeniya, Peradeniya, Sri Lanka

Abbreviations

CMV	Cucumber mosaic virus
Cryo-TEM	Cryogenic transmission electron microscopy
EB	Electron beam
EM	Electron microscopy
PBS	Phosphate buffered saline
PCR	Polymerase chain reaction
PTA	Phosphotungstic acid
PVX	Potato virus X
PVY	Potato virus Y
SEM	Scanning electron microscopy
TEM	Transmission electron microscopy
TMV	Tobacco mosaic virus
TSWV	Tomato spotted wilt virus
VLPs	Virus-like particles
ZYMV	Zucchini yellow mosaic virus

8.1 Introduction

Detection and diagnosis of viruses has become a vital requirement in medicine, pharmaceuticals, and food and agriculture industry. Virus detection includes direct and indirect approaches. Direct approaches include microscopy, cell culture methods, and

Viral Infections and Antiviral Therapies
DOI: https://doi.org/10.1016/B978-0-323-91814-5.00008-8

molecular and nucleic acid-based techniques. Also, direct fluorescent antibody staining, rapid enzyme immunoassay, enzyme-linked immunosorbent assays are used in clinical virology laboratories. Moreover, polymerase chain reaction (PCR) assays are used to detect different viruses in a single test [1].

Microscopic methods such as electron microscopy (EM) is one of the oldest technique that is still being applied in diagnosis of viruses in biological specimens. The EM is a catch-all technique that provides an open view of the target of analysis. This has been used in many applications especially in clinical diagnosis, veterinary, and plant pathology. In addition, EM is used as a diagnostic tool to ensure the quality of biosafety products. The EM facilitates to observe the target samples for viral particles in diagnostic applications. Furthermore, this is a basic tool to identify and classify viruses at least up to family and genus [2]. Therefore, EM has become a basic tool in virus particle analysis. The EM consists of two classes as Scanning Electron Microscopy (SEM) and Transmission Electron Microscopy (TEM).

SEM is a better tool to analyze the surface morphology and fine structures of biological samples. However, its application is less compared to TEM for virus characterization because of the struggle to prepare specimens to obtain high-resolution images [3]. Structural details of virus surfaces or the appearance of complete virus can be obtained from SEM, which will aid in the characterization of viruses [4]. The TEM is used to study the internal structures of samples analyzed. Still, these two techniques are prominent techniques in diagnostic virology. To better understand, Fig. 8.1 elaborates on the TEM image of coronavirus [5].

8.1.1 Electron microscopy as a diagnostic tool

The EM analysis is a predominant tool to investigate novel and evolving pathogens. This is imperative when scientists are lacking enough evidences to recognize the etiological agent [3]. The most recent and one of the best examples in clinical virology is the

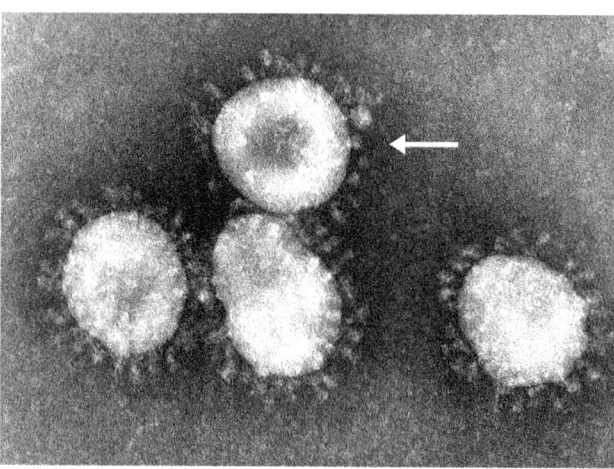

FIGURE 8.1 TEM micrograph of SARS-CoV, arrow head points a single virus particle [5].

detection of coronavirus disease. The EM was contributory technique to clarifying viral agents such as Ebola virus (Zaire in 1976), Ebola Reston infection (Reston in 1989) by filovirus, and the henipavirus epidemics in Australia and Asia [6] Furthermore, well-defined sample preparation methods enable researchers to march through the right path even in an emergency. Moreover, it does not require prior information, as in the case of molecular biology and nucleic acids-based assays [7].

The EM is a standout technique to observe viral pathogens directly from infected cell lines and from dirty clinical samples, such as plasma, serum, feces, urine, and vesicle fluids [2,8,9]. Hence, it does not require prior purification procedures as required in non-EM based methods. Sometimes long-term preservation of samples and environmental parameters such as temperature and relative humidity [10,11] can influence on virus viability. This can hinder the application of culture-dependent or molecular testings for virus diagnosis. Albeit, EM analyses are independent of culture-dependent methods where samples can be directly used for the assessment. Hence, it mitigates the limitations of culture-dependent methods as it does not require live specimens for diagnosis.

Immunological and molecular biology assays are highly specific and detection limits up to the species level. Molecular methods such as PCR require previous knowledge of genomic data of the organisms. Therefore, to assess and identify organisms it might require bioinformatics. In addition, primer designing required to amplify a target region of a sequence to obtain multiple copies of a segment of DNA [12,13]. However, these techniques can contribute to misidentifications or failures in diagnosis due to different antigenic determinants. Since viruses are highly prone to mutations this can lead to false-negative results. Although EM is a low-throughput technique it is the gold standard for detecting unknown agents where it does not require probes or primers for analysis [2]. Furthermore, molecular techniques might fail to identify dual infectious agents and EM can be used to diagnose such viruses rapidly. Therefore, EM is the backbone of detecting new and unknown viruses where it does not require virus specific reagents for the identification [6].

Scientists utilize EM to characterize viral pathogens and to study their virulence and distribution in the environment. Also, resolution level of EM has a significant impact on characterizing viruses to elaborate on distinct morphological components such as the nature of viral capsid and their nucleic acids [2,14,15]. In addition, EM analysis plays a vital role in understanding the mechanisms involved in viral pathogenesis. For instance, EM was used to identify the mode of endothelial infections by coronavirus particles for COVID-19 chilblains. Here, TEM played a vital role in understanding these mechanisms via histological evidence [16]. More importantly, EM helps study the mechanisms underlying the virus attachment and replication within the hosts. This is imperative in designing novel antiviral drugs and vaccines. Also, this is useful in clarifying the functions of different viral elements [17]. We always believe that a picture is worth a thousand words, therefore, EMs confirm the details concerning virus screening [2]. This is important for the development of novel antiviral strategies for unknown agents in an emergency [4].

The EM is also a basic tool in veterinary pathology for the detection of viruses infecting animals. Piñeyro and coworkers [18] reported the reasons for the sudden increase in the death of piglets using EM and immunohistochemistry. Here, EM was used to confirm transmissible gastroenteritis virus as the route cause for a higher mortality rate of pigs in Argentina. Therefore, this has become a basic tool in identifying etiologic agents in the veterinary field as well. Differential diagnosis and study of viruses make use EM to identify these agents. Barreto-Vieira et al. [19] proved that arboviruses such as dengue virus and its serotypes, Zika virus, Yellow fever virus, and Chikungunya virus with similar morphologies with different sizes are infecting the people in Brazil using TEM. Furthermore, Gibbs et al. [20] explained the use of EM in the rapid diagnosis of cowpox, pseudocowpox, and bovine herpes mammillitis in veterinary medicine.

The microbial contamination is a serious issue in animal cell cultures. It has become a major requirement to identify the potential contaminants. Different types of microbes including viruses cause adverse effects. Therefore, scientists have applied EM to detect the viral infections initially and the molecular and immunological methods to detect them. Ultrathin sectioning and staining procedures such as negative stains are commonly used sample preparations in recognizing viruses in growing cultures [21]. A case study from Fischer et al. [21] reported that the receivers of organ transplants were infected with lymphocytic choriomeningitis virus (family Arenaviridae) that was identified using thin section EM. Also, EMs are implemented as quality control steps especially to guarantee the biosafety in vaccines, test kits, etc. [2].

8.1.2 Electron microscopy in plant and animal virus diagnosis

Employing a virus detection method depends on the cost-effectiveness, sensitivity, rapidity, available instrumentation, and the stage of the disease or infection [22]. Conventional approaches include visual methods. For instance, plant viral detection procedures are based on symptomatology or indicator plants. Among the most widely used diagnosis methods of plant and animal viral diseases, serological and molecular biological methods possess a greater importance. Moreover, novel advanced methods include high-throughput imaging and high-throughput sequencing of nucleic acids. Albeit, EM is still in use for the discovery of plant and animal viruses.

The TEM is used as the main tool for the identification of plant viral particles. Also, this can be used in plant quarantine applications in detecting newly intruding viral pathogens to a country. For instance, TEM was used as a basic tool to diagnose cucumber mosaic virus (CMV) in pumpkins for the first time in Ukraine [23]. However, the sample preparation for plant virus detection using TEM is lengthy. The slow rate of chemical infiltration and polymerization of resins can take hours and days in sample preparations. Therefore, the application of TEM in plant virus analysis is comparatively low [24]. In this section, a selected number of plant and animal viruses that have been categorized as most disastrous at global scale are discussed. In addition to that some other cases have also been discussed where EM has been implemented for diagnosis. This section elaborates on some examples of animals and plant viruses where EM had been used as a diagnostic tool.

8.1.3 Dengue virus

Dengue virus is a mosquitoborne RNA virus that infects humans in various parts of the globe. This is a spherical-shaped viral particle with a diameter of 40−50 nm where EM has been used as a basic tool to characterize its morphology. This is also applied to explore the virus—host interactions. For instance, Noisakran et al. [25] stated that dengue viruses were observed within vacuoles or endosomal vesicle of platelets isolated from dengue patients. In addition, ultrathin sections using TEM showed dengue virus-like particles (VLPs) of size 50—60 nm were detected in specimens collected from spleen, liver, brain, and lungs for the first time from a fatal dengue cases.

8.1.4 Rabies virus

Rabies virus is a zoonotic virus belonging to genus *Lyssavirus*. Rabies reported to cause lethal neural infections on humans [26]. Typically, rabies can be diagnosed using the presence of special structures called Negri bodies on specimens. Also, EM has been used to study the inclusion bodies and virus particles. For instance, EM observations on Negri bodies showed that these are made of a matrix of granular or filamentous materials consisting of nucleoproteins of viruses [27].

8.1.5 Ebola virus

Ebola virus is deadly viral pathogen identified in 1976 from Southern Sudan and Northern Zaire [28]. Same as in other applications EM was used to characterize Ebola. The traditional gold standard to clarify the existence of Ebola virus is viral isolation in cell culture, using Vero E6 African Green monkey kidney cells. Therefore, propagated Ebola viruses in cell cultures can be directly be visualized by EM within a period of 1—5 days after inoculation [29]. Bukreyev et al. [30] utilized EM to characterize the chimeric human parainfluenza virus that contains the ebola virus glycoprotein to develop live attenuated vaccine against ebola virus.

8.1.5.1 Cucumber mosaic virus

Coexistence of CMV with tospoviruses on *Capsicum annuum* was documented in India. Infected plants had isometric virus particles of 24—52 nm (CMV) and quasispherical virus particles of 70—120 nm (Tospovirus particles) [31]. Icosahedral particles of CMV having 30 nm of diameter were detected through EM in CMV infected *Passiflora edulis* in China [32]. CMV particles having spherical shape in the range of 28—32 nm were observed in plant tissues of *Tylophora indica* in India [33]. Isometric CMV particles having 28—30 nm size were also observed in *Plectranthus barbatus* in India [34]. Natural occurrence of CMV in squash and a weed host, *Rumex hastatus* in India was confirmed by Kumar et al. [35]. The virus particles visualized by TEM showed isometric particles of CMV. TEM indicated drastic cellular level changes in two CMV isolates inoculated to *Nicotina glutinosa* in Egypt. Infected cells had abnormal and swollen like chloroplasts and intergrana lamellae were disorganized while grana appeared loose and lacking the form of integrity. Starch grains were very rare or not present at all. Ruptured chloroplasts were seen in some cells. Moreover, the cell walls of infected cells showed serrated appearance. Electron micrographs of purified CMV showed isometric particles of 28 nm in size [36]. Moreover, in a previous study CMV infected tobacco. Plasmolysis, collapse of the tonoplast, vesiculation in the interspace between the cell wall and the plasma membrane, abnormal chloroplast and hyperaccumulation of starch, and few thylakoid membranes were prominent features of CMV inoculated tobacco plants [37].

8.1.5.2 Tobacco mosaic virus

Thin sections of Tobacco mosaic virus (TMV) infected tobacco leaves labeled with tritiated uridine were observed under EM. Silver grains were observed in the nuclei and cytoplasmic ground substance but no grains were detected over mitochondria, vacuoles and intercellular spaces [38]. Viral particles are abundantly aggregated in outer tissues of necrotic zones, as compared to the necrotic center of tomato plants infected with TMV. Also, TMV particles were high in number in upper epidermis and palisade layers, but not in the lower epidermis of leaves [39]. TMV-inoculated plants nuclei possessed highly condensed and peripheral chromatin and revealed various time schedules for nuclei changes in different tissues. The number of mitochondrial cristae were observed and in chloroplasts, stacked grana, and stroma thylakoids remained normal. With time the mitochondria were degraded, but with intact outer membranes. Moreover, with time the infected plants had destroyed mesophyll cells in chloroplasts. Stacked nature of grana were destroyed in chloroplasts. Furthermore, invaginated tonoplasts, plasma memebranes retracted from the cell wall, fragments of endoplasmic reticulum [40]. In tomato roots coinfected with *Glomus* spp. and TMV intercellular hyphae associated with the cortical cells, showed a fimbriate cell wall structure separated by one or more osmiophilic bands. Also the adjoining cortical cells demonstrated high concentrations of virus particles [41]. EM evidences were presented for the infection of isolated tomato-fruit locule tissue protoplasts by TMV. The TMV particles were found attached to the protoplast surface form both invaginations and narrower end of plasmalemma [42].

Interestingly plant tissues with mixed infections of TMV in *Nicotiana tabacum* (280 × 17 nm in size, flexuous, rod-shaped) and Zucchini yellow mosaic virus (ZYMV) in *Cucurbita pepo* (707 × 12 nm in size, flexuous, rod-shaped) were observed under TEM [24]. Accordingly, dense cytosol with well-preserved organelles were seen in control and test samples. In all specimens, chloroplasts with dense stroma having thylakoids, starch grains, and plastoglobuli were observed. Moreover, mitochondria had a dense matrix and cristae. Nuclei had well-preserved heterochromatin and nucleoli. In TMV infected tissues, large areas having virions aligned parallel could be observed. Leaf cells infected with ZYMV showed cylindrical inclusion bodies in the cytosol. They appeared as bundles (when cut longitudinally) or as scrolls and pinwheels (when cut transversely). Moreover, proliferated and dilated endoplasmic reticula appeared frequently throughout the infected cells. Such modifications were not found in leaf cells of *Nicotiana* or *Cucurbita* control plants. In TMV-infected *N. tabacum* plants virions accumulated in parallel layers within the cytosol. In ZYMV infected *C. pepo* plants cylindrical inclusions including pinwheels and dilated endoplasmic reticula occurred in the cytosol.

8.1.5.3 Tomato spotted wilt virus

Infection of Tomato spotted wilt virus (TSWV) in *Peperomia obtusifolia, Humulus japonicas, Eustoma grandiflorum* were confirmed in South Korea [43–45]. Also, researchers have observed Tospovirus particles (80–100 nm in diameter) through TEM. Moreover, membrane-bounded, roundish Tospovirus particles of 80–100 nm in diameter were observed in cisternae of the endoplasmic reticulum in many epidermal and parenchymal cells of the tissues with lesions using TEM [46]. Cryogenic TEM (Cryo-TEM) was employed to study morphological changes related to extracellular vesicles upon infection by TSWV of certain aggregate structures that had no smooth contour, which were likely to be lipoprotein aggregates. Due to the thickness of ice layer (100 nm) larger vesicles were squeezed giving a shape of oblate ellipsoid, which appeared circular from the top. Bending of membrane to avoid contact with neighboring vesicles was also noted [47].

8.1.5.4 *Tomato yellow leaf curl virus*

The phenomenon of induction of autophagy in whiteflies due to infection of Tomato yellow leaf curl virus was demonstrated in a study. The TEM graph analysis showed that the infection induced the formation of autophagosomes [48].

8.1.5.5 *Potato virus X and potato virus Y*

Apoplast of *Nicotiana benthamiana* infected with Potato virus X (PVX) contained curved rod-shaped viral particles. It was observed that PVX particles were not associated with the exosomes [49]. Potato virus Y (PVY) infection in *Solanum tuberosum* L. loosen cell walls and created folds between necrotized and nonnecrotized mesophyll cells. Also, cell wall thickening with paramural bodies in phloem cells, paramural bodies connected with plasmodesmata in association with virus cytoplasmic inclusions, cell wall thickening with intense vesicles distribution were prominent morphological changes. Furthermore, PVY particles were aggregated in vesicles derived from cell wall with other paramural bodies at the vicinity of changed cell wall and PVY inclusions were observed in mesophyll cells. Moreover, PVY inclusions were found close to plasma membranes and paramural bodies along the deformed cell wall [50]. The PVY virions were detected in test plants inoculated with viruliferous cabbage aphids (*Brevicoryne brassicae* L.) under immunosorbent-TEM. Accordingly, the potyvirus-like virions from test plants trapped and decorated with homologous antiserum clumped with PVY antiserum were observed. The virus transmission efficiency was 11.25% [51].

8.2 Other plant viruses

Flexuous filamentous particles (580 × 13 nm) of Ligustrum Virus A were observed on *Syringa reticulata* var. mandshurica (Oleaceae) in China. The disease was first reported in *S. reticulata* with new yellow vein and malformed needle-shaped leaf symptoms [52]. *S. reticulata* is important as a plant widely used in urban home gardening. The TEM assay on a representative sample detected flexuous filament-shaped VLP of sweet potato feathery mottle virus (600−900 nm in length) infected *Chrysanthemum morifolium* in China. The plant is one of the eight well-known traditional medicinal herbs China. Telosma mosaic virus (TeMV) particles of length having nearly 700 nm and linear in shape were detected in virus-infected fruits of *P. edulis* using TEM [53]. The symptoms were visible as severe mosaic and distorted leaves and fruits with mosaic skin [54].

8.2.1 Sample preparation for electron microscopic analysis

Sample preparation is a critical step in EM analysis especially for biological specimens. Different methods have been applied based on the analysis requirement. Sample preparation for microscopic analysis has two main objectives. First, it is important to preserve the specimen with original characters. Second, the specimen should be visible with the method of imaging [55]. Therefore, sample preparations are imperative to enhance the quality of imaging in EM.

Sample preparation for virus diagnosis have several challenges and consideration. EM operate under high vacuum and samples interact with high energetic electron beam (EB). If underprepared specimens are placed within the sample holder for analysis, there is a greater possibility for the emission of gases compounds. These outgas molecules from the specimen tend to interact with both the primary EBs and the secondary or backscattered electrons emitted from the sample leading to low image quality. In addition, it can have a significant impact on creating artifacts in images generated from EM. Therefore, care should be taken to make specimens water-free and free of any organic contaminants that may potentially outgas in a high vacuum environment. Hence, biological samples should be prepared with the purpose of preserving the specimens with a minimum level of damage during the sample preparation and analysis [56].

Another challenge of imaging biological specimens is to differentiate between biological structures from artifacts created during image processing. Biotic components are largely composed of water and macromolecules, such as carbohydrates, proteins, lipids, and nucleic acids. These are majorly composed of low molecular weight elements, such as C, H, O, N, P, S, K, Ca, and Mg. Minute quantities of Cr, Co, Cu, Zn, Fe, Mn, and I may present and the contents may vary from specimen to specimen [55,56]. Therefore, images obtained from EM analysis of biological samples are very low in contrast because of their lower electron scattering power. Hence, staining procedures are implemented to enhance the quality of visualization. For that, specimens are infiltered with heavy metals to enhance the ultrastructural details [2].

Organelles such as lysosomes release enzymes such as hydrolases that tend to destroy fine structures within the specimens. Therefore, it is imperative to prevent the specimen from decomposing from reactive biological molecules. Generally, chemical and physical approaches are implemented to fix the specimen, which is the first step in sample preparations. Fixation ensures that structures will not be destroyed by enzymes and other reactive species. Chemicals used as fixatives are capable of stopping cellular processes that lead to decomposition. Fixatives acts on biological specimens either by denaturing and coagulating proteins or by cross-linking [55,57]. Therefore, the objective of utilizing fixatives is to stabilize the structure of the specimen from damages during consequent handling steps and analysis under the EM. In addition, critical point drying or using freeze-drying are used as physical methods of fixation [3]. Table 8.1 shows some of the commonly applied chemical fixatives and their principle of chemical fixation. Better fixative penetrates readily through the cells and reacts quickly and irreversible with target molecules. More importantly, they do not produce artifacts on the final images [55]. Sometimes if required postfixation steps are also applied [58] during sample preparations for EM.

TABLE 8.1 Commonly applied chemical fixative in samples and their mode of action [57].

Agent	Principle
Acetone	Acts as coagulants leading to protein denature and effective lipid solvent
Alcohols (ethanol/methanol)	Act as coagulants leading to protein denature and fixation occurs at 50%−60% for ethanol and >80% for methanol
Formaldehyde	Reacts with protein side chains present in protein molecules forming hydroxyl methyl groups. If calcium ions are available it reacts with some groups of unsaturated fatty acids
Glutaraldehyde	Slowly decomposes to form glutaric acids and polymerize to form cyclic and oligomer compounds by reacting with peptides
Osmium tetraoxide	Reacts with unsaturated fatty acids and phospholipids by reducing osmium tetraoxide to lower oxides which are black, insoluble, and deposited in tissue membranes
Protein cross-linking reagent Diimidoesters Diethylpyrocarbonate	Cross-link amino groups of proteins Reacts with tryptophan residue of proteins

Dehydration is a vital step in sample preparation because biotic components contain a greater proportion of water within their cells. Organic solvents such as acetone and ethanol are used to dehydrate the specimens. However, air drying is not recommended for dehydration since it can shrink the specimen [58,59]. Dehydration can collapse, shrink, and distort the specimen even after they are fixed with fixatives. Also, observation under a high vacuum can dry the specimen, thus reducing the contrast and resolution of the EM analysis [3]. Therefore, dehydration should be carefully performed for such preparations.

Based on EM techniques applied, samples may require specific requirements. For instance, specimens should be thin enough for TEM analysis to facilitate the electrons to travel through it. Hence, specimens should be thin as 300−500 nm for better imaging [56]. The EM requires conducting surfaces to obtain an acceptable level of contrast. Also, it is required to reduce the charging of organic materials to attain magnifications above 1000 times in EM. Charging leads to anomalous contrast for nonconductive specimens. Anomalous contrasts have different impacts on the image quality due to charge imbalance. Also, this has a significant impact on the secondary electrons generated from the specimen. In addition, image contrast might reduce leading to less topographic contrast. Therefore, images may appear with uncharacteristic darker or brighter regions. Hence, image distortion can ascend from deflections in the incident EB. Also, charging can damage the specimen forming several artifacts on the final image [60]. Surface coatings are necessary to prevent charging and to enhance the conductivity. Biological samples are required to have a minimum electrical conductivity of 100 mScm^{-1} for better imaging. If specimens are observed without surface coatings, the targets can be electrically charged from the uneven distribution of electrons on the specimen surface [3]. Primitive approach to make biological specimens to be electrically conductive by coating the surface with thin-film evaporation or sputtering of carbon or metals in a vacuum evaporator. In addition, infiltration with ionic liquids, such as 1-butyl-3-methylimidazolium tetrafluoroborate, can be implemented. These approaches make specimens electronically conductive preventing the risk charging. Conductive substrates such as indium-tin oxide, aluminum foil and metal-coated coverslips are used to fix biological specimens [3,61].

A considerable quantity of viral particles are required for EM analysis. For instance, a minimum viral quantity of 10^5 to 10^6 particles per milliliter are required for poxvirus or polyomaviruses in TEM. Filtration and centrifugation can enhance concentration of particles quantity in biological fluids, which can increase the sensitivity of EM [3]. In detail, viruses collected from dirty clinical samples such as serum, a low speed centrifugation can be used to get rid of cells and sucrose gradient ultra-centrifugation can be applied to obtained purified viruses [4,62].

Several methods have been utilized to prepare diagnostics samples. Staining procedures are common procedures, especially for EM. In clinical diagnostics using EM, staining, shadow castings, and thin sectioning of cellular and tissue samples are widely used. Negative staining for diagnostic virology has been the gold standard for TEM [7]. The following Tables 8.2 and 8.3 discuss some of the detailed aspects involved in sample preparation for plant and animal virus diagnosis.

8.2.2 Scanning and transmission electron microscopy: structure and functions

The SEM and TEM generate images using a highly focused EBs targeting the specimen in a vacuum chamber. Here, the signals produced by electron interaction with the specimen synchronize to form the final image. The SEM can deliver data on surface topography and crystalline structure of a material. Also, signals generated from SEMs even provide chemical composition and electrical behavior of the sample of analysis. In TEM, EB transmits through the specimen. It is important to note that specimen should be thin enough for the beam to penetrate the specimen. Detectors are used to collect all the signal generated and EB and the signal intensity controls the image quality. Compared to TEM, SEM can be used to assess larger specimens (>200 mm diameter) and a nondestructive technique with lesser damage during the analysis of the specimen. As TEM requires sample preparations specially for biological materials and this may be complex and time consuming technique [74,75].

The EMs have a higher resolution limit. Resolution is the smallest distance where two structures can be visualized as two separate objects. Resolution specifically depends on the wavelength of illumination source [74]. This can be mathematically defined using Abbe' equation. When the resolution of a microscopy exceeds the limit, magnified image starts to blur. Diffraction and interferences fails to focus the light as a perfect point which will form images with larger radii consisting of a disk composed of concentric circles having gradual decline in intensity. This is known as an Airy disk and may consist of several wave fronts. If the radii of an airy disk is small it defines the microscope is having a higher resolution. When the center of two primary peaks are separated by a distance equal to the radius of Airy disk, the two objects can be distinguished from each other [76]. Wave

TABLE 8.2   Sample preparation in plant virus diagnosis.

Plant/vector species	Virus	Sample preparation	References
N. tabacum and *Cucurbita pepo*	TMV and ZYMV	**Method 1**	[24]

Sections of leaves (1 mm²) from controls and plants infected with TMV or ZYMV were cut on a modeling wax plate in a drop of 3% glutaraldehyde in 0.06 M Sørensen phosphate buffer at pH 7.2. Samples were then transferred into glass vials and fixed for 90 min at room temperature in the medium. Samples were rinsed in buffer (4 times at 15 min each) and postfixed in 1% osmium tetroxide for 90 min at room temperature. The samples were dehydrated for 20 min each step in a graded series of increasing concentrations of acetone (50%, 70%, 90%, and 100%). Pure acetone was then exchanged by propylene oxide, and specimen were infiltrated with increasing concentrations (30%, 50%, 70% and 100%) of agar 100 epoxy resin mixed with propylene oxide for a minimum of 3 h per step. Samples were left overnight in a 1:1 mixture of Agar 100 epoxy resin and 100% propylene oxide. Infiltration was continued the next day and finally samples were embedded in pure, fresh agar 100 epoxy resin and polymerized at 60°C for 48 h. Ultrathin sections (80 nm) were cut with a ultramicrotome and poststained for 5 min with lead and for 15 min with uranyl acetate at room temperature before they were observed with TEM.

Method 2

Sample preparation was carried out with an automated microwave tissue processor. Small sections of leaves (1 mm²) were cut on a wax plate in a drop of 3% glutaraldehyde in 0.06 M Sørensen phosphate buffer (pH 7.2) and transferred immediately into small baskets with a mesh width of approximately 200 m. TMV-infected and ZYMV-infected samples and the controls were processed on two different days. Three baskets holding four sections each were filled with infected plant samples and the remaining two baskets with control samples. These baskets were then transferred into the chamber of the microwave processor which already contained a vial filled with the fixative solution. Sample preparation was started approximately 2 min after cutting of the samples by starting the programmed protocol. After the samples were infiltrated with pure agar 100 epoxy resin they were transferred manually into polymerization forms containing fresh agar 100 epoxy resin and then into the microwave chamber. Polymerization was carried out automatically for 85 min at a maximal temperature of 90°C. Sectioning and poststaining were undertaken as described for conventional sample preparation.

(Continued)

TABLE 8.2   (Continued)

Plant/vector species	Virus	Sample preparation	References
		Negative staining	
		Conducted on formvar coated copper grids (200-mesh) while fixation, dehydration and infiltration were carried out automatically with the microwave tissue processor. Crude sap of leaf material infected with TMV and ZYMV was applied to the top of the grid for 5 min. Then, the grid was washed twice in 0.06 M Sørensen phosphate buffer (pH 7.2). Deposits on the grid were stained with 2%-phosphotungstic acid (PTA) in 0.06 M Sørensen phosphate buffer (pH 6.5) for 2 min. The grids were air-dried and observed by TEM.	
Tobacco	TMV	For immunogold labeling, fragments of agroinoculated or TMV infected leaves of *Nicotiana benthamiana* were embedded in Epon 812 after fixation of samples with 2.5% glutaraldehyde. Ultrathin sections of 60 nm were prepared with an ultramicrotome and were placed on nickel grids covered with pioloform film. Grids were incubated for 15 min in 0.1 M nacacodylate buffer, pH 7.4, containing 1% of bovine serum albumin (cacodylate-buffer/bovine serum albumin), washed with cacodylate-buffer, and incubated with mast cell-specific monoclonal antibodies open reading frame 6 of TMV diluted to 2.05.0 µg/mL overnight at room temperature. After washing, the grids were incubated for 2 h with rabbit antimouse Immunoglobulin G conjugated to 15 nm colloidal gold particles. Then the grids were stained with 2% uranyl acetate for 30 min and analyzed by Immuno-EM.	[63]
Plectranthus barbatus	CMV	The suspension obtained by partial purification of samples from both healthy and infected plants for TEM. The formavar coated grids were immersed on purified suspension for 10 min. Then, the grids were stained with 2% PTA for 5 min and allowed to dry.	[34]
Tobacco	TMV	TMV particles were adsorbed for 10 min to Formvar-coated and carbon-sputtered 400 mesh copper grids, washed three times for 1 min on drops of double-distilled H_2O and negatively stained using three drops of uranyl acetate supplemented with bacitracin (2% (w/v) uranyl acetate; 250 mg/mL bacitracin), with 1 min incubation on each drop. TEM was carried out at 120 kV.	[64]
Tomato	TMV	Young roots, sampled from virus-infected and control plants, were vigorously washed in tap water and fixed for 2 h in 2% glutaraldehyde in 0.1 M phosphate buffer (pH 6.8). After further processing through excising unwanted components again washed in buffer and postfixed for 2 h in 1% osmium tetroxide in phosphate buffer (pH 6.8). Following dehydration in a graded acetone series, tissue segments were embedded in Spurr's resin. The sections were mounted in 51 µm mesh nickel grids, stained with uranyl acetate and Millonig's lead citrate, and examined in an EM (60 kV).	[41]

(Continued)

TABLE 8.2 (Continued)

Plant/vector species	Virus	Sample preparation	References
Tomato	TMV	Protoplasts were fixed in 6% glutaraldehyde in 0.025 M-sodium phosphate buffer, pH 7.0, containing 10% (w/v) sucrose for approximately 12 h at 4°C, washed well with 0.1 M-sodium phosphate buffer pH 7.0 and postfixed in 2% osmium tetroxide (0.1 M-sodium phosphate buffer pH 7.0 containing 10% sucrose) for 3 h. Protoplasts were stained during dehydration for 15 h in 1% uranyl acetate in 70% ethanol at 25° and subsequently embedded in the butylmethacrylatestyrene embedding medium. Gold sections were cut using glass knives on a ultramicrotome and collected on carbon-coated Athene type new 200 grids. They were poststained for 30 min with lead citrate and examined at 60 kv using a 50 μm. objective aperture.	[42]
Tomato	Not specified	**Cryo-TEM** Samples for the cryo-TEM were prepared by using the Vitrobot Mark IV. Quantifoil R 2/2, 200 or C-Flat 2/2, 200 mesh holey carbon grids were glow discharged for 60 s at 20 mA and positive polarity in air atmosphere. Vitrobot conditions were set to 4°C, 95% relative humidity, Blot time 3 s and Blot force 1. Then, 2 μL of the suspension was applied to the grid, blotted and vitrified in liquid ethane. Excess liquid was removed by filter paper. Samples were visualized with a 200 kV microscope Glacios with a Falcon 3EC detector. **SEM** Samples were incubated for two hours in 2% OsO_4 and dehydrated in a graded series of ethanol (30%−100%), followed by a graded series of hexamethyldisilazane (mixed with absolute ethanol; 30%, 50% and 100%), and finally air dried. The dehydrated samples were coated with gold and palladium and examined using Field Emission Scanning Electron Microscope.	[47]
Tomato	TMV	Tissue samples (1 mm³) of an uninoculated leaf, root-tip and stem-apex from healthy and infected plants were quickly excised and fixed overnight in 2.5% glutaraldehyde with 0.1 mol/L phosphate buffered saline (PBS) buffer (pH 7.2) at room temperature. Then samples were washed in the PBS buffer for 2 h and postfixed with 1.0% aqueous osmium tetroxide for 4 h at room temperature. After washing with PBS for 15 min, tissues were dehydrated in a graded ethanol series, transferred to acetone and filtered through acetone-Spurr resin mixtures. The filtered tissues were then embedded in pure Spurr resin, and polymerized at 37°C for 12 h, 45°C for 12 h, and 60°C for 48 h. Sections were cut on a ultramicrotome, collected onto formvar-coated grids, and stained with uranyl acetate and lead citrate.	[40,65]

(Continued)

II. Viral infections and transmission

TABLE 8.2   (Continued)

Plant/vector species	Virus	Sample preparation	References
Tobacco	Tomato spotted wilt virus	Samples (1 mm × 4 mm) were excised from leaves of *N. benthamiana* plants infected with TSWV rescued from the full-length infectious clones. The sample tissues were fixed in 2.5% v/v glutaraldehyde and 1% w/v osmium tetroxide in 100 mM phosphate buffer (pH 7.0) and then embedded in Epon 812 resin. Ultrathin sections (70 nm) were mounted on formvar-coated grids and then stained with uranyl acetate for 10 min, and later with lead citrate for 10 min. The stained sections were examined with a TEM.	[66]
Tobacco	Sonchus yellow net virus		[67]
Rice	Rice stripe virus		[68]
Chenopodium quinoa and *Vicia faba*	Broad bean wilt virus		[69]
Spider lily (*Hymenocallis littoralis*) plants	Tomato spotted wilt virus	Small sections from the lesions of collected leaf samples were fixed in 2.5% glutaraldehyde and 2% paraformaldehyde in 0.05 M cacodylate buffer, pH 7.2, and postfixed in 1% OsO$_4$, dehydrated in ethanol and embedded in low viscosity Spurr's epoxy resin. Thin sections from the embedded tissues were obtained using an ultramicrotome equipped with Diatome diamond knife, mounted on 300 mesh copper grids and stained with 3% uranyl acetate and Reynold's lead citrate.	[46]
Whiteflies (*Bemisia tabaci*)	Tomato yellow leaf curl virus	For TEM analyses, whitefly guts were first dissected and fixed with 2.5% glutaraldehyde in phosphate buffer (0.1 M, pH 7.0) for 4 h, and then postfixed with 1% OsO$_4$ in phosphate buffer for 2 h. After a series of ethanol dehydration, gut samples were transferred to acetone for 20 min and then embedded in Spurr resin. After staining with uranyl acetate and alkaline lead citrate, samples were sectioned and observed under TEM.	[48]

particle duality of electrons portrays that electron can behave as waves. Wavelengths of electrons depend on their momentum, which can be adjusted by the voltage supplied to the electron emitter. For instance, electrons with wavelengths of 40 pm down to 1 pm can be readily achieved using accelerating voltages of 1−300 kV [74]. Therefore, EM having higher resolution limits which will enable to visualize virus like nanoparticles.

Abbe's equation,

$$d = 0.6121/n\sin\alpha$$

where,

d = resolution
l = wavelength of the imaging radiation
n = index of refraction of medium between point source and lens, relative to free space
a = half the angle of the cone of light from specimen plane accepted by the objective (half aperture angle in radians)
$n \sin \alpha$ is often called numerical aperture

TABLE 8.3 Sample preparation in animal virus diagnosis.

Animal species/ specimens	Virus	Sample preparation	References
Capybara (*Hydrochoerus hydrochaeris*) i	Coronaviruses (Coronaviridae family) f	**Negative staining technique** (rapid preparation). For TEM analysis, feces and small intestine fragments from capybaras were suspended in 0.1 M phosphate buffer (pH 7.0). Specimens were introduced in to a copper grids with carbon stabilized supporting film of 0.5% collodium in amyl acetate in drop-wise. Excessive contents were removed with a filter paper and negatively stained using 2% ammonium molybdate, pH 5.0.	[70]
Human liver and suckling mouse brains	Dengue virus	The middle sections of the mouse cerebral hemispheres and human liver sections fixed overnight with 2.5% glutaraldehyde in 0.1 M sodium cacodylate buffer at 4°C. Then samples were postfixed using 1% osmium tetroxide in Millonig buffer for 1 h at 4°C. Fixed specimens were dehydrated using ethyl alcohol and shifted to a 50:50 mixture of ethanol:propylene oxide and to 100% propylene oxide before being infiltrated with epoxy resin and polymerized at 60°C for 48 h. Ultrathin sections were obtained using microtome. Thin sections were placed on grids and stained with 2% uranyl acetate and lead citrate solution. The samples were examined using an FEI Tecnai T20 TEM functioned at 200 kV.	[71]
Nasopharyngeal and sputum	Coronavirus Influenzavirus, Adenovirus, Respiratory Syncytial Virus (RSV)	**SEM** Specimens were fixed using 2.5% glutaraldehyde for 1 h. A carbon grid (glow discharge for 2 min). was placed on top of 10−30 μL of the fixed sample for 15 min. The grids stained with a 1% molybdate solution and 18 contrasted carbon grids were then loaded onto a multiwell glass slide using double-sided conductive adhesive tape and sputtered with a 5-μm-thick platinum layer.	[72]
Lung tissues	pH1N1 2009 influenzavirus	Lung tissue were fixed with 3% glutaraldehyde in 0.2 M sodium cacodylate buffer (pH 7.2). Fixed samples were then embedded in epoxy resin blocks. Ultrathin sections of the specimen was stained with uranyl acetate and contrasted with Reynolds lead citrate. Specimens thus prepared were observed in a TEM.	[73]
Platelets	Dengue virus	Fresh platelets samples were quickly fixed using 4% glutaraldehyde in PBS overnight at 4°C and washed once with PBS and thrice with phosphate buffer (pH 7.3). Then specimens were incubated with 2% phosphate-buffered osmium tetroxide (45 min) and washed with distilled water. Afterwards, specimens were stained with 2% aqueous uranyl acetate (30 min) and dehydrated using 70% and 80% ethanol, 90% and 95% absolute ethanol and propylene oxide. Then it was infiltrated with propylene oxide and epoxy resin (50:50) at 37°C (30 min). This was followed by an epoxy resin mixture alone for 2 h at 37°C. Finally, samples were embedded in a polypropylene capsule and allowed to polymerize at 70°C overnight.	[25]

8.2.3 Electron generator

Electron guns are used to provide a steady beam of electrons. Thermionic emitters and field emitters are used as electron guns in EMs. Tungsten or lanthanum hexaboride filaments are common gun types. Here, electrons are generated by thermal emission in thermionic emitters which is induced by passing a current through it. Electrostatic fields are used induce the emission of electrons via quantum mechanical tunneling from field emitters. Electrons emitted from field emitters have lower deviations in their wavelengths [75]. The EMs having field emission guns are expensive in comparison to EMs with thermionic.

8.2.4 Electron lenses

The EBs need to be properly focused for better imaging. Electric and magnetic fields are applied to control the path of an EB. In EM, electron lenses prepared using electromagnets where magnetic fields are used to adjust the trajectory of EB. Generally, electron lenses are used to control diameters of the EB.

The EBs beam coming out from guns are emitted in a dispersed manner and required to converge in to parallel and tiny beam. Condenser lenses are used for this purpose. Generally, condenser lenses are composed of two iron poles wound with a coil of copper to provide the required electromagnetic field. The tiny hole at the center of the lens allows the passage of EB. Lens gap of the condenser lens involves in the focusing of EB and lens current controls the position of the focal point. Objective lenses are applied to focus the EB in to a target point which interacts with the specimens [76]. Therefore, objective lenses used to demagnify the EB further to enhance the image resolution. Hence, condenser and objective lenses are employed to obtain variable focal lengths for EBs.

8.2.5 Signals and detectors

The interaction of EBs with specimens generate secondary, backscattered, Auger electrons, X-rays, and perhaps light. These signals are collected by detectors and assess to build up the image and other information from the specimen. The EMs consists of secondary electron detectors, in addition backscattered electron detectors and X-ray spectrometers are also available [74].

8.2.6 Vacuum system

Vacuum system is a principle component in EM. Vacuum systems are used to remove gaseous molecules from EM system to prevent EB scattering. This provides a free path for the EB to interact with the specimen. Also, it is imperative to prevent the specimen, the apertures and the parts of the electron gun becoming contaminated by contaminants such as hydrocarbons and water vapor [77].

8.3 Concluding remarks and future trends

The EM is a catch all technique that has been used to characterize micro- and nanostructures. These are used to elaborate on the surface morphology and internal structures of a given material where SEM and TEM are the common types. In biological specimens it requires different sample preparation to enhance the image quality and to preserve the specimens from any damage. Viruses are biological nanoparticles that require EM to visualize and understand them as a basic tool. Viruses are obligate parasites that can cause harmful or detrimental effects on humans, animals, and plants. Therefore, EM is used as a basic diagnostic tool in virology. This has been widely used in medical applications, veterinary science, and in plant pathology. In addition, modified applications such as immunosorbent EM and cryogenic EM like techniques have been introduced to analyze biological specimens. Immunosorbent electron microscopies are used to localize viruses by tagging them with specific antibodies. Cryoelectron microscopies are used to image specimens by freezing them at cryogenic temperatures. This will enable to preserve the biological specimens without applying any fixative on specimen. However, there are limitations in viral diagnosis using EM. These analyses require well-trained personals to handle the instrument. Also, it requires higher capital expenses and maintenance costs. Furthermore, acquisition and maintenance require a suitable working area for its operation [2]. Also, it is important to accurately explore the nature of viruses using EM, because certain cellular components such as perichromatin granules, improperly fixed chromatin, nuclear pores, melanosomes, cilia and microvilli, microtubules, granules and glycogen can imitate the structure of viral particles [78]. If the viral load is very low and specimens are heavily damaged due to sampling degradation or due to improper preservation procedures the sample might not be useful for detection via EM [2]. "A picture is worth a thousand words," therefore, EM provides the opening to understanding the structural characteristics of infectious virus particles [79]. Hence, EM finds numerous applications as a diagnostic tool in basic virology as well as different fields.

References

[1] Mahony JB, Petrich A, Smieja M. Molecular diagnosis of respiratory virus infections. Crit Rev Clin Lab Sci 2011;48(5−6):217−49.
[2] Vale FF, Correia AC, Matos B, Moura Nunes JF, Alves, de Matos AP. Applications of transmission electron microscopy to virus detection and identification. Microsc: Sci Technol Appl Educ 2010;1:128−36.
[3] Golding CG, Lamboo LL, Beniac DR, Booth TF. The scanning electron microscope in microbiology and diagnosis of infectious disease. Sci Rep 2016;6(1):1−8.
[4] Lin Y, Yan X, Cao W, Wang C, Feng J, Duan J, et al. Short communication Probing the structure of the SARS coronavirus using scanning electron microscopy. Antivir Ther 2004;9:287−9.
[5] Valencia DN. Brief review on COVID-19: the 2020 pandemic caused by SARS-CoV-2. Cureus 2020;12(3).
[6] Goldsmith CS, Miller SE. Modern uses of electron microscopy for detection of viruses. Clin Microbiol Rev 2009;22(4):552−63.
[7] Beniac DR, Siemens CG, Wright CJ, Booth TF. A filtration based technique for simultaneous SEM and TEM sample preparation for the rapid detection of pathogens. Viruses 2014;6(9):3458−71.
[8] Maillard P, Krawczynski K, Nitkiewicz J, Bronnert C, Sidorkiewicz M, Gounon P, et al. Nonenveloped nucleo-capsids of hepatitis C virus in the serum of infected patients. J Virol 2001;75(17):8240−50.

[9] Oberste MS, Gotuzzo E, Blair P, Nix WA, Ksiazek TG, Comer JA, et al. Human febrile illness caused by encephalomyocarditis virus infection, Peru. Emerg Infect Dis 2009;15(4):640.

[10] Olson MR, Axler RP, Hicks RE. Effects of freezing and storage temperature on MS2 viability. J Virol Methods 2004;122(2):147−52.

[11] Chan KH, Peiris JM, Lam SY, Poon LL, Yuen KY, Seto WH. The effects of temperature and relative humidity on the viability of the SARS coronavirus. Adv Virol 2011;2011.

[12] Kageyama T, Kojima S, Shinohara M, Uchida K, Fukushi S, Hoshino FB, et al. Broadly reactive and highly sensitive assay for Norwalk-like viruses based on real-time quantitative reverse transcription-PCR. J Clin Microbiol 2003;41(4):1548−57.

[13] Jothikumar N, Cromeans TL, Robertson BH, Meng XJ, Hill VR. A broadly reactive one-step real-time RT-PCR assay for rapid and sensitive detection of hepatitis E virus. J Virol Methods 2006;131(1):65−71.

[14] Taha BA, Al Mashhadany Y, Bachok NN, Bakar AA, Hafiz Mokhtar MH, Dzulkefly Bin Zan MS, et al. Detection of COVID-19 Virus on Surfaces Using Photonics: Challenges and Perspectives. Diagnostics 2021;11(6):1119.

[15] Guarner J. Three emerging coronaviruses in two decades: the story of SARS, MERS, and now COVID-19.

[16] Colmenero I, Santonja C, Alonso-Riaño M, Noguera-Morel L, Hernández-Martín A, Andina D, et al. SARS-CoV-2 endothelial infection causes COVID-19 chilblains: histopathological, immunohistochemical and ultra-structural study of seven paediatric cases. Br J Dermatol 2020;183(4):729−37.

[17] Scheller C, Krebs F, Minkner R, Astner I, Gil-Moles M, Wätzig H. Physicochemical properties of SARS-CoV-2 for drug targeting, virus inactivation and attenuation, vaccine formulation and quality control. Electrophoresis 2020;41(13−14):1137−51.

[18] Piñeyro PE, Lozada MI, Alarcón LV, Sanguinetti R, Cappuccio JA, Pérez EM, et al. First retrospective studies with etiological confirmation of porcine transmissible gastroenteritis virus infection in Argentina. BMC Vet Res 2018;14(1):1−8.

[19] Barreto-Vieira DF, Couto-Lima D, Jácome FC, Caldas GC, Barth OM. Dengue, Yellow Fever, Zika and Chikungunya epidemic arboviruses in Brazil: ultrastructural aspects. Mem Inst Oswaldo Cruz 2021;115.

[20] Gibbs EP, Johnson RH, Voyle CA. Differential diagnosis of virus infections of the bovine teat skin by electron microscopy. J Comp Pathol 1970;80(3):455−63.

[21] Mahmood A, Ali S. Microbial and viral contamination of animal and stem cell cultures: common contaminants, detection and elimination. J Stem Cell Res Therap 2017;2(5):1−8.

[22] Gachon C, Mingam A, Charrier B. Real-time PCR: what relevance to plant studies? J Exp Bot 2004;55 (402):1445−54.

[23] Zitikaitė I, Staniulis J, Urbanavičienė L, Žižytė M. Cucumber mosaic virus identification in pumpkin plants. Žemdirbystė = Agric 2011;98(4):421−6.

[24] Zechmann B, Zellnig G. Rapid diagnosis of plant virus diseases by transmission electron microscopy. J Virol Methods 2009;162(1−2):163−9.

[25] Noisakran S, Gibbons RV, Songprakhon P, Jairungsri A, Ajariyakhajorn C, Nisalak A, et al. Detection of dengue virus in platelets isolated from dengue patients. Southeast Asian J Trop Med Public Health 2009;40 (2):253.

[26] Singh R, Singh KP, Cherian S, Saminathan M, Kapoor S, Manjunatha Reddy GB, et al. Rabies−epidemiology, pathogenesis, public health concerns and advances in diagnosis and control: a comprehensive review. Vet Quart 2017;37(1):212−51.

[27] Lahaye X, Vidy A, Pomier C, Obiang L, Harper F, Gaudin Y, et al. Functional characterization of Negri bodies (NBs) in rabies virus-infected cells: evidence that NBs are sites of viral transcription and replication. J Virol 2009;83(16):7948−58.

[28] Le Guenno B, Formenty P, Wyers M, Gounon P, Walker F, Boesch C. Isolation and partial characterisation of a new strain of Ebola virus. Lancet 1995;345(8960):1271−4.

[29] Broadhurst MJ, Brooks TJ, Pollock NR. Diagnosis of Ebola virus disease: past, present, and future. Clin Microbiol Rev 2016;29(4):773−93.

[30] Bukreyev A, Marzi A, Feldmann F, Zhang L, Yang L, Ward JM, et al. Chimeric human parainfluenza virus bearing the Ebola virus glycoprotein as the sole surface protein is immunogenic and highly protective against Ebola virus challenge. Virology 2009;383(2):348−61.

[31] Vinodhini J, Rajendran L, Abirami R, Karthikeyan G. Co-existence of chlorosis inducing strain of Cucumber mosaic virus with tospoviruses on hot pepper (Capsicum annuum) in India. Sci Rep 2021;11(1):1−9.

[32] Lan H, Lai B, Zhao P, Dong X, Wei W, Ye Y, et al. Cucumber mosaic virus infection modulated the phyto-chemical contents of Passiflora edulis. Microb Pathog 2020;138:103828.

[33] Meena RP, Manivel P. First report of cucumber mosaic virus infecting antamul vine (Tylophora indica) in India. Virusdisease 2019;30(2):319−20.

[34] Pavithra BS, Govin K, Renuka HM, Krishnareddy M, Jalali S, Samuel DK, et al. Characterization of cucumber mosaic virus infecting coleus (*Plectranthus barbatus*) in Karnataka. Virusdisease 2019;30(3):403−12.

[35] Kumar A, Jain RK, Bhattarai A, Rathore AS, Watpade SG, Parkash C, et al. Occurrence of cucumber mosaic virus Subgroup I on different plant families in Himachal Pradesh and its infection to new hosts. Indian Phytopathol 2020;73(4):759−66.

[36] Wagih EE, Zalat MM, Kawanna MA. Cytological, histological and molecular characterization of two isolates of cucumber mosaic virus (CMV) in Egypt. Int J Phytopathol 2021;10(1):09−18.

[37] Mochizuki T, Ohki ST. Single amino acid substitutions at residue 129 in the coat protein of cucumber mosaic virus affect symptom expression and thylakoid structure. Arch Virol 2011;156(5):881−6.

[38] Hibino H, Matsui C. Electron microscopic autoradiography of leaf cells infected with tobacco mosaic virus I. Uridine-H3 uptake. Virology 1964;24(1):102−6.

[39] Stobbs LW, Manocha MS, Dias HF. Histological changes associated with virus localization in TMV-infected Pinto bean leaves. Physiol Plant Pathol 1977;11(1):87 IN13.

[40] Zhou S, Liu W, Kong L, Wang M. Systemic PCD occurs in TMV-tomato interaction. Sci China Ser C: Life Sci 2008;51(11):1009−19.

[41] Jabaji-Hare SH, Stobbs LW. Electron microscopic examination of tomato roots coinfected with Glomus sp. and tobacco mosaic virus. Phytopathology 1984;74(3):277−9.

[42] Cocking EC, Pojnar E. An electron microscopic study of the infection of isolated tomato fruit protoplasts by tobacco mosaic virus. J Gen Virol 1969;4(3):305−12.

[43] Yoon JY, Choi GS, Choi SK. First report of Tomato spotted wilt virus in Eustoma grandiflorum in Korea. Plant Dis 2017;101(3):515.

[44] Yoon JY, Choi GS, Jang SW, Park SH, Choi SK. First report of Tomato spotted wilt virus in *Humulus japonicus* in Korea. Plant Dis 2018;102(3):690.

[45] Yoon JY, Choi GS, Kwon SJ, Cho IS. First report of Tomato spotted wilt virus infecting *Peperomia obtusifolia* in South Korea. Plant Dis 2019;103(3):593.

[46] Dietzgen RG, Freitas-Astúa J, Salaroli RB, Kitajima EW. Tomato spotted wilt virus infects spider lily plants in Australia. Australas Plant Dis Notes 2018;13(1):1−3.

[47] Mammadova R, Fiume I, Bokka R, Kralj-Iglič V, Božič D, Kisovec M, et al. Identification of tomato infecting viruses that co-isolate with nanovesicles using a combined proteomics and electron-microscopic approach. Nanomaterials 2021;11(8):1922.

[48] Wang LL, Wang XR, Wei XM, Huang H, Wu JX, Chen XX, et al. The autophagy pathway participates in resistance to tomato yellow leaf curl virus infection in whiteflies. Autophagy 2016;12(9):1560−74.

[49] Hu S, Yin Y, Chen B, Lin Q, Tian Y, Song X, et al. Identification of viral particles in the apoplast of Nicotiana b enthamiana leaves infected by potato virus X. Mol Plant Pathol 2021;22(4):456−64.

[50] Otulak-Kozieł K, Kozieł E, Lockhart BE. Plant cell wall dynamics in compatible and incompatible potato response to infection caused by Potato virus Y (PVYNTN). Int J Mol Sci 2018;19(3):862.

[51] Sridhar J, Venkateswarlu V, Shah MA, Kumari N, Bhatnagar A, Raigond B, et al. Incidence of the cabbage aphid, *Brevicoryne brassicae* L. in potato crops in India and its efficiency for transmission of potato virus Y o. Int J Trop Insect Sci 2021;1−7.

[52] Han T, Yang CX, Fu JJ, Hou QS, Gang S, Chen S, et al. First report of Ligustrum virus A on *Syringa reticulata* var. mandshurica (Oleaceae) with a new yellow vein and malformed needle-shaped leaf disease in China. Plant Dis 2018;102(10):2053.

[53] Chen S, Yu N, Yang S, Zhong B, Lan H. Identification of Telosma mosaic virus infection in Passiflora edulis and its impact on phytochemical contents. Virol J 2018;15(1):1−8.

[54] Brenner S, Horne RW. A negative staining method for high resolution electron microscopy of viruses. Biochim Biophys Acta 1959;34:103−10.

[55] Mehdizadeh KA, Tahermanesh K, Chaichian S, Joghataei MT, Moradi F, Tavangar SM, et al. How to prepare biological samples and live tissues for scanning electron microscopy (SEM).

[56] Tizro P, Choi C, Khanlou N. Sample preparation for transmission electron microscopy. Biobanking 2019;417−24.

[57] Rolls G. Fixation and fixatives (3)—fixing agents other than the common aldehydes.

[58] Al Shehadat S, Gorduysus MO, Hamid SS, Abdullah NA, Samsudin AR, Ahmad A. Optimization of scanning electron microscope technique for amniotic membrane investigation: a preliminary study. Eur J Dent 2018;12(04):574—8.

[59] Zhang Y, Huang T, Jorgens DM, Nickerson A, Lin LJ, Pelz J, et al. Quantitating morphological changes in biological samples during scanning electron microscopy sample preparation with correlative super-resolution microscopy. PLoS ONE 2017;12(5):e0176839.

[60] Keywords "charging phenomenon", JEOL. Jeol.co.jp <https://www.jeol.co.jp/en/words/semterms/search_result.html?keyword = charging%20phenomenon>; 2021 [accessed 5.11.21].

[61] Torimoto T, Tsuda T, Okazaki KI, Kuwabata S. New frontiers in materials science opened by ionic liquids. Adv. Mater. 2010;22(11):1196—221.

[62] Hodoši R, Nováková E, Šupolíková M. Purification of murine gammaherpesvirus 68 with use of differential centrifugation. Eur Pharm J 2021;68(1):76—9.

[63] Erokhina TN, Lazareva EA, Richert-Pöggeler KR, Sheval EV, Solovyev AG, Morozov SY. Subcellular localization and detection of tobacco mosaic virus ORF6 protein by immunoelectron microscopy. Biochem (Moscow) 2017;82(1):60—6.

[64] Koch C, Wabbel K, Eber FJ, Krolla-Sidenstein P, Azucena C, Gliemann H, et al. Modified TMV particles as beneficial scaffolds to present sensor enzymes. Front Plant Sci 2015;6:1137.

[65] Zhou S, Hong Q, Li Y, Li Q, Wang M. Autophagy contributes to regulate the ROS levels and PCD progress in TMV-infected tomatoes. Plant Sci 2017;269:12—19.

[66] Li J, Feng Z, Wu J, Huang Y, Lu G, Zhu M, et al. Structure and function analysis of nucleocapsid protein of tomato spotted wilt virus interacting with RNA using homology modeling. J Biol Chem 2014;290(7):3950—61.

[67] Wang Q, Ma X, Qian S, Zhou X, Sun K, Chen X, et al. Rescue of a plant negative-strand RNA virus from cloned cDNA: insights into enveloped plant virus movement and morphogenesis. PLoS Pathog 2015;11(10):1005223.

[68] Kong L, Wu J, Lu L, Xu Y, Zhou X. Interaction between Rice stripe virus disease-specific protein and host PsbP enhances virus symptoms. Mol Plant 2014;7(4):691—708.

[69] Liu Y, Wang Z, Qian Y, Mu J, Shen L, Wang F, et al. Rapid detection of tobacco mosaic virus using the reverse transcription loop-mediated isothermal amplification method. Arch Virol 2010;155(10):1681—5.

[70] Catroxo MH, Araújo LB, Lavorenti A, Petrella SM, Melo NA, Martins AM. Detection of coronavirus in capybaras (Hydrochoeris hydrochaeris) by transmission electron microscopy in São Paulo, Brazil.

[71] Win MM, Charngkaew K, Punyadee N, Aye KS, Win N, Chaisri U, et al. Ultrastructural features of human liver specimens from patients who died of dengue hemorrhagic fever. Trop Med Infect Dis 2019;4(2):63.

[72] Haddad G, Bellali S, Fontanini A, Francis R, La Scola B, Levasseur A, et al. Rapid scanning electron microscopy detection and sequencing of severe acute respiratory syndrome coronavirus 2 and other respiratory viruses. Front Microbiol 2020;11:2883.

[73] Basu A, Shelke V, Chadha M, Kadam D, Sangle S, Gangodkar S, et al. Direct imaging of pH1N1 2009 influenza virus replication in alveolar pneumocytes in fatal cases by transmission electron microscopy. J Electron Microsc (Tokyo) 2011;60(1):89—93.

[74] Inkson BJ. Scanning electron microscopy (SEM) and transmission electron microscopy (TEM) for materials characterization. Materials characterization using nondestructive evaluation (NDE) methods. Woodhead Publishing; 2016. p. 17—43.

[75] Vernon-Parry KD. Scanning electron microscopy: an introduction. III—Vs Rev 2000;13(4):40—4.

[76] Aharinejad SH, Lametschwandtner A. Fundamentals of scanning electron microscopy. Microvascular corrosion casting in scanning electron microscopy. Vienna: Springer; 1992. p. 44—51.

[77] Crewe AV. Scanning electron microscopes: is high resolution possible? Science 1966;154(3750):729—38.

[78] Akilesh S, Nicosia RF, Alpers CE, Tretiakova M, Hsiang TY, Gale Jr M, et al. Characterizing viral infection by electron microscopy: lessons from the coronavirus disease 2019 pandemic. Am J Pathol 2021;.

[79] Gastaminza P, Dryden KA, Boyd B, Wood MR, Law M, Yeager M, et al. Ultrastructural and biophysical characterization of hepatitis C virus particles produced in cell culture. J Virol 2010;84(21):10999—1009.

Antiviral agents and therapeutics

Virotherapy

Sathiamoorthi Thangavelu[1], Saikishore Ramanthan[1], Palanivel Velmurugan[2] and Ranjithkumar Dhandapani[3]

[1]Medical Microbiology Unit, Department of Microbiology, Alagappa University, Karaikudi, Tamil Nadu, India [2]Center for Materials Engineering and Regenerative Medicine, Bharath Institute of Higher Education and Research, Chennai, Tamil Nadu, India [3]Chimertech Private Limited, Chennai, Tamil Nadu, India

9.1 Introduction

Cancer is listed as one of the major cause-specific Disability-Adjusted Life Years by WHO [1] with 19.3 million cases recorded worldwide and 10 million recorded deaths according to the GLOBACAN survey 2020 [2]. Having a very aggressive pathology it does not follow the conventional pathogenesis but rather a gradual destabilization of fundamental cellular processes resulting in tumor heterogeneity with unique spatio-temporal signatures. Thus, resulting in diverse cell types in solid tumors displaying increased complexity and chemotherapeutic resistance [3]. With personalized medicine tailored to each individual, research on multiple strategies to combat cancer is on the rise to improve Disease Free Survival where oncolytic viruses (virotherapy) could be considered a viable option in the field of cancer therapeutics. Viruses being diverse entities are ubiquitous, playing a key role in maintaining the ecology as a whole, forming symbiotic relationships with the host exhibiting both parasitic and mutualistic relationships enhancing genome plasticity [4] acting as a forefront of evolution. Despite the antagonistic potential of viruses infecting a multitude of organisms, they serve as brilliant therapeutic machines that could be used for cancer treatment through (1) oncolysis (*certain wild-type or engineered viruses with selective tumor-killing capacity through lysis owing to its high replicative efficiency*). More modern strategies involve oncolytic viruses with dual-action exhibiting oncolysis and expressing certain tumor-specific antigen resulting in (2) immunomodulation-inducing systemic antitumor immunity [5], and also by using (3) viral vectors for gene therapy/prodrug therapy by inserting certain edited transgenes and enhancing product availability by viral replication or by engineering the wild type virus exhibiting selective tropism by exploiting aberrant pathway-related genes improving the therapeutic effect [6]. With many comprehensive reviews available on virotherapy, we in this chapter solely focus on recent advancements and clinical trials underway in the past five years from 2016 to 2021 explaining various strategies employed by using oncolytic viruses as therapeutic sources to combat the most common forms of cancer.

9.2 Oncolytic virus in common cancers and molecular changes observed during infection

The practice of viruses being used as tumor-killing machines has been around for almost 100 years [7]. The innate property of viruses to favorably replicate in actively proliferating cells [8] is exploited by a subset of oncotropic viruses where tumor destruction by such viruses is achieved through a multitude of mechanisms. With many viruses now being tested for oncolysis below, we discuss a wide range of oncolytic viruses used against common forms of cancer along with different strategies and any associated molecular signatures during infection (Fig. 9.1).

Oncolytic viruses exhibit selective tropism conditionally replicating in cancer cells that might be controlled by vectors or the virus utilizing aberrantly regulated pathways but do not infect normal healthy cells, (1) OV gain entry into the cell by binding to specific receptors, (2) after entry they replicate in high numbers producing new viral particles, and (3) after a threshold cancer cell lysis occurs releasing progeny viruses into the tumor bed infecting another cancer cell resulting in tumor regression.

Note: The details of the control vectors, modifications, cell lines, and mouse models used are briefed in Table 9.1. Whereas the clinical trial numbers and their related information are listed briefly in Table 9.2. Below is a representation (Fig. 9.2) of the most commonly used viruses in oncolytic virotherapy (Fig. 9.3).

9.3 Breast cancer

Breast cancer is the most commonly encountered cancer type occurring 1 in 8 women listing second in mortality rates observed in ages between 45 and 55 years with treatment including removal of whole tissue, hormonal therapy, chemotherapy, and radiotherapy [48]. Despite reduced breast cancer mortality, there is an increased incidence in cancer rates [49] requiring alternatives to conventional treatment modality bringing about unconventional approaches such as oncolytic viruses in breast cancer where research has ongoing since more than two decades reflecting good potential. Abd-Aziz et al. [9] investigated the effects of the use of Newcastle disease virus (NDV) against MCF-7 breast cancer cell lines to monitor HIF-1α expression in hypoxic cancer cells. This study revealed that infection using NDV displayed HIF-1α degradation on a posttranslational level but the transcriptional level of HIF-1α was consistent. Interestingly they reported the degradation of VHL tumor suppressor protein which suggests that HIF-1α degradation is independent of VHL and is purely by NDV. The degradation initiated by HIF-1α was found to be due to NDV L-protein with SOCS box motif that functions as a substrate-recognition protein

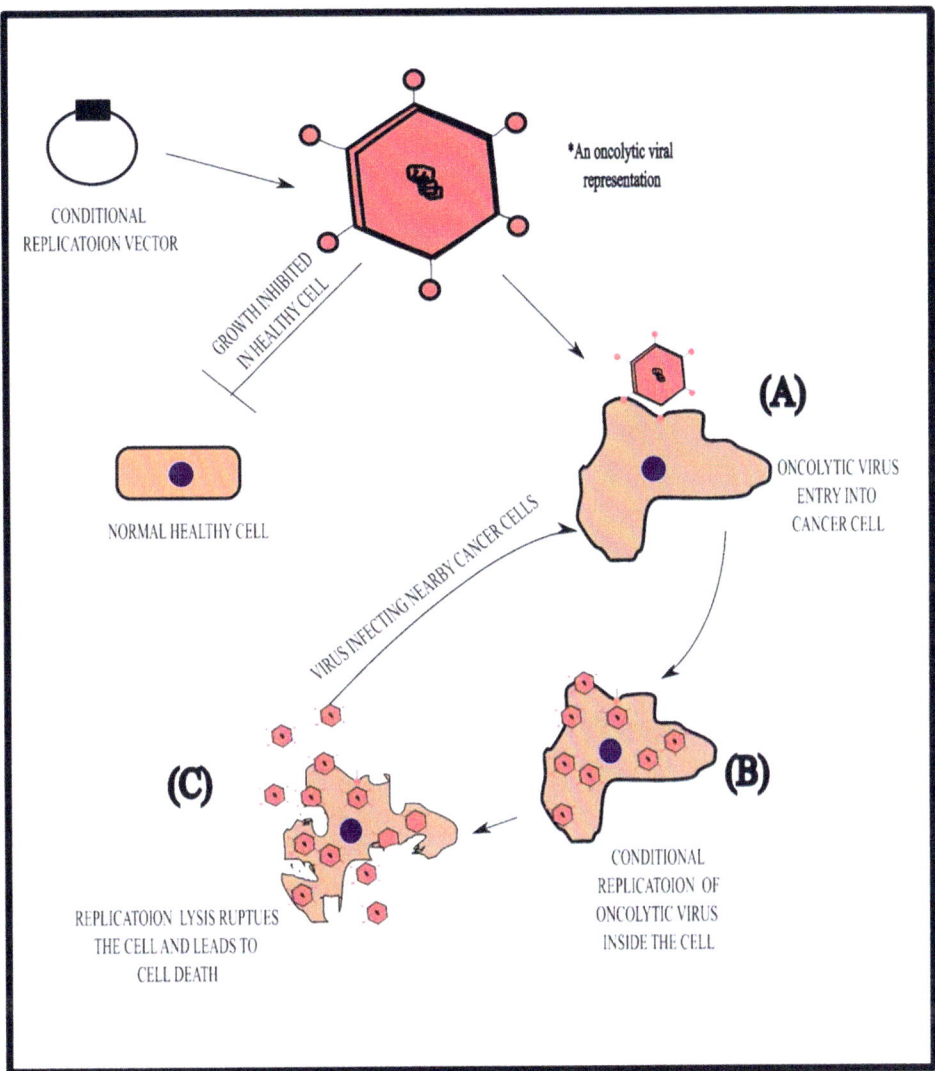

FIGURE 9.1   Generalized mechanism of oncolysis elicited by oncolytic virus. (A) Oncolytic virus enter into cancer cell; (B) conditional replication of oncolytic virus inside the cell; (C) replication lysis ruptures the cell and leads to cell death

involved in proteasome degradation pathways. NDV also caused a reduction in the CAIX expression, which could improve the prognosis for the patients treated with NDV. To improve the therapeutic potential of oncolytic therapy in breast cancers Mostafa et al. [10] combined anti-PD-L1 therapy with reovirus tested in vitro against T-47D, SK-BR-3, MDA-MB-468, MCF-7, Hs 578 T, EMT6, and 4T1 cell lines and in vivo using BALB/c mouse where reovirus was found to initiate oncolysis displaying superior tumor regression

TABLE 9.1 Brief information different viruses used for each type of cancer with their reported molecular signatures.

Cancer type	Virus/combined therapeutic strategy used	Viral modification/ plasmid	Tissue/cell line/model used	Observed immunomodulation/ molecular and metabolism signatures	References
Breast cancer	Newcastle disease virus (velogenic strain)	None	MCV-7	Degradation of HIF-1α, VHL protein and CAIX, VEGF, GLUT1 expression reduction	[9]
	Reovirus W/Anti PD-1 therapy	None	T-47D, SK-BR-3, MDA-MB-468, MCF-7, Hs 578 T, EMT6, 4T1 cell lines and BALB/c mouse	Increased expression of IL10, RANTES, IL2, IFNγ, CD8+, CD4+ and TNFα	[10]
	Adenovirus W/mcherry and Temozolomide	pSIΔ24- pIX-mCherry	HTB133, HTB 132, MCF7, HTB30	Conversion of LC3I and LC3II	[11]
	Reovirus w/doxorubicin conjugation	Not mentioned	MDA-MB-436, MDA-MB-231, L929, and 4T1 cell lines.	Increased expression of STAT1, STAT2, STAT3, IFNL1, IFNB1, and IFN-λ	[12]
Lung cancer	Reovirus rsT1L, rsT3D	Not mentioned	H661, H1299, H1437, H1563, H1975 cell lines.	TNF-α/CHX induced caspase 3/7 expression.	[13]
	Vesicular stomatitis virus (VSVΔ51) W/Sulphorapane	Not mentioned	A549	Nrf2 Axis, increased expression of ATM protein, inhibition of mTORC1	[14]
	Newcastle disease virus/FMN	Not Mentioned	H549, H1650, H460 cell lines and BALB/c Nude mice	Increased expression of CRT, HMGB1, HSP70/90	[15]
	Adenovirus (Ad5D24-CpG) W/pacitaxel encapsulated inside EV	CpG-shuttle plasmid recombination pTHSN-CpG1	H549, PNT2 cell line, BALB/c nude mice	Increased expression of E2F4 and decreased expression of USP37, SNx33, and POLE	[16]
Bladder cancer	CoxSackie virus A21 w/ mitomycin c	Engineered to target ICAM-1	T24, 5637, TCCSUP, KU19–19, VM-CUB1, RT-112, 253 J, VM-CUB2, and HCV29, C57BL/6 mice	Increases expression of caspase 7, caspase 3, PARP, ecto-calreticulin, HMGB1	[17]
	Adenovirus (Ad-VT) w/Rapamycin	TERTp controlled adenovirus w/CMV promoter	HEK-239, UM- UC-3, T24, 5637, SV-HUC1 and EJ-1	AMPK-RAPTOR-mTOR pathway, LC3-II, autophagosome, phosphorylated AMPK, Phosphorylated RAPTOR	[18]

Cancer type	Virus	Modification	Cell lines/models	Effects	Reference
Colorectal cancer	Measles Virus- GFP with starvation MV-GFP	Not mentioned	HT-29, HCT-15, and HCT-1	No gene signature reported	[19]
	Recombinant vaccinia virus VG9-IL-24	Disruption of the viral TK gene region based Strain VG9	CT26, HCT116, HT29, HCT8, 4T1, BSC-40 and BALB/c Nude mice	Increase in PARP Cleavage, Bad expression, PKR, JNK, and decrease in ERK and STAT3	[20]
Renal cancer	BlueTongue virus serotype 10 (BTV-10)	None	ACHN, CAKI-1, OS-PC-2, 786-O AND A498	Increased P53, BAX and increased apoptosis indicated by caspase 9, caspase 3, and PARP cleavage.	[21]
Prostate cancer	Vesicular stomatitis virus-hIFN3 (VSV-hIFN3) With Radiotherapy	ISRE luciferase	LNcaP, PC3, RM9, IMR-90 and athymic nude mouse C57BL/6	Increased caspase 3/7, EGFR, GADD45B, KLF-4, IL24, IFNγ, CXCL10, c-JUN and ATF-3 Decreased pro survival proteins pAKT and BCL-XL etc. and PKR	[22]
Endometrial cancer	Coxsackievirus- B3 strain 2035A (CV-B3 /2035A)	None	HEC-1-A, HEC-1-B, Ishikawa cells and BALB/c Nude mice	Increased CAR and DAF receptors	[23]
Leukemia	Measles virus- super cytosine deaminase MeV-SCD	Measels virus armed with Suicide death gene	MM-6, NOMO-1, SKM-1	Increased STAT 1 activation and IFIT1 regulation indicating IFN signaling	[24]
Hepatocellular carcinoma	Newcastle disease virus w/fludarabine	None	HCCLM3, H22, HepG2, Hepa 1-6, C57BL/6 nude mice	Increased ubiquitin- proteasomal degradation of STAT 3 and IDO1. Downregulation of STAT1 activation. Increased infiltration of CD8+ and NK cells with reduction in MDSC	[25]
	Adenovirus with PD1PVR and fludarabine	none	H22, HCC-LM3, Hepa1-6, HEK 293, C57Bl/6 mice	Increased CD8+ and NK cells infiltration, upregulation of TNF-a, COX-2, IL-1β, IL10, and TGF-3	[26]
Melanoma	Vaccinia virus JX-GFP; TG6002	Tyrosine kinase and ribonucleo tide reductase deletion and expression of suicide gene FCU1	SK-29 mel-1, SK-29 mel 1.22 cell lines		[27]

(Continued)

TABLE 9.1 (Continued)

Cancer type	Virus/combined therapeutic strategy used	Viral modification/ plasmid	Tissue/cell line/model used	Observed immunomodulation/ molecular and metabolism signatures	References
Brain Cancer	(Poliovirus:Rhi novirus Chimera) PVSRIPO	Not mentioned	D238, D341, 654, 2363 and patient derived tissue sections	Increased PARP cleavage observed after infection	[28]
	Recombinant NewCastle Disease Virus (NDV) NDV-LaSota- F3aa-GFP and rNDV-LaSota- NS1	Engineered to express NS1	HEP2, Vero, GBM18, GBM27, GBM38, GBM123, GBM128B, and GBM128D	Upregulation of RIG-I, MDA5, STAT1, Mx1, OAS1, ISG54, or ISG15 in type 1 competent cell lines and the vice versa for type 1 noncompetent cell lines	[29]

TABLE 9.2 Oncolytic viruses in current clinical trial.

Virus under trial	Clinical trial number	Purpose of trial	Genetic recombination with additional therapy	Number of patients	References
Vaccinia Virus Ankara (TG4023)	NCT00978107	Demonstrating PoC of MTD of TG4023 with systemic administration of 5-FC and tumor response monitoring for PoC for VDEPT	FUC1 gene, systemic administration of 5-FC	16	[30]
Herpes Simplex Virus (G207)	NCT02457845	Combined effect of conditionally replicating HSV and radiotherapy against pediatric recurrent malignant supratentorial brain tumors	Deletion of $\gamma^1 34.5$ region with lac z insertion at UL39	24	[31]
Dearing Type 3 Reovirus (Reolysin/ pelareorep)	NCT00984464	Investigation of therapeutic effect of reolysin with combination of carboplatin and placitaxel in treatment against metastatic melanoma	Carboplatin/ Placitaxel W/ Reolysin	14	[32]
Herpes Simplex Virus (HSV1716)	NCT00931931	Assessing safety of HSV1716 when administered against pediatric tumors against sarcomas, clival chordoma, malignant peripheral nerve sheath tumor	Not mentioned in current publication	9	[33]
Herpes Simplex Virus 1 (Talimogene Laherparepvec)	NCT02263508	Assessing the safety, tolerance of oncolytic potential of Talimogene Laherparepvec for increasing CD4+ and CD8+ improving pembrolizomab potential	Pembrolizomab	21	[34]
Herpes Simplex Virus 1 (Talimogene Laherparepvec)	NCT01740297	Evaluate the safety and efficiency of Taliomogene Laherparapvec in combination with ipilimumab	Ipilimumab	198	[35]
Herpes Simplex Virus 1 (Talimogene Laherparapvec)	NCT02014441	Evaluating biodistribution, shedding and viral transmission to healthcare workers during treatment	Not mentioned in current publication	60	[36]
Adenovirus (Enadenotucirev)	NCT02053220	Assessing viral delivery, replication and inflammation when delivered intravenously or through intratumoral injection against CRC, NSCLC, UCC, RCC	Not mentioned in current publication	17	[37]
Vaccinia Virus (GL-ONC1)	NCT01584284	Safety assessment of vaccinia virus when administered with chemotherapy and radiation	Not mentioned in current publication	19	[38]

(Continued)

TABLE 9.2 (Continued)

Virus under trial	Clinical trial number	Purpose of trial	Genetic recombination with additional therapy	Number of patients	References
Edmonston-lineage Measles Virus (MV-NIS)	NCT00450814	MTD assessment and safety profile. Monitoring viral gene expression, elimination, biodistribution, and humoral response	With or without cyclophosphamide	32	[39]
DNA-2401 O. Adenovirus	NCT00805376	Mechanism of action assessment of DNX-2401 through intratumoral administration	Not mentioned in current publication	37	[40]
HF-10 Herpes Simplex Virus-1	UMIN000010150	Safety assessment of HF-10 intratumoral administration	Erlotinib and gemcitabine	10	[41]
Vaccinia Virus	NCT01443260	Assessment of safety and antitumor activity of intraperitoneal administration of GL-ONC1 against peritoneal carcinomatosis	Not mentioned in current publication	9	[42]
Enadenotucirev O.Ad	NCT02028442	Establish MTD for enadenotucirev and identify suitable schedules for repeated cycles	Not mentioned in current publication	61	[43]
Vaccinia Virus ACAM2000	ISRCTN#10201650	Assessment of safety and pharmacodynamics against solid organ cancers	Not mentioned in current publication	26	[44]
Coxsackie Virus A21 (CVA21)	NCT02316171	Evaluating the use of ICAM-1 targeted Coxsackievirus A21 against bladder cancer	ICAM-1 targeted therapy w/mitomycin c	15	[45]
Seneca Valley virus (NTX-010)	NCT01017601	Evaluate the interim futility response for Progression free survival in Patients treated with NTX-010 vs Placebo patients	Not mentioned in current publication	50	[46]
Pelareorep (Oncolytic Reovirus)	NCT02620423	Determine safety and dose-limiting toxicities of pelareorep and its combination with pembrolizumab along with ORR and PFS assessment	Pembrolizumab	11	[47]

5-FC, Fluorocytosine; *5-FU*, fluorouracil; *MTD*, maximum tolerance dose; *O-Ad*, oncolytic adenovirus; *ORR*, overall response rate; *PFS*, progression free survival; *PoC*, proof of concept; *VDEPT*, virus directed enzyme therapy.

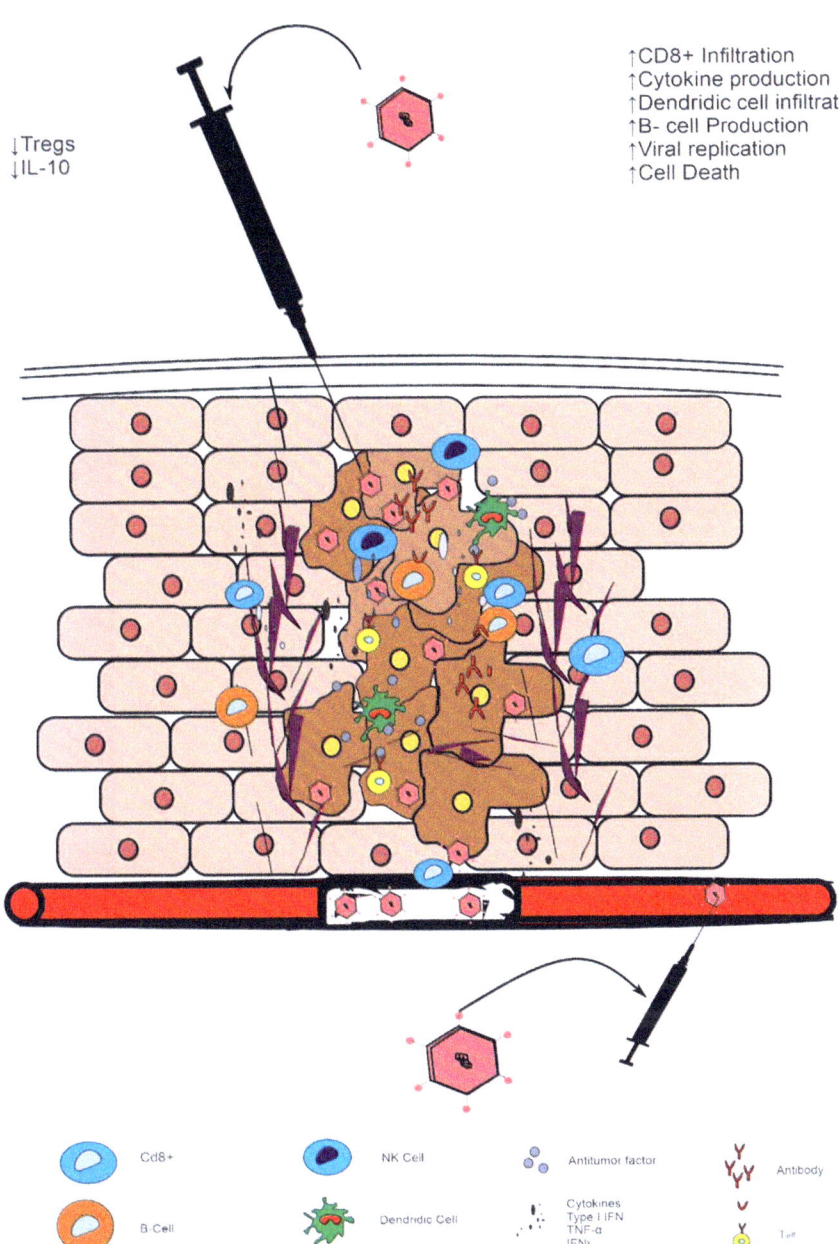

↓Tregs
↓IL-10

↑CD8+ Infiltration
↑Cytokine production
↑Dendridic cell infiltration
↑B- cell Production
↑Viral replication
↑Cell Death

FIGURE 9.2   Immunomodulation elicited by oncolytic virus through intratumoral and intravenous administration of oncolytic viruses.

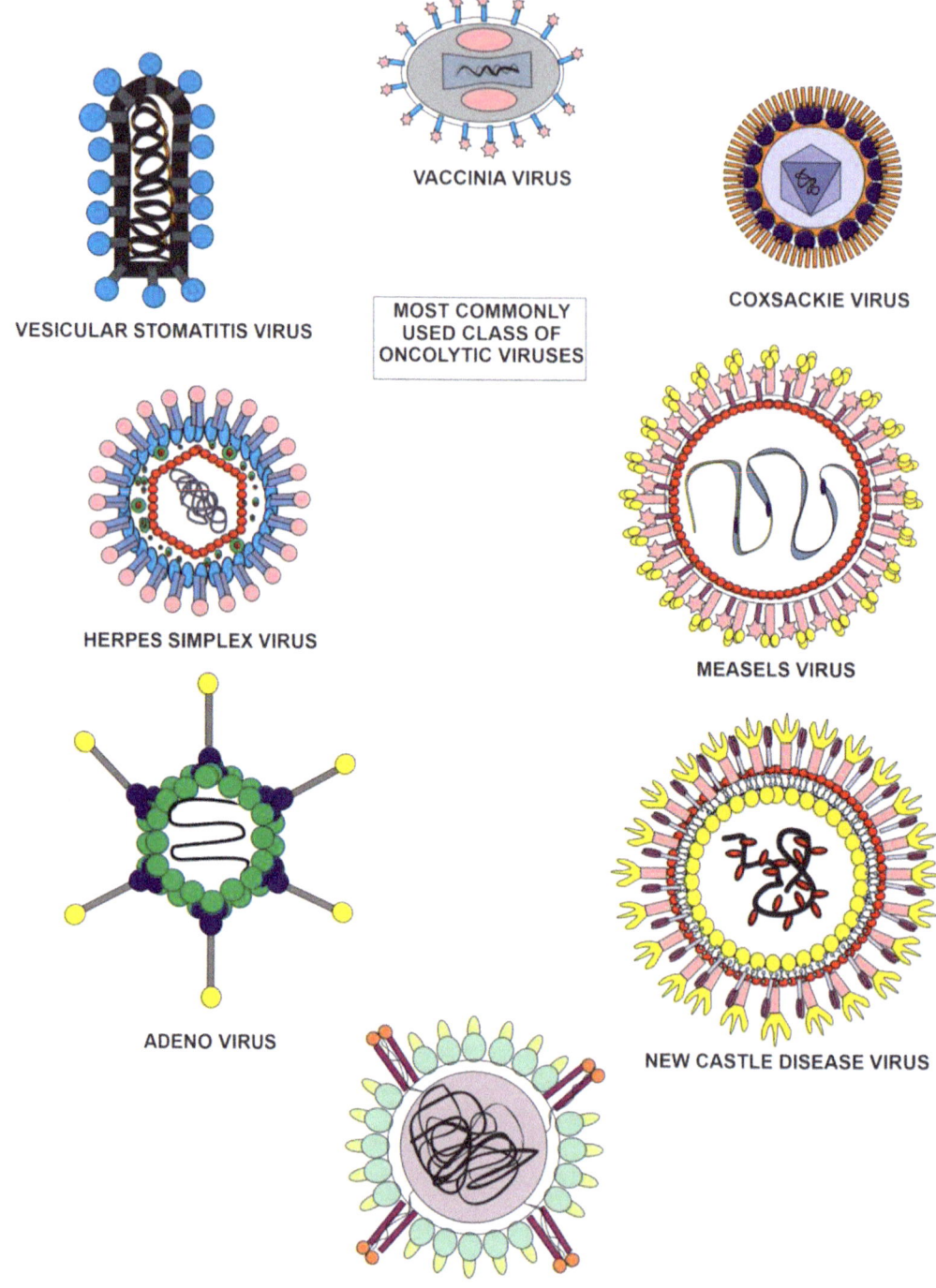

FIGURE 9.3 The most commonly selected viruses for oncolytic virotherapy.

III. Antiviral agents and therapeutics

in vitro and in vivo as a monotherapy when compared to anti-PD1 monotherapy additionally it was able to stimulate cytokines production in cells with most prominent being RANTES and IL10. Production of PD-L1 was observed to be increased, which was found to be due to secreted factor as a result of reovirus infection, and its combination with anti-PD1 therapy cured $\sim 70\%$ of the mice used in the study and an increased production of CD8+, CD4+, IFN-γ, TNF-α, and IL2 was observed. The mice treated with reovirus + anti-PD-L1 when rechallenged with introduction of tumors displayed protective immunity mediated by CD8+ cells, whereas depletion of CD8+ population decreased the protective immunity in the models. Berry et al. [12] demonstrated the use of functional drug delivery to tumor cells using reovirus embedded with chemotherapeutic target doxorubicin without altering the biology of the virus. The viral particle hosts around 3000 molecules of doxorubicin, where it does not interfere with the binding of the virus with the cell or interfere with the drug dynamics and elicits much more robust cytopathic effects. The reovirus-doxorubicin infection resulted in higher expression of IFNB1, IFNL1, STAT1, STAT2, STAT3 with IFN-γ providing a favorable environment for immunomodulatory cells to the site of infection. It also disrupts the mitochondrial membrane potential resulting in reduced tumor metastatic burden. Garza-Morales et al. [11] improved the therapeutic efficiency of adenovirus expressing mcherry (OAd-mcherry) in triple negative breast cancer (TNBC) along with temozolomide (TMZ). When tested in human-derived TNBC cell lines, TMZ tends to increase the infectivity of OAd-mcherry through enhancing the entry of the virus, which leads to 10-fold multiplication in the tested cell lines resulting in cytopathic effects. Their combination resulted in induction of autophagy through the conversion of LC3-I to LC3-II and reduced the clonogenic cell survival in TNBC cell lines to 10% compared to their monotherapy which showed a greater clonal survival of 45% for TMZ and 65% for OAd-mcherry.

9.4 Lung cancer

Lung cancer is a highly invasive and metastasizing cancer with a 5-year survival rate between 4% and 17% with high cancer associated deaths among both men and women where 90% of the disease arises from smoking and the use of tobacco products [50,51]. Lung cancer is differentiated into two major subtypes Non-Small Cell Lung Carcinoma (NSCLC) contributing to 80%−85% of lung cancers and small cell lung cancer (SCLC) contributing 15% of lung cancer [52]. NSCLC treatment is usually tailored for individuals with no standard treatment where targeted therapies are being developed [53] and treatment for SCLC involves chemotherapy or at times combined with radiotherapy [54]. Simon et al. [13] tested the oncolytic laboratory recombinant reovirus strain rsT1L and rsT3D engineered from wt-T1L & wt-T3D against NSCLC cell lines H661, H1299, H1437, H1563, H1975 (large cell carcinoma & adenovirus cell lines). They found that recombinant strains equally induced cell death as in wild type strain but large cell carcinoma killing capacity of rsT1D was more effective compared to rsT3D, which was found to be due to differences in L2, L3, and M1 gene segments indicating genetic complement of the virus. Despite the differences the viral strains were reported to readily infect and kill all NSCLC cell line panels. Olagnier et al. [14] increased the sensitivity of lung cancer cell line A549 to oncolytic vesicular stomatitis virus Δ51 infection by administration of sulforaphane, dampening the immune system by reducing IRF3 expression by blocking its binding to its cognate receptor and by activating the Nrf2 signaling increasing autophagy by inhibition of mTORC pathway improving viral replication of the VSVΔ51 improving therapeutic potential. Oncolytic viruses are not only limited to tumor lysis by replication but could also induce the immunogenic cell death which is demonstrated by Ye et al. [15] when they infected a panel of lung cancer cell lines H549, H1650, H460, and mouse model BALB/c nude mice with NDV strain FMW, wherein they observed an increase in calreticulin expression on the surface of lung cancer cells along with an increase in immunogenic cell death markers, such as HSP70/90, HMGB1, and ATP, where even after apoptosis and necroptosis suppression cell death was observed suggesting other possible pathway roles in cell death. Additionally, they found that NCD/FMW-induced cell death partially relies on autophagy as through suppressing Atg gene suppressed NDV induction of ICD. Effective administration strategies of oncolytic viruses are needed to improve tumor infection to circumvent some problems encountered during therapy, such as neutralizing antibodies initiated by the immune system and degradation of viral particles reducing efficiency. Garofalo et al. [16] utilized intracellular transportation potential of extracellular vesicles (EV) to increase paclitaxel and oncolytic virus delivery and efficiency. When they treated A549 xenograft established BALB/c nude mouse with EV coated virus and paclitaxel (EV-virus-PTX formulation) intravenously they found significant decrease in tumor growth compared with EV-virus alone and abraxane treatments but the same when administered intratumorally had no significant difference between them. Even when administered intravenously, no traces of adenoviral particles were found in the system indicating the specificity of the virus to replicate and infect tumor cells. Evaluation of cell death intro revealed that EV-virus-PTX formulations induced early and late apoptotic events. Even the transduction efficiency and viral titers were reported to be high in EV-coated formulations. They even established that the cell-killing effect of EV formulation is not cell line dependent by culturing viruses from prostate cancer cell lines PC3 and infecting them in A549 cell lines. When cells were treated with EV-formulations they observed an increase in the pathways related to antitumor mechanisms like increase in the E2F4 gene expression negatively regulates the G1 phase of cell cycle with downregulation of USP37, SNX33, and POLE reducing proliferation of tumor cells.

9.5 Bladder and endometrial cancer

Bladder cancer, the fourth most common cancer in men, is an invasive and aggressive disease with a 5-year survival rate of 70% where even post cystectomy the cancer relapse rate is around 50% [55,56]. The worldwide shortage in the firstline drug BCG vaccines for the past six years [57] has prompted scientists to look for alternative strategies to treat bladder carcinoma. Annels et al. [17] used coxsackievirus A21 (CVA21) against Nonmuscle invasive bladder cancers and enhanced its oncolysis by administering mitomycin c inducing an increase in ICAM-1 expression on the cell surface against a series of bladder cancer cell lines. Treatment with CVA21 and mitomycin induced cell death by apoptosis with increase in effector caspase 3, caspase 7, and complete PARP antibody cleavage. Monotherapy with mitomycin C did not have any profound effect but combinatorial therapy displayed superior therapeutic potential. But, mitomycin c only displayed increased ICAM-1 expression in ICAM-1+ cells but did not have any effect on ICAM-1cells even with combinatorial therapy. They observed upregulation of ICD markers calreticulin and HMGB1 whereas in murine models the mice treated with combined therapy displayed resistance to rechallenge of tumors due to the presence of CD4+ cells. The depletion of CD4+ resulted in outgrowth of tumors. Shang et al. [18] demonstrated autophagic death of tumor cells through treatment with adenovirus (Ad-VT) and rapamycin. Ad-VT when administered as monotherapy reduced the viability of tumors without any detrimental effects to the immortalized normal bladder cell line SV-HUC1 when compared against chemotherapeutic agents displaying cytotoxicity to normal cells. When tested in murine models Ad-VT displayed proliferation inhibition and apoptosis induction with an increase in LC3II levels and autophagosomes were reported. Autophagy was found to be induced by the inhibition of mTOR by increased levels of phosphorylated AMPK and Raptor. When Ad-VT administered with autophagy stimulating drug rapamycin improved tumor suppression with increased survival. Endometrial cancer being the fourth most common gynecological cancer in women has no particular screening test available only diagnosed through biopsy or sometimes with ultrasound when presented with abnormal symptoms like postmenopausal vaginal bleeding, endometrial stripe or anomalys like polyps where treatment options commonly include hormone therapy, adjuvant radiotherapy, surgery, and combinatorial chemotherapeutics, e.g., doxorubicin, paclitaxel, cisplatin [58]. Despite the available treatment options, recurrence is seen in three years with 55% median survival in case of cancer limited to pelvic region and 17% in case of metastases [59]. This prompts new therapeutic interventions in treating endometrial cancer. Lin et al. [23] assessed the effectiveness of coxsackie B3 strain 2035A against endometrial cancer cell lines and rhabdomyosarcoma cell lines HEC-1A, HEC-1-B, Ishikawa, and RD. They evaluated the expression of CAR and DAF, which are major Coxsackie virus internalization receptors and were found to be highly expressed in all the cell lines. When infected with CV-B3/2035A, all the cell lines exhibit oncolysis. To understand the in vivo effects of CV-B3/2053A, xenografts were established from the HEC-1-A, HEC-1-B, and Ishikawa cells. When treated with five consecutive doses of CV-B3/2035A, complete tumor elimination was observed with HEC-1-B and Ishikawa cells but not in HEC-1-A xenografts; however, it displayed significant tumor reduction. To check whether CV-B3/2035A exhibited systemic oncolytic effects they established bilateral xenografts of HEC-1B where with five consecutive administration of CV-B3/2035A in the right flank tumors suppressed their growth and the distant left flank tumor displaying significant reduction in size indicating systemic oncolytic effect through viremia and normal colon cell lines CCD-18 and CCD-841. MV-GFP infection worked the same in all conditions but with 120-hour glucose starvation there was an increased incidence in cell death. They also observed the effects of glucose and starvation on viral replication initially with shorter starvation time the viral replication was observed to be increased but not much difference was observed in long time starvation treatment contrastingly replication was impaired by long-term glucose starvation treatment. With long-term serum starvation treatment the oncolysis of MV-GFP was observed to be increased in HT-29 cells. They also assessed the effect of starvation of nonmalignant cells and even with higher doses of MV-GFP not much change was observed but a chromatin clumping and apoptotic body formation inducing cell cycle arrest on the G2/M phase. Increase in PARP cleavage, bad corresponding to apoptosis, and phosphorylation of PKR, JNK was observed but a decrease in ERK phosphorylation and STAT3 was also observed. When HCT116 xenograft tumor mice were infected with VG9-IL-24 and VG9-EGFP, VG9-IL-24 displayed strong tumor inhibition. Tumor rechallenge was performed by establishing CT26 xenograft tumors previously healed by VG9-IL-24 displayed resistance to CT26 but at the same time when 4T1 xenograft was established the mice developed tumors indicating the potential of VG9-IL-24 to develop antitumor immunity.

9.6 Renal and prostate cancer

Wang et al. [21] used Bluetongue virus serotype 10 (BTV-10) against a panel of renal cell carcinoma cell lines (ACHN, CAKI-1, OS-PC-2, 786-O, and A498) and human primary proximal tubular epithelial cell (HPTEC). Renal cell carcinoma cells displayed enhanced sensitivity towards BTV-10 when infected at different MOI whereas HPTEC cells were resistant even at higher viral MOI. The tumor cells treated with BTV-10 displayed increased chromatin condensation, expansion of endoplasmic reticulum, and mitochondrial membrane disintegration associated with decrease in matrix density but HPTEC cells displayed increased presence of organelles in the cytoplasm with nuclear membrane integrity indicating resistance of normal cells to BTV. An increase in apoptotic cells were also observed compared to HPTEC cells indicating BTV-10-induced apoptosis limited to cancer cells where they observed an increase of P53, BAX, PARP, and caspase-9, -3 cleavage. Six BALB/c mice with OS-PC-2 established xenografts when infected with BTV-10 had a survival of 6 months and 67% of the mice were cured of the disease and they started to gain weight and decrease in the tumor size and vasculature corresponding to good response to BTV. Even at a higher dose of $6*10^6$ BTV-10 it did not cause any premature death within the tested mice population and no local inflammation was observed to nearby vital organs emphasizing the safety of usage of BTV in treatment of renal cancer. Udayakumar et al. [22] used radiotherapy to sensitize resistant cells to oncolytic vesicular stomatitis virus infection. They observed PC3 and LNCaP cells displayed resistance to infection even when infected with high viral titers of VSV-hIFNβ. But with VSV-hIFNβ infection along with radiotherapy a rapid decrease in cell count was reported with increase in caspase 3/7 activity indicating apoptosis. But, heat inactivated VSV-hIFNβ infection did not cause any change in the caspase 3/7 activity. The reported changes were limited to tumor cells but not the fibroblasts. In the presence of a single dose of 5gyr radiation an increase of VSV-hIFNc replication was observed by monitoring the expression of VSV-encoding GFP. With VSV-hIFNβ infection IFN pathway was reported to be highly functional in PC3 and LNCaP cell lines, which was monitored by luciferase production by ISRE-luciferase plasmid and treatment of radiotherapy alone did not activate IFN pathway suggesting the virus-mediated activation of IFN pathway. With combined treatment with radiotherapy, they saw a decrease in antiviral pathway gene expression like the decrease of IFN-inducible RNA-dependent protein kinase supporting the role of sensitizing the cells to oncolysis along with gene involved in apoptosis, such as EGFR, GADD45B, KLF-4, IL24, IFNγ, CXCL10, c-JUN, and ATF-3. The in vivo experiments in the mice with PC3 and RM9 xenografts displayed similar effects as observed in vitro with complete tumor response in RM9 xenograft of C57BL/6 mice when treated with VSV-mIFNβ and 5 days of single dose of 5gyr radiotherapy with an increase in CD8+ and CD4+ cells indicating antitumor immunity. When rechallenged with the same tumor cell line in the cured mice no growth was observed and depletion of CD8+ cells re-established the tumor growth, which shows the role of antitumor immunity in cancer progression and by infection of the tumor cells with VSV promotes antitumor immunity.

9.7 Leukemia

Leukemia is a malignancy that occurs as a result of uncontrolled proliferation of hematopoietic stem cells in bone marrow with four subtypes: acute lymphoblastic leukemia, acute myelogenous leukemia, chronic lymphocytic leukemia, and chronic myelogenous leukemia. Common lab tests for evaluating leukemia are bone marrow aspirate, cytogenetic testing, flow cytometry with immunophenotyping, peripheral smear, and molecular testing where treatment for leukemia include chemotherapeutics, monoclonal antibodies, stem cell transplantation, and tyrosine kinase inhibitors. But with chemotherapy, tumor lysis syndrome is a possibility and immunosuppression for transplantation will result in secondary infections, hence limiting dosage reduces the effectiveness of the outcome [60]. Maurer et al. [24] utilized measles virus armed with suicide gene (super cytosine deaminse) along with prodrug 5-FC against acute myeloid leukemia. While initially infecting the cell line panels and patient-derived AML blasts with MeV-GFP they observed an decrease in cell viability and metabolic activity. This effect was more pronounced in MM6 cells followed by SKM-1 cells and NOMO-1 cells and in the case of patient-derived AML blasts one displayed reduction in viability and the other blast displayed resistance to MeV-GFP infections but did not have much degratory effects in blasts from healthy donors. With MeV-GFP infection an increase in STAT 1 activation and downstream regulation of IFIT-1 was observed, indicating the activation of IFN signaling cascade. The authors administered the prodrug 5-FC added super cytosine deaminase to measles virus Mev-SCD to convert 5-FC to 5-FU. They observed a huge decrease in viability of the cells after combined therapy with substantial induction of apoptosis.

9.8 Hepatocellular carcinoma

Hepatocellular carcinoma being the most common primary liver cancer with increasing incidence since 1980s is due to chronic liver disease, hepatitis B virus infections, hepatitis C virus, and nonalcoholic steatohepatitis with a 5-year survival rate of 18%. Diagnosis for hepatocellular carcinoma is done through imaging CT and MRI where the observed lesions are classified into five major categories. Treatment for HCC is recommended to have a multidisciplinary approach with surgical resection, liver transplantation, local ablative therapy, and chemo and radioembolization. Systemic therapies include sorafenib and lenvatinib combined with atezolizumab and bevacizumab [61]. Treatments for HCC being highly invasive alternate noninvasive effective strategies must be kept in place with a good therapeutic outcome. Meng et al. [25] explored the effects of using fludarabine as an adjuvant alongside NDV improving oncolytic activity. With fludarabine treatment the authors observed increased viral replication by two- to sixfold after 24 hours of infection by downregulating IFN-β and CXCL10 through inhibiting STAT 1 activation and promoted ubiquitin-proteasomal degradation of STAT-3 and IDO1 by increasing proteasome inhibitor MG132 and polyubiquitinated proteins. In xenografts-established mouse models, NDV increased the infiltration of CD8+ cells and NK cells but also increased MDSCs, respectively. When combined with fludarabine an increased NK cell infiltration was noted than CD8+ cells and reduction of MDSC was reported improving antitumor immunity. NDV with fludarabine increased overall survival (OS) in the mice models with no therapy-associated side effects. Zhang et al. [26] designed and armed the adenovirus with fusion protein sPD1PVR to induce long-term antitumor immunity. When infecting HCC H22 xenograft model mice intraperitoneally with Ad5sPD1PVR, they observed increased CD8+ infiltrations and only in the ascites there was a high infiltration of CD4+ and NK cell. When depleting CD8+ and NK cells the IFN-γ production was depleted in response to CD8+ depletion suggesting tumor specific long-term immune surveillance. Rechallenge studies of H22 xenografts had no tumor growth, whereas when challenged with a different cell line Hepa 1−6 displayed rapid tumor growth. During infection, Ad5sPD1PVR treatment significantly upregulated pro-inflammatory cytokines TNF-a, COX-2, IL-1β, and immunosuppressive factors such as IL10 and TGF-β. But this effect was abrogated by low doses of fludarabine.

9.9 Melanoma

Melanoma is the most aggressive cancer of the skin in all racial groups and has been listed as the third most common form of skin cancer [62], which always leads to poor prognosis for the patient where early interventional strategy includes surgery but often has higher chance of recurrence but when metastasized surgery is not an option which often requires isolated limb perfusion, chemotherapy, radiotherapy, targeted therapy, and immunotherapy [63,64]. The most utilized strategy to combat melanoma using oncolytic virus is by the modulation of the immune system resulting in immunogenic cell death of the tumor cells. Heinrich et al. [27] used vaccinia virus Jx-GFP and TG6002 along with prodrug 5-FC for oncolysis of melanoma cells with immunomodulatory activity. The cell lines SK29-MEL-1 and HLA-loss clone SK-29-MEL-1.22 when infected with the viruses displayed reduced viability and along with the combination of 5-FC only TG6002 synergistic response. Each virus displayed different apoptotic characters JX-GFP displaying necrosis and TG6002 displaying late apoptosis. Both JX-GFP and TG6002 displayed increased concentration of HMGB1 release upon infection. Calreticulin expression was high in TG6002+ 5-FC treated cells but JX-GFP treated cells calreticulin expression was not found to be significant. After viral infection (JX-GFP and TG6002) to the human immune cell line there was increased maturation in dendritic cells wherein CD83 and CD86 displayed high expression. Even when the virally induced TCL were cocultured with CTL to determine its activation where no significant activation of IFN-γ was observed. But treatment with 5-FU alone resulted in activation of IFN-γ.

This indicated the partial induction of adaptive simulating dendritic cells resulting in immuno-modulated cell death.

9.10 Brain cancer

Although brain cancers comprise only 2% of the cancers, they are particularly difficult to treat due to the location of tumor and genetics [65]. With Blood-Brian-Barrier in place exerting stringent control of passage of substances reduces the effective delivery of the drugs, a major rate limiting factor and also a barrier for tracers making it difficult for screening of tumors through PET, MRI, or contrast CT [66,67]. Treatment is based on type, location, malignancy of tumor along with age and current physical condition but is usually a combination of surgery, chemo, and radiotherapy [68]. Virotherapy has shown to display a good adjuvant effect through stimulating innate immunity along with oncolysis [69]. Thompson et al. [62] explored the tissue-wide expression of PVR/CD155 for the use of polio rhinovirus recombinant chimera PVSRIPO as oncolytic therapy in pediatric brain tumor tissues. Staining of the tissue sections revealed the robust expression of PVR in all pediatric brain tumor specimens with very high expression seen in Group 3-γ, WNTα, and WNTβ compared to other subtypes. As a proof of concept for PVSRIPO they infected a panel of low grade tumor (PXA 654 of BRAF V600E mutation and PXA 2363 of BRAF wild type) and malignant tumor (medulloblastoma D341 and D283) cell lines where they were able propagate and infect in both medulloblastoma and PXA equally where the cell lines displayed diminished growth and cell death was validated by increased expression of PARP cleavage after 24 and 48 hours of infection. Garcia-Romero et al. [29] bypassed the resistance of glioblastoma tumors against NDV by arming NDV with NS1 (rNDV-NS1) a potent negative regulator of IFN pathway, a key antiviral infection pathway thus successfully overcoming the resistance exhibited by tumor cells. In lieu of absence of CDKN2A and IFNβ1 the cells were reported to be sensitive to NDV infections. They infected a panel of glioblastoma cell lines where GBM27, GBM128B, and GBM128D having codeletion of CDKN2A and IFNβ1 deletion known as type 1 noncompetent cells whereas GBM18, GBM38, GBM123 are type 1 competent having the abovementioned genes intact and functional. During infection GBM27, GBM128B, and GBM128D displayed higher viral titers and replication compared to type 1 competent cell lines due to its impaired IFN pathway. The interferon stimulated genes RIG-I, MDA5, STAT1, Mx1, OAS1, ISG54, or ISG15 were upregulated in type 1 competent cells compared to type 1 noncompetent cells. MHC-I expression was upregulated in type 1 noncompetent cells upon NDV infection. IL6 and TNF-a expression was noted to be upregulated in all cell lines upon infection except GBM27. Type 1 noncompetent cells were found to be susceptible to NDV infections. As NDV infection resistance is displayed by type 1 competent cells, the NDV is armed with NS1, a strong antagonist of IFN signaling derived from influenza A. A 1000-fold increase of titer was observed in the type 1 competent cells indicating the overcoming of resistance displayed by cancer cells.

9.11 Oncolytic viruses under clinical trial

Owing to decades of research displaying the oncolytic potential and modification flexibility of oncolytic viruses, many clinical trials have been approved for testing assessing the safety profile and efficacy of the virus (Table 9.2). Husseini et al. [30] tested the maximal tolerant dose of engineered nonreplicating vaccinia virus Ankara named TG4023 armed with FCU1 gene along with systemic administration of 5-FC, a nontoxic prodrug fluorocytosine against 16 patients with nonresectable liver tumor. They assessed tumor response establishing the basis for virus-directed enzyme prodrug therapy (VDEPT) for the conversion of prodrug 5-FC to cytotoxic 5-FU by the action of FCU1 gene. The most commonly reported aggressive reactions were pyrexia, decreased appetite, nausea, and asthenia. Tumor assessment and examination revealed reduced CEA and CA 19−9 tumor markers. The level of 5-FU produced by enzymatic conversion in the tumor was equivalent to that of direct 5-FU treatment methods reflecting reduced cytotoxicity to nearby healthy cells reflecting the modification flexibility of oncolytic viruses. Contrast to monotherapy using oncolytic viruses as combinatorial therapy with chemotherapy and radiation has shown more promise in tumor regression as Waters et al. [31] proposed and designed a Phase 1 clinical trial for using conditionally replicating G207 HSV (deletion of $\gamma_1 34.5$ region with lac z insertion at UL39) with single radiation dose against pediatric supratentorial brain tumors in 24 participants aged between 3 and 18 years (results not available in clinicaltrials.gov). The above study was based upon the preclinical and three Phase 1 clinical trials in adults using G207 and single-dose radiation therapy where they displayed extended disease-free survival with increase in tumor infiltrating tumor lymphocytes and macrophages. Radiotherapy displayed synergistic effects by enhancing transcriptional mechanisms of G207 through activation of late HSV-1 promoters required for viral replication. It was also tested against pediatric patient-derived xenografts revealing higher sensitivity to therapy compared to adult glioblastoma. Mahalingam et al. [32] tested the combined effect of carboplatin/placitaxel with oncolytic reovirus named reolysin against metastatic melanoma where the combinatorial therapy displayed partial response in patients with metastatic melanoma who are resistant to firstline of chemotherapeutic agents with side effects pertaining to pyrexia, which was manageable. The 1-year survival was about 43% with progression free survival (PFS) of 5.2 months and OS of 10.9 months. Streby et al. [33] made a safety assessment for the use of HSV1716 in 9 pediatric patients with refractive sarcomas, clival chordoma, and malignant peripheral nerve sheath tumor where no objective response (no reduction in tumor size at the injected and noninjected site) and no dose-limiting toxicities were observed with the administered single doses but displayed stable disease with no progression with low standard uptake values in PET scan except two patients where sudden increase was in standard uptake value was observed but was reduced subsequently which was considered to be an effect due to inflammation elicited by the viral replication indicating the role of HSV in simulating antitumor immunity and one patient succumbed to the disease. Overall the virus was found to be well tolerated and the authors proposed further study with multiple doses and combinatorial therapy. Ribas et al. [34] administered oncolytic vector HSV1 also known as Talimogene Laherparepvec to increase the therapeutic effect of pembrolizumab (anti-PD-1) by increasing the influx of CD4+ and CD8+ cells initiated by oncolysis against metastatic melanoma. With a total of 21 participants under study, they did not observe any dose-limiting toxicities with an objective response rate of 62% and complete response rate of 33%. A higher influx of CD4+ and CD8+ T cells were observed with coexpression of PD-L1, PD-1, CD-56, and CD-20 expressing cells signifying tumor microenvironment modulation with regression in the remote lesions (visceral and nonvisceral) reporting minimal adverse effects related to pembrolizumab such as fatigue, rashes, arthralgia, fever, and chills highlighting the effect of therapy on remotely targeting metastatic cells improving the survival of the patients. A similar study by Chesney et al. [35] evaluated the efficacy and safety profile of combinatorial administration of Talimogene Laherparapvec with ipilimumab or monotherapy with ipilimumab consisting of 198 participants against advanced unresectable melanoma. The overall response rate was much better in patients with combinatorial therapy compared to ipilimumab or Talimogene Laherparepvec monotherapy and response to BRAF-wt tumors was greater (42%) than BRAF-mutant tumors (34%). Adverse effects were limited to flu-like symptoms related to Talimogene Laherparepvec and PFS was 8.2 months for combinatorial therapy compared to 6.4 months with ipilimumab monotherapy. Considering the treatment involving live oncolytic viruses Andtbacka et al. [36] observed the transmissibility of infection by Talimogene Laherparapvec to healthcare workers and also assessing its biodistribution and shedding in 60 patients with advanced melanoma. Results revealed that transmissibility was very low: about 1.1% (7 samples out of the 740 samples) involved surface injecting lesions with the infected individual displaying mild to no symptoms. Additionally, they monitored the adverse effects, viral DNA distribution, shedding and objective response rate in patients treated with T-VEC with most commonly reported adverse effects associated with flu-like symptoms but treatment in five patients was stopped during consecutive cycles due to pyrexia, delirium, atrial fibrillation, and posterior reversible encephalopathy syndrome. The viral DNA distribution and shedding was observed to be at peak in cycle 2 due to the higher dose administered and with each consecutive cycle of treatment clearance of virus was much faster due to humoral immunity. The objective response rate was 35% with a median response time of 3.1 months. Garcia-Carbonero et al. [37] assessed the effectiveness of IV compared to IT administration of enadenotucirev to 17 participants suffering from colorectal carcinoma, NSCLC, urothelial cell cancer, and renal cell carcinoma. The study assessed the viral delivery, replication, inflammation, and adverse effects caused by IV application and noted a marked increase in CD8+ cells signifying immune regulation with viral replication occurring in most tumor tissues. They also observe the Tregs being countered by infiltrating CTL phenotypes in nontumor healthy stroma challenging immunosuppression after administration of enadenotucirev. Intratumoral administration did not have any antiviral response compared to that of IV displaying mild antiviral immune response but it did not influence the administration through IV as no adverse effects and dose-limiting toxicities were observed. A similar intravenous administration strategy was performed by Mell et al. [38] using vaccinia virus GL-ONC1 combined with chemotherapy and radiation for HNSCC (head and neck squamous cell carcinoma) and assessing its safety profile against 19 patients. GL-ONC1 was found to be well tolerated in both single and multiple doses with side effects being hypotension, nausea, vomiting, and pox-like lesions. Other effects were associated with standard radiotherapy and cisplatin chemotherapy. But of the 19 they encountered seven treatment failures and seven deaths associated with p16-negative tumors with PFS and OS of 2 years was 51.8% and 58.6% and OS of but was much higher in p16-positive tumor patients with 64.1% and 69.2%. Dispenzieri et al. [39] intravenously administered Edmonston-lineage measles virus assessing its safety profile and dynamics targeting metastasis in patients with relapsed or refractory melanoma. They found no dose-limiting toxicity even at high doses with few significant reactions to the virus like fever, chills and gastrointestinal problems. Rapid administration was found not to be feasible over a period of 30 minutes resulting in headaches and much more tolerated with less rapid infusions. With higher dose infusions consistent viremia was observed while viral shedding was observed in the sputum and urine for 8–11 days. With treatment out of the 32 patients they reported one patient with hematological response, 4 patients with temporary reduction in serum immunoglobulin light chain with 4 weeks of therapy, 1 patient with complete response for a period of 9 months, and 1 patient with reductions in extramedullary plasmacytoma of the thighs and back. Of all the 32 participants, patients with 25% reduction in cFLC displayed lower levels of CD46 and higher SLAM expression. Lang et al. [40] observed the mechanism of action in tumors through dose escalation of DNX-2401 oncolytic adenovirus by intratumoral administration against recurrent high-grade glioma with 37 participants. They observed no dose-limiting toxicities and were found to vigorously replicate and kill recurrent glioblastoma cells resulting in >95% tumor regression with consistent treatment of 3 months in three patients with increased PFS and OS. DNX-2401 was detected in posttreatment biopsies taken after 2 weeks but not after a month suggesting its short-term survival in glioblastoma cells. Treatment with DXN-2401 did not have any effect modification on PD-L1 expression but reduced expression of TIM3 resulting in lower T-cell exhaustion eliciting antitumor response. Hirooka et al. [41] conducted a safety assessment of administration of EUS-guided intratumoral administration of HF-10 herpes simplex virus-1 against pancreatic cancer. Ten patients tolerating gemcitabine and erlotinib therapy were chosen for study and during HF-10 treatment five patients displayed myelosuppression due to chemotherapy, two with serious adverse effects, and one patient with duodenal stenosis and another patient experienced grade 4 hepatic dysfunction but the abovementioned reactions were confirmed not to be associated with HF-10 and found to be relatively safe. Through administration of HF-10 they observed three patients with partial response, four with stable disease, and two patients with progressive disease with a reported efficiency of 78%. After treatment the PFS was assessed to be around 6.3 months and OS of 15.5 months. Lauer et al. [42] tested the antitumour efficiency of GL-ONC1 vaccinia virus through intraperitoneal administration against peritoneal carcinomatosis that involved nine patients with cancers of gastric carcinoma, mesothelioma, peritoneal carcinoma, and cancer of primary unknown origin. No dose-limiting toxicities were observed and any adverse reaction observed were found to be dose independent with most common effects reported involving decrease in lymphocyte count, increase in CRP, pyrexia, and abdominal pain. The ascitic fluid monitoring for viral particles revealed to be increased fourfold from the administration of GL-ONC1 indicating splendid in-tumor replication where the virus titer was found even after 21 days posttreatment and oncolysis was verified through monitoring the high expression of β-Gluc and LDH1. Even with profound oncolytic ability of vaccinia virus significant elimination of tumors were not observed and long-term disease-free survival was not observed as all the patients under the study were diagnosed with advanced tumors refractory to all conventional therapies. Machiels et al. [43] evaluated the safety profile and efficacy of IV dose escalation of enadenotucirev in CRC patients and determined the dose schedule for metastatic CRC and UCC. The study included 61 patients with advanced invasive epithelial tumors. The maximal tolerant dose was determined at about 3×10^{12} viral particles with commonly reported adverse reactions pertaining to inflammatory reaction related to flu with increase in circulating cytokine levels IFN-γ, IL-6, MCP-1, and TNF-α after 24 hours but treatment related adverse effects reduced at day 3 to 5 due to immune desentization. Even with antibody response to the virus there was significant detection of viral particles but with each consecutive cycle of therapy an increase in the antibody level was reported displaying a neutralizing effect reducing effectiveness of oncolysis. An increase in infiltrating CD8+ cells was also observed in the microenvironment which in future could be exploited by combinatorial administration of immune checkpoint inhibitors improving therapy. Minev et al. [44] assessed the tolerance, safety profile, and effectiveness of oncolytic vaccinia virus ACAM2000 delivered by stromal vascular fraction cells against solid tumors and AML in 25 participants. No dose-limiting toxicities or cardiac complications were reported with rapid clearance of the administered virus within an hour but reappeared within one week displaying replication in tumors. The most commonly reported adverse effects were flu-like symptoms related to the administered vaccinia virus and viral DNA was not detected in remote areas signifying self limiting infection solely to the tumor without affecting the healthy cells. Even though Il-6 baseline levels were high it did not reach any toxic limits along with other cytokines and chemokines but the responses were delayed by 3 months, which correlates to no immediate toxicity that further supports the use of ACAM2000 in treatment of solid tumors. As an extension of previous work on bladder cancer [17], Annels et al. [45] conducted a window of opportunity trial consisting of 15 patients against bladder cancer using ICAM-1 targeted oncolytic Coxsackievirus A21 (CAVTAK) combined with low doses of mitomycin-c. Overall no major dose-limiting toxicities were observed but six patients experienced UTI due to the insertion of catheter resulting in displacement of bacteria but recovered with antibiotic treatment with oral amoxicillin 5-day course. Most patients displaced response in the form of surface hemorrhage, inflammation, and one patient was reported to have completed response. With increased expression of calreticulin, HMGB1, and RIPK3 it could be considered that CAVTAK treatment-induced necroptosis and ICD in the tumor. Through RNA-seq profile of the dissected tumors, increased levels of IFN-Υ induced chemokine genes CXCL-9, CXCL-10, CXCL-12 were observed associated with Th-1 tumor immunity but at the same instance increase levels of PD-L1, LAG-3, IPO were also observed, which might be the reason for tumor suppression leading to no marked infiltration of CD8+ cells into the environment compared to tissues prior to CAVTAK treatment. This opens new opportunities to administer immune checkpoint inhibitors like anti-PD-L1 in combination to CAVTAK improving tumor response to therapy. For effective CAVTAK treatment ICAM-1 expression was found to be crucial, which was in line with their preclinical trials. Schensk et al. [46] performed futility study for PFS in a Phase II trial administering Seneca Valley Virus NTX-010 and placebo in patients with SCLC not responding to more than four cycles of platinum-based chemotherapy. The group observed no significant PFS from the baseline control from placebo treatment. OS was found to be low in patients with detectable levels of virus on day 7 or 14 compared to patients with rapid viral clearance (3.3 months vs 9.9 months), which highlights the importance of viral persistence in patient prognosis after therapy. NTX-010 was found to be well tolerated among the treatment group with adverse effects limited to flu-like symptoms. Mahalingam et al. [47] determine safety and dose-limiting toxicities of pelareorep and its combination with pembrolizumab. A total of 11 patients with three treatment cycles displayed good tolerance with commonly reported adverse effects limited to fever, chills, fatigue, headache, emesis, flu like symptoms, hypotension, nausea, and neutropenia. No complete responses were achieved through the treatment with a median PFS of 2 months and OS of 3.1 months with 1 and 2 year survival of 35% and 23%. They found out that pelareorep replication was found in a higher level in the tumor tissues but had a significantly lower virus-mediated tumor lysis. However, an increase in CD8+ cells was observed in the biopsy with significant increases in PD-L1 and IO1 expression. An increase in caspase 3 activity was also observed by indicating increase in apoptosis. Assessment of T-cell clonality was performed to understand the influence of pelareorep in production of neoantigens, which in disregard of one patient there was a decrease in clonality and increase in diversity but for the one patient in which there was increased T-cell clonality had higher survival. Exploring the immune gene expression of the patient with high clonality revealed an increase in CCL3L1, which is a ligand for chemokine receptors. With treatment it was found out that production of early clones regardless of the oncolysis improved the survival of the patients with long-term outcomes.

9.12 Future directions

Intense research in the field of virotherapy has led to the emergence of multiple oncolytic drugs such as T-VEC, reolysin, H101, etc., which have shown promise in clinical trials. Yet, despite their potential problem persists when the selected patients for therapy display resistance to infection of such oncolytic viruses. So, strategies needed to overcome such barriers should be investigated. Increased studies exploring combinatorial therapy with existing radiotherapy and chemotherapy regimen sensitizing the resistant cells to oncolytic virus infection as shown by Udayakumar et al. [22] needs to be explored where they increased the sensitivity of tumor cells resistant to VSV infection by radiotherapy. More studies on chances of transmissibility of oncolytic virus from the treated patient to the administrator of such drugs must be performed to devise specific safety guidelines and regulations on personnel protection and environmental protection during administering of such drugs. Effective systemic administration strategies are required to improve the efficiency of treatment. With the advent of single-cell studies (both RNA and immunoprofiling), the effect of such therapy on cells could give more insights into regulatory pathways and immune cell landscape changes during infection in a superior resolution. Even with decades of research there is much in the field of virotherapy that needs to be explored.

Softwares used for images

Figs. 9.1 and 9.2 were drawn using Inkscape v1.1 and Fig. 9.3 using CorelDraw v2021 (Trial).

Author contribution

S.R and R.D drafted the manuscript. S.R designed the images. P.V and T.S designed and drafted the manuscript. All the authors read and approved the final manuscript.

Conflicts of interest

The authors declare no conflict of interest.

Funding statement

None.

Authors statement

Virotherapy being a very vast field, this book chapter has reviewed only a selected number of articles limited to the preclinical research and clinical trial conducted in the past 5 years.

References

[1] World Health Organization. Global Health Estimates 2016: Disease Burden by Cause, Age, by Country and by Region, 2000−2016; 2018.
[2] Sung H, Ferlay J, Siegel RL, Laversanne M, Soerjomataram I, Jemal A, et al. Global cancer statistics 2020: GLOBOCAN estimates of incidence and mortality worldwide for 36 cancers in 185 countries. CA Cancer J Clin 2021;71:209−49.
[3] Dagogo-Jack I, Shaw AT. Tumour heterogeneity and resistance to cancer therapies. Nat Rev Clin Oncol 2018;15:81−94.
[4] Roossinck MJ, Bazán ER. Symbiosis: viruses as intimate partners. Annu Rev Virol 2017;4:123−39.
[5] Kaufman HL, Kohlhapp FJ, Zloza A. Oncolytic viruses: a new class of immunotherapy drugs. Nat Rev Drug Discov 2015;14:642−62. Available from: https://doi.org/10.1038/nrd4663.
[6] Liu T-C, Thorne SH, Kirn DH. Oncolytic adenoviruses for cancer gene therapy. Gene therapy protocols. Springer; 2008. p. 243−58.
[7] Cao G-D, He X-B, Sun Q, Chen S, Wan K, Xu X, et al. The oncolytic virus in cancer diagnosis and treatment. Front Oncol 2020;10:1786. Available from: https://doi.org/10.3389/fonc.2020.01786.
[8] Motalleb G. Virotherapy in cancer. Iran J Cancer Prev 2013;6:101−7.
[9] Abd-Aziz N, Stanbridge EJ, Shafee N. Newcastle disease virus degrades HIF-1α through proteasomal pathways independent of VHL and p53. J Gen Virol 2016;97:3174.
[10] Mostafa AA, Meyers DE, Thirukkumaran CM, Liu PJ, Gratton K, Spurrell J, et al. Oncolytic reovirus and immune checkpoint inhibition as a novel immunotherapeutic strategy for breast cancer. Cancers 2018;10:205.
[11] Garza-Morales R, Gonzalez-Ramos R, Chiba A, Montes de Oca-Luna R, McNally LR, McMasters KM, et al. Temozolomide enhances triple-negative breast cancer virotherapy in vitro. Cancers 2018;10:144. Available from: https://doi.org/10.3390/cancers10050144.

[12] Berry JTL, Muñoz LE, Rodríguez Stewart RM, Selvaraj P, Mainou BA. Doxorubicin conjugation to reovirus improves oncolytic efficacy in triple-negative breast cancer. Mol Ther Oncolytics 2020;18:556–72. Available from: https://doi.org/10.1016/j.omto.2020.08.008.

[13] Simon EJ, Howells MA, Stuart JD, Boehme KW. Serotype-specific killing of large cell carcinoma cells by reovirus. Viruses 2017;9:140. Available from: https://doi.org/10.3390/v9060140.

[14] Olagnier D, Lababidi RR, Hadj SB, Sze A, Liu Y, Naidu SD, et al. Activation of Nrf2 signaling augments vesicular stomatitis virus oncolysis via autophagy-driven suppression of antiviral immunity. Mol Ther J Am Soc Gene Ther 2017;25:1900–16. Available from: https://doi.org/10.1016/j.ymthe.2017.04.022.

[15] Ye T, Jiang K, Wei L, Barr MP, Xu Q, Zhang G, et al. Oncolytic Newcastle disease virus induces autophagy-dependent immunogenic cell death in lung cancer cells. Am J Cancer Res 2018;8:1514–27.

[16] Garofalo M, Saari H, Somersalo P, Crescenti D, Kuryk L, Aksela L, et al. Antitumor effect of oncolytic virus and paclitaxel encapsulated in extracellular vesicles for lung cancer treatment. J Control Release 2018;283:223–34. Available from: https://doi.org/10.1016/j.jconrel.2018.05.015.

[17] Annels NE, Arif M, Simpson GR, Denyer M, Moller-Levet C, Mansfield D, et al. Oncolytic immunotherapy for bladder cancer using Coxsackie A21 virus. Mol Ther Oncolytics 2018;9:1–12. Available from: https://doi.org/10.1016/j.omto.2018.02.001.

[18] Shang C, Zhu Y-L, Li Y-Q, Song G-J, Ge C-C, Lu J, et al. Autophagy promotes oncolysis of an adenovirus expressing apoptin in human bladder cancer models. Invest N Drugs 2021;1–12.

[19] Scheubeck G, Berchtold S, Smirnow I, Schenk A, Beil J, Lauer UM. Starvation-induced differential virotherapy using an oncolytic measles vaccine virus. Viruses 2019;11:614. Available from: https://doi.org/10.3390/v11070614.

[20] Deng L, Yang X, Fan J, Ding Y, Peng Y, Xu D, et al. IL-24-armed oncolytic vaccinia virus exerts potent antitumor effects via multiple pathways in colorectal cancer. Oncol Res 2021;28:579–90. Available from: https://doi.org/10.3727/096504020X15942028641011.

[21] Wang H, Song L, Zhang X, Zhang X, Zhou X. Bluetongue viruses act as novel oncolytic viruses to effectively inhibit human renal cancer cell growth in vitro and in vivo. Med Sci Monit Int Med J Exp Clin Res 2021;27:e930634. Available from: https://doi.org/10.12659/MSM.930634.

[22] Udayakumar TS, Betancourt DM, Ahmad A, Tao W, Totiger TM, Patel M, et al. Radiation attenuates prostate tumor antiviral responses to vesicular stomatitis virus containing IFNβ, resulting in pronounced antitumor systemic immune responses. Mol Cancer Res 2020;18:1232–43.

[23] Lin Y, Wang W, Wan J, Yang Y, Fu W, Pan D, et al. Oncolytic activity of a coxsackievirus B3 strain in human endometrial cancer cell lines. Virol J 2018;15:65. Available from: https://doi.org/10.1186/s12985-018-0975-x.

[24] Maurer S, Salih HR, Smirnow I, Lauer UM, Berchtold S. Suicide gene-armed measles vaccine virus for the treatment of AML. Int J Oncol 2019;55:347–58. Available from: https://doi.org/10.3892/ijo.2019.4835.

[25] Meng G, Fei Z, Fang M, Li B, Chen A, Xu C, et al. Fludarabine as an adjuvant improves newcastle disease virus-mediated antitumor immunity in hepatocellular carcinoma. Mol Ther Oncolytics 2019;13:22–34. Available from: https://doi.org/10.1016/j.omto.2019.03.004.

[26] Zhang H, Zhang Y, Dong J, Zuo S, Meng G, Wu J, et al. Recombinant adenovirus expressing the fusion protein PD1PVR improves CD8+ T cell-mediated antitumor efficacy with long-term tumor-specific immune surveillance in hepatocellular carcinoma. Cell Oncol 2021;44:1243–55. Available from: https://doi.org/10.1007/s13402-021-00633-w.

[27] Heinrich B, Klein J, Delic M, Goepfert K, Engel V, Geberzahn L, et al. Immunogenicity of oncolytic vaccinia viruses JX-GFP and TG6002 in a human melanoma in vitro model: studying immunogenic cell death, dendritic cell maturation and interaction with cytotoxic T lymphocytes. OncoTargets Ther 2017;10:2389–401. Available from: https://doi.org/10.2147/OTT.S126320.

[28] Thompson EM, Brown M, Dobrikova E, Ramaswamy V, Taylor MD, McLendon R, et al. Poliovirus Receptor (CD155) expression in pediatric brain tumors mediates oncolysis of medulloblastoma and pleomorphic xanthoastrocytoma. J Neuropathol Exp Neurol 2018;77:696–702. Available from: https://doi.org/10.1093/jnen/nly045.

[29] García-Romero N, Palacín-Aliana I, Esteban-Rubio S, Madurga R, Rius-Rocabert S, Carrión-Navarro J, et al. Newcastle Disease Virus (NDV) oncolytic activity in human glioma tumors is dependent on CDKN2A-Type I IFN gene cluster codeletion. Cells 2020;9:1405. Available from: https://doi.org/10.3390/cells9061405.

[30] Husseini F, Delord J-P, Fournel-Federico C, Guitton J, Erbs P, Homerin M, et al. Vectorized gene therapy of liver tumors: proof-of-concept of TG4023 (MVA-FCU1) in combination with flucytosine. Ann Oncol 2017;28:169–74.

[31] Waters AM, Johnston JM, Reddy AT, Fiveash J, Madan-Swain A, Kachurak K, et al. Rationale and design of a phase 1 clinical trial to evaluate HSV G207 alone or with a single radiation dose in children with progressive or recurrent malignant supratentorial brain tumors. Hum Gene Ther Clin Dev 2017;28:7–16.

[32] Mahalingam D, Fountzilas C, Moseley J, Noronha N, Tran H, Chakrabarty R, et al. A phase II study of REOLYSIN®(pelareorep) in combination with carboplatin and paclitaxel for patients with advanced malignant melanoma. Cancer Chemother Pharmacol 2017;79:697–703.

[33] Streby KA, Geller JI, Currier MA, Warren PS, Racadio JM, Towbin AJ, et al. Intratumoral injection of HSV1716, an oncolytic herpes virus, is safe and shows evidence of immune response and viral replication in young cancer patients. Clin Cancer Res 2017;23:3566–74.

[34] Ribas A, Dummer R, Puzanov I, VanderWalde A, Andtbacka RHI, Michielin O, et al. Oncolytic virotherapy promotes intratumoral t cell infiltration and improves anti-PD-1 immunotherapy. Cell 2017;170:1109–1119. e10. Available from: https://doi.org/10.1016/j.cell.2017.08.027.

[35] Chesney J, Puzanov I, Collichio F, Singh P, Milhem MM, Glaspy J, et al. Randomized, Open-Label Phase II study evaluating the efficacy and safety of talimogene laherparepvec in combination with ipilimumab vs ipilimumab alone in patients with advanced, unresectable melanoma. J Clin Oncol J Am Soc Clin Oncol 2018;36:1658–67. Available from: https://doi.org/10.1200/JCO.2017.73.7379.

[36] Andtbacka RHI, Amatruda T, Nemunaitis J, Zager JS, Walker J, Chesney JA, et al. Biodistribution, shedding, and transmissibility of the oncolytic virus talimogene laherparepvec in patients with melanoma. EBioMedicine 2019;47:89–97. Available from: https://doi.org/10.1016/j.ebiom.2019.07.066.

[37] Garcia-Carbonero R, Salazar R, Duran I, Osman-Garcia I, Paz-Ares L, Bozada JM, et al. Phase 1 study of intravenous administration of the chimeric adenovirus enadenotucirev in patients undergoing primary tumor resection. J Immunother Cancer 2017;5:71. Available from: https://doi.org/10.1186/s40425-017-0277-7.

[38] Mell LK, Brumund KT, Daniels GA, Advani SJ, Zakeri K, Wright ME, et al. Phase I trial of intravenous oncolytic vaccinia virus (GL-ONC1) with cisplatin and radiotherapy in patients with locoregionally advanced head and neck carcinoma. Clin Cancer Res 2017;23:5696–702.

[39] Dispenzieri A, Tong C, LaPlant B, Lacy MQ, Laumann K, Dingli D, et al. Phase I trial of systemic administration of Edmonston strain of measles virus genetically engineered to express the sodium iodide symporter in patients with recurrent or refractory multiple myeloma. Leukemia 2017;31:2791–8. Available from: https://doi.org/10.1038/leu.2017.120.

[40] Lang FF, Conrad C, Gomez-Manzano C, Yung WKA, Sawaya R, Weinberg JS, et al. Phase I study of DNX-2401 (Delta-24-RGD) oncolytic adenovirus: replication and immunotherapeutic effects in recurrent malignant glioma. J Clin Oncol J Am Soc Clin Oncol 2018;36:1419–27. Available from: https://doi.org/10.1200/JCO.2017.75.8219.

[41] Hirooka Y, Kasuya H, Ishikawa T, Kawashima H, Ohno E, Villalobos IB, et al. A Phase I clinical trial of EUS-guided intratumoral injection of the oncolytic virus, HF10 for unresectable locally advanced pancreatic cancer. BMC Cancer 2018;18:596. Available from: https://doi.org/10.1186/s12885-018-4453-z.

[42] Lauer UM, Schell M, Beil J, Berchtold S, Koppenhöfer U, Glatzle J, et al. Phase I study of oncolytic vaccinia virus GL-ONC1 in patients with peritoneal carcinomatosis. Clin Cancer Res 2018;24:4388–98.

[43] Machiels J-P, Salazar R, Rottey S, Duran I, Dirix L, Geboes K, et al. A phase 1 dose escalation study of the oncolytic adenovirus enadenotucirev, administered intravenously to patients with epithelial solid tumors (EVOLVE). J Immunother Cancer 2019;7:20. Available from: https://doi.org/10.1186/s40425-019-0510-7.

[44] Minev BR, Lander E, Feller JF, Berman M, Greenwood BM, Minev I, et al. First-in-human study of TK-positive oncolytic vaccinia virus delivered by adipose stromal vascular fraction cells. J Transl Med 2019;17:271. Available from: https://doi.org/10.1186/s12967-019-2011-3.

[45] Annels NE, Mansfield D, Arif M, Ballesteros-Merino C, Simpson GR, Denyer M, et al. Phase I trial of an ICAM-1-targeted immunotherapeutic-coxsackievirus A21 (CVA21) as an oncolytic agent against non muscle-invasive bladder cancer. Clin Cancer Res 2019;25:5818–31.

[46] Schenk EL, Mandrekar SJ, Dy GK, Aubry MC, Tan AD, Dakhil SR, et al. A randomized double-blind phase II study of the seneca valley virus (NTX-010) vs placebo for patients with extensive-stage SCLC (ES SCLC) who were stable or responding after at least four cycles of platinum-based chemotherapy: north central cancer treatment group (alliance) N0923 study. J Thorac Oncol Off Publ Int Assoc Study Lung Cancer 2020;15:110–19. Available from: https://doi.org/10.1016/j.jtho.2019.09.083.

[47] Mahalingam D, Wilkinson GA, Eng KH, Fields P, Raber P, Moseley JL, et al. Pembrolizumab in combination with the oncolytic virus pelareorep and chemotherapy in patients with advanced pancreatic adenocarcinoma: a phase Ib study. Clin Cancer Res J Am Assoc Cancer Res 2020;26:71−81. Available from: https://doi.org/10.1158/1078-0432.CCR-19-2078.

[48] Ataollahi MR, Sharifi J, Paknahad MR, Paknahad A. Breast cancer and associated factors: a review. J Med Life 2015;8:6−11.

[49] Kuo Y-T, Liu C-H, Wong SH, Pan Y-C, Lin L-T. Small molecules baicalein and cinnamaldehyde are potentiators of measles virus-induced breast cancer oncolysis. Phytomedicine 2021;153611.

[50] Lemjabbar-Alaoui H, Hassan OU, Yang Y-W, Buchanan P. Lung cancer: biology and treatment options. Biochim Biophys Acta 2015;1856:189−210. Available from: https://doi.org/10.1016/j.bbcan.2015.08.002.

[51] Hirsch FR, Scagliotti GV, Mulshine JL, Kwon R, Curran Jr WJ, Wu Y-L, et al. Lung cancer: current therapies and new targeted treatments. Lancet 2017;389:299−311.

[52] Xie S, Wu Z, Qi Y, Wu B, Zhu X. The metastasizing mechanisms of lung cancer: recent advances and therapeutic challenges. Biomed Pharmacother 2021;138:111450. Available from: https://doi.org/10.1016/j.biopha.2021.111450.

[53] Nguyen THT, Pham XD, Dao KL, Vo TT. Response to a combination of full-dose osimertinib and ceritinib in a non-small cell lung cancer patient with EML4-ALK rearrangement and epidermal growth factor receptor co-mutation. Case Rep Oncol 2021;14:1085−91.

[54] Fan Z, Huang Z, Tong Y, Zhu Z, Huang X, Sun H. Sites of synchronous distant metastases, prognosis, and nomogram for small cell lung cancer patients with bone metastasis: a large cohort retrospective study. J Oncol 2021;2021:9949714. Available from: https://doi.org/10.1155/2021/9949714.

[55] Girardi DM, Ghatalia P, Singh P, Iyer G, Sridhar SS, Apolo AB. Systemic therapy in bladder preservation. Elsevier; 2020.

[56] Meeks JJ, Carneiro BA, Pai SG, Oberlin DT, Rademaker A, Fedorchak K, et al. Genomic characterization of high-risk non-muscle invasive bladder cancer. Oncotarget 2016;7:75176−84. Available from: https://doi.org/10.18632/oncotarget.12661.

[57] Guallar-Garrido S, Julián E. Bacillus Calmette-Guérin (BCG) therapy for bladder cancer: an update. ImmunoTargets Ther 2020;9:1−11. Available from: https://doi.org/10.2147/ITT.S202006.

[58] Leslie KK, Thiel KW, Goodheart MJ, De Geest K, Jia Y, Yang S. Endometrial cancer. Obstet Gynecol Clin North Am 2012;39:255−68. Available from: https://doi.org/10.1016/j.ogc.2012.04.001.

[59] Rütten H, Verhoef C, van Weelden WJ, Smits A, Dhanis J, Ottevanger N, et al. Recurrent endometrial cancer: local and systemic treatment options. Cancers 2021;13:6275. Available from: https://doi.org/10.3390/cancers13246275.

[60] Davis A, Viera AJ, Mead MD. Leukemia: an overview for primary care. Am Fam Physician 2014;89:731−8.

[61] Ferrante ND, Pillai A, Singal AG. Update on the diagnosis and treatment of hepatocellular carcinoma. Gastroenterol Hepatol 2020;16:506−16.

[62] Bradford PT. Skin cancer in skin of color. Dermatol Nurs 2009;21:170−8.

[63] Madamsetty VS, Paul MK, Mukherjee A, Mukherjee S. Functionalization of nanomaterials and their application in melanoma cancer theranostics. ACS Biomater Sci Eng 2019;6:167−81.

[64] Zhu C, Zhu Y, Pan H, Chen Z, Zhu Q. Current progresses of functional nanomaterials for imaging diagnosis and treatment of melanoma. Curr Top Med Chem 2019;19:2494−506.

[65] Gould J. Breaking down the epidemiology of brain cancer. Nature 2018;561:S40.

[66] Herholz K, Langen K-J, Schiepers C, Mountz JM. Brain tumors. Semin Nucl Med 2012;42:356−70. Available from: https://doi.org/10.1053/j.semnuclmed.2012.06.001.

[67] Arvanitis CD, Ferraro GB, Jain RK. The blood-brain barrier and blood-tumour barrier in brain tumours and metastases. Nat Rev Cancer 2020;20:26−41. Available from: https://doi.org/10.1038/s41568-019-0205-x.

[68] Perkins A, Liu G. Primary brain tumors in adults: diagnosis and treatment. Am Fam Physician 2016;93:211−17.

[69] Martikainen M, Essand M. Virus-based immunotherapy of glioblastoma. Cancers 2019;11:186.

Challenges in designing antiviral agents

Igor José dos Santos Nascimento, Leandro Rocha Silva and Edeildo Ferreira da Silva-Júnior

Institute of Chemistry and Biotechnology, Federal University of Alagoas, Maceió, Brazil

10.1 Introduction

Even after 56 years since the 1st Conference on Antiviral Substances took place, sponsored by the New York Academy of Science, several viruses still represent a meaningful threat to our modern society. In those years, viral replication was considered to be carried out by some cellular enzymes, which could be "promptly" blocked by using selective inhibitors. Posteriorly, the first viral protein was reported by Kates & McAuslan in 1967. It was an enzyme, the DNA-dependent RNA polymerase, isolated from Poxvirus (Poxviridae) [1], which was the first mechanistic basis for developing antiviral agents, followed by several other viral targets. The interest in developing new agents against the herpes simplex virus (HSV) allowed that the field of antiviral drugs to expand quickly, mainly with the development of acyclovir (Zovirax) (Fig. 10.1) [2]. However, the major interest for antiviral drugs emerged in face of the human immunodeficiency virus (HIV), in its first report in 1981 [3]; rapidly resulting in many anti-HIV agents capable of inhibiting different viral enzymes.

Similar to antibiotics for the treatment of bacterial diseases, almost all of today's antiviral drugs are developed targeting a specific virus-encoded gene or function [4]. However, continual use of these compounds intensifies the ability of viruses to acquire resistant mutations, which can result in the emergence of drug-resistant strains [5–7]. In this context, mutations associated with virus-encoded polymerases' errors during the RNA replication are frequently responsible for accumulating mutations and generating quasispecies [4].

Throughout this chapter, the reader will be guided into discussions involving the most relevant development of antiviral drugs against the most challenging viruses that humankind has faced. Also, some aspects associated with the perspective of medical needs, technical possibilities, and economic restrictions are covered in this actual piece of art.

Viral Infections and Antiviral Therapies
DOI: https://doi.org/10.1016/B978-0-323-91814-5.00017-9

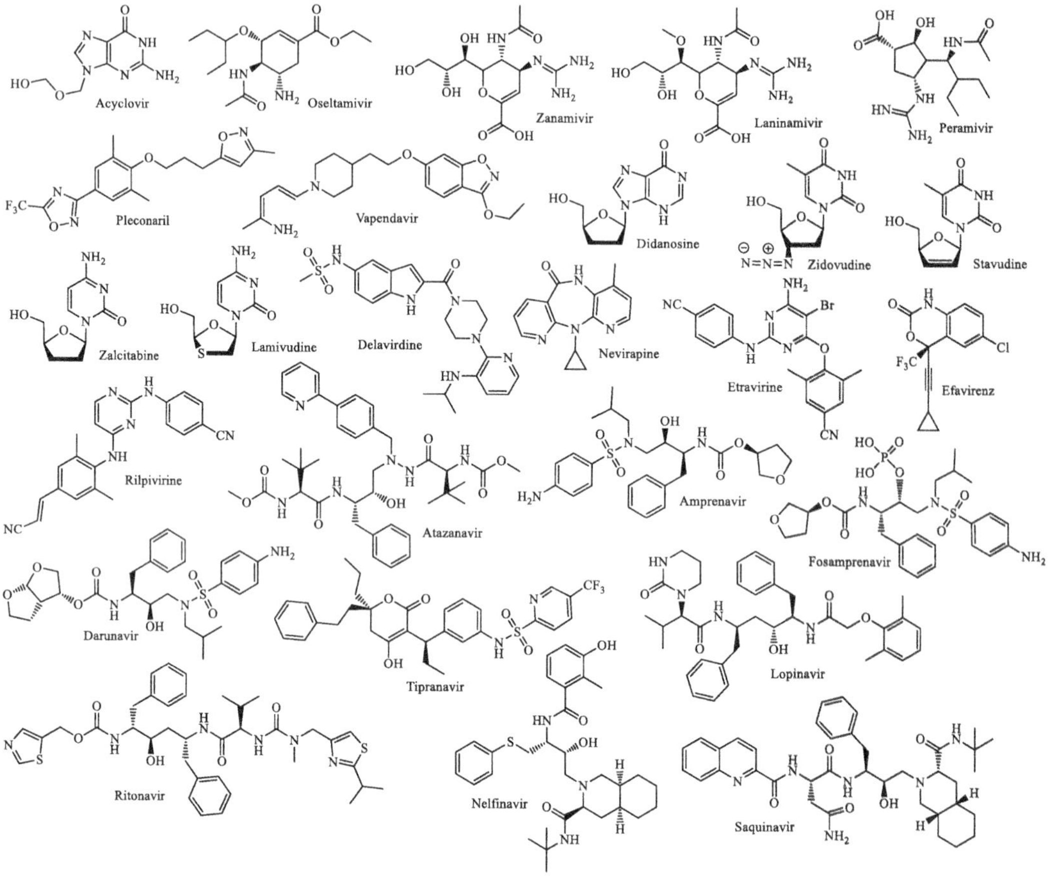

FIGURE 10.1 Current approved antiviral drugs utilized against several viruses.

10.2 Strategies for the design of antiviral agents

Over the past 30 years, antiviral drugs targeting host factors or viral proteins have been broadly developed, belonging to two groups: (i) inhibitors targeting the viruses themselves; or (ii) inhibitors targeting host-cell factors [8]. Concerning the antivirals that target the viruses, these can perform their effects by directly or indirectly inhibiting viral proteins, mainly enzymes [8]. In contrast, host-factor inhibitors act on host proteins that regulate functions in the viral life cycle, by improving the immune system or other cellular processes in host cells [8].

Currently, some issues are highlighted as the main reasons why antiviral pharmacotherapy is limited, such as: (1) the high cost of drug production has limited the market to some viral diseases (e.g. flu and herpes) since antiviral drugs must be highly specific for one single infectious agent, which its correct diagnoses should be made before to initiate the pharmacological therapy; (2) the existence of effective and safe vaccines, and the increasing number of vaccinated individuals, have led to delay in the development of new antiviral drugs [9]; (3) the prompt emergence of drug-resistant viral mutants (variants), which are required to be overcome by developing new antiviral classes having broad-spectrum antiviral efficacy [4]. Considering drug development an extremely prolonged and expensive process, the rapid emergence of drug-resistant viruses can render a previously effective antiviral drug useless within a short time, causing an enormous economic impact [4]. To curb the higher costs for producing a new antiviral addressed to a newly emerging virus, drugs targeting host-encoded functions or pathways can be an interesting alternative [4].

Notwithstanding all these facts, in the following subsections, we present the most relevant targets (viral adsorption, entry, polymerase, and proteases) used as alternatives to combat viral infections, allowing the readers to have an idea about the dimension of chemical classes of inhibitors and their mechanisms of action. Finally, the readers are invited to question themselves: *"Even with all these therapeutic alternatives, are there still viruses that can overcome them and cause outbreaks or pandemics?"* However, this is a question to be answered in Section 10.3 (*biggest challenging viruses*).

10.2.1 Virus attachment (or adsorption) inhibitors

The interaction of the virus with a functional receptor in host cells characterizes the first step in the virus life cycle, known as attachment [8]. In the case of enveloped viruses, some viral proteins placed on the outer envelope are responsible for recognizing receptors and attaching to host cells, as observed in HIV [10]. Examples of attachment inhibitors of enveloped viruses are oseltamivir (Tamiflu), zanamivir (Relenza) [11], laninamivir (Inavir) [12], and peramivir (Rapivab) [13] (Fig. 10.1). In contrast, nonenveloped viruses attach to functional receptors from host cells by using viral capsid proteins, as observed for picornavirus (Picornaviridae) [14]. Surprisingly, attachment inhibitors for nonenveloped viruses have not demonstrated clinical successes [8].

10.2.2 Virus entry inhibitors

Posteriorly to the virus attachment, it will release its genome into the cytoplasm via direct membrane fusion or endocytosis, being also a key early step in the viral life cycle and a promising target for designing drugs. Interestingly, fusion inhibitors are designed to prevent the conformational changes that are required for membrane fusion events [8]. In sense, enfuvirtide (Fuzeon) has been a clinically approved fusion inhibitor [15–17], which can inhibit a broad range of HIV strains, but its poor bioavailability makes its clinical utilization difficult [18,19]. To solve its pharmacokinetic problem, different peptides have been used, such as sifuvirtide [20], CP32M [17], and T2635 [21]. Regarding nonenveloped viruses, these adopt a different strategy for virus entry. After their attachment to host cells, their genetic material is released into the host cell cytoplasm due to a conformational change of their capsid protein, as verified for Enterovirus 71 (EV71), in which its capsid harbors 60 copies of a hydrophobic *pocket factor* (sphingosine), at the base of the canyon in the capsid protein VP1. The release of sphingosine and virus attachment triggers the opening of the capsid, releasing the viral genome [22,23]. However, pleconaril (Picovir) [24] and BTA798 (Vapendavir) (Fig. 10.1) are two hydrophobic molecules that can replace the natural pocket factor and inhibit picornaviral uncoating by generating resistance to the expulsion of this factor [25—27]. Additionally, pleconaril has been exhibited meaningful results against EV71 [28].

10.2.3 Viral polymerase inhibitors

During the virus life cycle, replication and transcription are the key steps, and these are the actions of polymerases [8]. In general, these polymerases are mainly responsible for replicating the virus genome, performing the three basic steps, being chain initiation, elongation, and termination [29]. In view of this, polymerases represent a promising alternative for designing new antiviral agents [8]. Concerning the structure and function of viral polymerases, there are two major types of polymerase inhibitors: (1) nucleoside/nucleotide substrate analogs, and (2) allosteric inhibitors. Thusly, nucleoside/nucleotide analogs play a central role in antiviral drugs targeting polymerases. These are firstly triphosphated by host cells to produce the active inhibitor and then act as a competitive inhibitor towards the natural nucleoside triphosphates and terminate the growing viral nucleic acids. In contrast, the main disadvantage of this type of inhibitor remains on the initial phosphorylation step, which is required for conversion of a monophosphorylated form into a triphosphated, and could not occur on the host cell correctly [8,30]. Therefore, monophosphate nucleotide analogs have been developed as polymerase inhibitors to prevent this obstacle [30—33]. Examples of approved antiviral drugs that used this mechanism include didanosine (Videx), zidovudine (Ziddivir), stavudine (Zerit), zalcitabine (Hivid), lamivudine (Duovir), among others [31—33]. On the other hand, during the polymerase cycle, the relative orientation of the polymerase domains undergoes a slight shift and it causes a conformational change of a specific binding site, the allosteric site [8]. Then, allosteric inhibitors block the structural movement of polymerase domains, inhibiting the viral polymerase activity. Finally, delavirdine (Rescriptor), nevirapine (Viramune), etravirine (Tibotec), efavirenz (Sustiva), and rilpivirine (Edurant) (Fig. 10.1) are examples of FDA-approved drugs that perform their activities via allosteric site [8,34].

10.2.4 Viral protease inhibitors

Typically, most viruses encode one or diverse proteases that are responsible for performing essential roles in their life cycle. These proteases carry out the proteolysis of a polyprotein precursor, which will originate and release functional viral proteins, allowing them to perform replication/transcription and maturation steps [35,36]. Still, their proteases effectively protect viral proteins by modulating host cells, including ISGylation and ubiquitination pathways [37—42]. In general, all viral proteases have a catalytic site, in which some inhibitors can interact to block them, varying in specificity and potency. In this context, endogenous protease substrates can be used for designing inhibitors, containing similar chemical groups to generate high-affinity binding and thus provide potent candidates for further drug discovery [8]. The most known protease inhibitors are indicated for the treatment of HIV-1 infection, being atazanavir, amprenavir, fosamprenavir (Lexiva), darunavir, tipranavir, lopinavir, ritonavir, nelfinavir, and saquinavir [43] (Fig. 10.1). All these compounds share similar chemical structures derived from their natural peptidic substrate and, therefore, posess cross-resistance to protease inhibitors at the active site of HIV-1 protease [8,44].

10.3 Biggest challenging viruses

10.3.1 *Herpes viruses* (HSV-1 and HSV-2)

HSV is an enveloped double-stranded DNA virus that belongs to the human Herpesviridae family and occurs in two different types: Herpes Simplex type 1 (HSV-1) and type 2 (HSV-2), which share up to 40% sequence homology of their genome structures. HSV-1 and HSV-2 clinically vary and present different degrees of severity [45,46]. Both viruses can infect orofacial areas, genital tract, and are known to cause central nervous system (CNS) infections, causing sporadically encephalitis [46,47]. Both of these diseases affect a significant number of individuals around the world, which can also cause encephalitis, eye disease, and generalized infections in immunodeficient patients or in newborns [9]. Furthermore, the major part of the symptoms observed in a cold sore is caused by inflammation produced by the action of the host immune system [48]. HSV-1 can cause herpes labialis, herpetic stomatitis, and herpetic stromal keratitis (HSK), whereas HSV-2 is mainly responsible for genital herpes [45]. HSV is also associated with visual impairment and blindness, which could be caused by acute retinal necrosis (ARN) [49].

Several glycoproteins (GPs) are involved in fusion machinery between host and viral membrane. HSV infection begins when a large number of virions are attached to heparan sulfate (HS), located on host cell surfaces by gB and/or gC, promoting gD or gE binding to relevant receptors known as heparin sulfate proteoglycans. Subsequently, gH/gL, gB, and gD are able to induce fusion of cellular membranes and are thus essential for the entry process. After adsorption gD, gH, gI, and gB, GPs synergically interact with the HSV entry medium, nectin-1, nectin-2, and 3-O-sulfated heparin proteoglycan, triggering conformational changes in the virus and plasma membrane, resulting in fusion and viral invasion. Then, gH is considered a fusion regulator and seems that the gH/gL complex can undergo dynamic rearrangements during viral fusion. Other ways of viral entry on the host include partial infection through pH-dependent endocytosis in certain types of host cells and phagocytosis-like mechanisms [50–52].

Most of the drugs addressed to fight against herpes viruses are mainly nucleoside inhibitors of DNA polymerase, which need to be phosphorylated by viral kinases [9]. In this context, acyclovir (see Fig. 10.1) and its prodrug have been found to be a prophylaxis alternative against episodes of genital herpes, avoiding painful episodes of genital sores [53].

10.3.1.1 *Promising compounds against herpes simplex virus*

Currently, there are three classes of licensed antiviral agents available for treat HSV infections, which are acyclic guanosine analogs, including acyclovir, which interferes with viral DNA replication through activation by viral thymidine kinase; penciclovir and ganciclovir (Fig. 10.2); acyclic nucleotide analogs like cidofovir (Fig. 10.2); and pyrophosphate

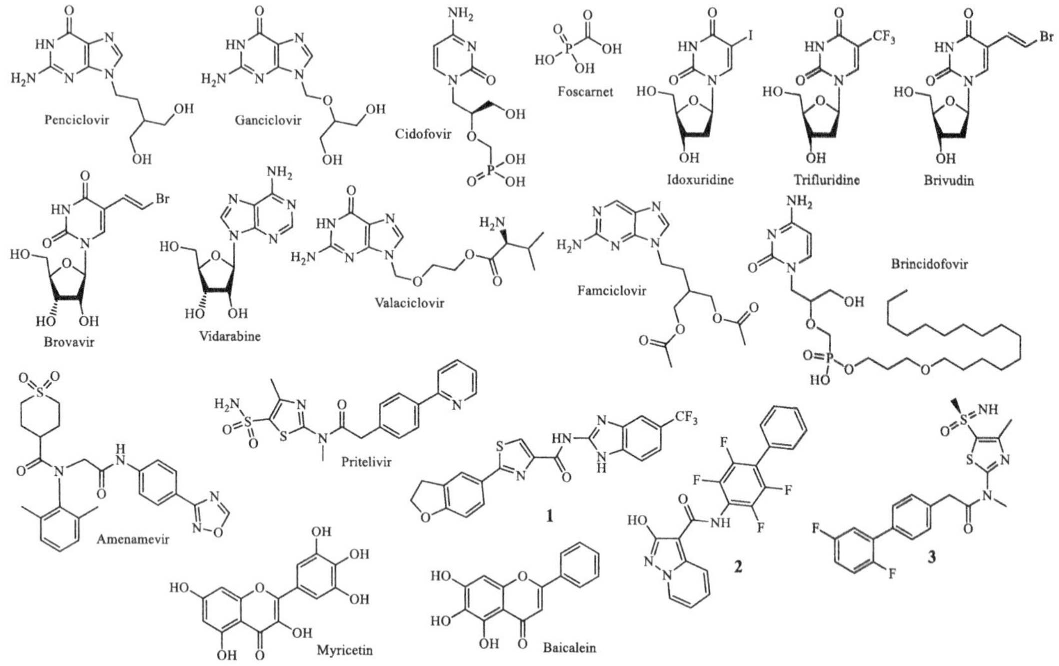

FIGURE 10.2   Approved antiviral drugs and promising active compounds against HSV.

analog foscarnet (Fig. 10.2), all of them targeting viral DNA replication [54,55]. After their stepwise activation by viral TK and cellular kinases, the phosphorylated compounds are incorporated by the DNA polymerase of HSV-1 into the nascent viral DNA chain, preventing further elongation [56]. Other licensed drugs targeting DNA polymerase include interferons associated with enzymatic degradation of mRNA; idoxuridine (Fig. 10.2), related nonselective activation by viral TK; trifluridine (Fig. 10.2), a competitive inhibitor of HSV DNA polymerase; brivudin (Fig. 10.2) related to activation to mono- and diphosphate by HSV-1; brovavir (Fig. 10.2) which acts in breakage of DNA strains and like competitive inhibitor; vidarabine (Fig. 10.2) activated to Ara-MP by cellular TK; foscarnet, a noncompetitive inhibitor of HSV DNA polymerase; valaciclovir (Valtrex) (Fig. 10.2), a prodrug of acyclovir; famciclovir (Fig. 10.2), a prodrug of penciclovir; brincidofovir (Fig. 10.2), a lipid conjugated prodrug of cidofovir [57,58].

HSK and ARN are treated from intravenous or oral therapy with acyclovir or newer prodrugs like famciclovir and valaciclovir, which have a better bioavailability in the CNS than acyclovir. Cidofovir and ganciclovir are alternatives antiviral drugs, which possess a poor oral bioavailability and are associated with severe side effects like nephrotoxicity that are used due to emerging of acyclovir-resistant HSV [49,55].

Viral resistance is defined as an acquisition of a gene mutation that reduces or alters TK activity, or alters viral DNA polymerase in HSV [59]. The most common cause of acyclovir resistance is mutations in viral TK, TK-gene (T66R), UL23 gene, polymerase gene (D672N), and UL30 gene. HSV-1-acyclovir-resistance exhibits a single-point mutation in the DNA polymerase gene, which leads to an amino acid exchange, important for the interaction between acyclovir and the protein. Due to mutations, viral replication cannot be blocked anymore by acyclovir [49,50,54]. In total, 95% of acyclovir-resistant HSV are associated with TK enzyme, while only 5% are associated with viral DNA polymerase [58]. The emergence of HSV-resistant strains is associated with mistakes in DNA polymerase, which appear during the replication process. Thus, constant suppression of the virus prevents the occurrence of mutations that may lead to resistance. In addition, cross-infections between HSV-1 and HSV-2 have always been the biggest challenge for the scientific community [52]. Antiviral strategies for HSV include directly blocking viral attachment factors such as HSV gB, gC, and gH, interfering with their conformational changes; phosphorylation of drug by TK to selectively inhibit viral DNA replication with low host cell toxicity, and interfering with virus receptor genes to block virus binding at the DNA level [50]. The mechanism of cell-to-cell spread is used by HSV to escape from the host's immune response and requires among others the viral GP gB. Thus, completely aborting the cell-to-cell transmission of the virus seems to be an attractive antiviral strategy [49]. Other effective target sites for limiting HSV productive and latent infections are modifications in essential genes like UL30, UL42, UL9, UL8, UL52, UL29, UL36, UL37, UL15, UL27, and UL54, which are involved in DNA replication. At the RNA level, synthetic siRNA is a potent treatment strategy for viral genome replication [50,60]. In addition to anti-HSV-1 drugs, antimicrobial photodynamic therapy (aPDT) is stated to inactivate the virus when use in the vesicle phase, by use a photosensitizer, a specific light source mainly when associated with photobiomodulation. aPDT is able to remove the viral load in lesions and even previous contaminations no-invasively and not cause resistance, being a therapy extremely important in the pandemic situations [61]. Acyclovir-resistant HSV strains and side effects associated to other licensed antiviral drugs have led to search for novel effective and well tolerable anti-HSV agents. Then, some promising candidates for the treatment of drug-resistant HSV infections are rising.

New drugs are required to be developed against HSV-resistant, while multidrug therapy is beneficial to prevent resistance. Schuhmacher et al. observed that peppermint oil (0.01%) inhibited plaque formation of HSV-1 and HSV-2 in a dose-dependent manner and HSV-1-acyclovir-resistance titer was reduced by 99%, probably due to direct interactions with the viral envelope and GPs [54]. Greeley et al. noted that combinations of acyclovir, cidofovir, and amenamevir (Fig. 10.2), which individually inhibit HSV-1 replication through different mechanisms of action, when combined leads to an additive relationship, suggesting that these drugs could be used as a viable combination therapy against HSV-1 [62]. Compound 1 (Fig. 10.2) has been identified as a new promising candidate for resistant HSV to acyclovir. Compound 1 interferes with a process that occurs during the early phase of HSV-1 replication, without toxic effect on Vero cells. In addition, compound 1 neutralized HSV-1 and HSV-2 irrespective of their resistance or cross-resistance toward standard antiviral drugs [55].

Antibody-based immunotherapy with mAb hu2c towards acyclovir-resistant ocular HSV infections in mice model has shown antiviral potency in the prevention of drug-resistant HSV-infections of the eye [49]. Wu et al. studied the effects of epigallocatechin-3-O-gallate (EGCG) against HSV-1 and observed that treatment of oral epithelial cells with EGCG may significantly prevent HSV-1-induced cell death and also reduce the viral release from oral epithelial cells. EGCG may affect viral life cycles at the early stage, to interfere with viral attachment by competing with HS binding HSV-1 and can interact with viral gB and gD proteins [63]. In addition to new anti-HSV agents, host-targeted small molecules able to modulate virus—host interactions may be an alternative to de novo development of virally targeted agents. Pyrimidines' availability into cells host is a critical factor for viral replication. Thus, compounds targeting the de novo pyrimidine biosynthetic pathway are potentials, host-acting antiviral agents, overcoming viral drug resistance. In this regard, compound 2 (Fig. 10.2) (MEDS433) has been identified as a potent inhibitor against in vitro replication of HSV-1 and HSV-2 via a mechanism that seems to selectively block the dihydroorotate dehydrogenase (DHODH) enzymatic activity [64]. Other strategies involve developing alternative therapeutics targeting non-TK dependents mechanisms of action. In this regard, anti-HSV agents that target the viral helicase primase (HP) complex may contribute to the effective therapy of HSV-resistant. HPIs inhibit the heterotrimeric protein complex in the initial stage of HSV DNA replication, which has multiple enzymatic activities including DNA unwinding at the replication fork, separating double-stranded DNA, DNA-dependent ATPase activity, and synthesis of RNA primers. Furthermore, HP inhibitors do not depend on the viral TK activation like nucleoside prodrugs, contributing to the effective therapy even to HSV nucleoside drug resistance. There are three classes of HP inhibitors: oxadiazolylphenyl, 2-amino-thiazole amide (BILS 179 BS), and thiazole amide [pritelivir (Fig. 10.2)], which can contribute to the suppression of viral recurrences and asymptomatic virus shedding, reducing person-to-person transmission [65]. Compound 3 (Fig. 10.2) presented antiherpetic potency as a novel HP inhibitor, with low cytotoxicity in vitro, high efficacy, and safety.

Myricetin and baicalein showed anti-HSV activities. Myricetin blocks HSV-1 and HSV-2 infection through direct interaction with virus gD protein to interfere with virus adsorption and membrane fusion. Whereas baicalein inhibits HSV-1 replication by two mechanisms, namely the inactivation of free viral particles to neutralize the infectivity and the suppression of NF-κB activation [66,67].

10.3.2 Respiratory viruses

10.3.2.1 Influenza A and B

10.3.2.1.1 Adamantane analogs and their limitations in drug design

For many years, INFV was responsible for several death cases worldwide, and the discovery of new drugs and vaccines was able to decrease its number of infections. The first treatments involved adamantanes, such as amantadine (Fig. 10.3), the first drug of this class used since 1964. Posteriorly, its analog, rimantadine (Fig. 10.3), was approved against INFV in 1993, in oral treatment [68]. These drugs act by inhibiting the M2 ion-channel activity, avoiding increasing the endosomal pH, inhibiting conformational changes of the viral HA protein [68,69]. One of the limitations of these agents is their restricted activity against INFV-A since INFV-B showed alterations at the M2 protein, limiting and motivating discoveries targeting other viral targets [68]. For many years, these drugs were the unique alternatives against INFV, with 70%—90% efficacy in prophylactic administration. However, the high incidence of side effects, including neurological disorders, and the rapid development of viral resistance limited their clinical use. Mutations at pore-facing residues, such as V27A, A30T, S31N, G43E, as well as, mutations at the N-terminal of M2

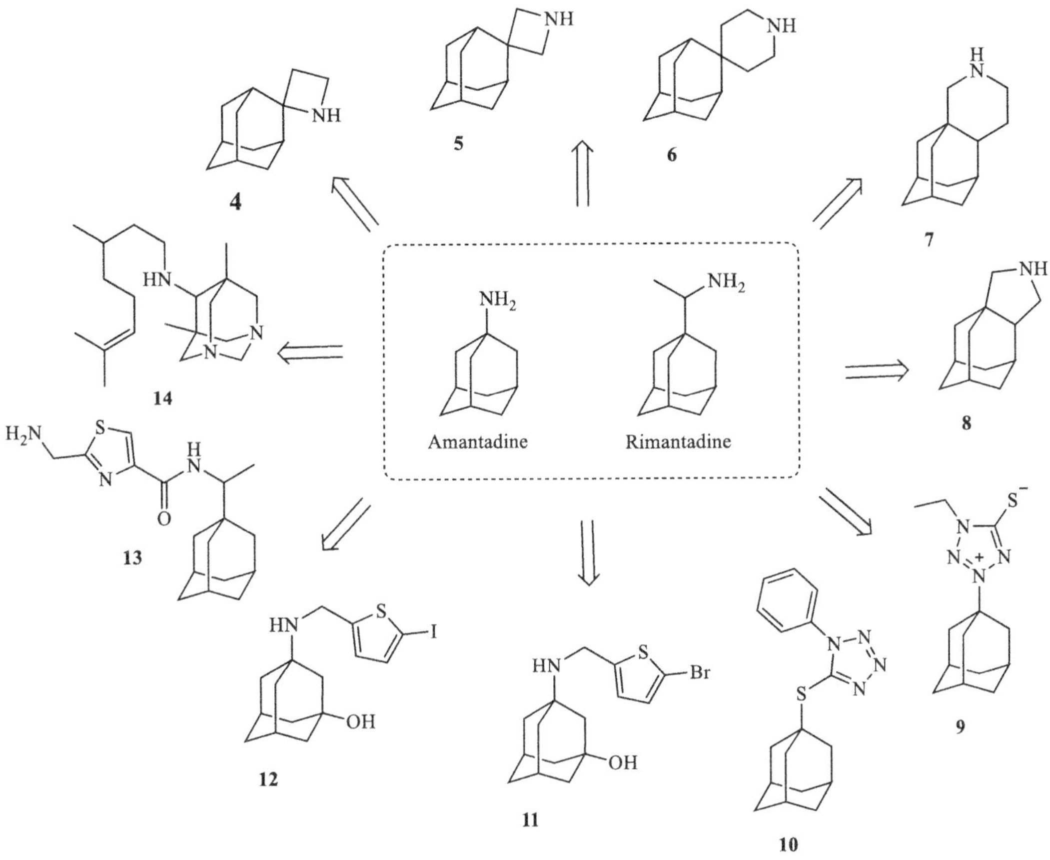

FIGURE 10.3   Chemical structures of main adamantane analogs capable of acting as M2 channel inhibitors.

ion-channel (L26F) and other in interhelical residues at C-terminus have been responsible for viral resistance cases [70]. For these reasons, the clinical use of adamantanes against INFV is not encouraged [71]. Thus, there are constant efforts to discover new analogs that can overcome these limitations, being therapeutically useful against resistant strains.

Heterocycle-fused adamantane analogs were synthesized by Kolocouris et al., showing high activity against INFV-A/H3N2, in which compounds 4, 5, and 6 (Fig. 10.3) were the most effective analogs, with EC_{50} values of 0.59, 0.33, and 0.16 μM, respectively [72]. The authors highlighted that the insertion of a methyl group at nitrogen atom led to potency reduction of their antiinfluenza activity. On the other hand, Zoidis et al. synthesized piper-idine analogs, where compounds 7 and 8 (Fig. 10.3) had the most activity against INFV-A/H3N2, exhibiting EC_{50} values of 0.6 ± 0.4 and 0.5 ± 0.3 μM, respectively [73]. SAR investigations of these analogs revealed that by replacing the five-membered cycle with a six-membered one the antiviral activity remains. Still, unsubstituted nitrogens result in the most active compounds, as in the study aforementioned. Finally, all compounds did not show activity against INFV-B due to differences in its M2 channel, highlighting the limita-tions of this chemical class to treat it.

Azolo-adamantanes were explored by Zarubaev and coworkers, aiming to overcome the viral resistance issue, mainly towards S31N mutation [74]. Analogs 9 and 10 (Fig. 10.3) had the most activities (EC_{50} values of 3 and 2 μg/mL, respectively), in which it was determined that their antiviral activities are related to tetrazole and adamantyl groups and amino replacement with heterocycles. Compounds were most effective in suppressing the viral replication than amantadine against rimantadine-resistant strains. The authors suggested that the use of meso-ionic adamantyl tetrazole analogs can be overcome the limitation related to viral resistance. This scaffold is not a natural occurrence, making it difficult for the recognition of mutant strains for developing resistance against it. Then, its optimization can be promising to discover new anti-INFV drugs. In another study, Hu and coworkers designed new analogs also against the S31N mutation, verifying that analogs 11 and 12 (Fig. 10.3) were the most promising candidates, with EC_{50} values of 1.1 ± 2.0, and 0.9 ± 0.1 μM against A/California/07/2009/H1N1, respectively [75]. In addition, these compounds were effective against other resistant strains, including oseltamivir-sensitive, and high microsomal ($T_{1/2}$ value of 145 min), and membrane stabilities (> 200 nm/s). The use of thiophen-adamantane analogs can overcome the limitations related to the resistance of this chemical class. However, Stankova et al. used thiazole-adamantane peptidomimetics to identify new anti-INFV active compounds targeting A/Hong Kong H1N1 strain [76]. Analog 13 (Fig. 10.3) was found to be the most effective, displaying IC_{50} and CC_{50} values of 0.11 and 50 μg/mL, respectively. Still, this study revealed that NH_2 protection is detrimental to antiviral activity.

Analogs explored by Suslov et al. contained monoterpenoid and hetero-adamantane moieties [77]. Compounds were evaluated against the rimantadine-resistant INFV-A/Puerto Rico/8/34 (H1N1) strain, in which analog 14 (Fig. 10.3) showed the best results, having IC_{50}, CC_{50}, and SI values of 8 ± 2, 239 ± 21, and 30, respectively. Molecular docking revealed that the interaction of it with Ala[30] and Asn[31] residues is essential for binding on the M2 channel. Thus, the use of diaza-adamantane analogs can be promising to identify new drugs that overcome the resistance of adamantane derivatives.

10.3.2.1.2 Neuraminidase inhibitors against influenza-A and influenza-B

The high prevalence of resistance to adamantanes motivated the discovery of new drugs that act in other targets related to INFV. This fact is supported because resistance to amantadine occurs in 33% of patients within 5 days of treatment [78]. Furthermore, adamantanes are not active against INFV-B, which instigated the investigation of NA as a drug target, leading to the discovery of zanamivir (see Fig. 10.1) as the first therapeutic agent of this class [78,79]. NA inhibitors appeared to be the solution to the problem related to virus resistance and posteriorly discovered analogs as oseltamivir, peramivir, and laninamivir (see Fig. 10.1), which showed great results in virus elimination [17,18]. On the other hand, the continuous utilization of NA inhibitors provided new virus mutations, such as H275Y. Interestingly, estimative data pointed that the 2009 H1N1 pandemic presented this mutation, which motivated the exploration and development of new analogs against these targets [80]. Main mutations against oseltamivir are related to modifications at R292K, N294S, and H274Y residues from NA. The excessive prophylactic use of oseltamivir is related to the high resistance to this drug compared to zanamivir [81]. In fact, the discovery of new anti-INVF drugs needs to overcome some challenges, such as for example, identifying a new drug for different targets from NA, and being active against both INFV-A and B [82]. Regarding these facts, the next topic will be focused on the latest developments in NA inhibitors and their main challenges faced.

To discover new effective NA inhibitors, Wang and coworkers used cyclohexene analogs and ligand- and structure-based drug design (SBDD) approaches [83]. In this way, compound 15 (Fig. 10.4) was identified to be the most active molecule against NA (IC$_{50}$: 39.6 ± 6.5 μM), and A/WSN/33 H1N1 strain, with IC$_{50}$ and CC$_{50}$ values of 1.72 ± 0.35 μM and 373 ± 19.5 μM, respectively. This compound interacted with key residues, Arg^{152}, Arg^{292}, Arg^{371}, and Tyr^{347} from NA, and showed stability within 20 ns of molecular dynamics simulations. In addition, it was most effective than oseltamivir (IC$_{50}$: 61.1 ± 6.3 μM against NA), suggesting that the N-substitutions can provide the most effective analogs. Still, Wang et al. explored new analogs without a basic group against NA from resistant INFV [84]. Compound 16 (Fig. 10.4) showed the best activity against H5N1-H274Y (IC$_{50}$: 2075 nM) and H1N1-H274Y (IC$_{50}$: 1382 nM) mutated NA, being more active than oseltamivir (IC$_{50}$: 6095 and 4071 nM for both mutations,

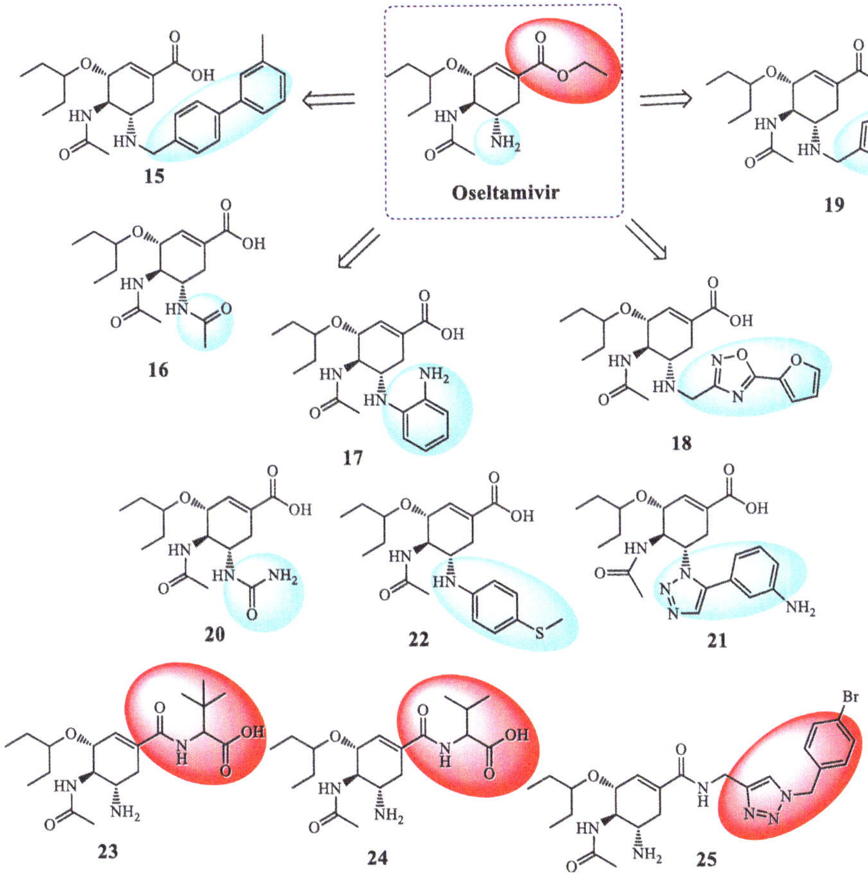

FIGURE 10.4 Neuraminidase inhibitors and mains oseltamivir analogs.

respectively). Molecular modeling studies suggested that the new chemical fragment introduced at this compound was able to interact with Glu[119] and Asp[151] residues, displaying a binding pose similar to oseltamivir. Thus, substitutions at the nitrogen group can be promising for designing new inhibitors against resistant INFV strains.

Oseltamivir analogs pyridyl based were synthesized by Wang et al., in which compound 17 (Fig. 10.4) showed high inhibitory potency against H5N1 (87% inhibition), with an IC_{50} value of 320 nM, similar to oseltamivir (IC_{50}: 210 nM) [85]. The cytopathic effect (CPE) inhibition assay on MDCK cells using A/Puerto Rico/8/1934 and A/LiaoNing-ZhenXing/1109/2010 strains demonstrated that it was the most promising analog, with EC_{50} values of 12.68 ± 8.96, and 14.31 ± 2.59 μM upon both strains, respectively. The great potential of this compound can be related to forming H-bonding interactions with Arg[118] and Tyr[406] residues. Despite its promising activity, compound 17 showed unsatisfactory pharmacokinetics properties. Ye et al. explored $C5\text{-}NHCH_2\text{-}Aryl$ groups against INFV-A NA [86]. In this way, compounds 18 and 19 (Fig. 10.4) were the most active against A/H3N2 NA, exhibiting IC_{50} values of 1.92 ± 0.24 and 1.63 ± 0.16 nM, respectively. The authors suggested that it can provide new solutions against resistant INFV strains. Posteriorly, a promising analog of it was synthesized by Zhang and coworkers by introducing a urea group at the C-5 position, providing compound 20 (Fig. 10.4) [87]. It showed potent activity against NA from H5N1 (IC_{50}: 0.254 ± 0.038 μM), H1N1 (IC_{50}: 8.7 ± 0.4 nM), and H3N2 (IC_{50}: 169 ± 15.5 nM), as well as, the mutants H5N1-H274Y (IC_{50}: 2005 ± 61 nM) and H1N1-H274Y (IC_{50}: 667 ± 70.5 nM). In addition, its pharmacokinetic profile in intravenous or oral administration was similar to oseltamivir. Still performing modifications at C-5 position, Wang and coworkers synthesized compound 21 (Fig. 10.4), with excellent activity against H12N5 (IC_{50}: 0.045 ± 0.016 μM), H3N2 (IC_{50}: 0.019 ± 0.0040 μM), and mutant H3N2-E119V (IC_{50}: 0.79 ± 0.074 μM) [88]. Molecular modeling suggested that the phenylamino group performing H-bonding interactions with Asp[151], surrounding the 150-cavity.

Other analogs modified at C-1 and $C\text{-}5\text{-}NH_2$ positions were evaluated by Jia et al. to discover that oseltamivir analogs were more resistant [89]. Thus, it was identified analog 22 (Fig. 10.4), with great activity against NA H5N1 (IC_{50}: 0.044 ± 0.0049 μM) and H5N1-H274Y (IC_{50}: 1.40 ± 0.061 μM), similar to oseltamivir (IC_{50}: 0.067 ± 0.0095 and 2.45 ± 0.31 μM) without toxicity ($CC_{50} > 200$ μM). In addition, it inhibited the enzyme from INFV-B (IC_{50}: 7.97 ± 1.27 μM), and high activity against H5N1 (EC_{50}: 0.66 μM), being most potent than oseltamivir (EC_{50}: 0.82 μM). Finally, in vivo assays in Kunming mice in intragastric administration demonstrated that this compound could be safely administered.

Modifications at the carbonyl-group of oseltamivir were performed by Wang et al. and evaluated against H5N1 [90]. Thus, compound 23 (Fig. 10.4) showed high inhibitory potency (IC_{50}: 1.30 ± 0.23 μM). Molecular modeling studies suggested that the tert-butyl group interacts with the 430-loop region. In addition, this compound showed high metabolic stability and lipophilicity, which improved permeability and oral absorption. On the other hand, Ju and coworkers synthesized the compound 24 (Fig. 10.4) modified at the carbonyl group, showing high potency against H5N1 (IC_{50}: 0.0883 ± 0.0133 μM) and H5N6 (IC_{50}: 0.0967 ± 0.0035 μM), and great EC_{50} values, being 4.26 and 1.31 μM, respectively [91]. In addition, its potency against H5N1-H274Y NA (IC_{50}: 37.83 ± 2.63 μM) was similar to oseltamivir (IC_{50}: 5.05 ± 0.19 μM). Notability, molecular modeling revealed that the elongation of the carbonyl group directing to 430-cavity is beneficial to the activity. Other substitutions at carbonyl group were performed by Ju et al., obtaining the compound 25 (Fig. 10.4) with high activity against H5N1, H5N2, and H5N6 strains, presenting IC_{50} values of 0.12, 0.049, and 0.16 μM, respectively; while EC_{50} values of 2.45, 0.43, and 2.8 μM, respectively [92]. In addition, molecular docking revealed that the triazole group occupies the 430-cavity. Finally, an experimental embryonated egg model revealed that this compound can exert a protective effect similar to oseltamivir.

10.3.2.2 Severe-acute respiratory syndrome coronavirus-2

Since December 31, 2019, Severe-Acute Respiratory Syndrome Coronavirus-2 (SARS-CoV-2), called COVID-19, has been causing health and economic damages, in which its eradication is a worldwide priority [93,94]. Despite the discovery of new vaccines, and some drugs can be used in emergencies or more serious cases, such as remdesivir or molnupiravir (Fig. 10.5). However, it is still necessary to identify the most effective agents [95,96].

Even after almost two years of the pandemic, several challenges are related to discovering new anti-SARS-CoV-2 drugs. One of them is the low financial investment, which results in the reduced availability of new antivirals. In addition, due to the limited therapeutic arsenal of antivirals' lack of a properly approved drug or even an effective molecule with a well-understood mechanism of action against this disease, there is a high need for molecular scaffolds as a starting point. This is highly limiting and justifies the delay in finding an effective agent against SARS-CoV-2 [93,97]. An ideal drug against this disease is the one that interferes in entry events, because blocking the transmission between the people, stopping the viral dissemination. Another strategy aiming the inhibition of viral proteases, such as 3CL^{Pro}, PL^{Pro}, and others like RdRp (RNA-dependent RNA polymerase)

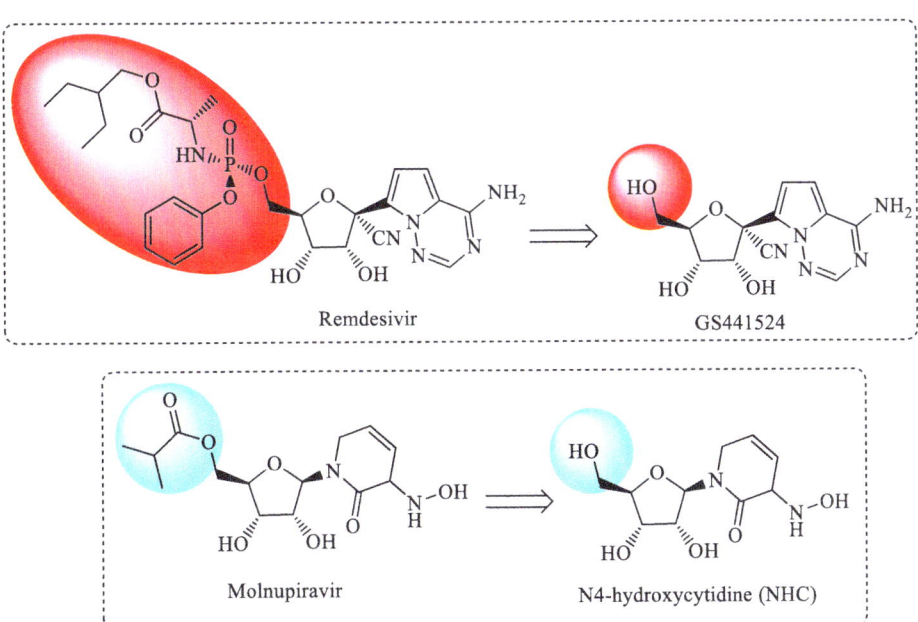

FIGURE 10.5 Chemical structures of nucleoside drugs active against SARS-CoV-2.

and helicase, or even human targets like TMPRSS2 (transmembrane protease serine 2) or cathepsin [98,99]. On the other hand, targets like cysteine or serine proteases face challenges related to selectivity for human enzymes, because the drugs must show high selectivity against the virus enzyme and low or none against host targets, representing one of the greatest challenges in the antiviral design [100,101]. In this way, we see in this topic some strategies in discovering new drugs against SARS-CoV-2, limitations, and how to overcome them.

10.3.2.2.1 Nucleoside analogs against severe-acute respiratory syndrome coronavirus-2

Several challenges must be overcome, mainly when choosing a compound in vitro for its subsequent evaluation in vivo. In vitro assays are performed against a specific target or in cell cultures, in which plasma proteins are absent. Basically, the interaction with plasma proteins decreases the drug available for its pharmacological effect. Neglecting this factor can lead to failures in vivo studies, in which IC_{50} values tend to increase in vivo assays due to the interaction of the drug with plasma proteins. Furthermore, different molecules can present different degrees of interaction with plasma proteins. Thus, a careful evaluation of preclinical candidates must be carried out to avoid further clinical failures [102]. In addition, the cell line or model assay must be adequate, prioritizing new hits and discarding compounds inactive [103].

Remdesivir was the first FDA-approved treatment against SARS-CoV-2. Several studies, among computational to clinical experiments proposed its efficacy against this virus [104–106]. This is a nucleoside analog prodrug, which is metabolized to GS-441524 (Fig. 10.5) and inhibits the RdRp as its mechanism of action [104,107]. The clinical use is limited for its administration form, in which the intravenous pathway showed limitation in its ambulatorial use [102]. In this way, searching for a new therapy for oral administrations is the bigger challenge against this virus. Recently, molnupiravir showed a similar mechanism of action, as RdRp inhibitor, being also a prodrug, in which β-D-N^4-hydroxycytidine (NHC) (Fig. 10.5) is its active form. The great advantage of this compound is its oral administration and low resistance probability [102,108], making it to be called the "COVID-19 pill."

Remdesivir is converted in GS-441524 monophosphate as its active form, acting as nonobligate RNA chain termination in the RdRp from SARS-CoV-2. The virus resistance is a problem faced to discover new drugs against SARS-CoV-2. Some studies revealed that mutations in this enzyme can affect the binding site of remdesivir, inducing resistance. Similar to other viruses, overcoming the resistance of SARS-CoV-2 is one of the most challenging tasks in drug discovery [109].

10.3.3 Human immunodeficiency virus

10.3.3.1 The virus and its some limitations in drug design

HIV, the etiological agent of the acquired immune deficiency syndrome, is a retrovirus identified around 1981, leading to several deaths worldwide until today. This virus targets T-cells from the immunological system, in which when infecting them, it remains latent and its activation destroys these cells, compromising the host's immune system. Unfortunately, there is no cure for this disease, only drugs capable of increasing patients' life expectancy. However, virus resistance is a common phenomenon, which limits the treatment, becoming a challenge in medicinal chemistry to discover new drugs [110,111]. In addition, the toxicity of current drugs instigates the discovery of new scaffolds, which act against other targets, overcoming the resistance problem induced by drugs with the most effective and security [112,113]. The first approved treatment against this virus was zidovudine (see Fig. 10.1) in 1987. Actually, there are 30 licensed drugs but the emergence of multi-resistant strains is still the problem [114,115]. Drugs acting in the main stages of the viral cycle, such as viral entry, reverse transcriptase (RT) (nucleosides and non-nucleosides), and maturation inhibitors (protease inhibitors) are used in combination to avoid viral resistance [115–117].

10.3.3.2 Reverse transcriptase inhibitors

To discover new non-nucleoside analogs as transcriptase reverse (NNRT) inhibitors that overcome the limitations related to viral resistance, Kankanala et al. synthesized and evaluated *N*-hydroxy hienopyrimidine-2,3-dione derivatives, identifying compound 26 (Fig. 10.6) was the most effective [118]. It showed a great result against RT-associated ribonuclease H (RNase H) (IC$_{50}$: 0.04 μM), in which crystallographic studies confirm the binding mode at the RNase H region. In addition, it presented high antiviral activity without toxicity (EC$_{50}$: 7.4 μM and CC$_{50}$ > 100 μM). Despite these great results, this compound did not inhibit the RT polymerase (pol).

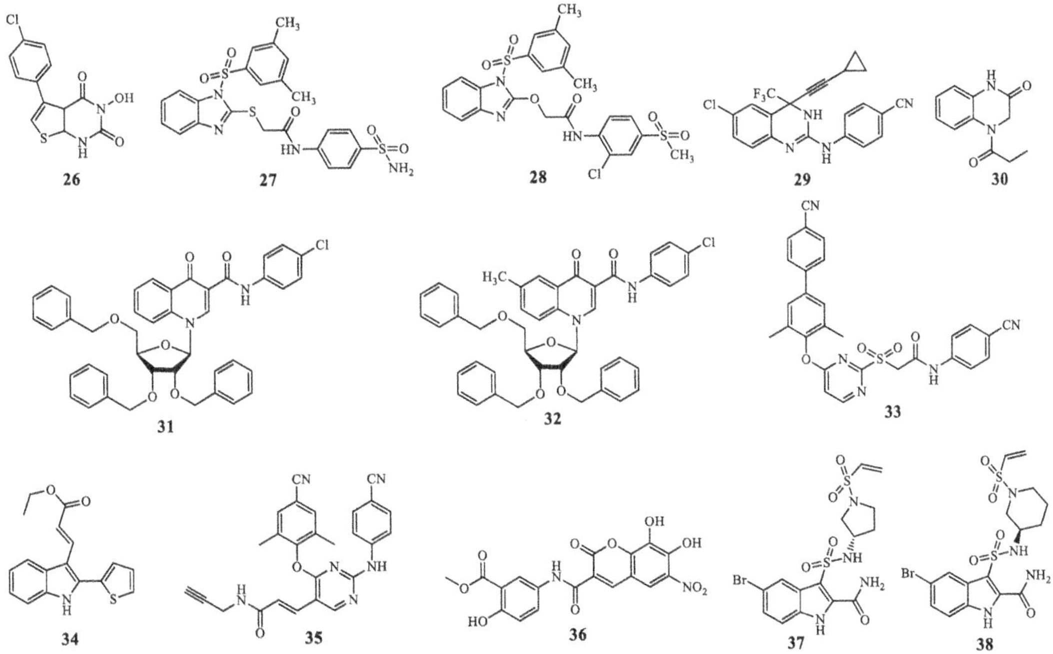

FIGURE 10.6 Chemical structures of anti-transcriptase reverse drugs and the most promising transcriptase reverse inhibitors active against several mutant strains.

The scaffold of N^1-aryl-2-arylthioacetamido-benzimidazoles was explored by Monforte et al. as RT inhibitors [119]. In this context, compounds 27 and 28 (Fig. 10.6) were the most effective against RT, with IC_{50} values of 0.05 ± 0.0005 and $0.76 \pm 0.19 \mu M$, respectively. Also, these presented good results in the cell-based infected assay (EC_{50}: 0.024 ± 0.01 and $0.12 \pm 0.08 \mu M$, respectively) without toxicity. In addition, compound 27 showed the best results against resistant strains (L100I, K103N, Y181C, Y188L, E138K, F227L/V106A, and K103N/Y181C). Molecular modeling reveals that its group 3,5-dimethylbenzyl interacts with Tyr^{181} and binds at the region localized between Tyr^{181}, Tyr^{188}, and Trp^{229}, in addition to a H-bonding interaction with Lys^{101}, justifying the activity against resistant strains. Thus, the use of these analogs could be interesting to overcome the limitations related to the viral resistance of RT. On other hand, Jin and coworkers explored dihydroquinazolin-2-amine analogs, showing compound 29 (Fig. 10.6) as the most promising against HIV-1IIIB strain (EC_{50}: 0.84 nM) [120]. This compound was similar to efavirenz and etravirine (see Fig. 10.1), inhibiting the WT RT, with an IC_{50} value of $0.01 \mu M$. Furthermore, molecular modeling studies suggested similar binding modes for compound 29 and these other drugs, in which interactions with Glu^{138} and Val^{179} residues can explain the best activity of this compound against the mutation E138K (EC_{50}: 3.5 nM) when compared to efavirenz (EC_{50}: 6 nM) and etravirine (EC_{50}: 4.3 nM).

SBDD strategies were applied by Fabian et al., in building a ligand database followed by a virtual screening based on molecular docking and 3D-QSAR [121]. In sense, quinoxaline derivatives were selected for synthesis and biological evaluation, in which compound 30 (Fig. 10.6) showed the best results in RT inhibition, with an IC_{50} value of $0.63 \mu M$, similar to than nevirapine (see Fig. 10.1) (IC_{50}: $0.13 \mu M$). In addition, compound 30 showed inhibition against K103N mutant strain (47%), also similar to nevirapine (50%), with best results in the cell-based infected assay (EC_{50}: 3.1 nM for compound 30 and 6.7 nM for nevirapine), without significant toxicity, and a great SI value of 31,798.

The antiviral properties of oxoquinoline moiety instigated Forezi et al. to develop new analogs targeting the RT inhibition [122]. Compounds 31 and 32 (Fig. 10.6) were found to be the best RT inhibition, exhibiting IC_{50} values of 1.4 and $1.6 \mu M$, respectively. Furthermore, these compounds showed less toxicity compared to atazanavir. Molecular modeling reveals interactions with Tyr^{318} residue for compound 31; while H-bonding interactions with Lys^{103} and Val^{106} residues for compound 32, suggesting that these interactions are essential to the activity, making them less susceptible to resistance.

By using pharmacophore modeling, Sang et al. developed rilpivirine analogs and screened them against several mutant strains [123]. Thusly, compound 33 (Fig. 10.6) showed high potency against the wild virus (EC_{50}: 6 nM) and also against other mutant strains, such as L100I, K103N, Y181C, Y188L, and E138K (EC_{50} values of 8, 6, 26, 122, and 26 nM, respectively). Docking results showed interactions with key residues (Tyr^{188}, Lys^{223}, and Lys^{103}) at the active site of RT, which are related to the activity against the mutant strains.

2-(Thiophen-2-yl)-1H-indole derivatives were synthesized by El-Hussieny et al., in which compound 34 (Fig. 10.6) was found to be the most effective (IC_{50}: 2.93 nM) than efavirenz (IC_{50}: 6.03 nM) against RT [124]. Molecular docking revealed the H-bonding interactions of indole ring with Lys^{101} and Asn^{103}, and stacking interactions with Leu^{100}, Val^{106}, and Tyr^{318} residues at the binding site of K103N mutant are relevant for its activity.

Diarylpyrimidines showing dual-site binding were explored by Feng et al. to discover new NNRT inhibitors [125]. Thus, compound 35 (Fig. 10.6) was the most potent molecule against several mutant strains, including E138K (EC$_{50}$: 10.6 nM), similar to etravirine (see Fig. 10.1) (EC$_{50}$ 9.8 nM). Furthermore, this compound inhibited the WT RT (IC$_{50}$: 0.0589 μM). Additionally, molecular dynamics simulations showed that compound 35 was stable during the whole simulation time (100 ns), interacting mainly with Tyr^{188} and Tyr^{181}, as well as, H-bonding interactions with Leu^{102}, Lys^{103}, and Lys^{223} residues, related to the best results against the mutant strains.

In another study, coumarin derivatives were evaluated by Kang et al. against RT-associated ribonuclease H [126]. In this way, compound 36 (Fig. 10.6) showed high potency in cell-based infection assay (EC$_{50}$: 3.94 ± 0.22 μM), and great inhibition against several mutant strains. Finally, the enzymatic inhibition assay obtained an IC$_{50}$ value of 12.3 μM, in which molecular docking suggested a binding mode similar to other RNase H inhibitors. Moreover, this target presented a magnesium ion chelating interaction, and H-bonding interactions with Lys^{540} residue, justifying the great results of this compound.

Instigated in overcoming the resistance induced by Y181C mutation, Gao et al. explored a series of indolylarylsulfone derivatives [127]. Thus, the most effective compounds 37 and 38 (Fig. 10.6) showed high activity to Y181C (EC$_{50}$: 2.36 ± 0.91 and 5.32 ± 1.02 μM, respectively) when compared with WT (EC$_{50}$: 9.74 ± 0.34 and 40.7 μM, respectively). Furthermore, interactions with Phe^{227} and Pro^{236} residues at the binding site are related to the best activity of the compounds against this mutant strain.

10.3.3.3 Anti-HIV protease inhibitors

Using a bis-THF structural template, Ghosh et al. synthesized a series of new compounds to interact with the S2 subsite of HIV-1 protease [128]. Compound 39 (Fig. 10.7) was designed based on 6 − 5 − 5 ring-fused in tetrahydropyranofuran at P2, and aminobenzothiazole at P2′ position, and (R)-hydroxyethylsulfonamide isoster. This compound was the most effective, mainly against protease (IC$_{50}$: 0.26 nM, and K$_i$: 0.04 nM) and DRV-resistant HIV-1 variants, with EC$_{50}$ varying between 0.17 to 35 nM. X-ray complex analysis showed H-bonding interactions with Asp^{29} and Asp^{30} residues, which are related to the drug resistance. In another study, Ghosh et al. synthesized compound 40 (Fig. 10.7) with a K$_i$ value of 0.025 nM, and promising antiviral activity, showing an EC$_{50}$ value of 69 nM [129]. Moreover, the X-ray structure of a complex with this compound showed interaction with Asp$^{29′}$, Arg$^{8′}$, Ile$^{50′}$, Val^{32}, Ile^{47}, and Ile^{84} residues. These interactions can be explored in the design of new inhibitors. Once again, Ghosh et al. explored new analogs against HIV-1 protease, obtaining compounds 41 (K$_i$: 0.38 nM and IC$_{50}$: 47 nM) and 42 (K$_i$: 0.06 nM and IC$_{50}$: 232 nM) (Fig. 10.7) [130].

Continuing the efforts to discover new protease inhibitors based on (R)-hydroxyethylsulfonamides, Zhu et al. synthesized compound 43 (Fig. 10.7) as the most promising molecule against the enzyme (K$_i$: 0.029 nM and IC$_{50}$: 0.13 ± 0.01 nM), around sixfold most active than darunavir [131] (see Fig. 10.1). On the other hand, the potency against DRV-resistant mutations and HIV-1NL4−3 was similar to darunavir, and great against HIV-C, showing EC$_{50}$ in the nanomolar range. Molecular modeling revealed that this compound forms H-bonding interaction with Ile^{50}, Ile$^{50′}$, and Asp$^{29′}$ residues and occupies similar

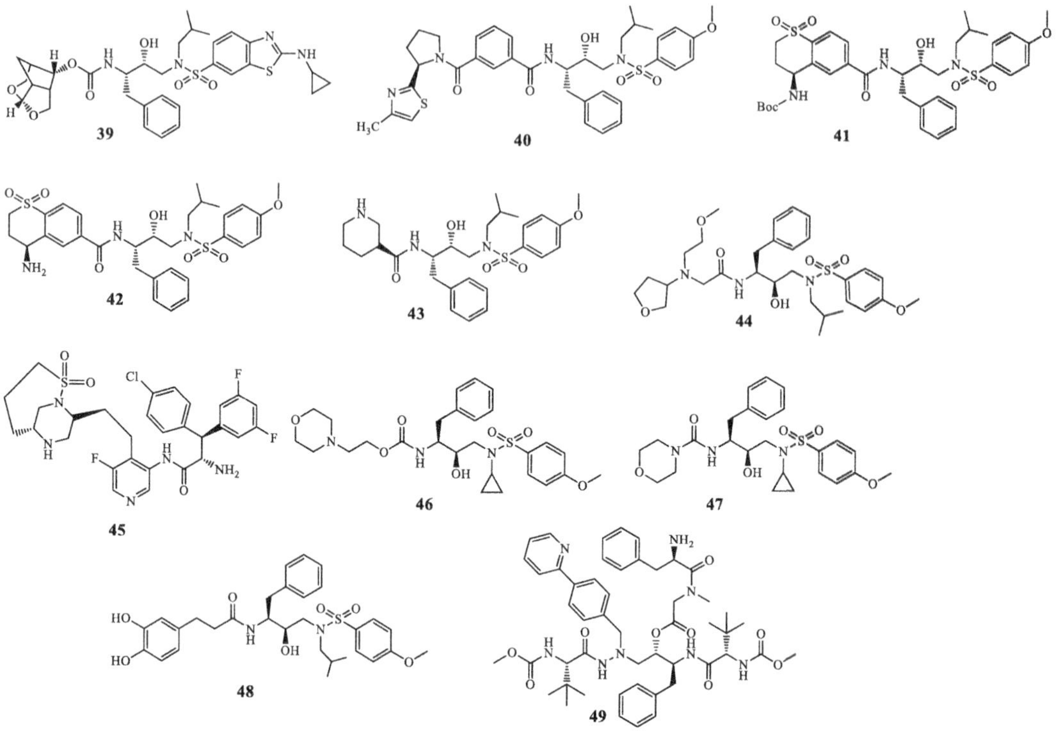

FIGURE 10.7   Chemical structures of promising anti-HIV protease inhibitors.

regions of darunavir. On other hand, Bai and coworkers synthesized an active THF deri-
vate, compound 44 (Fig. 10.7), which presented an IC_{50} value against the enzyme protease
of 0.35 nM and low toxicity [132]. The study highlighted that (S)-THF ring and flexibility
of P2 substituent are essential to the activity.

Encouraged by the previous studies, Bungard et al. synthesized new morpholine-based
piperazine sulfonamides that interact with Asp^{25A} and Asp^{25B}, and Ile^{50A} and Ile^{50B} [133].
Then, their efforts resulted in compound 45 (Fig. 10.7) with great activity in the enzyme
HIV protease (IC_{50}: 12 ± 1 pM) and antiviral (EC_{50}: 2.8 ± 0.4 nM), being most promising
than atazanavir and darunavir. Thus, the design of proteases inhibitors targeting key residues showed to be a great strategy in the development of new drugs. Other morpholine
derivatives were synthesized by Dou et al., in which compounds 46 and 47 (Fig. 10.7)
showed a great activity against protease, exhibiting IC_{50} values of 0.054 ± 0.004 and
0.047 ± 0.004 μM, respectively [134]. The crystal structure of compound 46 in complex with
the protease revealed H-bonding interactions with Ile^{50} and $Ile^{50'}$ residues.

To overcome limitations related to viral resistance, Zhu et al. developed dual inhibitors
of protease and RT based on cinnamic acids [135]. Thus, compound 48 (Fig. 10.7) was
identified as the best molecule against protease (IC_{50}: 0.081 nM), and promising results
against RT (IC_{50}: 2.5 μM). H-bonding interaction with $Asp^{30'}$ residue is mainly related to
the activity of compounds against protease, and molecular dynamics simulations showed

stability of the complex with both enzymes. Finally, their results indicated that this compound could act as a dual inhibitor.

Due to peptide character, protease inhibitors showed the bigger limitation related to solubility, which results in low bioavailability. Then, Subbaiah et al. produced atazanavir-based phosphate prodrugs presenting different spacer groups [136]. A derivative containing a methylene group as spacer presented the best results in drug release. In addition, other chemical groups were used as spacers, such as amino acids, in which the L-Phe ester provided high levels of circulating drug and low indices of its prodrug form. Their efforts resulted in the discovery of compound 49 (Fig. 10.7), an L-Phe-Sar dipeptide analog that showed the best levels of circulating drug when compared with atazanavir. These findings showed that this prodrug approach is powerful to increase the bioavailability and overcome the limitations related to solubility and drug delivery.

10.3.4 Emerging viruses

10.3.4.1 *Hepatitis B virus*

Hepatitis B virus (HBV) has shown high prevalence worldwide, although in the past years its occurrence has decreased due to the development of efficient vaccines. On the other hand, infected people around the world can act as a reservoir for it, which increases the chances to develop hepatocellular carcinoma, chronic liver disease, and cirrhosis. The treatment of patients with chronic HBV involves different classes of cytokines (Interferon) and nucleoside analogs, such as amivudine, emtricitabine, telbivudine, clevudine, adefovir, tenofovir, abacavir, entecavir, and besifovir (Fig. 10.8) [137,138].

10.3.4.1.1 Some limitations and challenges to identifying new anti-hepatitis B virus drugs

The limitation to using interferon remains in its administration pathway, in which its possible only intravenous. Nucleoside analogs, despite being effective to treat the disease

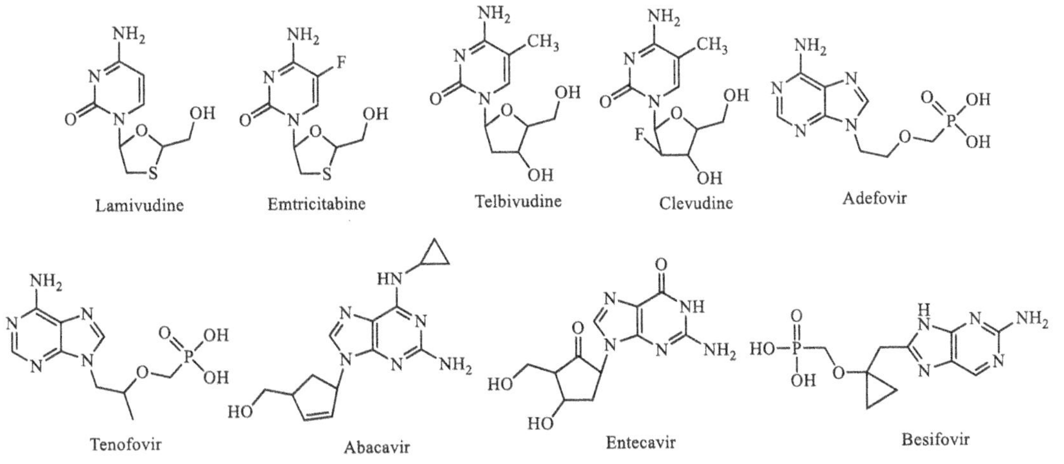

FIGURE 10.8   Nucleoside analogs used in HBV treatment.

and showing oral administration pathway, suffers from the same problem related to other viral diseases, mainly viral resistance. These drugs act inhibiting the viral RT, although its use is limited by enzyme mutation. Another limitation is that their use for long periods to observe their initial effects can increase the probability of resistance developing. In this context, to overcome these limitations, understanding the viral resistance mechanisms is essential to designing new drugs [137]. In fact, the discovery of new effective drug targets can aid during such difficult tasks [139,140]. In addition, combining drugs that act against different targets could be an alternative to discover new alternative treatments [141,142].

10.3.4.1.2 Promising compounds targeting anti-hepatitis B virus activity

Jia et al. used bioisosterism and hybrid pharmacophore-based as strategies to discover new anti-HBV agents [143]. Their efforts resulted in the discovery of compound 50 (Fig. 10.9), showing great activity against HBsAg (IC$_{50}$: 24.33 μM) and HBeAg (IC$_{50}$: 2.22 μM) proteins. The preliminary SAR study showed that thiazol-2-yl group is essential to the activity. Continuing their efforts, in another study Jia et al. synthesized compound 51 (Fig. 10.9), which displayed a promising activity against replication of HBV DNA (IC$_{50}$: of 3.4 μM), showing a new scaffold that could be further explored [144]. Still, Jia et al. use a similar approach to that previously cited [145]. Thus, compound 52 (Fig. 10.9) was synthesized and presented a promising anti-HBV DNA replication activity, being active against the HBV capsid protein with IC$_{50}$ and K$_D$ values of 2.2 ± 1.1 and 60.0 μM. Molecular docking simulations showed that interactions with Thr^{33}, Thr^{128}, Pro^{130}, and Pro^{129} are mainly related to the anti-HBV activity.

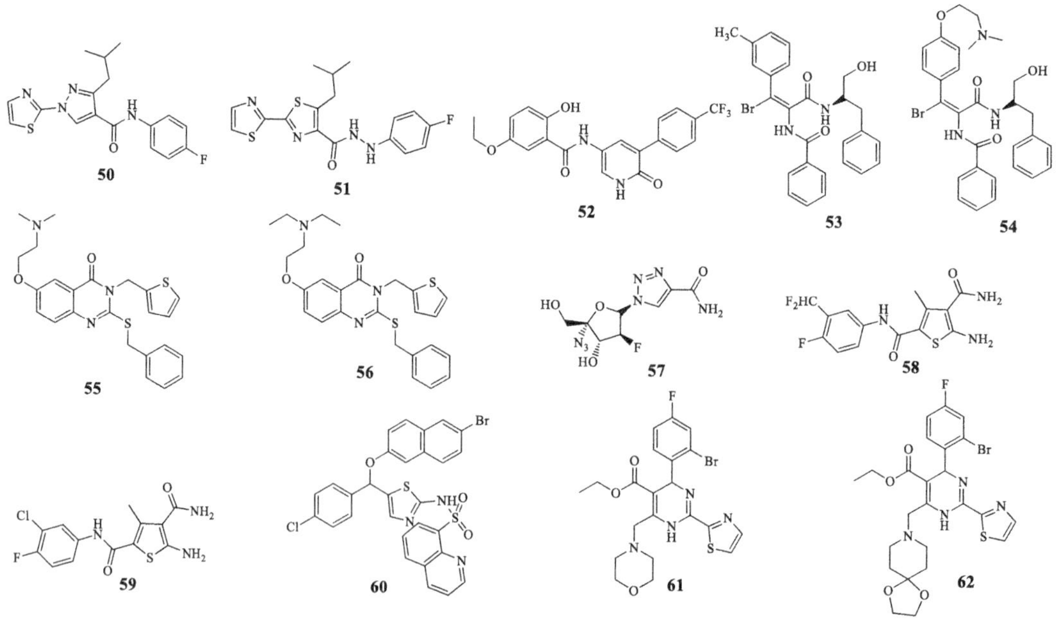

FIGURE 10.9   Chemical structures of the most promising compounds with anti-HBC activity.

A series of phenyl propionamide analogs were synthesized by Qiu et al., using a pharmacophore fusion approach [146]. This strategy was able to identify compounds 53 and 54 (Fig. 10.9) with the most excellent activity against HBV, showing great IC_{50} values against HBV DNA replication, being 0.46 and 0.14 μM, respectively. These compounds were promising antiviral activity against lamivudine- and entecavir-resistant mutant strains, displaying IC_{50} values of 0.77 and 0.32 μM, respectively. A mechanism of action study indicated that compound 53 could inhibit pgRNA expression, as well as, of the RT, been a compound with multiple mechanisms of action, and can be evaluated in posteriors assays as HBV drug. Qiu et al. explored quinazolinones for their anti-HBV activity [147]. Compounds 55 and 56 (Fig. 10.9) exhibited good results for inhibition of HBV DNA replication, with IC_{50} values of 1.54 and 0.71 μM, respectively. In addition, assays against lamivudine- and entecavir-resistant strains showed IC_{50} values of 1.90, and 0.84 μM, respectively. Thus, this study provides a new chemical class that could overcome the resistance of the actual drugs against HBV.

The study of Liu et al. explored 2′-deoxy-2′-β-fluoro-4′-azido-β-D-arabinofuranosyl 1,2,3-triazole nucleoside analogs to overcome the antiviral resistance of HBV [148]. They identified compound **57** (Fig. 10.9), with low cytotoxicity and great anti-HBV activity (85.7% inhibition). In addition, this compound showed activity against lamivudine-resistant strain (64.1% inhibition). Molecular docking and dynamics simulations predicted the interaction of its azide group with Va^{184}, Phe^{88}, Leu^{180}, and Met^{204} residues, essential to its activity. On the other hand, Tang et al. evaluated 5-amino-3-methylthiophene-2,4-dicarboxamide analogs, showing that compounds 58 and 59 (Fig. 10.9) were the most promising toward the HBV core protein assay, exhibiting EC_{50} values of 0.11 and 0.31 μM, respectively [149]. Molecular modeling revealed that the interaction of thiophene with Phe^{23}, Tyr^{118}, and Phe^{122} is related to the best activity of these compounds. Finally, ADME experiments proposed that these compounds showed a great oral bioavailability profile.

To discover a new capsid inhibitor from HBV, Pan and coworkers explored a series of aminothiazole-based [150]. In this way, compound 60 (Fig. 10.9) showed the best activity in anti-HBV replication assays (IC_{50}: 0.18 μM) and drug-likeness properties. Additionally, it interacts with Thr^{128} and Tyr^{118}. In addition, this compound showed a high affinity upon WT HBV capsid protein due to its H-bonding interaction with Ser^{121}, although presented a low affinity to S121A mutant.

To overcome the limitation related to the low solubility of compound 61 (Fig. 10.9), recently screened in phase II clinical trials, Ma et al. developed new analogs that improve its hydrophilicity and improvement of water solubility [151]. A morpholine group was replaced with spiro rings, in which compound 62 (Fig. 10.9) was the most promising in anti-HBV assays with low cytotoxicity (EC_{50}: 0.20 $\pm$ 0.0 μM, and CC_{50} > 87.03 μM), being more potent than lamivudine (EC_{50}: 0.37 $\pm$ 0.04 μM and CC_{50} > 100 μM). Finally, molecular docking at the capsid protein revealed that the spiro group may have the function of interacting as a hydrophobic tail with the solvent, where replacement of this ring with appropriate groups may improve aqueous solubility.

10.3.4.2 Dengue virus

Dengue virus (DENV) is a Flavivirus (Flaviviridae), transmitted by mosquitoes from the *Aedes* genus, responsible for several cases and deaths worldwide. The virus has four serotypes, DEN1−4, generating symptoms similar to flu, or even hemorrhagic dengue, in with its progression leads to dengue shock syndrome and death. This disease was endemic in nine countries before 1970, and today is an endemic disease in approximately 100 countries, producing several damages to public health [152,153]. Despite these facts, there are no approved drugs to treat this disease, where its treatment is based on the control of symptoms with palliative drugs and fluid therapy [154]. However, there are several studies and analogs presented in the literature that can be useful and explored in clinical trials [155]. Drug design targeting anti-DENV drugs advanced and various researchers have discovered very promising analogs [156]. However, several challenges can be overcome for identifying an innovative drug against DENV [156−158].

10.3.4.2.1 Promising compounds discovered against dengue virus

The first limitation in this field is related to clinical trials because the viremia is reduced very rapidly, and patients selected for essays must be in the phase of the infection, aiming to initiate the treatment as early as possible. However, there is a lack of diagnostic effectiveness of infection caused by DENV, which can be different from other diseases. There are impacts in the measurement of secondary clinical outcomes, making it difficult to assess the drug in the progression of the disease. In addition, the patients show a great varying of viral titers, which difficult for the selection and evaluation of a new efficient drug. Many dengue-infected patients are children under 15 years old, and adult patients infected with DENV often have other comorbidities. The physiology and immunological status of the disease in different patients influence the pharmacokinetics of the drug and make it difficult to evaluate the candidate from a safety point of view, requiring careful monitoring of the effective concentrations of the drug and evaluation of safety. Other challenges to overcome the preclinical assay, in which using mice is not acceptable to total events related to the disease, because it is not known the additional tissues or cells related to the viral pathogens and primate models are necessary for this. Also, the preclinical assay must consider the various strains of DENV1−4 to result in the best result in candidate selection. Finally, toxicity in several tissues, like the kidney, brain, and liver is responsible for clinical failures, and the use of in vitro assay by new techniques, such as 3D-organoids of the liver, cellular thermal shift assays, and cardiac patch microtissues should be considered in initial phases of drug design and development [159].

10.3.4.2.2 Promising active compounds against dengue virus

To discover new drugs against DENV, Weng et al. explored pyrrolidine and imidazolidinone as inhibitors targeting DENV-2 NS2B-NS3 protease [160]. Then, analogs 63 and 64 (Fig. 10.10) showed the great activity against the protease (IC_{50}: $1.2 \pm 0.4\ \mu M$, for both compounds) and WT DENV-2 (EC_{50}: 38.7 ± 5.4 and $39.4 \pm 6.2\ \mu M$, respectively) without cytotoxicity. The authors highlighted that the *p*-nitro substituent decreases the activity in presence of methylene at the side-chain, indicating that the suppression of methylene is an essential requisite for molecular recognition. Molecular docking showed similar binding modes for both compounds and interactions with a catalytic triad (His^{51}, Asp^{75}, and Ser^{135}). In another study, a high-throughput screening (HTS) was employed by Benmansour and coworkers to discover new non-nucleoside analogs to inhibit DENV RdRp [161]. The authors explored oxadiazole and thiophene moieties, resulting in the identification of compounds 65, 66, 67,

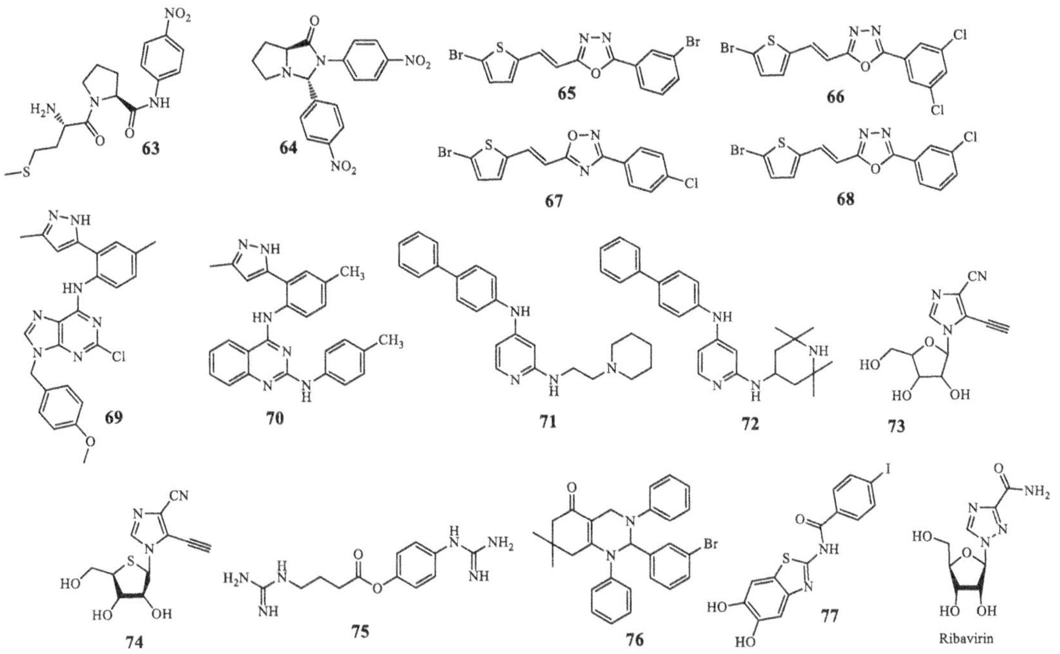

FIGURE 10.10 Chemical structures of main analogs identified against different DENV serotypes.

and 68 (Fig. 10.10) as the most promising against RdRp, exhibiting IC_{50} values of 4.0 ± 0.5, 11.0 ± 1.8, 2.2 ± 0.1, and $9.1 \pm 1.3 \mu M$, respectively. Finally, these compounds showed great activity against all four DENV serotypes (EC_{50} values ranging from 2 to 9.8 μM).

Venkatesham et al. explored 2,6-diaminopurine or 2,4-diaminoquinazoline analogs, which resulted in the identification of purine 69 and quinazoline 70 (Fig. 10.10) compounds as the most promising inhibitors [162]. The derivatives showed great EC_{50} values (1.9 and 2.6 μM, respectively). In addition, molecular docking simulations revealed that the binding pose for compound 69 is different from compound 70, which may be explain their different biological results.

Targeting the DENV envelope protein (E), Leal et al. perform a de novo drug design study to discover new drugs against DENV [163]. In this way, approximately 240,000 compounds were designed and screened, in which 2,4-pyrimidine analogs were synthesized, obtaining analogs 71 and 72 (Fig. 10.10) as the most promising molecules. Compounds showed the most promising EC_{50} value ($0.8 \pm 0.2 \mu M$, for both compounds), without cytotoxicity. In addition, these compounds were active against all four DENV serotypes. In vitro pharmacokinetics assays revealed the great solubility of these compounds and high stability.

Another chemical scaffold explored was imidazole nucleoside analogs, Okano et al. in a screening of 150 compounds [164]. Thus, compound 73 (Fig. 10.10) showed the most promising potential, with an IC_{50} value of 0.55 μM. Instigated to improve the activity and decrease the toxicity, new analogs were obtained, in which compound 74 (Fig. 10.10) showed the best activity without cytotoxicity (IC_{50}: 6.57 μM, and $CC_{50} > 50 \mu M$), better than control, ribavirin (Fig. 10.10) (IC_{50}: 6.66 μM and CC_{50}: 48.98 μM), conducting to new compounds that can be further optimized.

Based on previous studies which show 4-guanidinobenzoate moiety as a promising DENV inhibitor, Sundermann et al. explored this chemical class in antiviral assays against flaviviruses [165]. The main findings of the authors were that substitutions of the guanidine group with aromatic or ester are detrimental to their activities. 4-Guanidinobenzoate is essential to the activity, and compound 75 (Fig. 10.10) was the most effective (IC_{50}: 4.2 μM) against DENV. In addition, experiments suggest that this compound performed reversible covalent adducts. Furthermore, Wan et al. explored a series of octahydroquinazoline-5-ones and evaluated them against DENV-2 [166]. Thus, compound 76 (Fig. 10.10) was the most potent (EC_{50}: 1.31 $\pm$ 0.21 μM) without cytotoxicity (CC_{50} value > 20 μM). In addition, experimental assays with this compound suggest inhibition of the expression of E protein, reverting DENV-2 CPE-induced, without affecting MTase and RdRp activities.

Benzo[*d*]thiazole analogs were explored by Maus et al. such as allosteric NS2B/NS3 inhibitors [167]. The screening of these compounds revealed that analog 77 (Fig. 10.10) is the most promising molecule against DENV-2, with an IC_{50} value of 4.38 μM. Molecular docking simulations identified several interactions of the catechol group with Asn^{152}, Lys^{74}, and Gly^{148} residues, and interaction of benzothiazole with Leu^{76}, Gly^{148}, and Ile^{165} at the NS2B/NS3 allosteric site.

10.3.5 Hemorrhagic fever viruses

10.3.5.1 Ebola virus

The infection of Ebola virus (EBOV) is one the most threatening virus to humanity, with a mortality rate higher than 90%, occurring mainly in the African continent, especially in sub-Saharan Africa [168,169]. This pathogen is related to several outbreaks in the past few years, occurring in Guinea, Liberia, and Sierra Leone between the years of 2014 and 2016, leading to 28,616 cases and approximately 11,310 deaths [170–172]. Most recently, this disease was responsible for an outbreak in the Democratic Republic of Congo, reaching around 28% of children and adolescents, with 80% lethality, showing the high necessity of a new effective drug that can fight against this disease [170,173]. Despite its high lethality rate and its threatening potential, there is need for new effective drugs and diagnostic tests for this disease, and several challenges must be overcome for this. One of the main limitations is the classification of biosafety level 4 (BSL-4), and its manipulation must be in laboratories of the high-level expertise for biocontainment [174,175]. Due to this fact, for many years, its treatment involves non-pharmacological measures, such as maintaining electrolytes, blood pressure, oxygenation, and treating other infections classified by opportunistic [176–178]. The urgent need for new therapies encourages researchers and pharmaceutical companies around the world to discover new therapies. Their efforts resulted in the discovery of two monoclonal antibodies (mAbs), Ebanga and Inmazeb, approved for EBOV treatment in late 2020. In addition, the European Medicines Agency authorized the use of vaccines containing Zabdeno and Mvabea [168]. Furthermore, small molecules such as remdesivir (see Fig. 10.5) are under clinical trials against this agent [179–181]. In fact, there is high need for new effective treatments, as well as for several challenges to be overcome [180,181].

10.3.5.1.1 Some limitations to overcome in drug discovery targeting Ebola virus

In addition to be a BSL-4 pathogen, there are ethical conflicts related to the clinical trials involving the discovery of new drugs against EBOV, because placebo can never be used by humans due to the imminent threat of death of the patient, which implicates in clinical trials of low quality and unreliable [182]. In this way, the patients must be adequately informed about the risk of the experimental treatment, with data transparency and respecting freedom of choice, preserving their privacy [182,183]. Furthermore, the used animal infection models for EBOV in nonprimates, in many cases, do not reproduce the same results in other species. In addition, in vitro assays using human stomatitis virus infection yet need better validation. This finding confirms the need and importance of the development of new assays more robust and efficient, which could reproduce real outcomes [184]. Related to the drug design, the main failures concerning chemical factors associated with compounds, such as the inappropriate drug-likeness properties, or high toxicity in animal models, mechanisms of action were not identified. Among the difficult aspects involving the workflow with a BSL-4 virus, the manipulations of these types of viruses by pharmacologists represent one of the most significant challenges, reducing, even more, the chances to discover a new drug anti-EBOV [185–187].

10.3.5.1.2 Promising compounds against Ebola virus

To identify new drugs against EBOV, Capuzzi et al. performed a QSAR-based virtual screening in a dataset of FDA-approved drugs [188]. From 17 million compounds virtually screened, 102 molecules were selected for biological assays, in which 14 compounds showed the most promising results. During the experimental investigation, vindesine and BIX-01294 (Fig. 10.11) showed the most promising inhibition in the anti-EBOV assay (IC$_{50}$: 0.34 and 0.966 μM, respectively), without cytotoxicity. Mechanism of action studies reveals that these compounds can interact with NPC1 protein, cathepsin B/L, and lysosomal function.

Based on previous studies that showed compound 78 (Fig. 10.11), Liu and coworkers used this molecule as a starting point in an optimization study, targeting fewer hydrophobics analogs [189]. A series of analogs were synthesized and screened, in which compounds 79 and 80 (Fig. 10.11) were identified as the most potent in assay VSV (vesicular stomatitis virus) EBOV infection of Vero cells (IC$_{50}$: 19 and 21 nM, respectively). Besides, the coadministration of these compounds with ritonavir (CYP3A4 inhibitor) in mice showed an increase in its plasmatic concentrations, suggesting that these compounds can have oral administration. In another study, 4-(aminomethyl) benzamides were explored by Gaisina et al. against the filoviruses EBOV and Marburg virus (MARV) [190]. Their efforts resulted in the identification of compounds 81 and 82 (Fig. 10.11) with high anti-EBOV and anti-MARV activities, displaying EC$_{50}$ values of 0.11 and 0.31 μM; 1.25 and 0.82 μM for both of these compounds, respectively. In addition, the great ADME properties of these analogs suggested their high oral viability.

Adamantane carboxamides were investigated by Plewe et al. in a VSV/EBOV assay [191]. Then, compound 83 (Fig. 10.11) was found to be the most promising candidate,

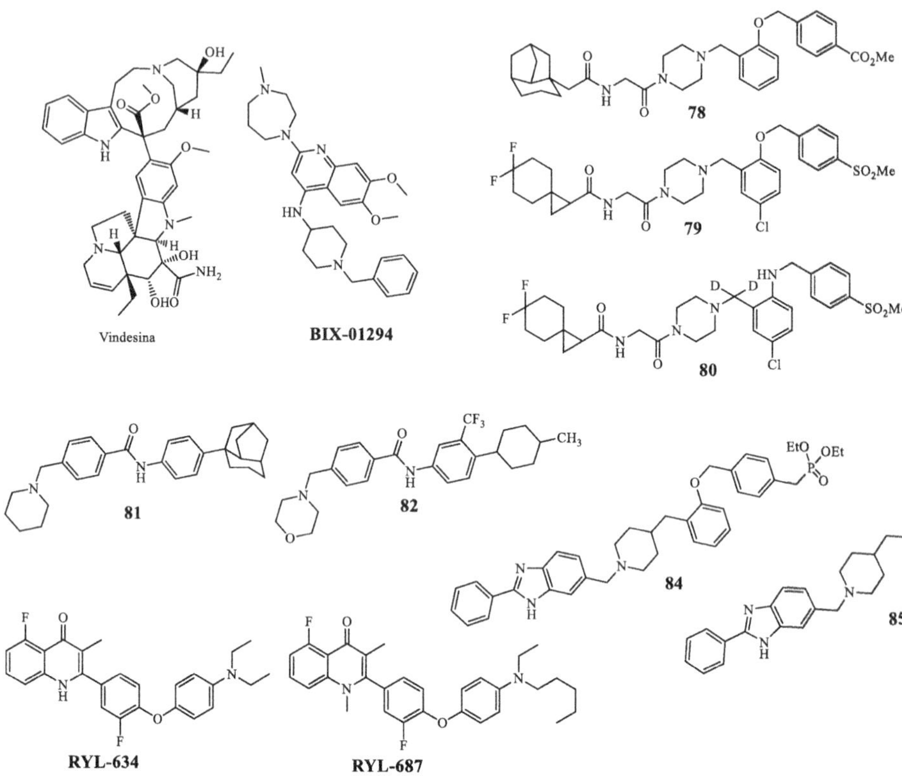

FIGURE 10.11   Chemical structures of main EBOV inhibitors presented in the last years.

showing a pEBOV EC_{50} value of 0.014 μM. In addition, the crystal structure of the compound in complex with GP protein showed hydrophobic interactions with residues Val^{66}, Leu^{184}, Leu^{186}, Leu^{515}, and Tyr^{517}, while H-bonding interaction was observed with Arg^{64}. These finds can facilitate the design of new adamantane-based inhibitors. On the other hand, Bessières et al. performed a study involving 2-substituted-6-[(4-substituted-1-piperidyl)methyl]-1H-benzimidazoles to design new EBOV inhibitors [192]. Compounds 84 and 85 (Fig. 10.11) presented the best activity in anti-EBOV assay, exhibiting EC_{50} values of 0.93 and 0.64 μM, respectively. Also, the mechanism of action assay reveals the inhibition of viral entry by NPC1 as mainly related. The molecular docking showed interactions with Gln^{421}, Tyr^{423}, Phe^{504}, and Phe^{503} residues, which are related to binging of NPC1 to GP protein. Finally, in silico ADME properties prediction revealed that these compounds have great drug-likeness properties.

Instigated in the potential of RYL-634 (Fig. 10.11), a quinolone derivative, as a DHODH inhibitor and useful against RNA viruses, Gong et al. performed optimizations in its structure aiming the anti-EBOV activity. In this way, some analogs were promising against in vitro EBOV infection, especially RYL-687 (Fig. 10.11), which showed an EC_{50} value of 7.469 nM, being more promising than remdesivir (EC_{50}: 40. 77 nM). The authors highlighted those preclinical assays are needed to confirm its efficacy.

10.3.5.2 *Lassa virus*

Lassa virus (LASV) is an RNA virus of a segmented negative-strand, belonging to Arenaviridae family [193]. LASV is related to the Lassa fever, in which this clinical manifestation generates symptoms such as slight fever or hemorrhagic fever that can be conducted to multiorgan failure and death in most cases [194]. Their natural reservoir is multimammate rats, *Mastomys natalensis*, and these infected rodents have been mainly reported in West African, in countries like Guinea, Nigeria, and Sierra Leone [194,195]. Its transmission is through ingestion of food contaminated by feces and urine of infected rodents, or even through inhalation of excreta. Among humans, this infection occurs by nosocomial pathways [195]. There is an estimation that this disease affects around 150,000 to 300,000 people per year, causing 5000 deaths [196,197]. The first case of LASV infection was identified in 1969 (Nigeria) by American nurses, which led later to identification in other countries in West Africa [197,198]. For many years, the treatment of this disease involved support therapy and broad-spectrum antivirals, such as ribavirin (see Fig. 10.10). Despite this, this nucleoside analog is not effective, justifying the necessity of discovering new drugs against this disease. Most recently, favipiravir (Fig. 10.12) is being tested in combination with ribavirin. In addition, several vaccines are in preclinical trials, highlighting the efforts to discover new drugs against this virus [199]. However, several challenges must be overcome, similar to other viruses cited in this chapter, such preclinical assays are most effective in stimulating the events associated with the infection. In addition, the

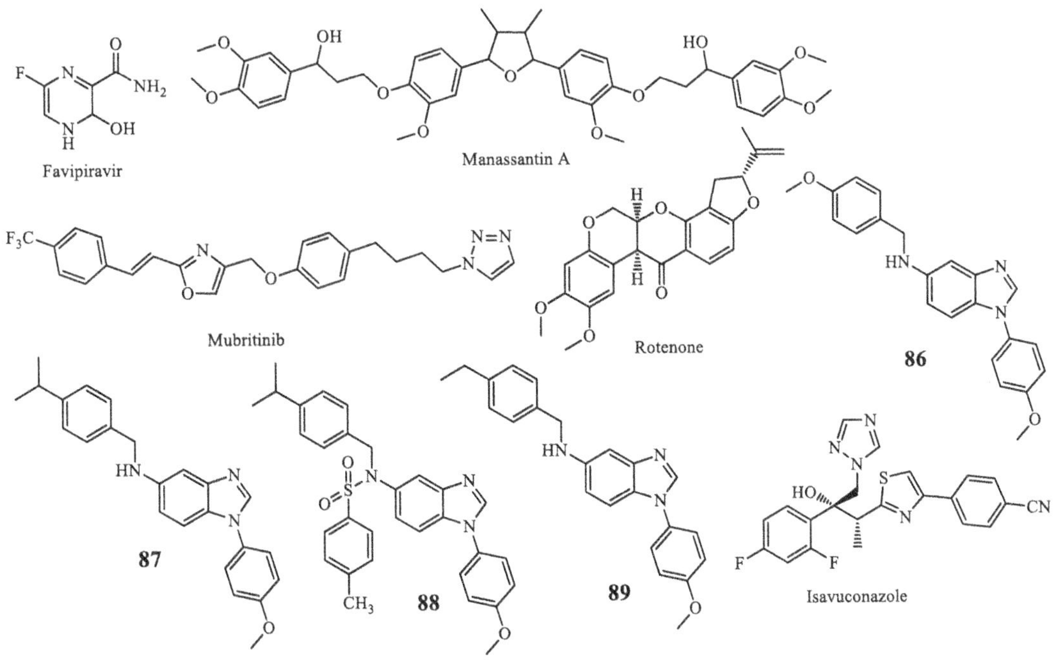

FIGURE 10.12 Chemical compounds as LASV inhibitors identified in the last years.

efficacy of ribavirin is constantly questioned, and new analogs are necessary to provide a new horizon for researchers to discover a cure against this viral disease [174,200].

10.3.5.2.1 Promising compounds against Lassa virus

Encouraged by the development of new in vitro assays against this virus, Cubitt et al. developed an assay on human cells that can express functions vRNPs of LASV, following to drug repurposing [201]. The authors highlighted that the expression LASV vRNP on human cells was able to overcome the limitations related to BSL-4 facilities and their methods could be used in the development of new chemical entities. Using their screening protocols, 16 repurposed drugs were identified, in which manassantin A, mubritinib, and rotenone (Fig. 10.12) showed the best results (EC$_{50}$ values of 0.02 μM for all of them), demonstrating the relevance of a new assay protocol, conducing to a new horizon in drug design against LASV.

To identify new chemical scaffolds useful against LASV, Dai et al. performed a HTS in a database containing 400,000 compounds in antiviral assays using pseudotyped-envelope GP gene incorporated of the LASV [202]. These efforts resulted in the identification of benzimidazole scaffold, and compound 86 (EC$_{50}$: 11 nM) (Fig. 10.12) as the most promising molecule. Thus, it was optimized, conducting to compounds 87, 88, and 89 (Fig. 10.12), with EC$_{50}$ values of 1.4, 11, and 110 nM, respectively. These compounds were active against other arenaviruses, such as MACV (Machupo virus), Junin (JUNV), Sabiá, and Guanarito.

New chemicals agents were identified by Zhang et al., which evaluated 56 broad-spectrum antifungals against LASV, using HIV-based pseudovirus platform [203]. Thus, isavuconazole (Fig. 10.12), a tetrazole that can inhibit the virus entry with an EC$_{50}$ value of 1.2 μM, was identified in their study. The mechanism of action of this compound involves the prevention of pH-mediated membrane fusion at the LASV GP. In addition, mutations in the residues Ser^{27} in the transmembrane region of N-terminal, and transmembrane domain of GP2 at residues Val^{431}, Phe^{434}, and Val^{435} led to decrease in isavuconazole activity.

10.4 New trends, challenges, and opportunities

As shown in this chapter, the design and discovery of new antivirals need to overcome several challenges currently. In fact, the development of new vaccines against some diseases, such as SARS-CoV-2, HIV, EBOV, LASV, DENV, and others, has unquestionably produced benefits since these could prevent and block viral transmission. Thus, these represent a great opportunity to be explored against these diseases. However, the development of vaccines needs to overcome the challenges related to genetic variations of the virus, which implies multiple resistance mechanisms to avoid the host's immune response. Furthermore, the lack of animal models to assess the effectiveness of vaccines remains a major challenge since current models are insufficient. Furthermore, the understanding concerning immunological events involved in viral infections may generate new opportunities for discovering new therapies related to vaccine development. Thus, vaccine development requires the integration of research in diverse fields to discover a revolutionary treatment [204−206].

An approach that has been growing in the discovery of new antiviral therapies is the development of a plant-based vaccine [207]. For more than 30 years, plants are used in folk culture, in which secondary metabolites as produced to be used as antibodies and antigens using genetic engineering [208,209]. The great advantage to using plant-based vaccines is the oral administration and their production on large scale, conducting to low financial cost [209]. This approach is constantly used against HBV, INFV, HIV, SARS-CoV-2, and ZIKV [208,210]. To find a cure that can finish the SARS-CoV-2 pandemic, the efforts to discover new vaccines are seeking plant-based approach, highlighting Medicago, Kentucky BioProcessing, and iBio, that present vaccines plant-based in clinical trials, showing that this approach can be promising in the design of new antivirals [211].

The development of new phenotypic screening assays is a great opportunity in antivirals drug design. This is notable that the targeted phenotypic screens are considered a great advance and a relevant means of discovering new drugs, being necessary for the development of new assays [212]. In addition, the innovative chemical class is necessary, leading to scaffolds that can be used as starting points in several optimizations, generating new most promising compounds. In this way, X-ray crystallography has provided several opportunities in the design of new antivirals, allowing the utilization of approaches such as fragment-based drug design or other virtual screening methods [212]. On the other hand, to increase the effectiveness of biological response and decrease the probability of mutations resistance, the combination of therapies can offer new opportunities in the discovery of new antivirals. Studies indicate that combined therapies decrease the appearance of mutations, and drugs can act by different targets, possibility new opportunities in antiviral development [213]. Thus, the discovery of a new effective antiviral must overcome some requisites, such as: (1) overcome the viral resistance; (2) acting against the virus incorporated into the DNA from the human genome; (3) drugs must act against different viruses, being broad-spectrum agents; (4) low side effects due to combination of drugs; and (5) without toxicity and low cost [214].

The use of human mAbs has been a tendency in the drug design and development against viral infections. The antibody variable gene sequencing techniques and B-cell isolation have contributed to the identification of several compounds useful against a range of pathogens, including viral diseases, such as HIV, INFV, EBOV, ZIKV, SARS-CoV-2, and others. Approximately 25 mAbs are been used and under evaluation in clinical trials against viral diseases, demonstrating the importance of this approach in antiviral drug design [215,216]. Antibody production is the body's natural reaction to block viral infection in host cells or through the production of viral antigens on host cells, in which effector cells are responsible for eliminating infected target cells by interacting with antibodies. In this way, mAbs have high target specificity, neutralize the virus by preventing its interaction with the host cell by direct binding to the virus, or by interacting with host cell in cell surface receptors, necessary for fusion with the membrane viral. Viral shedding can also occur through the presentation of infected cells to effector cells of the immune system. Thus, the design of antibodies represents a great opportunity in the discovery of new, more effective antivirals, being a field that should be better explored in upcoming years. This is justified by the discovery of mAbs palivizumab and ibalizumab approved for the treatment of various viral infections. Thus, using this type of strategy can minimize side effects and produce more selective drugs, also reducing the incidence of viral drug resistance, which is one of the biggest problems faced in the development of new antiviral therapies [217].

10.5 Conclusions

The discovery of new antiviral drugs is the need today, evidenced by the pandemic caused by SARS-CoV-2, as well as outbreaks of other viral agents in various parts of the world, such as the EBOV in Africa, and DENV and ZIKV in the Americas and Asia. Thus, the application of new approaches in drug and vaccine design are important in medicinal chemistry to discover efficient, new therapies against these threatening pathogens.

In fact, the greatest difficulties encountered in discovering new antivirals are mainly related to the high capacity of these pathogens to rapidly develop drug resistance. Furthermore, low bioavailability, difficulties in finding molecules to perform more efficiently in vitro and in vivo assays are common difficulties for all viruses presented in this manuscript. Overcoming these limitations is the first step forward to the efficient identification of new antivirals that could generate possible promising treatments. Further, investing in new approaches such as plant-based vaccine development, or even mAbs could be the next generation in antiviral research and development against emerging viruses.

Finally, we hope that this chapter can inspire medicinal chemists around the world and which the knowledge compiled here can be used in their researches. Thus, enabling the discovery of new antiviral drugs that can be new hopes against these concerning diseases, which have caused serious damages to the world population.

Acknowledgments

The authors thank Coordenação de Aperfeiçoamento Pessoal de Nível Superior (CAPES), National Council for Scientific and Technological Development (CNPq), Fundação de Amparo à Pesquisa do Estado de Alagoas (FAPEAL), and Financier of Studies and Projects (FINEP).

Conflict of interest

The authors confirm that there are no conflicts of interest.

Consent for publication

Not Applicable.

References

[1] Kates JR, McAuslan BR. Poxvirus DNA-dependent RNA polymerase. Proc Natl Acad Sci 1967;58:134–41. Available from: https://doi.org/10.1073/pnas.58.1.134.

[2] Schaeffer HJ, Beauchamp L, de Miranda P, Elion GB, Bauer DJ, Collins P. 9-(2-Hydroxyethoxymethyl)guanine activity against viruses of the herpes group. Nature 1978;272:583–5. Available from: https://doi.org/10.1038/272583a0.

[3] CDC. Pneumocystis pneumonia. MMWR Morb Mortal Wkly Rep 1981;250–2. https://www.cdc.gov/mmwr/preview/mmwrhtml/june_5.htm [accessed 17.10.21].

[4] Zhang X. Challenges and opportunities in the development of therapeutics for viral infectious diseases in the 21st century. Virol Mycol 2012;1:1000e101. Available from: https://doi.org/10.4172/2161-0517.1000e101.

[5] De Clercq E. HIV resistance to reverse transcriptase inhibitors. Biochem Pharmacol 1994;47:155−69. Available from: https://doi.org/10.1016/0006-2952(94)90001-9.

[6] Moore C, Galiano M, Lackenby A, Abdelrahman T, Barnes R, Evans MR, et al. Evidence of person-to-person transmission of oseltamivir-resistant pandemic influenza A(H1N1) 2009 virus in a hematology unit. J Infect Dis 2011;203:18−24. Available from: https://doi.org/10.1093/infdis/jiq007.

[7] Yarchoan R, Mitsuya H, Broder S. AIDS therapies. Sci Am 1988;259:110−19. Available from: https://doi.org/10.1038/scientificamerican1088-110.

[8] Lou Z, Sun Y, Rao Z. Current progress in antiviral strategies. Trends Pharmacol Sci 2014;35:86−102. Available from: https://doi.org/10.1016/j.tips.2013.11.006.

[9] Littler E, Oberg B. Achievements and challenges in antiviral drug discovery. Antivir Chem Chemother 2005;16:155−68. Available from: https://doi.org/10.1177/095632020501600302.

[10] Mao Y, Wang L, Gu C, Herschhorn A, Desormeaux A, Finzi A, et al. Molecular architecture of the uncleaved HIV-1 envelope glycoprotein trimer. Proc Natl Acad Sci 2013;110:12438−43. Available from: https://doi.org/10.1073/pnas.1307382110.

[11] Ison MG. Antivirals and resistance: influenza virus. Curr Opin Virol 2011;1:563−73. Available from: https://doi.org/10.1016/j.coviro.2011.09.002.

[12] Yamashita M. Laninamivir and its prodrug, CS-8958: long-acting neuraminidase inhibitors for the treatment of influenza. Antivir Chem Chemother 2010;21:71−84. Available from: https://doi.org/10.3851/IMP1688.

[13] Hernandez JE, Adiga R, Armstrong R, Bazan J, Bonilla H, Bradley J, et al. Clinical experience in adults and children treated with intravenous peramivir for 2009 influenza A (H1N1) under an emergency IND program in the United States. Clin Infect Dis 2011;52:695−706. Available from: https://doi.org/10.1093/cid/cir001.

[14] Porta C, Kotecha A, Burman A, Jackson T, Ren J, Loureiro S, et al. Rational engineering of recombinant picornavirus capsids to produce safe, protective vaccine antigen. PLoS Pathog 2013;9:e1003255. Available from: https://doi.org/10.1371/journal.ppat.1003255.

[15] Wild CT, Shugars DC, Greenwell TK, McDanal CB, Matthews TJ. Peptides corresponding to a predictive alpha-helical domain of human immunodeficiency virus type 1 gp41 are potent inhibitors of virus infection. Proc Natl Acad Sci 1994;91:9770−4. Available from: https://doi.org/10.1073/pnas.91.21.9770.

[16] Kilby JM, Hopkins S, Venetta TM, DiMassimo B, Cloud GA, Lee JY, et al. Potent suppression of HIV-1 replication in humans by T-20, a peptide inhibitor of gp41-mediated virus entry. Nat Med 1998;4:1302−7. Available from: https://doi.org/10.1038/3293.

[17] He Y, Cheng J, Lu H, Li J, Hu J, Qi Z, et al. Potent HIV fusion inhibitors against Enfuvirtide-resistant HIV-1 strains. Proc Natl Acad Sci 2008;105:16332−7. Available from: https://doi.org/10.1073/pnas.0807335105.

[18] Kilby JM, Lalezari JP, Eron JJ, Carlson M, Cohen C, Arduino RC, et al. The safety, plasma pharmacokinetics, and antiviral activity of subcutaneous enfuvirtide (T-20), a peptide inhibitor of gp41-mediated virus fusion, in HIV-infected adults. AIDS Res Hum Retroviruses 2002;18:685−93. Available from: https://doi.org/10.1089/088922202760072294.

[19] Lalezari JP, Eron JJ, Carlson M, Cohen C, DeJesus E, Arduino RC, et al. A phase II clinical study of the long-term safety and antiviral activity of enfuvirtide-based antiretroviral therapy. AIDS 2003;17:691−8. Available from: https://doi.org/10.1097/00002030-200303280-00007.

[20] He Y. Synthesized peptide inhibitors of HIV-1 gp41-dependent membrane fusion. Curr Pharm Des 2013;19:1800−9. Available from: https://doi.org/10.2174/1381612811319100004.

[21] Dwyer JJ, Wilson KL, Davison DK, Freel SA, Seedorff JE, Wring SA, et al. Design of helical, oligomeric HIV-1 fusion inhibitor peptides with potent activity against enfuvirtide-resistant virus. Proc Natl Acad Sci 2007;104:12772−7. Available from: https://doi.org/10.1073/pnas.0701478104.

[22] Ren J, Wang X, Hu Z, Gao Q, Sun Y, Li X, et al. Picornavirus uncoating intermediate captured in atomic detail. Nat Commun 2013;4:1929. Available from: https://doi.org/10.1038/ncomms2889.

[23] Harrison SC. Viral membrane fusion. Nat Struct Mol Biol 2008;15:690−8. Available from: https://doi.org/10.1038/nsmb.1456.

[24] Zhang G, Zhou F, Gu B, Ding C, Feng D, Xie F, et al. In vitro and in vivo evaluation of ribavirin and pleconaril antiviral activity against enterovirus 71 infection. Arch Virol 2012;157:669−79. Available from: https://doi.org/10.1007/s00705-011-1222-6.

[25] Rotbart HA. Treatment of picornavirus infections. Antivir Res 2002;53:83−98. Available from: https://doi.org/10.1016/S0166-3542(01)00206-6.

[26] Phelps DK, Post CB. A novel basis for capsid stabilization by antiviral compounds. J Mol Biol 1995;254:544−51. Available from: https://doi.org/10.1006/jmbi.1995.0637.

[27] Tsang SK, Danthi P, Chow M, Hogle JM. Stabilization of poliovirus by capsid-binding antiviral drugs is due to entropic effects 1 1. In: Wilson IA, Mol J, editors. Biol, 296. 2000. p. 335−40. Available from: https://doi.org/10.1006/jmbi.1999.3483.

[28] Shia K-S, Li W-T, Chang C-M, Hsu M-C, Chern J-H, Leong MK, et al. Design, synthesis, and structure-−activity relationship of pyridyl imidazolidinones: a novel class of potent and selective human enterovirus 71 inhibitors. J Med Chem 2002;45:1644−55. Available from: https://doi.org/10.1021/jm010536a.

[29] Nascimento IJ, dos S, Santos-Júnior PF, da S, Aquino TM, de, Araújo-Júnior JX, et al. Insights on dengue and Zika NS5 RNA-dependent RNA polymerase (RdRp) inhibitors. Eur J Med Chem 2021;224:113698. Available from: https://doi.org/10.1016/j.ejmech.2021.113698.

[30] Öberg B. Rational design of polymerase inhibitors as antiviral drugs. Antivir Res 2006;71:90−5. Available from: https://doi.org/10.1016/j.antiviral.2006.05.012.

[31] Andrei G, De Clercq E, Snoeck R. Drug targets in cytomegalovirus infection. Infect Disord—Drug Targets 2009;9:201−22. Available from: https://doi.org/10.2174/187152609787847758.

[32] Palumbo E. New drugs for chronic hepatitis B: a review. Am J Ther 2008;15:167−72. Available from: https://doi.org/10.1097/MJT.0b013e318155a191.

[33] Wegzyn CM, Wyles DL. Antiviral drug advances in the treatment of human immunodeficiency virus (HIV) and chronic hepatitis C virus (HCV). Curr Opin Pharmacol 2012;12:556−61. Available from: https://doi.org/10.1016/j.coph.2012.06.005.

[34] Beaulieu PL, Bös M, Bousquet Y, Fazal G, Gauthier J, Gillard J, et al. Non-nucleoside inhibitors of the hepatitis C virus NS5B polymerase: discovery and preliminary SAR of benzimidazole derivatives. Bioorg Med Chem Lett 2004;14:119−24. Available from: https://doi.org/10.1016/j.bmcl.2003.10.023.

[35] Summers DF, Maizel JV. Evidence for large precursor proteins in poliovirus synthesis. Proc Natl Acad Sci 1968;59:966−71. Available from: https://doi.org/10.1073/pnas.59.3.966.

[36] Pelham HRB. Translation of encephalomyocarditis virus RNA in vitro yields an active proteolytic processing enzyme. Eur J Biochem 1978;85:457−62. Available from: https://doi.org/10.1111/j.1432-1033.1978.tb12260.x.

[37] Thiel V, Ivanov KA, Putics Á, Hertzig T, Schelle B, Bayer S, et al. Mechanisms and enzymes involved in SARS coronavirus genome expression. J Gen Virol 2003;84:2305−15. Available from: https://doi.org/10.1099/vir.0.19424-0.

[38] Harcourt BH, Jukneliene D, Kanjanahaluethai A, Bechill J, Severson KM, Smith CM, et al. Identification of severe acute respiratory syndrome coronavirus replicase products and characterization of papain-like protease activity. J Virol 2004;78:13600−12. Available from: https://doi.org/10.1128/JVI.78.24.13600-13612.2004.

[39] Ratia K, Saikatendu KS, Santarsiero BD, Barreto N, Baker SC, Stevens RC, et al. Severe acute respiratory syndrome coronavirus papain-like-protease: structure of a viral deubiquitinating enzyme. Proc Natl Acad Sci U S A 2006;103:5717−22. Available from: https://doi.org/10.1073/pnas.0510851103.

[40] Schlieker C, Weihofen WA, Frijns E, Kattenhorn LM, Gaudet R, Ploegh HL. Structure of a herpesvirus-encoded cysteine protease reveals a unique class of deubiquitinating enzymes. Mol Cell 2007;25:677−87. Available from: https://doi.org/10.1016/j.molcel.2007.01.033.

[41] Balakirev MY, Jaquinod M, Haas AL, Chroboczek J. Deubiquitinating function of adenovirus proteinase. J Virol 2002;76:6323−31. Available from: https://doi.org/10.1128/JVI.76.12.6323-6331.2002.

[42] Sun Z, Chen Z, Lawson SR, Fang Y. The cysteine protease domain of porcine reproductive and respiratory syndrome virus nonstructural protein 2 possesses deubiquitinating and interferon antagonism functions. J Virol 2010;84:7832−46. Available from: https://doi.org/10.1128/JVI.00217-10.

[43] Llibre JM, Imaz A, Clotet B. From TMC114 to darunavir: five years of data on efficacy. AIDS Rev 2013;15:112−21.

[44] Nijhuis M, Deeks S, Boucher C. Implications of antiretroviral resistance on viral fitness. Curr Opin Infect Dis 2001;14:23−8. Available from: https://doi.org/10.1097/00001432-200102000-00005.

[45] Cole S. Herpes simplex virus: epidemiology, diagnosis, and treatment. Nurs Clin North Am 2020;55:337−45. Available from: https://doi.org/10.1016/j.cnur.2020.05.004.

[46] Hammad WAB, Konje JC. Herpes simplex virus infection in pregnancy—an update. Eur J Obstet Gynecol Reprod Biol 2021;259:38−45. Available from: https://doi.org/10.1016/j.ejogrb.2021.01.055.

[47] Fan TH, Khoury J, Cho SM, Bhimraj A, Shoskes A, Uchino K. Cerebrovascular complications and vasculopathy in patients with herpes simplex virus central nervous system infection. J Neurol Sci 2020;419:117200. Available from: https://doi.org/10.1016/j.jns.2020.117200.

[48] Spruance SL, Nett R, Marbury T, Wolff R, Johnson J. Spaulding for the acyclovir cream study group T. Acyclovir cream for treatment of herpes simplex labialis: results of two randomized, double-blind, vehicle-controlled, multicenter clinical trials. Antimicrob Agents Chemother 2002;46:2238–43. Available from: https://doi.org/10.1128/AAC.46.7.2238-2243.2002.

[49] Bauer D, Keller J, Alt M, Schubert A, Aufderhorst UW, Palapys V, et al. Antibody-based immunotherapy of aciclovir resistant ocular herpes simplex virus infections. Virology 2017;512:194–200. Available from: https://doi.org/10.1016/j.virol.2017.09.021.

[50] Dong H, Wang Z, Zhao D, Leng X, Zhao Y. Antiviral strategies targeting herpesviruses. J Virus Erad 2021;7:100047. Available from: https://doi.org/10.1016/j.jve.2021.100047.

[51] Lombardi L, Falanga A, Del Genio V, Palomba L, Galdiero M, Franci G, et al. A boost to the antiviral activity: cholesterol tagged peptides derived from glycoprotein B of Herpes Simplex virus type I. Int J Biol Macromol 2020;162:882–93. Available from: https://doi.org/10.1016/j.ijbiomac.2020.06.134.

[52] Liu Zhao, Niu Fju, Xie Yxin, Xie Smin, Liu Ynan, Yang Yying, et al. A review: natural polysaccharides from medicinal plants and microorganisms and their anti-herpetic mechanism. Biomed Pharmacother 2020;129:110469. Available from: https://doi.org/10.1016/j.biopha.2020.110469.

[53] Griffiths PD. Tomorrow's challenges for herpesvirus management: potential applications of valacyclovir. J Infect Dis 2002;186:S131–7. Available from: https://doi.org/10.1086/342960.

[54] Schuhmacher A, Reichling J, Schnitzler P. Virucidal effect of peppermint oil on the enveloped viruses herpes simplex virus type 1 and type 2 in vitro. Phytomedicine 2003;10:504–10. Available from: https://doi.org/10.1078/094471103322331467.

[55] Zinser E, Krawczyk A, Mühl-Zürbes P, Aufderhorst U, Draßner C, Stich L, et al. A new promising candidate to overcome drug resistant herpes simplex virus infections. Antivir Res 2018;149:202–10. Available from: https://doi.org/10.1016/j.antiviral.2017.11.012.

[56] Rose R, Brunnemann AK, Baukmann S, Bühler S, Fickenscher H, Sauerbrei A, et al. Antiviral susceptibility of recombinant Herpes simplex virus 1 strains with specific polymerase amino acid changes. Antivir Res 2021;195:105166. Available from: https://doi.org/10.1016/j.antiviral.2021.105166.

[57] Weber B, Cinatl J. Antiviral therapy of herpes simplex virus infection: recent developments. J Eur Acad Dermatol Venereol 1996;6:112–26. Available from: https://doi.org/10.1016/0926-9959(95)00152-2.

[58] Anton-Vazquez V, Mehra V, Mbisa JL, Bradshaw D, Basu TN, Daly ML, et al. Challenges of aciclovir-resistant HSV infection in allogeneic bone marrow transplant recipients. J Clin Virol 2020;128:104421. Available from: https://doi.org/10.1016/j.jcv.2020.104421.

[59] Akahoshi Y, Kanda J, Ohno A, Komiya Y, Gomyo A, Hayakawa J, et al. Acyclovir-resistant herpes simplex virus 1 infection early after allogeneic hematopoietic stem cell transplantation with T-cell depletion. J Infect Chemother 2017;23:485–7. Available from: https://doi.org/10.1016/j.jiac.2017.02.001.

[60] van Diemen FR, Kruse EM, Hooykaas MJG, Bruggeling CE, Schürch AC, van Ham PM, et al. CRISPR/Cas9-mediated genome editing of herpesviruses limits productive and latent infections. PLoS Pathog 2016;12:1–29. Available from: https://doi.org/10.1371/journal.ppat.1005701.

[61] Lago ADN, Fortes ABC, Furtado GS, Menezes CFS, Gonçalves LM. Association of antimicrobial photodynamic therapy and photobiomodulation for herpes simplex labialis resolution: case series. Photodiagnosis Photodyn Ther 2020;32:102070. Available from: https://doi.org/10.1016/j.pdpdt.2020.102070.

[62] Greeley ZW, Giannasca NJ, Porter MJ, Margulies BJ. Acyclovir, cidofovir, and amenamevir have additive antiviral effects on herpes simplex virus TYPE 1. Antivir Res 2020;176:104754. Available from: https://doi.org/10.1016/j.antiviral.2020.104754.

[63] Wu CY, Yu ZY, Chen YC, Hung SL. Effects of epigallocatechin-3-gallate and acyclovir on herpes simplex virus type 1 infection in oral epithelial cells. J Formos Med Assoc 2021;. Available from: https://doi.org/10.1016/j.jfma.2020.12.018.

[64] Luganini A, Sibille G, Mognetti B, Sainas S, Pippione AC, Giorgis M, et al. Effective deploying of a novel DHODH inhibitor against herpes simplex type 1 and type 2 replication. Antivir Res 2021;189:105057. Available from: https://doi.org/10.1016/j.antiviral.2021.105057.

[65] Uhlig N, Donner AK, Gege C, Lange F, Kleymann G, Grunwald T. Helicase primase inhibitors (HPIs) are efficacious for therapy of human herpes simplex virus (HSV) disease in an infection mouse model. Antivir Res 2021;195:105190. Available from: https://doi.org/10.1016/j.antiviral.2021.105190.

[66] Luo Z, Kuang XP, Zhou QQ, Yan CY, Li W, Gong HB, et al. Inhibitory effects of baicalein against herpes simplex virus type 1. Acta Pharm Sin B 2020;10:2323−38. Available from: https://doi.org/10.1016/j.apsb.2020.06.008.

[67] Li W, Xu C, Hao C, Zhang Y, Wang Z, Wang S, et al. Inhibition of herpes simplex virus by myricetin through targeting viral gD protein and cellular EGFR/PI3K/Akt pathway. Antivir Res 2020;177:104714. Available from: https://doi.org/10.1016/j.antiviral.2020.104714.

[68] Hurt AC, Ho H-T, Barr I. Resistance to anti-influenza drugs: adamantanes and neuraminidase inhibitors. Expert Rev Anti Infect Ther 2006;4:795−805. Available from: https://doi.org/10.1586/14787210.4.5.795.

[69] Shibnev VA, Garaev TM, Finogenova MP, Shevchenko ES, Burtseva EI. New adamantane derivatives can overcome resistance of influenza A(H1N1)pdm2009 and A(H3N2) viruses to remantadine. Bull Exp Biol Med 2012;153:233−5. Available from: https://doi.org/10.1007/s10517-012-1684-x.

[70] Gasparini R, Amicizia D, Lai PL, Bragazzi NL, Panatto D. Compounds with anti-influenza activity: present and future of strategies for the optimal treatment and management of influenza. Part I: influenza life-cycle and currently available drugs. J Prev Med Hyg 2014;55:69−85.

[71] Principi N, Camilloni B, Alunno A, Polinori I, Argentiero A, Esposito S. Drugs for influenza treatment: is there significant news? Front Med 2019;6. Available from: https://doi.org/10.3389/fmed.2019.00109.

[72] Kolocouris N, Zoidis G, Foscolos GB, Fytas G, Prathalingham SR, Kelly JM, et al. Design and synthesis of bioactive adamantane spiro heterocycles. Bioorg Med Chem Lett 2007;17:4358−62. Available from: https://doi.org/10.1016/j.bmcl.2007.04.108.

[73] Zoidis G, Kolocouris N, Naesens L, De Clercq E. Design and synthesis of 1,2-annulated adamantane piperidines with anti-influenza virus activity. Bioorg Med Chem 2009;17:1534−41. Available from: https://doi.org/10.1016/j.bmc.2009.01.009.

[74] Zarubaev VV, Golod EL, Anfimov PM, Shtro AA, Saraev VV, Gavrilov AS, et al. Synthesis and anti-viral activity of azolo-adamantanes against influenza A virus. Bioorg Med Chem 2010;18:839−48. Available from: https://doi.org/10.1016/j.bmc.2009.11.047.

[75] Hu Y, Hau RK, Wang Y, Tuohy P, Zhang Y, Xu S, et al. Structure−property relationship studies of influenza a virus AM2-S31N proton channel blockers. ACS Med Chem Lett 2018;9:1111−16. Available from: https://doi.org/10.1021/acsmedchemlett.8b00336.

[76] Stankova I, Chuchkov K, Chayrov R, Mukova L, Galabov A, Marinkova D, et al. Adamantane derivatives containing thiazole moiety: synthesis, antiviral and antibacterial activity. Int J Pept Res Ther 2020;26:1781−7. Available from: https://doi.org/10.1007/s10989-019-09983-4.

[77] Suslov E, Zarubaev VV, Slita AV, Ponomarev K, Korchagina D, Ayine-Tora DM, et al. Anti-influenza activity of diazaadamantanes combined with monoterpene moieties. Bioorg Med Chem Lett 2017;27:4531−5. Available from: https://doi.org/10.1016/j.bmcl.2017.08.062.

[78] Stiver G. The treatment of influenza with antiviral drugs. CMAJ 2003;168:49−56.

[79] Jefferson T, Doshi P. Multisystem failure: the story of anti-influenza drugs. BMJ 2014;348:g2263. Available from: https://doi.org/10.1136/bmj.g2263 −g2263.

[80] Boltz DA, Aldridge JR, Webster RG, Govorkova EA. Drugs in development for influenza. Drugs 2010;70:1349−62. Available from: https://doi.org/10.2165/11537960-000000000-00000.

[81] Patel A, Gorman SE. Stockpiling antiviral drugs for the next influenza pandemic. Clin Pharmacol Ther 2009;86:241−3. Available from: https://doi.org/10.1038/clpt.2009.142.

[82] Shaw ML. The next wave of influenza drugs. ACS Infect Dis 2017;3:691−4. Available from: https://doi.org/10.1021/acsinfecdis.7b00142.

[83] Wang Z, Cheng LP, Zhang XH, Pang W, Li L, Zhao JL. Design, synthesis and biological evaluation of novel oseltamivir derivatives as potent neuraminidase inhibitors. Bioorg Med Chem Lett 2017;27:5429−35. Available from: https://doi.org/10.1016/j.bmcl.2017.11.003.

[84] Wang K, Yang F, Wang L, Liu K, Sun L, Lin B, et al. Synthesis and biological evaluation of NH 2-acyl oseltamivir analogues as potent neuraminidase inhibitors. Eur J Med Chem 2017;141:648−56. Available from: https://doi.org/10.1016/j.ejmech.2017.10.004.

[85] Wang K, Lei Z, Zhao L, Chen B, Yang F, Liu K, et al. Design, synthesis and biological evaluation of oseltamivir derivatives containing pyridyl group as potent inhibitors of neuraminidase for influenza A. Eur J Med Chem 2020;185:111841. Available from: https://doi.org/10.1016/j.ejmech.2019.111841.

[86] Ye J, Yang X, Xu M, Chan PK, Ma C. Novel N-Substituted oseltamivir derivatives as potent influenza neuraminidase inhibitors: design, synthesis, biological evaluation, ADME prediction and molecular docking studies. Eur J Med Chem 2019;182:111635. Available from: https://doi.org/10.1016/j.ejmech.2019.111635.

[87] Zhang H, Wang K, Zhu H, Zhao X, Zhao H, Lei Z, et al. Discovery of a non-zwitterionic oseltamivir analogue as a potent influenza a neuraminidase inhibitor. Eur J Med Chem 2020;200:112423. Available from: https://doi.org/10.1016/j.ejmech.2020.112423.

[88] Wang P, Oladejo BO, Li C, Fu L, Zhang S, Qi J, et al. Structure-based design of 5′-substituted 1,2,3-triazolylated oseltamivir derivatives as potent influenza neuraminidase inhibitors. RSC Adv 2021;11:9528−41. Available from: https://doi.org/10.1039/D1RA00472G.

[89] Jia R, Zhang J, Ai W, Ding X, Desta S, Sun L, et al. Design, synthesis and biological evaluation of "Multi-Site"-binding influenza virus neuraminidase inhibitors. Eur J Med Chem 2019;178:64−80. Available from: https://doi.org/10.1016/j.ejmech.2019.05.076.

[90] Wang B, Wang K, Meng P, Hu Y, Yang F, Liu K, et al. Design, synthesis, and evaluation of carboxyl-modified oseltamivir derivatives with improved lipophilicity as neuraminidase inhibitors. Bioorg Med Chem Lett 2018;28:3477−82. Available from: https://doi.org/10.1016/j.bmcl.2018.09.014.

[91] Ju H, Zhang J, Sun Z, Huang Z, Qi W, Huang B, et al. Discovery of C-1 modified oseltamivir derivatives as potent influenza neuraminidase inhibitors. Eur J Med Chem 2018;146:220−31. Available from: https://doi.org/10.1016/j.ejmech.2018.01.050.

[92] Ju H, Xiu S, Ding X, Shang M, Jia R, Huang B, et al. Discovery of novel 1,2,3-triazole oseltamivir derivatives as potent influenza neuraminidase inhibitors targeting the 430-cavity. Eur J Med Chem 2020;187:111940. Available from: https://doi.org/10.1016/j.ejmech.2019.111940.

[93] Adamson CS, Chibale K, Goss RJM, Jaspars M, Newman DJ, Dorrington RA. Antiviral drug discovery: preparing for the next pandemic. Chem Soc Rev 2021;50:3647−55. Available from: https://doi.org/10.1039/D0CS01118E.

[94] Soufi GJ, Hekmatnia A, Nasrollahzadeh M, Shafiei N, Sajjadi M, Iravani P, et al. SARS-CoV-2 (COVID-19): new discoveries and current challenges. Appl Sci 2020;10:3641. Available from: https://doi.org/10.3390/app10103641.

[95] Humeniuk R, Mathias A, Kirby BJ, Lutz JD, Cao H, Osinusi A, et al. Pharmacokinetic, pharmacodynamic, and drug-interaction profile of remdesivir, a SARS-CoV-2 replication inhibitor. Clin Pharmacokinet 2021;60:569−83. Available from: https://doi.org/10.1007/s40262-021-00984-5.

[96] How antiviral pill molnupiravir shot ahead in the COVID drug hunt n.d.

[97] Chavda VP, Vora LK, Pandya AK, Patravale VB. Intranasal vaccines for SARS-CoV-2: from challenges to potential in COVID-19 management. Drug Discov Today 2021;. Available from: https://doi.org/10.1016/j.drudis.2021.07.021.

[98] Steuten K, Kim H, Widen JC, Babin BM, Onguka O, Lovell S, et al. Challenges for targeting SARS-CoV-2 proteases as a therapeutic strategy for COVID-19. ACS. Infect Dis 2021;7:1457−68. Available from: https://doi.org/10.1021/acsinfecdis.0c00815.

[99] Tiwari V, Beer JC, Sankaranarayanan NV, Swanson-Mungerson M, Desai UR. Discovering small-molecule therapeutics against SARS-CoV-2. Drug Discov Today 2020;25:1535−44. Available from: https://doi.org/10.1016/j.drudis.2020.06.017.

[100] Huggins DJ, Sherman W, Tidor B. Rational approaches to improving selectivity in drug design. J Med Chem 2012;55:1424−44. Available from: https://doi.org/10.1021/jm2010332.

[101] Siklos M, BenAissa M, Thatcher GRJ. Cysteine proteases as therapeutic targets: does selectivity matter? A systematic review of calpain and cathepsin inhibitors. Acta Pharm Sin B 2015;5:506−19. Available from: https://doi.org/10.1016/j.apsb.2015.08.001.

[102] Jonsson CB, Golden JE, Meibohm B. Time to 'Mind the Gap' in novel small molecule drug discovery for direct-acting antivirals for SARS-CoV-2. Curr Opin Virol 2021;50:1−7. Available from: https://doi.org/10.1016/j.coviro.2021.06.008.

[103] Murgolo N, Therien AG, Howell B, Klein D, Koeplinger K, Lieberman LA, et al. SARS-CoV-2 tropism, entry, replication, and propagation: considerations for drug discovery and development. PLOS Pathog 2021;17:e1009225. Available from: https://doi.org/10.1371/journal.ppat.1009225.

[104] Li Y, Cao L, Li G, Cong F, Li Y, Sun J, et al. Remdesivir metabolite GS-441524 effectively inhibits SARS-CoV-2 infection in mouse models. J Med Chem 2021;. Available from: https://doi.org/10.1021/acs.jmedchem.0c01929 acs.jmedchem.0c01929.

[105] Yu R, Chen L, Lan R, Shen R, Li P. Computational screening of antagonists against the SARS-CoV-2 (COVID-19) coronavirus by molecular docking. Int J Antimicrob Agents 2020;2:3−8. Available from: https://doi.org/10.1016/j.ijantimicag.2020.106012.

[106] Ko WC, Rolain JM, Lee NY, Chen PL, Huang CT, Lee PI, et al. Arguments in favour of remdesivir for treating SARS-CoV-2 infections. Int J Antimicrob Agents 2020;105933. Available from: https://doi.org/10.1016/j.ijantimicag.2020.105933.

[107] Kokic G, Hillen HS, Tegunov D, Dienemann C, Seitz F, Schmitzova J, et al. Mechanism of SARS-CoV-2 polymerase stalling by remdesivir. Nat Commun 2021;12:279. Available from: https://doi.org/10.1038/s41467-020-20542-0.

[108] Malone B, Campbell EA. Molnupiravir: coding for catastrophe. Nat Struct Mol Biol 2021;28:706−8. Available from: https://doi.org/10.1038/s41594-021-00657-8.

[109] Padhi AK, Shukla R, Saudagar P, Tripathi T. High-throughput rational design of the remdesivir binding site in the RdRp of SARS-CoV-2: implications for potential resistance. IScience 2021;24:101992. Available from: https://doi.org/10.1016/j.isci.2020.101992.

[110] Shukla H, Anti DM. HIV prolific drug discovery. J Curr Med Res Opin 2019;. Available from: https://doi.org/10.15520/jcmro.v2i11.230.

[111] Margolis DM, Archin NM. Proviral latency, persistent human immunodeficiency virus infection, and the development of latency reversing agents. J Infect Dis 2017;215:S111−18. Available from: https://doi.org/10.1093/infdis/jiw618.

[112] Puhl AC, Garzino Demo A, Makarov VA, Ekins S. New targets for HIV drug discovery. Drug Discov Today 2019;24:1139−47. Available from: https://doi.org/10.1016/j.drudis.2019.03.013.

[113] Zuo X, Huo Z, Kang D, Wu G, Zhou Z, Liu X, et al. Current insights into anti-HIV drug discovery and development: a review of recent patent literature (2014−2017). Expert Opin Ther Pat 2018;28:299−316. Available from: https://doi.org/10.1080/13543776.2018.1438410.

[114] Flexner C. Modern human immunodeficiency virus therapy: progress and prospects. Clin Pharmacol Ther 2019;105:61−70. Available from: https://doi.org/10.1002/cpt.1284.

[115] Menéndez-Arias L. Molecular basis of human immunodeficiency virus drug resistance: an update. Antivir Res 2010;85:210−31. Available from: https://doi.org/10.1016/j.antiviral.2009.07.006.

[116] Le Grice SFJ. Human immunodeficiency virus reverse transcriptase: 25 years of research, drug discovery, and promise. J Biol Chem 2012;287:40850−7. Available from: https://doi.org/10.1074/jbc.R112.389056.

[117] Menéndez-Arias L, Álvarez M. Antiretroviral therapy and drug resistance in human immunodeficiency virus type 2 infection. Antivir Res 2014;102:70−86. Available from: https://doi.org/10.1016/j.antiviral.2013.12.001.

[118] Kankanala J, Kirby KA, Huber AD, Casey MC, Wilson DJ, Sarafianos SG, et al. Design, synthesis and biological evaluations of N-Hydroxy thienopyrimidine-2,4-diones as inhibitors of HIV reverse transcriptase-associated RNase H. Eur J Med Chem 2017;141:149−61. Available from: https://doi.org/10.1016/j.ejmech.2017.09.054.

[119] Monforte AM, De Luca L, Buemi MR, Agharbaoui FE, Pannecouque C, Ferro S. Structural optimization of N1-aryl-benzimidazoles for the discovery of new non-nucleoside reverse transcriptase inhibitors active against wild-type and mutant HIV-1 strains. Bioorg Med Chem 2018;26:661−74. Available from: https://doi.org/10.1016/j.bmc.2017.12.033.

[120] Jin K, Sang Y, Han S, De Clercq E, Pannecouque C, Meng G, et al. Synthesis and biological evaluation of dihydroquinazoline-2-amines as potent non-nucleoside reverse transcriptase inhibitors of wild-type and mutant HIV-1 strains. Eur J Med Chem 2019;176:11−20. Available from: https://doi.org/10.1016/j.ejmech.2019.05.011.

[121] Fabian L, Taverna Porro M, Gómez N, Salvatori M, Turk G, Estrin D, et al. Design, synthesis and biological evaluation of quinoxaline compounds as anti-HIV agents targeting reverse transcriptase enzyme. Eur J Med Chem 2020;188:111987. Available from: https://doi.org/10.1016/j.ejmech.2019.111987.

[122] Forezi L, da SM, Ribeiro MMJ, Marttorelli A, Abrantes JL, Rodrigues CR, et al. Design, synthesis, in vitro and in silico studies of novel 4-oxoquinoline ribonucleoside derivatives as HIV-1 reverse transcriptase inhibitors. Eur J Med Chem 2020;194:112255. Available from: https://doi.org/10.1016/j.ejmech.2020.112255.

[123] Sang Y, Pannecouque C, De Clercq E, Zhuang C, Chen F. Pharmacophore-fusing design of pyrimidine sulfonylacetanilides as potent non-nucleoside inhibitors of HIV-1 reverse transcriptase. Bioorg Chem 2020;96:103595. Available from: https://doi.org/10.1016/j.bioorg.2020.103595.

[124] El-Hussieny M, El-Sayed NF, Ewies EF, Ibrahim NM, Mahran MRH, Fouad MA. Synthesis, molecular docking and biological evaluation of 2-(thiophen-2-yl)-1H-indoles as potent HIV-1 non-nucleoside reverse transcriptase inhibitors. Bioorg Chem 2020;95:103521. Available from: https://doi.org/10.1016/j.bioorg.2019.103521.

[125] Feng D, Zuo X, Jing L, Chen C-H, Olotu FA, Lin H, et al. Design, synthesis, and evaluation of "dual-site"-binding diarylpyrimidines targeting both NNIBP and the NNRTI adjacent site of the HIV-1 reverse transcriptase. Eur J Med Chem 2021;211:113063. Available from: https://doi.org/10.1016/j.ejmech.2020.113063.

[126] Kang D, Urhan Ç, Wei F, Frutos-Beltrán E, Sun L, Álvarez M, et al. Discovery, optimization, and target identification of novel coumarin derivatives as HIV-1 reverse transcriptase-associated ribonuclease H inhibitors. Eur J Med Chem 2021;225:113769. Available from: https://doi.org/10.1016/j.ejmech.2021.113769.

[127] Gao P, Song S, Frutos-Beltrán E, Li W, Sun B, Kang D, et al. Novel indolylarylsulfone derivatives as covalent HIV-1 reverse transcriptase inhibitors specifically targeting the drug-resistant mutant Y181C. Bioorg Med Chem 2021;30:115927. Available from: https://doi.org/10.1016/j.bmc.2020.115927.

[128] Ghosh AK, Rao KV, Nyalapatla PR, Osswald HL, Martyr CD, Aoki M, et al. Design and development of highly potent HIV-1 protease inhibitors with a crown-like oxotricyclic core as the P2-ligand to combat multidrug-resistant HIV variants. J Med Chem 2017;60:4267–78. Available from: https://doi.org/10.1021/acs.jmedchem.7b00172.

[129] Ghosh AK, Brindisi M, Nyalapatla PR, Takayama J, Ella-Menye J-R, Yashchuk S, et al. Design of novel HIV-1 protease inhibitors incorporating isophthalamide-derived P2-P3 ligands: synthesis, biological evaluation and x-ray structural studies of inhibitor-HIV-1 protease complex. Bioorg Med Chem 2017;25:5114–27. Available from: https://doi.org/10.1016/j.bmc.2017.04.005.

[130] Ghosh AK, Jadhav RD, Simpson H, Kovela S, Osswald H, Agniswamy J, et al. Design, synthesis, and X-ray studies of potent HIV-1 protease inhibitors incorporating aminothiochromane and aminotetrahydronaphthalene carboxamide derivatives as the P2 ligands. Eur J Med Chem 2018;160:171–82. Available from: https://doi.org/10.1016/j.ejmech.2018.09.046.

[131] Zhu M, Zhou H, Ma L, Dong B, Zhou J, Zhang G, et al. Design and evaluation of novel piperidine HIV-1 protease inhibitors with potency against DRV-resistant variants. Eur J Med Chem 2021;220:113450. Available from: https://doi.org/10.1016/j.ejmech.2021.113450.

[132] Bai X, Yang Z, Zhu M, Dong B, Zhou L, Zhang G, et al. Design and synthesis of potent HIV-1 protease inhibitors with (S)-tetrahydrofuran-tertiary amine-acetamide as P2 – ligand: structure – activity studies and biological evaluation. Eur J Med Chem 2017;137:30–44. Available from: https://doi.org/10.1016/j.ejmech.2017.05.024.

[133] Bungard CJ, Williams PD, Schulz J, Wiscount CM, Holloway MK, Loughran HM, et al. Design and synthesis of piperazine sulfonamide cores leading to highly potent HIV-1 protease inhibitors. ACS Med Chem Lett 2017;8:1292–7. Available from: https://doi.org/10.1021/acsmedchemlett.7b00386.

[134] Dou Y, Zhu M, Dong B, Wang J-X, Zhang G-N, Zhang F, et al. Design, synthesis and biological evaluation of HIV-1 protease inhibitors with morpholine derivatives as P2 ligands in combination with cyclopropyl as P1′ ligand. Bioorg Med Chem Lett 2020;30:127019. Available from: https://doi.org/10.1016/j.bmcl.2020.127019.

[135] Zhu M, Shan Q, Ma L, Wen J, Dong B, Zhang G, et al. Design and biological evaluation of cinnamic and phenylpropionic amide derivatives as novel dual inhibitors of HIV-1 protease and reverse transcriptase. Eur J Med Chem 2021;220:113498. Available from: https://doi.org/10.1016/j.ejmech.2021.113498.

[136] Subbaiah AM, Mandlekar M, Desikan S, Ramar S, Subramani T, Annadurai L, et al. Design, synthesis, and pharmacokinetic evaluation of phosphate and amino acid ester prodrugs for improving the oral bioavailability of the HIV-1 protease inhibitor atazanavir. J Med Chem 2019;62:3553–74. Available from: https://doi.org/10.1021/acs.jmedchem.9b00002.

[137] Ghany M, Liang TJ. Drug targets and molecular mechanisms of drug resistance in chronic hepatitis B. Gastroenterology 2007;132:1574–85. Available from: https://doi.org/10.1053/j.gastro.2007.02.039.

[138] Spyrou E, Smith CI, Ghany MG, Hepatitis B. Gastroenterol Clin North Am 2020;49:215–38. Available from: https://doi.org/10.1016/j.gtc.2020.01.003.

[139] Liang TJ, Block TM, McMahon BJ, Ghany MG, Urban S, Guo J-T, et al. Present and future therapies of hepatitis B: from discovery to cure. Hepatology 2015;62:1893–908. Available from: https://doi.org/10.1002/hep.28025.

[140] Testoni B, Durantel D, Zoulim F. Novel targets for hepatitis B virus therapy. Liver Int 2017;37:33–9. Available from: https://doi.org/10.1111/liv.13307.

[141] Block TM, Alter H, Brown N, Brownstein A, Brosgart C, Chang K-M, et al. Research priorities for the discovery of a cure for chronic hepatitis B: report of a workshop. Antivir Res 2018;150:93–100. Available from: https://doi.org/10.1016/j.antiviral.2017.12.006.

[142] Tang L, Zhao Q, Wu S, Cheng J, Chang J, Guo J-T. The current status and future directions of hepatitis B antiviral drug discovery. Expert Opin Drug Discov 2017;12:5–15. Available from: https://doi.org/10.1080/17460441.2017.1255195.

[143] Jia H, Bai F, Liu N, Liang X, Zhan P, Ma C, et al. Design, synthesis and evaluation of pyrazole derivatives as non-nucleoside hepatitis B virus inhibitors. Eur J Med Chem 2016;123:202–10. Available from: https://doi.org/10.1016/j.ejmech.2016.07.048.

[144] Jia H, Song Y, Yu J, Zhan P, Rai D, Liang X, et al. Design, synthesis and primary biological evaluation of the novel 2-pyridone derivatives as potent non-nucleoside HBV inhibitors. Eur J Med Chem 2017;136:144–53. Available from: https://doi.org/10.1016/j.ejmech.2017.04.048.

[145] Jia H, Yu J, Du X, Cherukupalli S, Zhan P, Liu X. Design, diversity-oriented synthesis and biological evaluation of novel heterocycle derivatives as non-nucleoside HBV capsid protein inhibitors. Eur J Med Chem 2020;202:112495. Available from: https://doi.org/10.1016/j.ejmech.2020.112495.

[146] Qiu J, Gong Q, Gao J, Chen W, Zhang Y, Gu X, et al. Design, synthesis and evaluation of novel phenyl propionamide derivatives as non-nucleoside hepatitis B virus inhibitors. Eur J Med Chem 2018;144:424–34. Available from: https://doi.org/10.1016/j.ejmech.2017.12.042.

[147] Qiu J, Chen W, Zhang Y, Zhou Q, Chen J, Yang L, et al. Assessment of quinazolinone derivatives as novel non-nucleoside hepatitis B virus inhibitors. Eur J Med Chem 2019;176:41–9. Available from: https://doi.org/10.1016/j.ejmech.2019.05.014.

[148] Liu Y, Peng Y, Lu J, Wang J, Ma H, Song C, et al. Design, synthesis, and biological evaluation of new 1,2,3-triazolo-2'-deoxy-2'-fluoro- 4'-azido nucleoside derivatives as potent anti-HBV agents. Eur J Med Chem 2018;143:137–49. Available from: https://doi.org/10.1016/j.ejmech.2017.11.028.

[149] Tang J, Huber AD, Pineda DL, Boschert KN, Wolf JJ, Kankanala J, et al. 5-Aminothiophene-2,4-dicarboxamide analogues as hepatitis B virus capsid assembly effectors. Eur J Med Chem 2019;164:179–92. Available from: https://doi.org/10.1016/j.ejmech.2018.12.047.

[150] Pan T, Ding Y, Wu L, Liang L, He X, Li Q, et al. Design and synthesis of aminothiazole based Hepatitis B Virus (HBV) capsid inhibitors. Eur J Med Chem 2019;166:480–501. Available from: https://doi.org/10.1016/j.ejmech.2019.01.059.

[151] Ma Y, Zhao S, Ren Y, Cherukupalli S, Li Q, Woodson ME, et al. Design, synthesis and evaluation of heteroaryldihydropyrimidine analogues bearing spiro ring as hepatitis B virus capsid protein inhibitors. Eur J Med Chem 2021;225:113780. Available from: https://doi.org/10.1016/j.ejmech.2021.113780.

[152] Chawla P, Yadav A, Chawla V. Clinical implications and treatment of dengue. Asian Pac J Trop Med 2014;7:169–78. Available from: https://doi.org/10.1016/S1995-7645(14)60016-X.

[153] Chan CY, Ooi EE. Dengue: an update on treatment options. Future Microbiol 2015;10:2017–31. Available from: https://doi.org/10.2217/fmb.15.105.

[154] Rajapakse S, Rodrigo C. Rajapakse. Treat Dengue Fever Infect Drug Resist 2012;103. Available from: https://doi.org/10.2147/IDR.S22613.

[155] Low JGH, Ooi EE, Vasudevan SG. Current status of dengue therapeutics research and development. J Infect Dis 2017;215:S96–102. Available from: https://doi.org/10.1093/infdis/jiw423.

[156] Behnam Ma M, Nitsche C, Boldescu V, Klein CD. The medicinal chemistry of dengue virus. J Med Chem 2016;59:5622–49. Available from: https://doi.org/10.1021/acs.jmedchem.5b01653.

[157] Ali F, Chorsiya A, Anjum V, Khasimbi S, Ali A. A systematic review on phytochemicals for the treatment of dengue. Phyther Res 2021;35:1782–816. Available from: https://doi.org/10.1002/ptr.6917.

[158] Wiwanitkit V. Dengue fever: diagnosis and treatment. Expert Rev Anti Infect Ther 2010;8:841–5. Available from: https://doi.org/10.1586/eri.10.53.

[159] Lim SP. Dengue drug discovery: progress, challenges and outlook. Antivir Res 2019;163:156–78. Available from: https://doi.org/10.1016/j.antiviral.2018.12.016.

[160] Weng Z, Shao X, Graf D, Wang C, Klein CD, Wang J, et al. Identification of fused bicyclic derivatives of pyr-rolidine and imidazolidinone as dengue virus-2 NS2B-NS3 protease inhibitors. Eur J Med Chem 2017;125:751−9. Available from: https://doi.org/10.1016/j.ejmech.2016.09.063.

[161] Benmansour F, Eydoux C, Querat G, De Lamballerie X, Canard B, Alvarez K, et al. Novel 2-phenyl-5-[(E)-2-(thiophen-2-yl)ethenyl]-1,3,4-oxadiazole and 3-phenyl-5-[(E)-2-(thiophen-2-yl)ethenyl]-1,2,4-oxadiazole deri-vatives as dengue virus inhibitors targeting NS5 polymerase. Eur J Med Chem 2016;109:146−56. Available from: https://doi.org/10.1016/j.ejmech.2015.12.046.

[162] Venkatesham A, Saudi M, Kaptein S, Neyts J, Rozenski J, Froeyen M, et al. Aminopurine and aminoquina-zoline scaffolds for development of potential dengue virus inhibitors. Eur J Med Chem 2017;126:101−9. Available from: https://doi.org/10.1016/j.ejmech.2016.10.008.

[163] Leal ES, Adler NS, Fernández GA, Gebhard LG, Battini L, Aucar MG, et al. De novo design approaches tar-geting an envelope protein pocket to identify small molecules against dengue virus. Eur J Med Chem 2019;182. Available from: https://doi.org/10.1016/j.ejmech.2019.111628.

[164] Okano Y, Saito-Tarashima N, Kurosawa M, Iwabu A, Ota M, Watanabe T, et al. Synthesis and biological evaluation of novel imidazole nucleosides as potential anti-dengue virus agents. Bioorg Med Chem 2019;27:2181−6. Available from: https://doi.org/10.1016/j.bmc.2019.04.015.

[165] Sundermann TR, Benzin CV, Dražić T, Klein CD. Synthesis and structure-activity relationships of small-molecular di-basic esters, amides and carbamates as flaviviral protease inhibitors. Eur J Med Chem 2019;176:187−94. Available from: https://doi.org/10.1016/j.ejmech.2019.05.025.

[166] Wan Y, Wu S, Zheng S, Liang E, Liu S, Yao X, et al. A series of octahydroquinazoline-5-ones as novel inhibi-tors against dengue virus. Eur J Med Chem 2020;200:112318. Available from: https://doi.org/10.1016/j.ejmech.2020.112318.

[167] Maus H, Barthels F, Hammerschmidt SJ, Kopp K, Millies B, Gellert A, et al. SAR of novel benzothiazoles targeting an allosteric pocket of DENV and ZIKV NS2B/NS3 proteases. Bioorg Med Chem 2021;47:116392. Available from: https://doi.org/10.1016/j.bmc.2021.116392.

[168] Ebola virus disease. n.d.

[169] Ebola (Ebola Virus Disease). n.d.

[170] Tapia MD, Sow SO, Mbaye KD, Thiongane A, Ndiaye BP, Ndour CT, et al. Safety, reactogenicity, and immunogenicity of a chimpanzee adenovirus vectored Ebola vaccine in children in Africa: a randomised, observer-blind, placebo-controlled, phase 2 trial. Lancet Infect Dis 2020;20:719−30. Available from: https://doi.org/10.1016/S1473-3099(20)30019-0.

[171] Berry DE, Bavinger JC, Fernandes A, Mattia JG, Mustapha J, Harrison-Williams L, et al. Posterior segment ophthalmic manifestations in ebola survivors, sierra leone. Ophthalmology 2021;. Available from: https://doi.org/10.1016/j.ophtha.2021.02.001.

[172] Agnihotri S, Alpren C, Bangura B, Bennett S, Gorina Y, Harding JD, et al. Building the Sierra Leone Ebola Database: organization and characteristics of data systematically collected during 2014-2015 Ebola epidemic. Ann Epidemiol 2021;. Available from: https://doi.org/10.1016/j.annepidem.2021.04.017.

[173] Tapia MD, Sow SO, Ndiaye BP, Mbaye KD, Thiongane A, Ndour CT, et al. Safety, reactogenicity, and immunogenicity of a chimpanzee adenovirus vectored Ebola vaccine in adults in Africa: a randomised, observer-blind, placebo-controlled, phase 2 trial. Lancet Infect Dis 2020;20:707−18. Available from: https://doi.org/10.1016/S1473-3099(20)30016-5.

[174] Iannetta M, Di Caro A, Nicastri E, Vairo F, Masanja H, Kobinger G, et al. Viral hemorrhagic fevers other than ebola and lassa. Infect Dis Clin North Am 2019;33:977−1002. Available from: https://doi.org/10.1016/j.idc.2019.08.003.

[175] Kiiza P, Adhikari NKJ, Mullin S, Teo K, Fowler RA. Principles and practices of establishing a hospital-based ebola treatment unit. Crit Care Clin 2019;35:697−710. Available from: https://doi.org/10.1016/j.ccc.2019.06.011.

[176] Rewar S, Mirdha D. Transmission of Ebola virus disease: an overview. Ann Glob Heal 2014;80:444−51. Available from: https://doi.org/10.1016/j.aogh.2015.02.005.

[177] Lachâtre M, Yazdanpanah Y. Maladie à virus Ebola: actualités thérapeutiques. J Des Anti-Infectieux 2016;18:117−25. Available from: https://doi.org/10.1016/j.antinf.2016.07.002.

[178] Schuit M, Dunning R, Freeburger D, Miller D, Hooper I, Faisca L, et al. The use of an Ebola virus reporter cell line in a semi-automated microtitration assay. J Virol Methods 2021;292:114116. Available from: https://doi.org/10.1016/j.jviromet.2021.114116.

[179] Nicastri E, Kobinger G, Vairo F, Montaldo C, Mboera LEG, Ansunama R, et al. Ebola virus disease. Infect Dis Clin North Am 2019;33:953−76. Available from: https://doi.org/10.1016/j.idc.2019.08.005.

[180] Lasala F, García-Rubia A, Requena C, Galindo I, Cuesta-Geijo MA, García-Dorival I, et al. Identification of potential inhibitors of protein-protein interaction useful to fight against Ebola and other highly pathogenic viruses. Antivir Res 2021;186:105011. Available from: https://doi.org/10.1016/j.antiviral.2021.105011.

[181] Mutters NT, Malek V, Agnandji ST, Günther F, Tacconelli E. Evaluation of the scientific impact of the Ebola epidemic: a systematic review. Clin Microbiol Infect 2018;24:573−6. Available from: https://doi.org/10.1016/j.cmi.2017.08.027.

[182] Sykes C, Reisman M. Ebola: working toward treatments and vaccines. P T 2015;40:521−5.

[183] Buseh AG, Stevens PE, Bromberg M, Kelber ST. The Ebola epidemic in West Africa: challenges, opportunities, and policy priority areas. Nurs Outlook 2015;63:30−40. Available from: https://doi.org/10.1016/j.outlook.2014.12.013.

[184] Picazo E, Giordanetto F. Small molecule inhibitors of ebola virus infection. Drug Discov Today 2015;20:277−86. Available from: https://doi.org/10.1016/j.drudis.2014.12.010.

[185] Mirza MU, Vanmeert M, Ali A, Iman K, Froeyen M, Idrees M. Perspect towards antiviral drug discovery Ebola virus. Journal of Medical Virology 2019;91. Available from: https://doi.org/10.1002/jmv.25357.

[186] Ekins S, Southan C, Coffee M. Finding small molecules for the 'next Ebola.'. F1000Research 2015;4:58. Available from: https://doi.org/10.12688/f1000research.6181.2.

[187] De Rycker M, Baragaña B, Duce SL, Gilbert IH. Challenges and recent progress in drug discovery for tropical diseases. Nature 2018;559:498−506. Available from: https://doi.org/10.1038/s41586-018-0327-4.

[188] Capuzzi SJ, Sun W, Muratov EN, Martínez-Romero C, He S, Zhu W, et al. Computer-aided discovery and characterization of novel ebola virus inhibitors. J Med Chem 2018;61:3582−94. Available from: https://doi.org/10.1021/acs.jmedchem.8b00035.

[189] Liu H, Tian Y, Lee K, Krishnan P, Wang MKM, Whelan S, et al. Identification of potent ebola virus entry inhibitors with suitable properties for in vivo studies. J Med Chem 2018;61:6293−307. Available from: https://doi.org/10.1021/acs.jmedchem.8b00704.

[190] Gaisina IN, Peet NP, Wong L, Schafer AM, Cheng H, Anantpadma M, et al. Discovery and structural optimization of 4-(aminomethyl)benzamides as potent entry inhibitors of ebola and marburg virus infections. J Med Chem 2020;63:7211−25. Available from: https://doi.org/10.1021/acs.jmedchem.0c00463.

[191] Plewe MB, Sokolova NV, Gantla VR, Brown ER, Naik S, Fetsko A, et al. Discovery of adamantane carboxamides as ebola virus cell entry and glycoprotein inhibitors. ACS Med Chem Lett 2020;11:1160−7. Available from: https://doi.org/10.1021/acsmedchemlett.0c00025.

[192] Bessières M, Plebanek E, Chatterjee P, Shrivastava-Ranjan P, Flint M, Spiropoulou CF, et al. Design, synthesis and biological evaluation of 2-substituted-6-[(4-substituted-1-piperidyl)methyl]-1H-benzimidazoles as inhibitors of ebola virus infection. Eur J Med Chem 2021;214:113211. Available from: https://doi.org/10.1016/j.ejmech.2021.113211.

[193] Günther S, Lenz O. Lassa virus. Crit Rev Clin Lab Sci 2004;41:339−90. Available from: https://doi.org/10.1080/10408360490497456.

[194] Safronetz D, Lopez JE, Sogoba N, Traore' SF, Raffel SJ, Fischer ER, et al. Detection of lassa virus, mali. Emerg Infect Dis 2010;16:1123−6. Available from: https://doi.org/10.3201/eid1607.100146.

[195] Salvato MS, Domi A, Guzmán-Cardozo C, Medina-Moreno S, Zapata JC, Hsu H, et al. A single dose of modified vaccinia ankara expressing lassa virus-like particles protects mice from lethal intra-cerebral virus challenge. Pathogens 2019;8:133. Available from: https://doi.org/10.3390/pathogens8030133.

[196] Warner BM, Safronetz D, Stein DR. Current research for a vaccine against Lassa hemorrhagic fever virus. Drug Des Devel Ther 2018;12:2519−27. Available from: https://doi.org/10.2147/DDDT.S147276 Volume.

[197] Olayemi A, Cadar D, Magassouba N, Obadare A, Kourouma F, Oyeyiola A, et al. New hosts of the lassa virus. Sci Rep 2016;6:25280. Available from: https://doi.org/10.1038/srep25280.

[198] Monath TP. A short history of Lassa fever: the first 10−15 years after discovery. Curr Opin Virol 2019;37:77−83. Available from: https://doi.org/10.1016/j.coviro.2019.06.005.

[199] Garnett LE, Strong JE. Lassa fever: with 50 years of study, hundreds of thousands of patients and an extremely high disease burden, what have we learned? Curr Opin Virol 2019;37:123−31. Available from: https://doi.org/10.1016/j.coviro.2019.07.009.

[200] Warner BM, Siragam V, Stein DR. Assessment of antiviral therapeutics in animal models of Lassa fever. Curr Opin Virol 2019;37:84−90. Available from: https://doi.org/10.1016/j.coviro.2019.06.010.

[201] Cubitt B, Ortiz-Riano E, Cheng BY, Kim Y-J, Yeh CD, Chen CZ, et al. A cell-based, infectious-free, platform to identify inhibitors of lassa virus ribonucleoprotein (vRNP) activity. Antivir Res 2020;173:104667. Available from: https://doi.org/10.1016/j.antiviral.2019.104667.

[202] Dai D, Burgeson JR, Gharaibeh DN, Moore AL, Larson RA, Cerruti NR, et al. Discovery and optimization of potent broad-spectrum arenavirus inhibitors derived from benzimidazole. Bioorg Med Chem Lett 2013;23:744−9. Available from: https://doi.org/10.1016/j.bmcl.2012.11.095.

[203] Zhang X, Tang K, Guo Y. The antifungal isavuconazole inhibits the entry of lassa virus by targeting the stable signal peptide-GP2 subunit interface of lassa virus glycoprotein. Antivir Res 2020;174:104701. Available from: https://doi.org/10.1016/j.antiviral.2019.104701.

[204] Ellebedy AH, Ahmed R. Antiviral vaccines: challenges and advances. Vaccine B 2016;283−310. Available from: https://doi.org/10.1016/B978-0-12-802174-3.00015-1.

[205] Baumert TF, Schuster C. Editorial overview: viral resistance and challenges for antiviral therapies and vaccines. Curr Opin Virol 2016;20:vi−vii. Available from: https://doi.org/10.1016/j.coviro.2016.10.003.

[206] Cruz-Teran C, Tiruthani K, McSweeney M, Ma A, Pickles R, Lai SK. Challenges and opportunities for antiviral monoclonal antibodies as COVID-19 therapy. Adv Drug Deliv Rev 2021;169:100−17. Available from: https://doi.org/10.1016/j.addr.2020.12.004.

[207] Nascimento IJ, dos S, Santos-Júnior PF, da S, Silva-Júnior EFda. Brief introduction of measles virus and its therapeutic strategies. Human viruses diseases treatments vaccines. Cham: Springer International Publishing; 2021. p. 503−30. Available from: https://doi.org/10.1007/978-3-030-71165-8_23.

[208] Venkataraman S, Hefferon K, Makhzoum A, Abouhaidar M. Combating human viral diseases: will plant-based vaccines be the answer? Vaccines 2021;9:761. Available from: https://doi.org/10.3390/vaccines9070761.

[209] Mahmood N, Nasir SB, Hefferon K. Plant-based drugs and vaccines for COVID-19. Vaccines 2020;9:15. Available from: https://doi.org/10.3390/vaccines9010015.

[210] Rybicki EP. Plant-based vaccines against viruses. Virol J 2014;11:205. Available from: https://doi.org/10.1186/s12985-014-0205-0.

[211] LeBlanc Z, Waterhouse P, Bally J. Plant-based vaccines: the way ahead? Viruses 2020;13:5. Available from: https://doi.org/10.3390/v13010005.

[212] Chaudhuri S, Symons JA, Deval J. Innovation and trends in the development and approval of antiviral medicines: 1987−2017 and beyond. Antivir Res 2018;155:76−88. Available from: https://doi.org/10.1016/j.antiviral.2018.05.005.

[213] Domingo E. Trends in antiviral strategies. Virus Popul 2020;301−39. Available from: https://doi.org/10.1016/B978-0-12-816331-3.00009-X.

[214] Tompa DR, Immanuel A, Srikanth S, Kadhirvel S. Trends and strategies to combat viral infections: a review on FDA approved antiviral drugs. Int J Biol Macromol 2021;172:524−41. Available from: https://doi.org/10.1016/j.ijbiomac.2021.01.076.

[215] Steinhardt JJ, Guenaga J, Turner HL, McKee K, Louder MK, O'Dell S, et al. Rational design of a trispecific antibody targeting the HIV-1 Env with elevated anti-viral activity. Nat Commun 2018;9:877. Available from: https://doi.org/10.1038/s41467-018-03335-4.

[216] Gilchuk P, Bombardi RG, Erasmus JH, Tan Q, Nargi R, Soto C, et al. Integrated pipeline for the accelerated discovery of antiviral antibody therapeutics. Nat Biomed Eng 2020;4:1030−43. Available from: https://doi.org/10.1038/s41551-020-0594-x.

[217] Ahangarzadeh S, Payandeh Z, Arezumand R, Shahzamani K, Yarian F, Alibakhshi A. An update on antiviral antibody-based biopharmaceuticals. Int Immunopharmacol 2020;86:106760. Available from: https://doi.org/10.1016/j.intimp.2020.106760.

Anti-influenza agents

Sambuddha Chakraborty and Ashwini Chauhan

Department of Microbiology, Tripura University (A Central University), Suryamaninagar, Tripura West, India

11.1 Introduction

In 412 BC, Hippocrates in his book "Book of Epidemics" described a disease outbreak in an ancient Greek town "Perinthus" ("fever of Perinthus or "cough of Perinthus"), which progressed with flu-like symptoms. Some historians advocate the event as the first-ever influenza epidemic, though some of them found diphtheria a better fit for the same [1]. However, the concrete evidence of the first influenza pandemic was reported in Modena, Italy, 1510 [2]. Tommasino de' Bianchi, a then resident of the town was the first person to write down the disease course of flu; he chronicled the course of the disease and recorded the manifestations as high fever, headache, and cough, he also noted that the fever could last for three days and the cough for up to eight days [3]. It was a misconception that *Haemophilus influenzae* is the causative agent of human epidemic influenza. In 1933, during an influenza epidemic in the United Kingdom, physician Wilson Smith, C. H. Andrew, and P. P. Laidlaw filtered patient throat washings through a bacterial impermeable membrane and inoculated ferrets with those bacteriologically sterile filtrates, the subjected ferrets developed fever and they concluded that etiological agent of influenza is a virus [4].

In the past 100 years, four overwhelming influenza pandemics occurred: 1918 Spanish flu, 1957 Asian flu, 1968 Hong Kong flu, and 2009 swine flu, which took approximately 56 million human lives [5] and caused a huge economical and psychological burden.

Influenza virus belongs to the family Orthomyxoviridae [6], a family of segmented [7], negative sense [7] RNA viruses where it is placed with other members like Thogoto Virus, Infectious Salmon Anemia Virus, and Quaranfil Virus [8]. There are four genera of influenza viruses: influenza A, influenza B, and influenza C, and a relatively new type isolated from diseased pigs, the influenza D virus [9]. Influenza A and influenza B can infect human beings and these two types of viruses spread routinely in the human population. Influenza A virus is responsible for both seasonal flu and pandemic influenza. The virus can infect a large number of hosts like birds, horses, ferrets, mice, guinea pigs, macaques, and marmosets [10].

Viral Infections and Antiviral Therapies
DOI: https://doi.org/10.1016/B978-0-323-91814-5.00006-4

Vaccines have always been the mainstay of viral disease prevention and prophylaxis, several approaches are used to develop a vaccine against influenza. It was in 1940 that for the first time Thomas "Tommy" Francis Jr. and Jonas Salk successfully developed a monovalent inactivated influenza vaccine against the H1N1 strain [11]. Currently, there are several inactivated and live attenuated influenza vaccines available in the market with the trivalent and quadrivalent formulations. The Centre for Disease Control (CDC) recommends a yearly vaccination as the primary preventive measure against flu. The vaccine is recommended for individuals older than 6 months. Due to the probability of the emergence of new influenza strains every year, the viral content of the vaccine formulations keeps changing. To formulate a vaccine, it is necessary to first predict and identify the probable viral strain(s), which may take up to 8 months. Vaccine formulations may become ineffective or may have reduced efficacy due to in-between mutations occurring due to antigenic drift and these occasional antigenic shifts may result in a new pandemic [12]. The uncertain efficacy of the vaccine upholds the importance of antivirals. Currently, there are four classes of antivirals, that is, Matrix Protein 2 (M2) ion channel inhibitors, neuraminidase (NA) inhibitors, RNA-dependent RNA polymerase (RdRp) inhibitors, and polymerase acidic protein (PA) inhibitors. Among these four classes of antiviral drugs, only eight drugs are approved by different Pharmaceutical Regulatory Agencies like US-FDA, MHLW, SFDA, SWISSMEDIC, viz., amantadine, rimantadine (Symmetrel, Flumadine), which are M2 ion channel inhibitors, peramivir, zanamivir (ZAN), laninamivir (LAN), oseltamivir (OS) (Rapivab, Relenza, Inavir, Tamiflu) are classified as NA inhibitors, favipiravir (Avigan) RdRp inhibitor, and baloxavir marboxil (Xofluza) as PA inhibitor [13–15]. The development of antiviral resistance and the side effects associated with the drugs are major drawbacks of synthetic drugs and hence, researchers are also trying to adapt traditional natural remedies to treat influenza infection [16].

This chapter focuses on different anti-influenza agents including approved drugs that are currently available in the market against flu and also discusses potential anti-influenza agents that are under clinical trials or in the basic experimental stage.

11.2 The virus

11.2.1 Virion structure

Influenza is an enveloped virus. The virus is observed mostly in a spherical shape or occasionally in the filamentous form [17]. Viral envelopes are usually derived from the infected host cells, either from the plasma membrane or from the other internal organelles like the nucleus, ER, or Golgi complexes [18,19]. Studies have shown that compositional difference in viral envelop alters viral infectivity. The envelopes are mainly composed of phospholipids and glycolipids as protein component glycoproteins are embedded in the lipid bilayer known as peplomers or spikes [20]. In 1934 Laidlaw, and Smith found that the influenza virus will pass through the 600 μm Gradocol collodion membrane [21]. Several ideas, techniques, and experiments were explored to measure the size of viral particles of influenza, such as centrifugation, diffusion, ultracentrifugation, or ultrafiltration and the data from these methods revealed the size between 6 and 160 μm [22]. Now by

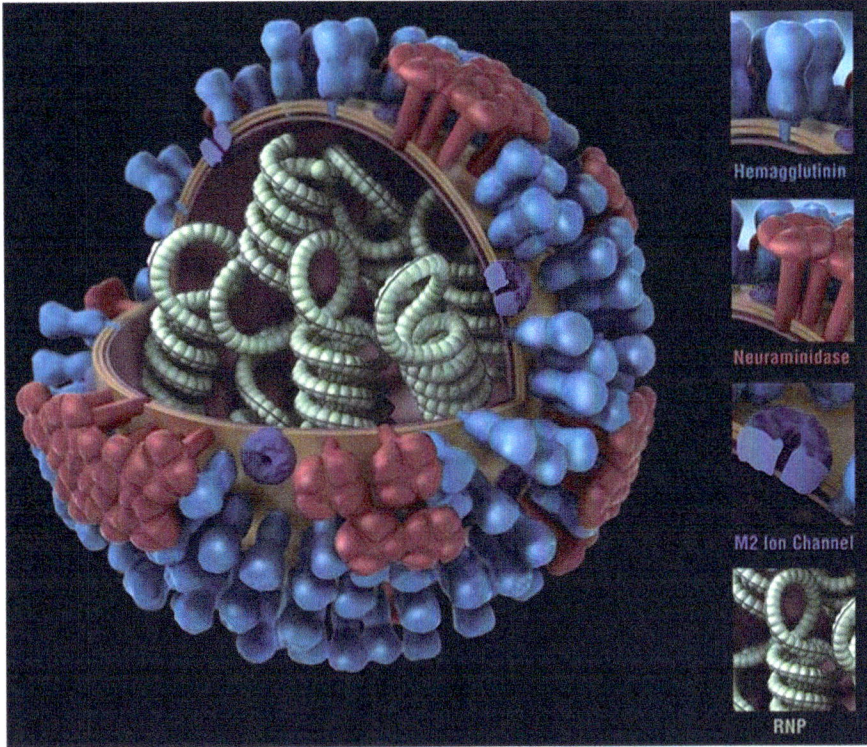

FIGURE 11.1 Structure of influenza virus. *From https://www.cdc.gov/flu/resource-center/freeresources/graphics/images.htm.*

using more advanced methods such as Negative Staining EM, TEM, or AFM, it is clear that the viral spheres range from 80 to 100 nm while the filamentous particles can reach more than 20 μm in size [23] (Fig. 11.1).

One of the most important structural features of any virus is the spike proteins, proteins that bind to the specific host cellular receptors and initiate infectivity. Traditional viral vaccines work by generating immunity against these spike proteins. Influenza spike protein is called hemagglutinin, or HA. HA is the most studied protein in the world and probably one of the deadliest proteins too as it has costed humanity an unprecedented number of lives. HA is a membrane fusion glycoprotein that binds to the host cell receptors and also helps in uncoating by fusing viral and host cell membranes [24]. Another surface glycoprotein of the influenza virus is NA a sialidase that cleaves the glycosidic linkages of sialic acid (SA) and helps the virus to bud out [25]. Although HA and NA together comprise 97% of total viral envelop protein, HA contributes to almost 80% of total envelop protein, and 16 to 18 M2 proteins are observed in one virion [26]. Another envelop transmembrane protein of influenza virus is Matrix 2 protein or M2 protein. The M2 protein is a proton channel that helps to equilibrate the pH necessary for virus entry and maturation [27–29]. Under the envelope another protein called Matrix 1 protein or M1 protein surrounds the internal viral interior. The M1 protein functions to communicate between envelop proteins and the viral genome during infection and also plays a role in virus budding out [30]. The core of the flu virus contains eight segmented negative-sense RNA and Nuclear Export Protein (NEP) [31]. Different RNA segments code for different specific proteins inside the infected host cells. NEP is a multipurpose protein of influenza virus it exports viral ribonuclear proteins (RNP) in the nucleus for replication and also triggers the budding of progeny viral particles from the infected cell.

11.2.2 Viral genes and viral proteins

Influenza A and B possess eight segments of negative-strand RNAs [32] whereas C and D have only seven segments of RNAs [17,33]. Negative strand RNA or antisense RNA requires primary transcription of negative-sense genome into mRNA or sense RNA for the translations of proteins [34]. The genes are conventionally numbered in descending order according to the genome size. Eight genetic segments code for 12 proteins among which the structural proteins help in viral replication and the nonstructural proteins protect the virus from host defense by antagonizing host antiviral response [35].

Influenza virus RdRp is a trimeric protein with a mass of 250 kDa, the protein is composed of three subunits namely PA, PB1, and PB2 [36]. The central role of RdRp is the translation and replication of influenza viral RNA. Mentioned three different subunits of RdRp has separate roles associated with virus infection. PB2 subunit of flu RdRp is an β interferon antagonist and it inhibits interferon expression by associating with the mitochondrial antiviral signaling [37,38]. The study revealed that in PB2 there are two functional domains where NP and PB1 can bind causing negative regulation of viral gene expression [39].

PB1 subunit of the flu is the minimal essential subunit of RdRp which can catalyze RNA synthesis in the absence of PB1 and PA also [40]. Apart from that PB1-F2 subunit of PB1, which is a protein composed of 90 amino acids can enhance pro-inflammatory response and promotes cell death [41]. PB1 N40 is a subunit of PB1 protein and it is found that PB1, PB1-F2, and PB1 N40 have an interdependent expression and PB1 N40 can modulate the ability of replication of the virus [42].

PA is one of the key proteins of the influenza virus which can be cleaved into two domains a small N terminal 25 kDa domain and a larger 55 kDa C-terminal domain (CTD). The major function of PA in viral RdRp is cap snitching as the N-terminal domain of PA is a cap-dependent endonuclease (CEN). PA and PB1 directly interact to form a stable complex which may inhibit viral replication [43,44]. HA protein is a trimeric glycoprotein [45] that binds to the host cell receptors and helps the virus enter the cell [46]. Influenza A virus is subtyped based on HA (H) and NA (N) proteins, till date there are 18 different HA (H18) subtypes and 11 different NA (N11) subtypes have been reported [47]. Primarily Nucleoprotein (NP) encapsidate the viral genome and forms RNP for the sake of replication, transcription, trafficking of RNP, and packaging, [48,49]. NA is a tetrameric sialidase enzyme, tetramers are made up of four identical polypeptide monomers consisting ~470 amino acids. Each monomer forms structural domains like cytoplasmic tails, transmembrane region, stalk, and catalytic head. It helps in the

TABLE 11.1 Viral genes and proteins.

Sr. no.	Gene segment	Encoded protein/s	Functions
1	PB2	Polymerase basic 2 protein	Viral polymerase subunit, antagonize mitochondrial interferon regulations, modulates replication, RNA synthesis,
2	PB1	i. PB1 (Polymerase basic 1 protein) ii. PB1-F2 iii. PB1-N40	i. Viral polymerase subunit, helps in viral replication, can catalyze RNA synthesis independently ii. Trigger apoptosis, interferon antagonist iii. Regulates expression of PB1 and PB1-F2
3	PA	Polymerase acidic protein	Viral polymerase subunit, cap snitching activity
4	HA	Hemagglutinin	Host cell receptor attachment, flu subtyping
5	NP	Nucleoprotein	RNA synthesis, RNP trafficking
6	NA	Neuraminidase	Release of viral progeny from host cell Helps in motility of the virus through the mucous lining of the respiratory tract.
7	M	i. M1 (Matrix protein) ii. M2 (Membrane protein)	i. Assembly and budding ii. Viral Entry, ion channel
8	NS1	i. NS1 (Nonstructural protein 1) ii. NEP/NS2 (Nuclear export protein)	i. IFN antagonist, regulation of RNA synthesis ii. Nuclear export of vRNP

movement of the virus but the major function is of course the catalytic activity where it cleaves the SA residue from the cell surface and lets the progeny virus release [50,51]. M gene of influenza viruses is 1027 nucleotides long and can code for two proteins M1 and M2 [52]. Both M1 and M2 are involved in many crucial functions in viral replication like viral entry, assembly, budding, and nuclear export of vRNP [27,53,54]. Nonstructural protein 1 and 2 of the influenza virus is coded by the 8th gene of the virus [55]. NS1 protein of influenza virus is constituted of 215 to 237 amino acids and bears a molecule weight of 26 kDa, it acts as an interferon antagonist by blocking the activation of the 2′-5′-oligoadenylate synthetase (OAS)/RNase L pathway [56]. It can selectively upregulate the synthesis of viral mRNA by regulating the activity of viral RdRp. NS1 is made up of RNA Binding Domain (RBD) and effector domain linked by a flexible linker region [57]. NS2 or NEP is involved in the export of vRNP outside the nucleus during the late stage of the infection [58] (Table 11.1).

11.2.3 Life cycle of influenza virus

Like most viruses, the steps of the flu life cycle can also be divided into six major categories Attachment—Penetration—Uncoating—Replication—Assembly, and Release [59]. Viruses are fastidious about their hosts; hence different viruses have different cell or tissue tropism. Like HIV replicates in CD4$^+$ T lymphocytes [60], hydrophobia causing rabies virus replicates inside the muscle cells and neurons, [61,62] and so on. The primary reason for the tissue tropism of any virus is the receptor (present on host cell surface) and antireceptor (spike proteins) compatibility between host and virus. Specifically, the RBD present in the spike proteins are the key players in virus attachment and entry [63]. Influenza virus mainly infects upper and lower respiratory tract epithelial cells [64], respiratory tract epithelial cells contain α2,6-linked or α2,3-linked SAs to the penultimate galactose of cell surface glycoconjugates (glycoproteins or glycolipids). It is known to us that the HA protein of flu binds to any molecule present with terminal sialosides [65]. Interestingly human influenza virus can recognize α2,6-linked SA moieties which are mostly present in the upper respiratory tract (trachea) of humans, whereas Avian influenza is more specific to α2,3-linked SAs present in the lower respiratory tract (type II alveolar cells) of humans or GI tract of birds [66]. The distribution of SA receptors in humans and swine are similar, as in α2,6 linked SA are predominant in the upper respiratory tract and lower respiratory tract are frequent with α2,3 SA and this physiology of swine is a suitable for reassortment or genetic shift of the virus and hence pigs are called as mixing vessels of Influenza [67]. Nevertheless, recent studies have reported sialoglycan-independent entries of influenza virus where the virus interacts with different receptors and coreceptors for the attachment and internalization to host cell [68,69]. Viral particles reach the receptor sites by crawling or gliding movement facilitated by NA and HA around the cell surface [70]. Adequate binding of HA and SA triggers clathrin-mediated or independent endocytic pathway of virus internalization by the host cell plasma membrane [71]. Besides endocytosis, another alternate mode of internalization of the receptor attached viral particles is macropinocytosis [72]. Being internalized the virus undergoes microtubule dependent transportation towards the perinuclear region consisting of the formation of early endosome, middle endosome, and late endosome before viral fusion [73]. Fusion of viral membrane with endosome acidifies the late endosome as well as the virus core which disassociates vRNPs from M1 and triggers the release of viral ribonucleic proteins, vRNP (vRNA + NP = vRNP), a more stabilized complex in the host cell cytoplasm [73,74] (Fig. 11.2).

Positive sense mRNA synthesis and RNA replication of influenza virus occurs inside the host nucleus [75,76]. vRNPs contain Nuclear Localization Sequences [77] which gets exposed during the dissociation from the M1 and finally gets translocated in the nucleus [78]. The genetic material of influenza is negative-strand RNA (antisense RNA), so the virus needs to synthesize mRNA (positive sense RNA/sense RNA) for further transcription and replication. PA subunit of heteromeric viral RdRp carries out the "cap snitching" from host cellular RNA Polymerase II (RNAPII) and RdRp performs the transcription and replication of the viral genome [79]. Nuclear export of the synthesized viral RNA to the cytoplasm is predominantly a NEP mediated activity, besides host factor, Nucleoporin 93 and viral M1 protein by CRM1 dependent pathway [80] are also found to be contributing to vRNA nuclear export. Protein synthesis in influenza virus is completely dependent on the host cellular translation machineries. Viral mRNA translation occurs in two different locations; PB1, PB2, PA, NP, NS1, NS2, and M1 are translated in cytosolic ribosomes, and envelop embedded proteins HA, NA, and M2 are translated in endoplasmic reticulum-associated ribosomes [31]. After protein synthesis virus orchestrates the assembly of viral proteins and vRNPs and prepares to bud out from the

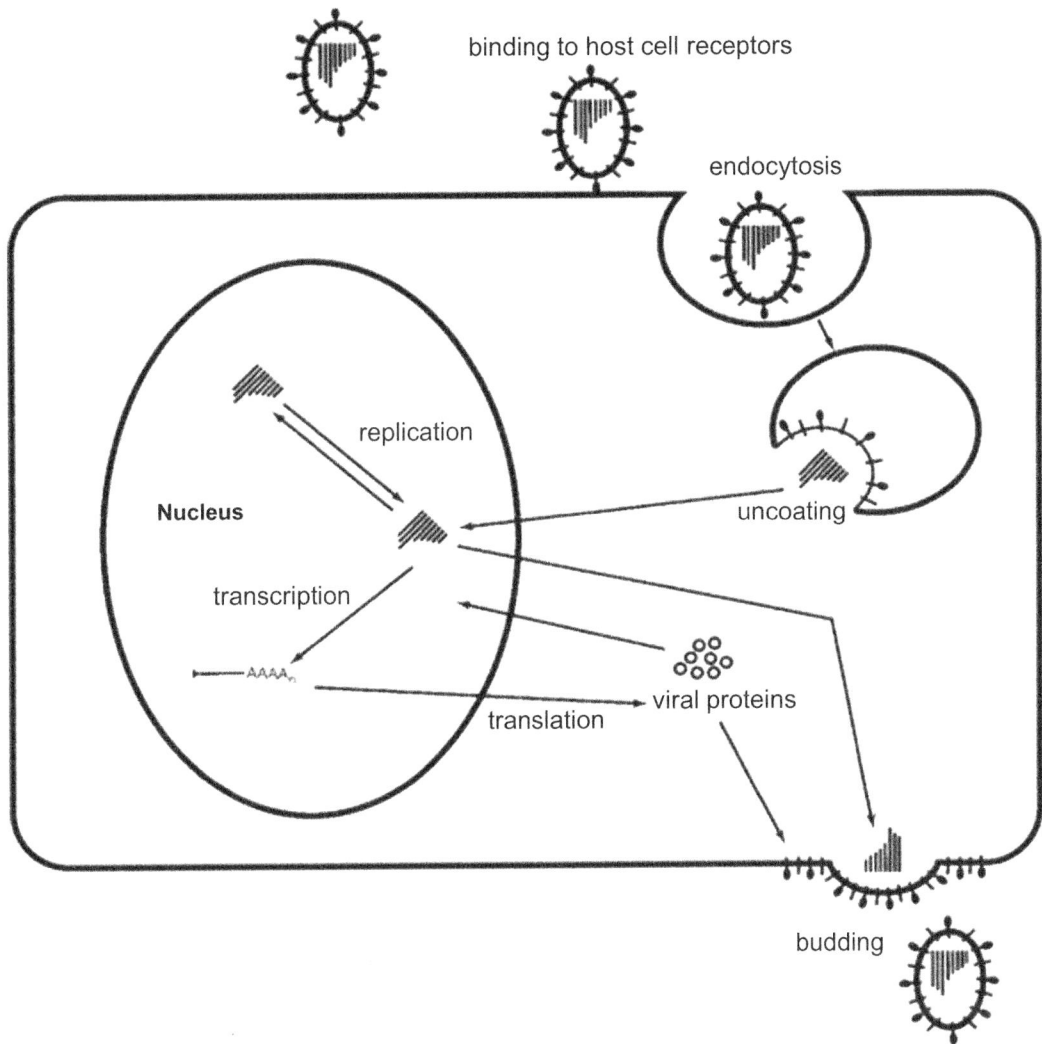

FIGURE 11.2   Structure of influenza virus replication.

cell. It is evident that the influenza virus utilizes cholesterol and sphingolipid-enriched lipid raft domains present in the plasma membrane of infected cells as sites of virus assembly and budding [81]. The vRNPs get delivered to the cell periphery by Ras-related protein 11 (Rab11) and localize to the budding site [82,83]. Influenza generally buds out from the apical plasma membrane of the cell [84]. The transmembrane proteins like HA, NA, M2, and the matrix protein M1 interact with the inner leaflet of the lipid bilayer and play an essential role in budding, M1 agglomeration results in the bending of plasma membrane and initiates budding [85]. Finally, the release of progeny virus requires sialidase activity of NA to prevent self-aggregation and reattachment to the infected cell [86]. After release, the progeny virus particles are ready to infect new cells and continue to replicate.

11.3 Anti-influenza agents

11.3.1 Regulatory authority-approved anti-influenza agents

11.3.1.1 Class I. Matrix protein 2 ion channel inhibitor

11.3.1.1.1 Amantadine

1-Adamantanamine or amantadine is a stable, colorless, crystalline amine synthesized from adamantine [87]. It is the first synthetic compound that exhibited antiviral activity against the influenza virus. Amantadine was discovered in the year 1960 and approved as prophylaxis of specifically Asian influenza, influenza-A (H2N2) infection in United States by October 1966 [88], and for prophylaxis and treatment of all influenza A infections in 1976 [89]. The M2 protein of influenza A virus is a tetramer composed of four α helices and acts as an ion channel that modulates the pH necessary for virus entry [27,90]. Amantadine acts on M2 ion channels and blocks the influx of H^+ ions resulting in inhibition of fusion between viral and host endosomal membrane [91]. Amantadine was accidentally discovered effective against Parkinson's disease. The compound can raise the concentration of dopamine in the synaptic cleft due to this neurological property. Amantadine was adopted as a drug against Parkinson's disease [89,92]. Amantadine resistance in treatment study was executed by Hall et al. in 1987 [93]. Among 37 children subjected to the study 10 children were shedding amantadine-resistant virus. Point mutation leading to a single amino acid change in positions 26, 27 (valine to alanine), 30 (alanine to threonine), 31 (serine to asparagine), or 34 (glycine to glutamic) of M2 protein made the virus resistant to amantadine [94,95]. After four decades of use in 2006, the CDC restricted the use of amantadine against flu [96].

11.3.1.1.2 Rimantadine

Rimantadine is a methyl derivative of amantadine, discovered in the year 1963 [97]. Former USSR started using Rimantadine as a drug of choice against influenza A in 1969 [98]. In 1993 rimantadine received US FDA approval and was indicated for both prophylactic and curative treatment of influenza A [99]. The mode of action of the rimantadine is similar to that of the M2 channel blockade of amantadine molecule. X-ray crystallography and NMR studies revealed that four rimantadine molecules bind to the amino acid residues of alpha helicase towards the cytoplasmic side of the membrane by hydrogen bonding [100] promoting inhibition of M2 ion channel activity. In 1987 Thompson et al. during a therapeutic trial in children, reported shedding of rimantadine resistant virus strain in the late stage of infection [101]. Similar observations were made by another group (Hall C. B., R. Dolin, C. L. 1987) and the viruses were genetically studied by Belshe et al. and found a point mutation in 30 (Ala to val) and 31 (Ser to Asn) number amino acids of M2 protein are responsible rimantadine resistance [102]. In a drug resistance survey study led by

Ziegler et al. 1999. 2017 isolates of flu were collected and 16 viruses were found resistant to rimantadine [103]. A point mutation of the transmembrane domain M2 gene in the 31st position (Ser to Asn) was reported for 15 viruses and the remaining one showed a point mutation in 30th position (ala to thr), the mutation pattern backed the study data of Belshe et al. [104]. An increased adamantane resistance against influenza A (H3N2) has been reported hence amantadine and rimantadine are not recommended by CDC for the treatment or prophylaxis against influenza infection [105].

11.3.1.2 Class II. Neuraminidase inhibitors

11.3.1.2.1 Zanamivir

Mark von Itzstein group of departments of pharmaceutical chemistry, Victorian college of pharmacy, Australia designed an influenza NA substrate; 4-guanidino-2,4-dideoxy-2,3dehydro-X-acetylneuraminic acid (4guanidino-Neu5Ac2en) [106] by using Grid a bioinformatics software [107]. The molecule was further subjected to in vivo studies in mice by the researchers at Glaxo Group Research Ltd., United Kingdom to evaluate its efficacy to prevent influenza infection. The molecule was found to be the potential anti-influenza agent. Further modifications to the compounds were made to increase their antiviral efficacy and finally as a more efficacious composition, 4-deoxy-4guanidino-Neu5Ac2en also called DANA was selected by Glaxo as a drug candidate under the generic name ZAN [108] and trade name as Relenza. ZAN inhibits the release of progeny virus by acting and binding with the viral NA as substrate [109]. In 1998 clinical study of the drug was completed on 6000 adult and adolescent patients in North America and showed a highly effective and safe result against influenza A and B virus [110]. In 1999, ZAN was approved as the first neuraminidase inhibitors (NAI) to use in the treatment of influenza A and influenza B by US FDA. Oral bioavailability of ZAN is poor hence the drug is administrated by oral inhalation [111]. More than 75% of the inhaled drug gets deposited in the oropharynx and swallowed, 13% of the drug gets distributed in the lungs and airways [112] and gets concentrated in the mucosa.

Prolonged treatment with ZAN can develop influenza resistance to the drug. In a study wherein an immunocompromised patient who was treated with OS for more than two months and then ZAN for more than two months, the prolonged treatment with ZAN caused point mutation in I223R position which was earlier shown to cause ZAN resistance [113,114]. In another case study with 18 years old immunocompromised patient with influenza B infection resistance against ZAN was reported to be developed due to prolonged exposure to the drug [115]. The study reported that 12 to 15 days of treatment with ZAN can cause a point mutation in the position 152 Arg to Lys that reduced the enzyme sensitivity to ZAN by 1000 folds.

11.3.1.2.2 Oseltamivir

This is the first oral anti-influenza antiviral medication of NAI class [116], famous as Tamiflu. The drug belongs to a big house; F. Hoffmann-La Roche AG, commonly known as Roche, a Swiss multinational healthcare company. The company claimed that Tamiflu cuts the hospital admission and reduces complications like sinusitis, bronchitis, and pneumoniae by 61% and 67% respectively, and reduces the number of deaths. On the contrary, in 1999, the approval committee of the FDA turned down the approval of Tamiflu because it had no impact on reducing death or pneumonia [117]. However, the notion was overruled by the FDA authority and Tamiflu received the approval for use against influenza A and B in the same year [118]. OS is a prodrug that needs to get hydrolyzed by hepatic esterase and convert into active drug OS carboxylate which further can inhibit the NA activity of influenza A and B viruses [119]. In 2014 British Medical Journal (BMJ) published a review of all clinical studies done on Tamiflu and concluded its insignificancy as an efficient and safe anti-influenza drug. BMJ countered WHO and CDC for listing Tamiflu as an essential drug and stockpiling the same [120]. The drug is currently recommended to clinicians to use against uncomplicated influenza infection by CDC. OS is a prodrug of OS carboxylate, whose twice 20 to 200 mg oral administration in a clinical study restricted influenza A and B viral load within 5 days [119]. Excessive use of OS increases the chance of OS-resistant viral population, a single amino acid change in H274Y position of NA has resulted into increased circulation of OS-resistant seasonal flu [121].

11.3.1.2.3 Peramivir

Peramivir was given emergency use authorization (EUA) during the 2009 H1N1 pandemic till 2010 July [122] and was approved by FDA in 2014 December 19th [123]. Peramivir was approved by the Japanese government in 2010 and it is licensed in South Korea also [124]. The drug is licensed to be sold under trade name Rapivab manufactured by BioCryst Pharmaceuticals, United States. Peramivir is a pentacyclic drug with a C4-guanidino substitution and hydrophobic side chain which can cause a strong interaction between peramivir and the enzyme active site [125,126]. It is indicated to use in uncomplicated influenza infections of adults and children above 2 years of age. In a study, it was shown that OS-resistant H274Y mutation also reduces the efficacy of peramivir in suppressing the viral replication [127,128].

11.3.1.2.4 Laninamivir

In 2008, a new compound was discovered by the researchers at Daiichi Sankyo Co. Ltd a Japanese pharmaceutical company, LAN octanoate, a prodrug of LAN, effective against influenza A and B viruses [129]. In 2010 Daiichi Sankyo Co., Ltd, Tokyo was granted license to manufacture the drug under trade name Inavir [130]. LAN is a long-acting NAI. The drug shares same binding sites in NA enzyme active site like other three NAIs OS, ZAN, and peramivir. It is an inhalation drug; it persists in lungs for a long time and the concentration of drug that resides in lungs is reported to be sufficient to suppress the influenza replication [131]. Moreover, LAN is found to be efficient against OS-resistant (His274Tyr) strain [132]. Core structural confirmation of LAN is like ZAN, is Neu5Ac2en with 4-guanidino group, in addition, LAN contains an additional 7methoxy group [132].

11.3.1.3 Class III. RNA-dependent RNA polymerase inhibitors

Favipiravir (T-705/6-fluoro-3-hydroxy-2-pyrazinecarboxamide) was discovered during a chemical screening library against influenza virus by Furuta and his team of

Toyama Chemicals, a Japanese pharmaceutical held by Fujifilm Corporation [133]. It is a prodrug molecule. The drug was approved by PMDA for use in Japan in February 2014 under the trade name Avigan [134] and the United States in 2018 [135]. It is a prodrug that needs to undergo an intracellular phosphoribosylation to be in active form, favipiravir-RTP (favipiravir ribofuranosyl-5B-triphosphate) [133]. Favipiravir directly inhibits viral replication by targeting viral RdRp [133]. Favipiravir is a broad-spectrum antiviral and is used against Lassa and Ebola virus [136–138]. Favipiravir is also considered to treat COVID-19 and the clinical trials are going on [139]. On 15th September 2021, drug maker Glenmark confirmed its safety in the study of the clinical trial with 1000 COVID-19 patients [140]. This pyrazine compound was found to be a potent anti-influenza agent. It targets the middle stage of virus replication. The compound inhibits the virus replication by targeting the RNA polymerase activity in GTP-competitive manner [136]. Furthermore, the addition of purine nucleoside attenuates the antiviral activity of the compound.

11.3.1.4 Class IV. Polymerase acidic protein inhibitor

Baloxavir (BXA) marboxil is a prodrug of baloxavir acid, which is an inhibitor of PA of the influenza virus RdRp complex [141]. Cap snatching is an important phenomenon in influenza viral replication where interaction between viral RdRp and CTD of cellular DNA dependent RNAPII occurs where 5′-capped transcripts produced by RNAPII get cleaved and then the cleaved and capped host cellular oligonucleotide primer is used to initiate viral RNA replication [142,143]. CEN is a structural part of PA (RdRp subunit [144]) which commits cap snatching from the host cellular RNA [145]. BXA inhibits CEN activity and suppresses viral replication [146]. The drug was given FDA approval under trade name Xofluza as a new anti-influenza antiviral in October, 2018 [147,148].

11.3.1.5 Hemagglutinin inhibitor

Umifenovir or Arbidol (ARB) is (ethyl-6-bromo-4-[(dimethylamino)methyl]-5-hydroxy-1-methyl-2-[(phenylthio)methyl]-indole-3-carboxylate hydrochloride monohydrate) is a broad-spectrum antiviral used against both enveloped and non-enveloped viruses like respiratory syncytial, adenovirus, coxsackie B5, parainfluenza, Ebola, and hepatitis B and C virus [149]. The drug was made by the Russian Re-search Chemical Pharmaceutical Institute in the year 1988 [150]. The medicine is used as a prophylaxis and treatment measure for respiratory illness since 1990. It is an indole-derived compound and synthesized from 1,2-dimethyl-5-hydroxyindole-3-acetic acid ethyl ester [151]. ARB has exhibited potent and broad anti-influenza activity against both human and avian influenza A virus strains (H1N1, H2N2, H3N2, H5N1, and H9N2) and influenza B virus and C virus [152]. Structural study of Arbidol–HA complexes revealed that ARB binds to a hydrophobic cavity at the interface of two protomers of HA trimer. ARB uses an induced-fit mechanism involving major and minor conformational changes in the binding pocket resulting in the breaking of preexisting salt bridges and the formation of new ones. The drug inhibits cell entry of enveloped viruses by attaching to the HA protein and blocking viral fusion with the host cell membrane. The drug is licensed in Russia (1990) and China (2006) [150] and was never submitted for FDA approval.

TABLE 11.2   Government regulatory agency-approved anti influenza agents.

Name of the compound	Trade name	Mode of action	Approval status
Amantadine	Symmetrel	M2 ion channel blockade	1966 US FDA Prophylaxis 1976 US FDA Prophylaxis and Treatment 2006 Restricted by CDC
Rimantadine	Flumadine	M2 ion channel blockade	1969 USSR 1993 US FDA
Zanamivir	Relenza	NAI inhibitor	1999 Australia 1999 US FDA
Oseltamivir	Tamiflu	NAI inhibitor	1999 Switzerland 1999 US FDA
Peramivir	Rapivab	NAI inhibitor	2010 JAPAN 2014 US FDA
Laninamivir	Inavir	NAI inhibitor	2010 US JAPAN
Favipiravir	Avigan	RdRp Inhibitor	2014 US JAPAN 2018 US FDA
Baloxavir marboxil	Xofluza	PA Inhibitor	2018 US FDA
Umifenovir	Arbidol	HA inhibitor/Fusion Inhibitor	1990 Russia 2006 China

Currently, the drug is considered to repurpose against COVID-19 and is currently under clinical trials [153,154] (Table 11.2, Fig. 11.3).

11.3.2 Anti-influenza agents under development

11.3.2.1 Anti-influenza agents under clinical trials

Influenza has been a health concern for a long time. Currently, there are four anti-influenza antivirals recommended by CDC to use against seasonal flu, oseltamivir, zanamivir, peramivir, and baloxavir marboxil, among these four drugs three drugs are NAI and baloxavir marboxil is an inhibitor of PA.

All viral pandemics have reminded us that we lack proper antiviral agents to deal with virulent viral pathogens. And as influenza is a rapidly mutating virus, excessive use of any drug will give rise to resistant variants due to selective pressure exerted by the drugs. Hence to deal with such highly unstable viruses like influenza we need a hot pipeline of antiviral agents against them. Three of the agents which can be acted as anti-influenza antivirals are in the clinical phases; a repurposed antiparasitic drug "Nitazoxanide" is currently in phase three clinical trials [155]. A Conjugated sialidase "DAS181" is in phase two clinical trial [156]. An oral endonuclease Inhibitor AL-794 is on phase one study [157] (Fig. 11.4).

11.3.2.1.1 Nitazoxanide

Nitazoxanide is a triazolide compound that was developed to use against protozoa and helminthes [158]. In 2002 the drug was approved by the US FDA for use in 1–11 years of

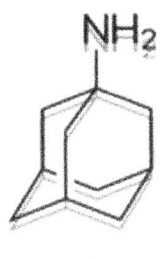

Amantadine

Rimantadine

Zanamivir

Oseltamivir

Peramivir

Laninamivir

Favipiravir

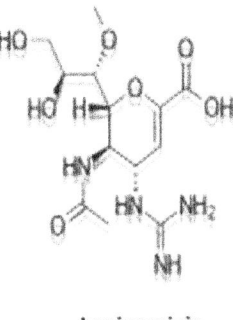

Baloxavir marboxil

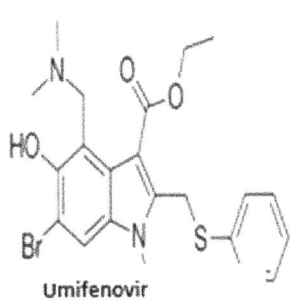

Umifenovir

FIGURE 11.3 Structures of agency approved anti-influenza agents.

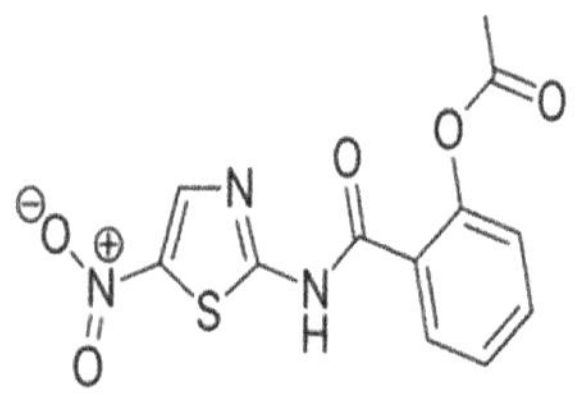

FIGURE 11.4 Nitazoxanide.

age against *Cryptosporidium* and *Giardia intestinalis* diarrhea patients. In 2006 it was approved to be used against G. intestinalis infections in adults. The mode of action of nitazoxanide is inhibition of pyruvate-ferredoxin oxidoreductase an enzyme essential for an anaerobic metabolic pathway of protozoa and bacteria [159] and it can derange mitochondrial respiration of parasites [160]. Rossignol et al. for the first time studied the anti-influenza activity of nitazoxanide, the group interpreted the effect and mechanism of nitazoxanide and its active metabolite tizoxanide. These compounds suppress the influenza replication by blocking the maturation of HA post translation and subsequently failing the assembly and release of progeny virus [161]. Danielle Tilmanis et al. in 2017 tested the susceptibility of the compound against influenza A (H1N1) and (H3N2), and influenza B viruses over 210 isolates and found positive [162]. In a double-blind, randomized, placebo-controlled, phase 2b/3 trial with 624 symptomatic participants, patients showed a significant reduction in symptom duration with the drug, and no significant side effects were presented in either the placebo or intervention group [163]. Phase three of the trial of the drug is currently underway, and if it receives approval for the intended use, it will surely be a major success.

11.3.2.1.2 DAS181

DAS181 is a chimeric sialidase protein constructed by fusing *Actinomyces viscous* derived sialidase catalytic domain and cell surface-anchor [164]. The construct is taken by inhalation to cleave the respiratory tract epithelial cell SA receptors. It can cleave both α2−6 and α2−3 linkage of SA. Due to the cleavage of SA residues from the cell surface, the viruses do not get a chance to get attached to the cell surface. It is reported that three days of treatment with the DAS181 reduced viral shedding. DAS181 is found effective against both influenza A and B virus [165]. In the first double-blind, placebo-controlled phase II clinical trial with 177 laboratory-confirmed influenza-infected participants for the efficacy and safety of this antiviral agent, the drug showed a well tolerance among the receivers and a significant reduction of viral load [166]. To examine the impact of higher dosing levels on virus load and tolerance further studies are required. DAS181 is a potential host-directed antiviral approach for treating resistant and novel influenza strains.

11.3.2.1.3 AL-794

AL-794 is a prodrug of ALS-033719 (structures are not revealed by the company). It inhibits the endonuclease activity of the influenza virus RdRp enzyme by binding to the endonuclease domain of influenza virus polymerase acidic protein. ALS-033719 has shown an inhibitory effect on different strains of both influenza virus A and B Virus. 61 individuals participated in the human challenge study of safety. The individuals were inoculated with 1 mL 5.5 $\log_{10}$/mL of good manufacturing practice–grade influenza A/Perth/16/2009[H3N2]. 150 mg of the drug reduced the viral load to 2.77 $\log_{10}$/mL.

Adverse events reported were mostly mild in severity, and there were no significant differences with the placebo group.

11.3.2.2 Anti-influenza agents under basic research

As we already discussed earlier, it is necessary to keep experimenting and learning about new possible drug targets for influenza-like viruses because the virus is unpredictable due to its patterns of mutations. So, it is always a matter of risk that an upcoming pandemic may outpace our anti-influenza arsenal. However, it is a matter of fact that understanding the probable outcome of the flu pandemic in the anti-influenza agent development research has not been overlooked in recent times. There are more than 2800 patents on anti-influenza therapeutics registered worldwide [167]. Different approaches for designing influenza antiviral are as follows [88] (Table 11.3).

11.3.3 Anti-influenza agents from traditional plants

To date, all the approved agents against flu are mainly synthetic chemical components. During 2009, H1N1 pandemic Chinese government officially recommended four antiflu TCM (Traditional Chinese Medicine) along with oseltamivir and zanamivir, for the treatment purpose of influenza [16]. The emergence of drug resistance variance in response to synthetic drug application is one of the serious public health issues. Identification of new antiviral agents as an alternative to synthetic therapeutics need intense investigation and strong experimental approach. "Ayurveda" is an ancient book of India where sages have written their understanding and findings on natural elements like herbs, stones, salts, etc. and their usefulness as medicines. Surprisingly the ways of Chinese medicine and Ayurveda are similar with both being scripted in their respective geographic regions [201]. Side effects of the natural products are very less to none, compared to the synthetic medicinal compounds, and the production is eco-friendly and cost-efficient. However, we cannot deny the fact that with the advancement of modern medical practices, the age-old traditional practices of medicines have been diluted. Nevertheless, it is necessary to establish a productive collaboration between ancient contemplation and modern techniques for the betterment of the human race.

Bioactive compounds like polyphenol, saponin, flavonoid, glucosides, and alkaloids derived from the medicinal herbs have been extensively studied for anti-influenza activity [202,203]. The derivatives were able to constrict viral replication by blocking different stages of the life cycle that is, attachment, penetration, synthesis of genetic material and maturation. As attachment inhibitors, a lysoganglioside/poly-L-glutamic acid conjugate was reported by the group of Kamitakahara [204]. Two types of flavonoids extracted from the elderberry were reported to possess anti-HA properties. Direct binding assay and DART (Direct Analysis in Real Time) revealed that these flavonoids bind to H1N1 virion and inhibit attachment to the cell surface receptor. Investigation on popular traditional treatment of flu and cold by an herb *Echinacea purpurea* has discovered proinflammatory response in epithelial cells and dose dependent killing of influenza H1N1, H3N2, H5N1,

TABLE 11.3 Agents under basic research.

Sr. no.	Approaches	Details
1.	Antisense, siRNA and miRNA	Selective gene silencing with RNA interference (RNAi) like Antisense RNA, siRNA, or miRNA been studied as potential anti-influenza therapeutics. PB1, PB2, PA, NP, M, NS genes of influenza have been targeted to silence both in vitro and in vivo (for PB1, PA, NP) and suppressed the flu replication [168]
2.	Drugs combined with Apoptosis inhibitors	Amantadine drugs combined with biological nanoparticles (NPs) like functionalized selenium nanoparticles, silver nanoparticles, sodium selenite inhibits infection of influenza virus to host cells and inhibits apoptosis [169–171]
3.	Immunomodulation by probiotics	Immunomodulatory probiotics like *Bifidobacterium bifidum*, heat killed *Lactobacillus planatarum* can suppress the respiratory viral infection severity by regulating the humoral immunity [172,173]
4.	Antagonists of Immune hyper activation	Cellular inhibitors of apoptosis proteins are crucial in regulation of cell death and deficiency of these proteins increase the mortality in influenza infection by death receptor modulated programmed apoptosis. Usage of necrostatin-1 (Nec-1) reduces the lethality by inhibiting receptor-interacting serine/threonine-protein kinase 1 (RIPK1) a modulator of inflammation and apoptosis [174]
5.	Small-molecule antagonists of cap-dependent viral endonuclease	Transcription starts with the PB2 and capped host cell pre-mRNAs binding which leads to, PA N-terminal (PAN) catalysation of endonuclease activity which results in cleavage of the host pre-mRNA and this "cap-snatching" mechanism produce short capped RNA oligoribonucleotides which is further used to prime the viral mRNA transcription by PB1. A small molecule that inhibits the activity of polymerase can be a potential drug target against influenza virus [175]. Peptide derived from influenza with a single influenza B-specific amino acid substitution facilities the binding of PA of both the viruses and blocks the viral polymerase activity and replication of both type of the virus [176].
6.	RAF/MEK/ERK inhibitors	Studies has shown that inhibition NF-kappaB and Raf/MEK/ERK activation during virus infection can reduce viral load and reduce cytokine storm. [177,178] (e.g., of the inhibitors Bay-7082 and U0126.)
7.	Kinase modulator compounds	Studies have revealed that flavonoids inhibit viral replications by modulating MAP kinase pathways [179].
8.	Aprotinin, leupaptin, camostat	The drug aprotinin is approved to use against pancreatitis and to reduce bleeding after operation. However, it can inhibit the serine protease enzyme by site directed structural modification by adding arginine. Serine protease helps in activation of influenza virus by cleaving subunits of HA. Similar anti influenza activity is exerted by other two chemicals leupaptin and camostat [180].
9.	FluPep	FluPep is a novel family of peptides that interacts with HA and blocks the SA binding site [181].

(Continued)

TABLE 11.3 (Continued)

Sr. no.	Approaches	Details
10.	HB80, HB36	HB80, HB36 are computer designed protein molecules having affinity to bind the surface patch on the stem of influenza HA protein, and have potential to replicate in experimental study and use of therapeutics [182].
11.	Cyanovirin-N	Cyanovirin-N (CV-N) is a novel antiviral protein and shown efficient antiviral activity against HIV. However further studies have revealed that it can interact with viral HA and suppress viral replication [183].
12.	Iota-Carrageenan	Study reported that iota-carrageenan can be use an alternative to NA inhibitors [184].
13.	Pentraxin PTX3	Pentraxins are a superfamily of conserved proteins and showed anti influenza activity by binding to hemagglutinin glycoproteins [185].
14.	6′ sialyl-N-acetyllactosamine	Sialoglycopeptide (SGP) is basically the glycopeptide found in hen's egg yolk, which can inhibit influenza hemaglutination [186]
15.	Recombinant human galectin-1	Gelactin-1 are S type lectins. Study shows that when cells are treated with recombinant human galectin-1, the viral replication gets suppressed as Galectin-1 could directly bind to the enveloped glycoproteins of influenza A/WSN/33 virus and inhibit its hemagglutination activity which results in lowering the infectivity [187]
16.	Viramidine	Viramidine is a derivative of ribavirin found effective against different strains of influenza [188]
17.	Specific on/off adapter hepatitis delta virus ribozymes (SOFA -HDV Rz-) ribozymes	SOFA-HDV-Ribozyme is derived from a natural motif found in HDV, and it can target and cleave target RNA molecules. SOFA -HDV Rz can be used against influenza virus as it suppresses the flu replication [189].
18.	EB peptide	It is a 20 amino acid long peptide derived from fibroblast growth factor 4 can bind to HA protein and exhibit broad spectrum antiviral activity against influenza viruses [190].
19.	Sphingosine mimics	Sphingolipids act as immunosuppressor. Combination treatment of oseltamivir and sphingosine reported an increased survival rate in mice [191].
20.	Statins	Cellular 3-hydroxy-3-methylglutaryl-coen-zyme A (HMG-CoA reductase) is an enzyme which performs biosynthesis of cholesterol. It was tested against influenza, it is possible that it can reduce the morbidity and mortality caused by cytokine storm [192].
21.	Defensins	Small peptide containing 6 cystrine residues. Due to the 6 cys residue a disulfide bond is formed which regulates the structure and function. There are three types of defensins depending upon their molecular weight α, β, and θ defensins. Studies reported that the defensins can inhibit influenza replication though the inhibitory mechanism that is still not clear [193]
22.	Collectins	Collectins are superfamily collagen-containing C-type lectins. The main function of this protein inside human body is to recognize pathogenic molecules. It binds to viral protein glycoproteins and inhibit virus entry inside cell [194].

(Continued)

III. Antiviral agents and therapeutics

TABLE 11.3   (Continued)

Sr. no.	Approaches	Details
23.	JNJ4796	JNJ4796 is a small molecule fuse inhibitor which can target the conserved stem region of hemagglutinin, and have shown effective neutralization of broad spectrum on influenza A virus (IAV) [195].
24.	Punicalagin	Punicalagin targets NA, and inhibits influenza replication. It can target both influenza A and B virus. Interestingly the component can inhibit the replication of oseltamivir NA/H274Y strain [196]
25.	Filamentous anti-influenza agents	Biologically replicable filamentous nano-scaffold M13 phage can serve as a potent plat-form for blocking the entry of influenza viruses as into the host cell by wrapping around the viral particle [197].
26.	H84T BanLec	Lectins can selectively bind to viral surface glycol proteins. A study identified a broad spectrum H84T BanLec with robust activity against all strains of influenza virus [198].
27.	MZ7465	MZ7465 is a derivative of salcomine. Studies reported that treatment with this compound reduced viral protein and RNA synthesis [199]
28.	D715−2441	1,3-dihydroxy-6-benzo[c] chromene or D715−2441 is a small molecule has reported to have antiviral activity against influenza virus. It is effective against different strains of influenza A virus, for example, H1N1, H5N1, H7N9, H3N. It is also effective against oseltamivir H274Y NA strain. The compound also blocks the PB2 activity of the virus [200]

and H7N7 [205]. Hemagglutination study directed that the extract of *Echinacea purpurea* can inhibit the cell binding capacity of the virus [206]. Phenolic compounds derived from widely distributed *Chaenomeles sinensis* plant fruit are reported to inhibit the hemagglutination activity of the influenza virus [207]. Flavonoid compounds are frequently found in plants and primarily inhibits the activity of NA and membrane fusion [208]. Structure Activity Relationship analysis disclosed the 4′-OH, 7-OH, C4=O, and C2=C3 groups present in the flavonoids are essential for NA inhibition [209]. *Citrus junos* plant fruit extract contained flavanone triglycoside, naringenin 7-O-(2′,6′-di-O-alpha-rhamnopyranosyl)-beta-glucopyranoside and hesperetin 7-O-(2′,6′-di-O-alpha rhamnopyranosyl)-beta-glucopyranoside hesperidin and narirutin which found to be effective against influenza [210]. Flavonoid derivatives apigenin, dinatin 2-((E)-4 hydroxyphenylidene)-6-hydroxy-2,3-dihydrobenzofuran-3-one was also reported effective against influenza viruses [209]. Total flavonoid extracts of *Bupleurum chinese* reported effective against influenza B [211]. Red fleshed potato (*Solanum tuberosum*) has been proven effective against both influenza A and B [212]. Bi flavonoids extracted from the *Ginkgo biloba* has anti-NA activities [213]. Pterocarpan and flavanones derived from popular TCM herb are found to have effective inhibitory effects on influenza by binding with the pocket adjacent to the active site [214]. *Cudrania tricuspidate* extracted flavanone and xanthone was also reported for their NA inhibition activity [215]. Saponins are reported to be an effective adjuvant for the pathogens having HA proteins [216]. In a collaborative study between Environment and Biotechnology Centre Swinburne University of Technology, Australia and Sarawak Biodiversity Centre, Malaysia on the medicinal plant extracts collected from tropical

TABLE 11.4 Traditional plant extracts as anti influenza agents.

Plants	Compound or active component	Target site
Sambucus nigra	Flavonoids 1 and 2	HA inhibition
Echinacea purpurea herb	Alkamides, glycoproteins	HA inhibition
Chaenomeles sinensis fruit extract	Polyphenols	NS2 protein inhibition
Elsholtzia rugulosa	Apigenin, luteolin	NA inhibitor
Ginkgo biloba	Biflavonoids (amentoflavone derivatives)	NA inhibitor
Cephalotaxus harringtonia	Biflavonoids (amentoflavone derivatives)	NA inhibitor
Aesculus chinensis	Flavanoids 10 and 11	NA inhibitor
Citrus junos	Pterocarpans and flavanones	NA inhibitor
Cudrania tricuspidata	Flavanone and xanthone	NA inhibitor
Rhodiola rosea	Gossypetin and kaempferol	NA inhibitor
Glycyrrhiza uralensis	polyphenol compounds	NA inhibitor
Sanicula europaea	Not detected	RdRp inhibitor
Thalictrum simplex	pavine alkaloid (-)-thalimonine (Thl)	Protein synthesis

rainforests of Borneo, all the plant extract showed anti-influenza effects either by NA inhibition or hemagglutination inhibition HI or both [217]. Interestingly the extracts which showed NA inhibition (NI) was able to reduce viral replication by 90%. A South African medicinal plant *Rapanea melanophloeos* extract was found to be effective against flu [218]. A study on South Indian traditional plants showed plant extract from *Wrightia tinctoria* possesses excellent anti-influenza activity in terms of an inhibitory concentration of 50% (IC_{50}). IC_{50} of Oseltamivir is 6.44 lg/mL whereas *Wrightia tinctoria* extract showed IC_{50} at 2.25 lg/mL [219] (Table 11.4).

11.4 Conclusion

For the treatment of influenza infection, we are primarily dependent on the antivirals. The current egg-dependent influenza vaccine production process is too slow and takes around 6 months. In addition to traditional egg-based vaccines, new vaccine technologies are coming into existence, including combination vaccines, vaccine vectors, DNA vaccines, and transgenic plants However, all the vaccines are prophylactic in nature and are only effective if they match the strain of the circulating virus. In addition, vaccines offer only modest protection to high-risk groups like immunocompromised, older adults, and infants. Hence, antiviral agents are a crucial defense strategy to combat seasonal and pandemic influenza strains. Currently, there are eight anti-influenza agents approved by regulatory agencies in different countries, and three are undergoing clinical trials to be approved by the US Food and Drug Administration. Additionally, many different approaches have been adopted to develop new antivirals that are expected to be registered for clinical trials with fewer side effects and greater efficacy.

The inability to combat viral outbreaks appropriately is another issue that needs to be addressed. Influenza also poses a high risk of causing a next worldwide pandemic and the primary reason for our incapacity is the lack of available antivirals that are ready for immediate use. It may be due to complete dependency on the synthetic drugs. An average of 800 million dollars and 12 years of time are needed for the introduction of a new drug from the bench to the bedside, and a single amino acid change in the target protein site can cause the virus to become resistant to that new drug. As an example, earlier we discussed that replacement of amino acid Histidine to Tyrosine at 274th position on NA protein which gave rise to oseltamivir resistance strains. So, interdisciplinary collaboration is essential in the development of new anti-influenza drugs and it is probably high time to identify and scientifically investigate traditional components to implement them along with modern practices to increase the overall effectiveness of the treatment. Nevertheless, to compete with highly mutating viruses like influenza, we must always be prepared with alternatives ready at hand.

Acknowledgments

Ashwini Chauhan would like to thank UGC-BSR (F.30-487/2019(BSR)), DST-Nanomission (DST/NM/NB/2018/203), ICMR (OMI/20/2020-ECD-1), and SERB-CRG (CRG/2021/001974) for the funding support. Research fellowship of S. C. is supported by ICMR (OMI/20/2020-ECD-1).

References

[1] Barberis I, Myles P, Ault SK, Bragazzi NL, Martini M. History and evolution of influenza control through vaccination: from the first monovalent vaccine to universal vaccines. J Prev Med Hyg 2016;57(3):E115−20.
[2] Morens DM, Taubenberger JK, Folkers GK, Fauci AS. Pandemic influenza's 500th anniversary. Clin Infect Dis 2010;51(12):1442−4.
[3] Morens DM, North M, Taubenberger JK. Eyewitness accounts of the 1510 influenza pandemic in Europe. Lancet 2010;376(9756):1894−5.
[4] Smith W, Andrewes CH, Laidlaw PP. A virus obtained from influenza patients. Lancet. 1933;222(5732):66−8.
[5] Saunders-Hastings PR, Krewski D. Reviewing the history of pandemic influenza: understanding patterns of emergence and transmission. Pathogens 2016;5(4).
[6] Baltimore D. Expression of animal virus genomes. Bacteriol Rev 1971;35(3):235−41.
[7] Virus OFI, Duesberg BYPH. The RNA's of influenza virus * by peter h duesberg. 1967; 930−937.
[8] Megan L, Shaw PP Orthomyxoviridae. Angew. Chem. Int. Ed. 1967; 6(11):951−52.
[9] Su S, Fu X, Li G, Kerlin F, Veit M. Novel influenza D virus: epidemiology, pathology, evolution and biological characteristics. Virulence [Internet] 2017;8(8):1580−91. Available from: https://doi.org/10.1080/21505594.2017.1365216.
[10] Long JS, Mistry B, Haslam SM, Barclay WS. Host and viral determinants of influenza A virus species specificity. Nat Rev Microbiol [Internet] 2019;17(2):67−81. Available from: https://doi.org/10.1038/s41579-018-0115-z.
[11] Nypaver C, Dehlinger C, Carter C. Influenza and influenza vaccine: a review. J Midwifery Women's Health 2021;66(1):45−53.
[12] Barberis I, Martini M, Iavarone F, Orsi A. Available influenza vaccines: immunization strategies, history and new tools for fighting the disease. J Prev Med Hyg 2016;57(1):E41−6.
[13] van der Vries E, Ison MG. Antiviral resistance in influenza viruses. Clin Epidemiol Asp Antimicrob Drug Resist 2017;1165−83.
[14] Abraham GM, Morton JB, Saravolatz D. Baloxavir: a novel antivir agent treat influenza. 2020; 508, 1−21.

[15] Reports BS. Influenza (Flu) antivir drugs related 2021;8—10.

[16] Ge H, Wang YF, Xu J, Gu Q, Liu HB, Xiao PG, et al. Anti-influenza agents from traditional Chinese medicine. Nat Prod Rep 2010;27(12):1758—80.

[17] Bouvier NM, Peter P. The biology of influenza viruses. Vaccine. 2008; 23(1), 1—7.

[18] Hoyle L. Structure of the influenza virus the relation between biological activity and chemical structure of virus fractions. J Hyg (Lond) 1952.

[19] Fenner F, Bachmann PA, Gibbs EPJ, Murphy FA, Studdert MJ, White DO. Structure and composition of viruses. Vet Virol 1987;3—19.

[20] Ivanova PT, Myers DS, Milne SB, McClaren JL, Thomas PG, Brown HA. Lipid composition of the viral envelope of three strains of influenza virus—not all viruses are created equal. ACS Infect Dis 2016;1(9):435—42.

[21] Burrell CJ, Howard CR, Murphy FA. Virion structure and composition. Fenner White's Med Virol 2017;27—37.

[22] Andrewes CH, Laidlaw PP, Smith W. The susceptibility of mice to the viruses of human and swine influenza. Lancet. 1934;224(5799):859—62.

[23] Stanley WM. The size of influenza virus. J Exp Med 1944;2:267—83.

[24] White J, Kartenbeck J, Helenius A. Membrane fusion activity of influenza virus. EMBO J 1982;1(2):217—22.

[25] Air GM. Influenza neuraminidase. Influenza Other Respi Viruses. 2012;6(4):245—56.

[26] Samji T. Influenza A: understanding the viral life cycle. Yale J Biol Med 2009;82(4):153—9.

[27] Pinto LH, Holsinger LJ, Lamb RA. Influenza virus M2 protein has ion channel activity. Cell. 1992;69(3):517—28.

[28] Rafal MPielak, James JChou1*. Influenza M2 proton channels Rafal. Biochim Biophys Acta 2009;23(1):1—7.

[29] Pinto LH, Lamb RA. The M2 proton channels of influenza A and B viruses. J Biol Chem [Internet] 2006;281 (14):8997—9000. Available from: https://doi.org/10.1074/jbc.R500020200.

[30] Hilsch M, Goldenbogen B, Sieben C, Höfer CT, Rabe JP, Klipp E, et al. Influenza a matrix protein m1 multimerizes upon binding to lipid membranes. Biophys J 2014;107(4):912—23.

[31] Paterson D, Fodor E. Emerging roles for the Influenza A virus nuclear export protein (NEP). PLoS Pathog 2012;8(12).

[32] McGeoch D, Fellner P, Newton C. Influenza virus genome consists of eight distinct RNA species. Proc Natl Acad Sci U S A 1976;73(9):3045—9.

[33] Jiang WM, Wang SC, Peng C, Yu JM, Zhuang QY, Hou GY, et al. Identification of a potential novel type of influenza virus in Bovine in China. Virus Genes 2014;49(3):493—6.

[34] Ahlquist P. Parallels among positive-strand RNA viruses, reverse-transcribing viruses and double-stranded RNA viruses. Nat Rev Microbiol 2006;4(5):371—82.

[35] Stephan P. Overview of influenza viruses article. Curr Top Microbiol Immunol 2013;(November):435. Available from: http://books.google.com/books?id = _DDwCqx6wpcC&printsec = frontcover&dq = unwritten + rules + of + phd + research&hl = &cd = 1&source = gbs_api%255Cnpapers2://publication/uuid/48967E01-55F9-4397-B941-310D9C5405FA%255Cnhttp://medcontent.metapress.com/index/A65RM03P4874243N.p.

[36] Chang S, Sun D, Liang H, Wang J, Li J, Guo L, et al. Cryo-EM structure of influenza virus RNA polymerase complex at 4.3Å resolution. Mol Cell 2015;57(5):925—35.

[37] Graef KM, Vreede FT, Lau Y-F, McCall AW, Carr SM, Subbarao K, et al. The PB2 subunit of the influenza virus RNA polymerase affects virulence by interacting with the mitochondrial antiviral signaling protein and inhibiting expression of beta interferon. J Virol 2010;84(17):8433—45.

[38] Zhao Z, Yi C, Zhao L, Wang S, Zhou L, Hu Y, et al. PB2—588I enhances 2009 H1N1 pandemic influenza virus virulence by increasing viral replication and exacerbating PB2 inhibition of beta interferon expression. J Virol 2014;88(4):2260—7.

[39] Poole E, Elton D, Medcalf L, Digard P. Functional domains of the influenza A virus PB2 protein: identification of NP- and PB1-binding sites. Virology 2004;321(1):120—33.

[40] Kobayashi M, Toyoda T, Ishihama A. Influenza virus PB1 protein is the minimal and essential subunit of RNA polymerase. Arch Virol 1996;141(3—4):525—39.

[41] Varga ZT, Palese, P. The influenza A virus protein PB1-F2 Killing two birds with one stone? Killing two birds with one stone. Circulation 2011; (November/December).

[42] Wang Q, Liu R, Li Q, Wang F, Zhu B, Zheng M, et al. Host cell interactome of PB1 N40 protein of H5N1 influenza A virus in chicken cells. J Proteom [Internet] 2019;197(January):34—41. Available from: https://doi.org/10.1016/j.jprot.2019.02.011.

[43] Perez DR, Donis RO. Functional analysis of PA binding by Influenza A virus PB1: effects on polymerase activity and viral infectivity. J Virol 2001;75(17):8127–36.

[44] Liu Y, Lou Z, Bartlam M, Rao Z. Structure-function studies of the influenza virus RNA polymerase PA subunit. Sci China, Ser C Life Sci 2009;52(5):450–8.

[45] Wilson IA, Skehel JJ, Wiley DC. Structure of the haemagglutinin membrane glycoprotein of influenza virus at 3 Å resolution. Nature 1981;289(5796):366–73.

[46] Wagner R, Matrosovich M, Klenk HD. Functional balance between haemagglutinin and neuraminidase in influenza virus infections. Rev Med Virol 2002;12(3):159–66.

[47] Centers for Disease Control. Influenza (flu): types of influenza viruses. 2020; 1–4. Available from: https://www.cdc.gov/flu/about/viruses/types.htm#:~:text = .

[48] Biswas SK, Boutz PL, Nayak DP. Influenza virus nucleoprotein interacts with influenza virus polymerase proteins. J Virol 1998;72(7):5493–501.

[49] Portela A, Digard P. The influenza virus nucleoprotein: a multifunctional RNA-binding protein pivotal to virus replication. J Gen Virol 2002;83(4):723–34.

[50] McAuley JL, Gilbertson BP, Trifkovic S, Brown LE, McKimm-Breschkin JL. Influenza virus neuraminidase structure and functions. Front Microbiol 2019;10(Jan).

[51] Zanin M, Baviskar P, Webster R, Webby R. The interaction between respiratory pathogens and mucus. Cell Host Microbe 2016;19(2):159–68.

[52] Ito T, Gorman OT, Kawaoka Y, Bean WJ, Webster RG. Evolutionary analysis of the influenza A virus M gene with comparison of the M1 and M2 proteins. J Virol 1991;65(10):5491–8.

[53] Gómez-Puertas P, Albo C, Pérez-Pastrana E, Vivo A, Portela A. Influenza virus matrix protein is the major driving force in virus budding. J Virol 2000;74(24):11538–47.

[54] Zvonarjev AY, Ghendon YZ. Influence of membrane (M) protein on influenza A virus virion transcriptase activity in vitro and its susceptibility to rimantadine. J Virol 1980;33(2):583–6.

[55] de Chassey B, Aublin-Gex A, Ruggieri A, Meyniel-Schicklin L, Pradezynski F, Davoust N, et al. The interactomes of influenza virus NS1 and NS2 proteins identify new host factors and provide insights for ADAR1 playing a supportive role in virus replication. PLoS Pathog 2013;9(7).

[56] Han CW, Jeong MS, Jang SB. Structure and function of the influenza a virus non-structural protein 1. J Microbiol Biotechnol 2019;29(8):1184–92.

[57] Marc D. Influenza virus non-structural protein NS1: interferon antagonism and beyond. J Gen Virol 2014;95:2594–611.

[58] O'Neill RE, Talon J, Palese P. The influenza virus NEP (NS2 protein) mediates the nuclear export of viral ribonucleoproteins. EMBO J 1998;17(1):288–96.

[59] Deeks SG, Overbaugh J, Phillips A, Buchbinder S. HIV infection. Nat Rev Dis Prim 2015;1(October).

[60] Jackson AC. Rabies virus. Encycl Neurol Sci 2014;3:1027–30.

[61] Lafon M. Modulation of the immune response in the nervous system by rabies virus. Curr Top Microbiol Immunol 2005;289:239–58.

[62] Yen HL, Aldridge JR, Boon ACM, Ilyushina NA, Salomon R, Hulse-Post DJ, et al. Changes in H5N1 influenza virus hemagglutinin receptor binding domain affect systemic spread. Proc Natl Acad Sci U S A 2009;106(1):286–91.

[63] Kalil AC, Thomas PG. Influenza virus-related critical illness: pathophysiology and epidemiology. Crit Care 2019;23(258):1–7.

[64] Zhang H. Tissue and host tropism of influenza viruses: importance of quantitative analysis. Sci China, Ser C Life Sci 2009;52(12):1101–10.

[65] García-Sastre A. Influenza virus receptor specificity: disease and transmission. Am J Pathol [Internet] 2010;176(4):1584–5. Available from: https://doi.org/10.2353/ajpath.2010.100066.

[66] Korteweg C, Gu J. Pathology, molecular biology, and pathogenesis of avian influenza A (H5N1) infection in humans. Am J Pathol 2008;172(5):1155–70.

[67] Sakai T, Nishimura SI, Naito T, Saito M. Influenza A virus hemagglutinin and neuraminidase act as novel motile machinery. Sci Rep [Internet] 2017;7(March):1–11. Available from: https://doi.org/10.1038/srep45043.

[68] Kajiwara N, Nomura N, Ukaji M, Yamamoto N, Kohara M, Yasui F, et al. Cell-penetrating peptide-mediated cell entry of H5N1 highly pathogenic avian influenza virus. Sci Rep [Internet] 2020;10(1):1–13. Available from: https://doi.org/10.1038/s41598-020-74604-w.

[69] Karakus U, Pohl, Marie OSS. Breaking the convention: sialoglycan variants, coreceptors, and alternative receptors for influenza A virus entry. J Virol 2020;1−9 (November 2019).

[70] Rajao DS, Vincent AL, Perez DR. Adaptation of human influenza viruses to swine. Front Vet Sci 2019;5(JAN):1−12.

[71] Melike Lakadamyali MXZ. Endocytosis of influenza viruses Melike. Microbes Infect [Internet] 2004. Available from: https://www.ncbi.nlm.nih.gov/pmc/articles/PMC3624763/pdf/nihms412728.pdf.

[72] de Vries E, Tscherne DM, Wienholts MJ, Cobos-Jiménez V, Scholte F, García-Sastre A, et al. Dissection of the influenza a virus endocytic routes reveals macropinocytosis as an alternative entry pathway. PLoS Pathog 2011;7(3).

[73] Lakadamyali M, Rust MJ, Babcock HP, Zhuang X. Visualizing infection of individual influenza viruses. Proc Natl Acad Sci U S A 2003;100(16):9280−5.

[74] Baudin F, Bach C, Cusack S, Ruigrok RWH. Structure of influenza virus RNP. I. Influenza virus nucleoprotein melts secondary structure in panhandle RNA and exposes the bases to the solvent. EMBO J 1994;13 (13):3158−65.

[75] Herz C, Stavnezer E, Krug RM, Gurney T. Influenza virus, an RNA virus, synthesizes its messenger RNA in the nucleus of infected cells. Cell 1981;26:391−400 (3 PART 1).

[76] Te Velthuis AJW, Fodor E. Influenza virus RNA polymerase: insights into the mechanisms of viral RNA synthesis. Nat Rev Microbiol 2016;14(8):479−93.

[77] Wu WWH, Sun YHB, Panté N. Nuclear import of influenza A viral ribonucleoprotein complexes is mediated by two nuclear localization sequences on viral nucleoprotein. Virol J 2007;4:1−12.

[78] Hutchinson EC, Fodor E. Nuclear import of the influenza A virus transcriptional machinery. Vaccine [Internet] 2012;30(51):7353−8. Available from: https://doi.org/10.1016/j.vaccine.2012.04.085.

[79] Eisfeld AJ, Neumann G, Kawaoka Y. At the centre: influenza A virus ribonucleoproteins. Nat Rev Microbiol 2015;13(1):28−41.

[80] Furusawa Y, Yamada S, Kawaoka Y. Host factor Nucleoporin 93 is involved in the nuclear export of influenza virus RNA. Front Microbiol 2018;9(Jul):1−9.

[81] Jeremy SR, Robert AL. Influenza virus assembly and budding. Virology 2011;229−36.

[82] Eisfeld AJ, Kawakami E, Watanabe T, Neumann G, Kawaoka Y. RAB11A is essential for transport of the influenza virus genome to the plasma membrane. J Virol 2011;85(13):6117−26.

[83] Momose F, Sekimoto T, Ohkura T, Jo S, Kawaguchi A, Nagata K, et al. Apical transport of influenza A virus ribonucleoprotein requires Rab11-positive recycling endosome. PLoS One 2011;6(6):1−15.

[84] Mora R, Rodriguez-Boulan E, Palese P, García-Sastre A. Apical budding of a recombinant influenza A virus expressing a hemagglutinin protein with a basolateral localization signal. J Virol 2002;76(7):3544−53.

[85] Nayak DP, Hui EKW, Barman S. Assembly budding influenza virus. Virus Res 2004;106:147−65 (2 SPEC.ISS.).

[86] Palese P, Tobita K, Ueda M, Compans RW. Characterization of temperature sensitive influenza virus mutants defective in neuraminidase. Virology 1974;61(2):397−410.

[87] Davies WL, Grunert RR, Haff RF, Mcgahen JW, Neumayer EM, Paulshock M, et al. Antiviral activity of 1-adamantanamine (amantadine). Science 1964;144(3620):862−3 (80).

[88] De Clercq E. Antiviral agents active against influenza A viruses. Nat Rev Drug Discov 2006;5(12):1015−25.

[89] Hubsher G, Haider M, Okun MS. Amantadine: the journey from fighting flu to treating Parkinson disease. Neurology 2012;78(14):1096−9.

[90] Sansom MSP, Kerr ID. Influenza virus m2 protein: a molecular modelling study of the ion channel. Protein Eng Des Sel 1993;6(1):65−74.

[91] Wang C, Takeuchi K, Pinto LH, Lamb RA. Ion channel activity of influenza A virus M2 protein: characterization of the amantadine block. J Virol 1993;67(9):5585−94.

[92] Tiwari P. Antiparkinsonian drugs. Methods Drug Eval 2018;108 −108.

[93] Hayden FG, Hay AJ. Emergence and transmission of influenza A viruses resistant to amantadine and rimantadine. Curr Top Microbiol Immunol 1992;176:119−30.

[94] Hay AJ, Wolstenholme AJ, Skehel JJ, Smith MH. The molecular basis of the specific anti-influenza action of amantadine. EMBO J 1985;4(11):3021−4.

[95] Nelson MI, Simonsen L, Viboud C, Miller MA, Holmes EC. The origin and global emergence of adamantane resistant A/H3N2 influenza viruses. Virology. 2009;388(2):270−8.

[96] CDC recommends against the use of amantadine and rimantadine for the treatment or prophylaxis of influenza in the United States during the 2005–06 influenza season. Available from: https://stacks.cdc.gov/view/cdc/25151.01/14/2006. CDC January;14:2006.

[97] Office USP. Rimantadine patent.pdf; 1967.

[98] Zlydnikov DM, Kubar OI, Kovaleva TP, Kamforin LE. Study of rimantadine in the USSR: a review of the literature. Rev Infect Dis 2021;3(3):408—21. Available from: https://www.jstor.org/stable/4452575.

[99] Wintenneyer SM, Nahata MC. Rimantadine: a clin perspect. 2016; 29, 299—310.

[100] Thomaston JL, Polizzi NF, Konstantinidi A, Wang J, Kolocouris A, Degrado WF. Inhibitors of the M2 proton channel engage and disrupt transmembrane networks of hydrogen-bonded waters. J Am Chem Soc 2018;140(45):15219—26.

[101] Thompson J, Fleet W, Lawrence E, Pierce E, Morris L, Wright P. A comparison of acetaminophen and rimantadine in the treatment of influenza A infection in children. J Med Virol 1987;21(3):249—55.

[102] Belshe RB, Burk B, Newman F, Cerruti RL, Sim IS. Resistance of influenza a virus to amantadine and rimantadine: results of one decade of surveillance. J Infect Dis 1989;159(3):430—5.

[103] Ziegler T, Hemphill ML, Ziegler ML, Perez-Oronoz G, Klimov AI, Hampson AW, et al. Low incidence of rimantadine resistance in field isolates of influenza A viruses. J Infect Dis 1999;180(4):935—9.

[104] Belshe RB, Smith H, Hall CB, Betts R, Hay AJ. Genetic basis of resistance to rimantadine emerging during treatment of influenza virus infection. Microbiology 1988;62(5):1508—12.

[105] Anthony EF, Fry A, Shay MD, Gubareva L, Bresee JSM, Uyeki TMM. Antiviral agents for the treatment and chemoprophylaxis of influenza recommendations of the advisory committee on immunization practices (ACIP) morbidity and mortality weekly report hemagglutinin neuraminidase M2 ion channel RNP centers for disease control. Recomm Reports [Internet]. 2011;60(1). Available from: http://www.cdc.gov/flu.

[106] Von Itzstein M, Wu WY, Kok GB, Pegg MS, Dyason JC, Jin B, et al. Rational design of potent sialidase-based inhibitors of influenza virus replication. Nature. 1993;363(6428):418—23.

[107] Jacq N, Blanchet C, Combet C, Cornillot E, Duret L, Kurata K, et al. Grid as a bioinformatic tool. Parallel Comput 2004;30(9—10):1093—107.

[108] von Itzstein M. The war against influenza: discovery and development of sialidase inhibitors. Nat Rev Drug Discov. 2007; 6(December):967—74.

[109] Elliott M. Zanamivir: from drug design to the clinic. Philos Trans R Soc B Biol Sci 2001;356(1416):1885—93.

[110] Freund B, Gravenstein S, Elliott M, Miller IZanamivir. A review of clinical safety. Drug Saf 1999;21 (4):267—81.

[111] Kimberlin DW. Antiviral agents. In: Principles and practice of pediatric infectious diseases. 5th ed. Elsevier Inc. 2018. 1551—1567.e6 p. Available from: https://doi.org/10.1016/B978-0-323-40181-4.00295-4

[112] Cass LMR, Brown J, Pickford M, Fayinka S, Newman SP, Johansson CJ, et al. Pharmacoscintigraphic evaluation of lung deposition of inhaled zanamivir in healthy volunteers. Clin Pharmacokinet 1999;36(Suppl. 1):21—31.

[113] Trebbien R, Pedersen SS, Vorborg K, Franck KT, Fischer TK. Development of oseltamivir and zanamivir resistance in influenza a(H1N1)pdm09 virus, Denmark, 2014. Eurosurveillance 2017;22(3):1—8.

[114] Pizzorno A, Abed Y, Bouhy X, Beaulieu É, Mallett C, Russell R, et al. Impact of mutations at residue I223 of the neuraminidase protein on the resistance profile, replication level, and virulence of the 2009 pandemic influenza virus. Antimicrob Agents Chemother 2012;56(3):1208—14.

[115] Gubareva LV. Evidence for zanamivir resistance in an immunocompromised child infected with influenza B virus. J Infect Dis 1998;178(5 Suppl.):1257—62.

[116] FDA. Tamiflu (oseltamivir phosphate) Information. 31. Available from: https://www.fda.gov/drugs/post-market-drug-safety-information-patients-and-providers/tamiflu-oseltamivir-phosphate-information

[117] Brownlee BS, Lenzer J. Tamiflu: myth and misconception when it decided to recommend tamiflu as a first line. Atl [Internet] 2013;1—7. Available from: https://www.theatlantic.com/health/archive/2013/02/tamiflu-myth-and-misconception/273167/.

[118] FDA. The FDA approves first generic version of widely used influenza drug, tamifluu on. Routledge; 2016. Available from: https://www.fda.gov/drugs/postmarket-drug-safety-information-patients-and-providers/fda-approves-first-generic-version-widely-used-influenza-dr.

[119] McClellan K, Perry CM. Oseltamivir: a review of its use in influenza. Drugs 2001;61(2):263—83.

[120] Heneghan CJ, Onakpoya I, Thompson M, Spencer EA, Jones M, Jefferson T. Oseltamivir for influenza in adults and children: systematic review of clinical study reports and summary of regulatory comments. BMJ 2014;348:1—27.

[121] Hurt AC, Holien JK, Parker MW, Barr IG. Oseltamivir resistance and the H274Y neuraminidase mutation in seasonal, pandemic and highly pathogenic influenza viruses. Drugs 2009;69(18):2523—31.

[122] Thorlund K, Awad T, Boivin G, Thabane L. Systematic review of influenza resistance to the neuraminidase inhibitors. BMC Infect Dis [Internet] 2011;11(1):134. Available from: http://www.biomedcentral.com/1471-2334/11/134.

[123] US FDA. FDA approves Rapivab to treat flu infection. 2017;8–10. Available from: https://www.fda.gov/NewsEvents/Newsroom/PressAnnouncements/ucm557102.htm

[124] Zaraket H, Saito R. Japanese surveillance systems and treatment for influenza. Curr Treat Options Infect Dis 2016;8(4):311–28.

[125] Kim CU, Lew W, Williams MA, Liu H, Zhang L, Swaminathan S, et al. Influenza neuraminidase inhibitors possessing a novel hydrophobic interaction in the enzyme active site: design, synthesis, and structural analysis of carbocyclic sialic acid analogues with potent anti-influenza activity. J Am Chem Soc 1997;119(4):681–90.

[126] Malaisree M, Rungrotmongkol T, Decha P, Intharathep P, Aruksakunwong O, Hannongbua S. Understanding of known drug-target interactions in the catalytic pocket of neuraminidase subtype N1. Proteins Struct Funct Genet 2008;71(4):1908–18.

[127] Okomo-Adhiambo M, Sleeman K, Ballenger K, Nguyen HT, Mishin VP, Sheu TG, et al. Neuraminidase inhibitor susceptibility testing in human influenza viruses: a laboratory urveillance perspective. Viruses 2010;2(10):2269–89.

[128] Takashita E, Ejima M, Itoh R, Miura M, Ohnishi A, Nishimura H, et al. A community cluster of influenza a (H1N1)pdm09 virus exhibiting cross-resistance to oseltamivir and peramivir in Japan, November to December 2013. Eurosurveillance [Internet] 2014;19(1):20666. Available from: https://doi.org/10.2807/1560-7917.ES2014.19.1.20666.

[129] Yamashita M, Tomozawa T, Kakuta M, Tokumitsu A, Nasu H, Kubo S. CS-8958, a prodrug of the new neuraminidase inhibitor R-125489, shows long-acting anti-influenza virus activity. Antimicrob Agents Chemother 2009;53(1):186–92.

[130] Yamashita M. Laninamivir and its prodrug, CS-8958: long-acting neuraminidase inhibitors for the treatment of influenza. Antivir Chem Chemother 2010;21(2):71–84.

[131] Ishizuka H, Yoshiba S, Okabe H, Yoshihara K. Clinical pharmacokinetics of Laninamivir, a novel long-acting neuraminidase inhibitor, after single and multiple inhaled doses of its prodrug, CS-8958, in healthy male volunteers. J Clin Pharmacol 2010;50(11):1319–29.

[132] Kiso M, Kubo S, Ozawa M, Le QM, Nidom CA, Yamashita M, et al. Efficacy of the new neuraminidase inhibitor CS-8958 against H5N1 influenza viruses. PLoS Pathog 2010;6(2):1–10.

[133] Yousuke Furuta, Brian B. Gowen, Kazumi Takahashi, Kimiyasu Shiraki, Donald F. Smee and DLB. Favipiravir (T-705), a novel viral RNA polymerase inhibitor Yousuke. Antiviral Res [Internet]. 2013. Available from: https://www.ncbi.nlm.nih.gov/pmc/articles/PMC3624763/pdf/nihms412728.pdf

[134] PMDA. Favipiravir report on the deliberation results; 2014.

[135] Fang Q, Wang D. Advanced researches on the inhibition of influenza virus by Favipiravir and Baloxavir. Biosaf Heal [Internet] 2020;2(2):64–70. Available from: https://doi.org/10.1016/j.bsheal.2020.04.004.

[136] Furuta Y, Takahashi K, Kuno-Maekawa M, Sangawa H, Uehara S, Kozaki K, et al. Mechanism of action of T-705 against influenza virus. Antimicrob Agents Chemother 2005;49(3):981–6.

[137] Rosenke K, Feldmann H, Westover JB, Hanley PW, Martellaro C, Feldmann F, et al. Use of favipiravir to treat lassa virus infection in Macaques. Emerg Infect Dis 2018;24(9):1696–9.

[138] Sissoko D, Laouenan C, Folkesson E, M'Lebing AB, Beavogui AH, Baize S, et al. Experimental treatment with favipiravir for ebola virus disease (the JIKI Trial): a historically controlled, single-arm proof-of-concept trial in Guinea. PLoS Med 2016;13(3):1–36.

[139] Seneviratne SL, Abeysuriya V, Mel SDe, Zoysa IDe, Niloofa R. Favipiravir in Covid-19. Int J Prog Sci Technol 2020;19(2):143–5.

[140] Hassanipour S, Arab-Zozani M, Amani B, Heidarzad F, Fathalipour M, Martinez-de-Hoyo R. The efficacy and safety of favipiravir in treatment of COVID-19: a systematic review and meta-analysis of clinical trials. Sci Rep [Internet] 2021;11(1):1–11. Available from: https://doi.org/10.1038/s41598-021-90551-6.

[141] Shirley M. Baloxavir marboxil: a review in acute uncomplicated influenza. Drugs [Internet] 2020;80 (11):1109–18. Available from: https://doi.org/10.1007/s40265-020-01350-8.

[142] De Vlugt C, Sikora D, Pelchat M. Insight into influenza: a virus cap-snatching. Viruses 2018;10:11.

[143] Moeller A, Kirchdoerfer RN, Potter CS, Carragher B, Wilson IA. Organization of the influenza virus replication machinery. Science 2012;338(6114):1631–4 (80).

[144] Stubbs TM, Te Velthuis AJW. The RNA-dependent RNA polymerase of the influenza A virus. Future Virol 2014;9(9):863−76.

[145] Yuan P, Bartlam M, Lou Z, Chen S, Zhou J, He X, et al. Crystal structure of an avian influenza polymerase PA N reveals an endonuclease active site. Nature 2009;458(7240):909−13.

[146] Noshi T, Kitano M, Taniguchi K, Yamamoto A, Omoto S, Baba K, et al. In vitro characterization of baloxavir acid, a first-in-class cap-dependent endonuclease inhibitor of the influenza virus polymerase PA subunit. Antivir Res [Internet] 2018;160(June):109−17. Available from: https://doi.org/10.1016/j.antiviral.2018.10.008.

[147] FDA approves new drug to treat influenza. Case Med Res. 2018;1−2.

[148] O'Hanlon R, Shaw ML. Baloxavir marboxil: the new influenza drug on the market. Curr Opin Virol [Internet] 2019;35(Table 1):14−18. Available from: https://doi.org/10.1016/j.coviro.2019.01.006.

[149] Kadam RU, Wilson IA. Structural basis of influenza virus fusion inhibition by the antiviral drug Arbidol. Proc Natl Acad Sci U S A 2017;114(2):206−14.

[150] Boriskin Y, Leneva I, Pecheur E-I, Polyak S. Arbidol: a broad-spectrum antiviral compound that blocks viral fusion. Curr Med Chem 2008;15(10):997−1005.

[151] Chemistry P, Vol J, Zhurnal K, Chemistry D, Radiology M, Ratnikova LI. Search for new drugs arbidol: a new domestic immunomodulant (a review). Pharm Chem J 1999;33(3):115−22.

[152] Teissier E, Zandomeneghi G, Loquet A, Lavillette D, Lavergne JP, Montserret R, et al. Mechanism of inhibition of enveloped virus membrane fusion by the antiviral drug arbidol. PLoS One 2011;6:1.

[153] Nojomi M, Yassin Z, Keyvani H, Makiani MJ, Roham M, Laali A, et al. Effect of arbidol (Umifenovir) on COVID-19: a randomized controlled trial. BMC Infect Dis 2020;20(1):1−10.

[154] Lian N, Xie H, Lin S, Huang J, Zhao J, Lin Q. Umifenovir treatment is not associated with improved outcomes in patients with coronavirus disease 2019: a retrospective study. Clin Microbiol Infect [Internet] 2020;26(7):917−21. Available from: https://doi.org/10.1016/j.watres.2021.117043.

[155] Rossignol JF. Nitazoxanide: a first-in-class broad-spectrum antiviral agent. Antivir Res 2014;110 (August):94−103.

[156] Hayden FG. Newer influenza antivirals, biotherapeutics and combinations. Influenza Other Respir Viruses 2013;7:63−75 (Suppl. 1).

[157] Yogaratnam J, Rito J, Kakuda TN, Fennema H, Gupta K, Jekle CA, et al. antiviral activity, safety, and pharmacokinetics of AL-794, a novel oral influenza endonuclease inhibitor: results of an influenza human challenge study. J Infect Dis 2019;219(2):177−85.

[158] Fox LM, Saravolatz LD. Nitazoxanide: a new thiazolide antiparasitic agent. Clin Infect Dis 2005;40(8):1173−80.

[159] Sisson G, Goodwin A, Raudonikiene A, Hughes NJ, Mukhopadhyay AK, Berg DE, et al. Enzymes associated with reductive activation and action of nitazoxanide, nitrofurans, and metronidazole in *Helicobacter pylori*. Antimicrob Agents Chemother 2002;46(7):2116−23.

[160] Mifsud EJ, Tilmanis D, Oh DY, Ming-Kay Tai C, Rossignol JF, Hurt AC. Prophylaxis of ferrets with nitazoxanide and oseltamivir combinations is more effective at reducing the impact of influenza a virus infection compared to oseltamivir monotherapy. Antivir Res [Internet] 2020;176:104751. Available from: https://doi.org/10.1016/j.antiviral.2020.104751.

[161] Rossignol JF, La Frazia S, Chiappa L, Ciucci A, Santoro MG. Thiazolides, a new class of anti-influenza molecules targeting viral hemagglutinin at the post-translational level. J Biol Chem 2009;284(43):29798−808.

[162] Tilmanis D, van Baalen C, Oh DY, Rossignol JF, Hurt AC. The susceptibility of circulating human influenza viruses to tizoxanide, the active metabolite of nitazoxanide. Antivir Res [Internet] 2017;147:142−8. Available from: https://doi.org/10.1016/j.antiviral.2017.10.002.

[163] Haffizulla J, Hartman A, Hoppers M, Resnick H, Samudrala S, Ginocchio C, et al. Effect of nitazoxanide in adults and adolescents with acute uncomplicated influenza: a double-blind, randomised, placebo-controlled, phase 2b/3 trial. Lancet Infect Dis 2014;14(7):609−18.

[164] Malakhov MP, Aschenbrenner LM, Smee DF, Wandersee MK, Sidwell RW, Gubareva LV, et al. Sialidase fusion protein as a novel broad-spectrum inhibitor of influenza virus infection. Antimicrob Agents Chemother 2006;50(4):1470−9.

[165] Belser JA, Lu X, Szretter KJ, Jin X, Aschenbrenner LM, Lee A, et al. DAS181, a novel sialidase fusion protein, protects mice from lethal avian influenza h5N1 virus infection. J Infect Dis 2007;196(10):1493−9.

[166] Moss RB, Hansen C, Sanders RL, Hawley S, Li T, Steigbigel RT. A phase II study of DAS181, a novel host directed antiviral for the treatment of influenza infection. J Infect Dis 2012;206(12):1844−51.

[167] Mayburd AL. Influenza antiviral therapeutics. Recent Pat Antiinfect Drug Discov 2010;5(1):64−75.

[168] Drugs Future. Antiviral compounds in the pipeline to tackle h1n1 influenza infection. Bone; 2010.

[169] Khanna M, Saxena L, Rajput R, Kumar B, Prasad R. Gene silencing: a therapeutic approach to combat influenza virus infections. Future Microbiol 2015;10(1):131−40.

[170] Li Y, Lin Z, Guo M, Zhao M, Xia Y, Wang C, et al. Inhibition of H1N1 influenza virus-induced apoptosis by functionalized selenium nanoparticles with amantadine through ROS-mediated AKT signaling pathways. Int J Nanomed 2018;13:2005−16.

[171] Lin Z, Li Y, Guo M, Xu T, Wang C, Zhao M, et al. The inhibition of H1N1 influenza virus-induced apoptosis by silver nanoparticles functionalized with zanamivir. RSC Adv 2017;7(2):742−50.

[172] Takeda S, Takeshita M, Kikuchi Y, Dashnyam B, Kawahara S, Yoshida H, et al. Efficacy of oral administration of heat-killed probiotics from Mongolian dairy products against influenza infection in mice: alleviation of influenza infection by its immunomodulatory activity through intestinal immunity. Int Immunopharmacol [Internet] 2011;11(12):1976−83. Available from: https://doi.org/10.1016/j.intimp.2011.08.007.

[173] Mahooti M, Abdolalipour E, Salehzadeh A, Mohebbi SR, Gorji A, Ghaemi A. Immunomodulatory and prophylactic effects of *Bifidobacterium bifidum* probiotic strain on influenza infection in mice. World J Microbiol Biotechnol [Internet] 2019;35(6):1−9. Available from: https://doi.org/10.1007/s11274-019-2667-0.

[174] Rodrigue-Gervais IG, Labbé K, Dagenais M, Dupaul-Chicoine J, Champagne C, Morizot A, et al. Cellular inhibitor of apoptosis protein cIAP2 protects against pulmonary tissue necrosis during influenza virus infection to promote host survival. Cell Host Microbe 2014;15(1):23−35.

[175] Baughman BM, Jake Slavish P, Dubois RM, Boyd VA, White SW, Webb TR. Identification of influenza endonuclease inhibitors using a novel fluorescence polarization assay. ACS Chem Biol 2012;7(3):526−34.

[176] Wunderlich K, Mayer D, Ranadheera C, Holler AS, Mänz B, Martin A, et al. Identification of a PA-binding peptide with inhibitory activity against influenza A and B virus replication. PLoS One 2009;4(10).

[177] Pinto R, Herold S, Cakarova L, Hoegner K, Lohmeyer J, Planz O, et al. Inhibition of influenza virus-induced NF-kappaB and Raf/MEK/ERK activation can reduce both virus titers and cytokine expression simultaneously in vitro and in vivo. Antivir Res [Internet] 2011;92(1):45−56. Available from: https://doi.org/10.1016/j.antiviral.2011.05.009.

[178] Planz O. Development of cellular signaling pathway inhibitors as new antivirals against influenza. Antivir Res [Internet] 2013;98(3):457−68. Available from: https://doi.org/10.1016/j.antiviral.2013.04.008.

[179] Dong W, Wei X, Zhang F, Hao J, Huang F, Zhang C, et al. A dual character of flavonoids in influenza A virus replication and spread through modulating cell-autonomous immunity by MAPK signaling pathways. Sci Rep 2014;4:1−12.

[180] Zhirnov OP, Klenk HD, Wright PF. Aprotinin and similar protease inhibitors as drugs against influenza. Antivir Res [Internet] 2011;92(1):27−36. Available from: https://doi.org/10.1016/j.antiviral.2011.07.014.

[181] Nicol MQ, Ligertwood Y, Bacon MN, Dutia BM, Nash AA. A novel family of peptides with potent activity against influenza a viruses. J Gen Virol 2012;93(5):980−6.

[182] Sarel JFleishman, Timothy AWhitehead, Damian CEkiert, Cyrille Dreyfus, Jacob ECorn, Eva-Maria Strauch, et al. Computational design of proteins targeting the conserved stem region of influenza hemagglutinin. Science [Internet] 2011. Available from: https://www.ncbi.nlm.nih.gov/pmc/articles/PMC3624763/pdf/nihms412728.pdf.

[183] O'Keefe BR, Smee DF, Turpin JA, Saucedo CJ, Gustafson KR, Mori T, et al. Potent anti-influenza activity of cyanovirin-N and interactions with viral hemagglutinin. Antimicrob Agents Chemother 2003;47(8):2518−25.

[184] Leibbrandt A, Meier C, König-Schuster M, Weinmüllner R, Kalthoff D, Pflugfelder B, et al. Iota-carrageenan is a potent inhibitor of influenza a virus infection. PLoS One 2010;5(12):1−11.

[185] Reading PC, Bozza S, Gilbertson B, Tate M, Moretti S, Job ER, et al. Antiviral activity of the long chain pentraxin PTX3 against influenza viruses. J Immunol 2008;180(5):3391−8.

[186] Makimura Y, Watanabe S, Suzuki T, Suzuki Y, Ishida H, Kiso M, et al. Chemoenzymatic synthesis and application of a sialoglycopolymer with a chitosan backbone as a potent inhibitor of human influenza virus hemagglutination. Carbohydr Res 2006;341(11):1803−8.

[187] Yang M-L, Chen Y-H, Wang S-W, Huang Y-J, Leu C-H, Yeh N-C, et al. Galectin-1 binds to influenza virus and ameliorates influenza virus pathogenesis. J Virol 2011;85(19):10010−20.

[188] Sidwell RW, Bailey KW, Wong MH, Barnard DL, Smee DF. In vitro and in vivo influenza virus-inhibitory effects of viramidine. Antivir Res 2005;68(1):10−17.

[189] Motard J, Rouxel R, Paun A, von Messling V, Bisaillon M, Perreault JP. A novel ribozyme-based prophylaxis inhibits influenza a virus replication and protects from severe disease. PLoS One 2011;6:11.

[190] Jones JC, Turpin EA, Bultmann H, Brandt CR, Schultz-Cherry S. Inhibition of influenza virus infection by a novel antiviral peptide that targets viral attachment to cells. J Virol 2006;80(24):11960−7.

[191] Walsh KB, Teijaro JR, Wilker PR, Jatzek A, Fremgen DM, Das SC, et al. Suppression of cytokine storm with a sphingosine analog provides protection against pathogenic influenza virus. Proc Natl Acad Sci U S A 2011;108(29):12018−23.

[192] Mehrbod P, Omar AR, Hair-Bejo M, Haghani A, Ideris A. Mechanisms of action and efficacy of statins against influenza. Biomed Res Int 2014;2014:11−14.

[193] Doss M, White MR, Tecle T, Gantz D, Crouch EC, Jung G, et al. Interactions of α-, β-, and θ-defensins with influenza A virus and surfactant protein D. J Immunol 2009;182(12):7878−87.

[194] Yang J, Li M, Shen X, Liu S. Influenza A virus entry inhibitors targeting the hemagglutinin. Viruses 2012;5 (1):352−73.

[195] Wang A, Li Y, Lv K, Gao R, Wang A, Yan H, et al. Optimization and SAR research at the piperazine and phenyl rings of JNJ4796 as new anti-influenza A virus agents, part 1. Eur J Med Chem [Internet] 2021;222:113591. Available from: https://doi.org/10.1016/j.ejmech.2021.113591.

[196] Li P, Du R, Chen Z, Wang Y, Zhan P, Liu X, et al. Punicalagin is a neuraminidase inhibitor of influenza viruses. J Med Virol 2021;93(6):3465−72.

[197] Chung J, Jung Y, Hong C, Kim S, Moon S, Kwak EA, et al. Filamentous anti-influenza agents wrapping around viruses. J Colloid Interface Sci [Internet] 2021;583:267−78. Available from: https://doi.org/10.1016/j.jcis.2020.09.012.

[198] Covés-Datson EM, King SR, Legendre M, Gupta A, Chan SM, Gitlin E, et al. A molecularly engineered antiviral banana lectin inhibits fusion and is efficacious against influenza virus infection in vivo. Proc Natl Acad Sci U S A 2020;117(4):2122−32.

[199] Takizawa N, Kimura T, Watanabe T, Shibasaki M. Anti-influenza virus activity of a salcomine derivative mediated by inhibition of viral RNA synthesis. Arch Virol [Internet] 2018;163(6):1607−14. Available from: https://doi.org/10.1007/s00705-018-3779-9.

[200] Liu T, Liu M, Chen F, Chen F, Tian Y, Huang Q, et al. A small-molecule compound has anti-influenza a virus activity by acting as a "pB2 inhibitor.". Mol Pharm 2018;15(9):4110−20.

[201] Patwardhan B, Warude D, Pushpangadan P, Bhatt N. Ayurveda and traditional Chinese medicine: a comparative overview. Evidence-based Complement Altern Med 2005;2(4):465−73.

[202] Wang X, Jia W, Zhao A, Wang X. Anti-influenza agents from plants and traditional Chinese medicine. Phyther Res 2006;20(5):335−41.

[203] Tungmunnithum D, Thongboonyou A, Pholboon A, Yangsabai A. Flavonoids and other phenolic compounds from medicinal plants for pharmaceutical and medical aspects: an overview. Medicines 2018;5(3):93.

[204] Kamitakahara H, Suzuki T, Nishigori N, Suzuki Y, Kanie O, Wong CH. A lysoganglioside/poly-L-glutamic acid conjugate as a picomolar inhibitor of influenza hemagglutinin. Angew Chem—Int (Ed.) 1998;37 (11):1524−8.

[205] Hudson J, Vimalanathan S. Echinacea-A source of potent antivirals for respiratory virus infections. Pharmaceuticals 2011;4(7):1019−31.

[206] Pleschka S, Stein M, Schoop R, Hudson JB. Anti-viral properties and mode of action of standardized Echinacea purpurea extract against highly pathogenic avian Influenza virus (H5N1, H7N7) and swine-origin H1N1 (S-OIV). Virol J 2009;6:1−9.

[207] Sawai-Kuroda R, Kikuchi S, Shimizu YK, Sasaki Y, Kuroda K, Tanaka T, et al. A polyphenol-rich extract from Chaenomeles sinensis (Chinese quince) inhibits influenza A virus infection by preventing primary transcription in vitro. J Ethnopharmacol [Internet] 2013;146(3):866−72. Available from: https://doi.org/10.1016/j.jep.2013.02.020.

[208] Nagai T, Miyaichi Y, Tomimori T, Suzuki Y, Yamada H. Inhibition of influenza virus sialidase and anti-influenza virus activity by plant flavonoids. Chem Pharm Bull. 1990;38(1):1329−32.

[209] Liu AL, Wang HDi, Lee SMY, Wang YT, Du GH. Structure-activity relationship of flavonoids as influenza virus neuraminidase inhibitors and their in vitro anti-viral activities. Bioorg Med Chem 2008;16(15):7141−7.

[210] Ho Kyoung Kim Won Kyung Jeon BSK. Flavanone glycosides from citrus junos and their anti-influenza virus. Planta Med 2001;67(2):192−6.

[211] Wang XG, Liu ZJ. Prevention and treatment of viral respiratory infections by traditional Chinese herbs. Chin Med J (Engl) 2014;127(7):1344−50.

[212] Hayashi K, Mori M, Knox YM, Suzutan T, Ogasawara M, Yoshida I, et al. Anti influenza virus activity of a red-fleshed potato anthocyanin. Food Sci Technol Res 2003;9(3):242−4.

[213] Ibrahim MA, Ramadan HH. Mohammed RN. Evidence that Ginkgo Biloba could use in the influenza and coronavirus COVID-19 infections. J Basic Clin Physiol Pharmacol 2021;32(3):131−43.

[214] Ryu YB, Curtis-Long MJ, Kim JH, Jeong SH, Yang MS, Lee KW, et al. Pterocarpans and flavanones from Sophora flavescens displaying potent neuraminidase inhibition. Bioorg Med Chem Lett [Internet] 2008;18 (23):6046−9. Available from: https://doi.org/10.1016/j.bmcl.2008.10.033.

[215] Ryu YB, Curtis-Long MJ, Lee JW, Kim JH, Kim JY, Kang KY, et al. Characteristic of neuraminidase inhibitory xanthones from Cudrania tricuspidata. Bioorg Med Chem [Internet] 2009;17(7):2744−50. Available from: http://doi.org/10.1016/j.bmc.2009.02.042.

[216] Rajput ZI, Hu Shua, Xiao Cwen, Arijo AG. Adjuvant effects of saponins on animal immune responses. J Zhejiang Univ Sci B 2007;8(3):153−61.

[217] Rajasekaran D, Palombo EA, Yeo TC, Ley DLS, Tu CL, Malherbe F, et al. Identification of traditional medicinal plant extracts with novel anti-influenza activity. PLoS One 2013;8(11):1−15.

[218] Mehrbod P, Abdalla MA, Njoya EM, Ahmed AS, Fotouhi F, Farahmand B, et al. South African medicinal plant extracts active against influenza A virus. BMC Complement Altern Med 2018;18(1):1−10.

[219] Maria John KM, Enkhtaivan G, Ayyanar M, Jin K, Yeon JB, Kim DH. Screening of ethnic medicinal plants of South India against influenza (H1N1) and their antioxidant activity. Saudi J Biol Sci [Internet] 2015;22 (2):191−7. Available from: https://doi.org/10.1016/j.sjbs.2014.09.009.

Anti-herpes virus agents

Joy Mondal[1,2], Debprasad Chattopadhyay[1,3] and Keshab C. Mandal[2]

[1]ICMR-NICED Virus Unit, ID and BG Hospital, Kolkata, West Bengal, India [2]Department of Microbiology, Vidyasagar University, Midnapore, West Bengal, India [3]ICMR-National Institute of Traditional Medicine, Belagavi, Karnataka, India

12.1 Herpes simplex: a DNA virus

Herpes simplex viruses (HSV) are one of the most important human pathogens causing diseases in a variety of different tissues and animal species. This is the leading cause of human viral disease, second to the infections caused by air-borne influenza viruses. Herpes viruses have distinct characteristics compared to other viruses which make them more virulent. They have the unique ability to cause explicit disease while they remain silent inside the patient body for many years and get reactivated from time to time or in suitable condition or immunocompromised state of the patient, for example, shingles. The word *"Herpes"* is a Latin term coined from the Greek word *"herpein"* which means "to creep," due to the creeping or spreading nature of the skin lesions caused by herpes viruses. After, a person gets infected by the herpes virus; the infection usually remains for life. The initial infection is followed by latency and with the subsequent appearance of disease upon reactivation. There are at least 25 viruses in the family of Herpesviridae, which is classified into three subfamilies: alpha (α), beta (β), and gamma (γ) herpes viruses, based on biological properties; currently there are nine herpes virus types that frequently infect human beings and are recognized as natural human pathogens [1−6] (Table 12.1). There are mainly two antigenic types, Herpes simplex virus-1 (HSV-1) and Herpes simplex virus-2 (HSV-2), with HSV-1 being most often transmitted nonsexually and HSV-2 most usually sexually transmitted [7,8]. Herpes viruses are larger viruses and their genome encodes at least 80 proteins. Many of these proteins are not directly involved in viral replication or structural modification but can contribute to the interaction with the host cell or host immune response.

All human herpesviruses, which belong to the "Herpesviridae" family, must contain a large double-stranded, linear DNA with 100−200 genes covered within an icosahedral protein

TABLE 12.1   Classification and disease profile of herpes viruses.

Type	Synonym	Subfamily	Pathophysiology
HHV-1	Herpes simplex virus-1 (HSV-1)	*Alpha herpesvirinae* (α-herpes virus)	Oral and/or genital herpes; usually orofacial
HHV-2	Herpes simplex virus-2 (HSV-2)	α (Alpha)	Oral and/or genital herpes; usually genital
HHV-3	Varicella zoster virus (VZV)	α (Alpha)	Chickenpox and Shingles
HHV-4	Epstein-Barr virus (EBV) Lymphocryptovirus (LCV)	*Gamma herpesvirinae* (γ-herpes virus)	Infectious mononucleosis, Burkitt's lymphoma, CNS lymphoma (in AIDS patients), Post-transplant lymphoproliferative syndrome (PTLD), Nasopharyngeal carcinoma.
HHV-5	Cytomegalovirus (CMV)	*Betaherpesvirinae* (β- herpes virus)	Infectious mononucleosis-like syndrome, retinitis etc.
HHV-6, 7	Roseolo virus	β (Beta)	Roseola infantum or *exanthem subitum*
HHV-8	Kaposi's sarcoma-associated herpesvirus (KSHV), a rhadino-virus	γ (Gamma)	Kaposi's sarcoma, primary effusion lymphoma, some multicentric Castleman's disease.

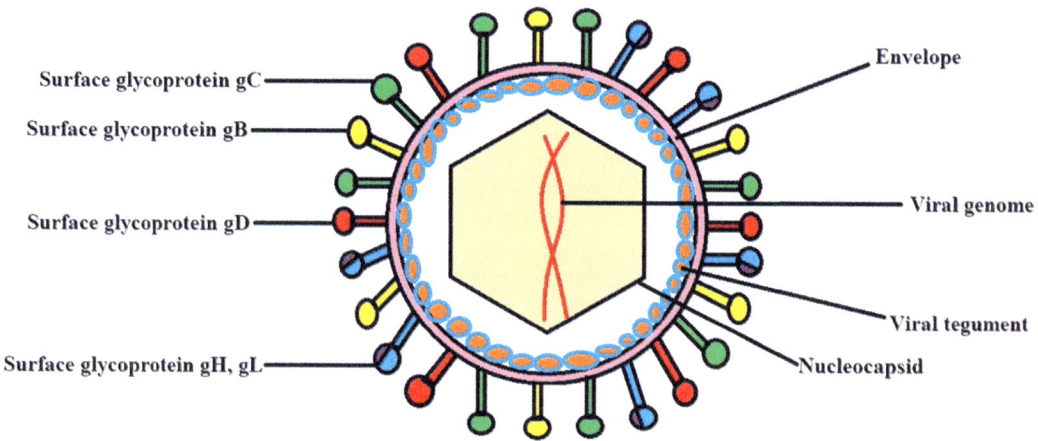

FIGURE 12.1   Diagram of herpes simplex virus particle.

capsid enfolded in a lipid bilayer envelope, called virion (Fig. 12.1). After successful binding of viral envelope glycoproteins with the host cell membrane receptors, the virion is internalized and dismantled, allowing its viral DNA to transfer into the nucleus of the host cell, where viral DNA replication and transcription arise. A successful replication cycle of HSV involves a several steps: (1) virion entry (attachment, fusion, and penetration), (2) gene expression in three phages: immediate-early (IE), early I, and late (L). The expression of IE (α) genes include infected cell protein (ICP) 0 and 4, (3) E gene (β_1, β_2) includes DNA polymerase and thymidine kinase (TK), (4) L genes (γ_1, γ_2) containing glycoprotein B (gB), C (gC), ICP5, and

(5) replication of unpaired DNA [9]. During symptomatic infection, infected cells transcribe lytic viral genes, but sometimes a small number of viral genes (latency-associated transcripts, LAT) accumulate, and the virus can persist in the host cell indefinitely. The primary infection is a self-limited period of illness, but long-term latency is symptom-free. Following reactivation, transcription of viral genes switches from LAT to multiple lytic genes and consequently leading to enhanced replication and virion production. Herpesviruses are common pathogens that cause localized skin infections of the mucosal epithelia of the oral cavity, the pharynx, the esophagus, and the eye, or genitals, depending upon the type involved. Moreover, herpes virus infections may also cause severe problems to infected individuals due to the properties such as (1) the virus establishes latent infections that can be periodically reactivated. (2) Under certain circumstances, the virus can produce serious infections of the central nervous system including acute necrotizing encephalitis and meningitis; the viruses may also produce fatal infections in patients with immune deficiencies. (3) The IE genes of HSV-1 can stimulate the activation of genes belonging to different viruses such as human immunodeficiency virus (HIV), varicella zoster virus (VZV), or human papillomavirus type 18. Additionally, HSV infections were reported to be a significant risk factor for transmission of HIV/AIDS. HSV-2 is also known as an oncogenic virus which can convert cells into tumor cells. Infection with a herpesvirus can also lead to scarification, a major cause of blindness in developing nations. Acute and recurrent HSV infections remains the most important health problem. The search for selective antiviral agents has been vigorous in recent years, but the need for new antiviral therapies still exists since many of the problems relating to the treatment of HSV infections remain unresolved, such as the generation of viral resistance and conflicting efficacy in recurrent infection and immunocompromised patients [10].

12.2 Clinical administration of viral infection

Viral infection management is a major challenge in the modern era of public health research. There are three basic strategies involved to control viral infections which are: (1) public health measures that minimize the risk of infection; (2) treatment of infected individuals with the anti-infective drugs, and (3) vaccination of the exposed or potentially vulnerable immunocompromised individuals. However, to provide the benefits of these three strategies to the people of the underdeveloped and developing world is a challenge as infection can and will occur despite the best effort of mankind. Effective public health measures can only protect the individual from water and airborne infections; but public health measure is never totally effective for the entire population, predominantly in the case of developing and underdeveloped world. The second strategy is to vaccinate vulnerable and exposed individuals. Although, to date there is no effective vaccine formulation available for all the viral infections, and one single vaccine cannot eradicate all viral infections; due to a number of factors such as the ability of the viruses to mutate rapidly, as well as emerging and re-emerging viral infections. Therefore, treatment of the infected population with suitable antiviral drugs is the only way to control and manage the viral diseases, though the development of suitable antiviral drug against newly emerging and re-emerging viral infections is a big challenge for the researchers.

To date there is no 100% effective drug development protocol available that can produce an antiviral drug that can kill a pathogenic virus without damaging the host cell.

Therefore, scientists are continuously looking into the features of an ideal antiviral drug. The ideal antiviral drug should cover four specific criteria: (1) should effectively target specific steps of viral life cycle or inhibit some essential viral processes, (2) should prevent the development of drug-resistant viral strains; (3) must have broad spectrum of activity, that is, single drug against any of the 100 + Rhinoviruses or common cold viruses; and (4) minimum or no effect on host cell or system.

To date the Food and Drug Administration (FDA), USA has approved only 40 antiviral drugs, and most of these antiviral drugs are either nucleoside or nucleotide analogs (Presented in Table 12.2). These nucleoside analogs incorporate into the replicating viral

TABLE 12.2 List of some FDA-approved antiviral drug and their mode of action.

Name of drug	Against DNA virus	Mode of action
Acyclovir	HSV, VZV, Cytomegalovirus (CMV)	Acyclovir-triphosphate serves as a competitive substrate for viral DNA polymerase, and its incorporation into the DNA chain results in termination of viral replication.
Brivudin	HSV, VZV	Competitive inhibitor of viral DNA polymerase
Cidofovir	CMV and other DNA viral infections.	Blocks viral DNA synthesis.
Famciclovir	HSV, VZV, Epstein-Barr virus (EBV)	Acts as a substrate for viral DNA polymerase, and thereby inhibit DNA synthesis.
Fomivirsen	CMV	Acts as a complementary strand to mRNA of the IE region in viral IE transcription.
Foscarnet	HSV, VZV, CMV	Inhibits pyrophosphate binding on viral DNA polymerases.
Ganciclovir	HSV, CMV, EBV	Competitively incorporated during viral DNA synthesis, leading to the DNA chain termination.
Penciclovir	HSV, VZV, EBV	Acts as a competitive inhibitor of deoxyguanosine triphosphate, to inhibit DNA polymerase activity.
Valacyclovir	CMV	Inhibits viral replication as a competitive substrate for viral DNA polymerase.
Vidarabine	HSV, VZV	Inhibits viral DNA polymerase.
Adefovir	HBV	Competitively inhibits viral DNA polymerase.
Amantadine	Influenza A virus	Inhibits viral replication by impairing the membrane protein M_2.
Boceprevir	HCV	Reversibly binds with the HCV NS3 to inhibit viral replication.
Entecavir	HBV	Priming (1) viral DNA polymerase, (2) reverse transcription of the negative strand from the pregenomic mRNA, and (3) synthesis of positive-strand HBV DNA.

(Continued)

III. Antiviral agents and therapeutics

TABLE 12.2　(Continued)

Name of drug	Against DNA virus	Mode of action
Interferons (IFNs)	HBV, HCV	
1. IFN-α, 2. IFN-β, 3. IFN-γ		1. Inhibition of virus replication 2. Suppression of cell proliferation 3. Enhanced phagocytic activity of macrophages
Lamivudine	HBV	Inhibits DNA chain termination step.
Oseltamivir	Influenza A and B viruses	Neuraminidase inhibitor of influenza virus.
Ribavirin	ADV, Poxvirus, HCV, Lassa virus, influenza, measles, mumps, RSV, HIV	Potent competitive inhibitor of inosine monophosphate dehydrogenase, viral RNA polymerase, and mRNA guanylyl transferase.
Rimantadine	Influenza virus	Inhibits the ion channel function of M_2, thereby the viral uncoating.
Telbivudine	HBV	Inhibits viral DNA polymerase.
Telaprevir	HCV	Inhibits HCV replication by binding reversibly to NS3 serine protease.
Zanamivir	Influenza A and B viruses	Inhibits neuraminidase activity of Influenza virus.

genome and either inhibit its replication or result in incomplete viral replication and reproduction. An example of one such widely used nucleoside analog is Acyclovir, the most effective and frequently prescribed antiviral drugs discovered so far since the 1970s. The HSV infections are usually managed with acyclovir, but extensive and long-term clinical use of acyclovir and related analogs such as ganciclovir, famciclovir, penciclovir, etc. cause severe side effects as well as the emergence of drug-resistant viruses [11,12]. Further, the increasing clinical use of acyclovir and related drugs against HSV, VZV, and cytomegalovirus (CMV), especially during long-term therapy results in the emergence of drug-resistant strains with severe side effects and thus, are completely unsafe for pregnant women and neonates [1–4]. Severe forms of disseminated HSV-2 infection, often seen in immunocompromised individuals with increased recurrence are found to increase the level of drug resistance. The problem of increasing drug resistance is further intensified by the wide dissemination of resistant clones [3, 13]. On the other hand, the major determinants of effective immunity against HSV infection are yet to be identified [14], and animal efficacy has not predicted success in humans [15]. Furthermore, the therapeutic vaccines failed to induce antibody-specific responses to protect recipients from recurrences [16]. Therefore, there is an unmated and urgent need for a cheap, readily available, and less toxic alternate agent to control and prevent HSV infection and its transmission.

12.2.1 Acyclovir

Acyclovir is the most commonly prescribed antiviral drug to treat HSV infections. Other oral medications include famciclovir, valacyclovir, penciclovir, foscarnet, etc. These prodrugs are converted to their active triphosphate-form to inhibit HSV replication. Oral therapy is effective for nonlife-threatening HSV infections (e.g., primary orolabial, genital); while intravenous acyclovir is used for the treatment of encephalitis, neonatal diseases, and severe infection in immunocompromised patients, and occasional cases of severe orolabial or genital disease. It is also useful in the suppression of recurrent genital HSV infections to reduce viral shedding and decrease rates of clinical recurrences. Studies have indicated that daily use of acyclovir and valacyclovir can reduce HSV-2 shedding by 60%−80% and HSV-2 transmission risk in half [17].

12.2.1.1 Mechanism of action

Acyclovir (acycloguanosine) (Fig. 12.2), after entering into the infected cell undergoes stepwise metabolic breakdown in three steps: first, it is converted to acyclovir monophosphate by the viral enzyme TK. Then cellular kinase enzymes of human cells add another phosphate in two sequential steps to form the active drug acyclovir-triphosphate. The acyclovir triphosphate competes with 2-deoxyguanosine triphosphate (dGTP), a substrate for viral DNA polymerase, and finally leads to the chain terminator. In actual infection, the HSV releases its naked capsid that delivers DNA to the nucleus of the host cell; and the active drug acyclovir triphosphate exerts its action on the viral DNA in the nucleus (Fig. 12.3).

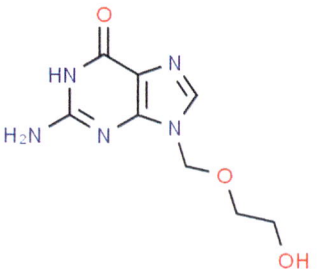

FIGURE 12.2   Chemical structure of acyclovir.

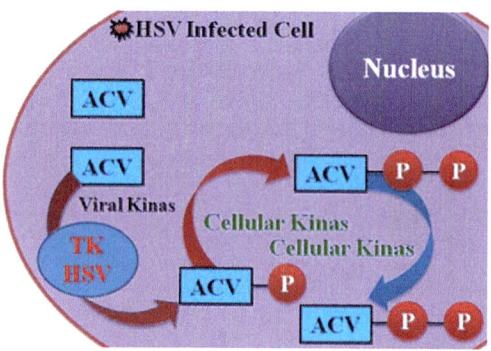

FIGURE 12.3   Mechanisms of action of acyclovir.

12.3 Disadvantages of acyclovir

The antiherpes virus drug acyclovir and its related analogs frequently develop drug-resistant strains [18,19] and adverse drug reactions during pregnancy [4], in neonates and children [20]. Moreover, acyclovir therapy is associated with few adverse effects. The most common effects associated with acyclovir therapy include nausea, diarrhea, and headache. Reversible crystalline nephropathy can result in case of rapid infusion of intravenous acyclovir. The rapid infusion of intravenous acyclovir may cause CNS disturbances, including agitation hallucinations, disorientation, tremors, and myoclonus [21].

12.3.1 Famciclovir/penciclovir

The oral form of penciclovir is termed famciclovir (Fig. 12.4). It is a synthetic acyclic guanine derivative and the inactive diacetyl prodrug of penciclovir. Penciclovir does not terminate any function in the viral DNA chain. Viral TK phosphorylates penciclovir to penciclovir monophosphate. Penciclovir achieves higher intracellular concentrations and for a longer duration than acyclovir. The adverse events and resistance of penciclovir and famciclovir are similar to those for acyclovir and valacyclovir [22].

12.3.1.1 Mechanism of action

In infected (HSV-1, HSV-2, or VZV) cells famciclovir is converted to a monophosphate form that is, penciclovir triphosphate with the help of viral TK or cellular killers, which has inhibitory activity against HSV types 1 (HSV-1) and 2 (HSV-2) and VZV. The result of in vitro studies proves that penciclovir triphosphate inhibits HSV-2 DNA polymerase competitively with deoxyguanosine triphosphate. Consequently, herpes viral DNA synthesis and, therefore, replication are selectively inhibited.

12.3.2 Ganciclovir

The chemical structure of ganciclovir (Fig. 12.5) is similar to the acyclovir with the addition of the 3'-hydroxy-methyl group. Similar to acyclovir, ganciclovir is also converted to ganciclovir monophosphate by the virus-specific TK followed by phosphorylation to the triphosphate analog by cellular enzymes [23]. It inhibits HCMV replication by

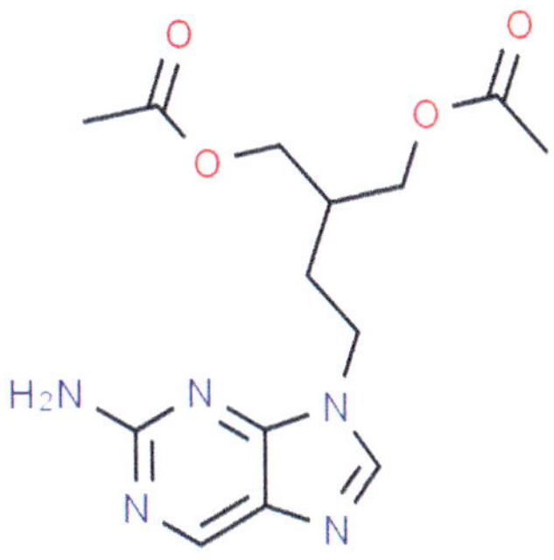

FIGURE 12.4 Chemical structure of famciclovir.

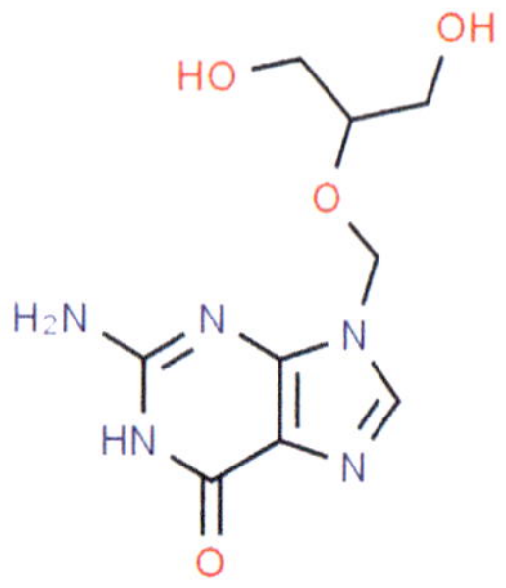

FIGURE 12.5   Chemical structure of ganciclovir.

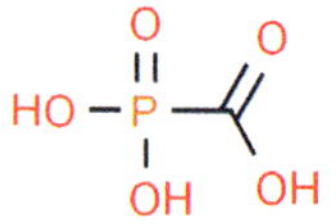

FIGURE 12.6   Chemical structure of foscarnet.

incorporating its triphosphate form into the viral DNA which causes termination of the viral chain by inhibiting DNA polymerase. The activity of ganciclovir against HSV-1, HSV-2, and VZV is the same as acyclovir. However, its greatest activity is found in HCMV-infected individuals, as its intracellular concentration is at least 10-fold higher than acyclovir and has an intracellular half-life of at least 24 hours [24].

12.3.2.1 *Mechanism of action*

The antiviral activity of ganciclovir is highly selective; it inhibits virus replication. A virus-encoded cellular enzyme, TK converted the drug to its active form by phosphorylation of ganciclovir to the monophosphate, which is then subsequently converted into the diphosphate by cellular guanylate kinase and into the triphosphate by several cellular enzymes. Inside the cell, ganciclovir triphosphate inhibits the replication of herpes viral DNA. It can also be used as a substrate for viral DNA polymerase; ganciclovir triphosphate competitively inhibits dATP leading to the formation of "faulty" DNA. Ganciclovir triphosphate is incorporated into the DNA strand by replacing many of the adenosine bases. This results in the prevention of DNA synthesis, as phosphodiester bridges can destabilize strand. Ganciclovir inhibits viral DNA polymerases more effectively than it does cellular polymerase, and chain elongation resumes when ganciclovir is removed.

12.3.3 Foscarnet

Foscarnet (Fig. 12.6) is an inorganic pyrophosphate analog also termed phosphonomethanoic acid. It is an antiherpetic drug with the brand name Foscavir. Like other antiviral drugs (acyclovir or ganciclovir) foscarnet does not require any activation through phosphorylation by TK or other kinases and therefore it is active in vitro against HSV TK deficient mutants and CMV UL97 mutants [25].

12.3.3.1 Mechanism of action

Foscarnet employs its antiviral activity by selectively acting at the pyrophosphate binding site of viral DNA polymerase and retroviral reverse transcriptase, to selectively inhibit the pyrophosphate binding site on virus-specific DNA polymerases at concentrations that do not affect cellular DNA polymerases and prevent elongation of DNA chain [25].

12.3.3.2 Cidofovir

Cidofovir (Fig. 12.7) is classified as a nucleotide analog, unlike ganciclovir it is a nucleoside monophosphate derivative. It is an injectable antiviral medication used to treat CMV. It downregulates CMV replication through selective inhibition of viral DNA synthesis. It was manufactured by Gilead and initially approved by the FDA in 1996 but has since been discontinued.

12.3.3.2.1 Mechanism of action

Cidofovir inhibits CMV replication by selectively suppressing the synthesis of viral DNA polymerase. Cidofovir diphosphate, the active intracellular metabolite of cidofovir selectively inhibits the synthesis of viral DNA polymerase. Cidofovir diphosphate inhibits herpesvirus polymerases at concentrations that are 8- to 600-fold lower than those needed to inhibit human cellular DNA polymerase alpha, beta, and gamma. Cidofovir acts as a competitive inhibitor of normal substrate dCTP, which results in reductions in the rate of viral DNA synthesis and finally termination of the CMV DNA chain [26].

12.3.3.3 Fomivirsen

Fomivirisen sodium (Vitravene) (Fig. 12.8) is an antiviral agent which contains 21 mer phosphorothioate oligonucleotide. It acts as a complementary nucleotide to the mRNA of the major IE region protein sequence of CMV. Fomivirsen binds with CMV RNA and inhibits CMV replication by an antisense mechanism.

12.3.3.4 Mechanism of action

Fomivirsen is a phosphorothioate oligonucleotide, complementary to a sequence in mRNA transcripts of the major IE region 2 (IE2) of human CMV, which is responsible for the synthesis of multiple proteins that regulates viral gene expression that is essential for viral replication. After successful binding of fomivirisen to the target mRNA, Fomivirsen inhibits the IE2 protein synthesis and disrupts viral replication [27].

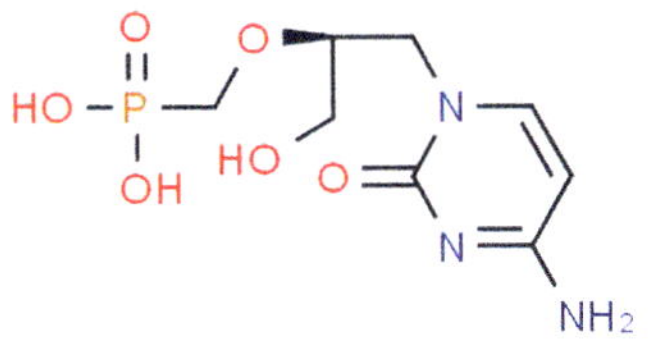

FIGURE 12.7   Chemical structure of cidofovir.

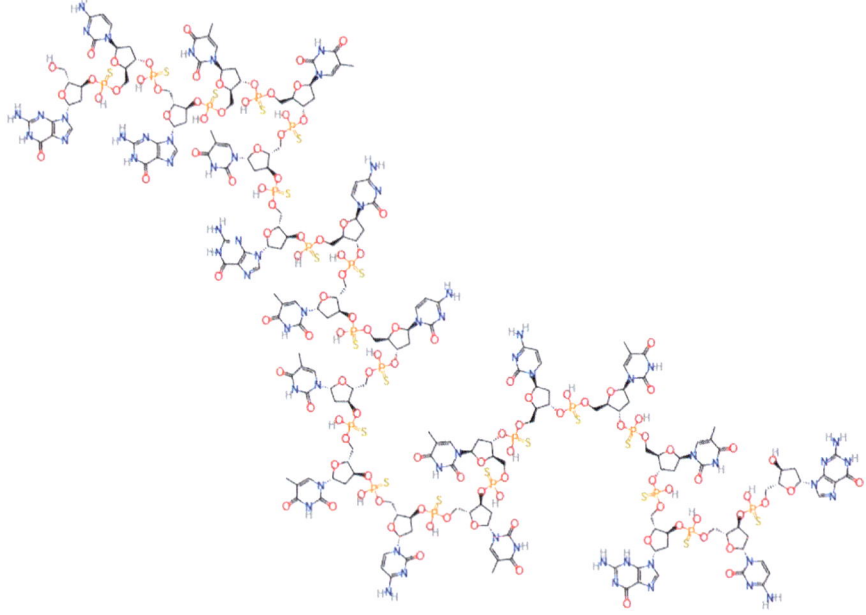

FIGURE 12.8   Chemical structure of fomivirsen.

12.3.3.5 Trifluridine

Trifluridine (Fig. 12.9) is an active antikeratoconjunctivitis drug, structurally related to idoxuridine. It is a fluorinated pyrimidine nucleoside. It is used in ophthalmic solutions and mainly used in the treatment of primary keratoconjunctivitis and recurrent epithelial keratitis due to HSV. It exhibits potent antiviral activity against both HSV types 1 and 2 [28].

12.3.3.6 Mechanism of action

The mechanism action of Trifluridine as an active antiviral agent has not been fully established, but it acts as an active molecule that inhibits viral replication. During viral replication, Trifluridine incorporates into the viral DNA which leads to the formation of defective proteins and increases the rate of mutation. Trifluridine is phosphorylated with help of viral TK and converted into its active monophosphate form [29].

12.3.4 Indoxuridine

Indoxuridine (Fig. 12.10) is chemically similar to the structure of thymidine, one of the four nucleosides or building blocks of DNA. It competitively replaces thymidine during the enzymatic step of viral replication which leads to the production of faulty DNA resulting in a pseudostructure that cannot infect or destroy tissue.

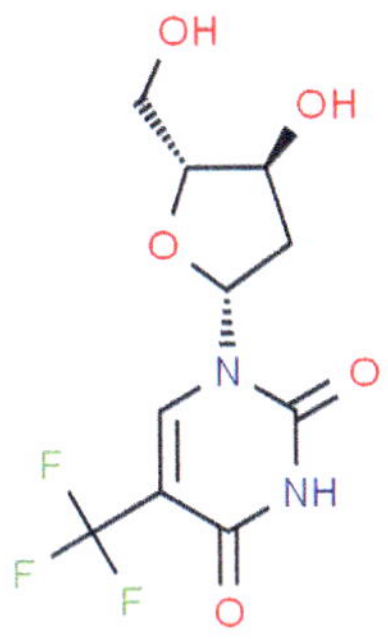

FIGURE 12.9   Chemical structure of trifluridine.

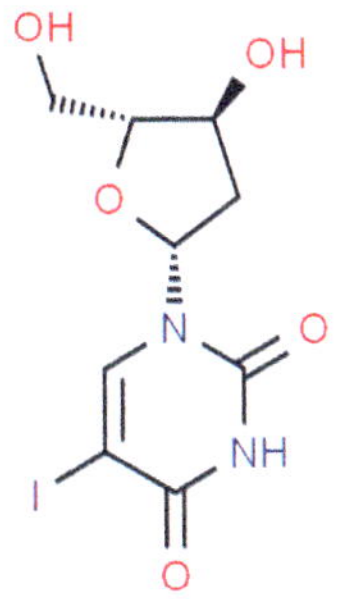

FIGURE 12.10   Chemical structure of indoxuridine.

12.3.4.1 *Mechanism of action*

Idoxuridine is a potent antiviral agent, which inhibits viral replication by competitively substituting itself in place of thymidine in the viral DNA, which results in inhibition of thymidylate phosphorylase and viral DNA polymerases from properly functioning. The application of Idoxuridine results in the inability of the virus to reproduce or to infect/destroy tissue [30].

12.4 Ethnomedicine: a gift of God to solve the problems of synthetic drugs

Nature provides us with all possible ways to combat with carnages of illness and diseases and protects the mankind from all threats of nature like viruses. The gift of God for the safety of mankind is kept in nature as a treasure chest, which is available as natural products or compounds with diverse structures and bioactivities, these bioactive molecules or agents are necessary to maintain good health as well as to fight the huge number of lifestyle disorders and infections including viral, bacterial, parasites, etc. The synthetic antivirals are creating problems of viral resistance, latency, and reactivation along with the shorter effective life span of most, these circumstances forced the scientists to look into the natural treasure chests to find out effective or potent secondary metabolites produced by the plants for the management of viral diseases.

Reports from the WHO-TRM (Traditional Medicinal Programme) centers spread across the world, show that a total of 122 compounds from 94 plant species were identified, 80%

of which were used for the same or related ethno medical purposes [31]. Since these compounds were derived only from 94 species of plants out of an estimate of 250,000 flowering plants, one might imagine the abundance of drugs remaining to be identified from these plants.

12.5 Mode of action of plant-derived anti-herpes virus agents

Ocimum basilicum or the sweet Basil a common herb found in India and China has a broad-spectrum of antiviral activity. The aqueous and ethanolic extract along with purified apigenin, linalool, and ursolic acid showed strong activity against HSV-1, adenovirus 8 (ADV-8), and CVB1 [32]. The Egyptian plant *Melaleuca armillaris* has a virucidal effect. Isoborneol, a monoterpene of essential oils isolated from *Melaleuca alternifolia* exhibited anti-HSV-1 activity by inactivating HSV-1 replication within 30 minutes of exposure [33]. Luteolin a bioflavonoid has been shown to have activity against HSV [34] (Fig. 12.14; Table 12.3).

12.6 Inhibition of virus replication

Resveratrol, which is a natural component of certain foods, such as grapes, has been shown to limit HSV-1 lesion formation by inhibiting viral replication [35]. Pentacyclic triterpenes isolated from birch bark extract can inhibit HSV-1 [36]. Proanthocyanidins a purified constituent and an extract named Oximacro isolated from a novel cranberry can inhibit HSV-1 and HSV-2 replications in vitro. It has also been reported that the compound and the extract can prevent virus adsorption and targets viral envelop protein gD and gB [37]. Fruit extract (dichloromethane) of *Angelica archangelica* L. is reported to be a potential antiviral agent against HSV type l by inhibiting its replication [38]. Oxyresveratrol, a compound purified from the Thai traditional medicinal plant *Artocarpus lakoocha*, has been reported to possess therapeutic effects in mice infected with HSV-1. It inhibits the early and late phases of viral DNA replication. The anti-HSV effects of soybean-derived isoflavonoids can be attributed to the inhibition of viral DNA replication [39]. These compounds can inhibit acyclovir-sensitive clinical isolates and acyclovir-resistant clinical isolates of HSV as well (Fig. 12.11).

12.7 Inhibition of herpes simplex viruses by immunomodulation

Odina wodier bark used by many tribes can prevent HSV by immunomodulation [40]. The methanolic extract from this plant acts by regulating COX-2 dependent prostaglandin-E2 and TLR 4 signaling pathway. *Boswellia serrata* oleo-gum-resin and β-boswellic acid inhibit HSV-1 infection in vitro through modulation of NF-κB and p38 MAP kinase signaling [41] (Fig. 12.12).

12.8 Interference with virus release

Thai medicinal plant *Dunbariabella Prain* interferes with HSV release [42].

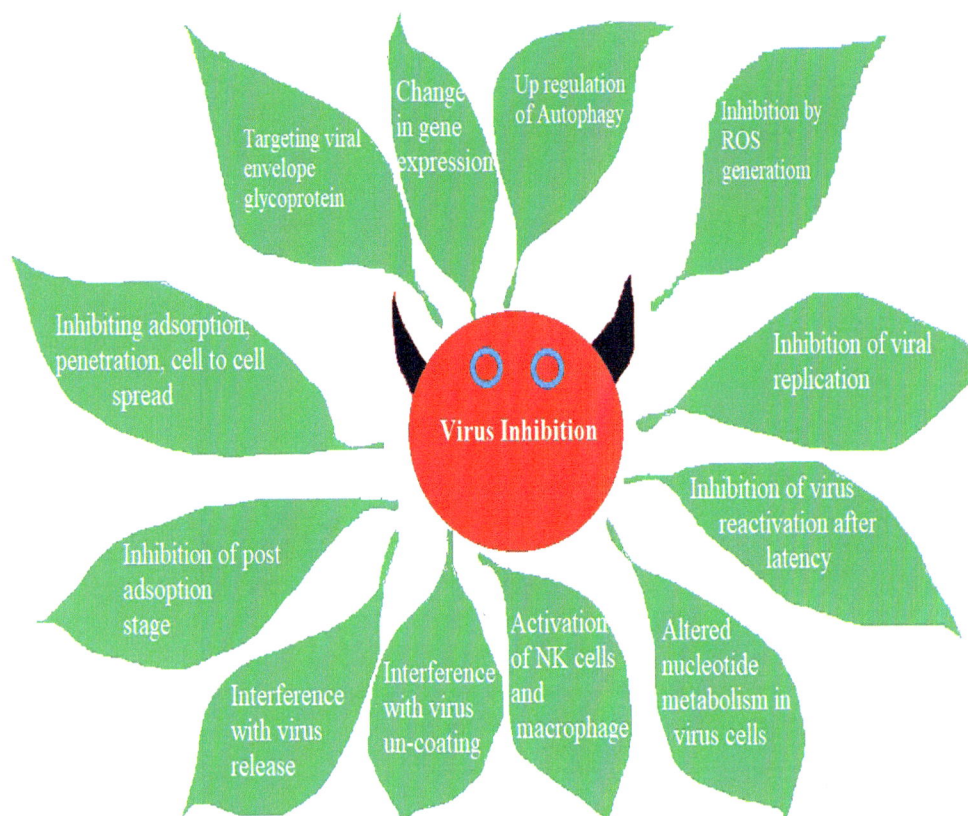

FIGURE 12.11   Twelve different modes of action of plant-derived antiviral agents.

12.9 Inhibition of herpes simplex viruses by autophagy

Pentagalloylglucose a hydrolyzable polyphenol mainly isolated from the branches and leaves of *Phyllanthus emblica*, effectively inhibit HSV-1 by induction of autophagosomes that engulfed HSV-1 virion [43]. Triterpene glycyrrhizic acid, isolated from the plant *Glycyrrhiza glabra*, is effective against HSV by promoting autophagy [44].

12.10 Inhibition of viral entry into the host cell

The anti-HSV-1 & 2 actions of sulfated galactans from the red seaweeds *Gymnogongrus griffithsiae* and *Cryptonemia crenulata* can be ascribed to the inhibition of the virus entry [45]. *Mentha suaveolens* essential oil and its main component piperitenone oxide exert an anti-HSV effect by inhibiting viral adsorption [46]. Moreover, piperitenone oxide could interfere with some redox-sensitive cellular pathways exploited for viral replication and thus inhibits viral growth and proliferation. The virucidal effect of peppermint oil, the

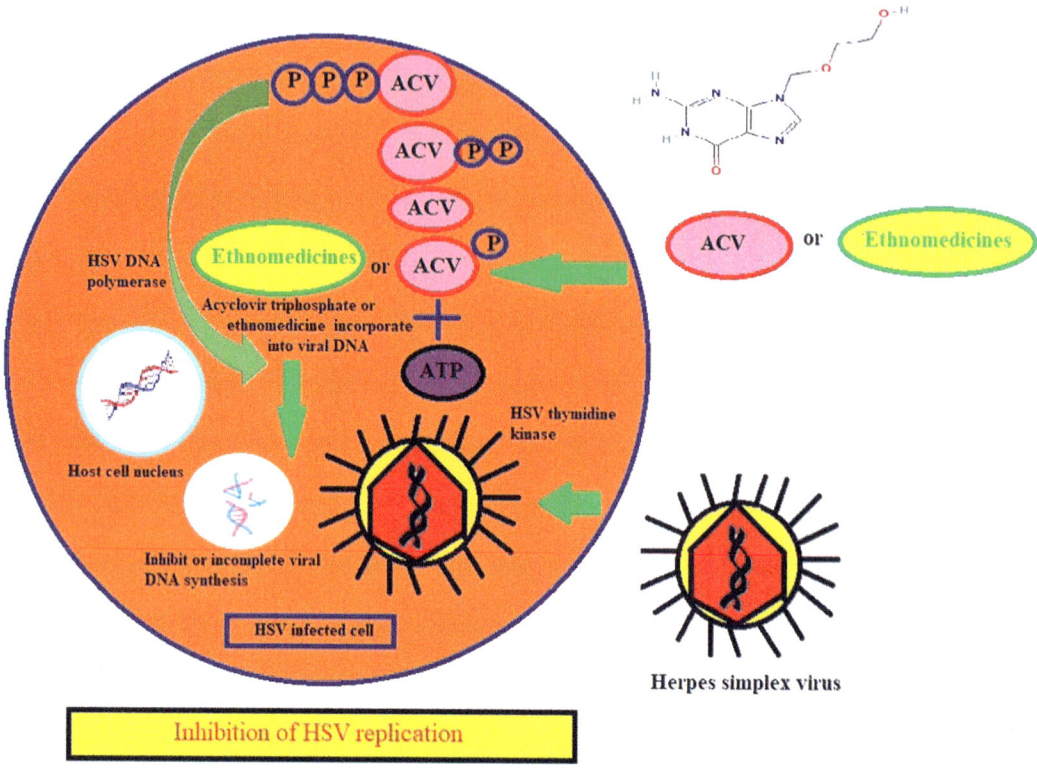

FIGURE 12.12   Inhibition of herpes virus replication by ethno medicines or synthetic drugs.

essential oil of *Mentha piperita*, against HSV, can be attributed to the interference of peppermint oil with viral adsorption to the host cell [47]. Chloroform extract of an ethnomedicinal plant *Stephania hernandifolia* inhibits HSV-1 entry by direct damage to the virion and by affecting the host cell to block virion uptake [48] (Fig. 12.13).

12.11 Conclusion

The management of herpetic infections is possible with the help of currently available FDA approved antiviral agents, which demonstrated antiviral efficacy and can be used to treat diseases caused by herpes viruses including HSV-1 and HSV-2, varicella zoster virus, and CMV. Researchers have continuously progressed their extensive research in the development of newer agents with enhanced activity against these viruses including resistant strains. The USFDA has approved 40 antiviral drugs so far, but most of these approved antivirals are either nucleoside or nucleotide analogs. These synthetic nucleoside analogs or drugs can incorporate into the replicating viral genome which results in inhibition of viral replication or incomplete viral

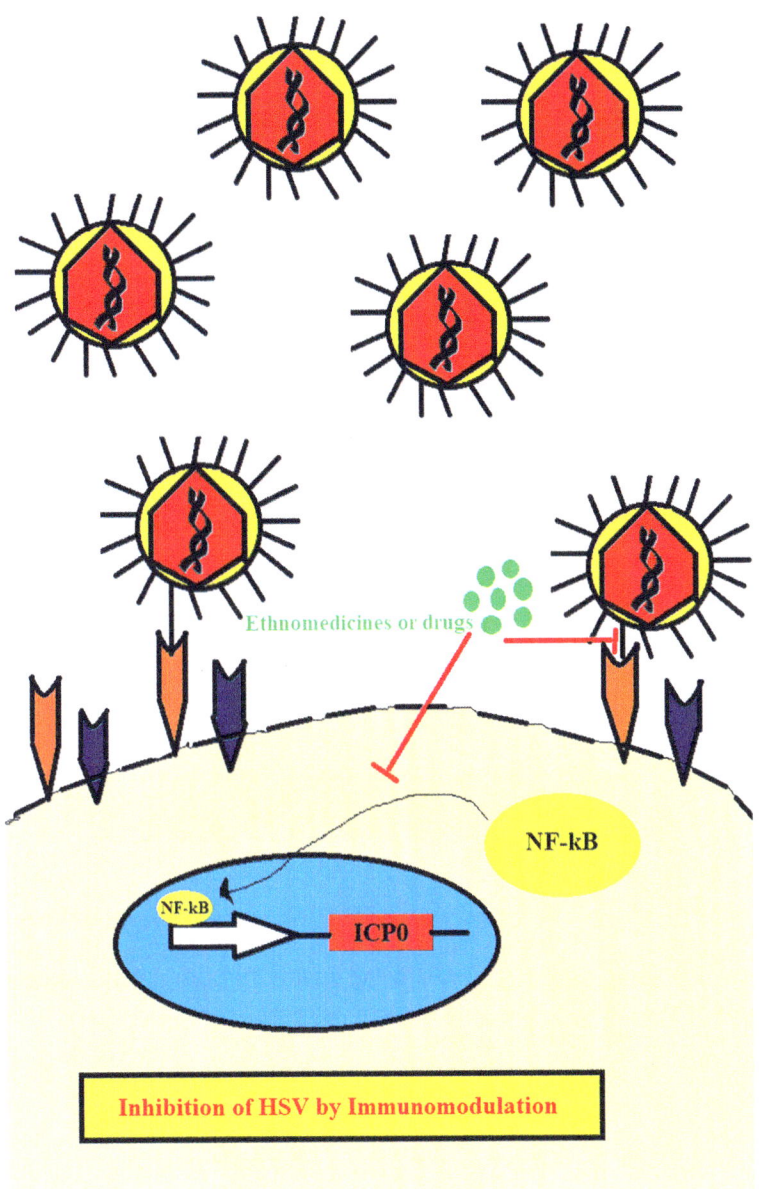

FIGURE   12.13   Inhibition of HSV by immunomodulation.

replication and reproduction. Acyclovir is one of the examples of such a widely used nucleoside analog, the most effective and frequently prescribed antiviral drug discovered so far since the 1970s. Ethnomedicines have been increasingly used by many communities from the ancient periods to till date, revealing a very important role of plants in maintaining the healthy status of

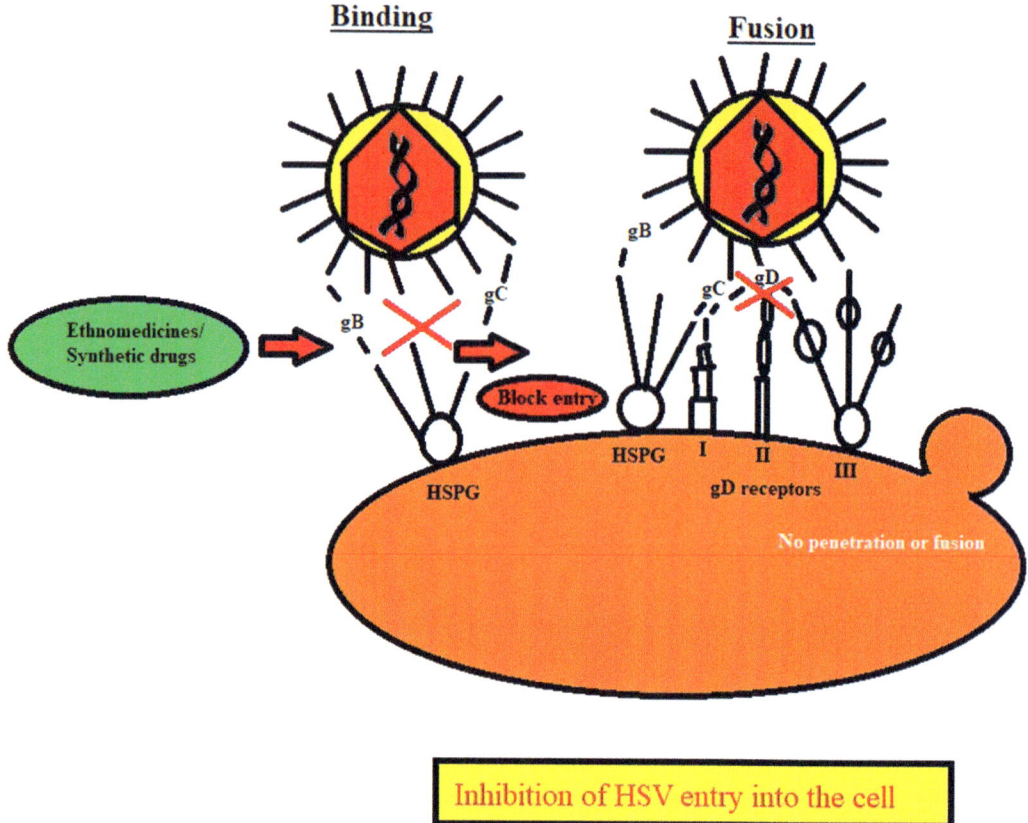

FIGURE 12.14   Inhibition of herpes virus entry into the cell.

the people. The World Health Organization conferring that, the goal of "health for all" cannot be achieved without the help of herbal medicines. The people of ancient civilization taught us a beautiful lesson, how to blend traditional knowledge and natural herbs to combat diseases and control public health. Researchers now use an increasing number of traditional plants as a source of new antiviral agents. The age-old medicinal knowledge is being revisited by the researchers to withstand the emerging public health issues. So, there is a big reason to look into the natural treasures of ethnomedicinal plants to improve drug discovery technology to find out natural products or natural antiviral agents, which can tackle the emerging diseases and protect our bodies. Innovative strategies would be required to reveal and contribute to the use of complex structures of phytochemicals that not only mimic as synthetic drugs but also use the full range of chemical diversity of these natural products as a unique lead molecule towards drug discovery process. Extensive research by the scientists on diverse chemical structures of phytochemicals alone or with the combination of synthetic drugs by using the advanced technology can made discovery of great antivirals from the treasure of ethnomedicines which can eradicate virus.

TABLE 12.3 Mode of action of plant-derived antiherpetic agents.

Serial no.	Extract/compound	Plant origin or source	Phytocomponents or metabolites	Virus	MOA	Reference
1	A Chinese herbal medicine, with the Japanese name hochu-ekki-to HET; Chinese name, bu-zhong-Ž yi-qi-tang, a mixture of ten plants.			MCMV	Induction of IFN-γ production and perforin-mediated cytotoxicity via activation of Natural Killer (NK) cells.	[49]
2.	Quercetin 3-O-rutinoside, kaempferol 3-O-rutinoside and kaempferol 3-O-robinobioside	Ethanolic extract of *Ficus benjamina* leaves		Herpes simplex virus-1 (HSV)		[50]
3.	1,2,4-Triazoloazine		Analogs of natural nucleic bases	HSV	Replication	[51]
4.	*Sophora subprostrata*		Propolis	HSV	Pre-treatment of this drug shows antiherpetic activity	[52]
5.	Chebulagic acid and Punicalagin	*Terminalia chebula Retz*	Hydrolyzable tannins	HSV, RSV, MCMV	Inactivating free virus particles by inhibiting early steps of virus entry	[53]
6.	3.4-Dihydroxybenzoic acid, p-OH-Phenylacetic acid, N-arginine	*Ficus carica* latex extracts		ECV-11 HSV-1, and ADV		[54]
7.	Ursolic acid	*Mallotus peltatus* from the rain forests of Middle and Southern Andaman		HSV-1 and HSV-2	Inhibits HSV multiplication at an early stage	[55]
8.	2-aminobenzamide				HSP90 inhibitors	[56]
9.	Oleanolic acid	*Achyranthes aspera*	Triterpene acid	HSV-1 and HSV-2	Inhibits the early stage of HSV replication and also modulates some early immunological parameters of infected host cell	[57]

(*Continued*)

TABLE 12.3 (Continued)

Serial no.	Extract/compound	Plant origin or source	Phytocomponents or metabolites	Virus	MOA	Reference
10.	methoxy-1-methyl-4,9-dihydro-3H-pyrido[3,4-b]indole	Ophiorrhiza nicobarica Balkr.	Indole	HSV-2	Immediate early transcriptional event of HSV-2 interfered with the recruitment of LSD-1 by HCF-1	[58]
11.	Sulfated polysaccharide	Caesalpinia ferrea		HSV	Inhibited virus adsorption and steps after penetration, and inhibited the synthesis of viral protein	[59]
12.	Berberine		Isoquinoline alkaloids	HSV	Modulating multiple cellular signaling pathways, including p53, nuclear factor κB (NF-κB), and mitogen-activated protein kinase	[60]
13.	Piperitenone oxide		Essential oils	HSV	Late phase	[46]
14.	Curcumin	Curcuma longa		Human cytomegalovirus (HCMV)	Effect on protein expression of Hsp90α	[61]
15.	3,19-isopropylidene andrographolide	Andrographis paniculata	Diterpenoid lactone	HSV	Early gene expression, inhibit postentry step, suppressed ICP8 transcription and translation as well as DNA replication and gD expression	[62]
16.	Xanthotoxin, bergapten, imperatorin, phellopterin, and isoimperatorin, and the mixture of imperatorin and phellopterin	Angelica archangelica	Isopentenyloxy moiety at C-8 position	HSV-1 and Coxsackievirus B3	Titer of virus in relation to the virus control	[38]
17.		Coptidis rhizoma extract		RSV	By the induction of type I interferon-related signaling	[63]

References

[1] Fatahzadeh M, Schwartz RA. Human herpes simplex virus infections: epidemiology, pathogenesis, symptomatology, diagnosis, and management. J Am Acad Dermatol 2007;57(5):737–63.

[2] Wald A. Synergistic interactions between herpes simplex virus type-2 and human immunodeficiency virus epidemics. Herpes J IHMF 2004;11(3):70–6.

[3] Mbopi-Kéou FX, Grésenguet G, Mayaud P, Weiss HA, Gopal R, Matta M, et al. Interactions between herpes simplex virus type 2 and human immunodeficiency virus type 1 infection in African women: opportunities for intervention. J Infect Dis 2000;182(4):1090–6.

[4] Narayana K. A purine nucleoside analogue-acyclovir [9-(2-hydroxyethoxymethyl)-9h–guanine] reversibly impairs testicular functions in mouse. J Toxicol Sci 2008;33(1):61–70.

[5] Prober CG. Antiviral agents and interferons. In: Lon SS, Pickering LK, Prober CG, editors. Principles and practice or pediatic infectious diseases. New York: Churchil Livingstone; 1997. p. 1682–703.

[6] Roizman B, Pellet PE. The family herpesviridae: a brief introduction. In. Fields virology 2001.

[7] Umene K, Kawana T. Molecular epidemiology of herpes simplex virus type 1 genital infection in association with clinical manifestations. Arch Virol 2000;145(3):505–22.

[8] Umene K, Kawana T. Divergence of reiterated sequences in a series of genital isolates of herpes simplex virus type 1 from individual patients. J Gen Virol 2003;84(4):917–23.

[9] Sandri-Goldin RM. Alpha herpesviruses: molecular and cellular biology. Caister Academic Press; 2006.

[10] Roizman B. The function of herpes simplex virus genes: a primer for genetic engineering of novel vectors. Proc Natl Acad Sci 1996;93(21):11307–12.

[11] De Clercq E. Hamao Umezawa memorial award Lecture: 'an odyssey in the viral chemotherapy field'. Int J Antimicrob Agents 2001;18(4):309–28.

[12] De Clercq E. Strategies in the design of antiviral drugs. Nat Rev Drug Discov 2002;1(1):13–25.

[13] Safrin S, Crumpacker C, Chatis P, Davis R, Hafner R, Rush J, et al. AIDS Clinical Trials Group*. A controlled trial comparing foscarnet with vidarabine for acyclovir-resistant mucocutaneous herpes simplex in the acquired immunodeficiency syndrome. N Engl J Med 1991;325(8):551–5.

[14] Stanberry LR. Clinical trials of prophylactic and therapeutic herpes simplex virus vaccines. Herpes: J IHMF 2004;11:161A–169AA.

[15] Koelle DM. Vaccines for herpes simplex virus infections,. Curr Opin Invest Drugs 2006;7:136–41.

[16] Dropulic LK, Cohen JI. The challenge of developing a herpes simplex virus 2 vaccine. Expert Rev Vaccines 2012;11(12):1429–40.

[17] Wald A, Corey L, Cone R, Hobson A, Davis G, Zeh J. Frequent genital herpes simplex virus 2 shedding in immunocompetent women. Effect of acyclovir treatment. J Clin Invest 1997;99(5):1092–7.

[18] Kleymann G. New antiviral drugs that target herpesvirus helicase primase enzymes. Herpes J IHMF 2003;10(2):46–52.

[19] Miserocchi E, Modorati G, Galli L, Rama P. Efficacy of valacyclovir vs acyclovir for the prevention of recurrent herpes simplex virus eye disease: a pilot study. Am J Ophthalmol 2007;144(4):547–51.

[20] Sawyer MH, Webb DE, Balow JE, Straus SE. Acyclovir-induced renal failure: clinical course and histology. Am J Med 1988;84(6):1067–71.

[21] Pistelli L, Giorgi I. Antimicrobial properties of flavonoids. Dietary phytochemicals and microbes. Dordrecht: Springer; 2012. p. 33–91.

[22] Grose C, Wiedeman J. Generic acyclovir vs. famciclovir and valacyclovir. Pediatr Infect Dis J 1997;16(9):838–41.

[23] Fan-Havard P, Nahata MC, Brady MT. Ganciclovir—a review of pharmacology, therapeutic efficacy and potential use for treatment of congenital cytomegalovirus infections. J Clin Pharm Therap 1989;14(5):329–40.

[24] Abdel-Haq NM, Asmar BI. Anti-herpes viruses agents. Indian J Pediatr 2001;68(7):649–54.

[25] Wagstaff AJ, Bryson HM. Foscarnet. Drugs 1994;48(2):199–226.

[26] Plosker GL, Noble S. Cidofovir: a review of its use in cytomegalovirus retinitis in patients with AIDS. Drugs 1999;58(2):325–45.

[27] Isomura H, Stinski MF. The human cytomegalovirus major immediate-early enhancer determines the efficiency of immediate-early gene transcription and viral replication in permissive cells at low multiplicity of infection. J Virol 2003;77(6):3602–14.

[28] Carmine AA, Brogden RN, Heel RC, Speight TM, Avery GS. Trifluridine: a review of its antiviral activity and therapeutic use in the topical treatment of viral eye infections. Drugs 1982;23(5):329–53.

[29] Burness CB, Duggan ST. Trifluridine/tipiracil: a review in metastatic colorectal cancer. Drugs 2016;76 (14):1393–402.

[30] Lalezari JP, Stagg RJ, Kuppermann BD, Holland GN, Kramer F, Ives DV, et al. Intravenous cidofovir for peripheral cytomegalovirus retinitis in patients with AIDS: a randomized, controlled trial. Ann Intern Med 1997;126(4):257–63.

[31] Fabricant DS, Farnsworth NR. The value of plants used in traditional medicine for drug discovery. Environ Health Perspect 2001;109(Suppl. 1):69–75.

[32] Chattopadhyay D, Naik TN. Antivirals of ethnomedicinal origin: structure-activity relationship and scope. Mini Rev Med Chem 2007;7(3):275–301.

[33] Armaka M, Papanikolaou E, Sivropoulou A, Arsenakis M. Antiviral properties of isoborneol, a potent inhibitor of herpes simplex virus type 1. Antivir Res 1999;43(2):79–92.

[34] Ojha D, Das R, Sobia P, Dwivedi V, Ghosh S, Samanta A, et al. Pedilanthus tithymaloides inhibits HSV infection by modulating NF-κB signaling. PLoS One 2015;10(9):e0139338.

[35] Docherty JJ, Fu MM, Hah JM, Sweet TJ, Faith SA, Booth T. Effect of resveratrol on herpes simplex virus vaginal infection in the mouse. Antivir Res 2005;67(3):155–62.

[36] Navid MH, Laszczyk-Lauer MN, Reichling J, Schnitzler P. Pentacyclic triterpenes in birch bark extract inhibit early step of herpes simplex virus type 1 replication. Phytomedicine 2014;21(11):1273–80.

[37] Terlizzi ME, Occhipinti A, Luganini A, Maffei ME, Gribaudo G. Inhibition of herpes simplex type 1 and type 2 infections by Oximacro®, a cranberry extract with a high content of A-type proanthocyanidins (PACs-A). Antivir Res 2016;132:154–64.

[38] Rajtar B, Skalicka-Woźniak K, Świątek Ł, Stec A, Boguszewska A, Polz-Dacewicz M. Antiviral effect of compounds derived from Angelica archangelica L. on Herpes simplex virus-1 and Coxsackievirus B3 infections. Food Chem Toxicol 2017;109:1026–31.

[39] Argenta DF, Bidone J, Koester LS, Bassani VL, Simões CM, Teixeira HF. Topical delivery of coumestrol from lipid nanoemulsions thickened with hydroxyethylcellulose for antiherpes treatment. AAPS Pharm 2018;19(1):192–200.

[40] Ojha D, Mukherjee H, Mondal S, Jena A, Dwivedi VP, Mondal KC, et al. Anti-inflammatory activity of Odina wodier Roxb, an Indian folk remedy, through inhibition of toll-like receptor 4 signaling pathway. PLoS One 2014;9(8):e104939.

[41] Goswami D, Mahapatra AD, Banerjee S, Kar A, Ojha D, Mukherjee PK, et al. Boswellia serrata oleo-gum-resin and β-boswellic acid inhibits HSV-1 infection in vitro through modulation of NF-κB and p38 MAP kinase signaling. Phytomedicine 2018;51:94–103.

[42] Akanitapichat P, Wangmaneerat A, Wilairat P, Bastow KF. Anti-herpes virus activity of Dunbaria bella Prain. J Ethnopharmacol 2006;105(1–2):64–8.

[43] Pei Y, Chen ZP, Ju HQ, Komatsu M, Ji YH, Liu G, et al. Autophagy is involved in anti-viral activity of pentagalloylglucose (PGG) against Herpes simplex virus type 1 infection in vitro. Biochem Biophys Res Commun 2011;405(2):186–91.

[44] Laconi S, Madeddu MA, Pompei R. Autophagy activation and antiviral activity by a licorice triterpene. Phytother Res 2014;28(12):1890–2.

[45] Talarico LB, Zibetti RG, Faria PC, Scolaro LA, Duarte ME, Noseda MD, et al. Anti-herpes simplex virus activity of sulfated galactans from the red seaweeds Gymnogongrus griffithsiae and Cryptonemia crenulata. Int J Biol Macromolecules 2004;34(1–2):63–71.

[46] Civitelli L, Panella S, Marcocci ME, De Petris A, Garzoli S, Pepi F, et al. In vitro inhibition of herpes simplex virus type 1 replication by Mentha suaveolens essential oil and its main component piperitenone oxide. Phytomedicine 2014;21(6):857–65.

[47] Schuhmacher A, Reichling J, Schnitzler PJ. Virucidal effect of peppermint oil on the enveloped viruses herpes simplex virus type 1 and type 2 in vitro. Phytomedicine 2003;10(6–7):504–10.

[48] Mondal J, Mahapatra AD, Mandal KC, Chattopadhyay D. An extract of Stephania hernandifolia, an ethnomedicinal plant, inhibits herpes simplex virus 1 entry. Arch Virol 2021;1–2 May 26.

[49] Hossain MS, Takimoto H, Hamano S, Yoshida H, Ninomiya T, Minamishima Y, et al. Protective effects of hochu-ekki-to, a Chinese traditional herbal medicine against murine cytomegalovirus infection. Immunopharmacology 1999;41(3):169–81.

[50] Yarmolinsky L, Zaccai M, Ben-Shabat S, Mills D, Huleihel M. Antiviral activity of ethanol extracts of Ficus binjamina and Lilium candidum in vitro. N Biotechnol 2009;26(6):307–13.

[51] Deev SL, Yasko MV, Karpenko IL, Korovina AN, Khandazhinskaya AL, Andronova VL, et al. 1, 2, 4-Triazoloazine derivatives as a new type of herpes simplex virus inhibitors. Bioorg Chem 2010;38(6):265—70.

[52] Nolkemper S, Reichling J, Sensch KH, Schnitzler P. Mechanism of herpes simplex virus type 2 suppression by propolis extracts. Phytomedicine 2010;17(2):132—8.

[53] Lin LT, Chen TY, Chung CY, Noyce RS, Grindley TB, McCormick C, et al. Hydrolyzable tannins (chebulagic acid and punicalagin) target viral glycoprotein-glycosaminoglycan interactions to inhibit herpes simplex virus 1 entry and cell-to-cell spread. J Virol 2011;85(9):4386—98.

[54] Lazreg Aref H, Gaaliche B, Fekih A, Mars M, Aouni M, Pierre Chaumon J, et al. In vitro cytotoxic and antiviral activities of *Ficus carica* latex extracts. Nat Prod Res 2011;25(3):310—19.

[55] Bag P, Chattopadhyay D, Mukherjee H, Ojha D, Mandal N, Sarkar MC, et al. Anti-herpes virus activities of bioactive fraction and isolated pure constituent of *Mallotus peltatus*: an ethnomedicine from Andaman Islands. Virol J 2012;9(1):1—2.

[56] Xiang YF, Qian CW, Xing GW, Hao J, Xia M, Wang YF. Anti-herpes simplex virus efficacies of 2-aminobenzamide derivatives as novel HSP90 inhibitors. Bioorg Med Chem Lett 2012;22(14):4703—6.

[57] Mukherjee H, Ojha D, Bag P, Chandel HS, Bhattacharyya S, Chatterjee TK, et al. Anti-herpes virus activities of Achyranthes aspera: an Indian ethnomedicine, and its triterpene acid. Microbiol Res 2013;168(4):238—44.

[58] Bag P, Ojha D, Mukherjee H, Halder UC, Mondal S, Chandra NS, et al. An indole alkaloid from a tribal folklore inhibits immediate early event in HSV-2 infected cells with therapeutic efficacy in vaginally infected mice. PLoS One 2013;8(10):e77937.

[59] Lopes N, Faccin-Galhardi LC, Espada SF, Pacheco AC, Ricardo NM, Linhares RE, et al. Sulfated polysaccharide of *Caesalpinia ferrea* inhibits herpes simplex virus and poliovirus. Int J Biol Macromol 2013;60:93—9.

[60] Song S, Qiu M, Chu Y, Chen D, Wang X, Su A, et al. Downregulation of cellular c-Jun N-terminal protein kinase and NF-κB activation by berberine may result in inhibition of herpes simplex virus replication. Antimicrob Agents Chemother 2014;58(9):5068—78.

[61] Lv Y, Gong L, Wang Z, Han F, Liu H, Lu X, et al. Curcumin inhibits human cytomegalovirus by downregulating heat shock protein 90. Mol Med Rep 2015;12(3):4789—93.

[62] Kongyingyoes B, Priengprom T, Pientong C, Aromdee C, Suebsasana S, Ekalaksananan T. 3, 19-isopropylideneandrographolide suppresses early gene expression of drug-resistant and wild type herpes simplex viruses. Antivir Res 2016;132:281—6.

[63] Lee BH, Chathuranga K, Uddin MB, Weeratunga P, Kim MS, Cho WK, et al. *Coptidis rhizoma* extract inhibits replication of respiratory syncytial virus in vitro and in vivo by inducing antiviral state. J Microbiol 2017;55(6):488—98.

CHAPTER

13

Antiretroviral therapy

Suman Ganguly[1] and Debjit Chakraborty[2]

[1]West Bengal State AIDS Prevention and Control Society, Kolkata, West Bengal, India
[2]Division of Epidemiology, ICMR-National Institute of Cholera and Enteric Disease, Kolkata, West Bengal, India

13.1 Introduction

The human immunodeficiency virus (HIV) causes acquired immunodeficiency syndrome (AIDS), the most advanced stage of HIV infection. Once HIV infection is acquired, it will remain forever. HIV per se has no cure but it is a chronic manageable disease. The treatment for HIV is called antiretroviral therapy (ART), which involves taking a combination of anti-HIV drugs. ART cannot cure HIV infection and there is no way to eradicate HIV from human body once infected. The virus tends to persist in various sanctuaries in the body, such as brain, liver, spleen, and other parts of the lympho-reticular system. The basic goal of ART is the highest and sustained reduction in plasma viral load and restoration of immunity of the body. Once an HIV infected person is put on ART, it is expected to reduce the plasma viral load to undetectable levels after some duration. "U" is equal to "U," that is, a person with undetectable viral load is untransmissible to HIV to others to a large extent. Therefore, ART does not only reduce the viral load in the body but also increases the quality of life of an HIV-infected person and reduces the chance of new HIV infection from an infected individual [1,2]. The goal of ART is as follows [3,4].

1. **Virological goal:** To sustain plasma viral load suppression
2. **Immunological goal:** To restore immunological status of the individual infected with HIV
3. **Clinical goal:** To increase survival and improvement in quality of life
4. **Preventive goal:** To prevent HIV transmission from an infected person to an uninfected one

During initial years of HIV pandemic, it was an eventual fatal disease. Soon after identification of this retrovirus, attempts were made to develop therapy for the new infection. During the first phase, necessary effort was made to protect $CD4^+$ T cells from being killed by HIV virus. As the pathogenesis of HIV virus was discovered, effort was undertaken to develop drug to prevent viral replication. During 1985, zidovudine was found to effective in an in vitro study and subsequent clinical trial started with this medicine. Thus zidovudine brought the first ray of hope towards containing HIV replication in the body. Subsequently, it was attempted to reach out for more avenues targeting different stages of pathogenesis of HIV. Currently, a series of antiretroviral drugs are in use and many more in pipeline. The currently available drugs directed at various steps of replication are as follow [5−10].

1. Fusion Inhibitor: Blocks binding of HIV to the target cell
2. Chemokine receptor blocker: Prevents binding of the virus inhibiting conformational changes of chemokine receptor
3. Reverse transcriptase inhibitor: May be nucleotide based or nucleoside based and blocks the viral RNA cleavage and inhibits reverse transcriptase enzyme
4. Integrase inhibitor: Blocks the enzyme integrase, which helps in the proviral DNA being incorporated into the host cell genome
5. Protease inhibitor: Blocks enzyme protease and prevents post-translational modification

Lifecycle of HIV is illustrated in Fig. 13.1 and commonly used ART drugs are presented in Table 13.1.

13.2 Formulation of antiretroviral treatment

Single antiretroviral medicine is not used in the formulation of ART regimen but generally three drugs are used in appropriate combination to develop a regimen for ART. To formulate ART regimen, two drugs are generally taken from one class of antiretroviral drug and one from other class [16−19]. For operational feasibility and patients'

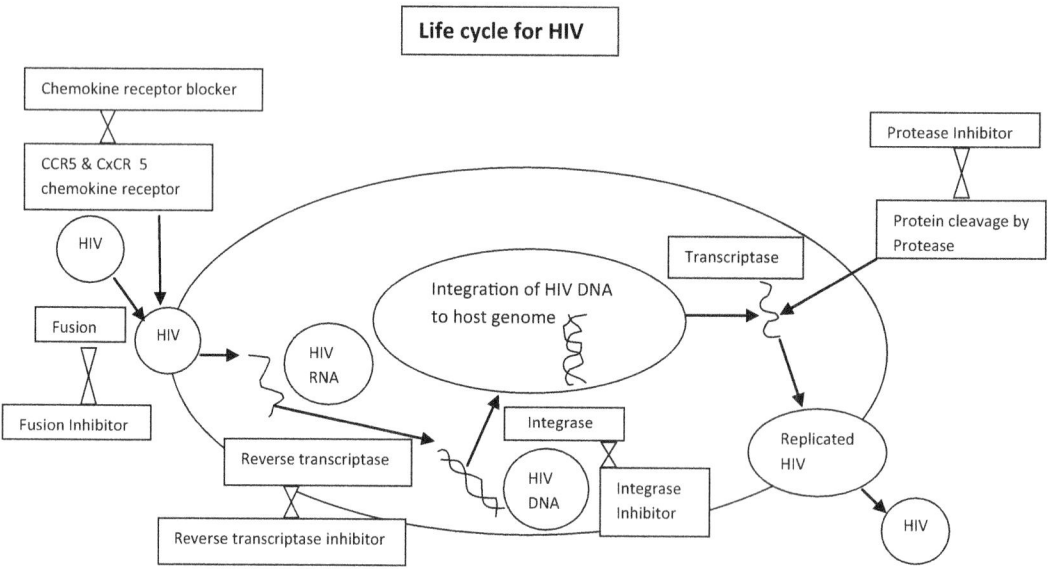

FIGURE 13.1   Lifecycle of human immunodeficiency virus.

TABLE 13.1 Commonly used antiretroviral therapy drugs [5–15].

Group of the drug	Name of the drug	Standard adult dosage	Mechanism of action	Adverse effects
Nucleoside Reverse Transcriptase Inhibitors (NRTIs)	Zidovudine (ZDV, AZT)	300 mg twice a day-oral	Nucleoside Reverse Transcriptase Inhibitors (NRTIs) inhibit reverse transcription by blocking the enzyme causing chain termination after they have been incorporated into viral DNA. The drug is activated by phosphorylating intracellularly. The resultant enzyme inhibition cause development of incomplete DNA and new virus formation is halted.	Anemia, neutropenia, bone marrow suppression, gastrointestinal intolerance, headache, insomnia, myopathy, lactic acidosis, skin and nail hyperpigmentation
	Lamivudine (3TC)	150 mg twice a day or 300 mg once a day-oral		Minimal toxicity, rash though very rare
	Abacavir (ABC)	300 mg twice a day or 600 mg once a day-oral		Most dangerous side effect:Hypersensitivity reaction in 3% to 5% (can be fatal), fever, rash, fatigue, nausea, vomiting, re-challenging after discontinuation can be fatal.
	Emtricitabine (FTC)	200 mg once a day-oral		Generally mild, rash, diarrhea, headache, nausea
	Stavudine(D4T)	40 mg twice a day (weight dependent)-oral		Lactic acidosis, pancreatitis, peripheral neuritis, lipodystrophy
Some other NRTIs are didanosine and zalcitabine, which are hardly used now a days.				
Nucleotide Reverse Transcriptase Inhibitors (NtRTIs)	Tenofovir (TDF)	300 mg once a day-oral	Nucleotide analog of reverse transcriptase inhibitor works similarly to nucleosides, but they have a nonpeptidic chemical structure	Renal toxicity, bone demineralization
Non-Nucleoside Reverse Transcriptase Inhibitors (NNRTIs)	Nevirapine (NVP)	200 mg twice a day (starting with 200 mg once a day for 2 weeks when ART is started with NVP)-oral	Nonnucleoside reverse transcriptase inhibitors (NNRTIs) stop viral replication by binding on the reverse transcriptase. These drugs are called "nonnucleoside" inhibitors because, even though they work at the same stage as nucleoside analogs, as chain terminators, they inhibit the HIV reverse transcriptase enzyme by directly binding to it	Hepatitis (usually within 12 weeks); sometime life-threatening hepatic toxicity. Skin rash occasionally progressing to severe conditions with mucosal and systemic involvement, leading to Stevens Johnson syndrome (SJS), and Toxic Epidermal Necrolysis (TEN).
	Efavirenz (EFV)	600 mg once a day oral		CNS symptoms (dizziness, somnolence, insomnia, confusion, hallucinations, agitation) and personality change even suicidal ideation. CNS symptoms generally go away by 2 to 6 weeks. Rash may sometimes occur. Avoid taking EFV after heavy fatty meals
	Etravirine	200 mg twice a day-oral		Rash, nausea, and other gastrointestinal symptoms. Mild to moderate hepatitis
	Rilpivirine	Oral as well as long acting intra muscular forms		Injection site pain, swelling, fever, rash, occasional CNS side effects
	Delavirdine	400 mg thrice a day-oral		Moderate to severe rash up to 20% of the patients, fatigue, nausea, vomiting, fever

(Continued)

TABLE 13.1 (Continued)

Group of the drug	Name of the drug	Standard adult dosage	Mechanism of action	Adverse effects
Protease inhibitors (PIs)	Atazanavir/ ritonavir (ATV/r)	300/100 mg once a day-oral	A protease enzyme catalyzes proteolysis, breaking down proteins into smaller polypeptides or single amino acids, and spurring the formation of new protein products. It is mainly important for posttranslational modification. Protease Inhibitors work at the last stage of the viral replication by preventing HIV from being successfully assembled and released from the infected host cell.	Hyper bilirubinaemia (cosmetic). Few lipid problems than LPV/r. Hyperglycemia, lipodystrophy, nephrolithiasis
	Lopinavir/ ritonavir (LPV/r)	200/100 mg twice a day-oral		Diarrhea, nausea, vomiting, abnormal lipid profiles, glucose intolerance.
	Darunavir/ Ritonavir(DRV/r)	600/100 mg twice a day-oral		Hepatotoxicity, rash, impaired glucose tolerance, diarrhea, lipodystrophy

There are some protease inhibitors that are not commonly used now days, such as saquinavir, nelfinavir, indinavir, fosamprenavir, tipranavir, and more. The drug ritonavir is not generally used singly but it is used in combination with other protease inhibitor for pharmacological boosting.

Group of the drug	Name of the drug	Standard adult dosage	Mechanism of action	Adverse effects
Integrase inhibitors	Raltegravir (RAL)	400 mg twice a day-oral	Integrase inhibitors block the action of the enzyme called integrase, a viral enzyme that facilitates insertion of the viral genome into the genome of the host cell	Rhabdomyolysis, myalgia, myopathy, fever, rash, hepatitis- rash may sometimes lead to Steven Jhonson Syndrome
	Dolutegravir (DTG)	50 mg once a day oral		Insomnia, headache, dyspepsia, hepatitis

Elvitegravir and bictegravir are also integrase inhibitors that are not used singly.

Group of the drug	Name of the drug	Standard adult dosage	Mechanism of action	Adverse effects
Fusion inhibitor	Enfuvirtide (T20)	90 mg twice a day-subcutaneously	Enfuvirtide disrupts the fusion of HIV-1 with the target cell, preventing uninfected cells from becoming infected. HIV binds to the host CD4 + T cell receptor via the viral protein gp120; gp41, a viral transmembrane protein, then undergoes a conformational change that assists in the fusion of the viral membrane to the host cell membrane. Enfuvirtide binds to gp41 preventing the creation of an entry pore for the capsid of the virus, keeping it out of the cell	Injection site reaction, insomnia, peripheral neuropathy, anorexia, arthralgia, depression
Chemokine inhibitor or CCR5 inhibitor	Maraviroc	300 mg twice a day-oral	The drug binds to CCR5 receptors expressed on the cell of human body resulting in interference with fusion of HIV protein gp120 and human cells like human macrophages and T cells	Hypersensitivity reaction, can be life threatening, hepatitis, arthralgia

Aplaviroc, cenicriviroc, and vicriviroc are other chemokine receptor blockers.

compliance, fixed dose combination of drug is preferred. The regimens are termed as first line, second line, alternate first line, and third line ART. The first line ART is offered to those who are ART naïve, that is, who have never been exposed to ART earlier. The recommended first line ART for adults and adolescents is a combination of tenofovir and lamivudin or emtricitabine and dolutegravir. This is a fixed dose combination and is given once a day. Alternatively the following regimen may be used when the first one is not available or contraindicated.

1. TDF+3TC (or FTC)+EFV
2. AZT+3TC+EFV (or NVP)
3. TDF+3TC (or FTC) +NVP

For adolescent, abacavir (ABC) may be used instead of TDF if weight of the adolescent is less than 35 kg. For children 3 years to less than 10 years old, the preferred first line of treatment is ABC+3TC +EFV. Alternatively the following regimen may be used.

1. ABC+3TC+NVP
2. AZT+3TC+EFV (or NVP)
3. TDF+3TC (or FTC) +EFV (or NVP)

For children less than 3 years, the first line of treatment is ABC (or AZT) +3TC +LPV/r. AZT should only be used if Hb% is more than 9 gm%. Alternative to this regimen, ABC +3TC +NVP may be used. In special circumstances, the regimen ABC or AZT+3TC +RAL may also be used.

13.3 General principles for antiretroviral therapy initiation

1. ART is to be started for all HIV-infected individuals after proper preparedness counseling and exclusion of opportunistic infections [20–23].
2. ART should be started with preferable first line of ART regimen as adopted by the country.
3. Alternate first line should be offered in case of absence of first line medicine or for individual experiencing side effects.
4. ART preparedness is to be assessed before initiating ART.
5. Patients need to assess clinically and microbiologically to rule out presence of any opportunistic infection.
6. The opportunistic infections are to be treated first before ART initiation. Once stabilized on the treatment for opportunistic infection, ART should be initiated.
7. Baseline clinical and biochemical assessment should be done as per prevailing norms and as per predictive adverse reactions for the ART regimen to be started.

13.4 Considerations before initiation of antiretroviral therapy

All people with confirmed HIV infection should be linked to the treatment center. ART initiation should be preceded by comprehensive clinical and laboratory evaluation to assess baseline status, treatment of preexisting opportunistic infections, treatment preparedness, counseling, and timely ART initiation. The following principles need to be exercised before initiation of ART:

1. The patient should be counseled and informed about HIV and its prognosis and probable treatment outcome.
2. The patient should be adequately prepared, and informed consent should be obtained from the patient or from the caregiver in case the patient is a minor, before initiating HIV care and ART.
3. Treatment should be started based on the person's informed decision and preparedness to initiate ART with information and understanding of the benefits of treatment, lifelong course of medication, issues related to adherence and positive prevention.
4. A caregiver should be identified for each person to provide adequate support. Caregivers must be oriented to support the people living with HIV on treatment adherence, followup visits and shared decision-making.
5. All patients with WHO clinical stage 3 and 4 and/or those with CD4 less than 350 cells/cmm need to be put on Co-trimoxazole Preventive Therapy. This is an important prophylaxis and it helps in prevention of bacterial diarrhea, bacterial respiratory tract infections, isospora diarrhea, pneumocystis jirovicci pneumonia, *Toxoplasma gondii* encephalitis, and sometimes tuberculosis with questionable efficacy. This prophylaxis should be continued till immunological status of the patient is restored.
6. All patients should be screened for tuberculosis. For screening, a strategy for 4-symptom-screening tool may be employed. Anybody with one of the four symptoms, that is, current cough, fever, night sweats, and weight loss, should be linked to tuberculosis testing center for microbiological confirmation.
7. Those HIV-infected individuals without any sign and symptom for tuberculosis should be started with Tubercuslosis Preventive Therapy with INH prophylaxis.
8. Patient should be clinically checked for other active or latent opportunistic infections and is to be treated for opportunistic infections first.

13.5 Monitoring on the patient on antiretroviral therapy

1. Adherence monitoring: The ART should be taken every day on time. Treatment compliance is the most important predictor for positive treatment outcome. Therefore, for each visit subsequent to initiation of ART, the adherence to medicine is to be monitored stringently by pill count, recollection method, or by any other method. Whenever there is lack in adherence, the reason for the same needs to identified and addressed as far as feasible [24−40].
2. Clinical monitoring: The clinical monitoring includes the status of patients' wellbeing, weight, sign of clinical improvement as compared to the last visit, any sign and symptom for adverse reaction of the ART medicines, any sign and symptom for immune reconstitution inflammatory syndrome, etc. It is also to be monitored whether new opportunistic infection is present or not.

3. Laboratory monitoring: It is important to identify any impending adverse reaction to the patient on ART.
4. Immunological monitoring: The major immunological marker that is followed up is CD4 + T cell count though this monitoring is gradually being phased out from all the countries due to its lack of precision and huge variability.
5. Virological monitoring: The major goal of ART is to reduce plasma viral load to such an extent so that the viral load becomes undetectable. Therefore, periodic plasma viral load testing in follow up visits play important role in monitoring the treatment outcome and early identification of treatment failure.

13.6 Immune reconstitution inflammatory syndrome

This is an atypical condition observed after initiating ART in some cases, especially when ART is initiated before opportunistic infection is stabilized or diagnosed or with latent opportunistic infection or patient with very low CD4 + T cell count at the time of ART initiation. This is worsening of signs and symptoms due to known infections after initiation of ART. This is a result of rapid restoration of immune status after ART initiation, which further results in inflammatory response. It is a spectrum of clinical signs and symptoms resulting from the body's ability to mount an inflammatory response associated with immune recovery. A sudden increase in inflammatory response produces nonspecific symptoms, such as fever and in some cases a paradoxical worsening of preexisting symptoms of infective or noninfective conditions. It happens in 10%–30% of patients initiating ART for the first time, usually within first 4–8 weeks. However, late IRIS can be observed upto 6 months after initiating ART [40]. Some risk factors for IRIS are as below:

1. People with CD4 + T counts below 100 cells/cm before initiating ART (lower CD4 counts at ART initiation) or lower CD4: CD8 ratio at ART initiation
2. People with rapid initial fall in HIV viral load due to therapy
3. Presence of latent opportunistic infections
4. Less gap between ATD and ART initiation
5. Higher HIV RNA at ART initiation
6. Male sex with younger age group
7. Genetic susceptibility

ART should not be stopped in IRIS and management of atypical inflammatory response may be done with NSAIDS, steroids, etc., along with the management of underlying disorder.

13.7 Antiretroviral failure

With antiretroviral treatment, it is expected that a person will become asymptomatic by 12 weeks. The viral load is expected to be undetectable and CD4 + T cell is expected to rise significantly by a maximum period of 6 months following ART initiation. This may be earlier in case of some regimen, especially DTG-based regimen. The most important aspect is to be adherent to ART treatment. With lack of adherence to ART treatment and with evolution of mutant-resistant strain against drug or drug regimen, the ongoing ART regimen may fail. Therefore, it is important to have a high index of suspicion for treatment failure [41−43]. Treatment failure is suspected among patients who are on first line ART for at least 6 months:

1. New OIs/recurrence/clinical events after 6 months on ART (after ruling out IRIS)
2. Progressive decline in CD4 + T counts
3. Slow/no clinical improvement over 6−12 months, associated with stationary CD4 + T cell count, despite good adherence
4. Failure to suppress plasma viral load below detectable level

The failures are of three types:

1. Clinical failure: Appearance of new or recurrent clinical event indicating severe immunodeficiency or new opportunistic infections (WHO clinical stage 4 condition) after 6 months of effective treatment.
2. Immunological failure: CD4 + T cell count falls to the baseline (or below) or Persistent CD4 + T cell level below 100 cells/cm or 50% fall from "on treatment" therapy peak level of CD4 + T cell count.
3. Virologic failure: Persistent plasma viral load above 1000 copies per ml after 6 months of ART intake with good ART adherence.

Chronologically, virologic failure occurs at first followed by immunological failure and then clinical failure. The clinical failure is the last to appear. ART regimen should be changed if there is treatment failure but identification of failure is important before taking decision to change the regimen. A thorough assessment is to be done regarding treatment adherence and drug interaction profile. Moreover, certain opportunistic infection may lead to temporary decline in CD4 + T cells. After correcting all these factors, necessary decision is to be taken to initiate new ART regimen. This is called ART switch, when the entire regimen is undergoing change (a change in single drug from ART regimen due to adverse effect is called substitution). Another important thing is to identify both genotypic and phenotypic resistance profile of the virus and based on which next ART regimen may be tailored.

The preferred second line of treatment is also constituted of two NRTIs, wherein at least one must be other than what were included in the first line regimen plus one boosted PI. If ABC + 3TC or TDF + 3TC (or FTC) was used in failing first line regimen, AZT + 3TC may be used in second line or vice versa. The preferred PI is ATV/r or LPV/r. Alternatively, DRV/r may also be used. For children less than 3 years, two NRTI plus LPV/r-based regimen may be switched to other two NRTI-based regimen plus RAL or two NRTI plus EFV. For children between 3 and 10 years, the NNRTI should be replaced by boosted PI or vice versa with change in drugs in the two NRTI backbones as stated earlier. In this case, the option of RAL may be kept for future use. When NRTI or NNRTI options cannot be considered, the regimen may be LPV/r or DRV/r plus RAL in special circumstances. The formulation of second and third line ART is a crucial task and the regimen is tailored after going through the entire treatment history of the patients based on resistance profile.

ATV/r is not to be used for pregnant women exposed to single dose-nevirapine in the past; they shall be provided with lopinavir/ritonavir-based regimen.

13.8 Drug interaction

ART drugs are associated with many drug-to-drug interactions. The most important interaction takes place with antitubercular drugs. Rifampicin is an inducer of many enzymes of the cytochrome P450 super family [44–48]. Possible interactions are noted with antiretroviral agents, atorvastatin, rosiglitazone/pioglitazone, celecoxib, clarithromycin, caspofungin, and lorazepam. Co-administration of ATD and ART, especially protease inhibitors and NNRTI, is associated with significant drug interactions. This may result in subtherapeutic plasma concentrations of antiretroviral drugs. For example, protease inhibitors must often be avoided if the potent CYP inducer like rifampicin is given concurrently. Alternatively, rifamycin and rifabutin, which have similar efficacy to rifampicin, can be used with dose modifications. Available clinical data suggest that, for most individuals, rifampicin-based regimens can be successfully combined with the nonnucleoside reverse transcriptase inhibitor, efavirenz, but not with nevirapine. Whenever protease inhibitor-based ART regimens are used, the comorbid TB must be treated with anti-TB regimen containing rifabutin instead of rifampicin. Regarding integrase inhibitor, when raltegravir is coadministered with ATD, rifampicin either should be substituted by rifamcin or rifabutin or the dosage of raltegravir may be increased to 800 mg twice a day. Similarly, the dose of DTG should be increased to 50 mg twice a day instead of 50 mg once a day.

In addition to ATD, the commonly used drugs like H2 blockers and Proton Pump Inhibitors (PPIs) interact with ART drugs. Boosted PI like ATV/r should not preferably be coprescribed with H2 blockers and PPIs. If PPI is to be prescribed for the patient on ATV/r-based ART regimen, the gap between both the drugs should be 12 hours. In that case the dosage should not exceed minimal dosage like omeprazole 20, rabeprazole 20, etc. So far as DTG is concerned, coadministration of anticonvulsant drugs like carbamazepine and phenytoin may preferably be avoided. So far as antidiabetic medicine is concerned, the dosage of metformin should not be increased beyond 1000 mg per day. The polyvalent cationic products like Al, Ca, Mg, Fe, Zn, antacids, multivitamins should be given either 2 hours before or 6 hours after DTG intake.

13.9 Antiretroviral drug resistance

At the end of the past decade, globally, more than 27.5 million people were receiving ART [49–54]. HIV drug resistance is the most crucial factor that can compromise the effectiveness of antiretroviral drugs in reducing HIV incidence and HIV-associated morbidity and mortality. So far we were concerned about development of resistance, which may lead to treatment failure. Currently, some studies related to pretreatment resistance, it was demonstrated that there may be more than 10% people living with HIV have acquired resistance to NVP and EFV. Pretreatment HIV drug resistance to the NNRTI drug class is up to three times more common in people with previous exposure to antiretroviral drugs. Nearly one half of infants born to mothers infected with HIV have HIV drug resistance to one or more NNRTIs. Therefore, the drug resistance is of basically two types:

1. Pre treatment resistance/Transmitted resistance: This is the drug resistance that appears before initiation of ART. Some countries are routinely undertaking pretreatment genotypic and phenotypic resistance profile and ART regimen is tailored based on the report. As NNRTI resistance is found to be pretty common, DTG-based ART regimen is being advocated nowadays.
2. Acquired resistance: This is the resistance to particular ART drug that develops in due course of treatment. Therefore, after initiation of ART routine viral load testing is recommended. If viral load is not suppressed or starts rising following achievement of viral load suppression, acquired drug resistance is to be suspected.

All the mutations do not lead to acquired drug resistance. M184V is the primary resistance mutation generated during HIV treatment with FTC and 3TC. This mutation renders a genetically unfit HIV strain. Therefore, keeping FTC and 3TC even in presence of M184V mutation increases the sensitivity to other thymidine analog ART drug. Moreover, single mutation does not result in acquired drug resistance. When multiple mutations are accumulated, it may result in acquired drug resistance. Thymidine Analog Mutations (TAMs) are nonpolymorphic mutations demonstrated in presence of AZT and d4T. They reduce NRTI susceptibility by facilitating primer unblocking. At least multiple levels of TAMs are required to develop genotypic resistance for thymidine analog ART drug. K103N is a nonpolymorphic mutation that causes high-level resistance to NVP and EFV. K103R is another polymorphic mutation that alone has no effect on NNRTI susceptibility. However, in combination with V179D, it reduces NVP and EFV susceptibility.

Some commonly occurring drug mutations are as follows, which may result in drug resistance (Table 13.2).

13.10 Preexposure prophylaxis

Preexposure prophylaxis is defined as the administration of antiretroviral drugs to an uninfected person before potential HIV exposure to reduce the risk of HIV infection. PrEP

TABLE 13.2 Some commonly occurring drug mutations that may result in drug resistance.

Drug class	Mutations
Nucleoside Reverse Transcriptase Inhibitors (NNRTI)	184V is the most common. Others are 70R,215Y,215F, 65R, 74V
Non-Nucleoside Reverse Transcriptase Inhibitors (NNRTIs)	103N is the most common mutation. The other mutations are 181C. 190A,188L, 103R, 103S, etc.
Protease Inhibitors (PIs)	M46I, V82A, I54V, L90M, I84V, M46L, and L76V

is for people who do not have HIV but are at risk of getting HIV through unprotected sexual activities or injectable drug use by multiple partners [1,3]. PrEP is indicated in the following situations:

1. PrEP is prescribed for the individuals who are HIV negative and
   1. Have sexual partner with HIV infection.
   2. Have anal or vaginal sex in the past 6 months.
   3. Have inconsistently used a condom.
   4. Have been diagnosed with an STD in the past 6 months.
2. PrEP may also be prescribed for Injectable drug users who have an HIV infected partner and share injection syringe, needle and related equipment.
3. People who have taken PEP but continue with risk behavior and have taken PrEP for several times.
4. If an HIV negative woman has an HIV infected partner and both are willing to have their baby, PrEP may be prescribed for the woman. PrEP may help to protect the woman and her baby from getting HIV while trying to get pregnant, during pregnancy, or while breastfeeding.

If PrEP is needed, necessary consultation is to be done with doctors.

PrEP reduces the risk of getting HIV from sex by about 99% when taken as prescribed. Among people who inject drugs, it reduces the risk by at least 74% when taken as prescribed. PrEP is much less effective when it is not taken consistently.

The PrEP may be a combination of oral drug or injectable. There are two oral formulations approved for daily use as PrEP. They are combinations of two ART drugs in a fixed dose combination:

1. The combination of emtricitabine and tenofovir disoproxil fumarate (TDF) is for people at risk of acquiring HIV sexual route or use of injectable drugs.
2. The combination of emtricitabine and tenofovir alafenamide (TAF) is for sexually active men and transgender women at risk of getting HIV. This combination is not used for HIV prevention for receptive vaginal sex, that is, for woman who may conceive.

A long-acting injectable form of PrEP is cabetogravir (CAB) for sexually active adults. It is administered by every 2 months instead of daily oral pills. This is prescribed for those who have problems in taking oral PrEP, who have got preexisting kidney disease, and those who prefer two monthly dosages instead of once a day dosage for oral formulations.

PrEP has been found to be safe so far. No significant health hazards have been seen in people who are HIV negative and have taken PrEP for up to 5 years. Some people taking PrEP may experience side effects, like nausea, diarrhea, headache, fatigue, and stomach ache. These side effects are transient and mild and go away with time.

PrEP reduces the risk of getting HIV from sexual route by around 99% when taken regularly and on time. Among the injectable drug user, it reduces the risk by at least 74% when taken as prescribed. Just like ART, PrEP is also less effective if not taken regularly as scheduled.

This is to be reiterated that PrEP protects against HIV but not against other sexually transmitted infections (STIs). Combining PrEP with use of condom will reduce your risk further along with reducing the chance of getting other STIs. While on oral PrEP, one must undergo HIV tests every 3 months and for injectable PrEP, HIV tests are to repeated every 2 months. If an individual is tested HIV positive while on PrEP, s/he should be linked to treatment center. On PrEP, the individual needs to be followed up for probable adverse effects in terms of monitoring creatinine clearance, lipid profile, urea, and creatinine. If PrEP is to be stopped by the individual, necessary consultation is to be done with the healthcare provider.

13.11 Postexposure prophylaxis

Postexposure prophylaxis (PEP) is meant for preventing HIV following accidental exposure [19]. The risks of HIV transmission through various routes are as follows (Table 13.3):

Therefore, it is evident that with accidental needle-prick injury, the chance of HIV transmission is around 0.3% while the risk after needle-stick injury for hepatitis B is 9% to 30% and for hepatitis C is 1 to1.8%. All the body fluids are not harmful in terms of transmitting HIV infection. Body fluid causing HIV transmission is illustrated in Fig. 13.2.

The body fluids like tears, sweat, saliva, sputum do not transmit the virus until and unless it is mixed with blood or other potentially infective body fluid. The healthcare providers are specifically at risk of needle-stick injury and the chance of transmission increases when exposed to large volume of the blood or infective materials, when there is deep injury, when there is visible blood on the device. To reduce the chance of transmission, PEP is indicated. PEP can effectively prevent HIV infection in an exposed individual when initiated within 2 hours but not later than 72 hours following an exposure.

Management following exposure:

TABLE 13.3   Risk of acquiring human immunodeficiency virus through different routes of transmission.

Routes of transmission	Risks
Blood transfusion	90%−95%
Vaginal sexual intercourse	0.05%−0.1%
Anal sexual route	0.065% to 0.5%
Through oral sex	0.005% to 0.01%
Through needle syringe sharing	0.67%
Through accidental needle prick	0.3%
Mucous membrane splash to eye, buccal, or nasal mucosa	0.09%

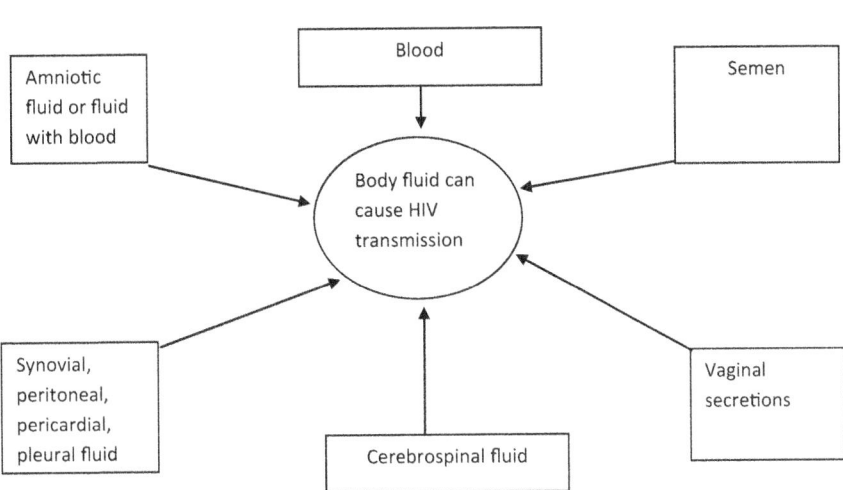

FIGURE 13.2 Body fluid causing human immunodeficiency virus transmission.

HIV exposure through any of the four routes is a medical emergency. Whenever someone has sustained a needle-stick injury, s/he should never panic. The injured area should never be squeezed or scrubbed. The injured area along with its adjoining areas should be washed with soap and water. When someone is exposed through splashing of infective material in eyes, the eyes should be thoroughly washed water or saline water. If the exposed person has contact lens, it should remain as it is till washing is complete. Then the lens is to be removed. If the exposure happens through splash in buccal mucosa, it is to be spit out immediately. Then s/he has to rinse the mouth with water or saline water several times and spit it again. Under any circumstances, antiseptic materials should not be used. After initial management, eligibility for the PEP is to be ascertained after consultation with physician. A proper counseling is often needed before initiating PEP to make the exposed person understand the risk and benefits of PEP and protocol for follow up. Subsequently, PEP should be initiated preferably within 2 hours of exposure.

The eligibility for PEP is ascertained based on the HIV status of the source and type and severity of the exposure. Under the following circumstances, PEP is not warranted.

1. When the exposed individual is already HIV positive
2. When the source is HIV negative
3. Exposures to body fluids that do not pose significant risk or not tinged with blood

When the source status is unknown, then local epidemiology, risk behavior of source is to be considered. In the setting with generalized epidemic or the source is a member of high-risk group, that is, sex workers, injectable drug user, men having sex with men, then exposure to a source with unknown HIV status may warrant PEP. After initiation of PEP,

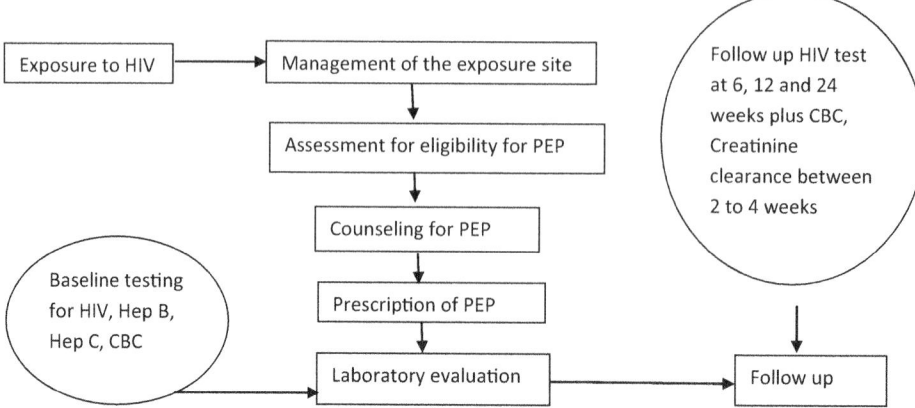

FIGURE 13.3   Flow chart indicative of management of postexposure prophylaxis.

if the source is found to be HIV negative, it may be discontinued. Ascertaining the source HIV may be difficult in some settings. PEP initiation should never be delayed due to unavailability of the source's HIV test results. The following flow chart (Fig. 13.3) is indicative of management of PEP.

Regimen for PEP:

As per current guideline, the regimen of PEP should essentially be triple drug ART regimen as advocated for ART initiation. The preferred PEP regimen for otherwise healthy adults and adolescents is TDF (300 mg) +emtricitabine (FTC) 200 mg or lamivudin (3TC) 300 mg once a day plus raltegravir (RAL) (400 mg) twice daily or dolutegravir (DTG) (50 mg) once daily. Alternatively, boosted PI, that is, lopinavir boosted by ritonavir (200 mg plus 100 mg twice a day) or efavirenz (600 mg once a day) may be used along with nucleoside and nucleotide reverse transcriptase inhibitor backbone.

PEP is to be given uninterruptedly for 28 days. Biochemical follow up should be done in line with the adverse effect of the ART drug used for PEP and management for adverse reaction is to be done accordingly. Follow up HIV testing is to be done 6, 12, and 24 weeks subsequent to exposure.

13.12 Prevention of mother child transmission

Mother to child transmission of HIV is an important route for HIV transmission, which can be prevented to a great extent with multiple interventions [1,2,16,17,55−60]. HIV transmission from mother to child occurs during:

1. Pregnancy
2. Labor and delivery
3. Breastfeeding

TABLE 13.4 Risk of mother to child transmission of human immunodeficiency virus through different routes.

Time of transmission	Risk of transmission
During pregnancy	5%−10%
During labor and delivery	10%−15%
During breastfeeding	5%−20%
Overall without breastfeeding	15%−25%
Overall with breastfeeding upto 6 months	20%−35%
Overall with breastfeeding for 18−24 months	30%−45%

The major factor that increases the risk of MTCT is the amount of HIV (viral load) in the mother's blood and other body fluid. The chance of transmission through vertical route is as follows (Table 13.4).

The risk of transmission can be prevented to an extent to even less that 1% with combination of PMTCT interventions. PMTCT is a four pronged strategy to make our future generation free from HIV. They are:

1. First prong: Primary prevention of HIV in the population within reproductive age bracket.
2. Second Prong: Prevention of unwanted pregnancy in the HIV infected women through promotion of barrier method of contraception and other contraceptive modalities and adoption of comprehension abortion care services.
3. Third prong: Medical intervention directed at mother to child transmission of HIV for mother baby pair.
4. Fourth prong: Continuum of care, support and treatment for both mother baby pair. The cascades of PMTCT services are as follows:
    1. Offering universal routine HIV counseling and testing services to all pregnant women preferably during the first trimester.
    2. Counsel HIV negative pregnant women for future risk reduction and issues related to window period.
    3. Repeat HIV testing of the at risk pregnant women (spouse of HIV infected person, women with high risk exposure or behavior, etc.)
    4. Ensure involvement of spouse & family members for better obstetric outcome.
    5. Provision of ART to HIV positive pregnant and breastfeeding women on priority basis for her own health and prevention of vertical transmission.
    6. Facilitation for medical termination services as per individual's will and medical indication.
    7. Facilitation for institutional deliveries of positive pregnant women.
    8. Provision of nutritional counseling, infant feeding counseling, and psychosocial support to positive pregnant women.

9. Spouse or partner testing for HIV and contact tracing.
10. Provision of care for other sexually transmitted infections and opportunistic infections.
11. Other package of antenatal services as applicable in the state and countries.
12. Standard intra natal care.
13. Standard postnatal care including immunization, ARV prophylaxis.
14. Followup of HIV-exposed infants through regular clinical check up.
15. Cotrimoxazole prophylactic therapy & early infant diagnosis to be started at 6 weeks and to be continued till confirmed negative.
16. Confirmatory HIV testing at 18 months using standard antibody testing.

In high prevalent setup, HIV screening may be repeated in third trimester, other-wise single HIV test during pregnancy is recommended in low prevalence settings unless there is any risk factor associated with it.

ART intervention for HIV infected pregnant women:

All HIV positive pregnant women including those presenting in labor and breastfeeding women should be initiated on a triple ART irrespective of maternal gestational age, plasma viral load, WHO clinical staging, or CD4 + T cell count for maternal health benefit as well as for preventing mother-to-child transmission risk and should continue lifelong ART. The ART regimen is standard for different countries and the current preferred regimen is a fixed-dose combination of TDF (300 mg) with emtricitabine (FTC) 200 mg or lamivudin (3TC) 300 mg once a day plus dolutegravir (DTG) (50 mg) once daily, which is also a preferred regimen for all HIV-infected adult and adolescent population. Like others, ART should be continued lifelong.

ARV prophylaxis for exposed children:

All the children born to HIV-infected mother are eligible for ARV prophylaxis. Currently, there are several ARV prophylaxis regimens in use across the world. Some countries prefer to use single drug whereas some countries prefer to use dual drug ARV prophylaxis. Some countries prefer to use triple drug ARV prophylaxis and in some cases combination of all are being used based on the quantum of probable risks associated for vertical transmission. This prophylaxis should be initiated at birth and the minimum duration for ARV prophylaxis should be 4–6 weeks and in some cases it may be extended as per the policy of the country. The various regimens are as follow:

1. Single-dose prophylaxis: It is generally done with either nevirapine or zidovudine. NVP is the most commonly used single drug prophylaxis used and it is given once a day. Zidovudine or AZT is given in some cases as first line ARV prophylaxis but in some cases where NVP is used as first line, AZT is reserved in the following conditions:
   a. Infants born to mothers infected with confirmed HIV-2 or HIV 1 & 2 mixed infections
   b. Infants born to a mother who received single-dose NVP during earlier pregnancy/delivery
   c. Infant born to a mother who is on PI based regimen due to treatment failure
   d. Infant born to mother with achieved NNRTI mutation

Other single drug that may be used is lopinavir boosted by ritonavir but it should not be given before 42 weeks of gestation due to probable cardiotoxicity. Another reserved category drug which coming up for ARV prophylaxis is raltegravir, which is only used for some special cases.

2. Dual drug prophylaxis: This is the combination of ARV drugs used for ARV prophylaxis. The commonly used drugs in dual drug ARV prophylaxis are AZT and NVP. Generally it is given separately as AZT is given twice a day and NVP once a day. If anyone of these two drugs cannot be given for medical reasons, other drug may be combined with lopinavir plus ritonavir suspension but in this case, initiation of lopinavir and ritonavir may be deferred for atleast 2 weeks post delivery.

3. Triple drug combination: In some countries, triple drug ARV prophylaxis rather presumptive ARV therapy is used. The recommended regimen is the combination of zidovudin (60 mg), lamivudin (30 mg), and nevirapine (50 mg), which is given twice a day. Alternatively, raltegravir may be used instead of nevirapine. Abacavir (ABC) cannot be used in infant less than 1 month of age and dolutegravir (DTG) and efavirenz (EFV) are not recommended for infants.

The dosage of the drugs and formulation used for ARV prophylaxis may either be per kilogram weight dependent or may be fixed dosage for particular weight bracket. Some countries tailor the ARV prophylaxis regimen based on the risk assessment. Infants who are at high risk of acquisition of HIV vertically may be initiated on presumptive HIV therapy as per policies of some countries. The inclusion criteria are:

1. Mother did not receive antepartum ART.
2. Mother received only intrapartum ART.
3. Mother received antepartum ART but did not achieve viral suppression (defined as a confirmed HIV RNA level <50 copies/mL) within 4 weeks of delivery. This is important to know that the cut off value of plasma viral load for determining suppression varies from machine to machine and also from country to country.
4. Mother with primary or acute HIV infection during pregnancy or breastfeeding.

Presumptive ART treatment is also offered for the HIV exposed babies with AIDS defining illness without confirmation of HIV status.

PMTCT interventions play the most crucial role in preventing HIV-related morbidity and mortality in children. In absence of PMTCT interventions and ART for children 30% of the HIV exposed children are likely to die by 1st year, 50% by 2nd year, and 80% by 5th year.

PMTCT program and its related interventions vary from country-to-country based on prevalence, availability of medicines and other logistics. A prototype of this intervention has been presented below, which is currently followed in India [1,2,16,17,59,60].

The PMTCT flow chart is presented in Fig. 13.4.

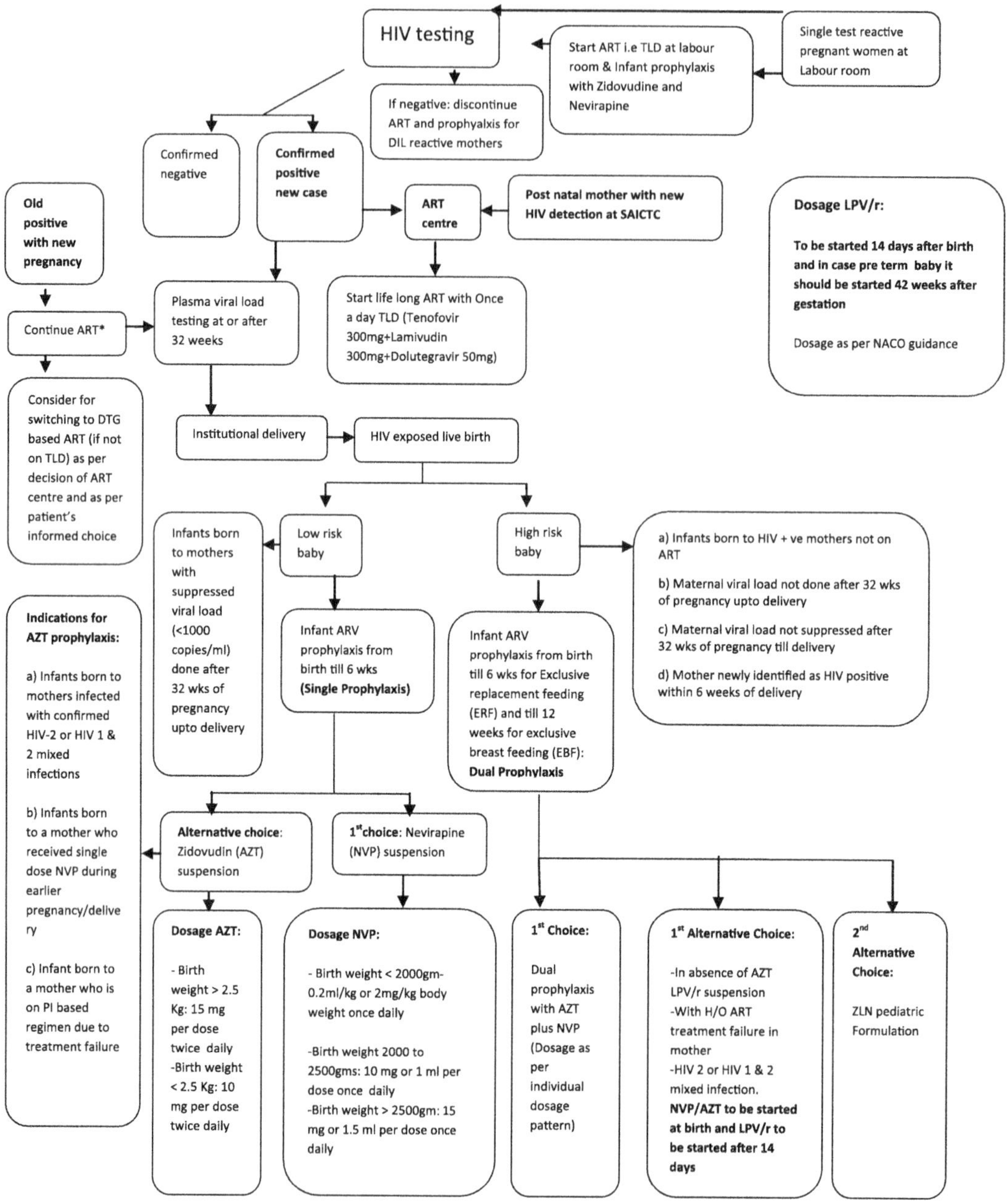

FIGURE 13.4 PMTCT flow chart.(All pregnant women should be screened for HIV preferably during first trimester with "Opt out" strategy: Repeat test for high risk individuals). *Assess viral suppression (PVL) soon after pregnancy detection if ART taken more than 6 months.

References

[1] <http://naco.gov.in/sites/default/files/NACO%20-%20National%20Technical%20Guidelines%20on%20ART_October%202018%20%281%29.pdf> [accessed 26.12.21].

[2] <https://www.nicd.ac.za/wp-content/uploads/2019/11/2019-ART-Clinical-Guidelines-25-Nov.pdf> [accessed 26.12.21].

[3] <https://www.hiv.gov/hiv-basics/hiv-prevention/using-hiv-medication-to-reduce-risk/pre-exposure-prophylaxis> [accessed 10.01.22].

[4] <https://www.cdc.gov/hiv/clinicians/prevention/prep.html> [accessed 12.01.22].

[5] Quinn TC. HIV epidemiology and the effects of antiviral therapy on long-term consequences. AIDS 2008;22 (Suppl. 3):S7−12.

[6] Flexner C. HIV drug development: the next 25 years. Nat Rev Drug Discov 2007;6(12):959−66.

[7] Gao Y, Kraft JC, Yu D, et al. Recent developments of nanotherapeutics for targeted and long-acting, combination HIV chemotherapy. Eur J Pharm Biopharm 2018. Available from: https://doi.org/10.1016/j.ejpb.2018.04.014 April 17.

[8] Tomkowicz B, Lee C, Ravyn V, et al. The Src kinase Lyn is required for CCR5 signaling in response to MIP-1beta and R5 HIV-1 gp120 in human macrophages. Blood 2006;108(4):1145−50 Epub 2006 Apr 18.

[9] Herbein G, Coaquette A, Perez-Bercoff D, et al. Macrophage activation and HIV infection: can the Trojan horse turn into a fortress? Curr Mol Med 2002;2(8):723−38.

[10] Aquaro S, Svicher V, Schols D, et al. Mechanisms underlying activity of antiretroviral drugs in HIV-1-infected macrophages: new therapeutic strategies. J Leukoc Biol 2006;80(5):1103−10.

[11] Roujeau J-C, Stern RS. Severe adverse cutaneous reactions to drugs. N Engl J Med 1994;331:1272−85.

[12] Coopman MA, Johnson RA, Platt R, Stern RS. Cutaneous disease and drug reactions in HIV infection. N Engl J Med 1993;328:1670−4.

[13] Race EM, Adelson-Mitty J, Kriegel GR, et al. Focal mycobacterial lymphadenitis following initiation of protease-inhibitor therapy in patients with advanced HIV-1 disease. Lancet 1998;351:252−5.

[14] Miller J, Carr A, Smith D, et al. Lipodystrophy following antiretroviral therapy of primary HIV infection. AIDS, 2000.

[15] Ledergerber B, Egger M, Opravil M, et al. Clinical progression and virological failure of highly active antiretroviral therapy in HIV-1 patients: a progressive cohort study. Lancet 1999;353:863−8.

[16] <https://www.who.int/hiv/pub/arv/chapter4.pdf>. [accessed 10.01.22].

[17] <https://www.who.int/news-room/fact-sheets/detail/hiv-drug-resistance> [accessed 11.01.22].

[18] <https://www.who.int/hiv/pub/guidelines/PEP/en>. [accessed 9.01.22].

[19] <https://www.cdc.gov/hiv/risk/pep/index.html>. [accessed 11.01.22].

[20] Lane HC, Masur H, Gelmann EP, et al. Correlation between immunologic function and clinical subpopulations of patients with the acquired immune deficiency syndrome. Am J Med 1985;78:417−22.

[21] Phillips AN, Lundgren JD. The CD4 lymphocyte count and risk of clinical progression. Curr Opin HIV AIDS 2006;1:43−9.

[22] Lundgren JD, Babiker A, El-Sadr W, et al. Inferior clinical outcome of the CD4 + cell count-guided antiretroviral treatment interruption strategy in the SMART study: role of CD4 + cell counts and HIV RNA levels during follow-up. J Infect Dis 2008;197:1145−55.

[23] De Cock KM, El-Sadr WM. When to start ART in Africa — an urgent research priority. N Engl J Med 2013;368:886−9.

[24] Kamya MR, Mayanja-Kizza H, Kambugu A, et al. Predictors of long-term viral failure among ugandan children and adults treated with antiretroviral therapy. J Acquir Immune Defic Syndr 2007;46:187−93.

[25] Rouet F, Fassinou P, Inwoley A, et al. Long-term survival and immuno-virological response of African HIV-1-infected children to highly active antiretroviral therapy regimens. AIDS 2006;20:2315−19.

[26] Wade AM, Ades AE. Incorporating correlations between measurements into the estimation of age-related reference ranges. Stat Med 1998;17:1989−2002.

[27] Stein DS, Korvick JA, Vermund SH. CD4 + lymphocyte cell enumeration for prediction of clinical course of human immunodeficiency virus disease: a review. J Infect Dis 1992;165:352−63.

[28] Hughes MD, Stein DS, Gundacker HM, Valentine FT, Phair JP, Volberding PA. Within-subject variation in CD4 lymphocyte count in asymptomatic human immunodeficiency virus infection: implications for patient monitoring. J Infect Dis 1994;169:28−36.

[29] Schooley RT. Viral load testing in resource-limited settings. Clin Infect Dis 2007;44:139−40.

[30] Cohen MS, Chen YQ, McCauley M, et al. Prevention of HIV-1 infection with early antiretroviral therapy. N Engl J Med 2011;365:493−505.

[31] Temprano ANRS Study Group. A trial of early antiretrovirals and isoniazid preventive therapy in Africa. N Engl J Med 2015;373:808−22.

[32] Insight START Study Group. Initiation of antiretroviral therapy in early asymptomatic HIV infection. N Engl J Med 2015;373:795−807.

[33] Anglaret X, Scott CA, Walensky RP, et al. Could early antiretroviral therapy entail more risks than benefits in sub-Saharan African HIV-infected adults? A model based analysis. Antivir Ther 2013;18:45−55.

[34] Freedberg KA, Losina E, Weinstein MC, et al. The cost effectiveness of combination antiretroviral therapy for HIV disease. N Engl J Med 2001;344:824−31.

[35] Ouattara EN, Ross EL, Yazdanpanah Y, et al. Clinical impact and cost-effectiveness of making third-line antiretroviral therapy available in sub-Saharan Africa: a model based analysis in Côte d'Ivoire. J Acquir Immune Defic Syndr 2014;66:294−302.

[36] Weinstein MC, Siegel JE, Gold MR, Kamlet MS, Russell LB. Recommendations of the panel on cost-effectiveness in health and medicine. JAMA 1996;276:1253−8.

[37] Anglaret X, Chêne G, Attia A, et al. Early chemoprophylaxis with trimethoprim-sulphamethoxazole for HIV-1-infected adults in Abidjan, Côte d'Ivoire: a randomised trial. Cotrimo-CI Study Group. Lancet 1999;353:1463−8.

[38] Anglaret X, Minga A, Gabillard D, et al. AIDS and non-AIDS morbidity and mortality across the spectrum of CD4 cell counts in HIV-infected adults before starting antiretroviral therapy in Côte d'Ivoire. Clin Infect Dis 2012;54:714−23.

[39] Messou E, Chaix ML, Gabillard D, et al. Association between medication possession ratio, virologic failure and drug resistance in HIV-1-infected adults on antiretroviral therapy in Côte d'Ivoire. J Acquir Immune Defic Syndr 2011;56:356−64.

[40] Fox MP, Ive P, Long L, Maskew M, Sanne I. High rates of survival, immune reconstitution, and virologic suppression on second-line antiretroviral therapy in South Africa. J Acquir Immune Defic Syndr 2010;53:500−6.

[41] Fethia KSM, Kassu K, Shewaminale Y, Zelalem T. National guidelines for comprehensive HIV prevention, care and treatment federal minstry of health, 5th ed. Ethiopia; 2017 <https://aidsfree.gov/ethiopia_art_guidelines_2017> [accessed 7.01.21].

[42] Kassa D, Gebremichael G, Alemayehu Y, Wolday D, Messele T, van Baarle D. Virologic and immunologic outcome of HAART in Human Immunodeficiency Virus (HIV)-1 infected patients with and without tuberculosis (TB) and latent TB infection (LTBI) in Addis Ababa, Ethiopia. AIDS Res Ther 2013;10(1):18.

[43] Roberts T, Cohn J, Bonner K, Hargreaves S. Scale-up of routine viral load testing in resource-poor settings: current and future implementation challenges. Clin Infect Dis 2016;62(8):1043−8.

[44] Marzolini C, Elzi L, Gibbons S, et al. Prevalence of comedications and effect of potential drug-drug interactions in the Swiss HIV Cohort Study. Antivir Ther 2010;15:413−23.

[45] Von Moltke LL, Greenblatt DJ, Grassi JM, et al. Protease inhibitors as inhibitors of human cytochromes P450: high risk associated with ritonavir. J Clin Pharmacol 1998;38:106−11.

[46] Minuesa G, Huber-Ruano I, Pastor-Anglada M, Koepsell H, Clotet B, Martinez-Picado J. Drug uptake transporters in antiretroviral therapy. Pharmacol Ther 2011;132:268−79.

[47] Kim RB. Drug transporters in HIV Therapy. Top HIV Med 2003;11:136−9.

[48] Belanger AS, Caron P, Harvey M, Zimmerman PA, Mehlotra RK, Guillemette C. Glucuronidation of the antiretroviral drug efavirenz by UGT2B7 and an in vitro investigation of drug-drug interaction with zidovudine. Drug Metab Dispos 2009;37:1793−6.

[49] World Health Organization HIV drug resistance report. <http://apps.whoint/iris/bitstream/10665/75183/1/9789241503938_eng.pdf>; 2012. [accessed 12.01.22].

[50] Frentz D, Boucher CA, van de Vijver DA. Temporal changes in the epidemiology of transmission of drug-resistant HIV-1 across the world. AIDS Rev 2012;14:17−27.

[51] Gupta RK, Jordan MR, Sultan BJ, et al. Global trends in antiretroviral resistance in treatment-naive individuals with HIV after rollout of antiretroviral treatment in resource-limited settings: a global collaborative study and *meta*-regression analysis in resource limited settings 2001−2011. Lancet 2012;380:1250−8.

[52] Wittkop L, Günthard HF, de Wolf F, et al. Effect of transmitted drug resistance on virological and immunological response to initial combination antiretroviral therapy for HIV (EuroCoord-CHAIN joint project): a European multicohort study. Lancet Infect Dis 2011;11:363−71.

[53] Stadeli KM, Richman DD. Rates of emergence of HIV drug resistance in resource limited settings: a systematic review. Antivir Ther 2013;18:115–23.

[54] Phillips AN, Dunn D, Sabin C, et al. Long term probability of detection of HIV-1 drug resistance after starting antiretroviral therapy in routine clinical practice. AIDS 2005;9:487–94.

[55] Marazzi MC, Palombi L, Nielsen-Saines K, Haswell J, Zimba I, Magid NA, et al. Extended antenatal use of triple antiretroviral therapy for prevention of mother-to-child transmission of HIV-1 correlates with favorable pregnancy outcomes. AIDS 2011;25(13):1611–18.

[56] Chen JY, Ribaudo HJ, Souda S, Parekh N, Ogwu A, Lockman S, et al. Highly active antiretroviral therapy and adverse birth outcomes among HIV-infected women in Botswana. J Infect Dis 2012;206(11):1695–705.

[57] Shapiro RL, Souda S, Parekh N, Binda K, Kayembe M, Lockman S, et al. High prevalence of hypertension and placental insufficiency, but no in utero HIV transmission, among women on HAART with stillbirths in Botswana. PLoS One 2012;7(2):e31580.

[58] Ford N, Mofenson L, Kranzer K, Medu L, Frigati L, Mills EJ, et al. Safety of efavirenz in first trimester of pregnancy: a systematic review and *meta*-analysis of outcomes from observational cohorts. AIDS 2010;24 (10):1461–70.

[59] Gibb DM, Kizito H, Russell EC, Chidziva E, Zalwango E, Nalumenya R, et al. Pregnancy and infant outcomes among HIV-infected women taking long-term ART with and without tenofovir in the DART trial. PLoS Med 2012;9(5):e1001217.

[60] Dryden-Peterson S, Shapiro RL, Hughes MD, Powis K, Ogwu A, Moffat C, et al. Increased risk of severe infant anemia after exposure to maternal HAART, Botswana. J Acquir Immune Defic Syndr 2011;56 (5):428–36.

Rotavirus and antirotaviral therapeutics: trends and advances

Ujjwal Kumar De[1], Yashpal Singh Malik[2], Gollahalli Eregowda Chethan[3], Babul Rudra Paul[1], Jitendra Singh Gandhar[1], Varun Kumar Sarkar[1], Srishti Soni[1] and Kuldeep Dhama[4]

[1]Division of Medicine, ICAR-Indian Veterinary Research Institute, Izatnagar, Uttar Pradesh, India [2]College of Animal Biotechnology, Guru Angad Dev Veterinary and Animal Sciences University, Ludhiana, Punjab, India [3]Department of Veterinary Medicine, College of Veterinary Science and Animal Husbandry, Central Agriculture University, Aizawl, Mizoram, India [4]Division of Pathology, ICAR-Indian Veterinary Research Institute, Izatnagar, Uttar Pradesh, India

14.1 Introduction

Rotavirus (RV), a contagious viral disease of infants and young children, is characterized by diarrhea, fever, vomiting, abdominal pain, and other intestinal symptoms. RV causes inflammation in the stomach and intestines. It is one of the important viral enteropathogens causing acute diarrhea in young animals usually affecting up to one month of age. The disease is zoonotic and has an economic impact on animal production [1]. The infection spreads rapidly and causes extensive damage to the gut mucosa, which results in rapid fluid loss and severe dehydration [2]. An accurate and quick diagnosis is of prime importance for the timely administration of correct therapeutic interventions to contain the spread of RV infections and help in mitigating the severity of clinical symptoms. Many significant advances made in this direction have been reviewed extensively elsewhere [3]. Though the treatment of RV is primarily aimed at restoring body fluid levels and electrolyte balance by fluid therapy, there are no specific antiviral drugs that are approved to treat RV infections so far. However, significant efforts have been put together by researchers worldwide in RV drugs and therapeutics development. In this chapter, we converse on the trends and advances in the specific therapeutic approaches being targeted against RV.

14.2 Supportive/symptomatic therapies

14.2.1 Fluid therapy

Symptomatic therapies are the most widely accepted and the unequivocal option offered to lessen the RV-associated morbidity and mortality due to the unavailability of specific treatment for RV infection [4]. Maintenance of hydration and oncotic support as well as correction of acid—base and electrolyte imbalances is of utmost importance in RV enteritis [5]. Hence, fluid and electrolyte therapy is considered one of the major aspects of the therapy to restore the electrolyte reserve and rehydration. It prevents the mortality rate from hypovolemic shock [6,7]. The application of oral rehydration therapy (ORT) along with an early nutritional supplementation realized to diminish the mortality in neonates suffering from acute RV diarrhea [8]. Intravenous fluids can be used in cases of severe dehydration, hyperemesis, ORT failure, or severe electrolyte imbalances [9].

14.2.2 Antibiotic treatment

The concurrent infection with secondary pathogens influences the severity of the RV diarrhea. Therefore, antimicrobial therapy has a key role in reducing RV-associated mortality. Typically, virus-associated diarrheal cases are self-limiting and the duration of diarrhea does not get affected by the use of antimicrobials, except in certain cases of secondary bacterial complications [10]. Although antibiotics play a major part in reducing mortality among severely ill patients, it must always be kept in mind that antimicrobial therapy should be reserved for severe, prolonged, or potentially complicated cases, as most patients respond fairly well to supportive therapy, and their indiscriminate use carries the threat of increasing antimicrobial resistance [11].

14.3 Antiviral drugs/mimetics

Keeping in view the limitations of rotaviral vaccines, effective antiviral drugs against RVs have become important, particularly for the treatment of acute-phase infections in nonimmune individuals [12]. However, in comparison to antibacterial or antibiotics, there is a shortage of potent antiviral agents against viral diseases and not many studies have been carried out to assess the efficacy of such drugs or synthetic mimetics against enteric virus infections. We have discussed RV-specific antiviral drugs and mimetics based on their mode of action.

14.3.1 Interference in attachment and entry of virus into host cells

Few of the lactose-based sialylmimetics have been biologically evaluated as anti-RV agents [13,14]. The results of in vitro studies showed modest inhibition of human RVA (strain Wa) but not of rhesus RV (RRV) strain. A novel sialylphospholipid was shown to hinder both simian and human RVs in MA-104 cells [15].

Donkey lactadherin-derived peptides carrying DGE (Asp-Gly-Glu) and RGD (Arg-Gly-Asp) motifs are shown to block the RV infection via integrin $\alpha2\beta1$ competition that is crucial during the early steps of RV host cell adhesion [16]. RV entry into host cells has been proposed to be dependent on the reductase activity of cell surface protein disulfide isomerase (PDIs) of a family of oxidoreductases [17,18]. PDIs on cell surfaces could create thiol groups in VP5 and/or VP6 proteins which help in interaction with other cell surface receptors or mediate virus disassembly [18]. Membrane impermeant thiol/disulfide-blockers like DTNB [5,5-dithio-bis-(2-nitrobenzoic acid)] and bacitracin have been reported to inhibit redox activity of PDI and the treatment of MA104 cells with such agents resulted in reducing the RV infectivity, however, these inhibitors have shown poor inhibition postentry [19].

14.3.2 Interefrence in host cell lipid metabolic pathways

The lipid homeostasis is vital during the replication of RV. An increase in the triglycerides levels is the major alteration in lipid metabolic pathways in the host cells due to RV infection [20,21]. Previous reports show the beneficial effects of bile acids and synthetic FXR agonists on virus replication in conjunction with proper cellular lipid homeostasis. The CDCA, deoxycholic acid, and synthetic FXR agonist (GW4064), markedly reduce the virus replication in cells by inhibiting the FXR pathway activation and declining the cellular triglyceride levels. Another study confirmed that oral administration of CDCA significantly reduced virus shedding in mice feces and this could be due to lipid synthesis downregulation [20,21]. Stilbenoids, the phenolic compounds found in peanuts, grapes, and berries have been reported to block RV infection in host cells through modulation of the cannabinoid receptors and lipid metabolism [22]. Recent in vitro studies have shown that certain compounds viz., 5-(tetradecyloxy)-2-furoic acid (TOFA), Triacsin C, isobutyl-methylxanthine (IBMX), and isoproterenol interfere with the homeostasis of LD inhibited RV replication [21,23,24]. Fatty acid synthesis inhibition could negatively affect the later stages of RV infection in host cells [21].

14.3.3 Inhibition of viroplasm formation

In the past two decades, few research trials have been conducted with racecadotril, an enkephalinase inhibitor, and nitazoxanide to examine their effects against viruses including RV [25–27]. The nitazoxanide and its active metabolite, tizoxanide, were reported to inhibit the replication of simian A/SA11-G3P[2] and human Wa-G1P[8] RV in different types of cells by altering the architecture of viroplasms and through blocking the NSP5 and NSP2 interactions [12]. Interestingly, a small molecule, ML-60218 (ML) previously known to inhibit an RNA polymerase III is found to disrupt viroplasms, the viral replication machinery by targeting the viral protein VP6 in double-layered particles [28].

14.3.4 Intererfence in viral RNA and protein synthesis

Genistein, an isoflavone, has been reported to block RV (Wa type) replication in vitro by upregulating aquaporin 4 (AQP4) mRNA and protein expression in RV-infected Caco-2 cells. Furthermore, it induced cAMP response element binding phosphorylation via PKA (protein kinase A) activation pathway and thus enhanced AQP4 gene transcription. Moreover, genistein proves to be a promising candidate in the formulation of a new anti-RV compound [29]. Phosphonoformic acid (foscarnet, PFA), a nonnucleoside pyrophosphate analog is a known inhibitor of viral polymerases, and it inhibits both (+) strand and (−) strand RNA synthesis [13,30]. An antineoplastic antibiotic known as Actinomycin D, an antineoplastic compound is reported to inhibit RNA translation by making a complex with viral RNA [31].

Immuno-suppressants like mycophenolic acid and 6-Thioguanine (6-TG) were shown to inhibit RV infection in Caco-2 cells and organoids via inhibition of inosine-5-monophosphate dehydrogenase enzyme and Rac1 activation and depletion of intracellular guanosine nucleotides [32,33]. Inosine pranobex (Isoprinosine R), an immunomodulating as well as an antiviral agent was shown to inhibit the synthesis of RV protein and double-stranded nucleic acid [34,35]. Recently, the methanol extracts of the cyanobacterial isolates namely *Arthrospira platensis* and *Oscillatoria* sp. have been reported to possess a potent antiviral effect against RV in vitro, making them promising sources of new safe antiviral [36].

14.3.5 Targeting RNA interference pathway

In a recent study, the anti-RV activity of RNA interference (RNAi) containing short hairpin RNA (shRNA) targeting enterotoxin nonstructural protein-4 (NSP4) gene of bovine RV, namely RNAi-351 and RNAi-492 has been explored in vivo and in vitro studies and found that RNAi-492 was more effective than RNAi-351 to keep NSP4 silenced, while synergistic effects exerted by both shRNA were shown to silence NSP4 gene expression completely. Besides, the in vivo study in suckling mice prevented the appearance of diarrhea when RNAi-492 or a combination of RNAi-492/ + RNAi-351 was used. Recently, it has been documented that miR-7, a microRNA inhibits viroplasm formation and RV replication by downregulating RV NSP5 expression in vitro and in mice model [37]. Thus, this shRNA could be exploited as an antiviral agent in clinical RV diarrhea cases [38].

14.4 Passive immunotherapy

Passive immunotherapy represents a prominent and effective way to counter gastrointestinal pathogens and can be achieved by oral administration of anti-RV immunoglobulin preparations [39].

The immunoglobulin derived from hyperimmune bovine colostrum had shown substantial benefits on childhood infectious diarrhea [40,41]. Immunoglobulin of bovine colostrum were found to show higher titer of neutralizing antibodies compared to egg yolk and pooled plasma of humans [42]. However, a combination treatment of colostrum-derived antibodies with probiotics (*Lactobacillus rhamnosus* strain GG) exhibited superior effects than the symptomatic therapies [43,44].

Heavy chain antibody fragments (VHHs) possess high degrees of solubility and robustness enabling the generation of multivalent constructs with increased avidity characteristics which makes them superior to other antibody fragments and monoclonal antibodies [45]. An anti-RV VHH, termed ARP1 (anti-RV protein), derived from a llama immunized with a rhesus-monkey RVA was found to effectively neutralize many RV strains in vitro [13]. The production of neutralizing antibodies in vegetables like transgenic purple tomatoes has also been targeted for RV treatment [46].

The placebo-controlled randomized clinical trials of hyperimmunized chicken egg yolk (IgY) immunoglobulin in humans and animals with RV infection demonstrated that treatment with IgY resulted in the reduction of intravenous administrations as well as virus shedding in feces resulted in early recovery [47–51]. Orally, milk supplementation with 6% bovine RV-immune egg yolk in experimentally RV-infected newborn calf effectively prevented bovine RV diarrhea. The bovine RV-immune egg treatment induces bovine RV specific antibody-secreting cells of the gut, indicating the modulation of mucosal immunity by egg yolk immunity against RV infection [52]. In the gnotobiotic neonatal piglet disease model, oral administration of RV-specific IgY antibodies as a milk supplement protected neonatal pigs against virulent human RV by inducing enhanced production of IgG in serum and local IgA and IgG antibody responses in gut contents [53].

14.5 Immunotherapeutics

Immunotherapeutic regimens on viral infections are a popular and attractive therapeutic approach because of reduced side-effects as compared to antiviral drugs and the rare generation of antiviral resistance during the treatment of viral infections. The immune-therapeutic approaches, in particular to RV treatment, have been categorized into three sections viz. immunomodulators, cytokines, and Toll-like receptors (TLRs) based therapeutics.

14.6 Immunomodulators

The stimulus to the host's innate immune system is of paramount importance to counteract the virus infection [54]. RV infection also triggers innate immune responses, which are critical to reducing RV replication [55–57]. An immunomodulating dose of *Saccharomyces cerevisiae* derived β-glucan in piglets suffering from RVA diarrhea was shown to increase the production of IFN-γ concentration and resulted in early recovery [4] and reduced intestinal injury as evident by reduced serum intestinal fatty acid-binding protein-2 concentration which is now considered as one of the biomarkers for intestinal injury [4]. Similar therapeutic effects were observed by using levamisole as an immunomodulator along with supportive therapy [58].

Ergoferon, an antiviral complex drug containing released-active forms of antibodies to interferon-gamma (anti-IFN-gamma), CD4-coreceptor, and histamine, exhibited high anti-RV activity in MA-104 cells [59]. Cyclosporin A, an established immunomodulatory agent, had shown anti-RV activity in a dose-dependent manner by increasing the mRNA expression of IFN-β and decreasing the mRNA expression of inflammatory cytokines viz. IL-8, IL-10, IFN-γ, and tumor necrosis factor-α (TNF-α) and a few of the inflammatory signaling pathways (p38, c-Jun N-terminal kinase, activator protein-1, and nuclear factor-kappa B) [60].

14.7 Cytokines-based therapeutics

Cytokines, as we know, are a diverse group of protein molecules bearing both pro- and antiinflammatory properties and play important roles in immunity and protection against viral diseases [61]. Pro-inflammatory cytokines such as TNF-α is a potent mediator of antimicrobial immunity [62]. A strong inhibitory effect of TNF-α has recently been reported against RV both at intracellular and extracellular levels, and this effect was mediated via TNF receptor 1 and classical nuclear factor-κB signaling pathways. Notably, the anti-RV effects of TNF-α were independent of type I IFN production [63].

The exogenously administered IFNs were also protective against RV diarrhea in calves and pigs [64,65]. Besides, IFN-α had shown significant inhibition of RV replication in organoids of both mouse and human gut tissues in an in vitro study [66].

Interleukin-22 (IL-22), a member of the IL-10 superfamily exhibited antiviral activity against porcine RV (PRV) [67]. However, the possible impact of cytokine therapies should be more systematically evaluated in the target hosts.

14.8 Toll-like receptors-based therapeutics

The TLRs establish antiviral innate immune responses, providing a long-lasting adaptive immunity to prevent the progress of viral pathogenesis [68]. TLR3, TLR7, and TLR9 have been identified for stimulating innate responses to RV infection [69,70].

Bacterial flagellin, the prime constituent of bacterial flagella, acts as a potential activator of local intestinal mucosal immunity and enhances the host defense gene expression in intestinal epithelial cells [71,72]. Stimulation of innate immunity using bacterial flagellin TLR5/NLRC4 mediated production of IL-22 and IL-18 alleviated even chronic RV infection in immune-compromised animals [73]. Similarly, the MyD88, a central molecule in innate immunity, was reported to show anti-RV function by activating the TLR7, TLR8, and TLR9 signaling pathway and increasing RV-specific Abs in the suckling mice model [74]. A lavandulylated flavanone, norkurarinol, extracted from the roots of *Sophora flavescens* restricted RV replication via inhibition of TLR3 mediated pro-inflammatory signaling pathway [75]. Recently, Melanoma Differentiation-Associated protein 5 (MDA5), an important cytoplasmic receptor and a member of pattern recognition receptors (PRR) is found to have antirotaviral function. It has been demonstrated that MDA5 effectively inhibits the RV replication through provoking a noncanonical IFN-like response, which is partially dependent on the JAK-STAT cascade in Caco2 cell culture [76]. The possible therapeutic efficacy of TLR agonists targeting the RV disease treatment and prevention needs to be explored as the same has been achieved successfully by pharmaceutical companies for human papillomavirus infection [77].

14.9 Herbal/medicinal plants

An interest in natural products including herbs, plants, and their extracts/metabolites as antiviral drug candidates has increased in the last few decades especially due to the rising emergence of antimicrobial resistance globally and the potential side effects of many antimicrobials [78]. In vitro studies using various plants such as *Pinus koraiensis*, *Lomatium dissectum*, *Artocarpus integrifolia*, *Myristica fragrans*, *Spongias lutea*, *Tylosema esculentum*, *Byrsonima verbascifolia*, *Myracrodruon urundeuva*, *Eugenia dysenterica*, *Hymenaeacour baril* and *Achillea kellalensis* exhibited anti-RV activities with no apparent cytotoxicity [79–85]. A considerable number of studies using plant food species like *P. koraiensis*, *Theobroma cacao*, *T. esculentum*, *Aegle marmelos*, *Psidium guajava*, *Vaccinium macrocarpon* and *Vitis labrusca* also showed inhibitory action against RV [82,86]. In addition, plant species like *Stevia rebaudiana*, *V. macrocarpon*, *V. labrusca* and. *M. urundeuva* were found to act on outer viral capsid proteins thereby reducing the infection of RV [82].

Extracts from *Glycyrrhiza uralensis* were seen to reduce the severity of RV diarrhea and reduces lesions in piglet's small intestine as it might exert anti-RV as well as antiinflammatory effects [87]. The tannin-rich *Persimmon* extracts, also possessed anti-RV activity through aggregating viral proteins [88]. Saponin extracts from *Quillaja saponaria* Molina and Pectic polysaccharides derived from *Panax ginseng* disrupted virus receptors on the host cell membranes thus blocking virus attachment [89,90]. The anti-RV activities of more than 50 flavones and flavonols had been reported [91]. A Thailand origin herbal product, Krisanaklan, reduced RV-induced diarrhea in mice by blocking the luminal cAMP-dependent chloride channel and Ca(2 +) activated Cl(−) channels in enterocytes [92]. Another herbal preparation Q-iwei-Bai-zhu powder, a Chinese traditional medicine, was shown to cure RV diarrhea in mice and also enhance absorption in the small intestine [93]. Methanolic extract of *Dodonaea viscosa* leaves showed beneficial activity against RV-induced gastroenteritis in mice [94]. *D. viscosa* plant extract reduced the mortality, viral titers in fecal specimens, duration of illness, and improved the healing of intestinal lesions caused by RV [94].

Dietary supplementation of resveratrol alleviated RV-induced diarrhea in pigs by its antioxidant and immunomodulatory activities [95]. Ginsenosides, a class of steroid glycosides, and triterpene saponins and its hydrolytic products from Korean red ginseng were found to have anti-RV effect in mice with RV strain SA11 infection [96].

14.10 Probiotics

Probiotics have become increasingly popular as a safe harmless substitute with minimum cost against various diseases including inflammatory and diarrheal complications due to RV infection [97–99]. Probiotics basically constitute live microorganisms in fermented foods which establish a balance in the intestinal microflora and promote good health [100,101]. Species that are commonly used as probiotics in treatment of RV infections include *Lactobacillus lactis*, *L. paracasei*, *L. rhamnosus*, *Bifidobacterium longum*, *Bifidobacterium infantis*, *Enterococcus faecalis*, and *Saccharomyces boulardii* [102–104]. Most of these probiotics work on the competitive-expulsion phenomenon and inhibit colonization of invading pathogenic bacteria with available receptors or intra-luminal nutrients, rather than producing antimicrobials that trigger host immunity [105,106]. In vitro studies on the pathogenesis of PRV OSU strain in porcine jejunum epithelial cell line (IPEC-J2) with *Lactobacillus acidophilus* or *L. rhamnosus* GG demonstrated that these did not reduce the virus replication but they showed immunomodulating and antiinflammatory effects [107]. Similarly, *S. boulardii* (Sb) was found to inhibit RV infection by increasing antioxidant defense and reducing chloride secretion [108]. Recently, exopolysaccharides from *Lactobacillus plantarum* LRCC5310 showed a decrease in RV-induced diarrhea duration and prevented enteric epithelium damage in mice [109].

Supplementation of *Bifidobacterium species* (*Bifidobacterium bifidum* and *B. infantis*) to the Balb/c mice with RRV infection stimulated local and humoral immune responses which resulted in mitigation of severity of RV induced diarrhea [110]. Lately, oral administration of *B. bifidum* G9−1 (BBG9−1) to suckling mice model alleviated RV-diarrhea by upregulating gene expression of small intestinal protective factors such as MUC2−4, TGFβ1, and TFF3 [111].

14.11 Advances in drug delivery: nanotechnology-based approach

Proteolytic degradation of drugs and reduced absorption via epithelial barriers are the main drawbacks of the oral route of drug administration for the management of enteric diseases. The acidic environment of the stomach inactivates the drug delivery systems (DDS), thereby, hampering normal absorption. Notably, researchers have come up with a novel DDS comprising of RV VP6 protein construct in conjunction with a small ubiquitin-like modifier (SUMO) [112]. Therefore, the major advantages of this nanotechnology-based DDS include small size, high monodispersity, higher stability, and uptake in the nucleus and cytoplasm of intestinal cells. These advantages propose SUMO-VP6 nanocarriers as potent DDS for the oral administration of low soluble drugs [112]. Immunization of mice with RV (strain SA11) after incorporating into a 4-arm star PELA microspheres induced a long-lasting IgA and IgG antibody response compared to the free RV antigen immunization [113]. In another study, silver nanoparticle, an antiviral and antiinflammatory agent, was also shown to reduce the viral load and augment the transcripts for TGF-β mRNA in RRV inoculated mice [114].

The incorporation of nanotechnology for the treatment of RV infection offers enormous potential for enhanced mechanisms of action of currently available therapeutics or the development of novel therapeutics.

14.12 Neutraceuticals

One of the most important consequences of RV enteritis is compromised gut health with increased intestinal permeability. Therefore, augmentation of gut health is of prime importance to reduce RV pathogenesis. Nutraceuticals are known to improve health benefits including the prevention and treatment of various diseases. The following neutra-ceuticals have demonstrated promising therapeutics to treat RV gastroenteritis.

14.12.1 Milk proteins

An earlier study identified the anti-RV activity of the milk proteins, probably achieved by blocking virus-integrin interactions and virus aggregations [115]. The bovine milk pro-tein, κ-casein was shown to bind to viral particles through glycan residues and inhibited human RV multiplication [116]. Additionally, different components of milk such as milk mucin, apolactoferrin, Fe(3 +)-lactoferrin, beta-lactoglobulin, human lactadherin, bovine IgG, and bovine kappa-casein were found to inhibit RV infection in human beings [117]. In a comparative study, the RV-specific inhibitory action was more pronounced with apo-lactoferrin followed by Fe3 + -lactoferrin, mucin, and β-lactoglobulin, respectively. But no such effect was noticed with α-lactalbumin [118−121].

14.12.2 Cholesterol

Lipid rafts present in plasma membranes play vital roles in RV infection [122]. Recently, anti-RV activity of 25-hydroxycholesterol (25HC) and 27-hydroxycholesterol (27HC) has been documented [123]. Such hydroxycholesterols exhibited antiviral activity by modulating endogenous production of oxysterols and are considered new therapeutic strategies against RV infection [124,125]. Reduction in cholesterol found to restrict host cell entry of RVs. However, inhibition reversed near to normal with supplementation of exoge-nous cholesterol [126]. The disruption of NSP4 (112−140)-N-caveolin-1(19−40) and choles-terol interactions could inhibit NSP4 intracellular transport thereby decreasing the enterotoxicity of RV infection [127].

14.12.3 L-isoleucine

Recently, the dietary L-isoleucine supplementation has shown a beneficial effect for RV treatment in infants and young animals. In an in vitro study, the L-isoleucine at the con-centration of 8 mM exhibited anti-RV activity in IPEC-J2 and 3D4/31cell culture systems by upregulation of β-defensins gene expressions and PRR signaling pathway. The in vivo study with crossbred weaned pigs revealed that dietary supplementation of L-isoleucine attenuated RV diarrhea and improved growth performance via enhancing the functional activity of immunoglobulins, antibodies, cytokines, and β-defensins and inhibiting the NSP4 concentration in ileal mucosa [128].

14.12.4 Vitamin D3

The supplementation of vitamin D3 was shown to be effective in mitigating the negative effects caused by RV challenges such as body weight gain, feed intake, villus height, and fecal consistency [129]. A study by Tian et al. [130] divulged that on the treatment of 25-hydroxyvitamin D3 to RV-infected IPEC-J2 cells markedly reduced the RV antigen load and NSP4 protein with an increased in Beclin 1 and PR-39 mRNA expression through autophagy signaling pathway [130]. Similarly, 25-hydroxyvitamin D3 also modified certain porcine cathelicidin gene expression in infected cells by decreasing p62 mRNA and increasing mRNA expression in porcine cathelicidins (PMAP23, PG1−5, and PR-39), leading to inhibition of RV infection [130,131]. Notably, the dietary vitamin D supplementation imparted anti-RV activity by the retinoic acid-inducible gene-1 and TANK-binding kinase 1/interferon regulatory factors 3 signaling pathways in RV challenged pigs [130,132].

14.12.5 Oligosaccharides

The dietary supplementation of pectic oligosaccharides (POS) exhibited beneficial effects in PRV enteritis [133]. The dietary intervention with POS to weaned piglets orally challenged with RV was shown to improve the growth performance and augmented serum IgA, lipase, and tryptase. An enhancement in the levels of sIgA, IL-4, and IFN-γ in jejunal and/or ileal mucosa was also prudent [134]. In another experimental study with porcine model, it was found that dietary supplementation of human milk oligosaccharides resulted in early recovery from PRV OSU strain-induced diarrhea by altering colonic microbiota and mucosal immune response [135]. Furthermore, dietary supplementation of apple pectic oligosaccharide enhanced the total antioxidant capacity and decreases malondialdehyde (MDA) levels in jejunal mucosa of RV-infected weaned piglets [136]. Recently, it is documented that dietary intervention of short-chain galactooligosaccharides [2'-fucosyllactose (2'-FL)] and long-chain fructooligosaccharides (scGOS/lcFOS) and their combination (scGOS/lcFOS/2'-FL) prevents RV-induced diarrhea by inhibiting intestinal dysbiosis and boosting TLR5, TLR7 and TLR9 in neonatal Lewis rats [137]. These accumulated pieces of evidence are suggestive of the beneficial effects of nutraceuticals in managing or treating the rotaviral diarrhea efficiently.

14.13 Antioxidants

Antioxidants have proven medicinal and therapeutic applications against various diseases and disorders including infectious diseases [138,139]. Therefore, understanding the role of oxidative stress in RV infection might help in formulating improved novel anti-RV therapeutic strategies [140,141]. The studies from piglets suffering from acute RV enteritis showed a marked decrease in total antioxidant capacity and antioxidant enzyme activities with a concomitant increase in oxidative stress indices like MDA and nitric oxide (NO) [141].

Inhibition of RV replication has been studied in HT29 cells using antioxidants like stilbenoids, trans-Piceatannol (t-PA, a hydroxylated analog of resveratrol), trans-Arachidin-1 (t-A1, a prenylated analog of piceatannol), and trans-arachidin-3 (t-A3. a prenylated analog of resveratrol) [22]. Antioxidant and antiinflammatory activities of resveratrol, a stilbene, and a naturally occurring phytoalexin produced in different plants, have been demonstrated in piglets infected with RV [95]. Resveratrol is found to alleviate the inflammation due to RV by decreasing TNF-α concentration and inhibiting RV infection via elevating

IFN-γ concentration [95]. Redox reactions have been significantly reduced by using another antioxidant known as N-acetyl cysteine treatment in cultured MA104 cells [142] and infected mice [143]. Recently, the antirotaviur activity of lentinan (LNT), a potent antioxidant available in the fruit body of shiitake mushroom, has been explored. Dietary LNT supplementation relieves RV-induced diarrhea through the increased antioxidant capacity, reduction in apoptosis, and improvement of the microbiota-increased gut barrier in the porcine model [144].

The understanding of oxidative stress-mediated RV pathogenesis offers an opportunity for the development of novel therapeutic strategies using antioxidants interfering with RV infection.

14.14 Combinational therapy

Most of the time, two or more agents in combinations exhibit better results in treating viral infectious diseases [145]. The major advantage of combination therapy is a reduction in toxicity as well as the duration of treatment [146]. Keeping in view the synergistic activity of a combination of herbs on therapeutic efficacy [147], it was found that the extracts of *Glycyrrhiza glabra* plus *Nelumbo nucifera* and *Urtica dioica* plus *N. nucifera* markedly inhibited RV infectivity [82]. In addition, stevioside from *S. rebaudiana* along with *S. flavescens* reduced intestinal lesions in RV infected piglets [148]. Synergisms among natural products such as a flavin and its gallate derivatives; isoflavones mixtures (genistin, acetylgenistin, daidzin, daidzein, acetyldaidzin, glycitin) had shown strong inhibitory action against RV infection [149,150]. Furthermore, isoflavones containing genistin suppressed the RV infectivity whereas no such effect was noticed when isoflavones were used alone. Synergistic effect of the two flavonoids [Epigallocatechin gallate of green tea and the neutraceutical CystiCranR-40 (containing 40% proanthocyanidins)] was reported in Coliphage T4II (phage T4) and the RV strain SA-11(RTV) model systems [151]. The activity-guided separation of most active compounds from crude extract will help to develop potent anti-RV drugs. Hence, more rigorous clinical trials are required on target species using these efficacious phytochemicals.

14.15 Other potential therapeutic approaches

The gut microbiota, a large diverse ecosystem of microorganisms, of a host plays critical roles in protecting against pathogens, mainly by preventing the colonization of pathogens. Recently, a specific class of bacteria, segmented filamentous bacteria conferred protection in mice against RV infection and associated diarrhea [152]. Similarly, Baicalin (5,6,7-trihydroxyflavone 7-O-beta-d-glucuronide or baicalein 7-O-β-d-glucuronic acid or 7-d-glucuronic acid-5,6-dihydroxyflavone), the most important flavonoid compounds of the roots of *Scutellaria baicalensis* from traditional Chinese medicinal plant were found to have potent anti-RV effect in mice with RV induced diarrhea [153].

14.16 Conclusion and future prospects

Non-availability of approved drugs to treat RV infection augments an additional burden on human and livestock health. However, symptomatic therapies including fluid therapy remain the primary choice to alleviate the RV-associated symptoms and thereby the treatment of RV infections. The global efforts of researchers put in to search for effective RV treatment have led to the elucidation of anti-RV activities of many chemical compounds, immuno-therapeutics, neutraceuticals, potent plant extracts, probiotics, antioxidants, metabolites of marine sponges and finally identifying a species-specific bacteria from intestinal microbiota. The authors believe that coordinated research efforts should be initiated toward a combinatorial therapy to find a potent solution for the RV treatment. Ultimately, the healthcare professionals and the Governments should use these findings to create therapeutics methodically to reduce the global burden of RV on human and animal health.

References

[1] Geletu US, Usmael MA, Bari FD. Rotavirus in calves and its zoonotic importance. Vet Med Int 2021;2021:6639701.

[2] Foster DM, Smith GW. Pathophysiology of diarrhea in calves. Vet Clin North Am Food An Pract 2009;25 (1):13.

[3] Malik YS, Verma AK, Kumar N, Touil N, Karthik K, Tiwari R, et al. Advances in diagnostic approaches for viral etiologies of diarrhea: from the lab to the field. Front Microbiol 2019;10:1957.

[4] Chethan GE, Garkhal J, Sircar S, Malik Y, Mukherjee R, Sahoo NR, et al. Immunomodulatory potential of β-glucan as supportive treatment in porcine rotavirus enteritis. Vet Immunol Immunopathol 2017;191:36−43.

[5] Wielgos K, Setkowicz W, Pasternak G, Lewandowicz-Uszyńska A. Postępowanie w ostrej biegunce infekcyjnej u dzieci [Management of acute gastroenteritis in children]. Pol Merku Lekarski 2019;47(278):76−9.

[6] Steele AD, Geyer A, Gerdes GH. Rotavirus infections. In: Coetzer JAW, Tustin RC, editors. Infectious diseases of livestock. Southern Africa: Oxford University Press; 2004, p. 1256−64.

[7] Murphy FA, Gibbs EPJ, Horzinek MC, Studdert MJ. Reoviridae. In: Murphy FA, Gibbs EPJ, Horzinek MC, Studdert MJ, editors. Veterinary virology. USA: Academic Press; 1999, p. 391−404.

[8] Duggan C, Santosham M, Glass RI. The management of acute diarrhea in children: oral rehydration, maintenance, and nutritional therapy. Centers for Disease Control and Prevention. MMWR Recomm Rep 1992;41 (RR-16):1−20.

[9] Parashar UD, Nelson EA, Kang G. Diagnosis, management, and prevention of rotavirus gastroenteritis in children. BMJ (Clinical Research Ed) 2013;347:7204.

[10] Bruzzese E, Giannattasio A, Guarino A. Antibiotic treatment of acute gastroenteritis in children. F1000 Res 2018;7:193.

[11] Efunshile AM, Ezeanosike O, Nwangwu CC, König B, Jokelainen P, Robertson LJ. Apparent overuse of antibiotics in the management of watery diarrhoea in children in Abakaliki, Nigeria. BMC Infect Dis 2019;19 (1):275.

[12] La Frazia S, Ciucci A, Arnoldi F, Coira M, Gianferretti P, Angelini M, et al. Thiazolides, a new class of antiviral agents effective against rotavirus infection, target viral morphogenesis, inhibiting viroplasm formation. J Virol 2013;87(20):11096−106.

[13] Ghosh S, Malik YS, Kobayashi N. Therapeutics and immunoprophylaxis against noroviruses and rotaviruses. Curr Drug Metab 2017;19(3):170−91.

[14] Liakatos A, Kiefel MJ, Fleming F, Coulson B, von Itzstein M. The synthesis and biological evaluation of lactose-based sialylmimetics as inhibitors of rotaviral infection. Bioorg Med Chem 2006;14(3):739−57.

[15] Koketsu M, Nitoda T, Sugino H, Juneja LR, Kim M, Yamamoto T, et al. Synthesis of a novel sialic acid derivative (sialylphospholipid) as an antirotaviral agent. J Med Chem 1997;40(21):3332−5.

[16] Civra A, Giuffrida MG, Donalisio M, Napolitano L, Takada Y, Coulson BS, et al. Identification of equine lactadherin-derived peptides that inhibit rotavirus infection via integrin receptor competition. J Biol Chem 2015;290(19):12403−14.

[17] Jordan PA, Gibbins JM. Extracellular disulfide exchange and the regulation of cellular function. Antioxid Redox Signal 2006;8(3−4):312−24.

[18] Moreno LY, Guerrero CA, Acosta O. Protein disulfide isomerase and heat shock cognate protein 70 interactions with rotavirus structural proteins using their purified recombinant versions. Rev Colomb Biotecnol 2016;18(1):33−48.

[19] Calderon MN, Guerrero CA, Acosta O, Lopez S, Arias CF. Inhibiting rotavirus infection by membrane-impermeant thiol/disulfide exchange blockers and antibodies against protein disulfide isomerase. Intervirology 2012;55(6):451−64.

[20] Kim Y, Chang KO. Inhibitory effects of bile acids and synthetic farnesoid X receptor agonists on rotavirus replication. J Virol 2011;85(23):12570−7.

[21] Lever A, Desselberger U. Rotavirus replication and the role of cellular lipid droplets: new therapeutic targets? J Formos Med Assoc 2016;115(6):389−94.

[22] Ball JM, Medina-Bolivar F, Defrates K, Hambleton E, Hurlburt ME, Fang L, et al. Investigation of stilbenoids as potential therapeutic agents for rotavirus gastroenteritis. Adv Virol 2015;2015:293524.

[23] Cheung W, Gill M, Esposito A, Kaminski CF, Courousse N, Chwetzoff S, et al. Rotaviruses associate with cellular lipid droplet components to replicate in viroplasms, and compounds disrupting or blocking lipid droplets inhibit viroplasm formation and viral replication. J Virol 2010;84(13):6782−98.

[24] Kim Y, George D, Prior AM, Prasain K, Hao S, Le DD, et al. Novel triacsin C analogs as potential antivirals against rotavirus infections. Eur J Med Chem 2012;50:311−18.

[25] Waddington CS, McLeod C, Morris P, Bowen A, Naunton M, Carapetis J, et al. The NICE-GUT trial protocol: a randomised, placebo controlled trial of oral nitazoxanide for the empiric treatment of acute gastroenteritis among Australian Aboriginal children. BMJ Open 2018;8(2):019632.

[26] Flerlage T, Hayden R, Cross SJ, Dallas R, Srinivasan A, Tang L, et al. Rotavirus infection in pediatric allogeneic hematopoietic cell transplant recipients: clinical course and experience using nitazoxanide and enterally administered immunoglobulins. Pediatr Inf Dis J 2018;37(2):176−81.

[27] Lehert P, Chéron G, Calatayud GA, Cézard JP, Castrellón PG, Garcia JM, et al. Racecadotril for childhood gastroenteritis: an individual patient data *meta*-analysis. Dig Liver Dis 2011;43(9):707−13.

[28] Eichwald C, De Lorenzo G, Schraner EM, Papa G, Bollati M, Swuec P, et al. Identification of a small molecule that compromises the structural integrity of viroplasms and rotavirus double-layered particles. J Virol 2018;92(3):01943 17.

[29] Huang H, Liao D, Liang L, Song L, Zhao W. Genistein inhibits rotavirus replication and upregulates AQP4 expression in rotavirus-infected Caco-2 cells. Arch Virol 2015;160(6):1421−33.

[30] Rios M, Munoz M, Spencer E. Antiviral activity of phosphonoformate on rotavirus transcription and replication. Antivir Res 1995;27(1−2):71−83.

[31] Stefanelli CC, Castilho JG, Botelho MV, Linhares RE, Nozawa CM. Effect of actinomycin D on simian rotavirus (SA11) replication in cell culture. Braz J Med Biol Res 2002;35(4):445−9.

[32] Yin Y, Chen S, Hakim MS, Wang W, Xu L, Dang W, et al. 6-Thioguanine inhibits rotavirus replication through suppression of Rac1 GDP/GTP. Antivir Res 2018;156:92−101.

[33] Yin Y, Wang Y, Dang W, Xu L, Su J, Zhou X, et al. Mycophenolic acid potently inhibits rotavirus infection with a high barrier to resistance development. Antivir Res 2016;133:41−9.

[34] Sliva J, Pantzartzi CN, Votava M. Inosine pranobex: a key player in the game against a wide range of viral infections and non-infectious diseases. Adv Ther 2019;36(8):1878−905.

[35] Pavlova EL, Simeonova LS, Gegova GA. Combined efficacy of oseltamivir, isoprinosine and ellagic acid in influenza A(H3N2)-infected mice. Biomed Pharmacother 2018;98:29−35.

[36] Deyab M, Mofeed J, El-Bilawy E, Ward F. Antiviral activity of five filamentous cyanobacteria against coxsackievirus B3 and rotavirus. Arch Microb 2020;202(2):213−23.

[37] Zhou Y, Chen L, Du J, Hu X, Xie Y, Wu J, et al. MicroRNA-7 inhibits rotavirus replication by targeting viral NSP5 *in vivo* and *in vitro*. Viruses. 2020;12(2):209.

[38] Chen F, Hongmei W, Hongbin H, Lingling S, Jianming W, Yundong G, et al. Short hairpin RNA-mediated silencing of bovine rotavirus NSP4 gene prevents diarrhoea in suckling mice. J Gen Virol 2011;92:945−51.

[39] Alexander E, Hommeida S, Stephens MC, Manini ML, Absah I. The role of oral administration of immunoglobulin in managing diarrheal illness in immunocompromised children. Pediatr Drugs 2020;22(3):331−4.

[40] Civra A, Altomare A, Francese R, Donalisio M, Aldini G, Lembo D. Colostrum from cows immunized with a veterinary vaccine against bovine rotavirus displays enhanced *In vitro* anti-human rotavirus activity. J Dairy Sci 2019;102(6):4857−69.

[41] Li YT, Xu H, Ye JZ, Wu WR, Shi D, Fang DQ, et al. Efficacy of *Lactobacillus rhamnosus* GG in treatment of acute pediatric diarrhea: a systematic review with *meta*-analysis. World J Gastroenterol 2019;25 (33):4999−5016.

[42] Bogstedt AK, Johansen K, Hatta H, Kim M, Casswall T, Svensson L, et al. Passive immunity against diarrhoea. Acta Paediatr 1996;85(2):125−8.

[43] Ghosh S, Malik YS, Kobayashi N. Therapeutics and immunoprophylaxis against noroviruses and rotaviruses: the past, present, and future. Curr Drug Metab 2018;19(3):170−91.

[44] Pant N, Marcotte H, Brüssow H, Svensson L, Hammarström L. Effective prophylaxis against rotavirus diarrhea using a combination of *Lactobacillus rhamnosus* GG and antibodies. BMC Microbiol 2007;7:86.

[45] Wilken L, McPherson A. Application of camelid heavy-chain variable domains (VHHs) in prevention and treatment of bacterial and viral infections. Int Rev Immunol 2018;37(1):69−76.

[46] Juárez P, Presa S, Espí J, Pineda B, Antón MT, Moreno V, et al. Neutralizing antibodies against rotavirus produced in transgenically labelled purple tomatoes. Plant Biotechnol J 2012;10(3):341−52.

[47] Abbas AT, El-Kafrawy SA, Sohrab SS, Azhar EIA. IgY antibodies for the immunoprophylaxis and therapy of respiratory infections. Hum Vaccine Immunother 2019;15(1):264−75.

[48] Rahman S, Higo-Moriguchi K, Htun KW, Taniguchi K, Icatlo Jr FC, Tsuji T, et al. Randomized placebo-controlled clinical trial of immunoglobulin Y as adjunct to standard supportive therapy for rotavirus-associated diarrhea among pediatric patients. Vaccine. 2012;30(31):4661−9.

[49] Thu HM, Myat TW, Win MM, Thant KZ, Rahman S, Umeda K, et al. Chicken egg yolk antibodies (IgY) for prophylaxis and treatment of rotavirus diarrhea in human and animal neonates: a concise review. Korean J Food Sci Ani Res 2017;37(1):1−9.

[50] Vega C, Bok M, Saif L, Fernandez F, Parreño V. Egg yolk IgY antibodies: a therapeutic intervention against group A rotavirus in calves. Res Vet Sci 2015;103:1−10.

[51] Wang X, Song L, Tan W, Zhao W. Clinical efficacy of oral immunoglobulin Y in infant rotavirus enteritis: systematic review and *meta*-analysis. Med (Baltim) 2019;98(27):16100.

[52] Vega C, Bok M, Chacana P, Saif L, Fernandez F, Parreño V. Egg yolk IgY: protection against rotavirus induced diarrhea and modulatory effect on the systemic and mucosal antibody responses in newborn calves. Vet Immunol Immunopathol 2011;142(3−4):156−69.

[53] Vega CG, Bok M, Vlasova AN, Chattha KS, Fernández FM, Wigdorovitz A, et al. IgY antibodies protect against human Rotavirus induced diarrhea in the neonatal gnotobiotic piglet disease model. PLoS One 2012;7:42788.

[54] Takeuchi O, Akira S. Innate immunity to virus infection. Immunol Rev 2009;227:75−86.

[55] Rollo EE, Kumar KP, Reich NC, Cohen J, Angel J, Greenberg HB, et al. The epithelial cell response to rotavirus infection. J Immunol 1999;163(8):4442−52.

[56] Azevedo MS, Yuan L, Jeong KI, Gonzalez A, Nguyen TV, Pouly S, et al. Viremia and nasal and rectal shedding of rotavirus in gnotobiotic pigs inoculated with Wa human rotavirus. J Virol 2005;79(9):5428−36.

[57] Holloway G, Coulson BS. Innate cellular responses to rotavirus infection. J Gen Virol 2013;94:1151−60.

[58] Chethan GE, Kumar De, U, Garkhal J, Sircar S, Malik YPS, Sahoo NR, et al. Immunomodulating dose of levamisole stimulates innate immune response and prevents intestinal damage in porcine rotavirus diarrhea: a restricted-randomized, single-blinded, and placebo-controlled clinical trial. Trop Anim Health Prod 2019;51 (6):1455−65.

[59] Emelianova AG, Shilovskii IP, Sundukova MS, Khaitov MR, Epshtein OI. Antiviral activity of ergoferon against group A rotavirus. Bull Exp Biol Med 2016;161(6):806−7.

[60] Shen Z, Tian Z, He H, Zhang J, Li J, Wu Y. Antiviral effects of cyclosporine A in neonatal mice with rotavirus-induced diarrhea. J Pediatr Gastroenterol Nutr 2015;60(1):11−17.

[61] Rojas JM, Avia M, Martín V, Sevilla N. IL-10: a multifunctional cytokine in viral infections. J Immunol Res 2017;6104054.

[62] Sedger LM, McDermott MF. TNF and TNF-receptors: from mediators of cell death and inflammation to therapeutic giants—past, present and future. Cytokine Growth Factor Rev 2014;25(4):453–72.

[63] Hakim MS, Ding S, Chen S, Yin Y, Su J, van der Woude CJ, et al. TNF-α exerts potent anti-rotavirus effects via the activation of classical NF-κB pathway. Virus Res 2018;253:28–37.

[64] Schwers A, Broecke CV, Maenhoudt M, Beduin JM, Werenne J, Pastoret PP. Experimental rotavirus diarrhoea in colostrum-deprived newborn calves: assay of treatment by administration of bacterially produced human interferon (Hu-IFNα2). AnnRech Vet 1985;16(3):213–18.

[65] Lecce JG, Cummins JM, Richards AB. Treatment of rotavirus infection in neonate and weanling pigs using natural human interferon alpha. Mol Biother 1990;2:211–16.

[66] Yin Y, Bijvelds M, Dang W, Xu L, van der Eijk AA, Knipping K, et al. Modeling rotavirus infection and antiviral therapy using primary intestinal organoids. Antivir Res 2015;123:120–31.

[67] Xue M, Zhao J, Ying L, Fu F, Li L, Ma Y, et al. IL-22 suppresses the infection of porcine enteric coronaviruses and rotavirus by activating STAT3 signal pathway. Antivir Res 2017;142:68–75.

[68] Patel MC, Shirey KA, Pletneva LM, Boukhvalova MS, Garzino-Demo A, Vogel SN, et al. Novel drugs targeting Toll-like receptors for antiviral therapy. Future Virol 2014;9(9):811–29.

[69] Pott J, Stockinger S, Torow N, Smoczek A, Lindner C, McInerney G, et al. Age-dependent TLR3 expression of the intestinal epithelium contributes to rotavirus susceptibility. PLoS Pathog 2012;8(5):1002670.

[70] Arnold MM, Sen A, Greenberg HB, Patton JT. The battle between rotavirus and its host for control of the interferon signalling pathway. PLoS Pathog 2013;9(1):1003064.

[71] Vijay-Kumar M, Gewirtz AT. Flagellin: key target of mucosal innate immunity. Mucosal Immunol 2009;2(3):197.

[72] Zeng H, Carlson AQ, Guo Y, Yu Y, Collier-Hyams LS, Madara JL, et al. Flagellin is the major proinflammatory determinant of enteropathogenic Salmonella. J Immunol 2003;171(7):3668–74.

[73] Zhang B, Chassaing B, Shi Z, Uchiyama R, Zhang Z, Denning TL, et al. Viral infection. Prevention and cure of rotavirus infection via TLR5/NLRC4-mediated production of IL-22 and IL-18. Science 2014;346(6211):861–5.

[74] Uchiyama R, Chassaing B, Zhang B, Gewirtz AT. MyD88-mediated TLR signaling protects against acute rotavirus infection while inflammasome cytokines direct Ab response. Innate Immun 2015;21(4):416–28.

[75] Oh HM, Lee SW, Park MH, Kim MH, Ryu YB, Kim MS, et al. Norkurarinol inhibits Toll-like receptor 3 (TLR3)-mediated pro-inflammatory signaling pathway and rotavirus replication. J Pharmacol Sci 2012;118(2):161–70.

[76] Li Y, Yu P, Qu C, Li P, Li Y, Ma Z, et al. MDA5 against enteric viruses through induction of interferon-like response partially via the JAK-STAT cascade. Antivir Res 2020;176:104743.

[77] Es-Saad S, Tremblay N, Baril M, Lamarre D. Regulators of innate immunity as novel targets for panviral therapeutics. Curr Opin Virol 2012;2(5):622–8.

[78] Dhama K, Karthik K, Khandia R, Munjal A, Tiwari R, Rana R, et al. Medicinal and therapeutic potential of herbs and plant metabolites/extracts countering viral pathogens-current knowledge and future prospects. Curr Drug Metab 2018;19(3):236–63.

[79] Taherkhani R, Farshadpour F, Makvandi M. In vitro anti-rotaviral activity of Achillea kellalensis. Jundishapur J Nat Pharm Prod 2013;8:138–43.

[80] cgqiílio AB, de Faria DB, de Carvalho Oliveira P, Caldas S, de Oliveira DA, Sobral MEG, et al. Screening of Brazilian medicinal plants for antiviral activity against rotavirus. J Ethnopharmacol 2012;141(3):975–81.

[81] Chingwaru W, Majinda RT, Yeboah SO, Jackson JC, Kapewangolo PT, Kandawa-Schulz M, et al. Tylosema esculentum (Marama) tuber and bean extracts are strong antiviral agents against rotavirus infection. Evid Based Complement Altern Med 2011;2011.

[82] Gandhi GR, Barreto PG, dos Santos Lima B, Quintans JDSS, de Souza Araujo AA, Narain N, et al. Medicinal plants and natural molecules with in vitro and in vivo activity against rotavirus: a systematic review. Phytomedicine 2016;23(14):1830–42.

[83] Gonçalves JL, Lopes RC, Oliveira DB, Costa SS, Miranda MM, Romanos MT, et al. In vitro anti-rotavirus activity of some medicinal plants used in Brazil against diarrhea. J Ethnopharmacol 2005;99:403–7.

[84] McCutcheon AR, Roberts TE, Gibbons E, Ellis SM, Babiuk LA, Hancock RE, et al. Antiviral screening of British Columbian medicinal plants. J Ethnopharmacol 1995;49(2):101–10.

[85] Mukoyama A, Ushijima H, Unten S, Nishimura S, Yoshihara M, Sakagami H. Effect of pine seed shell extract on rotavirus and enterovirus infections. Lett Appl Microbiol 1991;13:109–11.

[86] Sauer S. Amorfrutins: a promising class of natural products that are beneficial to health. ChemBioChem 2014;15(9):1231–8.

[87] Alfajaro MM, Kim HJ, Park JG, Ryu EH, Kim JY, Jeong YJ, et al. Anti-rotaviral effects of *Glycyrrhiza uralensis* extract in piglets with rotavirus diarrhea. Virol J 2012;9(1):1–10.

[88] Ueda K, Kawabata R, Irie T, Nakai Y, Tohya Y, Sakaguchi T. Inactivation of pathogenic viruses by plant-derived tannins: strong effects of extracts from persimmon (diospyros kaki) on a broad range of viruses. PLoS One 2013;8(1):55343.

[89] Baek SH, Lee JG, Park SY, Bae ON, Kim DH, Park JH. Pectic polysaccharides from panax ginseng as the antirotavirus principals in ginseng. Biomacromolecules. 2010;11(8):2044–52.

[90] Tam KI, Roner MR. Characterization of *in vivo* anti-rotavirus activities of saponin extracts from quillaja saponaria molina. Antivir Res 2011;90(3):231–41.

[91] Savi LA, Caon T, de Oliveira AP, Sobottka AM, Werner W, Reginatto FH, et al. Evaluation of antirotavirus activity of flavonoids. Fitoterapia 2010;81(8):1142–6.

[92] Tradtrantip L, Ko E, Verkman SA. Antidiarrheal efficacy and cellular mechanisms of a Thai herbal remedy. PLoS Negl Trop Dis 2014;8(2):1–11.

[93] He ST, He FZ, Wu CR, Li SX, Liu WX, Yang YF, et al. Treatment of rotaviral gastroenteritis with Qiwei Baizhu powder. World J Gastroenterol 2001;7(5):735–40.

[94] Shaheen M, Mostafa S, El-Esnawy N. Prevention of coxsackieviruses and rotaviruses infections. Hum Virol Retrovirol 2017;5(5):00171.

[95] Cui Q, Fu Q, Zhao X, Song X, Yu J, Yang Y, et al. Protective effects and immunomodulation on piglets infected with rotavirus following resveratrol supplementation. PLoS One 2018;13(2):0192692.

[96] Yang H, Oh KH, Kim HJ, Cho YH, Yoo YC. Ginsenoside-Rb2 and 20 (S)-Ginsenoside-Rg3 from Korean Red ginseng prevent rotavirus infection in newborn mice. J Microbiol Biotechnol 2018;28(3):391–6.

[97] Teran CG, Teran-Escalera CN, Villarroel P. Nitazoxanide vs. probiotics for the treatment of acute rotavirus diarrhea in children: a randomized, single-blind, controlled trial in Bolivian children. Int J Infect Dis 2009;13:518–23.

[98] Ahmadi E, Alizadeh-Navaei R, Rezai MS. Efficacy of probiotic use in acute rotavirus diarrhea in children: a systematic review and *meta*-analysis. Casp J Intern Med 2015;6(4):187–95.

[99] Liévin-Le Moal V, Servin AL. Anti-infective activities of lactobacillus strains in the human intestinal microbiota: from probiotics to gastrointestinal anti-infectious biotherapeutic agents. Clin Microbiol Rev 2014;27(2):167–99.

[100] Rigo-Adrover M, Knipping K, Garssen J, van Limpt K, Knol J, Franch À, et al. Prevention of rotavirus diarrhea in suckling rats by a specific fermented milk concentrate with prebiotic mixture. Nutrients 2019;11(1):189.

[101] Gibson GR, Roberfroid MB. Dietary modulation of the human colonic microbiota: introducing the concept of prebiotics. J Nutr 1995;125:1401–12.

[102] Yang Y, Pei J, Qin Z, Wei L. Efficacy of probiotics to prevent and/or alleviate childhood rotavirus infections. J Funct Foods 2019;52:90–9.

[103] Azagra-Boronat I, Massot-Cladera M, Knipping K, Garssen J, Ben Amor K, Knol J, et al. Strain-specific probiotic properties of bifidobacteria and lactobacilli for the prevention of diarrhea caused by rotavirus in a preclinical model. Nutrients 2020;12(2):498.

[104] Park MS, Kwon B, Ku S, Ji GE. The efficacy of bifidobacterium longum BORI and *Lactobacillus acidophilus* AD031 probiotic treatment in infants with rotavirus infection. Nutrients 2017;9(8):887.

[105] Walker WA. Role of nutrients and bacterial colonization in the development of intestinal host defense. J Pediatr Gastroenterol Nutr 2000;30(2):S2–7.

[106] Lu L, Walker WA. Pathologic and physiologic interactions of bacteria with the gastrointestinal epithelium. Am J Clin Nutr 2001;73:1124S–1130SS.

[107] Liu F, Li G, Wen K, Bui T, Cao D, Zhang Y, et al. Porcine small intestinal epithelial cell line (IPEC-J2) of rotavirus infection as a new model for the study of innate immune responses to rotaviruses and probiotics. Viral Immunol 2010;23(2):135–49.

[108] Buccigrossi V, Laudiero G, Russo C, Miele E, Sofia M, Monini M, et al. Chloride secretion induced by rotavirus is oxidative stress-dependent and inhibited by *Saccharomyces boulardii* in human enterocytes. PLoS One 2014;9(6):99830.

[109] Kim K, Lee G, Thanh HD, Kim JH, Konkit M, Yoon S, et al. Exopolysaccharide from *Lactobacillus plantarum* LRCC5310 offers protection against rotavirus-induced diarrhea and regulates inflammatory response. J Dairy Sci 2018;101(7):5702–12.

[110] Qiao H, Duffy LC, Griffiths E, Dryja D, Leavens A, Rossman J, et al. Immune responses in rhesus rotavirus-challenged BALB/c mice treated with bifidobacteria and prebiotic supplements. Pediatr Res 2002;51(6):750.

[111] Kawahara T, Makizaki Y, Oikawa Y, Tanaka Y, Maeda A, Shimakawa M, et al. Oral administration of *Bifidobacterium bifidum* G9-1 alleviates rotavirus gastroenteritis through regulation of intestinal homeostasis by inducing mucosal protective factors. PLoS One 2017;12(3):0173979.

[112] Palmieri V, Bugli F, Papi M, Ciasca G, Maulucci G, Galgano S. VP6-SUMO self-assembly as nanocarriers for gastrointestinal delivery. J Nanomater 2015;1–7.

[113] Qingcong L, Xiaoxia P, Hongli L, Minglong Y. Preparation of 4-arm star PELA and its encapsulation of rotavirus for drug delivery. Int J Pharm 2015;491(1–2):123–9.

[114] Zhang R, Lin Z, Lui V, Wong K, Tam P, Lee P, et al. Silver nanoparticle treatment ameliorates biliary atresia syndrome in rhesus rotavirus inoculated mice. Nanomedicine 2017;13(3):1041–50.

[115] Reading PC, Holmskov U, Anders EM. Antiviral activity of bovine collectins against rotaviruses. J Gen Virol 1998;79(9):255–2263.

[116] Inagaki M, Muranishi H, Yamada K, Kakehi K, Uchida K, Suzuki T, et al. Bovine κ-casein inhibits human rotavirus (HRV) infection via direct binding of glycans to HRV. J Dairy Sci 2014;97(5):2653–61.

[117] Ng TB, Cheung RC, Wong JH, Wang Y, Ip DT, Wan DC, et al. Antiviral activities of whey proteins. Appl Microbiol Biotechnol 2015;99(17):6997–7008.

[118] Superti F, Marziano ML, Tinari A, Donelli G. Effect of polyions on the infectivity of SA-11 rotavirus in LLC-MK2 cells. Comp Immunol Microbiol Infect Dis 1993;16:55–62.

[119] Yolken RH, Peterson JA, Vonderfecht SL, Fouts ET, Midthun K, Newburg DS. Human milk mucin inhibits rotavirus replication and prevents experimental gastroenteritis. J Clin Invest 1992;90:1984–91.

[120] Yolken RH, Willoughby R, Wee SB, Miskuff R, Vonderfecht S. Sialic acid glycoproteins inhibit *in vitro* and *in vivo* replication of rotaviruses. J Clin Invest 1987;79:148–54.

[121] Parrón JA, Ripollés D, Ramos SJ, Pérez MD, Semen Z, Rubio P, et al. Antirotaviral potential of lactoferrin from different origin: effect of thermal and high pressure treatments. Biometals 2018;31(3):343–55.

[122] Cui J, Fu X, Xie J, Gao M, Hong M, Chen Y, et al. Critical role of cellular cholesterol in bovine rotavirus infection. Virol J 2014;11:98.

[123] Civra A, Francese R, Gamba P, Testa G, Cagno V, Poli G, et al. 25-Hydroxycholesterol and 27-hydroxycholesterol inhibit human rotavirus infection by sequestering viral particles into late endosomes. Redox Biol 2018;19:318–30.

[124] Civra A, Cagno V, Donalisio M, Biasi F, Leonarduzzi G, Poli G, et al. Inhibition of pathogenic non-enveloped viruses by 25-hydroxycholesterol and 27-hydroxycholesterol. Sci Rep 2014;4(1):1–6.

[125] Civra A, Leoni V, Caccia C, Sottemano S, Tonetto P, Coscia A, et al. Antiviral oxysterols are present in human milk at diverse stages of lactation. J Steroid Biochem Mol Biol 2019;193:105424.

[126] Dou X, Li Y, Han J, Zarlenga DS, Zhu W, Ren X, et al. Cholesterol of lipid rafts is a key determinant for entry and post-entry control of porcine rotavirus infection. BMC Vet Res 2018;14(1):45.

[127] Schroeder ME, Hostetler HA, Schroeder F, Ball JM. Elucidation of the rotavirus NSP4-Caveolin-1 and -cholesterol interactions using synthetic peptides. J Amino Acids 2012;2012:575180.

[128] Mao X, Gu C, Ren M, Chen D, Yu B, He J, et al. l-Isoleucine administration alleviates rotavirus infection and immune response in the weaned piglet model. Front Immunol 2018;9:1654.

[129] Zhao Y, Yu B, Mao X, He J, Huang Z, Zheng P, et al. Dietary vitamin D supplementation attenuates immune responses of pigs challenged with rotavirus potentially through the retinoic acid-inducible gene I signalling pathway. Br J Nutr 2014;112(3):381–9.

[130] Tian G, Liang X, Chen D, Mao X, Yu J, Zheng P, et al. Vitamin D3 supplementation alleviates rotavirus infection in pigs and IPEC-J2 cells via regulating the autophagy signaling pathway. J Steroid Biochem Mol Biol 2016;163:157–63.

[131] Lee C. Controversial effects of vitamin D and related genes on viral infections, pathogenesis, and treatment outcomes. Nutrients 2020;12(4):962.

[132] Zhao Y, Ran Z, Jiang Q, Hu N, Yu B, Zhu L, et al. Vitamin D alleviates rotavirus infection through a Microrna-155-5p mediated regulation of the TBK1/IRF3 signaling pathway *in vivo* and *in vitro*. Int J Mol Sci 2019;20(14):3562.

III. Antiviral agents and therapeutics

[133] Hester SN, Chen X, Li M, Monaco MH, Comstock SS, Kuhlenschmidt TB, et al. Human milk oligosaccharides inhibit rotavirus infectivity *in vitro* and in acutely infected piglets. Br J Nutr 2013;110(7):1233–42.

[134] Chen H, Hu H, Chen D, Tang J, Yu B, Luo J, et al. Dietary pectic oligosaccharide administration improves growth performance and immunity in weaned pigs infected by rotavirus. J Agric Food Chem 2017;65(14):2923–9.

[135] Li M, Monaco MH, Wang M, Comstock SS, Kuhlenschmidt TB, Fahey Jr GC, et al. Human milk oligosaccharides shorten rotavirus-induced diarrhea and modulate piglet mucosal immunity and colonic microbiota. ISME J 2014;8(8):1609–20.

[136] Mao X, Xiao X, Chen D, Yu B, He J, Chen H, et al. Dietary apple pectic oligosaccharide improves gut barrier function of rotavirus-challenged weaned pigs by increasing antioxidant capacity of enterocytes. Oncotarget 2017;8(54):92420–30.

[137] Azagra-Boronat I, Massot-Cladera M, Knipping K, Van't Land B, Tims S, Stahl B, et al. Oligosaccharides modulate rotavirus-associated dysbiosis and TLR gene expression in neonatal rats. Cells 2019;8(8):876.

[138] Rahal A, Kumar A, Singh V, Yadav B, Tiwari R, Chakraborty S, et al. Oxidative stress, prooxidants, and antioxidants: the interplay. BioMed Res Int 2014;2014:761264.

[139] Mohamed AA, Ali SI, El-Baz FK, El-Senousy WM. New insights into antioxidant and antiviral activities of two wild medicinal plants: *Achillea fragrantissima* and *Nitraria retusa*. Int J Pharm Bio Sci 2015;6:708–22.

[140] Guerrero CA, Acosta O. Inflammatory and oxidative stress in rotavirus infection. World J Virol 2016;5(2):38.

[141] De UK, Mukherjee R, Nandi S, Patel BH, Dimri C, Ravishankar U, et al. Alterations in oxidant/antioxidant balance, high-mobility group box 1 protein and acute phase response in cross-bred suckling piglets suffering from rotaviral enteritis. Trop Anim Health Prod 2014;46:1127–33.

[142] Guerrero CA, Murillo A, Acosta O. Inhibition of rotavirus infection in cultured cells by N-acetylcysteine, PPARc agonists and NSAIDs. Antivir Res 2012;96:1–12.

[143] Guerrero CA, Pardo P, Rodriguez V, Guerrero R, Acosta O. Inhibition of rotavirus ECwt infection in ICR suckling mice by N-acetylcysteine, peroxisome proliferator-activated receptor gamma agonists and cyclooxygenase-2 inhibitors. Mem Inst Oswaldo Cruz 2013;108:741–54.

[144] Mao X, Hu H, Xiao X, Chen D, Yu B, He J, et al. Lentinan administration relieves gut barrier dysfunction induced by rotavirus in a weaned piglet model. Food Funct 2019;10(4):2094–101.

[145] Aiyegoro OA, Okoh AI. Use of bioactive plant products in combination with standard antibiotics: implications in antimicrobial chemotherapy. J Med Plants Res 2009;13:1147–52.

[146] Butler A, Keating R. Old herbal remedies and modern combination therapy. Scott Med J 2011;56:170–3.

[147] Che C, Wang ZJ, Chow MSS, Lam CWK. Herb-herb combination for therapeutic enhancement and advancement: theory, practice and future perspectives. Molecules 2013;18:5125–41.

[148] Alfajaro MM, Rho MC, Kim HJ, Park JG, Kim DS, Hosmillo M, et al. Anti-rotavirus effects by combination therapy of stevioside and Sophora flavescens extract. Res Vet Sci 2014;96(3):567–75.

[149] Clark KJ, Grant PG, Sarr AB, Belakere JR, Swaggerty CL, Phillips TD, et al. An *in vitro* study of theaflavins extracted from black tea to neutralize bovine rotavirus and bovine coronavirus infections. Vet Microbiol 1998;63(2–4):147–57.

[150] Andres A, Donovan SM, Kuhlenschmidt TB, Kuhlenschmidt MS. Isoflavones at concentrations present in soy infant formula inhibit rotavirus infection *in vitro*. J Nutr 2007;137:2068–73.

[151] Lipson SM, Karalis G, Karthikeyan L, Ozen FS, Gordon RE, Ponnala S, et al. Mechanism of anti-rotavirus synergistic activity by epigallocatechin gallate and a proanthocyanidin-containing nutraceutical. Food Environ Virol 2017;9(4):434–43.

[152] Shi Z, Zou J, Zhang Z, Zhao X, Noriega J, Zhang B, et al. Segmented filamentous bacteria prevent and cure rotavirus infection. Cell 2019;179(3):644–58.

[153] Shen J, Chen JJ, Zhang BM, Zhao J, Chen L, Ye QY, et al. Baicalin is curative against rotavirus damp heat diarrhea by tuning colonic mucosal barrier and lung immune function. Dig Dis Sci 2020;65(8):2234–45.

15

Current therapeutic strategies and novel antiviral compounds for the treatment of nonpolio enteroviruses

Angeline Jessika Suresh and Regina Sharmila Dass

Fungal Genetics and Mycotoxicology Laboratory, Department of Microbiology, School of Life Sciences, Pondicherry University, Pondicherry, India

15.1 Introduction

The enteroviruses have established a persistent presence across the globe, with periodic outbreaks occurring since the mid-20th century in both community and health-care settings [1,2]. In literature, they are often separated into polioviruses and nonpolio enteroviruses, due to the latter's versatility in their mechanisms of pathogenesis and the resultant range of neurological, respiratory, and systemic manifestations [3]. Endemicity and recurrence of nonpolio enteroviruses is particularly observed in countries of the Asia-Pacific region, such as China, Taiwan, Malaysia, and Japan, with young children and the immunocompromised constituting a significant portion of the susceptible population [4−10]. Recently, outbreaks of the numbered enteroviruses in Western nations, such as the United States of America and the Netherlands, have further brought attention to the rapid propagation of the nonpolio enteroviruses [11,12].

There are several reasons why nonpolio enteroviruses, particularly those responsible for eliciting neurological complications, are among the leading public health threats in numerous countries. Like most RNA viruses, the enteroviruses undergo events of recombination, such as template switching, and accumulate mutations at alarming speeds due to the poor fidelity of their RNA polymerases. This results in a wide range of genetic diversity and altered states of virulence among the viruses, posing a challenge in the development of specific antivirals [13]. Additionally, the involvement of the central nervous system and severity of symptoms often leads to lifelong debilitating consequences even after the eradication of infection [14]. Finally, enteroviruses are capable of transmitting to susceptible hosts through respiratory secretions [3]. Respiratory viruses can lead to deadly scenarios as they rapidly spread in overcrowded, population-dense communities and resist measures of control, as seen with the influenza H1N1, MERS, and COVID-19 pandemics of the 21st century.

The enteroviruses belong to the large and diverse viral family of the Picornaviridae. After undergoing several taxonomic and classification revisions, there are currently four recognized human enterovirus species, designated enteroviruses A-D, which are further divided into subtypes [9]. Notable human pathogens include the polioviruses, echoviruses, human rhinoviruses, coxsackie A and B viruses (CV-A and CV-B), EV-D68, and EV-A71. Except the conflict-ridden regions of Afghanistan and Pakistan inaccessible to vaccination campaign efforts, the WHO has declared the successful eradication of the poliovirus in the remaining countries. Therefore, in this chapter, we will cover research focusing on novel therapeutic interventions for nonpolio enteroviruses that have emerged as a growing concern in the past decade, with emphasis on EV-A71, EV-D68, and CV-A16.

15.2 Structure and life cycle

The enteroviruses are transmitted to susceptible populations via oropharyngeal secretions and by indirect and direct fecal—oral routes [3]. Like other members of the Picornaviridae, they are small, single-stranded, positive-sense RNA viruses that lack a viral envelope. Their genome encodes a single polyprotein that is proteolytically cleaved by viral-encoded proteases into eleven proteins, four structural and seven nonstructural [15]. The four structural proteins contribute to capsid formation and are designated VP0 (processed into VP1 and VP4 during later stages of viral release), VP2, and VP3. The nonstructural proteins are 2A, 2B, 2C, 3A, 3B, 3C, and 3D^{Pol}. These proteins are vital to viral synthesis and are all involved in a multitude roles including immune evasion, formation of replication organelles, and the trafficking of host resources into enterovirus replication. They have also been recently found to mediate apoptosis and other signaling pathways [3,15].

After entering the new host, the enteroviruses are internalized by their target cells through receptor-mediated endocytosis. Some host receptors are shared by the enteroviruses, but many nonpolio enteroviruses exhibit interactions with selected receptors that are possibly unique to them. EV-A71 employs predominantly two receptors, the human scavenger receptor class B, member 2 (hSCARB2), and the P-selectin glycoprotein ligand 1 [3]. These receptors are expressed heavily in specific neuronal cells, explaining the neurotrophic nature of the virus [9]. Other cellular receptors that interact with the enteroviruses include annexin-2, sialylated glycans, and heparin sulfate [3]. A concave depression in VP1, known as the canyon, appears to mediate receptor binding and uncoating, and is a favored target for therapeutic intervention. Following ligand binding, studies show that the virus undergoes modifications in structural conformation and forms a pore in the endosomal membrane. Through this channel, the viral genome is injected and enters the cytoplasm where it is readily transcribed due to its positive-sense conformation. Enterovirus replication occurs in designated, localized membranous domains formed with coordination between the nonstructural viral components and the host cell. An extensive, detailed review of the life cycle of nonpolio enteroviruses is covered by [3].

15.3 Clinical manifestations

Most infections by nonpolio enteroviruses are self-resolving, asymptomatic or mild, and brief. However, in individuals incapable of presenting a sufficient immune response, such as infants, young children, and those with a compromised immune system, the infection may progress into severe neurological disease. EV-A71, CV-A16, and recently CV-A6 are the etiological agents of hand, foot, and mouth disease, which is prevalent in neonates and infants. HMFD is characterized by febrile illness, maculopapular rashes or erythema of the extremities, and ulcerative lesions in the oral cavity [16,17]. As the viruses exhibit strong neurotropism, especially EV-A71, complications such as brainstem encephalitis, acute flaccid paralysis, and aseptic meningitis are present during the seasonal epidemics. The enteroviruses may cause fatal disease by disseminating to other surrounding tissues and cause pulmonary edema, septic shock in neonates, conjunctivitis, or cardiac dysfunction [9].

Infections caused by EV-D68 differ from those resulting from other members of the genus. The viruses are acid-intolerant and prefer cooler temperatures, thus colonizing the nasal passages of the upper respiratory tract rather than the highly acidic gastric environment [18]. They are associated with moderate to severe respiratory illnesses, including severe bronchitis and interstitial pneumonia. However, like the other human enteroviruses, it is capable of replicating in neurological tissues and is seen as a high threat pathogen after a series of outbreaks occurring in 2014 [12,18].

15.4 Antiviral agents

15.4.1 Capsid inhibitors

Capsid binders are a large group of antiviral drugs that target the surface proteins of viruses and inhibit viral entry [19]. Despite the drugs exhibiting a higher risk of inducing antiviral drug resistance, capsid inhibitors are extensively studied in the literature [20]. Pyridyl imidazolidinone and its derivatives exhibit high and selective potency against EV-A71 infection, possibly by targeting capsid protein VP1 [21]. Another compound that targets a highly conserved residue of VP1 is rosmarinic acid, which has shown promising results during in vitro cell culture and in vivo mouse studies [22,23] Attempting to fill the void in selective antivirals for EV-D68 infections [24], reported a novel triazole compound that inhibited EV-D68 uncoating by inhabiting the hydrophobic depression of VP1 protein. Pleconaril is a well-characterized, capsid inhibiting, antiviral drug for picornavirus infections but has not yet reached the mainstream market due to concerns about its toxicity [25]. Alternatively, antibodies blocking enterovirus receptors, for example, hSCARB2, are also a viable strategy for inhibiting capsid-receptor recognition [26].

15.4.2 Inhibitors of nonstructural viral components

There is very little literature on enteroviral 2A^{pro}, and 2B, 3A, and 3B proteins as potential targets for antienteroviral efforts [26]. The research of [27] supports the repurposing of anti-HCV drug Telaprevir for EV-D68 infections, as the pharmaceutical agent binds and inhibits EV-D68 2A protein. Telaprevir is ideally administered in a combination drug therapy with other drugs, such as enterovirus 2C inhibitor R523062, which not only provides additive protection but also reduces the probability of antiviral resistance [28].

Proteins 2C, 3C, and 3D^{pol} are popular, well-investigated drug targets and therefore may prove to be more suitable for drug development. The ATP-dependent RNA helicase 2C is a multifaceted protein involved in host membrane alteration, the formation of viral replication complexes, and the evasion of innate immunity pathways [15]. During wide-scale drug repurposing screenings, the S-enantiomer of fluoxetine was found to deter infections caused by enterovirus B and D by binding to the 2C protein, effectively halting viral genomic replication [29–31]. Fluoxetine, commercially known as Prozac, was further brought to attention for successfully treating enteroviral induced encephalitis in an immunocompromised five-year-old child [32]. The highly conserved nature of the protein is an attractive property for drug design as it is ideal for the development of broad-spectrum antienteroviral pharmaceuticals [29]. Other drugs identified as 2C inhibitors of the enterovirus species during repurposing screenings include dibucaine, pirlindole, and zuclopenthixol [33,34].

The cysteine protease 3C contributes to the virulence of EV-A71 species and is responsible for the proteolytic segmentation of the precursor polyprotein and modulation of host cell apoptosis. Similar to the 2C protein, it is capable of binding to RNA and plays a significant role in inactivating host innate immunity through cleavage of essential chemical mediators [15]. The 3C protease is particularly suitable as a target for infections caused by human rhinoviruses and enterovirus B [35]. The researchers [36] have designed potent inhibitors for protein 3C expressed by a circulating strain of EV-A71. The antiviral candidates were derived from peptidomimetic inhibitor AG7088 which was originally obtained to target rhinoviruses and showed no cytotoxicity during cellular assays. Rupintrivir and its analogs are well-known, nontoxic inhibitors of human rhinovirus 3C protease and have also been well received as potential leads for the treatment of EV-A71 infection [37,38]. Small molecules that directly bind to the substrate-binding cavity of 3C protease, such as quercetin, diminish the viral progeny load by deterring substrate recognition [39]. Perhaps the most promising candidates in this age of coexistence and cocirculation of numerous respiratory viral genera is the broad-spectrum alpha-ketoamides, designed by [40]. Two compounds, 11u and 11r, presented low cytotoxicity and possessed inhibitory action in vitro against enteroviruses, alphacoronaviruses, and beta-coronaviruses.

Viral polymerases have always been an excellent target in many antiviral strategies and the enteroviral RNA-dependent 3D^{pol} is no exception. Given that the Picornaviridae family houses several human pathogens, their polymerases have often been on the receiving end of broad-spectrum antiviral drug development [41]. Examples of 3D^{pol} inhibitors that show promise as broad-spectrum antivirals include the pyrimidine analog FNC and nonnucleoside compound GPC-N114 [41,42]. Aurintricarboxylic acid (ATA) is a well-known, broad-spectrum antiviral compound possessing inhibitory effects against a spectrum of infections, such as those caused by HIV, HAV, and influenza viruses. Hung et al. [43]

demonstrated that ATA interfered with the function of EV-A71 $3D^{pol}$, diminishing the cytopathic effects of EV-A71 infected vero cells.

15.4.3 Nucleoside analogs

Nucleoside analogs inhibit viral replication by competing with nucleotide substrates and have experienced great success in treating RNA viral infections, as seen with HIV infections. NITD008 is an adenosine analog that has shown success in halting flavivirus synthesis. Deng et al. [44] showed the compound induces similar results in EV-A71 infections, with an EC50 of $0.67\,\mu M$ and an absence of life-threatening symptoms in inoculated mice. The group also highlighted the need for combination therapy to prevent the occurrence of antiviral resistance by performing resistance analyses with mutated strains. Additionally, [45] noted the synergistic antiviral effects on CV-B3, human rhinoviruses, and EV-A71 exhibited by a combinational drug therapy involving gemcitabine (an inhibitor of pyrimidine biosynthesis) and ribavirin (an inhibitor of inosine monophosphate dehydrogenase). Recently, the [46] group demonstrated the antienteroviral activity of Remdesivir, a popular broad-spectrum antiviral pharmaceutical and was found to be a potent inhibitor of EV-A71 genomic synthesis.

15.4.4 Inhibition of host cellular components

The success of an enterovirus infection not only depends on viral protein activity but also on the actions of "hijacked" host biomolecules. While targeting host cellular responses eliminates the risk of the viral pathogen developing drug resistance, there is a larger risk of host cells experiencing toxicity and inappropriate signaling behavior. The host protein, phosphatidylinositol 4-kinase type IIIβ (PI4KB), is essential for the formation of viral replication organelles and thus the efficacy and toxicity of its inhibitors are covered extensively in research [47–49]. Other host factors that may serve as a potential site for antiviral efforts include oxysterol-binding protein and cyclophilins [19].

15.4.5 Other compounds

Traditional Chinese Medicine has served as both a source of inspiration and a repertoire of bioactive small molecules for modern pharmaceutical companies. The flavonoid baicalin, extracted from the root of *Scutellaria*, was found to inhibit the transcriptional and translation activities of $3D^{pol}$ [50]. Other flavonoids, such as apigenin, luteolin, kaempferol, and their derivatives, exhibited similar potent antiviral activities against a range of enteroviruses, although the mode of action for many of the compounds remains unelucidated [51].

15.5 Advances in vaccine development for HFMD and EV-D68 infections

Considering the heightened global emergence of the human numbered enteroviruses, the fatality of illness, and the highly contagious nature of the pathogens, there is an immediate requirement for nontoxic and efficient prophylactic measures against the nonpolio enteroviruses. Vaccine development for enteroviruses remains a challenge for countries located in the Asia-Pacific region, such as Taiwan, Malaysia, Japan, and China, due to their genetic variance. EV-A71 possesses three genotypes, A, B, and C, with the latter two containing their subtypes, with their distribution varying on a temporal and spatial scale [13]. The cocirculation of various genotypes and subgenotypes facilitates recombination between viral strains, known as inter- and intra-genotypic shifts respectively, resulting in the formation of novel enterovirus strains. Additionally, there is mounting evidence that coexistence with CV-A16 also produces new genotypes through recombination [13,52]. The periodic occurrence of novel strains and the continuous processes of antigenic drift and shift require consistent global surveillance, without which successful vaccine development cannot occur [53].

15.5.1 Inactivated whole vaccines

In the current market, there are three formalin-inactivated vaccines available against EV-A71 and are manufactured by Beijing Vigoo Biological Co., Ltd. (Vigoo), Sinovac Biotech Co., Ltd. (Sinovac), and the Chinese Academy of Medical Sciences (CAMS) [26]. Inactivated vaccines appear more accessible in terms of production and marketing as they are more cost-effective, safer, and manufactured using well-tested practices. The three vaccines, based on the C4-subgenotype, were deemed safe and effective in phase III clinical trials and received their licensure for production and marketing in 2015 by China's Food and Drug Administration [26]. The vaccines showed protection effectiveness of 89.7% against EV-A71 infection along with a 4.58% rate of reported adverse reaction in phase IV clinical trials held in China, spanning 14 months [54]. Additionally, they were found to exhibit cross-neutralizing effect against subgenotypes B1, B4, B5, and C4A [55]. Unfortunately, a major disadvantage of the CAM and Sinovac EV-A71 vaccines is the inability to offer prophylactic action against HFMD cases caused by CV-A16, which is a significant contributor to cyclic epidemics of the disease [26,56]. The National Health Research Institute, Taiwan, and Inviragen Inc, Singapore have entered their own inactivated vaccines into clinical trials, with the former estimated to finish phase III clinical trials by 2022 [26]. Unlike the Chinese vaccines, the Taiwan and Singapore vaccines are developed using strains derived from the subgenotypes B4 and B3, respectively [26]. Another promising vaccine candidate to enter clinical trials is the formalin-inactivated, C4-subgenotype based, EV-A71 strain manufactured by the Korea National Research Institute of Health [57,58].

15.5.2 Recombinant subunit vaccines

Researchers employed various expression systems to express virus-like particles (VLPs) containing the capsid antigens VP0, VP2, and VP3 of EV-A71, CV-A16, CV-A6, and CV-A10. These expression systems include the yeasts *Saccharomyces cerevisiae* and *Pichia pastoris*, and the baculovirus-insect system [54]. Yang et al. [59] have successfully expressed a high quantity of VLPs using the *P. pastoris* system and have demonstrated its safety and efficacy in preventing lethal EV-D68 infections during passive immunization of neonatal mice. The results also show that administration of sera containing viral antigen-specific antibodies in dams induced protection of suckling mice through passive transfer of maternal antibodies in breast milk. VLP-based EV-A71 vaccines were further found to be comparable in respect to nontoxicity and efficacy with the CAMS inactivated EV-A71 vaccine [60]. The promising results of [59] have led to the introduction of *P. pastoris* expressed EV-D68 VLPs into the clinical trials as a potential vaccine candidate for HFMD [54].

15.5.3 Multivalent and chimeric vaccines

Monovalent vaccines may not be effective in preventing epidemics as there are several etiological agents in HFMD disease, including EV-A71, CV-A16, and CV-A6. The development of a bivalent (EV-A71 and CV-A16) and trivalent (EV-A71, CV-A16, and CV-A6) has shown immense success in eliciting cross-protective immune responses in rhesus macaques and mice, respectively [61,62]. However, further progress may be hindered by the relatively expensive manufacturing process of simply "mixing" the inactivated viruses into one dose [54]. To overcome the cost and increase protective efficacy against a broad range of enteroviral agents, the production of novel chimeric vaccines has been investigated in the past few years [63]. For example, [64] and [65] created a bivalent vaccine candidate by replacing an epitope of EV-A71 with one of the CV-A16 to create a chimeric EV-A71/CV-A16 strain and VLP, respectively, that were capable of inducing a strong humoral immune response in mice.

15.5.4 Vaccine candidates for enterovirus D68

To our knowledge, there are currently no vaccine candidates for the respiratory pathogen EV-D68 approved for clinical trials. In the last decade, efforts have been focused on identifying appropriate animal models that mimic human enteroviral infections and those suitable for vaccine studies [66,67]. Raychoudhuri et al. [68] show that a formalin-inactivated strain of EV-D68 from the 2014 American outbreak is a promising candidate for vaccine development. The strain is vero cell adapted and injected into Balb/c mice, forming a high titer of neutralizing, viral antigen-specific, IgG antibodies that protect recipient neonatal mice from life-threatening symptoms. Another experimental vaccine employs the use of VLPs. The [69] group demonstrated protective serum antibody production in mice by injecting them with a recombinant baculovirus vector housing the viral P1 precursor (encoding V0, V2, V3) and 3CD protease adapted from insect cells. Not only did the sera protect against lethal infection in neonatal mice when administered as passive immunization, but also provided the same protection in suckling mice when the sera were administered to the maternal mice. Zhang et al. [70] provide further evidence of the protective efficacy of sera induced by VPL vaccines produced through the passage in *P. pastoris* yeast cells and the suitability of enteroviral VPLs as immunogens.

15.5.5 Live attenuated vaccines

When compared to inactivated viral vaccines, live attenuated vaccines offer a comparatively longer-lasting immune response, stimulating both the humoral and cellular arms. The increased effectiveness of attenuated vaccines offers an additional advantage as it reduces the requirement for multiple booster shots. microRNA (miRNA) based vaccine candidates may offer a solution to the growing necessity for enteroviral prophylaxis. Several types of endogenous miRNA control translational gene regulation in a tissue-specific manner. Inserting miRNA sites into the viral genome has been found to restrict viral tropisms by inhibiting replication in some cell types while allowing it in others [71]. This mode of synthesis mounts a sufficient immune response involving T-cells while preventing viral pathogenesis. Yee et al. [72] created an EV-A71 miRNA-based vaccine strain (pIY) with neuronal-specific miRNAs, let-7a and miR-124a. The experimental vaccine, which also encoded for a high-fidelity RNA-dependent RNA polymerase, retained genetic stability and conferred protection against hind limb paralysis in four-week mice. The group's findings demonstrate the potential of the pIY vaccine for further development as a live attenuated vaccine candidate.

15.6 Conclusion

Antienteroviral development has progressed immensely in the past decade, as seen by increasing amount of literature regarding novel therapeutics and their supportive clinical studies. Considering the risk of antiviral resistance, researchers have investigated a wide range of pharmaceutical candidates targeting host factors and structural and nonstructural components of the enterovirus itself, such as transcription factors and viral enzymes. Many of these compounds have entered clinical trials after experiencing success in both in vitro cell culture studies and in vivo mouse models. Additionally, research on prophylactic measures, particularly regarding monovalent and polyvalent vaccines, has seen a positive trajectory. However, the future of antienteroviral pharmaceuticals rests on three important actions.

One, the exact mechanisms of pathogenesis of the numbered viruses remain yet to be completely elucidated. Understanding the infection process will reveal new therapeutic targets, aiding in the development of safer, selective, and specific antiviral agents. Second, the number of pathogenesis studies remains insufficient due to the lack of appropriate animal models that mimic human enterovirus infections. Hence, future research should not neglect the development of suitable animal models as they not only sustain investigations on the clinical manifestation of enterovirus pathogenesis but also streamline the drug development pipeline. Finally, enterovirus surveillance predominantly occurs in countries that experience frequent outbreaks, therefore providing a global picture of enterovirus evolution and circulating serotypes in the remainder of the world. Global surveillance is pivotal to antienteroviral research because the distribution of serotypes is constantly changing and new strains emerge rapidly. Without understanding the current geography of enterovirus subtypes, it is impossible to produce effective antiviral agents against future epidemics.

References

[1] Holm-Hansen CC, Midgley SE, Fischer TK. Global emergence of enterovirus D68: a systematic review. Lancet Infect Dis 2016;16:e64−75. Available from: https://doi.org/10.1016/S1473-3099(15)00543-5.

[2] Fujimoto T, Iizuka S, Enomoto M, Abe K, Yamashita K, Hanaoka N, et al. Hand, foot, and mouth disease caused by coxsackievirus A6, Japan, 2011. Emerg Infect Dis 2012;18:337. Available from: https://doi.org/10.3201/eid1802.111147.

[3] Baggen J, Thibaut HJ, Strating JR, van Kuppeveld FJ. The life cycle of non-polio enteroviruses and how to target it. Nat Rev Microbiol 2018;16:368–81. Available from: https://doi.org/10.1038/s41579-018-0005-4.

[4] Ho M, Chen E-R, Hsu K-H, Twu S-J, Chen K-T, Tsai S-F, et al. An epidemic of enterovirus 71 infection in tai-wan. N Engl J Med 1999;341:929–35. Available from: https://doi.org/10.1056/nejm199909233411301.

[5] Zhao K. Circulating Coxsackievirus A16 identified as recombinant type A human enterovirus, China. Emerg Infect Dis 2011;. Available from: https://doi.org/10.3201/eid1708.101719.

[6] Yang F, Ren L, Xiong Z, Li J, Xiao Y, Zhao R, et al. Enterovirus 71 outbreak in the People's Republic of China in 2008. J Clin Microbiol 2009;47:2351–2. Available from: https://doi.org/10.1128/jcm.00563-09.

[7] Ikeda T, Mizuta K, Abiko C, Aoki Y, Itagaki T, Katsushima F, et al. Acute respiratory infections due to enterovirus 68 in Yamagata, Japan between 2005 and 2010. Microbiol Immunol 2012;56:139–43. Available from: https://doi.org/10.1111/j.1348-0421.2012.00411.x.

[8] Kaida A. Enterovirus 68 in children with acute respiratory tract infections, Osaka, Japan. Emerg Infect Dis 2011;. Available from: https://doi.org/10.3201/eid1708.110028.

[9] Solomon T, Lewthwaite P, Perera D, Cardosa MJ, McMinn P, Ooi MH. Virology, epidemiology, pathogenesis, and control of enterovirus 71. Lancet Infect Dis 2010;10:778–90. Available from: https://doi.org/10.1016/s1473-3099(10)70194-8.

[10] Herrero LJ, Lee CS, Hurrelbrink RJ, Chua BH, Chua KB, McMinn PC. Molecular epidemiology of enterovirus 71 in peninsular Malaysia, 1997–2000. Arch Virol 2003;148:1369–85. Available from: https://doi.org/10.1007/s00705-003-0100-2.

[11] Meijer A, van der Sanden S, Snijders BEP, Jaramillo-Gutierrez G, Bont L, van der Ent CK, et al. Emergence and epidemic occurrence of enterovirus 68 respiratory infections in The Netherlands in 2010. Virology 2012;423:49–57. Available from: https://doi.org/10.1016/j.virol.2011.11.021.

[12] Oermann CM, Schuster JE, Conners GP, Newland JG, Selvarangan R, Jackson MA. Enterovirus D68. A focused review and clinical highlights from the 2014 U.S. Outbreak. Ann ATS 2015;12:775–81. Available from: https://doi.org/10.1513/annalsats.201412-592fr.

[13] Yip CCY, Lau SKP, Woo PCY, Yuen K-Y. Human enterovirus 71 epidemics: what's next? Emerg Health Threats J 2013;6:19780. Available from: https://doi.org/10.3402/ehtj.v6i0.19780.

[14] McMinn PC. Recent advances in the molecular epidemiology and control of human enterovirus 71 infection. Curr Opin Virol 2012;2:199–205. Available from: https://doi.org/10.1016/j.coviro.2012.02.009.

[15] Yuan J, Shen L, Wu J, Zou X, Gu J, Chen J, et al. Enterovirus A71 proteins: structure and function. Front Microbiol 2018;9. Available from: https://doi.org/10.3389/fmicb.2018.00286.

[16] Lee T-C, Guo H-R, Su H-JJ, Yang Y-C, Chang H-L, Chen K-T. Diseases Caused by Enterovirus 71 infection. Pediatr Infect Dis J 2009;28:904–10. Available from: https://doi.org/10.1097/inf.0b013e3181a41d63.

[17] Cabrerizo M, Tarragó D, Muñoz-Almagro C, del Amo E, Domínguez-Gil M, Eiros JM, et al. Molecular epi-demiology of enterovirus 71, coxsackievirus A16 and A6 associated with hand, foot and mouth disease in Spain. Clin Microbiol Infect 2014;20:O150–6. Available from: https://doi.org/10.1111/1469-0691.12361.

[18] Sun J, Hu X-Y, Yu X-F. Current understanding of human enterovirus D68. Viruses 2019;11:490. Available from: https://doi.org/10.3390/v11060490.

[19] Bauer L, Lyoo H, van der Schaar HM, Strating JR, van Kuppeveld FJ. Direct-acting antivirals and host-targeting strategies to combat enterovirus infections. Curr Opin Virol 2017;24:1–8. Available from: https://doi.org/10.1016/j.coviro.2017.03.009.

[20] Li P, Yu J, Hao F, He H, Shi X, Hu J, et al. Discovery of potent EV71 capsid inhibitors for treatment of HFMD. ACS Med Chem Lett 2017;8:841–6. Available from: https://doi.org/10.1021/acsmedchemlett.7b00188.

[21] Chen T-C, Liu S-C, Huang P-N, Chang H-Y, Chern J-H, Shih S-R. Antiviral activity of pyridyl imidazolidi-nones against enterovirus 71 variants. J Biomed Sci 2008;15:291–300. Available from: https://doi.org/10.1007/s11373-007-9228-5.

[22] Hsieh C-F, Jheng J-R, Lin G-H, Chen Y-L, Ho J-Y, Liu C-J, et al. Rosmarinic acid exhibits broad anti-enterovirus A71 activity by inhibiting the interaction between the fivefold axis of capsid VP1 and cognate sulfated receptors. Emerg Microbes Infect 2020;9:1194–205. Available from: https://doi.org/10.1080/22221751.2020.1767512.

[23] Lin W-Y, Yu Y-J, Jinn T-R. Evaluation of the virucidal effects of rosmarinic acid against enterovirus 71 infection via in vitro and in vivo study. Virol J 2019;16. Available from: https://doi.org/10.1186/s12985-019-1203-z.

[24] Ma C, Hu Y, Zhang J, Musharrafieh R, Wang J. A novel capsid binding inhibitor displays potent antiviral activity against enterovirus D68. ACS Infect Dis 2019;5:1952—62. Available from: https://doi.org/10.1021/acsinfecdis.9b00284.

[25] Lacroix C, Laconi S, Angius F, Coluccia A, Silvestri R, Pompei R, et al. In vitro characterisation of a pleconaril/pirodavir-like compound with potent activity against rhinoviruses. Virol J 2015;12. Available from: https://doi.org/10.1186/s12985-015-0330-4.

[26] Lin J-Y, Kung Y-A, Shih S-R. Antivirals and vaccines for Enterovirus A71. J Biomed Sci 2019;26. Available from: https://doi.org/10.1186/s12929-019-0560-7.

[27] Musharrafieh R, Ma C, Zhang J, Hu Y, Diesing JM, Marty MT, et al. Validating enterovirus D68-2A pro as an antiviral drug target and the discovery of telaprevir as a potent D68-2A pro inhibitor. J Virol 2019;93. Available from: https://doi.org/10.1128/jvi.02221-18.

[28] Ma C, Hu Y, Zhang J, Wang J. Pharmacological characterization of the mechanism of action of R523062, a promising antiviral for enterovirus D68. ACS Infect Dis 2020;6:2260—70. Available from: https://doi.org/10.1021/acsinfecdis.0c00383.

[29] Bauer L, Manganaro R, Zonsics B, Strating JRPM, El Kazzi P, Lorenzo Lopez M, et al. Fluoxetine inhibits enterovirus replication by targeting the viral 2C protein in a stereospecific manner. ACS Infect Dis 2019;5:1609—23. Available from: https://doi.org/10.1021/acsinfecdis.9b00179.

[30] Ulferts R, van der Linden L, Thibaut HJ, Lanke KHW, Leyssen P, Coutard B, et al. Selective serotonin reuptake inhibitor fluoxetine inhibits replication of human enteroviruses B and D by targeting viral protein 2C. Antimicrob Agents Chemother 2013;57:1952—6. Available from: https://doi.org/10.1128/aac.02084-12.

[31] Zuo J, Quinn KK, Kye S, Cooper P, Damoiseaux R, Krogstad P. Fluoxetine is a potent inhibitor of coxsackievirus replication. Antimicrob Agents Chemother 2012;56:4838—44. Available from: https://doi.org/10.1128/aac.00983-12.

[32] Gofshteyn J, Cárdenas AM, Bearden D. Treatment of chronic enterovirus encephalitis with fluoxetine in a patient with X-linked agammaglobulinemia. Pediatr Neurol 2016;64:94—8. Available from: https://doi.org/10.1016/j.pediatrneurol.2016.06.014.

[33] Tang Q, Xu Z, Jin M, Shu T, Chen Y, Feng L, et al. Identification of dibucaine derivatives as novel potent enterovirus 2C helicase inhibitors: in vitro, in vivo, and combination therapy study. Eur J Med Chem 2020;202:112310. Available from: https://doi.org/10.1016/j.ejmech.2020.112310.

[34] Ulferts R, de Boer SM, van der Linden L, Bauer L, Lyoo HR, Maté MJ, et al. Screening of a library of FDA-approved drugs identifies several enterovirus replication inhibitors that target viral protein 2C. Antimicrob Agents Chemother 2016;60:2627—38. Available from: https://doi.org/10.1128/aac.02182-15.

[35] Diarimalala RO, Hu M, Wei Y, Hu K. Recent advances of enterovirus 71 3Cpro targeting Inhibitors. Virol J 2020;17. Available from: https://doi.org/10.1186/s12985-020-01430-x.

[36] Kuo C-J, Shie J-J, Fang J-M, Yen G-R, Hsu JT-A, Liu H-G, et al. Design, synthesis, and evaluation of 3C protease inhibitors as anti-enterovirus 71 agents. Bioorg Med Chem 2008;16:7388—98. Available from: https://doi.org/10.1016/j.bmc.2008.06.015.

[37] Tan YW, Ang MJY, Lau QY, Poulsen A, Ng FM, Then SW, et al. Antiviral activities of peptide-based covalent inhibitors of the Enterovirus 71 3C protease. Sci Rep 2016;6. Available from: https://doi.org/10.1038/srep33663.

[38] Zhang XN, Song ZG, Jiang T, Shi BS, Hu YW, Yuan ZH. Rupintrivir is a promising candidate for treating severe cases of Enterovirus-71 infection. WJG 2010;16:201. Available from: https://doi.org/10.3748/wjg.v16.i2.201.

[39] Yao C, Xi C, Hu K, Gao W, Cai X, Qin J, et al. Inhibition of enterovirus 71 replication and viral 3C protease by quercetin. Virol J 2018;15. Available from: https://doi.org/10.1186/s12985-018-1023-6.

[40] Zhang L, Lin D, Kusov Y, Nian Y, Ma Q, Wang J, et al. α-Ketoamides as broad-spectrum inhibitors of coronavirus and enterovirus replication: structure-based design, synthesis, and activity assessment. J Med Chem 2020;63:4562—78. Available from: https://doi.org/10.1021/acs.jmedchem.9b01828.

[41] van der Linden L, Vives-Adrián L, Selisko B, Ferrer-Orta C, Liu X, Lanke K, et al. The RNA template channel of the RNA-dependent RNA polymerase as a target for development of antiviral therapy of multiple genera within a virus family. PLoS Pathog 2015;11:e1004733. Available from: https://doi.org/10.1371/journal.ppat.1004733.

[42] Xu N, Yang J, Zheng B, Zhang Y, Cao Y, Huan C, et al. The pyrimidine analog FNC potently inhibits the replication of multiple enteroviruses. J Virol 2020;94. Available from: https://doi.org/10.1128/jvi.00204-20.

[43] Hung H-C, Chen T-C, Fang M-Y, Yen K-J, Shih S-R, Hsu JT-A, et al. Inhibition of enterovirus 71 replication and the viral 3D polymerase by aurintricarboxylic acid. J Antimicrob Chemother 2010;65:676–83. Available from: https://doi.org/10.1093/jac/dkp502.

[44] Deng C-L, Yeo H, Ye H-Q, Liu S-Q, Shang B-D, Gong P, et al. Inhibition of Enterovirus 71 by Adenosine analog NITD008. J Virol 2014;88:11915–23. Available from: https://doi.org/10.1128/jvi.01207-14.

[45] Kang H, Kim C, Kim D, Song J-H, Choi M, Choi K, et al. Synergistic antiviral activity of gemcitabine and ribavirin against enteroviruses. Antivir Res 2015;124:1–10. Available from: https://doi.org/10.1016/j.antiviral.2015.10.011.

[46] Ye W, Yao M, Dong Y, Ye C, Wang D, Liu H, et al. Remdesivir (GS-5734) impedes enterovirus replication through viral RNA synthesis inhibition. Front Microbiol 2020;11. Available from: https://doi.org/10.3389/fmicb.2020.01105.

[47] LaMarche MJ, Borawski J, Bose A, Capacci-Daniel C, Colvin R, Dennehy M, et al. Anti-hepatitis C virus activity and toxicity of Type III phosphatidylinositol-4-kinase beta inhibitors. Antimicrob Agents Chemother 2012;56:5149–56. Available from: https://doi.org/10.1128/aac.00946-12.

[48] van der Linden L, Wolthers K, van Kuppeveld F. Replication and inhibitors of enteroviruses and parechoviruses. Viruses 2015;7:4529–62. Available from: https://doi.org/10.3390/v7082832.

[49] van der Schaar HM, van der Linden L, Lanke KHW, Strating JRPM, Pürstinger G, de Vries E, et al. Coxsackievirus mutants that can bypass host factor PI4KIIIβ and the need for high levels of PI4P lipids for replication. Cell Res 2012;22:1576–92. Available from: https://doi.org/10.1038/cr.2012.129.

[50] Li X, Liu Y, Wu T, Jin Y, Cheng J, Wan C, et al. The antiviral effect of baicalin on enterovirus 71 in vitro. Viruses 2015;7:4756–71. Available from: https://doi.org/10.3390/v7082841.

[51] Lalani S, Poh CL. Flavonoids as antiviral agents for enterovirus A71 (EV-A71). Viruses 2020;12:184. Available from: https://doi.org/10.3390/v12020184.

[52] Huang S-W, Cheng H-L, Hsieh H-Y, Chang C-L, Tsai H-P, Kuo P-H, et al. Mutations in the non-structural protein region contribute to intra-genotypic evolution of enterovirus 71. J Biomed Sci 2014;21. Available from: https://doi.org/10.1186/1423-0127-21-33.

[53] Chong P, Liu C-C, Chow Y-H, Chou A-H, Klein M. Review of enterovirus 71 vaccines. Clin Infect Dis 2014;60:797–803. Available from: https://doi.org/10.1093/cid/ciu852.

[54] He X, Zhang M, Zhao C, Zheng P, Zhang X, Xu J. From monovalent to multivalent vaccines, the exploration for potential preventive strategies against hand, foot, and mouth disease (HFMD). Virol Sin 2020;36:167–75. Available from: https://doi.org/10.1007/s12250-020-00294-3.

[55] Chou A-H, Liu C-C, Chang J-Y, Jiang R, Hsieh Y-C, Tsao A, et al. Formalin-inactivated EV71 vaccine candidate induced cross-neutralizing antibody against subgenotypes B1, B4, B5 and C4A in adult volunteers. PLoS ONE 2013;8:e79783. Available from: https://doi.org/10.1371/journal.pone.0079783.

[56] Mao Q, Wang Y, Yao X, Bian L, Wu X, Xu M, et al. Coxsackievirus A16. Hum Vaccin Immunother 2013;10:360–7. Available from: https://doi.org/10.4161/hv.27087.

[57] In HJ, Lim H, Lee J-A, Kim HJ, Kim J-W, Hyeon J-Y, et al. An inactivated hand-foot-and-mouth disease vaccine using the enterovirus 71 (C4a) strain isolated from a Korean patient induces a strong immunogenic response in mice. PLoS ONE 2017;12:e0178259. Available from: https://doi.org/10.1371/journal.pone.0178259.

[58] Li M-L, Shih S-R, Tolbert BS, Brewer G. Enterovirus A71 vaccines. Vaccines 2021;9:199. Available from: https://doi.org/10.3390/vaccines9030199.

[59] Yang Z, Gao F, Wang X, Shi L, Zhou Z, Jiang Y, et al. Development and characterization of an enterovirus 71 (EV71) virus-like particles (VLPs) vaccine produced in Pichia pastoris. Hum Vaccin Immunother 2019;16:1602–10. Available from: https://doi.org/10.1080/21645515.2019.1649554.

[60] Wang Z, Zhou C, Gao F, Zhu Q, Jiang Y, Ma X, et al. Preclinical evaluation of recombinant HFMD vaccine based on enterovirus 71 (EV71) virus-like particles (VLP): immunogenicity, efficacy and toxicology. Vaccine 2021;39:4296–305. Available from: https://doi.org/10.1016/j.vaccine.2021.06.031.

[61] Fan S, Liao Y, Jiang G, Jiang L, Wang L, Xu X, et al. Study of integrated protective immunity induced in rhesus macaques by the intradermal administration of a bivalent EV71-CA16 inactivated vaccine. Vaccine 2020;38:2034–44. Available from: https://doi.org/10.1016/j.vaccine.2019.12.057.

[62] Caine E, Fuchs J, Das S, Partidos C, Osorio J. Efficacy of a trivalent hand, foot, and mouth disease vaccine against enterovirus 71 and coxsackieviruses A16 and A6 in mice. Viruses 2015;7:5919–32. Available from: https://doi.org/10.3390/v7112916.

[63] Luo J, Huo C, Qin H, Hu J, Lei L, Pan Z. Chimeric enterovirus 71 virus-like particle displaying conserved coxsackievirus A16 epitopes elicits potent immune responses and protects mice against lethal EV71 and CA16 infection. Vaccine 2021;39:4135–43. Available from: https://doi.org/10.1016/j.vaccine.2021.05.093.

[64] Yang L, Liu Y, Li S, Zhao H, Lin Q, Yu H, et al. A novel inactivated enterovirus 71 vaccine can elicit cross-protective immunity against coxsackievirus A16 in mice. Vaccine 2016;34:5938–45. Available from: https://doi.org/10.1016/j.vaccine.2016.10.018.

[65] Zhao H, Li H-Y, Han J-F, Deng Y-Q, Zhu S-Y, Li X-F, et al. Novel recombinant chimeric virus-like particle is immunogenic and protective against both enterovirus 71 and coxsackievirus A16 in mice. Sci Rep 2015;5. Available from: https://doi.org/10.1038/srep07878.

[66] Sun S, Bian L, Gao F, Du R, Hu Y, Fu Y, et al. A neonatal mouse model of Enterovirus D68 infection induces both interstitial pneumonia and acute flaccid myelitis. Antivir Res 2019;161:108–15. Available from: https://doi.org/10.1016/j.antiviral.2018.11.013.

[67] Zhang C, Zhang X, Dai W, Liu Q, Xiong P, Wang S, et al. A mouse model of enterovirus d68 infection for assessment of the efficacy of inactivated vaccine. Viruses 2018;10:58. Available from: https://doi.org/10.3390/v10020058.

[68] Raychoudhuri A, Naru AK, Kanubothula SR, Uddala R. Development of an experimental inactivated vaccine from vero cell adapted Enterovirus D68. Virus Res 2021;304:198528. Available from: https://doi.org/10.1016/j.virusres.2021.198528.

[69] Dai W, Zhang C, Zhang X, Xiong P, Liu Q, Gong S, et al. A virus-like particle vaccine confers protection against enterovirus D68 lethal challenge in mice. Vaccine 2018;36:653–9. Available from: https://doi.org/10.1016/j.vaccine.2017.12.057.

[70] Zhang C, Zhang X, Zhang W, Dai W, Xie J, Ye L, et al. Enterovirus D68 virus-like particles expressed in *Pichia pastoris* potently induce neutralizing antibody responses and confer protection against lethal viral infection in mice. Emerg Microbes Infect 2018;7:1–22. Available from: https://doi.org/10.1038/s41426-017-0005-x.

[71] Fay E, Langlois R. MicroRNA-attenuated virus vaccines. ncRNA 2018;4:25. Available from: https://doi.org/10.3390/ncrna4040025.

[72] Yee PTI, Tan SH, Ong KC, Tan KO, Wong KT, Hassan SS, et al. Development of live attenuated Enterovirus 71 vaccine strains that confer protection against lethal challenge in mice. Sci Rep 2019;9. Available from: https://doi.org/10.1038/s41598-019-41285-z.

Antiviral agents against flaviviruses

Érica Erlanny S. Rodrigues[1,2],
Ana Beatriz Souza Flor dos Santos[2],
Manuele Figueiredo da Silva[2],
João Xavier de Araújo-Júnior[1,2] and
Edeildo Ferreira da Silva-Júnior[1]

[1]Institute of Chemistry and Biotechnology, Federal University of Alagoas, Maceió, Brazil
[2]Institute of Pharmaceutical Sciences, Federal University of Alagoas, Maceió, Brazil

16.1 Introduction

Infectious diseases include a group of pathologies, caused by microorganisms such as fungi, protozoa, bacteria, and viruses, of which 73% are responsible for emerging zoonotic infections. These can cause acute or chronic conditions, which can lead to lethal cases. These agents are more prevalent in underdeveloped countries, where there are low incentives to health policies and unsatisfactory sanitization conditions [1]. Among these pathogenic agents, viruses have been perpetuated in society since antiquity. Regarding the risk of viral infections and their set of symptoms, these depend on viral characteristics and host innate and acquired resistance [2]. In this context, human immunodeficiency virus, SARS-CoV-2 (also known as COVID-19), Ebola, Chikungunya (CHIKV), Dengue (DENV), Zika (ZIKV), and Yellow Fever (YFV) are examples of viruses responsible for causing high social and public health damages. Among these, DENV, ZIKV, and YFV belong to Flaviviridae family, and these are capable of generating high rates of morbidity and mortality, especially DENV, which is endemic in more than 100 countries, according to the WHO [3,4]. However, there are no specific drugs to treat these infections. Thus, constant investments in drug development are needed to overcome this issue.

In addition, DENV, ZIKV, YFV, West Nile (WNV), and Japanese encephalitis (JEV) viruses belong to the Flaviviridae family, and represent a major threat to human health, causing global impacts, with up to 400 million people infected annually [5,6]. Flaviviruses infect arthropods and mammals, causing human diseases. They are mainly distributed in tropical areas with high population density, which reflects in the number of people susceptible to infections by such viruses. These viruses can cause clinical complications, such as ZIKV

Viral Infections and Antiviral Therapies
DOI: https://doi.org/10.1016/B978-0-323-91814-5.00012-X

which can cause neurological problems, including encephalitis, Guillain-Barré Syndrome, and microcephaly in newborns [7]. However, DENV can establish a more severe infection, causing hemorrhagic fever or dengue shock syndrome (DSS). It is estimated that from 100 to 400 million people worldwide have symptomatic infections very similar to the fever caused by ZIKV [8]. On the other hand, although most cases of WNV infections are asymptomatic, in which one in every 150 infected individuals develops meningoencephalitis, with a 4%–14% mortality rate [9,10]. Besides, JEV is responsible for 10,000 to 15,000 deaths each year, which is a reflection of the aggressiveness associated with Japanese fever [11]. Complementarily, other flaviviruses such as the Hepatitis C virus (HCV) is responsible for almost 170 million infections globally [12].

Concerning all these facts is possible to verify the importance of flaviviruses in the world scenario, even more considering the lack of effective chemotherapeutic agents for controlling and treating them. Hence, there is an unmet need for constant investments in studies addressed to the development of drugs and effective therapeutic alternatives. Thus, this chapter will cover the main points about flaviviruses, such as biological characteristics, mechanism of action, clinical manifestations, and epidemiology, aiming to provide an updated and concise knowledge about this group of viral pathogens.

16.2 Flaviviruses

Flaviviridae family comprises impacting arthropod-borne viruses, including (I) *Flavivirus*, (II) *Hepacivirus*, (III) *Pegivirus*, and (IV) *Pestivirus* genera [13,14]. Moreover, these present similar structural and genomic organizations, although the Flaviviridae members have a distinct set of biological characteristics [14]. The *flavivirus* genus is represented by more than 70 viruses (53 species), including DENV, ZIKV, YFV, JEV, Tick-borne Encephalitis (TBEV), WNV, Kunjin (KUNV), HCV viruses, and others [12,13]. Interestingly, Hepatitis G virus is a unique species from *Pegivirus* genus, which is transmitted to humans, while *Pestivirus* genus infects only animals [14].

The transmission of flaviviruses occurs by different pathways since their infections can be spread by mosquitoes, ticks, vertical transmission (mother to child in pregnant women), or even person-to-person [6,14,15]. In general, infected female mosquitoes from *Aedes* genus (e.g., *Aedes aegypti* and *Aedes albopictus*) are associated with the transmission of DENV, ZIKV, and YFV [11,13], while *Culex* spp., *Culex tritaeniorhynchus*, and *Culex annulirostris* mosquitoes are responsible for transmitting WNV and JEV, respectively. In contrast, TBEV transmission is mediated mainly by bites of infected *Ixodes* spp. ticks (e.g., *Ixodes ricinus* and *Ixodes persulcatus*), while *Haemagogus* mosquitoes have already been associated with YFV transmission [14,16]. In addition to vector-mediated ZIKV transmission [17], it also can be vertically [18–20] and sexually [21] transmitted, as well as, for blood transfusions [22].

Dengue fever is one of the 20 Neglected Tropical Diseases according to the WHO's ranking. In addition, it has also been considered the fastest-spreading mosquito-borne viral disease [23–25]. Infections and endemics caused by DENV occur primarily in Sub- and Tropical countries/territories, involving all four Dengue serotypes (DENV-1, DENV-2, DENV-3, and DENV-4). These serotypes are further divided based on their antigenic characteristic of the envelope (E) protein. Then, infections caused by DENV can be primary or secondary [26]. In general, infections caused by each different serotype can generate clinical symptoms ranging from mild fever (Dengue Fever) or even more severe cases, such as Dengue Hemorrhagic Fever, and DSS [27,28]. Secondary infections, which are considered when an individual is infected by a different serotype after a previous infection, are a risk factor for severe dengue cases [6,29]. Since 2016, WHO has recommended the use of Dengvaxia (or CYD-TDV, by Sanofi Pasteur), a vaccine formed by combining the YFV 17D genes with those of all DENV serotypes, addressed to the immunization of individuals older than 9 years, living in endemic regions. However, the Dengvaxia efficacy against infections associated with DENV-1 and DENV-2 has been discussed. Also, there is a discussion on the higher risk of development of the severe form of the disease in immune-naïve individuals and vaccines who are later infected [30−33].

ZIKV, another arbovirus, was first found in *Rhesus* monkeys (*Macaca mulata*) in 1947, in Uganda in the Zika forest [34]. The first identification of infection in humans occurred in 1952, in Uganda and Nigeria [35]. Although sporadic outbreaks were later reported, the first ZIKV outbreak occurred in Micronesia in 2007 [36,37]. In general, symptoms presented by individuals infected with ZIKV are similar to those developed by patients with DENV and CHIKV, a fact that makes their differential diagnosis difficult [38]. Its symptoms include fever accompanied by polyarthralgia, headaches, conjunctivitis, edema, and maculopapular rash [36,39]. However, it is estimated that ZIKV infection in humans results in low hospitalization numbers, although ZIKV sequelae can be severe [39]. These sequelae were well-evidenced during the ZIKV outbreak in Brazil since then it has been associated with microcephaly in newborns and fetal malformations in pregnant women [40]. Thus, the congenital ZIKA syndrome manifests itself beyond microcephaly, by cerebral palsy, intellectual disability, epilepsy, as well as, visual, auditive, and behavioral disturbances [41,42]. Furthermore, it has been responsible for Guillain-Barré syndrome in adult individuals [40,43]. In addition, ZIKV, TBEV, WNV, and JEV represent neurotropic agents associated with encephalitis [6].

Since its identification, WNV has circulated among birds but has also caused infections in humans, horses, and other mammals. Despite causing sporadic outbreaks, the scenario after the 90 began to transform, neurological conditions began to gain prominence and currently cases associated with WNV emerged in Europe. To date, no vaccine is available for prophylaxis of human WNV infections but three vaccines are available for horses [44]. Overall, it is estimated that 80% of infections caused by WNV are asymptomatic and 20% of individuals can progress to West Nile fever (characterized by headaches, myalgia, fever, rash, lymphadenopathy, and gastrointestinal symptoms) or severe neurological diseases, such as meningitis, encephalitis or flaccid paralysis (similar to polio) [45,46].

JEV is one of the major encephalitis-causing viral diseases in rural and suburban areas with the highest prevalence in many Asian countries [47,48]. Regarding this, around 68,000 cases are estimated every year in Asia. This virus is classified as a single serotype with five distinct genotypes that are responsible most often for mild clinical conditions characterized by fever, headaches, neuralgia, and others. However, when developing severe cases characterized by JEV, meningitis, or paralysis, in which approximately 30% of cases are fatal or still cause significant neurological sequelae [26,47,49]. Interestingly, it is a disease with a prophylactic vaccine available, although JEV is still one of the main causes of encephalitis in the world, having approximately 80% of infection cases occur in areas with vaccination policies [48].

16.2.1 *Flavivirus* genus

Flaviviruses comprise arthropod-borne viruses that present a positive-sense RNA genome with a size of approximately 11 kb, consisting of a single open-reading frame (ORF) and two untranslated regions at both 5' and 3' terminals. Structurally, these viruses appear as small spherical particles ($\sim$50 nm) with a tightly adherent lipid envelope [12,14,26].

In general, flaviviruses have a similarity in their process of infection and replication [15]. Thus, the infection begins with the release of viral particles in the host, as a consequence of its interaction with arthropods that feed on the hosts' blood. The viral particles then infect susceptible host cells through interactions between their E protein and host attachment factors. These interactions are mediated by factors such as the dendritic cell (DC)-specific intercellular adhesion molecule-3-grabbing nonintegrin (DC-SIGN), AXL, TIM-1, Tyro3, and MertK [50]. Fixation is followed by endocytosis by clathrin-dependent mechanisms. In the endosome, a low pH ($\sim$6.0) allows the trimerization of the E protein, resulting in the fusion of cell and viral membranes. This condition allows the decapsulation of the nucleocapsid to occur and then the viral genome is released into the host cell cytoplasm. In the ribosome, positive-stranded RNA is translated into a polyprotein with the endoplasmic reticulum (ER)-localization signal, promoting the association of the ribosome to the ER membrane. The polyprotein is then cleaved into non and structural proteins (discussed below). Finally, viral particles formed are transported to the Golgi apparatus and matured [51] after the action of host furin protease and signalase, as well as, viral NS2B/NS3 protease, and these are then released from the cell via exocytosis [6,15,52−54], as shown in Fig. 16.1.

16.2.2 Flaviviruses proteins

Flaviviruses encode a single ORF that is translated at the ER into a polyprotein [6]. The processing of this polyprotein by viral proteases and host cells results in ten functional proteins, being three structural proteins: Capsid (C), pre-Membrane (prM), and Envelope (E); and seven nonstructural (NS) proteins, named NS1, NS2A, NS2B, NS3, NS4A, NS4B, and NS5. In general, structural proteins are involved in the process of virus entry, assembly, and budding virions from the host cells, while NS is essential for genomic RNA replication, regulation of the host immune pathways to viral infection, and viral maturation [55].

16.2.2.1 Structural proteins

Capsid (C) protein is relatively small and participates in the viral assembly process, in RNA replication and also aids to form the virus envelope and enables the structure of the E protein [26]. By the action of the NS2B-NS3 protease, this protein is cleaved from the viral polyprotein between C and prM proteins. Moreover, E protein is involved in the virus entry and membrane fusion. Several studies investigated molecules capable of interacting with E protein, especially in a binding pocket occupied by a detergent molecule, *n*-octyl-β-*D*-glucoside (β-OG). This pocket is placed between the I and II domains from the E protein [56,57]. prM protein plays an essential role in virion assembly. Basically, prM-E heterodimers are located on the surface of immature particles and, when these reach the Golgi apparatus, the peptide pr is cleaved by furins, releasing the M protein [26,54].

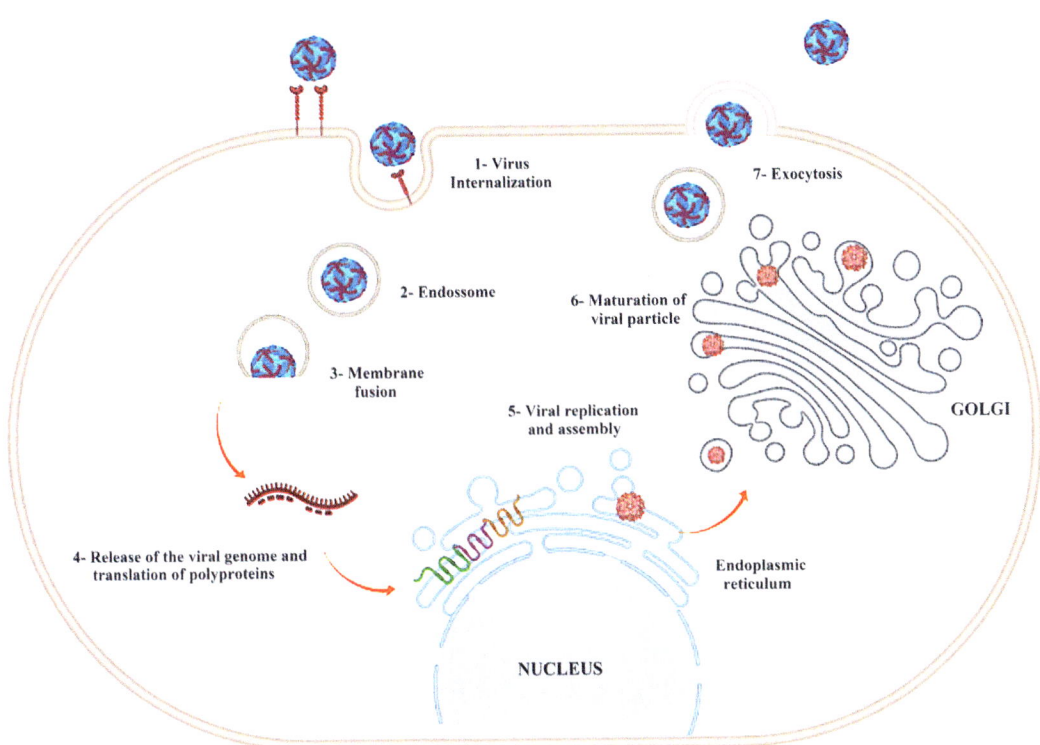

FIGURE 16.1 **Flaviviruses' life cycle.** **(1)** Internalization of viral particles mediated by the interaction between viral E protein and host cell receptors; **(2)** Acidic endosome pH causes trimerization of flavivirus E protein; **(3)** Disassembly of viral particles; **(4)** Release of viral RNA into the cytoplasm. RNA is translated into a polyprotein, which is then processed and cleaved into structural and nonstructural proteins; **(5)** Viral replication and assembly of immature viral particles in the endoplasmic reticulum; **(6)** Viral maturation into the Golgi complex by the action of furin protease; **(7)** Mature viral particles released by exocytosis. Illustration elaborated by using the web server BioRender, available at https://biorender.com.

16.2.2.2 Nonstructural proteins

These proteins are constantly targeted in studies for developing antiviral drugs. NS1 (46 kDa) is essential for the inhibition of complement-mediated immune response and viral replication. Similarly, NS2A (22 kDa) has been associated with participation within the process of replication, virion assembly, and immune system evasion. The helicase activity is attributed to NS2B protease. However, its prominent role occurs as a cofactor of NS3 protease (NS3^{pro}) during viral polyprotein processing. On the other hand, NS3 contains serine protease, RNA helicase, 5′ RNA triphosphatase (RTPase), and Nucleoside 5′ triphosphatase (NTPase) activities [13,55,58,59]. Thus, due to its critical role in viral replication, NS2B-NS3^{pro} has been targeted in several research projects focused on the development of new antiviral agents, in which its inhibition leads to nonfunctional proteins [60]. Effects in the induction of membrane rearrangement and autophagy response to viral infection of host cells have been speculated under NS4A, while the anchoring of the replication complex to the endoplasmic reticulum is attributed to NS4B activity. NS5 is the largest NS protein and since it is not present in the host cell, it has been associated with studies for drug discovery. This protein has an RNA-dependent RNA polymerase (RdRp) domain, as well as, guanylyltransferase, and methyltransferase domains [13,55,58,59].

16.3 Why develop novel antiviral drugs?

Flavivirus-associated infections are marked by recurrent patterns of outbreaks and severe clinical symptoms, such as microcephaly, congenital anomalies, and paralysis. Its wide distribution characterizes a global public health problem not only in developing countries [6]. The increase in the number of cases of infections caused by flaviviruses can be related to the unintentional vectors' transport to nonendemic areas [5]. It is possible due to the climate changes, which are responsible for vector distribution, feeding a new environment adaptation by arthropods, such as mosquitoes and ticks [61]. Furthermore, travels involving natives from endemic areas are one of the main factors for dissemination, as well, through the geographical expansion of vectors [62]. On the other hand, the unavailability of vaccines that can effectively control flaviviruses or their mutations has limited their prophylaxis and/or treatment nowadays [26]. Still, associated with all these factors, the absence of antiviral drugs for most of these infections justifies current efforts to develop drugs capable of fighting against these pathogens.

16.4 Recent advances in inhibitors targeting flaviviruses

16.4.1 Hepatitis C virus

The treatment of chronic infections caused by HCV has been revolutionized by the development of direct-acting antivirals (DAAs) [63]. A great deal of success has been attributed to combination therapies with DAAs, as these have significantly improved the HCV patients' prognosis [63]. DAAs target three major HCV proteins required for viral replication: (I) NS3/4A, (II) NS5A, and (III) NS5B polymerase. HCV NS3/4A protease represents a promising target in the development of pan-genotypic DAAs with an improved resistance profile for the treatment of HCV infections [64]. However, the ineffectiveness of NS3/4A protease inhibitors against all genotypes and preexisting resistance-associated substitutions (RASs) limit their clinical uses. In addition, factors such as RASs at positions A156, R155, and D168 or polymorphism between genotypes are then associated with reduced anti-HCV effects [65].

Approved pan-genotypic HCV NS3/4A protease inhibitors glecaprevir (**1**), voxilaprevir (**2**), and grazoprevir (**3**) (Fig. 16.2) present a tripeptide structure with a macrocyclic ring that is built between the P4 terminus of the molecule and the P2* quinoxaline heterocycle. Linear HCV NS3/4A protease inhibitors without the macrocyclic linker and greater conformational flexibility have already been explored and, although less potent than macrocyclic analogs, these were less susceptible to resistance [64].

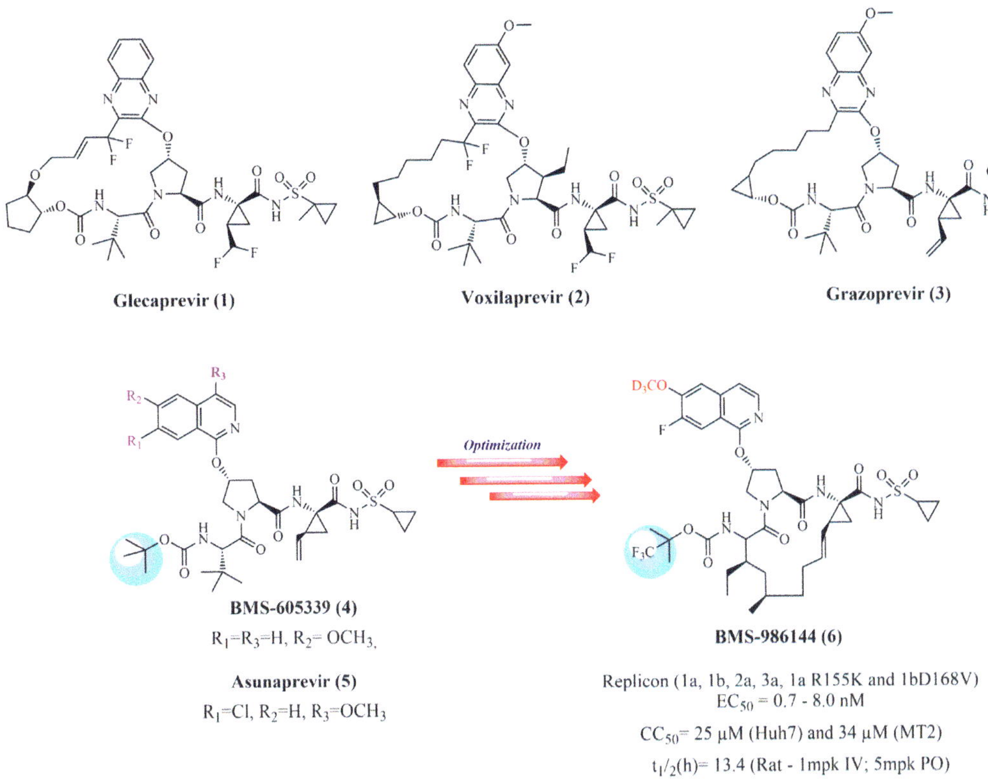

FIGURE 16.2 **Inhibitors of HCV NS3/4A.** Structures of approved drugs glecaprevir, voxilaprevir, grazoprevir, and clinical candidates BMS-605339 (4), asunaprevir (5), and BMS-986144 (6).

With a proposal to extend the antiviral effects of known NS3/4A protease inhibitors BMS-605339 **(4)** and asunaprevir **(5)** (Fig. 16.2) [66,67], new inhibitors were designed by Sun et al. [68]. For this, macrocyclization with a tether linked P1 − P3 subsites was performed, allowing a series of additional structural modifications. An exploration of structure − activity relationship analysis involving 18 new compounds, among them the most promising BMS-986144 **(6)** (Fig. 16.2), exhibited characteristics of a clinical candidate. BMS-986144 **(6)**, a new HCV NS3/4A inhibitor tripeptide features an all-carbon tether linking P1 − P3 subsites of its tripeptide structure, and a deuterated substituent at C-6 position from the isoquinoline ring. GT-1a, GT-1b, GT-2a, and GT-3a, and resistant variants (Arg155Lys and Asp168Val) replicon assays showed that macrocycle BMS-986144 **(6)** exhibited similar effects to acyclic compound **(5)** only on GT-1a and GT-1b. Investigating its preclinical pharmacokinetic properties, BMS-986144 **(6)** did not exhibit adverse clinical effects and significant histopathological alterations in rats. In addition, in rat and rabbit models, it was observed to have minimal risks of finding cardiovascular effects. Several in vitro assays show evidences of CYP3A4 induction and clastogenicity (in vitro micronucleus assay in Chinese hamster ovary cells) or mutagenicity (Ames assay). Furthermore, this pan-genotype is found to be a better inhibitor than acyclic derivative (5), exhibiting lower renal clearance in rats and dogs, while oral bioavailability and C_{24} plasma levels were found elevated [68].

New NS5A inhibitors for composition of oral treatment regimens against HCV have been pursued [69,70]. In recent years, the discovery of MK-4882 (7) (Fig. 16.3) and its cyclically restricted derivative, MK-8742, the latter currently in clinical trials, supported the development of new compounds targeting NS5A [70]. Further studies developed by Nair et al. [71] motivated by the optimization of effects of MK-4882 (7) resulted in new silyl proline-containing HCV NS5A inhibitors. Among them, MK-8325 (8) and analog (9) (Fig. 16.3) demonstrated promising effects with improved pan-genotypic activity, and their rat PK profiles, when compared to MK-4882 (7) [71]. The structure of these derivatives presents silyl and difluoro substituted proline on either side. In this study, MK-8325 (8) and compound (9) exhibited in HCV replicon assays (GT1a, GT1b, GT2a, and GT3a) EC_{50} values of 0.015, 0.001, 0.003, and 0.25 nM, and Rat IV $T_{1/2}$ of 7.6 hours for compound MK-8325 (8), while compound (9) exhibited EC_{50} values of 0.009, 0.001, 0.002, and 0.2 nM, and Rat IV $T_{1/2}$ of 6.7 hours [71]. A more detailed description of MK-8325 (8) and compound (9) was further developed by Nair et al. [72]. In this, replicon assays corroborated their previous findings, although potencies of these compounds against GT2b encoding the M31 variant were reduced. Moreover, these derivatives exhibited a better overall PK profiles in rats, dogs, and monkeys compared to MK-4882 (7). Additionally, in vitro and in vivo tests have shown that MK-8325 (8) is orally bioavailable with adequate pharmacokinetics for once-daily dosing in patients [72].

Sofosbuvir (Sovaldi) (10) (Fig. 16.4), the unique FDA- and EU-approved DAA nucleoside inhibitor that targets NS5B polymerase, is used to treat infections caused by HCV genotypes 1−4. This prodrug of nucleoside 5′-phosphate phosphoramidate 2′-deoxy-2′-α-fluoro-2′-β-methyl uridine has its antiviral effects attributed to 5′-phosphoramidate, as this makes 5′ possible -phosphorylation of uridine nucleosides in hepatocytes, becoming active in its 5′-triphosphate form [73,74]. In the search for new active dihalogenated nucleoside inhibitors, a prodrug of 2′-deoxy-2′-α-bromo-2′-β-chloro uridine 5′-phosphoramidate was identified. Compared to sofosbuvir, compound (11) (Fig. 16.4) exhibited a potent effect against HCV

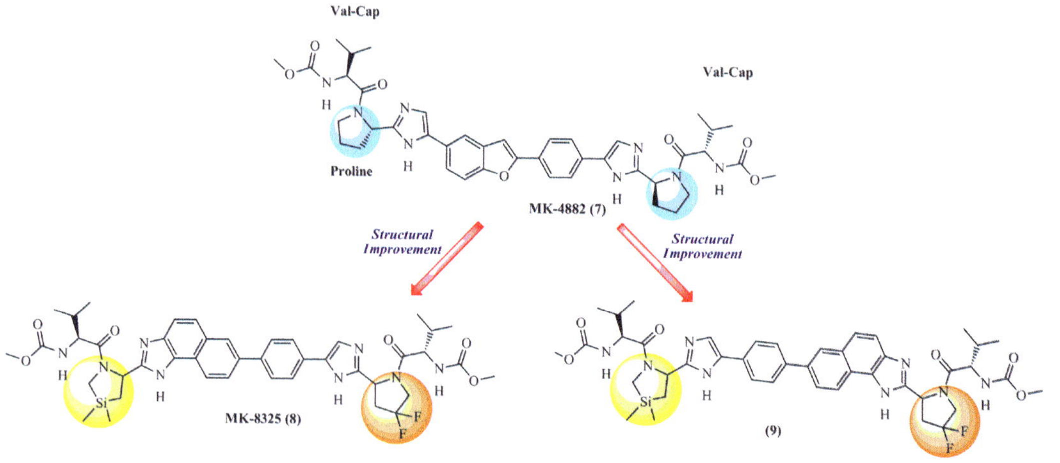

FIGURE 16.3   **HCV NS5A inhibitors.** Structure of MK-4882 and new silyl proline derivatives.

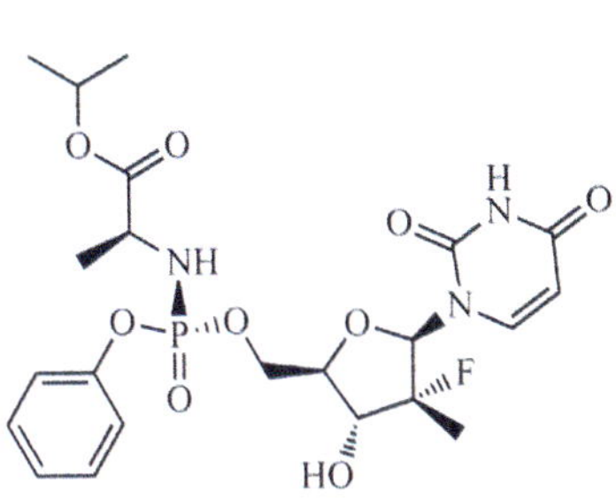

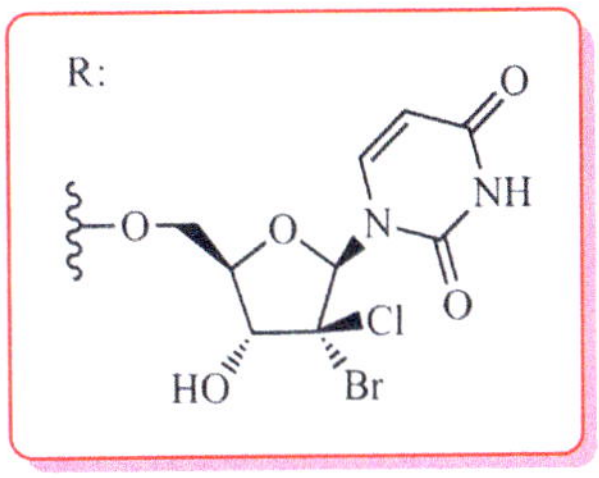

FIGURE 16.4 Sofosbuvir, a prodrug inhibitor of HCV NS5B and analogs.

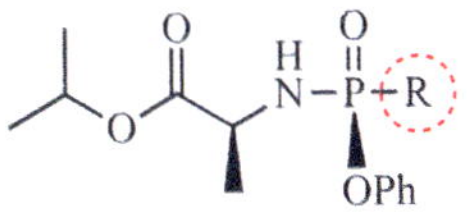

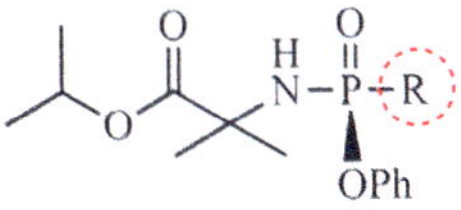

genotype 1. However, tests in dogs (exposed to 5 mg of both compounds) showed that the active triphosphate concentration of compound (11) measured after liver biopsy w lower than sofosbuvir (10). These observations indicated that the clinical dose of compound (11) should be higher than that for sofosbuvir (400 mg/day). However, compared to sofosbuvir, compound (11) exhibited better permeability in PAMPA assay. Due to the limitation of compound (11), modifications of its 5'-phosphoramidate prodrug portion were performed and led to the identification of aminoisobutyric acid ethyl ester (AIBEE) 5'-phosphoramidate, compound (12) (Fig. 16.4). This new prodrug exhibited EC_{50} values of 0.96 and 1.17 μM in replicon assays for GT-1a and GT-1b, respectively. In addition to being permeable in PAMPA assays (value of 2.36), the analysis of the concentration of its triphosphate form in dog liver was superior to the triphosphate produced by sofosbuvir and 100-fold greater than that produced by its precursor compound (11) [75].

16.4.2 Dengue and Zika viruses

Niclosamide (13) (Fig. 16.5) is an oral anthelmintic drug included in the World Health Organization's Essential Medicines List [76]. Current advances have shown that it exhibits effects on diseases such as cancer, metabolic diseases, bacterial, and viral infections, among others. In the antiviral field, effects against Severe Acute Respiratory Syndrome Coronavirus (SARS-CoV) and SARS-CoV-2, CHIKV [77], and ZIKV [78] have been also reported. However, previous studies have shown that niclosamide (13) interferes with the process of viral entry and endosomal acidification. In contrast, its limited effect against ZIKV has been attributed to its pharmacokinetic properties.

Notwithstanding the antiviral effects of niclosamide (13), a small library of analogs was investigated against. In vitro and in vivo assays have shown that JMX0207 (14) (Fig. 16.5) has promising antiviral effects and improved pharmacokinetic profile. SLC-based NS2B − NS3 interaction (DENV-2), NS2B − NS3 protease (DENV-2) and cell-based antiviral assays demonstrated that JMX0207 (14) exhibited antiviral effects superior to those observed for niclosamide (13). In addition, JMX0207 (14) effectively inhibited ZIKV protein expression and RNA synthesis in human-induced pluripotent stem cell-derived neural progenitors and also significantly reduced ZIKV infection in an organoid miniature model.

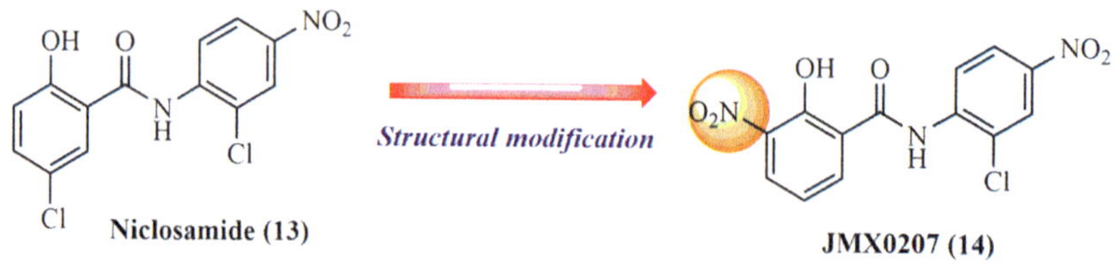

FIGURE 16.5   JMX0207, a new niclosamide-based antiviral agent against DENV and ZIKV.

Moreover, it did present toxic effects on 3D-brain derived from pluripotent neural stem cells at $1.5\,\mu M$ concentration. Its pharmacokinetic profile after oral administration at $40\,mg/kg$ in B6 mice was excellent, unlike niclosamide (13) which exhibited low C_{max} and short $T_{1/2}$. Furthermore, JMX0207 (14) did not exhibit any toxicity in rats and when treated with $20\,mg/kg/day$, while a reduction in viremia was observed [79].

Yao et al. [80] screened pyrazine analogs, in which they identified some potential inhibitors against NS2B-NS3^{Pro}. Among them, compound (15) (Fig. 16.6) exhibited a promising antiviral effect against ZIKV. This molecule showed activity against ZIKV NS2B-NS3 inhibition, with an IC_{50} value of $0.20 \pm 0.01\,\mu M$. Furthermore, this compound also showed inhibition against WNV NS2B-NS3 with DENV-2 NS2B-NS3 an IC_{50} value of $0.59 \pm 0.02\,\mu M$; DENV-3 NS2B-NS3 with an IC_{50} value of $0.52 \pm 0.06\,\mu M$, and WNV NS2B-NS3 with an IC_{50} value of $0.78 \pm 0.02\,\mu M$, showing that this compound has active inhibition with high selectivity against flavivirus proteases. X-ray studies of DENV-2 NS2B-NS3 showed that the surface of NS3 is able to remodel itself and, consequently, create a deep pocket in the L-form. This pocket is responsible to accommodate inhibitors. Furthermore, the interactions between the protein and the inhibitors were also elucidated, in which it was possible to verify that the pyrazine ring and the furanylphenyl groups of the compound (15) are related to the pocket. Substituents R1 and R2 are mainly responsible for hydrophobic interactions, and positively charged NH_2 performs electrostatic and hydrogen-bonding interactions with Asp^{75} residue present at the catalytic triad of this protease. It is noteworthy that the similarity for DENV-2 and ZIKV NS2B-NS3 led the authors to suggest that compound (15) binds at ZIKV NS2B-NS3 similarly to DENV-2 NS2B-NS3. Still, U87 glioma cells were used to evaluate the activity of compound (15) against ZIKV. From these tests, it is found that compound (15) reduced the ZIKV RNA. Compound (15), at $5\,\mu M$ concentration, inhibited 97% of DENV-2 replication on Vero cells. Furthermore, this compound also inhibited NS3 and NS5 proteins [80,81]. Additionally, in vivo evaluation of compound (15) against ZIKV was performed in mice. These analyses showed that administration of compound (15) at doses of $15\,mg/kg$ every 12/24 hours was able to reduce ZIKV RNA copies by 96 and 98%. Furthermore, administration of 30 and $20\,mg/kg$ *per* day within 3 days increased the survival of ZIKV-infected mice. Lastly, the authors reported that this compound can effectively inhibit ZIKV replication [80].

Li et al. [81] evaluated the antiviral profile of cyanohydrazones and their ability of binding to E protein. Among the cyanohydrazones analyzed, compound 3−110−22 (16) (Fig. 16.7) exhibited a significant antiviral activity against JEV, WNV, and ZIKV, verified from in vitro assays. Thus, compound 3−110−22 (16) has been shown to be a fusion inhibitor in liposome tests. However, further investigations were performed for compound

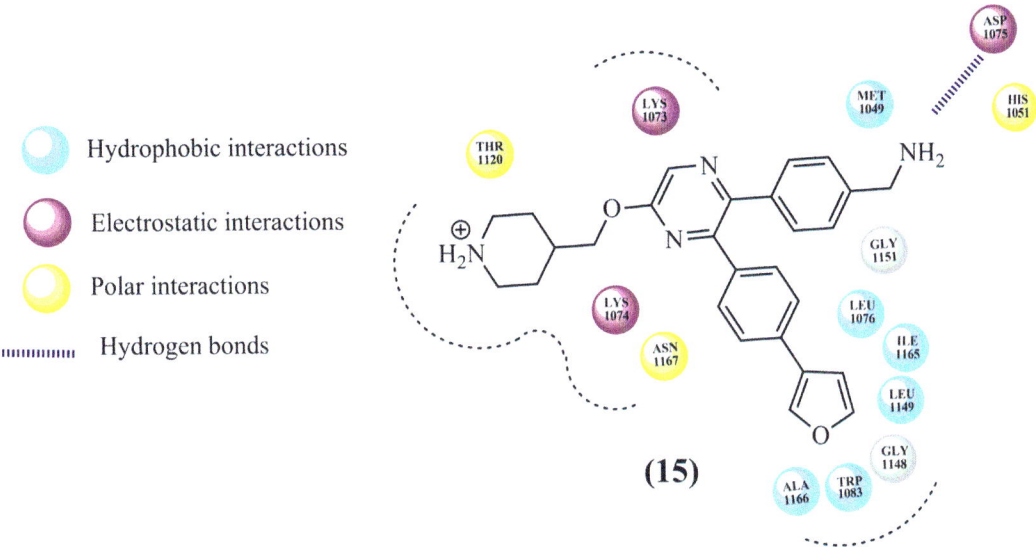

FIGURE 16.6 A novel pyrazine-based NS2B-NS3 inhibitor of DENV and ZIKV. Molecular docking for compound **(15)** shows its main interactions.

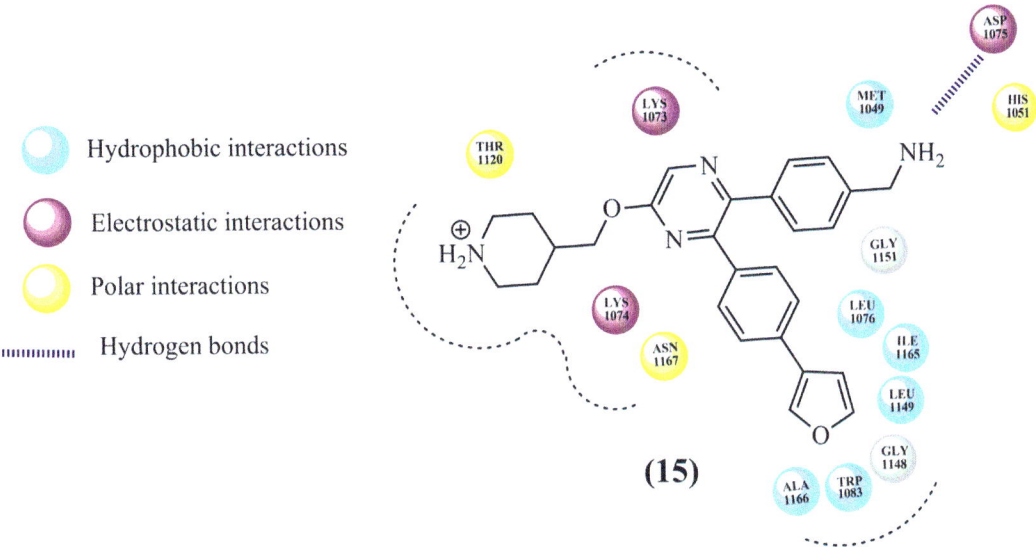

3-110-22 (16)   JBJ-01-162-04 (17)

FIGURE 16.7 Cyanohydrazones as fusion inhibitors targeting DENV, ZIKV, and other flaviviruses.

3−110−22 **(16)**, in which the authors noticed that it had low stability against in vitro mouse microsomes, having a $T_{1/2}$ value of 13.1 min. Furthermore, compound 3−110−22 **(16)** exhibited acute toxicity in mice. Finally, optimizations performed on structure 3−110−22 **(16)** resulted in the discovery of JBJ-01−162−04 **(17)** (Fig. 16.7) [81].

Compound JBJ-01−162−04 **(17)** had a $T_{1/2}$ value of 64.8 min, and activity against DENV with an IC_{90} value of $1.5 \pm 0.4\,\mu M$, in addition to reduced cell cytotoxicity. Furthermore, JBJ-01−162−04 **(17)** has a CC_{50} value of $49.1 \pm 2.6\,\mu M$, which indicates that its antiviral activity is not associated with general cytotoxicity. Also, compound JBJ-01−162−04 **(17)** was found to have an equilibrium dissociation constant (K_D) of $1.4 \pm 0.5\,\mu M$ against the interaction of DENV-2 E (sE2) and E protein, in addition, and viral inhibition obtained by liposome assays. In vivo assays involving JBJ-01−162−04 **(17)** demonstrated that it inhibits DENV-2 S221 strain with an IC_{90} value of $1.4 \pm 0.6\,\mu M$. Furthermore, when analyzing supernatants collected at 48 and 72 hours after infection, it was possible to observe that JBJ-01−162−04

(17) reduced viral titers by more than 95%, at 5 µM concentration. Its pharmacokinetic evaluation showed that intraperitoneal administration is the best pathway for it. Additionally, administration of 40 mg/kg twice a day managed to maintain its plasma concentration with IC_{90} greater than 1.4 µM. It was noteworthy that the administration in mice started one day before inoculation resulted in a reduction on the 3rd day postinfection. The authors reported that the modest effect of this compound may be associated with a high level of plasma protein binding. Thus, they suggested that studies focusing on the reduction of JBJ-01–162–04 **(17)** plasma protein binding could promote better results [81].

16.4.3 Japanese encephalitis virus

E protein of flaviviruses is an antiparallel dimer of inducer-binding domain. Moreover, domain II (DII) is placed into a cavity that is formed by DI and DIII in the opposite subunit, which has a functional relevance to the viral cycle. However, for JEV, the dimer is mainly considered by loop fusion-loop pocket interaction and DII-DII is deficient in contact. Thus, residues W101 and F108 were identified as important for fusion-loop functionality. Based on this, a recent work performed by Ye et al. [82] explored the pocket of JEV-based fusion loop to identify active compounds against this flavivirus [82]. Initially, a pharmacophore model based on the structures of W101 and F108 residues from the JEV (PDBID:5YWP) was created and used to screen 1000 compounds from the ChemDiv database. According to the free binding energy assessed by AutoDock 4.2, only five promising compounds were chosen for in vitro testing. These initial analyses led to the discovery of ChemDiv-3 **(18)** (Fig. 16.8), responsible for reducing the viral titer on BHK-21 cells, at 20 µM concentration [82].

From the qRT-PCR analysis of JEV-infected BHK-21 cells, it was seen that ChemDiv-3 **(18)** significantly reduced RNA on cells and virus particles in the supernatant, reducing the PFU/mL. In vivo assays showed that the intraperitoneal administration of ChemDiv-3 **(18)** in mice after 2 days of infection produced a survival of 3 days longer when compared to the control group. These results suggested that ChemDiv-3 **(18)** can provide certain protection for these animals. However, an improvement in the antiviral activity should be explored more [82]. In addition, molecular docking revealed hydrogen-bonding interactions between ChemDiv-3 **(18)** and Ala[315], Asn[313], Pro[314], and Val[323] residues from

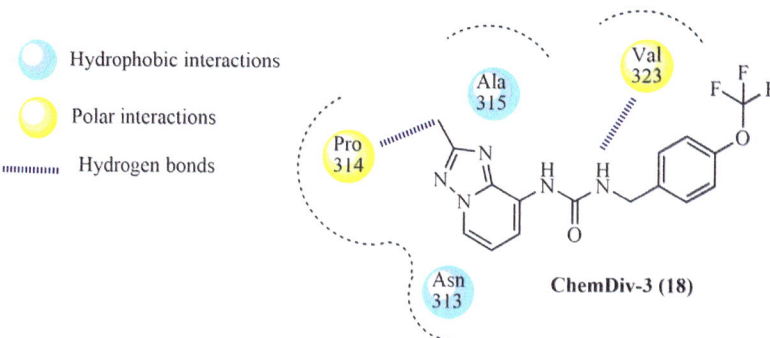

FIGURE 16.8   ChemDiv-3, a new in vitro and in vivo anti-JEV agent.

JEV E protein. A comparison of flavivirus fusion-loop pockets demonstrated a similarity between ZIKV, DENV-2, and JEV. Therefore, the fusion-loop pocket, which is formed by Lys ^{312}, Thr ^{317}, and Ala ^{513} and which may is a path in the process of developing antiviral therapies for flaviviruses [82].

Recent work has identified new JEV E protein inhibitors using phage display screening. For this, a screening of a phage 12-peptide library against E protein of JEV evidenced five promising peptide sequences [P1 (TPDCTRWWCPLT); P2 (YPICTDTLCRLS); P3 (RQRTPPTRTIRS); P4 (RPRTRLHTHRNR); and P5 (TQTTKSISTNRI)]. Then, these were identified, synthesized, and screened for their antiviral potential. These peptides did not show toxicity on BHK-21 cells up to 500 μM concentration. Plaque formation assays and qRT-PCR showed that these peptides were able to inhibit JEV replication on BHK-21 cells. However, P1 had best antiviral effects, being further investigated. Thus, P1 was able to inhibit JEV N28 infection on BHK-21 cells with an IC_{50} value of 35.9 μM. Moreover, it reduced JEV-induced cytopathic effects and inhibited JEV-infected Vero cells (SA14). P1 effects were more significant when it was added before or during the infection on BHK-21 cells. Thus, this peptide reduced 27.2% $\pm$ 4.6% and 38.6% $\pm$ 3.7% of JEV replication before and during the infection, respectively. Therefore, the authors stated that P1 is associated with the inhibition of JEV infection during the onset of infection. BiFC assays were performed to confirm the interaction between P1 and JEV E protein. From these assays, it was possible to verify that P1 and JEV E protein interact directly, which is a crucial factor for the in vitro inhibition of JEV infection [83].

In vivo tests performed with C57BL/6 mice showed that P1 did not exhibit toxic effects at 5.0 mg/kg dose. It was shown that pre-incubation of JEV followed by treatment with P1 resulted in a higher survival rate compared to the control group treated with another peptide. However, the intraperitoneal injection of P1 after infection with JEV was not effective. Furthermore, viremia was reduced in treated mice. Finally, histopathological evaluations of mice brains revealed that P1 does not exhibit signs of alterations associated with JEV, in preincubation experiments [83].

16.4.4 West Nile virus

WNV E protein has three distinct domains (DI, DII, and DIII), in which DIII has great relevance in the mediation of the connection between the virus and cellular receptor [84]. Therefore, Mertinková et al. [85] investigated 7-*mer* cyclic or 12-*mer* linear peptides that were able to block the interaction between human brain microvascular endothelial cells (hBMECs) and WNV E protein DIII. ELISA assays showed that all 7-*mer* cyclic peptides [CP2 (CTKTDVHFC); CP4 (CIHSSTRAC); CP6 (CMQTQRAHC); CP12 (CTYENHRTC); CP15 (CDPRHSKFC); and CP16 (CLAQSHPLC)] had binding potential with DIII, since they had absorbance > 1.9. However, among 12-*mer* linear peptides (LP3 and LP19), only LP3 could bind DIII with $A_{450} = 2.564$ [85].

Thus, peptides that showed potential binding to DIII were submitted to new ELISA tests, now to assess their ability to block the binding between hBMECs and recombinant DIII (rDIII). Among the analyzed peptides, CP2 showed better inhibition with 89% and $A_{450} = 0.117$. Furthermore, CP2, CP4, CP12, CP15, CP16, and CP6 showed satisfactory results with inhibitions ranging from 47% to 67%. Then, western blot was applied to analyze their ability to block the binding interaction between low molecular weight receptors (∼15 kDa) of hBMECs and rDIII. For this, assays performed with the preincubation at 2.5 µg (0.19 nM) of rDIII along with 1.9 nM of each peptide, showed that CP2, CP4, CP12, and C16 had significative inhibitions. These compounds were further investigated by immunocytochemical tests, confirming that they were able to preblock rDIII in endothelial cells, at 3.9 nM concentration [4]. Finally, CP2, CP4, CP12, and C16 were also submitted to neutralization tests, where it was possible to verify the blocking of the infection of each one. The results showed that CP2 had better infection neutralization with 2.43 nM, followed by CP12 (10.4 nM), CP16 (14.8 nM), and CP4 (17 nM). Furthermore, these peptides showed good cytotoxic results, since none of them showed toxicity in eukaryotic cells, at 6 µM concentration [4]. Thus, Mertinková et al. [85] suggested that CP2, CP4, CP12, and C16 can be considered models in the development of new antivirals that can inhibit the entry of neuroinvasion promoted by WNV [85].

Brai et al. [86] investigated a new group of selective inhibitors of the human adenosine-triphosphatase/RNA helicase X-linked DEAD-box polypeptide 3 (DDX3X), targeting WNV infection [86]. A virtual screening based on homology modeling identified compounds (19) and (20) (Fig. 16.9) as the most promising DDX3X inhibitors. Additional molecular docking simulations with hybrid compounds, built from fragments of compounds (19) (triazole) and (20) (sulfonamide) resulted in the identification of derivatives capable of interacting with the most important amino acid residues of DDX3X (Arg[276], Pro[274], and Arg[480]) [86]. Enzymatic assays under DDX3X, antiviral assays in Huh-7 cells, and analysis of their pharmacokinetic parameters resulted in obtaining a more potent compound (21) (Fig. 16.9), exhibiting an EC_{50} value of 2.3 µM (Fig. 16.7). Furthermore, the time

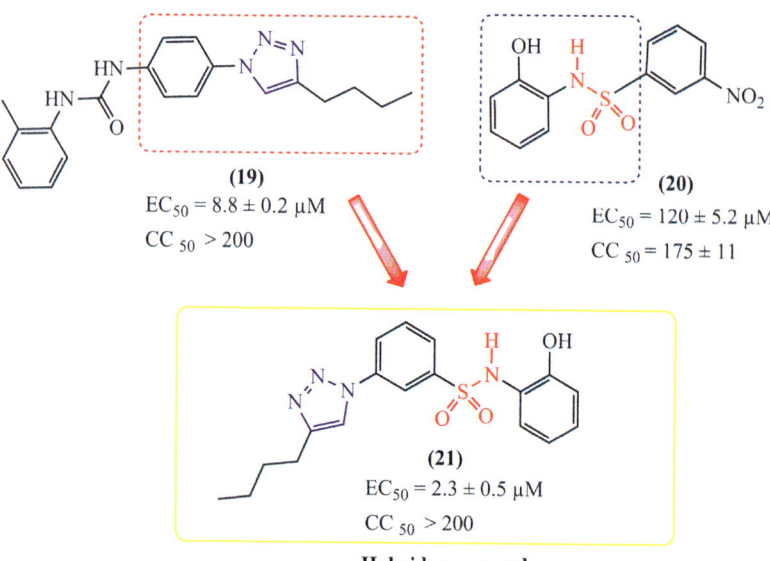

FIGURE 16.9 New DDX3X inhibitors as anti-WNV agents.

of drug addition experiments showed that these molecules act, probably, in the first stages of WNV replication [86].

Dragoni et al. [87] determined the antiviral activity of sofosbuvir **(10)** (see Fig. 16.4) against WNV, using plate assay and immunodetection assay in human cell lines and by enzymatic RdRp assay [87]. The plate assay showed that sofosbuvir **(10)** exhibited significant anti-WNV effects with IC_{50} values of 1.2 ± 0.3, 5.3 ± 0.9, 7.8 ± 2.5, and 63.4 ± 14.1 μM on Huh-7, neuronal U87, LN-18, and A549 cells, respectively. On the other hand, immunodetection assay results revealed the anti-WNV activity, with IC_{50} values of 1.7 ± 0.5 and 7.3 ± 2.0 μM on Huh-7 and U87 cells [87].

In addition, molecular docking was performed for sofosbuvir **(10)**, docked in both wild models for WNV RdRp and its S604T and M479K/A483G/L721M WNV RdRp variants. Thus, docking results evidenced that the bioactive triphosphate form of sofosbuvir **(10)** binds to the wild-type WNV RdRp catalytic site by pairing the RNA template uracil in a wobble-like conformation. It was possible to identify that the hydroxyl group of this antiviral establishes two hydrogen-bonding interactions with Asp^{541} and Asn^{613}, while the phosphate groups interact with Arg^{474} and coordinated with catalytic ions of Mg^{2+} [87]. Therefore, the set of findings in molecular docking suggests that mutations may occur within the catalytic site, which may affect pharmacophoric characteristics and the general form of this protein. As a consequence of this, the binding mode of sofosbuvir **(10)** may be impaired, especially about base-pairing with uracil from the RNA template [87].

Konkolova et al. [88] conducted studies involving remdesivir **(22)** (Fig. 16.10), an antiviral known for its inhibitory activity against different coronaviruses and Ebola RdRp [88]. The active triphosphate form of remdesivir, compound **(23)** (Fig. 16.10), was tested directly in an in vitro polymerase assay and an alkaline phosphatase-coupled polymerase assay (FAPA). The results revealed that remdesivir **(22)** can inhibit all flaviviral polymerases (DENV-3, TBEV, JEV, WNV, ZIKV, and YFV), with IC_{50} values ranging from 0.26 to 2.2 μM, being the best result against YFV polymerase [88].

16.4.5 Tick-born encephalitis virus

Akaberi et al. [89] predicted the antiviral activity of the compound **(24)** (Fig. 16.11), a potential tripeptide inhibitor of TBEV NS2B-NS3 protease [89]. Compound **(24)** was analyzed by in silico approaches involving molecular docking and molecular dynamics

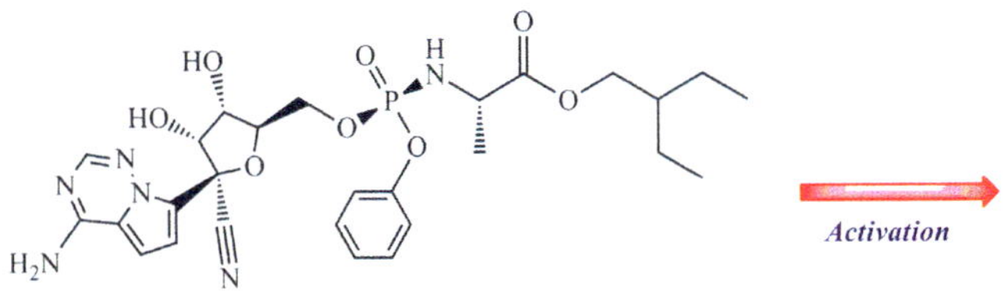

Remdesivir **(22)**
Pro-drug

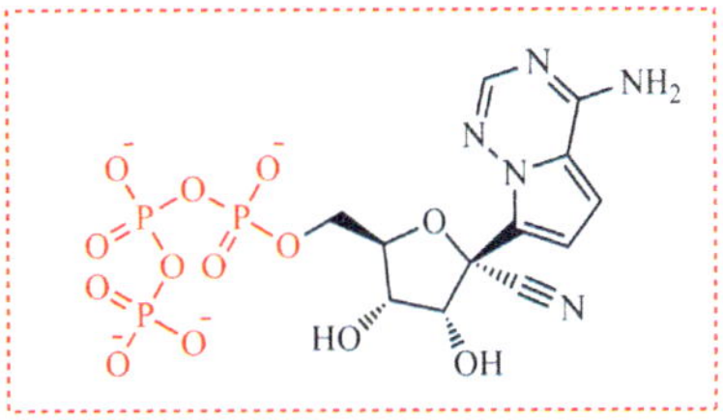

Remdesivir triphosphate **(23)**
Active triphosphate form

FIGURE 16.10 Remdesevir, a prodrug polymerase inhibitor (NS5) of flaviviruses.

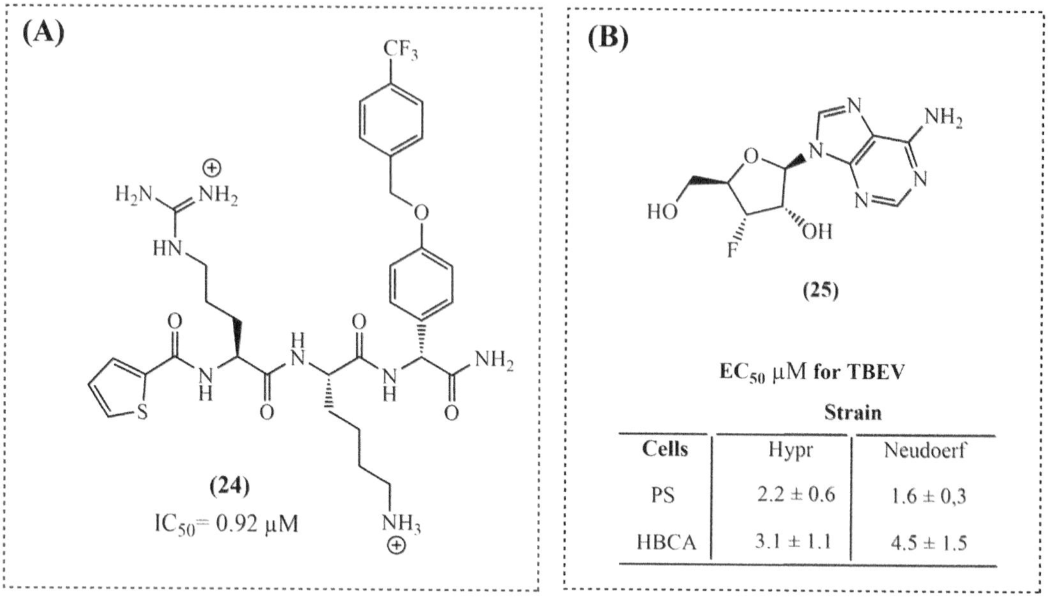

EC_{50} μM for TBEV

Cells	Strain: Hypr	Neudoerf
PS	2.2 ± 0.6	1.6 ± 0,3
HBCA	3.1 ± 1.1	4.5 ± 1.5

FIGURE 16.11 New agents against TBEV. (A) Tripeptide NS2B-NS3pro inhibitor of TBEV. (B) Nucleoside derivative active against TBEV and other flaviviruses.

simulations. Molecular docking results demonstrated that compound **(24)** binds to the active site of the TBEV NS2B-NS3 serine protease, which showed a free binding energy of −6.91 Kcal/mol. Compound **(24)** remained stable in the complex with TBEV protease, during 50 ns of MD simulations. FRET-based enzymatic assay revealed that compound **(24)** inhibited the TBEV protease with an IC_{50} value of 0.92 μM, being the first TBEV protease inhibitor described to date [89].

Eyer et al. [90] analyzed the antiviral activity of 28 nucleoside analogs that have a fluorine substituent at various positions at the ribose ring. However, most of these compounds were toxic or had no antiviral effects. Thus, among them, only 3′-deoxy-3′-fluoroadenosine **(25)** (Fig. 16.11) had a relevant antiviral effect [90].

The antiviral activity of 3′-deoxy-3′-fluoroadenosine **(25)** was evaluated against ZIKV (strains MR-766 and Paraiba_01), WNV (strains Eg-101 and 13–104), and TBEV (strains Hypr and Neudoerf). To analyze the antiviral effect of this compound, two cell lines, PS and HBCA were used. In TBEV, the treatment was divided into a pretreatment (24 hours before), simultaneous treatment, and posttreatment (2 hours later), with a strong effect in all treatments. However, the pretreatment was found to be the best one since concentrations above 10 μM of 3′-deoxy-3′-fluoroadenosine **(25)** were able to inhibit 100% of viral replication. The Anti-TBEV effects on PS cells were observed with an EC_{50} value of 2.2 ± 0.6 μM for the Hypr strain, whille 1.6 ± 0.3 μM for the Neudoerf strain. These authors also conducted studies in mice, which allowed them to assess the toxicity exhibited by 3′-deoxy-3′-fluoroadenosine after administration of 25 mg/kg twice a day for 6 days intraperitoneally. At first, it was possible to identify mild side effects that disappeared after discontinuing the administration of this compound. Then, it was possible to verify that infected mice with WNV and treated with

3'-deoxy-3'-fluoroadenosine had protection from the disease with a survival rate of 70%. Furthermore, no clinical signs of infection were observed in 70% of treated mice, while more than 30% of mice had prolonged survival. For mice infected with TBEV, this treatment promoted a deceleration of clinical symptoms related to neuroinfection, in addition to an increase in survival time, factors that should be considered, even though all mice died [90].

16.4.6 Yellow fever virus

The search for effective antivirals against YFV is a reality among many research groups. Thus, Mendes et al. [91] carried out a study that aimed to identify an effective antiviral against YFV, based on drug repurposing [91]. For this, DAAs, drugs known for their role in inhibiting HCV, were investigated against infections by flaviviruses, including sofosbuvir (10) (see Fig. 16.4), a nucleotide triphosphate prodrug, which acts as an inhibitor of HCV NS5B RdRp [92].

Sofosbuvir (10) underwent a reliable and rapid cell-based screening assay using Vero or Huh-7 cells. The results showed that sofosbuvir (10) alone presented an EC_{50} value of $0.4 \pm 0.1\,\mu M$ and $SI > 2,56.6$. When it was combined with ledipasvir (26) (Fig. 16.12), it was proved to be even more effective in inhibiting YFV replication, when compared to other DAAs, presenting an EC_{50} value of $0.36 \pm 0.03\,\mu M$. In addition, sofosbuvir (10) demonstrated superior results in the YFV high-content screening compared to Interferon-α-2a (IFN-α2a). Therefore, from a 16-point dose—response assay, and experimental plates, sofosbuvir (10) was found to be the most potent and effective compound, exhibiting a high index of selectivity against Huh-7 cells [91].

Based on its promising results from in vitro studies, aiming to evaluate the effectiveness of sofosbuvir (10) in humans, two patients received a single daily dose of 400 mg of this drug, orally, in an off-label treatment against YFV. Consequently, after one week of treatment, there was a reduction in viremia and an improvement in the clinical status associated with YFV infection in both patients. These effects were also observed in tests with neonatal Swiss mice, which confirmed the efficiency of this antiviral in controlling YFV viremia in vivo [91].

Srivastava et al. [93] proposed the design of benzimidazole derivatives, containing substituents at 6, 5, and N-1 positions, and then were tested for their anti-YFV properties [93]. Among the newly investigated benzimidazole derivatives, compound (27) (Fig. 16.12) had

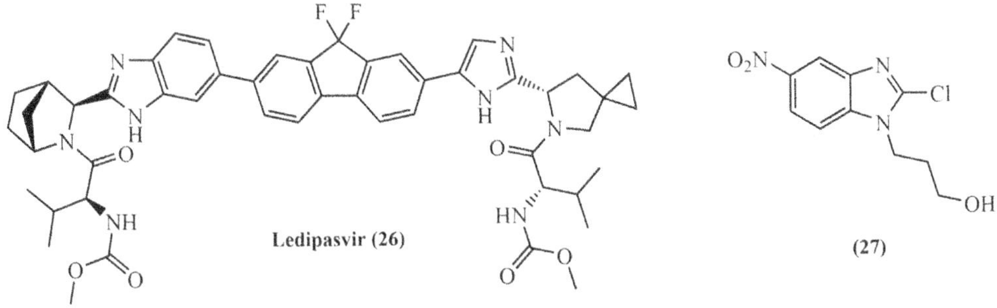

FIGURE 16.12 Structures of lepidasvir and novel benzimidazole derivative as agents against YFV.

promising anti-YFV effects. This compound showed inhibitory activity against YFV replication on Vero cells, with an EC_{50} value of 0.007824 µM and an SI of 50 [93].

Gawrijuk et al. [94] built pharmacophore models capable of predicting drug candidates against YFV to select the best ones for in vitro testing. For that, they used several machine learning methods, intending to create a compound library. In total, 7865 compounds were virtually screened for their activities. Among them, 142 compounds were found to be active [94]. After screening, only five compounds remained (Fig. 16.13) suitable for the next stages of the study and then were submitted for in vitro testing. These molecules were chosen according to some criteria, such as FDA-approved drug: paroxetine **(28)** (Fig. 16.13); molecules with previously reported antiviral activity ezetimibe **(29)**, and SB202190 **(30)** (Fig. 16.13); a randomly selected compound cinoxacin **(31)** (Fig. 16.13); and a new synthetic compound 11626003 **(32)** (Fig. 16.13) [94].

It can be noted that the five molecules chosen share similar chemical groups or structures, such as sulfonamides or 1,3-benzodioxoles. However, these compounds do not share the same pharmacophore, they present active fragments, as established by the algorithms used in the computational search. Additionally, pyrazolesulfonamide (11626003) **(32)** was the most promising, with an EC_{50} value of 3.2 µM. However, it was shown to be slightly cytotoxic, with CC_{50} values of 24 and 11 µM for visual and neutral red assays, respectively. In this context, compound 11626003 **(32)** may be a new scaffold for further

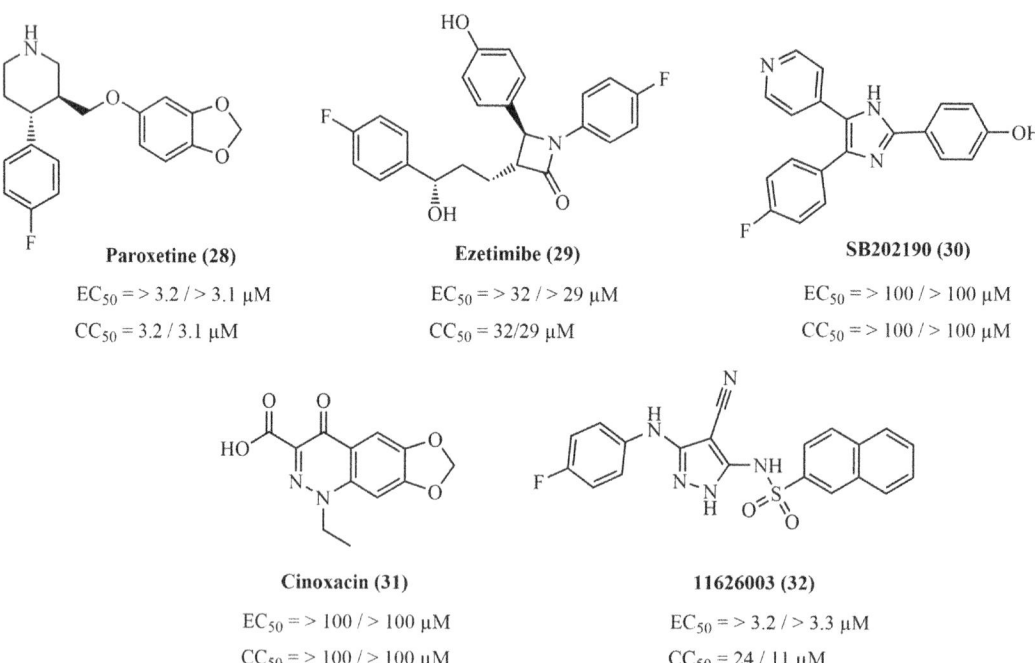

Paroxetine (28)

EC_{50} = > 3.2 / > 3.1 µM

CC_{50} = 3.2 / 3.1 µM

Ezetimibe (29)

EC_{50} = > 32 / > 29 µM

CC_{50} = 32/29 µM

SB202190 (30)

EC_{50} = > 100 / > 100 µM

CC_{50} = > 100 / > 100 µM

Cinoxacin (31)

EC_{50} = > 100 / > 100 µM

CC_{50} = > 100 / > 100 µM

11626003 (32)

EC_{50} = > 3.2 / > 3.3 µM

CC_{50} = 24 / 11 µM

FIGURE 16.13 Drugs repositioned as new scaffolds against YFV, showing EC_{50} and CC_{50} values for visual and neutral red assays.

hit-to-lead optimization studies, which can actively contribute to the discovery of candidates against YFV [94].

16.5 Conclusion

Even several decades after the discovery of flaviviruses as causative agents of a variety of human diseases, infection control has faced numerous challenges. Although the development of vaccines has made progress, limitations associated with prophylactic agents have contributed to viral dissemination or worsening of clinical conditions, as in the case of Dengvaxia. In this sense, despite efforts to develop drugs, the current evolution of flaviviruses continues to be marked by recurrent outbreaks, high prevalence, and the appearance of new severe symptoms, such as those associated with ZIKV. The current therapeutic arsenal shows a more promising scenario for the treatment of HCV, in which there are already approved drugs and, in which DAADs have been the most explored agents recently. However, other flaviviruses continue to be addressed to treat signs and symptoms. In the search for new viral agents, in silico approaches have been useful tools for drug design and development. These techniques have been essential for understanding the essential targets for a viral infection that have been recently investigated. Thus, molecules targeting E, NS2B-NS3, NS5, and NS3/4A proteins represent the main targets currently studied. Special attention has been given to NS5 (RdRp) since it is a viral target absent in hosts. The active chemical scaffolds against flaviviruses are quite diversified, although peptides and/or peptidomimetics have been promising compounds against several flaviviruses. Furthermore, drug repurposing remains one of the main strategies for the discovery of antiviral agents. Advances are ascending but pharmacokinetic parameters such as half-life, binding to plasma proteins, and toxicity are factors associated with in vivo models that limit the progress of active substances to clinical trials.

Acknowledgments

The authors thank to Coordenação de Aperfeiçoamento Pessoal de Nível Superior (CAPES), National Council for Scientific and Technological Development (CNPq), Fundação de Amparo à Pesquisa do Estado de Alagoas (FAPEAL), and Financier of Studies and Projects (FINEP).

Conflict of interest

The authors confirm that there are no conflicts of interest.

References

[1] Sims N, Kasprzyk-Hordern B. Future perspectives of wastewater-based epidemiology: monitoring infectious disease spread and resistance to the community level. Env Int 2020;139:105689. Available from: https://doi.org/10.1016/j.envint.2020.105689.

[2] Burrell CJ, Howard CR, Murphy FA. Epidemiology of viral infections. Fenner White's medical virology. Elsevier; 2017. p. 185−203. Available from: https://doi.org/10.1016/B978-0-12-375156-0.00013-8.

[3] Chong HY, Leow CY, Abdul Majeed AB, Leow CH. Flavivirus infection—a review of immunopathogenesis, immunological response, and immunodiagnosis. Virus Res 2019;274:197770. Available from: https://doi.org/10.1016/j.virusres.2019.197770.

[4] Mao K, Zhang K, Du W, Ali W, Feng X, Zhang H. The potential of wastewater-based epidemiology as surveillance and early warning of infectious disease outbreaks. Curr Opin Env Sci Heal 2020;17:1−7. Available from: https://doi.org/10.1016/j.coesh.2020.04.006.

[5] Cuevas-Juárez E, Pando-Robles V, Palomares LA. Flavivirus vaccines: virus-like particles and single-round infectious particles as promising alternatives. Vaccine 2021. Available from: https://doi.org/10.1016/j.vaccine.2021.10.049.

[6] Pierson TC, Diamond MS. The continued threat of emerging flaviviruses. Nat Microbiol 2020;5:796−812. Available from: https://doi.org/10.1038/s41564-020-0714-0.

[7] de St Maurice A, Ervin E, Chu A. Ebola, dengue, chikungunya, and Zika infections in neonates and infants. Clin Perinatol 2021;48:311−29. Available from: https://doi.org/10.1016/j.clp.2021.03.006.

[8] Messina JP, Brady OJ, Golding N, Kraemer MUG, Wint GRW, Ray SE, et al. The current and future global distribution and population at risk of dengue. Nat Microbiol 2019;4:1508−15. Available from: https://doi.org/10.1038/s41564-019-0476-8.

[9] Petersen LR, Marfin AA. West Nile virus: a primer for the clinician. Ann Intern Med 2002;137:173. Available from: https://doi.org/10.7326/0003-4819-137-3-200208060-00009.

[10] Ozawa T, Masaki H, Takasaki T, Aoyama I, Yumisashi T, Yamanaka A, et al. Human monoclonal antibodies against West Nile virus from Japanese encephalitis-vaccinated volunteers. Antivir Res 2018;154:58−65. Available from: https://doi.org/10.1016/j.antiviral.2018.04.011.

[11] Hong SH, Joung I, Flores-Canales JC, Manavalan B, Cheng Q, Heo S, et al. Protein structure modeling and refinement by global optimization in CASP12. Proteins Struct Funct Bioinforma 2018;86:122−35. Available from: https://doi.org/10.1002/prot.25426.

[12] Kümmerer BM. Dengue and Zika: control and antiviral treatment strategies, Vol. 1062. Singapore: Springer Singapore; 2018. Available from: https://doi.org/10.1007/978-981-10-8727-1.

[13] Apte-Sengupta S, Sirohi D, Kuhn RJ. Coupling of replication and assembly in flaviviruses. Curr Opin Virol 2014;9:134−42. Available from: https://doi.org/10.1016/j.coviro.2014.09.020.

[14] Burrell CJ, Howard CR, Murphy FA. Flaviviruses. Fenner and White's medical virology. Elsevier; 2017. p. 493−518. Available from: https://doi.org/10.1016/B978-0-12-375156-0.00036-9.

[15] Silva-Júnior EF, Schirmeister T, Araújo-Júnior JX. Recent advances in inhibitors of flavivirus NS2B-NS3 protease from Dengue, Zika, and West Nile viruses. Vector-Borne Diseases Treatment. 1st ed. Las Vegas: Open Access eBooks;; 2018. p. 1−25.

[16] Ruzek D, Avšič Županc T, Borde J, Chrdle A, Eyer L, Karganova G, et al. Tick-borne encephalitis in Europe and Russia: review of pathogenesis, clinical features, therapy, and vaccines. Antivir Res 2019;164:23−51. Available from: https://doi.org/10.1016/j.antiviral.2019.01.014.

[17] Guerbois M, Fernandez-Salas I, Azar SR, Danis-Lozano R, Alpuche-Aranda CM, Leal G, et al. Outbreak of Zika virus infection, Chiapas State, Mexico, 2015, and first confirmed transmission by *Aedes aegypti* mosquitoes in the Americas. J Infect Dis 2016;214:1349−56. Available from: https://doi.org/10.1093/infdis/jiw302.

[18] Mlakar J, Korva M, Tul N, Popović M, Poljšak-Prijatelj M, Mraz J, et al. Zika virus associated with microcephaly. N Engl J Med 2016;374:951−8. Available from: https://doi.org/10.1056/NEJMoa1600651.

[19] Oliveira Melo AS, Malinger G, Ximenes R, Szejnfeld PO, Alves Sampaio S, Bispo de Filippis AM. Zika virus intrauterine infection causes fetal brain abnormality and microcephaly: tip of the iceberg? Ultrasound Obstet Gynecol 2016;47:6−7. Available from: https://doi.org/10.1002/uog.15831.

[20] Calvet G, Aguiarv RS, Melo AS, Sampaio SA, de Filippis I, Fabri A, et al. Case report of detection of Zika virus genome in amniotic fluid of affected fetuses: association with microcephaly outbreak in Brazil. Lancet Infect Dis 2016;. Available from: https://doi.org/10.1017/CBO9781107415324.004.

[21] Moreira J, Peixoto TM, Siqueira AM, Lamas CC. Sexually acquired Zika virus: a systematic review. Clin Microbiol Infect 2017;23:296−305. Available from: https://doi.org/10.1016/j.cmi.2016.12.027.

[22] Motta IJF, Spencer BR, Cordeiro da Silva SG, Arruda MB, Dobbin JA, Gonzaga YBM, et al. Evidence for transmission of Zika virus by platelet transfusion. N Engl J Med 2016;375:1101−3. Available from: https://doi.org/10.1056/NEJMc1607262.

[23] Barnett R. Dengue. Lancet 2017;390:1941. Available from: https://doi.org/10.1016/S0140-6736(17)32651-X.

[24] Moloo A. Neglected tropical diseases. World Health Organization; 2020.

[25] Guzman MG, Harris E. Dengue. Lancet 2015;385:453−65. Available from: https://doi.org/10.1016/S0140-6736(14)60572-9.

[26] Zhao R, Wang M, Cao J, Shen J, Zhou X, Wang D, et al. Flavivirus: from structure to therapeutics development. Life 2021;11:615. Available from: https://doi.org/10.3390/life11070615.

[27] Rigau-Pérez JG, Clark GG, Gubler DJ, Reiter P, Sanders EJ, Vance Vorndam A. Dengue and dengue haemorrhagic fever. Lancet 1998;352:971−7. Available from: https://doi.org/10.1016/S0140-6736(97)12483-7.

[28] Murgue B. Severe dengue: questioning the paradigm. Microbes Infect 2010;12:113−18. Available from: https://doi.org/10.1016/j.micinf.2009.11.006.

[29] Sangkawibha N, Rojanasuphot S, Ahandrik S, Viriyapongse S, Sujartijatanasen, Salttul V, et al. Risk factors in dengue shock syndrome: a prospective epidemiologic study in Rayong, Thailand. I. The 1980 outbreak. Am J Epidemiol 1984;120. Available from: https://doi.org/10.1093/oxfordjournals.aje.a113932.

[30] Vannice KS, Wilder-smith A, Barrett ADT, Carrijo K, Cavaleri M, de Silva A, et al. Clinical development and regulatory points for consideration for second-generation live attenuated dengue vaccines. Vaccine 2018;36:3411−17. Available from: https://doi.org/10.1016/j.vaccine.2018.02.062.

[31] Guy B, Noriega F, Ochiai RL, Maïna L, Skipetrova A, Verdier F, et al. A recombinant live attenuated tetravalent vaccine for the prevention of dengue. Expert Rev Vaccines 2017;00:1−13. Available from: https://doi.org/10.1080/14760584.2017.1335201.

[32] Halstead SB. Critique of World Health Organization recommendation of a dengue vaccine. J Infect Dis 2016;214:1793−5. Available from: https://doi.org/10.1093/infdis/jiw340.

[33] Katzelnick LC, Gresh L, Halloran ME, Mercado JC, Kuan G, Gordon A, et al. Antibody-dependent enhancement of severe dengue disease in humans. Science 2017;358:929−32. Available from: https://doi.org/10.1126/science.aan6836 80−.

[34] Dick G. Zika isolation and serological specificity. Trans R Soc Trop Med Hyg 1952;46:509−20.

[35] MacNamara FN. Zika virus: a report on three cases of human infection during an epidemic of jaundice in Nigeria. Trans R Soc Trop Med Hyg 1954;48:139−45. Available from: https://doi.org/10.1016/0035-9203(54)90006-1.

[36] Hayes EB. Zika virus outside Africa. Emerg Infect Dis 2009;15:1347−50. Available from: https://doi.org/10.3201/eid1509.090442.

[37] Lanciotti RS, Kosoy OL, Laven JJ, Velez JO, Lambert AJ, Johnson AJ, et al. Genetic and serologic properties of Zika virus associated with an epidemic, Yap State, Micronesia, 2007. Emerg Infect Dis 2008;14:1232−9. Available from: https://doi.org/10.3201/eid1408.080287.

[38] Aubry M, Finke J, Teissier A, Roche C, Broult J, Paulous S, et al. Seroprevalence of arboviruses among blood donors in French Polynesia, 2011−2013. Int J Infect Dis 2015;41:11−12. Available from: https://doi.org/10.1016/j.ijid.2015.10.005.

[39] Duffy MR, Chen T-H, Hancock WT, Powers AM, Kool JL, Lanciotti RS, et al. Zika virus outbreak on Yap Island, Federated States of Micronesia. N Engl J Med 2009;360:2536−43. Available from: https://doi.org/10.1056/NEJMoa0805715.

[40] Plourde AR, Bloch EM. A literature review of Zika virus. Emerg Infect Dis 2016;22:1185−92. Available from: https://doi.org/10.3201/eid2207.151990.

[41] Pan American Health Organisation. Neurological syndrome, congenital malformations, and Zika virus infection. Implications for public health in the Americas. Pan Am Heal Organ; 2015.

[42] Ashwal S, Michelson D, Plawner L, Dobyns WB. Practice parameter: evaluation of the child with microcephaly (an evidence-based review): report of the quality standards subcommittee of the American academy of neurology and the practice committee of the child neurology society. Neurology 2009. Available from: https://doi.org/10.1212/WNL.0b013e3181b783f7.

[43] WHO. Zika virus; 2018.

[44] De Filette M, Ulbert S, Diamond M, Sanders NN. Recent progress in West Nile virus diagnosis and vaccination. Vet Res 2012;43:16. Available from: https://doi.org/10.1186/1297-9716-43-16.

[45] Donadieu E, Bahuon C, Lowenski S, Zientara S, Coulpier M, Lecollinet S. Differential virulence and pathogenesis of West Nile viruses. Viruses 2013;5:2856−80. Available from: https://doi.org/10.3390/v5112856.

[46] Sejvar J. Clinical manifestations and outcomes of West Nile virus infection. Viruses 2014;6:606−23. Available from: https://doi.org/10.3390/v6020606.

[47] WHO. Japanese encephalitis; 2019.

[48] Campbell G, Hills S, Fischer M, Jacobson J, Hoke C, Hombach J, et al. Estimated global incidence of Japanese encephalitis. Bull World Health Organ 2011;89:766—74. Available from: https://doi.org/10.2471/BLT.10.085233.

[49] Hameed M, Wahaab A, Nawaz M, Khan S, Nazir J, Liu K, et al. Potential role of birds in Japanese encephalitis virus zoonotic transmission and genotype shift. Viruses 2021;13:357. Available from: https://doi.org/10.3390/v13030357.

[50] Perera-lecoin M, Meertens L, Carnec X, Amara A. Flavivirus entry receptors: an update; 2014:69—88. https://doi.org/10.3390/v6010069.

[51] Kurz M, Stefan N, Zhu J, Skern T. NS2B/3 proteolysis at the C-prM junction of the tick-borne encephalitis virus polyprotein is highly membrane dependent. Virus Res 2012;168:48—55. Available from: https://doi.org/10.1016/j.virusres.2012.06.012.

[52] Lindenbach BD, Rice CM. Molecular biology of flaviviruses. Adv Virus Res 2003;59:23—61.

[53] Preugschat F, Yao CW, Strauss JH. In vitro processing of dengue virus type 2 nonstructural proteins NS2A, NS2B, and NS3. J Virol 1990;64:4364—74. Available from: https://doi.org/10.1128/JVI.64.9.4364-4374.1990.

[54] Rodrigues ÉE, da S, Maus H, Hammerschmidt SJ, Ruggieri A, dos Santos EC, et al. The medicinal chemistry of Zika virus. In: Ahmad SI, editor. Human viruses: diseases, treatments and vaccines new insights. Cham: Springer International Publishing; 2021. p. 233—95. Available from: https://doi.org/10.1007/978-3-030-71165-8_13.

[55] Botting C, Kuhn RJ, Lafayette W. Novel approaches to flavivirus drug discovery. Expert Opin Drug Discov 2018;7:417—28. Available from: https://doi.org/10.1517/17460441.2012.673579.Novel.

[56] de Wispelaere M, Lian W, Potisopon S, Li PC, Jang J, Ficarro SB, et al. Inhibition of flaviviruses by targeting a conserved pocket on the viral envelope protein. Cell Chem Biol 2018;25:1006—1016.e8. Available from: https://doi.org/10.1016/j.chembiol.2018.05.011.

[57] Modis Y, Ogata S, Clements D, Harrison SC. A ligand-binding pocket in the dengue virus envelope glycoprotein; 2003.

[58] Nascimento IJ, dos S, Santos-Júnior PF, da S, Aquino TM, Araújo-Júnior JX, et al. Chemistry insights on Dengue and Zika NS5 RNA-dependent RNA polymerase (RdRp) inhibitors. Eur J Med Chem 2021;224. Available from: https://doi.org/10.1016/j.ejmech.2021.113698.

[59] Sampath A, Padmanabhan R. Molecular targets for flavivirus drug discovery. Antivir Res 2009;81:6—15. Available from: https://doi.org/10.1016/j.antiviral.2008.08.004.

[60] da Silva-Júnior EF, de Araújo-Júnior JX. Peptide derivatives as inhibitors of NS2B-NS3 protease from Dengue, West Nile, and Zika flaviviruses. Bioorg Med Chem 2019;27:3963—78. Available from: https://doi.org/10.1016/j.bmc.2019.07.038.

[61] Tabachnick WJ. Climate change and the arboviruses: lessons from the evolution of the dengue and yellow fever viruses. Annu Rev Virol 2016;3:125—45. Available from: https://doi.org/10.1146/annurev-virology-110615-035630.

[62] Young PR. Arboviruses: a family move. Springer; 2018. p. 1—10. Available from: https://doi.org/10.1007/978-981-10-8727-1_1.

[63] Vermehren J, Park JS, Jacobson IM, Zeuzem S. Challenges and perspectives of direct antivirals for the treatment of hepatitis C virus infection. J Hepatol 2018;69:1178—87. Available from: https://doi.org/10.1016/j.jhep.2018.07.002.

[64] Rusere LN, Matthew AN, Lockbaum GJ, Jahangir M, Newton A, Petropoulos CJ, et al. Quinoxaline-based linear HCV NS3/4A protease inhibitors exhibit potent activity against drug resistant variants. ACS Med Chem Lett 2018;9:691—6. Available from: https://doi.org/10.1021/acsmedchemlett.8b00150.

[65] Rao DN, Zephyr J, Henes M, Chan ET, Matthew AN, Hedger AK, et al. Discovery of Quinoxaline-based P1—P3 macrocyclic NS3/4A protease inhibitors with potent activity against drug-resistant hepatitis C virus variants; 2021. https://doi.org/10.1021/acs.jmedchem.1c00554.

[66] Scola PM, Sun LQ, Wang AX, Chen J, Sin N, Venables BL, et al. The discovery of asunaprevir (BMS-650032), an orally efficacious NS3 protease inhibitor for the treatment of hepatitis C virus infection. J Med Chem 2014;57:1730—52. Available from: https://doi.org/10.1021/jm500297k.

[67] Scola PM, Wang AX, Good AC, Sun L-Q, Combrink KD, Campbell JA, et al. Discovery and early clinical evaluation of BMS-605339, a potent and orally efficacious tripeptidic acylsulfonamide NS3 protease inhibitor for the treatment of hepatitis C virus infection. J Med Chem 2014;57:1708—29. Available from: https://doi.org/10.1021/jm401840s.

[68] Sun L, Mull E, Andrea SD, Zheng B, Hiebert S, Gillis E, et al. Discovery of BMS-986144, a third-generation, pan-genotype ns3/4a protease inhibitor for the treatment of hepatitis C virus infection; 2020:605339. https://doi.org/10.1021/acs.jmedchem.0c01296.

[69] Tong L, Yu W, Chen L, Selyutin O, Dwyer MP, Nair AG, et al. Discovery of Ruzasvir (MK-8408): a potent, pan-genotype HCV NS5A inhibitor with optimized activity against common resistance-associated polymorphisms; 2017. https://doi.org/10.1021/acs.jmedchem.6b01310.

[70] Coburn CA, Meinke PT, Chang W, Fandozzi CM, Graham DJ, Hu B, et al. Discovery of MK-8742: an HCV NS5A inhibitor with broad genotype activity; 2013:1930—1940. https://doi.org/10.1002/cmdc.201300343.

[71] Nair AG, Zeng Q, Selyutin O, Rosenblum SB, Jiang Y, Yang D, et al. Discovery of silyl proline containing HCV NS5A inhibitors with pan-genotype activity: SAR development. Bioorg Med Chem Lett 2016;26:1475—9. Available from: https://doi.org/10.1016/j.bmcl.2016.01.050.

[72] Nair AG, Zeng Q, Selyutin O, Rosenblum SB, Jiang Y, Yang D, et al. MK-8325: a silyl proline-containing NS5A inhibitor with pan-genotype activity for treatment of HCV. Bioorg Med Chem Lett 2018;28:1954—7.

[73] Sofia MJ. Beyond sofosbuvir: What opportunity exists for a better nucleoside/nucleotide to treat hepatitis C? Antivir Res 2014;107:119—24. Available from: https://doi.org/10.1016/j.antiviral.2014.04.008.

[74] Bhatia HK, Singh H, Grewal N, Natt NK. Molecules of the millennium sofosbuvir: a novel treatment option for chronic hepatitis C infection; 2014:5. https://doi.org/10.4103/0976-500X.142464.

[75] Randolph JT, Li T, Chris Krueger A, Heyman HR, Chen H-J, Bow DAJ, et al. Discovery of 2-aminoisobutyric acid ethyl ester (AIBEE) phosphoramidate prodrugs for delivering nucleoside HCV NS5B polymerase inhibitors. Bioorg Med Chem Lett 2020;30:126986. Available from: https://doi.org/10.1016/j.bmcl.2020.126986.

[76] Chen W, Mook RA, Premont RT, Wang J. Niclosamide: beyond an antihelminthic drug. Cell Signal 2018;41:89—96. Available from: https://doi.org/10.1016/j.cellsig.2017.04.001.

[77] Wang YM, Lu JW, Lin CC, Chin YF, Wu TY, Lin LI, et al. Antiviral activities of niclosamide and nitazoxanide against chikungunya virus entry and transmission. Antivir Res 2016;135:81—90. Available from: https://doi.org/10.1016/j.antiviral.2016.10.003.

[78] Xu M, Lee EM, Wen Z, Cheng Y, Huang WK, Qian X, et al. Identification of small-molecule inhibitors of Zika virus infection and induced neural cell death via a drug repurposing screen. Nat Med 2016;22:1101—7. Available from: https://doi.org/10.1038/nm.4184.

[79] Li Z, Xu J, Lang Y, Fan X, Kuo L, Brant LD, et al. JMX0207, a niclosamide derivative with improved pharmacokinetics, suppresses Zika virus infection both in vitro and in vivo; 2020. https://doi.org/10.1021/acsinfecdis.0c00217.

[80] Yao Y, Huo T, Lin Y-L, Nie S, Wu F, Hua Y, et al. Discovery, X-ray crystallography and antiviral activity of allosteric inhibitors of flavivirus NS2B-NS3 protease. J Am Chem Soc 2019;141:6832—6. Available from: https://doi.org/10.1021/jacs.9b02505.

[81] Li P-C, Jang J, Hsia C-Y, Groomes PV, Lian W, de Wispelaere M, et al. Small Molecules targeting the Flavivirus E protein with broad-spectrum activity and antiviral efficacy in vivo. ACS Infect Dis 2019;5:460—72. Available from: https://doi.org/10.1021/acsinfecdis.8b00322.

[82] Ye C, Bian P, Zhang J, Xiao H, Zhang L, Ye W, et al. Structure-based discovery of antiviral inhibitors targeting the E dimer interface of Japanese encephalitis virus. Biochem Biophys Res Commun 2019;515:366—71. Available from: https://doi.org/10.1016/j.bbrc.2019.05.148.

[83] Wei J, Hameed M, Wang X, Zhang J, Guo S, Anwar MN, et al. Antiviral activity of phage display-selected peptides against Japanese encephalitis virus infection in vitro and in vivo. Antivir Res 2020;174:104673. Available from: https://doi.org/10.1016/j.antiviral.2019.104673.

[84] Chu JJH, Rajamanonmani R, Li J, Bhuvanakantham R, Lescar J, Ng M-L. Inhibition of West Nile virus entry by using a recombinant domain III from the envelope glycoprotein. J Gen Virol 2005;86:405—12. Available from: https://doi.org/10.1099/vir.0.80411-0.

[85] Mertinková P, Mochnáčová E, Bhide K, Kulkarni A, Tkáčová Z, Hruškovicová J, et al. Development of peptides targeting receptor binding site of the envelope glycoprotein to contain the West Nile virus infection. Sci Rep 2021;11:20131. Available from: https://doi.org/10.1038/s41598-021-99696-w.

[86] Brai A, Martelli F, Riva V, Garbelli A, Fazi R, Zamperini C, et al. DDX3X helicase inhibitors as a new strategy to fight the West Nile virus infection. J Med Chem 2019;62:2333—47. Available from: https://doi.org/10.1021/acs.jmedchem.8b01403.

[87] Dragoni F, Boccuto A, Picarazzi F, Giannini A, Giammarino F, Saladini F, et al. Evaluation of sofosbuvir activity and resistance profile against West Nile virus in vitro. Antivir Res 2020;175:104708. Available from: https://doi.org/10.1016/j.antiviral.2020.104708.

[88] Konkolova E, Dejmek M, Hřebabecký H, Šála M, Böserle J, Nencka R, et al. Remdesivir triphosphate can efficiently inhibit the RNA-dependent RNA polymerase from various flaviviruses. Antivir Res 2020;182:104899. Available from: https://doi.org/10.1016/j.antiviral.2020.104899.

[89] Akaberi D, Båhlström A, Chinthakindi PK, Nyman T, Sandström A, Järhult JD, et al. Targeting the NS2B-NS3 protease of tick-borne encephalitis virus with pan-flaviviral protease inhibitors. Antivir Res 2021;190:105074. Available from: https://doi.org/10.1016/j.antiviral.2021.105074.

[90] Eyer L, Svoboda P, Balvan J, Vičar T, Raudenská M, Štefánik M, et al. Broad-spectrum antiviral activity of 3'-deoxy-3'-fluoroadenosine against emerging flaviviruses. Antimicrob Agents Chemother 2021;65. Available from: https://doi.org/10.1128/AAC.01522-20.

[91] Mendes EA, Pilger DRB, de, Santos Nastri AC, de S, Malta F, de M, et al. Sofosbuvir inhibits yellow fever virus in vitro and in patients with acute liver failure. Ann Hepatol 2019;18:816−24. Available from: https://doi.org/10.1016/j.aohep.2019.09.001.

[92] Ansari U, Henderson LI, Stott G, Parr K. Treatment with ledipasvir-sofosbuvir for hepatitis C resulting in improvement of lichen planus. JAAD Case Rep 2017;3:67−9. Available from: https://doi.org/10.1016/j.jdcr.2016.12.005.

[93] Srivastava R, Gupta SK, Naaz F, Sen Gupta PS, Yadav M, Singh VK, et al. Alkylated benzimidazoles: design, synthesis, docking, DFT analysis, ADMET property, molecular dynamics and activity against HIV and YFV. Comput Biol Chem 2020;89:107400. Available from: https://doi.org/10.1016/j.compbiolchem.2020.107400.

[94] Gawriljuk VO, Foil DH, Puhl AC, Zorn KM, Lane TR, Riabova O, et al. Development of machine learning models and the discovery of a new antiviral compound against yellow fever virus. J Chem Inf Model 2021;61:3804−13. Available from: https://doi.org/10.1021/acs.jcim.1c00460.

Pathophysiology of HIV and strategies to eliminate AIDS as a public health threat

Omar Sued[1] and Tomás M. Grosso[2,3]

[1]Pan American Health Organization, Washington, DC, United States [2]Laboratory of Immunology, National University of Luján, Buenos Aires, Argentina [3]Clinical Trials Unit, Huésped Foundation, Buenos Aires, Argentina

17.1 Background

17.1.1 Human immunodeficiency virus

The human immunodeficiency virus type 1 (HIV-1) belongs to the Retroviridae family, Orthoretrovirinae subfamily, within the genus *Lentivirus* (international classification of viral taxonomy). These viruses have an RNA genome that a catalyst enzyme—reverse transcriptase—uses as a template to synthesize proviral DNA in a process called retrotranscription. Viral DNA is then integrated to the host cell genome via an integrase enzyme action.

The HIV-1 virion (Fig. 17.1A) is a spherical particle of approximately 100–120 nm of diameter, formed by an external, host-derived lipid envelope (Env), with a conical capsid and the virus genome inside. The capsid consists of 60 triangular faces formed by protein p24 where the viral genome, the structural nucleocapsid proteins, and the nonstructural proteins (integrase, protease, and reverse transcriptase) are located. Two surface glycoproteins—namely, transmembrane gp41 and external gp120—are expressed within the Env and are essential for the entry of the virus into the host cell [1]. This process is mediated by a primary gp120-CD4 interaction, followed by gp120 binding to a coreceptor (CCR5 or CXCR4), and gp41 attachment to the host cell membrane [2].

The HIV-1 genome (Fig. 17.1B) consists of two nearly identical molecules of single-chain ribonucleic acid with positive polarity (cSRN +), of 9.2 kilobases each, encoding

Viral Infections and Antiviral Therapies
DOI: https://doi.org/10.1016/B978-0-323-91814-5.00023-4

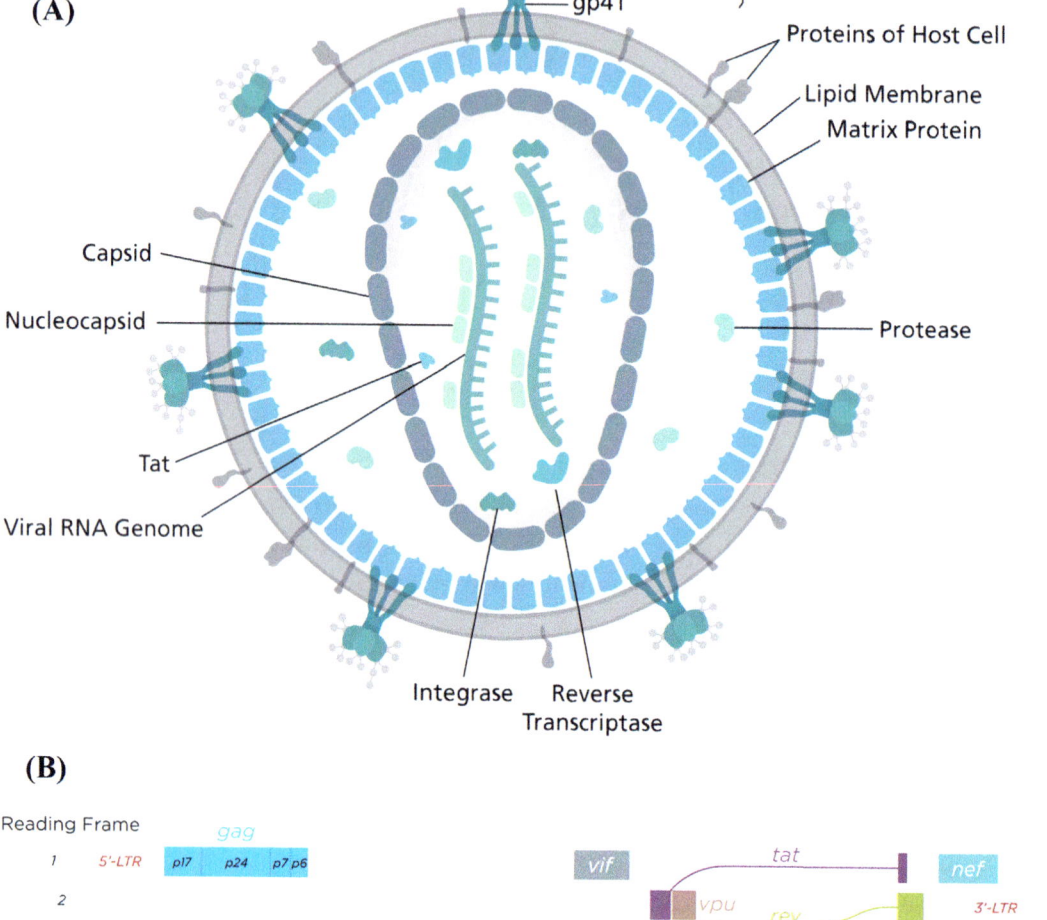

FIGURE 17.1 (A) HIV-1 virion diagram. Its main structures are detailed. (B) Organization of the viral genome. Source: (A) Thomas Splettstoesser (www.scistyle.com), CC BY-SA 4.0 , via Wikimedia Commons: https://commons.wikimedia.org/wiki/File:HI-virion-structure_en.svg; (B) Thomas Splettstoesser (www.scistyle.com), CC BY-SA 3.0 , via Wikimedia Commons: https://commons.wikimedia.org/wiki/File:HIV-genome.png.

3 genes that are common to all retroviruses (gag, pol, and env) along with 6 other genes that code for regulatory (tat and tev), and accessory (nef, vif, vpr, and vpu) activities. The gag gene encodes p24, p6, p7, and p17. The pol gen encodes viral enzymes including reverse transcriptase integrase and protease. The env region encodes the glycoprotein

TABLE 17.1 HIV genes and their encoded proteins.

Gene		Encoded protein(s)	Function
gag		Pr55Gag	Precursor of structural proteins
	p24	CA (capsid protein)	Structural—formation of viral capsid
	p6		Structural—viral release
	p7	NC (nucleoprotein)	Structural—formation NC/RNA complex
	p17	MA (matrix protein)	Structural—formation inner membrane
pol		Pr160GagPol	Precursor of viral enzymes
	P51	RT (reverse transcriptase)	Retro transcription of HIV RNA in proviral DNA
	P32	IN (integrase)	Proviral DNA integration into the host genome
	P10	PR (protease)	Cleaves viral proteins into their final functional units
	P15	RNase H	Degradation of viral RNA in the complex RNA/DNA
env		PrGp160	Precursor of gp41 and gp120 − host cell interaction
	Gp120	Surface glycoprotein	Attachment to CD4
	Gp41	Transmembrane protein	Gp120 anchorage and fusion
tat	P14	Tat (trans activator protein)	Upregulates viral transcription
rev	P19	Rev (RNA splicing regulator)	Regulates splicing and export of mRNA
nef	P27	Nef (negative regulating factor)	Affects viral replication and infectivity
vif	P23	Vif (viral infectivity protein)	Viral assemble
vpr	P15	Vpr (viral protein R)	Proviral DNA transport to cell nucleus
vpu	P16	Vpu (virus protein unique)	Particle release, intracellular trafficking
tev	P26	Tat-rev protein	Regulates the activity of Tat and Rev in nucleus

Each protein has its main function detailed in this table.
IN, Integrase, *RT*, reverse transcriptase, *PR*, protease.

gp160, which is the gp120 and gp41 precursor, that form a protein complex which interacts specifically with the host cellular receptor. These characteristics are summarized in Table 17.1.

Once integrated into the cellular genome, proviral DNA is flanked by Long Terminal Repeats (LTRs) generated during the retro-transcription process. LTRs are composed of the U3, R, and U5 regions and are responsible for regulating, at least in part, the expression of viral genes [3].

Two HIV viral types have been described, HIV-1, of universal distribution, and HIV-2, isolated in West Africa in 1985, where it is endemic. HIV-1 has a high degree of diversity and variability, with 4 groups (M, N, O, and P) recognized so far. Group M is responsible for the global pandemic and is subdivided into 11 subtypes (from A to K) that can combine with each other to create other viral variants. In Europe, the Americas and Oceania, subtype B is predominant, except in Brazil, Uruguay, and Argentina, where almost half of infected individuals harbor subtype F virus, or some recombinant B/F forms [4]. Subtype A is the most prevalent group in East Africa, Russia, and former Soviet Union countries; subtype C in Southern Africa and India; and circulating forms CRF01_AE in Asia and CRF02_AG in Western Africa [5].

The virus can also be classified according to the type of coreceptor it uses when the gp120 glycoprotein attaches to the CD4 receptor. Hence, R5 viruses use C-C receptor type 5 (CCR5), whereas X4 viruses use the C-X-C receptor type 4 (CXCR4). However, some strains have dual tropism. CCR5 receptors are expressed on various cells such as T lymphocytes, dendritic cells (DCs), and macrophages. R5 viruses predominate during the early stages of infection while X4 viruses emerge during later stages due to a small number of mutations within the highly variable region of the V3 domain [6].

17.1.2 Acquired immunodeficiency syndrome

Acquired immunodeficiency syndrome (AIDS) represents the last clinical stage of HIV infection. It was first described in the United States in 1981 from a report of a cluster of *Pneumocystis jiroveci* pneumonia cases and an unexpected increase in cases of Kaposi's sarcoma in men who had sex with men [7,8].

The exponential increase in cases triggered a frantic search for etiology. In 1983, Dr Françoise Barré-Sinoussi, from the retrovirus laboratory of the Pasteur Institute of Paris, led by Dr. Luc Montagnier, described a new retrovirus [9], a finding for which they were distinguished in 2008 with the Nobel Prize in Medicine.

However, the origin of the HIV pandemic dates back to the early 20th century. Studying the probability of mutations in the viral sequence of the first identified virus, it was possible to identify the origin of the crossover from monkeys to humans in Kinshasa (now Democratic Republic of Congo) around 1920 [10]. Significant sociodemographic changes in this area at that time, including a significant shift in residency patterns within different cities due to the opening of new trade routes, associated migrations, an increase in sex work, and the expansion of medical procedures without sterilized materials, facilitated the slow growth and local dissemination of cases within sub-Saharan Africa. In the 1980s, the generalization of intercontinental travel, sex tourism, and explosion of intravenous drug use made possible the establishment of the HIV pandemic [11].

17.1.3 Epidemiology

Since its discovery, it is estimated that HIV has infected more than 80 million people and caused more than 36 million deaths. At the end of 2020, an estimated 37.7 million people were living with HIV/AIDS, with 1.5 million new cases during this year and 680,000 associated deaths. This represents an impressive advance compared with the peak of 3 million new cases registered in 1997 and with 2.1 million deaths reported in 2004. These achievements were only possible due to unprecedented global efforts and huge investments that resulted in providing treatment to 27.5 million individuals (73% of people living with HIV), and extensive prevention activities [12].

The global prevalence of HIV infection stands at 0.8% of the total population, but this proportion varies significantly, from extremely low prevalence to 2-digit figures in some countries in sub-Saharan Africa. Out of all new cases, 62% are concentrated among specific key populations—namely, men who have sex with men, people who inject drugs, transgender people, and sexual workers and their clients—and this percentage increases to 93% when excluding sub-Saharan Africa. Women in Africa suffer a greater impact, with girls between 15 and 24 showing twice as high an HIV prevalence compared to men of the same age. Many causes explain this disproportion, such as social and economic inequities, exacerbated by political violence, racism, poverty, and male chauvinism [13].

17.1.4 Transmission and establishment of infection

T helper lymphocytes (CD4 + T cells) represent the main target cell in which HIV-1 can reproduce efficiently. The sequence of events that occurs between viral exposure and productive infection of these cells involves complex virus-mucosae interactions that are not fully known yet. The acquisition of HIV through a sexual route might hide important differences in terms of susceptibility and resistance to infection of mucosal membranes. Most of the information about initial events comes either from in vitro studies using human tissue explants, or animal model studies (i.e., *rhesus* macaques). To initiate sustained replication, HIV must cross the anatomical mucosal barrier, must evade the innate defense system and infect CD4 + T cells to initiate viral replication [14]. Mucosal erosions, ulcers, hormonal changes, or pregnancy can alter tissue integrity and facilitate infection. The first part of this process, which involves infecting intraepithelial CD4 + T cells, Langerhans cells, or submucosal DCs located in the lamina propria, may take a few hours. The route (intravenous, rectal, or vaginal) and quantity of the inoculum influences the peak of viremia: the larger the inoculum, the faster infection occurs, therefore, viremia peak is delayed when the transmission route is by intra-rectal inoculation compared to the intravenous route because of the time it takes for the virus to pass through the columnar colonic epithelium [15].

Mucosal barriers also limit access to the different viral variables creating a "bottleneck" that favors viral infection by a homogeneous viral population, which is observed in approximately 80% of cases of heterosexual transmission, and to a lesser extent during homosexual transmission (60%) or by use of intravenous drugs (40%) [16].

DCs are key to facilitating the infection. They migrate from the epithelium to the regional lymph patrolling mucosal territories to detect invader pathogens [17], and are among the earliest cells to encounter the HIV-1 virions transmitted through sexual contact, bringing them to the regional lymph nodes and passing them directly to the CD4 + T lymphocytes in the parafollicular area [18]. HIV-1 exploits the antigen-presenting cell functions of DCs to be transferred to CD4 + T-cells [19–21], through the so-called virological synapse. This transfer, known as trans-infection, constitutes a mechanism of HIV-1 spreading and occurs even in the absence of productive infection of DCs. In fact, immature DCs are highly resistant to productive HIV-1 infection, while mature DCs are even more resistant, although, intriguingly, much more effective to transfer HIV-1 to CD4 + T-cells [20,22].

Due to the large concentration of target cells (CD4 + T cells) in the lymph node, this step results in a massive infection of lymphocytes and consequent explosion of viral replication that is released into the systemic circulation [23], further leading to colonizing all tissues and establishing the viral reservoir.

During sexual transmission, the virus must overcome two major limitations to reach CD4 + T cells: the low density of target cells within the mucous membranes, and the host innate immune system. For the former, the virus induces a local inflammatory reaction that initially recruits DCs, which in turn generates more local inflammation and T cell recruitment. The virus does not productively infect DCs, but joins lectins on its surface (DC-SIGN and others) to be transported, and that favors local inflammation and lymphocyte recruitment [24]. The participation of DCs in the initial phases is one of the possible explanations for the positive selection of R5 viruses in the early stages of infection, taking into account that most people with acute infection present this tropism [25]. Another possible explanation would be that the production of stromal cell-derived factor-1 by DCs blocks the CXCR4 receptor, making it unavailable for X4 viruses [26].

17.1.5 Human immunodeficiency virus life cycle

To successfully bind to the cell, the viral glycoprotein gp120 must encounter the CD4 cell receptor. This binding induces a series of conformational changes that expose the V3 domain and adjacent regions, which form the binding domain of gp120 to chemokine receptors. This second interaction induces changes in the structure of gp41, which exposes, in the N-terminal region, a highly hydrophobic domain that is anchored in the plasma membrane, generating a binding movement of gp41 domains. During this process, the host and viral membranes approach and fuse [27]. Subsequently, the nucleocapsid is internalized and the decapsidation and release of the viral genome take place. The virus must inhibit the cellular protein TRIM5α, a restriction factor that disrupts capsid stability to carry out this process [28]. At this point, reverse transcriptase initiates DNA from RNA. The virus can infect activated or resting cells, although in resting lymphocytes retro transcription occurs incompletely [29]. The lack of proofreading control activity of reverse transcriptase facilitates the constant generation of mutants or variants called quasispecies. The synthesized proviral DNA is coupled to a series of cellular factors and viral proteins (IN, RT, p17, and Vpr) [27] defining the preintegration complex that is transported to the nucleus, where it is integrated, via the viral integrase, into the genome, giving rise to the proviral form of HIV. To begin replication requires that transcription of the viral genome be initiated [30], a process that is influenced by multiple factors such as the cellular proteins that regulate the expression of multiple cellular genes involved in immune recognition; and the viral protein Tat, that increases transcription; or the viral protein Rev, which regulates processing, transport and coupling of mRNA to ribosomes. Once the transcript is synthesized, it is encapsulated and transported to the regions where the virus is budding. It is of note that recent reports are shifting this model to a different one. It was seen that viral cores are translocated within the nucleus, where retrotranscription occurs, and then uncoated near the site of integration [31].

Viral proteins Vif and Vpu are involved in the assembly of viral proteins to form mature particles. Vif impedes the correct functioning of the APOBEC3G protein, one of the mechanisms of human innate antiviral immunity against retroviruses [32]. The viral protease processes Gag and Gag-Pol polyproteins to form mature particles that sprout across the cell membrane and are then released into the extracellular space. The role of Vpu in this scenario is to inhibit tetherin protein [33].

In the absence of antiretrovirals and cellular factors that prevent transcription, the activation and cell proliferation of infected CD4 + T cells determines a cycle of uncontrolled viral replication, which will infect new cells leading to disseminated infection and occupation of what in the future will be called a viral reservoir. Each of these steps are targets for antiretroviral drugs (Fig. 17.2).

17.1.6 Physiopathogenesis

17.1.6.1 *CD4 + T cell depletion*

CD4 + lymphocytes are indispensable to maintain adaptive responses, which gives HIV a unique trait — it infects the cells that should control the infection, thus evading its control and elimination. The progressive decrease of CD4 + lymphocytes is the distinctive feature of the infection and that is produced by different mechanisms: pyroptosis, apoptosis, redistribution, sequestration in peripheral nodes and a blockade in lymphocyte regeneration, the latter accomplished by a mechanism not yet completely clarified [34,35].

Activated and proliferating lymphocytes are especially susceptible to infection and to allowing viral replication because of the high levels of CCR5 receptors on their surface, high levels of nucleotides and ATP, and they have activated the transcription factors that HIV needs for its replication. Moreover, HIV preferentially infects memory CD4 + cells [35−37].

Once cell infection occurs, three situations may arise: (1) the viral cycle is interrupted before integration into the genome, usually in cells that are not permissive to infection (for example due to the presence of restriction factors); (2) in permissive cells, the cycle is completed and the cell begins to produce viruses; (3) some cells that sustained replication can revert to the state of latency and become part of the viral reservoir [36]. In-cell death can occur due to different mechanisms: in the first case, cells die from caspase1/3-dependent pyroptosis. In the second case, integration stimulates the phosphorylation and activation of DNA protein kinase, which activates p-53 and its cell death pathways. After integration, cell death mechanisms can also be activated in response to the expression of the HIV protease. This triggers the Casp8p41 system, which in turn activates the NF-κB factor and other mechanisms that favor cell death [36].

17.1.6.2 *Immunoactivation*

Inflammation and cellular activation are normal events, typically driven by the interaction between a pathogen and the immune system. If the infection or damage is not eliminated, the inflammation persists and evolves into chronic inflammation, with persistent cell activation, continuous cytokine production, mononuclear cell infiltration, granuloma formation, and attempts at tissue reconstruction leading, over time, to fibrosis. This is a well-known, heterogeneous process in both quantity and quality, associated with changes in plasma levels of different cytokines, and with the activation of different cell types, and

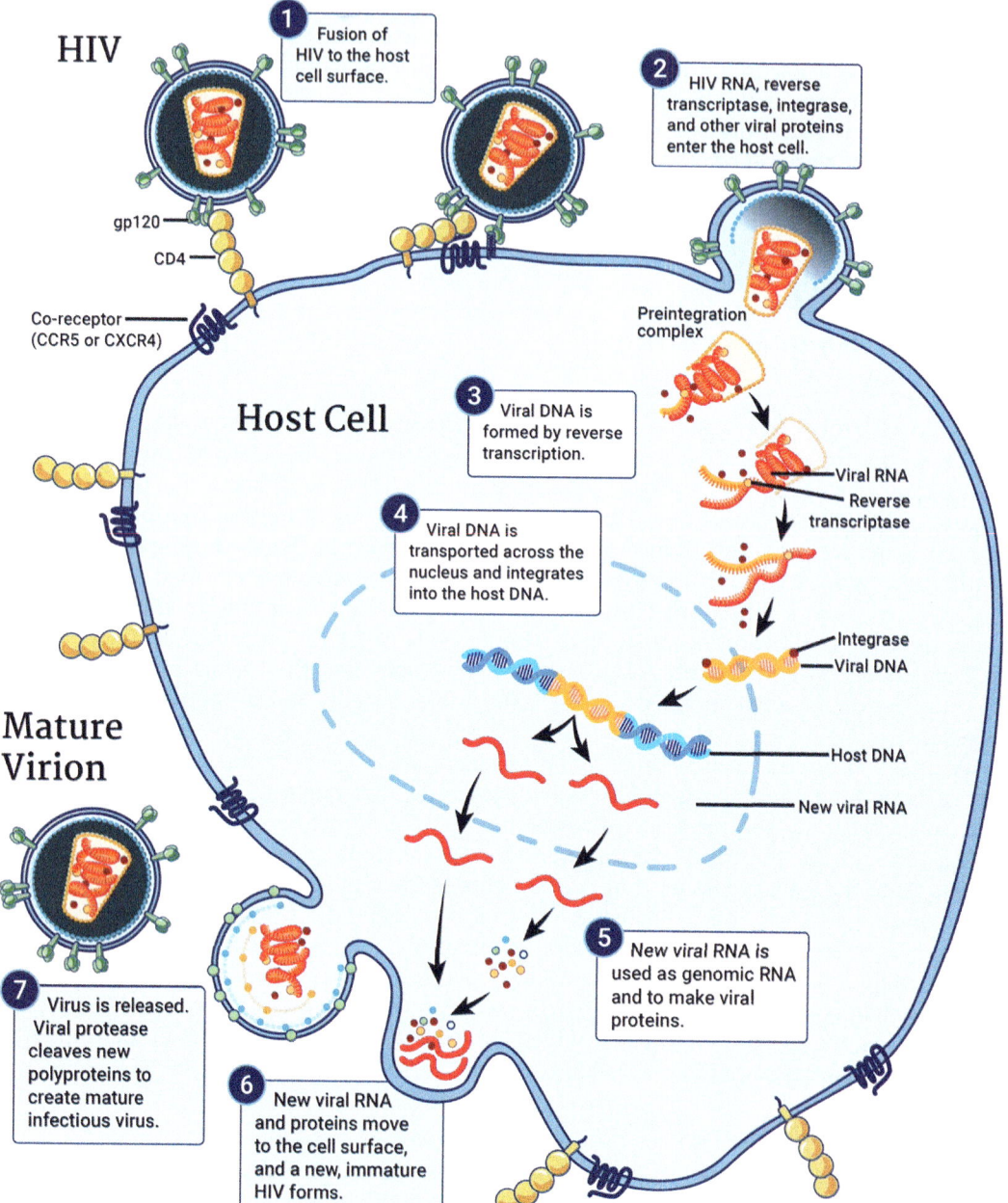

FIGURE 17.2 Simplified diagram of the classical model of HIV-1 replication cycle. Source: *NIAID, CC BY 2.0 , via Wikimedia Commons: https://commons.wikimedia.org/wiki/File:HI-Virus_Replication_Cycle_(5057022555).jpg.*

that has been studied in various ways. In the case of HIV infection, massive inflammation and immunoactivation occur during acute infection, followed by a persistent inflammatory state during the chronic phase, that can be partially restored with treatment [38,39]. Increased immune activation can predict further disease progression [40]. The persistence of this immune activation, produced by dysregulation of a normal defense process, has been related to various clinical conditions such as atherosclerosis, development of cancer, or neurocognitive disorders, and currently it is recognized as one of the most important factors that influence the aging process and its associated diseases [41].

With the increase in life expectancy (due to the availability of safer and more effective drugs), aging and prevention, diagnosis, and management of chronic diseases have become important pillars of HIV patient management, as it has been seen that HIV-positive individuals have a higher prevalence of cardiovascular diseases, osteoporosis, cancer, and other chronic diseases [42,43].

Immune activation is characterized by an increased rate of activation and apoptosis of CD4 + and CD8 + (cytotoxic) T cells as well as natural killer (NK) cells; activation of polyclonal B cells with increased immunoglobulin levels; and elevated production of proinflammatory cytokines [44]. It should be noted that these alterations only describe what happens in peripheral blood, although there are changes in all compartments. Intestinal tissue and the events that lead to its alteration are crucial to establishing generalized immunoactivation. Within this context, cytotoxic T cells, which should control viral replication, are affected by generalized immunoactivation and by the antigenic load produced by chronic stimulation that alters their functions [45]. An early observation that synthesizes the impact of immunoactivation is the association between activated CD4 + cells (CD4 + CD38 + HLA-Dr +) with higher HIV viral load (VL) and faster depletion of CD4 + cells [46]. On top of that, expression of CD38 might influence immune cells to HIV infection and even accentuate viral replication [47].

Therefore, early control of immune activation should be one of the goals of the management of patients with acute infection. Cells that regulate immunoactivation include natural regulatory T cells (Tregs) (CD4$^+$CD25^{high}FoxP3$^+$), regulatory CD8 + cells (CD8 + FoxP3 +), and T cell precursor cells or double negative cells (CD3 + CD4 − CD8 −) and could be targeted for new therapeutic approaches. In a small longitudinal study, no relationship could be evidenced between the proportion of Treg and activated CD4 + or CD8 + cells, but an inverse relationship was found in the baseline situation between the proportion of double negative cells with the proportion of activated CD8 + cells (CD38 + and HLA-Dr +) and CD8 + cells in apoptosis (Ki-67 +), and between these double negative cells with VL values as well. This inverse relationship between double negative cells and activation predicted the possibility of maintaining immunoactivation at 6 months. The conclusion is that these double-negative cells could play an important role in controlling the widespread activation of T cells through the production of IL-10 and TGF-β and that they could be an easily applicable marker [48].

Another parameter valued as a marker of immunoactivation is the specific soluble receptor of monocytes and macrophages (sCD163). Its function is to recycle extracellular iron, plus it has bactericidal activity and inhibits the proliferation and activation of T lymphocytes. Its increase is associated with the expansion and activation of monocytes, in particular CD14 + CD16 + monocytes, and this correlates with increased VL and progression to AIDS [38]. Antiretroviral therapy (ART) decreases sCD163 although not to normal values, and its level correlates with the proportion of CD8 + HLA-Dr + CD38 + lymphocytes [38].

Increased inflammation presents with increased cytokine production. The acute phase of the infection has been described as a "cytokine storm", in which significant increases in pro-inflammatory cytokines such as IL-1β, IL-6, IL-18, TNF-α, IL-10, and interferon-induced protein 10 have been seen [49,50]. A subanalysis of 75 patients not receiving treatment in the SPARTAC study showed that baseline IL-6 was associated with progression. The mean value was 1.45 pg/mL and it was shown that for each additional pg/mL the risk of progression to less than 350 CD4 + cells increased by 38% during followup [51]. ART tends to reduce elevated cytokines, although not to normal levels, in particular with IL-18, IL-6, IL-10, and IL-12 [52–54].

Inflammation and immunoactivation favors the prevalence of comorbidities, particularly noncommunicable diseases. Lately, a plasma proteomic study yielded that the complement protein C5 strongly correlated with non-AIDS comorbidities in HIV-positive patients, outscoring classic markers such as IL-6 and D-dimer [55].

17.1.6.3 Depletion of intestinal lymphoid tissue

DCs and lymphocytes located in the lamina propria of the intestinal mucosa constitute the *GALT system (gut-lymphoid associated tissue)* that represents 50% of the lymphatic tissue and it is the main defense mechanism against microbes that penetrate the digestive tract [56]. Most lymphocytes have an activated memory phenotype and an overregulated CCR5 receptor, given the increased exposure to antigens. HIV infection and replication in CD4 + cells and macrophages of the GALT system produces rapid depletion of these cells, in particular of CCR5 + CCR6 + Th17 cells. These cells are critical for controlling the bacterial proliferation and maintaining enterocyte homeostasis, secreting IL-17, IL-22, and recruiting NK cells, and their deficit or excess can produce several pathologies [57]. HIV-induced disruptions in tight junctions leads to loss of enterocytes, a decrease of Th17 lymphocytes, alteration of the hepatic architecture and depletion of myelomonocytic cells [58], and eventually causes *bacterial translocation*; understood as the dissemination of microbial products from the intestinal lumen to systemic circulation [58]. These products include lipopolysaccharides (LPS), peptidoglycans, lipoteic acid, and bacterial DNA. The involvement of enterocytes can be directly by the viral proteins Tat and Gp120, which favor their apoptosis, or by alteration of the gut microbiota, affecting the integrity of the membrane. HIV interferes through multiple mechanisms with the usual process of clearing microbial products that enter circulation and travel through the portal vein until they are recognized by toll-like receptors in Kupffer cells and hepatocytes [59,60]. The virus also induces an increase in fibrosis, probably because it stimulates the secretion of TGF-β1, which would favor the disposition of collagen in tissues [61]. A vicious circle is thus established, where the initial direct HIV infection causes depletion of lymphocytes of the GALT system, favors tissue damage and bacterial translocation, leading to generalized and persistent immune activation, further affecting the immune system and the functioning of the gastrointestinal tract. Therefore, the disruption of the intestinal barrier integrity and bacterial translocation are fundamental characteristics of viral pathogenesis, associated with persistent inflammation and sustained systemic immunoactivation that favors the development of noninfectious complications [62–64].

Conspicuous translocation biomarkers include direct determination of LPS and increase of a soluble marker of LPS-induced macrophage activation (sCD14) [65,66]. This increase is accompanied by increased IL-1β, IL-10, IFN-gamma, and TNF-α, and correlates with increased VL. Some patients may also have significant increases in IL-6, C-reactive protein, neopterin and β2 microglobulin, and others may show atypical patterns of inflammation [67].

ART has been shown to reduce sCD14 and LPS levels, but these values do not normalize. ART better restores Th17 and Kupffer cell when it is started at early stages [68].

17.1.6.4 Metabolic alterations

Inflammation and immune activation are closely linked to metabolism, since establishing an immune response requires an important change in metabolic processes due to high energy demand for the biosynthesis of multiple molecules (pro-inflammatory cytokines, antigen processing, phagocytosis, etc.). On the other hand, patients with HIV infection have metabolic changes secondary to immunoactivation which in turn worsens the metabolic situation [69]. Immunosuppression, altered immunological status, and rapid decrease of VL and restoration of CD4 + T cells can trigger the immune reconstitution inflammatory syndrome [70].

Reactive oxygen species (ROS) are highly reactive molecules that are produced by oxygen metabolism, such as hydrogen peroxide (H_2O_2), superoxide anion, or hydroxyl radical. They play a role in cell signaling, homeostasis, and antimicrobial/antitumor protection and have also been implicated in chronic inflammation [71]. Increased production of these metabolites increases oxidative stress, which damages cellular structures and promotes inflammation by increasing the production of inflammatory cytokines (IL-1β, IL-6, INF, TNF-α) [72,73]. The body regulates these processes through antioxidant enzymes such as superoxide dismutase, peroxidase/reductase catalase glutathione, vitamins A, C, and E and small redox proteins, such as glutathione and thioredoxin. Patients with HIV show high oxygen consumption, increased hydroperoxide, and reduced protective enzymes, probably by direct induction of the viral proteins Tat, Vpr, Nef, and Gp120. During infection, T cells increase the production of ROS, which reduces the response to the cytokines IL2, IL4, and promotes cell dysfunction and apoptosis by inducing PD-1 expression and reducing the vitamin D receptor [74–77].

Tryptophan is an essential amino acid whose catabolism generates products such as kineurin, a precursor to several molecules involved in energy production. In people with HIV infection, tryptophan metabolism is increased, which is evidenced by an increase in IDO-1. This correlates with immunoactivation and can be partially reversed with ART [78]. Early induction of IDO-1 activity in macrophages and DCs limits the antiretroviral response and contributes to disease progression [79–81].

Immunoactivation caused by HIV also results in increased glucose consumption to address cell proliferation, survival, and immune function [82]. Infection of CD4 + T cells causes an increase in glycolytic flow to its maximum capacity and facilitates apoptosis. In fact, the expression of GLUT-1 (the glucose transporter in T cells that is increased in CD4 + of HIV patients) is associated with immunoactivation and cell depletion that does not improve with ART [83]. Likewise, infected macrophages have a significant reduction in glucose uptake and lower generation of intermediate metabolites [84]. Vpr inhibits hexokinase 1, which is the enzyme responsible for converting glucose into glucose 6 phosphate and promotes mitochondrial integrity by a nonmetabolic route [85,86], which could have consequences in eradication efforts. Together these data support the negative impact of HIV infection on glucose metabolism.

17.1.7 Response to human immunodeficiency virus infection

During the acute phase of infection, interactions between the virus and the organism determine whether the virus is eliminated in the mucosae or if it is established as an infecting agent.

The first response to the virus is innate and consists of identifying the pathogen and activating the signaling pathways of the innate system. Identification is achieved by the recognition of viral pathogen-associated molecular patterns. This system induces viral restriction factors that will suppress or limit viral replication, produce IFN and pro-inflammatory cytokines and chemokines that recruit and activate innate immune cells such as macrophages and NK cells [87]. Exaggerated response or persistence of innate system activation can also have deleterious consequences, as it promotes viral replication and increases CD4+ cell loss, for example, a large activation of DCs can exacerbate pro-apoptotic effects on CD4+ cells [24]. Currently, there is great interest in identifying the regulatory mechanisms of the innate response in HIV to try to identify potential therapeutic targets.

The Major Histocompatibility Complex is the largest genetic determinant of immune responses (in humans, HLA), and its polymorphisms can influence innate and adaptive responses to HIV. Individuals with the HLA-B57 allele have less progression, less frequently present symptoms during acute infection, and achieve better spontaneous control over viremia [88]. Other alleles have been described with beneficial effects (e.g., KIR3DS1/HLA-Bw480I) or with negative effects (HLA-B35-Px) on infection progression [89].

DCs regulate the balance between tolerance and immune protection in both the innate and adaptive systems. One of the most important functions of DCs is the regulation, activation, and survival of B-cell development by producing B-cell growth factor (BLyS) *lymphocyte stimulator*. During acute infection, the amount of mature myeloid DCs decrease and persist below normal levels in the subsequent phases of infection even after successful treatment, although it is speculated that this is due to an active recruitment of these cells into peripheral tissues [90].

Another regulation mechanism between the innate and adaptive systems is the suppressive and stimulating signals mediated by galectins. Galectins are a family of 15 highly conserved lectins that bind to β-galactoside that share a consensus sequence in their carbohydrate recognition domain and that binds to bacterial membranes or activated cells [91]. Galectin 1 promotes cellular HIV infection and promotes replication by facilitating the binding between gp120 and the CD4 molecule. Galectin 9 (Gal-9) is a lectin, Tim-3 ligand, a specific surface receptor on CD4+ lymphocytes that induces peripheral tolerance by generating IL-17-secreting CD4+ T cells (Th17), and inducing Tregs. Gal-9 levels increase very significantly during the first weeks of acute HIV infection, proportionally to viremia and markers of inflammation such as IL-10, TNF-α, and IL-1β [92]. In fact, Gal-9 in serum is a promising marker for the degree of potential damage of acute and chronic infectious diseases [93], its levels decrease with treatment [94], but remain higher than in the HIV-free population, even in those that spontaneously control viral replication (elite controllers, or ECs). Some experts suggest that Gal-9 is responsible for the CD8+ cell dysfunction seen in HIV infection [92].

The adaptive response is generated approximately 12 weeks after infection and generates both specific antibodies and CD8+ lymphocytes with cytotoxic activity against HIV.

The implementation of these mechanisms of specific immunity achieves a partial control of viral replication and causes a sharp drop in viremia that is subsequently maintained at a relatively stable level and that varies in each patient (viral set-point). This basal viremia ranges from undetectable levels to more than hundreds of thousands of copies, and works as a parameter that reflects the new balance between viral replication and immune control and that represents a prognostic marker of the speed of evolution to AIDS [95]. Specific immune responses are, however, unable to eradicate the virus that has become cantoned in different reservoirs in which it persistently replicates. Throughout the course of infection, the immune system maintains a pressure against replication, gradually producing humoral and cellular responses specific to the circulating virus, and stimulating the continuous generation of escape variants to evade these responses, on which specific cellular and humoral responses will be induced again [96].

After several years of infection, the immune system progressively loses its power. The high rate of cell turnover affects homeostasis, clonal depletion of T cells associated with memory cell loss begins, chronic inflammation and collagen deposition produce permanent damage in lymphocyte production centers and increased immunoactivation provides more substrate to the virus, leading to the impossibility of the body to maintain CD4 + T cell levels, with the gradual fall of their number and the development of the immune catastrophe [45]. Depletion and immune dysfunction is characterized by increased expression of inhibitory receptors of PD-1 [97] and CTLA-4 [97,98].

17.1.7.1 *Humoral response*

HIV infection induces a humoral response to virtually all of HIV's regulatory and structural proteins, including Env antibodies, matrix proteins, viral nucleocapsid proteins, and virus regulatory proteins. Antibodies appear after the reduction of the initial viral replication peak, so they are considered to have a limited role in the initial control and the establishment of viral set-point. The first antibodies target nonneutralizing epitopes in the Env. During the following months, a minority of individuals are able to develop neutralizing antibodies [99] but the virus quickly escapes their inhibition [100]. Some antibodies, such as those directed against gp41 and against certain domains of gp120, have neutralizing capacity in vitro [101].

So far, it has not been possible to produce an immune response that allows creation of an effective vaccine. Probably the most important explanation is the great variability of protein gp160 exposed parts [102]. The Env is a trimeric structure that hides conserved domains; besides, interaction with CD4 is needed to expose interaction domains [103]. In recent years some findings have reactivated interest in neutralizing protective antibodies with the aim of being able to induce them with vaccines or as a passive transfer strategy. In animal models passive transfer of antibodies prevents infection or suppresses VL [104], something that has also been observed in initial studies with infected individuals [105].

Ideally, if it were possible to produce neutralizing antibodies against the gp120 gp120—the first component to contact cells—these antibodies could neutralize the infection or stimulate antibody-dependent cytotoxicity. However, the production of these has been a huge challenge because of the hypervariability of this region. The glycan shield that protects it and the folding of proteins on the conserved regions prevent these antibodies from being produced [106]. In addition, significant alteration in the production of antibody-producing B cells is seen, due to the loss of more than 80% of the germinal centers of the gastrointestinal lymphoid tissue, along with severe alteration of the environment that supports the maturation of B cells in the terminal ileon [107].

The identification, in serum of ECs, of antibodies that could neutralize a broad spectrum of viral variants permitted to proposing them as therapeutics [108]. Among the first identified, two targeting the CD4 interaction domain—VRC01 and VRC02—neutralize more than 90% of HIV-1 isolates [109]. Current research is moving towards how to identify broad-amplitude antibodies and evaluating combinations of several antibodies directed against different sites to avoid the emergence of resistance [110,111].

17.1.7.2 Cellular response

In HIV infection, the antiviral response is produced by different cell populations: CD4 + helper lymphocytes, cytotoxic CD8 + lymphocytes (CTLs), and NK cells [112]. The expansion of cytotoxic cell response is a remarkable phenomenon in the modulation of infection, the establishment of viral set-point and risk of progression. Studies carried out in Simian Immunodeficiency Virus infected macaques, whose CD8 + lymphocytes were eliminated at early stages, demonstrated the critical importance of these cells for the early control of viral replication as their presence is reduced by ten times, compared with their set point value, over a 21-day period [113]. Therefore, it has been suggested that HIV-specific cytotoxic CD8 + T cells and, to a lesser extent, CD4 + responses and target cell depletion, are the causes of decreased peak viremia, disappearance of acute symptoms, and establishment of a viral set-point [114]. The role of specific CD4 + cells has not been fully described. These cells, apart from being collaborators, recognize and destroy infected cells. In a longitudinal study, it was observed that people who controlled the infection spontaneously had a greater expansion of specific CD4 + HIV cells, but not of the CD8 + cytotoxic response, which was characterized by cytolytic activity and expression of perforins and granzymes from the onset of infection, suggesting that specific CD4 + responses are a predictor of, viral set-point and evolution without treatment [115].

The first cellular responses to the virus target a limited spectrum of epitopes, and are able to slow down initial replication [116]. However, HIV escapes immune control by CD8 + cells by selecting mutations. This explains why, despite the importance of CD8 responses in containing the initial viremia, neither their amplitude, nor their magnitude, nor their function, can predict the sustained control of replication. Paradoxically, some patients with high levels of response present high set-points, suggesting that these responses somehow reflect the level of viremia [116].

Antiviral functions of CD8 + T cells are characterized by not having a single cellular function, but each cell can produce a broad spectrum of effector protein molecules. Cellular polyfunctionality is described as the ability of a T cell to produce at least three markers, among which are CD107A/B, MIP-1β, IFN-γ, IL-2, and TNF-α, as the most typical. Polyfunctionality of the response has been associated with the ability to mediate viral control [117–119] and high levels of perforins [120]. ECs restrain HIV viremia without treatment, do not progress in the long term, and demonstrate increased CD8 + -specific suppressive capacity [121,122]. Early in the HIV infection, a large polyfunctionality of T cells, a strong specific CD4 + response, and a lack of CD8 + with suppressive capacity is seen [123]. CD8 + cell dysfunction is a common feature of chronic viral infections, including hepatitis C. The mechanisms are complex and partly mediated by a set of inhibitory control points (Tim-3, PDI, and CD160). A study has shown the presence of HIV-specific CD4 + CD8 + double cells during the acute phase that would be responsible for 16% of the proliferative response and more than 70% of the multifunctional anti-HIV response [124], although there are cases in which double-positive T cells do not exert enhanced antiviral activity [125]. ECs orient us in the search for substitutes of protection in vaccine trials and potential HIV control mechanisms. However, there is some consensus that there is a significant heterogeneity in terms of virological, immunological, and clinical outcomes in this population, aside from the traditional definition of elite controller status [126].

17.1.7.3 *Viral escape mechanism*

The most relevant mechanism of HIV viral escape is the very high rate of variability due to the raised error rate of reverse transcriptase (a substitution by 10,000–100,000 nucleotides and copy rounds). Mutations are responsible for a large proportion of defective viruses and the great variability of proteins that favor viral escape from the specific immune response [101,102]. Viral variants appear within a few days by mutations in the Env region, allowing them to evade the humoral response established against the virus a few months after the development of antibodies [127].

17.2 Natural history of human immunodeficiency virus infection

Viral replication and cell damage caused by HIV in immune cells are the most important determinants of the clinical progression of untreated HIV infection. Three different stages may be listed: primary or acute infection, asymptomatic phase (clinical latency), and the clinical phase of AIDS (Fig. 17.3).

Primary infection occurs immediately after the virus enters the body. Once the virus crosses the mucosae, DCs capture it and migrate to the regional ganglia, where they arrive 24–72 hours later to present antigens. Inside the ganglion there is an explosion of viral replication that produces high levels of VL and spreading of the virus to all future reservoirs (gastrointestinal mucosa, central nervous system, kidney, etc.). This peak of viremia is associated with some symptoms in 80% of cases, giving rise to a syndrome called "acute retroviral syndrome", which consists of fever pharyngitis, adenomegalies, headache, diarrhea, and night sweats that usually last between 2 and 3 weeks [129].

The improvement of symptoms is concomitant with a decrease in VL due to the generation of a specific immune response that partially controls the infection and modulates the viral set-point. CD4 + lymphocyte count, which prior to infection is normally between 500 and 1500 cells/mm^3, decreases during the acute phase, then recovers to levels close to normal during the first months after infection, and subsequently decreases in individuals without ART to, on average, from 60 to 100 cells/mm^3/year. This phenomenon constitutes the pathognomonic pattern of HIV infection and the main explanation for the pathogenesis of associated infections.

Once the viral set-point is established, the stable clinical phase begins. Its duration depends on the VL value that the individual maintains during the subsequent months and

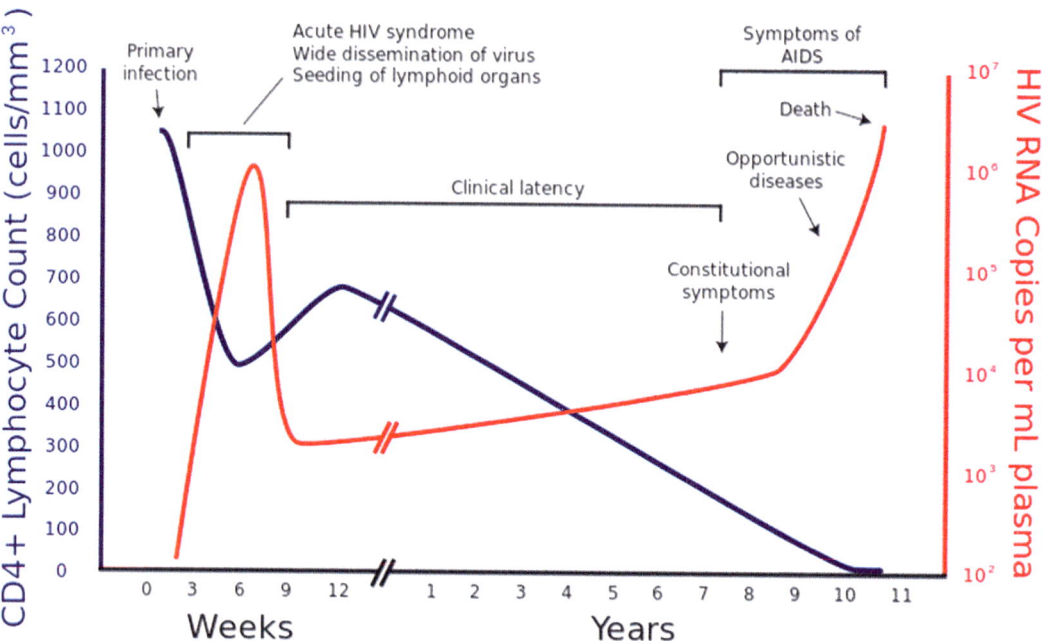

FIGURE 17.3 Natural history of HIV infection from primary infection to death [128]. Source: *Modified from Pantaleo G, Graziosi C, Fauci AS. New concepts in the immunopathogenesis of human immunodeficiency virus infection. N Engl J Med 1993;328:327–35. Sigve, CC0, via Wikimedia Commons: https://commons.wikimedia.org/wiki/File:Hiv-time-course_copy.svg.*

years, how rapidly the CD4+ lymphocyte count falls, and possible intercurrences that alter this balance. After 8–10 years of this stable clinical phase, the number of CD4+ T cells is so low that serious opportunistic infections (initially called AIDS-defining diseases) begin to appear [130]. If the person is not treated for both these infections and HIV, the risk of death is extremely high.

17.2.1 Transmission route

The virus is transmitted between humans by three main routes:

1. Sexually, through semen and cervicovaginal infected secretions.
2. Parenterally, through blood or blood products (by transfusions, sharing needles between drug users, or accidentally with contaminated sharp instruments during health procedures).
3. The perinatal route or transmission from the infected mother to the child, either intrauterine or during passage through the birth canal, and subsequently through breastmilk.

Different social characteristics influence transmission and must be considered when establishing health prevention, diagnosis and treatment programs. Well-known factors remain important and responsible for new acute infections, such as unprotected sexual activity and attendance at anonymous sexual encounter sites [131,132]. This expresses the importance of implementing innovative strategies to reduce new infections among people at higher risk, particularly those interventions aimed at reducing social factors and transmission networks, as well as other individual interventions of proven efficacy such as pre-exposure prophylaxis [133–135].

Pregnancy implies a greater susceptibility to infection, not only because of the biological changes that happen but also because of the low frequency of diagnosed male partners. In a cohort of 1200 pregnant women in Mozambique, for instance, there were 14 seroconversions but only 19% knew the situation of their companion [136]. In places of high incidence, up to 25% of transmissions can be linked to seroconversion during pregnancy [137]. For these reasons, more and more countries are incorporating couple HIV testing and the repetition of the HIV test during the third trimester of pregnancy among pregnant women [138,139].

The VL of the source individual is the factor more relevant in determining the risk of HIV transmission. In a randomized study, ART reduced the risk of HIV transmission more than 96%, with the remaining 4% events produced during few days after treatment initiation [140]. Several other studies show that HIV-treated individuals do not transmit HIV and are undetectable [141,142], and this is known by the acronym U = U (undetectable = untransmissible). This information has a significant potential to reduce the stigma and discrimination.

During acute infection, viruses with high replicative capacity are most likely selected. Moreover, the absence of neutralizing antibodies that can block or inactivate circulating virus and specific cellular responses explains extremely high values of VLs that are associated with greater infectivity. It is estimated that for every 10-fold VL increase, there is a 2.5-fold increase in transmission risk [143].

Several *meta*analyses reviewed the estimated transmission rates for each exposure event between serodiscordant couples, including *meta*analyses transmission via sexual [144,145], oral [146], and intravenous [147] means (Table 17.2). These estimates may not reflect actual infectivity as they do not distinguish the contribution of different cofactors in transmission risk that can have a significant impact on transmission. These cofactors are detailed in Table 17.3.

Estimates may vary depending on the age of the population studied, with slightly higher values for younger people, as well as higher values in people with greater sexual activity [148]. It should also be mentioned that these risks are substantially attenuated by the use of condoms and antiretroviral treatment [149].

17.2.2 Manifestation of acute human immunodeficiency virus infection

In 1985, Cooper et al. published the first description of symptoms attributable to acute HIV infection in 12 men having sex with men (MSM) who presented a picture compatible with "infectious mononucleosis" (fever, pharyngitis, and rash), that tested negative serology for the Epstein-Barr virus, and in which HIV infection was finally confirmed. Since

TABLE 17.2　Risk per act of exposure.

Contact type	Risk per act (%)	CI95%
Blood transfusion	92.50	89−96.10
Drug use sharing syringes	0.62	0.41−0.93
Stabbing accident	0.23	0−0.46
Receptive anal sex	1.38	1.02−1.86
Insertive anal sex	0.11	0.04−0.28
Receptive vaginal sex	0.08	0.06−0.11
Insertive vaginal sex	0.04	0.01−0.14
Receptive oral sex	ND	0−0.04
Insertive oral sex	ND	0−0.04
Mother-to-child transmission	22.6	17−29

Compiled from references [144−147].

TABLE 17.3　Cofactors that increase the risk of transmission.

Cofactor	Relative risk	CI95%
Acute versus asymptomatic infection	7.25	3.05−17.3
Final versus asymptomatic stage	5.81	3.00−11.4
High VL (per log)	2.89	2.19−3.89
Genital ulcer	2.65	1.35−5.19

Compiled from references [144−147].

then, this syndrome has been known as "acute retroviral syndrome" or "acute HIV sero-conversion syndrome".

The incubation period of HIV infection varies between 1 and 3 weeks (typically 14 days) and the duration of the symptomatic period is 7−14 days, rarely more than 2 weeks. In different studies, the prevalence of symptoms during acute HIV infection varies between 40% and 90%. In prospective studies of high-risk and potential HIV-exposed patients (e.g., patients treated in sexually transmitted disease care facilities), compared with a control group, between 53% and 88% had symptoms that were temporarily thought to be associated with infection. However, up to 50% of patients who did not acquire the infection suffered similar symptoms, demonstrating its low specificity [150,151].

The clinical spectrum of acute HIV infection (Table 17.4) varies from banal differential diagnoses, which are confused with a nonspecific virus, to severe cases with neurological involvement. The most common symptoms include fever, rash, oral and/or genital ulcers, lymphadenopathies, marked asthenia, arthro-myalgias, and aseptic meningitis [152] ().

Body temperature is often very high at the beginning, it drops slowly, and the fever can persist for several days. It is usually associated with night sweats and significant asthenia,

TABLE 17.4 Most common symptoms [153].

Symptom	Frequency
Fever	53%−87.5%
Exanthema	9%−57.5%
Oral ulcers	7.5%−37%
Arthro-myalgias	24%−54%
Pharyngitis	14%−44%
Weight loss	32%
Asthenia	68%−72.5%
Headache	54%−55%
Night sweats	50%−51%
Adenomegalies	7%−37.5%
Diarrhea	24%

Reported in Miro JM, Sued O, Plana M, Pumarola T, Gallart T. Advances in the diagnosis and treatment of acute human immunodeficiency virus type 1 (HIV-1) infection. Enferm Infect Microbiol Clin 2004;22:643−59.

forcing prolonged rest [154]. The rash usually appears in the 24−48 hours after the onset of fever, with a frequency ranging from 21% to 57% of cases. It can be generalized, or affect only the face and trunk, and has characteristics of papules or erythematous macules and sometimes presents as hives, flaking of palms, and soles or alopecia. It is described much less frequently in non-Caucasians [155].

Less commonly, cervical, axillary, occipital or generalized lymphadenopathies have been reported, which in some cases may persist for several months, even years after sero-conversion. Gastrointestinal manifestations are mostly diarrhea, nausea or vomiting. Cases of acute infection with chronic genital ulcerations have been described as the only symptom [156]. Rare cases of pancreatitis [151,157], acute renal failure [158,159], rhabdomyolisis [160−162], acute urinary retention [163], autoimmune hemolytic anemia [164,165], splenic rupture [166,167], and thrombocytopenia [168,169] have been reported.

The most serious clinical manifestations of acute infection are neurological conditions. Globally, these conditions affect up to 20% of patients and are the most frequent cause of hospital admission.

HIV is a neurotropic virus, and has been shown to invade multinucleated giant cells, mononuclear cells, endothelial cells, and microglia. Infection of macrophages and glia can result in toxicity and release of TNF-α, and IL-1β, which promote the release of arachidonic acid metabolites in astrocytes. During infection there is disruption of the blood-brain barrier [170], which correlates with an increase in neopterin, increased production of glutamate [171], increased activation of CD4 + and CD8 + lymphocytes [172] and macrophages [173]. The presence of neurological symptoms correlates with VL in cerebrospinal fluid (CSF), which is usually lower than in plasma [174]. Severe cases of acute encephalitis have been reported, particularly with affectation of the limbic region [175−177]. Neurological alterations are usually associated with radiological changes, and more recently noted, spectroscopic alterations in MRI suggested a rapid turnover of membrane lipids, which might mean glial inflammation without a large change in metabolites, indicative of gliosis or neuronal damage, demonstrated by an increased choline-creatine ratio (CHO/CR) in the basal ganglia and occipital gray matter [178]. As monocytes are cells susceptible to HIV infection and are capable of crossing the blood-brain barrier, the expansion of infected lymphocytes could be one of the main mechanisms of brain reservoir establishment and HIV-associated neurological injury. High choline levels also correlate with macrophagic activation of monocytes in the periphery (CD16 +) [179].

Peripheral nervous system involvement is also common, occurring in up to one-third of people with acute infection [180]. Guillain-Barre syndrome has been a recognized complication since 1989, sometimes fatal [181–183]. Depression, measured by the Beck Depression scale, can affect up to half of HIV-positive people, and in a third severely [184], and psychosis has been described in which brain biopsy showed HIV encephalitis with microglial nodules [185].

Furthermore, some patients may have simultaneous infections related to the local epidemiology, their risk behavior and, less frequently, an opportunistic infection in those who present a profound immunosuppression during the acute period.

With regard to HIV-related infections during the acute stage, these are generally episodes of herpes simplex or zoster, oral, or esophageal candidiasis [186,187] but cases of tuberculosis [188,189] cerebral toxoplasmosis [190], *P. jiroveci* [191], or cryptosporidiosis [192] have also been reported. Most of these cases were observed in patients with acute HIV infection who had a significant decrease in CD4 + lymphocytes [193].

17.2.3 Laboratory diagnosis

The diagnosis of acute infection has always been difficult due to several factors:

1. Low risk perception by patients
2. Lack of clinical suspicion
3. Presence of a window period for serological diagnosis
4. Low frequency of periodic testing in at-risk populations

The varied symptomatology and the lack of knowledge among patients and health workers make the clinical picture, although characteristic, go unnoticed. Even if the patient consults emergency personnel, the diagnosis is not usually hierarchized due to either lack of experience or omission of collecting information about possible sexual exposures in recent weeks. Therefore, the frequency of patients detected at this stage is relatively low, with less than 3% of patients currently under followup diagnosed during this phase [194].

Improving the capacity to diagnose acute infection is of fundamental importance. It is necessary to increase the level of alertness in exposed populations, provide greater training of health personnel to recognize these symptoms and, on top of that, institute changes at the national level to modify diagnostic algorithms. To facilitate the understanding of the required tests, it is useful to use the Fiebig stages (Fig. 17.4), in which the time when different testing modalities are positivized are clearly graphed. Fiebig carried out this work by studying plasma panels constructed with samples collected in a serial way in people with seroconversion whom he gradually observed, analyzing the moment when each test is positivet [195].

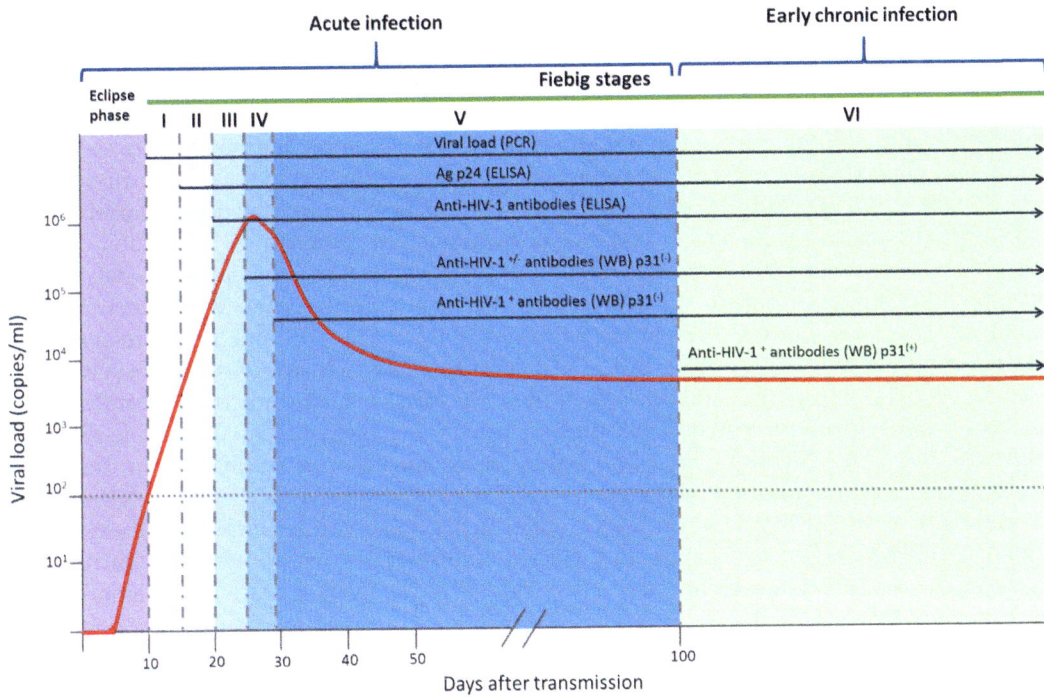

FIGURE 17.4 Fiebig stages to establish the time from infection to diagnosis. WB: Western Blot. *Source: Modified from McMichael A.J., Borrow P., Tomaras G.D., Goonetilleke N., Haynes B.F., The immune response during acute HIV-1 infection: clues for vaccine development. Nat Rev Immunol 2010; 11-23 [196].*

Tests that are performed to detect HIV infection include, then (sorted according to the time they become positive):

1. VL or other nucleic acid test
2. p24 antigen
3. Serology (4th generation ELISA)
4. 3rd or 2nd generation ELISA serology
5. Western blot

Once the virus gets into the body, an initial phase (i.e., eclipse phase) is established. This period lasts approximately 7 days, and all laboratory tests, both antibodies and antigens identification are negative. After the seventh day, the virus can already be detected in blood by qualitative and quantitative molecular techniques, while p24 antigen appears 4–5 days after VL. From 20 to 25 days after initial infection, most of the tests that detect antibodies are already positive [197]. This means that, if an acute infection is being searched, some other factors are needed to confirm it, namely—documentation of a recent negative test or examining the pattern of the western blot test where a negative or indeterminate result can still be obtained.

The sensitivity and specificity of molecular tests when the patient has symptoms (∼14 days after infection) is 100% and 97%, respectively [198]. Given their high cost, they have not been incorporated into individual diagnosis routinely, but in centers like blood banks, laboratories for the care of sexually transmitted diseases and in public laboratories, they have proven to be cost-effective. In these situations, samples are gathered in *pools* as a means to reduce the cost [199–202]. The implementation of this technique results in a significant increase in case detection. For example, in an experience limited to MSM, 11.5% more cases were diagnosed than via the 3rd-generation ELISA [203]. In other studies in high-risk individuals, the detection rate was 39 cases per 10,000 samples processed at a cost of USD 4535 per case detected [204].

When there is no access to *VL* pooling, p24 antigen can be used in individual cases. It can also be detected in blood by ELISA, and is positivized a few days before the onset of symptoms, disappearing with the increase in the level of antibodies in serum [205]. The sensitivity and specificity of p24 antigenemia in plasma are 89% and 100%, respectively. The p24 antigen is now included in the 4th-generation ELISA samples, which have been used since 2011, in Europe, and facilitates the detection of acute infection cases. Its introduction was delayed in the US because it was not approved by the FDA. From its approval, the CDC incorporated it and issued recommendations to change the algorithm to simplify the HIV diagnostic process and facilitate the detection of acute infection. It was concluded that the 4th-generation tests, which in the case of being reactive must be followed by an IgG-based test capable of discriminating HIV 1 or 2, allow for establishing most diagnoses. Molecular techniques are reserved for discordant cases [206–208].

Serological tests are highly sensitive and specific to detect general infection, but obviously the sensitivity in cases of acute infection decreases as you approach the date of exposure. The ARCHITECT test has a sensitivity of 99.94% and a specificity of 98.78%, and in cases of people with acute infection the sensitivity decreases to 83% [209]. Most 4th-generation tests can fail to identify some group O and group P and HIV-2 strains [210]. Since FDA approval, almost all countries are implementing 4th-generation ELISA, which has allowed HIV diagnosis to expand and increase cases detected in the acute infection phase [200,211].

Western blot detects antibodies against specific proteins of HIV-1 and HIV-2 and, according to the CDC, it requires the presence of at least two out of four bands (gp160, gp120, gp41, gp24) to be considered positive. Results with a single band with or without antibodies against additional bands are considered indeterminate and require a subsequent study with the same criteria to be positive. During the acute phase, the test yields a negative or indeterminate result, but from the fourth week of infection forward, it is usually positive and during the following weeks, additional bands will be added. The gp31 band usually appears at 12 weeks and its absence makes a recent infection suspect [212].

Innovation is clearly required to expand the diagnosis of acute infection, and recent advances allow to improve it [213]. Innovations need to be implemented, not only in a technological manner, but also the inclusion of rapid tests, self-testing, and complementary approaches [214,215], such as testing in pharmacies [216] or emergency departments [217].

In short, the key to diagnosis is based on clinical suspicion and performing laboratory tests at the right time. Early detection is important for patient management as allows for taking advantage of the potential benefits of early treatment and for epidemiological purposes, and because that allows for better estimates of the incidence, and potentially avoiding secondary transmissions [218] and detecting early changes in the epidemic [219]. Some cities have incorporated the active search for these cases as a tool for the control of the HIV epidemic [220,221]. Clearly, to expand treatment, new diagnostic approaches are needed and the advancement of technology is creating opportunities to achieve this expansion of testing.

17.2.4 Antiretroviral treatment

Since 1987, when zidovudine was first used for the treatment of HIV, we have witnessed one of the fastest advances in pharmaceutical drug development, resulting in more than 40 different drugs and coformulations approved for treating HIV today. In 1996, the demonstration that a triple-drug therapy could suppress viral replication in plasma established a new paradigm in the treatment of HIV. Two nucleoside reverse transcriptase inhibitors, combined with a nonnucleoside reverse transcriptase inhibitor, or one protease inhibitor, were defined as the cornerstone of ART. However, high pill burden, acute and long-term toxicities, drug-drug interactions, and specific monitoring requirements made simplified management difficult. The incorporation of new antiretroviral families, such as entry inhibitors [this group includes chemokine receptor antagonists CCR5 (maraviroc), fusion inhibitors (enfuvirtide) and binding inhibitors], but predominantly, the incorporation of integrase inhibitors expanded the option of treatment for patients harboring resistant virus.

Faced with the impossibility of achieving eradication, the objectives of treatment remain the same as in recent decades: to suppress the VL in a sustained way to improve immune functions and preserve future treatment options, improve quality of life, eliminate onward transmission, and reduce morbidity and mortality.

Integrase inhibitors of second generation (dolutegravir and bictegravir) are the most potent antiretroviral drugs, with the highest resistance barrier, and a safer profile, and are considered as the preferred treatment in all the international guidelines. Dolutegravir (in combination with lamivudine and tenofovir), is the WHO recommended therapy for all populations. Dolutegravir with lamivudine is also accepted by international guidelines and has demonstrated the same efficacy as a three-drug regimen.

Novel delivery mechanisms for antiretrovirals like injections and implants is an issue of interest among people living with HIV, since it means not taking a daily pill [222]. In this sense, ATLAS and FLAIR clinical trials yielded that the intramuscular injection of a long-acting second generation integrase inhibitor cabotegravir, in combination with the a long-acting nonnuclesoside transcriptase reverse inhibitor rilpivirine was noninferior to regular oral ART for suppressing VL [223,224]. This combination is currently approved in the United States, Canada, and Europe. Cabotegravir is also approved as a preventive strategy (see PrEP).

Novel antiretroviral drugs are in an advanced stage in clinical trials. Islatravir, a nucleoside reverse transcriptase translocation inhibitor that has shown to inhibit HIV-1 replication with higher potency, including on drug-resistant strains and long half-life. These characteristics makes islatravir particularly interesting for long-acting formulations, such as injections or subdermal drug-eluting implant that can last up to one year [225].

Lenacapavir (GS-6207), a capsid binding inhibitor, is effective in T cells and PBMCs at picomolar concentrations with no cross-resistance with protease, reverse transcriptase, or integrase inhibitors. Its pharmacokinetics also make this drug a great candidate for long-acting treatment, as it can sustain adequate levels during 6 months after a subcutaneous injection [222].

These both compounds are the promise for new combinations that can be used for treatment or prevention. The ART field is continuously evolving and new generation of long acting drugs with stronger potency, multimodal mechanism of action, and minimal side-effects will continue to arise. In the same way, new drug delivery technologies such as injectables and implants will certainly become more important in the upcoming years.

17.3 Strategies to eliminate human immunodeficiency virus as a public health threat

17.3.1 Community participation

The involvement of vulnerable populations in the planning, communication, and provision of HIV interventions is a crucial strategy for establishing a health response according to their expectations and needs. Discrimination and stigma continue to be the main barrier in terms of access to care for key populations [226]. That is why plans for community engagement and specific training curricula are urgently needed in separate care for MSM and transgender women. In this context, community-delivered services may be a promising approach, looking for higher reach, uptake, and adherence. Online technologies have been demonstrated to be effective targeting MSM, particularly those not self-identified as gay [227]. Regarding transgender people, programs including integrated care, harm reduction, mental health support, STI screening, treatment, and gender affirmation should improve their healthcare [228].

17.3.2 Expansion of testing

Reinforcement of HIV testing volume is crucial. Physicians routinely deal with the need to discuss sexual activity and preferences with their patients. Incorporation of HIV testing into routine care may overcome this barrier. Considering that MSM is a heterogeneous population, facilitated access for them is recommended. Greater comprehension of the various visions and contexts of patients vision will improve the implementation of different options and the frequency of HIV testing [226].

Another interesting way of addressing this issue is to expand self-testing of key populations, which has been shown to increase the frequency of testing and results in earlier identification of positive cases and helps to reach first-time testers [229]. For example, a multidisciplinary study in Argentina of a transgender women cohort compared four home-based HIV testing interventions and yielded that it is feasible to perform rapid tests followed by molecular diagnoses with improved effectiveness. Moreover, a history of STIs and of never being tested for HIV were highly correlated with positive diagnoses, showing the importance of reaching this population [230].

Elimination of structural barriers can also increase testing, such as modifying punitive laws that criminalize sexual and gender minorities, reducing pretest counseling time, and allow testing without a signed informed consent.

17.3.3 Preexposure prophylaxis

Along with male circumcision and condom use, PrEP is a novel instrument for preventing HIV transmission, particularly in vulnerable populations. Current research is promising on PrEP efficacy, but it is highly dependent on adherence. Hence, other possibilities are being tested such as intermittent, sex-driven PrEP (i.e. usage only before and after high-risk sexual encounters), but with inconclusive outcomes [43].

Latin America is now a cutting-edge research center on PrEP. High enrollment rates in iPrEX, MTN-017, ACTG A5175, HPTN-077, HPTN-083, and Discover trials demonstrates its high acceptability. The ImPrEP study is an ongoing, same-day oral PrEP study which preliminary reports showed 0.26% baseline prevalence of acute HIV infection, and a high prevalence of active syphilis, chlamydia, and gonorrhea with 9.9%, 11.7% and 7.4%, respectively, high uptake, and adherence [226].

A significant decline in HIV transmission is seen in places where treatment, testing, and PrEP uptake rates are high. However, only a handful of countries provide access to PrEP, and most low- & middle-income countries rely on either programs funded by international donors, or small demonstration projects targeting people at risk. Therefore, a more ambitious approach to PrEP politics is needed. This includes normalizing PrEP, making it available for a wider range of people, offering PrEP outside the boundaries of an HIV consultation, and a better PrEP promoting focusing on positive motivators, such as decreased anxiety and increased pleasure or intimacy [231].

New PrEP modalities are a promising way of increasing its acceptancy, since daily oral administration is not suitable for everyone, especially considering that adherence is fundamental. For instance, a monthly dapivirine vaginal ring and long-acting injectables are being developed. In this line, on December 2021, the FDA approved the administration of bimonthly injectable cabotegravir (Apretude), after showing superiority compared to oral tenofovir (Truvada) in randomized clinical trials. Injection site reactions were common, but declines along the time, and were associated with study drug discontinuation in very few individuals [232]. The only limitation to these approaches is the access and the unknown price policy for developing countries.

Advances in both investigation and programmatic implementation of PrEP will certainly help to reduce new HIV diagnoses. However, as the use of PrEP grows, addressing other STI transmissions will be necessary.

17.3.4 Antiretroviral treatment

The beneficial effect of ART in reducing HIV transmission is well documented in the publication of the ACTG 076 study, which showed that a course of zidovudine in pregnant HIV-positive women reduced mother-to-child transmission from 22.6% to 7.6% [233]. During sexual intercourse, transmission is reduced to almost negligible values when the

HIV-positive individual is on effective ART [140,141,234]. At the community level, several ecological studies conducted in Canada [235], China [236], and South Africa [237] show how the expansion of treatment programs is associated with a reduction in the incidence of new HIV infections.

These studies provide the evidence base for the "treatment as prevention" strategy, where reducing community VL through widespread expansion of diagnosis and immediate treatment of positive cases could reduce HIV transmission in the community to levels that allow the elimination of the epidemic. The implementation of this strategy would require a significant investment of resources and exceptionally large efforts [238].

The acute phase of infection is a period of very high transmissibility and, in this line, it is estimated that between 20% and 50% of new infections are acquired from a person with acute or recent HIV infection [239,240]. Therefore, a high rate of people in the seroconversion phase within a community could limit the effectiveness of these interventions. In a mathematical model in Malawi, where it is estimated that 38% of new infections are associated with a case of recent infection, it was established that if cases of acute infection are not considered, long-term elimination of HIV is practically impossible and would require the diagnosis and treatment of all cases of chronic infection in the whole community. The addition of interventions to identify and treat people in early stages of infection dramatically improves the model with a significant reduction in the incidence of new cases and the possibility of the elimination of transmission at 30 years [241].

The main goal in identifying individuals at early stages of infection is to improve the rate of diagnosis. Barriers to testing access should be identified and lowered, particularly for those at higher risk of infection. The perception of a lack of risk, the need to attend several times to perform the test and return for the results, and the absence of innovation such as the promotion of self-testing are the most frequently mentioned barriers [242]. In spite of this, even if patients consult doctors in time, we usually miss the opportunity to diagnose 70% of cases, despite the presence of classic symptoms or our recognizing a risky exposure [243]. Therefore, it is very important to raise awareness and improve knowledge about acute retroviral syndrome among doctors and populations at high risk, since many studies show that more than half of the people with the highest exposure are unaware of the symptoms of acute infection [244,245]. Another difficulty is the low priority for health systems to identify cases of acute infection [246], due to the high cost of technologies that can detect acute HIV [207,247].

These concepts justify the need to implement integrated diagnostic systems from reference laboratories with epidemiological surveillance systems. This should include case definition, appropriate identification, verification of laboratory information and collection of clinical information, active communication with the individual to ensure that the necessary information is provided and that treatment begins quickly, and other services such as notification and contact management [221]. When implementing these programs, it should be considered that there are successful experiences in some cities that can serve as guidance for these efforts [248].

17.3.5 Human immunodeficiency virus cure

A recent report by the International AIDS Society highlighted the importance of several research goals to be addressed within 2021−26, which were better understanding HIV reservoirs and its measurements, mechanisms of virus control, targeting the provirus and the immune system, cell and gene therapy, pediatric remission and cure, and social, behavioral and ethical aspects if the HIV cure [249].

It could be said that antiretroviral treatment initiated early allows for reestablishing the damage caused to the immune system, and in a sustained way can achieve an immuno-virological situation similar to that of EC individuals, both in terms of residual replication, volume of reservoirs and the functional profile of CD4 + and CD8 + lymphocytes [250]. That is why these patients are in a privileged place when it comes to selecting patients to evaluate future eradication strategies.

In recent years, there has been renewed interest in the cure agenda. There have been very significant advances, such as the identification of different cell populations that make up the viral reservoir [251], a better description of the mechanisms that regulate immune responses [252], the mechanisms of elimination of infected CD4 + cells [36], the mechanisms of latency and how to evade them [253] and, strategies to reduce reservoirs [254], among many others.

One of the biggest challenges today is to be able to standardize laboratory studies to measure reservoirs. More than 90% of the viral DNA that can be measurable by standard studies does not replicate and therefore should not be considered part of the reservoir. Methods that can better describe and explain the discrepancy between viral culture-based and molecular methods are needed. Eriksson compared different methods against the laborious method of viral growth of infected CD4 + cells in 30 people under treatment in which he was able to show large differences in the frequencies of infected cells, due to the poor sensitivity of PCR studies to detect active replication and by the detection of cells infected with defective viruses [255]. That is why the standardization and clinical valida-tion of methods to quantify the viral reservoir is a priority, being necessary methods that can measure not only the virus within CD4 + cells but also in other cells and tissues (macrophages, monocytes, gastrointestinal tissue, etc.).

Therapeutic approaches require additional measures beyond mere ART. Combination treatment [256], even started at very early stages [257] has not been sufficient to achieve infection control without treatment. The virus integrated into latently infected cells con-tinues to be a barrier to eradication. Histone deacetylase inhibitors (HDACi) are able to reactivate latency in infected cells but have not been able to reduce the total amount of integrated DNA [254]. Vorinostat, romdepsin, and panobinostat have been evaluated with positive results but drugs that can kill cells that restart viral transcription should be identified.

These cells could be eliminated by the immune system, so improving specific immune cell responses, innate immunity and neutralizing antibodies are attractive approaches for researchers. Some authors propose to improve cytotoxic responses before stimulating latent virus reactivation based on in vitro studies that have been shown to be effective [258]. In this sense, combined approaches appear to be the most promising strategies. The combination of HDACi with therapeutic vaccines (based on DNA vectors, viral or DC vec-tors), cytokines, or immunomodulators such as PD-1 inhibitors offers a possible additional tool [259]. Autologous DC vaccines were shown to generate a strong specific response, positioning them as an element to be combined in functional cure approaches [252]. Monoclonal neutralizing antibodies are other candidates to be used in immunotherapy

TABLE 17.5   Strategies exploring cure in patients with acute HIV-1 infection.

Identification of the study in clinical trial database	Place	Phase	Intervention
NCT02231281	China	III	HIV-specific autologous T cells + ART versus ART alone
NCT02018510	USED	I	Neutralizing monoclonal ac (3BNC117)
NCT01950325	USED	I	Neutralizing monoclonal ac (VRC601)
NCT02028403	USED	I	BMS936559 (anti-PD1) in suppressed patients
CHERUB001	UK	I	IV immunoglobulin in acute patients
NCT01365065	Australia	II	Vorinostat in suppressed patients
NCT01319383	USED	I/II	Vorinostat in suppressed patients
RIVER	UK	II	Vorinostat + vaccine (ChAdV63. HIV + MVA. HIVconsv)
NCT01933594	USED	I/II	Romdepsin in suppressed patients
REDUCED	Denmark	I/II	Romdepsin + vaccine (Vacc-4x) + GM-CSF
NCT01944371	USED	I/II	Disulfiram in suppressed patients
NCT02071095	USED	I/II	Poli-ICLC (TLR-3 agonist) + ART in acute pac.

Modified from Thornhill J, Fidler S, Frater, J. Advancing the HIV cure agenda: the next 5 years. Curr Opin Infect Dis 2015;28:1−9.

and have been tested in animal models with good response. Antibodies directed to CD4 binding or the V3 region reduced viremia and proviral DNA although viral escape situations were detected [260,261]. Table 17.5 mentions some strategies currently under study.

In summary, the search for combinations that can enable a functional cure in people with acute infection is broad. Moreover, most HIV research is performed within high-income countries where viral burden is comparatively low to low- & middle-income ones. This means that more basic research is needed since HIV infection differs by strains and so do host immunity, influenced by ethnicity, biological sex and geographical location. With the availability of safe and less toxic drugs, the evidence of a long life expectancy and the unknown risks of some proposed interventions, in particular those that require discontinuation of treatment, the need to discuss the ethical limits of the incorporation of humans into some of these strategies is imposed [249,262]. The road is difficult and surely still long, but it is important to keep pace and hope.

References

[1] Zhu P, et al. Distribution and three-dimensional structure of AIDS virus envelope spikes. Nature 2006;441:847−52.
[2] Seitz R. Human immunodeficiency virus (HIV). Transfus Med Hemotherapy 2016;43:203−22.
[3] Pereira LA, Bentley K, Peeters A, Churchill MJ, Deacon NJ. A compilation of cellular transcription factor interactions with the HIV-1 LTR promoter. Nucleic Acids Res 2000;28:663−8.
[4] Dilernia DA, et al. Analysis of HIV type 1 BF recombinant sequences from south america dates the origin of CRF12-BF to a recombination event in the 1970s. AIDS Res Hum Retroviruses 2011;27:569−78.
[5] Bbosa N, Kaleebu P, Ssemwanga D. HIV subtype diversity worldwide. Curr Opin HIV AIDS 2019;14:153−60.

[6] Lindemann D, Steffen I, Pöhlmann S. Cellular entry of retroviruses. Adv Exp Med Biol 2013;790:128−49.

[7] Gottlieb GJ, et al. A preliminary communication on extensively disseminated Kaposi's sarcoma in young homosexual men. Am J Dermatopathol 1981;3:111−14.

[8] Centers for Disease Control (CDC). Pneumocystis pneumonia−Los Angeles. MMWR Morb Mortal Wkly Rep 1981;30:250−2.

[9] Barré-Sinoussi F, et al. Isolation of a T-lymphotropic retrovirus from a patient at risk for acquired immune deficiency syndrome (AIDS). Science 1983;220:868−71.

[10] Hooper E, et al. Search for the origin of HIV and AIDS. Science 2000;289:1140−1.

[11] Agarwal-Jans S. Timeline: HIV. Cell 2020;183:550.

[12] UNAIDS. Confronting inequalities-lessons for pandemic responses from 40 years of AIDS. Global AIDS update. <https://www.unaids.org/sites/default/files/media_asset/2021-global-aids-update_en.pdf>; 2021.

[13] Richardson ET, et al. Gender inequality and HIV transmission: a global analysis. J Int AIDS Soc 2014;17:1−5.

[14] Keele BF, Estes JD. Barriers to mucosal transmission of immunodeficiency viruses. Blood 2011;118:839−46.

[15] Nishimura Y, Martin MA. The acute HIV infection: implications for intervention, prevention and development of an effective AIDS vaccine. Curr Opin Virol 2011;1:204−10.

[16] Shaw GM, Hunter E. HIV transmission. Cold Spring Harb Perspect Med 2012;2.

[17] Steinman RM. Decisions about dendritic cells: past, present, and future. Annu Rev Immunol 2012;30:1−22.

[18] Harman AN, Kim M, Nasr N, Sandgren KJ, Cameron PU. Tissue dendritic cells as portals for HIV entry. Rev Med Virol 2013;23:319−33.

[19] Teleshova N, Frank I, Pope M. Immunodeficiency virus exploitation of dendritic cells in the early steps of infection. J Leukoc Biol 2003;74:683−90.

[20] Piguet V, Steinman RM. The interaction of HIV with dendritic cells: outcomes and pathways. Trends Immunol 2007;28:503−10.

[21] Wu L, KewalRamani VN. Dendritic-cell interactions with HIV: infection and viral dissemination. Nat Rev Immunol 2006;6:859−68.

[22] Wang J-H, Janas AM, Olson WJ, Wu L. Functionally distinct transmission of human immunodeficiency virus Type 1 mediated by immature and mature dendritic cells. J Virol 2007;81:8933−43.

[23] Miller CJ, et al. Propagation and dissemination of infection after vaginal transmission of simian immunodeficiency virus. J Virol 2005;79:9217−27.

[24] Borrow P. Innate immunity in acute HIV-1 infection. Curr Opin HIV AIDS 2011;6:353−63.

[25] Yandrapally S, Mohareer K, Arekuti G, Vadankula GR, Banerjee S. HIV co-receptor-tropism: cellular and molecular events behind the enigmatic co-receptor switching. Crit Rev Microbiol 2021;47:499−516.

[26] González N, et al. SDF-1/CXCL12 production by mature dendritic cells inhibits the propagation of X4-tropic HIV-1 isolates at the dendritic cell-T-cell infectious synapse. J Virol 2010;84:4341−51.

[27] Matreyek KA, Engelman A. Viral and cellular requirements for the nuclear entry of retroviral preintegration nucleoprotein complexes. Viruses 2013;5:2483−511.

[28] Yang H, et al. Structural insight into HIV-1 capsid recognition by rhesus TRIM5α. Proc Natl Acad Sci U S A 2012;109:18372−7.

[29] Coiras M, López-Huertas MR, Pérez-Olmeda M, Alcamí J. Understanding HIV-1 latency provides clues for the eradication of long-term reservoirs. Nat Rev Microbiol 2009;7:798−812.

[30] Lever AML, Jeang K-T. Replication of human immunodeficiency virus type 1 from entry to exit. Int J Hematol 2006;84:23−30.

[31] Burdick Ryan C, Li Chenglei, Munshi MohamedHusen, Rawson Jonathan M O, Nagashima Kunio, Hu Wei-Shau, et al. HIV-1 uncoats in the nucleus near sites of integration. Proc Natl Acad Sci U S A 2020;117 (10):5486−93. Available from: https://doi.org/10.1073/pnas.1920631117.

[32] Feng Y, Baig TT, Love RP, Chelico L. Suppression of APOBEC3-mediated restriction of HIV-1 by Vif. Front Microbiol 2014;5:450.

[33] Neil SJD, Zang T, Bieniasz PD. Tetherin inhibits retrovirus release and is antagonized by HIV-1 Vpu. Nature 2008;451:425−30.

[34] Cummins NW, Badley AD. Mechanisms of HIV-associated lymphocyte apoptosis: 2010. Cell Death Dis 2010;1:e99.

[35] Douek DC, et al. HIV preferentially infects HIV-specific CD4 + T cells. Nature 2002;417:95−8.

[36] Cummins NW, Badley AD. Making sense of how HIV kills infected CD4T cells: implications for HIV cure. Mol Cell Ther 2014;2:20.

[37] Card CM, et al. Reduced cellular susceptibility to in vitro HIV infection is associated with CD4 + T cell quiescence. PLoS One 2012;7.

[38] Burdo TH, et al. Soluble CD163 made by monocyte/macrophages is a novel marker of HIV activity in early and chronic infection prior to and after antiretroviral therapy. J Infect Dis 2011;204:154−63.

[39] Vinikoor MJ, et al. Antiretroviral therapy initiated during acute HIV infection fails to prevent persistent T-cell activation. J Acquir Immune Defic Syndr 2013;62:505−8.

[40] Giorgi JV, et al. Shorter survival in advanced human immunodeficiency virus type 1 infection is more closely associated with T lymphocyte activation than with plasma virus burden or virus chemokine coreceptor usage. J Infect Dis 1999;179:859−70.

[41] Younas M, Psomas C, Reynes J, Corbeau P. Immune activation in the course of HIV-1 infection: causes, phenotypes and persistence under therapy. HIV Med n/a-n/a 2015;. Available from: https://doi.org/10.1111/hiv.12310.

[42] High KP, et al. HIV and aging: state of knowledge and areas of critical need for research. A report to the NIH Office of AIDS Research by the HIV and Aging Working Group. J Acquir Immune Defic Syndr 2015;60:1−18.

[43] Ghosn J, Taiwo B, Seedat S, Autran B, Katlama C. Seminar: HIV. Lancet 2018;392:685−97.

[44] Vrisekoop N, Mandl JN, Germain RN. Life and death as a T lymphocyte: from immune protection to HIV pathogenesis. J Biol 2009;8:91.

[45] Douek DC, Roederer M, Koup RA. Emerging concepts in the immunopathogenesis of AIDS. Annu Rev Med 2009;60:471−84.

[46] Zhang Z, et al. CD4 + CD38 + HLA-DR + cells: a predictor of viral set point in Chinese men with primary HIV infection who have sex with men. Jpn J Infect Dis 2011;64:423−5.

[47] Lu L, Wang J, Yang Q, Xie X, Huang Y. The role of CD38 in HIV infection. AIDS Res Ther 2021;18.

[48] Petitjean G, et al. Level of double negative T cells, which produce TGF-β and IL-10, predicts CD8 T-cell activation in primary HIV-1 infection. AIDS 2012;26:139−48.

[49] Biancotto A, et al. Abnormal activation and cytokine spectra in lymph nodes of people chronically infected with HIV-1. Blood 2007;109:4272−9.

[50] Decrion AZ, Dichamp I, Varin A, Herbein G. HIV and inflammation. Curr HIV Res 2005;3:243−59.

[51] Hamlyn E, et al. Interleukin-6 and D-dimer levels at seroconversion as predictors of HIV-1 disease progression. AIDS 2014;28:869−74.

[52] Leeansyah E, Malone DFG, Anthony DD, Sandberg JK. Soluble biomarkers of HIV transmission, disease progression and comorbidities. Curr Opin HIV AIDS 2013;8:117−24.

[53] Gay C, et al. Cross-sectional detection of acute HIV infection: timing of transmission, inflammation and antiretroviral therapy. PLoS One 2011;6:e19617.

[54] Gutierrez M, del M, Mateo MG, Vidal F, Domingo P. Does choice of antiretroviral drugs matter for inflammation? Expert Rev Clin Pharmacol 2019;12:389−96.

[55] Vujkovic-Cvijin I, et al. The complement pathway is activated in people with HIV and is associated with non-AIDS comorbidities. J Infect Dis 2021;224:1405−9.

[56] Forchielli ML, Walker WA. The role of gut-associated lymphoid tissues and mucosal defence. Br J Nutr 2005;93(Suppl. 1):S41−8.

[57] Liu JZ, Pezeshki M, Raffatellu M. Th17 cytokines and host-pathogen interactions at the mucosa: dichotomies of help and harm. Cytokine 2009;48:156−60.

[58] Brenchley JM, Douek DC. Microbial translocation across the GI tract. Annu Rev Immunol 2012;30:149−73.

[59] Maresca M, et al. The virotoxin model of HIV-1 enteropathy: involvement of GPR15/Bob and galactosylceramide in the cytopathic effects induced by HIV-1 gp120 in the HT-29-D4 intestinal cell line. J Biomed Sci 2003;10:156−66.

[60] Balagopal A, et al. Kupffer cells are depleted with HIV immunodeficiency and partially recovered with antiretroviral immune reconstitution. AIDS 2009;23:2397−404.

[61] Schacker TW, et al. Collagen deposition in HIV-1 infected lymphatic tissues and T cell homeostasis. J Clin Invest 2002;110:1133−9.

[62] Litzman J, et al. Chronic immune activation in common variable immunodeficiency (CVID) is associated with elevated serum levels of soluble CD14 and CD25 but not endotoxaemia. Clin Exp Immunol 2012;170:321−32.

[63] Sandler NG, et al. Plasma levels of soluble CD14 independently predict mortality in HIV infection. J Infect Dis 2011;203:780–90.

[64] Somsouk M, et al. Gut epithelial barrier and systemic inflammation during chronic HIV infection. AIDS 2015;29:43–51.

[65] Romero-Sánchez M, et al. Different biological significance of sCD14 and LPS in HIV-infection: importance of the immunovirology stage and association with HIV-disease progression markers. J Infect 2012;65:431–8.

[66] Nockher WA, Bergmann L, Scherberich JE. Increased soluble CD14 serum levels and altered CD14 expression of peripheral blood monocytes in HIV-infected patients. Clin Exp Immunol 1994;98:369–74.

[67] Aziz N, et al. Acute HIV-1 Seroconversion with an Unusual Plasma Biomarker Profile. Clin Vaccine Immunol 2013;20:1774–7.

[68] Guadalupe M, et al. Severe CD4 + T-cell depletion in gut lymphoid tissue during primary human immunodeficiency virus type 1 infection and substantial delay in restoration following highly active antiretroviral therapy. J Virol 2003;77:11708–17.

[69] Dagenais-Lussier X, et al. Current topics in HIV-1 pathogenesis: the emergence of deregulated immunometabolism in HIV-infected subjects. Cytokine Growth Factor Rev 2015;. Available from: https://doi.org/10.1016/j.cytogfr.2015.09.001.

[70] Pei L, et al. Plasma metabolomics reveals dysregulated metabolic signatures in HIV-associated immune reconstitution inflammatory syndrome. Front Immunol 2021;12:693074.

[71] Dröge W. Free radicals in the physiological control of cell function. Physiol Rev 2002;82:47–95.

[72] Valko M, et al. Free radicals and antioxidants in normal physiological functions and human disease. Int J Biochem Cell Biol 2007;39:44–84.

[73] Fang FC. Antimicrobial actions of reactive oxygen species. MBio 2011;2.

[74] Chandel N, et al. VDR hypermethylation and HIV-induced T cell loss. J Leukoc Biol 2013;93:623–31.

[75] Petrovas C, et al. PD-1 is a regulator of virus-specific CD8 + T cell survival in HIV infection. J Exp Med 2006;203:2281–92.

[76] Tkachev V, et al. Programmed death-1 controls T cell survival by regulating oxidative metabolism. J Immunol 2015;194:5789–800.

[77] Pulliam L, Calosing C, Sun B, Grunfeld C, Rempel H. Monocyte activation from interferon-α in HIV infection increases acetylated LDL uptake and ROS production. J Interferon Cytokine Res 2014;34:822–8.

[78] Neurauter G, et al. Effective antiretroviral therapy reduces degradation of tryptophan in patients with HIV-1 infection. Adv Exp Med Biol 2003;527:317–23.

[79] Routy JP, et al. Clinical relevance of kynurenine pathway in HIV/AIDS: an immune checkpoint at the crossroads of metabolism and inflammation. AIDS Rev 2015;17:96–106.

[80] Gaardbo JC, et al. Increased tryptophan catabolism is associated with increased frequency of CD161 + Tc17/MAIT Cells and Lower CD4 + T-cell count in HIV-1 infected patients on cART after 2 years of follow-up. J Acquir Immune Defic Syndr 2015;70:228–35.

[81] Wang X, Mehra S, Kaushal D, Veazey RS, Xu H. Abnormal tryptophan metabolism in HIV and *Mycobacterium tuberculosis* Infection. Front Microbiol 2021;12.

[82] Calder PC, Dimitriadis G, Newsholme P. Glucose metabolism in lymphoid and inflammatory cells and tissues. Curr Opin Clin Nutr Metab Care 2007;10:531–40.

[83] Palmer CS, et al. Increased glucose metabolic activity is associated with CD4 + T-cell activation and depletion during chronic HIV infection. AIDS 2014;28:297–309.

[84] Hollenbaugh JA, Munger J, Kim B. Metabolite profiles of human immunodeficiency virus infected CD4 + T cells and macrophages using LC–MS/MS analysis. Virology 2011;415:153–9.

[85] Azoulay-Zohar H, Israelson A, Abu-Hamad S, Shoshan-Barmatz V. In self-defence: hexokinase promotes voltage-dependent anion channel closure and prevents mitochondria-mediated apoptotic cell death. Biochem J 2004;377:347–55.

[86] Sen S, et al. Role of Hexokinase-1 in the survival of HIV-1-infected macrophages. Cell Cycle 2015;14:980–9.

[87] Altfeld M, Gale M. Innate immunity against HIV-1 infection. Nat Immunol 2015;16:554–62.

[88] Altfeld M, et al. Influence of HLA-B57 on clinical presentation and viral control during acute HIV-1 infection. AIDS 2003;17:2581–91.

[89] Willberg CB, et al. Rapid progressing allele HLA-B35 Px restricted anti-HIV-1 CD8 + T cells recognize vestigial CTL epitopes. PLoS One 2010;5:e10249.

[90] Fontaine J, Chagnon-Choquet J, Valcke HS, Poudrier J, Roger M. High expression levels of B lymphocyte stimulator (BLyS) by dendritic cells correlate with HIV-related B-cell disease progression in humans. Blood 2011;117:145–55.

[91] Liu F, Bevins CL. A sweet target for innate immunity. Nat Med 2010;16:263–4.

[92] Tandon R, et al. Galectin-9 is rapidly released during acute HIV-1 infection and remains sustained at high levels despite viral suppression even in elite controllers. AIDS Res Hum Retroviruses 2014;30:654–64.

[93] Iwasaki-Hozumi H, Chagan-Yasutan H, Ashino Y, Hattori T. Blood levels of galectin-9, an immuno-regulating molecule, reflect the severity for the acute and chronic infectious diseases. Biomol 2021;11:430.

[94] Saitoh H, et al. Rapid decrease of plasma galectin-9 levels in patients with acute HIV infection after therapy. Tohoku J Exp Med 2012;228:157–61.

[95] Mellors JW, et al. Prognosis in HIV-1 infection predicted by the quantity of virus in plasma. Science 1996;272:1167–70.

[96] Turnbull EL, et al. Escape is a more common mechanism than avidity reduction for evasion of CD8 + T cell responses in primary human immunodeficiency virus type 1 infection. Retrovirology 2011;8:41.

[97] Day CL, et al. PD-1 expression on HIV-specific T cells is associated with T-cell exhaustion and disease progression. Nature 2006;443:350–4.

[98] Cecchinato V, et al. Immune activation driven by CTLA-4 blockade augments viral replication at mucosal sites in simian immunodeficiency virus infection. J Immunol 2008;180:5439–47.

[99] Walker LM, et al. Broad neutralization coverage of HIV by multiple highly potent antibodies. Nature 2011;477:466–70.

[100] Liao H-X, et al. Co-evolution of a broadly neutralizing HIV-1 antibody and founder virus. Nature 2013;496:469–76.

[101] Wei X, et al. Antibody neutralization and escape by HIV-1. Nature 2003;422:307–12.

[102] Burton DR, et al. HIV vaccine design and the neutralizing antibody problem. Nat Immunol 2004;5:233–6.

[103] González N, Alvarez A, Alcamí J. Broadly neutralizing antibodies and their significance for HIV-1 vaccines. Curr HIV Res 2010;8:602–12.

[104] Shingai M, et al. Passive transfer of modest titers of potent and broadly neutralizing anti-HIV monoclonal antibodies block SHIV infection in macaques. J Exp Med 2014;211:2061–74.

[105] Caskey M, et al. Viraemia suppressed in HIV-1-infected humans by broadly neutralizing antibody 3BNC117. Nature 2015;522:487–91.

[106] Lewis GK, DeVico AL, Gallo RC. Antibody persistence and T-cell balance: two key factors confronting HIV vaccine development. Proc Natl Acad Sci U S A 2014;111:15614–21.

[107] Alter G, Moody MA. The humoral response to HIV-1: new insights, renewed focus. J Infect Dis 2010;202 (Suppl):S315–22.

[108] Simek MD, et al. Human immunodeficiency virus type 1 elite neutralizers: individuals with broad and potent neutralizing activity identified by using a high-throughput neutralization assay together with an analytical selection algorithm. J Virol 2009;83:7337–48.

[109] Zhou T, et al. Structural basis for broad and potent neutralization of HIV-1 by antibody VRC01. Science (80-) 2010;329:811–17.

[110] McCoy LE, Weiss RA. Neutralizing antibodies to HIV-1 induced by immunization. J Exp Med 2013;210:209–23.

[111] Liu Y, Cao W, Sun M, Li T. Broadly neutralizing antibodies for HIV-1: efficacies, challenges and opportunities. Emerg Microbes Infect 2020;9:194–206.

[112] McMichael AJ, Rowland-Jones SL. Cellular immune responses to HIV. Nature 2001;410:980–7.

[113] Schmitz JE, et al. Control of viremia in simian immunodeficiency virus infection by CD8 + lymphocytes. Science 1999;283:857–60.

[114] Addo MM, et al. Fully Differentiated HIV-1 specific CD8 + T effector cells are more frequently detectable in controlled than in progressive HIV-1 infection. PLoS One 2007;2:e321.

[115] Soghoian DZ, et al. HIV-specific cytolytic CD4T cell responses during acute HIV infection predict disease outcome. Sci Transl Med 2012;4:123–5.

[116] Streeck H, et al. Emergence of individual HIV-specific CD8T cell responses during primary HIV-1 infection can determine long-term disease outcome. J Virol 2014;88:12793–801.

[117] Betts MR, et al. HIV nonprogressors preferentially maintain highly functional HIV-specific CD8 + T cells. Blood 2006;107:4781–9.

[118] Almeida JR, et al. Antigen sensitivity is a major determinant of CD8 + T-cell polyfunctionality and HIV-suppressive activity. Blood 2009;113:6351−60.

[119] Ferre AL, et al. Mucosal immune responses to HIV-1 in elite controllers: a potential correlate of immune control. Blood 2008;113:3978−89.

[120] Hersperger AR, et al. Perforin expression directly ex vivo by HIV-specific CD8 T-cells is a correlate of HIV elite control. PLoS Pathog 2010;6:e1000917.

[121] Kloosterboer N, et al. Natural controlled HIV infection: preserved HIV-specific immunity despite undetectable replication competent virus. Virology 2005;339:70−80.

[122] Lambotte O, Delfraissy J-F. HIV controllers: a homogeneous group of HIV-1 infected patients with a spontaneous control of viral replication. Pathol Biol (Paris) 2006;54:566−71.

[123] Lécuroux C, et al. CD8 T-cells from most HIV-infected patients lack ex vivo HIV-suppressive capacity during acute and early infection. PLoS One 2013;8:e59767.

[124] Frahm MA, et al. CD4 + CD8 + T cells represent a significant portion of the anti-HIV T cell response to acute HIV infection. J Immunol 2012;188:4289−96.

[125] Durand CM et al. A human immunodeficiency virus controller with a large population of CD4 + CD8 + double-positive T cells. 2015. Available from: https://doi.org/10.1093/ofid/ofv039.

[126] Navarrete-Muñoz MA, Restrepo C, Benito JM, Rallón N. Elite controllers: a heterogeneous group of HIV-infected patients. 2020. Available from: https://doi.org/10.1080/21505594.2020.1788887.

[127] Sanchez-Merino V, et al. Detection of broadly neutralizing activity within the first months of HIV-1 infection. J Virol 2016;90:5231−45.

[128] Pantaleo G, Graziosi C, Fauci AS. New concepts in the immunopathogenesis of human immunodeficiency virus infection. N Engl J Med 1993;328:327−35.

[129] Miró JM, et al. Avances en el diagnóstico y tratamiento de la infección aguda por el VIH-1^ies. Enferm Infect Microbiol Clín (Ed) 2004;22:643−59.

[130] Simon V, Ho DD, Abdool Karim Q. HIV/AIDS epidemiology, pathogenesis, prevention, and treatment. Lancet 2006;368:489−504.

[131] Chen Y-J, et al. Risk factors for HIV-1 seroconversion among Taiwanese men visiting gay saunas who have sex with men. BMC Infect Dis 2011;11:334.

[132] Silva AP, Greco M, Fausto MA, Greco DB, Carneiro M. Risk factors associated with HIV infection among male homosexuals and bisexuals followed in an open cohort study: Project Horizonte, Brazil (1994−2010). PLoS One 2014;9:e109390.

[133] Beyrer C. Strategies to manage the HIV epidemic in gay, bisexual, and other men who have sex with men. Curr Opin Infect Dis 2014;27:1−8.

[134] Burns DN, Grossman C, Turpin J, Elharrar V, Veronese F. Role of oral pre-exposure prophylaxis (PrEP) in current and future HIV prevention strategies. Curr HIV/AIDS Rep 2014;11:393−403.

[135] Phanuphak N, Gulick RM. HIV treatment and prevention 2019: current standards of care. Curr Opin HIV AIDS 2020;15:4−12.

[136] De Schacht C, et al. High rates of HIV seroconversion in pregnant women and low reported levels of HIV testing among male partners in southern mozambique: results from a mixed methods study. PLoS One 2014;9:e115014.

[137] Johnson LF, et al. The contribution of maternal HIV seroconversion during late pregnancy and breastfeeding to mother-to-child transmission of HIV. J Acquir Immune Defic Syndr 2012;59:417−25.

[138] Williams B, et al. Repeat antenatal HIV testing in the third trimester: a study of feasibility and maternal uptake rates. HIV Med 2014;15:362−6.

[139] Yeganeh N, et al. HIV testing of male partners of pregnant women in Porto Alegre, Brazil: a potential strategy for reduction of HIV seroconversion during pregnancy. AIDS Care, 26. 2014. p. 790−4.

[140] Cohen MS, et al. Prevention of HIV-1 infection with early antiretroviral therapy. N Engl J Med 2011;. Available from: https://doi.org/10.1056/NEJMoa1105243.

[141] Rodger A, et al. HIV transmission risk through condomless sex if the HIV positive partner is on suppressive ART: PARTNER study, CROI 153LB, 2014.

[142] Kumi Smith M, Jewell BL, Hallett TB, Cohen MS. Treatment of HIV for the prevention of transmission in discordant couples and at the population level. in. Adv Exp Med Biol 2018;1075:125−62.

[143] Quinn TC, et al. Viral load and heterosexual transmission of human immunodeficiency virus type 1. Rakai Project Study Group. N Engl J Med 2000;342:921−9.

[144] Boily M-C, et al. Heterosexual risk of HIV-1 infection per sexual act: systematic review and *meta*-analysis of observational studies. Lancet Infect Dis 2009;9:118−29.

[145] Baggaley RF, White RG, Boily M-C. HIV transmission risk through anal intercourse: systematic review, *meta*-analysis and implications for HIV prevention. Int J Epidemiol 2010;39:1048−63.

[146] Baggaley RF, White RG, Boily M-C. Systematic review of orogenital HIV-1 transmission probabilities. Int J Epidemiol 2008;37:1255−65.

[147] Patel P, et al. Estimating per-act HIV transmission risk: a systematic review. AIDS 2014;28:1509−19.

[148] Scott HM, et al. Age, race/ethnicity, and behavioral risk factors associated with per contact risk of HIV infection among men who have sex with men in the United States. J Acquir Immune Defic Syndr 2014;65:115−21.

[149] Liu H, et al. Effectiveness of ART and condom use for prevention of sexual HIV transmission in serodiscordant couples: a systematic review and *meta*-analysis. PLoS One 2014;9:e111175.

[150] Hecht FM, et al. Use of laboratory tests and clinical symptoms for identification of primary HIV infection. AIDS 2002;16:1119−29.

[151] Vanhems P, et al. Comprehensive classification of symptoms and signs reported among 218 patients with acute HIV-1 infection. J Acquir Immune Defic Syndr 1999;21:99−106.

[152] Kassutto S, Rosenberg ES. Primary HIV type 1 infection. Clin Infect Dis 2004;38:1447−53.

[153] Miro JM, Sued O, Plana M, Pumarola T, Gallart T. Advances in the diagnosis and treatment of acute human immunodeficiency virus type 1 (HIV-1) infection. Enferm Infect Microbiol Clin 2004;22:643−59.

[154] Schacker T, Collier AC, Hughes J, Shea T, Corey L. Clinical and epidemiologic features of primary HIV infection. Ann Intern Med 1996;125:257−64.

[155] Bollinger RC, et al. Risk factors and clinical presentation of acute primary HIV infection in India. JAMA 1997;278:2085−9.

[156] Gentile M, et al. Acute HIV infection: a misleading presentation. Int J STD AIDS 2011;22:766−7.

[157] Bitar A, Altaf M, Sferra TJ. Acute pancreatitis: manifestation of acute HIV infection in an adolescent. Am. J Case Rep 2012;13:17−18.

[158] Ananworanich J, et al. Acute tubular nephropathy in a patient with acute HIV infection: review of the literature. AIDS Res Ther 2014;11:34.

[159] Gameiro J, Jorge S, Lopes JA. HIV and renal disease: a contemporary review. Int J STD AIDS 2018;29:714−19.

[160] Takahashi T. A case of primary human immunodeficiency virus infection with severe rhabdomyolysis without acute renal failure. Kansenshogaku Zasshi 2011;85:268−71.

[161] Babiker ZOE, Wingfield T, Galloway J, Snowden N, Ustianowski A. Extreme elevation of ferritin and creatine kinase in primary infection with HIV-1. Int J STD AIDS 2015;26:68−71.

[162] Same RG, McAleese S, Agwu AL, Arav-Boger R. Acute HIV in an Adolescent Male With Fever and Rhabdomyolysis. J Adolesc Health 2019;65:567−9.

[163] Abbass K, Qazi A, Markert RJ, Gul W. Primary HIV infection associated with acute urinary retention. J Int Assoc Physicians AIDS Care 2011;10:133−4.

[164] Sherry NL, Woolley IJ, Korman TM. Autoimmune haemolytic anaemia: an unusual presentation of HIV seroconversion disease. AIDS 2010;24:1968−70.

[165] Yen YF, et al. Human immunodeficiency virus infection increases the risk of incident autoimmune hemolytic anemia: a population-based cohort study in Taiwan. J Infect Dis 2017;216:1000−7.

[166] Yamashita T, Tokushige S, Maekawa R, Shiio Y. Reversible splenial lesion associated with acute HIV infection. Intern Med 2012;51:1643.

[167] Vallabhaneni S, Scott H, Carter J, Treseler P, Machtinger EL. Atraumatic splenic rupture: an unusual manifestation of acute HIV infection. AIDS Patient Care STDS 2011;25:461−4.

[168] Ghosn J, et al. Thrombocytopenia during primary HIV-1 infection predicts the risk of recurrence during chronic infection. JAIDS J Acquir Immune Defic Syndr 2012;60:e112−15.

[169] Marchionatti A, Parisi MM. Anemia and thrombocytopenia in people living with HIV/AIDS: a narrative literature review. Int Health 2021;13:98−109.

[170] Wright PW, et al. Cerebral white matter integrity during primary HIV infection. AIDS 2015;29:433−42.

[171] Joshi SG, Cho TA. Pathophysiological mechanisms of headache in patients with HIV. Headache 2014;54:946−50.

[172] Suh J, et al. Progressive increase in central nervous system immune activation in untreated primary HIV-1 infection. J Neuroinflam 2014;11:199.

[173] Spudich S, et al. Central nervous system immune activation characterizes primary human immunodeficiency virus 1 infection even in participants with minimal cerebrospinal fluid viral burden. J Infect Dis 2011;204:753–60.

[174] Tambussi G, et al. Neurological symptoms during primary human immunodeficiency virus (HIV) infection correlate with high levels of HIV RNA in cerebrospinal fluid. Clin Infect Dis 2000;30:962–5.

[175] Ferrada MA, Xie Y, Nuermberger E. Primary HIV infection presenting as limbic encephalitis and rhabdomyolysis. Int J STD AIDS 2014;26:835–6.

[176] Scriven J, Davies S, Banerjee AK, Jenkins N, Watson J. Limbic encephalitis secondary to HIV seroconversion. Int J STD AIDS 2011;22:236–7.

[177] Nzwalo H, Añón RP, Àguas MJ. Acute encephalitis as initial presentation of primary HIV infection. BMJ Case Rep 2012;2012.

[178] Young AC, et al. Cerebral metabolite changes prior to and after antiretroviral therapy in primary HIV infection. Neurology 2014;83:1592–600.

[179] Lentz MR, et al. Alterations in brain metabolism during the first year of HIV infection. J Neurovirol 2011;17:220–9.

[180] Wang SXY, et al. Peripheral neuropathy in primary HIV infection associates with systemic and central nervous system immune activation. J Acquir Immune Defic Syndr 2014;66:303–10.

[181] de Castro G, Bastos PG, Martinez R, de Castro Figueiredo JF. Episodes of Guillain-Barré syndrome associated with the acute phase of HIV-1 infection and with recurrence of viremia. Arq Neuropsiquiatr 2006;64:606–8.

[182] Pontali E, Feasi M, Crisalli MP, Cassola G. Guillain-Barré syndrome with fatal outcome during HIV-1-seroconversion: a case report. Case Rep Infect Dis 2011;2011:972096.

[183] Varshney AN, et al. HIV seroconversion manifesting as Guillian-Barre syndrome. Chin Med J (Engl) 2014;127:396.

[184] Gold JA, et al. Longitudinal characterization of depression and mood states beginning in primary HIV infection. AIDS Behav 2014;18:1124–32.

[185] Helleberg M, Kirk O. Encephalitis in primary HIV infection: challenges in diagnosis and treatment. Int J STD AIDS 2013;24:489–93.

[186] Lafeuillade A, et al. Oesophageal candidiasis in primary HIV infection. Eur J Med 1992;1:126.

[187] Pena JM, Martinez-Lopez MA, Arnalich F, Barbado FJ, Vazquez JJ. Esophageal candidiasis associated with acute infection due to human immunodeficiency virus: case report and review. Rev Infect Dis 1991;13:872–5.

[188] Sued O, et al. Acute HIV seroconversion presenting with active tuberculosis and associated with high levels of T-regulatory cells. Viral Immunol 2011;24:347–9.

[189] Yoo KM, et al. Dissemination of multidrug-resistant tuberculosis in a patient with acute HIV infection. BMC Infect Dis 2014;14:462.

[190] Mateos Rodríguez F, Fuertes Martín A, Marcos Toledano M, Jiménez López A. Primary HIV infection with esophageal candidiasis and acute toxoplasmosis. An Med Interna 1998;15:50–1.

[191] Vento S, Di Perri G, Garofano T, Concia E, Bassetti D. Pneumocystis carinii pneumonia during primary HIV-1 infection. Lancet (London, Engl) 1993;342:24–5.

[192] Moss PJ, Read RC, Kudesia G, McKendrick MW. Prolonged cryptosporidiosis during primary HIV infection. J Infect 1995;30:51–3.

[193] Szabo S, James CW, Telford G. Unusual presentations of primary human immunodeficiency virus infection. AIDS Patient Care STDS 2002;16:251–4.

[194] Sued O, et al. Primary human immunodeficiency virus type 1 infection: clinical, virological and immunological characteristics of 75 patients (1997–2003). Enferm Infecc Microbiol Clin 2006;24:238–44.

[195] Fiebig EW, et al. Dynamics of HIV viremia and antibody seroconversion in plasma donors: implications for diagnosis and staging of primary HIV infection. AIDS 2003;17:1871–9.

[196] McMichael AJ, Borrow P, Tomaras GD, Goonetilleke N, Haynes BF. The immune response during acute HIV-1 infection: clues for vaccine development. Nat Rev Immunol 2010;10:11–23.

[197] Thorstensson R, et al. Evaluation of 14 commercial HIV-1/HIV-2 antibody assays using serum panels of different geographical origin and clinical stage including a unique seroconversion panel. J Virol Methods 1998;70:139–51.

[198] Daar ES, et al. Diagnosis of primary HIV-1 infection. Los Angeles County primary HIV infection recruitment network. Ann Intern Med 2001;134:25−9.

[199] Sullivan TJ, Patel P, Hutchinson A, Ethridge SF, Parker MM. Evaluation of pooling strategies for acute HIV-1 infection screening using nucleic acid amplification testing. J Clin Microbiol 2011;49:3667−8.

[200] Pilcher CD, et al. Detection of acute infections during HIV testing in North Carolina. N Engl J Med 2005;352:1873−83.

[201] Muthukumar A, et al. Comparison of 4th-generation HIV antigen/antibody combination assay with 3rd-generation HIV antibody assays for the occurrence of false-positive and false-negative results. Lab Med 2015;46:84−9.

[202] Long EF. HIV screening via fourth-generation immunoassay or nucleic acid amplification test in the United States: a cost-effectiveness analysis. PLoS One 2011;6:e27625.

[203] Gilbert M, et al. Targeting screening and social marketing to increase detection of acute HIV infection in men who have sex with men in Vancouver, British Columbia. AIDS 2013;27:2649−54.

[204] Borges CM, Pathela P, Pirillo R, Blank S. Targeting the use of pooled hiv rna screening to reduce cost in health department std clinics: New York City, 2009−2011. Public Health Rep 2015;130:81−6.

[205] Kessler HA, et al. Diagnosis of human immunodeficiency virus infection in seronegative homosexuals presenting with an acute viral syndrome. JAMA 1987;258:1196−9.

[206] Masciotra S, et al. Evaluation of an alternative HIV diagnostic algorithm using specimens from seroconversion panels and persons with established HIV infections. J Clin Virol 2011;52(Suppl. 1):S17−22.

[207] Branson BM, Stekler JD. Detection of acute HIV infection: we can't close the window. J Infect Dis 2012;205:521−4.

[208] Branson BM, Mermin J. Establishing the diagnosis of HIV infection: new tests and a new algorithm for the United States. J Clin Virol 2011;52:S3−4.

[209] Chavez P, Wesolowski L, Patel P, Delaney K, Owen SM. Evaluation of the performance of the Abbott ARCHITECT HIV Ag/Ab Combo Assay. J Clin Virol 2011;52(Suppl. 1):S51−5.

[210] Ly TD, et al. The variable sensitivity of HIV Ag/Ab combination assays in the detection of p24Ag according to genotype could compromise the diagnosis of early HIV infection. J Clin Virol 2012;55:121−7.

[211] Patel P, et al. Detection of acute HIV infections in high-risk patients in California. JAIDS J Acquir Immune Defic Syndr 2006;42:75−9.

[212] Lindbäck S, et al. Diagnosis of primary HIV-1 infection and duration of follow-up after HIV exposure. Karolinska Institute Primary HIV Infection Study Group. AIDS 2000;14:2333−9.

[213] Rosenberg NE, Pilcher CD, Busch MP, Cohen MS. How can we better identify early HIV infections? Curr Opin HIV AIDS 2015;10:61−8.

[214] World Health Organization. Consolidated guidelines on HIV testing services 2015, 2015.

[215] Cherutich P, Bunnell R, Mermin J. HIV testing: current practice and future directions. Curr HIV/AIDS Rep 2013;10:134−41.

[216] Mugo PM, et al. Engaging young adult clients of community pharmacies for HIV screening in Coastal Kenya: a cross-sectional study: Table 1. Sex Transm Infect 2015;91:257−9.

[217] Geren KI, et al. Identification of acute HIV infection using fourth-generation testing in an opt-out emergency department screening program. Ann Emerg Med 2014;64:537−46.

[218] Ambrosioni J, et al. Impact of highly active antiretroviral therapy on the molecular epidemiology of newly diagnosed HIV infections. AIDS 2012;26:2079−86.

[219] Petroll AE, Pinkerton SD. A conceptual model of interventions to increase diagnosis of acute HIV infection and reduce forward transmission. AIDS Behav 2011;15:1715−20.

[220] Sabharwal CJ, Bodach S, Braunstein SL, Sepkowitz K, Shepard C. Entry into care and clinician management of acute HIV infection in New York City. AIDS Patient Care STDS 2012;26:129−31.

[221] Bodach S, et al. Integrating acute HIV infection within routine public health surveillance practices in New York City. Public Health Rep 2012;127:451−9.

[222] Thornhill J, Orkin C. Long-acting injectable HIV therapies: the next frontier. Curr Opin Infect Dis 2021;34:8−15.

[223] Swindells S, et al. Long-acting cabotegravir and rilpivirine for maintenance of HIV-1 suppression. N Engl J Med 2020;382:1112−23.

[224] Orkin C, et al. Long-acting cabotegravir and rilpivirine after oral induction for HIV-1 infection. N Engl J Med 2020;382:1124–35.

[225] Menéndez-Arias L, Delgado R. Update and latest advances in antiretroviral therapy. Trends Pharmacol Sci 2022;43:16–29.

[226] Sued O, Cahn P. Latin America Priorities after 40 years of the beginning of the HIV pandemic. Lancet Reg Heal—Am 2021;1:100024.

[227] Rebe K, Hoosen N, McIntyre JA. Strategies to improve access for MSM in low-income and middle-income countries. Curr Opin HIV AIDS 2019;14:387–92.

[228] Radusky PD, et al. Reduction of gender identity stigma and improvements in mental health among transgender women initiating hiv treatment in a trans-sensitive clinic in Argentina. Transgender Heal 2020;5:216–24.

[229] De Boni RB, et al. An internet-based hiv self-testing program to increase HIV testing uptake among men who have sex with men in brazil: descriptive cross-sectional analysis. J Med Internet Res 2019;21.

[230] Frola CE, et al. Home-based HIV testing: using different strategies among transgender women in Argentina. PLoS One 2020;15:1–13.

[231] Bavinton BR, Grulich AE. HIV pre-exposure prophylaxis: scaling up for impact now and in the future. Lancet Public Heal 2021;6:e528–33.

[232] Tantibanchachai C. FDA approves first injectable drug for HIV treatment. FDA News Release. <https://www.fda.gov/news-events/press-announcements/fda-approves-first-injectable-treatment-hiv-pre-exposure-prevention>; 2021.

[233] Sperling RS, et al. Maternal viral load, zidovudine treatment, and the risk of transmission of human immunodeficiency virus type 1 from mother to infant. N Engl J Med 1996;335:1621–9.

[234] Del Romero J, et al. Absence of transmission from HIV-infected individuals with HAART to their heterosexual serodiscordant partners. Enferm Infecc Microbiol Clin 2014;. Available from: https://doi.org/10.1016/j.eimc.2014.10.020.

[235] Montaner JSG, et al. Expansion of HAART coverage is associated with sustained decreases in HIV/AIDS morbidity, mortality and HIV transmission: the 'HIV Treatment as Prevention' experience in a Canadian setting. PLoS One 2014;9:e87872.

[236] Smith MK, et al. Treatment to prevent HIV transmission in serodiscordant couples in Henan, China, 2006 to 2012. Clin Infect Dis 2015;61:111–19.

[237] Vandormael A, Newell M-L, Bärnighausen T, Tanser F. Use of antiretroviral therapy in households and risk of HIV acquisition in rural KwaZulu-Natal, South Africa, 2004–12: a prospective cohort study. Lancet Glob Heal 2014;2:e209–15.

[238] Kretzschmar ME, Schim van der Loeff MF, Birrell PJ, De Angelis D, Coutinho RA. Prospects of elimination of HIV with test-and-treat strategy. Proc Natl Acad Sci U S A 2013;110:15538–43.

[239] Cope AB, et al. Ongoing HIV transmission and the HIV care continuum in North Carolina. PLoS One 2015;10:e0127950.

[240] Brenner BG, et al. High rates of forward transmission events after acute/early HIV-1 infection. J Infect Dis 2007;195:951–9.

[241] Powers KA, et al. The role of acute and early HIV infection in the spread of HIV and implications for transmission prevention strategies in Lilongwe, Malawi: a modelling study. Lancet (London, Engl) 2011;378:256–68.

[242] Prestage G, Brown G, Keen P. Barriers to HIV testing among Australian gay men. Sex Health 2012;9:453.

[243] Sharrocks K, et al. Missed opportunities for identifying primary HIV within genitourinary medical/HIV services. Int J STD AIDS 2012;23:540–3.

[244] Grin B, Chan PA, Operario D. Knowledge of acute human immnuodeficiency virus infection among gay and bisexual male college students. J Am Coll Heal 2013;61:232–41.

[245] Siegler AJ, et al. Knowledge and awareness of acute human immunodeficiency virus infection among mobile app-using men who have sex with men: a missed public health opportunity. Open Forum Infect Dis 2015;2 ofv016.

[246] McNairy ML, El-Sadr WM. Antiretroviral therapy for the prevention of HIV transmission: what will it take? Clin Infect Dis 2014;58:1003–11.

[247] Owen SM. Testing for acute HIV infection: implications for treatment as prevention. Curr Opin HIV AIDS 2012;7:125–30.

[248] Buskin SE, Fida NG, Bennett AB, Golden MR, Stekler JD. Evaluating new definitions of acute and early HIV infection from HIV surveillance data. Open AIDS J 2014;8:45−9.

[249] Deeks SG, et al. Research priorities for an HIV cure: International AIDS Society Global Scientific Strategy 2021. Nat Med 2021;27:2085−98.

[250] Cellerai C, et al. Early and prolonged antiretroviral therapy is associated with an HIV-1-specific T-cell profile comparable to that of long-term non-progressors. PLoS One 2011;6:e18164.

[251] Chéret A, et al. Combined ART started during acute HIV infection protects central memory CD4 + T cells and can induce remission. J Antimicrob Chemother 2015;70:2108−20.

[252] Mylvaganam GH, Silvestri G, Amara RR. HIV therapeutic vaccines: moving towards a functional cure. Curr Opin Immunol 2015;35:1−8.

[253] Dahabieh MS, Battivelli E, Verdin E. Understanding HIV Latency: the road to an HIV cure. Annu Rev Med 2015;66:407−21.

[254] Archin NM, Margolis DM. Emerging strategies to deplete the HIV reservoir. Curr Opin Infect Dis 2014;27:29−35.

[255] Eriksson S, et al. Comparative analysis of measures of viral reservoirs in HIV-1 eradication studies. PLoS Pathog 2013;9:e1003174.

[256] Chéret A, et al. Intensive five-drug antiretroviral therapy regimen vs standard triple-drug therapy during primary HIV-1 infection (OPTIPRIM-ANRS 147): a randomised, open-label, phase 3 trial. Lancet Infect Dis 2015;15:387−96.

[257] Ananworanich J, et al. Impact of multi-targeted antiretroviral treatment on gut T cell depletion and HIV reservoir seeding during acute HIV infection. PLoS One 2012;7:e33948.

[258] Shan L, et al. Stimulation of HIV-1-specific cytolytic T lymphocytes facilitates elimination of latent viral reservoir after virus reactivation. Immunity 2012;36:491−501.

[259] Kaufmann DE, Walker BD. PD-1 and CTLA-4 inhibitory cosignaling pathways in HIV infection and the potential for therapeutic intervention. J Immunol 2009;182:5891−7.

[260] Barouch DH, et al. Therapeutic efficacy of potent neutralizing HIV-1-specific monoclonal antibodies in SHIV-infected rhesus monkeys. Nature 2013;503:224−8.

[261] Shingai M, et al. Antibody-mediated immunotherapy of macaques chronically infected with SHIV suppresses viraemia. Nature 2013;503:277−80.

[262] Thornhill J, Fidler S, Frater J. Advancing the HIV cure agenda: the next 5 years. Curr Opin Infect Dis 2015;28:1−9.

Herbal drugs to combat viruses

Benil P.B.[1], Rajakrishnan Rajagopal[2], Ahmed Alfarhan[2] and Jacob Thomas[2]

[1]Department of Agadatantra, Vaidyaratnam P.S. Varier Ayurveda College, Kottakkal, Kerala, India [2]Department of Botany and Microbiology, College of Science, King Saud University, Riyadh, Saudi Arabia

18.1 Introduction

Emerging diseases are causing a very high impact on the social, cultural, and economic circles of people around the globe. The definitions of health, healthy living standards, and health priorities have been changed drastically with the recent pandemic of COVID-19. The unpreparedness felt at the insurgency of the new viral strain in the environment has created panic among medical fraternity and the treasure trove of all the medicaments available till date fall short. There were heated debates on the ethical aspects regarding the use of antiviral agents that have not passed through a Phase III double blinded placebo controlled randomized clinical trial being put to use for this new disease [1]. Even under this turmoil there were sincere efforts to sought for newer alternatives against this menace. To make the situation grimmer, the virus underwent several mutations that incapacitated the activity of certain medicines [2]. Traditionally, humans depended on the natural resources for all their needs and especially for medical needs. Owing to this same fact, a large percentage of drugs currently in use have been derived from plant sources or their modified forms [3]. These precious phytochemicals derived from the medicinal plants world over were extensively researched to find a cure or as an adjuvant therapy or as an immune-booster to combat viral infections [4]. These studies were carried out on the basis of the positive results already obtained for similar studies conducted on other viral infections. Studies ranging from in silico, in vitro, in vivo, and clinical trials were performed to establish an effective treatment against several viral infections including COVID-19. The strategies of drug targeting against viral infections are intricately woven with the pathology of establishing an infection. The drug targets against viruses involves molecules interacting with the adhesion of the virus on to the host cell, those inhibiting the transport of the virus in to the host cell, those interfering with the replication of the nucleic acids and those inhibiting the viral assembly and release from the host cell. The advantages of phytochemicals in this regard are that they can exert their influence on multiple targets at a time. This advantageous effect makes them ideal candidate to be used as adjuvant therapies against mainstay antiviral agents. Several of these phytochemicals have exhibited very strong viricidal activity comparable in dose with those of conventional antivirals. This chapter concentrates on the important drug targets for curbing viral infections with phytochemicals derived from medicinal plants.

18.2 Phytochemicals preventing attachment of virus to host cell

Viruses generally are classified into enveloped and nonenveloped based on their acquisition of a lipid membrane derived from the host cell or not. The initiation of a viral infection is with the process of attachment of a virus to receptors on cell surface and their subsequent delivery of the viral genome to the cell cytoplasm [5]. The interaction of the virus with the host cell is complex and serves multiples purposes like binding to several receptors of the host cell [6] (Fig. 18.1). The initial interaction with host cell receptors serves for attachment of the virus followed by the several other cellular receptors for irreversible binding providing redundancy and in invading to establish infection by undergoing latency in the host tissues [7,8]. The receptors favored by viruses can be either proteins, carbohydrates, or lipids. A particular virus may exhibit specificity regarding cellular receptors while different viruses may utilize the same receptors and vice versa [9,10]. Viruses mostly employs receptors that are fundamental to the cell's function because they are highly conserved over time.

Phytochemicals binds to the carbohydrate moiety and inhibits the cell entry of the viruses [11]. This event prevents the viral penetration, resists its coating and inhibits the growth of viruses. Plant-derived compounds like lignans, saponins, alkaloids, kaempferol, luteolin, apigenin, baicalin, quercetin, catechol, flavonoids, and sulphated polysaccharides have reported roles in preventing the viral entry, disrupting the nucleocapsid and genetic material and inhibiting replication of the viruses like dengue, herpes simplex virus, hepatitis C virus, influenza, chikungunya, SARS, and more. Terpenes are generally shown to disrupt the attachment of herpes simplex virus - 1 (HSV-1) to the cell membrane and interferes with the replication of the virus. Saikoponins, a terpenoid from *Bupleurum kaoi* root inhibited hepatitis C cell cycle at the early stage of viral entry and fusion [12]. Loliolide, a monoterpene from *Phyllanthus urinaria* inhibited the attachment of hepatitis C virus on to the host cell membrane and prevented its cell entry [13,14]. Viruses like SARS-CoV-2 binds to the angiotensin-converting-enzyme - 2 (ACE2) receptor to gain entry into the host cell with the help of TMPRSS2 enzyme [15,16]. The virus then gets rid of its envelope and starts replication of its genome. In silico investigations showed that glyasperin-A, a compound isolated from *Glycyrrhiza glabra* showed its ability to inhibit nonstructural protein-15 endoribonuclease and glycyrrhizic acid showed ability to bind to the spike glycoprotein and inhibited the entry of the virus [17]. ACE-2 facilitates the entry of SARS-CoV-2 and activates the antiinflammatory pathway that evades the body's immune response against invading virus. Glycyrrhizin and glycyrrhetinic acid, an active metabolite

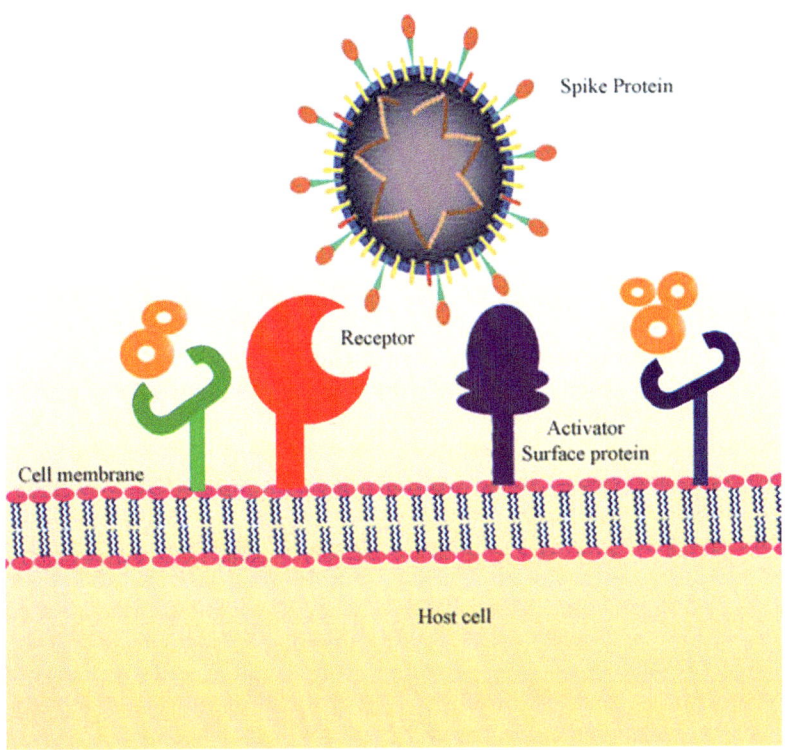

FIGURE 18.1 Interaction of virus with the membrane receptors for cell entry.

of glycyrrhizin exhibits antiinflammatory activity with the help of toll-like receptor 4 antagonism, and activates the inflammatory response against the virus. These two phytochemicals actively inhibit the expression of type 2 transmembrane serine proteases, that helps in the uptake of virus particles [18].

Grape seed extracts showed the decreasing binding ability of human norovirus (NoVs) GII.4 and human NoVs GII.4.P particles in a dose dependent manner [19]. Studies conducted on various berries have shown that Cranberries are most efficient than blueberry, blackberry, raspberry, and strawberry in lowering the attachment of NoV P-particles [19]. Pomegranate juice also showed efficacy equivalent to that of cranberries to disrupt NoV attachment [19]. Resveratrol, a polyphenolic compound isolated from cranberries exhibited antiviral activity against reovirus, enterovirus, and influenza viruses by inhibiting the process of attachment of viruses to the host cell. Among different investigations conducted on phytochemicals, flavonoids, especially from grape seeds, pomegranates, mulberries, black raspberries, cranberries, persimmons, and green tea extracts showed the highest anti-NoV activity. Glycoside ginsenoside derived from *Panax quinquefolium* have showed activity against influenza A H1N1 viruses by preventing their attachments with α 2−3′ sialic acid receptors [20]. Punicalagin, a hydrolysable tannin obtained from *Punica granatum* showed inhibiting viral replication by targeting its attachment on influenza-A H1N1, H3N2, and influenza B viruses [21]. Curcumin demonstrated a significance in plaque reduction assay in H1N1 and H6N1 viruses indicating its early stage inhibitory activity from preventing the attachment of the viruses [22–24]. Elderberry extracts containing a flavonoid 5,7,30,40-tetra-O-methylquercetin showed activity against influenza-A H1N1 virus infection at low doses equivalent to the antiviral agent oseltamivir and prevented the viral attachment to the host cell [25]. 5,7,3′, 4′-tetra-O-methylquercetin also shown its capability to prevent the attachment and entry of H1N1 virus by blocking the viral surface glycol-proteins in in vitro studies [25]. Quercetagetin, another flavonoid in vitro activity against Chikungunya virus (CHIKV) mainly by blocking its attachment to host cells [26]. Myrciatrin V, showed activity against Ebola virus in *in silico* studies by exhibiting strong interaction and attachment with the envelop glycol-proteins and thus prevents its entry and attachment to the host cell [27]. Dihydromyricetin showed in vitro activity against influenza virus H1N1 by blocking the viral surface proteins attachment to host cells [25]. A related polyphenol, catechin inhibits viral entry and attachment in West Nile virus (WNV), hepatitis C and B viruses, herpes simplex virus, influenza A virus, dengue virus, adenovirus, reovirus, and ZIKV infections [28]. In silico analysis of cyanidin-3-(p-coumaroyil)-rutinoside-5-glucoside showed strong interaction and attachment with the envelop glycol-proteins of Sudan Ebolavirus (SEBOV). Similarly, 7-O-(6-feruoylglucosyl) isoorientin also showed to be preventing the attachment and entry of SEBOV. *Azadirachta indica* bark extracts inhibited viral attachment and blocked viral glycoprotein mediated cell-cell fusion in HSV-1 viruses. *Ilex paraguariensis* leaf hydroethanolic extracts showed antiviral activity against HSV-1 and HSV-2 viruses by preventing viral attachment and penetration. *Melissa officinalis* leaf aqueous extract inhibited the HSV-1 attachment and inactivated the virions in an in vitro study conducted on infected RC-37 cell lines. *Rhododendron ferrugineum* aerial part aqueous extract also inhibited HSV-1 attachment and penetration. *Rheum officinale, Paeonia suffruticosa*, and *Melia toosendan* aqueous, methanolic, and ethanolic extracts showed inhibition of viral attachment against HSV-1. These plants are widely used in Chinese traditional medicine [29]. Hydromethanolic extract of *Indigofera heterantha* roots, a western Himalayan plant inhibited the viral replication cycle of HSV-2 by preventing its attachment into host cells [30]. The aqueous and chloroform extracts of *Ficus religiosa* showed activity against HSV-2 and acyclovir-resistant HSV-2 viruses by inactivating the virions prior to infection and preventing viral attachment respectively [31]. Ethanolic extracts of the fruits of *Terminalia chebula* on pretreating HSV-2 viruses inhibited viral attachment and penetration into host cells [32]. Hydromethanolic extract of *Hemidesmus indicus* root inhibited HSV-1 and HSV-2 infections throughout their infection cycle. But their effects are more potent during the preinfection stage as they specifically inhibit viral attachment [33]. *Arisaema tortuosum* leaves have reported antiherpetic activity. Their chloroform extract showed direct virucidal activity by inhibiting viral attachment, adsorption, and replication [34]. Green fresh water algae spirogyra aqueous, ethanolic, and methanolic extracts showed antiviral activity against ASV-1 (Strain F) and HSV-2 (Strain G). The extracts were tested for their activity before, during and after viral attachment and showed highest inhibition when added to cells during attachment. HSV-1 showed susceptibility to ethanolic extract while HSV-2 showed activity with methanolic extract. *G. glabra* is yet another medicinal plant that showed significant antiviral activity against HSV-1.

The antiviral activity of *G. glabra* is from the direct viral inhibition and by preventing viral attachment.

Yupingfeng powder, a traditional Chinese medicine formulation containing *Atractylodes macrocephala*, *Atractylodes lancea*, *G. glabra*, *Lonicera japonica*, and *Saposhnikovia divaricata* was found to be effective against SARS-CoV-2 virus [35]. *Sambucus javanica* prevented the viral attachment to host cells in human coronavirus. The three phenolic acids; caffeic, gallic, and chlorogenic acids significantly contributed to the inhibition of replication and the attachment of the virus [36]. In silico studies showed the efficacy of phytoconstituents from *A. indica*; nimbin, berberine, mangiferin and tinosporin from *Tinospora cordifolia* showed significant binding affinity towards SARS-CoV-2 spike proteins and ACE2 receptor. Garlic and onion are good sources of organosulfur compounds like allicin and quercetin that inhibits viral infections by inhibiting virus attachment to host cell. Anthocyanin delphinidin has been proved as an antiviral agent and it prevents the adsorption and attachment of hepatitis C virus (HCV) in the inoculation phase. Even phytochemicals obtained from lower organisms also showed significant antiviral properties. Griffithsin, a protein obtained from red algae *Griffithsia* inhibits MERS-CoV and other human coronaviruses by preventing their attachment to host cells by binding on to glycans of protein spikes of viruses with its three carbohydrate binding domains. Caffeic acid derived from *Sambucus formosana* inhibits the attachment of HCoV and hepatitis B viruses to host cell by binding to S proteins. Agglutinin (a mannose-binding lectin) derived from *Galanthus nivalis* inhibited feline coronavirus (FCoV) by binding to membrane spike proteins and prevents attachment to host cells. The common Indian spice *Zingiber officinalis* has been studied for their action against the Human respiratory syncytial virus (RSV) and was found to inhibit the plaque formation on the epithelium of the airways by preventing viral attachment and internalization. Similarly, *Pelargonium sidoides* extracts show activity against rhinovirus by inhibiting viral binding. Honokiol, a lignin from the Magnolia tree, widely used in Chinese medicine inhibited the endocytic entry of dengue virus into the host cells.

18.3 Phytochemicals preventing penetration and uncoating of viruses

Mammalian cells, in general, utilizes the receptor-mediated endocytosis for transporting macromolecules across the plasma membrane. Both enveloped and nonenveloped viruses make use of two important routes to get internalized into the host cells. The endocytic route in which the clathrin-coated vesicles and pits, nonclathrin-coated pits, micropinocytosis, or caveolae are engaged. Virion attachment to the receptors clustered at clathrin-coated pits gets endocytosed into clathrin-coated vesicles. As the vesicles enters the cytoplasm, the clathrin coat is removed and the virion fuses with the endosomes (acidic prelysosomal vacuoles). Lower pH in the vesicle produces changes in the surface protein of the virions leading to fusion of viral envelope with the endosomal membrane and the subsequent release of the viral nucleocapsid into the cytoplasm (Fig. 18.2).

The nonendocytic route utilizes the direct transport across the plasma membrane at a neutral pH. The F (Fusion) glycoprotein of certain viruses like paramyxovirus causes the viral envelope to directly fuse with the plasma membrane at neutral pH of 7 and allows the viral nucleocapsid to be released directly into the cytoplasm. Some viruses which

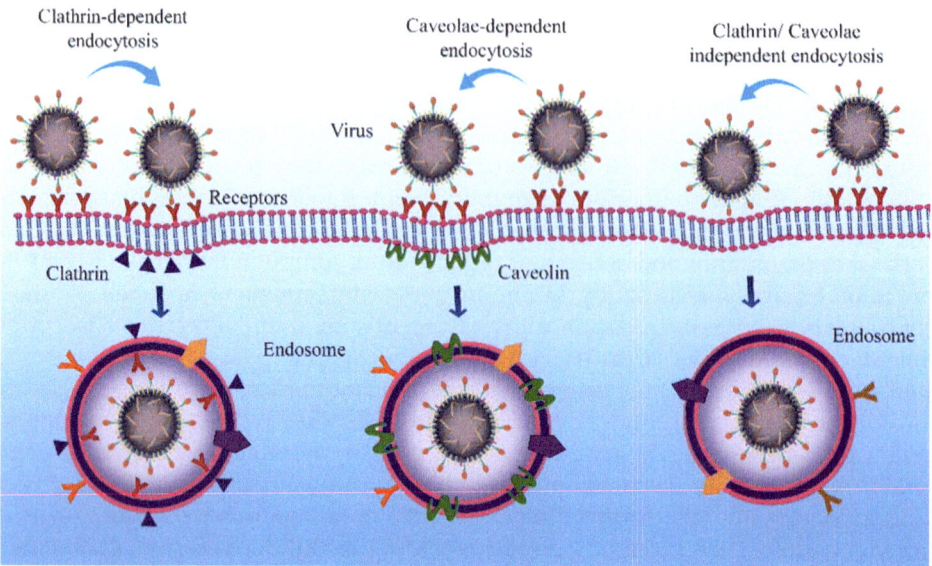

FIGURE 18.2   Different mechanisms of internalization of viruses.

predominantly uses the nonendocytic routes may sometimes utilize the endocytic pathway for internalization. In this process, the virus binds on to the cell-surface receptors associating with the endosomal membranes followed by conformational changes brought about in the viral capsid which exposes target regions that interact with the endosomal membrane resulting in the opening up of specific transport channels for carrying viral genome across the plasma membrane. In all these processes, membrane fusion is the key event that leads to the phagocytosis, pinocytosis and vesicular trafficking. This process is regulated by membrane proteins and is associated with the conformational changes in the viral proteins as well as the host-cell receptor proteins mainly from the low endosomal pH. The nonenveloped viruses gain entry to the host cell by penetration while the enveloped virus fuse to the cell membrane to gain entry.

With the entry of the virus into the host cell, they are transported across the cytoplasm as a nucleoprotein complex. Specific surface-exposed nuclear localization signals give them entry into the nucleus and establishes the infection. Several other viruses utilize another mechanism with the help of lipid microdomains or lipid rafts which helps in transporting sufficient number of viruses from one cell to the other. Direct cell-to-cell entry of viruses by a process called "virological synapse" are also possible with the help of interaction between infected cells and polarized cytoskeleton, adhesion molecules, and viral proteins. Transcytosis, another specialized method of transmission of viruses without actual crossing of membranes. Transcytosis is a method of transport of materials utilizing the vesicular transport from one side of the cell to the other.

Upon entry, the virions need to be uncoated to initiate the transcription of the viral genes. In enveloped RNA viruses, the nucleocapsid is discharged into the cytoplasm after the fusion of the virion and the process of transcription of viral genome in initiated while it is still associated with the proteins of nucleocapsid. Certain viruses require complete uncoating before the initiation of the transcription. Viruses also show diversity in the place of uncoating like the endosomes while in cytoplasm or inside the nucleus.

Glycyrrhizic acid is a strong antiinflammatory and immune-active agent exhibiting membrane bound and cytoplasmic effects. Glycyrrhizic acid favors the entry of the virus into the host cell by disorganizing the lipid cores by a cholesterol dependent pathway [35]. Several flavonoids have shown promising results in preventing the viral entry in case of SARS-CoV-2 infections. Luteolin, inhibits the entry of SARS-CoV-2 virus entry and fusion with human receptors [37]. Saponins are also investigated for their antiviral activity and evidences shows that they inhibit cellular attachment, entry adsorption, and penetration of a virion. Hydrolysable tannins from *T. chebula*, chebulagic acid and punicalagin, inhibit the entry and transport of HSV-1 virus [38]. Apigenin is found to inhibit the entry of picrona viral protein entry. Apigenin inhibits enterovirus A71 (EV-A71) infection by inhibiting internal ribosome entry site (IRES) activity [39]. Withanone from *Withania somnifera* interrupts the electrostatic interaction between the ACE2-RBD complex and weakens the entry of SARS-CoV-2 viral entry [40]. Phytochemicals widely used in Chinese traditional medicine like Betulinic aicd, coumaroyltyramine, kaempferol, lignan, desmethoxyreserpine, cryptotanshinone, dihomo-γ-linolenic acid, dihydrotanshinone, moupinamide, N-cis-feruloyiltyramine, quercetin, and sugiol inhibits 3CL^{Pro} and PL^{Pro} and prevents cellular entry of human coronaviruses [41]. Oleanene derivatives existing as glycosides named saikosaponin A, B2, C, and D derived from the plant *Scrophularia scorodonia*, *Heteromorpha* spp., and *Bupleurum* spp. Showed inhibitory activity on viral attachment and penetration against (human coronavirus 229E) HCoV-229E, measles virus, hepatitis B virus (HBV), and hepatitis C virus (HCV) [42]. Saikosaponin B2 showed specific inhibition of cellular entry by hepatitis C virus [12,43]. Several phytochemicals from the *Cinnamomi* spp. Inhibits the clathrin-based endocytosis pathways and prevents the virus entry [44]. Polysaccharide mixture from the ethanolic extract of *Ginko biloba* prevents the attachment and entry of viruses [45]. Lectins inhibit virus penetration. Pterocarnin A, a constituent from *Pterocarya stenoptera* and *Prunella vulgaris* lignin-carbohydrate complex was shown effective against HSV-2 binding and penetration of host cells [46]. Polyphenols inhibit the viral cell entry by modulating surface protein expression. Spiroketalenol and its derivatives mainly isolated from the fresh rhizomes of *Tanacetum vulgare* and samarangenin B from the roots of *Limonium sinense* inhibited the process of host cell entry by HSV-1 [47,48]. Jatrophane esters inhibited the viral entry and expression of receptors on host cell surface by HIV [49]. *Saururus chinensis* yielded manassantin B, which was proved beneficial in preventing the entry of Epstein-Barr virus (EBV) into the host cell [50]. Tetranotriterpenoid 1-cinnamoyl-3,11-dihydroxymeliacarpin (CDM) from *Melia azedarach* blocked vesicular stomatitis virus (VSV) from entering host cell and also prevented its intracellular transport [51]. Flavonoids from *Galla chinensis* or *Veronica linrrifolia* extract inactivated the spike proteins of the SARS virus by binding on to them and prevented the virus particles from penetrating the cell [52]. Tetra-O-galoyl-β-d-glucose (TGG) luteoline isolated from traditional Chinese herbs inhibited SARS-CoV surface protein interaction to prevent their entry into the host cells [53]. *Pogostemon cabilin* dried aerial parts yielded a sesquiterpene Patchouli alcohol which showed promising activity against Influenza-A H2N2 viruses by inhibiting viral penetration [19]. Polysaccharides from the roots of *Isatis indigotica* showed cell protection of Madin-Darby canine kidney (MDCK) cells infected with swine influenza virus (SIV) and inhibited the adsorption, penetration of SIV [54]. Quercetin flavonoids interacted with the surface glycoprotein hemagglutinin (HA) of influenza-A viruses, which plays a crucial step in the fusion and entry of the virus to the host cell. Quercetin has demonstrated strong inhibition of HA and has also shown significant reduction in the transcription of HA mRNA [55]. Baicalein and quercetagetin inhibited the entry of chikungunya virus (CHIKV) into Vero cell cultures in in vitro studies. Baicalein also showed activity against human cytomegalovirus (HCMV) by blocking the kinase activity of the epidermal growth factor receptor (EGFR) gene and thus prevented the entry of the virus [56]. Genistein, another flavonoid also showed 90% inhibition of HCMV DNA synthesis [57]. Ladanein, a flavonoid isolated from *Marrubium peregrinum* showed minimal activity on preventing the attachment of HCV to host cells, but it inhibited its cell entry. Ladanein showed postattachment entry inhibition against several genotype strains of HCV like 1a, 1b, 2b, 3a, 4a, and 5a [58]. investigations on two other flavonoids isolated from *Pterogyne nitens*; sorbifolin and pedalitin also showed inhibition of HCV infection. Among these flavonoids, pedalitin showed 78.7% inhibition of virus entry into host cells [59]. Kaempferol showed notable antiviral activity against EV-A71 up to 80% by inhibiting IRES [60]. Epigallocatechin gallate (EGCG) inhibited zika virus MR-766 strain which was found to be due to the inhibition of viral entry [61]. Quercetin-3-β-O-D-glucoside on treating with Vero E6 cells showed no toxicity and inhibited the Ebola virus wild type strain entry into the host cells [62]. Sodium Rutin sulfate inhibited human immunodeficiency virus (HIV-1) virus entry by inhibiting the glycoproteins of HIV-1 envelop [63]. Pinocembrin prevented zika virus replication and entry [64]. Myrciatrin V showed inhibition of viral entry in Ebola viruses in in silico studies [27]. Ladanein inhibited the entry of HCV [58]. Isoquercetin showed activity against zika virus by preventing its entry in in vitro studies [65].

Whole plant aqueous extracts of *Achyrocline flaccida* prevented HSV-1 adsorption and penetration [66]. *Ficus carica* extracts also showed inhibition of viral penetration and adsorption [67]. *Ilex paraguariensis* leaf hydroalcoholic extracts prevented viral attachment and penetration of HSV-1 [68]. *Pelargonum sidoides* root aqueous extract prevented the penetration of both HSV-1 and HSV-2 viruses [69]. *P. suffruticosa* methanolic and ethanolic extracts, *R. officinale* and *M. toosendan* ethanolic extracts inhibited viral penetration in HSV-1 infected cell cultures [29]. Pretreatment with the methanolic extracts of *Ophiorrhiza nicobarica* showed the prevention of viral prevention and adsorption in HSV-1 and HSV-2 viruses [70]. *T. chebula* ethanolic extract inhibited viral attachment and penetration of HSV-1 [71]. *Phyllanthus orbicularis* butanolic and acetic acid extracts reduced viral adsorption, attachment and penetration into host cells with HSV-1 infection [72,73]. *F. carica* hexane and hexane-ethyl acetate extracts showed inhibition of HSV-1 viral penetration on host cells [67]. Aqueous extract of *R. ferrugineum* prevented HSV-1 attachment and penetration [74]. Tea tree oil from *Melaleuca alternifolia* and eucalyptus oil from *Eucalptus globulus* inhibited viral penetration of HSV-1 and HSV-2 [75,76]. *Cordia salicifolia* ethanolic extract from its dried leaves and twigs inhibited viral penetration in HSV-1 virions [77]. *A. flaccida* aqueous extract showed to prevent the penetration of HSV-1 virions in radiolabeled virion study [66]. *I. paraguariensis* hydroethanolic extract of the leaves fractionated with butanol and ethyl acetate and investigated for antiherpetic activity [78]. The ethyl acetate fraction exhibited prevented significant viral attachment and penetration into host cells. *Schinus terebinthifolia* crude hydroethanolic extract exhibited inhibiting viral attachment and penetration [79]. *Strychnos pseudoquina* ethyl acetate extract exhibited viral adsorption and penetration against HSV-1 and HSV-2 viral strains [80]. *Acacia catechu, Lagerstroemia speciosa, T. chebula,* and *Phyllanthus emblica* in a gel form used on open ulcers due to HSV-1 infection showed significant prevention of penetration by the virus [32]. *A. indica* bark aqueous extract prevented the entry of HSV-1 virus into the host cells. *Houttuynia cordata,* a Chinese herb, aqueous extract inhibited three strains of HSV by interfering with viral binding, penetration, and blocking viral entry [81,82]. *I. heterantha,* a western Himalayan plant showed inhibition of HSV-2 entry into host cells [30]. *F. religiosa* bark chloroform extract inhibited HSV-2 viral entry into cells [31]. *H. indicus,* the Indian sarsaparilla root inhibited HSV-1 and HSV-2 viral infections by preventing their entry at higher concentrations [33]. *P. orbicularis,* a native of Cuba, showed antiherpetic activity through cell entry inhibition [72]. Mushroom species *Inonotus obliquus* aqueous extract showed antiviral activity against HSV-1 by preventing viral entry to host cells [83]. *I. heterantha* hydromethanolic extract showed in vivo antiherpetic activity in BALB/c mice against HSV-2 by preventing their attachment, adsorption and entry into host cells [30].

18.4 Phytochemicals inhibiting replication of viral nucleic acids

Viruses use diverse strategies to replicate their nucleic acids and this largely depends on the families they belong to. Papillomaviruses, polyomaviruses, and herpesviruses synthesize daughter strands via a replication fork, while adenoviruses, parvoviruses, and poxviruses replicate by strand displacement. DNA viruses face a problem in replicating their DNA as the cellular DNA polymerases are unable to initiate synthesis of new DNA strand but can only perform the extension of a short primer. To tackle this adversity, one end of the newly synthesized viral DNA is retained as a short single-stranded region by some viruses or certain viruses had evolved a circular genome or by retaining complementary single stranded DNA primers at their termini or by incorporating a protein primer attached covalently to the 5′ terminus of the DNA strand. Viruses encode enzymes necessary for the successful replication of the viral DNA. Helicases to unwind the double helix, helix-destabilizing protein to separate the unwinded strands separate, DNA polymerase to copy the strand from the 5′ to 3′ direction, RNAase to degrade the RNA primer after the replication and DNA ligase to paste the Okazaki fragments together. The viral genome makes use of the host DNA polymerase α-primase to synthesize RNA primer for genome replication. Certain viruses require DNA replication initiation by the binding of an antigen in the form of large ′T′ binding to the regulatory sequence. This is exhibited by polyomaviruses and papillomavirus E1 and E2. Replication of the circular DNA proceeds from the replication fork in both directions commencing from the unique palindromic sequence. Replication can continue as either continuous or discontinuous pattern from the two expanding forks. In the discontinuous replication, the lagging strand showcases repeated short oligoribonucleotide primers giving rise to short nascent DNA strands called Okazaki fragments which by the action of DNA ligases covalently join to form a single long strand. In case of viruses with linear strand of DNA like adenoviruses, the replication starts from the 5′ end which being identical and covalently linked to the precursor of the primer. Replication proceeds from both ends continuously and synchronously with the help of DNA polymerase and forming a continuous long strand of DNA.

RNA viruses utilize a unique mechanism of replication by utilizing an RNA template which is not seen in any other organisms. For replication they require an RNA-dependent RNA polymerase which is coded by the virus and is not seen in uninfected cells. Viral RNA replication requires the synthesis of a complementary RNA chain which serves as its template. In viruses with negative sense RNA, the complementary RNA is positive sense and vice versa. The polymerase enzyme performs the same function as in the mRNA transcription.

EGCG inhibits viral enzymes like RNA polymerase, protease, and reverse transcriptase to produce antiviral activity [84]. Replication of HIV is restricted via inhibiting the phosphorylation of protein also. Baicalein inhibited the replication of dengue virus while Fisetin and quercetagetin, both belongs to the group of flavonols, inhibited the replication of dengue virus and enterovirus A71 in in vitro studies. Baicalein, quercetagenin, and fisetin are together reported to inhibit the replication of CHIKV replication at early stages [85]. Luteolin, a flavone widely present in chamomile tea, perilla leaf, green pepper and celery is a good virucidal agent, which inhibited the replication of *Japanese encephalitis* virus at postentry phase [86]. Pinocembrin, a constituent of honey, tea and red wine inhibited zika virus replication at postentry stage [64]. Genistein, has antiviral activity against hepatitis B virus and it targets the viral replication stage. This compound in combination with antiviral agents like acyclovir and ganciclovir significantly reduced the effective dose of drug. Genistein has been proved as an antiviral agent against a large group of viruses including HSV-1, cytomegalovirus, bovine herpesvirus-1, SV40, human papilloma virus, porcine reproductive and respiratory syndrome virus, African swine fever virus, and HIV [87]. Two flavonoids isolated from *Salvia plebeian* showed inhibition of replication cycle in H1N1 virus. They also proved inhibition of neuraminidase of H1N1 and thus inhibited the growth of the virus as well [88]. Researches from Indonesia reported antidengue activity of *Myristica fatua*, *Cymbopogon citratus*, and *Acorus calamus* probably by inhibiting replication [89]. Alkaloids, the most studied secondary metabolite from plants, are also widely studied for their anticancer, antibacterial, antiviral, and antiasthma activities. Alkaloids as antiviral agent against influenza viruses are probably due to the activation of interferons and the resultant immune system activation. Macrophage activation and thereby the escalated phagocytosis is also attributed to the action of alkaloids against viruses. Alkaloids inhibit the replication of influenza viruses by interfering at different stages of replication and viral protein synthesis. Homonojirimycin, an alkaloid from *Commelina communis* showed a strong activity in inhibiting influenza viral infection [90]. Dendrobine from *Dendrobium nobile* inhibited the replication of influenza viruses in the host cells. Dedrobine has been shown to bind to the nucleoprotein of the influenza virus and restrict its replication and export. A synergistic activation of antiinfluenza A activity is also exhibited by dendrobine along with the antiviral agent zanamivir in combating influenza infection [91]. Among the secondary metabolites from plant sources, terpenes form a major group of antiviral agents. 3, 4-secodammarane, a triterpenoid inhibited the replication of the viral genome as evidenced by its inhibition of S and G2/M phases of cell cycle [92]. Terpenes derived from *Aglaia* sp. inhibited HIV-1 infection in humans from exhibiting strong cytotoxic activity along with the inhibition of the viral proliferation. Terpenes have also shown activity against simian immunodeficiency virus and murine leukemic virus. Terpenes from *Marrubium vulgare* and oleanolic terpenoids from *Camellia japonica* flowers interrupted the replication of HSV-1 and porcine epidemic diarrhea virus (PEDV) respectively [93,94]. These terpenes inhibited viral replication by inhibiting the genes coding for GP2 spike, GP5 membrane protein and GP6 nucleocapsid and interrupting the viral assembly. Lignans are another important class of secondary metabolites exhibiting varied biological activity like antihepatotoxic, antiinflammatory, and antidepressant activities along with antiviral activity against HIV and influenza viruses. Honokiol, a lignin bisphenol from Magnolia tree inhibited the formation of dsRNA along with the nonstructural proteins ns1, ns3, and other intermediate proteins during replication in dengue viruses [95]. The traditional use of this tree in Chinese medicine is well substantiated through this finding. Phorbol-type and ingenane-type diterpene compounds with antiviral activity which potentially inhibits the viral replication has been given US patents assigned to Korean Research Institute of Bioscience. Lignans isolated from the Chinese medicinal plant *Radix isatidis* has shown antiviral activity against influenza virus, HSV, respiratory syncytial virus (RSV), mumps virus, coxsackie virus, and hepatitis B viruses. The mode of action is attributed to the disruption of the replication machinery in case of RSV [96]. Similarly, lignans from *Forsythia suspense* exhibited antiinfluenza A (H5N1) activity probably with similar mechanism of action [97]. *Swietenia macrophylla* yielded a lignan named 3-hydroxy caruilignan C (3-HCL-C) exhibited antiviral activity against chronic HCV in coadministration with interferon-α (IFN-α), 2'-C-methylcytidine or telaprevir by reducing viral replication at both translation and transcription levels [98].

Lectins are proved to agglutinate virions and inhibit them from binding on to the host cell surface and inhibits replication of viruses. *Boerhaavia diffusa*, *Eclipta alba* are considered hepatoprotective and their action in infective inflammatory viral infections like hepatitis including viral hepatitis is well documented [99]. Isoflavones isolated from *Psorothamus arborescens*, (2 S)-Eriodictyol 7-O-(6″-O-galloyl)-beta-d-glucopyranoside from *Phyllanthus emblica*, 3,5,7,3′,4′,5′-hexahydroxy flavanone-3-O-beta-d-glucopyranoside from *Phaseolus vulgaris*, methyl rosmarinate from *Hyptis atrorubens*, myricitin from *Myrica cerifera*, myricetin 3-O-beta-d-glucopyranoside from *Camellia sinesnsis*, amaranthin from *Amaranthus tricolor* and licoleafol from *Glycyrrhiza uralensis* are some of the probable phytochemicals that inhibits the replication of SARS-CoV-2 virus by targeting the 3-chymotrypsin-like cysteine protease (3CL^{pro}) essential for the replication of coronaviruses [100]. Plants from *Veronica* species are used for treating influenza as they inhibit the intracellular replication of viruses and provided symptomatic relief in HSV-1 infections [101]. Quercetin prevents HCV replication by inhibiting the HCV NS3 protease which ends the replication of HCV in the subgenomic RNA replicon cell system. Quercetin also inhibits the heat shock protein activity in HCV which is a crucial step in nonstructural protein 5A mediated viral entry site. In rhinoviruses, quercetin influences different stages of the life cycle of virus like inhibiting endocytosis, protein synthesis, and viral genome transcription [102−104]. Quercetin, myricetin, quercetagetin, and baicalin are found to be effective in preventing the growth and replication of Rauscher murine leukemia virus [105]. Quercetin showed a synergistic action against COVID-19 along with ascorbic acid. Coadministration of ascorbic acid was found to be beneficial in COVID-19 patients owing to the synergistic antiviral, antioxidant, and immunomodulatory activities along with the capability of ascorbic acid to recycle quercetin to provide more sustained effectiveness. Apigenin inhibits the replication of HCV by decreasing the liver specific microRNA 122 expression [106]. Baicalin, another flavonoid, inhibits the neuraminidase activity and prevents the replication of H5N1 viruses in human lung and monocyte-derived macrophages. Luteolin is found to be highly effective against the replication of enterovirus-71 and coxsackievirus A 16 due to the viral RNA disruption [107]. Tormentic acid glucosyl ester, a triterpenoid inhibited HSV-1 viruses by inhibiting the capsid protein synthesis and DNA replication [108]. Niranthin from *Phyllanthus amarus* inhibits HBV virus by downregulating the DNA replication [109]. Nordihydroguaiaretic acid, a lignan from the leaves of *Larrea tridentata* inhibited the genome replication and disrupted the viral assembly in HCV, dengue, influenza A, and zika virus infections [109–111]. Terameprocol, a lignin derived from the leaves of *Larrea tridentata* with semisynthetic modifications inhibited the replication of WNV, pox virus, HSV, and HIV viruses by interfering with the cell-to-cell transfer and by inhibiting the binding of host transcription factor [112]. Arctigenin showed antiviral properties against influenza A virus and HIV virus by IFN secretion induction and inhibiting protein expression [113]. Dried leaves of *Chamaecyparis obtuse* yields yatein which inhibited the DNA synthesis in HSV-1ICP0 and ICP4 strains [114]. Diphyllin from *Haplophyllum* inhibited vacuolar ATPase in zika virus and inhibited influenza A infection by downregulating genome replication [115]. Silymarin, a popular hepatoprotective agent obtained from the seeds of *Silybum marianum* inhibited HCV infection by inducing antiinflammatory and antiproliferative genes. Silymarin showcases other activities like regulation of lipid metabolism, apoptosis, protein, and cytokine expression along with viral replication inhibition [116]. Tannins are another important group of phytochemicals specifically studied for antimicrobial studies. They have potent capability to interfere with the different stages of viral life cycle like attachment, replication, assembly and protein transport [117]. Ellagitannins derived from *P. amarus* with the names Corilagin and geraniin interfered with HIV replication. Combinations of tannins with antiviral agents like acyclovir were investigated for their escalated action [118]. Castalagin, vescalagin, and grandinin were investigated for such an action and it was found that the action of castalagin and vescalagin showed equivalent action as acyclovir against HSV-1 viruses. These two tannins also showed high activity against alphaherpevirus-1 and showed good level of inhibition of replication of HSV-1 [119]. Triterpenoid saponin TS21 blocks the HSV viral capsid protein synthesis and replication. Niranthin inhibits DNA replication in duck HBV infection. Nordihydroguairetic acid affects the viral proliferation by inhibiting genome replication and viral assembly. It also inhibits the replication of influenza A virus [109,120]. Terameprocol shows its efficacy against WNV by arresting its replication. Curcumin has been subjected to extensive studies to establish its antiviral mechanisms. They directly interfere with viral replication machinery and suppresses cellular signaling pathways like PI3K/Akt and NF-kB [121]. Glycyrrhizin has shown inhibition of SARS-CoV replication in Vero cells. Apart from inhibition of viral replication they also act at multiple levels like preventing adsorption and penetration of the virus. The mechanism of action of glycyrrhizin inhibiting the viral cell replication is by inducing nitrous oxide synthase levels in the host cells [122]. Silvestrol derived from *Aglaia* sp. Inhibited replication of MERS-CoV by inhibiting the enzyme RNA helicase eIF4A and prevents the establishment of replication and transcription complexes [123]. Cepharanthine, tetrandrine, and fangchinoline isolated from *Stephania tetrandra* and related species showed antiviral activity against HCoV-OC43 by suppressing virus-induced host response, virus replication and viral S protein and nucleocapsid protein (N protein) expression [124]. Trytanthrin derived from the leaves of *Strobilanthes cusia* showed antiviral activity against HCoV-NL63 by preventing early and late replication periods. By inhibiting the postentry replication stage by blocking papain-like protease two activity and the viral RNA genome synthesis [125]. *Nigella sativa, Anthemis hyaline, Citrus sinensis*, and *Ziziphus jujuba* ethanolic extracts inhibited replication in mouse hepatitis virus (MHV) [126]. Avian infectious bronchitis virus infection was subdued with treatment with terpenoid compounds (-)-β-pinene, (-)-α-pinene and 1,8-cineole by breaking the replication cycle of the virus by binding to its N protein. Oxyresveratrol from *Artocarpus lakoocha* inhibited the early and late phases of HSV-I and HSV-2 replications [127]. *Ruta angustifolia* leaves yield two important antiviral agents in the form of Chalepin and Pseudane IX and they are reported to act in multiple ways by inhibiting HCV at postentry stages of RNA replication and viral protein synthesis [46]. Another important phytochemical from *Liriope platyphilla*, LPRP-97543 showed antiviral activity against HBV at gene expression, replication and viral promoter activity levels by hampering the binding activity of NF-kB to HBV surface gene CS1 element [48]. Polymethoxylated flavones isolated from the pericarp of *Citrus reticulate*, Tangeretin and Nobiletin, inhibited intracellular replication of RSV. Tangeretin is shown to inhibit the virus-to-cell fusion and fusion during the early stage of replication and cell-cell fusion at the end of replication cycle [49]. Dicaffeoylquinic acid isolated from *Schefflera hetraphylla* also showed inhibition of RSV [128]. Saururus chinensis derived Manassantin B also showed inhibition of EBV replication. Several studies related to the mechanism of action of phytochemicals have been done. Traditional Chinese medicine herbs *Cibotium barometz, Gentiana scabrea, Dioscorea batatas, Cassia tora*, and *Taxillus chinensis* inhibited SARS-CoV replication [129]. *Torreya nucifera* yielded a biflavonoid havind strong inhibition of the replication of SARS-CoV 3CLpro [130]. Quercetin-3-β-galactoside showed inhibition of protease activity in SARS-CoV while quercetin-7-rhamnoside flavonoid reduced replication of epidemic diarrhea virus in pigs and PEDV [131].

Among the plethora of phytochemicals, flavonoids inhibit Asn-130 glycosylation site where the nonstructural 1 (NS1) binds for the initiation of propagation of dengue virus. An escalation in the amount of NS1 will increase the triple-N-linked glycosylation as a crucial step in the attachment of cells. This event can control the replication of virus by its control over the functions of protein. Thus, preventing the Asn-130 binding with NS1 stops the replication of dengue virus. NS1 was has high affinity towards flavonoids with a strong bond-effect. Flavonoids from plants such as grape seeds, pomegranates, mulberries, black raspberries, cranberries, green tea extracts, persimmons, ginseng oregano essential oils, chitosan, citric acid, and proanthocyanins reduces NoV replication. Emodin, a polyphenol isolated from the roots, bark, and leaves of aloe vera, cascara, rhubarb, and senna can inhibit replication of influenza A virus by modulating signal pathways [132]. Chrysin, a natural flavonoid, from the bee propolis and honey is preventing the replication of coxsackievirus B3 (CVB3). A mannose-binding lectin from *Narcissus tazetta* inhibited H1N1, H3N2, and influenza B viruses and showed a promising activity against the replication of H1N1A early phase by interacting with surface glycoproteins and also prevents the adherence and fusion of virus. A fetulin-binding nonspecific lipid transfer protein named NTP from the same plant inhibited the neuraminidase of H1N1 and inhibited its replication.

Anthraquinones (2',3', 4', 6'-Tetra-O-acetyl-b-D-glucopyranosyl-aloesaponarin-I) and 3-(2',3',4',6'-Tetra-O-aetyl-β-D-glucopyranosyl-aloesaponarin-II) derived from aloe vera inhibited replication of H1N1. *Arctium lappa* fruits yielded two lignans with the names arctiin and arctigenin which inhibited with the early replication of H1N1 viruses and inhibited virion release [133]. *D. nobile* derived alkaloid, dendrobine inhibited the early stage of viral replication in H1N1 and H3N2 viruses [91]. The flavone, Quercetin 3-rhamnoside from *Houttuynia cordata* reduced viral mRNA synthesis in H1N1 viruses. *I. indigotica* from Brassicaceae family yielded clemastanin A which inhibited the early replication stages in H1N1 [134]. Hydrolysable tannin, punicalagin from *P. granatum* fruit inhibited H1N1, H3N2 and influenza B virus replication by preventing target attachment [38]. *Scutelaria baicalensis* leaves and roots yielded flavones named baicalin, 5,7,4'-trihydroy-8-methoxyflavone (F36), Isoscutellarein (5,7,8,4'-tetrahydroxyflavone), wogonin, baicalein. Among these F36 inhibited infection and replication of Influenza-A H1N1, H3N2, and influenza B viruses. Isoscutellarein inhibited sialidase enzyme prevented the replication of influenza A H1N1 virus [135]. Wagonin also exhibited significant suppression of H1Ni virus replication. Baicalein also inhibited influenza H1N1 virus replication by interacting with the NA1 active sites [107]. *Aronia melanocarpa* is a rich source of polyphenolic compounds with very good antiviral properties. Myricetin, a flavone and ellagic acid inhibited the replication of H1N1 and H3N2 viruses [136].

Hydrolyzable tannins from *Camellia sinensis* named strictinin inhibited influenza A and B virus replications [137]. Curcumin exhibited dose dependent inhibition of viral replication of H1N1 and H6N1 viruses [22]. Quercetin was also investigated for their activity against DENV-2 viruses and found to inhibit their replication significantly [138]. Baicalin also showed significant reduction in the replication of DENV-2 and CHIKV viruses [135]. Apigenin on the other hand inhibited the replication of HCV. Kaempferol interfered with the replication of IRES in EV-A71 infected RD cells. Rhinovirus replication and transport was inhibited by quercetin. Myricetin and other quercetin flavonoids targeted the replication of zika virus by blocking the RNA production. Researches have also shown that naringenin, a flavonoid inhibited the viral NS2B-NS3 protease and prevented the replication of ZIKV. Baicalein, another flavonoid, also inhibited cytomegalovirus replication in human embryonic lung fibroblasts cells [139]. Genistein also inhibited CMV replication [140]. These compounds targeted the immediate early proteins IE1-72 and IE2-86, which are essential for replication of viruses. Pinocembrin, belonging to flavonoid group also produced significant antiviral activity in JEG-3 cells infected with ZIKV at low concentrations. The replication inhibition is by targeting the protein synthesis. Among flavonoids, 3G (Quercetin-3-β-O-D-glucoside) inhibited viral replication in EBOV strains EBOV-kikwit and EBOV-Makona. Taxifolin interacted and inhibited VP35 protein of Ebola virus in in silico studies. VP35 proteins in Ebola virus is vital for the formation of replication complex. Dihydroquercetin inhibited Coxsackievirus B4 (CVB4) in a mouse model by suppressing replication process. Luteolin inhibited coxsackie A16 strain grown on cultured cells and found to affect viral-RNA replication and also inhibited postattachment stage of the virus. Quercetin-7-rhamnoside inhibited PEDV by inhibiting the replication process at an early phase [141].

Alkaloids isolated from the seeds of *Peganum harmala*, a member of Nitrariaceae inhibited viral RNA replication and polymerase activity in influenza A virus [142]. Lycorine, an alkaloid from *Lycoris radiata* inhibited propagation and replication of HCoV-OC43 [143]. Total alkaloid fraction from *Alstonia scholaris* inhibited viral replication, mRNA inhibition of cytokines and proteins and increased survival rates along with deactivation of pattern recognition receptor and interferon-activated signal transduction pathways. In animal models these alkaloids were capable of decreasing innate immune cell infiltration and improved lung histopathology [144]. *Cryptolepis sanguinolenta* alkaloids cryptomisrine, cryptospirolepine and cryptoquindoline potent interactions with Mpro/RdRp of ARS-CoV-2 virus [145]. Ipecac alkaloid emetine inhibited replication of SARS-CoV-2 virus through the inhibition of viral polymerases in chicken coronavirus [146]. Envelope (E) protein is another drug target in SARS-CoV-2 virus, which helps in the assembly of the virus during replication. Berbamine, an alkaloid and its derivatives exhibited anti-SARS-CoV-2 activity through inhibition of E protein [147]. Nonstructural proteins Nsp15 also plays a role in the replication of SARS-CoV-2. Alkaloids like ajmalicine, reserpine, berberine, and taspine produces anti-SARS-CoV-2 activity by inhibiting Nsp15. Berberine showed promising results in reducing the complications of COVID-19 and also provides additional protection against many other viruses owing to its reported antioxidant, antiinflammatory, immunomodulatory, neuroprotective, and antilung-injury effects [148–150]. *Avicenna marina* inhibited HSV-2 replication with its methanolic extract of leaves. *Pongamia pinnata* aqueous extract showed a complete inhibition of HSV-1 AC strain and HSV-2 HV-219 strain replications. *H. cordata*, traditionally used by Japanese and Chinese medicines, inhibited replication of HSV-1 HF strain. *H. cordata* was also subjected to study their effect against HSV-2 replication and found that they were effective when added prior to infection. These results showed the efficacy of these drugs to be used as agents that prevent viral infection [81,82]. *P. suffruticosa*, *M. toosendan* and *R. officinale* ethanolic and methanolic extracts acted at different stages of viral replication. Different species of *Phyllanthus* like *P. amarus*, *P. niruri*, *P. urinaria*, and *P. watsonii* aqueous extracts shown activity against HSV-1 and HSV-2 and was found to significantly inhibit their replication [72]. *Lobelia chinensis* methanolic extract inhibited HSV-1 KOS strain in HeLa cells by blocking its replication by impairing DNA synthesis. A Chinese herb *Viola yedoensis* aqueous extract inhibited HSV-1 replication in human neuroblastoma cells. The leguminous plant *Cajanus cajan*, pigeon pea exhibited antiviral effects against HSV-1 and HSV-2. The antiviral activity was compared with that of acyclovir and it was found that the *C. cajan* elicits a direct antiviral effect on the HSV virions [151]. *Pedilanthus tithymaloides* showed strong antiviral activity against HSV-2 G strain by inhibiting NF-κB, which is required for viral replication [152]. *I. heterantha* hydromethanolic extract showed antiviral activity against HSV-2 G strain by inhibiting viral replication cycle at an early stage. *Euphorbia spinidens* methanolic extract inhibited HSV-1 replication [153]. *H. indicus* root hydromethanolic extract inhibit both HSV-1 and HSV-2 replication to a similar degree. The chloroform fraction of the ethanolic extract of *A. tortuosum* leaves exhibited antiherpetic activity by direct virucidal agent and inhibited replication [34]. The fresh water algae Spirogyra methanolic extract inhibited viral replication in HSV-1 F strain. *Mentha piperita* hot water extract pretreatment with HSV-1 inactivated the virus. *F. carica* hexane, ethyl acetate, methanol, chloroform and hexane-ethyl acetate extracts were studied for their antiherpetic activities. The hexane and hexane-ethyl acetate extracts inhibited HSV-1 intracellular replication [67]. *Lentinus edodes*, a mushroom water extract inhibited HSV-1 replication in a dose-dependent manner.

Haematococcus pluvialis and *Dunaliella salina*, microalgae were extracted with hexane, ethanol and water under pressurized liquid inhibited intracellular replication of HSV-1 KOS strain virus [154]. *G. glabra* inhibited HSV-1 infection by interfering with gene expression and delays viral replication [155]. *Opuntia streptacantha* ether, acetone and chloroform extracts of dried powder inhibited HSV-2 3345 strain by reducing viral replication and the replication seems to be more pronounced when the extracts were pretreated of the cells. *C. salicifolia* chloroform and ethanolic extracts of the dried leaves inhibited penetration and replication of HSV-1 virus [156]. *Baccharis trinervis, Heisteria acuminate*, and *Eupatorium articulatum* aqueous and ethanolic extracts showed replication inhibition in HSV-1 infected cells [157]. *Cecropia glaziovii* leaves hydroethanolic extract inhibited replication of HSV-1 stain 29-R/acyclovir resistant and of HSV-2 replication on pretreatment. Hydroethanolic extracts of *Equisetum giganteum, Croton lechleri, Uncaria tomentosa*, and *Copaifera reticulata* inhibited HSV-2 333 strain replication on pretreatment of cells [158]. *Tanacetum parthenium* aerial parts where screened for antiviral activity and their hydroethanolic extract inhibited HSV-1 replication in Vero cells [159]. *Fomes fomentarius, Pleurotus ostreatus, Auriporia aurea* and *Trametes versicolor*, four mushroom species as aqueous extracts inhibited HSV-2 replication. *Eugenia caryophyllus* flower bud ethanolic extract inhibited the replication of HSV-1 and HSV-2 [160]. *T. vulgare* ethyl acetate extract prevented HSV-1 replication [161]. Ethanolic extract from berries of *Rubus eubatus* prevented adsorption and replication of HSV-1 strain 17 + in cultured epithelial cells derived from oral cavity (OKF6). They also produced virucidal activity [162]. *L. chinensis* methanolic extract demonstrated antiherpetic activity against HSV-1 virus in infected BALB/c mice model. *L. chinensis* blocked the DNA synthesis in in vitro studies and thus prevented the replication of HSV-1 [163]. A Chinese herbal drug combination named Lianhuaqingwen containing menthol demonstrated anti-inflammatory activity and also inhibited SARS-CoV-2 virus replication [164]. Yupingfeng powder, a polyherbal formulation from Chinese medicine containing *A. lancea, A. macrocephala, Astragalus mongholicus, G. glabra, L. japonica*, and *S. divaricate* and ethanolic extract of *S. javanica* stem inhibited virus attachment and also prevented the replication of SARS-CoV-2 and H1N1 viruses [165]. *Dioscorea polystachya* tubers and *C. barometz* rhizomes inhibited SARS-CoV 3CL protease, which is the major protein for the genome replication of SARS-CoV [129]. A study from Iran showed the efficacy of four medicinal plants; *Quercus infectoria, Onopordum acanthium, Crataegus laevigata*, and *Berberis integerrima* inhibited ACE receptors in SARS-CoV infection [166]. Nimbin, berberine, and mangiferin isolated from the Indian medicinal plants *A. indica* and *Tinospora cordifolia* showed significant binding affinity to ACE2 receptors of SARS-CoV-2 in in silico studies. The study concluded that the phytoconstituents of *A. indica* and *T. cordifolia* could be utilized in the prophylaxis as well as the therapy of COVID-19 as they are proved to restrict viral attachment and prevents the replication of viral RNA [167]. Allicin, a bioactive compound from garlic inhibits viral replication and acts as an immunomodulator. The presence of selenium enhances the virucidal activity of garlic as selenium is proved to be an inhibitor of replication of coxsackie virus [168,169]. Polyphenolic compound, resveratrol, derived from berries like blueberries and cranberries exhibit strong virucidal activity by inhibiting viral replication and exhibiting antioxidant effects [170]. Blueberries are in-addition considered as a rich source of anthocyanin which has shown inhibition of influenza A(H3N2) virus. Anthocyanins from *Solanum paniculatum* extract inhibited the replication of Human herpes virus (HHV-1) virus [171]. Alkaloid, dendrobine from *D. nobile* shown high degree of influenza A virus replication [91]. Flavonoids isolated from the aerial parts of *Salvia plebeian* also inhibited replication of influenza A viruses in a dose-dependent manner. Flavonoid, genistein present in soy-based products are proved to be inhibitors of herpes B virus [87]. 11-dihydroxymeliacarpin extracted from the leaves of *M. azedarach* and a lignin-carbohydrate complex isolated from *P. vulgaris* targeted HSV-1 and HSV-2 viral infections by inhibiting the late phase of their replications [172].

18.5 Phytochemicals preventing assembly and release of virus

The structural proteins formed associate to form the structural units called capsomers which self-assemble later to form capsids followed by the insertion of viral nucleic acid and form the virion which is released. Viral genome gets attached to a nucleotide sequence known as packaging sequences at one end which guides the viral DNA to enter the procapsid bound to basic core proteins followed by the enzymatic cleaving of some of the capsid proteins to produce mature virion. Generally, nonenveloped viruses attain an icosahedral structure and assemble themselves in the cytoplasm or the nucleus of the cell and releases with the lysis of the cell. Some viruses on the other hand mature after acquiring an envelope by the process of budding through cellular membranes. Enveloped viruses after acquiring the envelope within the cell are transported within vesicles to the cell surface and they bud from the plasma membrane or the internal cytoplasmic membranes or the nuclear membrane. Budding from the surface of the membranes occurs through the incorporation of viral glycoproteins being inserted into the lipid bilayer by lateral displacement of cellular proteins. The single viral glycoproteins are inserted as rod-shaped or club-shaped peplomers with a hydrophilic protruding domain to the external surface of the cell and a hydrophobic transmembrane anchor domain and another hydrophilic cytoplasmic domain. During budding, protein molecules of the nucleocapsid binds to the cytoplasmic domain of viral glycoprotein and molds them on to the exterior of the nucleocapsid. Budding and release of the enveloped virion does not affect the physical integrity of the plasma membrane. Hence, several virions can bud off from the plasma membrane over a period of time without damaging the plasma membrane (Fig. 18.3).

Nordihydroguaiaretic acid isolated from the leaves of *Larrea tridentata* showed antiviral property against HCV, dengue virus, influenza A, and zika virus by inhibiting the viral assembly [109–111]. Tannins like ellagitannins and geraniin from *P. urinaria* and *P. amarus*, punicalagin and chebulagic acid from *T. chebula* inhibited viral assembly and transport [21,32,109]. Many plant-derived flavonoids are proved to be inhibitors of formation of viral envelop glycoproteins and viral release. Rhamnose residue containing kaempferol isolated from *Ficus benjamina* inhibited HSV-1, HSV-2, and corona virus infections by affecting their release by via 3a channel inhibition [173]. *Ribes nigrum* crude extracts inhibited virus release in MDCK cells infected with H1N1 when treated post infection showing complete clearance of virus titers in culture fluids [174]. Anthocyanins isolated from the same plant; 3-O-alpha-L-rhamnopyranosyl-beta-D-glucopyranosyl-cyanidin, 3-O-beta-D-glucopyranosyl-cyanidin, 3-O-alpha-L-rhamnopyranosyl-beta-D-glucopyranosyl-delphinidin, and 3-O-beta-D-glucopyranosyl-delphinidin inhibited the release of influenza viruses [175]. A lignan, arctigenin, isolated from *A. lappa*, showed inhibition of viral replication and progeny release in MDCK cells infected with H1N1 virus [176]. The flavonoid,

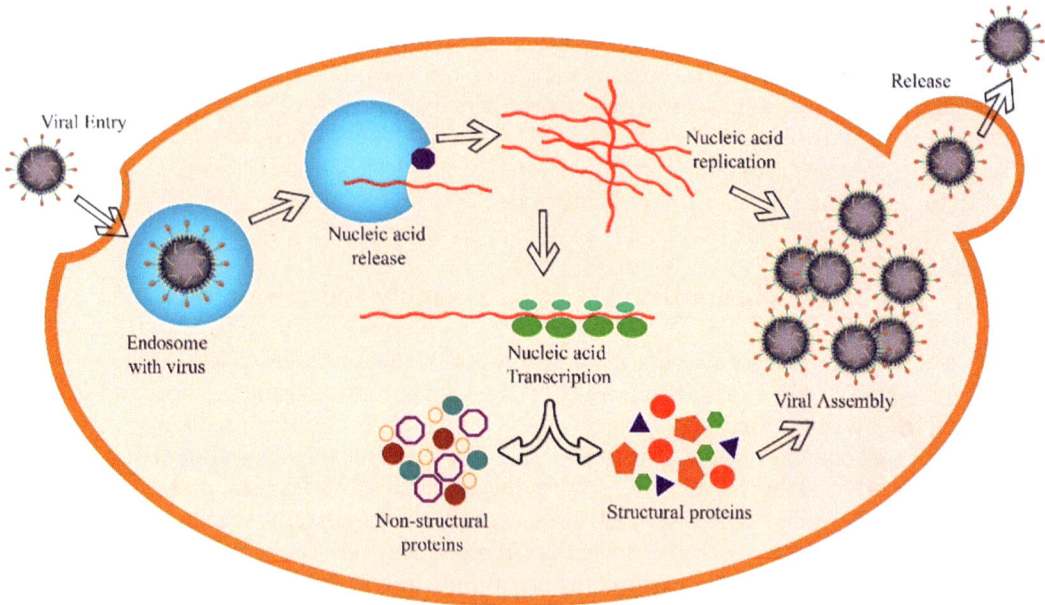

FIGURE 18.3   Nucleic acid replication, viral assembly, and release from the host cell.

TABLE 18.1   Medicinal plants with antiviral activity.

Scientific name	Part used	Antiviral activity	References
Acorus calamus L.	Rhizome	HSV-1 and HSV-2	Umamaheshwari and Rekha, [177].
Aegle marmelos (L.) Correa	Root bark	Ranikhet disease virus	Dhar et al. (1973) [178]
Allium sativum L.	Bulb	Influenza A and B Cytomegalovirus Rhinovirus HIV herpes simplex virus 1, herpes simplex virus 2, viral pneumonia, rotavirus, decrease expression of proinflammatory cytokines SARS-CoV-s Hepatits A virusInfluenza A (H1N1) pdm09 virus—inhition of of viral nucleoprotein synthesis and polymerase activity	Ankri and Mirelman [179], Chavan et al. (2016) [214]
Alpinia galanga (L.) Willd.	Rhizome	HIV type 1 blocking reverse transport Immune-stimulating activity in mice	Ye and Li [180], Bendjeddou et al. [181]
Anisomeles indica (L.) Kuntze	Whole plant	HIV-1 virus—ovatodiolide	Alam et al. [182]
Annona reticulata L.	Fruit	HIV	Pathak and Zaman [183]
Artemisia vulgaris L.	Whole plant	Yellow fever virus	Meneses et al. [184]

(Continued)

TABLE 18.1 (Continued)

Scientific name	Part used	Antiviral activity	References
Artemisia parviflora Buch. Ham. Ex D. Don	Aerial parts	Ranikhet disease virus Vaccinia virus	Dhar et al. (1973) [178]
Azadirachta indica A. Juss	Whole plant	HIV virus	Badam et al. [185], Tiwari et al. [186]
Baccaurea ramiflora Lour	Fruit, leaves, bark, root, seeds	Semliki forest virus	Das and Das [187]
Bergenia ciliata (Haw.) Sternb.	Whole plant	Influenza A virus	Bhandari et al. [188]
Boehmeria nivea (L.) Gaudich	Roots and leaves	Hepatitis B virus	Chang et al. [189]
Capparis multiflora Hook.f. & Thomson	Whole plant	Ranikhet disease virus	
Castanea sativa Mill.	Bark, seeds, nuts, leaves	Reo virus Avian metapneumovirus	Lupini et al. [190]
Cissampelos pareira L. var. hirsuta (Buch. Ham. Ex DC.)	Whole plant	Dengue virus	Bhatnagar and Popli [191]
Citrus hystrix DC.	Fruit, leaves and rind of fruits	Herpes simplex virus type-1	Fortin et al. [192]
Curcuma longa L.	Rhizome	Influenza virus, Respiratory syncytial virus, Herpes simplex-1, Human Papilloma virus, Human norovirus	Li et al. [97], Singh et al. [193], Pécheur [194]
Dillenia pentagyna Roxb.	Stem bark	Raniket disease virus	Boparai et al. [195]
Dioscorea bulbifera L.	Bulbils and tuber	HIV	Chaniad et al. [196], Lebot et al. [197], Ahmed et al. [198]
Elephantopus scaber L.	Whole plant	HIV-1, Rhinovirus	Wang et al. [199]
Houttuynia cordata Thunb.	Whole plant	Influenza, HIV, Porcine epidemic diarrhea virus, SARS-CoV	Kumar et al. [200]
Leea indica (Burm.f.) Merr.	Leaves	Herpes simplex virus type-1	Hamidi et al. [201]
Nyctanthes arbor-tristis L.	Aerial part	Encephalomyocarditis virus Semliki Forest virus	Saxena et al. [202]
Psidium guajava L.	Whole plant	Dengue virus serotype-2, Influenza H1N1, Newcastle disease virus	Sriwilaijaroen et al. [203], Chollom et al. [204]

(Continued)

TABLE 18.1   (Continued)

Scientific name	Part used	Antiviral activity	References
Punica granatum L.	Whole plant	Human herpes virus-3, Adenovirus, Herpes simplex virus type-1, Human respiratory syncytial virus, Influenza virus	Angamuthu et al. [205], Moradi et al. [206], Haidari et al. [21]
Rhus chinensis Mill.	Whole plant	HIV-1, Herpes simplex virus type-1,	Wang et al. [207]
Scoparia dulcis L.	Whole plant	Herpes simplex virus type 1	Murti et al. [208]
Siegesbeckia orientalis L.	Whole plant	Ranikhet disease virus	Chang et al. [209]
Solanum virginianum L.	Whole plant	Arbovirus	Karthik et al. [210]
Strobilanthus cusia (Nees) Kuntze	Whole plant	Influenza, Human coronavirus NL63	Tsai et al. [125]
Syzygium jambos (L.) Alston	Whole plant	Herpes Simplex Virus type-1 and type-2, Vesicular stomatitis virus, Avian influenza virus (H5N1)	Teixeira et al. [211]
Terminalia chebula Retz.	Fruit	HIV, AIDS, HSV-2	Kesharwani et al. [71]
Urtica dioica L.	Whole plant	SARS-CoV, HIV-1, HIV-2, CMV, RSV, Influenza A, Dengue virus	Kumaki et al. [212]
Zingiber zerumbet (L.) Roscoe ex Smith.	Whole plant	Epstein Barr Virus, HIV	Murakami et al. [213]

naringenin impaired the virion assembly of zika virus in infected A549 cell lines. Viruses have genes coding for ion-selective channels which later gets integrated to the host plasma membrane. These channels help in the viral release [105]. Genistein, a flavonoid has shown inhibition of the viral-protein-U (Vpu), an ion channel expressed by HIV-1 viruses for their release [87]. Flavonoids kaempferol and juglanin inhibited 3a-channel protein, which is used to release viruses in SARS-CoV-1 infections [60].

Apart from these there are several researches being conducted on phytochemicals to combat viral infections. However, their molecular mechanisms have not been clearly elucidated. (Table 18.1).

Viruses have evolved several mechanisms to evade the immune mechanisms of the host cell to establish the infection. This evolutionary advantage gained by the viruses provide them the strength to overcome the most powerful antiviral agents. Powerful, potent and tactful treatment modalities to be adopted to overcome these trick strategies adapted by viruses. A multiprong approach is the need of the hour wherein drugs acting at multiple drug targets needs to be promoted. Phytochemicals derived from medicinal plants are ideal for this purpose as there are ample evidence for their antiviral activity. The vast research findings help in developing new antiviral agents as well as adjuvant therapies that aid in overcoming the obstacles in treatment like resistance development.

References

[1] Hsu NS, Hendriks S, Ramos KM, Grady C. Ethical considerations of COVID-19-related adjustments to clinical research. Nat Med 2021;27(2):191−3.

[2] Cascella M, Rajnik M, Aleem A, Dulebohn S, Di Napoli R. Features, evaluation, and treatment of coronavirus (COVID-19). StatPearls. 2021 Apr 20.

[3] Jansi S, Khusro A, Agastian P, Alfarhan A, Al-Dhabi NA, Arasu MV, et al. Emerging paradigms of viral diseases and paramount role of natural resources as antiviral agents. Sci Total Environ 2021;759:14539.

[4] Ghildiyal R, Prakash V, Chaudhary VK, Gupta V, Gabrani R. Phytochemicals as antiviral agents: recent updates. Plant-derived bioactives. Singapore: Springer; 2020. p. 279−95.

[5] Sieczkarski SB, Whittaker GR. Dissecting virus entry via endocytosis. J Gen Virol 2002;83(7):1535−45.

[6] Maginnis MS. Virus−receptor interactions: the key to cellular invasion. J Mol Biol 2018;430(17):2590−611.

[7] Dermody TS, Kirchner E, Guglielmi KM, Stehle T. Immunoglobulin superfamily virus receptors and the evolution of adaptive immunity. PLoS Pathog 2009;5(11):e1000481.

[8] Yamauchi Y, Helenius A. Virus entry at a glance. J Cell Sci 2013;126(6):1289−95.

[9] Barton ES, Forrest JC, Connolly JL, Chappell JD, Liu Y, Schnell FJ, et al. Junction adhesion molecule is a receptor for reovirus. Cell. 2001;104(3):441−51.

[10] Konopka-Anstadt JL, Mainou BA, Sutherland DM, Sekine Y, Strittmatter SM, Dermody TS. The Nogo receptor NgR1 mediates infection by mammalian reovirus. Cell Host Microbe 2014;15(6):681−91.

[11] Idris F, Muharram SH, Diah S. Glycosylation of dengue virus glycoproteins and their interactions with carbohydrate receptors: possible targets for antiviral therapy. Arch Virol 2016;161(7):1751−60.

[12] Lin LT, Chung CY, Hsu WC, Chang SP, Hung TC, Shields J, et al. Saikosaponin b2 is a naturally occurring terpenoid that efficiently inhibits hepatitis C virus entry. J Hepatol 2015;62(3):541−8.

[13] Chung CY, Liu CH, Burnouf T, Wang GH, Chang SP, Jassey A, et al. Activity-based and fraction-guided analysis of *Phyllanthus urinaria* identifies loliolide as a potent inhibitor of hepatitis C virus entry. Antivir Res 2016;130:58−68.

[14] Corlay N, Delang L, Girard-Valenciennes E, Neyts J, Clerc P, Smadja J, et al. Tigliane diterpenes from *Croton mauritianus* as inhibitors of chikungunya virus replication. Fitoterapia. 2014;97:87−91.

[15] Hoffmann M, Kleine-Weber H, Schroeder S, Krüger N, Herrler T, Erichsen S, et al. SARS-CoV-2 cell entry depends on ACE2 and TMPRSS2 and is blocked by a clinically proven protease inhibitor. Cell. 2020;181 (2):271−80.

[16] South AM, Brady TM, Flynn JT. ACE2 (angiotensin-converting enzyme 2), COVID-19, and ACE inhibitor and Ang II (Angiotensin II) receptor blocker use during the pandemic: the pediatric perspective. Hypertension. 2020;76(1):16−22.

[17] Umesh KD, Selvaraj C, Singh SK, Dubey VK. Identification of new anti-nCoV drug chemical compounds from Indian spices exploiting SARS-CoV-2 main protease as target. J Biomol Struct Dyn 2021;39(9):3428−34.

[18] Murck H. Symptomatic protective action of glycyrrhizin (licorice) in COVID-19 infection? Front Immunol 2020;11:1239.

[19] Li D, Baert L, Zhang D, Xia M, Zhong W, Van Coillie E, et al. Effect of grape seed extract on human norovirus GII. 4 and murine norovirus 1 in viral suspensions, on stainless steel discs, and in lettuce wash water. Appl Environ Microbiol 2012;78(21):7572−8.

[20] Dong W, Farooqui A, Leon AJ, Kelvin DJ. Inhibition of influenza A virus infection by ginsenosides. PLoS One 2017;12(2):e0171936.

[21] Haidari M, Ali M, Casscells III SW, Madjid M. Pomegranate (*Punica granatum*) purified polyphenol extract inhibits influenza virus and has a synergistic effect with oseltamivir. Phytomedicine. 2009;16(12):1127−36.

[22] Chen DY, Shien JH, Tiley L, Chiou SS, Wang SY, Chang TJ, et al. Curcumin inhibits influenza virus infection and haemagglutination activity. Food Chem 2010;119(4):1346−51.

[23] Dao TT, Nguyen PH, Won HK, Kim EH, Park J, Won BY, et al. Curcuminoids from *Curcuma longa* and their inhibitory activities on influenza A neuraminidases. Food Chem 2012;134(1):21−8.

[24] Liao Q, Qian Z, Liu R, An L, Chen X. Germacrone inhibits early stages of influenza virus infection. Antivir Res 2013;100(3):578−88.

[25] Roschek Jr B, Fink RC, McMichael MD, Li D, Alberte RS. Elderberry flavonoids bind to and prevent H1N1 infection *in vitro*. Phytochemistry. 2009;70(10):1255−61.

[26] Lani R, Hassandarvish P, Shu MH, Phoon WH, Chu JJ, Higgs S, et al. Antiviral activity of selected flavonoids against Chikungunya virus. Antivir Res 2016;133:50−61.

[27] Putra RP, Alkaff AH, Nasution MA, Corona A, Kantale B, Tambunan US. Searching of flavonoid compounds as a new antiviral for Sudan Ebolavirus glycoprotein using *in silico* methods. Int J 2018;15(49):78−84.

[28] Vázquez-Calvo Á, Jiménez de Oya N, Martín-Acebes MA, Garcia-Moruno E, Saiz JC. Antiviral properties of the natural polyphenols delphinidin and epigallocatechin gallate against the flaviviruses West Nile virus, Zika virus, and dengue virus. Front microbiology 2017;8:1314.

[29] Hsiang CY, Hsieh CL, Wu SL, Lai IL, Ho TY. Inhibitory effect of anti-pyretic and anti-inflammatory herbs on herpes simplex virus replication. Am J Chin Med 2001;29(03n04):459−67.

[30] Kaushik NK, Guha R, Saravanabalaji S, Saklani A, Singh R, Sivaramakrishnan H, et al. Antiviral activity of *Indigofera heterantha* Wall. ex brandis against herpes simplex virus−type 2 (HSV-2). Int J Adv Res 2015;3:1365−76.

[31] Ghosh M, Civra A, Rittà M, Cagno V, Mavuduru SG, Awasthi P, et al. *Ficus religiosa* L. bark extracts inhibit infection by herpes simplex virus type 2 vitro. Arch Virol 2016;161(12):3509−14.

[32] Mishra NN, Kesharwani A, Agarwal A, Polachira SK, Nair R, Gupta SK. Herbal gel formulation developed for anti-human immunodeficiency virus (HIV)-1 activity also inhibits *in vitro* HSV-2 infection. Viruses 2018;10(11):580.

[33] Bonvicini F, Lianza M, Mandrone M, Poli F, Gentilomi GA, Antognoni F. Hemidesmus indicus (L.) R. Br. extract inhibits the early step of herpes simplex type 1 and type 2 replication. N Microbiol 2018;41(3):187−94.

[34] Rittà M, Marengo A, Civra A, Lembo D, Cagliero C, Kant K, et al. Antiviral activity of a Arisaema tortuosum leaf extract and some of its constituents against herpes simplex virus type 2. Planta Med 2020;86(04):267−75.

[35] Bailly C, Vergoten G. Glycyrrhizin: an alternative drug for the treatment of COVID-19 infection and the associated respiratory syndrome? Pharmacol Therap 2020;214:107618.

[36] Weng JR, Lin CS, Lai HC, Lin YP, Wang CY, Tsai YC, et al. Antiviral activity of Sambucus Formosana Nakai ethanol extract and related phenolic acid constituents against human coronavirus NL63. Virus Res 2019;273:197767.

[37] Ansari WA, Ahamad T, Khan MA, Khan ZA, Khan MF. Luteolin: a dietary molecule as potential anti-COVID-19 agent. Comput Chem 2020;.

[38] Lin LT, Chen TY, Chung CY, Noyce RS, Grindley TB, McCormick C, et al. Hydrolyzable tannins (chebulagic acid and punicalagin) target viral glycoprotein-glycosaminoglycan interactions to inhibit herpes simplex virus 1 entry and cell-to-cell spread. J Virol 2011;85(9):4386−98.

[39] Lv X, Qiu M, Chen D, Zheng N, Jin Y, Wu Z. Apigenin inhibits enterovirus 71 replication through suppressing viral IRES activity and modulating cellular JNK pathway. Antivir Res 2014;109:30−41.

[40] Balkrishna A, Pokhrel S, Singh J, Varshney A. 2020. Withanone from Withania somnifera may inhibit novel coronavirus (COVID-19) entry by disrupting interactions between viral S-protein receptor binding domain and host ACE2 receptor.

[41] Zhang DH, Wu KL, Zhang X, Deng SQ, Peng B. *In silico* screening of Chinese herbal medicines with the potential to directly inhibit 2019 novel coronavirus. J Integr Med 2020;18(2):152−8.

[42] Zorofchian Moghadamtousi S, Hajrezaei M, Abdul Kadir H, Zandi K. *Loranthus micranthus* Linn.: biological activities and phytochemistry. Evid. Based Complement Alternat Med 2013;2013.

[43] Zhao Y, Feng L, Liu L, Zhao R. Saikosaponin b2 enhances the hepatotargeting effect of anticancer drugs through inhibition of multidrug resistance-associated drug transporters. Life Sci 2019;231:116557.

[44] Zhuang M, Jiang H, Suzuki Y, Li X, Xiao P, Tanaka T, et al. Procyanidins and butanol extract of Cinnamomi cortex inhibit SARS-CoV infection. Antivir Res 2009;82(1):73−81.

[45] Lee JH, Park JS, Lee SW, Hwang SY, Young BE, Choi HJ. Porcine epidemic diarrhea virus infection: inhibition by polysaccharide from *Ginkgo biloba* exocarp and mode of its action. Virus Res 2015;195:148−52.

[46] Wahyuni TS, Widyawaruyanti A, Lusida MI, Fuad A, Fuchino H, Kawahara N, et al. Inhibition of hepatitis C virus replication by chalepin and pseudane IX isolated from Ruta angustifolia leaves. Fitoterapia. 2014;99:276−83.

[47] Álvarez ÁL, Habtemariam S, Moneim AE, Melón S, Dalton KP, Parra F. A spiroketal-enol ether derivative from *Tanacetum vulgare* selectively inhibits HSV-1 and HSV-2 glycoprotein accumulation in Vero cells. Antivir Res 2015;119:8−18.

[48] Huang TJ, Tsai YC, Chiang SY, Wang GJ, Kuo YC, Chang YC, et al. Anti-viral effect of a compound isolated from *Liriope platyphylla* against hepatitis B virus *in vitro*. Virus Res 2014;192:16−24.

[49] Nothias-Scaglia LF, Retailleau P, Paolini J, Pannecouque C, Neyts J, Dumontet V, et al. Jatrophane diterpenes as inhibitors of chikungunya virus replication: structure—activity relationship and discovery of a potent lead. J Nat Prod 2014;77(6):1505—12.

[50] Cui H, Xu B, Wu T, Xu J, Yuan Y, Gu Q. Potential antiviral lignans from the roots of *Saururus chinensis* with activity against Epstein—Barr virus lytic replication. J Nat products 2014;77(1):100—10.

[51] Barquero AA, Michelini FM, Alché LE. 1-Cinnamoyl-3, 11-dihydroxymeliacarpin is a natural bioactive compound with antiviral and nuclear factor-κB modulating properties. Biochem Biophys Res Commun 2006;344 (3):955—62.

[52] Campani ST, Moreira JD, Tietbohel CN. Pulmonary tuberculosis treatment regimen recommended by the Brazilian National Ministry of Health: predictors of treatment noncompliance in the city of Porto Alegre, Brazil. J Bras Pneumol 2011;37:776—82.

[53] Yi L, Li Z, Yuan K, Qu X, Chen J, Wang G, et al. Small molecules blocking the entry of severe acute respiratory syndrome coronavirus into host cells. J Virol 2004;78(20):11334—9.

[54] Yang Z, Wang Y, Zheng Z, Zhao S, Zhao JI, Lin Q, et al. Antiviral activity of Isatis indigotica root-derived clemastanin B against human and avian influenza A and B viruses *in vitro*. Int J Mol Med 2013;31(4):867—73.

[55] Wu W, Li R, Li X, He J, Jiang S, Liu S, et al. Quercetin as an antiviral agent inhibits influenza A virus (IAV) entry. Viruses. 2016;8(1):6.

[56] Evers DL, Chao CF, Wang X, Zhang Z, Huong SM, Huang ES. Human cytomegalovirus-inhibitory flavonoids: studies on antiviral activity and mechanism of action. Antivir Res 2005;68(3):124—34.

[57] Sauter D, Schwarz S, Wang K, Zhang R, Sun B, Schwarz W. Genistein as antiviral drug against HIV ion channel. Planta Med 2014;80(08/09):682—7.

[58] Haid S, Novodomská A, Gentzsch J, Grethe C, Geuenich S, Bankwitz D, et al. A plant-derived flavonoid inhibits entry of all HCV genotypes into human hepatocytes. Gastroenterology. 2012;143(1):213—22.

[59] Shimizu JF, Lima CS, Pereira CM, Bittar C, Batista MN, Nazaré AC, et al. Flavonoids from Pterogyne nitens inhibit hepatitis C virus entry. Sci Rep 2017;7(1):1—9.

[60] Tsai FJ, Lin CW, Lai CC, Lan YC, Lai CH, Hung CH, et al. Kaempferol inhibits enterovirus 71 replication and internal ribosome entry site (IRES) activity through FUBP and HNRP proteins. Food Chem 2011;128 (2):312—22.

[61] Imanishi N, Tuji Y, Katada Y, Maruhashi M, Konosu S, Mantani N, et al. Additional inhibitory effect of tea extract on the growth of influenza A and B viruses in MDCK cells. Microbiol Immunol 2002;46(7):491—4.

[62] Fanunza E, Iampietro M, Distinto S, Corona A, Quartu M, Maccioni E, et al. Quercetin blocks Ebola virus infection by counteracting the VP24 interferon-inhibitory function. Antimicrobial Agents Chemother 2020;64 (7):e00530 20.

[63] Tao J, Hu Q, Yang J, Li R, Li X, Lu C, et al. *In vitro* anti-HIV and-HSV activity and safety of sodium rutin sulfate as a microbicide candidate. Antivir Res 2007;75(3):227—33.

[64] Le Lee J, Loe MW, Lee RC, Chu JJ. Antiviral activity of pinocembrin against Zika virus replication. Antivir Res 2019;167:13—24.

[65] Gaudry A, Bos S, Viranaicken W, Roche M, Krejbich-Trotot P, Gadea G, et al. The flavonoid isoquercitrin precludes initiation of Zika virus infection in human cells. Int J Mol Sci 2018;19(4):1093.

[66] García G, Cavallaro L, Broussalis A, Ferraro G, Martino V, Campos R. Biological and chemical characterization of the fraction with antiherpetic activity from *Achyrocline flaccida*. Planta Med 1999;65(04):343—6.

[67] Lazreg Aref H, Gaaliche B, Fekih A, Mars M, Aouni M, Pierre Chaumon J, et al. *In vitro* cytotoxic and antiviral activities of *Ficus carica* latex extracts. Nat Prod Res 2011;25(3):310—19.

[68] Lückemeyer DD, Müller VD, Moritz MI, Stoco PH, Schenkel EP, Barardi CR, et al. Effects of Ilex paraguariensis A. St. Hil.(yerba mate) on herpes simplex virus types 1 and 2 replication. Phytother Res 2012;26 (4):535—40.

[69] Schneider S, Reichling J, Stintzing FC, Carle R, Schnitzler P. Efficacy of an aqueous Pelargonium sidoides extract against herpesvirus. Planta Med 2008;74(09):PA322.

[70] Chattopadhyay D, Arunachalam G, Mandal AB, Bhattacharya SK. Dose-dependent therapeutic antiinfectives from ethnomedicines of bay islands. Chemotherapy. 2006;52(3):151—7.

[71] Kesharwani A, Polachira SK, Nair R, Agarwal A, Mishra NN, Gupta SK. Anti-HSV-2 activity of Terminalia chebula Retz extract and its constituents, chebulagic and chebulinic acids. BMC Complement Altern Med 2017;17(1):1.

[72] Del Barrio G, Parra F. Evaluation of the antiviral activity of an aqueous extract from *Phyllanthus orbicularis*. J Ethnopharmacol 2000;72(1−2):317−22.

[73] Fernández Romero JA, Del Barrio Alonso G, Romeu Alvarez B, Gutierrez Y, Valdés VS, Parra F. *In vitro* antiviral activity of *Phyllanthus orbicularis* extracts against herpes simplex virus type 1. Phytother Res 2003;17 (8):980−2.

[74] Gescher K, Kühn J, Hafezi W, Louis A, Derksen A, Deters A, et al. Inhibition of viral adsorption and penetration by an aqueous extract from *Rhododendron ferrugineum* L. as antiviral principle against herpes simplex virus type-1. Fitoterapia. 2011;82(3):408−13.

[75] Carson CF, Ashton L, Dry L, Smith DW, Riley TV. Melaleuca alternifolia (tea tree) oil gel (6%) for the treatment of recurrent herpes labialis. J Antimicrob Chemother 2001;48(3):450−1.

[76] Schnitzler P, Schön K, Reichling J. Antiviral activity of Australian tea tree oil and eucalyptus oil against herpes simplex virus in cell culture. Die Pharmazie 2001;56(4):343−7.

[77] Hayashi K, Hayashi T, Morita N, Niwayama S. Antiviral activity of an extract of *Cordia salicifolia* on herpes simplex virus type 1. Planta Med 1990;56(05):439−43.

[78] Müller V, Chávez JH, Reginatto FH, Zucolotto SM, Niero R, Navarro D, et al. Evaluation of antiviral activity of South American plant extracts against herpes simplex virus type 1 and rabies virus. Phytother Res 2007;21 (10):970−4.

[79] Nocchi SR, Companhoni MV, de Mello JC, Dias Filho BP, Nakamura CV, Carollo CA, et al. Antiviral activity of crude hydroethanolic extract from *Schinus terebinthifolia* against herpes simplex virus type 1. Planta Med 2017;234(06):509−18.

[80] Boff L, Silva IT, Argenta DF, Farias LM, Alvarenga LF, Pádua RM, et al. *Strychnos pseudoquina* A. St. Hil.: a Brazilian medicinal plant with promising *in vitro* antiherpes activity. J Appl Microbiol 2016;121(6):1519−29.

[81] Chiow KH, Phoon MC, Putti T, Tan BK, Chow VT. Evaluation of antiviral activities of *Houttuynia cordata* Thunb. extract, quercetin, quercetrin and cinanserin on murine coronavirus and dengue virus infection. Asian Pac J Trop Med 2016;9(1):1−7.

[82] Liang W, He L, Ning P, Lin J, Li H, Lin Z, et al. (+)-Catechin inhibition of transmissible gastroenteritis coronavirus in swine testicular cells is involved its antioxidation. Res Vet Sci 2015;103:28−33.

[83] Pan HH, Yu XT, Li T, Wu HL, Jiao CW, Cai MH, et al. Aqueous extract from a Chaga medicinal mushroom, Inonotus obliquus (higher basidiomyetes), prevents herpes simplex virus entry through inhibition of viral-induced membrane fusion. Int J Med Mushrooms 2013;15(1).

[84] Lipson SM, Karalis G, Karthikeyan L, Ozen FS, Gordon RE, Ponnala S, et al. Mechanism of anti-rotavirus synergistic activity by epigallocatechin gallate and a proanthocyanidin-containing nutraceutical. Food Environ Virol 2017;9(4):434−43.

[85] Zakaryan H, Arabyan E, Oo A, Zandi K. Flavonoids: promising natural compounds against viral infections. Arch Virol 2017;162(9):2539−51.

[86] Fan W, Qian S, Qian P, Li X. Antiviral activity of luteolin against *Japanese encephalitis* virus. Virus Res 2016;220:112−16.

[87] LeCher JC, Diep N, Krug PW, Hilliard JK. Genistein has antiviral activity against herpes B virus and acts synergistically with antiviral treatments to reduce effective dose. Viruses. 2019;11(6):499.

[88] Bang S, Ha TK, Lee C, Li W, Oh WK, Shim SH. Antiviral activities of compounds from aerial parts of *Salvia plebeia* R. Br. J Ethnopharmacol 2016;192:398−405.

[89] Rosmalena R, Elya B, Dewi BE, Fithriyah F, Desti H, Angelina M, et al. The antiviral effect of indonesian medicinal plant extracts against dengue virus *in vitro* and *in silico*. Pathogens 2019;8(2):85.

[90] Zhang G, Zhang B, Zhang X. Bing F. Homonojirimycin, an alkaloid from dayflower inhibits the growth of influenza A virus *in vitro*. Acta Virol 2013;57(1):85−6.

[91] Li R, Liu T, Liu M, Chen F, Liu S, Yang J. Anti-influenza A virus activity of dendrobine and its mechanism of action. J Agric Food Chem 2017;65(18):3665−74.

[92] Esimone CO, Eck G, Nworu CS, Hoffmann D, Überla K, Proksch P. Dammarenolic acid, a secodammarane triterpenoid from Aglaia sp. shows potent anti-retroviral activity *in vitro*. Phytomedicine. 2010;17(7):540−7.

[93] Fayyad AG, Ibrahim N, Yaakob WA. Phytochemical screening and antiviral activity of *Marrubium vulgare*. Malays J Microbiol 2014;10(2):106−11.

[94] Amal Gaber SF, Ibrahim N, Yaakob WA. Phytochemical screening and antiviral activity of Marrubium vulgare. Malays J Microbiol 2014;10(2):106−11.

[95] Fang CY, Chen SJ, Wu HN, Ping YH, Lin CY, Shiuan D, et al. Honokiol, a lignan biphenol derived from the magnolia tree, inhibits dengue virus type 2 infection. Viruses. 2015;7(9):4894—910.

[96] Hui QX, He LW, Ying YD, et al. S. Extraction of total lignans from *Radix isatidis* and its ant-RSV virus effect. Clin J Pharmacol Pharmacother 2019;1(1):1005.

[97] Li C, Wei Q, Zou ZH, Sun CZ, Wang Q, Zhao G, et al. A lignan and a lignan derivative from the fruit of Forsythia suspensa. Phytochem Lett 2019;32:115—18.

[98] Wu SF, Lin CK, Chuang YS, Chang FR, Tseng CK, Wu YC, et al. Anti-hepatitis C virus activity of 3-hydroxy caruilignan C from *Swietenia macrophylla* stems. J Viral Hepat 2012;19(5):364—70.

[99] Verma HN, Awasthi LP. Occurrence of a highly antiviral agent in plants treated with *Boerhaavia diffusa* inhibitor. Can J Bot 1980;58(20):2141—4.

[100] Ul Qamar MT, Alqahtani SM, Alamri MA, Chen LL. Structural basis of SARS-CoV-2 3CLpro and anti-COVID-19 drug discovery from medicinal plants. J Pharm Anal 2020;10(4):313—19.

[101] Salehi B, Shivaprasad Shetty M, Anil Kumar NV, Živković J, Calina D, Oana Docea G, et al. Veronica plants—drifting from farm to traditional healing, food application, and phytopharmacology. Molecules. 2019;24(13):2454.

[102] Bachmetov L, Gal-Tanamy M, Shapira A, Vorobeychik M, Giterman-Galam T, Sathiyamoorthy P, et al. Suppression of hepatitis C virus by the flavonoid quercetin is mediated by inhibition of NS3 protease activity. J Viral Hepat 2012;19(2):e81—8.

[103] Ganesan S, Faris AN, Comstock AT, Wang Q, Nanua S, Hershenson MB, et al. Quercetin inhibits rhinovirus replication *in vitro* and *in vivo*. Antivir Res 2012;94(3):258—71.

[104] Gonzalez O, Fontanes V, Raychaudhuri S, Loo R, Loo J, Arumugaswami V, et al. The heat shock protein inhibitor Quercetin attenuates hepatitis C virus production. Hepatology. 2009;50(6):1756—64.

[105] Zandi K, Teoh BT, Sam SS, Wong PF, Mustafa MR, AbuBakar S. Antiviral activity of four types of bioflavonoid against dengue virus type-2. Virol J 2011;8(1):1.

[106] Shibata C, Ohno M, Otsuka M, Kishikawa T, Goto K, Muroyama R, et al. The flavonoid apigenin inhibits hepatitis C virus replication by decreasing mature microRNA122 levels. Virology. 2014;462:42—8.

[107] Chu M, Xu L, Zhang MB, Chu ZY, Wang YD. Role of Baicalin in anti-influenza virus A as a potent inducer of IFN-gamma. BioMed Res Int 2015;2015 Dec 10.

[108] Simões CM, Amoros M, Girre L. Mechanism of antiviral activity of triterpenoid saponins. Phytother Res 1999;13(4):323—8.

[109] Liu S, Wei W, Shi K, Cao X, Zhou M, Liu Z. *In vitro* and *in vivo* anti-hepatitis B virus activities of the lignan niranthin isolated from *Phyllanthus niruri* L. J Ethnopharmacol 2014;155(2):1061—7.

[110] Merino-Ramos T, Jiménez de Oya N, Saiz JC, Martín-Acebes MA. Antiviral activity of nordihydroguaiaretic acid and its derivative tetra-O-methyl nordihydroguaiaretic acid against West Nile virus and Zika virus. Antimicrob Agents Chemother 2017;61(8):e00376 17.

[111] Soto-Acosta R, Bautista-Carbajal P, Syed GH, Siddiqui A, Del Angel RM. Nordihydroguaiaretic acid (NDGA) inhibits replication and viral morphogenesis of dengue virus. Antivir Res 2014;109:132—40.

[112] Oyegunwa AO, Sikes ML, Wilson JR, Scholle F, Laster SM. Tetra-O-methyl nordihydroguaiaretic acid (Terameprocol) inhibits the NF-κB-dependent transcription of TNF-α and MCP-1/CCL2 genes by preventing RelA from binding its cognate sites on DNA. J Inflamm 2010;7(1):1.

[113] Fu L, Xu P, Liu N, Yang Z, Zhang F, Hu Y. Antiviral effect of arctigenin compound on influenza virus. Tradit Chin Drug Res Clin Pharmacol 2008;19(4).

[114] Kuo YC, Kuo YH, Lin YL, Tsai WJ. Yatein from *Chamaecyparis obtusa* suppresses herpes simplex virus type 1 replication in HeLa cells by interruption the immediate-early gene expression. Antivir Res 2006;70 (3):112—20.

[115] Martinez-Lopez A, Persaud M, Chavez MP, Zhang H, Rong L, Liu S, et al. Glycosylated diphyllin as a broad-spectrum antiviral agent against Zika virus. EBioMedicine. 2019;47:269—83.

[116] Liu CH, Jassey A, Hsu HY, Lin LT. Antiviral activities of silymarin and derivatives. Molecules. 2019;24 (8):1552.

[117] Aires A, editor. Tannins: Structural Properties, Biological Properties and Current Knowledge. BoD—Books on Demand; 2020 Jan 8.

[118] Vilhelmova N, Jacquet R, Quideau S, Stoyanova A, Galabov AS. Three-dimensional analysis of combination effect of ellagitannins and acyclovir on herpes simplex virus types 1 and 2. Antivir Res 2011;89(2):174—81.

[119] Vilhelmova-Ilieva N, Deffieux D, Quideau S. Castalagin: some aspects of the mode of anti-herpes virus activity. Ann Antivir Antiretrovirals 2018;2(1):004–7.

[120] Huang RL, Huang YL, Ou JC, Chen CC, Hsu FL, Chang C. Screening of 25 compounds isolated from Phyllanthus species for anti-human hepatitis B virus in vitro. Phytother Res 2003;17(5):449–53.

[121] Pollara JJ, Laster SM, Petty IT. Inhibition of poxvirus growth by Terameprocol, a methylated derivative of nordihydroguaiaretic acid. Antivir Res 2010;88(3):287–95.

[122] Michaelis M, Geiler J, Naczk P, Sithisarn P, Leutz A, Doerr HW, et al. Glycyrrhizin exerts antioxidative effects in H5N1 influenza A virus-infected cells and inhibits virus replication and pro-inflammatory gene expression. PLoS One 2011;6(5):e19705.

[123] Müller C, Schulte FW, Lange-Grünweller K, Obermann W, Madhugiri R, Pleschka S, et al. Broad-spectrum antiviral activity of the eIF4A inhibitor silvestrol against corona-and picornaviruses. Antivir Res 2018;150:123–9. Available from: https://doi.org/10.1016/j.antiviral.2017.12.010.

[124] Kim DE, Min JS, Jang MS, Lee JY, Shin YS, Park CM, et al. Natural bis-benzylisoquinoline alkaloids-tetrandrine, fangchinoline, and cepharanthine, inhibit human coronavirus OC43 infection of MRC-5 human lung cells. Biomolecules. 2019;9(11):696.

[125] Tsai YC, Lee CL, Yen HR, Chang YS, Lin YP, Huang SH, et al. Antiviral action of tryptanthrin isolated from Strobilanthes cusia leaf against human coronavirus NL63. Biomolecules. 2020;10(3):366.

[126] Ulasli M, Gurses SA, Bayraktar R, Yumrutas O, Oztuzcu S, Igci M, et al. The effects of Nigella sativa (Ns), Anthemis hyalina (Ah) and Citrus sinensis (Cs) extracts on the replication of coronavirus and the expression of TRP genes family. Mol Biol Rep 2014;41(3):1703–11.

[127] Chuanasa T, Phromjai J, Lipipun V, Likhitwitayawuid K, Suzuki M, Pramyothin P, et al. Anti-herpes simplex virus (HSV-1) activity of oxyresveratrol derived from Thai medicinal plant: mechanism of action and therapeutic efficacy on cutaneous HSV-1 infection in mice. Antivir Res 2008;80(1):62–70.

[128] Hayashi K, Niwayama S, Hayashi T, Nago R, Ochiai H, Morita N. In vitro and in vivo antiviral activity of scopadulcic acid B from Scoparia dulcis, Scrophulariaceae, against herpes simplex virus type 1. Antivir Res 1988;9(6):345–54.

[129] Wen CC, Shyur LF, Jan JT, Liang PH, Kuo CJ, Arulselvan P, et al. Traditional Chinese medicine herbal extracts of Cibotium barometz, Gentiana scabra, Dioscorea batatas, Cassia tora, and Taxillus chinensis inhibit SARS-CoV replication. J Tradit Complement Med 2011;1(1):41–50.

[130] Ryu YB, Jeong HJ, Kim JH, Kim YM, Park JY, Kim D, et al. Biflavonoids from Torreya nucifera displaying SARS-CoV 3CLpro inhibition. Bioorg Med Chem 2010;18(22):7940–7. Available from: https://doi.org/10.1016/j.bmc.2010.09.035.

[131] Song JH, Shim JK, Choi HJ. Quercetin 7-rhamnoside reduces porcine epidemic diarrhea virus replication via independent pathway of viral induced reactive oxygen species. Virol J 2011;8(1):1–6.

[132] Ho TY, Wu SL, Chen JC, Li CC, Hsiang CY. Emodin blocks the SARS coronavirus spike protein and angiotensin-converting enzyme 2 interaction. Antivir Res 2007;74(2):92–101.

[133] Hayashi K, Narutaki K, Nagaoka Y, Hayashi T, Uesato S. Therapeutic effect of arctiin and arctigenin in immunocompetent and immunocompromised mice infected with influenza A virus. Bio Pharm Bull 2010;33 (7):1199–205. Available from: https://doi.org/10.1248/bpb.33.1199.

[134] Yoshikawa TT, High KP. Nutritional strategies to boost immunity and prevent infection in elderly individuals. Clin Infect Dis 2001;33(11):1892–900.

[135] Steinfeld B, Scott J, Vilander G, Marx L, Quirk M, Lindberg J, et al. The role of lean process improvement in implementation of evidence-based practices in behavioral health care. J Behav Health Serv Res 2015;42 (4):504–18.

[136] Lu ZF, Doulabi BZ, Wuisman PI, Bank RA, Helder MN. Differentiation of adipose stem cells by nucleus pulposus cells: configuration effect. Biochem Biophys Res Commun 2007;359(4):991–6.

[137] Redfearn DP, Trim GM, Skanes AC, Petrellis B, Krahn AD, Yee R, et al. Esophageal temperature monitoring during radiofrequency ablation of atrial fibrillation. J Cardiovas Electrophysiol 2005;16(6):589–93.

[138] Choi HJ, Song JH, Park KS, Kwon DH. Inhibitory effects of quercetin 3-rhamnoside on influenza A virus replication. Eur J Pharm Sci 2009;37(3–4):329–33.

[139] Hour MJ, Huang SH, Chang CY, Lin YK, Wang CY, Chang YS, et al. Baicalein, ethyl acetate, and chloroform extracts of Scutellaria baicalensis inhibit the neuraminidase activity of pandemic 2009 H1N1 and seasonal influenza A viruses. Evid-Based Complement Alt Med 2013;2013.

[140] Malla A, Ramalingam S. Health perspectives of an isoflavonoid genistein and its quantification in economically important plants. Role of materials science in food bioengineering. Academic Press; 2018. p. 353−79. Jan 1.

[141] Liu AL, Liu B, Qin HL, Lee SM, Wang YT, Du GH. Anti-influenza virus activities of flavonoids from the medicinal plant *Elsholtzia rugulosa*. Planta medica 2008;74(08):847−51.

[142] Moradi MT, Karimi A, Rafieian-Kopaei M, Fotouhi F. *In vitro* antiviral effects of *Peganum harmala* seed extract and its total alkaloids against influenza virus. Microb pathogenesis 2017;110:42−9.

[143] Li X, Yu HY, Wang ZY, Pi HF, Zhang P, Ruan HL. Neuroprotective compounds from the bulbs of *Lycoris radiata*. Fitoterapia. 2013;88:82−90.

[144] Zhou HX, Li RF, Wang YF, Shen LH, Cai LH, Weng YC, et al. Total alkaloids from *Alstonia scholaris* inhibit influenza a virus replication and lung immunopathology by regulating the innate immune response. Phytomedicine. 2020;77:153272.

[145] Borquaye LS, Gasu EN, Ampomah GB, Kyei LK, Amarh MA, Mensah CN, et al. Alkaloids from Cryptolepis sanguinolenta as potential inhibitors of SARS-CoV-2 viral proteins: an *in silico* study. BioMed Res Int 2020;2020 Sep 22.

[146] Bleasel MD, Peterson GM. Emetine, ipecac, ipecac alkaloids and analogues as potential antiviral agents for coronaviruses. Pharmaceuticals. 2020;13(3):51.

[147] Huang L, Li H, Yuen TT, Ye Z, Fu Q, Sun W, et al. 2020. Berbamine inhibits the infection of SARS-CoV-2 and flaviviruses by compromising TPRMLs-mediated endolysosomal trafficking of viral receptors.

[148] Kuo CL, Chi CW, Liu TY. The anti-inflammatory potential of berberine *in vitro* and *in vivo*. Cancer Lett 2004;203(2):127−37.

[149] Majnooni MB, Fakhri S, Shokoohinia Y, Kiyani N, Stage K, Mohammadi P, et al. Phytochemicals: potential therapeutic interventions against coronavirus-associated lung injury. Front Pharmacol 2020;11:1744.

[150] Neag MA, Mocan A, Echeverría J, Pop RM, Bocsan CI, Crişan G, et al. Berberine: botanical occurrence, traditional uses, extraction methods, and relevance in cardiovascular, metabolic, hepatic, and renal disorders. Front Pharmacol 2018;9:557.

[151] Zu Y, Fu Y, Wang W, Wu N, Liu W, Kong Y, et al. Comparative study on the antiherpetic activity of aqueous and ethanolic extracts derived from *Cajanus cajan* (L.) Millsp. Complement Med Res 2010;17(1):15−20.

[152] Ojha D, Das R, Sobia P, Dwivedi V, Ghosh S, Samanta A, et al. Pedilanthus tithymaloides inhibits HSV infection by modulating NF-κB signaling. PLoS One 2015;10(9):e0139338.

[153] Karimi A, Mohammadi-Kamalabadi M, Rafieian-Kopaei M, Amjad L. Determination of antioxidant activity, phenolic contents and antiviral potential of methanol extract of *Euphorbia spinidens* Bornm (Euphorbiaceae). Trop J Pharm Res 2016;15(4):759−64.

[154] Binns SE, Hudson J, Merali S, Arnason JT. Antiviral activity of characterized extracts from Echinacea spp. (Heliantheae: Asteraceae) against herpes simplex virus (HSV-I). Planta Med 2002;68(09):780−3.

[155] Ghannad MS, Mohammadi A, Safiallahy S, Faradmal J, Azizi M, Ahmadvand Z. The effect of aqueous extract of *Glycyrrhiza glabra* on herpes simplex virus 1. Jundishapur J Microbiol 2014;7(7).

[156] Ahmad A, Davies J, Randall S, Skinner GR. Antiviral properties of extract of *Opuntia streptacantha*. Antivir Res 1996;30(2−3):75−85.

[157] Abad MJ, Bermejo P, Gonzales E, Iglesias I, Irurzun A, Carrasco L. Antiviral activity of Bolivian plant extracts. Gen Pharmacol: Vasc Syst 1999;32(4):499−503.

[158] Churqui MP, Lind L, Thörn K, Svensson A, Savolainen O, Aranda KT, et al. Extracts of *Equisetum giganteum* L and *Copaifera reticulata* Ducke show strong antiviral activity against the sexually transmitted pathogen herpes simplex virus type 2. J Ethnopharmacol 2018;210:192−7.

[159] Benassi-Zanqueta É, Marques CF, Valone LM, Pellegrini BL, Bauermeister A, Ferreira IC, et al. Evaluation of anti-HSV-1 activity and toxicity of hydroethanolic extract of *Tanacetum parthenium* (L.) Sch. Bip. (Asteraceae). Phytomedicine. 2019;55:249−54.

[160] Tragoolpua Y, Jatisatienr A. Anti-herpes simplex virus activities of Eugenia caryophyllus (Spreng.) Bullock & SG Harrison and essential oil, eugenol. Phytother Res 2007;21(12):1153−8.

[161] Onozato T, Nakamura CV, Cortez DA, Filho BP, Ueda-Nakamura T. Tanacetum vulgare: antiherpes virus activity of crude extract and the purified compound parthenolide. Phytother Res 2009;23(6):791−6.

[162] Danaher RJ, Wang C, Dai J, Mumper RJ, Miller CS. Antiviral effects of blackberry extract against herpes simplex virus type 1. Oral Surg, Oral Med, Oral Pathol, Oral Radiol, Endodontol 2011;112(3):e31−5.

[163] Kuo YC, Lee YC, Leu YL, Tsai WJ, Chang SC. Efficacy of orally administered *Lobelia chinensis* extracts on herpes simplex virus type 1 infection in BALB/c mice. Antivir Res 2008;80(2):206—12.

[164] Runfeng L, Yunlong H, Jicheng H, et al. Lianhuaqingwen exerts anti-viral and anti-inflammatory activity against novel coronavirus (SARS-CoV-2). Pharmacol Res 2020;156:104761. Available from: https://doi.org/10.1016/j.phrs.2020.104761.

[165] Luo H, Tang QL, Shang YX, Liang SB, Yang M, Robinson N, et al. Can Chinese medicine be used for prevention of corona virus disease 2019 (COVID-19)? A review of historical classics, research evidence and current prevention programs. Chin J Integr Med 2020;26(4):243—50.

[166] Tayel AA, El-Sedfy MA, Ibrahim AI, Moussa SH. Application of *Quercus infectoria* extract as a natural antimicrobial agent for chicken egg decontamination. Rev Argent Microbiol 2018;50(4):391—7.

[167] Maurya VK, Kumar S, Prasad AK, Bhatt ML, Saxena SK. Structure-based drug designing for potential antiviral activity of selected natural products from Ayurveda against SARS-CoV-2 spike glycoprotein and its cellular receptor. Virusdisease. 2020;31(2):179—93.

[168] Lanzotti V. The analysis of onion and garlic. J Chromatogr A 2006;1112(1—2):3—22.

[169] Cermelli C, Vinceti M, Scaltriti E, Bazzani E, Beretti F, Vivoli G, et al. Selenite inhibition of Coxsackie virus B5 replication: implications on the etiology of Keshan disease. J Trace Elem Med Biol 2002;16(1):41—6.

[170] Abba Y, Hassim H, Hamzah H, Noordin MM. Antiviral activity of resveratrol against human and animal viruses. Adv Virol 2015;2015.

[171] Javed T, Ashfaq UA, Riaz S, Rehman S, Riazuddin S. In-vitro antiviral activity of *Solanum nigrum* against hepatitis C virus. Virol J 2011;8(1):1—7.

[172] Kumar KS, Ganguly S, Veerasamy R, De Clercq E. Synthesis, antiviral activity and cytotoxicity evaluation of Schiff bases of some 2-phenyl quinazoline-4 (3) H-ones. Eur J Med Chem 2010;45(11):5474—9.

[173] Yarmolinsky L, Zaccai M, Ben-Shabat S, Mills D, Huleihel M. Antiviral activity of ethanol extracts of *Ficus binjamina* and *Lilium candidum in vitro*. N Biotechnol 2009;26(6):307—13.

[174] Suzutani T, Ogasawara M, Yoshida I, Azuma M, Knox YM. Anti-herpesvirus activity of an extract of *Ribes nigrum* L. Phytother Res 2003;17(6):609—13.

[175] Castañeda-Ovando A, de Lourdes Pacheco-Hernández M, Páez-Hernández ME, Rodríguez JA, Galán-Vidal CA. Chemical studies of anthocyanins: a review. Food Chem 2009;113(4):859—71.

[176] Kyoko H, Kazuto N, Yasuo N, Shinichi U. Therapeutic effect of arctiin and arctigenin in immunocompetent and immunocompromised mice infected with influenza. Bio Pharm Bull 2010;33(7):1199—205.

[177] Umamaheshwari N, Rekha A. Sweet flag:(*Acorus calamus*)—An incredible medicinal herb. J Pharmacogn Phytochem 2018;7(6):15—22.

[178] Dhar ML, Dhar MM, Dhawan BN, Mehrotra BN, Srimal RC, Tandon JS. Screening of indian plants for biological activity: Part I V. Indian J Exp Biol 1973;11:43—54.

[179] Ankri S, Mirelman D. Antimicrobial properties of allicin from garlic. Microbes Infect 1999;1(2):125—9.

[180] Ye Y, Li B. 1′ S-1′-acetoxychavicol acetate isolated from *Alpinia galanga* inhibits human immunodeficiency virus type 1 replication by blocking Rev transport. J Gen Virol 2006;87(7):2047—53.

[181] Bendjeddou D, Lalaoui K, Satta D. Immunostimulating activity of the hot water-soluble polysaccharide extracts of Anacyclus pyrethrum, *Alpinia galanga* and *Citrullus colocynthis*. J Ethnopharmacol 2003;88 (2—3):155—60.

[182] Alam MS, Quader MA, Rashid MA. HIV-inhibitory diterpenoid from *Anisomeles indica*. Fitoterapia. 2000;71 (5):574—6.

[183] Pathak K, Zaman K. An overview on medicinally important plant-*Annona reticulata* Linn. Int J Pharmacogn Phytochem Res 2013;5(4):299—301.

[184] Meneses R, Ocazionez RE, Martínez JR, Stashenko EE. Inhibitory effect of essential oils obtained from plants grown in Colombia on yellow fever virus replication *in vitro*. Ann Clin Microbiol Antimicrob 2009;8(1):1—6.

[185] Badam L, Joshi SP, Bedekar SS. 'In vitro' antiviral activity of neem (*Azadirachta indica* A. Juss) leaf extract against group B coxsackieviruses. J Commun Dis 1999;31(2):79—90.

[186] Tiwari D, Mishra SP, Mishra M, Dubey RS. Biosorptive behaviour of Mango (*Mangifera indica*) and Neem (*Azadirachta indica*) bark for Hg2 + , Cr3 + and Cd2 + toxic ions from aqueous solutions: a radiotracer study. Appl Radiat Isotopes 1999;50(4):631—42.

[187] Das T, Das AK. Inventorying plant biodiversity in homegardens: a case study in Barak Valley, Assam, North East India. Curr Sci 2005;155—63.

[188] Bhandari MR, Jong-Anurakkun N, Hong G, Kawabata J. α-Glucosidase and α-amylase inhibitory activities of Nepalese medicinal herb Pakhanbhed (Bergenia ciliata, Haw.). Food Chem 2008;106(1):247−52.

[189] Chang JM, Huang KL, Yuan TT, Lai YK, Hung LM. The anti-hepatitis B virus activity of *Boehmeria nivea* extract in HBV-viremia SCID mice. Evid-Based Complement Altern Med 2010;7(2):189−95.

[190] Lupini C, Cecchinato M, Scagliarini A, Graziani R, Catelli E. *In vitro* antiviral activity of chestnut and quebracho woods extracts against avian reovirus and metapneumovirus. Res Vet Sci 2009;87(3):482−7.

[191] Bhatnagar AK, Popli SP. Chemical examination of the roots of *Cissampelos pareira* Linn. Part V. structure and stereochemistry of hayatidin. Experientia. 1967;23(4):242−3.

[192] Fortin H, Vigor C, Lohézic-Le Dévéhat F, Robin V, Le Bossé B, Boustie J, et al. *In vitro* antiviral activity of thirty-six plants from La Réunion Island. Fitoterapia 2002;73(4):346−50.

[193] Singh D, Rathod V, Ninganagouda S, Hiremath J, Singh AK, Mathew J. Optimization and characterization of silver nanoparticle by endophytic fungi *Penicillium* sp. isolated from *Curcuma longa* (turmeric) and application studies against MDR *E. coli* and *S. aureus*. Bioinorg Chem Appl 2014;2014.

[194] Pécheur EI. Curcumin against hepatitis C virus infection: spicing up antiviral therapies with 'nutraceuticals'? Gut. 2014;63(7):1035−7.

[195] Boparai A, Niazi J, Bajwa N, Singh PA. A review update on *Dillenia indica* f. elongata (MIQ.) MIQ. J Drug Deliv Therap 2016;6(2):62−70.

[196] Chaniad P, Tewtrakul S, Sudsai T, Langyanai S, Kaewdana K. Anti-inflammatory, wound healing and antioxidant potential of compounds from *Dioscorea bulbifera* L. bulbils. PLoS One 2020;15(12):e0243632.

[197] Lebot V, Faloye B, Okon E, Gueye B. Simultaneous quantification of allantoin and steroidal saponins in yam (*Dioscorea* spp.) powders. J Appl Res Med Aromat Plants 2019;13:100200.

[198] Ahmed Z, Chishti MZ, Johri RK, Bhagat A, Gupta KK, Ram G. Antihyperglycemic and antidyslipidemic activity of aqueous extract of *Dioscorea bulbifera* tubers. Diabetol Croat 2009;38(3):63−72.

[199] Wang J, Li P, Li B, Guo Z, Kennelly EJ, Long C. Bioactivities of compounds from *Elephantopus scaber*, an ethnomedicinal plant from Southwest China. Evid-Based Complement Altern Med 2014;2014.

[200] Kumar M, Prasad SK, Hemalatha S. A current update on the phytopharmacological aspects of *Houttuynia cordata* Thunb. Pharmacogn Rev 2014;8(15):22.

[201] Hamidi JA, Ismaili NH, Ahmadi FB, Lajisi NH. Antiviral and cytotoxic activities of some plants used in Malaysian indigenous medicine. Pertanika J Trop Agric Sci 1996;19(2/3):129−36.

[202] Saxena RS, Gupta B, Lata S. Tranquilizing, antihistaminic and purgative activity of *Nyctanthes arbor tristis* leaf extract. J Ethnopharmacol 2002;81(3):321−5.

[203] Sriwilaijaroen N, Fukumoto S, Kumagai K, Hiramatsu H, Odagiri T, Tashiro M, et al. Antiviral effects of *Psidium guajava* Linn.(guava) tea on the growth of clinical isolated H1N1 viruses: its role in viral hemagglutination and neuraminidase inhibition. Antivir Res 2012;94(2):139−46.

[204] Chollom SC, Agada GO, Bot DY, Okolo MO, Dantong DD, Choji TP, et al. Phytochemical analysis and antiviral potential of aqueous leaf extract of *Psidium guajava* against newcastle disease virus in ovo, 2012.

[205] Angamuthu D, Purushothaman I, Kothandan S, Swaminathan R. Antiviral study on *Punica granatum* L., *Momordica charantia* L., *Andrographis paniculata* Nees, and *Melia azedarach* L., to human herpes virus-3. Eur J Integr Med 2019;28:98−108.

[206] Moradi MT, Karimi A, Shahrani M, Hashemi L, Ghaffari-Goosheh MS. Anti-influenza virus activity and phenolic content of pomegranate (Punica granatum L.) peel extract and fractions. Avicenna J Med Biotechnol 2019;11(4):285.

[207] Wang RR, Gu Q, Wang YH, Zhang XM, Yang LM, Zhou J, et al. Anti-HIV-1 activities of compounds isolated from the medicinal plant Rhus chinensis. J Ethnopharmacol 2008;117(2):249−56.

[208] Murti K, Lambole V, Panchal M, Shah M, Gajera V. Evaluation of wound healing activity of polyherbal formulation in rats. Res J Pharmacogn Phytochem 2011;3(3):112−15.

[209] Chang CC, Hsu HF, Huang KH, Wu JM, Kuo SM, Ling XH, et al. Anti-proliferative effects of Siegesbeckia orientalis ethanol extract on human endometrial RL-95 cancer cells. Molecules. 2014;19(12):19980−94.

[210] Karthik KN, Ankith GN, Avinash HC, Rajesh MR, Kekuda PT, Raghavendra HL. Antifungal activity of some botanicals against seed-borne fungi. J Agric Food Nat Resour 2017;1(1):40−3.

[211] Teixeira CC, Weinert LS, Barbosa DC, Ricken C, Esteves JF, Fuchs FD. *Syzygium cumini* (L.) Skeels in the treatment of type 2 diabetes: results of a randomized, double-blind, double-dummy, controlled trial. Diabetes Care 2004;27(12):3019−20.

[212] Kumaki Y, Wandersee MK, Smith AJ, Zhou Y, Simmons G, Nelson NM, et al. Inhibition of severe acute respiratory syndrome coronavirus replication in a lethal SARS-CoV BALB/c mouse model by stinging nettle lectin, *Urtica dioica* agglutinin. Antivir Res 2011;90(1):22–32.

[213] Murakami A, Takahashi M, Jiwajinda S, Koshimizu K, Ohigashi H. Identification of zerumbone in *Zingiber zerumbet* Smith as a potent inhibitor of 12-O-tetradecanoylphorbol-13-acetate-induced Epstein-Barr virus activation. Biosci, Biotechnol, Biochem 1999;63(10):1811–12.

[214] Chavan R.D., Shinde P., Girkar K., Madage R., Chowdhary A. Assessment of anti-influenza activity and hemagglutination inhibition of Plumbago indica and Allium sativum extracts. Pharmacognosy Research. 2016;8(2):105.6

Strategies for delivery of antiviral agents

Vuyolwethu Khwaza, Buhle Buyana, Xhamla Nqoro, Sijongesonke Peter, Zintle Mbese, Zizo Feketshane, Sibusiso Alven and Blessing A. Aderibigbe

Department of Chemistry, University of Fort Hare, Alice, Eastern Cape, South Africa

19.1 Introduction

Viral infections are the main health problem worldwide and they are a major cause of high mortality along with an adverse continuously amplified socioeconomic impact [1]. Viruses are small infectious agents which cause infectious diseases and they have an effect on human health which could be mild or life-threatening [2]. The understanding of the global economic effect of viral diseases is poorly studied, regardless of the increasing transmission of viral infections, making it challenging to evaluate the cost-effectiveness and the societal costs of preventative efforts [3]. The problem of estimating a general effect of viral infections includes numerous factors that hamper the technique such as the incidence of associated comorbidities, the variety of viral infections, psychological and social problems having economic repercussions, the diversity of medications and vaccinations that can be utilized in the evaluation [3].

Viral infections are classified as single-stranded (ss) DNA viruses, double-stranded (ds) DNA viruses, dsRNA viruses, positive (+)ssRNA viruses, negative(−)ssRNA viruses, ssRNA-reverse transcriptase (RT) viruses, and dsDNA-RT viruses [4]. Over 6590 virus species have been identified and more than 10^{31} viruses were estimated on Earth. However, only 50 antiviral drugs have been permitted for human use [5]. Researchers have estimated that infectious viruses are accountable for approximately two million deaths per year. Many different types of viruses exist, however only about five thousand of them have been identified and characterized. These kinds of viruses can enter the human body through different routes such as mouth, skin, eyes and nose. Major pathogenic viruses that are responsible for a huge number of human mortality and morbidity are human immunodeficiency virus (HIV), hepatitis viruses, herpes simplex virus (HSV), norovirus, human papillomavirus (HPV) and coronaviruses and COVID-19 [6,7]. Several infectious diseases have high infection rates. Viruses including the Middle East respiratory syndrome (MERS), severe acute respiratory syndrome (SARS), COVID-19, and Ebola virus are major threats to the public health sector [8].

In the past eras, several infectious viruses have emerged or re-emerged, producing severe threats to public health and the economic system globally. Examples of viral infections which are transmitted from animals to humans include (MERS, Ebola, and Marburg hemorrhagic fevers, West Nile fever, Lassa fever, Yellow fever, SARS, and COVID-19) [9]. Hepatitis C (HCV) and HIV are chronic viral infections that do not have effective vaccines for the prevention of these infections. Numerous viral illnesses also require therapeutic vaccines to prevent their recurrences [10]. Many viruses cause chronic diseases and the lack of powerful prophylactic vaccines that patients could additionally benefit from making the prevention of these infections challenging [10]. Vaccination is one effective approach to empower the immune system defensive responses against pathogens, typically decreasing infectious disease burdens and reducing mortality and morbidity. Under health emergency circumstances, alternative and new procedures in vaccine development and design are vital for rapid and huge vaccination coverage, to manage a disease outbreak and limit the epidemic spread [9]. Nonetheless, despite global efforts, getting a vaccine to the public takes time, dosing issues, side effects and manufacturing problems can all cause delays. Moreover, the time and costs for faster development, the lack of suitable experimental animal models and universally issuing active vaccine applicants are some of the main challenges to overcome in case of pandemic threats. Hence, alternative and/or new methods in vaccine development and design are necessary to quickly act against outbreak conditions [9]. It is essential to have effective means of inhibiting viral transmission and decreasing its devastating outcomes on animal and human health.

Antiviral drugs are effective against viruses. Even though several antivirals are already accessible, their efficacy is frequently limited due to factors that include poor oral bioavailability, adverse side effects, poor solubility, untargeted release, low permeability and antiviral resistance [8]. The aforementioned complications can be overcome using advanced antiviral delivery systems developed using nanotechnology principles. These delivery systems loaded with antiviral drugs are prepared from synthetic or natural materials. Nanomaterials provide unique physicochemical properties which have related advantages for drug delivery as perfect tools for the treatment of viral infections. Presently, different kinds of nanomaterials specifically nanoparticles, nanospheres, nanosuspensions, nanovesicles, nanoemulsions, nanogels, liposomes, dendrimers and polymeric micelles had been studied either in vitro or in vivo for drug delivery of antiviral agents with prospects to be transformed in clinical practice [1,11]. Moreover, there is increasing emphasis on the development of antiviral delivery systems from natural substances including phospholipids, lipids, proteins, surfactants and polysaccharides, because of health and environmental issues. The dimensions, morphology, composition and interfacial characteristics of nanoparticles can be manipulated to enhance the stability, handling and potency of antivirals [12]. More studies on the in vivo metabolism and pharmacokinetic characteristics of antiviral drugs may additionally assist to resolve the issues related to antiviral drugs and improve their efficacy [8]. This chapter will focus on the efficacy of hybrid compounds, drug delivery systems, and nanoparticles containing antiviral drugs for the treatment of viral infections such as HIV, herpes, hepatitis, Ebola, HPV, viral pneumonia, common cold, COVID-19 and MERS.

19.2 Classes of antiviral drugs

19.2.1 Antihuman immunodeficiency virus drugs

In the treatment of HIV infection, **Zidovudine** was the first approved drug and is still part of the first line routine in Highly Active Antiretroviral Therapy. Regardless of its effectiveness, some factors limit its clinical use such as suboptimal bioavailability, toxicity and pharmacokinetics that includes inhibition of mitochondrial machinery, bone marrow aplasia, high hepatic first-pass metabolism and short plasma half-life [13,14]. **Stavudine** is one of the initial antiretroviral drugs (ARVs) from the class of nucleoside reverse transcriptase inhibitors (NRTIs). It was approved for the treatment of HIV infection and has been extensively utilized as a part of initial combination antiretroviral therapy (cART) [15]. It is highly effective in treating HIV, although it is associated with serious long-term side effects. Due to the severe side effects associated with stavudine, in 2009, World Health Organization (WHO) suggested the replacement of stavudine with zidovudine and/or tenofovir for primary treatment of HIV, since there are fewer side effects and drug toxicities associated with zidovudine and tenofovir [16].

Tenofovir is an effective antiviral inhibitor belonging to the acyclic nucleoside phosphonate family. It displays additive and synergistic activity when combined with some ARVs such as zidovudine, abacavir, lamivudine and ritonavir. It exhibits no antagonistic interactions when used in combination with some ARVs [17]. Tenofovir exhibits immunomodulatory effects with minimal cytotoxicity. Data from both murine and human cell lines in HIV-1 infected peripheral blood mononuclear cells (PBMC) and T-lymphocytes revealed tenofovir concentrations required to kill 50% of cells was 29 and 22 μmol/L, respectively [17]. **Tenofovir disoproxil fumarate** (TDF) is a prodrug of tenofovir and acts as a powerful competitive inhibitor of hepatitis B virus (HBV) and HIV-1 reverse transcription. TDF is utilized alone in mono-infected patients with HBV but in combination with other ARVs (emtricitabine, rilpivirine, etc.) for the treatment of HIV-1 [18]. TDF is the most widely utilized ARV since it has an extended plasma and intracellular half-life with a high barrier against the development of viral resistance mutations [19]. **Emtricitabine/tenofovir** is a fixed-dose combination (FDC) of emtricitabine (200 mg) and tenofovir (300 mg) that was initially approved to treat and prevent HIV-1 infection in patients at high risk. Severe side effects associated with the use of emtricitabine/tenofovir in studies of preexposure prophylaxis were limited [20]. Saravolatz et al. evaluated in vitro the efficacy of **Emtricitabine** against clinical isolates of HIV-1 and laboratory strains of HIV-1 and HIV-2. The EC_{50} ranged from 0.002 to 1.5 mmol/L, but depending on the cell line used and viral isolate. Emtricitabine exhibits in vitro synergy with stavudine and zidovudine and exhibited additive in vitro activity when combined with didanosine and/or zalcitabine against HIV-1 and HIV-2. Moreover, it also showed anti-HBV activity in vitro with EC_{50} values in the range of 0.01−0.04 mmol/L that is comparable to the anti-HBV activity of lamivudine [21].

Abacavir is a NRTI that is utilized for the treatment of HIV infection. The infection of HIV is typically treated with ARVs agents that include three or more different drugs used in combination. ARVss utilized in these agents include protease inhibitors, NRTIs, non-NRTIs and integrase strand inhibitors [22]. Abacavir makes an ideal addition to these types of combination drugs due to its dosing flexibility. It may be administered once or twice a day to match the dosing pattern of other drugs and can also be administered as tablets that contain other ARV drugs (zidovudine and lamivudine), allowing for a reduction in pill count. The common side effects of abacavir include headache, nausea and diarrhea [22–24]. Abacavir exhibits good bioavailability and is widely distributed and rapidly absorbed [25]. **Rilpivirine** is a novel NNRTI that is approved for HIV-1 treatment in combination with other ARVs [26]. It prevents HIV-1 replication by noncompetitive inhibition of HIV-1 RT and has high activity against wild-type and mutant virus strains, including K103N [27]. When rilpivirine is coadministered with raltegravir or NRTIs (emtricitabine, stavudine, abacavir, zidovudine and lamivudine) no clinically relevant drug–drug interactions are reported [26]. Rilpivirine exhibits in vitro activity against a variety of resistant HIV isolates and retains an EC_{50} of 1 nM against mutations K103N, V106A, L100I, G190A and G190S. Moreover, it showed antiviral activity against a broad panel of HIV-1 group M primary isolates, with EC_{50} values ranging from 0.07 to 1.01 nM. However, it was less effective against group O primary isolates, with EC_{50} values ranging from 2.88 to 8.45 nM. Rilpivirine has limited activity in cell culture against HIV-2 [28]. It causes a much lower increase in blood triglycerides, low-density lipoprotein cholesterol and high-density lipoprotein cholesterol [29].

Dolutegravir is a second-generation integrase strand transfer inhibitor (INSTI), which is used in different countries including the UK [30]. It is a relatively novel class of ARVs to treat HIV-1 infection. It prevents the combination of HIV-1 proviral DNA into the host cell genome, a step that is essential for viral replication. HIV-1 integrase core domain is a 32 kDa protein that consists of three distinct structural and functional domains; that is, the catalytic core domain, the N-terminal domain and C-terminal domain [30]. Dolutegravir is an effective integrase inhibitor that employs divalent cations (Mg^{2+}) to couple with the enzymatic active site of the viral integrase. Its structure permits it to enter the active and recently vacated enzymatic pocket where it binds farther in than prior drugs in its class. This provides a more stable and lasting bond compared to other precursor integrase inhibitors such that its dissociation constant is slower compared to either raltegravir or elvitegravir. Moreover, it also has a lower IC_{50} value of 1.6 nM for HIV-1 compared to raltegravir with IC_{50} value of 3.3 nM and elvitegravir with IC_{50} value of 6 nM [31]. The pharmacokinetic/pharmacodynamics features of dolutegravir include the absence of negative interactions with other antiretroviral drugs, prolonged intracellular half-life and effective activity against HIV-1 strains that are resistant to other INSTIs. Dolutegravir has a very good safety profile moreover; it can be administered once a day [32]. Its antiviral activity is via preventing HIV type-1 integrase-catalyzed strand transfer into host cell DNA. The mechanism of action of dolutegravir has been established through several measurements including in vitro resistance passage experiments, integrase enzyme assays and mechanistic cellular assays [33].

Dolutegravir/rilpivirine was approved by the US Food and Drug Administration (FDA) as the first dual ARV single-tablet combination regimens (STR) for the management of HIV-1 infection. The safety, effectiveness, tolerability and high resistance barrier of dolutegravir and rilpivirine including the favorable pharmacokinetic profiles of both drugs make them good candidates for combination therapy. The dolutegravir/rilpivirine presents a novel therapeutic approach for the treatment of the HIV-1 infected people [34]. Dowers et al. reported in vitro studies of dolutegravir/rilpivirine against HIV-1 and HIV-2. The results revealed that dolutegravir inhibited WT HIV-1 with an EC_{50} value of 1.6 nM, which retained high effectiveness against the N155H, Y143R and G140S/Q148H mutants, with less than fourfold reduction inactivity. While rilpivirine revealed activity against laboratory strains of WT HIV-1 in a highly infected T-cell line with an EC_{50} value of 0.73 nM for HIV-1. Moreover, rilpivirine exhibited antiviral action against a broad panel of HIV-1 group M and was less effective against group O primary isolates with EC_{50} values ranging from 2.88 to 8.45 nM. Nonetheless, limited activity in cell culture against HIV-2 with a median EC_{50} value of 5220 nM was observed [34]. Therefore, an in vitro study showed synergistic effects of dolutegravir and rilpivirine when combined. This synergy has contributed to their ARV effectiveness in combination therapy [35]. **Cabotegravir** is an HIV-1 INSTI and has the unique feature of a long half-life. The long-acting injectable formulations of cabotegravir are being studied for clinical development. It is known that carbotegravir has superior ARV activity with IC_{50} values of 0.22 and 0.34 nmol/L against HIV-1Bal and HIV-1NL-43, respectively [36]. Hence, cabotegravir has the potential to act as part of parenterally delivered long-acting treatment for the prevention of HIV infection. Moreover, it may be utilized as single-agent PrEP as a long-acting injection maintenance treatment, which can avoid NRTI-linked toxicity and increase its therapeutic options. It also has activity against all common clades of HIV-1 in vitro and good effectiveness at low concentrations, with 5 or 30 mg daily oral dosing revealed to decrease HIV-1 RNA by 2.2–2.3 log10 in a 10-day monotherapy trial and has a protein binding-adjusted IC_{90} value of 166 ng/mL. Furthermore, it exhibits high potency, low water solubility and could be formulated as a nano-suspension for long-acting injection at a concentration of 200 mg/mL [36,37]. **Lamivudine** is an NRTI and one of the first approved antiviral agents that are utilized for the treatment of HIV and HBV infections. It is safe and effective for the treatment of chronic HBV infection. However, it is associated with high drug resistance [38,39]. It is an orally administered NRTI that is used in combination with other ARVs to treat HIV-1 infection in patients with AIDS and is used as one dose in the treatment of HBC infection [40]. The effective form is lamivudine triphosphate, which is made through an intracellular triple phosphorylation process. Lamivudine triphosphate competitively prevents viral RT by causing termination of DNA replication, thus, interrupting HIV replication [40]. It shows strong activity against HIV-1 and HIV-2 and in the last decade, it has been one of the most effective remedies for the management of HIV as well as chronic hepatitis B. This strong NRTI is one of the preferred agents used in combination therapy for the first-line treatment of HIV [41]. Fig. 19.1 shows the structures of HIV antiviral drugs. The mode of action of the anti-HIV drugs are shown in Table 19.1.

19.2.2 Antiviral drugs used for the treatment of herpes

Acyclovir (Fig. 19.2) is a synthetic guanosine analog utilized for treating HSV, varicella-zoster virus (VZV) infections, chickenpox, and shingles [42]. Intravenous acyclovir provides

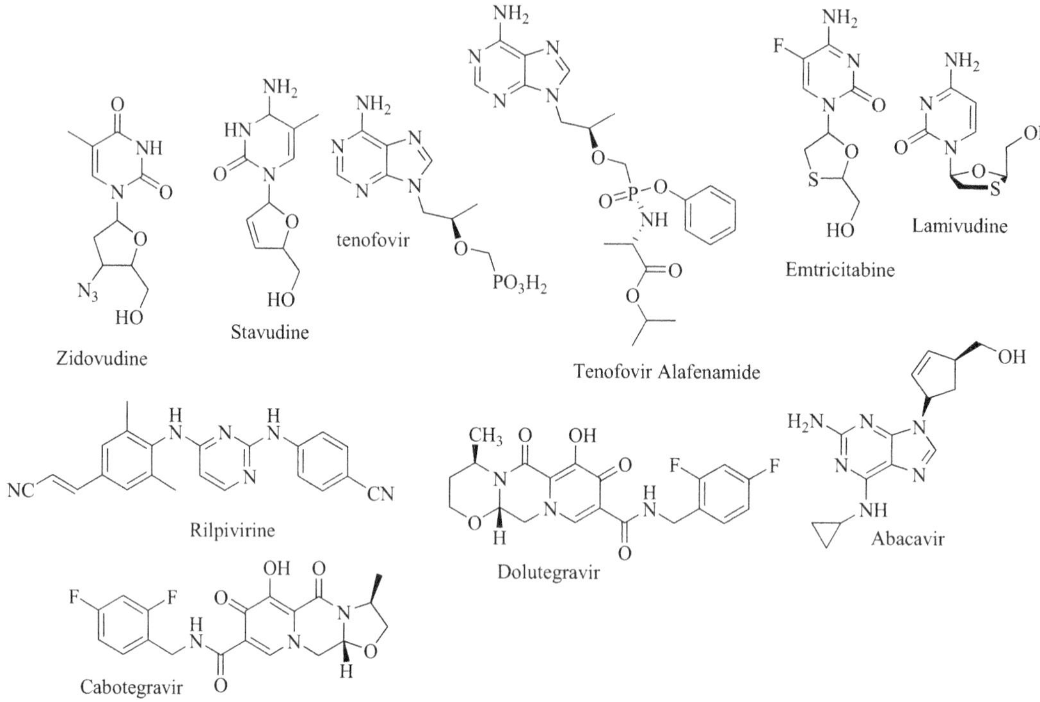

FIGURE 19.1   HIV antiviral drugs.

fluid penetration while oral acyclovir provides modest bioavailability. Herpesviruses have varying degrees of susceptibility to acyclovir, with HSV type 1 (HSV-1) being the most susceptible, followed by HSV type 2 (HSV-2) and VZV, and to a lesser extent Epstein-Barr virus [43]. The solubility of acyclovir in water and their oral bioavailability is limited, which require relatively large doses and frequent administration to maintain plasma levels of acyclovir high enough to achieve viral inhibition [44]. **Valacyclovir** (a valine ester of acyclovir) is a safe and efficient prodrug with three to five times higher oral bioavailability than acyclovir [45,46]. Many clinical studies showed that valacyclovir has a safety profile compared to acyclovir. It became a better option in the treatment of VZV infections since it requires less frequent dosing treatment than acyclovir, contributing to increased patient adherence to remedy [46]. Valacyclovir also exhibits increased oral bioavailability and this outcome prompted the design of prodrugs of other antivirals as well [47]. **Ganciclovir** is reported to have antiviral activity against HSV-1 and HSV-2 and it has also been used for the treatment of herpetic keratitis. However, it has also been reported to produce side effects in a high percentage of patients including neutropenia, nephrotoxicity, myelosuppression, anxiety, fever, altered mental status, confusion, tec [48]. **Valganciclovir** is the L-valyl ester prodrug of ganciclovir. Oral valganciclovir is well absorbed and converted to ganciclovir by first-pass intestinal or hepatic metabolism [49]. It exhibits its antiviral activity in the form of ganciclovir-triphosphate by hindering viral replication via helping as a competitive substrate for cytomegalovirus (CMV) DNA polymerase. The most common side effects of

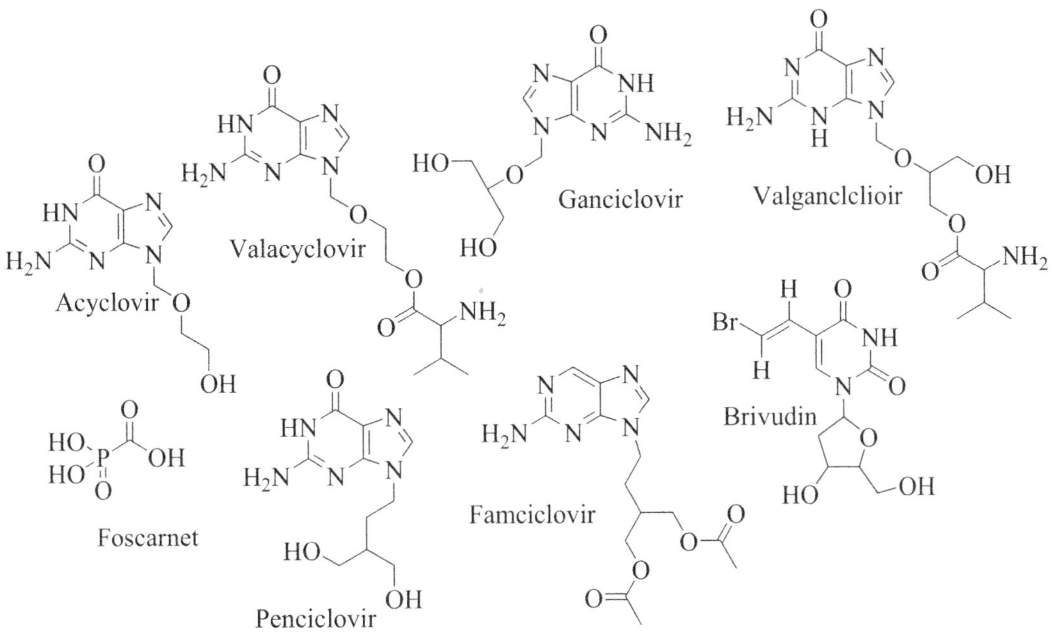

FIGURE 19.2   Antiviral drugs for the treatment of herpes.

valganciclovir are vomiting, diarrhea and nausea. Resistance to valganciclovir occurs via the *UL54* gene, which encodes for CMV DNA polymerase and mutations in the *UL97* gene, which encodes for CMV kinase [43].

Foscarnet is a pyrophosphate analog that reversibly binds to the viral DNA polymerase and hence does not require activation by TK. The binding of foscarnet does not result in chain termination just like other antiherpetic treatment, which binds to the viral DNA polymerase; the remedy binds at the pyrophosphate binding site, within the effective site of the herpesvirus DNA polymerase, inhibiting nucleotides from binding to the effective site and from being linked into the growing DNA strand [50]. Moreover, foscarnet is different from nucleoside-based remedies; since it occurs in its effective form and needs no additional adjustments to prevent herpesviruses. It could be utilized to prevent acyclovir-resistant herpesviruses [51]. However, its side effects include hypocalcemia, penile ulcerations, acute nephrotoxicity, electrolyte disturbances, metabolic disturbances, nausea and seizures [52,53]. Though foscarnet might prevent host DNA replication, the greater inhibitory impact on viral DNA replication dictates that foscarnet could still be utilized remedially for acyclovir-resistant HSVs [51]. It has a significant role to play in the management of HSV and CMV infections. It does not only prevent the proliferation of all herpes viruses but also of influenza viruses, HIV and HBV. The oral bioavailability of foscarnet is low; therefore, it must be administered intravenously and is excreted in over 90%–95% unchanged form through the kidneys [54].

Penciclovir is a guanosine analog antiviral drug utilized for the management of various herpesvirus infections. It is a nucleoside analog that shows good selectivity and low toxicity. It is very poorly absorbed when given orally and therefore, famciclovir, the diace-tylester of 6-deoxypenciclovir, was developed as the oral prodrug, which has an oral bio-availability of 77% [44]. Famciclovir is quickly and widely absorbed. It is proficiently converted to penciclovir in two steps: which is (i) removal of the two acetyl groups (the first one by esterases in the intestinal wall and the second one on the liver) and (ii) oxidation at the six position catalyzed by aldehyde oxidase that accounts for the con-version of 6-deoxypenciclovir to penciclovir. Famciclovir is well tolerated in patients and is active against HSV-1 and HSV-2 and VZV [44]. **Famciclovir** is a prodrug that is in the family of nucleoside analog antiviral agent. It is a drug that is utilized for the treatment and management of HSV and VZV infections. It is also used to treat HIV-infected patients and lowers HBV DNA levels in patients when utilized as a long-term treatment [55]. A study of 419 patients with Herpes Zoster (HZ) given either famciclovir or placebo exhib-ited that famciclovir management accelerated lesion healing and decreased the median duration of postherpetic neuralgia (PHN) resolution [56]. It has also confirmed efficiency in acute HZ, decreasing the duration of acute pain and enhancing cutaneous remedial. The median duration of pain was half as many days in people who received famciclovir when compared with those receiving placebo, which results in a 3.5-month decrease in the average duration of pain [56]. Famciclovir is a well-absorbed first-line option for the management of HZ and utilized for handling HZ in immunocompetent adults and immu-nosuppressed people older than 25 years, but not approved in childhood. In a study with 148 patients who were immunocompromised following bone marrow or solid organ trans-plantation or oncology treatment, the efficiency and well-being of famciclovir were evalu-ated. The results showed that oral famciclovir is suitable, active, and well-tolerated in immunocompromised people with HZ [57].

Brivudin is a thymidine nucleoside analog with strong and selective activity against VZV and HSV. It is considered safe and has not been linked with symptomatic hepato-toxicity [58]. It is approved for the management of HZ in many countries. The clinical trials confirmed that orally administered famciclovir, brivudin, acyclovir and valacyclovir decreased the duration of viral shedding and novel lesion formation and accelerates rash remedial in patients with HZ. The aforementioned drugs reduce the severity, duration of acute pain and antiviral therapy decreases the side effects of the acute phase of HZ on the quality of life [59]. Brivudine is orally administered and approximately 70% of the oral dos-age is quickly changed to bromovinyluracil during the first passage through the liver. It is effective in the management of HZ for both the formation of new lesions and the prevention of PHN impacts. Therefore, brivudine can be considered as the first choice to treat acute HZ cases considering that it is administered once daily and can control pain earlier [60,61].

19.2.3 Antihepatitis drugs

Adefovir is an antiviral drug that is used for the treatment of HBV infections. It can be used as a primary treatment in patients with hepatitis B e Antigen (HBeAg)-negative or HBeAg-positive chronic hepatitis B. Moreover, it can be used as an additional treatment for those with lamivudine-resistant HBV. Drug resistance associated with adefovir is less frequent compared to lamivudine [62]. Adefovir has an effective in vitro activity against herpes viruses and retroviruses. Adefovir diphosphate, an intracellular compound acts as a good inhibitor and chain-terminator of HBV replication mediated by HBV DNA polymerase. Moreover, it prevents HBV polymerase by direct binding in competition with the endogenous substrate and after incorporation into viral DNA, results in chain termination of DNA synthesis [63]. Low potential for resistance development with adefovir can be associated with its close structural relationship with the natural substrate which limits the potential for steric hindrance as a mechanism of resistance. The IC_{50} of adefovir diphosphate for HBV polymerase in HBV core particles isolated from transfected HepG2 cells is 0.2 mm. Adefovir is effective in vitro against all known emtricitabine, lamivudine, famciclovir and HBIG resistant HBV, using both cell culture and in vitro enzymatic assays. Mitochondrial DNA in skeletal muscle is unchanged [63–65].

Entecavir is a primary remedy in the treatment of chronic hepatitis B. It has a higher genetic barrier and higher antiviral potency than other antiviral drugs [66]. Entecavir displayed a high virologic response of approximately 93% and a rare genotypic resistance of only 1.2% during the 5-year followup in a cohort study [66]. However, few reports compare the short-term efficiency directly between TDF and entecavir in the treatment of chronic hepatitis B. The long-term efficiency in clinical practice is still limited [66]. Entecavir showed superior antiviral efficiency and a lower resistance rate than lamivudine [39]. Entecavir and adefovir are nucleoside analogs that have been approved for the treatment of chronic HBV infections for many years. Entecavir display delayed development of resistance and a low rate of resistance in the treatment of patients with chronic hepatitis B. Even though entecavir display some degree of cross-resistance, it shows antiviral activity against lamivudine-resistant HBV. Several clinical trials have compared the efficiency and side effects of entecavir and adefovir for the treatment of chronic hepatitis B [67–69].

Entecavir and Tenofovir are recently recommended as the primary treatment in patients with chronic hepatitis B. Entecavir and Tenofovir exhibit high rates of viral suppression and high genetic barriers with very low rates of resistance. The normalization of alanine aminotransferase, HBeAg seroconversion and the rates of HBV DNA suppression are reported to be comparable in treatment for patients with chronic hepatitis B and immune-active disease [67,70]. **Telbivudine** is a synthetic thymidine nucleoside analog that is approved as a pregnancy category B drug for treating chronic HBV infection. Many studies have reported that telbivudine can successfully block mother-to-infant transmission in late pregnancy with safety and good tolerance [71]. But the safety and efficiency of telbivudine in treating HBV-positive women during early pregnancy has not been well-known [71]. Some clinical trials demonstrated that telbivudine is more effective than adefovir and lamivudine in the treatment of patients with chronic hepatitis B, irrespective of the HBeAg detection. The side effects that are associated with telbivudine are peripheral neuropathy and muscle toxicity [72]. Telbivudine demonstrated enhanced eGFR and higher cumulative rates of HBeAg seroconversion than entecavir in patients with HBeAg-positive chronic hepatitis B. The CK elevation and the viral breakthrough were more common in the telbivudine group [73]. Therefore, the balance should be considered between high rate of antiviral drug resistance and the potential benefit of telbivudine on HBeAg seroconversion rate and also side effects when choosing telbivudine as a primary drug. Moreover, all associated conditions, such as pregnancy, renal function, age, myopathy and

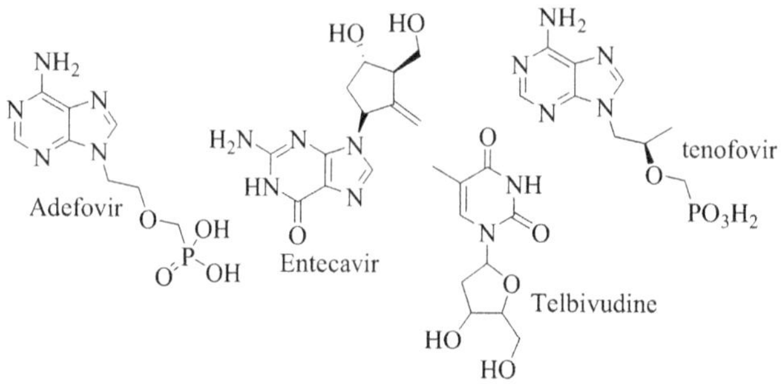

FIGURE 19.3 Antiviral drugs for the treatment of hepatitis.

economic circumstances need to be considered to enhance the therapeutic choice with telbivudine in patients with HBeAg-positive chronic hepatitis B [73]. Fig. 19.3 represent the structures of antiviral drugs for the treatment of hepatitis.

19.2.4 Antiviral drugs for the treatment of Ebola

Favipiravir is an antiviral drug that is utilized in treating Lassa fever, pandemic influenza H1N1 virus, Ebola, and Argentine hemorrhagic fever. Many concerns exist concerning favipiravir even though it is one of the choices for the remedy of patients with Ebola virus [74]. Guedj et al. evaluated the efficiency achieved by increasing the dosage of favipiravir in Ebola-virus-infected nonhuman primates. The results showed that the survival and virology results are supplemented by genomic and PK analysis to show new insights into the antiviral mechanism of action of favipiravir against Ebola virus and it supports the evaluation of high dosages of favipiravir in future human therapeutic interventions in Ebola virus disease [75]. Sissoko et al. evaluated the efficacy and safety of favipiravir in decreasing mortality and viral load in patients with Ebola virus disease. The results revealed that favipiravir monotherapy might not provide a benefit in patients with very high viral load, but proposes that more studies in the same context will be very unlikely to display that monotherapy with favipiravir reduces mortality in patients with very high viral load. As the changes that occur in patients with acute sepsis might influence pharmacokinetic parameters, it can also be of interest to evaluate higher dosages than those used in the JIKI trial [76]. Oestereich et al. evaluated the effectiveness of the pyrazinecarboxamide derivative (T-705) (favipiravir) against Zaire Ebola virus in vitro and in vivo. The outcome reveals that T-705 is an effective drug for improved Zaire Ebola virus infection in an animal model. It decreases ameliorates clinical, viremia and biochemical signs of disease and inhibits lethal results in 100% of the animals when the medication was administered 6 days after infection. T-705 has been revealed to be active at preventing replication of RNA viruses in various small animal models [77]. Pharmacokinetics information for patients with Ebola virus disease enrolled in the JIKI trial shown that treatment concentrations were lower than targeted levels. Administering higher dosages of favipiravir, modulating its metabolism and increasing its bioavailability may improve its therapeutic effect [78]. In conclusion, the observations regarding favipiravir there is a need for further preclinical and clinical studies to better understanding pharmacodynamics, side effects at several dosing, pharmacokinetics and pharmaceutical techniques [78]. **Remdesivir** also known as (GS-5734) is a novel antiviral drug which was developed by Gilead Sciences for the treatment of Ebola virus and Marburg virus infections [42]. It is a monophosphate prodrug that undergoes metabolism intracellularly to an active C-adenosine nucleoside triphosphate analog, preventing the viral RNA polymerases [42]. The agent was found among a screening process for antimicrobial activities against RNA viruses including Flaviviridae and Coronaviridae. Studies revealed that remdesivir was an effective agent during the time of the Ebola virus outbreak because of its low EC_{50} value and host polymerase selectivity against the Ebola virus [79]. Presently, remdesivir shows promising potential for the treatment of COVID-19 because of its broad-spectrum, effective in vitro activity against various nCoVs such as, SARS-CoV-2 with EC_{50} value of 0.77 μM and EC_{90} value of 1.76 μM, respectively. Moreover, in murine lung infection models with MERS-CoV, it prevented lung hemorrhage and reduced viral lung titers [79]. Remdesivir has been biochemically revealed to prevent the activity of Ebola virus large (L) RNA-dependent RNA polymerase (RdRp) as a nonobligate delayed chain terminator. Comparative structural modeling of the Ebola virus and SARS-CoV-2 RdRp domains shows that remdesivir targets the polymerases of EBOVs and CoVs equally, showing the effects of remdesivir drug on COVID-19 patients. Remdesivir prevents the replication of other pathogenic RNA viruses while having low cytotoxicity in a wide range of human primary cell lines [80,81]. Between 2013 and 2016, no antiviral drug and/or vaccines were approved for the treatment of Ebola virus disease.

Amiodarone has been identified as a powerful inhibitor of Ebola virus in epithelial and numerous endothelial cell lines at concentrations that are usually reached in humans treated with this remedy. Its mechanism of action appears to rely on the induction of a Niemann-Pick C-like phenotype that prevents late endosomal filovirus entry [82]. Amiodarone modifies the cell distribution of glycosphingolipids, sphingomyelin, cholesterol and its transporter NPC-1. Moreover, this molecule blocks the progression of fluid-phase endocytosis affecting the late endosomes [83]. The in vitro anti-Ebola virus activity of amiodarone was confirmed in many different cell types such as human macrophages. However, no anti-Ebola virus activity was observed for amiodarone when tested in the guinea pig model of Ebola virus disease [84]. **Clomiphene and Toremifene** act as effective inhibitors of Ebola virus infection. The estrogen receptor antagonists, clomiphene and toremifene are approved FDA drugs with oral availability, tolerability profiles, good safety and a long history of use. They have bioavailability and good plasma exposure, making them good candidates for the treatment of Ebola virus infection [85]. The availability of clomiphene and toremifene are useful in the resource inhibited geographical regions where outbreaks of filoviral infection occur frequently. Current efforts are focused on the improvement and development of these drugs as a therapeutic countermeasure alone or through synergistic combinations with other antiviral drugs [86]. Structures of antiviral drugs for the treatment of Ebola are shown in Fig. 19.4.

19.2.5 Antiviral drugs used for the treatment of human papillomavirus

Cidofovir is an antiviral that is used for the treatment of CMV retinitis in people with HIV. It has been shown to decrease E6 and E7 expression and to decrease the metastatic

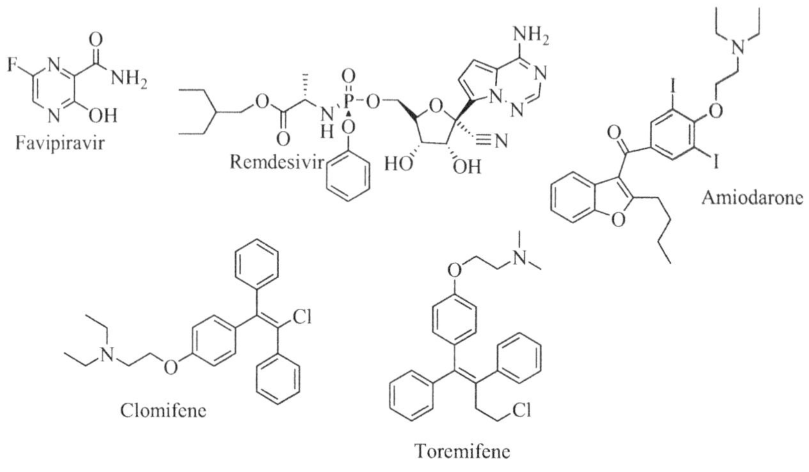

FIGURE 19.4   Antiviral drugs for the treatment of Ebola.

activities of HPV-positive tumor cells [87]. Cidofovir is a cytidine nucleoside analog with in vitro and in vivo activity against a broad spectrum of herpes viruses as well as HPVs, adenoviruses, human poxviruses and polyomaviruses. It blocks the proliferation of HPV-positive over HPV-negative cell lines [88]. A topical formulation of cidofovir for the treatment of recurrent herpes viruses, molluscum contagiosum, Kaposi sarcoma, and HPV associated lesions has been described. The topical formulation of cidofovir seems to cause major shrinkage and resolution of gingival HPV recalcitrant to traditional treatments. The topical application of cidofovir showed no side effects in the short-term use. However, the local side effects of topical cidofovir can include reversible alopecia, hyperpigmentation, and irritant dermatitis [89]. The topical formulation of cidofovir is effective and safe for the treatment of epithelial hyperplasia in patients with HPV infection. Moreover, chemotherapy is known as the standard therapy for those with advanced cervical cancer. Cisplatin, a platinum-based anticancer drug treat the disease effectively [90]. Yang et al. investigated the inhibitory effect of cidofovir on the proliferation of HPV 18-positive HeLa cells in cervical cancer using cisplatin as a positive control. The results revealed that the treatment of cidofovir with cisplatin prevented the proliferation of HeLa cells in a concentration- and time-dependent manner [90]. Cidofovir and cisplatin affect the expression levels of E6 and p53 proteins in HeLa cells, and consequently regulate cell cycle arrest and cell apoptosis. Furthermore, it may induce cell apoptosis similar to cisplatin and prevent HeLa cell proliferation, which indicates that Cidofovir might be a possible drug for cervical cancer therapy [90,91].

Ribavirin is an artificial guanosine nucleoside analog with broad-spectrum antiviral activities that was initially synthesized in 1972. It has been effectively utilized in humans to treat several pathologies such as Lassa fever, respiratory syncytial virus (RSV) and hepatitis C virus infections [92]. There are reports on the usage of ribavirin for the management of HPV infections. The advance of HPV infection utilizing peg-Interferon alfa-2b and ribavirin in the management of chronic HCV infection revealed that this association can be

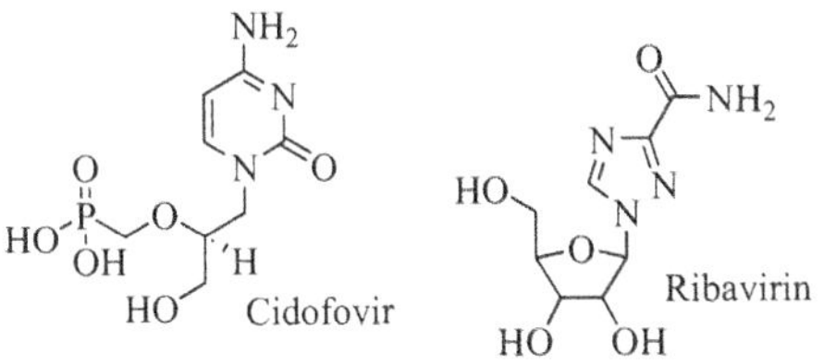

FIGURE 19.5   HPV antiviral drugs.

of great value for cases that present more resistance to the conventional treatment [93,94]. Fig. 19.5 presents the HPV antiviral drugs.

19.2.6 Viral pneumonia antiviral drugs

Ribavirin exhibit antiviral activity aginst a broad spectrum of RNA and DNA viruses. It is approved by FDA for the treatment of RSV in pediatric patients. Oral ribavirin has been utilized as a therapeutic agent in the management of noninfluenza respiratory viral infections (NIRVIs) [95]. Clinical trials were done to determine the efficiency of oral ribavirin for NIRVI. The evidence recommended that some benefit is derived from the usage of oral ribavirin for people with NIRVI. Therefore, oral ribavirin can be considered for the management of NIRVI in immunocompromised adults and patients with MERS-CoV [96,97]. The use of ribavirin for the treatment of RSV pneumonia is limited. The treatment efficacy is influenced by the immune status of patients and the benefit derived by ribavirin therapy is poor, also when specific immunoglobulins are administered [98]. Pneumonia is the one of the most common severe complications of influenza virus infection in all age groups.

Oseltamivir has been studied by some researchers. Chen et al. assessed the efficiencies of oseltamivir and peramivir in the management of severe influenza A patients with primary viral pneumonia. The results showed no major difference in the periods of influenza virus nucleic acid positivity, the remission time of cough symptoms and the remission times of clinical symptoms between the peramivir group and the oseltamivir group [99,100]. Therefore, oseltamivir did not induce a high efficacy than peramivir in the treatment of severe influenza A patients with primary viral pneumonia. Those who receive peramivir intravenously displayed significantly shorter remission times of fever symptoms than those treated with oral oseltamivir [100]. Oseltamivir treatment was recognized as an important protective factor against subsequent development of radiographic pneumonia, shorter viral RNA shedding and faster fever clearance times. More research is needed to define the effectiveness of antiviral treatment, pharmacodynamics and pharmacokinetic properties, viral shedding patterns and risk factors for an improved severity of illness, which will allow for enhancement in public health guidance and clinical treatment [101].

Zanamivir is a primary neuraminidase inhibitor approved for the treatment and prevention of influenza. Zanamivir exhibited efficiency for decreasing lung consolidation, pulmonary virus titers and mortality and morbidity scores in Influenza A virus infected mice and also for alleviating clinical symptoms of influenza-infected humans. Significantly, it was found to suppress the production of nitric oxide in Influenza A virus infected and

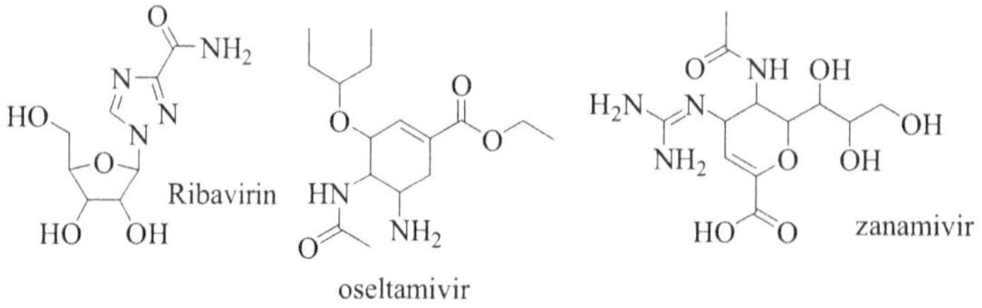

FIGURE 19.6 Antiviral drugs used for the treatment of viral pneumonia.

IFN-γ-activated RAW 264.7 macrophages in vitro. It has healing potential to decrease pulmonary damage in Influenza A virus-infected mice through a mechanism involving the suppression of nitric oxide synthesis. These results are relevant for the improvement of novel treatment methods for influenza infection and might also support a beneficial role of zanamivir in the early initiation of influenza treatment for inhibiting severe pneumonia [102,103]. The structures of viral pneumonia antiviral drugs are displayed in Fig. 19.6.

19.2.7 Antiviral drugs used for the treatment of respiratory infection

Ribavirin is one of the most significant nucleoside antivirals which were discovered in the 1970s [104]. Its activity against other nCoVs makes it a candidate for COVID-19 treatment. The in vitro activity of ribavirin against SARS-CoV is limited and a high concentration is required to prevent viral replication, requiring combination therapy and high dose [79]. **Umifenovir** is a nucleoside antiviral targeting the hemagglutinin envelope glycoprotein (HA) in the fusion machinery of influenza virus. A current study showed that administering umifenovir to COVID-19 patients in China resulted in negative viral conversion where the virus was not detected in 14 days. Umifenovir has displayed inhibitory effects on SARS virus-replication in vitro. However, clinical trials are in the process of evaluating the efficiency of umifenovir in China [79,105]. There is no effective treatment for the management of COVID-19. Many clinical trials are being studied using different drugs, vaccines and biologics on COVID-19 [106]. Okumus et al. evaluated the safety and efficiency of **ivermectin**. The results suggest that ivermectin is an alternative remedy that can be utilized in the management of COVID-19 [107]. Even when utilized in severe COVID-19 patients, it can reduce the mortality rates, provide an increase in clinical recovery and improvement in prognostic laboratory techniques. It can cause severe side effects in those without MDR-1/ABCB1 and/or CYP3A4 gene mutation; furthermore, the developing side effects may be removed with suitable treatment. Therefore, the results suggest that ivermectin can be used for the treatment of COVID-19 [107].

Favipiravir is an antiviral drug previously studied for the treatment of Ebola, Lassa fever and influenza virus. Its therapeutic efficacy has been well-known documented for these diseases [108]. Currently, it has potent effects in early clinical trials for the treatment of COVID-19. The safety data for favipiravir against COVID-19 are limited [109]. The EC_{50} value of favipiravir against SARS-CoV-2 was 61.88 μM/L in Vero E6 cells. Therefore, favipiravir have high potential in the treatment of patients with COVID-19. It improves viral clearance within 7 days and clinical enhancement within 14 days, particularly in those with mild-to-moderate COVID-19. However, the studies on the safety and effectiveness of favipiravir in those with COVID-19 are still limited [110,111]. Therefore, there is a need to confirm the efficiency and safety of favipiravir in viral infections as a randomized, multicentre and double-blind clinical trial on a large group of people. Furthermore, more studies at higher doses, in combination with other antivirals drugs are required to determine the efficacy and safety of favipiravir in high-risk patients [112]. The primary mechanism of action of favipiravir is via binding to and prevention of RdRp, which ultimately inhibits viral genome RNA transcription and replication. Since RdRp domains are not currently in human cells and are conserved among RNA viruses, this distinct specific mechanism targeting RNA viral polymerases makes favipiravir an attractive drug candidate [112]. **Remdesivir**, a nucleoside analog prodrug, has inhibitory effects on pathogenic animal and human coronaviruses such as SARS coronavirus 2 (SARS-CoV-2) in vitro, and inhibits MERS coronavirus (MERS)-CoV, SARS-CoV-1, and SARS-CoV-2 replication in animal models [113]. The active form of remdesivir acts as a nucleoside analog and inhibits the RNA-dependent RdRp of coronaviruses including SARS-CoV-2. Remdesivir is incorporated by the RdRp into the growing RNA product and allows for the addition of three more nucleotides before RNA synthesis stalls [114]. SARS-CoV-2 is a beta-coronavirus that belong to the same genus as MERS-CoV and SARS-CoV. Presently, there are no confirmed antiviral agents for COVID-19. Remdesivir is a promising drug for treating COVID-19 based on in vitro activity against SARS-CoV-2, limited data from clinical trials and uncontrolled clinical reports [115,116]. Studies in vitro and in vivo have revealed the effectiveness of remdesivir against coronaviruses [117]. There are some side effects associated with remdesivir, even though it has a specific therapeutic effect on COVID-19. Currently, several countries are actively developing drugs and vaccines to fight against COVID-19 [118,119].

Lopinavir is utilized in combination with ritonavir for the prevention and treatment of HIV infection. Lopinavir has been reported to prevent SARS-CoV-2 at an IC_{50} of (26.36 μM), the level of a drug that prompts a response halfway between the baseline and maximum after a specified exposure time. Administration of lopinavir as an emergency drug in China increased the eosinophil count among COVID-19 patients [105]. **Lopinavir-ritonavir** has been proposed as a monotherapy treatment for COVID-19 on the basis of preclinical studies, in vitro activity and observational studies. The results reveal that lopinavir−ritonavir monotherapy is not an active treatment for COVID−19 patients. It does not improve the clinical results in COVID-19 patients [120]. Alhumaid et al. evaluated the safety and efficiency lopinavir−ritonavir in COVID-19 patients. The results did not show any statistically major advantage in the efficiency of lopinavir−ritonavir in COVID-19 patients. In terms of safety, the study found a greater number of side effects in lopinavir−ritonavir. There is a need to conduct large randomized clinical trials to determine the safety and efficiency of lopinavir−ritonavir in the treatment of COVID-19 patients. Preferably, these studies should be double-blinded and conducted in a wide range of settings [121,122]. Lopinavir−ritonavir treatment did not significantly decrease mortality, accelerate clinical enhancement and diminish throat viral RNA detectability in COVID-19 patients [123]. Fig. 19.7 shows the structures of antiviral agents for the treatment of respiratory infections.

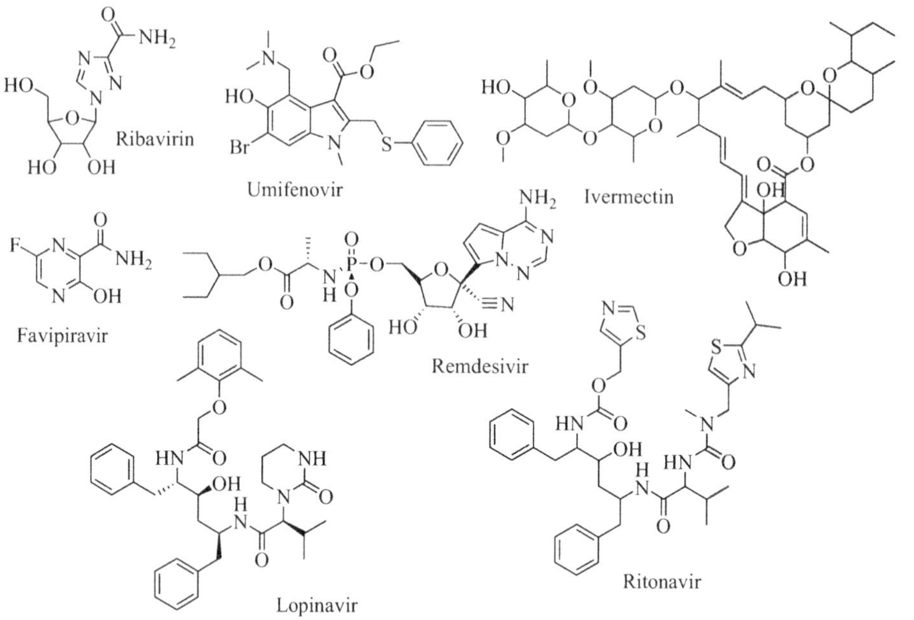

FIGURE 19.7 Antiviral drugs for the treatment of respiratory infections.

19.3 The general mechanism of viral infections

Infection begins when viruses enter the host cell surface and binds to at least one cell receptors, such as carbohydrates, proteins and lipids [124]. Some viruses enter cells via direct cell-to-cell contacts, utilizing structures that are shaped by the polarized cytoskeleton, viral proteins and adhesion molecules at the contaminated cell junction [125]. Viral infection includes the entry of viral DNA into a host cell, duplication of that DNA and releasing the novel infections, for example, the infection attaches to a host cell injecting its genetic material into the host cell through penetration stage and attachment [42]. The viral RNA and/or DNA is itself joined into the genetic material of the host cell prompting it to duplicate the viral genome and the host cell releases the recently made viruses, both by the breakage of the cell, waiting for cell death and through budding off by the cell membrane [126,127]. The study conducted by Dimitrov showed that the attachment of the virus to the host cell is the initial phase in infection [125]. A typical viral life cycle includes entry of the virus into the host cell, translation of viral proteins, replication of viral genome, assembly of viral particles, and final release of the mature virions into the extracellular milieu. All viruses have on their outside a receptor-binding protein [128]. When attached, the virus can enter the cell through combination with the cell layer, receptor intervened endocytosis or nonclathrin-mediated endocytosis. Ultimately, the viral genome is released into the cytoplasm where it will be interpreted and replicated or reach the nucleus of the cell [128].

TABLE 19.1 Mechanism of action of viral infections.

Viral infections	Mechanism of action	References
HIV	HIV targets and destroys important constituents of the human immune system	[129]
	HIV-1 mainly targets CD4 receptors and contaminates CD4+ T lymphocytes and also cells of the monocyte and/or macrophage lineages. Furthermore, it contaminates CD4- cells such as astrocytes	[130]
	HIV-1 may enter and exit cells except by crossing membranes using a process called transcytosis	[131]
	HIV-1 contaminates cells of the immune system and taking advantage of their precise particularities and functions	[132]
	HIV enters target cells with the assistance of two envelope glycoproteins (gp120 and gp41)	[133]
	The virus attaches to the cellular membrane via interaction of the viral envelope protein gp120 and the first extracellular domain of the cellular CD4 receptor	[134,135]
Herpes	HSV introduces membrane disruptions through developing pores and fragmentations in the membrane to prompt endocytosis	[136,137]
	- The virus enters and transport to the nucleus back to the cell membrane to spread out to the other cells with the assist of the rearrangement in the cytoskeleton components - The HSV viral proteins are able of restraining the usage of the translational and transcriptional tools for the cell itself so that its pathways remain unconstrained	[138]
	HSV attaches to the host's cell surface receptor through its viral glycoproteins such as gB and gC. It then slides on the cell surface and reaches the cell body. Furthermore, it binds with cell membrane receptors utilizing glycoproteins such as gD, gH/gL, and gB that triggers direct membrane cell	[139]
	HSV-1 gB is a fusogen that mediates the membrane cell to assist the penetration of the HSV into the host cell	[140]
	HSV-1 blocks host cell transcription termination via the viral immediate-early protein ICP27	[141]
	HSV enters the epithelial cells through the endocytic pathway and the neuronal cells through the membrane fusion pathway	[142]
	- HSV-1 prevents the production and blocks the accumulation of IFN-β mediated by the IRF-3 signaling pathway - HSV-1 infection blocks the nuclear accumulation of activated IRF-3 but does not block the initial virus-induced phosphorylation of IRF-3	[143]
	HSV-1 infection blocks the signaling impacts of α/β-IFNs through reversing the impacts of the double-stranded RNA-activated protein kinase and by blocking the phosphorylation of STAT1 and STAT2	[144]
Hepatitis	HAV enters the bloodstream through a poorly understood process to have access to the target organ, the liver and for its propagation	[145]
	HBV utilizes several pathways to harness host innate immunity to improve its duplication	[146]

(Continued)

TABLE 19.1   (Continued)

Viral infections	Mechanism of action	References
	The cell entry process of HBV infection involves a noncell-type precise main attachment to the cell-connected heparin-sulfate proteoglycans	[147]
	HCV particles interact with numerous receptors to prompt conformational changes and proceeds to enter the cell through clathrin-mediated endocytosis	[148]
Ebola	Ebola infection leads to host T cell apoptosis, neutrophilia, cytokine storm, disseminated intravascular coagulation, humoral immune response, cell-mediated, liver necrosis and lymphoid tissue necrosis	[149–152]
	The Ebola infection results in the activation of type-I effector mechanisms as well as strong T cell activation and differentiation	[153]
	Ebola virus can affect the mononuclear phagocytes differentiation that might interfere with their capacity to identify and present the antigen	[154]
	The Ebola virus primary attacks dendritic and macrophages immune cells. It attacks the kidneys and spleen, by killing cells that help the body to regulate its fluid and chemical balance, which help the blood to clot. Furthermore, the Ebola virus causes the lungs, kidneys and liver to shut down their functions and the blood vessels to leak fluid into surrounding tissues	[155–157]
	Ebola virus enters the human body through mucosal surfaces and injuries in the skin or by direct parental transmission. Moreover, it enters the target cell by utilizing different uptake mechanisms which include receptor-mediated endocytosis, lipid raft and micropinocytosis	[158,159]
	Ebola virus prevents IFN signaling through its VP24 protein that blocks the nuclear accumulation of tyrosine phosphorylated STAT1	[160]
	The cell entry of Ebola virus includes virus binding to the cell surface receptors followed by internalization via processing by endosomal proteases, macropinocytosis and transport to Niemann-Pick C1 (which is an internal receptor for Ebola virus) containing endolysosomes	[161]
HPV	Several HPV types enters the cell through a clathrin-dependent endocytic mechanism	[162]
	HPV protects the viral nucleic acid and provides the initial interaction site of the viral particle with the host cell. The virus is internalized and its coat is disassembled to allow the encapsidated genome access to the cellular transcription and replication machinery	[163]
	HPV may infect cells via damaged skin tissue. When the damage is deep enough and reaches the basement membrane, the virus can infect keratinocytes	[164]
	HPV induce immune evasion of the infected cells, which allow the virus to be untraceable for a long time	[165]
	HPV primarily infect the basal layers of the epithelium via micro-wounds and enter cells through interaction with certain receptors including α-6 integrin for HPV16	[166]
	E6 and E7 are among several proteins that HPV encodes They prevent the cell cycle arrest mechanisms, support carcinogenesis as they drive cellular proliferation and block apoptosis. The mechanisms of E6 and E7 prevent the normal regulations on the cell-cycle-progression system and accelerate the rate of cell division, thus leading to unchecked cell growth that can lead to cancer	[167]

(Continued)

TABLE 19.1 (Continued)

Viral infections	Mechanism of action	References
Viral pneumonia	Viral pneumonia infects the lung epithelium and prompts an inflammatory cascade	[168]
	Viruses may infect the low respiratory tract, causing bronchitis, bronchiolitis and pneumonia	[169]
Common cold	Respiratory infections affect the respiratory tract by a direct viral infection and/or by damage from the immune system response	[170]
	Common cold viruses are primarily cytolytic and destroy the cells of the host in which they multiply. They are highly infective and readily transmit to susceptible new hosts	[171]
COVID-19	Transmission of the virus can occur through direct or indirect contact with an infected individual or spread through inhalation and/or respiratory droplets released during coughing, sneezing, touching or talking	[172,173]
	The primary step in infection is virus binding to a host cell via its target receptor	[174]
	The S-protein is integrated over the surface of the virus; it mediates attachment of the virus to the host cell surface receptors and fusion between the viral and host cell membranes to facilitate viral entry into the host cell. It primary binds to a receptor on the host cell surface via its S1 subunit and then fuses viral and host membranes via its S2 subunit	[175,176]
	SARS-CoV is known to be able to enter host cells through directly fusing with the host membrane and the endosomal pathway through cathepsin B and L	[177]
MERS	The virus infects epithelial cells of the upper and lowers respiratory tract via endocytosis, as the disease gets worse	[178,179]
	Viruses may damage ciliated cells, which results in ciliostasis and thus, deterioration of mucociliary clearance	[180]
	MERS-CoV may enter the cell via an auxiliary pathway on the cell surface through transmembrane proteases	[181]

19.4 Challenges in the treatment of viral infections

Viral infections causes a burden in the global public health system and there is a rapid increase in the number of cases and deaths caused by these infections. Recently, a concerning and uncontrollable COVID-19 outbreak caused havoc (economically and social destruction) to the world and showed that there is a pressing need to be a step ahead and develop novel and effective viral therapies that can eradicate or at least control these viral infections. To develop better and effective models for the development of antiviral therapeutic agents, an overview of challenges hampering the treatment of viral infections need to be evaluated. Generally, there are several factors such as toxicity leading to side effects, short half-life, low bioavailability, poor solubility, poor delivery systems, drug-drug interaction, drug resistance, dose-dependent agents, etc. hinders the efficacy of most of the antiviral agents [182−186].

Drug resistance is a major challenge in the treatment of viral infections especially in patients with compromised immune systems. The prolonged use of antiviral agents, genetic variability (genotypes are different from patient to patient and region to region), short half-life, etc. results in treatment failure [187,188]. For instance, some therapies used such as antiretroviral therapy for HIV, direct-acting antivirals for HCV, etc. suffer from drug resistance due to change in viral mutation strains resulting in treatment failure [188–190]. Therefore, the development of multidrug agents to overcome drug resistance must be utilized urgently. However, these multidrug agents can develop multidrug resistance, drug toxicity due to high dose and dosing frequency leading to severe side effects [188–191]. Furthermore, drug toxicity is commonly caused by the use of drugs with a short half-life resulting in an increase in dose, the duration of treatment (long-term) which sometimes results in patients stopping following their daily medical prescriptions, adverse drug-drug interactions which are as a result of antiviral drugs interacting with regularly prescribed medicines [191].

In addition, the drug delivery system is another limiting factor when it comes to the treatment of viral infection as it also results in some organ toxicity and treatment failure [185]. The low bioavailability of antiviral agents which result in poor solubility, poor permeability compromised their efficacy as the amount of drug reaching the bloodstream decreases. Moreover, it is difficult to treat some viruses such as Ebola, HIV, etc. which are found in distal sites such as the lymphatic system, central nervous system (CNS), synovial fluid, etc. since the amount of drugs reaching those regions is limited due to the poor permeability of some antiviral. These aforementioned drug delivery incidents result in increased doses and dosing frequency which can lead to some side effects and drug resistance [185,190].

The other flaw in the treatment of viral infection is the lack of multitargeting drugs and selectivity of antiviral drugs as each virus consists of unique mechanisms (unique viral cycle) [192,193]. The lack of enough resources for clinical trials for some viral infections is one of the limitations hindering the development of novel antiviral agents. The lack of animal models for the CNS also leads to treatment failure and adverse side effects as a result of human studies which are conducted without preclinical animal efficacy testing and safety [194]. Inventing new therapies is quite expensive and time-consuming resulting in fewer therapeutic agents reaching the market especially in developing countries with fewer resources [195,196].

19.5 Combination therapy (fixed-dose combination) for the treatment of viral infections

One of the public health problems is the resistance of viral diseases to monotherapy. However, combination therapy (FDC) has been approved and reported by several scientists as one of the promising strategies which can prolong the life expectancy of people living with viral infections such as HIV, HCV, HPV, respiratory infections, etc. [197–200]. Since FDC is a combination of two or more antiviral drugs in a single dose form, it offers several advantages which include reduced resistance as drugs with different modifications are combined without changing their pharmacokinetics, they can target different sites, affordability, better tolerability and adherence as a person can take one pill per day instead of many pills [199,201,202]. Thus, several antiviral drugs were developed using this strategy and approved by FDA for the treatment of viral infections. However, in some cases, FDC can increase the prevalence of age, cause drug-drug interactions, and can be challenging on patients with some comorbidities and can result in larger pills which might reduce tolerability in some patients [198,203–205].

FDC is beneficial for people living with HIV and single tablet regimes (STRs) are recommended over multiple tablet regimes (MTRs) [198,204,205]. Furthermore, with the use of cART in people living with HIV, the socio-economic burden to the healthcare facilities has been reduced thus developed countries are focused on HIV related issues such as lifestyle-related comorbidities such as, osteoporosis, cardiovascular disease, diabetes mellitus, frailty and renal impairment HIV-related neurological disease, or associated opportunistic infections, which include coinfections, and primarily age [198,203–205].

When HIV mono-therapeutic inhibitors such as zidovudine, lamivudine, stavudine, etc. suffer from drug resistance, several FDC compounds were developed and approved. The first developed FDC anti-HIV drug was Combivir in 1997 a combination of zidovudine and lamivudine. Kaletra (a combination of ritonavir and lopinavir) was also developed as an FDC compound [206]. A combination of zidovudine, lamivudine, and abacavir sulfate (Trizivir) was also developed and these compounds were responsible for the inhibition of viral infection replication [206]. Promising evidence from these compounds resulted in several developments of anti-HIV agents through FDC therapy, specifically for HIV-1 infection [206]. Furthermore, an HIV drug, ibalizumab was approved in 2018 as a monotherapy drug and it was also used in combination therapy to treat multidrug resistance HIV-1 infection [206]. Moreover, other drugs namely cimduo, delstrigo, biktarvy, symtuza, symfi and symfi Lo, dovato, and fulyzaq were all developed and approved as STRs to treat and inhibit the multiplication of HIV-1 infection with fulyzqa responsible for the treatment of diarrhea in patients living with HIV and AIDS (as shown in Table 19.1) [206]. Other developed and approved drugs through FDC therapy for the treatment of HIV infections include epzicom, Truvada, Atripla, complera, stribild, triumeq, dutrebis, evotaz, genvoya, prezcobix, descovy, odefsey, juluca, and delstrigo (as shown in Table 19.2) [206].

Increasing cases of hepatitis with two common hepatitis types (HBV and HCV) which are responsible for the rapidly increasing numbers. Most HBV cases are being treated using monotherapy drugs and HCV cases are treated using combination therapy direct-acting antivirals (DAAs). The current concerning outbreaks that occur in 2015 were caused by HCV [207]. Thus, the use of DAAs with a different mode of actions and targets was implemented to combat this HC virus. Furthermore, the DAAs are reported as highly effective, have good tolerability, reduced side effects, short duration of treatment, are safe, and improve patient adherence. Thus, there are several developed and FDA-approved combination drugs for HCV that were reported since 2011, and the DAAs approach is considered as a promising strategy to reach the goal of eliminating HCV by 2030 according to the WHO [206–209]. Harvoni, Vieira Pak, daklinza/sunvepra, vanihep, technivie, daklinza/sovaldi, zepatier, epclusa, vosevi, and mavyret are the FDA approved drugs which are developed using the monotherapy approved drugs (as shown in Table 19.3) for the treatment of HCV. Furthermore, the advantage of these DAAs is that they can treat different

TABLE 19.2   Representing anti- HIV FDC regimens.

FDC drugs	Combined agents
Combivir	Zidovudine + Lamivudine
Kaletra	Ritonavir + Liponavir
Trizivir	Combivir + Abacavir sulfate
Cimduo	Lamivudine + Tenofovir Disoproxil Fumarate
Delstrigo	Doravirine + Cimduo
Biktarvy	Emtricitabine + Bictegravir + Tenofovir Alafenamide
Symtuza	Tenofovir Alafenamide + Darunavir + Emtricitabine + Cobicistat
Symfi	Efavirenz + Lamivudine + Tenofovir Disoproxil Fumarate
Fulyzaq	Catechin + Gallocatechin + Epicatechin + Epigallocatechin
Dovato	Dolutegravir + Lamivudine
Symfi Lo	Tenofovir Disoproxil Fumarate + Efavirenz + Lamivudine
Epzicom	Abacavir sulfate + Lamivudine
Truvada	Emtricitabine + Tenofovir Disoproxil Fumarate
Genvoya	Emtricitabine + Cobicistat + Tenofovir Alafenamide Fumarate + Elvitegravir
Complera	Emtricitabine + Rilpivirine Hydrochloride + Tenofovir Disoproxil Fumarate
Stribild	Cobicistat + Elvitegravir + Emtricitabine + Tenofovir Disoproxil Fumarate
Triumeq	Abacavir sulfate + Dolutegravir Sodium + Lamivudine
Dutrebis	Lamivudine + Raltegravir
Evotaz	Atazanavir sulfate + Cobicistat
Atripla	Tenofovir Disoproxil Fumarate + Efavirenz + Emtricitabine
Prezcobix	Cobicistat + Darunavir Ethanolate
Descovy	Emtricitabine + Tenofovir Alafenamide Fumarate
Odefsey	Emtricitabine + Rilpivirine hydrochloride + Tenofovir Alafenamide Fumarate
Juluca	Dolutegravir + Rilpivirine

genotypes as a single drug consisting of different modes of action as HCV contains seven genotypes [206,208,209].

The search for effective treatments of COVID-19 is required. The treatment emergency combination of available drugs seems like a possible approach [210,211]. Purwati et al. reported potential anti-COVID-19 drugs using FDC therapy [210]. Moreover, seven drugs including Hydroxychloroquine—Azithromycin (HCA), Hydroxychloroquine—Doxycycline (HCD), Lopinavir—Ritonavir-Clarithromycin (LRC), Hydroxychloroquine—Favipiravir (HCF), Lopinavir—Ritonavir—Azithromycin (LRA), Favipiravir (FAVI)-AZI,

TABLE 19.3 Representing anti- hepatitis FDC regimens.

FDC drugs	Combined agents
Harvoni	Ledipasvir + Sofosbuvir
Vieira Pak	Dasabuvir sodium + Ombitasvir + Paritaprevir + Ritonavir
Daklinza/sunvepra	Daclatasvir Dihydrochloride + Asunaprevir
Vanihep	Vaniprevir + Peginterferon alpha-2a + Ribavirin
Technivie	Ombitasvir + Paritaprevir + Ritonavir
Daklinza/sovaldi	Daclatasvir Dihydrochloride + Sofosbuvir
Zepatier	Elbasvir + Grazoprevir
Epclusa	Sofosbuvir + Velpatasvir
Vosevi	Sofosbuvir + Velpatasvir + Voxilaprevir
Mavyret	Glecaprevir + Pibrentasvir

Hydroxychloroquine—Lopinavir—Ritonavir (HCLR), and Lopinavir—Ritonavir—Doxycycline (LRD) were reported in this study. These aforementioned drugs display low toxicity than their parent drugs (single drugs) and were effective against coronavirus in the following order LRA> LRD> HCA> HCF> LRC> HCD, displaying a potential COVID-19 treatment [210]. Another report by Han et al. showed two FDC drugs where sofosbuvir is combined daclatasvir and lopinavir combined with ritonavir, respectively for the treatment of COVID-19 and sofosbuvir/daclatasvir was found to be effective with less duration of hospitalization and higher hospital discharge whereas lopinavir/ritonavir exhibited no clinical recovery or viral clearance [211].

FDC regimes Xerese is developed by the combination of acyclovir and hydrocortisone and is used to treat herpes infections [206]. Another FDC drug (a combination of celecoxib and famciclovir) was reported by Marchenkova et al. for the treatment of fibromyalgia which is caused by the reactivation of HSV type 1. This drug is safe with an improved tolerability profile and it is still in phase IIA of clinical practice [212]. Moreover, most of the viral infections such as Ebola, HBV, HPV, viral pneumonia, and some respiratory infections are still treated using mono-therapeutic agents [206]. However, there is limited literature on FDC therapy for the aforementioned therapies. In essence, these regimens showed that FDC is a promising approach in the treatment of viral infections as they improve the efficacy of ARV agents, can overcome resistance caused by mutations, reduce toxicity and costs, and increase tolerability and adherence as it is easy to take a single pill than multiple pills. However, more studies are needed to be conducted to develop more STRs for other viral infections.

19.6 Hybrid compounds designed for the treatment of viral infections

Combinational therapy is utilized to overcome the emergence of drug resistance in the treatment of complex diseases caused by viral pathogens. However, due to the differences in stability, solubility, and pharmacokinetic profile of the combined drugs, selecting drugs and doses for best combinational therapy can be difficult. Furthermore, clinical development of optimal combination therapy is expensive and also drug-drug interactions may result in additive toxic side effects. In the field of medicinal chemistry, the hybridization of active chemical scaffolds is one of the most promising and fundamentally unique techniques for the design of new lead structures and the development of new and effective drugs [213]. Combining two or more pharmacophores in one molecule (i.e. hybrid), have the potential to overcome these drawbacks of combinational therapy [214,215]. Hybrid molecules are tested in clinical trials for a variety of diseases, including those caused by drug-resistant organisms, indicating that hybridization is a potential technique for developing novel therapeutic drugs [216–218]. Hybrid compounds are categorized based on their chemical compositions, which may include the whole chemical structure(s) of the parent compound(s) or simply a portion of the parent compound(s). Direct conjugation via linking arms or combining various haptophoric moieties of two or more different drugs can all be used to combine two or more compounds. Hybrid compounds can be categorized in the following ways (Fig. 19.8) [219,220]: (i) **Cleavable hybrids**: are usually made up of amides, esters, or carbamates that are either directly linked to each other or linked with spacers that are usually enzymatically hydrolysable amides, esters, or carbamates. These hybrid compounds are expected to have two distinct mechanisms of action. (ii) **Non-cleavable hybrids**: are linked together by an enzymatically noncleavable linker, they also retain the pharmacological properties (such as affinity for targets) of two or more pharmacophores into a single chemical structure. (iii) **Merged or Overlapping Hybrids**: are created by overlapping two pharmacophores' common structural motifs. The functional features of the parental compounds may or may not be retained in these hybrids [219].

The synthesis of these novel hybrid molecules is described below, as well as their antiviral effectiveness against HIV, herpes, hepatitis, Ebola, HPV, viral pneumonia, respiratory infections (common cold, COVID-19, MERS).

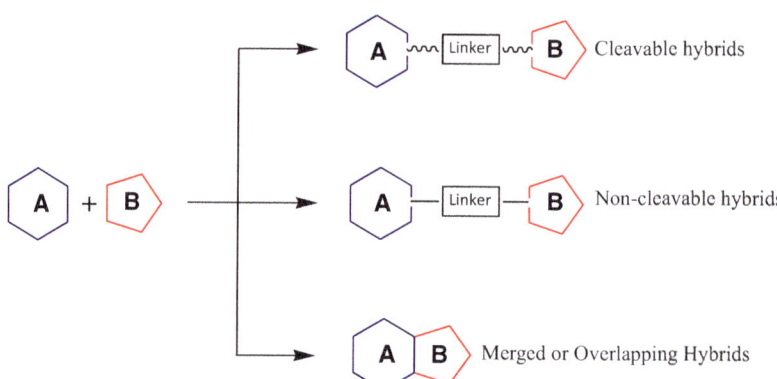

FIGURE 19.8   Classification of different types of hybrids.

19.6.1 Anti-human immunodeficiency virus hybrids

According to the literature, various hybrid compounds have shown excellent anti-HIV potential (Fig. 19.9). 4-phenyl-1H-1,2,3-triazole phenylalanine hybrid 1 (EC_{50} = 3.13 μM) demonstrated potent activity against HIV-1. This hybrid exhibits the effects in both the early and late stages of HIV-1 replication, according to the mechanism of action analysis [221]. Wu et al. also synthesized 1,2,3-triazole-containing phenylalanine hybrid compounds via CuAAC reaction. Hybrid 2 (EC_{50} = 4.33 μM, SI > 13.33), showed the best anti-HIV-1 activity among the synthesized hybrids and was better than the lead PF-74 (EC_{50} = 5.95 M, SI > 11.85) [222]. Jana et al. used a multicomponent organocatalytic procedure to produce a range of novel functionalized artemisinin derivatives. In a single step, a library of dihydroartemisinin fused and 1,5-disubstituted 1,2,3-triazole derivatives were developed and evaluated for their anti-HIV potential against MT-4 cells. Three of the dihydroartemisinin triazole compounds (compounds 3–5, Fig. 19.2) produced had half maximum IC_{50} values ranging from 1.34 to 2.65 μM [223]. Novel hybrids N'-benzylidene-2-(5-(4-chlorophenyl)-3-phenyl-4,5-dihydropyrazol-1-yl)thiazole-4-carbohydrazide Schiff bases containing 1,3-thiazole and 2-pyrazoline (compounds 6 and 7) possess potent activity against HIV1 replication with IC_{50} values of 0.50 and 0.45 μM, SI = 3 and 5, respectively. The SAR revealed that the 4-fluoro- or 2-hydroxy-4-chlorobenzylidine substituents at C-4 of the thiazole ring, were well tolerated in the hydrophobic binding pocket of HIV-1 RT and showed potent activity [224]. Chander et al. designed and synthesized a series of 5-benzoyl-4-methyl-1,3,4,5-tetrahydro-2H-1,5-benzodiazepin-2-one hybrid compounds for the search of potential HIV-1 RT inhibitors. Hybrids 8 and 9 (IC_{50}: 8.62 and 6.87 μM) indicated the most potent activity and according to SAR analysis of the studied hybrid compounds, the benzoyl ring with no substitution, methyl at the ortho, and chloro substitution at ortho and meta positions all favored the RT inhibitory potency. The docking analysis demonstrated that hybrid 9 had more prominent hydrophobic and hydrogen bonding interaction than hybrid 8 which could explain its superior in vitro potency [225].

The anti-HIV-1 activity of a series of 2-benzoxazolinones-based hybrid compounds as a potential scaffold for INSTI activity was examined by Safakish et al. and all the hybrids were active against HIV-1 at 100 μM. The most active hybrid (10, 84% at 100 μM) interacts with two Mg^{2+} cations at the IN active site *via* its 2-aminophenyl-1, 3, 4-thiadozole ring. According to the results of the docking analysis, the benzoxazolinone scaffold interacts with viral DNA in the IN active site in π–π stacking interactions [226]. Bielenica et al. have worked with a number of diaryl hybrids containing 1,3-thiazolidin-4-one and 1H-tetrazol-5-yl scaffolds to evaluate the anti-HIV-1 and Cytotoxicity activity. The hybrid N-(4-nitrophenyl)-1H-tetrazol-5-amine 11 (HIV-1: EC_{50} = > 3.2 mM) was the most promising compound and was more active than the origin thiourea. The MTT assay was further used to assess its cytotoxicity against HaCaT, HTB-140 and A549 cancer cell lines. The hybrid has a strong effect on tumor cells, but not on normal cells [227].

Two novel betulinic acid-nucleoside hybrids 12 (IC_{50} = 0.0078 μM, CC_{50} = 9.6 M) and 13 (IC_{50} = 0.020 μM, CC_{50} = 23.8 M) were very effective against HIV but displayed lesser cytotoxicity when compared to AZT and DSB [228]. Gao et al. designed novel diarylpyrimidine hybrids and tested them in human T-lymphocyte (MT-4) cells for anti-HIV activity. Out of the designed hybrids, compounds 14 and 15 were the most potent compounds with

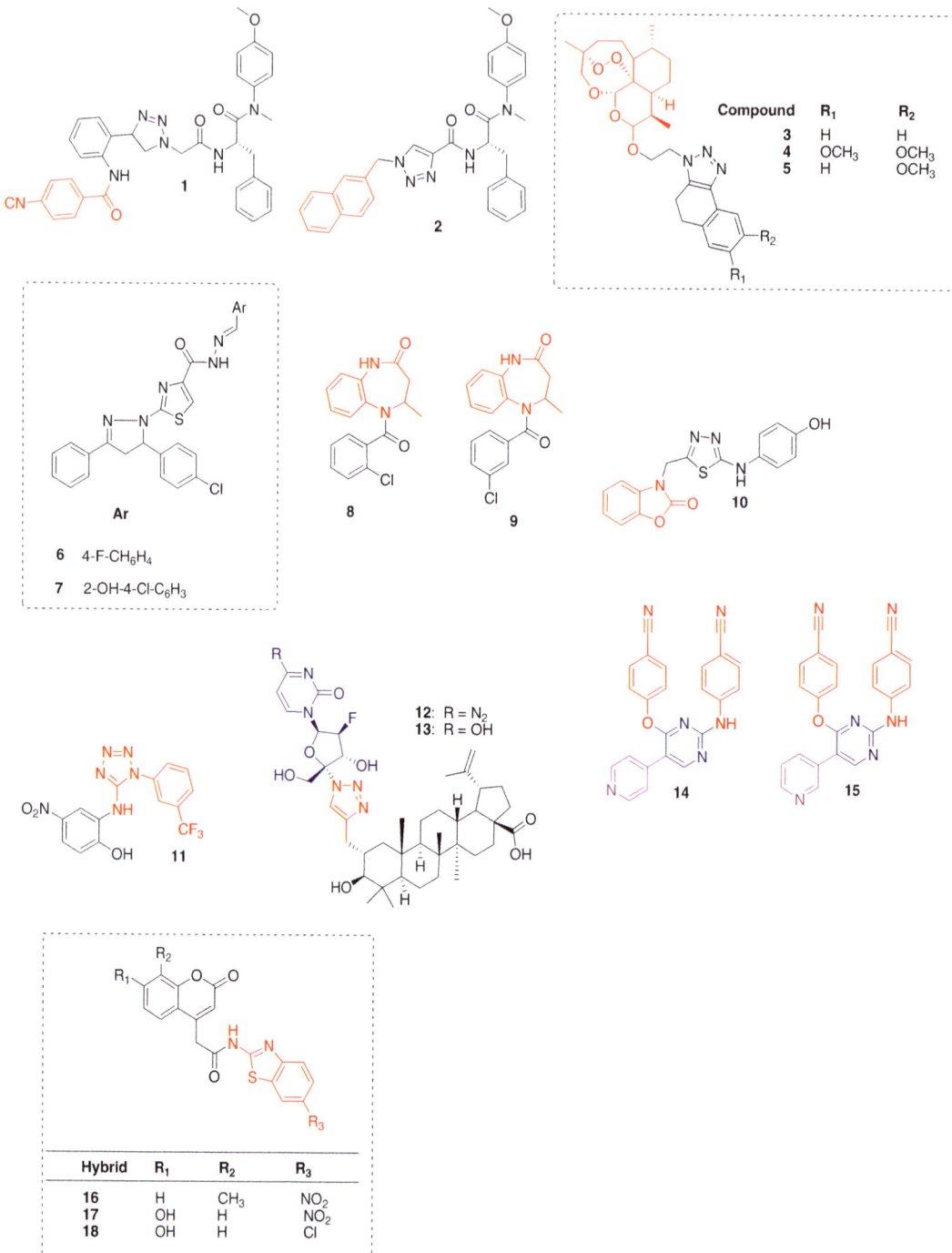

FIGURE 19.9 Hybrid compounds with potential anti-HIV activity.

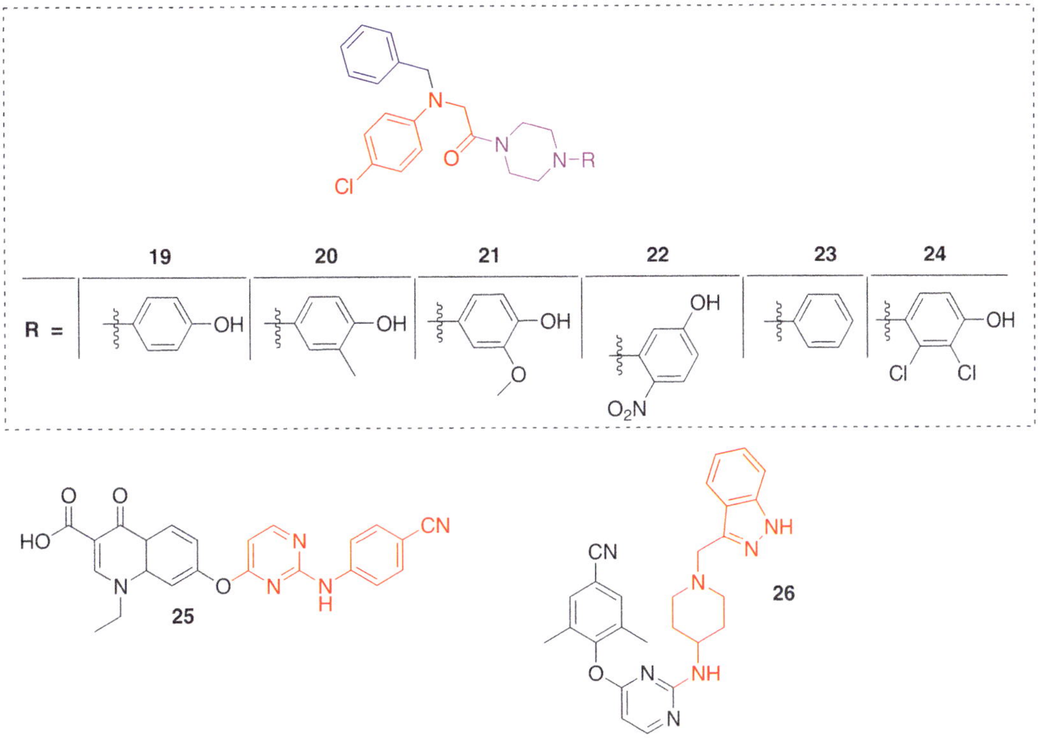

FIGURE 19.9   Continued.

anti-HIV-$_{1IIIB}$ efficacy equivalent to Efavirenz and Etravirine and superior to Nevirapine. Compound **14** showed potential RT inhibition [229]. In another similar study conducted by Shirvani and colleagues, molecular docking experiments were used to generate a new series of novel 4-nitroimidazole hybrid compounds containing a 5-hydroxy-4-pyridinone moiety. However, no activity was reported in an anti-HIV-1 assay of the produced hybrid compounds. Despite this, some hybrids indicated high cytotoxicity against MT-4 cells and were chosen for an antiproliferative test. They showed exceptional anticancer efficacy against leukemia MOLM-13 and K562 cells with low IC$_{50}$ values. These newly designed hybrids are likely to block tyrosine kinase enzymes, and they deserve further investigation in the hopes of generating novel promising tyrosine kinase inhibitors [230]. Bhavsar et al. generated a simple and effective synthetic method to design (N-1,3-benzo[d]thiazol-2-yl-2-(2-oxo-2H-chromen-4-yl)acetamide) hybrid compounds that show in vitro anti-HIV potential using MTT method. Three hybrid compounds, **16**, **17**, and **18**, have been reported to be potently active against wild-type HIV-1 cells with the EC$_{50}$ values of 9, 8 and 7, respectively and the activity was significantly lower than that of zidovudine. This technique is significant because it produces high yields and is simple to set up for biologically active compounds [231].

Chander et al. developed 14 new 2-(benzyl(4-chlorophenyl)amino)-1-(piperazin-1-yl) ethanone hybrid compounds as HIV-1 RT inhibitors in which hybrid **19–24** showed substantial inhibition of HIV-1 RT ($IC_{50} \leq 25\,\mu M$). Furthermore, with the exception of hybrid **21**, all of the top five RT inhibitors assessed for anti-HIV-1 activity retained high anti-HIV-1 efficacy while maintaining a strong safety index. Docking analysis of hybrid **24** against wild HIV-1 RT reveal extensive pi-pi stacking and hydrophobic contacts within NNIBP of a selected RT strain, which may be responsible for its great receptor binding affinity and, as a result, significant HIV-1 RT inhibitory efficacy [232].

Diarylpyrimidine—quinolone hybrids were developed and tested against both wild-type and mutant HIV-1 strains by Mao et al. Against HIV-1 III$_B$, the most active hybrid **25** had an EC_{50} value of $0.28 \pm 0.07\,\mu M$. A RT-targeted mechanism of action was identified by a couple of enzyme-based tests. Despite the bulky and polar features of a quinolone 3-carboxylic acid moiety in the molecules, docking analysis demonstrated that these hybrids may be well localized in the NNIBP of HIV-1 RT [233].

Indazolyl-substituted piperidin-4-yl-aminopyrimidines were developed and produced in a range of structurally varied hybrids. The biological activity data demonstrated that all of the target compounds had good to exceptional activity against the WT HIV-1 strain. Compound **26** was found to be the most effective hybrid against WT HIV-1 ($EC_{50} = 6.4$ nM, $SI = 2500$). It also had strong activity against K103 N ($EC_{50} = 0.077\,\mu M$), Y181C ($EC_{50} = 0.11\,\mu M$), and E138K ($EC_{50} = 0.057\,\mu M$), as well as moderate activity against RES056 ($EC_{50} = 8.7$ M) [234].

19.6.2 Anti-herpes simplex virus hybrids

HSV is an infectious disease that belongs to the Herpesviridae family. Humans are usually infected with HSV-1 and HSV-2. HSV 1 causes infections in the mouth, pharynx, face, and lips, whereas HSV-2 is linked to anogenital infections [235]. When the virus begins to shed from the human body, it appears to be the most contagious. Antiviral drugs such as acyclovir, famiclovir, and valacyclovir have been licensed for individuals with HSV infections. However, there is still a desire to discover novel anti-HSV compounds. Currently, Cunha et al. used spectroscopic methods to synthesis and characterize a series of novel hybrid compounds known as nitroxide-1H-1,2,3-triazoles, using single-crystal X-Ray diffraction (XRD) data to determine the crystal structure of several of them, which were then tested for antiviral HSV-1 activity. Compounds **27**, **28**, **29**, and **30** had the best anti-HSV-1 activity among these hybrid compounds, with compound **29** having the best antiviral HSV-1 activity with an IC_{50} of $0.80\,\mu M$ (Fig. 19.10) [236]. Pandey prepared benzimidazole derivatives containing triazoloquinolinyl, such as 7-hydroxy-4-methyl-8-(amino benzimidazolyl)-quinolinyl-1,5c-2-mercaptotriazoles. Compound **31** demonstrated promising results against the Japanese encephalitis virus and the HSV-1 [237]. Kharitonova and colleagues described the synthesis of nucleoside-containing benzimidazoles hybrid compounds. When compared to acyclovir drug (SI > 256), 2-amino-5,6-difluoro-benzimidazole riboside hybrid **32** showed good anti-HSV action (SI > 32). Hybrid **32** could be utilized in some circumstances where acyclovir proved ineffective in treating viral infections [238].

Jordão et al. developed and tested a new set of arysulfonylhydrazide-1,2,3-triazole hybrid compounds for antiviral activity against HSV-1 replication in cell culture. The most promising hybrid compounds with potent anti-HSV-1 activity were compounds **33** and **34**.

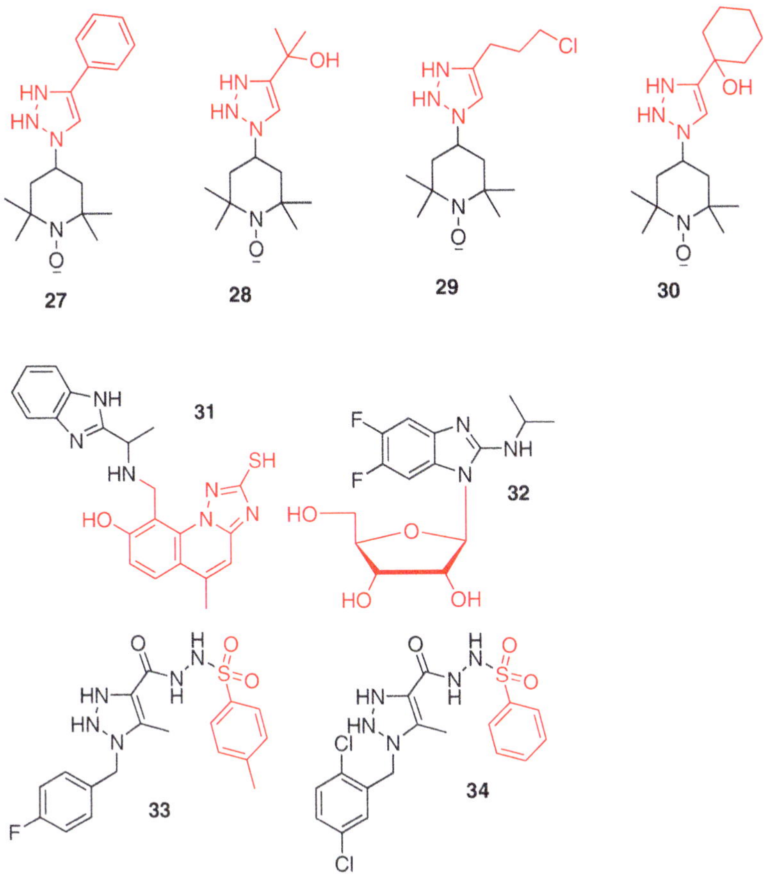

FIGURE 19.10 Hybrid compounds with potential anti-herpes simplex viral activity.

In comparison to the reference chemical acyclovir, these compounds displayed decreased cytotoxicity and greater selectivity indices [239].

19.6.3 Anti-hepatitis

Inflammatory liver disease is caused by the hepatitis virus. HAV, HBV, HCV, Hepatitis D (HDV), and hepatitis E virus (HEV) viruses are among the five types of hepatitis viruses. HAV and HEV are transmitted through the feces or orally. Both viruses are capable of causing severe sickness. HBV and HCV are hepatitis viruses that are transferred through the bloodstream and cause chronic hepatitis [240]. Hepatitis is divided into different families. HBA is a Piconaviridae virus, while HBV is a Hepadnaviridae virus, HCV is a Flaviviridae virus, and HEV is a Hepeviridae virus [241]. In 2015, the hepatitis virus was responsible for 1.34 million deaths, caused by HCV (30%), HAV (0.8%), HEV (3.3%) and HBV (66%) [242].

Wang et al. synthesized 46 hybrids of caudatin with substituted cinnamic acids and evaluated their anti-HBV activity in HepG 2.2.15 cells. The majority of the compounds showed significant anti-HBV activity, particularly blocking HBV DNA replication with IC_{50} values ranging from 2.44 to 22.89 μM. According to the acute toxicity investigation, the hybrid with the most potent activity was Compound **35**, which inhibited the secretion of HBsAg, HBeAg, and HBV DNA replication with IC_{50} values of 5.52, 5.52, 2.44 μM, respectively, and had good safety ($LD_{50} > 1250$ mg/kg) (Fig. 19.11) [243]. Zhang et al. synthesized flavonoid-triazolyl hybrids which were tested as potential anti-HCV agents.

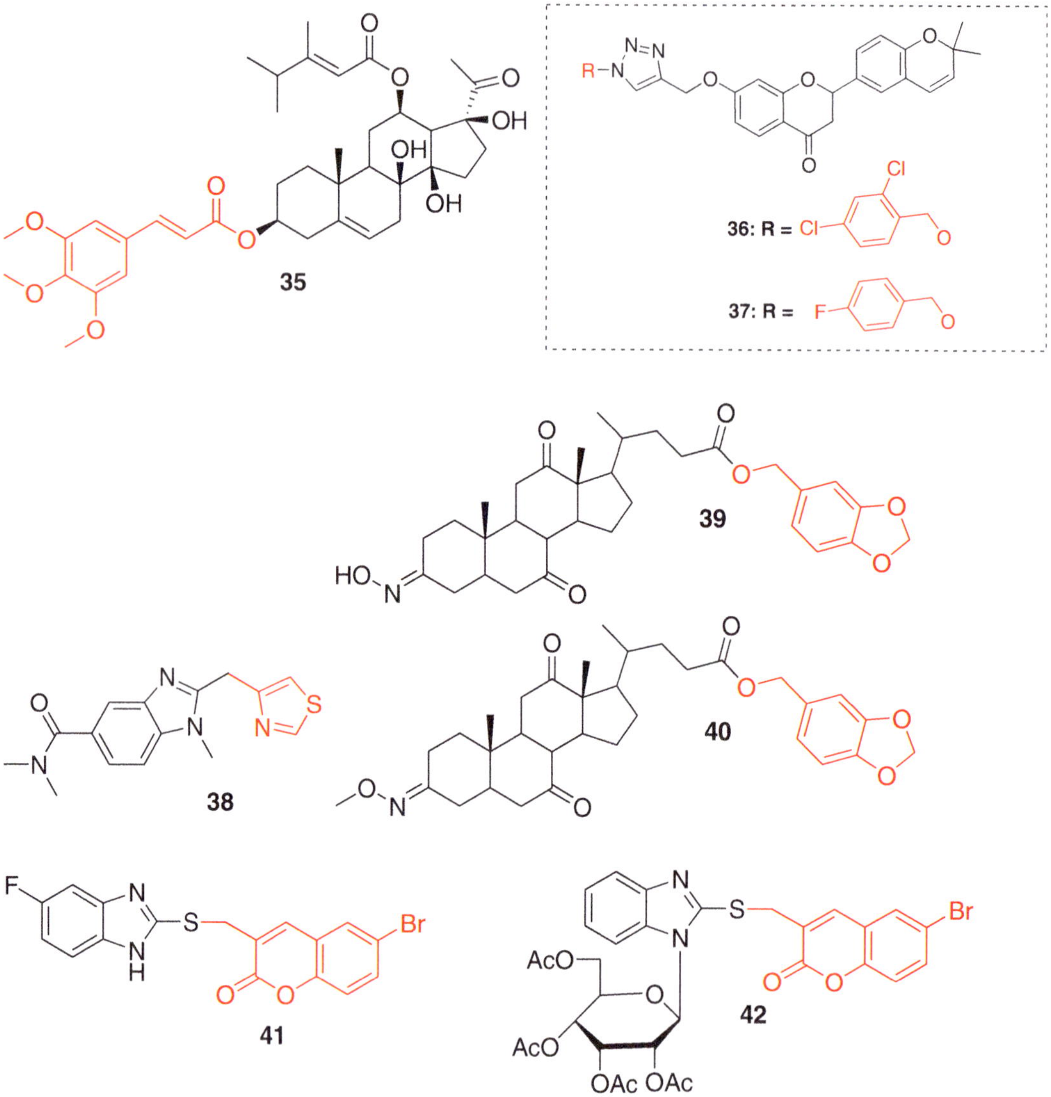

FIGURE 19.11   Hybrid compounds with potential Anti- Anti-hepatitis viral activity.

The majority of the synthesized compounds suppressed the formation of progeny virus at concentrations of 100 mg/mL. The anti-HCV activity of **36** and **37** was the strongest among these derivatives, and it suppressed HCV production in a dose-dependent manner. Additional action mechanism tests revealed that **36**, the most effective compounds. Compounds **36** and **37** blocked viral entrance by 34.0% and 52.0%, respectively [244].

Luo et al. designed and synthesized a variety of novel thiazolylbenzimidazole compounds and tested them on the HepG2.2.15 cell line for HBV activity and cytotoxicity. The compound indicating the most potent activity has been depicted as Compound **38** with $IC_{50} = 1.1 \mu M$ and SI > 90.9 [245].

Wei et al. synthesized twelve compounds from dehydrocholic acid, and their anti-HBV actions were tested in HepG 2.2.15 cells. The results revealed that five compounds inhibited HBeAg more effectively than the positive control, with **39** ($IC_{50} = 49.39 \pm 12.78 \mu M$, SI = 11.03) and **40** ($IC_{50} = 96.64 \pm 28.99 \mu M$, SI = 10.35) showing strong anti-HBV effects on decreasing HBeAg secretion compared to Entecavir ($IC_{50} = 161.24 \mu M$, SI = 3.72). Most of these compounds interacted with protein residues of heparan sulfate proteoglycan in host hepatocytes and bile acid receptors, according to molecular docking studies [246].

Hwu et al. successfully synthesized nineteen novel hybrid compounds from coumarin and benzimidazole derivatives using a one-flask method. To join these two types of derivatives, a methylenethio linker was used. Two of the most effective compounds **41** and **42** had EC_{50} values of 3.4 and 4.1 M, respectively, when tested against the HCV. Compound **42** reduced HCV RNA replication by 90% at a dose of 5.0 M and did not affect cell proliferation [247].

19.6.4 Hybrid compounds with anti-COVID-19

Verma et al. prepared a variety of indolo [3,2-c]isoquinoline (δ-carboline) derivatives with two pyrimidine and piperizine ring frameworks. The synthesized hybrids were evaluated for anticancer, antituberculosis, antimicrobial, antioxidant, effects, the replacement of fluorine, methyl, and methoxy of synthetic compounds with fluorine, methyl, and methoxy of synthetic compounds was proposed. Among these compounds, **43**, **44**, **45**, **46**, and **47** showed strong interactions with 6LZE (COVID-19) and 6XFN (SARS-CoV-2) at active sites, according to the molecular docking studies (Fig. 19.12) [248].

19.6.5 Ebola

Bessieres et al. designed and synthesized novel 2-substituted-6-[(4-substituted-1-piperidyl) methyl]-1H-benzimidazoles hybrids as Ebola virus inhibitors. Based on their spectral data and CHN investigations, the hypothesized structures of the newly produced benzimidazole-piperidine hybrids were confirmed. The anti-Ebola activity of the target compounds was tested in vitro. Compounds **48** ($EC_{50} = 0.93 \mu M$, SI = 10) and **49** (EC_{50} 140.64 μM, SI = 20) were equally powerful and selective against cell lines as Toremifene reference drug ($EC_{50} = 0.38 \mu M$, SI = 7) (Fig. 19.13). According to the data, **48** and **49** prevent EBOV infection by inhibiting viral entrance at the NPC1 level [249].

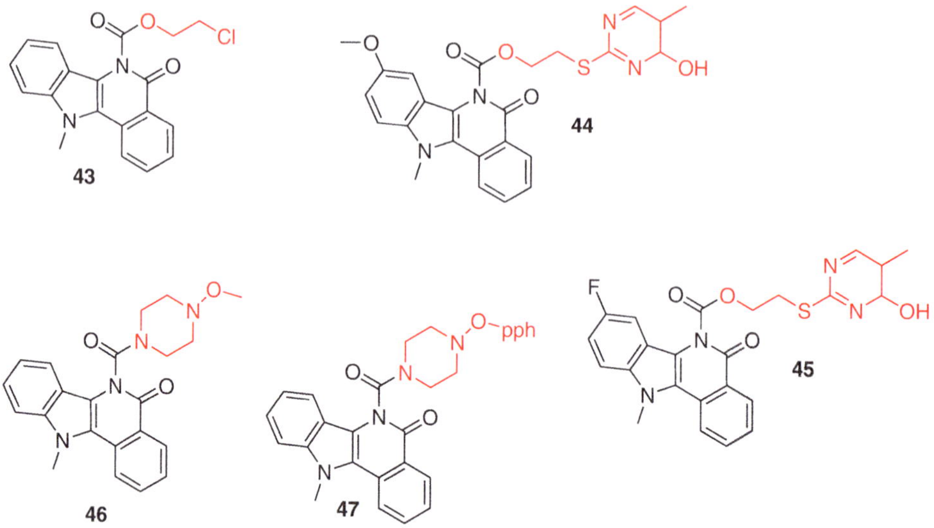

FIGURE 19.12   Hybrid compounds with potential anti-COVID-19 activity.

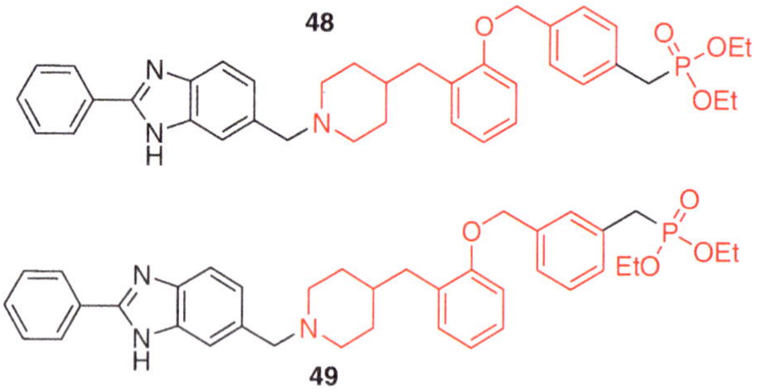

FIGURE 19.13   Hybrid compounds with potential Anti-ebola activity.

19.6.6 Human papilloma virus

Kazakova et al. synthesized a novel botulin-based hybrid **50** fused with nicotinoyl at positions C-3 and C28. The antiviral screening of hybrid **50** against HPV type 11 demonstrated the selectivity index of 35 with no cellular cytotoxicity and was also active against HCV replicon with the EC_{50} of 1. 32 μM (Fig. 19.14) [250]. Novel 1,3-oxazole compounds were synthesized by Kachaeva et al. and determine the in vitro antiviral activity of the compounds against HPV. The researchers used cytotoxicity tests to determine HPV-11 transient replication in transfected HEK293 cells and HPV-18 DNA amplification in an organotypic squamous epithelial raft culture of primary human keratinocytes (HK). In a transient DNA replication assay, compound **51** (EC_{50} = 6.13 μM) showed significant antiviral activity against low-risk HPV-11 and low cytotoxicity in HEK293 cells when compared to Cidofovir (IC_{50} = 148.00 μM), a clinically used antiviral drug [251].

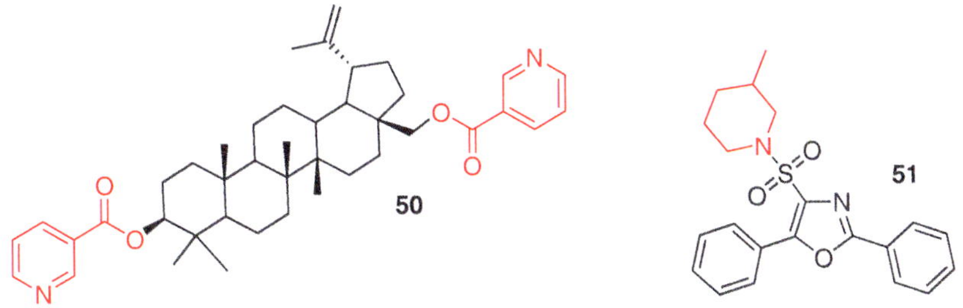

FIGURE 19.14 Hybrid compounds with potential Anti-HPV activity.

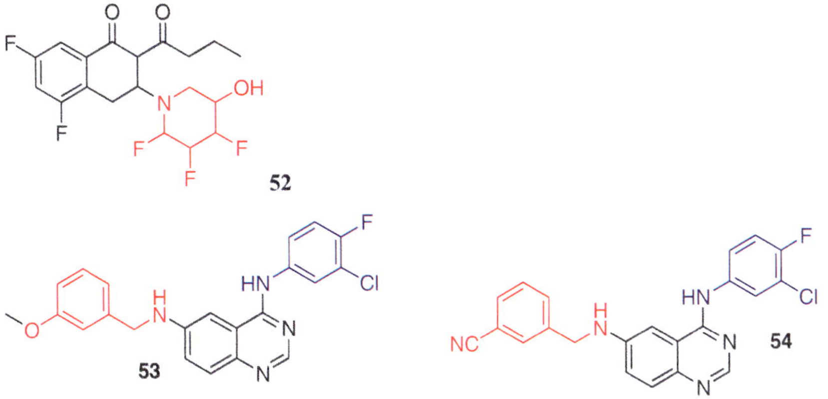

FIGURE 19.15 Hybrid compounds with potential anti-MERS activity.

19.6.7 Middle East respiratory syndrome

The anti-MERS-CoV inhibitory activity of synthesized 3-acyl-2-phenylamino-1,4-dihydroquinolin-4(1H)-one derivatives were evaluated. 6,8-difluoro-3-isobutyryl-2-((2,3,4-trifluorophenyl)amino) quinolin-4(1H)-one (**52**) displayed high inhibitory action ($IC_{50} = 0.086 \pm 0.041 \, \mu M$) and low toxicity ($CC_{50} > 25 \, \mu M$). It also displayed significant metabolic stability, a low hERG binding affinity, is noncytotoxic, and has excellent in vivo PK characteristics [252]. Lee et al. synthesized 4-anilino-6-aminoquinazoline derivatives and tested their anti-MERS-CoV activity. A hit drug for suppressing MERS-CoV infection was identified in a random screen, N-4-(3-Chloro-4-fluorophenyl)-N-6-(3-methoxybenzyl)quinazoline-4,6-diamine (**53**). Compound **54** was discovered to have a high inhibitory effect ($IC_{50} = 0.157 \, \mu M$, SI = 25) with no cytotoxicity and moderate in vivo PK characteristics during the optimization process (Fig. 19.15) [253] (Fig. 19.16).

19.7 Lipid-based drug delivery systems

19.7.1 Emulsion

Lipid-based drug delivery systems have been studied extensively as potential therapeutics for the treatment of viral infections (16). Emulsions, lipid-based drug delivery systems

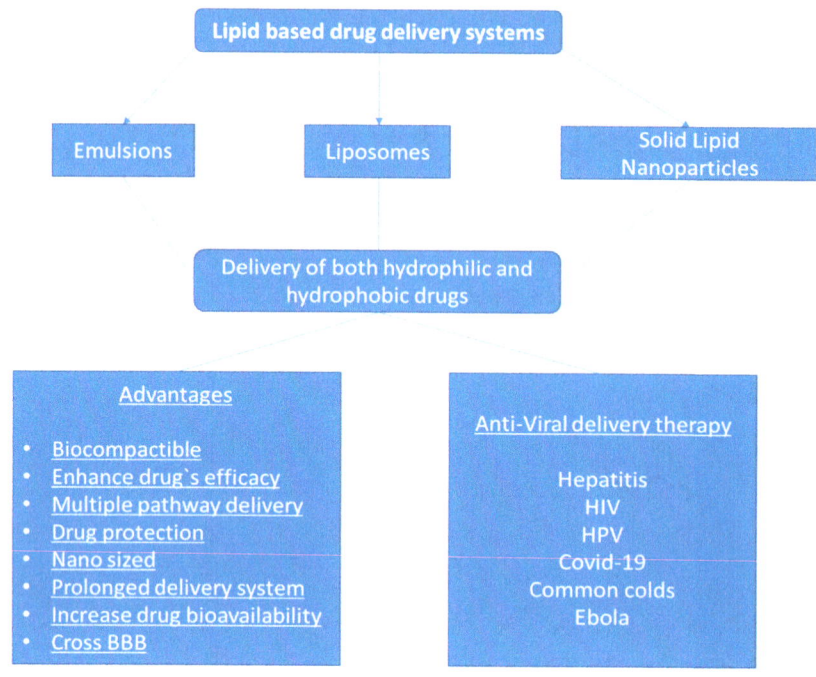

FIGURE 19.16   Lipid-based drug delivery systems.

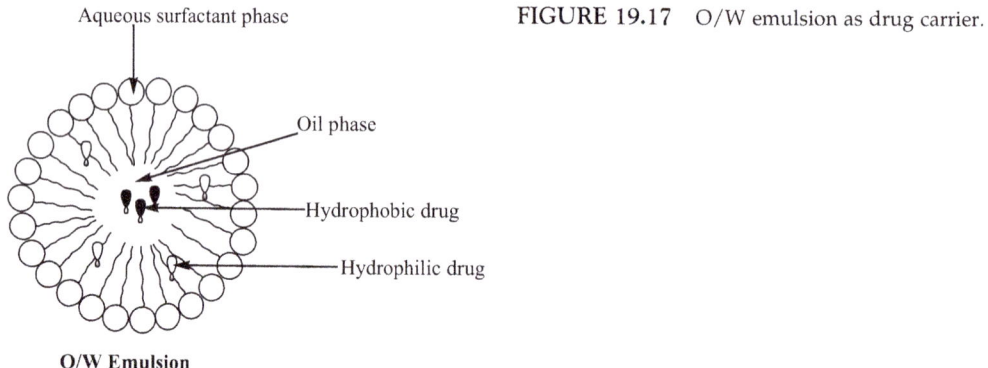

FIGURE 19.17   O/W emulsion as drug carrier.

can be categorized into different forms such as nano-emulsion, micro-emulsion, etc. depending on their nanoparticle size (Fig. 19.17). These formulations offer different advantages. Tamilvanan et al. reported micro-emulsion as self-micro emulsifying drug delivery systems which are clear and thermodynamically unstable colloidal systems [254] with the capacity to load lipophilic and hydrophilic drugs [255]. Nanoemulsions are drug delivery systems that are thermodynamically stable, obtainable as a single phase containing surfactant, water, oil, and cosurfactant [256]. Nanoemulsions have great potential in protecting, encapsulating, and delivering antiviral drugs to the targeted site. Emulsions can be primarily prepared at room temperature using (water, oil and emulsifiers) and are reported as oil in water (O/W) emulsion or water in oil emulsion (W/O). Furthermore, they can also be formulated as double solvent emulsions as water in oil in water (W/O/W) or oil in water in oil (O/W/O) emulsions, which are often unstable and require lipophilic and hydrophilic emulsifiers with the presence of high-pressure and high energy approaches systems [257]. Kotta et al. encapsulated efavirenz into nanoemulsions to improve its bioavailability in HIV therapy [258]. They prepared formulations with varying percentage compositions of surfactants (geucire and transcutol) and drug into oil (propylene glycol monocaprylate). The mean globule size of the formulations was between $65.483 \pm 1.978 - 26.427 \pm 1.960$ nm, with polydispersity index (PI) at the range of $0.517 \pm 0.0034 - 0.117 \pm 0.0034$. In vitro drug release studies revealed an initial release of approximately 40% within the 1st 35 minutes in all the formulations, which was later sustained until the 8th hour with F5 [transcutol HP (3.432% v/v), gelucire (13.728% v/v), capryol 90 (2.89% v/v), and double-distilled water (13.98% v/v)] exhibiting the highest drug release (90.02 ± 3.2). F5 was kept for further analysis due to its outstanding characteristics (least globule size ± 26 nm, PI of ± 0.117, and lowest viscosity ± 30 cP). Albino Wistar rats were used to check the plasma concentration-time profile of F5 versus efavirenz following an oral administration. A high concentration peak was observed for F5 compared to the efavirenz suspension (ES) in 4 hours, and the $AUC_{0 \to 24}$ was extremely significant for F5. In vivo bioavailability check revealed a greater extent of absorption of efavirenz from F5 formulation than in efavirenza suspension, showing a two-fold increase in bioavailability. The reported nanoemulsions in this study mostly F5 are promising scaffolds for the delivery of efavirenz for HIV therapy. However, further studies like cytotoxicity, anti-HIV activity, the mechanism involved in oral delivery of the drug and its protection from immune response still need to be discussed.

Saraf et al. prepared lipid microparticles (LM) loaded with hepatitis B surface antigen (HBsAg) for mucosal immunization using double solvent emulsion W/O/W and considerable used intranasal (IN) route [259]. The LM formulations were noted as LMST-HBsAg for LM with stearylamine, LM-HBsAg without stearylamine, alum-HBsAg with alum, and HBsAg for plain emulsion. The mean particle diameter (μm) was recorded as 1.61 ± 0.07 for LMST-HBsAg and 1.59 ± 0.06, with percentage entrapment efficiency of 67.90% and 58.17% respectively. LMST-HBsAg revealed a cumulative drug release % of 29.41 in 24 hours with mucoadhesion% of 42.21 ± 0.40, whereas LM-HBsAg presented a slightly higher release of 32.19% and 3.03 ± 0.033 mucoadhesin% which is significantly lower than that of LMST-HBsAg. The slow drug release of LMST formulation at pH 7.4 was linked to the ionic interaction of the cationic lipid in LMST and the protein antigen, and the higher release of LM formulation was linked to its particle size that may have resulted in an increased surface area [259]. Tracheal mucosa of the she-goat was used, and the % phagocytosis by alveolar macrophage displayed $64.67\% \pm 3.32\%$ and $73.5\% \pm 3.39\%$ in 1 hour of incubation for LM-HBsAg and LMST-HBsAg formulations, which increased to $74.6\% \pm 2.73\%$ and $88.16\% \pm 3.43\%$ on the 3rd hour, respectively. Following these percentages, the cell membrane as observed after 3 hours was slightly broken at certain areas with some visible granules. The in vivo drug delivery system was performed using Sprague-Dawley female albino rats and LM-HBsAg formulations were more distributed to the target site than LMST-HBsAg formulations, revealing localized uptake. The mucosal immune responses were in the following order: HBsAg < alum-HBsAg < LM-HBsAg < LMST-HBsAg and the same trend was observed for the systematic immune response. Further investigation still needs to be done to evaluate the significant efficacy of these formulations into the viral cultured cells and how delivering the drugs using double solvent emulsion improves the activity of the drug, or if there are any limitations.

Argenta et al. formulated W/O nanoemulsions containing isoflavone genistein for treatment of HSV using (dioleylphosphaditylcholine [DOPC] and distearoylphosphatidylcholine [DSPC]) as phospholipids and isopropyl myristate (IM) or castor oil (CO). They prepared 8 formulations (IM/DSPC/OLA = NE1, IM/DSPC/OA = NE2, IM/DOPC/OLA = N3, IM/DOPC/OA = N4, CO/DSPC/OLA = N5, CO/DSPC/OA = N6, CO/DOPC/OLA = N7, and CO/DOPC/OA = N8) of W/O nanoemulsion with cosurfactants (oleic acid [OA] and oleylamine [OLA]). The formulations exhibited varying mean droplet sizes with NE6 (426.40 ± 82.94 nm) being the highest and NE3 (161.19 ± 06.64 nm) significantly exhibiting the lowest mean size. Regarding PI formulations exhibited PI lower than 0.3, except for NE6 (0.58 ± 0.08) with NE1 (0.07 ± 00.02) exhibiting the lowest value. With regards to skin retention of genistein, formulations prepared from IM showed high retention in the skin with NE3 exhibiting $1.86 ± 0.32 \, \mu g/cm^2$. NE1 and NE3 were selected for further studies and Nile Red dye skin distribution results showed that the skin maintained a structure like that of untreated skin regardless of the two formulation's treatment, revealing little detachment on stratum corneum, cytoplasm with no vacuole, and the dermal-epidermal junction with no visible separation of the two layers. Nanoemulsion application exhibited higher intensity of the fluorescence than the control. Epidermis and dermis revealed a higher amount of genistein when it was incorporated into nanoemulsions, and this is linked to their cationic character which can interact with epithelial cell's negatively charged protein residues, resulting in increased drug permeability [260]. The antiherpetic activity (anti-HSV-1 and anti-HSV-2) of the two nanoemulsions decreased the IC_{50} values exhibiting significant inhibition of the tested viral plaque forms. Cytotoxicity results using Vero cells HSV-1 CC_{50} ($\mu g/mL$) showed that genistein (23.91 ± 1.12) exhibited higher cytotoxicity followed by NE3 (17.55 ± 2.82) and NE1 (16.59 ± 0.94). A similar trend was also observed using green monkey kidney (GMK AH1) cells HSV-2 CC_{50} ($\mu g/mL$) with genistein exhibiting (50.36 ± 7.97), NE3 (21.23 ± 0.61), and NE1 (19.88 ± 2.21). Cytotoxicity inhibition activity showed that the nanoemulsions exhibited higher activity against HSV-2 with IC_{50} values lower than 06.00 as compared to HSV-1 with IC_{50} values of ± 06.38. However, nanoemulsion without genistein exhibited no viral activity. These results suggest that the delivery of genistein using nanoemulsion significantly increased its activity revealing the synergistic effect of delivery the drug using a nanoemulsion and enhanced intracellular uptake of genistein due to their cationic character which may have increased the membrane permeability. Overall DOPC and IM nanoemulsion with cosurfactant OLA exhibited good results than IM, DSPC, and OA nanoemulsion, showing that the presence of cosurfactant OLA significantly played a vital role in drug delivery of these double solvent nanoemulsions. These results suggest that W/O/W nanoemulsions significantly enhance drug delivery to the targeted site improving its antiviral activity. However, the mechanism of action of the nanoemulsion still needs to be evaluated. Nanoemulsions are promising scaffolds for the treatment of herpes and the delivery of antiviral drugs, but it would be advisable to further evaluate them in vivo to compare them with the obtained in vitro results.

Kelmann et al. prepared O/W nanoemulsion for topical delivery of pentyl gallate (PG) for treatment of human herpes labialis [261]. The nanoemulsions were composed of 1:1 medium-chain triglyceride: castor oil mixture, soybean lecithin (SL), polysorbate 80 (PL80), and vitamin E. Nanoemulsion formulations were coded as F1 (CO: MCT/SL/PG), F2 (CO: MCT/SL/vitamin E/PG), F3 (CO: MCT/SL/PL80/PG), and F4 (CO: MCT/SL/PL80/vitamin E/PG). The PI of the formulations ranged between 0.129 ± 0.06 and 0.155 ± 0.02 with mean droplet size ranging between 143.7 ± 33.2 nm and 124.8 ± 15.1 nm, showing monodisperse distribution and nanometric size. The PG content was recorded between 103.6% and 94.3% and was observed in the oil part of the nanoemulsion as it was insoluble in water. In vitro penetration studies using the full-thickness skin of a pig ear at phosphate buffer of pH 5.5, and PG was diluted in propylene glycol and used as a control to compare against the nanoemulsion formulations. High PG concentration at the stratum corneum was observed using the control ($\pm 15.00 \, \mu g/cm^2$) followed by F2 and F1 ($\geq 10.00 \, \mu g/cm^2$), with F3 and F4 showing less than $5.00 \, \mu g/cm^2$. The epidermis showed a higher concentration of PG using F4 ($\geq 10.00 \, \mu g/cm^2$), followed by the control, F1 and F3 had no significant difference, and lastly F2 with concentration ($\leq 4.44 \, \mu g/cm^2$). Lastly in the dermis F4 and F3 formulations exhibited higher PG concentration as compared to other formulations and the control. Antiviral activity using HSV-1 (KOS strain) the nanoemulsion formulations exhibited IC_{50} values ranging between $19.26 \pm 3.67 \, \mu g/mL$ and $9.80 \pm 3.68 \, \mu g/mL$, while PG exhibited $11.08 \pm 0.84 \, \mu g/mL$, F4 had the highest inhibition activity than PG and all other formulations and hence was selected for further investigation. In vivo studies for skin sensitization and acute dermal toxicity were performed in 15 days using male and female Wistar rats with temperatures controlled between $22°C-25°C$. These studies showed no systematic toxicity signs (change in behavior patterns, circulatory, eyes, fur, CNS, mucous membrane, respiratory, or bodyweight) observed in the treated animals in both propylene glycol-diluted PG and F4 formulation. HE staining results revealed normal skin patterns of dermis, epidermis, and stratum corneum for the untreated group which was contra to those of the control group, where epidermis showed thinner aspect including the presence of inflammatory cells in both epidermis and dermis. Skin treated with F4 formulation exhibited no histological changes causing no erythema responses or other effects, with skin that resembles that of the untreated group. The presence of both PL80 and vitamin E in the nanoemulsion synergistically increased the biological activity of the formulation and enhanced the herpetic activity of PG. however, different formulations of F4 changing the concentration of the surfactants and PG still need to be practice and to check the mechanism behind the influence of both vitamin E and PL80 on the activity of PG. Moreover, F4 formulation is a promising antiherpetic scaffold, thus clinical trial will be proposed for this study.

Bidone et al. fabricated O/W nanoemulsions incorporated with *Achyrocline satureioides* extract (ASE) for skin/mucosa distribution to treat herpes [262]. These nanoemulsions were composed of egg lecithin, triglyceride, vitamin E, and polysorbate 80, nanoemulsions without ASE (Blank-NE) and quercetin-loaded nanoemulsions (QCT-NE) were used as control. The average mean droplet size was recorded at 207.29 ± 19.11 for Blank-NE,

ASE-NE (237.35 ± 12.71), and 223.88 ± 7.19 for QCT-NE, with PI, recorded at the range of 0.05 ± 0.04 to 0.09 ± 0.04 ASE-NE exhibiting the high values. In vitro antiherpes activity was performed by comparing the nanoemulsions and their controls. ASE exhibited 15.71 ± 8.61 CC_{50} (μg/mL) values and 14.07 ± 3.46 IC_{50} (μg/mL) values which were relatively higher than those of ASE-NE with 11.25 ± 0.48 CC_{50} (μg/mL) and 1.40 ± 0.88 IC_{50} (μg/mL) values. QCT showed 25.34 ± 6.88 CC_{50} values and 7.93 ± 1.53 IC_{50} values showing great potential than QCT-NE with 138.52 ± 33.36 CC_{50} and 9.51 ± 5.98 IC_{50} values. Encapsulation of ASE into the nanoemulsion significantly increased its herpetic activity 8 folds, showing that the nanoemulsion delivery system enhanced the intracellular uptake of this flavonoid. The upper layers of the esophageal mucosa and porcine ear skin epidermis showed high retention of the flavonoid, mostly in injured tissues. The nanoemulsions developed in this study are promising drug delivery systems for antiherpetic drugs or hydroethanolic extracts. Nanoemulsion interacts with intracellular pathways and protects the encapsulated antiviral molecule while in circulation. Further studies as to how the produced nanoemulsion interact with the intracellular pathways still need to be evaluated and discussed for future purposes.

Argenta et al. formulated genistein loaded cationic O/W nanoemulsions and evaluated the in vitro antiherpes activity [263]. Nanoemulsions were divided into two groups: dioleylphosphocholine (DOPC) and distearoylphosphocholine (DSPC), and formulated as NE-DSPC, HNE-DSPC, NE-DOPC, and HNE-DOPC depending on the presence of hydroxyethyl cellulose (HEC). DOPC formulations exhibited the smallest droplet size NE-DOPC (153.27 ± 08.67) and HNE-DOPC (160.40 ± 22) compared to DSPC formulation with 240.27 ± 11.99 and 225.06 ± 2.8, HEC significantly increase droplet size in both formulations. All formulations exhibited PI lower than 0.25 with NE-DSPC showing the lowest PI values of 0.08 ± 0.02, zeta potential showed values less than 33 mV in all formulations. Drug release profile showed high drug release ($103.34\% \pm 5.22\%$) of genistein using the control (propylene glycol) in 3 hours, and a prolonged release of up to 10 hours using NE-DOPC ($97.82\% \pm 1.66\%$), and NE-DSPC (61.47 ± 2.87). Lower drug release was observed for HEC containing formulations HNE-DSPC ($40.86\% \pm 5.25\%$) and HNE-DOPC ($49.44\% \pm 3.54\%$). Mucosa permeation/retention studies revealed approximately twofold reduction values using HEC formulations after 8 hours, with NE-DOPC and NE-DSPC exhibiting approximately 50 μg/cm^2. Nanoemulsion formulations exhibited higher amount of genistein retention 9.72 ± 1.29 μg/cm^2 for NE-DPOC, 8.04 ± 0.89 μg/cm^2 for NE-DSPC, and 6.85 ± 0.99 μg/cm^2 for HNE-DOPC compared to the control with 4.24 ± 1.08 μg/cm^2. histological and confocal analysis showed less compacted cells of superficial stratum surrounded by mucus, larger and compacted cells of the middle layer, more darkly stained basal layer. Esophageal mucosa showed a distribution of the Nile red after retention studies compared to propylene glycol treatment. Antiherpatic activity using Vero cells (anti-HSV-1 29R) revealed more cytotoxicity effet of the nanoemulsion formulation (IC_{50}: 5.4 μg/mL NE = DOPC and 5.3 μg/mL NE-DSPC) while free drug exhibited 14.70 μg/mL. Nanoemulsion formulations exhibited increased anti-HIV inhibition than the blank nanoemulsions by twofold.

Mahajan et al. developed O/W nanoemulsions encapsulated with saquinavir mesylate for brain targeting through IN drug delivery as neuro-AIDS treatment [264]. They prepared 9 formulations composed of oil (capmul MCM), surfactant (tween 80), and cosurfactant

(PEG 400) with different percentage compositions. Nano-formulation 4 (F4) was selected for further studies due to its stability during thermodynamic stability testing compared to all the formulations that were thermodynamically unstable. F4 was characterized for drug content/entrapment, pH, refractive index, PI, globule size, and zeta potential, the results were as follows 96.76% ± 1.16%, 5.8, 1.412 ± 0.018, 0.078 ± 0.01, 176.3 ± 4.21 nm, and −10.3 ± 1.67 mV, respectively. Ex vivo permeation studies between plain drug suspension (PDS) and F4 after 4 hours displayed 26.73% ± 3.60% and 76.96% ± 1.99% drug permeation, respectively. F4 exhibited high percentage diffusion and a faster rate of drug diffusion across nasal mucosa compared to PDS. This is an advantage in drug delivery because the drug is rapidly removed from the nasal mucosa by mucocilliary clearance. Permeability coefficient (P) was 0.51 cm/h for F4 and 0.17 cm/h for PDS. Nasal cilio toxicity test showed an unchanged mucosal structure (normal nasal mucosa) when treated with F4 than both positive and negative control. In vivo biodistribution evaluation of saquinavir mesylate into the brain was done comparing IN versus intravenous (IV) administration and higher drug concentrations in plasma and brain were detected through IN administration using drug-loaded nanoemulsion than in IV administration of PDS.

Maximum drug concentration in plasma and brain were observed after about 1 hours of IN delivery which was 62.29 times higher than IV delivered PDS. These results signify that the nanoemulsions improved the delivery of saquinavir mesylate through the blood-brain barrier using the nose-to-brain direct pathway, which hampered PDS accumulation into the brain through IV delivery. The possible mechanism of the nanoemulsions resulting from their small size was assumed to be via transcellularly delivery through olfactory neurons to the brain by means of several endocytic pathways of sustentacular. Drug transportation via the olfactory pathway into the brain was evaluated and recorded as ±96% direct transport percent (%DTP), and drug targeting efficiency was 2919.261 ± 5.68 suggesting that the nanoemulsions have better brain targeting, and can be used in the development of scaffolds for HIV therapy. Further studies to determine the biocompatibility of these nanoemulsions including their safety are required.

Bonfim et al. formulated curcumin encapsulated nanoemulsions and evaluated their activity using cell lines transducing different variants of HPV (HPV-16) for photodynamic therapy (PDT) in vulvar cell lines [265]. These nanoemulsions exhibited 180 ± 0.5 nm for nanoemulsion without curcumin (NE) and 195 ± 3.1 with curcumin (NE-Cur), 0.19 NE and 0.18 NE-Cur PI, and −46.3 ± 0.1 NE and −53.7 ± 5.84 NE-Cur zeta potential. Cytotoxicity and cellular uptake results using NE-Cur, NE, and free curcumin against A431 cells and human immortalized keratinocytes (HaCat) cells after 24 hours treatment, showed that free curcumin was very toxic to the cell mostly in higher dosages (80 μM) with 10% cell viability. NE exhibited no significant interference with cell viability of the tested cells despite the dose, while NE-Cur a slight drop of cell viability was observed again the dose had no notable influence. The internalization of (80 μM) NE-Cur by A431 cells and HaCat cells after 3 hours of incubation using fluorescence microscopy displayed cytoplasmic green fluorescence. Comparable levels of NE-Cur uptake to that of nontransduced parental cells was observed in different HPV-16 variants. Curcumin was used as a photosensitizer in PDT to check the phototoxicity of the nanoemulsions, against HaCat and A431 HPV-16 E6 transduced cells. The cell viability showed PDT of NE treated cells was not affected, even after two doses of irradiation, while NE-Cur associated with PDT strongly affected cell viability of all cell lines with over 85% drop after two irradiation doses. These results suggest that photoactivation potentiate curcumin effect strongly. Apoptosis analyses of NE-Cur in cells expressing HPV-16 E6 variants showed a noteworthy increase in the efficacy of caspases 3 and 7. Nanoemulsions are very promising scaffolds for PDT.

19.7.2 Liposomes

Liposomes (fat bodies) are small-sized colloidal spherical vesicles composed of an aqueous phase enclosed by phospholipid bilayer for the delivery of hydrophilic and lipophilic drugs to the targeted site (Fig. 19.18) [266—268]. They are classified into three types: small uni-lamellar vesicles, large uni-lamellar vesicles, and multilamellar vesicles owing to their number of bilayers and size [169,269]. Their drug encapsulation efficiency and drug release/circulation time characteristics are influenced by their membrane lamellarity and size [270]. The average size of liposomes can be between 15 asnd 1000 nm offering the advantage of good permeability [256,271]. Liposomes upon entering the living organisms are quickly absorbed by macrophages and cleared from circulation, making them beneficial as carriers of anti-HIV drugs into the macrophages [271—273].

Nayak et al. prepared gelatin-based liposomes loaded with stavudine for the treatment of HIV [274]. Three formulations SG-1-LP, SG-2-LP, and SG-3-LP were prepared with different concentrations of stavudine: 1, 5, and 10 mg, respectively. The average mean size was 248.2 ± 1.3, 230 ± 3.1, and 232.9 ± 1.5, zeta potential: -30.1 ± 1.54 mV, -43.7 ± 3.1 mV, and -44.6 ± 1.36 mV for SG-1-LP, SG-2-LP, and SG-3-LP, respectively with polydispersity less than 0.5 and pH kept at 8 for all the formulation. Encapsulation efficacy of the formulations was $38.7\% \pm 2.01\%$, $44.08\% \pm 1.01\%$, and $55.1\% \pm 2.07\%$ for SG-1-LP, SG-2-LP, and SG-3-LP, respectively. In vitro drug release studies showed by the formulations at 37°C after 12 hours was less than 80% and SG-1-LP exhibited highest (77.108%) and SG-3-LP showing the lowest (72.97%) controlled stavudine release. Hemolysis studies revealed that all formulations exhibited a safe hemolytic ratio with less than 5% hemolysis percentage after 30 minutes of incubation. At various doses of drug content, they formed a "button like" structure confirming their biocompatibility with hemoglobin. The 3 formulation showed good cell viability against Raw 264.7 cells, however SG-3-LP was highly effective

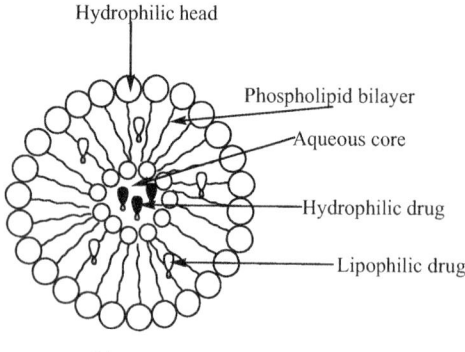

FIGURE 19.18 A schematic diagram of a liposome drug carrier.

exhibiting the least IC_{50} values. The cell viability studies reveal a dose dependent activity for these nanoformulations. The results are promising for HIV-1 therapy. However, in vivo evaluations still needs to be performed on these nanoformulations.

Zhen et al. designed pH-sensitive cationic liposomes to deliver HPV oncogene manipulation utilizing clustered regularly interspersed short palindromic repeats/Cas9 (CRISPR/Cas9) [275]. Off-target effect and endogenous shear activity of CRISPR/Cas9 gRNA-HPV16 E6/E7 showed double-strand breaks in E6/E7 locus of SiHa cells. Corresponding sites in CRISPR/Cas9 gRNA-HPV 16 E7 and CRISPR/Cas9 gRNA-HPV 16 E6 displayed the efficiency of mutations to be 69% and 72%, respectively. Two possible off-target sites were revealed using bioinformatics prediction for E6 and E7. The developed CRISPR/Cas9 gRNA-HPV 16 E6/E7 systems had no off-target effects in SiHa cells as the off-target region showed no T7E1 cleavage band. Expression levels of E6/E7 were significantly high in the control group than in CRISPR/Cas9 gRNA-HPV 16 E6/E7 treated cells. The control showed a significantly lower percentage of apoptotic cells and apoptosis rate than CRISPR/Cas9 gRNA-HPV 16 E6/E7 exhibiting a higher rate. CRISPR/Cas9 gRNA-HPV 16 E6/E7 loaded nano-liposomes showed a mean size of 145.7 ± 2.6 nm and zeta potential of 51.33 ± 1.8 mV, and pH-sensitive results displayed an increase of surface charge of the nano-liposomes as pH decreased. This signifies that the developed nano-liposomes have great potential to enter the cytoplasm and achieve lysosomal escape.

Drug entrapment efficiency of the nano-liposomes was recorded at $96.87\% \pm 0.13\%$ displaying that these scaffolds exhibited high drug encapsulation. In vitro drug release profiles showed that the nano-liposomes exhibited more than 90% drug released over 35 hours, with a recorded initial burst of 37.74% in the 1st 3 hours. CCK8 cells were used to detect cytotoxicity of these nano-liposomes, and the control phosphate buffer solution (PBS) exhibited the highest cell viability together with empty vector gRNA greater than 90% while nano-liposomes displayed slightly less than 80% viability. These nano-liposomes revealed lower levels of E6/E7 mRNA than of lipofectamine-2000-gRNA empty vector and naked CRISPR/Cas9 control groups, signifying high delivery ability into cells to perform gene knockout. In vivo results showed no significant toxicity of the nano-liposomes targeted to splicing HPV 16 E6/E7 with significant inhibition of the tumor growth in nude mice. These results suggest that the developed nano-liposomes promote the delivery of nonviral gene editing to hamper HPV related cancer.

Kulkarni et al. developed a vaccine candidate for mice immunized with liposomes encapsulated recombinant neutralizing epitopes (NE) protein of HEV checking effector memory T cells, participation of $CD4^+$ central, and specific memory B cell response [276]. Strong anti-HEV IgG response was produced by 5 μg dose of vaccine candidate immunized mice and was estimated to continue for a lifetime. The frequency of memory B cells postimmunization at day 120 was significantly higher in immunized mice than those treated with only adjuvant. Antibody response in mice with low (≤ 100 geometric mean titers, 1 μg) and high (> 800 geometric mean titers, 5 μg) anti-HEV IgG titers was heightened by a booster dose. Mice showing high anti-HEV IgG titers displayed high frequencies of effector memory T cells and $CD4^+$ central, whereas 50% and 100% mice with low and high anti-HEV IgG titers, respectively, displayed detection of HEV-specific antibody-secreting plasma cells 6 days' post booster dose. The liposome-based vaccine candidate produced in this study is a promising approach in HEV vaccine trials in humans to explore the longevity of HEV-Specific memory response.

Tiwari et al. designed a vaccine candidate-based liposome for the delivery of HBsAg administered intranasally [277]. Two major formulations were prepared, liposome-encapsulated HBsAg (Lipo-HBsAg), and influenza virus complexed recombinant protein haemagglutinin (HA) liposome encapsulated with HBsAg (HA-Lipo-HBsAg). The average vesicle size was 643 ± 10 nm Lipo-HBsAg, 712 ± 12 nm HA-Lipo-HBsAg, with zeta potential -31.17 ± 0.72 mV and -42.32 ± 0.91 mV, drug entrapment% $53.2\% \pm 1.2\%$ and $51.2\% \pm 0.8\%$, and PI 0.147 ± 0.019 and 0.159 ± 0.027, respectively. Drug release profiles displayed higher percentages for Lipo-HBsAg ($90.23\% \pm 2.10\%$) and lower percentages for HA-Lipo-HBsAg ($83.13\% \pm 1.72\%$) after 144 hours. Haemagglutination assay showed positive RCB's agglutination with a slight difference between liposomes coupling HA and plain HA in a buffer. HA-Lipo-HBsAg adhered to the mucosal surface more than in plain liposomes, and they also displayed a high immune response. The higher cellular response of HA-Lipo-HBsAg is linked to the pH-dependent fusion property of HA protein.

Akhtar et al. developed optimized liposomes for delivery of arsenic trioxide (ATO) to HPV-positive cancer cells [278]. They reported cell viability and uptake of ATO-encapsulated liposomes (Lipo-ATO) in negative and positive-HPV cancer cell lines. MTT assay showed a similar survival rate of the cells after 72 hours in both cervical cells after treatment with Lipo-ATO, however, free ATO treated cells exhibited a lower cell survival rate. This shows that liposomes improved the cytotoxicity of ATO against both cancer cell lines. A similar trend of LIpo-ATO was observed in apoptotic evaluation of the liposomes between the two cancer cell lines, showing biocompatibility of these liposomes scaffolds against either negative or positive-HPV cervical cancer cell lines. Moreover, the free ATO exhibited similar results to those of ATO loaded in liposomes. In vitro inhibition activity of Lipo-ATO exhibited no significant difference in both cell lines (IC_{50} values: 14.2 ± 0.2 μM HeLa and 12.2 ± 0.6 μM HT-3), however, it is evident that it is significantly more active against HT-3 cell lines after 48 hours of incubation. Further studies to evaluate Lipo-ATO formulations activity on noncancerous cells were performed using human colon cells (CRL-1790) and HK. Cytotoxicity investigation revealed higher uptake of Lipo-ATO in HK cells compared to CRL-1790 and HeLa cells. In vitro inhibition activity against these cells showed that Lipo-ATO exhibited IC_{50} values of 49.7 ± 1.2 μM, 48.7 ± 1.2 μM, and 29.7 ± 2.6 μM of HK, CRL-1790, and HeLa cells respectively. Results reported in this study show a promising approach for the effective management of HPV cervical cancers.

19.7.3 Solid lipid nanoparticles

Drug delivery systems are designed to protect and improve the transport of poorly water/lipid insoluble drugs including poorly bioavailable drugs. Solid Lipid Nanoparticles (SLNs) (Fig. 19.19) possess a solid lipid phase rather than a liquid phase inside the nanoparticle [279,280]. SLNs are designed to overcome common limitations such as drug leakage, promote long-acting drug release, are cost-effective, and provide drug protection from degradation, etc. [279,281,282]. Cationic SLNs are reported as considerable scaffolds for the delivery of nucleic acid [283]. These systems have great potential for the encapsulation of

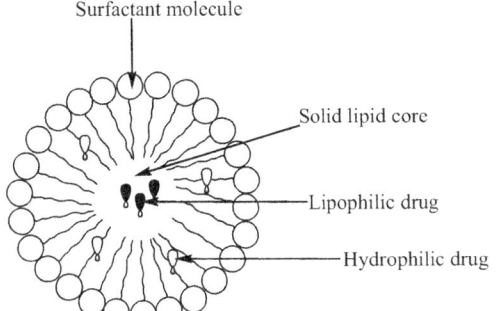

Surfactant molecule

Solid lipid core

Lipophilic drug

Hydrophilic drug

Solid Lipid Nanoparticles

FIGURE 19.19 Schematic diagram of SLN drug carrier.

both hydrophilic and lipophilic drugs and deliver the drugs to the targeted tissues with controlled release [284]. Hassan et al. prepared SLNs encapsulated with acyclovir (herpes simplex viral drug) to evaluate its pharmacokinetic profile when delivered orally [285]. Biogapress Vegetal 297 ATO was used to prepare the SLNs and Tween 80. The mean particle size of the SLNs was 122.72 ± 2.15 nm, PI 0.23 ± 0.01, and zeta potential was observed at -24.37 v 1.07. Encapsulation of acyclovir into the SLNs did not influence the physical appearance of the prepared SLNs. The drug entrapment efficacy showed a relatively high, $86.53\% \pm 0.49\%$ average percent of acyclovir-loaded SLNs.

The melting peak of pure acyclovir compound was recorded at 249.3°C, and its encapsulation to SLNs reduced it to 56.37°C, and blank SLNs recorded 56.13°C, which is a good temperature for oral delivery of drugs as it is greater than the room and body temperatures to maintain solid state of the SLNs. The melting enthalpy of bulk counterpart was relatively high (207.95 J/g), while the acyclovir-loaded SLNs exhibited 4.34 J/g with blank (3.93 J/g). Acyclovir-loaded SLNs exhibited a biphasic release profile, however within the first hour a 46% drug release was observed and a cumulative, sustained release was observed until the 24th hour with 100% drug released. The 46% initial burst could be linked to the acyclovir that was entrapped on the surface of SLNs. In vivo oral bioavailability of acyclovir-loaded SLNs were compared to that of commercially available acyclovir suspension. Acyclovir-loaded SLNs reached the maximum acyclovir plasma concentration in 1.25 hours whereas it was observed at the 1st hour postoral administration for the commercial acyclovir suspension, such results can be linked to the burst release of acyclovir-loaded SLNs. Acyclovir's bioavailability increased to 423.61% when loaded to SLNs supporting the aforementioned characteristics of SLNs as carriers that are developed to enhance the bioavailability of the drug encapsulated unto them. SLNs reported in this study are proved to be great carriers of acyclovir and its oral delivery, they did not only improve its half-life limitation but it also enhanced its intestinal permeability. This study is of great importance however, in vitro cytotoxicity and degradation studies are outstanding. Acyclovir-loaded SLNs are promising scaffolds for delivery and improvement of acyclovir limitation (bioavailability and intestinal permeability). Nonetheless, antiherpetic activity still needs to be performed and to further evaluate the mechanism that influences intestinal drug uptake.

Kondel et al. SLNs encapsulated with acyclovir for the treatment of HSV-1 infection [286]. The mean particle size (nm), PI, and zeta potential (mV) of the acyclovir-SLNs was reported at 131 ± 41.44, 0.30 ± 0.014, and -16 ± 1.90, respectively. In vitro drug release was determined using PBS pH 6.8, simulated gastric fluid (SGF) pH 1.2, and simulated intestinal fluid (SIF) pH 7.8 and the acyclovir-SLNs exhibited a biphasic drug release with an initial burst release in 24 hours followed by a controlled release that lasted until the 7th day. The initial burst release was recorded at 45% PBS, 51% SIF, and 52% SGF, and 100% at SIF, 95% at SGF, and 87% at PBS of the entrapped drug was released until day 7. Acyclovir-SLNs exhibited a plasma concentration for 7 days as compared to the free acyclovir group which only maintained it for 2.5 hours max. The EC_{50} against HSV-1 isolates ranged from $(0.02$ to $13.5 \mu g/mL)$ and the time $> EC_{50}$ was 4 hours for free acyclovir drug (36 mg/kg) and 168 hours for acyclovir-SLNs (equivalent to 36 mg/kg). Cytotoxicity studies were investigated using Vero cells and expressed in percentages. Free acyclovir and acyclovir-SLNs showed no significant difference. The cell viability was influence by the concentration of acyclovir at high concentration $(250 \mu g/mL)$ exhibited 19.1% and lower concentration $(100 \mu g/mL)$ exhibited 60.5% cell viability, suggesting that acyclovir concentrations $\geq 100 \mu g/mL$ are toxic to Vero cells. Plaque reduction assay was used to investigate the antiviral efficiency of acyclovir-SLNs against HSV-1. The lower concentration $(0.25 \mu M)$ of free acyclovir showed a 61% plaque count which decreased to 2% with an increase in concentration $(8 \mu M)$, suggesting that acyclovir plaque efficacy is dependent on concentration.

Contrary to acyclovir, acyclovir-SLNs exhibited a uniform efficacy against plaque count regardless of the concentration from 0.25 to $8 \mu M$ as no plaque formation was observed. In vivo evaluation was investigated using female BALB/c mice in 11 days, and acyclovir (400 mg) was given 3 times a day in 5 days while acyclovir-SLNs (400 mg) was administered only once. Saline was used as a control, and the lesion score was significantly reduced in acyclovir and acyclovir-SLNs treated groups as compared to the saline-treated group, including the healing time of lesions. The healing time for acyclovir and acyclovir-SLNs was recorded at day 4, whereas for saline it was at day 9. Results produced in this study show that a single dose of acyclovir-SLNs is effective as multiple doses of acyclovir, showing that acyclovir-SLNs are promising scaffolds with prolonged drug release.

Gaur et al. formulated SLNs encapsulated with efavirenz (HIV treatment drug) and checked their pharmacokinetic studies [287]. Formulations that only vary in concentration of surfactant ESLN-0 (SLN, 1:3 drug: lipid "glyceryl monostearate," and 0.5% surfactant "tween 80"), ESLN-1 (0.75% surfactant), ESLN-2 (1.0%), and ESLN-3 (1.25%) were prepared. These formulations presented physical characterization that range between $362 \pm 2.1 - 124.5 \pm 3.2$ nm particle size ESLN-0 highest and ESLN-3 lowest, $0.455 - 0.234$ PI, -15.9 to -22.1 zeta potential, and $86 \pm 1.03 - 46.28 \pm 1.05\%$ drug entrapment. Higher tween 80 concentration promoted smaller particle size and high drug entrapment percentage as ESLN-3 exhibited $86\% \pm 1.03\%$. ESLN-3 was selected for further investigations, and the in vitro drug release kinetics were done at pH 7.4. ESLN-3 was compared to ES, commercial formulation (EMF). It exhibited a cumulative, sustained drug release of 60.6%–98.22% in 24 hours, while ES and EMF exhibited 61.705% to $\pm 93\%$ and 86.705% to $\pm 93\%$ drug release in 16 hours, respectively. Pharmacokinetic studies using albino rats showed a 5.32-fold increase of ESLN-3 in peak plasma concentration and 10.98-fold increase in AUC, with AUC$_{0\rightarrow24}$ (76.4 µg/mL · h), C$_{max}$ (4.21 µg/mL) compared to ES with AUC$_{0\rightarrow24}$ (6.958 µg/mL · h), and C$_{max}$ (0.791 µg/mL). Moreover, temperature seemingly affected the physical characteristics of the formulation however, an increase in the concentration of the surfactant increased the stability of the ESLN-3 formulation as no drastic change was observed at different temperatures varying from 4°C to 25°C. The presence of surfactants in the formulation of SLNs is important for sustained drug release and enhanced activity of the encapsulated antiviral drug. However, more investigation is proposed regarding drug delivery, mechanism related to drug enhancement by the surfactant, in vitro cytotoxicity and in vitro anti-HIV activity of these SLNs.

Torrecilla et al. prepared nonviral vectors of SLNs for the treatment of chronic hepatitis C by RNA interference [288]. Dextran (DX), hyaluronic acid (HA), short-hairpin RNA expression plasmid (shRNA74), protamine (P), and SLNs were used to prepare the vector formulations. The vector formulations were divided into two, those containing DX and the ones with HA using different ratios of protamine, and an HA-SLN2 (0.5:2:1:2 of PA:P: shRNA74:SLN), HA-SLN5 (0.5:2:1:50), DX-SLN2 (1:2:1:2), DX-SLN5 (1:2:1:5). The average particle size for these vectors was between 242 ± 16–208 ± 9.9 nm, PI in the range of 0.23 ± 0.06–0.21 ± 0.012, and zeta potential +38.87 ± 0.64 to +29 ± 0.46. Agarose gel electrophoresis revealed that all formulations protected shRNA74 against inhibition by nuclease when treated with DNase I. however, DX formulated vectors showed a light tail which reveal a slight plasmid degradation. The successful release of shRNA was observed after treatment with SDS, and the cellular uptake revealed high intensity to vectors of 1:5 shRNA: SLN regardless of the polymer HA or DX with no significant difference between the two formulations. In vitro inhibition of internal ribosome entry site (IRES-GFP) revealed that both polymer vectors inhibited the expression of GFP using HepG2 cells. A range of 19%–67% silencing capacity was observed for HA vectors and it was directly proportional to shRNA74 dosage. Even though the increase in shRNA74 dose increase the silencing of the two formulations, HA-SLN5 exhibited the highest silencing percentage against all other formulations. DX vectors exhibited the highest silencing percentage of 32%, showing that overall HA vectors induced a twofold high rate of silencing than DX vectors. The cell viability studies showed that low ratios of SLN (SLN2) irrespective of the polymer or dose of shRNA 90% of HepG2 cells was observed, meanwhile, SLN5 ratios significantly decreased cell viability of HepG2 cell to 64% for HA-SLN5 and 76% for DX-SLN5. Non-viral SLN-based vectors produced in this study are promising with HA formulations exhibiting the most effective results against the treatment of hepatitis C virus through the IRES. However, more investigation is required regarding the influence of HA on the delivery of the nano-carriers. Different formulations using variations of surfactants, polymer are suggested, and to check the mechanisms involved in the inhibition of HCV replicon.

Torrecilla et al. shRNA encapsulated a nonviral vector based on SLNs for treatment of hepatitis C virus, targeted to the IRES [289]. The nano-formulations were coded with varying ratios of shRNA74: SLNs, where HA-SLN2 was 1:2 and HA-SLN5 was 1:5. Around 90% cell viability of Huh-7 cells was observed after 48 hours treatment with the HA-SLN vectors bearing different doses of shRNA74, naked shRNA74, and empty SLNs. The HA-SLN vectors have the potential to silence the expression of GFP in Huh-7 cells. There was no significant inhibition observed with cells treated with shRNA scrable, while 4%–50%

was noted on HA-SLN vector depending on concentration/dose of shRNA74 as higher concentrations exhibited higher inhibition. HA-SLN5 vector exhibited higher silencing than HA-SLN2 at each shRNA74 dose level. Inhibition of the hepatitis C virus RNA replication studies was similar to that of silencing studies with higher doses exhibiting higher inhibition, however, there was no significant difference between HA-SLN2 and HA-SLN5 despite the dose. Cellular uptake presented that both vectors were proficiently able to enter the Huh-7 cells, though in higher drug extension for HA-SLN5. Inhibition of endocytosis shows that lower temperatures (4°C) exhibited lower cell uptake, meanwhile higher temperatures (37°C) exhibited high cell uptake. Lower temperature (4°C) exhibited endocytosis inhibition of 16% while at 37°C inhibition of endocytosis was recorded at 32%. Interaction of both vectors with erythrocytes induced red cells agglutination, presenting a lack of hemolytic effect for these HA-SLN vectors. These results proved the biocompatibility of the HA-SLN vectors. A similar level of hemolysis to that of untreated blood was observed for both HA-SLN vectors. Internalization of HA-SLN vectors using CD44 receptor was also achieved, though this entry route was of less productivity for silencing as compared to that endocytosis.

Javan et al. prepared SLNs encapsulated with ritonavir for in vitro anti-HIV-1 activity. The SLNs were prepared via solvent emulsification (SE) evaporation and double emulsion (DE) method with varying surfactant (poloxamer 188 "P188" and tween 80 "T80") and drug concentration. Formulations were labeled as R_1 (DE:20 mg ritonavir: P188), R_2 (DE:20 mg ritonavir: T80), R_3 (DE:10 mg ritonavir: P188), R_4 (DE:10 mg ritonavir: T80), same trend was used to formulate SE-based SLNs formulations from R_5–R_8. Physical characteristics of these formulation were in the range of 178.1 ± 4.5–254.3 ± 16.6 nm, 0.225 ± 0.05–0.282 ± 0.05, -50.80 ± 4.8 to -39.35 ± 1.2 mV, and $21.44\% \pm 6.8\%$–$53.28\% \pm 12.8\%$, for particle size, PI, zeta potential, and EE%, respectively. The prepared SLNs displayed low drug entrapment efficacy, however, formulations containing poloxamer 188 as surfactant exhibited a higher percentage than the one with tween 80. This was likened to tween 80s ability to increase drug solubility in an aqueous phase, resulting in smaller drug entrapment by the lipid phase [290]. Although tween 80 exhibited poor drug encapsulation it significantly improved the reduction of the SLNs size. R_1 was selected for further studies due to its high drug entrapment percentage. In vitro drug release investigation displayed cumulative 40% drug release of ritonavir over 9 days with no initial burst, showing a great potential of these SLNs to prolonged treatment. In vitro antiviral experiments using lentiviral-based HIV-1 particles pseudo-type by VSV-G showed that both free drug and SLNs encapsulated drug maintained inhibition of virus production. This study shows that the produced SLNs are good drug carriers with prolonged drug release kinetics. In vitro cytotoxicity of the SLNs need to be reported, and to further do in vivo analysis. Poor drug entrapment of the SLNs needs to be improved.

19.8 Polymer-based drug delivery system for viral infections

19.8.1 Micelles

Micelles are colloidal or self-aggregated drug delivery systems that possess an average particle size of less than 100 nm. They are composed of amphiphiles or surfactants and are

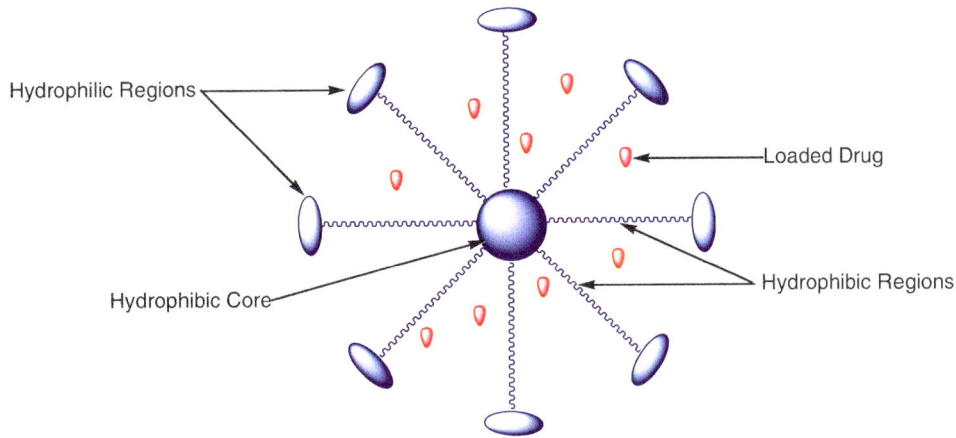

FIGURE 19.20 Schematic diagram showing micelles loaded with drugs.

made up of two parts: hydrophilic head and hydrophobic tails (Fig. 19.20) [291]. Micelles are formed or produced at a concentration called critical micelle concentration (CMC). Various factors affect the formulation of the micelles including the used solvent system, temperature, concentration of amphiphiles, and size of the hydrophobic domain in the amphiphilic molecule [292]. Micelles exhibit interesting advantages that make them useful in the field of drug delivery such as high drug encapsulation and loading capacity, high cellular uptake because of their nanosized range, reduced drug toxicity, disease site targeting of several drugs, easily eliminated from the biological environment after biodegradation, enhanced pharmacokinetic parameters of loaded drugs, they can be used in drug combination therapy, and improved drug stability and bioavailability [293]. These drug delivery systems are frequently synthesized from synthetic polymers: poly (vinyl alcohol) (PVA), poloxamer, poly (lactic-coglycolic acid) (PLGA), polylactide (PLA), poly (hydroxyethyl methacrylate) (PHEMA), polyglycolic acid (PGA), polyurethanes (PUs), poly (ethylene oxide) (PEO)/poly (ethylene glycol) (PEG), polycaprolactone (PCL), and poly (vinyl pyrrolidone). They can be also synthesized from natural polymers: chitosan, alginate, cellulose, chitin, gelatin, collagen, etc. Several research reports discuss the application of polymer-based micelles for the treatment of viral infections.

19.8.1.1 Micelles for loaded with antihuman immunodeficiency virus drugs

Efavirenz (EFV)(Sustiva) is a highly lipophilic nonnucleoside RT inhibitor classified in class II of the Biopharmaceutic Classification System. EFV is a first-choice ARV in pediatric pharmacotherapy and adult. The low solubility of the EFV in an aqueous medium hinders the biodistribution and absorption of the drug from the gastrointestinal (GI) tract. This limitation can be overcome by polymeric micelles. Chiappetta et al. formulated polymeric micelles for the delivery of an anti-HIV drug, efavirenz using N-alkylated poloxamines as nanocarriers [294]. The particle size analysis of efavirenz-loaded micelles using DLS displayed hydrodynamic radius of about 8 nm and a PDI of 0.30 that are suitable in the field of drug delivery. The in vitro drug release profile of micelles exhibited that the

N-alkylated poloxamines released efavirenz more rapidly when compared to micelles that were not N-alkylated. The in vitro cytocompatibility studies using methylthiazole tetrazolium (MTT) assay demonstrated that good viability of murine fibroblast cell lines ranging between 70% and 100% when were incubated with efavirenz-loaded micelles at a concentration of 0.01%, indicating excellent biocompatibility that is essential in the treatment of HIV and other viral infections [294].

Chiappetta et al. investigated the synergistic performance of combined polymer-based micelles composed of branched and linear poly (ethylene oxide)-poly (propylene oxide) (poloxamer or Pluronic F127) for drug delivery of efavirenz. The polymeric micelles exhibited monomodal size distributions at 37°C with small PI values ranging between 0.18 and 0.25 that are fixed with a full micelle aggregation. The particle size analysis of micelles displayed a particle size of 26.9 nm while TEM exhibited spherical morphology [295]. Other similar studies of efavirenz-loaded poloxamer micelles conducted by Chiappetta et al. demonstrated that these drug delivery systems are potential systems for improved oral pharmacokinetic and can be very useful in pediatric HIV/AIDS pharmacotherapy [296]. Another study showed that poly (ethylene oxide)-poly (propylene oxide) micelles significantly increase drug bioavailability of efavirenz fourfold in the CNS when compared to plain efavirenz, showing that these polymeric micelles can be a potential system for the management of HIV-related neurocognitive disorder [297].

Chiappetta et al. (formulated poly (ethylene oxide)−poly (propylene oxide)) block copolymer micelles loaded with efavirenz for pediatric anti-HIV pharmacotherapy. The in vitro drug release experiments of efavirenz-loaded micelles at intestine-like release medium showed a linear curve that was constant with zero-order kinetics. The pharmacokinetic experiments exhibited that block copolymer micelles significantly enhance the oral bioavailability of efavirenz and decrease the interindividual variability demonstrating their promising potential in pediatric anti-HIV pharmacotherapy [298]. Furthermore, the efavirenz-loaded poloxamer micelles prepared by Chiappetta et al. demonstrated rapid drug release at an intestine-like release medium. The taste tests achieved by adult healthy volunteers demonstrated that the unique combination of sweeteners and flavors utilized significantly (i) reduced the intensity of the Burning Mouth Syndrome (which is the major cause of unplanned disruption of the ARV pharmacotherapy) and (ii) significantly shortened its duration [299].

Chaudhari and Handge formulated polymer-based micelles loaded with an anti-HIV drug called lopinavir using Pluronic (F68 &F127) as the polymer or carrier. The successful formulation of lopinavir-loaded micelles was confirmed by Fourier transform infrared (FTIR), Differential scanning calorimetry (DSC), and XRD spectroscopy. The %Entrapment of lopinavir in the Pluronic-based micelles was found ranging between 12% and 28% [300]. Mahajan and Patil designed Vitamin E-TPGS micelles to improve the pharmacokinetic parameters of lopinavir for the treatment of HIV. The characterizations exhibited that formulated lopinavir-loaded micelles displayed particle size of about 91.71 nm, surface charge of −24.8 mV, PDI 0.129, drug entrapment efficiency 99.36% ± 1.06% and drug loading capacity of 20.83% ± 1.23%. The pharmacokinetic experiments demonstrated that the relative bioavailability of Vitamin E-TPGS micelles loaded with lopinavir was boosted by 3.17-folds when compared to plain lopinavir, suggesting that these micelles can be useful in the treatment of HIV [301].

19.8.1.2 Micelles for herpes management

Most of the micelles that are reported for the treatment of herpes are loaded with Acyclovir (ACV). ACV is the regularly utilized antiviral drug for the treatment of herpes infections and other related viral mucosal infections. But its application is hampered by poor water solubility that limits both its antiviral activity and drug bioavailability. That's where the polymeric micelles can be used to overcome these limitations. Sawdon and Peng formulated polymer-based micelles incorporated with ACV for the treatment of herpes using PCL and methoxy poly (ethylene glycol) (MPEG) or chitosan as polymeric nanocarriers [302]. The successful formulation and physicochemical properties of ACV-loaded micelles were confirmed by FTIR, proton nuclear magnetic resonance (1HNMR) spectroscope and gel permeation chromatography (GPC). The TEM analysis of ACV-loaded micelles displayed an average particle size of about 200 nm while the CMCs of ACV-PCL-chitosan and ACV-PCL-MPEG were and 6.6 mg/L and 2.0 mg/L, respectively. The in vitro drug release studies at 25°C exhibited that the ACV was initial burst release from within the first 2 hours with ~50% accumulative release from PCL-MPEG or PCL-chitosan micelles followed by a sustained drug release mechanism up to 2 days. The cytotoxicity experiments using MTT assay showed high cell viability of HT29 colorectal cells when were incubated with ACV-loaded micelles for 48 hours, suggesting that these micelles are nontoxic and possess excellent biocompatibility. These results demonstrated that ACV-loaded polymeric micelles can be potential candidates for the treatment of herpes viral infection without exacerbating any toxicity to normal cells [302].

Varela-Garcia et al. investigated sclera and cornea permeability of Solutol or Soluplus polymeric micelles loaded with ACV because almost 20% of people that suffer from HSV can have vision problems. The particle size analysis of Solutol and Soluplus micelles incorporated with ACV exhibited 134.9 nm (with PDI of 0.26) and 137 nm (with PDI of 0.30), respectively. Also, the zeta potentials of polymeric micelles were slightly negative, which can result in high eye tissue penetration. The bovine sclera and cornea permeability experiments exhibited that the Solutol and Soluplus polymeric micelles significantly facilitated the penetration of the ACV through sclera and cornea and led to an accumulation of the drug in both eye tissues, indicating that these micelles can be effective drug delivery systems in the treatment of herpes that affect vision [303].

Accardo et al. designed mixed peptide amphiphile micelles for the treatment of herpes simplex viral infections. The DLS results of micelles exhibited hydrodynamic radii that range between 50 and 80 nm and negative zeta potential of approximately 40 mV while CMC values were around 4.0×10^{-7} mol/Kg, demonstrating that these micelles can result in high cell uptake. The in vitro cytotoxicity results demonstrated that these micelles aggregates did not show any noticeable signs of cellular toxicity when were seeded with mouse and human macrophage cell lines. Other in vitro experiments displayed that peptide amphiphile micelle at a concentration of about 10 μM, activated RAW 264.7 and U937cells to release considerable levels of cytokines that can result in improved herpes treatment [304].

19.8.1.3 Micelles for hepatitis treatment

The antiviral drugs that are mostly used for the treatment of hepatitis suffer from several shortcomings such as high dosage, relapse after drug withdrawal, long course of treatment, drug resistance, and low efficacy. Therefore, genetic bioactive agents including deoxyribozyme 10−23 DNAzyme have been designed recently to directly target the life cycle of the HBV. Several biomedical researchers reported the polymer-based micelles loaded with 10−23 DNAzyme. Miao et al. reported galactosylated chitosan-oligosaccharide- SS-octadecylamine micelles encapsulated with 10−23 DNAzyme for hepatitis management [305]. These micelles exhibited the %encapsulation efficiency and %drug loading of 94.85% ± 0.31% and 1.555% ± 0.005%, respectively. The DLS results of 10−23 DNAzyme-loaded micelles displayed size diameter of about 246.73 ± 6.73 nm, zeta potential of 13.93 ± 1.05, and PDI of 0.22 ± 0.06. The in vitro cytotoxicity results using MTT assay showed more than 80% cell viability of HepG2.2.15 cells when were incubated with was 200 μg/mL 10−23 DNAzyme-loaded micelles, indicating nontoxicity and good biocompatibility of these polymeric micelles. The in vivo experiments using rAAV8−1.3HBV-infected male BALB/c mice demonstrated that 10−23 DNAzyme-loaded micelles significantly inhibited HBV antigen secretion when compared to the control group (Lipofectamine™ 2000), suggesting that10−23 DNAzyme-loaded micelles are potential systems that can be used for the treatment of hepatitis [305].

Hong et al. formulated chitosan-g-stearic acid micelles for intracellular delivery of 10−23 DNAzyme to HBV [306]. The mean size of micelles was around 158.0 ± 2.5 nm and the surface charge were 39.1 ± 3.3 mV in water. The TEM images indicate that micelles possessed a well-formed shape and compact structure. The in vitro cellular uptake experiments of 10−23 DNAzyme using HepG2.2.15 cell lines exhibited that cellular uptake is time-dependent, indicating that these micelles were potential intracellular delivery systems and could enhance the cellular uptake of loaded bioactive agents into HepG2.2.15 cells. The 10−23 DNAzyme-loaded micelles exhibited a significant HBsAg inhibition when compared with that of 10−23 DNAzyme-loaded Lipofectamine™ 2000. These results demonstrate that 10−23 DNAzyme-loaded micelles are safe and possess potential as an anti-HBV gene therapy [306]. Another study conducted by Miaoa et al. demonstrated that10−23 DNAzyme-loaded Lipofectamine 2000 complex displayed maximum inhibition rate on HBsAg expression of 46.53% ± 2.00% at 48 hours, and then decreased quickly, while 10−23 DNAzyme-loaded chitosan oligosaccharide−stearic acid micelles displayed maximum inhibition rate of 82.51% ± 1.28% at 72 hours and hold on inhibition rate above 70% until 96 hours, suggesting that DNAzyme-loaded chitosan oligosaccharide−stearic acid micelles are effective systems for hepatitis gene therapy [307].

Layek et al. formulated mannosylated phenylalanine grafted chitosan micelles as a transdermal delivery system of DNA vaccine for hepatitis treatment. The in vitro cytotoxicity studies of polymeric micelles using RAW 264.7 mouse macrophage and mouse dendritic cells (DC 2.4) exhibited high cell viability at concentrations that range between 0.1 and 1 mg/mL, indicating excellent biocompatibility and nontoxicity. The cellular uptake results displayed maximum cellular internalization for DNA-loaded mannosylated phenylalanine grafted chitosan micelles when compared with DNA-loaded chitosan micelles. The in vivo studies demonstrated that intradermal immunization of BALB/c mice showed that hepatitis B DNA vaccine-loaded polymeric micelles did not only stimulate multifold higher serum antibody titer but also significantly induced T-cell proliferation and skewed T helper toward Th1 polarization [308].

The shortcomings of antiviral drugs that are commonly used for the treatment of hepatitis, which is mentioned above can be overcome by micelle drug delivery systems. These antiviral drugs include Lamivudine, daclatasvir, sofosbuvir, ombitasvir, ritonavir, simeprevir, paritaprevir, acyclovir, etc. Li and coworkers formulated stearic acid-g-chitosan oligosaccharide polymeric micelles for drug delivery of lamivudine stearate [309]. The high-pressure liquid chromatography (HPLC) results of lamivudine stearate-loaded micelles showed a high encapsulation efficiency of more than 97%. The in vitro drug release experiments at pH 7.4 exhibited that approximately 90% lamivudine from plain lamivudine stearate was released in 24 hours, while around 32% of lamivudine was released from the polymeric micelles, indicating the potential of the stearic acid-g-chitosan oligosaccharide micelles as a sustained drug delivery system. The in vitro cytotoxicity assay of lamivudine stearate-loaded micelles employing MTT assay displayed low cytotoxicity when were cultured with HepG2.2.15 cells, demonstrating good biocompatibility. Other in vitro results showed that lamivudine stearate-loaded micelles possessed more noticeable anti-HBV activities when compared with plain lamivudine and lamivudine stearate solution [309].

Huang et al. formulated stearic acid-g-chitosan oligosaccharide micelles incorporated with acyclovir for HBV treatment. The physicochemical properties and successful formulation of acyclovir-incorporated micelles were confirmed by ^{1}H NMR. The cellular uptake experiments exhibited that even after the polymeric micelles were loaded with acyclovir could be also highly uptaken by HepG2 cells, and no significant difference was found between unloaded micelles and acyclovir-loaded micelles. The in vitro anti-HBV studies of the plain acyclovir and acyclovir-loaded micelles with acyclovir concentration of about 0.044 μM/mL showed that the inhibition of acyclovir on HBsAg was increased from 12.7% to 22.3% from 5th day to 9th day, while the inhibition of acyclovir-loaded micelles was increased from 58.2% to 80.3% from 5th day to 9th day, suggesting superior anti-HBV activity of acyclovir-loaded micelles that plain acyclovir [310].

19.8.1.4 Micelles for the treatment of Human papilloma virus

Genital warts are usually caused by HPV-11 viruses, whereby they penetrate the skin affecting epithelial cells in the urethra, vagina, cervix, and perianal skin. Wang et al. formulated glycyrrhizic acid micelles for transdermal delivery of podophyllotoxin to treat HPV caused warts. The DLS analysis and TEM micrographs displayed that podophyllotoxin-loaded micelles had a spherical shape with a particle size of about 10 nm. The in vitro drug release studies at simulated skin conditions demonstrated slowly sustained drug release of podophyllotoxin from the micelles within 12 hours when compared to the free drug, suggesting that these micelles delivery systems could be potential for transdermal delivery of drugs. In vivo skin distribution and drug-deposition experiments using anaesthetized rats showed that podophyllotoxin was released in a sustained manner from the glycyrrhizic acid micelles [311].

Nishida et al. prepared polyion complex micelles incorporated with siRNA for treatment of HPV-associated cervical cancer. The In vitro cell viability experiments demonstrated that HPV type 16 and 18 E6/E7 siRNAs inhibited proliferation of cervical cancer cells in an HPV type-specific phenomena. Also, the fluorescence imaging biodistribution test exhibited that fluorescence dye-labeled siRNA-encapsulated micelles significantly accumulated within the cervix tumor post the systemic administration. The in vivo anti-cancer studies using the mice model displayed that intravenous injection of siRNA-loaded polymeric micelles effectively suppressed the growth of subcutaneous HeLa and SiHa tumors, respectively. These results revealed that drug-loaded polymeric micelles could be the potential candidates for the treatment of HPV-related cervical cancer [312]. The PPO-pluronic micelles loaded with *Natronobacterium gregoryi* Argonaute or Cas9 plasmid designed by Lao et al. showed significant inhibition of HPV-stimulated tumor activity both in vivo and in vitro, while plain *N. gregoryi* Argonaute did not display any significant E7 inhibition on the mouse model [313].

19.8.1.5 Micelles for the treatment of respiratory infections (common cold, COVID-19, and Middle East respiratory syndrome)

The research studies on polymer-based micelles for the treatment of viral respiratory infections are very scarce. Nevertheless, Chauhan et al. designed b-casein micelles for combination therapy of tipranavir: efavirenz and darunavir: efavirenz: ritonavir and their physicochemical properties were confirmed by FTIR and XRD spectroscopy [314]. The in vitro drug release studies at gastric-like pH conditions exhibited initial burst release of coloaded ARV drugs from b-casein micelles followed by sustained drug released. These micelles demonstrated the ability to protect loaded drugs from gastric enzymes, indicating that ARV drugs can reach the targeted biological site without their therapeutic activity being destroyed. These results revealed that b-casein micelles are promising systems for FDCs of antiviral agents for the treatment of viral respiratory infections and other viral infections [314].

19.8.2 Dendrimers

Dendrimers are drug delivery systems that possess 3-dimensions, spherical, hyper-branched structures (Fig. 19.21). These delivery systems are used in various biomedical applications for the delivery of therapeutic agents such as antiviral, anticancer, antituber-cular, antibacterial, antimalarial drugs, etc [315]. The well-distinct surface functional groups and globular nanosized structure that range between 1 and 100 nm caused them to be very useful in the field of drug delivery. Polymer-based dendrimers such as polyami-doamine (PAMAM) dendrimers have been attracting great attention in biomedical applications because of their low toxicity and good biocompatibility [316]. The other advantages of polymeric dendrimers that make them useful in the treatment of viral infections include improved drug biocompatibility, enhanced drug efficacy, reduced drug toxicity, and sustained and controlled drug release profiles [316].

19.8.2.1 Dendrimers for the treatment of human immunodeficiency virus

Perisé-Barrios et al. designed second-generation carbosilane dendrimers as gene delivery systems for the treatment of HIV. The in vitro cytotoxicity studies of carbosilane dendrimers using MTT assay exhibited about 80% cell viability of CD4T lymphocytes, indicating their nontoxicity and good cytocompatibility. Other experiments demonstrated

Dendrimer Chains

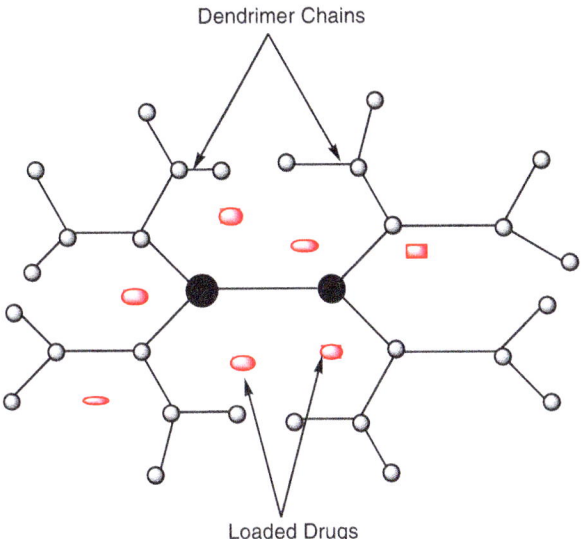

Loaded Drugs

FIGURE 19.21 Schematic Diagram of dendrimers.

that these dendrimers reduced the expression of the CCR2 receptor in macrophages, decreasing the possibility of HIV-1 infection [317]. Gutierrez-Ulloa et al. synthesized carbosilane dendrimers for the treatment of HIV infections. The in vitro antiviral experiments displayed 100% viral inhibition at 1 µM of carbosilane dendrimers for the X4 strain whereby in the case of the R5 strain 100% and 98% inhibition was observed at 20 and 100 µM, respectively, suggesting excellent HIV antiviral activity [318]. Sepúlveda-Crespo et al. formulated G2-STE16 carbosilane dendrimers combined with other carbosilane dendrimers, and the results demonstrated that these combinations resulted in 100% inhibition and exhibited a synergistic effect against various HIV-1 isolates of TZM.bl cells [319]. Furthermore, the G2-NF16 and G3-S16 carbosilane dendrimers reported by Vacas-Córdoba et al. exhibited an additive or synergistic efficacy profile with efavirenz, tenofovir, and zidovudine in most combinations investigated against the R5 and X4 tropic HIV-1 in cell lines, revealing promising strategy for the treatment of HIV [320]. Also, the carbosilane dendrimers studied by Cordoba demonstrated synergistic profile MRV against CCR5 and dual tropic HIV-1 when combined with maraviroc [321].

Ceña-Diez et al. formulated polyanionic carbosilane dendrimer G2-S16 for antiviral applications. The in vitro biological studies exhibited that G2-S16 dendrimer possesses anti-HIV-1 efficacy at an initial stage of viral replication, inhibiting the gp120/CD4/CCR5 interaction and offering a barrier to infection for long periods, revealing its nonspecific and multifactorial ability [322]. Briz et al. designed second- and third-generation polyanionic carbosilane dendrimers for combination therapy of antiviral drugs, tenofovir and raltegravir. The in vitro antiviral experiments demonstrated that G2-S16, G2-NS16 and G3-Sh16 exert anti-HIV-2 efficacy at an initial phase of viral replication inhibiting the virus, hindering cell-to-cell HIV-2 transmission, and hindering the binding of gp120 to CD4, and the HIV-2 entry. Triple combinations with raltegravir and tenofovir increased the anti-HIV-2 efficacy, constant with synergistic interactions [323]. The thiol-ene carbosilane dendrimers

Sánchez-Rodríguez and coworkers demonstrated these dendrimers did not only possess a higher ability to hinder the entry of various R5-HIV-1 and X4-HIV-1 isolates into epithelial cells but also avoid the HIV-1 infection of stimulated PBMCs [324]. Kandi et al. formulated anionic citrate-PEG-citrate dendrimer for HIV infection therapy. The DLS results displayed that the size of the polymeric dendrimers was about 90 nm with the negative surface charge of 2.46 mv, which may be due to the existence of free carboxylic acid functional groups. These dendrimers demonstrated that the CC50 (50% cytotoxic concentrations) of about 0.6 mM and IC50 (50% inhibitory concentration) of 2.4 mM on HIV infected cells [325].

Kumar et al. prepared PEGylated PAMAM dendrimers loaded with anti-HIV drug lamivudine. The successful preparation and physicochemical properties of lamivudine-loaded dendrimers were confirmed by TEM, IR, UV, and DSC. The % drug entrapment of PEGylated PAMAM dendrimers was high ranging between $28.21\% \pm 1.45\%$ and $71.54\% \pm 1.56\%$. the in vitro drug release profile exhibited that the release of lamivudine from PEGylated PAMAM dendrimer was significantly slower than non-PEGylated PAMAM dendrimers, revealing that PEGylated PAMAM dendrimers are potential systems for controlled and prolonged delivery of lamivudine in the treatment [326]. The SPL7013 dendrimers reported by Telwatte and coworkers demonstrated that these dendrimers are potential virucidal agents against HIV-1 strains that use the CXCR4 coreceptor that uniquely employ CCR5 by a mode that is distinct from virion loss or disruption of gp120 [327]. Maciel et al. formulated poly(alkylideneamine) dendrimers for HIV therapy. These polymeric dendrimers were successfully characterized by mass spectroscopy (Ms) ^{13}C- and ^{1}H-NMR, FTIR, and zeta potential techniques. The in vitro cytocompatibility studies using MTT assay displayed high cell viability of TZM-bl cells when were incubated with poly(alkylideneamine) dendrimers, suggesting excellent biocompatibility and nontoxicity. The antiviral studies of poly(alkylideneamine) dendrimers exhibited good antiviral activity when the TZM-bl cells were infected with R5-HIV-1$_{NLAD8}$ isolates [328].

19.8.2.2 Dendrimers for herpes treatment

Falanga et al. formulated polyamide-based dendrimers for herpes treatment. The physicochemical properties of polymeric dendrimers were confirmed by FTIR spectroscopy. The in vitro cytotoxicity studies using MTT assay demonstrated high cell viability of Vero cells when were incubated with dendrimers with various concentrations (5.5, 55 nM, 0.28, 0.55, 1.1, 2.8 μM) up to 48 hours, suggesting good biocompatibility and nontoxicity. The in vitro antiviral experiments exhibited that the inhibition of HSV-1 replication with the dendrimers possess an ability to reach 90% already at a concentration of 55 nM, indicating excellent virucidal activity against herpes [329]. Carberry et al. prepared polyamide-based dendrimers functioned with peptide gH (625−644) (gH625). The cellular uptake studies displayed high cellular internalization of gH625 functionalized dendrimers within the HeLa cells via a nonactive translocation mechanism, demonstrating their ability to deliver cargo inside the cells for the treatment of herpes infections [330].

Luganini et al. formulated peptide-derivatized dendrimers for the treatment of herpes. The in vitro experiments showed that the peptide dendrimers possessed their ability to hinder HSV adsorption at acidic conditions and in the presence of 10% human serum proteins, mimicking the physiological conditions of the vagina, a possible therapeutic site for such bioactive agents [331]. Lastly, when these dendrimers are combined with acyclovir resulted in highly synergistic efficacy. Beltrán and coworkers polyanionic carbosilane dendrimers for inhibition of herpes simplex type 2 infection. The in vivo studies exhibited that the G2-S16 dendrimers were able to completely inhibit the HSV-2 infection in normal conditions. Remarkably, when HSV-2 was cultured with semen and treated with these dendrimers, they significantly inhibited increased infection with similar actions, which in the absence of semen, hindered more than 99% of HSV-2 infection [332]. Gonga et al. formulated SPL7013 dendrimers and their plaque reduction experiments exhibited that the 50% effective concentrations (EC50) were about 2.0 and 0.5 μg/mL for HSV-1 and HSV-2, respectively. Inhibitory effects were also examined on HSV-infected cells with EC50s of 6.1 and 3.8 μg/mL for HSV-1 and HSV-2, respectively, revealing that these dendrimers are promising systems for the treatment of herpes [333]. The SPL-2999 dendrimers formulated by Gong et al. demonstrated EC50 values of 0.5 μg/mL (30 nM) and 1 μg/mL (60 nM) for HSV-2 and HSV-1, respectively [334].

19.8.2.3 Dendrimers for hepatitis treatment

Lancelota et al. prepared ammonium-terminated amphiphilic Janus dendrimers for drug delivery of camptothecin with antiviral efficacy [335]. The physiochemical properties and successful formulation of dendrimers were confirmed by FTIR, ^{1}H and ^{13}C NMR, and Ms spectroscopy. The TEM micrographs demonstrated that these polymeric dendrimers had spherical shapes with mean diameters that range between 14.5 ± 3.9 nm and 15.5 ± 4.0 nm. The %EE of dendrimers for camptothecin was more than 10%. The in vitro cytotoxicity and antiviral studies demonstrated concentration-dependent, for instance, cell viabilities were higher than 95% and virus replication values were above 80% at low concentrations of dendrimers. Furthermore, the CC50 and EC50 values demonstrated that camptothecin preserved its activity against hepatitis C viral replication when loaded within the dendrimer systems [335]. Sepúlveda-Crespo et al. formulated polyanionic carbosilane dendrimers for the treatment of hepatitis C virus infections [336]. The antiviral studies showed that the incorporation of an antiviral drug, sofosbuvir into the dendrimers significantly resulted in improved antiviral activity against hepatitis C viral strains when compared to the plain dendrimers or the free sofosbuvir [336].

Akao et al. fabricated dendrimer/α-cyclodextrin conjugates appended with fucose for the treatment of fulminant hepatitis in vivo. The particle size of these systems was ranging between 240 and 280 nm with PDI that is about 0.5. The cytotoxicity experiments demonstrated good cytocompatibility and nontoxicity of fucose-appended dendrimers when were incubated with NR8383 cells. The intravenous injection of dendrimer/α-cyclodextrin conjugates significantly prolonged the survival of lipopolysaccharide-induced fulminant hepatitis model mice. Furthermore, fucose-appended dendrimers intravenously administered highly accumulated in the liver tissue when compared to pristine dendrimers. These results demonstrated that dendrimer/α-cyclodextrin conjugates appended with fucose are potential candidates for the treatment of hepatitis [337]. Anselmo et al. formulated Janus dendrimers for drug delivery of iopanoic acid and tiratricol. The antiviral experiments exhibited that dual drug-loaded dendrimers possessed improved antiviral activity against hepatitis C when compared to the free drugs [338].

19.8.2.4 Dendrimers for Ebola treatment

Karpenko et al. formulated PAMAM dendrimers encapsulated with DNA vaccine constructs for Ebola management. The particle size analysis of PAMAM dendrimers displayed particle sizes that are less than 100 nm, suggesting their ability to deliver DNA vaccines to target cells. It was revealed that the loading of DNA vaccine constructs in the PAMAM dendrimers results in an increase in their immunogenicity than the group of rats immunized with the vector plasmid pcDNA3.1 (a negative control) [339]. Chahala et al. prepared methoxyPEG-based dendrimers incorporated with mRNA replicons for protective immunity against Ebola. The in vivo experiments demonstrated that the mice inoculated with about 40 μg of dendrimers showed robust GP-specific T-cell responses 9 days after inoculation, as investigated by IL-2 and IFN-γ expression by CD4+ and CD8+ splenocytes in feedback to ex vivo treatment with the EBOV GP-derived WE15 peptide. These outcomes demonstrated that the dendrimers are promising delivery systems for DNA vaccine constructs in the immunization against Ebola [340].

19.8.2.5 Dendrimers for Human papilloma virus treatment

Donalisio et al. formulated peptide dendrimers SB105-A10 loaded with clusters of basic amino acids for treatment of HPVs. The in vitro cytotoxicity experiments exhibited that the cell lines cultured with peptide dendrimers indicated no evidence of toxicity. The pseudovirus-based neutralization assays showed that these dendrimers were observed to be an effective inhibitor of genital HPV types (i.e., types 6, 16, and 18). The IC_{50} values were between 2.8 and 4.2 μg/mL (0.59 and 0.88 μM), demonstrating good antiviral activity against HPV types [341]. Lee et al. synthesized PAMAM dendrimers peptide derived from HPV type 11 E2 protein. The successful formulation and physicochemical properties of dendrimers were confirmed by FTIR and ^{1}H NMR. PAMAM dendrimers demonstrated high transfection efficiencies in NIH3T3 and Neuro-2A cell lines and low cytotoxicity, suggesting their safety to be used in HPV treatment [342].

19.8.2.6 Dendrimers efficacy against respiratory infections: COVID-19 and Middle East respiratory syndrome

Khaitov et al. formulated peptide dendrimer KK-46 encapsulated with siRNA for COVID-19 management. The in vitro studies demonstrated that siRNA-loaded dendrimers were effective to inhibit SARS-CoV-2 virus replication. The in vivo experiments exhibited an important reduction of virus titer and entire lung in inflammation in the animals visible by inhalation of siRNA-loaded dendrimers [343]. Furthermore, Bohr et al. fabricated PAMAM dendrimers loaded siRNA for the treatment of acute lung inflammation. The in vitro cellular uptake experiments demonstrated high cellular internalization of siRNA-loaded dendrimers inside the macrophage cells when compared with plain siRNA, which was caused by their nanosized range of about 127–153 nm. The in vivo studies showed that PAMAM dendrimers significantly induced effective TNF-α siRNA inhibition when compared with siRNA, upon pulmonary administration to mice with lipopolysaccharide-induced lung inflammation. These results demonstrated that these dendrimers are potential candidates for the treatment of viral respiratory infections such as the common cold, COVID-19, and MERS [344].

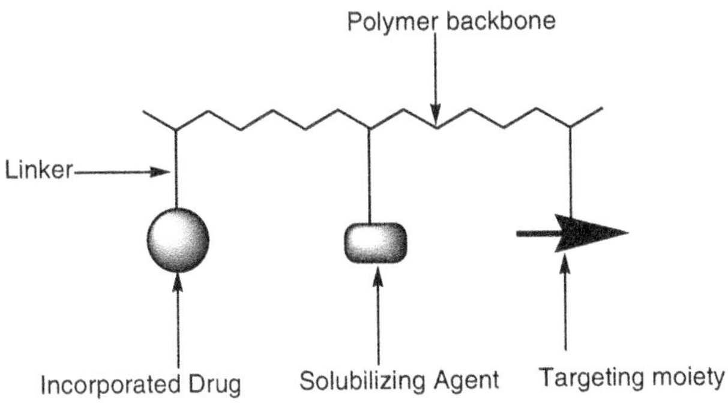

FIGURE 19.22 Schematic diagram of polymer-drug conjugates.

19.8.3 Polymer-drug conjugates

Polymer-drug conjugates that are called polymer prodrugs are drug delivery systems composed of three constituents: the drug, targeting moiety, and solubilising agents. The model of polymer-drug conjugates was firstly proposed in 1975 by Helmut Ringsdorf (Fig. 19.22) [345]. The drug, targeting moiety, and solubilising agents are covalently linked into the polymeric backbone via selected linkers, such as esters, amine, amides, and alcohols. The solubilizing agent and targeting moiety are linked into polymer-drug conjugates to enhance therapeutic or pharmacokinetic outcomes of the incorporated drugs [346]. Several polymers are utilized to formulate these delivery systems including polyamidoamine, polyglutamic acid, PEG, PGA, PCL, PVA, etc. The advantages of polymer-drug conjugates include reduced drug toxicity, good drug bioavailability and biodegradability, improved drug solubility, and improved pharmacological and pharmacokinetic parameters. In addition, they can preserve and protect the activity of incorporated drug(s) during circulation for attacks of enzymes and they deliver loaded drugs to the target site [347]. There some research reports that demonstrate the effectiveness of polymer-drug conjugates in the management of viral infections.

19.8.3.1 Polymer-drug conjugates for human immunodeficiency virus treatment

Aremu and coworkers formulated PEG succinate conjugates drug delivery of an anti-HIV drug, lopinavir. The physiochemical properties and successful formulation of PEG-lopinavir conjugates were confirmed by ^{1}H NMR, FTIR, XRD, and DSC. The solubility experiments of polymer-drug conjugates demonstrated an enhanced solubility of sixfold in water, fourfold phosphate-buffer saline, threefold in 0.1N of HCl, compared to plain lopinavir. The toxicity studies employing *Danio rerio* demonstrated that these conjugates possess low toxicity when compared with free lopinavir. These results demonstrated that PEG-based conjugates significantly improved water solubility of lopinavir with good cytocompatibility, suggesting their potential for HIV treatment [348]. The chitosan-based conjugates incorporated with stavudine via phosporamide linkers were formulated by Zeng et al. and they demonstrated good antiviral activity against HIV-1 in MT4 cells with low cytotoxicity [349].

Roy et al. fabricated polymer-based Pluronic conjugated loaded with anti-HIV drug efavirenz targeting Microfold cells (M-cells) in the gut-associated lymphoid tissue. DLS characterization displayed that the conjugates were of 140 nm size, PDI 0.3, and a negative surface charge of 19.38 ± 2.2 mv. The in vitro drug release profile exhibited sustained release of efavirenz from the polymeric prodrugs when compared with free efavirenz. The anti-HIV efficacy of the conjugates was significantly higher than that of pristine efavirenz drug. These outcomes demonstrated that these formulations were able to display sustained drug release and hinder the HIV-1 infection in the gut-associated lymphoid tissue when compared to the plain drug [350]. Dang et al. fabricated PCL-based conjugates coloaded with tenofovir and nevirapine for HIV treatment. The in vitro experiments of dual drug-loaded conjugates revealed a potential additive/synergistic effect of the released antiviral drugs on HIV-1 infection of HeLa cells, suggesting promising efficacy against HIV [351].

19.8.3.2 Polymer-drug conjugates for herpes and hepatitis treatment

Stegman et al. developed PCL-based conjugates incorporated with antiviral drugs acyclovir for HCV treatment. The physicochemical properties of conjugates were analyzed by 1H NMR, DSC, and IR spectroscope. The in vitro experiments showed that these polymer-drug conjugates with high enough acyclovir loads reduce primary HSV-1 infection in Vero cells as the same as a single dose of acyclovir. The in vivo studies of acyclovir-loaded conjugates exhibited activity in inhibiting reactivation of these HSV viruses with a single intervention [352]. Wohl and coworkers formulated Poly (acrylic acid)-based conjugates loaded with ribavirin for hepatitis C virus treatment. These conjugates were confirmed as potential inhibitors of replication of the hepatitis C viral RNA utilizing a subgenomic viral replicon system. Furthermore, negatively charged conjugates were shown to possess an intracellular efficacy against hepatitis C virus replication [353].

19.8.3.3 Polymer-drug conjugates for COVID-19 treatment

Thirumalaisamy et al. developed HA-based polymeric prodrugs incorporated with 2-Deoxy-D-Glucose for COVID-19 treatment [354]. The in silico molecular docking experiments of HA-2-Deoxy-D-Glucose conjugates against four various SARS-CoV-2 viral protein (S protein, PLpro, RdRp, and Mpro,) demonstrated that these conjugates possessed superior binding affinity (6.4, -7.0, -7.2, and -6.2 Kcal/mol) with all four screened SASR-CoV-2 viral targets when compared to plain 2-Deoxy-D-Glucose (4.7, -4.6, -4.9, and -4.8 Kcal/mol), respectively. The cytocompatibility studies exhibited that HA-2-Deoxy-D-Glucose conjugates possessed low toxicity than free2-Deoxy-D-Glucose. The study also demonstrated that the HA-2-Deoxy-D-Glucose conjugates have multiple advantages of effective drug delivery to its CD44 variant isoform receptors of the lower respiratory tract, highest interactive binding affinity with SARS-CoV-2 protein targets. Furthermore, the HA-2-Deoxy-D-Glucose conjugates possess additional advantages of biocompatibility, good biodegradability, nonimmunogenicity, and no toxicity. These outcomes revealed that the HA-based polymeric prodrugs are potential delivery systems for the treatment of COVID-19 [354].

19.8.4 Nanocapsules

Nanocapsules are drug delivery systems that are composed of a protective shell a core where drugs are loaded (Fig. 19.23) [355]. Various procedures are employed to produce

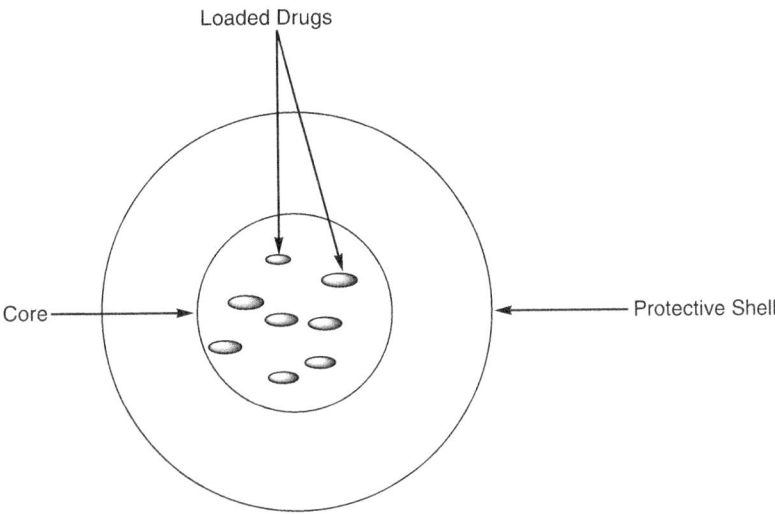

FIGURE 19.23 Nanocapsules.

polymeric nanocapsules, such as self-assembly of block copolymers, nanoemulsion polymerization, and solidification of droplet shells. These drug delivery systems possess a size that ranges between 10 and 1000 nm [356]. The advantages of polymer-based nanocapsules include reduced drug toxicity, enhanced drug bioavailability, high drug loading capacity, and they promote sustained and controlled drug release profile [357]. Several polymeric nanocapsules were reported for the treatment of viral infections in preclinical studies.

19.8.4.1 Nanocapsules for human immunodeficiency virus treatment

Hillaireau et al. fabricated poly(iso-butylcyanoacrylate)/poly(ethyleneimine) hybrid aqueous-cored nanocapsules encapsulated with an anti-HIV drug azidothymidine-triphosphate [358]. The hybrid polymeric systems resulted in high entrapment efficiency of azidothymidine-triphosphate in nanocapsules that reach 90%. TEM micrographs exhibited bimodal distribution and displayed polymeric hybrid nanocapsules possessing a central cavity enclosed by a polymer membrane, confirming the successful fabrication of nanocapsules. The in vitro drug release studies displayed that azidothymidine-triphosphate was initially rapid released (about 90% after 8 hours) from polymeric nanocapsules and significantly followed by slow sustained drug release. The in vitro experiments demonstrated that hybrid poly(iso-butylcyanoacrylate)/poly(ethyleneimine) nanocapsules efficiently deliver azidothymidine-triphosphate to macrophages: the cellular uptake was increased by 30-fold than the plain drug, reaching appropriate cellular concentrations for therapeutic determinations for HIV treatment [358]. The poly(iso-butylcyanoacrylate) nanocapsules coloaded with azidothymidine-triphosphate and cidofovir were formulated by Hillaireau et al. and demonstrated a gradual release of antiviral drugs followed by a sustained release profile [359].

Wen et al. formulated Poly(DL-lactide)-b-Poly(ethylene glycol)-b-Poly(DL-lactide)-diacrylate triblock nanocapsules loaded with broadly neutralizing antibodies for HIV treatment in the CNS. The results from this study revealed that a single dose of broadly neutralizing antibodies PGT121 encapsulated in polymeric nanocapsules when transported at 48 hours postinfection retarded early acute infection with $SHIV_{SF162P3}$ in infants, with one of four mice showing viral clearance. Significantly, the polymeric nanocapsules delivery of broadly neutralizing antibodies PGT121 promoted suppression of SHIV infection in the CNS relative to controls [360].

19.8.4.2 Nanocapsules for hepatitis treatment

Fichter and coworkers formulated block copolymer poly (ethylene-cobutylene)-b-(ethylene oxide) nanocapsules functionalized with hepatitis C virus nonstructural protein 5A (NS5A) and the adjuvant monophosphoryl lipid A (MPLA) for the treatment of hepatitis. The in vitro cytotoxicity studies showed low toxicity of functionalized nanocapsules when were cultured with murine nonparenchymal liver cells. The in vivo studies showed a preferential deposition of NS5A-MPLA functionalized nanocapsules in the liver that was visualized in a 3D reconstruction utilizing fluorescence imaging tomography, suggesting that these polymeric nanocapsules are potential candidates for the treatment of hepatitis [361]. Vicente et al. developed chitosan-based nanocapsules for codelivery of recombinant HBsAg and TLR7 agonist. The DLS analysis exhibited that nanocapsules possessed a nanometric size of about 200 nm with a high positive surface charge of 45 mV. The in vivo studies demonstrated the positive effect of the codelivery of recombinant HBsAg and TLR7 agonist from the nanocapsules was proved upon IN administration to mice [362].

Somiya et al. reported polyethyleneimine bio-nanocapsules loaded with HBV surface antigen. These nanocapsules displayed excellent transfection efficiency in human hepatic cells, revealing their ability to be used in gene delivery therapy for hepatitis treatment [363]. The squalene-loaded polyglucosamine nanocapsules designed by Vicente and coworkers exhibited particle sizes that range between 200 and 250 nm with a positive zeta-potential about +60 mV, while the in vivo studies demonstrated that these nanocapsules were able to significantly modulate and potentiate the immune response to the particular antigens when were intramuscular injected to mice [364]. The Monophosphoryl lipid A-coated polyglycerolpolyricinoleate nanocapsules encapsulated with HBV surface antigen were formulated by Nguyen et al. and they showed that cellular Uptake of nanocapsules by monocyte-derived neonatal and adult dendritic cells was dose-dependent. Monophosphoryl lipid A-coating significantly improved the cellular uptake of polymeric nanocapsules up to about 76% at a concentration of 100 μg/mL than the noncoated nanocapsules (43.7% cellular uptake) [365].

19.8.5 Polymeric nanoparticles and nanospheres

Polymer-based nanoparticles are drug delivery systems that are particulate dispersions and solid colloidal particles that possess a size diameter that range between 1 and 1000 nm [76]. These nanocarriers are usually synthesized from synthetic or semisynthetic polymers and natural polymers whereby the drugs are loaded, encapsulated, incorporated, adsorbed, or entrapped. These nanoparticles display fascinating properties such as excellent biocompatibility, good biodegradability, and versatility during their application.

The polymers that are frequently employed for the formulation of nanoparticles are PLGA, PLA, polyalkylcyanoacrylates, PGA, chitosan, cellulose, etc [366]. The advantages of polymeric nanoparticles in biomedical applications include controlled and sustained drug release profile, enhanced water solubility, the improved biological activity of loaded drugs, and the ability to be loaded with more than one drug [367].

19.8.5.1 Nanoparticles and nanospheres for human immunodeficiency virus treatment

Elkateb et al. developed PLGA-SL hybrid nanoparticles for combination therapy of ARV drugs; darunavir and ritonavir. The SEM micrographs of hybrid nanoparticles displayed spherical particles with smooth surfaces and a mean diameter of about 50 nm with narrow size distributions. The encapsulation efficiency of PLGA-based hybrid nanoparticles for the ARV drugs was significantly higher ranging between 62% and 90.8%. The in vitro drug release studies exhibited initial burst release of darunavir and ritonavir from the hybrid nanoparticles followed by a sustained drug release mechanism, suggesting PLGA-SL hybrid nanoparticles for drug delivery of ARV agents in the treatment of HIV [368]. The dolutegravir-loaded chitosan nanoparticles Dharshini and coworkers exhibited superior therapeutic efficiency in T-lymphatic cell line-infected with HIV_{IIIB} viral strain and lower cytotoxicity than free dolutegravir [369].

Dev et al. prepared PLA-chitosan hybrid nanoparticles incorporated ARV drug Lamivudine using the emulsion method. The FTIR results confirmed the successful formulation of lamivudine-loaded nanoparticles. The SEM pictures displayed that the hybrid nanoparticles possessed spherical morphology for both loaded and unloaded nanoparticles. The in vitro cytocompatibility studies showed high cell viability of mouse fibroblast cells when were incubated with lamivudine-loaded nanoparticles, indicating nontoxicity and good biocompatibility of nanoparticles. The in vitro drug release profile exhibited that lamivudine release rate was lower in the acidic pH condition than in basic pH condition, suggesting that these hybrid nanoparticles are an auspicious drug delivery system for controlled delivery of anti-HIV agents [370]. The dapivirine-encapsulated PLGA-based nanoparticles developed by das Neves and Sarmento exhibited a rapid release for the first 4 hours, followed by a sustained release for 24 hours at pH 4.2 and 7.4. Furthermore, the fluorescence microscopy results of polymeric nanoparticles were readily taken up by diverse macrophage and genital cell lines [371].

Zhang et al. developed PLGA-methacrylic acid copolymer (Eudragit) hybrid nanoparticles entrapped tenofovir or tenofovir for HIV treatment. The in vitro cytotoxicity experiments demonstrated nontoxicity of hybrid nanoparticles at a concentration of 10 mg/mL for 2 days to vaginal endocervical/epithelial cells. The cellular uptake studies exhibited high cellular uptake approximately 50% in 24 hours by these vaginal cell lines generally happened through a caveolin-mediated pathway. These results demonstrated that nanoparticles are alternative controlled drug delivery candidates in the intravaginal delivery of anti-HIV drugs [372]. Ogunwuyi et al. fabricated PEG-based nanoparticles loaded with various ARV drugs (lamivudine, zidovudine, raltegravir, and nevirapine). These polymeric nanoparticles effectively hindered HIV-1 infection in PBMCs and CEM T cells; they grasp the potential for HIV/AIDS treatment [373]. Sneha et al. fabricated carboxy methyl cellulose-polyvinyl pyrrolidone hybrid nanoparticles loaded with Lamivudine.

The in vitro drug release studies displayed initial burst release of lamivudine from hybrid nanoparticles followed by sustained drug release with good cytocompatibility when were incubated with liver cells [374].

The efavirenz-loaded PGLA nanoparticles developed by Chaowanachan et al. displayed a particle size of about 200 nm and a negative surface of -25 mV. The in vitro antiviral studies revealed that the HIV inhibitory result of efavirenz-loaded nanoparticles demonstrated up to a 50-fold reduction in the 50% inhibitory concentration when compared to plain efavirenz [375]. Leporati et al. synthesized methoxypoly(ethylene glycol) nanoparticles loaded efavirenz. These nanoparticles displayed particle sizes that range between 23 and 30 nm. The efavirenz-loaded nanoparticles showed a notable improvement of EC50 of efavirenz by 20-folds in the A17 strain [376]. The amyloid-binding polyacrylate-based nanoparticles synthesized by Sheik and coworkers demonstrated excellent capability to reduce semen-derived enhancer of virus infection-mediated improvement of HIV infection, with the polymeric nanoparticles showing the greatest efficacy with an IC_{50} value of approximately 4 μg/mL, suggesting outstanding antiviral activity [377].

19.8.5.2 Polymeric nanoparticles and nanospheres for the treatment of herpes

Tavares and coworkers formulated N, N, N-trimethylchitosan-poly (n-butylcyanoacrylate) core-shell nanoparticles loaded with acyclovir for herpes treatment. The DLS analysis of polymeric nanoparticles displayed particle size, zeta potential, and PDI of 296.2 ± 4.6 nm, $+36.5 \pm 2.7$ mV, and $0.245 \pm 0,125$, respectively. The in vitro cytotoxicity experiments displayed high cell viability of Caco-2 cells when were seeded with acyclovir-loaded nanoparticles at a concentration of about 8.0 μg/mL, suggesting excellent cytocompatibility [378]. Gourdon et al. designed PLA-PEG hybrid nanoparticles loaded acyclovir. The in vivo pharmacokinetic studies using mice demonstrated that acyclovir was eliminated more slowly in its loaded form as its half-time was 1.3 and 1.7 times higher when compared with elimination times seen for plain acyclovir and free valacyclovir forms, respectively [379]. Ramyadevi and Rajan developed acyclovir-loaded PVP/ethylcellulose/ Eudragit nanoparticles for herpes treatment. The hybrid nanoparticles displayed high 80% drug entrapment with a particle size of 100 nm and a zeta potential of $+26$ mV. SEM and XRD results of nanoparticles demonstrated spherical morphology and amorphous nature, respectively. The in vitro drug release profile showed initial rapid release of acyclovir from nanoparticles followed by a sustained release for more than 12 hours. The cellular uptake results exhibited high cell uptake acyclovir-loaded hybrid nanoparticles by corneal epithelial cells with high cell viability [380].

The acyclovir-loaded Eudragit nanoparticles formulated by Gandhi et al. exhibited sustained drug release for a prolonged time in vitro [381]. Shahsavari and coworkers synthesized chitosan nanoparticles loaded with acyclovir. The physicochemical properties of nanoparticles were confirmed by FTIR, TGA, and XRD. The DLS results displayed particle size of about 132 ± 24.3 nm, the surface charge of 32 ± 2.87 mV and PDI of 0.159 ± 0.05 with a % encapsulation efficiency of $85\% \pm 4.38\%$. the in vitro drug release profile showed sustained drug release of acyclovir from the nanoparticles [382]. Bhosale and coworkers formulated acyclovir-loaded PLGA nanoparticles and the in vivo studies exhibited that these nanoparticles clearly showed 2 to threefold improvement in bioavailability of acyclovir when compared to commercial available Zovirax tablets [383].

Steinbach et al. developed PLGA-based nanoparticles loaded siRNA for treatment of HSV-2 genital infection. The in vivo studies exhibited that the mice intravaginally injected with a poisonous dose of HSV-2, and treated with siRNA-loaded nanoparticles, demonstrated increased survival from ~9 days (in untreated mice) to more than 28 days (in nanoparticle treated mice), suggesting safety and excellent biocompatibility [384]. Lima and coworkers developed PLA nanoparticles loaded with chloroquine for the treatment of HSV-1. The SEM and DLS analysis showed that spherical nanoparticles were formulated with a modal diameter of less than 300 nm, the surface charge of −20 mv and %encapsulation efficiency of 64.1%. Furthermore, the antiviral efficacy investigated by the plaque reduction assay demonstrated greater activity for chloroquine-loaded nanoparticles when compared to plain chloroquine at concentrations lower than 20 μg/mL [385].

Al-Dhubiab developed PGLA-based nanospheres loaded with acyclovir and impregnated in HEC-eudragit films. The SEM micrographs exhibited that the acyclovir-loaded nanospheres were uniformly embedded in the polymeric films. The in vivo experiments of acyclovir-loaded nanospheres using rabbits demonstrated a significantly prolonged duration of about 6 hours, absorption of approximately 360.93 ng/mL, and greater AUC0-α of fivefold with buccal films when compared with oral solution [386]. Donalisio et al. formulated chitosan nanospheres loaded with acyclovir for herpes treatment. The acyclovir-loaded PGLA-based nanospheres exhibited higher antiviral activity against both the HSV-1 and the HSV-2 strain when compared to plain acyclovir [387]. Al-Dhubiab et al. fabricated acyclovir-loaded PLGA nanospheres embedded in films for drug delivery. The in vitro drug release experiments exhibited the potential of designed films to offer controlled release of acyclovir over a prolonged period. The ex vivo permeation experiments displayed the prospect of drug-loaded nanospheres to permeate through buccal film at a controlled rate [388].

19.8.5.3 Polymeric nanoparticles and nanospheres for the treatment of hepatitis

Wang et al. reported mPEG−PLA−chitosan-based nanoparticles loaded with siRNA for the treatment of hepatitis B. The DLS results displayed an average size diameter of about 226.7 nm, the zeta potential of +19.44, and PDI of 0.152. These results showed that these nanoparticles can possess high cellular uptake. The mPEG−PLA−chitosan-based nanoparticles loaded with siRNA significantly reduced the expression of HBsAg, indicating that the S_2RNA was efficiently transported into the cells so that they could effectively hinder the secretion of the HBsAg by the PLC/PRF/5 cells and resulting in good antihepatitis activity [389]. Jain and coworkers PEG−PLA−PEG block copolymeric nanoparticles loaded HBsAg. The results from this study showed that the in vivo and in vitro experiments of hybrid nanoparticles depict greater mucosal uptake resulting in actual immune response [390].

Saraf and coworkers developed chitosan-based nanoparticles loaded with lipopolysaccharide derived alginate coated hepatitis B antigen. SEM and DLS Results demonstrated that the prepared chitosan nanoparticles were spherical shape with an average particle size of approximately 605.23 nm, PDI 0.234 and surface charge −26.2 mV and could effectively protect loaded antigen at GIT in acidic conditions [391]. The chitosan-based nanoparticles loaded with recombinant HBsAg formulated by Pregoa et al. demonstrated sustained release way of antigen from nanoparticles without compromising its antigenicity, showing that these nanoparticles are auspicious candidates for vaccine delivery of antigens in the treatment of hepatitis [392]. Zhu et al. formulated PLGA-based nanoparticles loaded with HBsAg and the in vitro uptake experiments exhibited that these nanoparticles could lead to high internalization in RAW 264.7 cells and bone marrow-derived dendritic cells [393].

Antigen-loaded PLGA-based nanoparticles reported by Mishra et al. demonstrated a strongly systemic and mucosal response, revealing that they could be an auspicious carrier system for the M cell-targeted oral mucosal immunization against hepatitis B [394]. Shawky and coworkers developed chitosan-PVA nanoparticles loaded with Sofosbuvir drug antihepatitis C virus. The XRD and FTIR spectrums confirmed the successful formulation of hybrid nanoparticles. The in vitro drug release studies exhibited sustained and controlled drug release of sofosbuvir from the nanoparticles [395].

Gohar et al. formulated chitosan nanospheres for the treatment of hepatitis. The DLS results showed that chitosan nanospheres possessed a mean size of approximately 110 nm and a positive surface charge of 27.8 mv. The high cellular uptake of chitosan nanospheres by Hep G2 cell lines demonstrated that these nanospheres can be a suitable system for the inhibition of hepatitis C virus replication [396]. Zeng et al. prepared Chitosan-modified PLGA nanospheres for delivery plasmid and silencing DNA hepatitis C virus gene. chitosan modified PLGA nanospheres displayed much higher loading efficiency when compared to unmodified PLGA nanospheres. The EGFP expression experiments using observation with confocal laser scanning microscopy demonstrated that pDNA-encapsulated chitosan-modified PLGA nanospheres were more effectively taken up by the cells than unmodified PLGA nanospheres [397].

19.8.5.4 Polymeric nanoparticles and nanospheres for COVID-19 treatment

Khater et al. formulated PLGA-lipid hybrid nanoparticles loaded with various antiviral drugs (such as Atomoxetine, Fluoxetine hydrochloride, Paroxetine, Nisoxteine, Repoxteine SS and Repoxteine RR) for the treatment of COVID-19 [398]. The DLS results of polymer lipid nanoparticles displayed particle size of about 98.5 ± 3.5 nm and a negative surface charge of 10.5 ± 0.45 mV, showing their ability to promote cellular uptake. The in vitro cytocompatibility studies using MTT assay demonstrated a high cell viability of about 80% when nanoparticles were cultured with Human lung fibroblast (CCD-19Lu) Cells, suggesting excellent biocompatibility and nontoxicity that can be ideal for the treatment of COVID-19. Furthermore, the in vitro drug release profile exhibited sustained drug release of antiviral drugs from polymer lipid hybrid nanoparticles [398].

19.8.6 Hydrogels and nanogels

Hydrogels are drug delivery systems with 3-dimensional networks (Fig. 19.24) usually fabricated from natural and synthetic polymers. These delivery systems possessed the capability to engulf and preserve large volumes of water and biological fluids [399]. The porosity degree of polymer-based hydrogels is influenced by various factors such as formulation procedure, polymer composition, and the materials from which they are derived, etc. Hydrogels can be manipulated into different forms such as microparticles, slabs,

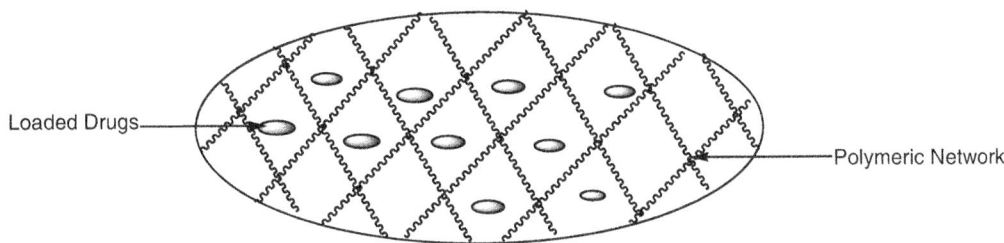

FIGURE 19.24 Criss-linked hydrogels loaded with drugs.

nanoparticles, and films. Hydrogels exhibit unique advantages such as nontoxicity, excellent biocompatibility, nonimmunogenicity, affordability, environmental sensitivity (e.g., pH, electric field, and temperature), and their drug release mode can be tailored [399].

19.8.6.1 Hydrogels and nanogels for human immunodeficiency virus treatment

Tian and coworkers fabricated poloxamer 407-based hydrogels for drug delivery of antiviral drugs; theaflavin and Nifeviroc [400]. The rheological studies demonstrated that the loss modulus (G00) and storage modulus (G0) was gradually increased with the increase in temperature and reached the gelation temperature at 25°C. The pharmacokinetic results demonstrated that the poloxamer 407-based hydrogels possessed an excellently controlled release profile. The ex vivo studies revealed that the polymeric hydrogel was able to bind to coreceptor CCR5 of DCs cells, indicating that poloxamer 407-based hydrogels could be an auspicious drug delivery system for intravaginal delivery of anti-HIV agents [400]. Macchione et al. reported poly(N-vinylcaprolactam)-based nanogels for HIV treatment. The TEM micrographs of the nanogels displayed a particle size of approximately 40 nm. The in vitro cytotoxicity experiments using MTT assay showed high cell viability of more than 80% for cervicovaginal epithelial cell lines when were incubated with poly(N-vinylcaprolactam)-based nanogels for 2 days. The in vitro antiviral analysis of the nanogels demonstrated an inhibitory effect against the R5-HIV-1 isolate with inhibition rates of more70% at the concentration of 10 μg/mL, showing excellent antiviral activity [401].

The chitosan-carboxyl methylcellulose nanogels loaded with Nevirapine nanoparticles formulated by Rahman and Ahmed exhibited good spreadability, homogeneity, content uniformity, and physical appearance. The pH of the chitosan-based hybrid nanogels was in the range of 3.70 to 4.56, which lies in the standard pH range of the vaginal fluid, indicating their ability to be used vaginal application for the HIV infection treatment [402]. Town et al. poly(N-isopropylacrylamide) nanogels for drug delivery of anti-HIV agent lopinavir. The DLS analysis of nanogels demonstrated particle sizes that range between 65 and 450 nm, demonstrating their ability to be used as potential nanocarriers. The in vitro drug release profile displayed prolonged sustained drug release of lopinavir from the polymeric nanogels [403].

19.8.6.2 Hydrogels and nanogels for herpes treatment

Sabbagh and Muhamad developed polyacrylamide-based hydrogels loaded with antiherpes drug acyclovir. The physicochemical properties of hydrogels were characterized by

FTIR. The in vitro drug release studies showed fast release of acyclovir from hydrogels at pH 4.2 and pH 7.4, demonstrating that these hydrogels are potential systems to deliver acyclovir. The acyclovir-loaded β-cyclodextrin-g-poly (AMPS) hydrogels demonstrated good pharmacokinetic parameters, suggesting their ability to be used in the treatment of herpes [404]. Houston et al. formulated hydroxypropyl methylcellulose hydrogels coloaded with punicalagin and zinc (II) ions for the treatment of herpes. The in vitro antiviral studies showed superior virucidal activity against HSV-1 infected Vero cells when treated with dual drug-loaded hydrogels compared with plain hydrogels [405].

Malik et al. fabricated chitosan/xanthan gum hybrid hydrogels for drug delivery of acyclovir. The TGA, FTIR, XRD results confirmed the successful formulation of acyclovir-loaded hydrogels. SEM images displayed the porous structure of fabricated hydrogels. The in vitro drug release experiments demonstrated that chitosan/xanthan gum hybrid hydrogels have revealed very less drug release at acidic pH, but it significantly increased as the pH of the dissolution medium increased to pH 7.4, demonstrating controlled drug release [406]. The β-cyclodextrin chitosan-based hydrogels loaded with acyclovir were formulated by Malik and exhibited no significant changes in clinical, behavioral, or histopathological parameters of rats. Furthermore, the pharmacokinetic experiment showed that chitosan-based hydrogels resulted in an important increase in oral bioavailability of acyclovir in rabbit plasma than oral suspension [407]. The acyclovir-loaded sericin hydrogels were formulated by Al-Tabakha et al. demonstrated the release of sericin from the sericin matrix happened via a diffusion-controlled mechanism, and they were finely tolerable up to 3800 mg/kg body weight of rabbits [408].

19.8.6.3 Hydrogels and nanogels for hepatitis and Human papilloma virus treatment

Tran et al. formulated alginate-based hydrogels loaded with Hepatic HuH-7 Cells for the treatment of hepatitis treatment. The SEM micrographs displayed that these hydrogels possessed high porosity. Alginate-based hydrogel formulations demonstrated the capability to block hepatitis C virus particle release [409]. The microneedle arrays-incorporated hydrogels formulated by Guo demonstrated good capacity to stimulate more potent immune responses when compared to the hydrogels without microneedle arrays, showing that they are a potential vaccination method against hepatitis infections [410]. Bae et al. formulated DX hydrogels for drug delivery of PEGylated protein drugs for the treatment of hepatitis. The physiochemical properties of hydrogels were characterized by FTIR spectroscopy. The pharmacokinetic experiments exhibited that these polymeric hydrogels significantly prolonged the circulation half-life of loaded PEGylated protein drugs, permitting for less regular dosing in a humanized mouse model of hepatitis C [411].

19.8.6.4 Hydrogels and nanogels for Ebola treatment

Wua et al. fabricated chitosan-based hydrogels loaded with Zaire Ebola virus glycoprotein antigen for Ebola management. The cytocompatibility experiments demonstrated that these hydrogels are nontoxic and safe for the treatment of Ebola. Other experiments exhibited that chitosan hydrogels loaded with Ebola antigen significantly stimulated the highest IgG2a IgG1, and IgG, antibody titers in serum and mucosal IgA responses in lung wash, which may cause by the extended antigen residence period because of the thermal response of these polymeric hydrogels [412]. Furthermore, the chitosan-based hydrogels encapsulated by the H5N1 vaccine reported by Fan and coworkers showed similar results. We applied in vivo imaging system and found that the moderate DQ 41% of these hydrogels resulted in extended antigen residence time in the nasal cavity, leading to the most effective systemic responses (HI, IgG, IgG2a, and IgG1), suggesting that these hydrogels are potential candidates for the vaccinations against Ebola [413].

19.9 Conclusion and future perspective

The challenges associated with the treatment of viral infections have been identified. The challenges have been highlighted and the solutions have been reported. Several researchers have developed different systems and therapeutics for the treatment of viral infections. However, the number of cases and deaths due to these viral infections are still increasing and these diseases are evolving into new genetic variants that become resistant to the currently used therapies. Therefore, collaborative research by physicians, chemists, doctors, private sponsors, academic institutes, pharmaceutical companies, media, communities, etc. is needed so that more research can be conducted to find new effective therapies.

Combination therapy has been considered as an effective approach for the treatment of infectious diseases and appropriate to overcome multidrug resistance by targeting different sites and pathogens, especially in severe infections. Different synthetic approach has been employed to design therapeutics containing two or more antiviral drugs such as hybrid compounds synthesis and the use of drug delivery systems. These approaches offer several advantages such as improved drug solubility, bioavailability, toxicity, biodegradability, overcome drug resistance, etc. The results obtained so far indicates that further research on antiviral agents will result in potent antiviral therapeutics.

Acknowledgments

The financial support of Govan Mbeki Research and Development Centre, University of Fort Hare, South Africa Medical Research Council, and National Research Foundation, South Africa towards this research is hereby acknowledged.

References

[1] Cojocaru FD, Botezat D, Gardikiotis I, Uritu CM, Dodi G, Trandafir L, et al. Nanomaterials designed for antiviral drug delivery transport across biological barriers. Pharmaceutics. 2020;12(2):171.

[2] Barker J, Stevens D, Bloomfield SF. Spread and prevention of some common viral infections in community facilities and domestic homes. J Appl Microbiol 2001;91(1):7.

[3] Szucs PA, Richman PB, Mandell M. Triage nurse application of the Ottawa knee rule. Acad Emerg Med 2001;8:112−16.

[4] Sander WJ, O'Neill HG, Pohl CH. Prostaglandin E2 as a modulator of viral infections. Front Physiol 2017;8:89.

[5] Yang KC, Lin JC, Tsai HH, Hsu CY, Shih V, Hu CM. Nanotechnology advances in pathogen-and host-targeted antiviral delivery: multipronged therapeutic intervention for pandemic control. Drug Deliv Transl Res 2021;1−8.

[6] Colpitts CC, Verrier ER, Baumert TF. Targeting viral entry for treatment of hepatitis B and C virus infections. ACS Infect Dis 2015;1(9):420−7.

[7] Rivera A, Messaoudi I. Pathophysiology of Ebola virus infection: current challenges and future hopes. ACS Infect Dis 2015;1(5):186−97.

[8] Chen R, Wang T, Song J, Pu D, He D, Li J, et al. Antiviral drug delivery system for enhanced bioactivity, better metabolism and pharmacokinetic characteristics. Int J Nanomed 2021;16:4959.

[9] Trovato M, Sartorius R, D'Apice L, Manco R, De Berardinis P. Viral emerging diseases: challenges in developing vaccination strategies. Front Immunol 2020;11.

[10] Rouse BT, Lukacher AE. Some unmet challenges in the immunology of viral infections. Discov Med 2010;10 (53):363.

[11] Sailaja I, Baghel MK, Shaker IA. Nanotechnology based drug delivery for HIV-AIDS treatment, 2021.

[12] Delshadi R, Bahrami A, Mcclements DJ, Moore MD, Williams L. Development of nanoparticle-delivery systems for antiviral agents: a review. J Control Release 2021.

[13] Kumar P, Lakshmi YS, Golla K, Kondapi AK. Improved safety, bioavailability and pharmacokinetics of zidovudine through lactoferrin nanoparticles during oral administration in rats. PLoS one 2015;10(10):e0140399.

[14] Phe T, Thai S, Veng C, Sok S, Lynen L, van Griensven J. Risk factors of treatment-limiting anemia after substitution of zidovudine for stavudine in HIV-infected adult patients on antiretroviral treatment. PLoS One 2013;8(3):e60206.

[15] Podlekareva D, Grint D, Karpov I, Rakmanova A, Mansinho K, Chentsova N, et al. Changing utilization of Stavudine (d4T) in HIV-positive people in 2006−2013 in the EuroSIDA study. HIV Med 2015;16(9):533−43.

[16] Brennan AT, Davies MA, Bor J, Wandeler G, Stinson K, Wood R, et al. Has the phasing out of stavudine in accordance with changes in WHO guidelines led to a decrease in single-drug substitutions in first-line antiretroviral therapy for HIV in sub-Saharan Africa? AIDS (London, Engl) 2017;31(1):147.

[17] James AM, Ofotokun I, Sheth A, Acosta EP, King JR. Tenofovir: once-daily dosage in the management of HIV infection. Clin Med Insights: Therap 2012;4 CMT-S8316.

[18] Nelson MR, Katlama C, Montaner JS, Cooper DA, Gazzard B, Clotet B, et al. The safety of tenofovir disoproxil fumarate for the treatment of HIV infection in adults: the first 4 years. AIDS. 2007;21(10):1273−81.

[19] Atta MG, De Seigneux S, Lucas GM. Clinical pharmacology in HIV therapy. Clin J Am Soc Nephrol 2019;14 (3):435−44.

[20] Coutinho B, Prasad R. Emtricitabine/tenofovir (Truvada) for HIV prophylaxis. Am family physician 2013;88 (8):535−40.

[21] Saravolatz LD, Saag MS. Emtricitabine, a new antiretroviral agent with activity against HIV and hepatitis B virus. Clin Infect Dis 2006;42(1):126−31.

[22] Barbarino JM, Kroetz DL, Altman RB, Klein TE. PharmGKB summary: abacavir pathway. Pharmacogenet. Genomics 2014;24(5):276.

[23] Adetokunboh OO, Schoonees A, Balogun TA, Wiysonge CS. Efficacy and safety of abacavir-containing combination antiretroviral therapy as first-line treatment of HIV infected children and adolescents: a systematic review and meta-analysis. BMC Infect Dis 2015;15(1):1−3.

[24] Mega TA, Usamo FB, Negera GZ. Abacavir vs zidovudine-based regimens for treatment of HIV-infected children in resource limited settings: a retrospective cohort study. BMC pediatrics 2020;20(1):1−9.

[25] Sivasubramanian G, Frempong-Manso E, MacArthur RD. Abacavir/lamivudine combination in the treatment of HIV: a review. Ther. Clin. Risk Manag. 2010;6:83.

[26] Verloes R, Deleu S, Niemeijer N, Crauwels H, Meyvisch P. Williams P. Safety, tolerability and pharmacokinetics of rilpivirine following administration of a long-acting formulation in healthy volunteers. HIV Med 2015;16(8):477−84.

[27] McGowan I, Dezzutti CS, Siegel A, Engstrom J, Nikiforov A, Duffill K, et al. Long-acting rilpivirine as potential pre-exposure prophylaxis for HIV-1 prevention (the MWRI-01 study): an open-label, phase 1, compartmental, pharmacokinetic and pharmacodynamic assessment. Lancet HIV 2016;3(12):e569−78.

[28] Sharma M, Saravolatz LD. Rilpivirine: a new non-nucleoside reverse transcriptase inhibitor. J Antimicrobial Chemother 2013;68(2):250−6.

[29] Díaz-Delfín J, Domingo P, Mateo MG, Gutierrez MD, Domingo JC, Giralt M, et al. Effects of rilpivirine on human adipocyte differentiation, gene expression, and release of adipokines and cytokines. Antimicrob. Agents Chemother 2012;56(6):3369−75.

[30] Taha H, Das A, Das S. Clinical effectiveness of dolutegravir in the treatment of HIV/AIDS. Infect Drug Resist 2015;8:339.

[31] Zamora FJ, Dowers E, Yasin F, Ogbuagu O. Dolutegravir and lamivudine combination for the treatment of HIV-1 infection. HIV/AIDS (Auckland, NZ) 2019;11:255.

[32] Bruzzese E, Vecchio AL, Smarrazzo A, Tambaro O, Palmiero G, Bonadies G, et al. Dolutegravir-based anti-retroviral therapy is effective and safe in HIV−infected paediatric patients. Ital J Pediatr 2018;44(1):1−5.

[33] Osterholzer DA, Goldman M. Dolutegravir: a next-generation integrase inhibitor for treatment of HIV infection. Clin Infect Dis 2014;59(2):265−71.

[34] Dowers E, Zamora F, Barakat LA, Ogbuagu O. Dolutegravir/rilpivirine for the treatment of HIV-1 infection. HIV/AIDS (Auckland, NZ) 2018;10:215.

[35] Hester EK, Astle K. Dolutegravir-rilpivirine, dual antiretroviral therapy for the treatment of HIV-1 infection. Ann Pharmacother 2019;53(8):860−6.

[36] Zhang X. Anti-retroviral drugs: current state and development in the next decade. Acta Pharm Sin B 2018;8(2):131−6.

[37] Whitfield T, Torkington A, van Halsema C. Profile of cabotegravir and its potential in the treatment and prevention of HIV-1 infection: evidence to date. HIV/Aids (Auckland, NZ) 2016;8:157.

[38] Colgan R, Michocki R, Greisman L, Moore TA. Antiviral drugs in the immunocompetent host: part I. Treatment of hepatitis, cytomegalovirus, and herpes infections. Am. Fam. Physician 2003;67(4):757−62.

[39] Tsai WL, Chiang PH, Chan HH, Lin HS, Lai KH, Cheng JS, et al. Early entecavir treatment for chronic hepatitis B with severe acute exacerbation. Antimicrob. Agents Chemother 2014;58(4):1918−21.

[40] Strauch S, Jantratid E, Dressman JB, Junginger HE, Kopp S, Midha KK, et al. COMMENTARY: biowaiver monographs for immediate release solid oral dosage forms: lamivudine. J Pharm Sci 2011;100(6):2054−63.

[41] Mandala D, Chada S, Watts P. Semi-continuous multi-step synthesis of lamivudine. Org Biomol Chem 2017;15(16):3444−54.

[42] Kausar S, Said Khan F, Ishaq Mujeeb Ur Rehman M, Akram M, Riaz M, Rasool G, et al. A review: mechanism of action of antiviral drugs. Int J Immunopathol Pharmacol 2021;35 20587384211002621.

[43] Razonable RR. Antiviral drugs for viruses other than human immunodeficiency virus. Mayo ClProc 2011;86(10):1009−26 Elsevier.

[44] Andrei G, Snoeck R. Advances and perspectives in the management of varicella-zoster virus infections. Molecules. 2021;26(4):1132.

[45] Vigil KJ, Chemaly RF. Valacyclovir: approved and off-label uses for the treatment of herpes virus infections in immunocompetent and immunocompromised adults. Expert Opin Pharmacother 2010;11:1901−13. Available from: https://doi.org/10.1517/14656566.2010.49417946.

[46] Rajalakshmi R, Kumari R, Thappa DM. Acyclovir vs valacyclovir. Indian J Dermatol Venereol Leprol 2010;76:439−44. Available from: https://doi.org/10.4103/0378-6323.66577.

[47] De Clercq E. Selective anti-herpesvirus agents. Antivir Chem Chemother 2013;23(3):93−101.

[48] Álvarez DM, Castillo E, Duarte LF, Arriagada J, Corrales N, Farías MA, et al. Current antivirals and novel botanical molecules interfering with herpes simplex virus infection. Front Microbiol 2020;11:139.

[49] Razonable RR, Paya CV. Valganciclovir for the prevention and treatment of cytomegalovirus disease in immunocompromised hosts. Expert Rev Anti Infect Ther 2004;2:27−41 YQFSU 3FW "OUJ *OGFDU 5IFS.

[50] Vashishtha AK, Kuchta RD. Effects of acyclovir, foscarnet, and ribonucleotides on herpes simplex virus-1 DNA polymerase: mechanistic insights and a novel mechanism for preventing stable incorporation of ribonucleotides into DNA. Biochemistry 2016;55:1168−77.

[51] Sadowski LA, Upadhyay R, Greeley ZW, Margulies BJ. Current drugs to treat infections with herpes simplex viruses-1 and-2. Viruses. 2021;13(7):1228.

[52] Leowattana W. Antiviral drugs and acute kidney injury (AKI). Infect Disord Drug Targets 2019;19:375−82.

[53] Mareri A, Lasorella S, Iapadre G, Maresca M, Tambucci R, Nigro G. Anti-viral therapy for congenital cytomegalovirus infection: pharmacokinetics, efficacy and side effects. J Matern Fetal Neonatal Med 2016;29:1657−64.

[54] Adalsteinsson JA, Pan M, Kaushik S, Ungar J. Foscarnet-induced genital lesions: an overview with a case report. Dermatology Rep 2018;10(1).

[55] Semaan JR, Parmar M. Famciclovir. StatPearls [Internet]. 2020 Dec 2.

[56] Whitley RJ, Volpi A, McKendrick M, van Wijck A, Oaklander AL. Management of herpes zoster and postherpetic neuralgia now and in the future. J Clin Virol 2010;48:S20−8.

[57] Wang L, Verschuuren EA, van Leer-Buter CC, Bakker SJ, de Joode AA, Westra J, et al. Herpes zoster and immunogenicity and safety of zoster vaccines in transplant patients: a narrative review of the literature. Front Immunol 2018;9:1632.

[58] Mottu A, Rubbia-Brandt L, Bihl F, Hadengue A, Spahr L. Acute hepatitis due to brivudin: a case report. J Hepatol 2009;51(5):967–9.

[59] Dworkin RH, Johnson RW, Breuer J, Gnann JW, Levin MJ, Backonja M, et al. Recommendations for the management of herpes zoster. Clin Infect Dis 2007;44(Suppl 1):S1–26.

[60] De Clercq E. Discovery and development of BVDU (brivudin) as a therapeutic for the treatment of herpes zoster. Biochem Pharmacol 2004;68:2301–15.

[61] Yaldiz M, Solak B, Kara RO, Cosansu N, Erdem MT. Comparison of famciclovir, valaciclovir, and brivudine treatments in adult immunocompetent patients with herpes zoster. Am J Ther 2018;25:e626–34.

[62] Osborn MK, Lok AS. Antiviral options for the treatment of chronic hepatitis B. J Antimicrobial Chemother 2006;57(6):1030–4.

[63] Dusheiko G. Adefovir dipivoxil for the treatment of HBeAg-positive chronic hepatitis B: a review of the major clinical studies. J Hepatol 2003;39:116–23.

[64] Izzedine H, Hulot JS, Launay-Vacher V, Marcellini P, Hadziyannis SJ, Currie G, et al. Renal safety of adefovir dipivoxil in patients with chronic hepatitis B: two double-blind, randomized, placebo-controlled studies. Kidney Int 2004;66(3):1153–8.

[65] Dey KK, Ghosh M. Investigation of the structure and dynamics of antiviral drug adefovir dipivoxil by site-specific spin–lattice relaxation time measurements and chemical shift anisotropy tensor measurements. ACS Omega 2020;5(45):29373–81.

[66] Park JW, Kwak KM, Kim SE, Jang MK, Suk KT, Kim DJ, et al. Comparison of the long-term efficacy between entecavir and tenofovir in treatment-naïve chronic hepatitis B patients. BMC Gastroenterol 2017; 17(1):1–9.

[67] Yim HJ, Lee HJ, Suh SJ, Seo YS, Kim CW, Lee CD, et al. Adefovir and lamivudine combination therapy in patients with entecavir-resistant chronic hepatitis B: antiviral responses and evolution of mutations. Intervirology. 2014;57(5):239–47.

[68] Zhao SH, Liu EQ, Cheng DX, Li YF, Wang YL, Chen YL, et al. Comparison of entecavir and adefovir for the treatment of chronic hepatitis B. Braz J Infect Dis 2012;16(4):366–72.

[69] Zhao SS, Tang LH, Dai XH, Wang W, Zhou RR, Chen LZ, et al. Comparison of the efficacy of tenofovir and adefovir in the treatment of chronic hepatitis B: a systematic review. Virol J 2011;8(1):1–9.

[70] Lee SW, Kwon JH, Lee HL, Yoo SH, Nam HC, Sung PS, et al. Comparison of tenofovir and entecavir on the risk of hepatocellular carcinoma and mortality in treatment-naïve patients with chronic hepatitis B in Korea: a large-scale, propensity score analysis. Gut. 2020;69(7):1301–8.

[71] Sun W, Zhao S, Ma L, Hao A, Zhao B, Zhou L, et al. Telbivudine treatment started in early and middle pregnancy completely blocks HBV vertical transmission. BMC Gastroenterol 2017;17(1):1–5.

[72] Prifti GM, Moianos D, Giannakopoulou E, Pardali V, Tavis JE, Zoidis G. Recent Advances in Hepatitis B Treatment. Pharmaceuticals. 2021;14(5):417.

[73] Zhang Y, Hu P, Qi X, Ren H, Mao RC, Zhang JM. A comparison of telbivudine and entecavir in the treatment of hepatitis B e antigen-positive patients: a prospective cohort study in China. Clin Microbiol Infect 2016;22(3):287 e1.

[74] Nagata T, Lefor AK, Hasegawa M, Ishii M. Favipiravir: a new medication for the Ebola virus disease pandemic. Disaster Med Public Health Prep 2015;9(1):79–81.

[75] Guedj J, Piorkowski G, Jacquot F, Madelain V, Nguyen TH, Rodallec A, et al. Antiviral efficacy of favipiravir against Ebola virus: a translational study in cynomolgus macaques. PLoS Med 2018;15(3):e1002535.

[76] Sissoko D, Laouenan C, Folkesson E, M'lebing AB, Beavogui AH, Baize S, et al. Experimental treatment with favipiravir for Ebola virus disease (the JIKI Trial): a historically controlled, single-arm proof-of-concept trial in Guinea. PLoS Med 2016;13(3):e1001967.

[77] Oestereich L, Lüdtke A, Wurr S, Rieger T, Muñoz-Fontela C, Günther S. Successful treatment of advanced Ebola virus infection with T-705 (favipiravir) in a small animal model. Antivir Res 2014;105:17–21.

[78] Kerber R, Lorenz E, Duraffour S, Sissoko D, Rudolf M, Jaeger A, et al. Laboratory findings, compassionate use of favipiravir, and outcome in patients with Ebola virus disease, Guinea, 2015—a retrospective observational study. J Infect Dis 2019;220(2):195–202.

[79] Sanders JM, Monogue ML, Jodlowski TZ, Cutrell JB. Pharmacologic treatments for coronavirus disease 2019 (COVID-19): a review. Jama 2020;323(18):1824−36.

[80] Lo MK, Albariño CG, Perry JK, Chang S, Tchesnokov EP, Guerrero L, et al. Remdesivir targets a structurally analogous region of the Ebola virus and SARS-CoV-2 polymerases. Proc Natl Acad Sci 2020;117(43): 26946−54.

[81] Malin JJ, Suárez I, Priesner V, Fätkenheuer G, Rybniker J. Remdesivir against COVID-19 and other viral diseases. Clin Microbiol Rev 2020;34(1):e00162 20.

[82] Sweiti H, Ekwunife O, Jaschinski T, Lhachimi SK. Repurposed therapeutic agents targeting the Ebola virus: a systematic review. Curr Therap Res 2017;84:10−21.

[83] Salata C, Munegato D, Martelli F, Parolin C, Calistri A, Baritussio A, et al. Amiodarone affects Ebola virus binding and entry into target cells. N Microbiol 2018;41(2):162−4.

[84] Dyall J, Johnson JC, Postnikova E, Cong Y, Zhou H, Gerhardt DM, et al. In vitro and in vivo activity of amiodarone against Ebola virus. J Infect Dis 2018;218(Suppl 5):S592−6.

[85] Johansen LM, Brannan JM, Delos SE, Shoemaker CJ, Stossel A, Lear C, et al. FDA-approved selective estrogen receptor modulators inhibit Ebola virus infection. Sci Transl Med 2013;5(190):190 79.

[86] Morello KC, Wurz GT, DeGregorio MW. Pharmacokinetics of selective estrogen receptor modulators. Clin Pharmacokinet 2003;42:361−72.

[87] Stern PL, van der Burg SH, Hampson IN, Broker TR, Fiander A, Lacey CJ, et al. Therapy of human papillomavirus-related disease. Vaccine. 2012;30:F71−82.

[88] Muffarrej D, Khattab E, Najjar R. Successful treatment of genital warts with cidofovir cream in a pediatric patient with Fanconi anemia. J Oncol Pharm Pract 2020;26(5):1234−6.

[89] DeRossi SS, Laudenbach J. The management of oral human papillomavirus with topical cidofovir: a case report. Cutis-New Y 2004;73(3):191−6.

[90] Yang J, Dai LX, Chen M, Li B, Ding N, Li G, et al. Inhibition of antiviral drug cidofovir on proliferation of human papillomavirus infected cervical cancer cells. Exp Therap Med 2016;12(5):2965−73.

[91] Gröne D, Treudler R, De Villiers EM, Husak R, Orfanos CE, Zouboulis C. Intravenous cidofovir treatment for recalcitrant warts in the setting of a patient with myelodysplastic syndrome. J Eur Acad Dermatol Venereol 2006;20(2):202−5.

[92] Casaos J, Gorelick NL, Huq S, Choi J, Xia Y, Serra R, et al. The use of ribavirin as an anticancer therapeutic: will it go viral? Mol Cancer therapeutics 2019;18(7):1185−94.

[93] Mosa C, Trizzino A, Trizzino A, Di Marco F, D'Angelo P, Farruggia P. Treatment of human papillomavirus infection with interferon alpha and ribavirin in a patient with acquired aplastic anemia. Int J Infect Dis 2014;23:25−7.

[94] Pavan MH, Velho PE, Vigani AG, Gonçalves Jr FL, Aoki FH. Treatment of human papillomavirus with peg-interferon alfa-2b and ribavirin. Braz J Infect Dis 2007;11(3):383−4.

[95] Gross AE, Bryson ML. Oral ribavirin for the treatment of noninfluenza respiratory viral infections: a systematic review. Ann Pharmacother 2015;49(10):1125−35.

[96] Fuehner T, Dierich M, Duesberg C, et al. Single-centre experience with oral ribavirin in lung transplant recipients with paramyxovirus infections. Antivir Ther 2011;16:733−40.

[97] Lehners N, Schnitzler P, Geis S, et al. Risk factors and containment of respiratory syncytial virus outbreak in a hematology and transplant unit. Bone Barrow Transpl 2013;48:1548−53.

[98] Pagliano P, Sellitto C, Conti V, Ascione T, Esposito S. Characteristics of viral pneumonia in the COVID-19 era: an update. Infection. 2021;1 0.

[99] Oboho IK, Bramley A, Finelli L, Fry A, Ampofo K, Arnold SR, et al. Oseltamivir use among children and adults hospitalized with community-acquired pneumonia. InOpen forum infectious diseases, Vol. 4. US: Oxford University Press; 2017. p. ofw254. No. 1.

[100] Chen HD, Wang X, Yu SL, Ding YH, Wang ML, Wang JN. Clinical effectiveness of intravenous peramivir compared with oseltamivir in patients with severe Influenza A with primary viral pneumonia: a randomized controlled study. InOpen Forum Infectious Diseases, Vol. 8. US: Oxford University Press; 2021. p. ofaa562. No. 1.

[101] Yu H, Liao Q, Yuan Y, Zhou L, Xiang N, Huai Y, et al. Effectiveness of oseltamivir on disease progression and viral RNA shedding in patients with mild pandemic 2009 influenza A H1N1: opportunistic retrospective study of medical charts in China. Bmj. 2010;341.

[102] Heneghan CJ, Onakpoya I, Thompson M, Spencer EA, Jones M, Jefferson T. Zanamivir for influenza in adults and children: systematic review of clinical study reports and summary of regulatory comments. BMJ 2014;348.

[103] Zablockienė B, Kačergius T, Ambrozaitis A, Žurauskas E, Bratchikov M, Jurgauskienė L, et al. Zanamivir diminishes lung damage in influenza a virus-infected mice by inhibiting nitric oxide production. vivo 2018;32(3):473−8.

[104] Geraghty RJ, Aliota MT, Bonnac LF. Broad-spectrum antiviral strategies and nucleoside analogues. Viruses. 2021;13(4):667.

[105] Frediansyah A, Tiwari R, Sharun K, Dhama K, Harapan H. Antivirals for COVID-19: a critical review. Clin Epidemiol Glob Health 2021;9:90−8.

[106] Kaur H, Shekhar N, Sharma S, Sarma P, Prakash A, Medhi B. Ivermectin as a potential drug for treatment of COVID-19: an in-sync review with clinical and computational attributes. Pharmacol Rep 2021;1−4.

[107] Okumuş N, Demirtürk N, Çetinkaya RA, Güner R, Avcı İY, Orhan S, et al. Evaluation of the effectiveness and safety of adding ivermectin to treatment in severe COVID-19 patients. BMC Infect Dis 2021;21(1):1.

[108] Dabbous HM, Abd-Elsalam S, El-Sayed MH, Sherief AF, Ebeid FF, et al. Efficacy of favipiravir in COVID-19 treatment: a multi-center randomized study. Arch Virol 2021;166(3):949−54.

[109] Pilkington V, Pepperrell T, Hill A. A review of the safety of favipiravir−a potential treatment in the COVID-19 pandemic? J Virus Erad 2020;6(2):45−51.

[110] Manabe T, Kambayashi D, Akatsu H, Kudo K. Favipiravir for the treatment of patients with COVID-19: a systematic review and meta-analysis. BMC Infect Dis 2021;21(1):1−3.

[111] Kaur RJ, Charan J, Dutta S, Sharma P, Bhardwaj P, Sharma P, et al. Favipiravir use in COVID-19: analysis of suspected adverse drug events reported in the WHO database. Infect Drug Resist 2020;13:4427.

[112] Łagocka R, Dziedziejko V, Kłos P, Pawlik A. Favipiravir in therapy of viral infections. J Clin Med 2021; 10(2):273.

[113] Wang Y, Zhang D, Du G, Du R, Zhao J, Jin Y, et al. Remdesivir in adults with severe COVID-19: a randomised, double-blind, placebo-controlled, multicentre trial. Lancet 2020;395(10236):1569−78.

[114] Kokic G, Hillen HS, Tegunov D, Dienemann C, Seitz F, Schmitzova J, et al. Mechanism of SARS-CoV-2 polymerase stalling by remdesivir. Nat Commun 2021;12(1):1−7.

[115] Pardo J, Shukla AM, Chamarthi G, Gupte A. The journey of remdesivir: from Ebola to COVID-19. Drugs Context 2020;9.

[116] Lin HX, Cho S, Aravamudan VM, Sanda HY, Palraj R, Molton JS, et al. Remdesivir in Coronavirus Disease 2019 (COVID-19) treatment: a review of evidence. Infection. 2021;401−10.

[117] Hashemian SM, Farhadi T, Velayati AA. A review on remdesivir: a possible promising agent for the treatment of COVID-19. Drug Design, Dev Ther 2020;14:3215.

[118] Hong YN, Xu J, Sasa GB, Zhou KX, Ding XF. Remdesivir as a broad-spectrum antiviral drug against COVID-19. Eur Rev Med Pharmacol Sci 2021;25(1):541−8.

[119] Al-Tannak NF, Novotny L, Alhunayan A. Remdesivir-bringing hope for COVID-19 treatment. Sci Pharm 2020;88(2):29.

[120] Horby PW, Mafham M, Bell JL, Linsell L, Staplin N, Emberson J, et al. Lopinavir−ritonavir in patients admitted to hospital with COVID-19 (RECOVERY): a randomised, controlled, open-label, platform trial. Lancet 2020;396(10259):1345−52.

[121] Alhumaid S, Al Mutair A, Al Alawi Z, Alhmeed N, Zaidi AR, Tobaiqy M. Efficacy and safety of lopinavir/ritonavir for treatment of COVID-19: a systematic review and meta-analysis. Trop Med Infect Dis 2020;5(4):180.

[122] Patel TK, Patel PB, Barvaliya M, Saurabh MK, Bhalla HL, Khosla PP. Efficacy and safety of Lopinavir-Ritonavir in COVID-19: a systematic review of randomized controlled trials. J Infect public health 2021.

[123] Cao B, Wang Y, Wen D, Liu W, Wang J, Fan G, et al. A trial of lopinavir−ritonavir in adults hospitalized with severe Covid-19. N Engl J Med 2020.

[124] Lozach PY. Cell biology of viral infections. Cells. 2020;9:2431.

[125] Dimitrov DS. Virus entry: molecular mechanisms and biomedical applications. Nat Rev Microbiol 2004;2 (2):109−22.

[126] Ryu W-S. Virus life cycle. Molecular Virology of Human Pathogenic Viruses 2017;2017:31−45 10.

[127] Connolly SA, Jackson JO, Jardetzky TS, et al. Fusing structure and function: a structural view of the herpesvirus entry machinery. Nat Rev Microbiol 2011;9:369−81.

[128] Pascutti MF, Erkelens MN, Nolte MA. Impact of viral infections on hematopoiesis: from beneficial to detrimental effects on bone marrow output. Front Immunol 2016;7:364.

[129] Wilkins T. HIV 1: epidemiology, pathophysiology and transmission. Nurs Times 2020;116(7):40−2.

[130] Tan IL, Smith BR, von Geldern G, Mateen FJ, McArthur JC. HIV-associated opportunistic infections of the CNS. Lancet Neurol 2012;11(7):605−17.

[131] Bomsel M, Alfsen A. Entry of viruses through the epithelial barrier: pathogenic trickery. Nat Rev Mol Cell Biol 2003;4:57−68.

[132] Bracq L, Xie M, Benichou S, Bouchet J. Mechanisms for cell-to-cell transmission of HIV-1. Front Immunol 2018;9:260.

[133] Sierra S, Walter H. Targets for inhibition of HIV replication: entry, enzyme action, release and maturation. Intervirology. 2012;55(2):84−97.

[134] Clapham PR, McKnight A. Cell surface receptors, virus entry and tropism of primate lentiviruses. J Gen Virol 2002;83:1809−29.

[135] Moore JP, Kitchen SG, Pugach P, Zack JA. The CCR5 and CXCR4 coreceptors—central to understanding the transmission and pathogenesis of human immunodeficiency virus type 1 infection. AIDS Res Hum Retroviruses 2004;20:111−26.

[136] Nicola AV, Hou J, Major EO, Straus SE. Herpes simplex virus type 1 enters human epidermal keratinocytes, but not neurons, via a pH-dependent endocytic pathway. J Virol 2005;79:7609−16. Available from: https://doi.org/10.1128/JVI.79.

[137] Miranda-Saksena M, Denes CE, Diefenbach RJ, Cunningham AL. Infection and transport of herpes simplex virus type 1 in neurons: role of the cytoskeleton. Viruses 2018;10:E92. Available from: https://doi.org/10.3390/v10020092.

[138] Banerjee A, Kulkarni S, Mukherjee A. Herpes simplex virus: the hostile guest that takes over your home. Front Microbiol 2020;11:733.

[139] Madavaraju K, Koganti R, Volety I, Yadavalli T, Shukla D. Herpes simplex virus cell entry mechanisms: an update. Front Cell Infect Microbiol. 2021;10:852.

[140] Atanasiu D, Cairns TM, Whitbeck JC, Saw WT, Rao S, Eisenberg RJ, et al. Regulation of herpes simplex virus gB-induced cell-cell fusion by mutant forms of gH/gL in the absence of gD and cellular receptors. mBio 2013;4:e00046−13. Available from: https://doi.org/10.1128/mBio.00046-13.

[141] Wang X, Liu L, Whisnant AW, Hennig T, Djakovic L, Haque N, et al. Mechanism and consequences of herpes simplex virus 1-mediated regulation of host mRNA alternative polyadenylation. PLoS Genet 2021;17(3):e1009263.

[142] Karasneh GA, Shukla D. Herpes simplex virus infects most cell types in vitro: clues to its success. Virol J 2011;8(1):1.

[143] Melroe GT, DeLuca NA, Knipe DM. Herpes simplex virus 1 has multiple mechanisms for blocking virus-induced interferon production. J virology 2004;78(16):8411−20.

[144] Yokota S, Yokosawa N, Kubota T, Suzutani T, Yoshida I, Miura S, et al. Herpes simplex virus type 1 suppresses the interferon signaling pathway by inhibiting phosphorylation of STATs and Janus kinases during an early infection stage. Virology286 2001;119−24.

[145] Wang M, Feng Z. Mechanisms of hepatocellular injury in hepatitis A. Viruses. 2021;13(5):861.

[146] Tsai KN, Kuo CF, Ou JH. Mechanisms of hepatitis B virus persistence. Trends Microbiol 2018;26(1):33−42.

[147] Dandri M, Petersen J. Mechanism of hepatitis B virus persistence in hepatocytes and its carcinogenic potential. Clin Infect Dis 2016;62(Suppl 4):S281−8.

[148] D'souza S, Lau KC, Coffin CS, Patel TR. Molecular mechanisms of viral hepatitis induced hepatocellular carcinoma. World J Gastroenterol 2020;26(38):5759.

[149] Jain S, Khaiboullina SF, Baranwal M. Immunological perspective for ebola virus infection and various treatment measures taken to fight the disease. Pathogens. 2020;9(10):850.

[150] Feldmann H, Sprecher A, Geisbert TW. Ebola. N Engl J Med 2020;382(19):1832−42.

[151] Goeijenbier M, van Kampen JJ, Reusken CB, Koopmans MP, van Gorp EC. Ebola virus disease: a review on epidemiology, symptoms, treatment and pathogenesis. Neth J Med 2014;72:442−8.

[152] Marcinkiewicz J, Bryniarski K, Nazimek K. Ebola haemorrhagic fever virus: pathogenesis, immune responses, potential prevention. Folia Med Crac 2014;54:39−48.

[153] Perdomo-Celis F, Salvato MS, Medina-Moreno S, Zapata JC. T-cell response to viral hemorrhagic fevers. Vaccines 2019;7:11.

[154] Geisbert TW, Hensley LE, Larsen T, Young HA, Reed DS, Geisbert JB, et al. Pathogenesis of Ebola hemorrhagic fever in cynomolgus macaques: evidence that dendritic cells are early and sustained targets of infection. Am J Pathol 2003;163:2347−70.

[155] Falasca L, Agrati C, Petrosillo N, Di Caro A, Capobianchi MR, Ippolito G, et al. Molecular mechanisms of Ebola virus pathogenesis: focus on cell death. Cell Death Differ 2015;22(8):1250−9.

[156] Saeed MF, Kolokoltsov AA, Albrecht T, Davey RA. Cellular entry of ebola virus involves uptake by a macropinocytosis-like mechanism and subsequent trafficking through early and late endosomes. PLoS Pathog 2010;6:e1001110.

[157] Patel D, Delgado E, Nasr P. Ebola vrius disease: the biology, pathology, treatments, and advancements. Kean Quest 2020;3(1):6.

[158] Nanbo A, Imai M, Watanabe S, Noda T, Takahashi K, Neumann G, et al. Ebolavirus is internalized into host cells via macropinocytosis in a viral glycoprotein-dependent manner. PLoS Pathog 2010;6: e1001121.

[159] Hunt CL, Kolokoltsov AA, Davey RA, Maury W. The Tyro3 receptor kinase Axl enhances macropinocytosis of Zaire ebolavirus. J Virol 2011;85:334−47.

[160] Valmas C, Grosch MN, Schümann M, Olejnik J, Martinez O, Best SM, et al. Marburg virus evades interferon responses by a mechanism distinct from ebola virus. PLoS Pathog 2010;6(1):e1000721.

[161] Dhama K, Karthik K, Khandia R, Chakraborty S, Munjal A, Latheef SK, et al. Advances in designing and developing vaccines, drugs, and therapies to counter Ebola virus. Front Immunol 2018;9:1803.

[162] Horvath CA, Boulet GA, Renoux VM, Delvenne PO, Bogers JP. Mechanisms of cell entry by human papillomaviruses: an overview. Virol J 2010;7(1):1−7.

[163] Richards RM, Lowy DR, Schiller JT, Day PM. Cleavage of the papillomavirus minor capsid protein, L2, at a furin consensus site is necessary for infection. Proc Natl Acad Sci 2006;103:1522−7.

[164] Tong Q, Zheng L, Zhao R, Xing T, Li Y, Lin T, et al. Human papillomavirus infection mechanism and vaccine of vulva carcinoma. Open Life Sci 2016;11(1):185−90.

[165] Song D, Li H, Li H, Dai J. Effect of human papillomavirus infection on the immune system and its role in the course of cervical cancer. Oncol Lett 2015;10(2):600−6.

[166] Narisawa-Saito M, Kiyono T. Basic mechanisms of high-risk human papillomavirus-induced carcinogenesis: roles of E6 and E7 proteins. Cancer Sci 2007;98(10):1505−11.

[167] Shang Z, Kouznetsova VL, Tsigelny IF. Human papillomavirus (HPV) viral proteins substitute for the impact of somatic mutations by affecting cancer-related genes: *meta*-analysis and perspectives. J Infectiol 2020;3(1).

[168] Koo HJ, Lim S, Choe J, Choi SH, Sung H, Do KH. Radiographic and CT features of viral pneumonia. Radiographics. 2018;38(3):719−39.

[169] Figueiredo LT. Viral pneumonia: epidemiological, clinical, pathophysiological and therapeutic aspects. J Bras Pneumol 2009;35:899−906.

[170] Kalil AC, Thomas PG. Influenza virus-related critical illness: pathophysiology and epidemiology. Crit care 2019;23(1):1−7.

[171] Hilleman MR. Strategies and mechanisms for host and pathogen survival in acute and persistent viral infections. Proc Natl Acad Sci 2004;101(Suppl 2):14560−6.

[172] Audi A, Allbrahim M, Kaddoura M, Hijazi G, Yassine HM, Zaraket H. Seasonality of respiratory viral infections: will COVID-19 follow suit? Front Public Health 2020;8:576.

[173] Boopathi S, Poma AB, Kolandaivel P. Novel 2019 coronavirus structure, mechanism of action, antiviral drug promises and rule out against its treatment. J Biomol Struct Dyn 2021;39(9):3409−18.

[174] Cevik M, Kuppalli K, Kindrachuk J, Peiris M. Virology, transmission, and pathogenesis of SARS-CoV-2. bmj. 2020;371.

[175] Kirchdoerfer RN, Cottrell CA, Wang N, Pallesen J, Yassine HM, Turner HL, et al. Pre-fusion structure of a human coronavirus spike protein. Nature 2016;531(7592):118−21.

[176] Li F. Structure, function, and evolution of coronavirus spike proteins. Annu Rev virology 2016;3:237−61.

[177] Shyr ZA, Gorshkov K, Chen CZ, Zheng W. Drug discovery strategies for SARS-CoV-2. J Pharmacol Exp Therap 2020;375(1):127−38.

[178] Gu Y, Zuo X, Zhang S, Ouyang Z, Jiang S, Wang F, et al. The mechanism behind influenza virus cytokine storm. Viruses. 2021;13(7):1362.

[179] Herold S, Becker C, Ridge KM, Budinger GS. Influenza virus-induced lung injury: pathogenesis and implications for treatment. Eur Respir J 2015;45(5):1463–78.

[180] Hendaus MA, Jomha FA, Alhammadi AH. Virus-induced secondary bacterial infection: a concise review. Therap Clin Risk Manag 2015;11:1265.

[181] Skariyachan S, Challapilli SB, Packirisamy S, Kumargowda ST, Sridhar VS. Recent aspects on the pathogenesis mechanism, animal models and novel therapeutic interventions for Middle East respiratory syndrome coronavirus infections. Front Microbiol 2019;10:569.

[182] Dawre S, Maru S. Human respiratory viral infections: current status and future prospects of nanotechnology-based approaches for prophylaxis and treatment. Life Sci 2021;278:119561. Available from: https://doi.org/10.1016/j.lfs.2021.119561.

[183] Hussain M, Galvin H, Haw TY, Nutsford A, Husain M. Drug resistance in influenza A virus: the epidemiology and management. Infect Drug Resist 2017;10:121–34. Available from: https://doi.org/10.2147/IDR.S105473.

[184] Zheng W, Sun W, Simeonov A. Drug repurposing screens and synergistic drug-combinations for infectious diseases. Br J Pharmacol 2018;175:181–91. Available from: https://doi.org/10.1111/bph.13895.

[185] Chakravarty M, Vora A. Nanotechnology-based antiviral therapeutics. Drug Deliv Transl Res 2021;11:748–87. Available from: https://doi.org/10.1007/s13346-020-00818-0.

[186] Lembo D, Donalisio M, Civra A, Argenziano M, Cavalli R. Nanomedicine formulations for the delivery of antiviral drugs: a promising solution for the treatment of viral infections. Expert Opin Drug Deliv 2018;15:93–114. Available from: https://doi.org/10.1080/17425247.2017.1360863 Taylor & Francis.

[187] Vermehren J, James SP, Ira MJ, Stefan Z. Challenges and perspectives of direct antivirals for the treatment of hepatitis C virus infection. J Hepatol 2018;69:1178–87. Available from: https://doi.org/10.1016/j.jhep.2018.07.002.

[188] Duncan JD, Urbanowicz RA, Tarr AW, Ball JK. Hepatitis C virus vaccine: challenges and prospects. Vaccines 2020;8:1–23. Available from: https://doi.org/10.3390/vaccines8010090.

[189] Li G, De Clercq E. Current therapy for chronic hepatitis C: the role of direct-acting antivirals. Antivir Res 2017;142:83–122. Available from: https://doi.org/10.1016/j.antiviral.2017.02.014.

[190] Shah S, Chougule MB, Kotha AK, Kashika R, Godugu C, Raghuvanshi RS, et al. Nanomedicine based approaches for combating viral infections. J Control Release 2021;338:80–104. Available from: https://doi.org/10.1016/j.jconrel.2021.08.011.

[191] Oshikoya KA, Oreagba LA, Ogunleye OO, Lawal S, Senbanjo LO. Clinically significant interactions between antiretroviral and co-prescribed drugs for HIV-infected children: profiling and comparison of two drug databases. Ther Clin Risk Manag 2013;9:215–21. Available from: https://doi.org/10.2147/TCRM.S44205.

[192] Bule M, Khan F, Niaz K. Antivirals: past, present and future. Recent Adv Anim Virol 2019;425–46. Available from: https://doi.org/10.1007/978-981-13-9073-9_22.

[193] Adalja A, Inglesby T. Broad-spectrum antiviral agents: a crucial pandemic tool. Expert Rev Anti-Infect Ther 2019;17:467–70. Available from: https://doi.org/10.1080/14787210.2019.1635009.

[194] Nath A, Tyler KL. Novel approaches and challenges to treatment of central nervous system viral infections. Ann Neurol 2013;74:412–22. Available from: https://doi.org/10.1002/ana.23988.

[195] Chang R, Sun WZ. Repositioning chloroquine as antiviral prophylaxis against COVID-19: potential and challenges. Drug Discov Today 2020;25:1786–92. Available from: https://doi.org/10.1016/j.drudis.2020.06.030.

[196] DiMasi JA, Grabowski HG, Hansen RW. Innovation in the pharmaceutical industry: new estimates of R&D costs. J Health Econ 2016;47:20–33. Available from: https://doi.org/10.1016/j.jhealeco.2016.01.012.

[197] Brunaugh AD, Sharma S, Smyth H. Inhaled fixed-dose combination powders for the treatment of respiratory infections. Expert Opin Drug Deliv 2021;18:1101–15. Available from: https://doi.org/10.1080/17425247.2021.1886074.

[198] Stellbrink HJ, Lazzarin A, Woolley I, Llibre JM. The potential role of bictegravir/emtricitabine/tenofovir alafenamide (BIC/FTC/TAF) single-tablet regimen in the expanding spectrum of fixed-dose combination therapy for HIV. HIV Med 2020;21:3–16. Available from: https://doi.org/10.1111/hiv.12833.

[199] Caplan MR, Daar ES, Corado KC. Next generation fixed dose combination pharmacotherapies for treating HIV. Expert Opin Pharmacother 2018;19:589–96. Available from: https://doi.org/10.1080/14656566.2018.1450866.

[200] Aygen B, Demirtürk N, Yıldız O, Çelen MK, Çelik I, Barut S, et al. Real-world efficacy, safety, and clinical outcomes of ombitasvir/paritaprevir/ritonavir ± dasabuvir ± ribavirin combination therapy in patients with hepatitis C virus genotype 1 or 4 infection: the Turkey experience experience. Turkish J Gastroenterol 2020;31:305–17. Available from: https://doi.org/10.5152/TJG.2020.19197.

[201] Costa JO, Ceccato MDGB, Silveira MR, Bonolo PF, Reis EA, Acurcio FA. Effectiveness of antiretroviral therapy in the single-tablet regimen era. Rev Saude Publica 2018;52:87.
[202] Aldir I, Horta A, Serrado M. Single-tablet regimens in HIV: does it really make a difference. Curr Med Res Opin 2014;30:89−97.
[203] Imaz A, Podzamczer D. Tenofovir alafenamide, emtricitabine, elvitegravir, and cobicistat combination therapy for the treatment of HIV. Expert Rev Anti-Infect Ther 2017;15:195−209. Available from: https://doi.org/10.1080/14787210.2017.1286736.
[204] Fields J, Go JT, Schulze KS. Pill properties that cause dysphagia and treatment failure. Curr Ther Res Clin Exp 2015;77:79−82. Available from: https://doi.org/10.1016/j.curtheres.2015.08.002.
[205] Sutton SS, Magagnoli J, Hardin JW. Odds of viral suppresion by single-tablet regimens, multiple-tablet regimens, and adherence level in HIV/AIDS patients receiving antiretroviral therapy. Pharmacotherapy 2017;37:204−13.
[206] Tompa DR, Immanue A, Srikanth S, Kadhirvel S. Trends and strategies to combat viral infections: a review on FDA approved antiviral drugs. Int J Biol Macromol 2021;172:524−41. Available from: https://doi.org/10.1016/j.ijbiomac.2021.01.076.
[207] World Health Organization. Global hepatitis report. <http://apps.who.int/iris/bitstream/10665/255016/1/9789241565455-eng.pdf?ua = 1>; 2017 [accessed 21.09.21].
[208] Dehghan Manshadi SA, Merat S, Mohraz M, Rasoolinejad M, Sali S, Mardani M, et al. Single-pill sofosbuvir and daclatasvir for treating hepatis C in patients co-infected with human immunodeficiency virus. Int J Clin Pract 2021;75(2021):0−2. Available from: https://doi.org/10.1111/ijcp.14304.
[209] Asselah T, Marcellin P, Schinazi RF. Treatment of hepatitis C virus infection with direct-acting antiviral agents: 100% cure? Liver Int 2018;38:7−13. Available from: https://doi.org/10.1111/liv.13673.
[210] Purwati AM, Nasronudin E, Hendrianto D, Karsari A, Dinaryanti N, Ertanti IS, et al. An in vitro study of dual drug combinations of anti-viral agents, antibiotics, and/or hydroxychloroquine against the SARS-CoV-2 virus isolated from hospitalized patients in Surabaya, Indonesia. PLoS ONE 2021;16:1−27. Available from: https://doi.org/10.1371/journal.pone.0252302.
[211] Han YJ, Lee KH, Yoon S, Nam SW, Ryu S, Seong D, et al. Treatment of severe acute respiratory syndrome (SARS), Middle East respiratory syndrome (MERS), and coronavirus disease 2019 (COVID-19): a systematic review of in vitro, in vivo, and clinical trials. Theranostics 2021;11:1207−31. Available from: https://doi.org/10.7150/thno.48342.
[212] Marchenkova L, Vasileva V, Eryomushkin M. Pos0016 the evaluation of functional abilities of patients with osteoporotic vertebral fractures as a basis for rehabilitation programs developing. Ann Rheum Dis 2021;80:209. Available from: https://doi.org/10.1136/annrheumdis-2021-eular.4260.
[213] Karagöz AÇ, et al. Synthesis of new betulinic acid/betulin-derived dimers and hybrids with potent antimalarial and antiviral activities. Bioorg Med Chem 2018;27(1):110−15.
[214] Battini L, Bollini M. Challenges and approaches in the discovery of human immunodeficiency virus type-1 non-nucleoside reverse transcriptase inhibitors. Med Res Rev 2019;39(4):1235−73.
[215] Feng LS, et al. Hybrid molecules with potential in vitro antiplasmodial and in vivo antimalarial activity against drug-resistant Plasmodium falciparum. Med Res Rev 2020;40(3):931−71.
[216] Gao F, Zhang X, Wang T, Xiao J. Quinolone hybrids and their anti-cancer activities: an overview. Eur J Med Chem 2019;165:59−79.
[217] Hu YQ, Zhang S, Xu Z, Lv ZS, Liu ML, Feng LS. 4-Quinolone hybrids and their antibacterial activities. Eur J Med Chem 2017;141:335−45.
[218] Nqoro X, Tobeka N, Aderibigbe BA. Quinoline-based hybrid compounds with antimalarial activity. Molecules 2017;22(12):2268.
[219] Srivastava V, Lee H. Chloroquine-based hybrid molecules as promising novel chemotherapeutic agents. Eur J Pharmacol 2015;762:472−86.
[220] Morphy R, Kay C, Rankovic Z. From magic bullets to designed multiple ligands. Drug Discov Today 2004;9(15):641−51.
[221] Sun L, et al. Design, synthesis and structure-activity relationships of 4-phenyl-1H-1,2,3-triazole phenylalanine derivatives as novel HIV-1 capsid inhibitors with promising antiviral activities. Eur J Med Chem 2020;190:112085.
[222] Wu G, et al. Discovery of phenylalanine derivatives as potent HIV-1 capsid inhibitors from click chemistry-based compound library. Eur J Med Chem 2018;158:478−92.

[223] Jana S, Iram S, Thomas J, Hayat MQ, Pannecouque C, Dehaen W. Application of the triazolization reaction to afford dihydroartemisinin derivatives with anti-HIV activity. Molecules 2017;22(2):303.

[224] Madni M, et al. Synthesis, crystal structure, anti-HIV, and antiproliferative activity of new pyrazolylthiazole derivatives. Med Chem Res 2017;26(10):2653−65.

[225] Chander S, et al. Synthesis and study of anti-HIV-1 RT activity of 5-benzoyl-4-methyl- 1,3,4,5-tetrahydro-2H-1,5-benzodiazepin-2-one derivatives. Bioor Chem 2017;72:74−9.

[226] Safakish M, Hajimahdi Z, Zabihollahi R, Aghasadeghi MR, Vahabpour R, Zarghi A. Design, synthesis, and docking studies of new 2-benzoxazolinone derivatives as anti-HIV-1 agents. Med Chem Res 2017;26 (11):2718−26.

[227] Bielenica A, et al. 1 H -Tetrazol-5-amine and 1, 3-thiazolidin-4-one derivatives containing 3- (tri fl uoro-methyl) phenyl scaffold: synthesis, cytotoxic and anti-HIV studies. Biomed Pharmacother 2017;94:804−12.

[228] Wang Q, et al. Novel betulinic acid-nucleoside hybrids with potent anti-HIV activity. ACS Med Chem Lett 2020;11(11):2290−3.

[229] Gao P, et al. Design, synthesis and anti-HIV evaluation of novel 5-substituted diarylpyrimidine derivatives as potent HIV-1 NNRTIs. Bioorg Med Chem 2021;40:116195.

[230] Shirvani P, Fassihi A, Saghaie L, Van Belle S, Debyser Z, Christ F. Synthesis, anti-HIV-1 and antiprolifera-tive evaluation of novel 4-nitroimidazole derivatives combined with 5-hydroxy-4-pyridinone moiety. J Mol Struct 2020;1202:127344.

[231] Bhavsar D, et al. Synthesis and in vitro anti-HIV activity of N-1,3-benzo[d]thiazol-2-yl-2- (2-oxo-2H-chro-men-4-yl)acetamide derivatives using MTT method. Bioorg Med Chem Lett 2011;21(11):3443−6.

[232] Chander S, Wang P, Ashok P, Yang LM, Zheng YT, Sankaranarayanan M. Design, synthesis and anti-HIV-1 RT evaluation of 2-(benzyl(4-chlorophenyl)amino)-1-(piperazin-1-yl)ethanone derivatives. Bioorg Med Chem Lett 2017;27(1):61−5.

[233] Mao TQ, et al. Anti-HIV diarylpyrimidine-quinolone hybrids and their mode of action. Bioorg Med Chem 2015;23(13):3860−8.

[234] Xiao T, et al. Indazolyl-substituted piperidin-4-yl-aminopyrimidines as HIV-1 NNRTIs: design, synthesis and biological activities. Eur J Med Chem 2020;186:111864.

[235] Kaur R, Kumar K. Synthetic and medicinal perspective of quinolines as antiviral agents. Eur J Med Chem 2021;113220.

[236] Cunha AC, et al. Chemistry and anti-herpes simplex virus type 1 evaluation of 4-substituted-1H-1,2,3-tria-zole-nitroxyl-linked hybrids. Mol Divers 2020;1−9.

[237] Pandey VK, Upadhyay M, Upadhyay M, Gupta VD, Tandon M. Benzimidazolyl quinolinyl mercaptotria-zoles as potential antimicrobial and antiviral agents. Acta Pharm 2005;55(2):47−56.

[238] Kharitonova MI, et al. New modified 2-aminobenzimidazole nucleosides:synthesis and evaluation of their activity against herpes simplex virus type 1. Bioorg Med Chem Lett 2017;27(11):2484−7.

[239] Jordão AK, et al. Synthesis and anti-HSV-1 activity of new 1,2,3-triazole derivatives. Bioorg Med Chem 2011;19(6):1860−5.

[240] Lefkowitch JH. Acute viral hepatitis. Scheuer's liver biopsy interpretation 2021;89.

[241] Ryu WS. Molecular virology of human pathogenic viruses. Academic Press; 2016.

[242] World Health Organization, Global hepatitis report, 2017.

[243] Wang LJ, et al. Design, synthesis, and molecular hybrids of caudatin and cinnamic acids as novel anti-hepatitis B virus agents. Eur J Med Chem 2012;54:352−65.

[244] Zhang H, et al. Flavonoid-triazolyl hybrids as potential anti-hepatitis C virus agents: synthesis and biologi-cal evaluation. Eur J Med Chem 2021;218:113395.

[245] Luo Y, et al. Synthesis and anti-Hepatitis B virus activity of a novel class of thiazolylbenzimidazole deriva-tives. Arch Pharm 2011;344(2):78−83.

[246] Wei Z, et al. Design, synthesis and bioactive evaluation of oxime derivatives of dehydrocholic acid as anti-hepatitis B virus agents. Molecules 2020;25(15):3359.

[247] Hwu JR, et al. Synthesis of new benzimidazole-coumarin conjugates as anti-hepatitis C virus agents. Antivir Res 2008;77(2):157−62.

[248] Verma VA, Saundane AR, Meti RS, Vennapu DR. Synthesis of novel indolo[3,2-c]isoquinoline derivatives bearing pyrimidine, piperazine rings and their biological evaluation and docking studies against COVID-19 virus main protease. J Mol Struct 2021;1229:129829.

[249] Bessières M, et al. Design, synthesis and biological evaluation of 2-substituted-6-[(4-substituted-1-piperidyl) methyl]-1H-benzimidazoles as inhibitors of ebola virus infection. Eur J Med Chem 2021;214:113211.

[250] Kazakova OB, Giniyatullina GV, Yamansarov EY, Tolstikov GA. Betulin and ursolic acid synthetic derivatives as inhibitors of Papilloma virus. Bioorg Med Chem Lett 2010;20(14):4088−90.

[251] Kachaeva M, et al. In vitro activity of novel 1,3-oxazole derivatives against human papillomavirus. Ibnosina J Med Biomed Sci 2017;9(4):111.

[252] Yoon JH, et al. Synthesis and biological evaluation of 3-acyl-2-phenylamino-1,4-dihydroquinolin-4(1H)-one derivatives as potential MERS-CoV inhibitors. Bioorg Med Chem Lett 2019;29(23):126727.

[253] Lee JY, et al. Identification of 4-anilino-6-aminoquinazoline derivatives as potential MERS-CoV inhibitors. Bioorg Med Chem Lett 2020;30(20):127472.

[254] Tamilvanan S, Benita S. The potential of lipid emulsion for ocular delivery of lipophilic drugs. Eur J Pharm Biopharm 2004;58(2):357−68. Available from: https://doi.org/10.1016/j.ejpb.2004.03.033.

[255] Tiwari R, Pandey V, Asati S, Soni V, Jain D. Therapeutic challenges in ocular delivery of lipid based emulsion. Egypt J Basic Appl Sci 2018;5(2):121−9. Available from: https://doi.org/10.1016/j.ejbas.2018.04.001.

[256] Maus A, Strait L, Zhu D. Nanoparticles as delivery vehicles for antiviral therapeutic drugs. Eng Regen 2020;2:31−46. Available from: https://doi.org/10.1016/j.engreg.2021.03.001.

[257] Saffarionpour S. One-step preparation of double emulsions stabilized with amphiphilic and stimuli-responsive block copolymers and nanoparticles for nutraceuticals and drug delivery. JCIS Open 2021;3. Available from: https://doi.org/10.1016/j.jciso.2021.100020.

[258] Kotta S, Khan AW, Ansari SH, Sharma RK, Ali J. Anti HIV nanoemulsion formulation: optimization and in vitro-in vivo evaluation. Int J Pharm 2014;462(1−2):129−34. Available from: https://doi.org/10.1016/j.ijpharm.2013.12.038.

[259] Saraf S, Mishra D, Asthana A, Jain R, Singh S, Jain NK. Lipid microparticles for mucosal immunization against hepatitis B. Vaccine 2006;24(1):45−56. Available from: https://doi.org/10.1016/j.vaccine.2005.07.053.

[260] Oliveira CM, Factorial design applied to the optimization of lipid composition of topical antiherpetic nanoemulsions containing isoflavone genistein, 2014; 4737−4747.

[261] Lopes DA et al., Pentyl gallate nanoemulsions as potential topical treatment of herpes labialis, 2016;6:1−10, Available from: https://doi.org/10.1016/j.xphs.2016.04.028.

[262] Bidone J, et al. Antiherpes activity and skin/mucosa distribution of flavonoids from achyrocline satureioides extract incorporated into topical nanoemulsions. BioMed Res Int 2015;2015. Available from: https://doi.org/10.1155/2015/238010.

[263] Nanoemulsions C, In vitro evaluation of mucosa permeation/retention and antiherpes activity of genistein from, 2016, Available from: https://doi.org/10.1166/jnn.2016.11676.

[264] Mahajan HS, Mahajan MS, Nerkar PP, Agrawal A. Nanoemulsion-based intranasal drug delivery system of saquinavir mesylate for brain targeting. Drug Deliv 2014;21(2):148−54. Available from: https://doi.org/10.3109/10717544.2013.838014.

[265] do Bonfim CM, et al. Antiviral activity of curcumin-nanoemulsion associated with photodynamic therapy in vulvar cell lines transducing different variants of HPV-16. Artif Cells, Nanomed Biotechnol 2020;48(1):515−24. Available from: https://doi.org/10.1080/21691401.2020.1725023.

[266] Nisini R, Poerio N, Mariotti S, De Santis F, Fraziano M. The multirole of liposomes in therapy and prevention of infectious diseases. Front Immunol 2018;9. Available from: https://doi.org/10.3389/fimmu.2018.00155 no. FEB.

[267] Magar KT, Boafo GF, Li X, Chen Z, He W. Liposome-based delivery of biological drugs. Chin Chem Lett 2021;. Available from: https://doi.org/10.1016/j.cclet.2021.08.020.

[268] Guimarães D, Cavaco-Paulo A, Nogueira E. Design of liposomes as drug delivery system for therapeutic applications. Int J Pharm 2021;601. Available from: https://doi.org/10.1016/j.ijpharm.2021.120571 no. February.

[269] Salunkhe SA, Chitkara D, Mahato RI, Mittal A. Lipid based nanocarriers for effective drug delivery and treatment of diabetes associated liver fibrosis. Adv Drug Deliv Rev 2021;173:394−415. Available from: https://doi.org/10.1016/j.addr.2021.04.003.

[270] Large DE, Abdelmessih RG, Fink EA, Auguste DT. Liposome composition in drug delivery design, synthesis, characterization, and clinical application. Adv Drug Deliv Rev 2021;113851. Available from: https://doi.org/10.1016/j.addr.2021.113851.

[271] Access O. We are IntechOpen, the world's leading publisher of Open Access books Built by scientists, for scientists TOP 1% nanotechnology based drug delivery for HIV-AIDS treatment.

[272] Rohan LC. Progress in antiretroviral drug delivery using nanotechnology; 2010: 533−547.

[273] Faria MJ, Lopes CM. Lipid nanocarriers for anti-HIV therapeutics: a focus on physicochemical properties and biotechnological advances, 2021.

[274] Nayak D, Boxi A, Ashe S, Thathapudi NC, Nayak B. Stavudine loaded gelatin liposomes for HIV therapy: preparation, characterization and in vitro cytotoxic evaluation. Mater Sci Eng C 2017;73:406−16. Available from: https://doi.org/10.1016/j.msec.2016.12.073.

[275] Yang X, Chen H, Chen W. Human papillomavirus oncogene manipulation using clustered regularly interspersed short palindromic repeats/Cas9 delivered by pH-sensitive cationic liposomes. Hum Gene Ther 2020;31. Available from: https://doi.org/10.1089/hum.2019.312.

[276] Kulkarni SP, Thanapati S, Arankalle VA, Tripathy AS. Specific memory B cell response and participation of CD4 + central and effector memory T cells in mice immunized with liposome encapsulated recombinant NE protein based Hepatitis E vaccine candidate. Vaccine 2016;34(48):5895−902. Available from: https://doi.org/10.1016/j.vaccine.2016.10.046.

[277] Tiwari S, Verma SK, Agrawal GP, Vyas SP. Viral protein complexed liposomes for intranasal delivery of hepatitis B surface antigen. Int J Pharm 2011;413(1−2):211−19. Available from: https://doi.org/10.1016/j.ijpharm.2011.04.029.

[278] Akhtar A, Wang SX, Ghali L, Bell C, Wen X. Effective delivery of arsenic trioxide to HPV-positive cervical cancer cells using optimised liposomes: a size and charge study. Int J Mol Sci 2018. Available from: https://doi.org/10.3390/ijms19041081.

[279] Mirchandani Y, Patravale VB, Brijesh S. Solid lipid nanoparticles for hydrophilic drugs. J Control Release 2021;335:457−64. Available from: https://doi.org/10.1016/j.jconrel.2021.05.032.

[280] Delshadi R, Bahrami A, McClements DJ, Moore MD, Williams L. Development of nanoparticle-delivery systems for antiviral agents: a review. J Control Release 2021;331:30−44. Available from: https://doi.org/10.1016/j.jconrel.2021.01.017.

[281] Basha SK, Dhandayuthabani R, Muzammil MS, Kumari VS. Solid lipid nanoparticles for oral drug delivery. Mater Today Proc 2019;36:313−24. Available from: https://doi.org/10.1016/j.matpr.2020.04.109.

[282] Elbrink K, et al. Application of solid lipid nanoparticles as a long-term drug delivery platform for intramuscular and subcutaneous administration: in vitro and in vivo evaluation. Eur J Pharm Biopharm 2021;163:158−70. Available from: https://doi.org/10.1016/j.ejpb.2021.04.004.

[283] Limeres MJ, et al. Development and characterization of an improved formulation of cholesteryl oleate-loaded cationic solid-lipid nanoparticles as an efficient non-viral gene delivery system. Colloids Surf B Biointerfaces 2019;184:110533. Available from: https://doi.org/10.1016/j.colsurfb.2019.110533.

[284] Yaghmur A, Mu H. Recent advances in drug delivery applications of cubosomes, hexosomes, and solid lipid nanoparticles. Acta Pharm Sin B 2021;11(4):871−85. Available from: https://doi.org/10.1016/j.apsb.2021.02.013.

[285] Hassan H, Bello R O, Adam S K, Alias E. Acyclovir-loaded solid lipid nanoparticles: optimization, characterization and evaluation of its pharmacokinetic profile.

[286] Kondel R, et al. Effect of acyclovir solid lipid nanoparticles for the treatment of herpes simplex virus (HSV) infection in an animal model of HSV-1 infection. Pharm Nanotechnol 2019;1−15. Available from: https://doi.org/10.2174/2211738507666190829161737.

[287] Gaur P K, Mishra S, Bajpai M, Mishra A. Enhanced oral bioavailability of efavirenz by solid lipid nanoparticles: in vitro drug release and pharmacokinetics studies, 2014.

[288] Torrecilla J, Del Pozo-Rodríguez A, Apaolaza PS, Solinís MÁ, Rodríguez-Gascón A. Solid lipid nanoparticles as non-viral vector for the treatment of chronic hepatitis C by RNA interference. Int J Pharm 2015;479(1):181−8. Available from: https://doi.org/10.1016/j.ijpharm.2014.12.047.

[289] Torrecilla J, et al. Silencing of hepatitis C virus replication by a non-viral vector based on solid lipid nanoparticles containing a shRNA targeted to the internal ribosome entry site (IRES). Colloids Surf B Biointerfaces 2016;146:808−17. Available from: https://doi.org/10.1016/j.colsurfb.2016.07.026.

[290] Javan F, Vatanara A. Encapsulation of ritonavir in solid lipid nanoparticles: in-vitro anti-HIV-1 activity using lentiviral particles. J Pharm Pharmacol 2017;1−8. Available from: https://doi.org/10.1111/jphp.12737.

III. Antiviral agents and therapeutics

[291] Mahmud A, Lavasanifar A. The effect of block copolymer structure on the internalization of polymeric micelles by human breast cancer cells. Colloids Surf B Biointerfaces 2005;45:82−9. Available from: https://doi.org/10.1016/j.colsurfb.2005.07.008.

[292] Choi J, Hyuk N, Kim D, Park J. Comparison of paclitaxel solid dispersion and polymeric micelles for improved oral bioavailability and in vitro anti-cancer effects. Mater Sci Eng C 2019;100:247−59. Available from: https://doi.org/10.1016/j.msec.2019.03.002.

[293] Alven S, Aderibigbe BA. The therapeutic efficacy of dendrimer and micelle formulations for breast cancer treatment. Pharmaceutics 2020;12:1−49. Available from: https://doi.org/10.3390/pharmaceutics12121212.

[294] Chiappetta DA, Alvarez-Lorenzo C, Rey-Rico A, Taboada P, Concheiro A, Sosnik A. N-alkylation of poloxamines modulates micellar assembly and encapsulation and release of the antiretroviral efavirenz. Eur J Pharm Biopharm 2010;76(1):24−37. Available from: https://doi.org/10.1016/j.ejpb.2010.05.007.

[295] Chiappetta DA, Facorro G, Rubin de Celis E, Sosnik A. Synergistic encapsulation of the anti-HIV agent efavirenz within mixed poloxamine/poloxamer polymeric micelles. Nanomed Nanotechnol Biol Med 2011;7(5):624−37. Available from: https://doi.org/10.1016/j.nano.2011.01.017.

[296] Chiappetta DA, Hocht C, Taira C, Sosnik A. Oral pharmacokinetics of the anti-HIV efavirenz encapsulated within polymeric micelles. Biomaterials 2011;32(9):2379−87. Available from: https://doi.org/10.1016/j.biomaterials.2010.11.082.

[297] Chiappetta DA, Hocht C, Opezzo JAW, Sosnik A. Intranasal administration of antiretroviral-loaded micelles for anatomical targeting to the brain in HIV. Nanomedicine 2013;8(2):223−37. Available from: https://doi.org/10.2217/nnm.12.104.

[298] Chiappetta DA, Hocht C, Taira C, Sosnik A. Efavirenz-loaded polymeric micelles for pediatric anti-HIV pharmacotherapy with significantly higher oral bioavailaibility. Nanomedicine 2010;5(1):11−23. Available from: https://doi.org/10.2217/nnm.09.90.

[299] Chiappetta DA, Hocht C, Sosnik A. A highly concentrated and taste-improved aqueous formulation of efavirenz for a more appropriate pediatric management of the anti-HIV therapy. Curr HIV Res 2010;8(3):223−31. Available from: https://doi.org/10.2174/157016210791111142.

[300] Chaudhari SP, Handge NM, Collegeofpharmacy DYP. Formulation, Development and Evaluation of Lopinavir Loaded Polymeric Micelles. J Sci Technol 2020;5(4):173−87. Available from: https://doi.org/10.46243/jst.2020.v5.i4.pp173-187.

[301] Mahajan HS, Patil PH. Central composite design-based optimization of lopinavir vitamin E-TPGS micelle: in vitro characterization and in vivo pharmacokinetic study. Colloids Surf B Biointerfaces 2020;194:111149. Available from: https://doi.org/10.1016/j.colsurfb.2020.111149.

[302] Sawdon AJ, Peng CA. Polymeric micelles for acyclovir drug delivery. Colloids Surf B Biointerfaces 2014;122:738−45. Available from: https://doi.org/10.1016/j.colsurfb.2014.08.011.

[303] Varela-Garcia A, Concheiro A, Alvarez-Lorenzo C. Soluplus micelles for acyclovir ocular delivery: formulation and cornea and sclera permeability. Int J Pharm 2018;552:39−47. Available from: https://doi.org/10.1016/j.ijpharm.2018.09.053.

[304] Accardo A, et al. Self-assembled or mixed peptide amphiphile micelles from herpes simplex virus glycoproteins as potential immunomodulatory treatment. Int J Nanomed 2014;9(1):2137−48. Available from: https://doi.org/10.2147/IJN.S57656.

[305] Miao J, et al. Redox-responsive chitosan oligosaccharide-SS-Octadecylamine polymeric carrier for efficient anti-Hepatitis B Virus gene therapy. Carbohydr Polym 2019;212:215−21. Available from: https://doi.org/10.1016/j.carbpol.2019.02.047.

[306] Hong Y, et al. Hepatitis B virus S gene therapy with 10-23 DNAzyme delivered by chitosan-: G-stearic acid micelles. RSC Adv 2019;9(27):15196−204. Available from: https://doi.org/10.1039/c9ra00330d.

[307] Miao J, et al. Inhibition on hepatitis B virus e-gene expression of 10-23 DNAzyme delivered by novel chitosan oligosaccharide-stearic acid micelles. Carbohydr Polym 2012;87(2):1342−7. Available from: https://doi.org/10.1016/j.carbpol.2011.09.022.

[308] Layek B, Lipp L, Singh J. APC targeted micelle for enhanced intradermal delivery of hepatitis B DNA vaccine. J Control Release 2015;207:143−53. Available from: https://doi.org/10.1016/j.jconrel.2015.04.014.

[309] Li Q, et al. Synthesis of Lamivudine stearate and antiviral activity of stearic acid-g-chitosan oligosaccharide polymeric micelles delivery system. Eur J Pharm Sci 2010;41(3−4):498−507. Available from: https://doi.org/10.1016/j.ejps.2010.08.004.

[310] Huang ST, et al. Synthesis and anti-hepatitis B virus activity of acyclovir conjugated stearic acid-g-chitosan oligosaccharide micelle. Carbohydr Polym 2011;83(4):1715–22. Available from: https://doi.org/10.1016/j.carbpol.2010.10.032.

[311] Wang Y, et al. Formulation and evaluation of novel glycyrrhizic acid micelles for transdermal delivery of podophyllotoxin. Drug Deliv 2016;23(5):1623–35. Available from: https://doi.org/10.3109/10717544.2015.1135489.

[312] Nishida H, et al. Systemic delivery of siRNA by actively targeted polyion complex micelles for silencing the E6 and E7 human papillomavirus oncogenes. J Control Release 2016;231:29–37. Available from: https://doi.org/10.1016/j.jconrel.2016.03.016.

[313] Lao YH, et al. HPV oncogene manipulation using nonvirally delivered CRISPR/Cas9 or natronobacterium gregoryi argonaute. Adv Sci 2018;5(7):1–12. Available from: https://doi.org/10.1002/advs.201700540.

[314] Singh Chauhan P, Abutbul Ionita I, Moshe Halamish H, Sosnik A, Danino D. Multidomain drug delivery systems of β-casein micelles for the local oral co-administration of antiretroviral combinations. J Colloid Interface Sci 2021;592:156–66. Available from: https://doi.org/10.1016/j.jcis.2020.12.021.

[315] Mhlwatika Z, Aderibigbe BA. Application of dendrimers for the treatment of infectious diseases. Molecules 2018;23(9). Available from: https://doi.org/10.3390/molecules23092205.

[316] Dias AP, et al. Dendrimers in the context of nanomedicine. Int J Pharm 2020;573:118814. Available from: https://doi.org/10.1016/j.ijpharm.2019.118814.

[317] Perisé-Barrios AJ, et al. Carbosilane dendrimers as gene delivery agents for the treatment of HIV infection. J Control Release 2014;184(1):51–7. Available from: https://doi.org/10.1016/j.jconrel.2014.03.048.

[318] Gutierrez-Ulloa CE, et al. Synthesis of bow-tie carbosilane dendrimers and their HIV antiviral capacity: a comparison of the dendritic topology on the biological process. Eur Polym J 2019;119:200–12. Available from: https://doi.org/10.1016/j.eurpolymj.2019.07.034.

[319] Sepúlveda-Crespo D, et al. Synergistic activity profile of carbosilane dendrimer G2-STE16 in combination with other dendrimers and antiretrovirals as topical anti-HIV-1 microbicide. Nanomed Nanotechnol Biol Med 2014;10(3):609–18. Available from: https://doi.org/10.1016/j.nano.2013.10.002.

[320] Vacas-Córdoba E, Galán M, de la Mata FJ, Gómez R, Pion M, Muñoz-Fernández MÁ. Enhanced activity of carbosilane dendrimers against HIV when combined with reverse transcriptase inhibitor drugs: searching for more potent microbicides. Int J Nanomed 2014;9(1):3591–600. Available from: https://doi.org/10.2147/IJN.S62673.

[321] Córdoba EV, et al. Synergistic activity of carbosilane dendrimers in combination with maraviroc against HIV in vitro. Aids 2013;27(13):2053–8. Available from: https://doi.org/10.1097/QAD.0b013e328361fa4a.

[322] Ceña-Diez R, García-Broncano P, de la Mata FJ, Gómez R, Muñoz-Fernández MÁ. Efficacy of HIV antiviral polyanionic carbosilane dendrimer G2-S16 in the presence of semen. Int J Nanomed 2016;11:2443–50. Available from: https://doi.org/10.2147/IJN.S104292.

[323] Briz V, et al. Development of water-soluble polyanionic carbosilane dendrimers as novel and highly potent topical anti-HIV-2 microbicides. Nanoscale 2015;7(35):14669–83. Available from: https://doi.org/10.1039/c5nr03644e.

[324] Sanchez-Rodriguez J, et al. Anti-human immunodeficiency virus activity of thiol-ene carbosilane dendrimers and their potential development as a topical microbicide. J Biomed Nanotechnol 2015;11(10):1783–98. Available from: https://doi.org/10.1166/jbn.2015.2109.

[325] Kandi MR, et al. Inherent anti-HIV activity of biocompatible anionic citrate-PEG-citrate dendrimer. Mol Biol Rep 2019;46(1):143–9. Available from: https://doi.org/10.1007/s11033-018-4455-6.

[326] Kumar PD, Kumar PV, Selvam TP, Rao KRSS. Prolonged drug delivery system of PEGylated PAMAM dendrimers with a anti-HIV drug. Res Pharm 2013;3(2):8–17.

[327] Telwatte S, et al. Virucidal activity of the dendrimer microbicide SPL7013 against HIV-1. Antivir Res 2011;90(3):195–9. Available from: https://doi.org/10.1016/j.antiviral.2011.03.186.

[328] Maciel D, Guerrero-Beltrán C, Ceña-Diez R, Tomás H, Muñoz-Fernández MÁ, Rodrigues J. New anionic poly(alkylideneamine) dendrimers as microbicide agents against HIV-1 infection. Nanoscale 2019;11(19):9679–90. Available from: https://doi.org/10.1039/c9nr00303g.

[329] Falanga A, et al. Engineering of janus-like dendrimers with peptides derived from glycoproteins of herpes simplex virus type 1: toward a versatile and novel antiviral platform. Int J Mol Sci 2021;22(12):6488. Available from: https://doi.org/10.3390/ijms22126488.

[330] Carberry TP, et al. Dendrimer functionalization with a membrane-interacting domain of herpes simplex virus type 1: towards intracellular delivery. Chem—A Eur J 2012;18(43):13678—85. Available from: https://doi.org/10.1002/chem.201202358.

[331] Luganini A, et al. Inhibition of herpes simplex virus type 1 and type 2 infections by peptide-derivatized dendrimers. Antimicrob Agents Chemother 2011;55(7):3231—9. Available from: https://doi.org/10.1128/AAC.00149-11.

[332] Guerrero-Beltrán C, et al. Cationic dendrimer g2-s16 inhibits herpes simplex type 2 infection and protects mice vaginal microbiome. Pharmaceutics 2020;12(6):1—14. Available from: https://doi.org/10.3390/pharmaceutics12060515.

[333] Gong E, et al. Evaluation of dendrimer SPL7013, a lead microbicide candidate against herpes simplex viruses. Antivir Res 2005;68(3):139—46. Available from: https://doi.org/10.1016/j.antiviral.2005.08.004.

[334] Gong Y, et al. Evidence of dual sites of action of dendrimers: SPL-2999 inhibits both virus entry and late stages of herpes simplex virus replication. Antivir Res 2002;55(2):319—29. Available from: https://doi.org/10.1016/S0166-3542(02)00054-2.

[335] Lancelot A, Clavería-Gimeno R, Velázquez-Campoy A, Abian O, Serrano JL, Sierra T. Nanostructures based on ammonium-terminated amphiphilic Janus dendrimers as camptothecin carriers with antiviral activity. Eur Polym J 2017;90:136—49. Available from: https://doi.org/10.1016/j.eurpolymj.2017.03.012 no. February.

[336] Sepúlveda-Crespo D, et al. Polyanionic carbosilane dendrimers prevent hepatitis C virus infection in cell culture. Nanomed Nanotechnol Biol Med 2017;13(1):49—58. Available from: https://doi.org/10.1016/j.nano.2016.08.018.

[337] Akao C, et al. Potential use of fucose-appended dendrimer/α-cyclodextrin conjugates as NF-κB decoy carriers for the treatment of lipopolysaccharide-induced fulminant hepatitis in mice. J Control Release 2014;193:35—41. Available from: https://doi.org/10.1016/j.jconrel.2014.07.004.

[338] Anselmo MS, et al. Janus dendrimers to assess the anti-hcv activity of molecules in cell-assays. Pharmaceutics 2020;12(11):1—24. Available from: https://doi.org/10.3390/pharmaceutics12111062.

[339] Karpenko LI, et al. Cationic polymers for the delivery of the ebola dna vaccine encoding artificial t-cell immunogen. Vaccines 2020;8(4):1—14. Available from: https://doi.org/10.3390/vaccines8040718.

[340] Chahal JS, et al. Dendrimer-RNA nanoparticles generate protective immunity against lethal Ebola, H1N1 influenza, and Toxoplasma gondii challenges with a single dose. Proc Natl Acad Sci U S A 2016;113(35):E5250. Available from: https://doi.org/10.1073/pnas.1612792113.

[341] Donalisio M, et al. Identification of a dendrimeric heparan sulfate-binding peptide that inhibits infectivity of genital types of human papillomaviruses. Antimicrob Agents Chemother 2010;54(10):4290—9. Available from: https://doi.org/10.1128/AAC.00471-10.

[342] Lee J, et al. Nonviral gene delivery using PAMAM dendrimer conjugated with the nuclear localization signal peptide derived from human papillomavirus type 11 E2 protein. J Biomater Sci Polym Ed 2021;32(9):1140—60. Available from: https://doi.org/10.1080/09205063.2021.1909411.

[343] Khaitov M, et al. Silencing of SARS-CoV-2 with modified siRNA-peptide dendrimer formulation. Allergy Eur J Allergy Clin Immunol 2021;76(9):2840—54. Available from: https://doi.org/10.1111/all.14850.

[344] Bohr A, Tsapis N, Foged C, Andreana I, Yang M, Fattal E. Treatment of acute lung inflammation by pulmonary delivery of anti-TNF-α siRNA with PAMAM dendrimers in a murine model. Eur J Pharm Biopharm 2020;156:114—20. Available from: https://doi.org/10.1016/j.ejpb.2020.08.009.

[345] Pang ZG, Yang X. Polymer-drug conjugates: recent progress on administration routes. Expert Opin Drug Deliv 2014;11:75—86.

[346] [56]Elvira C, Gallardo A, San Roman J, Cifuentes A. Covalent polymer-drug conjugates. Molecules 2005;10:114—25. Available from: https://doi.org/10.3390/10010114.

[347] Alven S, Aderibigbe B. Combination therapy strategies for the treatment of malaria. Molecules 2019;24:3601.

[348] Aremu OS, Katata-Seru L, Mkhize Z, Botha TL, Wepener V. Polyethylene glycol (5,000) succinate conjugate of lopinavir and its associated toxicity using Danio rerio as a model organism. Sci Rep 2020;10(1):1—9. Available from: https://doi.org/10.1038/s41598-020-68666-z.

[349] Zeng R, et al. Effect of bond linkage on in vitro drug release and anti-HIV activity of chitosan-stavudine conjugates. Macromol Res 2012;20(4):358—65. Available from: https://doi.org/10.1007/s13233-012-0022-5.

[350] Roy U, et al. Preparation and characterization of anti-HIV nanodrug targeted to microfold cell of gut-associated lymphoid tissue. Int J Nanomed 2015;10:5819—35. Available from: https://doi.org/10.2147/IJN.S68348.

[351] Dang NTT, Sivakumaran H, Harrich D, Shaw PN, Davis-Poynter N, Coombes AGA. Synergistic activity of tenofovir and nevirapine combinations released from polycaprolactone matrices for potential enhanced prevention of HIV infection through the vaginal route. Eur J Pharm Biopharm 2014;88(2):406–14. Available from: https://doi.org/10.1016/j.ejpb.2014.05.018.

[352] Stegman JR, et al. Volatile acid-solvent evaporation (VASE): molecularly homogeneous distribution of acyclovir in a bioerodable polymer matrix for long-term treatment of herpes simplex virus-1 infections. J Drug Deliv 2018;2018:1–13. Available from: https://doi.org/10.1155/2018/6161230.

[353] Wohl BM, Smith AAA, Jensen BEB, Zelikin AN. Macromolecular (pro)drugs with concurrent direct activity against the hepatitis C virus and inflammation. J Control Release 2014;196:197–207. Available from: https://doi.org/10.1016/j.jconrel.2014.09.032.

[354] Thirumalaisamy R, et al. Hyaluronic acid- 2-deoxy-D-glucose conjugate act as a promising targeted drug delivery option for the treatment of COVID-19. Int J Adv Sci Eng 2021;7(4):2–13. Available from: https://doi.org/10.29294/ijase.7.4.2021.1995-2005.

[355] Kothamasu P, Kanumur H, Ravur N, Maddu C, Parasuramrajam R, Thangavel S. Nanocapsules: the weapons for novel drug delivery systems. BioImpacts 2012;2(2):71–81. Available from: https://doi.org/10.5681/bi.2012.011.

[356] Ariga K, Lvov YM, Kawakami K, Ji Q, Hill JP. Layer-by-layer self-assembled shells for drug delivery. Adv Drug Deliv Rev 2011;63:762–71. Available from: https://doi.org/10.1016/j.addr.2011.03.016.

[357] Singh M, Hemant K, Ram M, Shivakumar H. Microencapsulation: a promising technique for controlled drug delivery. Res Pharm Sci 2010;5(2):67–77.

[358] Hillaireau H, Le Doan T, Appel M, Couvreur P. Hybrid polymer nanocapsules enhance in vitro delivery of azidothymidine-triphosphate to macrophages. J Control Release 2006;116(3):346–52. Available from: https://doi.org/10.1016/j.jconrel.2006.09.016.

[359] Hillaireau H, Le Doan T, Besnard M, Chacun H, Janin J, Couvreur P. Encapsulation of antiviral nucleotide analogues azidothymidine-triphosphate and cidofovir in poly(iso-butylcyanoacrylate) nanocapsules. Int J Pharm 2006;324(1):37–42. Available from: https://doi.org/10.1016/j.ijpharm.2006.07.006.

[360] Wen J, et al. Improved delivery of broadly neutralizing antibodies by nanocapsules suppresses SHIV infection in the CNS of infant rhesus macaques. PLoS Pathog 2021;17:1–27. Available from: https://doi.org/10.1371/journal.ppat.1009738.

[361] Fichter M, et al. Polymeric hepatitis C virus non-structural protein 5A nanocapsules induce intrahepatic antigen-specific immune responses. Biomaterials 2016;108:1–12. Available from: https://doi.org/10.1016/j.biomaterials.2016.08.046.

[362] Vicente S, Peleteiro M, Díaz-Freitas B, Sanchez A, González-Fernández Á, Alonso MJ. Co-delivery of viral proteins and a TLR7 agonist from polysaccharide nanocapsules: a needle-free vaccination strategy. J Control Release 2013;172(3):773–81. Available from: https://doi.org/10.1016/j.jconrel.2013.09.012.

[363] Somiya M, et al. Targeting of polyplex to human hepatic cells by bio-nanocapsules, hepatitis B virus surface antigen L protein particles. Bioorg Med Chem 2012;20(12):3873–9. Available from: https://doi.org/10.1016/j.bmc.2012.04.031.

[364] Vicente S, et al. Highly versatile immunostimulating nanocapsules for specific immune potentiation. Nanomedicine 2014;9(15):2273–89. Available from: https://doi.org/10.2217/nnm.14.10.

[365] Pietrzak-Nguyen A, et al. MPLA-coated hepatitis B virus surface antigen (HBsAg) nanocapsules induce vigorous T cell responses in cord blood derived human T cells. Nanomed Nanotechnol Biol Med 2016;12(8):2383–94. Available from: https://doi.org/10.1016/j.nano.2016.07.010.

[366] Alven S, Aderibigbe BA. Nanoparticles formulations of artemisinin and derivatives as potential therapeutics for the treatment of cancer, leishmaniasis and malaria. Pharmaceutics 2020;12(8):1–34. Available from: https://doi.org/10.3390/pharmaceutics12080748.

[367] Crucho CIC, Barros MT. Polymeric nanoparticles: a study on the preparation variables and characterization methods. Mater Sci Eng C 2017;80(2017):771–84. Available from: https://doi.org/10.1016/j.msec.2017.06.004.

[368] Elkateb H, et al. Optimization of the synthetic parameters of lipid polymer hybrid nanoparticles dual loaded with darunavir and ritonavir for the treatment of HIV. Int J Pharm 2020;588:119794. Available from: https://doi.org/10.1016/j.ijpharm.2020.119794.

[369] Priya Dharshini K, et al. pH-sensitive chitosan nanoparticles loaded with dolutegravir as milk and food admixture for paediatric anti-HIV therapy. Carbohydr Polym 2021;256:117440. Available from: https://doi.org/10.1016/j.carbpol.2020.117440.

[370] Dev A, et al. Preparation of poly(lactic acid)/chitosan nanoparticles for anti-HIV drug delivery applications. Carbohydr Polym 2010;80(3):833–8. Available from: https://doi.org/10.1016/j.carbpol.2009.12.040.

[371] Das Neves J, Sarmento B. Precise engineering of dapivirine-loaded nanoparticles for the development of anti-HIV vaginal microbicides. Acta Biomater 2015;18:77–87. Available from: https://doi.org/10.1016/j.actbio.2015.02.007.

[372] Zhang T, Sturgis TF, Youan BBC. PH-responsive nanoparticles releasing tenofovir intended for the prevention of HIV transmission. Eur J Pharm Biopharm 2011;79(3):526–36. Available from: https://doi.org/10.1016/j.ejpb.2011.06.007.

[373] Ogunwuyi O, et al. Antiretroviral drugs-loaded nanoparticles fabricated by dispersion polymerization with potential for HIV/AIDS treatment. Infect Dis Res Treat 2016;9. Available from: https://doi.org/10.4137/idrt.s38108.

[374] Sneha R, Vedha Hari BN, Ramya Devi D. Design of antiretroviral drug-polymeric nanoparticles laden buccal films for chronic HIV therapy in paediatrics. Colloids Interface Sci Commun 2018;27:49–59. Available from: https://doi.org/10.1016/j.colcom.2018.10.004.

[375] Chaowanachan T, Krogstad E, Ball C, Woodrow KA. Drug synergy of tenofovir and nanoparticle-based antiretrovirals for HIV prophylaxis. PLoS One 2013;8(4):e61416. Available from: https://doi.org/10.1371/journal.pone.0061416.

[376] Leporati A, et al. Antiretroviral hydrophobic core graft-copolymer nanoparticles: the effectiveness against mutant HIV-1 strains and in vivo distribution after topical application. Pharm Res 2019;36(5):1–12. Available from: https://doi.org/10.1007/s11095-019-2604-9.

[377] Heik DA, Brooks L, Frantzen K, Dewhurst S, Yang J. Inhibition of the enhancement of infection of human immunodeficiency virus by semen-derived enhancer of virus infection using amyloid-targeting polymeric nanoparticles. Physiol Behav 2015;9(2):1829–36. Available from: https://doi.org/10.1021/nn5067254. Inhibition.

[378] Tavares GD, et al. N, N, N-trimethylchitosan-poly (n-butylcyanoacrylate) core-shell nanoparticles as a potential oral delivery system for acyclovir. Colloids Surf B Biointerfaces 2020;196:111336. Available from: https://doi.org/10.1016/j.colsurfb.2020.111336.

[379] Gourdon B, et al. Functionalized PLA-PEG nanoparticles targeting intestinal transporter PepT1 for oral delivery of acyclovir. Int J Pharm 2017;529(1–2):357–70. Available from: https://doi.org/10.1016/j.ijpharm.2017.07.024.

[380] Ramyadevi D, Rajan KS. Interaction and release kinetics study of hybrid polymer blend nanoparticles for pH independent controlled release of an anti-viral drug. J Taiwan Inst Chem Eng 2015;50:1–11. Available from: https://doi.org/10.1016/j.jtice.2014.12.036.

[381] Gandhi A, Jana S, Sen KK. In-vitro release of acyclovir loaded Eudragit RLPO® nanoparticles for sustained drug delivery. Int J Biol Macromol 2014;67:478–82. Available from: https://doi.org/10.1016/j.ijbiomac.2014.04.019.

[382] Seifirad S, Karami H, Shahsavari S, Mirabbasi F, Dorkoosh FA. Design and characterization of mesalamine loaded nanoparticles for controlled delivery system. Nanomed Res J 2016;1(2):97–106. Available from: https://doi.org/10.7508/NMRJ.2016.02.006.

[383] Bhosale UV, Devi K, Choudhary S. Development and in vitro-in vivo evaluation of oral drug delivery system of acyclovir loaded PLGA nanoparticles. Int J Drug Deliv 2013;5(3):331–43. Available from: https://doi.org/10.5138/ijdd.v5i3.1038.

[384] Steinbach JM, Weller CE, Booth CJ, Saltzman WM. Polymer nanoparticles encapsulating siRNA for treatment of HSV-2 genital infection. J Control Release 2012;162(1):102–10. Available from: https://doi.org/10.1016/j.jconrel.2012.06.008.

[385] Lima TLC, et al. Improving encapsulation of hydrophilic chloroquine diphosphate into biodegradable nanoparticles: a promising approach against herpes virus simplex-1 infection. Pharmaceutics 2018;10(4):1–18. Available from: https://doi.org/10.3390/pharmaceutics10040255.

[386] Al-Dhubiab BE. Mucoadhesive buccal films embedded with antiviral drug loaded nanospheres. Turkish J Pharm Sci 2016;13(2):213–24. Available from: https://doi.org/10.5505/tjps.2016.05706.

[387] Donalisio M, et al. Acyclovir-loaded chitosan nanospheres from nano-emulsion templating for the topical treatment of herpesviruses infections. Pharmaceutics 2018;10(2):1–12. Available from: https://doi.org/10.3390/pharmaceutics10020046.

[388] Al-Dhubiab BE, Nair AB, Kumria R, Attimarad M, Harsha S. Formulation and evaluation of nano based drug delivery system for the buccal delivery of acyclovir. Colloids Surf B Biointerfaces 2015;136:878−84. Available from: https://doi.org/10.1016/j.colsurfb.2015.10.045.

[389] Wang J, Feng SS, Wang S, Ying Chen Z. Evaluation of cationic nanoparticles of biodegradable copolymers as siRNA delivery system for hepatitis B treatment. Int J Pharm 2010;400(1−2):194−200. Available from: https://doi.org/10.1016/j.ijpharm.2010.08.026.

[390] Jain AK, Goyal AK, Mishra N, Vaidya B, Mangal S, Vyas SP. PEG-PLA-PEG block copolymeric nanoparticles for oral immunization against hepatitis B. Int J Pharm 2010;387(1−2):253−62. Available from: https://doi.org/10.1016/j.ijpharm.2009.12.013.

[391] Saraf S, Jain S, Sahoo RN, Mallick S. Lipopolysaccharide derived alginate coated Hepatitis B antigen loaded chitosan nanoparticles for oral mucosal immunization. Int J Biol Macromol 2020;154:466−76. Available from: https://doi.org/10.1016/j.ijbiomac.2020.03.124.

[392] Prego C, et al. Chitosan-based nanoparticles for improving immunization against hepatitis B infection. Vaccine 2010;28(14):2607−14. Available from: https://doi.org/10.1016/j.vaccine.2010.01.011.

[393] Zhu J, et al. Mannose-modified PLGA nanoparticles for sustained and targeted delivery in hepatitis B virus immunoprophylaxis. AAPS PharmSciTech 2020;21(1):1−9. Available from: https://doi.org/10.1208/s12249-019-1526-5.

[394] Mishra N, Tiwari S, Vaidya B, Agrawal GP, Vyas SP. Lectin anchored PLGA nanoparticles for oral mucosal immunization against hepatitis B. J Drug Target 2011;19(1):67−78. Available from: https://doi.org/10.3109/10611861003733946.

[395] Shawky S, El-Shafai NM, El-Mehasseb IM, Shoueir KR, El-Kemary MA. Spectroscopic study of self-assembly of anti-hepatitis C virus sofosbuvir drug with bio-polymeric nanoparticles for improving the drug release effect. Spectrochim Acta—Part A Mol Biomol Spectrosc 2021;261:120008. Available from: https://doi.org/10.1016/j.saa.2021.120008.

[396] Gohar YM, et al. Cellular uptake of chitosan nanospheres by HEP G2 cells phagocytosis. J Chem Pharm Res 2016;8(4):499−505.

[397] Zeng P, Xu Y, Zeng C, Ren H, Peng M. Chitosan-modified poly(d, l-lactide-co-glycolide) nanospheres for plasmid DNA delivery and HBV gene-silencing. Int J Pharm 2011;415(1−2):259−66. Available from: https://doi.org/10.1016/j.ijpharm.2011.05.053.

[398] Khater SE, El-khouly A, Abdel-Bar HM, Al-mahallawi AM, Ghorab DM. Fluoxetine hydrochloride loaded lipid polymer hybrid nanoparticles showed possible efficiency against SARS-CoV-2 infection. Int J Pharm 2021;607:121023. Available from: https://doi.org/10.1016/j.ijpharm.2021.121023.

[399] Kopecek J. Hydrogel biomaterials: a smart future? Biomaterials 2007;28:5185−92.

[400] Tian W, et al. LDH hybrid thermosensitive hydrogel for intravaginal delivery of anti-HIV drugs. Artif Cells Nanomed Biotechnol 2019;47(1):1234−40. Available from: https://doi.org/10.1080/21691401.2019.1596935.

[401] Macchione MA, et al. Poly(N-vinylcaprolactam) nanogels with antiviral behavior against HIV-1 Infection. Sci Rep 2019;9(1):1−10. Available from: https://doi.org/10.1038/s41598-019-42150-9.

[402] Rahman SS, Ahmed AB. Development and evaluation of mucoadhesive nanogel of nevirapine for vaginal application. Int J Appl Pharm 2019;11(3):144−9. Available from: https://doi.org/10.22159/ijap.2019v11i3.32353.

[403] Town AR, et al. Tuning HIV drug release from a nanogel-based in situ forming implant by changing nanogel size. J Mater Chem B 2019;7(3):373−83. Available from: https://doi.org/10.1039/C8TB01597J.

[404] Sabbagh F, Muhamad II. Acrylamide-based hydrogel drug delivery systems: release of acyclovir from MgO nanocomposite hydrogel. J Taiwan Inst Chem Eng 2017;72:182−93. Available from: https://doi.org/10.1016/j.jtice.2016.11.032.

[405] Houston DMJ, Robins B, Bugert JJ, Denyer SP, Heard CM. In vitro permeation and biological activity of punicalagin and zinc (II) across skin and mucous membranes prone to Herpes simplex virus infection. Eur J Pharm Sci 2017;96:99−106. Available from: https://doi.org/10.1016/j.ejps.2016.08.013.

[406] Malik NS, et al. Chitosan/xanthan gum based hydrogels as potential carrier for an antiviral drug: fabrication, characterization, and safety evaluation. Front Chem 2020;8:1−16. Available from: https://doi.org/10.3389/fchem.2020.00050.

[407] Malik NS, et al. β-cyclodextrin chitosan-based hydrogels with tunable pH-responsive properties for controlled release of acyclovir: design, characterization, safety, and pharmacokinetic evaluation. Drug Deliv 2021;28(1):1093−108. Available from: https://doi.org/10.1080/10717544.2021.1921074.

[408] Al-Tabakha MM, et al. Synthesis, characterization and safety evaluation of sericin-based hydrogels for controlled delivery of acyclovir. Pharmaceuticals 2021;14(3):234. Available from: https://doi.org/10.3390/ph14030234.

[409] Tran NM, et al. Alginate hydrogel protects encapsulated hepatic HuH-7 cells against hepatitis C virus and other viral infections. PLoS One 2014;9(10):16−17. Available from: https://doi.org/10.1371/journal.pone.0109969.

[410] Guo L, Qiu Y, Chen J, Zhang S, Xu B, Gao Y. Effective transcutaneous immunization against hepatitis B virus by a combined approach of hydrogel patch formulation and microneedle arrays. Biomed Microdevices 2013;15(6):1077−85. Available from: https://doi.org/10.1007/s10544-013-9799-z.

[411] Bae KH, et al. Microstructured dextran hydrogels for burst-free sustained release of PEGylated protein drugs. Biomaterials 2015;63:146−57. Available from: https://doi.org/10.1016/j.biomaterials.2015.06.008.

[412] Wu Y, et al. Novel thermal-sensitive hydrogel enhances both humoral and cell-mediated immune responses by intranasal vaccine delivery. Eur J Pharm Biopharm 2012;81(3):486−97. Available from: https://doi.org/10.1016/j.ejpb.2012.03.021.

[413] Fan Q, et al. Hydroxypropyltrimethyl ammonium chloride chitosan-based hydrogel as the split H5N1 mucosal adjuvant: structure-activity relationship. Carbohydr Polym 2021;266(1):118139. Available from: https://doi.org/10.1016/j.carbpol.2021.118139.

Nanovesicles for delivery of antiviral agents

Yasmine Radwan, Ali H. Karaly and Ibrahim M. El-Sherbiny

Nanomedicine Research Laboratories, Center for Materials Science, Zewail City of Science and Technology, 6 of October City, Giza, Egypt

Abbreviation

AIDS	Acquired immune deficiency syndrome
BCS	Biopharmaceutical classification system
ddC	2′,3′-dideoxycytidine
ddCTP	dideoxycytidine-5′-triphosphate
DDS	Drug delivery system
DLS	Dynamic light scattering
DSC	Differential scanning calorimetry
EM	Erythrocyte membrane
hep-AIII	Heparin active serine antithrombin III
IFN	Interferon-induced
LPH-NPs	Lipid-polymer hybrid nanoparticles
MBSA	Maleylated bovine serum albumin
O-SAP	O-stearoyl amylopectin
PEG	Polyethylene glycol
PEVs	Penetration enhancer vesicles
RSV	Respiratory syncytial virus
SEM	Scanning electron microscope
TEM	Transmission electron microscope
VLPs	Virus-like particles

20.1 Introduction

The current antiviral therapies are based on the interferon effect. The interferon effect is basically the use of small molecular weight drugs or proteins that have the ability to stimulate the innate immune response. There are several antiviral agents currently in the market and in clinical practice [1]. For example, Table 20.1 shows the approved antiviral drugs for HIV

TABLE 20.1 Approved antiviral drugs for HIV infections [3].

Drug	Dosage form
Abacavir: 2-amino-6-cyclopropylaminopurin-9-yl-2-cyclopentene	Oral
Didanosine: 2′,3′-dideoxyinosine	Oral
Emtricitabine: (-)-β-l-3′-thia-2′,3′-dideoxy-5-fluorocytidine	Oral
Lamivudine: (-)-β-l-3′-thia-2′,3′-dideoxycytidine	Oral
Stavudine: 2′,3′-dideoxy-2′,3′-didehydrothymidine	Oral
Zalcitabine: 2′,3′-dideoxycytidine	Oral
Zidovudine: 3′-azido-2′,3′-dideoxythymidine	Oral
Tenofovir disoproxil fumarate: bis(isopropoxycarbonyloxymethyl)ester of (R)-9-(2-phosphonylmethoxypropyl)adenine	Oral
Delavirdine	Oral
Efavirenz	Oral
Etravirine	Oral
Nevirapine	Oral
Raltegravir	Oral
Amprenavir	Oral
Atazanavir	Oral
Darunavir	Oral
Fosamprenavir (a prodrug of amprenavir)	Oral
Indinavir	Oral
Lopinavir	Oral
Nelfinavir	Oral
Ritonavir	Oral
Saquinavir	Oral
Tipranavir	Oral
Enfuvirtide (T-20)	Subcutaneous
Maraviroc	Oral
Abacavir: 2-amino-6-cyclopropylaminopurin-9-yl-2-cyclopentene	Oral

infections. It can be noticed that most of them are administered orally. However, several antivirals have low bioavailability upon oral administration such as the antiretroviral acyclovir and ganciclovir. Nevertheless, the antiviral bioavailability is essential for its efficacy along with other properties such as good solubility, proper half-life, and high absorption. So, to achieve the required efficacy, high doses and frequent administration are applied, which can result in severe side effects as well as negatively affect the patients compliance [2].

Improving the delivery of antiviral agents can be challenging due to several reasons. One of which is related to the nature of the virus itself. As the viral infection mechanism depends on the nature of the host cell and the virus as well, it is challenging to synthesize a specific drug that can target the virus without harming the host cell. Theoretically, the drug should target the viral proteins responsible for its replication as well as the different pathogenesis from the host cell proteins to make the drug selective for a specific virus. Given that each virus has its own pathogenesis, hence, it is difficult to develop a broad-spectrum antiviral agent that is active against several viruses cause similar symptoms [3].

Another challenging aspect of the development process is the drug formulation development, which includes modifications in the physicochemical and biopharmaceutical properties of antiviral molecules. The enhancement of the bioavailability and pharmacokinetics was performed by reformulating the drug or by improving the delivery systems for antiviral administration using nanotechnology. The later has led to the development of various nanocarriers for the delivery of the antiviral drugs, and the evolving to nanopharmaceuticals [3].

20.2 Overcoming the challenges of traditional delivery of antiviral agents

Nanotechnology is the downsizing of materials to the nano level, which is 1×10^{-9} of meter, and utilizing it afterwards. The utilization of nanostructures in diagnostic and therapeutic fields is referred to as nanomedicine [4]. At a later estimation of the drug delivery industry, it worth about USD 80 billion, and most of it is devoted to the design of new controlled release and targeting biological systems [5].

Nanopharmaceuticals represent a relatively new class of therapeutic nanomaterials which displays the advantages of high surface-to-volume ratio, small size and rich surface chemistry. The rich surface chemistry allows the nanomaterials to be functionalized with small biological molecules such as protein and nucleic acids, which broadening their targeting spectrum, as well as increasing their compatibility with the immune system via modifying the surface with hydrophilic polymers such as polyethylene glycol (PEG) [6,7].

The downsizing of nanocarrier systems is imparted to several physicochemical changes such as increasing the surface-to-volume ratio, which reflects on the reactivity of molecules. Nanocarriers are promising to overcome several problems of antiviral drug delivery in the conventional dosage forms such as low bioavailability, high toxicity, drug resistance, unspecific targeting, and short half-life of the drug, as shown in Fig. 20.1. Nanodelivery systems can include several kinds of nanocarriers such as nanovesicles, nanoparticles, nanomicelles, dendrimers, etc. Fig. 20.2 shows the several types of nanocarriers and their advantages over conventional delivery methods. In addition, the antiviral nanocarriers in the case of intravenous administration can have a safe route without the danger of retaining by the pulmonary capillaries due to their small size [3].

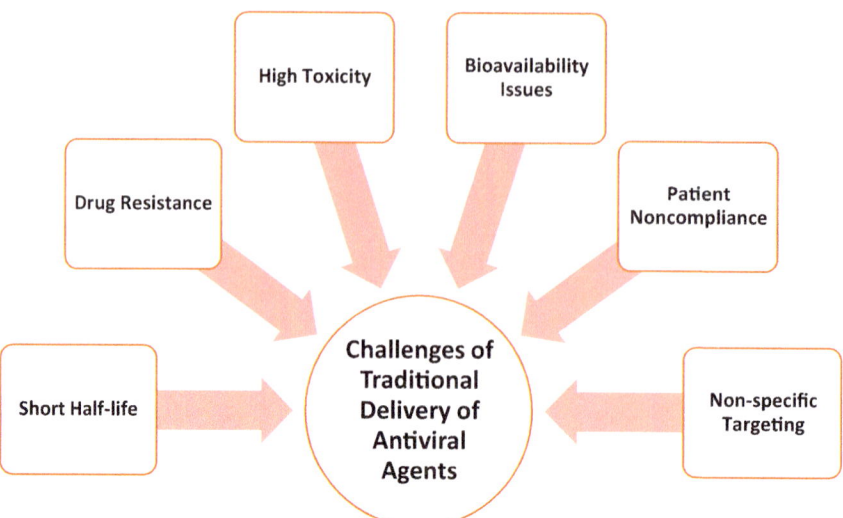

FIGURE 20.1 Challenges of traditional delivery of antiviral agents.

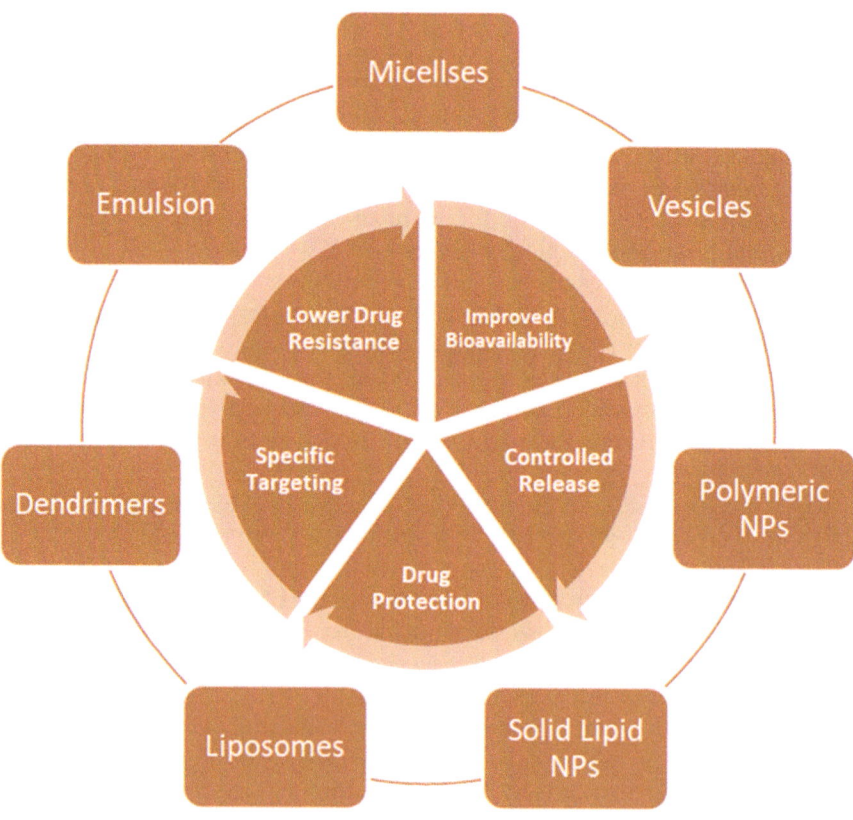

FIGURE 20.2 Types of nanocarriers for drug delivery, and their advantages over conventional drug delivery.

20.3 Nanovesicles

20.3.1 Nanovesicles composition

Vesicles are self-assembled structures that are produced naturally and can be also produced artificially. They are generally formed as spherical capsules composed of one or more lipid layers with an aqueous content core [8]. In nature, nanovesicles can be described as tiny sacs carrying chemical compounds released by cells as a way of communication [9]. There are several synthetic types of nanovesicles which include; liposomes, niosomes, transferosomes, and ethosomes. Each of these synthetic types has its own features of composition that will be discussed in this section.

Liposomes are spherical vesicles consisting of aqueous core enclosed by one or concentrically arranged bilayer membranes of natural or synthetic phospholipids. Their unique structure enables them to incorporate both hydrophilic and hydrophobic (lipophilic) drugs. The hydrophilic molecules can be loaded in their aqueous core, while the hydrophobic drugs are stored in the lipid bilayer.

Phospholipids have a ranger of a head and tail structures as shown in Fig. 20.3. The head is responsible for the surface charge of the liposomes. For example, phosphatidylcholine/phosphatidylethanolamine are neutral while phosphatidylglycerol and phosphatidylserine are negative. On the other hand, the tail is responsible for the physicochemical properties such as melting point and permeability. For example, dimyristoylphosphatidylcholine DMPC has a transition temperature of 23°C while distearoylphosphatidylcholine DSPC has a transition temperature 55°C.

Niosomes are vesicles composed of bilayer of non-ionic surfactants (e.g., Span 40 and Span 60) as shown in Fig. 20.4. They are more stable than the liposomes, and their lipid membrane can be either unilamellar or multilamellar.

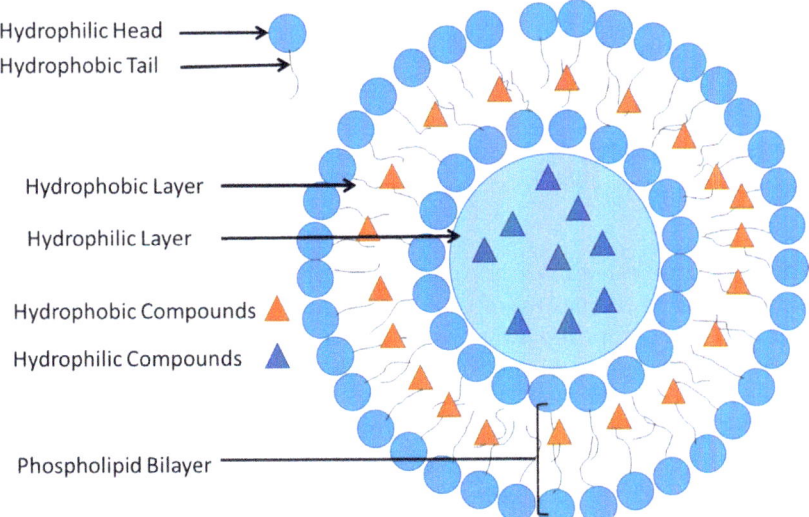

FIGURE 20.3 Schematic representation of liposomes structure.

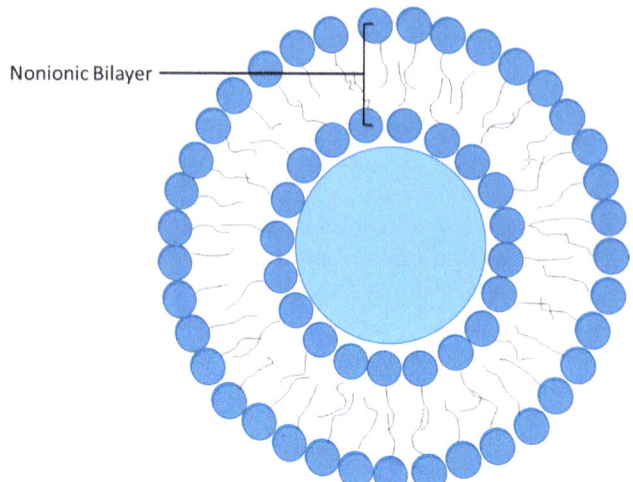

Nonionic Bilayer

FIGURE 20.4   Schematic representation of niosomes.

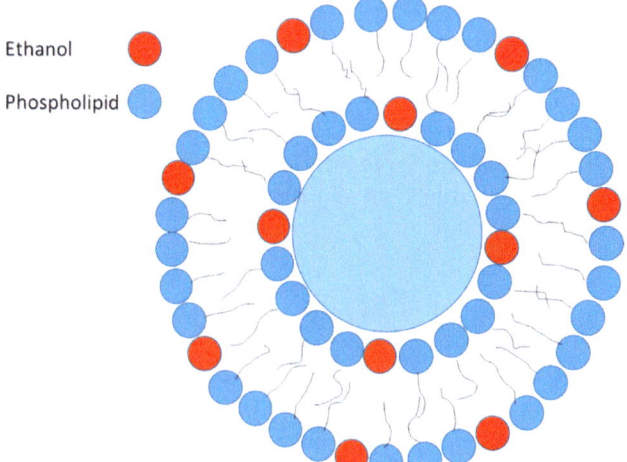

Ethanol

Phospholipid

FIGURE 20.5   Schematic representation of ethosomes.

Another type of nanovesicles is ethosomes. Ethosomes are more commonly used in cosmetics application due to the nature of their composition. As shown in Fig. 20.5, the membrane of ethosomes is composed of phospholipids with ethanol. This unique structure enables them to fluidize the stratum corneum lipids by virtue of their alcohol and increase their ability to reach deep skin layers. However, their ethanolic content can cause skin irritation, and their dose should be adjusted accordingly.

Transferosomes are another class of nanovesicles. They are composed of phospholipids and edge activators which increase the elasticity of the membrane (Fig. 20.6) as they are membrane softening agents (e.g., Tween 80, Span 80, and sodium cholate). Owing to their high elasticity, transferosomes can overcome the difficulty in skin penetration due to their

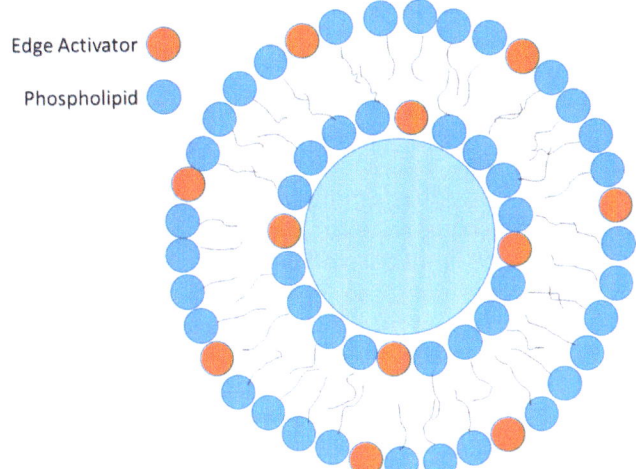

FIGURE 20.6 Schematic representation of transferosomes.

Edge Activator

Phospholipid

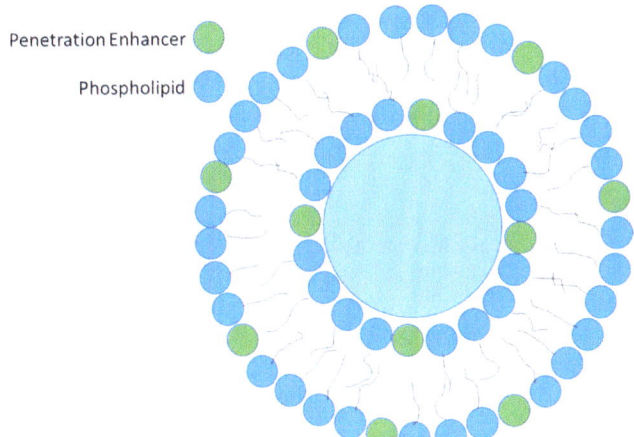

FIGURE 20.7 Schematic representation of penetration enhancer vesicles (PEVs).

Penetration Enhancer

Phospholipid

ability of penetrating different ranges of pore sizes, which makes them more advantageous than the nondeformable liposomes and niosomes.

Another class of nanovesicles is the Penetration Enhancer Vesicles (PEVs). They are overcoming the disadvantages of ethosomes by replacing the ethanol in the phospholipid bilayer by penetration enhancers such as labrasol, transcutol, and PEG 400 within their phospholipid bilayer as shown in Fig. 20.7.

20.3.2 Nanovesicles fabrication methods

Nanovesicles as biofunctional materials should have certain characteristics to preform optimally. In other words, some of the nanovesicles physicochemical parameters such as size, morphology, and complexity should be optimized to serve their target application. Those parameters are strogly affected by the synthesis approach and system composition during the prodution process. For example, in the conventional methods of nanovesicles synthesis, the formation of vesicles desertion in the aqueous phase during the production step by self-assembly can result in heterogeneity due to the production system composition and dimensions. So, to control size, morphology, lamerallity, and other desired structural characteristics, post formation steps can be introduced to the production process [4].

Generally, there are four main stages in synthesizing nanovesicles. First, the lipids need to be dried from its organic solvent, dispersed in aqueous media, purified, and finally characterized. To load the nanovesicles with active molecules, there are two main routes: passive loading and active loading. In the passive loading, the drug is loaded during the synthesis procedure, while in the active loading, the drug is loaded after the vesicle formation. Passive encapsulation efficiency of hydrophilic active molecules depends on the nanovesicles ability to trap aqueous phase during formation. On the other hand, active encapsulation efficiency depends on the ability of the drug to penetrate the nanovesicle membrane as well as the membrane-drug interaction. There are several methods for passive loading of drug during the nanovesicles synthesis. They can be categorized into three classes: mechanical dispersion, solvent dispersion, and detergent removal. The mechanical dispersion can be attained by several ways. Mainly the dried lipids are mixed with aqueous phase and the mechanical dispersion allows the assembly of the nanovesicles. The mechanical dispersion can be performed via sonication, extrusion, microemulsification, thin film hydration, etc. [10]. Some of the fabrication methods of nanovesicles are described below.

20.3.2.1 Thin film hydration method

Thin film hydration method is one of the simplest methods for nanovesicles preparation. It includes preparing a thin hydrophobic layer in a round bottom flask simply by removal of the solvent, followed by addition of dispersion medium in which nanovesicles will be formed upon agitation. With continuous agitation the nanovesicles start to form, however, in a heterogenous manner. To make it homogenous, extrusion through polycarbonate membrane is required [11].

20.3.2.2 Reverse phase evaporation method

It is a solvent dispersion method with the ability to increase the aqueous space to lipid ratio in the nanovesicle composition, which can be useful for higher entrapment of hydrophilic materials. This method is based on the synthesis of inverted micelles which are formed during sonication of the mixture of buffered aqueous phase that contains the hydrophilic materials to be encapsulated into the liposome and amphiphilic materials in the organic phase. The organic solvent phase is then eliminated slowly, which leads to the conversion of inverted micelles into viscous state and gel form. At this stage, excess of phospholipids are critical to aid in completion of the bilayer around the micelles and to create the vesicles. Otherwise, the gel state can reach to a critical point and collapse and eventually disrupt the inverted micelles formation. This method enables the nanovesicles synthesis with a relatively high aqueous volume to lipid ratio with a wide range of lipid formulations [10].

20.3.2.3 *Detergent removal by dialysis method*

Detergents are used to solubilize hydrophobic materials and lipids at their critical micelle concentrations, at which the concentrations of lipids and hydrophobic materials are at their maximum, and above which they start to aggregate and forms micelles. When detergent is detached, the micelles begin to aggregate and forms large unilamellar vesicles, and these detergents can be removed mainly by dialysis [10].

20.3.3 Nanovesicles characterization

There are several characteristics in nanovesicles that need to be tested to ensure their efficacy and structural integrity along with other desired properties. The main aspects that need to be tested are the entrapment efficiency of loaded cargo, morphological visualization, dynamic size and zeta potential, vesicle stability, transition temperature, drug release profile, penetration and permeation, as well as elasticity or deformability [12].

Entrapment efficiency is the percentage of the entrapped drug. It is estimated by the separation of the entrapped drug and measure its concentration, then disrupting the nanovesicles and measuring it again. The amount of free drug is determined and subtracted from the total amount of the added drug to estimate the actual loaded amount of drug. Measuring and quantification of the drug depends particularly on its nature, and it can be measured by spectrometry, florescence spectroscopy, enzyme-based methods, and electrochemical techniques. Similarly, drug content can be quantified by analytical methods such as high-pressure liquid chromatography (HPLC) and liquid chromatography coupled with mass spectrometry [9,11−13].

Drug release study could be done in vitro using artificial membrane such as dialysis bag, while the ex vivo studies are done using animal skin. The drug release studies can be done also with the aid of Franz diffusion cell, which has a donor compartment in which the drug is kept and a receptor compartment that contains the dissolution media [12,14].

Morphological visualization of nanovesicles can be performed using transmission electron microscopy (TEM) and scanning electron microscopy (SEM) to examine their topology and morphology of their surface [12,13,15−19]. The dynamic size average and distribution (poly-dispersity index) are critical parameters to asses, especially when using the nanovesicles for therapeutic applications, and dynamic light scattering (DLS) is one of the most common methods used for this purpose. DLS uses a computerized analysis system accompanied with photon correlation spectroscopy to estimate the dynamic size of particles. In addition, surface charge of nanovesicles (Zeta potential) is used to estimate the ability of the vesicles to stay suspend in a dispersion by Van Der Waals repulsion force and also can be assessed with Zetasizer machine [12,17,20−22].

The stability of the nanovesicles can be assessed by determining the size and structure of the vesicle over allocated time intervals with the aid of DLS and TEM. Transition temperature is defined as the temperature required to induce a change in the lipid physical state. It can be determined by differential scanning calorimetry (DSC). Moreover, it is important to study the vesicles penetration ability as well as the penetration mechanism through intact skin. This can be performed using confocal laser scanning microscopy (CLSM) [12,23]. One of the characteristics of nanovesicles, mainly "transferosomes" is their elasticity, which can be checked using the extrusion method. In this method, nanovesicles are allowed to pass through polycarbonate membranes of known size at a constant pressure over allocated time intervals followed by measuring the size of passed vesicles [12,24,25].

20.3.4 Nanovesicles applications in nanomedicine

Nanovesicular systems are widely used in different biomedical applications, including targeted drug delivery and diagnosis. To elaborate, nanovesicles are extensively used for the delivery of various classes of drugs and pharmaceuticals. They are also exploited in several drug delivery routes such as topical, oral, and transdermal delivery [26].

The most delivered therapeutics with the aid of nanovesicles as nanocarriers are anticancer drugs, where they are extensively investigated and are the most clinically approved nanoformulated products [4,27]. Various anticancer drugs such as cisplatin, doxorubicin, lapatinib, paclitaxel, vinorelbine, and 5-fluorouracil are being loaded into nanovesicles. While some of the developed nanoformulations are currently available in the market, many are still under clinical evaluations [4]. Currently, the delivery of anticancer drugs loaded into nanovesicular systems via topical administration is being extensively assessed. This is due to the high safety index and efficacy of theses drug-loaded nanovesicular systems in the treatment of skin cancer as compared to conventional treatments. In addition, using drug delivery systems (DDSs) for topical delivery increase the therapeutic index, and decrease systemic side effects of drugs [26].

Infectious diseases impose a challenge to the available conventional treatments. Thus, nanovesicular systems are being used for the delivery of various anti-infectious drugs as an effective alternative to enhance the therapeutic efficacy of these drugs [26].

Fig. 20.8 shows various nanovesicular systems intended for drug delivery for the treatment of various diseases, some of them; particularly liposomes, niosomes, transfersomes, and ethosomes have been described earlier and their nanomedicine applications will be covered in the below section.

20.3.4.1 Liposomes

Liposomes are the most studied nanocarriers for several biomedical applications including oral and topical drug delivery, diagnostic imaging, cosmetics, and delivery of proteins and genes [26]. Shedding the light on drug delivery applications via liposomes will highlight already marketed formulations or preparations under clinical evaluation. As an example, the first formulation that was commercialized in 1995 was named Doxil/Caelyxs, composed of doxorubicin encapsulated into stealth non-targeted liposomes for treatment of ovarian cancer and HIV-related Kaposi's sarcoma. Doxil/Caelyxs prolonged the half-life of the drug, enhanced its biodistribution, improved the therapeutic efficacy, and decreased the side effects including neutropenia and cardiotoxicity. The fabrication technique used to prepare Doxil/Caelyxs contributed to its high drug loading and stability on shelf and in blood. This smart drug loading technique actively loaded the amphipathic weak base drug post-production of liposomes via using transmembrane ammonium sulfate gradients. The development of this nano-DDS encouraged the progress of other nano-DDSs in clinical settings [28,29]. Lipo-Doxs is another marketed vesicular formulation, it is composed of doxorubicin encapsulated in liposomes. A study developed a

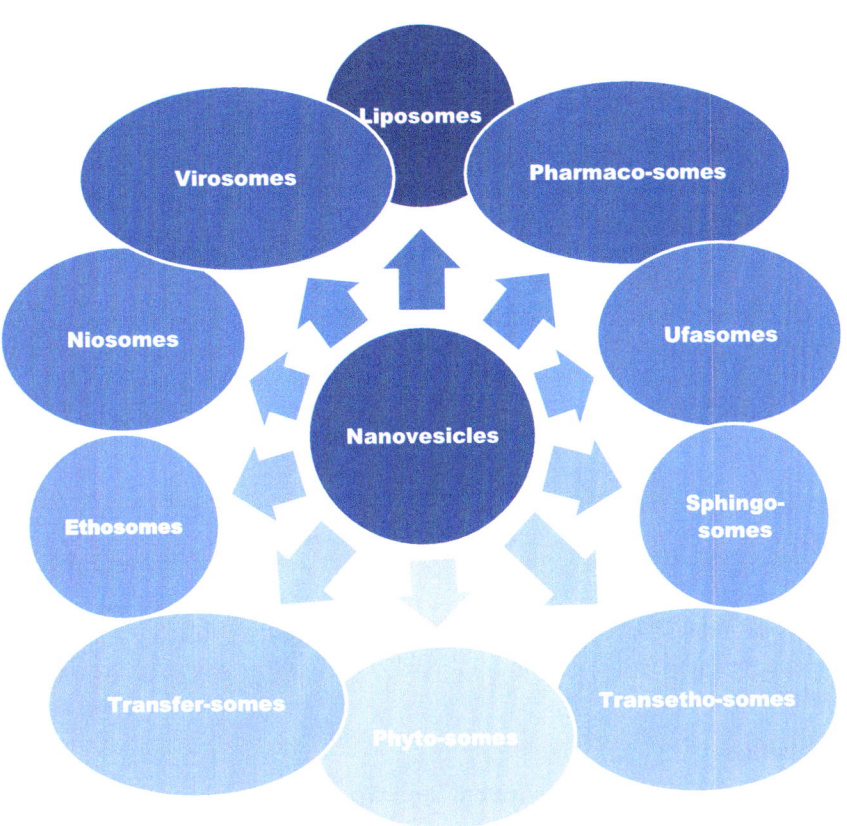

FIGURE 20.8   Visual representation of nanovesicular systems used in nanomedicine.

liposomal system coated with poly(allylamine hydrochloride), and poly-(acrylic acid) to deliver doxorubicin. The results revealed a significant enhancement in the oral bioavailability of the drug, and antitumor activity when administering the formulation orally in a multiple-dose regimen, as compared to single administration of Lipo-Doxs intravenously [30]. Another study functionalized the surface of liposomes carrying doxorubicin with antiHER2/neu peptide for targeting breast cancer. The targeted DDS enhanced the in-vivo anticancer activity and therapeutic efficacy comparable to Doxil/Caelyxs which is a non-targeted system [31].

Liposomes were also used by different research groups for delivering anti-infectious drugs. For example, a recent study intended to encapsulate ciprofloxacin, an anti-infectious drug, in liposomes. The study used an animal model with lethal tularemia. The results displayed that more than 85% animal survival rate was reached after 2 weeks from treatment [32]. In another work, tobramycin encapsulated in liposomes was used to treat infected rat model with *Pseudomonas aeruginosa* and *Burkholderia cepacia*. The results demonstrated that the developed nano-DDS enhanced the therapeutic efficacy of the drug as compared to the free drug [33].

Some challenges faced by conventional formulations include the obstacles in reaching therapeutic targets for treatment of infectious diseases like HIV, legionella, or tuberculosis.

This is due to their evading from the phagocytic system by infecting the cells internally [34,35]. Liposomes have been used to effectively deliver drugs into cells and enhance cellular uptake. Different studies have involved the encapsulation of drugs such as amikacin, kanamycin, gentamicin and streptomycin into liposomal DDSs [4]. The results confirmed enhanced therapeutic efficacy in-vivo. In addition, the results confirmed that using DDSs for drug delivery diminished the counts of viable pathogenic bacteria like *Klebsiella pneumonia* and *Mycobacterium tuberculosis*.

Another study aimed at targeting macrophages via loading rifampicin into functionalized liposomes. Results demonstrated that *Mycobacterium smegmatis* viable counts significantly reduced when treated with the liposomal DDS as compared to the free drug. Added to this, functionalizing the liposomes with surface antigens such as o-stearoyl amylopectin (O-SAP) and maleylated bovine serum albumin (MBSA) imposed a significant enhancement in the efficacy of the delivered drug due to the active targeting to the macrophages [36].

In spite of the extensive use of liposomes for the delivery of various drugs, their usage for delivery of biopharmaceuticals such as proteins and nucleic acid is still limited. This is due to the poor transfection efficiency, low stability and poor therapeutic efficacy. Most of the biopharmaceuticals loaded on liposomes nowadays are still under early stage of clinical evaluation. Several proteins and enzymes are being investigated for the delivery via liposomes because of the recent studies that confirmed the enhanced absorption and bioavailability of proteins loaded in liposomes [4]. However, further investigations and assessments are still needed along with clinical trials for transferring this technology into clinical settings.

20.3.4.2 Niosomes

Niosomes are being used in transdermal, topical, and ophthalmic drug delivery applications. They are more stable and cost-effective than liposomes [37]. In addition, they are used as adjuvants to immunological therapies [26]. In a recent study, niosomes were used to deliver a member of biopharmaceutical classification system (BCS) class IV drugs, named paclitaxel, which is suffering from poor characteristics impairing its oral delivery, and the results demonstrated that niosomes enhanced the oral bioavailability of paclitaxel [4,38,39].

For ophthalmic drug delivery, niosomes are widely used. For example, in a study niosomes loaded with tacrolimus were developed to enhance the drug biocompatibility with the cornea. Results indicated that using niosomes has minimized the corneal allograft rejection when transplanted as compared to the free drug and classic CsA treatment [4,40]. In another study, gatifloxacin was encapsulated in chitosan-coated niosomes. The formulation was used for the treatment of ocular infections, and developed to reduce the dosing frequency of the drug. Results displayed that the developed formula prolonged the retention time of the drug in the eyes, enhanced the transcorneal permeation of the drug, and had improved ocular tolerability [41].

For the treatment of inflammation, niosomal formulations are found to enhance anti-inflammatory activity as compared to free drugs. For instance, serratiopeptidase was encapsulated in niosomes to alleviate the systemic side effects caused by administering the free drug orally. Ex vivo and in vivo experiments were held to assess the efficacy of the gel niosomal formulation loaded with serratiopeptidase when applied topically. Results of the ex vivo experiments revealed that the developed formulation enhanced the skin permeation of the drug. While the results of the in vivo experiments demonstrated that the developed formula had a comparable anti-inflammatory activity to diclofenac gel [42].

20.3.4.3 *Transfersomes*

Transfersomes with their ultra-flexibility provide improved topical and transdermal delivery of different drugs including proteins and peptides [26]. Transfersomes are being used as drug delivery systems for various applications including treatment of cancer, hypertension, depression, leishmaniasis, and fungal infections [4].

A study involved the development of three nanovesicular systems for the topical delivery of 5-fluorouracil; liposomes, niosomes, and transfersomes. The results showed that the nanocarriers enhanced the therapeutic efficacy of the drug. However, transfersomes had the best penetration capacity thus it provided the highest anticancer activity [43]. In another study, the use of loaded transfersomes was investigated for the treatment of rheumatoid arthritis. Meloxicam was encapsulated in transfersomes for dermal drug delivery, and the results indicated that transfersomes allowed a better skin permeation of the drug as compared to free drug and drug-loaded liposomes [44]. In another study, transfersomes were used for the delivery of sertraline for treating depression. The nanocarrier system was investigated for the transdermal drug delivery to escape hepatic first-pass metabolism normally faced when the drug is orally administered. Results exhibited that using transferosomes for sertraline transdermal delivery decreased the required dose, enhanced the permeation in both in vivo and ex vivo experiments, and improved sertraline therapeutic efficacy as compared to the free drug [45].

A study explored the use of transferosomes for transdermal delivery of an antihypertension agent named felodipine. The in vitro assessments revealed that skin permeation was significantly improved as compared to the control formulation. In addition, pharmacokinetic results indicated that using transferosomes for the transdermal delivery maintained the drug concentration in blood with low plasma fluctuation. Finally, the bioavailability of the drug was significantly increased as compared to orally administered felodipine, due to prompt permeation of felodipine across skin layers [46].

Another study encapsulated amphotericin B in transferosomes for treating leishmaniasis. Results demonstrated an enhancement in skin permeation, and an improvement in the therapeutic efficacy of the drug against *L. donovani* after comparison with liposome-based formulation [47].

20.3.4.4 *Ethosomes*

Ethosomes are used mainly in transdermal and topical drug delivery [26]. This is due to the high concentration of ethanol that is forming ethosomes enhancing its penetration into deep dermal layers due to the ethosomal effect, increased plasticity of the lipid membranes, as illustrated in Fig. 20.9. Ethanol decreases the lipids' density in the cell membrane, and rises the membrane's fluidity leading to a partial dissolution of the intercellular matrix of lipids, this effect is known as the ethanolic effect [48].

In a study, ethosomes were used to encapsulate mitoxantrone for melanoma tumors treatment. The results exhibited an enhancement in the anticancer activity, initiation of the anticancer immune response, and a decrease of tumor size [49]. In another study,

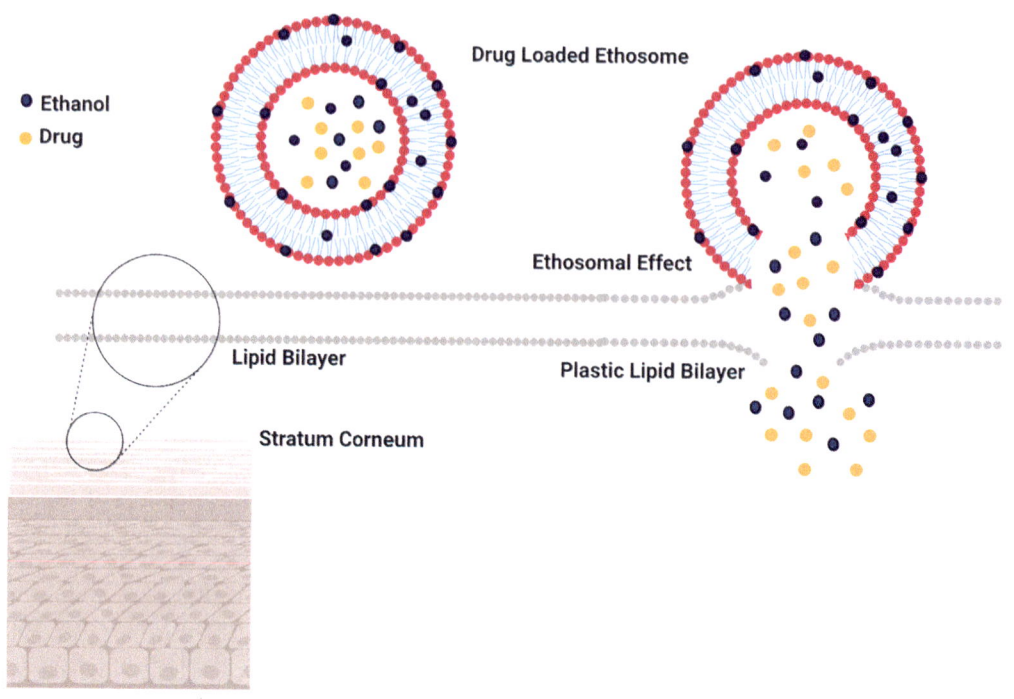

FIGURE 20.9   The ethosomal effect: ethosomes enhancing drug permeation into deep dermal layers.

glimepiride was encapsulated in ethosomes formed of ethanol, propylene glycol, Chol, and PC for treating diabetes. The developed system was delivered via a transdermal patch. Results of the study displayed that ethosomes enhanced the skin permeation of the drug, and increased its bioavailability in vivo, when compared to the patch loaded with the free drug. The results also highlighted the potential of using ethosomes for the transdermal delivery of the drug as an alternative to oral delivery [50].

In other study proposed the ethosomes as a nasal drug delivery system for delivery of zolmitriptan for migraine treatment. Zolmitriptan has low oral bioavailability, thus intranasal delivery will bypass this challenge. In addition, using a thermo-reversible gel-based formula helps in increasing the time of residence of the drug at the nasal cavity, hence decreasing the frequent dosing of the treatment. Added to this, the results of the study demonstrated that the developed formula has no toxicity on columnar epithelial cells [51].

In other work, transethosomes were used for the delivery of piroxicam for treating some inflammatory diseases such as osteoarthritis and rheumatoid arthritis. Transdermal delivery was investigated to mitigate the side effects caused due to oral delivery, such as peptic ulcer, ulcerative colitis, and gastrointestinal irritation. The loaded transethosomes were incorporated into a hydrogel, and the results showed that the developed system enhanced the drug permeation and retention in the skin. In addition, it provided the highest elasticity and stability as compared to other formulations such as liposomes and transferosomes [52].

20.3.5 Challenges of nanovesicles for nanomedicine applications

Using nanovesicles for various therapeutic applications has been extensively investigated. As an example, liposomes are being studied as drug delivery systems in humans and animals. In addition, several companies are investing in developing liposomes-based products for delivering antifungal and anticancer drugs to mitigate their cytotoxicity and adverse effects. However, using liposomal formulations for drug delivery in commercial pharmaceutical products is still limited [26].

Likewise, nanovesicles-based formulations are not widely available for commercial purposes due to challenges that hinder their development. However, some obstacles were resolved, such as controlling particle size, short half-life in circulation, drug entrapment, and batch-to-batch reproducibility. Yet, some other challenges are not resolved, such as stability issues, and identifying an appropriate sterilization method, especially for formulations including phospholipids as they are unstable to radiation, heat, and chemical sterilization. Filtration through sterile membranes is the alternative used for the sterilization of nanovesicles. Another challenge is the lack of mass production methods which is due to several reasons. One of which is the high cost of production in addition to the lack of economic feasibility to produce large-scale batches. For example, liposomes-based drug formulations already found in the market, are produced in small-scale production, thus increasing the manufacturing costs. In addition, lack of infrastructure and competent skilled researchers that are able to develop new nanovesicle systems or scale-up the production process for large batch sizes while cutting costs, hinder the commercialization of nanovesicles-based products. Finally, another posed challenge is the high maintenance cost of instruments and the high cost of research, development, and analysis [26]. Some of these challenges and limitations of the various types of nanovesicles are shown in Table 20.2.

TABLE 20.2 An overview of challenges and limitations of several types of nanovesicles.

Formulation	Challenges	References
Liposomes	• Drug leakage • Rapid clearance • Stability issues	[53]
Transfersomes	• Chemical instability	[53]
Niosomes	• Short half-life due to aggregation, hydrolysis, and leakage • Low bioavailability • Time-consuming manufacturing process	[53]
Phytosomes	• Short duration of action • Rapid elimination of phytoconstituent	[54]
Transethosomes	• Stability issues	[54]
Pharmacosomes	• Short half-life due to aggregation, hydrolysis, and fusion	[54]
Sphingosomes	• Low entrapment efficiency • Costly manufacturing process	[26]
Virosomes	• Expensive manufacturing process	[55]
Ufasomes	• Stability issues • Sensitivity to pH and ionic strength of medium	[26]

20.4 Nanovesicles and biomimetic nanovesicles for delivery of antiviral agents

In the advent era of nanobiotechnology, the development of drug delivery technologies is being revolutionized. Nanobiotechnology allowed a great enhancement in pharmacodynamics and pharmacokinetics of the drugs through the improvement of specific targeting of drug and prolongation of drug bioavailability in the body as compared to conventional therapy. Special focus on delivering antiviral agents via different nanocarriers is being given. The major reason is the unique characteristics imposed on the drug in the nanoscale governed by the chemical and physical properties of the nanocarrier [56,57]. Hence, enhancing the therapeutic index, increasing the half-life of the drugs, as well as its precision to reach the targeted cells without damaging healthy cells. In addition, using appropriate nanocarriers allowed the delivery of poorly-soluble antiviral drugs, enhanced the drug stability, improved ADME of drugs, and reduced undesired immune activation [58–61]. Nanovesicles are extensively studied as promising nanocarriers for the delivery of antiviral agents. Some of the most promising nanovesicular formulations loaded with antiviral drugs will be described below.

20.4.1 Liposomes for delivery of antiviral agents

The development of liposomes as drug delivery systems had contributed to the treatment of several viral diseases. For instance, various studies had developed liposome-based formulations aiming to eradicate HIV. The empirical application of liposomes for delivering antiretroviral drugs has been extended due to the modulated pharmacokinetic profile of the loaded drug. In addition, the liposomal delivery system provides enhanced stability, bioavailability, permeability, and bioaccumulation of the drug at HIV reservoir sites. Liposomes are also easily surface-modified and conjugated, making them a perfect fit for targeted drug delivery.

Jin, et al. encapsulated zidovudine myristate prodrug in liposomes to enhance its anti-HIV effect. Pharmacokinetic assessments revealed that the loading of zidovudine enhanced its pharmacokinetic profile as well as reduced its clearance and prolonged its half-life as compared to the zidovudine solution. In addition, the drug distribution in tissue was improved with an increased accumulation level of the drug-loaded liposome formulation in the reticuloendothelial system and brain. Results highlight the potential of using the liposomal prodrug system for enhancing zidovudine therapeutic efficacy and diminishing toxicity [62].

In another study, in vitro and in vivo investigations have been conducted to enhance the permeation of antiretroviral agents by encapsulating them in liposomes. In this study, a rat model was used to assess the skin permeation of zidovudine loaded into liposomes. Results demonstrated that the antiretroviral agents encapsulated in liposomes significantly enhanced the permeation of the drug to blood plasma as compared to the free drug. This is due to enhanced transdermal flux, and improved deposition of the drug in the reticuloendothelial organ after administration of the drug-loaded liposomes [63].

Some studies modified the liposomes' surface via ligands or charges, for selective lymphatic targeting. In one study, dicetyphosphate, stearyl amine, and mannose conjugate were used to coat liposomes loaded with zidovudine. The liposomes coated with mannose caused a significant decrease in drug release as compared to conventional liposomes. In addition, the results displayed that selective targeting was greatly achieved when using mannose-coated liposomes rather than positive or negative liposomes, as the concentration of drug in spleen and lymph nodes was significantly higher than in serum, and the results were supported by fluorescence imaging of spleen and lymph nodes. These outcomes highlight the efficacy of selective targeting of zidovudine to the lymphatic system for AIDS treatment [64].

A study investigated the effect of encapsulating 2',3'-dideoxycytidine (ddC) in liposomes. The formulation was assessed on a mouse macrophage cell line. Results displayed an improved intracellular absorption of the drug [65]. Another study investigated the effect of encapsulating dideoxycytidine-5'-triphosphate (ddCTP), a phosphorylated form of ddC, in liposomes by using a murine-acquired immunodeficiency syndrome model. The developed nanosystem showed good stability and high retention of the loaded drug. Moreover, the results demonstrated a reduction in viral load and proviral DNA in the mononuclear phagocyte system in bone marrow and spleen [66].

In one work, anti-HLA-Dr immunoliposomes, loaded with protease inhibitor named heparin active serine antithrombin III (hep-AIII) was assessed in vivo, in a nonhuman adult primate model. Results revealed that a steady decline in the plasma viral load. This study concluded that using hep-AIII is a good alternative to antiretroviral drugs for the treatment of resistant strains of HIV [67].

Another study tackled the challenge of macrophages' phagocytosis of liposomes by coating liposomes with polyethylene glycol (PEG) to make it stealth. The developed nanosystem was targeted by conjugating it to gp120 ligand, and was encapsulated with protease inhibitor for the treatment of HIV. Results revealed that the developed system selectively targeted HIV-1 infected cells, and the drug demonstrated higher and longer antiviral activity as compared to the free drug or the drug-loaded into conventional liposomes. This study highlighted the potential of using targeted delivery for antiretroviral drugs with enhanced therapeutic efficacy and decreased toxicity [68].

In another work, PEGylated liposomes loaded with saquinavir were assessed in regards to release rate in mammalian cells. In vitro studies indicated that the developed system has a sustained release profile and a lower cytotoxicity compared to non-PEGylated liposomes [69].

Other studies investigated the liposomal delivery of different antiviral agents. One study assessed charged and neutral liposomes for delivery of acyclovir to the eyes. Acyclovir is a 2/-deoxyguanosine analog, used for the treatment of Epstein Barr virus, cytomegalovirus, varicella-zoster virus and herpes virus [70]. The results indicated that only charged liposomes interacted with the drug. In addition, the highest entrapment efficiency was achieved with the negatively charged liposomes. However, the highest ocular bioavailability of the drug in the aqueous humor was achieved when using positively charged liposomes. The reason could be the increased permeability of the cornea due to using the positively charged liposomes, hence increasing the concentration of drug delivered [71]. Other studies confirmed the finding of the previous study, where the positively charged liposomes significantly enhanced acyclovir absorption into the cornea, and prolonged the residence time of the drug in the aqueous humor of the rabbit model. This is due to increased corneal adhesion of the positively charged liposomes relative to the free drug ointment. In conclusion, the surface charge and rate of corneal permeation could enhance the liposome-based ophthalmic delivery of acyclovir through the negatively charged corneal surface [72–74].

20.4.2 Niosomes for delivery of antiviral agents

In a study, a comparison between acyclovir-loaded liposomes and niosomes was held. In vitro assessments investigated the entrapment efficiency and release profile of the drug from both systems, and the residence time of the drug. The findings demonstrated that niosomes had a higher percentage of the loaded drug relative to liposomes. Moreover, the release profile of the drug from niosomes was better as only 50% of the drug was released in 200 minutes while liposomes released around 90% of the drug in only 150 minutes. The niosomes also provided a longer residence time in blood, hence would require less frequent dosing. In regards of stability, niosomes were more stable than liposomes. The study concluded that niosomes are a better fit for acyclovir intravenous delivery. Yet, further in vivo studies are required to authenticate the results [70]. In another work, zidovudine was encapsulated in niosomes for treatment of acquired immune deficiency syndrome (AIDS). In vivo studies were held on rabbit model, and the niosomal formulations were compared to liposomal-based ones. The results demonstrated a prolonged half-life of the drug from the niosomal formulations in the rabbit serum [75].

20.4.3 Ethosomes for delivery of antiviral agents

Ethosomes are used for antiviral agents' delivery as they provide better penetration of the drugs and prevent skin irritation. Several studies investigated the delivery of antiviral agents-loaded ethosomes [48]. For example, zidovudine was encapsulated in ethosomes and its skin penetration rate was assessed in comparison with zidovudine-loaded liposomes, zidovudine water-ethanol solution, and zidovudine ethanol solution. Fluorescence microscopy was used to evaluate the degree of skin permeability of each system. Results showed that ethosomes had the highest penetration rate as compared to all other groups [76].

In one study, acyclovir and acyclovir prodrug were encapsulated in ethosomes for transdermal delivery for comparison. The penetration capacity into epidermal layers was investigated for the encapsulated ethosomes, and free prodrug. Results demonstrated that ethosomes encapsulating acyclovir prodrug showed significantly higher concentration in the epidermal layers relative to the free prodrug, and ethosomes encapsulating acyclovir, 5.3-times higher and 3.4-times higher, respectively [77]. Another study explored the use of ethosomes for the delivery of lamivudine. In-vivo studies using rat model were performed to compare the lamivudine-loaded ethosomes and liposomes. Results indicated that lamivudine-loaded ethosomes had higher transdermal flux across rat skin, which was around 25-times more than the free drug solution. Moreover, higher penetration of the drug into rat skin layers was obtained by using ethosomes relative to liposomes. Scanning electron microscopy and transmission electron microscopy determined the change of stratum corneum structure influenced by ethosomes. Besides, cellular uptake results displayed that ethosomes provided higher intracellular uptake than the free drug solution. In conclusion, the ethosomal effect causing perturbation in the stratum corneum, along with the elasticity of ethosomes are the key reasons for enhanced skin permeation [78].

20.4.4 Biomimetic nanovesicles for delivery of antiviral agents

One of the strategies used to deliver antiviral agents and vaccines efficiently is using biomimetic nanovesicles. Biomimetic nanovesicles are formed by decorating the nanovesicle's surface with ligands mimicking cell surface proteins [79,80]. Biomimetic nanotechnology allowed the development of new therapeutic strategies for pathogen interception. One common strategy is blocking the interactions between the virus and host via using antibodies. However, this strategy is challenged by the high mutagenic nature of viruses, where the mutations have the ability to decrease the binding affinity of the virus to the antibody drugs [81]. One very recent case is the COVID-19 in which it has several mutants and variants leading to their escape from neutralizing antibodies [82]. Hence, the development of biomimetic nanoparticles having cell-like surface functionalities allowed its agonistic activity towards many virus variants, unlike antibodies that are only specific to one viral variant [81].

Biomimetic lipid-polymer hybrid nanoparticles (LPH NPs) are designed by incorporating ligands onto the surface of the nanosystem. These ligands mimic cell surface proteins and provide various advantages. For example, biomimetic LPH NPs provide selective targeting, enhanced efficacy, prolonged circulation time, and reduced drug resistance [79,83]. Two classes of nanovesicles fall under the biomimetic category, they are virosomes and virus-like particles (VLPs). Virosomes are nanovesicle with a functionalized phospholipid bilayer with viral envelope proteins such as NA and HA, however, they lack the genetic material of the viral source [84]. VLPs are self-assembled nanoparticles obtained from incorporating proteins such as encapsulin, ferritin, and lumazine synthase into virus envelope proteins or virus-derived capsids. VLPs have the advantage of tailoring specific structures, and easy surface functionalization. These platforms uphold various properties, one of which is the ability of loading different cargo into them such as proteins, drugs, contrast agents, and antibodies. In addition, their biomimetic nature allows their successful entry to the cell, prevents their endolysosomal entrapment, and provides selective delivery of the cargo [79,85]. Moreover, these systems show a low risk of immunogenic activation, and could be further enhanced by surface coatings such as PEG [86].

Many attempts were done to develop vaccines against influenza virus. One study attempted to develop a VLP to be used as an influenza vaccine. In the study, naturally occurring ferritin was genetically fused with influenza HA protein, that fusion self-assembled into VLPs tagged with HA trimeric spikes on the surface. Results demonstrated that the developed vaccine provided a broader and more potent response relative to the commercialized influenza vaccine [87]. In another study, for targeting and inhibiting influenza, sialic acid incorporated on nanoparticles' surface with a spacing of 3 nm between each residue was developed. Results demonstrated that this specific spacing provided the best binding affinity between the virus and nanoparticles. In-vivo studies performed on H1N1 infected mice, exhibited that the sialic acid conjugated nanoparticles decreased the viral load in lungs and enhanced survival [88].

In one work, the topography of influenza A virus was mimicked geometrically by spiky nanostructures. The study exhibited that spacing between spikes ranging from 5 to 10 nm provided the best binding affinity to the virions relative to smooth nanostructures. Afterward, the nanostructures were coated with erythrocyte membrane (EM) and the developed system prevented the virus from binding to the host cells and inhibited the infection. A post-infection study showed that EM-coated spiky nanostructures diminished more than 99.9% replication of the virus at a nontoxic cellular dose [89].

Marcandalli, et al. developed a potential vaccine for respiratory syncytial virus (RSV). The methodology was based on developing self-assembled protein nanoparticles surface-tagged with 20 antigenic prefusion-stabilized F glycoprotein trimer (DS-Cav1). In vivo studies performed on mouse model and nonhuman primate model, displayed that the designed system induced the production of neutralizing antibodies 10 times higher than single trimeric DS-Cav1. Added to this, the study provided that the developed system had optimum stability and immunogenicity, however further development is still needed [90].

Some efforts have been directed for the development of biomimetic nanovesicles for treating HIV. One study developed coated polymeric nanoparticles with CD + 4 T cell membrane. This aimed to target HIV as the T cell membrane encloses CCR5 and CXCR4 antigens that are critical for HIV targeting. Results confirmed that the developed T cell membrane-coated nanoparticles neutralized HIV by acting as a decoy for the virus and preventing it from attacking host cells [91].

Another study developed a T cell membrane-coated nanosystem to neutralize HIV-1 strains. Results demonstrated that the developed biomimetic system could neutralize a broad range of the HIV-1 viral strains, and be effective on all the 125 pseudotyped viruses assessed in the study. In addition, results exhibited that the nanosystem had the ability to selectively bind to HIV-infected cells. Finally, the nanosystem suppressed the viral replication via an autophagy-dependent mechanism and caused no cytotoxicity to neighbor cells [92].

In one important study, Bronshtein, et al., used cell-derived liposomes containing CCR5 surface ligands that target gp-120 expressing cells, these cells were used to mimic HIV-infected cells. Results were a proof of concept that the CCR5 tagged liposomes specifically targeted gp120- expressing cells. Loading the tagged liposomes with EDTA showed a significant cytotoxic effect on gp120-expressing cells relative to normal control cells. Hence, this study is real proof of principle that the biomimetic drug delivery approach has a significant ability to selectively target infected cells [93].

In the light of the most recent COVID-19 pandemic, the development of biomimetic nanosystems targeting SARS-CoV-2 has been extensively progressed. In one work, cellular nanosponges tagged with ACE2 originating from human macrophages or lung epithelial cells were assessed for their ability to neutralize SARS-CoV-2. The findings demonstrated the ability of the cellular nanosponges in neutralizing the virus and preventing it from infecting targeted cells [94]. Another study that confirmed the results of the previous one, used ACE-2 tagged nanovesicles derived from 293T cells that are genetically engineered, to neutralize SARS-CoV-2. This approach is specifically used to provide an abundant source of the derived vesicles to be used in future clinical applications [95].

One more advancement was made in treating and managing SARS-CoV-2 by developing hybrid nanovesicles. This was achieved by incorporating ACE-2 tagged nanovesicles with nanovesicles obtained from macrophage cells. The derived nanovesicles from macrophages have the ability to neutralize cytokine factors and suppress the inflammatory response. The findings of this study confirmed that the hybrid nanovesicles have the ability to neutralize the viral infection and suppress inflammatory response induced by the lung injury [96].

Intensive research held on developing virosomes for biopharmaceuticals delivery had led to the emerging of two FDA-approved intramuscular administered vaccine commercialized products. These liposome-based vaccines are named Inflexal Vs and Epaxal. Inflexel Vs is a vaccine against influenza virus, developed by incorporating inactive hemagglutinin of influenza virus strains A and B on the surface of virosomes. While Epaxal is a vaccine for hepatitis A virus, developed by conjugating inactive hepatitis A antigen on the virosomes surface. These vaccines are proved to be safe, highly effective, and robust stimulant for the immune system against the viruses [4,28].

20.4.5 Exosomes for delivery of antiviral agents

Exosomes are extracellular lipid bilayer nanovesicles produced from cells carrying cell-specific cargo such as genetic materials, lipids, and proteins. Exosomes are taken up by distant cells reprogramming them upon the specific cargo carried, and are abundantly found in body fluids such as saliva, blood, and urine [80]. Exosomes secreted from eukaryotic cells are known for normally carrying cargo between distant cells as an intercellular communicator, and play a major role during viral infections. For instance, during influenza virus infection, exosomes makeup is changed to express antiviral proteins aiding in inducing inflammatory response and inhibiting virus interaction with pulmonary cells [97,98]. However, exosomes might act in an undesired way by depending on the type of virus infecting the cell, for example, hepatitis virus-infected cells produce exosomes that provoke the infection rather than inhibit it.

Another favorable scenario is exosomes expressing interferon-induced (IFN) antiviral proteins interrupting viral replication in host cells and activating the defense of cells that could be potentially infected. A profound example is Type I IFN which, activates viral defense against Dengue and hepatitis B viruses [97].

Exosomes also have a unique capability to interact with recipient cells and have a high selective homing ability. Hence, exosomes have a great potential in antivirals delivery applications due to their innate response to viral infections and viruses, their innate function as bio-information transporters, and their promising properties that could easily be tailored to match the drug delivery application in the body.

It is of huge importance to assess the effectiveness of exosomes in vivo as the reported results are based mainly on in vitro studies. To address this issue, Bedford, et al., performed an in vivo study using a mouse model with influenza viral infection. The results highlighted that exosomes' protein composition was altered during the infection. In addition, intranasally inserting exosomes from infected mice to healthy mice, pulmonary inflammation was induced, and healthy cells were protected from infection as exosomes carrying antigens were attached to their surface. This study demonstrated the potential of using exosomes for treating influenza in humans as the results from mice studies were very promising [98].

One profound study used exosomes secreted from CD4 + T cells as nanodecoys to prevent HIV infection of host cells. The exosomes are surface-labeled with proteasomes similar to CD4 + T cells, hence, they can bind effectively to HIV-1 and impede the viral binding to healthy host cells. A further interesting finding was that exosomes released from the infected host cells had a role in preventing the viral invasion into other host cells [99].

20.5 Conclusion and future prospects

Viral diseases eradication is becoming more possible after the nanomedicine approach was incorporated into therapeutic strategies. Nanomedicine has provided various innovative solutions that improved treatment efficiency and success rate. Nanomedicine plays a major role in antiviral therapy as it enhances the bioavailability, allows specific targeting of the drugs, diminishes adverse side effects, decreases treatment costs, and improves antiviral delivery in general. These properties are being very useful for the success of viral diseases treatment as compared conventional treatment approaches, which are very expensive, require a high dose, and depend on the patients' adherence to the treatment regimen. In addition, nanomedicine enhances the patients' compliance as it shortens the treatment regimen, decreases intake frequency, and thus becoming a more cost-effective approach. Besides, nanomedicine is being employed in early viral diagnosis, prevention therapy, and personalized therapy [3,100]. Despite the various advantages the nanomedicine upholds in the delivery of antiviral agents, further critical aspects should be investigated before translating the nano-based systems into effective and safe formulations in markets [3].

The success of nanovesicles in the delivery of topical and systemic drugs provided a huge potential for their application in the delivery of antiviral agents. Nanovesicles uphold several advantages including being a noninvasive and painless delivery system. However, further research is still required for the development and commercialization of nanovesicles as delivery systems for antivirals [26]. One challenge is nanotoxicology, where toxicity and bio-elimination features of nanovesicles require an assessment to ensure the safety of the nanomaterials used in manufacturing [3,101].

Future prospects should include virologists in the development of nanovesicles for the delivery of antiviral agents. Their involvement is crucial for assessing the risk-benefit ratio of the developed formulations. Added to this, virologists would design novel antiviral agents to be incorporated into nanovesicles. Moreover, the selective targeting of nanovesicles should be furtherly investigated to deliver the desired drug into the specific infected tissues based on differential expression of molecules or specific functions held by the infected cells, hence, reducing off-site toxicity. One more issue that needs to be addressed by virologists is the passive targeting of antiviral agents loaded into nanocarriers, thus susceptibility of infected tissues to nanovesicles should be studied. In addition, nanoscientists along with virologists would maximize their efforts to develop more biocompatible, biodegradable, and nontoxic nanovesicle systems for delivery of antiviral agents. Finally, the development of nanomaterials that have intrinsic antiviral activity and as well act as the nano-vehicles will enhance viral infection treatments.

Future directions should also include a solid plan for cost-effective mass production of the developed nanoformulations following good manufacturing practices. As a final note, antiviral nanovesicles developed should be safe with high quality and therapeutic efficacy as well as being available to the developing countries. This could only be achieved by addressing the scientific and ethical challenges facing the antiviral nanomedicine research field.

References

[1] Ketzinel-Gilad M, Shaul Y, Galun E. RNA interference for antiviral therapy. J Gene Med 2006;8(8):933–50.
[2] Amidon GL, Lennernäs H, Shah VP, Crison JR. A theoretical basis for a biopharmaceutic drug classification: the correlation of in vitro drug product dissolution and in vivo bioavailability. Pharm Res 1995;12(3):413–20.
[3] Lembo D, Cavalli R. Nanoparticulate delivery systems for antiviral drugs. Antivir Chem Chemother 2010;21 (2):53–70.
[4] Grimaldi N, Andrade F, Segovia N, Ferrer-Tasies L, Sala S, Veciana J, et al. Lipid-based nanovesicles for nanomedicine. Chem Soc Rev 2016;45(23):6520–45.
[5] Arshady R, Kono K. Smart nanoparticles in nanomedicine: Kentus; 2006.

[6] Sosnik A, Amiji M. Nanotechnology solutions for infectious diseases in developing nations. Adv Drug Deliv Rev 2010;62:4–5.

[7] Mishra B, Patel BB, Tiwari S. Colloidal nanocarriers: a review on formulation technology, types and applications toward targeted drug delivery. Nanomed: Nanotechnol Biol Med 2010;6(1):9–24.

[8] Has C, Pan S. Vesicle formation mechanisms: an overview. J Liposome Res 2021;31(1):90–111.

[9] Piffoux M, Silva AK, Wilhelm C, Gazeau F, Tareste D. Modification of extracellular vesicles by fusion with liposomes for the design of personalized biogenic drug delivery systems. ACS Nano 2018;12(7):6830–42.

[10] Akbarzadeh A, Rezaei-Sadabady R, Davaran S, Joo SW, Zarghami N, Hanifehpour Y, et al. Liposome: classification, preparation, and applications. Nanoscale Res Lett 2013;8(1):1–9.

[11] Zhang H. Thin-film hydration followed by extrusion method for liposome preparation. Liposomes. Springer; 2017. p. 17–22.

[12] Wadhwa S, Garg V, Gulati M, Kapoor B, Singh SK, Mittal N. Nanovesicles for nanomedicine: theory and practices. Pharm Nanotechnol 2019;1–17.

[13] Lau KG, Hattori Y, Chopra S, O'Toole EA, Storey A, Nagai T, et al. Ultra-deformable liposomes containing bleomycin: in vitro stability and toxicity on human cutaneous keratinocyte cell lines. Int J Pharm 2005;300 (1–2):4–12.

[14] Song Y-K, Kim C-K. Topical delivery of low-molecular-weight heparin with surface-charged flexible liposomes. Biomaterials. 2006;27(2):271–80.

[15] Trotta M, Peira E, Debernardi F, Gallarate M. Elastic liposomes for skin delivery of dipotassium glycyrrhizinate. Int J Pharm 2002;241(2):319–27.

[16] Trotta M, Peira E, Carlotti ME, Gallarate M. Deformable liposomes for dermal administration of methotrexate. Int J Pharm 2004;270(1–2):119–25.

[17] New RRC. Liposomes: a practical approach. Oxford; New York: IRL Press; Oxford University Press; 1990.

[18] Agronskaia AV, Valentijn JA, van Driel LF, Schneijdenberg CT, Humbel BM, Henegouwen PMvB, et al. Integrated fluorescence and transmission electron microscopy. J Struct Biol 2008;164(2):183–9.

[19] Vernon-Parry K. Scanning electron microscopy: an introduction. III–Vs Rev 2000;13(4):40–4.

[20] Dragovic RA, Gardiner C, Brooks AS, Tannetta DS, Ferguson DJ, Hole P, et al. Sizing and phenotyping of cellular vesicles using nanoparticle tracking analysis. Nanomed Nanotechnol Biol Med 2011;7(6):780–8.

[21] Laouini A, Jaafar-Maalej C, Limayem-Blouza I, Sfar S, Charcosset C, Fessi H. Preparation, characterization and applications of liposomes: state of the art. J Colloid Sci Biotechnol 2012;1(2):147–68.

[22] Marsalek R. Particle size and zeta potential of ZnO. APCBEE Procedia 2014;9:13–17.

[23] Cevc G, Schätzlein A, Richardsen H. Ultradeformable lipid vesicles can penetrate the skin and other semipermeable barriers unfragmented. Evidence from double label CLSM experiments and direct size measurements. Biochim Biophys Acta (BBA)-Biomembranes 2002;1564(1):21–30.

[24] Gillet A, Lecomte F, Hubert P, Ducat E, Evrard B, Piel G. Skin penetration behaviour of liposomes as a function of their composition. Eur J Pharm Biopharm 2011;79(1):43–53.

[25] Cevc G, Blume G. Lipid vesicles penetrate into intact skin owing to the transdermal osmotic gradients and hydration force. Biochim Biophys Acta (BBA)—Biomembranes 1992;1104(1):226–32.

[26] Pharmaceutical Nanotechnology: Basic Protocols. Pharmaceutical Nanotechnology: Basic Protocols. 2019;2000:1–398.

[27] Wicki A, Witzigmann D, Balasubramanian V, Huwyler J. Nanomedicine in cancer therapy: challenges, opportunities, and clinical applications. J Control Release 2015;200:138–57.

[28] Chang HI, Yeh MK. Clinical development of liposome-based drugs: formulation, characterization, and therapeutic efficacy. Int J Nanomed 2012;7:49–60.

[29] Haran G, Cohen R, Bar LK, Barenholz Y. Transmembrane ammonium sulfate gradients in liposomes produce efficient and stable entrapment of amphipathic weak bases. Biochim Biophys Acta 1993;1151(2):201–15.

[30] Jain S, Patil SR, Swarnakar NK, Agrawal AK. Oral delivery of doxorubicin using novel polyelectrolyte-stabilized liposomes (layersomes). Mol Pharm 2012;9(9):2626–35.

[31] Zahmatkeshan M, Gheybi F, Rezayat SM, Jaafari MR. Improved drug delivery and therapeutic efficacy of PEgylated liposomal doxorubicin by targeting anti-HER2 peptide in murine breast tumor model. Eur J Pharm Sci 2016;86:125–35.

[32] Wong JP, Yang H, Blasetti KL, Schnell G, Conley J, Schofield LN. Liposome delivery of ciprofloxacin against intracellular *Francisella tularensis* infection. J Control Release 2003;92(3):265–73.

[33] Marier JF, Brazier JL, Lavigne J, Ducharme MP. Liposomal tobramycin against pulmonary infections of Pseudomonas aeruginosa: a pharmacokinetic and efficacy study following single and multiple intratracheal administrations in rats. J Antimicrob Chemother 2003;52(2):247—52.

[34] Pinto-Alphandary H, Andremont A, Couvreur P. Targeted delivery of antibiotics using liposomes and nanoparticles: research and applications. Int J Antimicrob Agents 2000;13(3):155—68.

[35] Khorsandi K, Hosseinzadeh R, Sadat Esfahani H, Keyvani-Ghamsari S, Ur Rahman S. Nanomaterials as drug delivery systems with antibacterial properties: current trends and future priorities. Expert Rev Anti Infect Ther 2021;19(10):1299—323.

[36] Vyas SP, Kannan ME, Jain S, Mishra V, Singh P. Design of liposomal aerosols for improved delivery of rifampicin to alveolar macrophages. Int J Pharm 2004;269(1):37—49.

[37] Yeo PL, Lim CL, Chye SM, Ling APK, Koh RY. Niosomes: a review of their structure, properties, methods of preparation, and medical applications. Asian Biomed 2017;11(4):301—13.

[38] Sezgin-Bayindir Z, Onay-Besikci A, Vural N, Yuksel N. Niosomes encapsulating paclitaxel for oral bioavailability enhancement: preparation, characterization, pharmacokinetics and biodistribution. J Microencapsul 2013;30(8):796—804.

[39] Jain S, Kumar D, Swarnakar NK, Thanki K. Polyelectrolyte stabilized multilayered liposomes for oral delivery of paclitaxel. Biomaterials. 2012;33(28):6758—68.

[40] Li Q, Li Z, Zeng W, Ge S, Lu H, Wu C, et al. Proniosome-derived niosomes for tacrolimus topical ocular delivery: in vitro cornea permeation, ocular irritation, and in vivo anti-allograft rejection. Eur J Pharm Sci 2014;62:115—23.

[41] Zubairu Y, Negi LM, Iqbal Z, Talegaonkar S. Design and development of novel bioadhesive niosomal formulation for the transcorneal delivery of anti-infective agent: In-vitro and ex-vivo investigations. Asian J Pharm Sci 2015;10(4):322—30.

[42] Shinde UA, Kanojiya SS. Serratiopeptidase niosomal gel with potential in topical delivery. J Pharm (Cairo) 2014;2014:382959.

[43] Alvi IA, Madan J, Kaushik D, Sardana S, Pandey RS, Ali A. Comparative study of transfersomes, liposomes, and niosomes for topical delivery of 5-fluorouracil to skin cancer cells: preparation, characterization, in-vitro release, and cytotoxicity analysis. Anticancer Drugs 2011;22(8):774—82.

[44] Duangjit S, Opanasopit P, Rojanarata T, Ngawhirunpat T. Evaluation of meloxicam-loaded cationic transfersomes as transdermal drug delivery carriers. AAPS PharmSciTech 2013;14(1):133—40.

[45] Gupta A, Aggarwal G, Singla S, Arora R. Transfersomes: a novel vesicular carrier for enhanced transdermal delivery of sertraline: development, characterization, and performance evaluation. Sci Pharm 2012;80(4):1061—80.

[46] Yusuf M, Sharma V, Pathak K. Nanovesicles for transdermal delivery of felodipine: Development, characterization, and pharmacokinetics. Int J Pharm Investig 2014;4(3):119—30.

[47] Singodia D, Gupta GK, Verma A, Singh V, Shukla P, Misra P, et al. Development and performance evaluation of amphotericin B transfersomes against resistant and sensitive clinical isolates of visceral leishmaniasis. J Biomed Nanotechnol 2010;6(3):293—302.

[48] Pilch E, Musiał W. Liposomes with an ethanol fraction as an application for drug delivery. Int J Mol Sci 2018;19(12).

[49] Yu X, Du L, Li Y, Fu G, Jin Y. Improved anti-melanoma effect of a transdermal mitoxantrone ethosome gel. Biomed Pharmacother 2015;73:6—11.

[50] Ahmed TA, El-Say KM, Aljaeid BM, Fahmy UA, Abd-Allah FI. Transdermal glimepiride delivery system based on optimized ethosomal nano-vesicles: preparation, characterization, in vitro, ex vivo and clinical evaluation. Int J Pharm 2016;500(1—2):245—54.

[51] Shelke S, Shahi S, Jalalpure S, Dhamecha D. Poloxamer 407-based intranasal thermoreversible gel of zolmitriptan-loaded nanoethosomes: formulation, optimization, evaluation and permeation studies. J Liposome Res 2016;26(4):313—23.

[52] Garg V, Singh H, Bhatia A, Raza K, Singh SK, Singh B, et al. Systematic development of transethosomal gel system of piroxicam: formulation optimization, in vitro evaluation, and ex vivo assessment. AAPS PharmSciTech 2017;18(1):58—71.

[53] Ascenso A, Raposo S, Batista C, Cardoso P, Mendes T, Praça FG, et al. Development, characterization, and skin delivery studies of related ultradeformable vesicles: transfersomes, ethosomes, and transethosomes. Int J Nanomed 2015;10:5837—51.

[54] Garg V, Singh H, Bimbrawh S, Singh SK, Gulati M, Vaidya Y, et al. Ethosomes and transfersomes: principles, perspectives and practices. Curr Drug Deliv 2017;14(5):613–33.

[55] Saroja C, Lakshmi P, Bhaskaran S. Recent trends in vaccine delivery systems: a review. Int J Pharm Investig 2011;1(2):64–74.

[56] Li SD, Huang L. Pharmacokinetics and biodistribution of nanoparticles. Mol Pharm 2008;5(4):496–504.

[57] LaVan DA, Lynn DM, Langer R. Moving smaller in drug discovery and delivery. Nat Rev Drug Discov 2002;1(1):77–84.

[58] Amiji MM, Vyas TK, Shah LK. Role of nanotechnology in HIV/AIDS treatment: potential to overcome the viral reservoir challenge. Discov Med 2006;6(34):157–62.

[59] Kumar L, Verma S, Prasad DN, Bhardwaj A, Vaidya B, Jain AK. Nanotechnology: a magic bullet for HIV AIDS treatment. Artif Cell Nanomed Biotechnol 2015;43(2):71–86.

[60] Sharma P, Garg S. Pure drug and polymer based nanotechnologies for the improved solubility, stability, bio-availability and targeting of anti-HIV drugs. Adv Drug Deliv Rev 2010;62(4–5):491–502.

[61] Takeuchi H, Yamamoto H, Kawashima Y. Mucoadhesive nanoparticulate systems for peptide drug delivery. Adv Drug Deliv Rev 2001;47(1):39–54.

[62] Jin SX, Bi DZ, Wang J, Wang YZ, Hu HG, Deng YH. Pharmacokinetics and tissue distribution of zidovudine in rats following intravenous administration of zidovudine myristate loaded liposomes. Pharmazie. 2005;60(11):840–3.

[63] Jain S, Tiwary AK, Jain NK. Sustained and targeted delivery of an anti-HIV agent using elastic liposomal formulation: mechanism of action. Curr Drug Deliv 2006;3(2):157–66.

[64] Kaur CD, Nahar M, Jain NK. Lymphatic targeting of zidovudine using surface-engineered liposomes. J Drug Target 2008;16(10):798–805.

[65] Makabi-Panzu B, Gourde P, Désormeaux A, Bergeron MG. Intracellular and serum stability of liposomal 2′,3′-dideoxycytidine. Effect of lipid composition. Cell Mol Biol (Noisy-le-grand 1998;44(2):277–84.

[66] Oussoren C, Magnani M, Fraternale A, Casabianca A, Chiarantini L, Ingebrigsten R, et al. Liposomes as carriers of the antiretroviral agent dideoxycytidine-5′-triphosphate. Int J Pharm 1999;180(2):261–70.

[67] Asmal M, Whitney JB, Luedemann C, Carville A, Steen R, Letvin NL, et al. In vivo anti-HIV activity of the heparin-activated serine protease inhibitor antithrombin III encapsulated in lymph-targeting immunoliposomes. PLoS One 2012;7(11):e48234.

[68] Clayton R, Ohagen A, Nicol F, Del Vecchio AM, Jonckers TH, Goethals O, et al. Sustained and specific in vitro inhibition of HIV-1 replication by a protease inhibitor encapsulated in gp120-targeted liposomes. Antivir Res 2009;84(2):142–9.

[69] Ramana LN, Sharma S, Sethuraman S, Ranga U, Krishnan UM. Investigation on the stability of saquinavir loaded liposomes: implication on stealth, release characteristics and cytotoxicity. Int J Pharm 2012;431 (1–2):120–9.

[70] Mukherjee B, Patra B, Layek B, Mukherjee A. Sustained release of acyclovir from nano-liposomes and nano-niosomes: an in vitro study. Int J Nanomed 2007;2(2):213–25.

[71] Fresta M, Panico AM, Bucolo C, Giannavola C, Puglisi G. Characterization and in-vivo ocular absorption of liposome-encapsulated acyclovir. J Pharm Pharmacol 1999;51(5):565–76.

[72] Law SL, Huang KJ, Chiang CH. Acyclovir-containing liposomes for potential ocular delivery. Corneal penetration and absorption. J Control Release 2000;63(1–2):135–40.

[73] Chetoni P, Rossi S, Burgalassi S, Monti D, Mariotti S, Saettone MF. Comparison of liposome-encapsulated acyclovir with acyclovir ointment: ocular pharmacokinetics in rabbits. J Ocul Pharmacol. Therap. 2004;20 (2):169–77.

[74] Hassan H, Adam SK, Othman F, Shamsuddin AF, Basir R. Antiviral nanodelivery systems: current trends in acyclovir administration. J Nanomaterials 2016;2016.

[75] Devaraj GN, Parakh SR, Devraj R, Apte SS, Rao BR, Rambhau D. Release studies on niosomes containing fatty alcohols as bilayer stabilizers instead of cholesterol. J Colloid Interface Sci 2002;251(2):360–5.

[76] Jain SK, Umamaheshwari RB, Bhadra D, Jain N. Ethosomes: a novel vesicular carrier for enhanced transdermal delivery of an AntiHIV agent. Indian J Pharm Sci 2004;66:72–81.

[77] Zhou Y, Wei YH, Zhang GQ, Wu XA. Synergistic penetration of ethosomes and lipophilic prodrug on the transdermal delivery of acyclovir. Arch Pharm Res 2010;33(4):567–74.

[78] Jain S, Tiwary AK, Sapra B, Jain NK. Formulation and evaluation of ethosomes for transdermal delivery of lamivudine. AAPS PharmSciTech 2007;8(4):E111.

[79] Chakravarty M, Vora A. Nanotechnology-based antiviral therapeutics. Drug Deliv Transl Res 2021;11 (3):748–87.

[80] Zhang Y, Liu Y, Liu H, Tang WH. Exosomes: biogenesis, biologic function and clinical potential. Cell Biosci 2019;9:19.

[81] Yang KC, Lin JC, Tsai HH, Hsu CY, Shih V, Hu CJ. Nanotechnology advances in pathogen- and host-targeted antiviral delivery: multipronged therapeutic intervention for pandemic control. Drug Deliv Transl Res 2021;11(4):1420–37.

[82] Li Q, Nie J, Wu J, Zhang L, Ding R, Wang H, et al. SARS-CoV-2 501Y. V2 variants lack higher infectivity but do have immune escape. Cell. 2021;184(9):2362–71 e9.

[83] Jin K, Luo Z, Zhang B, Pang Z. Biomimetic nanoparticles for inflammation targeting. Acta Pharm Sin B 2018;8(1):23–33.

[84] Abdoli A, Soleimanjahi H, Kheiri MT, Jamali A, Sohani H, Abdoli M, et al. Reconstruction of H3N2 influenza virus based virosome in-vitro. Iran J Microbiol 2013;5(2):166–71.

[85] Parodi A, Molinaro R, Sushnitha M, Evangelopoulos M, Martinez JO, Arrighetti N, et al. Bio-inspired engineering of cell- and virus-like nanoparticles for drug delivery. Biomaterials. 2017;147:155–68.

[86] Yang G, Chen S, Zhang J. Bioinspired and biomimetic nanotherapies for the treatment of infectious diseases. Front Pharmacol 2019;10:751.

[87] Kanekiyo M, Wei CJ, Yassine HM, McTamney PM, Boyington JC, Whittle JR, et al. Self-assembling influenza nanoparticle vaccines elicit broadly neutralizing H1N1 antibodies. Nature. 2013;499(7456):102–6.

[88] Chen HW, Fang ZS, Chen YT, Chen YI, Yao BY, Cheng JY, et al. Targeting and enrichment of viral pathogen by cell membrane cloaked magnetic nanoparticles for enhanced detection. ACS Appl Mater Interfaces 2017;9(46):39953–61.

[89] Nie C, Stadtmüller M, Yang H, Xia Y, Wolff T, Cheng C, et al. Spiky nanostructures with geometry-matching topography for virus inhibition. Nano Lett 2020;20(7):5367–75.

[90] Marcandalli J, Fiala B, Ols S, Perotti M, de van der Schueren W, Snijder J, et al. Induction of potent neutralizing antibody responses by a designed protein nanoparticle vaccine for respiratory syncytial virus. Cell. 2019;176(6):1420–31 e17.

[91] Wei X, Zhang G, Ran D, Krishnan N, Fang RH, Gao W, et al. T-Cell-Mimicking Nanoparticles can neutralize HIV infectivity. Adv Mater 2018;30(45):e1802233.

[92] Zhang G, Campbell GR, Zhang Q, Maule E, Hanna J, Gao W, et al. CD4. mBio 2020;11(5).

[93] Bronshtein T, Toledano N, Danino D, Pollack S, Machluf M. Cell derived liposomes expressing CCR5 as a new targeted drug-delivery system for HIV infected cells. J Control Release 2011;151(2):139–48.

[94] Zhang Q, Honko A, Zhou J, Gong H, Downs SN, Vasquez JH, et al. Cellular nanosponges inhibit SARS-CoV-2 infectivity. Nano Lett 2020;20(7):5570–4.

[95] Rao L, Xia S, Xu W, Tian R, Yu G, Gu C, et al. Decoy nanoparticles protect against COVID-19 by concurrently adsorbing viruses and inflammatory cytokines. Proc Natl Acad Sci U S A 2020;117(44):27141–7.

[96] Thamphiwatana S, Angsantikul P, Escajadillo T, Zhang Q, Olson J, Luk BT, et al. Macrophage-like nanoparticles concurrently absorbing endotoxins and proinflammatory cytokines for sepsis management. Proc Natl Acad Sci U S A 2017;114(43):11488–93.

[97] Maus A, Strait L, Zhu D. Nanoparticles as delivery vehicles for antiviral therapeutic drugs. Eng Regen 2021;2:31–46.

[98] Bedford JG, Infusini G, Dagley LF, Villalon-Letelier F, Zheng MZM, Bennett-Wood V, et al. Airway exosomes released during influenza virus infection serve as a key component of the antiviral innate immune response. Front Immunol 2020;11.

[99] de Carvalho JV, de Castro RO, da Silva EZ, Silveira PP, da Silva-Januário ME, Arruda E, et al. Nef neutralizes the ability of exosomes from CD4 + T cells to act as decoys during HIV-1 infection. PLoS One 2014;9 (11):e113691.

[100] De Jong WH, Borm PJ. Drug delivery and nanoparticles:applications and hazards. Int J Nanomed 2008;3 (2):133–49.

[101] Aillon KL, Xie Y, El-Gendy N, Berkland CJ, Forrest ML. Effects of nanomaterial physicochemical properties on in vivo toxicity Adv Drug Deliv Rev 2009;61(6):457–66Created with. Available from: BioRender.com.

Antiviral biomaterials

Sandhya Khunger

Department of Microbiology, Faculty of Allied Health Sciences, Shree Guru Gobind Singh
Tricentenary University, Gurugram, Haryana, India

21.1 Introduction to antiviral biomaterials

Viral infections pose a major and formidable challenge to public health worldwide. Infections including influenza, pneumonia, and other respiratory tract infections are contagious [1]. Therefore, such infections have a catastrophic impact causing millions of deaths worldwide and levying an enormous burden on the global economy [1,2]. In recent years, new infectious pathogens such as Zika virus, Ebola virus, Nipah virus, and severe acute respiratory syndrome (SARS) have emerged that constituted a severe health crisis due to their epidemic or pandemic status [3]. Recently, the outbreak of coronavirus disease 2019 (COVID-19) caused by SARS-coronavirus-2 (SARS-CoV-2) is classified as a public health emergency of international concern, which resulted in 23, 04,18,451 confirmed cases and 47,24,876 deaths worldwide [4]. The world has witnessed the most dramatic and deadly changes due to the lack of preventive measures for these epidemics and pandemics. Thus, the current challenge is to innovate and discover a therapeutic concept to prevent and control such viral infections. Vaccination remains an effective method to protect public health against this uncontrolled infection [5]. Conventional vaccines are derived from inactivated viruses, live attenuated pathogens, and recombinant proteins that elicit protective immune responses [5]. Data in the literature suggests that the mortality due to common disease including diptheria, polio, tetanus, and smallpox has significantly reduced by 97%−99% with successful vaccination [6]. However, several drawbacks such as low stability and poor effectiveness against infection, and adverse side effects are associated with the conventional vaccines [5]. Despite several vaccination programs, researchers are lagging behind and still working on therapeutic approaches to mitigate the severity of the infections. In this scenario, the concept of biomaterials needs particular attention for the potential management of emerging threats of viral diseases.

Biomaterials are the engineered or programmed components derived naturally or synthetically that interact with the biological systems for diagnostic and therapeutic purposes [7].

These are the effective platforms for drug screening, drug delivery, fostering antiviral strategies, and reinforcing vaccine efficacy [8]. Plenty of biomaterials have been designed and developed to combat viral and bacterial infections and have achieved promising results. For instance, VivaGel (SPL7013 Gel), a product of Starpharma is available in the market for the treatment of human immunodeficiency disease (HIV) and herpes simplex virus (HSV) [9]. Historically, Egyptians used animal sinew as a biomaterial for suturing purposes [10]. The biomaterial field has diversified and combined with other fields including medicine, nanotechnology, tissue engineering, and material science over the years [7]. With the advent of viral infections at the global level, the biomaterial communities have participated and engaged themselves in antiviral discipline.

21.1.1 Types of biomaterials

Ample of biomaterials have been obtained which are of particular relevance. Such biomaterials include:

21.1.1.1 Hydrogels

Hydrogels are three-dimensional (3D) cross-linked hydrophilic polymers that can be prepared using natural or synthetic polymers [11]. Natural polymers include chitosan, hyaluronic acid, heparin, alginate, and fibrin whereas synthetic polymers include polyvinyl alcohol, sodium polyacrylate, polyethylene glycol, and acrylate polymers [11,12]. Based upon a structure, hydrogels exist in several forms: amorphous, crystalline, semicrystalline, and hydrocolloid aggregates. The cross-links in hydrogel can be generated either by physical or chemical interaction. Physical cross-links in the hydrogel can be formed by hydrogen bonds, hydrophobic interactions, and chain entanglements. Such hydrogels are sometimes known as "reversible" hydrogels. On the other hand, chemical cross-links entail covalent bonds between polymer strands, and these are called "permanent" hydrogels [11–13]. Hydrogels do not dissolve in water and they can mimic the 3D microenvironment of cells. Biocompatibility and well-defined structures are the characteristic features of hydrogels due to their high water content. Moreover, they are highly absorbent and flexible like natural tissues [12,14]. Hydrogels are considerably used in biomedical applications including tissue engineering and drug delivery [15]. Other characteristic features of hydrogels for biomedical applications include biodegradability, biocompatibility, and antitoxicity. Polymers such as collagen, chitosan, cellulose, and poly (lactic-co-glycolic acid) have been used extensively for drug delivery. The most interesting feature of hydrogels is they undergo a volume phase transition or gel-sol phase transition for physical and chemical stimuli for example, temperature, solvent composition, pressure, and pH ions [15].

21.1.1.2 Cryogels

Cryogels are considered a subclass of hydrogels with exceptional features such as macroporous network, shape memory, and syringe injectability [16]. These are supermacroporous gel networking fabricated at subzero temperatures and developed by cryogelation of apposite monomers or polymeric precursors. Cryogelation process occurs in following steps:

- Phase separation with the ice crystal formations

- Cross-linking
- Polymerization
- Thawing of the ice crystals for the formation of interconnected porous cryogel network [16,17]

Cryogels are subjected to different modifications by coupling differential ligands or grafting polymeric chains to their surfaces. For instance, the pore size of thermoreversible PVA cryogels without a cross-linker varies between 100 nm and 1 μm, whereas it varies from 1 μm up to 150 μm in chemically cross-linked PVA cryogels [18]. The properties of cryogels depend on the pore structure including size, distribution, thickness, density, and interconnectivity. Several bio-related applications including bioseparation, tissue engineering, and immunotherapy are associated with the usage of cryogels due to their highly porous features with sufficient osmotic stability and mechanical strength [19].

21.1.1.3 *Nanoparticles*

Nanoparticles (NPs) are small sized particles ranging between 1–100 nm in size. They exhibit differential physical and chemical properties such as a larger surface area to volume ratio, electron confinement, and quantum effects. NPs may occur naturally or can be generated through engineering or as byproducts of combustion reactions. NPs such as liposomes, are nanoscale biomaterials that have been used as nanocarriers for drug and gene delivery [20].

21.2 Mechanism of action

Since the virus is an obligate intracellular parasite that uses the biosynthetic machinery of the host cell for its replication, the antiviral biomaterials have been engineered in such a way that attacks the target without causing any effect on the host cell [21].

21.2.1 Structure of virus

A virion is a fully assembled virus particle [22]. The basic structure of virion is composed of two major components: nucleic acid [single or double-stranded deoxyribonucleic acid (DNA) or ribonucleic acid (RNA)] and a capsid. The capsid is a protein layer that protects the viral genome and binds to the specific receptors in the host cell. Viruses are inert or "dead" outside the cell and depend completely on the host's machinery [23]. Enveloped viruses derive an additional coverage from the portions of the modified host cell membrane. The derived envelopes consist of lipid bilayers decorated with glycoproteins [24]. Glycoproteins identify and bind to the host receptor sites for the activation of the viral infectious cycle. The major steps of the viral infection cycle are:

1. Attachment of virion onto the host cell membrane,
2. Penetration of virion particles into the cell,
3. Uncoating of the capsid that releases nucleic acid,
4. Biosynthesis of viral nucleic acid and proteins for the formation of virus assembly, and
5. Virion release from the infected cells [25].

21.2.2 Action mechanism of biomaterials

The viral life cycle represents several potential targets for antiviral intervention. The canonical targets for therapeutics include viral enzymes such as polymerases and proteases. The antiviral action mechanism of biomaterials is based on the interference in the viral infection cycle (Fig. 21.1), which is detailed in the following sections.

21.2.2.1 Physical adsorption of viruses

The first step of the viral infection cycle is an attachment of the virus onto the host cell membrane. In this scenario, one such property of antiviral biomaterial is associated with the absorbance of a virus. For instance, absorbent materials such as cotton or cardboard have liquid absorption properties which offer better protection than fluid-repelling surfaces [26]. In a study, Lai et al. compared the survival of SARS-CoV on a fluid-repelling disposable gown and a cotton gown in the hospital setting. The authors found that cotton material efficiently absorbs concentrated virus droplets with no viable viruses after 1 hour [27]. Further, existing antiviral therapies become nonefficient due to frequent viral mutations. Therefore, designing and development of high-affinity biomaterials is needed as an alternative treatment approach for capturing viruses. As the virus is composed of capsid with abundant surface proteins, for example, hemagglutinin proteins, high-affinity biomaterials can adhere to the surface proteins and capture the virus by rendering its ability to infect cells [28]. Thus, hemagglutinin can be considered an attractive target for designing antiviral biomaterial. For instance, Li et al. synthesized sialyl lactose-incorporated chitosan-based biomaterials by grafting lactoside onto chitosan followed by enzymatic hydrolysis with sialyltransferase which effectively captured the influenza virus due to its high affinity to hemagglutinin [29]. Overall, the virus capture strategy using functional biomaterials imposes an efficient method for virus clearance.

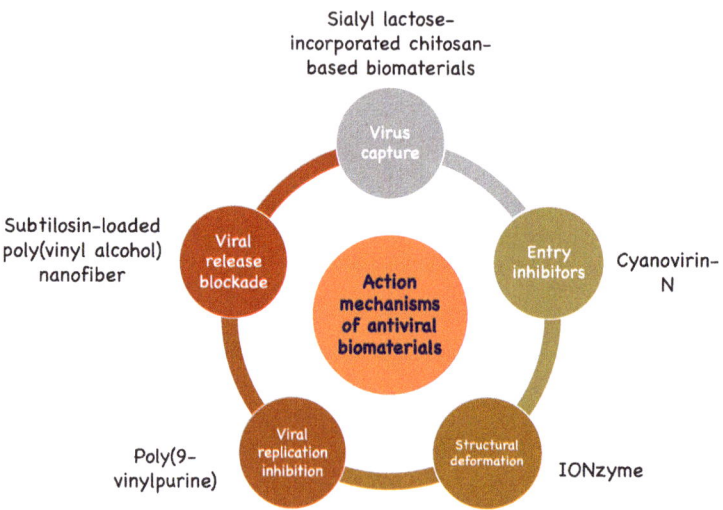

FIGURE 21.1 Schematic illustration of action mechanisms of antiviral biomaterials with examples.

21.2.2.2 Entry inhibitors

The second step of viral infection cycle is the penetration or entry of viruses into the cell. Entry of a viral particle is a multistep process comprised of viral-host interactions. Therefore, inhibition of virus entry pathways represents a promising target for antiviral biomaterials by targeting attachment onto the membrane, coreceptor binding, and virus—cell fusion processes. Antivirals as entry inhibitors provide a rational basis and play a significant role in preventing viral transmission. In most of the cases, carbohydrate- (lectins) or charge-based (glycosaminoglycans) interactions persist for viral attachment and soluble polysaccharides (heparin or microbicide carrageenan) can potentially inhibit these interactions [30]. For instance, cyanovirin-N (CV-N), a naturally-occurring lectin has the potential to inhibit attachment of HIV onto the membrane by binding to N-linked high-mannose oligosaccharides of the viral glycoprotein. Moreover, Barrientos et al. [31] reported CV-N as an effective entry inhibitor for Ebola virus using in vitro assays and in vivo mouse models [31]. Another promising targets are the receptors, for example, angiotensin-converting enzyme 2 and ephrinB2 are the functional or attachment receptors for SARS-CoV and Nipah virus, respectively [32].

21.2.2.3 Induction of irreversible viral deformation

Viruses maintain their integrity to infect the host cells. A virus can be inactivated by the destruction of viral structures or by causing irreversible viral damage. The inactivated virus permanently loses its infection ability. Several methods including temperature, pH, and ultraviolet radiation can destroy the virus structure quickly however, these methods are cytotoxic for host cells [33]. Antiviral biomaterials assume importance by exhibiting viricidal abilities and destroying the viral structure. For example, silver and gold NPs display viricidal properties by destroying viral capsids and neutralizing infectivity [34]. Qin and colleagues developed iron oxide nanozyme (IONzyme) that causes lipid peroxidation of the envelope of influenza A virus and thereby, protein disintegration. The authors examined the effect of IONzyme-pretreated virus on the lipid envelopes in H5N1-infected mice and found weight reduction and mortalities in infected mice. IONzyme-pretreated virus treatment caused no weight reduction, negligible pathological changes, and no mortality [35]. Recent reports stated that iron oxide NPs could inactivate the SARS-CoV-2 virus by causing conformational changes in the spike protein receptor-binding domain of the virus [36]. Apart from the viricidal efficacy of biomaterials, the decorated ligands on their surface also destroy viral structures. Cagno et al. developed antiviral AuNPs using long and flexible linkers that mimicked heparan sulfate proteoglycans. These NPs with strong multivalent binding forces led to irreversible deformation of viruses such as HSV, lentivirus, HPV, and dengue virus [37]. The ligand repeating units in the linkers for viral attachment can be considered promising strategies for developing future antiviral biomaterials.

21.2.2.4 Interference in nucleic acid replication

Nucleic acid replication is the core step in the viral infection cycle thus, researchers focus on the development of the antiviral agents that target this step. Biomaterials such as Poly(9-vinylpurine), PEG-coated zinc oxide NPs, and humic acid inhibit viral replication of influenza virus (HSV and H1N1) by interfering with polymerase activity [38]. In a clinical setting, reverse transcriptase (RT) inhibitors such as zidovudine and lamivudine have been used for the inhibition of nucleic acid replication in the HIV virus [39]. However, these agents were found to cause hepato-toxicity and neuro-toxicity. Therefore, another strategy of drug-loaded biomaterial has been developed that encapsulates RT inhibitors. Xu et al. developed a new class of hydrogels that sustained the release of RT inhibitors in combination with nonsteroidal antiinflammatory drugs [40]. These hydrogelator serves as matrices of hydrogels and act as potential anti-HIV agent. Several delivery systems such as alginate NPs, gelatin-modified NPs, and polyvinylpyrrolidone/stearic acid—PEG NPs have been developed for zidovudine carriers [41] that act as promising drug delivery vectors in the treatment of viral infections like AIDS.

21.2.2.5 Blockage of virion release from infected cells

The last step of the viral infection cycle is the release of virion particles from the infected cells. For instance, oseltamivir-based biomaterials, selenium NP surface modified with oseltamivir, subtilosin, and subtilosin-loaded poly(vinyl alcohol) nanofiber prevent the virion release by blocking chromatin condensation [42]. Further, nucleic acid-based polymer such as REP 2055 has the potential to block virus surface antigen and prevent the release of viral particles from the infected cells, thereby resulting in control of viral infection [43]. Overall, targeting the last step can effectively alleviate long-term viral infections.

Besides this, the combination of biomaterials and drugs shows synergistic interactions and provides better therapeutic or antiviral effects by inhibiting the virus with different levels of simultaneity. For example, dendritic sialopolymersome encapsulated with zanamivir was developed by Nazemi et al. against the influenza virus. This antiviral biomaterial interferes with the infectious cycle at two levels, that is, blocking the viral entry by inhibiting the binding of hemagglutinin to sialic acid on the host cells and preventing the release of progeny virus from host cells [44]. The combinatory antiviral drugs with excellent potential used for the treatment of viral infections are listed in Table 21.1.

21.3 Applications of antiviral biomaterials

Multidisciplinary approaches are involved with the usage of biomaterials for the protection and management of viral diseases. Antiviral biomaterials have both diagnostic and therapeutic applications, which are:

21.3.1 Diagnostics

21.3.1.1 Nucleic acid testing

The diagnostic method of nucleic acid testing has been known for decades however, it was limited to researchers and medical workers. With the outbreak of SARS-CoV-2 infection as a pandemic, RT-polymerase chain reaction (PCR) has emerged as the most common means for the screening of infected patient samples for the viral infection and is now known to all [60]. The PCR-based assay is a useful clinical tool for the detection of viral

TABLE 21.1 Different types of antiviral biomaterials.

Material	Mechanism of action	Target virus	References
Copper	Copper-linked (cross-linking) RNA damage Nucleic acid damage by cross-linking Damage to the viral capsid, RNA degradation	Infuenza A Bacteriophage Murine norovirus MNV-1	[45–47]
Polyethylenimines (PEI)	Charged "tentacle" fragments of polyanionic chains Destruction of viral membrane	Human and avian infuenza A Herpes simplex virus	[48]
Polyacrylic hydrogel containing monocarpin	Interference in viral replication	Herpes simplex virus	[49]
SL-chitosan	Blocking of host-virus interaction by binding to capsid protein (hemmaglutin)	Influenza	[50]
Fucoidan	Interference in viral replication	HIV	[51]
Glycopolymers with sialic acid on lactosamine repeats Mucin	Blocking of host-virus interaction by binding to hemmaglutin Entrapment of viral particles in mucin polymeric matrix	Human influenza viruses Swine influenza virus HPV-16 Merkel cell polyomavirus (MCV) influenza A	[52,53]
Cathelicidin	Prevention of endosomal cell-entry by impairing cathepsin B-mediated processing of viral glycoproteinDisruption of viral membrane integrity	West Africa Ebola virus (EBOV) Influenza A Respiratory syncytial virus ZIKV	[53–55]
Defensin	Interacting with viral envelope	Influenza A Herpes simplex virus	[56]
Hepcidin	Sequestering iron	Hepatitis B & C HIV Infectious pancreatic necrosis virus (IPNV)	[49,57,58]
Silica/30 nm silver particles	Interaction with viral membrane components	Influenza A	[59]

nucleic acids with high specificity and sensitivity. The detection method encompasses three steps: extraction and purification of RNA by lysing viral membranes, transcription of RNA to complementary DNA (cDNA) using RT enzymes, and followed by amplification of the target sequence of cDNA using thermocycling protocol [61]. Thermocycling is comprised of three phases: denaturation, annealing, and primer extension. In this way, amplification of target cDNA can detect viruses even with low detection limits. Further, Wei et al. developed a magnetic NP-based technique and a magnetic NP-based chemiluminescent reporter system for rapid isolation of RNA and DNA from the biological samples and detection of target sequences in the sample respectively [62].

21.3.1.2 Point-of-care tests

Point-of-care (POC) methods are the alternative strategies used in clinical settings. POC tests are used for rapid diagnoses in real-time without processing samples in the laboratory. It also eliminates the need for specialized instruments and equipment. Such tests are portable and can be self-conducted at home [63]. Biomaterial laboratories help in manufacturing and developing such tests. For instance, 3D printed nasal swabs are developed by Formlabs for the collection of respiratory samples [64]. Moreover, biosensor technology is another promising strategy for the detection of viral particles. For example, Quidel developed a fluorescent immunoassay-based antigen test for the detection of viral N proteins [65].

21.3.2 Antiviral therapies

The traditional antiviral therapies such as remdesivir, and chloroquine, are highly nonspecific targeted inhibitors, for example, HIV protease inhibitors, and RNAi that act on specific viral elements [66]. These therapies have tremendous shortcomings due to noneffectiveness against all viruses, limited specificity, and enormous side effects. Designing and developing new therapies require a considerable amount of time therefore, repurposing the existing therapies is preferred for better results. For example, hydroxychloroquine is an antimalarial and anti-inflammatory drug, used for the treatment of several rheumatological diseases [67]. Compelling evidence in the literature suggests that hydroxychloroquine acts well in the treatment of COVID-19 by blocking the entry of virion particles into the host cells and inhibiting replication [68]. In vitro studies in the literature revealed that Remdesivir, another antiviral prodrug is effective against Ebola, SARS-CoV, and MERS [69]. Despite this, a little progress has been made and hence, biomaterials offer several opportunities in antiviral therapeutics:

21.3.2.1 Drug delivery

An ample evidences in the literature reported that several drugs are available that have shown antiviral efficacy in vitro but did not acquiesce to clinical significance. The reason may include short half-lives, hydrophobicity, low accumulation in tissues, and may not be suitable for oral, intravenous, or systemic administration [70]. In this scenario, drug delivery systems can be usefully designed to directly deliver drugs to the affected tissue by reducing systemic exposure. This may resolve drug-related issues such as poor stability, low bioavailability, or severe side effects of high doses. For instance, self-assembling tyrosine-derived polymeric nanospheres (Tyrospheres) are used in cancer therapeutics for minimizing the toxic side effects and maintaining bioactivity [71].

21.3.2.2 Vaccination

Vaccination is an effective tool to reduce infection susceptibility at the population level. The recent development of the SARS-CoV-2 vaccine for attenuation and alleviation of

COVID-19 spread is the most potential research. Live attenuated viruses and S protein subunits moreover, inactivated viruses are usually preferred for vaccine development [72]. However, in vivo studies in the literature revealed that the live virus-based vaccination is associated with hypersensitive-type lung pathology [73]. Based on these, biomaterials-based approaches can be considered precise tools for immune modulation, characterized by sustained antibody production. For instance, microneedles have been proposed and applied topically for pain-free antigen delivery in the derma region [74].

21.3.3 Other antiviral strategies

21.3.3.1 *Surface inactivation*

Viral transmission occurs through aerosolized particles where large droplets of size >20 μm, fall on the nearer surfaces [75]. On the other hand, small droplets travel by air to a distance of few meters from the emission point. The viruses remain inactive outside the body however, these can remain active on the surfaces such as doorknobs, tables, and utensils [76]. These viruses then transfer from the contaminated surface to healthy humans and lead to infection. Sterilization is a valuable method through which contamination can be altered however, it is difficult to maintain in highly susceptible areas [77]. A recent study revealed that SARS-CoV-2 stays active on the surfaces like plastic, stainless steel, and cardboard for more than 4 hours approximately at a normal humid temperature [78]. Based on this, several strategies such as antiviral surfaces and antiviral coatings have been incorporated to mitigate the viral spread and minimize disease propagation. Biomaterials coatings made up of natural products, metals, and surfactants can be used [79]. Metals such as silver, gold, and copper have antiviral properties. However, colloidal silver is not intended for internal ingestion, as per NIH and FDA's guidelines [80]. Another example, N,N-dodecyl methyl-polyethylenimine (DM-PEI), a cationic surfactant is a quaternary ammonium compound that exhibits antiviral polymer properties. DM-PEI ruptures the cell membranes of the viral particle by interacting with polycationic chains and may act as an adhesive that entraps the virus [81]. In case of naturally-derived products, cellulose, pectin, alginate, and other antiviral coatings have been approved for coatings on fruits and vegetables and for packaging purposes to prevent the spread of viruses [82].

21.3.3.2 *Viral filtration*

In addition to metals, antiviral gloves and face masks are of particular interest for the safety of public and hospital care workers and to prevent human-to-human transmission. [83] Surgical masks are competent to filter 98% of large-sized droplets (3 μm) whereas N95 masks are designed to filter aerosolized particles of size, 0.1 μm which are transmitted by coughs or sneezes [75]. Most of the available masks are non-biodegradable and non-renewable which contribute to environmental pollution worldwide. Therefore, research in the field of chemically modified masks made up of biodegradable material with viral deactivation properties is underway. Tiliket et al. designed a virus-functionalized mask that can trap airborne influenza virus A (H5N2). This mask is developed by incorporating DM-PEI functionalized Kimwipes Lite layers into a commercial mask [84]. Other strategies such as the incorporation of green tea extracts, impregnating copper oxide and incorporating sialic acid successfully block the entry and route of the virus [85].

The porosity of the surface is another important feature that may prevent viral transmission. In a study, Konda et al. compared the efficacy of different fabrics for the prevention of aerosol particle inhalation. The authors observed that tighter weaves of the fabric lead to low porosity, which is considered ideal for mechanical filtration. Fabrics like natural silk, and chiffon have high electrostatic filtering capabilities whereas the fusion of cotton with silk or chiffon augments the filtration efficiency [86]. Thus, a face mask can be considered a good medium for the protecting humans from deadly viral infections.

21.4 Recent advancements

21.4.1 Biomaterials and nanotechnology

Combining biomaterials with nanotechnology has provided multifaceted benefits in the drug delivery discipline and for the development of new generation vaccines. Biomaterials-cum-drug cargos have an ample number of benefits including protection from enzymatic degradation, improve stability, and elicit bioavailability due to the distinctive physicochemical properties of biomaterials [87]. The advantages of vaccine nanotechnology are illustrated in Fig. 21.2. Several biomaterials have been designed that exert their antiviral activities against infections, which are discussed in the following sections.

21.4.1.1 Nanoparticles

Metal NPs have a significant role in viral inactivation such as HIV, influenza, and herpes. For example, silver NPs inactivate HIV and influenza viruses by binding to the sulfur-bearing residues present in the envelope of the virus particle [88]. Both silver and gold NPs directly inhibit the attachment of virion to the host by blocking the receptors on the cell and may act as physical barriers [88,89]. Conversely, copper oxide NPs damage virus complexes of both enveloped and nonenveloped viruses [90]. In a study, Mohammadyari, et al. reported

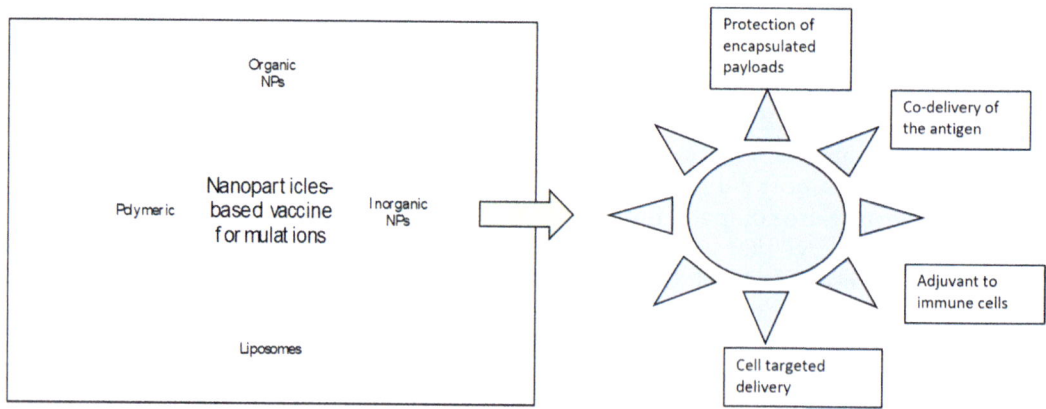

FIGURE 21.2   Schematic illustration of vaccine nanotechnology.

that <400 ppm dose of copper oxide NPs has no adverse effects on the organs of the rat such as kidney or liver, whereas the dose of >400 ppm was found to be toxic [91]. Further research on metal NPs with antiviral properties is needed to explore their clinical applications against particular viruses.

21.4.1.2 Nanodecoy

In alternative strategies, nanostructures are designed and developed to mimic the living cells, known as nanodecoys [92]. Nanodecoys are formulated from cell membrane-derived materials that entrap and capture the viral particles. Based on this technology, Rao et al. constructed a biomimetic nanodecoy by fusing cell membrane-derived vesicle from Aedes albopictus (C6/36) onto a gelatin NP to capture the Zika virus. The efficacy of nanodecoy against Zika virus was checked in A129 mice model and the data revealed significant reduction in infection with the usage of nanodecoy [93]. Similarly, Wei et al. successfully prepared a type of nanodecoy based on T cell membrane-coated poly(lactic-*co*-glycolic acid) (PLGA) NPs to capture HIV [94].

21.4.1.3 Nanosponges

Nanosponges are a novel class of drug delivery systems. They are tiny sponges of a size equivalent to a virus size. Nanosponges are filled with an extensive drug range for targeted delivery. Once entered into the body, these sponges circulate throughout till they reach the target and then predictably release the drug. They are highly efficient for the drugs with poor solubility [95]. Based on the nanosponge technology, Zhang et al. coated the PLGA core on the plasma membrane of human lung epithelial type II cells and human macrophages by developing a cellular nanosponge that targets SARS-CoV-2. These nanosponges efficiently neutralize the SARS-CoV-2 viruses and associated infection [96]. Overall, nanosponges are being used as decoys for the entrapment of pathogens.

21.4.2 Adjuvants

An adjuvant is a substance used for enhancing the antigenicity of an antigen. In a study, Park et al. have shown that the differentiation of dendritic cells can be upregulated with certain biomaterials by antigen presentation [97]. Such polymeric biomaterials act as adjuvants. An adjuvant is formulated from plant-derived polysaccharide inulin by incorporating delta inulin particles called AdvaxTM. AdvaxTM has the potential to enhance the vaccine efficacy against hepatitis B, influenza, Nile Virus, SARS, HIV, and several other epidemic and pandemic viral infections [98]. Adjuvants exist in organic or inorganic forms. The characteristic features of organic biomaterials are biodegradability, biocompatibility, and nontoxicity whereas inorganic biomaterials are highly stable compounds with tunable compositions, [99]. Moreover, inorganic biomaterials have a high surface area-volume ratio.

Further, liposomes can also be used as adjuvants. They are spherical-shaped vesicles that are prepared from biodegradable phospholipids that form a lipid bilayer upon hydration. In the past decade, liposome formulations have been approved clinically as therapeutic vaccines against infectious diseases [100] for example, EpaxalVR virosomes. EpaxalVR

TABLE 21.2   List of biomaterial-based nanoparticles.

Material	Mechanism of action	Target virus	References
PLGA nanosponge	PLGA coating on the membranes of targeted cells	SARS-CoV-2	[105]
PLGA nanodecoy	Capture the pathogens	HIV	[106]
Gelatin nanodecoy	Camouflaged by the host cell membrane	Zika virus	[107]
Iron oxide nanoparticle	Binding to the protein knob	Influenza virus	[108]
Cuprous oxide nanoparticle	Inhibit attachment and entry to hepatocytes	HCV	[109]
AgNP	Blocking of virus- cell interaction Damage viral coat protein	Influenza virus, RSV, HSV, norovirus, phage FX174, adenovirus, parainfluenza, virus type 3	[110−113]
PEG-coated zinc oxide nanoparticle	Interfere with viral nucleic acid replication	Influenza virus, HSV	[114]

contains antigens against the lipid components of hepatitis A, that is, dioleoylphosphatidy-lethanolamine and 1,2-dioleoyl-sn-glycero-3-phosphocholine [101]. The list of various biomaterials and their mechanism of action for target virus has been shown in Table 21.2.

21.4.3 Challenges associated with antiviral biomaterials

NPs are considered excellent supporters for targeted drug delivery or simultaneous co-delivering of different antigens and proteins. However, several challenges associated with biomaterials remain unaddressed. Most of the studies on NPs, hydrogel-based vaccines, and immunotherapies have been carried out in vitro only [102]. Accordingly, the translational applications of NPs would take considerable time to establish in the clinical setting. Therefore, a deeper understanding and knowledge are required to illustrate the impact of NPs on human health.

21.5 Summary/conclusion

The catastrophic outbreaks of viral infections and a spike in death rates have driven researchers to design and develop new antiviral therapeutic approaches and preventive strategies for the management of viral infections. In recent times, the field of biomaterial has emerged due to several ambiguities persisted in current antiviral therapies. Biomaterials in their nano forms are recognized as excellent candidates to combat viral infection by targeting viral infection pathways. [103] Antiviral biomaterials are used for vaccine generation, drug delivery, and gene delivery. Overall, these biomaterial-based systems are highly stable, biocompatible, and low cytotoxic [104]. Besides these achievements, antiviral biomaterials still lack recognition. The fate of biomaterials in vivo is stagnant. Further research is needed to be carried out to use antiviral biomaterials in clinical settings.

Conflict of interest

None declared.

References

[1] Thomas M, Bomar PA. Upper respiratory tract infection [Updated 2021 Jun 30] StatPearls [Internet]. Treasure Island, FL: StatPearls Publishing; 2021Jan-. Available from. Available from: https://www.ncbi.nlm.nih.gov/books/NBK532961/.

[2] Bloom DE, Cadarette D. Infectious disease threats in the twenty-first century: strengthening the global response. Front Immunol 2019;10:549. Available from: https://doi.org/10.3389/fimmu.2019.00549.

[3] Foreman KJ, Marquez N, Dolgert A, Fukutaki K, Fullman N, McGaughey M, et al. Forecasting life expectancy, years of life lost, and all-cause and cause-specific mortality for 250 causes of death: reference and alternative scenarios for 2016-40 for 195 countries and territories. Lancet. 2018;392:2052−90.

[4] WHO. COVID-19 Public Health Emergency of International Concern (PHEIC) Global research and innovation forum (who.int). < https://www.who.int/publications/m/item/covid-19-public-health-emergency-of-international-concern-(pheic)-global-research-and-innovation-forum >.

[5] Pollard AJ, Bijker EM. A guide to vaccinology: from basic principles to new developments. Nat Rev Immunol 2021;21:83−100. Available from: https://doi.org/10.1038/s41577-020-00479-7.

[6] Rodrigues CMC, Plotkin SA. Impact of vaccines; health, economic and social perspectives. Front Microbiol 2020;11:1526. Available from: https://doi.org/10.3389/fmicb.2020.01526.

[7] Chen FM, Liu X. Advancing biomaterials of human origin for tissue engineering. Prog Polym Sci 2016;53:86−168. Available from: https://doi.org/10.1016/j.progpolymsci.2015.02.004.

[8] Colombani T, Rogers ZJ, Eggermont LJ, Bencherif SA. Harnessing biomaterials for therapeutic strategies against COVID-19. Emerg Mater 2021;1−10. Available from: https://doi.org/10.1007/s42247-021-00171-z.

[9] Rupp R, Rosenthal SL, Stanberry LR. VivaGel (SPL7013 Gel): a candidate dendrimer−microbicide for the prevention of HIV and HSV infection. Int J Nanomed 2007;2(4):561−6.

[10] NIBIB. Biomaterials (nih.gov). < https://www.nibib.nih.gov/science-education/science-topics/biomaterials >.

[11] Chai Q, Jiao Y, Yu X. Hydrogels for biomedical applications: their characteristics and the mechanisms behind them. Gels 2017;3(1):6. Available from: https://doi.org/10.3390/gels3010006.

[12] Ahmed EM. Hydrogel: preparation, characterization, and applications: a review. J Adv Res 2015;6:105−21. Available from: https://doi.org/10.1016/j.jare.2013.07.006.

[13] Zhu J, Marchant RE. Design properties of hydrogel tissue-engineering scaffolds. Expert Rev Med Devices 2011;8(5):607−26. Available from: https://doi.org/10.1586/erd.11.27.

[14] Malpure PS, Patil SS, More YM, Nikam PP. A review on- hydrogel. Am J PharmTech Res 2018;8.

[15] Mantha S, Pillai S, Khayambashi P, et al. Smart hydrogels in tissue engineering and regenerative medicine. Materials 2019;12(20):3323. Available from: https://doi.org/10.3390/ma12203323 Basel.

[16] Eggermont LJ, Rogers ZJ, Colombani T, Memic A, Bencherif SA. Injectable cryogels for biomedical applications. Trends Biotechnol 2020;38(4):418−31. Available from: https://doi.org/10.1016/j.tibtech.2019.09.008.

[17] Reichelt S. Introduction to macroporous cryogels. Methods Mol Biol 2015;1286:173−81. Available from: https://doi.org/10.1007/978-1-4939-2447-9_14.

[18] Gun'ko VM, Savina IN, Mikhalovsky SV. Cryogels: morphological, structural and adsorption characterisation. Adv Colloid Interface Sci 2013;187−188:1−46. Available from: https://doi.org/10.1016/j.cis.2012.11.001.

[19] Çimen D, Özbek MA, Bereli N, Mattiasson B, Denizli A. Injectable cryogels in biomedicine. Gels 2021;7(2):38. Available from: https://doi.org/10.3390/gels7020038.

[20] Din FU, Aman W, Ullah I, et al. Effective use of nanocarriers as drug delivery systems for the treatment of selected tumors. Int J Nanomed 2017;12:7291−309. Available from: https://doi.org/10.2147/IJN.S146315 Published 2017 Oct 5.

[21] Szunerits S, Barras A, Khanal M, Pagneux Q, Boukherroub R. Nanostructures for the inhibition of viral infections. Molecules. 2015;20(8):14051−81. Available from: https://doi.org/10.3390/molecules200814051 Published 2015 Aug 3.

[22] Gelderblom HR. Structure and classification of viruses In: Baron S, editor. Medical microbiology. 4th ed Galveston, TX: University of Texas Medical Branch at Galveston; 1996Available from. Available from: https://www.ncbi.nlm.nih.gov/books/NBK8174/.

[23] Koonin EV, Starokadomskyy P. Are viruses alive? The replicator paradigm sheds decisive light on an old but misguided question. Stud Hist Philos Biol Biomed Sci 2016;59:125−34. Available from: https://doi.org/10.1016/j.shpsc.2016.02.016.

[24] Navaratnarajah CK, Warrier R, Kuhn RJ. Assembly of viruses: enveloped particles. Encycl Virology 2008;193−200. Available from: https://doi.org/10.1016/B978-012374410-4.00667-1.

[25] Chinchar VG. Replication of viruses. Encycl Virology 1999;1471−8. Available from: https://doi.org/10.1006/rwvi.1999.0245.

[26] Rakowska PD, Tiddia M, Faruqui N, et al. Antiviral surfaces and coatings and their mechanisms of action. Commun Mater 2021;2. Available from: https://doi.org/10.1038/s43246-021-00153-y.

[27] Lai MY, Cheng PK, Lim WW. Survival of severe acute respiratory syndrome coronavirus. Clin Infect Dis 2005;41(7):e67−71. Available from: https://doi.org/10.1086/433186.

[28] Navarre WW, Schneewind O. Surface proteins of gram-positive bacteria and mechanisms of their targeting to the cell wall envelope. Microbiol Mol Biol Rev 1999;63(1):174−229. Available from: https://doi.org/10.1128/MMBR.63.1.174-229.1999.

[29] Li X, Wu P, Gao GF, Cheng S. Carbohydrate-functionalized chitosan fiber for influenza virus capture. Biomacromolecules 2011;12(11):3962−9. Available from: https://doi.org/10.1021/bm200970x.

[30] Walker SJ, Pizzato M, Takeuchi Y, Devereux S. Heparin binds to murine leukemia virus and inhibits Env-independent attachment and infection. J Virol 2002;76(14):6909−18. Available from: https://doi.org/10.1128/jvi.76.14.6909-6918.2002.

[31] Barrientos LG, Lasala F, Otero JR, Sanchez A, Delgado R. In vitro evaluation of cyanovirin-N antiviral activity, by use of lentiviral vectors pseudotyped with filovirus envelope glycoproteins. J Infect Dis 2004;189 (8):1440−3. Available from: https://doi.org/10.1086/382658.

[32] Li W, Moore MJ, Vasilieva N, et al. Angiotensin-converting enzyme 2 is a functional receptor for the SARS coronavirus. Nature 2003;426(6965):450−4. Available from: https://doi.org/10.1038/nature02145.

[33] Sanders B, Koldijk M, Schuitemaker H. Inactivated viral vaccines. Vaccine analysis: strategies, principles, control 2014;45−80. Available from: https://doi.org/10.1007/978-3-662-45024-6_2 Published 2014 Nov 28.

[34] Galdiero S, Falanga A, Vitiello M, Cantisani M, Marra V, Galdiero M. Silver nanoparticles as potential antiviral agents. Molecules 2011;16(10):8894−918. Available from: https://doi.org/10.3390/molecules16108894 Published 2011 Oct 24.

[35] Qin T, Ma R, Yin Y, et al. Catalytic inactivation of influenza virus by iron oxide nanozyme. Theranostics 2019;9(23):6920−35. Available from: https://doi.org/10.7150/thno.35826.

[36] Khare S, Azevedo M, Parajuli P, Gokulan K. Conformational changes of the receptor binding domain of SARS-CoV-2 spike protein and prediction of a B-cell antigenic epitope using structural data. Front Artif Intell 2021;4:630955. Available from: https://doi.org/10.3389/frai.2021.630955.

[37] Jampílek J, Kráľová K. Nanoformulations: a valuable tool in the therapy of viral diseases attacking humans and animals. Nanotheranostics. 2019;137−78. Available from: https://doi.org/10.1007/978-3-030-29768-8_7 Published 2019 Nov 23.

[38] Ouyang K, Yu X-Y, Zhu Y, Gao C, Huang Q, Cai P. Effects of humic acid on the interactions between zinc oxide nanoparticles and bacterial biofilms. United States. < https://doi.org/10.1016/j.envpol.2017.07.003 >.

[39] Patel PH, Zulfiqar H. Reverse transcriptase inhibitors. [Updated 2021 Jun 14] StatPearls [Internet]. Treasure Island, FL: StatPearls Publishing; 2021Jan-. Available from. Available from: https://www.ncbi.nlm.nih.gov/books/NBK551504/.

[40] Li J, Kuang Y, Gao Y, Du X, Shi J, Xu B. D-amino acids boost the selectivity and confer supramolecular hydrogels of a nonsteroidal anti-inflammatory drug (NSAID). J Am Chem Soc 2013;135(2):542−5. Available from: https://doi.org/10.1021/ja310019x.

[41] Kerry RG, Malik S, Redda YT, Sahoo S, Patra JK, Majhi S. Nano-based approach to combat emerging viral (NIPAH virus) infection. Nanomedicine. 2019;18:196−220. Available from: https://doi.org/10.1016/j.nano.2019.03.004.

[42] Li Y, Lin Z, Guo M, et al. Inhibitory activity of selenium nanoparticles functionalized with oseltamivir on H1N1 influenza virus. Int J Nanomed 2017;12:5733−43. Available from: https://doi.org/10.2147/IJN.S140939.

[43] Noordeen F, Scougall CA, Grosse A, et al. Therapeutic antiviral effect of the nucleic acid polymer REP 2055 against persistent duck hepatitis B virus infection. PLoS One 2015;10(11):e0140909. Available from: https://doi.org/10.1371/journal.pone.0140909.

[44] McAuley JL, Gilbertson BP, Trifkovic S, Brown LE, McKimm-Breschkin JL. Influenza virus neuraminidase structure and functions. Front Microbiol 2019;10:39. Available from: https://doi.org/10.3389/fmicb.2019.00039.

[45] Warnes SL, Keevil CW. Inactivation of norovirus on dry copper alloy surfaces. PLoS One 2013;8:e75017.

[46] Warnes SL, Summersgill EN, Keevil CW. Inactivation of murine norovirus on a range of copper alloy surfaces is accompanied by loss of capsid integrity. Appl Environ Microbiol 2015;81:1085−91.

[47] Haldar J, Weight AK, Klibanov AM. Preparation, application and testing of permanent antibacterial and antiviral coatings. Nat Protoc 2007;2(10):2412−17. Available from: https://doi.org/10.1038/nprot.2007.353.

[48] Thorgeirsdóttir TO, Thormar H, Kristmundsdóttir T. Effects of polysorbates on antiviral and antibacterial activity of monoglyceride in pharmaceutical formulations. Pharmazie. 2003;58(4):286−7.

[49] Wang XH, Cheng PP, Jiang F, Jiao XY. The effect of hepatitis B virus infection on hepcidin expression in hepatitis B patients. Ann Clin Lab Sci 2013;43(2):126−34.

[50] Ponce NM, Pujol CA, Damonte EB, Flores ML, Stortz CA. Fucoidans from the brown seaweed *Adenocystis utricularis*: extraction methods, antiviral activity and structural studies. Carbohydr Res 2003;338(2):153−65. Available from: https://doi.org/10.1016/s0008-6215(02)00403-2.

[51] Hidari KI, Murata T, Yoshida K, et al. Chemoenzymatic synthesis, characterization, and application of glycopolymers carrying lactosamine repeats as entry inhibitors against influenza virus infection. Glycobiology 2008;18(10):779−88. Available from: https://doi.org/10.1093/glycob/cwn067.

[52] Lieleg O, Lieleg C, Bloom J, Buck CB, Ribbeck K. Mucin biopolymers as broad-spectrum antiviral agents. Biomacromolecules 2012;13(6):1724−32. Available from: https://doi.org/10.1021/bm3001292.

[53] Yu Y, Cooper CL, Wang G, et al. Engineered human cathelicidin antimicrobial peptides inhibit ebola virus infection. iScience 2020;23(4):100999. Available from: https://doi.org/10.1016/j.isci.2020.100999.

[54] Barlow PG, Svoboda P, Mackellar A, et al. Antiviral activity and increased host defense against influenza infection elicited by the human cathelicidin LL-37. PLoS One 2011;6(10):e25333. Available from: https://doi.org/10.1371/journal.pone.0025333.

[55] Currie SM, Gwyer Findlay E, McFarlane AJ, et al. Cathelicidins have direct antiviral activity against respiratory syncytial virus in vitro and protective function in vivo in mice and humans. J Immunol 2016;196(6):2699−710. Available from: https://doi.org/10.4049/jimmunol.1502478.

[56] Daher KA, Selsted ME, Lehrer RI. Direct inactivation of viruses by human granulocyte defensins. J Virol 1986;60(3):1068−74. Available from: https://doi.org/10.1128/JVI.60.3.1068-1074.1986.

[57] Armitage AE, Stacey AR, Giannoulatou E, et al. Distinct patterns of hepcidin and iron regulation during HIV-1, HBV, and HCV infections. Proc Natl Acad Sci U S A 2014;111(33):12187−92. Available from: https://doi.org/10.1073/pnas.1402351111.

[58] Rajanbabu V, Chen JY. Antiviral function of tilapia hepcidin 1-5 and its modulation of immune-related gene expressions against infectious pancreatic necrosis virus (IPNV) in Chinook salmon embryo (CHSE)-214 cells. Fish Shellfish Immunol 2011;30(1):39−44. Available from: https://doi.org/10.1016/j.fsi.2010.09.005.

[59] Park S, Ko YS, Lee SJ, Lee C, Woo K, Ko G. Inactivation of influenza A virus via exposure to silver nanoparticle-decorated silica hybrid composites. Environ Sci Pollut Res Int 2018;25(27):27021−30. Available from: https://doi.org/10.1007/s11356-018-2620-z.

[60] Emery SL, Erdman DD, Bowen MD, et al. Real-time reverse transcription-polymerase chain reaction assay for SARS-associated coronavirus. Emerg Infect Dis 2004;10(2):311−16. Available from: https://doi.org/10.3201/eid1002.030759.

[61] Udugama B, Kadhiresan P, Kozlowski HN, et al. Diagnosing COVID-19: the disease and tools for detection. ACS Nano 2020;14(4):3822−35. Available from: https://doi.org/10.1021/acsnano.0c02624.

[62] Wei X, Xia Y, Shen M, et al. Magnetic nanoparticle-based automatic chemiluminescent enzyme immunoassay for Golgi protein 73 and the clinical assessment. J Nanosci Nanotechnol 2019;19(4):1971−7. Available from: https://doi.org/10.1166/jnn.2019.16485.

[63] Mehta S, Dhawan V. Importance and feasibility of point-of-care testing in Takayasu's arteritis. INNOSC Theranos Pharmacol Sci 2020;3:3−14.

[64] Formlabs. 3D Printed COVID-19 Test Swabs (formlabs.com). < https://formlabs.com/asia/covid-19-response/covid-test-swabs/ > .

[65] Carter LJ, Garner LV, Smoot JW, et al. Assay techniques and test development for COVID-19 diagnosis. ACS Cent Sci 2020;6(5):591−605. Available from: https://doi.org/10.1021/acscentsci.0c00501.

[66] Khan M, Adil SF, Alkhathlan HZ, et al. COVID-19: a global challenge with old history, epidemiology and progress so far. Molecules. 2020;26(1):39. Available from: https://doi.org/10.3390/molecules26010039 Published 2020 Dec 23.

[67] Hu C, Lu L, Wan JP, Wen C. The pharmacological mechanisms and therapeutic activities of hydroxychloroquine in rheumatic and related diseases. Curr Med Chem 2017;24(20):2241−9. Available from: https://doi.org/10.2174/0929867324666170316115938.

[68] White NJ, Watson JA, Hoglund RM, Chan XHS, Cheah PY, Tarning J. COVID-19 prevention and treatment: a critical analysis of chloroquine and hydroxychloroquine clinical pharmacology. PLoS Med 2020;17(9): e1003252. Available from: https://doi.org/10.1371/journal.pmed.1003252.

[69] Eastman RT, Roth JS, Brimacombe KR, et al. Remdesivir: a review of its discovery and development leading to emergency use authorization for treatment of COVID-19 published correction appears in ACS Cent Sci. 2020;6(6):1009ACS Cent Sci 2020;6(5):672−83. Available from: https://doi.org/10.1021/acscentsci.0c00489.

[70] Agrahari V, Mandal A, Agrahari V, et al. A comprehensive insight on ocular pharmacokinetics. Drug Deliv Transl Res 2016;6(6):735−54. Available from: https://doi.org/10.1007/s13346-016-0339-2.

[71] Zhang Z, Tsai PC, Ramezanli T, Michniak-Kohn BB. Polymeric nanoparticles-based topical delivery systems for the treatment of dermatological diseases. Wiley Interdiscip Rev Nanomed Nanobiotechnol 2013;5 (3):205−18. Available from: https://doi.org/10.1002/wnan.1211.

[72] Dai L, Gao GF. Viral targets for vaccines against COVID-19. Nat Rev Immunol 2021;21:73−82. Available from: https://doi.org/10.1038/s41577-020-00480-0.

[73] McNeil MM, DeStefano F. Vaccine-associated hypersensitivity. J Allergy Clin Immunol 2018;141(2):463−72. Available from: https://doi.org/10.1016/j.jaci.2017.12.971.

[74] Kim YC, Park JH, Prausnitz MR. Microneedles for drug and vaccine delivery. Adv Drug Deliv Rev 2012;64 (14):1547−68. Available from: https://doi.org/10.1016/j.addr.2012.04.005.

[75] Jayaweera M, Perera H, Gunawardana B, Manatunge J. Transmission of COVID-19 virus by droplets and aerosols: a critical review on the unresolved dichotomy. Environ Res 2020;188:109819. Available from: https://doi.org/10.1016/j.envres.2020.109819.

[76] Lewis D. COVID-19 rarely spreads through surfaces. So why are we still deep cleaning? Nature 2021;590 (7844):26−8. Available from: https://doi.org/10.1038/d41586-021-00251-4.

[77] CDC. Guideline for Disinfection and Sterilization in Healthcare Facilities, 2008 (cdc.gov). < https://www.cdc.gov/infectioncontrol/pdf/guidelines/disinfection-guidelines-H.pdf > .

[78] Riddell S, Goldie S, Hill A, Eagles D, Drew TW. The effect of temperature on persistence of SARS-CoV-2 on common surfaces. Virol J 2020;17(1):145. Available from: https://doi.org/10.1186/s12985-020-01418-7.

[79] Lam MT, Wu JC. Biomaterial applications in cardiovascular tissue repair and regeneration. Expert Rev Cardiovasc Ther 2012;10(8):1039−49. Available from: https://doi.org/10.1586/erc.12.99.

[80] NIH. Colloidal Silver | NCCIH (nih.gov). < https://www.nccih.nih.gov/health/colloidal-silver > .

[81] Jiang X, Li Z, Young DJ, et al. Toward the prevention of coronavirus infection: what role can polymers play? Mater Today Adv 2021;10:100140. Available from: https://doi.org/10.1016/j.mtadv.2021.100140.

[82] Martău GA, Mihai M, Vodnar DC. The Use of chitosan, alginate, and pectin in the biomedical and food sector-biocompatibility, bioadhesiveness, and biodegradability. Polymers 2019;11(11):1837. Available from: https://doi.org/10.3390/polym11111837.

[83] Howard J, Huang A, Li Z, et al. An evidence review of face masks against COVID-19. Proc Natl Acad Sci U S A 2021;118(4). Available from: https://doi.org/10.1073/pnas.2014564118 e2014564118.

[84] Chakhalian D, Shultz RB, Miles CE, Kohn J. Opportunities for biomaterials to address the challenges of COVID-19. J Biomed Mater Res A 2020;108(10):1974−90. Available from: https://doi.org/10.1002/jbm.a.37059.

[85] Balasubramaniam B, Prateek, Ranjan S, et al. Antibacterial and antiviral functional materials: chemistry and biological activity toward tackling COVID-19-like pandemics. ACS Pharmacol Transl Sci 2020;4(1):8−54. Available from: https://doi.org/10.1021/acsptsci.0c00174 Published 2020 Dec 29.

[86] Konda A, Prakash A, Moss GA, Schmoldt M, Grant GD, Guha S. Aerosol filtration efficiency of common fabrics used in respiratory cloth masks published correction appears in ACS Nano. 2020 Aug 25;14(8): 10742-10743ACS Nano 2020;14(5):6339−47. Available from: https://doi.org/10.1021/acsnano.0c03252.

[87] Yin L, Yuvienco C, Montclare JK. Protein based therapeutic delivery agents: contemporary developments and challenges. Biomaterials 2017;134:91−116. Available from: https://doi.org/10.1016/j.biomaterials.2017.04.036.

[88] Lin N, Verma D, Saini N, et al. Antiviral nanoparticles for sanitizing surfaces: a roadmap to self-sterilizing against COVID-19. Nano Today 2021;40:101267. Available from: https://doi.org/10.1016/j.nantod.2021.101267.

[89] Singh L, Kruger HG, Maguire GEM, Govender T, Parboosing R. The role of nanotechnology in the treatment of viral infections. Ther Adv Infect Dis 2017;4(4):105−31. Available from: https://doi.org/10.1177/2049936117713593.

[90] Gurunathan S, Qasim M, Choi Y, et al. Antiviral potential of nanoparticles-can nanoparticles fight against coronaviruses? Nanomaterials 2020;10(9):1645. Available from: https://doi.org/10.3390/nano10091645.

[91] Mohammadyari A, Razavipour ST, Mohammadbeigi M, Negahdary M, Ajdary M. Exploring vivo toxicity assessment of copper oxide nanoparticle in Wistar rats. J Biol Today's World 2014;3(6):124−8.

[92] Kamat S, Kumari M, Jayabaskaran C. Nano-engineered tools in the diagnosis, therapeutics, prevention, and mitigation of SARS-CoV-2. J Control Release 2021;338:813−36. Available from: https://doi.org/10.1016/j.jconrel.2021.08.046.

[93] Rossi SL, Tesh RB, Azar SR, et al. Characterization of a novel murine model to study Zika virus. Am J Trop Med Hyg 2016;94(6):1362−9. Available from: https://doi.org/10.4269/ajtmh.16-0111.

[94] Wei X, Zhang G, Ran D, et al. T-cell-mimicking nanoparticles can neutralize HIV infectivity. Adv Mater 2018;30(45):e1802233. Available from: https://doi.org/10.1002/adma.201802233.

[95] S S, S A, Krishnamoorthy K, Rajappan M. Nanosponges: a novel class of drug delivery system−review. J Pharm Pharm Sci 2012;15(1):103−11. Available from: https://doi.org/10.18433/j3k308.

[96] Mulay A, Konda B, Garcia Jr G, et al. SARS-CoV-2 infection of primary human lung epithelium for COVID-19 modeling and drug discovery. Cell Rep 2021;35(5):109055. Available from: https://doi.org/10.1016/j.celrep.2021.109055.

[97] Park J, Babensee JE. Differential functional effects of biomaterials on dendritic cell maturation. Acta Biomater 2012;8(10):3606−17. Available from: https://doi.org/10.1016/j.actbio.2012.06.006.

[98] Kumari S, Chatterjee K. Biomaterials-based formulations and surfaces to combat viral infectious diseases. APL Bioeng 2021;5(1):011503. Available from: https://doi.org/10.1063/5.0029486 Published 2021 Feb 9.

[99] Bose RJ, Kim M, Chang JH, et al. Biodegradable polymers for modern vaccine development. J Ind Eng Chem 2019;77:12−24. Available from: https://doi.org/10.1016/j.jiec.2019.04.044.

[100] Nisini R, Poerio N, Mariotti S, De Santis F, Fraziano M. The multirole of liposomes in therapy and prevention of infectious diseases. Front Immunol 2018;9:155. Available from: https://doi.org/10.3389/fimmu.2018.00155.

[101] De Serrano LO, Burkhart DJ. Liposomal vaccine formulations as prophylactic agents: design considerations for modern vaccines. J Nanobiotechnol 2017;15(1):83. Available from: https://doi.org/10.1186/s12951-017-0319-9.

[102] Han L, Peng K, Qiu LY, et al. Hitchhiking on controlled-release drug delivery systems: opportunities and challenges for cancer vaccines. Front Pharmacol 2021;12:679602. Available from: https://doi.org/10.3389/fphar.2021.679602.

[103] Li S, Guo X, Gao R, Sun M, Xu L, Xu C, et al. Recent progress on biomaterials fighting against viruses. Adv Mater 2021;33:2005424. Available from: https://doi.org/10.1002/adma.202005424.

[104] Nikolova MP, Chavali MS. Recent advances in biomaterials for 3D scaffolds: a review. Bioact Mater 2019;4:271−92. Available from: https://doi.org/10.1016/j.bioactmat.2019.10.005.

[105] Zhang Q, Honko A, Zhou J, et al. Cellular nanosponges inhibit SARS-CoV-2 infectivity. Nano Lett 2020;20 (7):5570−4. Available from: https://doi.org/10.1021/acs.nanolett.0c02278.

[106] Zhang G, Campbell GR, Zhang Q, et al. CD4$^+$ T cell-mimicking nanoparticles broadly neutralize HIV-1 and suppress viral replication through autophagy. mBio 2020;11(5):e00903−20. Available from: https://doi.org/10.1128/mBio.00903-20.

[107] Rao L, Wang W, Meng QF, et al. A biomimetic nanodecoy traps Zika virus to prevent viral infection and fetal microcephaly development. Nano Lett 2019;19(4):2215–22. Available from: https://doi.org/10.1021/acs.nanolett.8b03913.

[108] Kumar R, Nayak M, Sahoo GC, et al. Iron oxide nanoparticles based antiviral activity of H1N1 influenza A virus. J Infect Chemother 2019;25(5):325–9. Available from: https://doi.org/10.1016/j.jiac.2018.12.006.

[109] Hang X, Peng H, Song H, Qi Z, Miao X, Xu W. Antiviral activity of cuprous oxide nanoparticles against Hepatitis C Virus in vitro. J Virol Methods 2015;222:150–7. Available from: https://doi.org/10.1016/j.jviromet.2015.06.010.

[110] Elechiguerra JL, Burt JL, Morones JR, et al. Interaction of silver nanoparticles with HIV-1. J Nanobiotechnol 2005;3:6. Available from: https://doi.org/10.1186/1477-3155-3-6 Published 2005 Jun 29.

[111] Gaikwad S, Ingle A, Gade A, et al. Antiviral activity of mycosynthesized silver nanoparticles against herpes simplex virus and human parainfluenza virus type 3. Int J Nanomed 2013;8:4303–14. Available from: https://doi.org/10.2147/IJN.S50070.

[112] Bekele AZ, Gokulan K, Williams KM, Khare S. Dose and size-dependent antiviral effects of silver nanoparticles on feline calicivirus, a human norovirus surrogate. Foodborne Pathog Dis 2016;13(5):239–44. Available from: https://doi.org/10.1089/fpd.2015.2054.

[113] Yang XX, Li CM, Huang CZ. Curcumin modified silver nanoparticles for highly efficient inhibition of respiratory syncytial virus infection. Nanoscale 2016;8(5):3040–8. Available from: https://doi.org/10.1039/c5nr07918g.

[114] Ghaffari H, Tavakoli A, Moradi A, et al. Inhibition of H1N1 influenza virus infection by zinc oxide nanoparticles: another emerging application of nanomedicine. J Biomed Sci 2019;26(1):70. Available from: https://doi.org/10.1186/s12929-019-0563-4.

Antiviral biomolecules from marine inhabitants

Ishwarya Ayyanar, Subidsha Suyambu Krishnan, Akila Ravindran, Sunandha Jeeva Bharathi Gunasekaran and Balasubramanian Vellaisamy

Department of Microbiology, Alagappa University, Science Block, Karaikudi, Tamil Nadu, India

22.1 Introduction

Every day, as technology advances, new infections emerge on a regular basis. The emergence might be caused by mutations or by exogenous substances that come into touch with the organisms. Pathogens that are evolving are becoming incredibly contagious and have a very unpleasant effect on humans, perhaps leading to a global pandemic. The most recent example is the recently discovered Severe Acute Respiratory Syndrome Coronavirus (SARS-CoV-2), which has resulted in significant fatality rates globally [1]. According to the World Health Organization (WHO), millions of people died as a result of the epidemic, and the mortality rate will continue to rise as the outbreak continues. Emerging viruses such as Zika, Ebola, and SARS-CoV-2 are developing resistance to antiviral vaccinations and antiviral medicines. When antiviral medications are consumed, some of them have negative effects [2].

Infectious illnesses can be caused by a variety of microorganisms, including bacteria, fungi, viruses, and parasites. Viruses, an obligatory intracellular parasite that kills millions of people globally each year, are among the many microorganisms. Despite their nano size, they cause significant problems and mortality in humans. It is seen with an electron microscope [2,3]. Viral surface proteins are responsible for viral attachment to host cell receptors. This interaction influences the virus's host specificity and organ specificity. Surface proteins are antibody targets, which means that antibodies attached to these surface proteins inhibit the virus from binding to the cell receptor. This effectively neutralizes viral replication. Internal proteins, some of which are DNA or RNA polymerases, are also found in viruses. Some viruses create antigenic variations of their surface proteins, allowing them to avoid detection by our host defences. Antibodies against one antigenic variation (serotype) will not neutralize antibodies against another serotype. Some viruses have only one serotype, whereas others have many [3].

The human immune system has evolved to identify and eradicate a large variety of diseases; yet, certain pathogens have gained the capacity to evade immune response [3,4]. Pathogens that exhibit resistance are treated with antiviral medications for viral illness control and cure, and there is also a vaccine therapy that is used to prevent viral infection. Antiviral agents are medications that have been authorized by the Food and Drug Administration to treat or control viral infections. Antiviral medications primarily target different phases of the viral life cycle. Viral attachment to host cell, uncoating, synthesis of viral mRNA, translation of mRNA, replication of viral RNA and DNA, maturation of new viral proteins, budding, release of freshly manufactured virus, and free virus in bodily fluids are the target phases in the viral life cycle. Fifty percent of current antiviral medications are used to treat human immunodeficiency virus (HIV) infections. Others are mostly used to treat herpes viruses, hepatitis B virus (HBV), hepatitis C virus (HCV), and respiratory viruses [4]. An ideal antiviral treatment should be active against both rapidly replicating and dormant viruses; nevertheless, the majority of antiviral drugs on the market are only effective against replicating viruses. The goal of antiviral medicine treatment in immunocompromised individuals is to lower the severity of the sickness and its consequences, as well as the rate of virus transmission [5].

However, the medicine has both benefits and drawbacks. There are two major issues that restrict the utility of antiviral medications: (a) toxicity and (b) the virus's development of resistance to antiviral treatments. Furthermore, host phenotypic responses to antiviral medications caused by genomic or epigenetic variables may restrict an antiviral agent's potency in an individual [4].

Vaccination is another method of preventing viral diseases. Vaccinations derived from viruses are known as viral vaccines. Inactivated or attenuated viruses are used in viral vaccinations. MMR (mumps, measles, and rubella) vaccination is a popular example of a viral vaccine [6,7]. Vaccination has been shown to be quite successful in activating protective immune responses against illnesses since its inception. Vaccination against common illnesses such as diphtheria, measles, mumps, pertussis, polio, tetanus, rubella, hepatitis B, meningitis, and smallpox has considerably reduced human morbidity and death by 97%−99% [7,8]. To induce protective immune responses against diseases, traditional vaccinations made from live attenuated pathogens and inactivated viruses, recombinant proteins, and synthetic peptides are typically utilized [7].

Despite their many advantages, these vaccines have numerous drawbacks, including limited bloodstream stability and poor efficacy against infectious illnesses. To address these constraints, antigenic components, either natural or recombinant proteins, are combined with immunologic adjuvants to increase and improve antigen immunogenicity [9,10]. For example, aluminum hydroxide (alum) based on inorganic salts is a widely used adjuvant on the market, but its usage is restricted by allergic responses at the injection site with the development of subcutaneous nodules and moderate toxicity in high dosages [11].

As a result, both antiviral medications and viral vaccines have certain negative effects. To combat all of these impacts, we may now work together with natural antiviral biomaterials or biomaterials. Recent discoveries in virology have enhanced our understanding of viral pathophysiology, improved immunization techniques, and the development of novel, more effective medicines for patients worldwide [12]. Aside from all of this treatment, there have been advancements in development of antiviral biomaterials being able to work against the virus.

The use of biomaterials into antiviral treatment provides diverse benefits and processes. Physical adsorption of viruses, binding to viruses as entrance inhibitors, production of irreversible viral deformation, interference with viral nucleic acid synthesis, and prevention of viral discharge from infected cells are some of the ways antiviral biomaterials act [13]. Noval natural biomolecule research have sparked interest in the development of biomaterial-based antiviral drug development and formulations of next-generation vaccines with natural biomaterial as adjuvant for the viral diseases. Biomaterials utilized in antiviral treatment are both synthetic and spontaneously produced. Innovative technologies that entail the creation of novel biomaterial-based formulations and surfaces endowed with broad-spectrum antiviral characteristics have recently emerged [13,14]. Biomaterials can be used in conjunction with other therapies such as vaccination to increase vaccine composition, distribution, and effectiveness. Biomaterials are employed in medicine because they are nontoxic, nonimmunogenic, and chemically inert [14,15].

The rising demand for novel, complex, multifunctional biomaterials has drawn attention to natural structural composites, which have undergone significant optimization over extended evolutionary selection pressure and adaption processes. Today's scientists believe that the marine biological materials are essential sources of inspiration for biomimetics as well as raw materials for practical applications in technology and healthcare. In contemporary materials research, the utilization of marine natural resources as multifunctional biomaterials is experiencing a revival. Marine resources are desirable since they are ecologically benign, low in pollutants, and metabolically compatible [16,17]. Marine resources, due to their abundance and diversity, contain antibacterial, anticancerous, antiinflammatory, and antiviral capabilities, making them important sources of pharmaceutical targets [16,18]. The antiviral ability of marine creatures makes them intriguing for therapeutic applications even in COVID-19 therapy [1,16,18−21]. Studies on the antiviral properties of marine natural materials, particularly marine polysaccharides (PSs) and marine peptides, have recently gained prominence across the world. A range of antiviral actions have been demonstrated for marine-derived PSs and their reduced molecular weight oligosaccharide derivatives and marine-derived antiviral peptides (AVPs) [16,18,22−26]. This chapter highlights research findings on the antiviral activity of marine-derived biomaterials such as PSs and peptides. This chapter mainly focuses on the antiviral benefits of most commonly reported marine PSs and their derivatives such as chitin, chitosan, carrageenan, alginate, etc. Many researchers have also turned their interest to marine-based peptides for the development of new antiviral therapies, and the results of such studies are highlighted in this chapter.

22.2 Marine polysaccharides

The field of marine-derived natural PSs is already huge and increasing. Seaweeds, such as alginates, agar, and agarose, as well as carrageenans, are the most prevalent source of PSs. Even cellulose and amylose have been extracted from the macroalga ULVA, which is found around the Mediterranean Sea's shores and in numerous lagoons, including Venice's. Chitin and chitosan are obtained from marine crustacean exoskeletons [16,22]. PSs generated from seawater have gained interest as a possible new class of biomaterials. Because of the abundance of marine species in the ocean, PSs taken from marine animals can be extracted at a lower cost than those recovered from plants [27–29]. Biotechnology advancements have also resulted in the in vitro manufacture of several marine PSs, with production yields considerably enhanced by the adjustment of growing conditions [16,29]. Because marine PSs have a wide range of structures and are yet underutilized, they should be explored as a potential source of natural chemicals for drug development [16]. Marine PSs are classified into three categories based on their source: marine animal PSs, plant PSs, and microbial PSs [30]. Antitumor, antiviral, anticoagulant, antioxidant, immuno-inflammatory, and other therapeutic characteristics have been demonstrated for marine-derived PSs [16,31]. Studies on the antiviral activities of marine PSs and their galacton derivatives, in particular, are gaining traction, and marine PSs are leading the opportunity for a better trend in antiviral medications.

22.2.1 Chitin, chitosan, and their derivatives

Chitin is a component found in the cell walls of the basidiomycete fungi. It is also considered a biodegradable component and the second most plentiful polysaccharide on earth, after cellulose [32,33]. Chitosan is a versatile and valuable natural biopolymer that can be obtained from various terrestrial and marine sources [33]. Chitosan is a low acetyl-substituted form of chitin (See Fig. 22.1). Chitin, chitosan, and their derivatives are organic materials that can be commonly found as pure chitin. Its properties include its ability to resist alkaline conditions and life form as associated chitin with a minute level of proteins, pigments, and minerals [17]. It has three different functional groups, which are an amino group at C-2, primary and secondary hydroxyl groups at C-3 and C-6, respectively. Although naturally occurring, chitin has a 30% functional group on the C-2 decarboxylated copolymer. This makes it a linear copolymer with −(1−4)-2-amino-2-deoxy-C-glucan [33,34]. Chitin and chitosan derivatives have various biological functions. They are known to have the following properties such as antioxidant, antimicrobial, blood anticoagulant, and besides efficient antiviral activity [35]. Baaten et al.'s [36] study showed that prophylactic nasal mucosal administration of chitin microparticles (CMP) dramatically lowered lung viral titers and clinical symptoms. By attracting innate cells such as neutrophils and boosting inflammatory cytokines, pretreatment with CMPs improved the innate immune response to future infection [36]. In another study, Hasegawa et al. [37] revealed that immunization of the influenza vaccine along with CMP induced primary and secondary anti-HA IgA reactions in the nasal wash and anti-HA IgG responses in the serum that was considerably greater than those of nasal vaccination without CMP, and presented adequate immunity against a homologous influenza virus challenge. Furthermore, CMP-based vaccination with A/Yamagata (H1N1) and A/Guizhou (H3N2) elicited PR8 HA-reactive IgA in nasal washes as well as specific-IgG in blood. In A/Yamagata (H1N1)-vaccinated mice, immunization with A/Yamagata and CMP resulted in complete protection against a PR8 (H1N1) challenge, whereas immunization with A/Guizhou (H3N2) and CMP resulted in a 100-fold reduction of nasal virus titer, demonstrating the

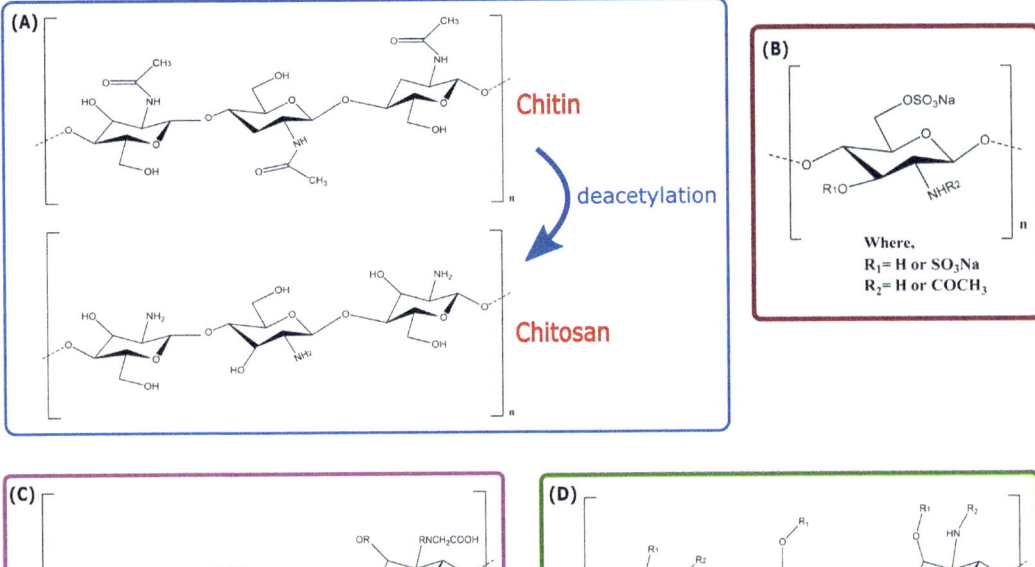

FIGURE 22.1   Basic chemical structure of Chitin and some of their derivatives (A) Chitin and chitosan [17], (B) 3,6-O-sulfated chitosan [43], (C) N-carboxymethylchitosan N, O, sulfate (NCMCS) [40,55], (D) Sulfated chitooligosaccharides (SCOS) [45].

cross-protective effect of CMP and influenza vaccine. And so Hasegawa et al., suggested that CMP may serve as a safe and effective adjuvant for nasal immunization with inactivated influenza vaccine [37]. It has been shown that the antiviral activity of low molecular weight chitosans generated by enzymatic hydrolysis improves dramatically with decreasing polymerization degree [38,39]. Sosa et al. stated through their study that the N-carboxymethylchitosan N, O, sulfate, a polysaccharide derivative of chitin, could inhibit the spread of HIV type 1 in CD4+ cells. They noted that this effect was due to the blocking of the interactions between the CD4 and viral surface proteins [40]. Nishimura et al. discovered that chitin sulfate has a good inhibitory impact on HIV-1 infection and that its inhibition action on HIV-1 is highly dependent on the locations of sulfation [30,41]. Similarly, it was revealed that aminoethyl-chitosan, which is made up of 50% deacetylated chitosan, had strong inhibitory action against HIV-1 in vitro [30,35]. Another study suggested that the effects of prophylactic intranasal administration of CMPs were found to stimulate the local accumulation of killer cells and suppress the secretion of cytokines, which could prevent the pathogenesis of H5N1 flu [42]. There is a robust investigation that reported the counter HPV (human papillomavirus) impact of 3,6-O-sulfated chitosan could be expected to targeting on the viral capsid protein, and to the guideline of the host

PI3K/Akt/mTOR pathways [43]. Low-subatomic subsidiaries of the chitosan polysaccharide with subatomic masses from 17 to 2 kDa were investigated for their antiviral movement against the tobacco mosaic virus (TMV) infection, and the results showed that these examples repressed the development of nearby putrefactions instigated by the infection for 50%—90%. The antiviral action of the Low Molecular derivative of chitosan essentially expanded with the bringing down of their polymerization degree [38]. Chitosans were discovered to successfully diminish the infection rate of the feline calicivirus FCV-F9 as well as the bacteriophages Ms2 and phiX174. Chitosan reduces Ms2 infectivity as its molecular weight increases, with high molecular weight chitosan (above 200 kDa) being able to entirely diminish Ms2 titers [30,44]. Sulfated chitooligosaccharides (SCOS), which are made by random sulfation of chitooligosaccharides, have been shown to have strong anti-HIV activity at the low molecular weight (3—5 kDa). SCOS inhibits viral entry and virus-cell fusion by interfering with the interaction of HIV-1 gp120 with CD4+ cytoplasmic membrane receptors [45—47]. Chirkov's [22] prior review shows that chitosan suppresses viral infections in mammalian cells, but provides limited data. Several investigations, however, have found that chitosan exhibits antiviral action against human cytomegalovirus (CMV) strain AD169 and H1N1 influenza A virus [47—49]. BALB/c mice were given an intranasal formulation of chitosan to boost the host immune responses against H7N9 influenza virus infection, which is generally very harmful to humans [47,50]. Chitosan was shown to be beneficial in shielding mice against this influenza strain and three other viral strains were examined. These studies demonstrate chitosan's great potentials as an antiinfluenza agent [47,51,52]. Hassan et al. [53] investigated chitosan's in vitro antiviral efficacy against Rift Valley Fever (RVFV), Herpes Simplex-1 (HSV-1), and Coxsackie viruses on mammalian cells. The results revealed a decrease in the viral load of these [53]. The patented invention of Zhu Yonghong et al. suggested that chitin, chitosan, and its derivatives can be made into antiviral or disinfecting preparations. In vitro, in vivo antiviral experiments, and histopathological inspection showed that chitin, chitosan, and their derivatives have broad-spectrum antiviral activity against RNA viruses, DNA viruses, and retroviruses, as well as an excellent effect in preventing and treating viral diseases with no adverse reactions. As a result, chitin, chitosan, and their derivatives can be made into antiviral or disinfecting preparations [23]. Existing research on the role of chitin and its derivatives suggests that this cationic polymer may be employed as an effective biodegradable immunotherapeutic adjuvant, and therefore this property may be relevant to SARS-CoV-2. Chitosan molecules were employed to transfer plasmid DNA encoding the nucleocapsid SARS-CoV-2 into the nose. This causes the SARS-CoV-2 spike protein to be secreted, which competes with active coronavirus for interacting with human ACE2 receptors. Chitosan inhibits DNA vaccine breakdown and, due to its cationic nature, promotes binding to negatively charged DNA. Furthermore, the mucoadhesive characteristics of chitosan stimulate its usage as a mucosal delivery adjuvant [54].

22.2.2 Carrageenan

Carrageenans are a prominent component of red seaweed cell walls, accounting for 30%—75% of the algal dry weight [21,56]. Carrageenan is derived primarily from specific genera of red seaweeds, including Chondrus, Gigartina, Hypnea, and Eucheuma, and is the most extensively researched red algal polysaccharide [57]. According to their structural properties, sulfation patterns, and the presence or absence of 3,6-anhydro bridges in -linked galactose residues, they are categorized into distinct categories denoted by Greek letters (iota, kappa, lambda) (see Fig. 22.2). Galactose units form lambda-carrageenans, whereas iota- and kappa-carrageenans contain equal quantities of galactose and 3,6 anhydrogalactose. At axial locations, kappa-carrageenan has one sulfate group, iota-carrageenan has two, and lambda-carrageenan has close to three equatorial sulfates. Natural carrageenans are typically found as mixes of several hybrid types, such as k-/i-hybrids, k-/μ-hybrids, or μ-/i-hybrids, which frequently generate cyclized derivatives. Kappa- and iota-carrageenans contain varying levels of their biological progenitors, mu- and nu-carrageenans, respectively [58]. k-, i-, and λ-carrageenans are the three most commercially utilized carrageenans [21].

Carrageenans are particularly efficient against enveloped viruses. Members of the Herpesviridae family, which includes herpes simplex virus types 1 and 2 (HSV-1 and 2), CMV, and varicella zoster virus (VZV), were the most researched viruses that caused diseases to humans, Dengue virus type 2 (DENV-2), HIV, Sindbis virus (SINV), Influenza virus, human metapneumovirus (HPNV), rabies virus (RABV), RVFV, Semliki forest virus, and hantaviruses were among the enveloped viruses impacted by carrageenans [21,59]. Some researchers have observed action against nonenveloped viruses such as human

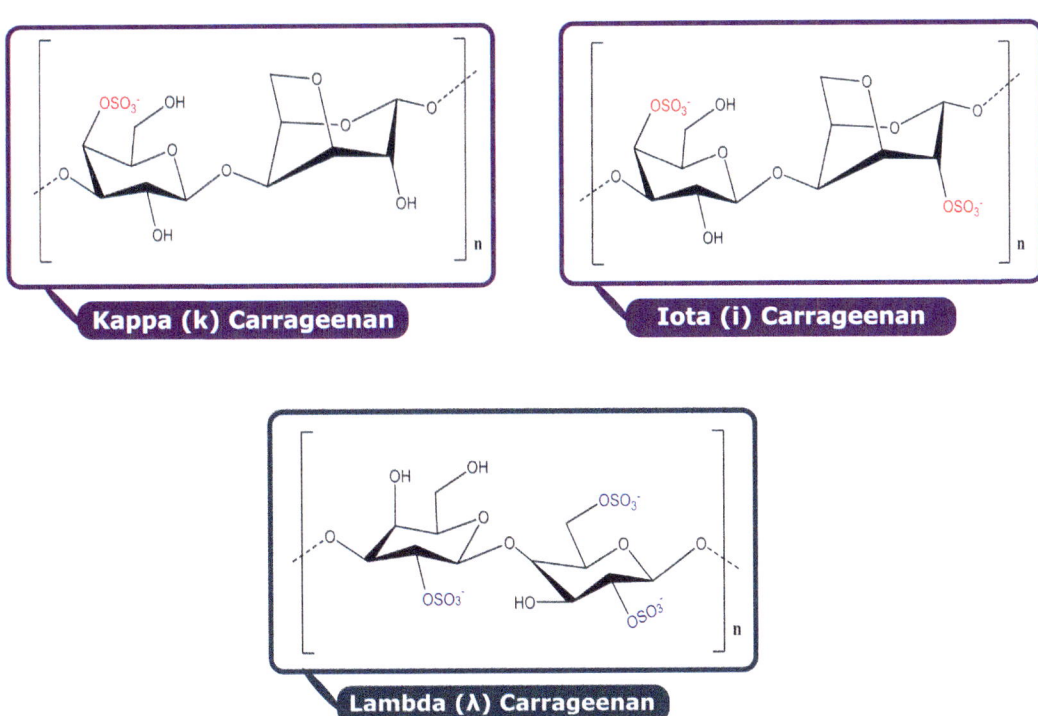

FIGURE 22.2   Chemical structure of three most common carrageenan molecules [17,21].

rhinovirus (HRV), enterovirus 71 (EV 71), and papillomavirus type 16. (HPV-16) [21]. The study of Talarico et al. revealed that the λ- and i-carrageenans, sulfated PSs containing linear chains of galactopyranosyl residues, were shown to be powerful inhibitors of DENV-2 and 3 (DENV-3) replication in Vero and HepG2 cells, with effective concentrations ranging from 0.14 to 4.1 g/mL. Plaque reduction, viral yield inhibition, and antigen expression studies revealed that this effect was independent of infection multiplicity in the range 0.001−1. The inhibitory activity of -carrageenan, a heparan sulfate (HS)-imitative substance, was exhibited through a combined interference with viral adsorption and nucleocapsid internalization into the cytoplasm [60]. Another study found that k-carrageenan has substantial anti-EV 71 action, including the ability to minimize plaque formation, block viral reproduction before or during viral adsorption, and inhibit EV 71-induced apoptosis. In a viral binding test, k-carrageenan was found to strongly bind EV 71, creating carrageenan-virus complexes and presumably disrupting the virus-receptor interaction. When all of these factors are considered, kappa-carrageenan may be an excellent option for development of anti-EV 71 drugs [61]. According to Girond et al., sulfated PSs such as iota-(ι), lambda(λ)-, and kappa(k)-carrageenans inhibited the replication of the hepatitis A virus (HAV) in the human hepatoma cell line PLC/PRF/5. These carrageenans reduced HAV-antigen generation and infectivity in a concentration-dependent manner, and no cytotoxic effects were seen at doses of up to 200 μg/mL. According to their findings, lota- and lambda-carrageenan are interesting options for acute hepatitis A chemotherapy [26]. In another study, carrageenan, at 5 μg/mL, protected the cell monolayer from HSV-1 development. Carrageenan, at 10 μg/mL, decreased the generation of new infectious HSV-1 by over five logs. When 10 μg/mL of carrageenan were introduced at the start of HSV-1 infection of HeLa cells, viral protein synthesis was significantly inhibited, while the cells continued to synthesize cellular proteins. These findings showed that this sulfated polysaccharide suppresses a stage in virus multiplication without any toxicity that occurs after viral internalization but before late viral protein synthesis begins [25,62]. Abu-Galiyun et al. used a plaque test to investigate the inhibitory effects of natural PSs derived from renewable sources that differed in structure and charge on VZV infection. The study findings, which demonstrated a substantial suppression of VZV infection whether cells were treated with iota-carrageenan (ι-carrageenan) only during infection or just after infection, might point to a multistep inhibitory action. It appears that may obstruct several phases of the viral reproduction cycle, including early steps such as absorption and penetration into host cells, as well as late processes following penetration into host cells [63]. Qiang et al. found that -carrageenan can considerably reduce SW731 replication by interfering with a few replication processes in the SW731 life cycle, including as adsorption, transcription, and viral protein production, particularly interactions between HA and cells. As a result, carrageenan may be a good alternative route to therapy for anti-IAV, as it has a HA similar to that of SW731 [64]. Yamada and colleagues discovered that iota-carrageenan and kappa-carrageenan had varied anti-HIV potencies than dextran sulfate and all other preparations of carrageenan even when their sulfate concentration was equal, suggesting that structural factors were also significantly influential on anti-HIV activity. The number of sulfate groups, as well as the molar mass of the biopolymer, play a significant role in the anti-HIV activity of sulfated carrageenans, indicating that the actions are not solely dependent on sulfate content; the positions and densities of the sulfate groups on the sugar chains can also be significant considerations [21,65]. Enomoto and cocreators proposed a condom with prophylactic activity against AIDS contamination covered with acidic sulfated PSs, among them carrageenan [21,66]. Grassauer and cocreators utilized i-, κ-, or λ-carrageenan or their blends for the production of a drug organization for the prophylaxis or treatment of rhinovirus contamination, i-carrageenan being the most effective to lessen the danger of disease. Additionally, they clearly mentioned that the antiviral efficacy of i- and/or k-carrageenan has been used in antiviral compositions for the prevention and cure of pulmonary viral disease (orthomyxovirus, paramyxovirus, adenovirus, and coronavirus). Carrageenans have been reported by Grassauer and coauthors to be useful for decongesting a stuffy nose as well as for preventative or therapeutic intervention in viral infections of the upper respiratory tract [21,31,67].

Yejin Jang et al. discovered through their investigation that the λ-carrageenan could be a promising antiviral agent for preventing infection with several respiratory viruses. Cell culture experiments found that the λ-carrageenan suppressed both influenza A and B viruses with EC50 values ranging from 0.3 to 1.4 μg/mL, as well as the presently spreading SARS-CoV-2 with an EC50 value of 0.9 1.1 μg/mL. At doses up to 300 μg/mL, no harm to host cells was found. Plaque titration and western blot analysis confirmed that λ-carrageenan inhibited viral protein expression in cell lysates and progeny virus formation in culture supernatants in a dose-dependent manner [68]. Bansal et al. mentioned that the nasal cavity and rhinopharynx are the locations of early replication of SARS-CoV-2. As a result, a nasal spray may be an appropriate dosage form for this purpose. To achieve this they evaluated the antiviral activity of three prospective nasal spray preparations against SARS-CoV-2 and discovered that iota-carrageenan at concentrations as low as 6 g/mL suppresses SARS-CoV-2 infection in Vero cell cultures. The doses demonstrated to be efficacious in vitro against SARS-CoV-2 may be easily reached by using nasal sprays that are currently on the market in various countries [69]. Also, the in vitro experimental findings of Morokutti-Kurz et al. on SARS-CoV-2 spike pseudotyped lentivirus and replication-competent SARS-CoV-2 showed that iota-carrageenan administration may be an effective and safe preventive or therapy for SARS-CoV-2 infections [24]. There was no evidence of toxicity for any of the carrageenan types tested. Carrageenan therapy did not have a negative impact on metabolic activity in many situations, and the lack of morphological changes in distinct cells at low doses was verified [20,64,70]. In vivo tests also revealed no signs of immunogenicity or immunotoxicity [71,72].

22.2.3 Alginates

Alginate is a biopolymer that is biodegradable, renewable, biocompatible, and water-soluble, having efficient antiviral and other biomedical uses. The term "alginate" is commonly used to refer to alginic acid salts, although it can also apply to all alginic acid derivatives and alginic acid itself. Alginates are natural anionic polymers derived from brown seaweed of the family *Phaeophyceae*, primarily the species *Laminaria hyperborea*, *Laminaria digitata*, *Macrocystis pyrifera*, and *Ascophyllum nodosum* [73]. *A. nodosum* has an alginate concentration of 22%−30% of its dry weight, but *L. hyperborea* does have a concentration ranging from 17%−33% to 25%−30% based on the portion of the algae from which the alginate is isolated [74]. Alginates are linear PSs comprised of (1-4)-d-mannuronic acid (M) blocks and C-5 epimer-l-glucuronic acid (G) blocks that may be arranged in a variety of ways that impact the physical characteristics of the alginate [75] (see Fig. 22.3). Aside from the mannuronate-to-guluronate (M/G) ratio, other parameters of alginate, such as molecular weight and degree of acetylation, influence its rheological qualities [76]. Diverse sources create alginate with varying G and M concentrations and block lengths, resulting in a plethora of different structures with varying characteristics. The species *L. digitata*, for example, has an M-block concentration of 49%, while other available alginates range from 15% to 43% [74].

The use of alginate salts in viral therapy is unavoidable nowadays and make it as a natural antiviral biomaterial choice for the researchers. Alginic acid was evaluated against the RABV in chicken embryo-related cells, with the antiviral action of alginate affecting the early stage [79]. The inhibitory impact of alginate on the RABV was shown to be dose-dependent at doses ranging from 1 to 100 g/mL. Alginic acid has also been demonstrated to have antiviral action against enveloped Baltimore group IV rubella virus (RV) infection in Vero cells [80]. An anti-HSV-1 action was discovered in a guluronic acid-rich sodium alginate (SA) produced from *Sargassum tenerrimum* [81]. Its antiviral effectiveness increased as the sulfate ester level rose. Another research, however, found that mannuronic-acid-rich alginate (M/G ratio = 1.88) isolated from *Sargassum trichophyllum* brown algae had no

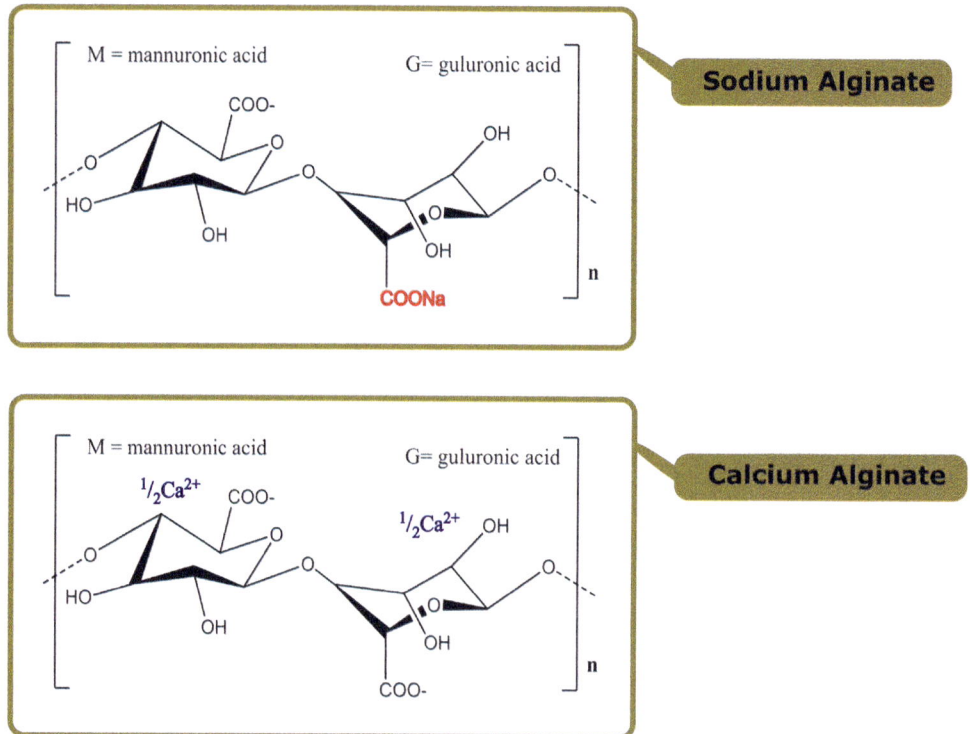

FIGURE 22.3   Basic chemical structure of sodium alginate [77] and calcium alginate [77,78].

impact on HSV-2 [82]. SA and its sulfated derivatives inhibited HSV-1 replication by interfering with virions and inhibiting viral adsorption/attachment to cells [83]. The antiviral potential of alginate hydrogels against HSV-1 when utilized as a sulfated substance was dependent on the sulfate concentrations of the PSs and the chemical characteristics of the sulfated alginate [84]. The in vitro properties demonstrated in this work showed that sulfated alginate may be a promising option for further antiviral investigation. SA, on the other hand, was less effective against HSV-1 than other sulfated PSs, as evidenced by its half-maximal inhibitory compound concentration (IC50), which ranged from 10 to 15 mg/mL, which is 10-fold higher than that of fucoidans and sulfated PSs [84,85]. Once calcium alginate microspheres were introduced to the supernatant of a human liver cell line (HuH-7 cells) as an encapsulating strategy, they displayed antiviral efficacy against numerous viruses, including SINV, poliovirus type 1 (PV-1), and HSV-1 [86]. In another study, Calcium alginate-based hydrogels have also shown antiviral activity against the influenza virus [87]. The antiviral studies with guluronic acid-rich SA (26 5 kDa) recommended that the antiviral actions of these biological molecules were exerted directly by interrupting with anti-HSV virion envelope structures or masking viral structures, which are actually needed for cell adsorption, thus trying to block entry of pathogens, as had previously been observed for diverse compounds, although more clarification is needed [81]. The antiviral activity of SA isolated from Sphacelaria indica, constituted of 41% G and 59% M blocks, and two sulfated versions (BS1 and BS2) against HSV-1 was examined, and the antiviral mechanism was explored in cells pretreated with these substances before infection [84,85]. According to Xin et al., a marine polysaccharide drug 911 produced from alginate may strongly suppress HIV-1 acute infection of MT-4 cells and chronic infection of H9 cells [88]. And this has been shown to strongly reduce HIV reproduction in vitro and in vivo, with its inhibitory impact attributable to the inhibition of viral reverse transcriptase (RT), interference with viral adsorption, and augmentation of immune function [88,89]. Geng et al. discovered that sulfated polymannuroguluronate (SPMG) inhibited HIV adsorption primarily through interfering with the interaction of the virus gp120 protein with the CD4 molecule on the surface of T cells [30,90,91]. In addition, their research found that the octasaccharide unit is the most potent SPMG fragment, preventing syncytium formation and reducing P24 antigen levels in HIV-IIIB-infected CEM cells [30,92]. Furthermore, 911 can boost host cell immune function and limit the activity of HBV DNA polymerase, therefore 911 can also be utilized to reduce HBV replication [93]. S. Wang et al. suggested that since brown algal-derived bioactive chemicals, particularly alginate PSs and fucans, show strong antiviral or antitumor properties, sulfated PSs from brown algae have the potential to be novel resources for the development of anti-HPV and associated cancer treatments [94]. Son et al. discovered that high mannuronic acid containing alginate can stimulate the creation of a variety of cytokine like IFN-α/IFN-β, many of which play a role in macrophagemediated cytostasis. However, only IFN-β is implicated in macrophage intrinsic anti-HSV-1 activity, despite the fact that several of these factors may have antiviral action in other cells [95]. The another brief review of Serrano-Aroca et al. demonstrated that alginate-based materials had antiviral efficacy against a diverse set of 17 viruses, including double-stranded DNA viruses, positive-sense or negative-sense single-stranded RNA viruses, and single-stranded RNA viruses with a DNA intermediary in their life cycle. HIV type 1, hepatitis A, B, and C viruses, SINV, HSV-1 and 2, PV-1, RABV, RV, and influenza virus may all infect various creatures. The author also highlighted that many of these viruses are enveloped viruses from the same Baltimore group IV as SARS-CoV-2, and so the author suggested to the research community that this alginate-based biomaterial holds a lot of potential for treating the quickly growing COVID-19 sickness [85]. As a result, the majority of the investigations included in this review demonstrated extremely low or no cytotoxicity of alginate-based biomaterials.

22.2.4 Fucans, fucoidans

Fucans are a kind of sulfated polysaccharide with a high molecular weight that can be divided into three primary groups: fucoidans, xylofucoglycuronans, and glycuronogalacto-fucans. Fucoidan is a fucose-enriched and sulfated polysaccharide derived mostly from the extracellular matrix of brown algae and it constitutes 25%−30% of the dry algal weigh [30,96]. Fucoidan is abundant in the leaves of *L. digitata, A. nodosum, M. pyrifera*, and *Fucus vesiculosus*, additionally Many algae and invertebrates have been studied for fucoidan content, including *Sargassum stenophyllum, Chorda filum, Dictyota menstrualis, Fucus evanescens, Fucus serratus, Fucus distichus, Caulerpa racemosa, Hizikia fusiforme, Padina gymnospora, Kjellmaniella crassifolia, Analipus japonicus*, and *L. hyperborean* [97]. Fucoidan is negatively charged and very hygroscopic polysaccharide and it may be dissolved in both water and acid solutions [97,98]. In a species of brown algae, fucoidan is composed of l-fucose, sulfate groups, and one or more tiny quantities of xylose, mannose, galactose, rhamnose, arabinose, glucose, glucuronic acid, and acetyl groups [97]. The structure of fucoidan varies depending on the type of seaweed used. Nonetheless, fucoidan often contains two kinds of homofucose. The first kind (I) includes repeated $(1 \rightarrow 3)$-l-fucopyranose, whereas the second form (II) has alternating and repeated $(1 \rightarrow 3)$- and $(1 \rightarrow 4)$-l-fucopyranose [99] (see Fig. 22.4). The rationale for the rise in research is that fucoidan possesses antiviral, antitumor, anticoagulant, and antioxidant properties, as well as being important in regulating glucose and cholesterol metabolism [97,100−102]. Wang et colleagues discovered that a fucoidan KW produced from the brown algae *K. crassifolia* successfully prevented IAV infection in vitro while being nontoxic. Fucoidan KW has the ability to inactivate virus particles before to infection and to prevent some steps after adsorption. Additionally, fucoidan KW also suppressed EGFR, PKC, NF-B, and Akt activation, as well as IAV endocytosis and EGFR internalization in IAV-infected cells, indicating that fucoidan KW may also block the cellular EGFR pathway. Furthermore, intranasal injection of fucoidan KW significantly increased lifespan and reduced viral titers in IAV-infected mice. And so, the author suggested that, fucoidan KW has the potential to be developed into a new nasal drop or spray for influenza prevention and treatment in the future [100]. Trejo-Avila and colleagues investigated the potential toxicity and antiviral efficacy of fucoidan from the seaweed *Cladosiphon okamuranus* against Newcastle disease virus (NDV), one of the world's most significant dangers to the chicken industry. Semi-quantification of viral RNA in pooled liver and small intestine of embryos treated with 4and 16 g fucoidan 1 hour before infection revealed decreases in virus replication of 60% and 99.8%, respectively. Because this strong anti-NDV activity in ovo was produced with very low dosages, the author recommended that the fucoidan from *C. okamuranus* might be a promising low-toxicity antiviral agent for usage in NDV-affected areas [101]. Similarly,

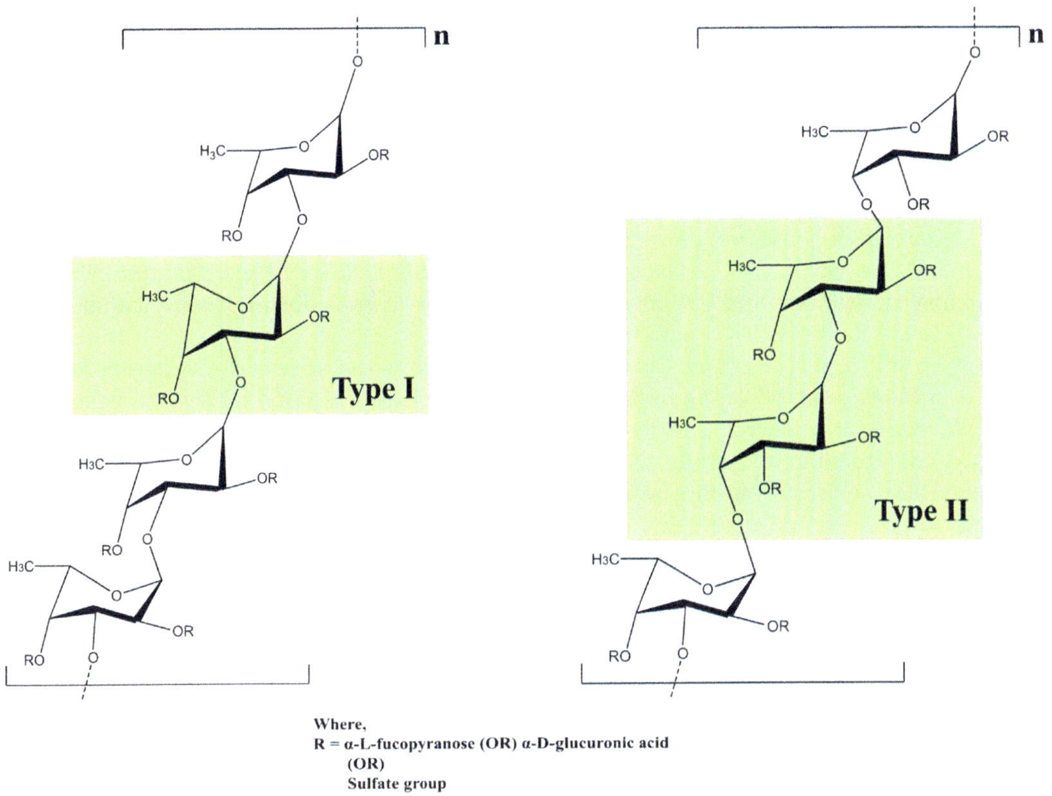

Type I

Type II

Where,
R = α-L-fucopyranose (OR) α-D-glucuronic acid
(OR)
Sulfate group

FIGURE 22.4 Two different structural backbone of fucoidans [97,108].

Elizondo-Gonzalez et al. proposed that fucoidan from *C. okamuranus* is a possible low-toxicity antiviral agent for the chicken industry against Newcastle virus (NDV), and the authors' work also provides a better knowledge of the method of action of sulfated PSs [102]. Another group of researchers tested two fucoidan fractions with low molecular weight and varying sulfate content from *Laminaria japonica* (LMW fucoidans) for antiviral activity in vitro and in vivo, as well as the influence on the immune system. In vitro, Hep-2, Hela, and MDCK cells were infected with type-I influenza virus, adenovirus, and Parainfluenza virus I, respectively. The author successfully demonstrated that, when compared to the control group, two types of LMW fucoidans had remarkable antiviral activity in vitro at middle and high doses, whereas at low doses, the antiviral activity of the two types of LMW fucoidans was not statistically different from that of the blank control group. And also the author concluded that both LMW fucoidans (LF1, LF2) could play an antiviral role by improving immune organ quality, immune cell phagocytosis, and humoral immunity [103]. Furthermore, Akamatsu et al. showed that "MC26," a novel form of fucose polysaccharide derived from brown seaweed, had outstanding antiinfluenza A virus activity in vitro and in vivo [104]. Another study demonstrated that, the sulfated fucans from the seaweeds *Dictyota mertensii, Lobophora variegata, Spatoglossum schroederi,* and *F. vesiculosus* have been shown to suppress HIV RT activity, which is essential for the virus RNA synthesis [105]. Hideri KI et al. determined in their study that a fucan polysaccharide from Cladosiphon okamuranus comprised of glucuronic acid and sulfated fucose units can actively prevent DENV-2 infection on BHK-21 cells but has little effect on the virus's other three serotypes (DENV-1, DENV-3 and DENV-4) [106]. In another study, methylthiazolyltetrazolium bromide and cytopathic effect (CPE) reduction assays were used to assess the cytotoxicity and antiviral activities of native fucoidan (FeF) and modified with enzyme (FeHMP) fucoidans from *F. evanescens* against herpes viruses (HSV-1, HSV-2) and enterovirus (ECHO-1) in Vero and human MT-4 cell lines. The scientists found that FeF and FeHMP exhibit equal effectiveness against many DNA and RNA viruses, allowing us to consider the examined fucoidans as prospective broad-spectrum antivirals [107].

Kwon et colleagues explored fucoidan, derived from the seaweed *Saccharina japonica*, heparan sulfates, and other highly sulfated PSs, and the findings of their investigation indicated excellent antiviral activity in vitro and minimal cytotoxicity. Thus, the author proposed that fucoidans, nebulized heparin, or potentially TriS-heparin, in conjunction with or without existing antiviral medications, be tested first in human primary epithelial cells and subsequently in human COVID-19 patients [109]. Sulfated fucoidan and crude PSs isolated from six different seaweed species were tested for their ability to block SARS-CoV-2 viral entrance. The above crude PSs have properties like high molecular weight (>800 kDa), high total carbohydrate content (62.799.1%), high fucose content (37.366.2%), and are extensively branched. The study authors suggest with their results that crude PSs derived from seaweeds can effectively block SARS-CoV-2 entrance [110]. According to many researchers, a recent in vitro studies found that a sulfated polysaccharide, a kind of fucoidan, had significant antiviral action against SARS-CoV-2 [111,112]. Fucoidan polysaccharide testing in vivo and in vitro found no substantial or extremely minimal immunotoxicity [104−107,109,110].

22.3 Other marine polysaccharides as antiviral biomaterials

Gerber and colleagues' [113] work that demonstrated suppression of mumps and influenza B virus by PSs produced from marine algae established algae-derived PSs as a viable source of antiviral medicines [113]. Over the next two decades, antiviral activity of various polysaccharide fractions derived from red algae were found against HSV and other viruses. Numerous research have now been published on the antiviral potential of different algae-derived PSs as well as their underlying mechanism of action [114−117]. Laminaran (or laminarin) was found initially in the laminaria species and seems being the food reservoir of all brown algae. Laminaran is a water-soluble polysaccharide with 20−25 glucose units made up of (1,3)β-d-glucan with β-(1,6) branching [30]. Laminaran, derived from brown algae, has a strong inhibitory impact on viral replication. At a concentration of 50 g/mL, laminaran PSs extracted from kelp were shown to effectively block HIV adsorption on lymphocytes and the activity of HIV RT, indicating that laminaran PSs had a good inhibitory impact on HIV propagation [30,118]. The marine heparinoid PSs are structurally related to heparin and have GAG-like biological characteristics, containing ulvans and their sulfated derivatives, as well as dextran sulfate and chitosan sulfate [30].

Recent research has revealed that heparinoid sulfate proteoglycans on the cell surface are the first receptors of human CMV, and bovine herpes virus in their infection processes [119–121]. Heparinoid PSs can engage with the positive charge areas of cell surface glycoproteins, resulting in a quenching effect on these surface areas and blocking viral attachment to the cell surface [30,119]. According to Lüscher-Mattli et al., dextran sulfate can block influenza virus fusion with cytoplasmic membrane and reduce influenza virus reproduction in vivo [122].

Witvrouw et al., discovered that dextran sulfate mostly slows the reproduction of enveloped viruses but has no effect on nonenveloped viruses. In addition, low molecular weight dextran sulfate (DS 1000) is practically ineffective against HSV, influenza virus, and certain other viruses, and its inhibitory action against HIV varies greatly depending on the viral strain and cell type [123]. Furthermore, Ivanova et al. discovered that ulvan PSs derived from green algae had a significant inhibitory impact on influenza A virus, which is dose-dependent and strain-specific [124]. Sulfated galactans are the primary extracellular PSs of red algae. With a few exceptions, they are composed of linear galactose chains; a chain of alternating 3−D-galactopyranose (G units) and 4−D-galactopyranose residues or 4−3,6-anhydrogalactopyranose residues completes their structural backbone, with the presence of D-series (D unit) in carrageenans and L-series (L unit) in agarans [125,126]. The different structural forms of these galactan PSs have demonstrated potent antiviral activity against a variety of enveloped viruses, including HSV-1 and HSV-2, DENV, HIV-1 and HIV-2, and the HAV [70,108,127]. Naviculan is a sulfated polysaccharide derived from the diatom *Navicula directa*. This polymer has a large molecular weight and is composed of multiple sugars and sulfate. According to Lee and colleagues, naviculan shows significant antiviral action against HSV-1 and HSV-2, as well as influenza virus, by suppressing the early phases of viral replication and perhaps limiting viral internalization into host cells proposed naviculan as a new antiviral sulfated polysaccharide with antienveloped virus activity. Furthermore, naviculan inhibited fusion among cells that express the CD4 receptor and the HIV gp160-producing HeLa cell line [82,108]. *Cochlodinium polykrikoides*, a marine microalga, has extracellular sulfated PSs A1 and A2. These PSs are made up of sugars and uronic acid, with sulfate groups distributed throughout. A1 and A2 suppressed the cytopathogenic effects of HIV-1 in MT-4 cells, influenza virus types A and B in MDCK cells, and respiratory syncytial virus types A and B in Hep-2 cells. A1 PSs were shown to be effective against HSV-1, whereas A2 PSs were found to be beneficial against parainfluenza virus type 2. At 100 g/mL, both A1 and A2 PSs are noncytotoxic and have a modest (10%) inhibitory impact on blood coagulation at precise amounts that produce viral inhibition [108,128].

Nostoflan (NSF) is an acidic polysaccharide from *Nostoc flagelliforme*, a nutritious edible blue-green alga. NSF's structural analysis indicated that it is primarily composed by sequences of (→4)-β-D-Glcp-(1→4)-D-Xylp-(1→4)-[β-D-GlcAp-(1→6)-]-β-D-Glcp-(1→4)-D-Galp-(1→). NSF, discovered by Kanekiyo and colleagues, has a powerful inhibitory impact on various enveloped viruses, including HSV-1 and HSV-2, human CMV, and influenza A virus. Further research verified that the antiherpetic activity of NSF was caused by the suppression of viral binding to host cells, and their findings indicated NSF as a promising antiherpes option [108,129]. The basic chemical structure of these PSs are shown in Fig. 22.5.

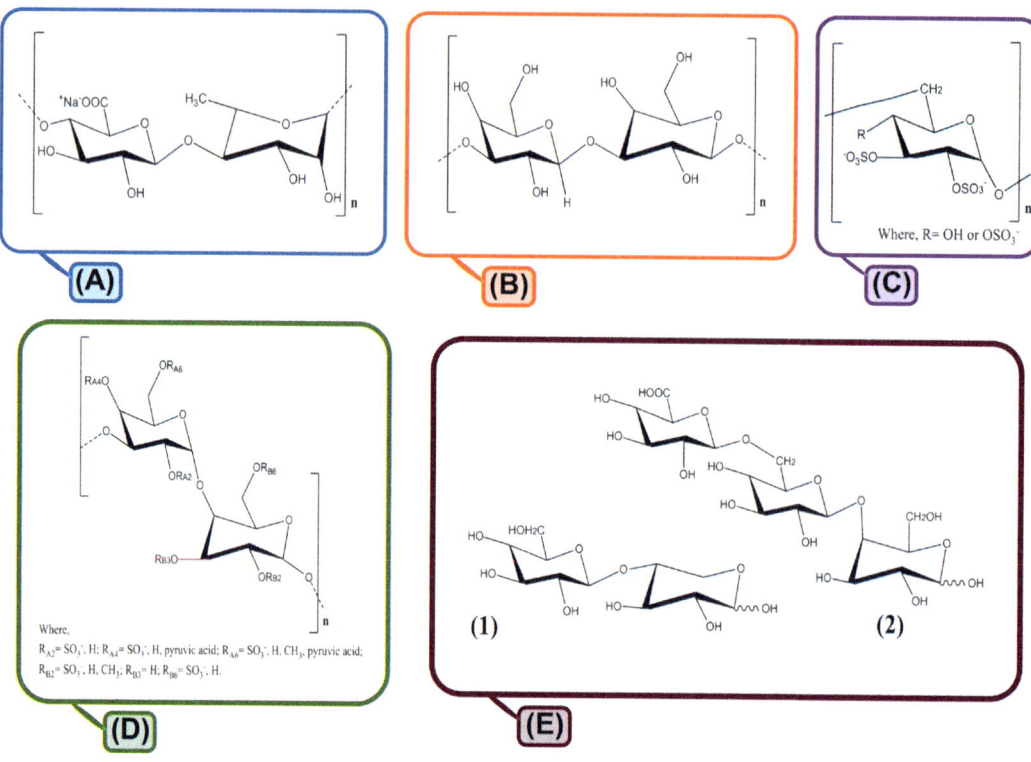

FIGURE 22.5 Basic chemical forms of (A) Ulvanobiouronic acid (Ulvan) [130], (B) Laminaran [108], (C) Dextran sulfate [131], (D) Galacton [108], and (E) two different chemical forms of Nostoflan backbone [132].

As mentioned earlier in this chapter, a variety of marine active PSs had novel antiviral therapeutic functions (Table 22.1). Since the marine-derived PSs and their derivatives have strong antiviral properties, they have the potential to be developed into a new class of antiviral drugs with high efficacy and low toxicity.

22.4 Marine peptides as antiviral biomaterials

Many molecules have been described so far to combat viral infections using new techniques, and most recently, the description of antimicrobial peptides has gained attention. Recent evidence emphasizes the role of antiviral proteinaceous molecules as a protective shield, and it is being displayed that some antimicrobial peptides may also exhibit activity against a wide range of viral infections, thus being referred to as AVPs [133]. Nowadays, natural AVPs are being studied more than synthetic AVPs, and scientists are venturing into the marine environment in search of natural AVPs. Because of their broad spectrum of bioactivities, these marine peptides have high potential nutraceutical and medicinal values [18]. Neurotoxin, cardiotonic peptide, antiviral and antitumor peptide, cardiotoxin,

TABLE 22.1 Short details on Marine Antimicrobial peptides as Antiviral Biomaterials.

S. no.	Marine polysaccharide	Source	Antiviral activity	References
1.	Chitin Microparticle	Chitin derivative (chitin extracted from shells of crustaceans and some fungal sp.)	H1N1, H3N2, HN1 FLU	[37,42]
2.	chitosan	Alkaline treatment and deacetylation of chitin	Calicivirus, Human cytomegalovirus strain AD169, H1N1 influenza A virus, H7N9 influenza virus, Rift Valley Fever virus (RVFV), Herpes Simplex-1 (HSV-1), and Coxsackie viruses	[30,44,47−50,53]
3.	N-carboxymethylchitosanN, O, Sulfate (NCMCS)	Chitosan derivative	HIV-1	[40]
4.	aminoethyl-chitosan	Chitosan derivative	HIV-1	[30,35]
5.	3,6-O-sulfated chitosan	Chitosan derivative	HPV	[43]
6.	Sulfated chitooligosaccharides (SCOS),	Chitin/Chitosan derivative	HIV-1	[45−47]
7.	i-carrageenans	*Meristiella gelidium*	DENV-2, DENV-3, HAV, VZV, HIV, orthomyxovirus, paramyxovirus, adenovirus, Rhinovirus, and coronavirus, SARS-CoV-2	[21,26,31,60,63,65,67,69]
8.	k-carrageenan	*Meristiella gelidium*	EV 71, HAV, HIV, orthomyxovirus, paramyxovirus, adenovirus, and coronavirus.	[21,26,31,61,65,67]
9.	λ-carrageenans	Red seaweed	DENV-2, DENV-3, influenza A and B viruses	[21,68]
10.	Other carrageenans derivatives	*Meristiella gelidium* and other red seaweed	HSV-1 and 2, CMV, and VZV, DENV-2, HIV, Sindbis virus, Influenza virus, HPNV, RABV, RVFV, Semliki forest virus, and hantaviruses and HRV, EV 71, and HPV-16	[21,25,59,62]
11.	Alginates	*Sargassum tenerrimum*, *Sphacelaria indica* and other brown seaweed	rabies virus, Baltimore group IV rubella virus, HIV-1, hepatitis A, B, and C viruses, Sindbis virus, HSV-1 AND HSV-2, poliovirus type 1, and influenza virus	[79,80,85]

(Continued)

TABLE 22.1 (Continued)

S. no.	Marine polysaccharide	Source	Antiviral activity	References
12.	Guluronic acid-rich sodium alginate (SA)	*Sargassum tenerrimum, Sphacelaria indica, Laminaria angustata*	HSV-1	[81,83–85]
13.	calcium alginate	Alginate derivative	Sindbis virus (SINV), Poliovirus type 1 (PV-1), HSV-1, Influenza virus	[86,87]
14.	sulfated polymannuroguluronate (SPMG)	Brown seaweed	HIV, HBV	[30,90,91,93]
15.	Fucose polysaccharide	Brown seaweed	Influenza A virus	[104]
16.	sulfated fucans	*Dictyota mertensii, Lobophora variegata, Spatoglossum schroederi,* and *Fucus vesiculosus*	HIV	[105]
17.	Fucoidans	*Kjellmaniella crassifolia, Cladosiphon okamuranus Undaria pinnatifida, Laminaria japonica, Hizikia fusiforme, Sargassum horneri, Codium fragile, Porphyra tenera, Haliotis discus hannai* & other brown seaweed	IAV, Newcastle disease virus (NDV), DENV-2, SARS-CoV-2	[100,101,106,111,112]
18.	Enzyme modified fucoidans (FeHMP)	*Fucus evanescens*	HSV-1, HSV-2 and enterovirus (ECHO-1)	[107]
19.	Crude extract of Algal polysaccharides	*Cryptosyphonia woodii, Farlowia mollis, Constantinea simplex* marine algae, Red algae	Mumps and Influenza B virus, HSV-1 & 2, Coxsackie Bs virus, Vesicular stomatitis virus (VSV), Vaccinia virus	[113–115,117]
20.	Laminaran	Kelp (large brown algae)	HIV	[30,118]
21.	Ulvans (ulvabiouronic acid)	*Ulva lactuca* (green algae)	Influenza A virus,	[124,151]
22.	Galactan polysaccharides	Red algae	HSV-1 and HSV-2, DENV, HIV-1 and HIV-2, and the hepatitis A virus	[70,108,127]
23.	Naviculan	*Navicula directa*	HSV-1, HSV-2, Influenza virus	[82,108]
24.	Extracellular sulfated polysaccharides A1 and A2	*Cochlodinium polykrikoides*	HIV-1, Influenza virus type A and B, HSV-1, Parainfluenza virus type 2	[108,128]
25.	Nostoflan	*Nostoc flagelliforme*	HSV-1 and HSV-2, human cytomegalovirus, and influenza A virus	[108,129]

and antimicrobial peptide were the first bioactive peptides discovered and isolated from marine species. Since then, research on marine bioactive peptides has continued with the goal of determining their applications [18,134]. Marine AVPs are typically found in sponges, ascidians, mollusks, sea anemones, and seaweeds, all of which are well-known for their unique bioactive metabolites. These natural marine-derived peptides are typically cyclic or linear peptides containing atypical amino acids. Peptides are frequently formed by unusual condensation between amino acids, which results in structures that are unique and rarely found in other terrestrial animals and microbes [135]. They combat virus infection in a variety of ways, including virus entry blocking, CPE inhibition, viral neutralization, fusion, and entry [18,135,136]. The molecular diversity of some marine peptides which can act as a potential antiviral biomaterials are highlighted in this chapter (Table 22.2).

Ma et al. displayed AVPs from marine gorgonian-derived fungus *Aspergillus* sp. SCSIO 41501, a new cyclic pentapeptide and three new linear peptides, aspergillipeptides D–G, were identified. Compounds 1 and 2 exhibited antiviral activity against HSV-1 with IC50 values of 9.5 and 19.8 mM against a Vero cell line, respectively, and 1 had also antiviral activity against acyclovir-resistant clinical strains of HSV-1 at noncytotoxic levels [137]. Another study used naturally occurring cyclicpeptide derivatives which are alkaloids associated indole diketopiperazine, including 15 novel rubrumlines A-O produced by the marine fungus Eurotium rubrum, to test the influenza A/WSN/33 virus. One among them, Neoechinulin B inhibited the H1N1 virus in MDCK cells and has been shown to be effective against a variety of influenza virus strains [138]. Similarly, another alkaloids associated thiodiketopiperazine-type peptides, eutypellazines A−M

TABLE 22.2 Short details on Marine Antimicrobial peptides as Antiviral Biomaterials.

S. no	Marine peptides	Source	Antiviral effects	References
1.	Aspergillipeptides D–G,	*Aspergillus* sp. SCSIO 41501	HSV-1	[137]
2.	Neoechinulin B	*Eurotium rubrum*	H1NI & other Influenza virus	[138]
3.	Eutypellazine E	*Eutypella* sp.	HIV-1	[139]
4.	Halovirs A−E	*Scytalidium sp.*	HSV-1, HSV-2	[140]
5.	Asperterrestide A	*Aspergillus terreus* SCSGAF0162	H1N1, H3N2	[141]
6.	Sansalvamide A	*Fusarium* sp.	MCV (Poxvirus)	[142,152]
7.	Kelletinin A	*Buccinulum corneum*	HTLV-1	[144]
8.	Microspinosamide	*Sidonops microspinosa*	HIV-1	[145]
9.	Mirabamides A-D & Mirabamides E-H	*Siliquariaspongia mirabilis, Stelletta clavosa*	HIV-1	[146,147]
10.	Tachyplesin I	*Tachypleus tridentatus*	HIV, VSV	[148,149]
11.	Polyphemusin II	*Limulus polyphemus*	HIV	[150]

derived from *Eutypella* sp., a deep-sea fungus, demonstrated low cytotoxicity and efficient antiviral activity. Eutypellazine E had the greatest anti-HIV-1 inhibitory effect, with an IC50 of 3.2 + 0.4 μM, while the others showed decent anti-HIV activity [139]. Rowley et al. demonstrated a marine AVP, Halovirs A—E synthesized by a marine-derived fungus genus *Scytalidium* during saline fermentation. The authors reported that these lipophilic, linear peptides suppress the HSV-1 and 2 in vitro. The peptide, halovirs directly inactivate herpes viruses, a method of action that might be useful in preventing HSV transmission. The halovir hexapeptides have a nitrogen terminus that has been acylated by myristic (C14) or lauric (C12) acid, a unique Aib-Hyp dipeptide sequence, and a carboxyl terminus that has been reduced to a primary alcohol. A qualitative examination of these compounds' secondary structures utilizing variable temperature NMR studies and NOE investigations is also presented by the author [140]. In another study, a noval antiviral cyclic tetrapeptide Asperterrestide A, demonstrated for its antiviral efficiency. This AVP has an unusual 3-OH-N-CH3-Phe residue obtained from the marine-derived fungus *Aspergillus terreus* SCSGAF0162. The authors of this study reported that the peptide Asperterrestide A, suppressed influenza virus strains H1N1 and H3N2 and exhibited cytotoxicity in human cancer cell lines U937 and MOLT4 [141]. Hwang et al. discovered and described an inhibitor of the virus-encoded type 1 topoisomerase, an enzyme that is likely necessary for the poxvirus molluscum contagiosum virus (MCV) replication. In vitro MCV topoisomerase tests were used to examine a library of marine extracts and natural compounds derived from microorganisms and the cyclic depsipeptide, Sansalvamide A was discovered to prevent topoisomerase-catalyzed DNA relaxing. In a counter-screen, Sansalvamide A was inactive against two additional DNA-modifying enzymes. Yet, Sansalvamide A also inhibits DNA binding by the separated catalytic domain, identifying the region of the protein that is susceptible to sansalvamide A. And so, the authors of this study concluded that sansalvamide A inhibits MCV topoisomerase and also stated that Sansalvamide A cyclic depsipeptides constitute a potentially interesting chemical family for the development of anti-MCV drug [142]. Another study briefly reviewed antiviral molecules from marine mollusks and found that antiviral compounds such as antiviral proteins and AVPs are a critical component of molluscan defences against viruses and have diverse mechanisms of action against a wide variety of viruses, including many human pathogens such as herpes viruses HSV and Epstein-Barr virus [143]. Silvestri and coworkers displayed Kelletinin A (KA), a bioactive peptide derived from the sea gastropod *Buccinulum corneum*, exhibited antiviral and antimitotic action against the human T-cell leukemia virus type 1 (HTLV-1) infested MT2 cells. The author also elaborated that KA reduced cellular DNA and RNA synthesis while having no effect on protein synthesis, and also interfered with viral transcription by decreasing the levels of high molecular mass transcripts. Finally, in vitro, the biomaterial KA blocked HTLV-1 RT [144]. In another study, Microspinosamide, a new cyclic depsipeptide containing 13 amino acid residues, was isolated from extracts of the marine sponge *Sidonops microspinosa* from an Indonesian collection, and its structural bio framework was elucidated through chemical degradation, extensive NMR and mass spectral analyses. The Microspinosamide peptide has a number of unusual amino acids, including the first naturally occurring peptide with a beta-hydroxy-p-bromophenylalanine residue. Microspinosamide shows antiviral efficacy, according to the scientists, because it decreased the cytopathic impact of HIV-1 infection in an XTT-based in vitro experiment [145]. Mirabamides A-D, isolated by Plaza et al. from the marine sponge *Siliquariaspongia mirabilis*, are four new natural marine depsipeptides. The authors used 1D and 2D NMR and ESIMS to identify the planar structures of the peptides, and advanced Marfey's technique, NMR, and GC-Ms to determine the absolute configurations. Mirabamides contain two novel entities: 4-chlorohomoproline (in 1–3) and an unusual glycosylated amino acid, beta-methoxytyrosine 4'-O-alpha-L-rhamnopyranoside (in 1, 2, and 4) as well as a rare N-terminal aliphatic hydroxy acid. These components were found to be beneficial in anti-HIV structure-activity relationship studies. Mirabamide A, as well as mirabamides C and D, inhibited HIV-1 in neutralization and fusion assays [146]. Similarly, Lu et al. isolated four new depsipeptides, mirabamides E-H, as well as the known depsipeptide mirabamide C, from the sponge *Stelletta clavosa*, which was obtained in the Torres Strait, Pacific Ocean. The authors did peptide determination techniques same as Plaza et al. Lu and colleagues also effectively demonstrated in vitro that all four novel compounds inhibited HIV-1 in a neutralization experiment [147]. Morimoto et al. investigated the inhibitory effects of an antimicrobial peptide, tachyplesin I, isolated from hemocytes of the Japanese horseshoe crab (*Tachypleus tridentatus*), on HIV infection in vitro. Tahyplesin I, at a concentration of 7.5 micrograms/mL, inhibited the development of CPE in HIV-infected MT-4 cells by more than 70%. (lymphadenopathy-associated virus). This inhibitory effect was only observed when the drug was added during the virus's adsorption to the cells [148]. Similarly, Murakami et al. tested the antimicrobial peptides tachyplesin I and related isopeptides isolated from the hemocytes of the horseshoe crab (*T. tridentatus* and *Limulus polyphemus*) against several viruses. Incubation with tachyplesin I and its isopeptides inactivated the Vesicular stomatitis virus (VSV), and the test results revealed that tachyplesin I directly inactivates the VSV by destroying its envelope subunits [149]. Likewise, Masuda et al. discovered that polyphemusin II, a marine-based antimicrobial peptide isolated from the hemocytes of the same horseshoe crab species, had extremely high anti-HIV activity, indicating that polyphemusin II could be a potential candidate for HIV infection therapy [150].

With the inclusion of marine peptides in the current preclinical and clinical pipelines, their contribution to the future pharmacopoeia appears promising. More research into marine diversity in search of therapeutic biomaterials will be required to ensure the future success of novel AVPs and other biomolecules that can contribute significantly to the treatment or prevention of various diseases. Some AVPs were shown in this chapter, but active AVPs from marine bacteria, fungi, and algae have yet to be discovered. As a need of natural therapeutic biomolecule, more AVPs must be found in the marine environment to combat emerging viral infections.

22.5 Conclusion

In conclusion, marine PSs and antimicrobial peptides have numerous advantages, including low production costs, a broad array of antiviral properties, nontoxicity, and wide ranging acceptability, indicating that these biomolecules justify further investigation as promising antivirals that can be used alone or in combination with existing drugs and vaccine adjuvants. The majority of the time, these marine biomolecules' antiviral activity is exerted by preventing viral replication and suppressing virus adhesion to host cells. Throughout our review of the literature, we discovered that, despite numerous research on marine-derived biomolecules being studied for their antiviral potential, there are still many gaps. The oceans are teeming with microorganisms, macroorganisms, plants, and animals, all of which contain large amounts of therapeutic molecules, the majority of which have yet to be discovered. More in vitro, in vivo, and clinical research on these marine-based molecules will provide us with novel and promising antiviral molecules to combat potentially lethal viral infections. This chapter will serve as a reminder to researchers in order to investigate more on natural organic biomaterials, as well as to explore the marine diversity for novel therapeutic molecules.

Acknowledgment

The authors are grateful to RUSA—phase 2.0, Alagappa University, Karaikudi, Tamil nadu, India.

References

[1] Wiersinga WJ, Rhodes A, Cheng AC, Peacock SJ, Prescott HC. Pathophysiology, transmission, diagnosis, and treatment of Coronavirus Disease 2019 (COVID-19): a review. JAMA [Internet] 2020;324(8):782–93, Aug 25. Available from: https://jamanetwork.com/journals/jama/fullarticle/2768391.

[2] Huang X, Zhao B. A permanent virtual memorial for a whistleblower of the COVID-19 pandemic: a case study of crypto place on the blockchain. Springer; 2021. p. 47–59. Available from: https://link.springer.com/chapter/10.1007/978-3-030-72808-3_4 [accessed 16.11.21].

[3] Allam A. Simplified medical microbiology and immunology for physicians antibiotic resistance view project Immunology of Bronchial Asthma View project. 2015; <https://www.researchgate.net/publication/326723347> [accessed 16.11.21].

[4] Paintsil E, Cheng YC. Antiviral agents. Encycl Microbiol [Internet]. 2019 Jan 1;176. <https://pmc/articles/PMC7150273/> [accessed 16.11.21].

[5] Whitley RJ, Gnann JW. Viral encephalitis: familiar infections and emerging pathogens. Lancet. 2002;359 (9305):507–13.

[6] Chattha KS, Roth JA, Saif LJ. Strategies for design and application of enteric viral vaccines. 2015 Feb 17;3:375–95. <https://www.annualreviews.org/doi/abs/10.1146/annurev-animal-022114-111038> [accessed 16.11.21].

[7] Rappuoli R, Miller HI, Falkow S. Medicine: the intangible value of vaccination. Science (80−) [Internet] 2002;297(5583):937–9, Aug 9. Available from: https://www.science.org/doi/abs/10.1126/science.1075173 [accessed 16.11.21].

[8] Ruiz-Palacios GM, Pérez-Schael I, Velázquez FR, Abate H, Breuer T, Clemens SC, et al. Safety and efficacy of an attenuated vaccine against severe rotavirus gastroenteritis. N Engl J Med [Internet] 2006;354(1):11–22, Jan 5. Available from: https://www.nejm.org/doi/full/10.1056/nejmoa052434 [accessed 16.11.21].

[9] Imler JL. Adenovirus vectors as recombinant viral vaccines. Vaccine. 1995;13(13):1143–51.

[10] Krammer F, Weir JP, Engelhardt O, Katz JM, Cox RJ. Meeting report and review: Immunological assays and correlates of protection for next-generation influenza vaccines. Influenza Other Respi Viruses [Internet] 2020;14(2):237–43, Mar 1. Available from: https://onlinelibrary.wiley.com/doi/full/10.1111/irv.12706 [accessed 16.11.21].

[11] He P, Zou Y, Hu Z. Advances in aluminum hydroxide-based adjuvant research and its mechanism. Hum Vaccin Immunother [Internet] 2015;11(2):477. Available from: https://pmc/articles/PMC4514166/ [accessed 16.11.21].

[12] Herrington CS, Coates PJ, Duprex WP. Viruses and disease: emerging concepts for prevention, diagnosis and treatment. J Pathol [Internet] 2015;235(2):149−52, Jan 1. Available from: https://onlinelibrary.wiley.com/doi/full/10.1002/path.4476 [accessed 16.11.21].

[13] Huang X, Xu W, Li M, Zhang P, Zhang YS, Ding J, et al. Antiviral biomaterials. Matter [Internet] 2021;4(6):1892−918, Jun 2. Available from: http://www.cell.com/article/S2590238521001235/fulltext [accessed 16.11.21].

[14] Kumari S, Chatterjee K. Biomaterials-based formulations and surfaces to combat viral infectious diseases. APL Bioeng [Internet] 2021;5(1), Mar 1. Available from: https://pubmed.ncbi.nlm.nih.gov/33598595/ [accessed 16.11.21].

[15] Sanina N. Vaccine adjuvants derived from marine organisms. Biomolecules [Internet] 2019;9(8), Aug 1. Available from: http://pmc/articles/PMC6723903/ [accessed 16.11.21].

[16] Laurienzo P. Marine polysaccharides in pharmaceutical applications: an overview. Mar Drugs 2010;8(9):2435−65.

[17] Wan MC, Qin W, Lei C, Li QH, Meng M, Fang M, et al. Biomaterials from the sea: Future building blocks for biomedical applications. Bioact Mater 2021;6(12):4255−85.

[18] Cheung RCF, Ng TB, Wong JH. Marine peptides: bioactivities and applications. Mar Drugs [Internet] 2015;13(7):4006, Jul 1. Available from: http://pmc/articles/PMC4515606/ [accessed 16.11.21].

[19] Geahchan S, Ehrlich H, Rahman MA. The anti-viral applications of marine resources for COVID-19 treatment: an overview. Mar Drugs 2021;19(8):409, 19, Page 409 [Internet]. 2021 Jul 23. Available from: https://www.mdpi.com/1660-3397/19/8/409/htm.

[20] Harden EA, Falshaw R, Carnachan SM, Kern ER, Prichard MN. Virucidal activity of polysaccharide extracts from four algal species against herpes simplex virus. Antivir Res 2009;83(3):282−9.

[21] Álvarez-Viñas M, Souto S, Flórez-Fernández N, Torres MD, Bandín I, Domínguez H. Antiviral activity of carrageenans and processing implications. Mar Drugs 2021;19(8):437, [Internet]. 2021 Jul 30. Available from: https://www.mdpi.com/1660-3397/19/8/437/htm [accessed 12.11.21].

[22] Chirkov SN. The antiviral activity of chitosan (review). Appl Biochem Microbiol 2002;38(1):1−8.

[23] Zhu Y., Ming F, Fenglan C, Yuewu Y, Xu L. Application of chitin, chitosan and their derivatives in preparing antivirotic [Internet]. 2003 <https://patents.google.com/patent/CN100577179C/en> [accessed 11.11.21].

[24] Morokutti-Kurz M, Fröba M, Graf P, Große M, Grassauer A, Auth J, et al. Iota-carrageenan neutralizes SARS-CoV-2 and inhibits viral replication in vitro. PLoS One [Internet] 2021;16(2):e0237480, Feb 1. Available from: https://journals.plos.org/plosone/article?id = 10.1371/journal.pone.0237480 [accessed 12.11.21].

[25] Gonzalez ME, Alarcon B, Carrasco L. Polysaccharides as antiviral agents: antiviral activity of carrageenan. Antimicrob Agents Chemother [Internet] 1987;31(9):1388. Available from: http://pmc/articles/PMC174948/?report = abstract.

[26] Girond S, Crance JM, Van Cuyck-Gandre H, Renaudet J, Deloince R. Antiviral activity of carrageenan on hepatitis A virus replication in cell culture. Res Virol [Internet] 1991;142(4):261−70. Available from: https://pubmed.ncbi.nlm.nih.gov/1665574/.

[27] Manivasagan P, Oh J. Marine polysaccharide-based nanomaterials as a novel source of nanobiotechnological applications. Int J Biol Macromol 2016;82:315−27.

[28] Cardoso MJ, Costa RR, Mano JF. Marine origin polysaccharides in drug delivery systems. Mar Drugs 2016;14(2):1−27.

[29] Lee YE, Kim H, Seo C, Park T, Lee KB, Yoo SY, et al. Marine polysaccharides: therapeutic efficacy and biomedical applications. Arch Pharmacal Res 2017;40(9):1006−20, 409 [Internet]. 2017 Sep 16. Available from: https://link.springer.com/article/10.1007/s12272-017-0958-2 [accessed 17.11.21].

[30] Wang W, Wang S-X, Guan H-S. The antiviral activities and mechanisms of marine polysaccharides: an overview. Mar Drugs [Internet] 2012;10(12):2795. Available from: http://pmc/articles/PMC3528127/ [accessed 27.10.21].

[31] Grassauer A, E Prieschl-Grassauer—US Patent 10 342,820, 2019 undefined. Antiviral composition comprising a sulfated polysaccharide. Google Patents [Internet]. 2019; <https://patents.google.com/patent/US10342820B2/en> [accessed 12.11.21].

[32] Muzzarelli RAA. Chitin. 1977;309.

[33] Hayes M. Chitin, chitosan and their derivatives from marine rest raw materials: potential food and pharmaceutical applications. Mar Bioact Compd Sources, Charact Appl [Internet] 2012;9781461412472:115−28, Jan 1. Available from: https://link.springer.com/chapter/10.1007/978-1-4614-1247-2_4.

[34] Furusaki E, Ueno Y, Sakairi N, Nishi N, Tokura S. Facile preparation and inclusion ability of a chitosan derivative bearing carboxymethyl-/I-cyclodextrin. Carbohydr Polym 1996;29(1):29–34.

[35] Vo TS, Kim SK. Potential anti-HIV agents from marine resources: an overview. Mar Drugs [Internet] 2010;8 (12):2871–92. Available from: https://pubmed.ncbi.nlm.nih.gov/21339954/ [accessed 27.10.21].

[36] Baaten BJG, Clarke B, Strong P, Hou S. Nasal mucosal administration of chitin microparticles boosts innate immunity against influenza A virus in the local pulmonary tissue. Vaccine. 2010;28(25):4130–7.

[37] Hasegawa H, Ichinohe T, Strong P, Watanabe I, Ito S, Tamura SI, et al. Protection against influenza virus infection by intranasal administration of hemagglutinin vaccine with chitin microparticles as an adjuvant. J Med Virol 2005;75(1):130–6.

[38] Davydova VN, Nagorskaya VP, Gorbach VI, Kalitnik AA, Reunov AV, Solov'eva TF, et al. Chitosan antiviral activity: dependence on structure and depolymerization method. Appl Biochem Microbiol 2011;47(1):103–8. Available from: https://link.springer.com/article/10.1134/S0003683811010042.

[39] Kulikov SN, Chirkov SN, Il'ina AV, Lopatin SA, Varlamov VP. Effect of the molecular weight of chitosan on its antiviral activity in plants. Appl Biochem Microbiol 2006;42(2):200–3, 422. Available from: https://link.springer.com/article/10.1134/S0003683806020165.

[40] Sosa MAG, Fazely F, Koch JA, Vercellotti SV, Ruprecht RM. N-carboxymethylchitosan-N, O-sulfate as an anti-HIV-1 agent. Biochem Biophys Res Commun [Internet] 1991;174(2):489–96. Available from: https://pubmed.ncbi.nlm.nih.gov/1704225/ [accessed 27.10.21].

[41] Niishimura SI, Kai H, Shinada K, Yoshida T, Tokura S, Kurita K, et al. Regioselective syntheses of sulfated polysaccharides: specific anti-HIV-1 activity of novel chitin sulfates. Carbohydr Res 1998;306(3):427–33.

[42] Ichinohe T, Nagata N, Strong P, Tamura SI, Takahashi H, Ninomiya A, et al. Prophylactic effects of chitin microparticles on highly pathogenic H5N1 influenza virus J Med Virol [Internet] 2007;79(6):811–19, Jun. Available from: https://pubmed.ncbi.nlm.nih.gov/17457919/ [accessed 24.10.21].

[43] Gao Y, Liu W, Wang W, Zhang X, Zhao X. The inhibitory effects and mechanisms of 3,6-O-sulfated chitosan against human papillomavirus infection. Carbohydr Polym [Internet] 2018;198:329–38. Available from: https://pubmed.ncbi.nlm.nih.gov/30093007/.

[44] Davis R, Zivanovic S, D'Souza DH, Davidson PM. Effectiveness of chitosan on the inactivation of enteric viral surrogates. Food Microbiol 2012;32(1):57–62.

[45] Artan M, Karadeniz F, Karagozlu MZ, Kim MM, Kim SK. Anti-HIV-1 activity of low molecular weight sulfated chitooligosaccharides. Carbohydr Res 2010;345(5):656–62.

[46] Pospieszny H, Atabekov JG. Effect of chitosan on the hypersensitive reaction of bean to alfalfa mosaic virus. Plant Sci 1989;62(1):29–31.

[47] Jaber N, Al-Remawi M, Al-Akayleh F, Al-Muhtaseb N, Al-Adham ISI, Collier PJ. A review of the antiviral activity of chitosan, including patented applications and its potential use against COVID-19. J Appl Microbiol 2021.

[48] Divya K, Vijayan S, George TK, Jisha MS. Antimicrobial properties of chitosan nanoparticles: mode of action and factors affecting activity. Fibers Polym. 2017;18(2):221–30, 182 [Internet]. 2017 Feb 28. Available from: https://link.springer.com/article/10.1007/s12221-017-6690-1 [accessed 11.11.21].

[49] Jarach N, Dodiuk H, Kenig S. Polymers in the medical antiviral front-line. Polym 2020;12(8):1727. Available from: https://www.mdpi.com/2073-4360/12/8/1727/htm.

[50] Zheng M, Qu D, Wang H, Sun Z, Liu X, Chen J, et al. Intranasal administration of chitosan against influenza A (H7N9) virus infection in a mouse model. Undefined; Jun 29, 2016;6.

[51] Khan MTH, Orhan I, Şenol FS, Kartal M, Şener B, Dvorská M, et al. Cholinesterase inhibitory activities of some flavonoid derivatives and chosen xanthone and their molecular docking studies. Chem Biol Interact 2009.

[52] Popescu R, Ghica MV, Dinu-Pîrvu CE, Anuţa V, Lupuliasa D, Popa L. New opportunity to formulate intranasal vaccines and drug delivery systems based on chitosan. Int J Mol Sci [Internet] 2020;21(14):1–23, Jul 2. Available from: https://pubmed.ncbi.nlm.nih.gov/32708704/ [accessed 11.11.21].

[53] Hassan MI, Mohamed AF, Taher FA, Kamel MR. Antimicrobial activities of chitosan nanoparticles preparedfrom lucila cuprina maggots (diptera: calliphoridae). J Egypt Soc Parasitol 2016;46(3):563–70.

[54] Tatlow D, Tatlow C, Tatlow S, Tatlow S. A novel concept for treatment and vaccination against Covid-19 with an inhaled chitosan-coated DNA vaccine encoding a secreted spike protein portion. Clin Exp Pharmacol Physiol [Internet] 2020;47(11):1874–8. Available from: http://pmc/articles/PMC7436441/?report = abstract [accessed 11.11.21].

[55] Raisi A, Asefnejad A, Shahali M, Sadat Kazerouni Z, Kolooshani A, Saber-Samandari S, et al. Preparation, characterization, and antibacterial studies of N, O-carboxymethyl chitosan as a wound dressing for bedsore application. Arch Trauma Res 2020;9(4):181.

[56] McCandless EL, Craigie JS, Walter JA. Carrageenans in the gametophytic and sporophytic stages of *Chondrus crispus*. Planta [Internet] 1973;112(3):201–12. Available from: https://pubmed.ncbi.nlm.nih.gov/24468729/ [accessed 12.11.21].

[57] Lahaye M. Developments on gelling algal galactans, their structure and physico-chemistry. J Appl Phycol 2001;13(2):173–84.

[58] Paula PC, Talarico LB, Noseda MD, Silvia SM, Damonte EB, Duarte MER. Chemical structure and antiviral activity of carrageenans from *Meristiella gelidium* against herpes simplex and dengue virus. Carbohydr Polym 2006;63(4):459–65.

[59] Shi Q, Wang A, Lu Z, Qin C, Hu J, Yin J. Overview on the antiviral activities and mechanisms of marine polysaccharides from seaweeds. Carbohydr Res 2017;453–454:1–9.

[60] Talarico LB, Damonte EB. Interference in dengue virus adsorption and uncoating by carrageenans. Virology. 2007;363(2):473–85.

[61] Chiu YH, Chan YL, Tsai LW, Li TL, Wu CJ. Prevention of human enterovirus 71 infection by kappa carrageenan. Antivir Res 2012;95(2):128–34.

[62] Talarico LB, Zibetti RGM, Faria PCS, Scolaro LA, Duarte MER, Noseda MD, et al. Anti-herpes simplex virus activity of sulfated galactans from the red seaweeds *Gymnogongrus griffithsiae* and *Cryptonemia crenulata*. Int J Biol Macromol 2004;34(1–2):63–71.

[63] Abu-Galiyun E, Huleihel M, Levy-Ontman O. Antiviral bioactivity of renewable polysaccharides against Varicella Zoster. Cell Cycle [Internet] 2019;18(24):3540. Available from: http://pmc/articles/PMC6927708/ [accessed 12.11.21].

[64] Shao Q, Guo Q, Xu WP, Li Z, Zhao TT. Specific inhibitory effect of κ-carrageenan polysaccharide on Swine pandemic 2009 H1N1 influenza virus. PLoS One [Internet] 2015;10(5):e0126577, May 13. Available from: https://journals.plos.org/plosone/article?id = 10.1371/journal.pone.0126577 [accessed 15.11.21].

[65] Yamada T, Ogamo A, Saito T, Watanabe J, Uchiyama H, Nakagawa Y. Preparation and anti-HIV activity of low-molecular-weight carrageenans and their sulfated derivatives. Carbohydr Polym 1997;32(1):51–5.

[66] Enomoto Y, Fujii M, Furusho T, Yamamoto N, US Patent 5 878,747, 1999 undefined. Condom coated with acidic polysaccharides. Google Patents [Internet]. 1999; <https://patents.google.com/patent/US5878747A/en> [accessed 12.11.21].

[67] Grassauer A, Weinmuellner R, Meier C, Pretsch A, Prieschl-Grassauer E, Unger H. Iota-Carrageenan is a potent inhibitor of rhinovirus infection. Virol J [Internet] 2008;5. Available from: https://pubmed.ncbi.nlm.nih.gov/18817582/ [accessed 12.11.21].

[68] Jang Y, Shin H, Lee MK, Kwon OS, Shin JS, Kim Yil, et al. Antiviral activity of lambda-carrageenan against influenza viruses and severe acute respiratory syndrome coronavirus 2. Sci Rep 2021;11(1):1–12. Available from: https://www.nature.com/articles/s41598-020-80896-9 [accessed 12.11.21].

[69] Bansal S, Jonsson CB, Taylor SL, Figueroa JM, Dugour AV, Palacios CA, et al. Iota-carrageenan and Xylitol inhibit SARS-CoV-2 in cell culture 2020. Available from: https://www.scienceopen.com/document?vid = 1a9f963a-02df-4dbb-8b8c-00fb2e51e71a [accessed 12.11.21].

[70] Carlucci MJ, Ciancia M, Matulewicz MC, Cerezo AS, Damonte EB. Antiherpetic activity and mode of action of natural carrageenans of diverse structural types. Antivir Res 1999;43(2):93–102.

[71] Hebar A, Koller C, Seifert JM, Chabicovsky M, Bodenteich A, Bernkop-Schnürch A, et al. Non-clinical safety evaluation of intranasal iota-carrageenan. PLoS One 2015;10(4).

[72] Pujol CA, Scolaro LA, Ciancia M, Matulewicz MC, Cerezo AS, Damonte EB. Antiviral activity of a carrageenan from gigartina skottsbergii against intraperitoneal murine herpes simplex virus infection. Planta Med [Internet] 2005;72(02):121–5, Nov 10. Available from: http://www.thieme-connect.com/products/ejournals/html/10.1055/s-2005-373168.

[73] Smidsrød O, Skjåk-Bræk G. Alginate as immobilization matrix for cells. Trends Biotechnol 1990;8(C):71–8.

[74] Qin Y. Alginate fibres: An overview of the production processes and applications in wound management. Polym Int 2008;57(2):171–80.

[75] Lee KY, Mooney DJ. Alginate: properties and biomedical applications. Prog Polym Sci 2012;37(1):106–26.

[76] Urtuvia V, Maturana N, Acevedo F, Peña C, Díaz-Barrera A. Bacterial alginate production: an overview of its biosynthesis and potential industrial production. World J Microbiol Biotechnol 2017;33(11).

[77] Homayouni A, Ehsani MR, Azizi A, Saeid Yarmand M, Razavi SH. Effect of lecithin and calcium chloride solution on the microencapsulation process yield of calcium alginate beads [Internet]. Iran Polym J (Engl) 2007;597−606. Available from: https://www.sid.ir/en/journal/ViewPaper.aspx?id = 95979 [accessed 21.11.21].

[78] Calcium alginate—Wikipedia [Internet]. <https://en.wikipedia.org/wiki/Calcium_alginate> [accessed 21.11.21].

[79] Pietropaolo V, Seganti L, Marchetti M, Sinibaldi L, Orsi N, Nicoletti R. Effect of natural and semisynthetic polymers on rabies virus infection in CER cells. Res Virol [Internet] 1993;144(2):151−8. Available from: https://pubmed.ncbi.nlm.nih.gov/8511399/ [accessed 18.11.21].

[80] Mastromarino P, Petruzziello R, Macchia S, Rieti S, Nicoletti R, Orsi N. Antiviral activity of natural and semi-synthetic polysaccharides on the early steps of rubella virus infection. J Antimicrob Chemother [Internet] 1997;39(3):339−45. Available from: https://pubmed.ncbi.nlm.nih.gov/9096183/ [accessed 18.11.21].

[81] Sinha S, Astani A, Ghosh T, Schnitzler P, Ray B. Polysaccharides from Sargassum tenerrimum: structural features, chemical modification and anti-viral activity. Phytochemistry. 2010;71(2−3):235−42.

[82] Lee JB, Hayashi K, Hirata M, Kuroda E, Suzuki E, Kubo Y, et al. Antiviral sulfated polysaccharide from Navicula directa, a diatom collected from deep-sea water in Toyama Bay. Biol Pharm Bull 2006;29(10):2135−9.

[83] Saha S, Navid M, … SB-C. undefined. Sulfated polysaccharides from Laminaria angustata: structural features and in vitro antiviral activities. Elsevier [Internet] 2012. Available from: https://www.sciencedirect.com/science/article/pii/S0144861711006163 [accessed 18.11.21].

[84] Bandyopadhyay SS, Navid MH, Ghosh T, Schnitzler P, Ray B. Structural features and in vitro antiviral activities of sulfated polysaccharides from Sphacelaria indica. Phytochemistry. 2011;72(2−3):276−83.

[85] Serrano-Aroca Á, Ferrandis-Montesinos M, Wang R. Antiviral properties of alginate-based biomaterials: promising antiviral agents against SARS-CoV-2. ACS Appl Bio Mater [Internet] 2021;4(8):5897−907, Aug 16. Available from: https://pubs.acs.org/doi/full/10.1021/acsabm.1c00523 [accessed 18.11.21].

[86] Tran NM, Dufresne M, Helle F, Hoffmann TW, Francois C, Brochot E, et al. Alginate hydrogel protects encapsulated hepatic HuH-7 cells against hepatitis C virus and other viral infections. PLoS One 2014;9(10).

[87] Gong Y, Han G, Li X, Wu Y. Cytotoxicity and antiviral activity of calcium alginate fibers and zinc alginate fibers. Trans Tech Publ [Internet]. <https://www.scientific.net/AMR.152-153.1475>; 2011 undefined [accessed 18.11.21].

[88] Xianliang X, Meiyu G, Huashi G, Zelin L. Study on the mechanism of inhibitory action of 911 on replication of HIV-1 in vitro. europepmc.org [Internet]. <https://europepmc.org/article/cba/339491>; 2000 undefined [accessed 18.11.21].

[89] Xianliang X, Hua D, Meiyu G, Huashi G, Zelin L. Studies of the anti-AIDS effects of marine polysaccharide drug 911 and its related mechanisms of action. europepmc.org [Internet]. <https://europepmc.org/article/cba/342376>; 2000 undefined [accessed 18.11.21].

[90] Meiyu G, Fuchuan L, Xianliang X, Jing L. The potential molecular targets of marine sulfated polymannuroguluronate interfering with HIV-1 entry: interaction between SPMG and HIV-1 rgp120 and CD4. Elsevier [Internet]. <https://www.sciencedirect.com/science/article/pii/S0166354203000688>; 2003 undefined [accessed 18.11.21].

[91] Miao B, Geng M, Li J, Li F, Chen H, Guan H, et al. Sulfated polymannuroguluronate, a novel anti-acquired immune deficiency syndrome (AIDS) drug candidate, targeting CD4 in lymphocytes. Biochem Pharmacol 2004;68(4):641−9.

[92] Liu H, Geng M, Xin X, Li F, Zhang Z, Li J, et al. Multiple and multivalent interactions of novel anti-AIDS drug candidates, sulfated polymannuronate (SPMG)-derived oligosaccharides, with gp120 and their anti-HIV activities. Glycobiology 2005;1815(5):501-510. https://academic.oup.com/glycob/article-abstract/15/5/501/602548 [Internet].

[93] Jiang B, Xu X, Li L, Medicine WY-MP. Study on—911∥ anti-HBV effect in HepG2. 2.15 cells culture. en.cnki.com.cn [Internet]. <https://en.cnki.com.cn/Article_en/CJFDTotal-XDYF200304031.htm>; 2003 undefined [accessed 18.11.21].

[94] Wang S, Zhang X, Guan H. Drugs WW-M. Potential anti-HPV and related cancer agents from marine resources: an overview. mdpi.com [Internet]. 2014;12(4):2019−35. <https://www.mdpi.com/69414>; 2014 undefined [accessed 18.11.21].

[95] Son EW, Rhee DK, Pyo S. Antiviral and tumoricidal activities of alginate-stimulated macrophages are mediated by different mechanisms. Arch Pharm Res 2003;26(11):960−6.

[96] Hans N, Malik A, Naik S. Antiviral activity of sulfated polysaccharides from marine algae and its application in combating COVID-19: Mini review. Bioresour Technol Rep 2021;13.

[97] Luthuli S, Wu S, Cheng Y, Zheng X, Wu M, Tong H. Therapeutic effects of fucoidan: a review on recent studies. Mar Drugs [Internet] 2019;17(9). Available from: http://pmc/articles/PMC6780838/ [accessed 18.11.21].

[98] Cui K, Tai W, Shan X, Hao J, Li G, Yu G. Structural characterization and anti-thrombotic properties of fucoidan from Nemacystus decipiens. Int J Biol Macromol 2018;120:1817−22.

[99] Usoltseva RV, Shevchenko NM, Malyarenko OS, Anastyuk SD, Kasprik AE, Zvyagintsev NV, et al. Fucoidans from brown algae *Laminaria longipes* and *Saccharina cichorioides*: structural characteristics, anticancer and radiosensitizing activity in vitro. Carbohydr Polym [Internet] 2019;221:157−65, Oct 1. Available from: https://pubmed.ncbi.nlm.nih.gov/31227154/ [accessed 18.11.21].

[100] Wang W, Wu J, Zhang X, Hao C, Zhao X, Jiao G, et al. Inhibition of Influenza A virus infection by Fucoidan targeting viral neuraminidase and cellular EGFR pathway. Sci Reports 2017;7(1):1−14, Jan 17. Available from: https://www.nature.com/articles/srep40760 [accessed 18.11.21].

[101] Trejo-Avila LM, Elizondo-Gonzalez R, Rodriguez-Santillan P, Alberto Aguilar-Briseño J, Ricque-Marie D, Rodriguez-Padilla C, et al. Innocuity and anti-Newcastle-virus-activity of *Cladosiphon okamuranus* fucoidan in chicken embryos. Poult Sci 2016;95(12):2795−802.

[102] Elizondo-Gonzalez R, Cruz-Suarez LE, Ricque-Marie D, Mendoza-Gamboa E, Rodriguez-Padilla C, Trejo-Avila LM. In vitro characterization of the antiviral activity of fucoidan from *Cladosiphon okamuranus* against Newcastle Disease Virus. Virol J [Internet] 2012;9. Available from: https://pubmed.ncbi.nlm.nih.gov/23234372/ [accessed 18.11.21].

[103] Sun T, Zhang X, Miao Y, Zhou Y, Shi J, Yan M, et al. Studies on antiviral and immuno-regulation activity of low molecular weight fucoidan from *Laminaria japonica*. J Ocean Univ China 2018;17(3):705−11, 173 [Internet]. 2018 May 9. Available from: https://link.springer.com/article/10.1007/s11802-018-3794-1 [accessed 18.11.21].

[104] Akamatsu E, Shimanaga M, Bioenvironment-Saga YK-C, 2003 undefined. Isolation of an anti-influenza virus substance, MC26 from a marine brown alga, Sargassum piluliferum and its antiviral activity against influenza virus. agris.fao.org [Internet]. <https://agris.fao.org/agris-search/search.do?recordID = JP2004002376> [accessed 18.11.21].

[105] Queiroz KCS, Medeiros VP, Queiroz LS, Abreu LRD, Rocha HAO, Ferreira CV, et al. Inhibition of reverse transcriptase activity of HIV by polysaccharides of brown algae. Biomed Pharmacother [Internet] 2008;62(5):303−7, Jun. Available from: https://pubmed.ncbi.nlm.nih.gov/18455359/ [accessed 18.11.21].

[106] Hidari KIPJ, Takahashi N, Arihara M, Nagaoka M, Morita K, Suzuki T. Structure and anti-dengue virus activity of sulfated polysaccharide from a marine alga. Biochem Biophys Res Commun [Internet] 2008;376(1):91−5. Available from: https://pubmed.ncbi.nlm.nih.gov/18762172/ [accessed 18.11.21].

[107] Krylova NV, Ermakova SP, Lavrov VF, Leneva IA, Kompanets GG, Iunikhina OV, et al. The comparative analysis of antiviral activity of native and modified fucoidans from brown algae fucus evanescens in vitro and in vivo. Mar Drugs [Internet] 2020;18(4), Apr 1. Available from: http://pmc/articles/PMC7230360/ [accessed 18.11.21].

[108] Ahmadi A, Zorofchian Moghadamtousi S, Abubakar S, Zandi K. Antiviral potential of algae polysaccharides isolated from marine sources: a review. Biomed Res Int 2015;2015.

[109] Kwon PS, Oh H, Kwon SJ, Jin W, Zhang F, Fraser K, et al. Sulfated polysaccharides effectively inhibit SARS-CoV-2 in vitro. Cell Discov 2020;6(1):1−4, 61 [Internet]. 2020 Jul 24. Available from: https://www.nature.com/articles/s41421-020-00192-8 [accessed 18.11.21].

[110] Yim SK, Kim K, Kim I, Chun SH, Oh TH, Kim JU, et al. Inhibition of SARS-CoV-2 virus entry by the crude polysaccharides of seaweeds and abalone viscera in vitro. Mar Drugs 2021;19(4):219, 19, Page 219 [Internet]. 2021 Apr 15. Available from: https://www.mdpi.com/1660-3397/19/4/219/htm [accessed 18.11.21].

[111] Itzhaki RF. Antivirals against SARS-CoV2: relevance to the treatment of Alzheimer's disease. J Alzheimer's Dis [Internet] 2020;78(3):905−6, Jan 1. Available from: https://clinicaltrials.gov/ct2/show/NCT03282916 [accessed 18.11.21].

[112] Pereira L, Critchley AT. The COVID 19 novel coronavirus pandemic 2020: seaweeds to the rescue? Why does substantial, supporting research about the antiviral properties of seaweed polysaccharides seem to go

unrecognized by the pharmaceutical community in these desperate times? J Appl Phycol [Internet] 2020;32 (3):1, Jun 1. Available from: http://pmc/articles/PMC7263178/ [accessed 18.11.21].

[113] Gerber P, Dutcher JD, Adams EV, Sherman JH. Protective effect of seaweed extracts for chicken embryos infected with Influenza B or mumps virus. Proc Soc Exp Biol Med 1958;99(3):590–3.

[114] Deig EF, Ehresmann DW, Hatch MT, Riedlinger DJ. Inhibition of herpesvirus replication by marine algae extracts. Antimicrob Agents Chemother [Internet] 1974;6(4):524–5. Available from: https://journals.asm.org/journal/aac [accessed 18.11.21].

[115] Ehresmann DW, Deig EF, Hatch MT, DiSalvo LH, Vedros NA. Antiviral substances from california marine algae. J Phycol 1977;13(1):37–40.

[116] Burkholder P, Lloydia GS. Antimicrobial agents from the sea. pubmed.ncbi.nlm.nih.gov [Internet]. <https://pubmed.ncbi.nlm.nih.gov/4984864/>; 1969 undefiend [accessed 18.11.21].

[117] Richards JT, Kern ER, Glasgow LA, Overall JC, Deign EF, Hatch MT. Antiviral activity of extracts from marine algae. Antimicrob Agents Chemother 1978;14(1):24–30.

[118] S. Muto, K. Niimura, M. Oohara, Y. Oguchi, K. Matsunaga, K. Hirose, et al. Polysaccharides from marine algae and antiviral drugs containing the same as active ingredient [Internet]. 1988;13 pp. <https://patents.google.com/patent/US5089481A/en> [accessed 18.11.21].

[119] Neyts J, Snoeck R, Schols D, Balzarini J, Esko JD, Van Schepdael A, et al. Sulfated polymers inhibit the interaction of human cytomegalovirus with cell surface heparan sulfate. Virology [Internet] 1992;189(1):48–58. Available from: https://pubmed.ncbi.nlm.nih.gov/1376540/ [accessed 18.11.21].

[120] Okazaki K, Matsuzaki T, Sugahara Y, Okada J, Hasebe M, Iwamura Y, et al. BHV-1 adsorption is mediated by the interaction of glycoprotein gIII with heparinlike moiety on the cell surface. Virology [Internet] 1991;181(2):666–70. Available from: https://pubmed.ncbi.nlm.nih.gov/2014642/ [accessed 18.11.21].

[121] B K, R G. A human cytomegalovirus glycoprotein complex designated gC-II is a major heparin-binding component of the envelope. J Virol [Internet] 1992;66(3):1761–4, Mar. Available from: https://pubmed.ncbi.nlm.nih.gov/1310777/ [accessed 18.11.21].

[122] Lüscher-Mattli M, Glück R. Dextran sulfate inhibits the fusion of influenza virus with model membranes, and suppresses influenza virus replication in vivo. Antivir Res [Internet] 1990;14(1):39–50. Available from: https://pubmed.ncbi.nlm.nih.gov/2080868/ [accessed 18.11.21].

[123] Witvrouw M, Schols D, Andrei G, Snoeck R, Hosoya M, Pauwels R, et al. Antiviral activity of low-MW dextran sulphate (derived from dextran MW 1000) compared to dextran sulphate samples of higher MW. Antivir Chem Chemother 1991;2(3):171–9.

[124] Ivanova V, Rouseva R, Kolarova M, Serkedjieva J, Rachev R, Manolova N. Isolation of a polysaccharide with antiviral effect from *Ulva lactuca*. Prep Biochem [Internet] 1994;24(2):83–97, May 1. Available from: https://pubmed.ncbi.nlm.nih.gov/8072958/ [accessed 18.11.21].

[125] McCandless EL, Craigie JS. Sulfated polysaccharides in red and brown algae. Annu Rev Plant Physiol 1979;30(1):41–53.

[126] Delattre C, Fenoradosoa T, Biology PM-BA of, 2011 undefined. Galactans: an overview of their most important sourcing and applications as natural polysaccharides. SciELO Bras [Internet]. <https://www.scielo.br/j/babt/a/XWMhGh9G6TTjyNCdhZ54Qrf/abstract/?lang = en> [accessed 18.11.21].

[127] Witvrouw M, Este JA, Mateu MQ, Reymen D, Andrei G, Snoeck R, et al. Activity of a sulfated polysaccharide extracted from the red seaweed *Aghardhiella tenera* against human immunodeficiency virus and other enveloped viruses. Antivir Chem Chemother 1994;5(5):297–303.

[128] Hasui M, Matsuda M, Okutani K. In vitro antiviral activities of sulfated polysaccharides from a marine microalga (*Cochlodinium polykrikoides*) against human immunodeficiency virus and other. Elsevier [Internet]. <https://www.sciencedirect.com/science/article/pii/014181309598157T>; 1995 undefined [accessed 18.11.21].

[129] Kanekiyo K, Hayashi K, Takenaka H, Lee J-B, Hayashi T. Anti-herpes simplex virus target of an acidic polysaccharide, nostoflan, from the edible blue-green alga Nostoc flagelliforme. jstagejstgojp [Internet] 2007;30 (8):1573–5. Available from: https://www.jstage.jst.go.jp/article/bpb/30/8/30_8_1573/_article/-char/ja/.

[130] Lahaye M, Robic A. Structure and functional properties of Ulvan, a polysaccharide from green seaweeds. Biomacromolecules [Internet] 2007;8(6):1765–74. Available from: https://pubs.acs.org/doi/full/10.1021/bm061185q [accessed 21.11.21].

[131] Delair T. Colloidal polyelectrolyte complexes of chitosan and dextran sulfate towards versatile nanocarriers of bioactive molecules. Eur J Pharm Biopharm 2011;78(1):10–18.

[132] Kanekiyo K, Lee JB, Hayashi K, Takenaka H, Hayakawa Y, Endo S, et al. Isolation of an Antiviral Polysaccharide, Nostoflan, from a Terrestrial Cyanobacterium, Nostoc flagelliforme. J Nat Prod [Internet] 2005;68(7):1037–41. Available from: https://pubs.acs.org/doi/full/10.1021/np050056c.

[133] Vilas Boas LCP, Campos ML, Berlanda RLA, de Carvalho Neves N, Franco OL. Antiviral peptides as promising therapeutic drugs. Cell Mol Life Sci 2019;76(18):3525–42. Available from: https://link.springer.com/article/10.1007/s00018-019-03138-w [accessed 19.11.21].

[134] Rinehart KL, Gloer JB, Cook JC, Mizsak SA, Scahill TA. Structures of the didemnins, antiviral and cytotoxic depsipeptides from a caribbean tunicate. J Am Chem Soc 1981;103(7):1857–9.

[135] Aneiros A, Garateix A. Bioactive peptides from marine sources: pharmacological properties and isolation procedures. J Chromatogr B Anal Technol Biomed Life Sci [Internet] 2004;803(1):41–53. Available from: https://pubmed.ncbi.nlm.nih.gov/15025997/ [accessed 19.11.21].

[136] Kang HK, Seo CH, Park Y. Marine peptides and their anti-infective activities. Mar Drugs [Internet] 2015;13(1):618. Available from: http://pmc/articles/PMC4306955/ [accessed 19.11.21].

[137] Ma X, Nong XH, Ren Z, Wang J, Liang X, Wang L, et al. Antiviral peptides from marine gorgonian-derived fungus Aspergillus sp. SCSIO 41501. Tetrahedron Lett 2017;58(12):1151–5.

[138] Chen X, Si L, Liu D, Proksch P, Zhang L, Zhou D, et al. Neoechinulin B and its analogues as potential entry inhibitors of influenza viruses, targeting viral hemagglutinin. Eur J Med Chem [Internet] 2015;93:182–95. Available from: https://pubmed.ncbi.nlm.nih.gov/25681711/ [accessed 19.11.21].

[139] Niu S, Liu D, Shao Z, Proksch P, Lin W. Eutypellazines A–M, thiodiketopiperazine-type alkaloids from deep sea derived fungus Eutypella sp. MCCC 3A00281. RSC Adv [Internet] 2017;7(53):33580–90. Available from: https://pubs.rsc.org/en/content/articlehtml/2017/ra/c7ra05774a [accessed 19.11.21].

[140] Rowley DC, Kelly S, Kauffman CA, Jensen PR, Fenical W, Halovirs A-E. New antiviral agents from a marine-derived fungus of the genus Scytalidium. Bioorg Med Chem [Internet] 2003;11(19):4263–74. Available from: https://pubmed.ncbi.nlm.nih.gov/12951157/.

[141] He F, Bao J, Zhang XY, Tu ZC, Shi YM, Qi SH. Asperterrestide A, a cytotoxic cyclic tetrapeptide from the marine-derived fungus *Aspergillus terreus* SCSGAF0162. J Nat Prod [Internet] 2013;76(6):1182–6. Available from: https://pubs.acs.org/doi/abs/10.1021/np300897v.

[142] Hwang Y, Rowley D, Rhodes D, Gertsch J, Fenical W, Bushman F. Mechanism of inhibition of a poxvirus topoisomerase by the marine natural product sansalvamide A. Mol Pharmacol [Internet] 1999;55(6):1049–53. Available from: https://pubmed.ncbi.nlm.nih.gov/10347247/ [accessed 19.11.21].

[143] Dang VT, Benkendorff K, Green T, Speck P. Marine snails and slugs: a great place to look for antiviral drugs. J Virol [Internet] 2015;89(16):8114. Available from: http://pmc/articles/PMC4524231/.

[144] Silvestri I, Albonici L, Ciotti M, Lombardi MP, Sinibaldi P, Manzari V, et al. Antimitotic and antiviral activities of Kelletinin A in HTLV-1 infected MT2 cells. Experientia [Internet] 1995;51(11):1076–80. Available from: https://pubmed.ncbi.nlm.nih.gov/7498449/.

[145] Rashid MA, Gustafson KR, Cartner LK, Shigematsu N, Pannell LK, Boyd MR. Microspinosamide, a new HIV-inhibitory cyclic depsipeptide from the marine sponge *Sidonops microspinosa*. J Nat Prod [Internet] 2001;64(1):117–21. Available from: https://pubmed.ncbi.nlm.nih.gov/11170684/ [accessed 19.11.21].

[146] Plaza A, Gustchina E, Baker HL, Kelly M, Bewley CA. Mirabamides A-D, depsipeptides from the sponge *Siliquariaspongia mirabilis* that inhibit HIV-1 fusion. J Nat Prod [Internet] 2007;70(11):1753–60. Available from: https://pubs.acs.org/doi/abs/10.1021/np070306k.

[147] Lu Z, Van Wagoner RM, Harper MK, Baker HL, Hooper JNA, Bewley CA, et al. Mirabamides E-H, HIV-inhibitory depsipeptides from the sponge *Stelletta clavosa*. J Nat Prod [Internet] 2011;74(2):185–93. Available from: https://pubmed.ncbi.nlm.nih.gov/21280591/.

[148] Morimoto M, Mori H, Otake T, Ueba N, Kunita N, Niwa M, et al. Inhibitory effect of tachyplesin I on the proliferation of human immunodeficiency virus in vitro. Chemotherapy [Internet] 1991;37(3):206–11. Available from: https://pubmed.ncbi.nlm.nih.gov/1889308/ [accessed 19.11.21].

[149] Murakami T, Niwa M, Tokunaga F, Miyata T, Iwanaga S. Direct virus inactivation of tachyplesin I and its isopeptides from horseshoe crab hemocytes. Chemotherapy [Internet] 1991;37(5):327–34. Available from: https://pubmed.ncbi.nlm.nih.gov/1666545/ [accessed 19.11.21].

[150] Masuda M, Nakashima H, Ueda T, Naba H, Ikoma R, Otaka A, et al. A novel anti-HIV synthetic peptide, T-22 ([Tyr5,12, Lys7]-polyphemusin II). Biochem Biophys Res Commun [Internet] 1992;189(2):845−50, Dec 15. Available from: https://pubmed.ncbi.nlm.nih.gov/1472056/.

[151] Sirisha VL, D'Souza J. Algal polysaccharides and their biological applications | Request PDF. Marine Algae Extracts [Internet]. Wiley-Blackwell; 2015. Available from: https://www.researchgate.net/publication/344295766_Algal_polysaccharides_and_their_biological_applications [accessed 22.11.21].

[152] Belofsky GN, Jensen PR, Fenical W. Sansalvamide: A new cytotoxic cyclic depsipeptide produced by a marine fungus of the genus Fusarium. Tetrahedron Lett 1999;40(15):2913−16.

Plant polysaccharides as antiviral agents

Bulu Mohanta[1], Amit Kumar Nayak[2] and Amal Kumar Dhara[3]

[1]Department of Pharmacology, Seemanta Institute of Pharmaceutical Sciences, Jharpokharia, Mayurbhanj, Odisha, India [2]Department of Pharmaceutics, Seemanta Institute of Pharmaceutical Sciences, Jharpokharia, Odisha, India [3]Department of Pharmacy, Contai Polytechnic, Purba Medinipur, West Bengal, India

23.1 Introduction

Viral infections have long been a global threat to human health and are now recognized as the most common natural diseases [1,2]. This is due to the capability of viruses to transmit, transform, and advance, just like other living organisms. These are highly pathogenic and fully reliant on the nutrition and physiological systems of target cells [1]. Certain viruses can also promote malignant cells to access the molecules and resources consumed for living activities [3]. However, one of the most essential characteristics of such an antiretroviral therapy is that it can block the internal events within organisms without influencing the regular physiology of cells [4−6]. A substantial number of antiviral therapies have serious complications, making it challenging to design efficient antiviral therapies with minimal adverse effects [7,8]. Polysaccharides are carbohydrate-based biological macromolecules present in a wide variety of plants [9,10]. They are crucial to the survival of organisms. Plant polysaccharides and their chemically modified derivatives have already been demonstrated in multiple studies to have a considerable inhibitory impact against the human immunodeficiency virus (HIV), herpes simplex virus (HSV), coxsackievirus B3, cytomegalovirus, influenza virus, hepatitis virus, etc [11−18]. The reasons behind the search for antiviral polysaccharides are availability of a wide range of plant resources in nature, easy and economic extraction, less adverse effects, etc. [12,13]. Plant polysaccharides can be artificially be modified to boost their pharmacological activities [17−20]. The antiviral mechanisms of polysaccharides(of different sources like plant, animal, marine, etc.)

Viral Infections and Antiviral Therapies
DOI: https://doi.org/10.1016/B978-0-323-91814-5.00026-X

and their derivatives have been summarized in the current chapter with the goal of providing a sound platform for future research on the antiviral properties of plant polysaccharides and their chemically modified derivatives.

23.2 Antiviral mechanisms in polysaccharides

23.2.1 Directly interacting with virus

Polysaccharides usually interfere with the membrane of viruses via their negative charge, limiting the spreading potential of viruses or killing viruses instantly. Pathogens engage glycosaminoglycans (GAGs) to encourage their adherence and penetration of infected cells, to migrate from one cell to another as well as defend themselves against the immune response at practically every primary access portal [21]. In the laboratory, fucosylated chondroitin sulfate inhibited HIV-1IIIB entrance and replication (4.26 and 0.73 g/mL, respectively), HIV-1KM018 and HIV-1TC-2 infection (31.35% and 31.86 g/mL, respectively) and suppressed HIV-1 drug-resistant virus (23.75 and 31.86 g/mL, respectively) [22]. A series of experiments have shown that the fucosylated chondroitin sulfate may effectively bind to the recombinant HIV-1 gp120 protein and then, block numerous strains of HIV-1 infection. It has been shown that N-(2-hydroxypropyl)-3-trimethyl ammoniumchitosan chloride (HTCC), a modified chitosan derivative, may effectively suppress the reproduction of the HCoV-NL63 virus. When HTCC polymer and a recombinant S protein ectodomain from CoV were studied together, it was found that the two molecules bound and formed protein-polymer complexes [23]. It is reasonable to presume that the virus will be rendered inactive as a consequence of this binding. Carrageenan works by stopping virions from attaching to or entering cells in the first place [24]. Sulfated polysaccharide, iota-carrageenan, is reported for antiviral therapy that interacts with viral surface molecules. Iota-carrageenan binds and inactivates viral particles, quickly and efficiently [25]. It has also been established in animal studies that iota-carrageenan can inhibit influenza virus transmission across the epithelial membranes of infected animals.

23.2.2 Inhibiting virus adsorption and invasion

Heparan sulfated proteoglycan on the cell surface is an example of a receptor that may be used by viruses to infect cells [26]. Polysaccharides, which have significant polyanionic characteristics, may block the positive charge on the cell surface to hinder viral adsorption or invasion. Endocytosis, fusion of viruses with cell membranes, and translocation of viruses are often related to viral invasion. Compounds that replicate the cell surface carbohydrates responsible for viral attachment have been designed to mimic heparin or heparin-like materials [27]. The polysaccharide produced from marine microalgae demonstrated substantial prevention against influenza A viruses (IAV) infection through the viral adsorption and internalization stages [28]. During dengue virus (DENV-2) adsorption and internalization, sulfated polysaccharides from seaweeds displayed their antiviral activity, primarily. In primary human nasal epithelial cells cultured with iota-carrageenan, human retrovirus replication is successfully inhibited [25]. Carrageenan seems to largely block virions from attaching to cells or entering them. To prevent IAV from being released, fucoidan binds to the IAV neuraminidase and inhibits the activity of an enzyme. As an additional benefit to the infected cells, fucoidan has been shown to suppress IAV endocytosis as well as epidermal growth factor receptor internalization by interfering with the activation of both pathways [29]. Fucoidans may have an antiviral effect by preventing HSV-2 adsorption into host cells by directly targeting the viral capsid protein and the host PI3K/Akt/mTOR pathway to prevent cell autophagy [30]. It has been shown that HTCC prevented the interaction between S protein and cell receptor, therefore, restricting viral entrance into cells and preventing virus infection by using HCoV-NL63 as an example system [31].

23.2.3 Inhibiting viral transcription and replication

Viral replication-related enzymes and host cell targets can directly be disrupted by polysaccharides, particularly by sulfated polysaccharides [32]. In both Marc-145 cells and swine alveolar macrophages, carrageenan efficiently inhibits reproductive and respiratory syndrome virus replication at the mRNA and protein levels [33]. Interferon does not seem to be necessary for the inhibitory effects of carrageenan oligosaccharide or its sulphated derivative on IAV replication, in vitro or in vivo. Polysaccharide from *Gracilaria lemaneiformis* was reported to inhibit viral adsorption and multiplication on host cells, in vitro, demonstrating antiinfluenza virus activity [34]. Enterovirus 71, (a positive-stranded RNA virus), was found resistant to polysaccharides derived from *Grifola frondosa*, which blocked the viral replication and inhibited the viral protein production and genomic RNA synthesis [35]. As little as 10 mg/kg (a 26% inhibitory dosage) of an *Angelica sinensis* sulfated polysaccharide (a frequently used traditional Chinese herbal treatment) suppressed the viral multiplication [36].

23.2.4 Activating host antiviral immunomodulatory system

Antiviral effects can be achieved by stimulating the innate immune system, which regulates the host NK (natural killer) cells as well as macrophage cells and induces the release of immune cytokines in response to the virus. In mice, chitosan was reported to reverse the Th2-skewed immune responses elicited by the inactivated respiratory syncytial virus vaccine elicited immune responses [37]. A red alga sulphated-carrageenan demonstrated a substantial influence on tobacco mosaic virus (TMV) infection by decreasing viral accumulation/infectivity and boosting plant immunity at the site of TMV infection [35]. After infection with the infectious bursa disease virus, astragalus polysaccharides were reported to improve the immune activity of chicken erythrocytes (IBDV) [38]. The H9N2 AIV replication was reduced by using astragalus polysaccharides with the development of early humoral immune responses in young chickens. A sulfated alginate derivative, polyguluronate sulfate (PGS), was reported to effectively block the production and secretion of HBsAg and HBeAg in HepG2.2.15 cells, preserving the interferon system and disrupting hepatitis B virus (HBV) transcription linked to the ability of PGS to activate NF-B and Raf/MEK/ERK signaling pathways [39]. Using PGS as a new anti-HBV drug, this work

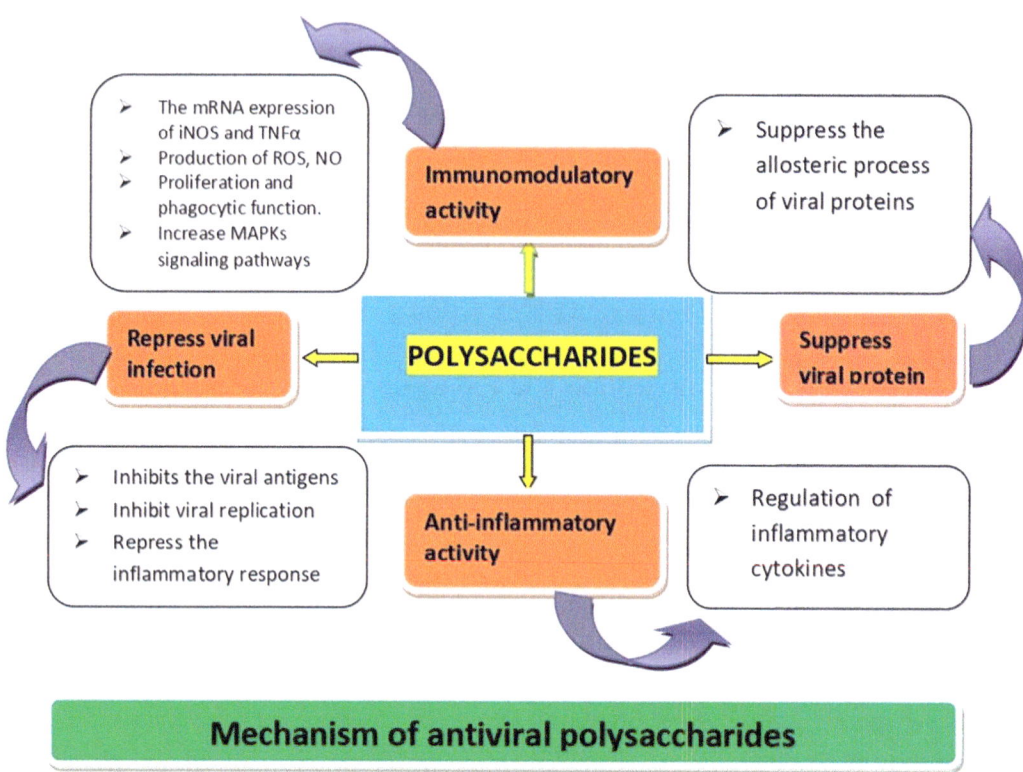

FIGURE 23.1　Mechanism of antiviral activity of polysaccharides.

suggested that the host innate immune system should be modulated in the future. Current anti-novel coronavirus therapy may be beneficial from these investigations.

The mechanism of antiviral activity of polysaccharides is schematically described in Fig. 23.1.

23.3 Plant polysaccharides

Plant polysaccharides are made up of two or more similar or dissimilar monosaccharides linked by α- or β-glycosidic linkages [40–42]. Even though plant polysaccharides are extracted from a variety of plant sources, the structural components as well as molecular structures of plant polysaccharides, vary from species to species [43]. Plant polysaccharides possess a variety of biological functions, including antioxidant, anticancer, antiviral, hypoglycemic, antiulcer, wound healing, immunological modulation, etc. [44–49]. Day-by-day, the studies on plant polysaccharides are gaining attention in health research. Plant polysaccharides have been reported to show antiviral properties, in vitro [11–13]. For example, different administrations of astragalus polysaccharides have already shown the alleviation of pathological damages caused by influenza virus-induced pneumonia by mediating TLR-4/7-MyD88-NF-B signal transduction pathways to reduce the inflammatory response and thus, can be used as an antiviral agent against antiinfluenza virus infection [50,51].

23.4 Antiviral activities of plant polysaccharides

23.4.1 Effect on hepatitis viruses

According to the reported research, plant polysaccharides appear to be capable of greatly suppressing the spread of hepatitis viruses [52]. Based on other recent experimental findings, the molecular mode of inhibitory action of plant polysaccharides is likely to interfere with DNA and/or RNA replications, which gives direct evidences of antiviral activities of *Platycladus orientalis*, *Codonopsis pilosula*, and *Astragali radix* polysaccharides [18,52,53]. Duck hepatitis A virus penetration may result in antiviral activity that is inversely proportional to the amount of viral proteins. Using phosphorylated *C. pilosula* polysaccharide, *Chrysanthemum indicum* polysaccharide and its phosphate derivative to treat duck embryonic hepatocytes after duck hepatitis A virus infection have been reported to lower the cytokine production [53]. Considering these facts, it is reasonable to assume that the phosphorylated *C. pilosula* polysaccharides could significantly decrease duck hepatitis A virus toxicity. Phosphorylated and sulfated *A. radix* polysaccharides reduced the release of duck hepatitis A virus from duck embryonic hepatocytes, in addition to pathogenicity, in vitro. Polysaccharides are generally considered to have antiviral properties by preventing viruses from adhering to cell surfaces. However, some recent research studies have suggested conflicting results, when it comes to the influence of certain polysaccharides on viral adhesion; both phosphorylated and sulfated *A. radix* polysaccharides reduced duck hepatitis A virus adhesion [53]. The viral genome, DNA polymerase, hepatitis B core antigen, and hepatitis B envelope antigen make up the inner core of the HBV [54]. Hepatitis B surface antigen has mostly been found on the surface of the virus. HBV antigens, such as HBsAg, HBcAg, and HBeAg are the significant indicators for the growth of hepatitis B in humans [55]. The polysaccharide component extracted from the leaves of *P. orientalis* was found to inhibit not only the generation of HBsAg and HBeAg, but also the propagation of HBV DNA [52]. In addition, a study conducted in HBV mouse models to assess the antiviral activity of emodin with *A. radix* polysaccharide found that somehow these polysaccharides could dramatically reduce the presentation of HBcAg, HBeAg, and HBsAg in the blood as well as the organs [18]. Liver cell expression levels are a form of protective mediator that can both prevent hepatic sclerosis and enhance hepatic cell proliferation. According to the studies, liver cell stimulant levels increase dramatically in both acute and chronic hepatitis C virus infections [56]. Furthermore, serum liver cell growth factor expression was found to be significantly linked to the plasma levels of liver enzymes in patients with chronic liver disease as well as three indexes (aspartate aminotransferase, alkaline phosphatase, and lactate dehydrogenase), explicitly or implicitly accounting for the degree of liver toxicity. The severity of liver toxicity was found to be significantly reduced after treatment using polysaccharides from *A. radix* and *C. indicum* along with their phosphate and sulfate derivatives for Duck hepatitis A virus infected animals [57]. In addition to enhanced serological index values, supplementary studies have revealed that the intramuscular injection of *A. radix* polysaccharide along with their sulfate derivatives enhanced liver cell markers, such as catalase, superoxide dismutase, glutathione reductase, and peroxidation in duck hepatitis A virus infected animals. It has been reported that the polysaccharide fraction derived from the leaves of *P. orientalis* could produce beneficial effects by enhancing macrophage nitric oxide, tumor necrosis factor, interleukin-6, and interleukin-12 signal transduction as well as establishing associated mRNA expression [52]. Furthermore, because there is conflicting data about the naturally resistant activity of *A. radix* polysaccharide itself and antiduck hepatitis A virus inhibitory activity, different studies showed that *Artemisia annua* polysaccharide could improve the immune response to the HCV/NS3 DNA vaccination [58].

23.4.2 Effect on influenza viruses

Several processes whereby influenza viruses enter, acquire, and mobilize the replication capacity of such a host genome have been described, in vivo and clinical experiments over the past decades, presenting new possibilities in the investigation of antiviral remedies [59]. Polysaccharides have a significant influence on improving influenza virus survival rates, which may be partially related to the signal transduction inhibition. *Panax ginseng* polysaccharides, among the polysaccharides isolated from different herbal resources, have been demonstrated to have a direct influence on the influenza virus mortality, which may be partially linked to signal transduction inhibition [60]. Additional animal research revealed that administering *P. ginseng* polysaccharides to a mouse model before infectious disease conferred tolerance to intranasal poisonous inflammation by various influenza strains [61]. In antiviral treatment, recent research on the benefits of an orally bioavailable whole extract of *P. ginseng*, terpenoids, and polysaccharide fractionation was studied. According to the results of this research, *P. ginseng* polysaccharide was found to be the most beneficial in reducing the symptoms of influenza virus infection [62]. Furthermore, in vitro investigations revealed more light on the different mechanisms of action for influenza prevention. Moreover, *Houttuynia cordata* polysaccharides were found to reduce pneumonia incidence and gastrointestinal damage in mice with H1N1 influenza by inhibiting TLR-4 and p-nuclear factor expression [63]. The kappa B p65, as well as the pathways, were also linked to the restoration of a compromised immunological and physical gastrointestinal barrier.

23.4.3 Effect on herpes simplex viruses

During viral symptoms, the HSV multiplies in the skin or mucosal epithelium, penetrates the peripheral neurons, and is carried to the implicated sensory neurons [64]. Acyclovir is one of several nucleoside analogs of guanine that are used in the treatment of HSV infections. However, the long-term use of acyclovir might lead to unpleasant side effects. This caused resistance to evolve via mutations as well. Where the DNA polymerase or thymidine kinase genes are located depends on whether an intensive search for medical anti-HSV compounds has been conducted. Heliantnus was also found to have antiviral polysaccharides. The tubers of *Helianthus tuberosus* contain a large amount of inulin-type carbohydrates. However, carboxymethylated derivatives of this polysaccharide material exhibited antiherpetic action [65].

Four acidic polysaccharides, containing two pectic polysaccharides and two pectin-type polysaccharides, have been tested in vitro for antiviral activity [66]. Both the pectins type polysaccharides showed modest anti-HSV-2 activity. It was also found that another polysaccharide, which contained homogalacturonan regions in its molecular structure, but not pectic areas, showed potential anti-HSV-2 action from *Portulaca oleracea* [67]. The results of the research suggested that the homogalacturonan region might contribute to the interference of interactions between positively charged viral glycoproteins and negatively sulfated heparin chains of the cell surface glycoprotein receptor, while high branching characteristics in type I rhamnogalacturonan regions might shield the electrostatic interactions. In another research, polysaccharides from *Echinacea purpurea* decreased the development of recurrent HSV-1 illness by increasing the immune response, lowering latency, and inhibiting wait time [68]. In a study, an arabinogalactan type II and a heteropolysaccharide extracted from Acanthopanaxsciadophylloides juvenile buds were examined, in vitro. Both the polysaccharides have exhibited indirect anti-HSV-2 actions by the way of their ability to stimulate the immune system [69].

23.4.4 Effect on human immunodeficiency viruses

Sulfated *Cynomorium songaricum* polysaccharides exhibited suppressive activities on HIV with an EC50 of 0.3–0.4 g/mL in in vitro experiments using the human leukemic and lymphomic cell line MT-4 [70]. Sulfated *Angelica sinensis* polysaccharides did not just effectively block the multiplication of the murine leukemia viruses (a genetic mutation similar to HIV), but they also increased the proportion of CD4 + lymphocytes and the CD4 + /CD8 + proportion of the immune response. In further in vivo trials, combivir and sulfated *A. sinensis* polysaccharides were administered together and exhibited a synergistic antiviral effect [36]. While the hematologic toxicity of combivir was found to be much decreased, the percentage and CD4 + /CD8 ratio of CD4 + cells and the CD4 + /CD8 ratio of peripheral blood cells were both dramatically raised in the optimal combination therapy.

23.4.5 Effect on enterovirus

The poliovirus is a tiny, nonenveloped, productive RNA virus that belongs to the Picornaviridae family of enteroviruses. According to the previously reported research pieces, 42 Egyptian medicinal plants led to the discovery of antipoliovirus agents from herbal resources [71]. The polysaccharides that occurred in the *Azadirachta indica*, and *Adenanthes pavonina* plants showed antipoliovirus activity [16,72]. HEp-2 cells were used to test the major mechanisms in the early stages of viral infection, in which they showed poliovirus type 1 (PV-1) inhibitory action with an IC50 value of 1.18 g/mL and a high selectivity index of 423. In in vitro study, when the polysaccharide was administered postinfection, a 100% suppression of PV-1 replication was likewise observed. Both the *A. indica* polysaccharide and their sulfated derivatives demonstrated a strong PV-1 inhibitory capability by suppressing the earliest phases of viral replication in HEp-2 cells. These polysaccharides had better inhibitory effects with a dose-dependent curve inhibition when they were added concurrently with virus infection. In the same study, fewer effects were observed when these preparations were added after viral infection with the least effects at pretreatment being the least effects of all. At all tested doses, the two native polysaccharides exhibited higher virucidal activity than their sulfated derivatives. Another study found that polysaccharides from *A. sinensis* and *Astragalus membranaceus* combined with ribavirin had higher antiviral effects on EV71-infected Vero cells [73,74]. Cardiomyopathy and viral myocarditis caused by coxsackievirus B3 are complicated by inflammation. A study found that the polysaccharides of *A. membranaceus* effectively inhibited the development of coxsackievirus B3-induced cardiac dysfunction, dilated cardiomyopathy, and fibrosis; the suppressive impact of which was proved to be tumor necrosis factor-α-independent.

23.4.6 Effect of Newcastle disease virus

Polysaccharides extracted from *Lycium barbarum*, and *C. pilosula*, have been shown to suppress the action of Newcastle disease virus (NDV) [75,76]. When tested in vivo, it was found that Chuanminshenviolaceum polysaccharides increased erythrocyte-C3b receptor rosette rate, elimination rate of circulating immune complexes as well as antibody titer, peripheral lymphocyte proliferation, and peripheral CD4 + /CD8 + ratio in chickens immunized with Newcastle disease live vaccine [77]. In vitro studies with chicken embryo fibroblasts showed that polysaccharides obtained from *L. barbarum* and their sulfates significantly inhibited the infectivity of NDV [75]. However, the sulfated derivatives of these polysaccharides exhibited stronger anti-NDV activity than the native form of polysaccharides. In a separate study, polygonatum polysaccharide and sulfated *C. pilosula* polysaccharide have shown their activity in combination to combat NDV [76]. The anti-NDV activities of five combinations are examined in this report. These include the combination of Polygonatum polysaccharide with sulfated *C. pilosula* polysaccharide, the combination of *C. pilosula* polysaccharide with sulfated *A. sinensis* polysaccharide, and the combination of *Epimedium polysaccharide* with sulfated *Astragalus membranaceous*.

23.4.7 Effect on rotavirus

The antidiarrheal effect of ginseng polysaccharides was evaluated in an in vitro model of rotavirus infection based on ethnopharmacological evidence about the treatment of diarrhea with *P. ginseng* [78]. A variety of mechanisms, including the prevention of virus attachment or penetration or viral propagation, have been discovered through in vitro experiments with two pectic ginseng polysaccharides with a major homogalacturonan backbone and hairy regions of rhamnogalacturonan I and arabinose-rich side chains. Additional investigations have shown that the hairy parts of the plants are not homogalacturonan, but rather the functional locations of the antirotavirus action. Many bioactive pectin-type polysaccharides possess hairy areas as a functional site. The incubation of ginseng polysaccharides for 30 min in the acidic medium (pH 2) did not alter the antirotavirus effect, indicating that the ginseng polysaccharides are resistant to digestion or absorption in the gastrointestinal tract and pass through the upper gastrointestinal tract with intact therapeutic effects.

23.4.8 Effect on other viruses

A. radix polysaccharides have been reported to suppress the porcine circovirus type 2 replication in PK15 cells in a variety of ways, including reducing oxidative stress and decreasing the nuclear factor kappa B signaling pathway [79]. In research, *Cyathula officinalis* polysaccharides are found to significantly inhibit canine parvovirus infection in F81 cells, and the potential antiviral mechanism might be related for involving the formation of a stable virion-polysaccharide complex, which could account for the occupancy of sites on the viral envelope required for viral infection, and thus preventing canine parvovirus from adhering to surfaces of host cells [80]. According to this hypothesis (attachment and penetration), the molecular method of action of *Ginkgo biloba* polysaccharides in Vero cells infected with swine epidemic diarrheaviruses, might be likely to entail the blocking of the early event of viral infection [81]. According to the experimental work performed in in vitro, *A. radix* polysaccharides extracted via water decoction and one-step or step-wise ethanol precipitation methods after modification by chlorosulfonic acid-pyridine have been shown to significantly inhibit bursal disease virus infectivity to the chicken embryo fibroblasts [82]. The inhibition of duck enteritis virus by sulfated Chuanminshenviolaceum polysaccharide has been shown in studies using direct immunofluorescence assays and transmission electron microscopy [83]. Fluorescent quantitative polymerase chain reaction results showed that the treatment with sulfated Chuanminshenviolaceum polysaccharide reduced the quantity of duck enteritis virus adsorption. In research, the sulfated polysaccharide of Chuanminshenviolaceum has been reported to block the propagation of the virus from cell to cell. Studies in animals have demonstrated the positive effects of *A. radix* and Chuanminshenviolaceum polysaccharides on the immune response to foot-and-mouth disease vaccine via the TLR-2 and TLR-4 signaling pathways.

23.5 Plant polysaccharide adjuvant for COVID-19 vaccine

The antiviral activities of plant polysaccharides are dependent not only on overall ion concentration and degree of polymerization but also on other precise structural features [12,13]. SARS-CoV, MERS-CoV, and novel SARS-CoV-2 all cause high morbidity and present a threat to human and environmental life demanding the development of effective antagonists. Because of their broad-spectrum antiviral activity and specific antiviral pathways, plant polysaccharides are extensively used as active components in herbal medicine, have a promising future in the research and therapy of coronavirus. Coronaviruses, such as the transmissible gastroenteritis virus, infectious bronchitis virus, mouse hepatitis virus, and feline coronaviruses serotypes I and II, etc., are all substantially inhibited by carbohydrate-binding compounds.

23.6 Conclusions and future perspectives

Although antiviral polysaccharide medications have a limited number of side effects, they have become more vital as viruses evolve into more resistant and mutated forms, which need more effective, low-toxicity antiviral treatments. In the near future, antiviral polysaccharide research for resistant viruses is expected to take place. Multicomponent, multipathway, multitarget impact polysaccharide offers unique benefits in antiviral therapy, making it difficult to develop drug resistance. Polysaccharides have a unique ability to work directly on viruses, stop them from growing and change the host's immune response while also reducing inflammation. Finding polysaccharides is an important first step in developing new antiviral drugs that are both highly effective and low in toxicity in the body. Investigations are needed to confirm the antiviral effects of this agent. Variations in how polysaccharides are absorbed and distributed in the body have been found. For the reasons mentioned, antiviral polysaccharides are only at the laboratory stage: The polysaccharide structure-antiviral connection has not yet been analyzed in most cases. The quality of each batch cannot be controlled; limiting the amount of industrial manufacturing that can be done. The majority of current research on antiviral polysaccharides takes place at the cellular or animal level. Preclinical and clinical contexts are little studied in terms of antiviral effects and mechanisms of various plant polysaccharides. Thus, there is still a long way to travel from the research laboratory to the retail market. Hence, more research is needed to figure out how it works and why it is important to study it more.

References

[1] Trovato M, Sartorius R, D'Apice L, Manco R, De Berardinis P. Viral emerging diseases: challenges in developing vaccination strategies. Front Immunol 2020;11:2130.
[2] Choi YK. Emerging and re-emerging fatal viral diseases. Exp Mol Med 2021;53(5):711−12.
[3] Krump NA, You J. Molecular mechanisms of viral oncogenesis in humans. Nat Rev Microbiol 2018;16 (11):684−98.
[4] Simon V, Ho DD, Abdool Karim Q. HIV/AIDS epidemiology, pathogenesis, prevention, and treatment. Lancet. 2006;368(9534):489−504.
[5] Boyd MA, Boffito M, Castagna A, Estrada V. Rapid initiation of antiretroviral therapy at HIV diagnosis: definition, process, knowledge gaps. HIV Med 2019;20(Suppl 1):3−11.
[6] Menéndez-Arias L, Delgado R. Update and latest advances in antiretroviral therapy. Trends Pharmacol Sci 2022;43(1):16−29.
[7] Antonelli G, Turriziani O. Antiviral therapy: old and current issues. Int J Antimicrob Agents 2012;40 (2):95−102.
[8] Reusser P. Antiviral therapy: current options and challenges. Schweiz Med Wochenschr 2000;130(4):101−12.
[9] Nayak AK, Hasnain MS, Dhara AK, Mandal SC. Herbal biopolysaccharides in drug delivery. In: Mandal SC, Nayak AK, Dhara AK, editors. Herbal biomolecules in healthcare applications. United States: Academic Press, Elsevier Inc.; 2021. p. 613−42.
[10] Mohanta B, Sen DJ, Mahanti B, Nayak AK. Antioxidant potential of herbal polysaccharides: an overview on recent researches. Sens Int 2022;3:100158.
[11] Martinez MJA, Olmo LMBD, Benito PB. Antiviral activities of polysaccharides from natural sources. Stud Nat Products Chem 2005;30:393−418.
[12] Sen IK, Chakraborty I, Mandal AK, Bhanja SK, Patra S, Maity P. A review on antiviral and immunomodulatory polysaccharides from Indian medicinal plants, which may be beneficial to COVID-19 infected patients. Int J Biol Macromol 2021;181:462−70.

[13] He X, Fang J, Guo Q, Wang M, Li Y, Meng Y, et al. Advances in antiviral polysaccharides derived from edible and medicinal plants and mushrooms. Carbohydr Polym 2020;229:115548.

[14] Claus-Desbonnet H, Nikly E, Nalbantova V, Karcheva-Bahchevanska D, Ivanova S, Pierre G, et al. Polysaccharides and their derivatives as potential antiviral molecules. Viruses. 2022;14(2):426.

[15] Lu W, Yang Z, Chen J, Wang D, Zhang Y. Recent advances in antiviral activities and potential mechanisms of sulfated polysaccharides. Carbohydr Polym 2021;272:118526.

[16] Faccin-Galhardi LC, Yamamoto KA, Ray S, Ray B, Carvalho Linhares RE, Nozawa C. The in vitro antiviral property of *Azadirachta indica* polysaccharides for poliovirus. J Ethnopharmacol 2012;142(1):86−90.

[17] Faccin-Galhardi LC, Ray S, Lopes N, Ali I, Espada SF, Dos Santos JP, et al. Assessment of antiherpetic activity of nonsulfated and sulfated polysaccharides from *Azadirachta indica*. Int J Biol Macromol 2019;137:54−61.

[18] Wang Y, Chen Y, Du H, Yang J, Ming K, Song M, et al. Comparison of the anti-duck hepatitis A virus activities of phosphorylated and sulfated Astragalus polysaccharides. Exp Biol Med (Maywood) 2017;242(3):344−53.

[19] Chen F, Huang G, Yang Z, Hou Y. Antioxidant activity of *Momordica charantia* polysaccharide and its derivatives. Int J Biol Macromol 2019;138:673−80.

[20] Fiorito S, Epifano F, Preziuso F, Taddeo VA, Genovese S. Selenylated plant polysaccharides: a survey of their chemical and pharmacological properties. Phytochemistry. 2018;153:1−10.

[21] Aquino RS, Park PW. Glycosaminoglycans and infection. Front Biosci (Landmark Ed) 2016;21(6):1260−77.

[22] Huang N, Wu MY, Zheng CB, Zhu L, Zhao JH, Zheng YT. The depolymerized fucosylated chondroitin sulfate from sea cucumber potently inhibits HIV replication via interfering with virus entry. Carbohydr Res 2013;380:64−9.

[23] Milewska A, Ciejka J, Kaminski K, Karewicz A, Bielska D, Zeglen S, et al. Novel polymeric inhibitors of HCoV-NL63. Antivir Res 2013;97(2):112−21.

[24] Buck CB, Thompson CD, Roberts JN, Müller M, Lowy DR, Schiller JT. Carrageenan is a potent inhibitor of papillomavirus infection. PLoS Pathog 2006;2(7):e69.

[25] Grassauer A, Weinmuellner R, Meier C, Pretsch A, Prieschl-Grassauer E, Unger H. Iota-Carrageenan is a potent inhibitor of rhinovirus infection. Virol J 2008;5:107 Sep 26.

[26] Chen L, Huang G. The antiviral activity of polysaccharides and their derivatives. Int J Biol Macromol 2018;115:77−82.

[27] Baram-Pinto D, Shukla S, Gedanken A, Sarid R. Inhibition of HSV-1 attachment, entry, and cell-to-cell spread by functionalized multivalent gold nanoparticles. Small. 2010;6(9):1044−50.

[28] Scordi-Bello IA, Mosoian A, He C, Chen Y, Cheng Y, Jarvis GA, et al. Candidate sulfonated and sulfated topical microbicides: comparison of anti-human immunodeficiency virus activities and mechanisms of action. Antimicrob Agents Chemother 2005;49(9):3607−15.

[29] Wang W, Wu J, Zhang X, Hao C, Zhao X, Jiao G, et al. Inhibition of influenza A virus infection by fucoidan targeting viral neuraminidase and cellular EGFR pathway. Sci Rep 2017;7:40760.

[30] Gao Y, Liu W, Wang W, Zhang X, Zhao X. The inhibitory effects and mechanisms of 3,6-O-sulfated chitosan against human papillomavirus infection. Carbohydr Polym 2018;198:329−38.

[31] Milewska A, Kaminski K, Ciejka J, Kosowicz K, Zeglen S, Wojarski J, et al. HTCC: broad range inhibitor of coronavirus entry. PLoS One 2016;11(6):e0156552.

[32] Guo C, Zhu Z, Yu P, Zhang X, Dong W, Wang X, et al. Inhibitory effect of iota-carrageenan on porcine reproductive and respiratory syndrome virus in vitro. AntivirTher. 2019;24(4):261−70.

[33] Wang W, Wang SX, Guan HS. The antiviral activities and mechanisms of marine polysaccharides: an overview. Mar Drugs 2012;10(12):2795−816.

[34] Chen MZ, Xie HG, Yang LW, Liao ZH, Yu J. In vitro anti-influenza virus activities of sulfated polysaccharide fractions from *Gracilarialemaneiformis*. Virol Sin 2010;25(5):341−51.

[35] Zhao C, Gao L, Wang C, Liu B, Jin Y, Xing Z. Structural characterization and antiviral activity of a novel heteropolysaccharide isolated from Grifolafrondosa against enterovirus 71. Carbohydr Polym 2016;144:382−9.

[36] Yang T, Jia M, Zhou S, Pan F, Mei Q. Antivirus and immune enhancement activities of sulfated polysaccharide from *Angelica sinensis*. Int J Biol Macromol 2012;50(3):768−72.

[37] Muralidharan A, Russell MS, Larocque L, Gravel C, Sauvé S, Chen Z, et al. Chitosan alters inactivated respiratory syncytial virus vaccine elicited immune responses without affecting lung histopathology in mice. Vaccine. 2019;37(30):4031−9.

[38] Jiang J, Wu C, Gao H, Song J, Li H. Effects of astragalus polysaccharides on immunologic function of erythrocyte in chickens infected with infectious bursa disease virus. Vaccine. 2010;28(34):5614–16.

[39] Wu L, Wang W, Zhang X, Zhao X, Yu G. Anti-HBV activity and mechanism of marine-derived polyguluronate sulfate (PGS) in vitro. Carbohydr Polym 2016;143:139–48.

[40] Nayak AK, Hasnain MS, Dhara AK, Pal D. Plant polysaccharides in pharmaceutical applications. In: Pal D, Nayak AK, editors. Bioactive natural products for pharmaceutical applications, advanced structured materials, Vol. 140. Cham: Springer; 2021. p. 93–125.

[41] Nayak AK, Hasnain MS. Plant polysaccharides in drug delivery applications. In: Nayak AK, Hasnain MS, editors. Plant polysaccharides-based multiple-unit systems for oral drug delivery. Singapore: Springer; 2019. p. 19–23.

[42] Pal D, Nayak AK. Plant polysaccharides-blended ionotropically-gelled alginate multiple-unit systems for sustained drug release. In: Thakur VK, Thakur MK, Kessler MR, editors. Handbook of composites from renewable materials, volume 6, polymeric composites. USA: WILEY-Scrivener; 2017. p. 399–400.

[43] Nayak AK, Bera H, Hasnain MS, Pal D. Synthesis and characterization of graft-copolymers of plant polysaccharides. In: Thakur VK, editor. Biopolymer grafting, synthesis and properties. Netherlands: Elsevier Inc; 2018. p. 1–62.

[44] Yarley OPN, Kojo AB, Zhou C, Yu X, Gideon A, Kwadwo HH, et al. Reviews on mechanisms of in vitro antioxidant, antibacterial and anticancer activities of water-soluble plant polysaccharides. Int J Biol Macromol 2021;183:2262–71.

[45] Albuquerque PBS, de Oliveira WF, Dos Santos Silva PM, Dos Santos Correia MT, Kennedy JF, Coelho LCBB. Skincare application of medicinal plant polysaccharides—a review. Carbohydr Polym 2022;277:118824.

[46] Xie JH, Jin ML, Morris GA, Zha XQ, Chen HQ, Yi Y, et al. Advances on bioactive polysaccharides from medicinal plants. Crit Rev Food Sci Nutr 2016;56(Suppl 1):S60–84.

[47] Yang W, Zhao P, Li X, Guo L, Gao W. The potential roles of natural plant polysaccharides in inflammatory bowel disease: a review. Carbohydr Polym 2022;277:118821.

[48] Huang R, Xie J, Yu Y, Shen M. Recent progress in the research of yam mucilage polysaccharides: isolation, structure and bioactivities. Int J Biol Macromol 2020;155:1262–9.

[49] Wong CK, Leung KN, Fung KP, Choy YM. Immunomodulatory and anti-tumour polysaccharides from medicinal plants. J Int Med Res 1994;22(6):299–312.

[50] Kallon S, Li X, Ji J, Chen C, Xi Q, Chang S, et al. Astragalus polysaccharide enhances immunity and inhibits H9N2 avian influenza virus in vitro and in vivo. J Anim Sci Biotechnol 2013;4(1):22.

[51] Zhang P, Liu X, Liu H, Wang W, Liu X, Li X, et al. Astragalus polysaccharides inhibit avian infectious bronchitis virus infection by regulating viral replication. MicrobPathog 2018;114:124–8.

[52] Lin Z, Liao W, Ren J. Physicochemical characterization of a polysaccharide fraction from *Platycladusorientalis* (L.) Franco and its macrophage immunomodulatory and anti-hepatitis B virus activities. J Agric Food Chem 2016;64(29):5813–23.

[53] Ming K, Chen Y, Yao F, Shi J, Yang J, Du H, et al. Phosphorylated *Codonopsispilosula* polysaccharide could inhibit the virulence of duck hepatitis A virus compared with *Codonopsispilosula* polysaccharide. Int J Biol Macromol 2017;94(Pt A):28–35.

[54] Datta S, Chatterjee S, Veer V, Chakravarty R. Molecular biology of the hepatitis B virus for clinicians. J Clin Exp Hepatol 2012;2(4):353–65.

[55] Liang TJ. Hepatitis B: the virus and disease. Hepatology. 2009;49(5 Suppl.):S13–21.

[56] Shepard CW, Simard EP, Finelli L, Fiore AE, Bell BP. Hepatitis B virus infection: epidemiology and vaccination. Epidemiol Rev 2006;28:112–25.

[57] Dang SS, Jia XL, Song P, Cheng YA, Zhang X, Sun MZ, et al. Inhibitory effect of emodin and Astragalus polysaccharide on the replication of HBV. World J Gastroenterol 2009;15(45):5669–73.

[58] Bao LD, Ren XH, Ma RL, Wang Y, Yuan HW, Lv HJ. Efficacy of *Artemisia annua* polysaccharides as an adjuvant to hepatitis C vaccination. Genet Mol Res 2015;14(2):4957–65.

[59] Yoo DG, Kim MC, Park MK, Park KM, Quan FS, Song JM, et al. Protective effect of ginseng polysaccharides on influenza viral infection. PLoS One 2012;7(3):e33678.

[60] Armstrong L, Hughes O, Yung S, Hyslop L, Stewart R, Wappler I, et al. The role of PI3K/AKT, MAPK/ERK and NFkappabeta signalling in the maintenance of human embryonic stem cell pluripotency and viability highlighted by transcriptional profiling and functional analysis. Hum Mol Genet 2006;15(11):1894–913.

[61] Chen PY, Sun JS, Tsuang YH, Chen MH, Weng PW, Lin FH. Simvastatin promotes osteoblast viability and differentiation via Ras/Smad/Erk/BMP-2 signaling pathway. Nutr Res 2010;30(3):191–9.

[62] Yin SY, Kim HJ, Kim HJ. A comparative study of the effects of whole red ginseng extract and polysaccharide and saponin fractions on influenza A (H1N1) virus infection. Biol Pharm Bull 2013;36(6):1002–7.

[63] Zhu H, Lu X, Ling L, Li H, Ou Y, Shi X, et al. *Houttuynia cordata* polysaccharides ameliorate pneumonia severity and intestinal injury in mice with influenza virus infection. J Ethnopharmacol 2018;218:90–9.

[64] Arduino PG, Porter SR. Herpes Simplex Virus Type 1 infection: overview on relevant clinico-pathological features. J Oral Pathol Med 2008;37(2):107–21.

[65] Rubel IA, Iraporda C, Novosad R, Cabrera FA, Genovese DB, Manrique GD. Inulin rich carbohydrates extraction from *Jerusalem artichoke* (*Helianthus tuberosus* L.) tubers and application of different drying methods. Food Res Int 2018;103:226–33.

[66] Ma FW, Kong SY, Tan HS, Wu R, Xia B, Zhou Y, et al. Structural characterization and antiviral effect of a novel polysaccharide PSP-2B from *Prunellae Spica*. Carbohydr Polym 2016;152:699–709.

[67] Dong CX, Hayashi K, Lee JB, Hayashi T. Characterization of structures and antiviral effects of polysaccharides from *Portulaca oleracea* L. Chem Pharm Bull (Tokyo) 2010;58(4):507–10.

[68] Ghaemi A, Soleimanjahi H, Gill P, Arefian E, Soudi S, Hassan Z. *Echinacea purpurea* polysaccharide reduces the latency rate in herpes simplex virus type-1 infections. Intervirology. 2009;52(1):29–34.

[69] Lee JB, Tanikawa T, Hayashi K, Asagi M, Kasahara Y, Hayashi T. Characterization and biological effects of two polysaccharides isolated from *Acanthopanax sciadophylloides*. Carbohydr Polym 2015;116:159–66.

[70] Tuvaanjav S, Shuqin H, Komata M, Ma C, Kanamoto T, Nakashima H, et al. Isolation and antiviral activity of water-soluble *Cynomoriumsongaricum* Rupr. polysaccharides. J Asian Nat Prod Res 2016;18(2):159–71.

[71] Soltan MM, Zaki AK. Antiviral screening of forty-two Egyptian medicinal plants. J Ethnopharmacol 2009;126 (1):102–7.

[72] de Godoi AM, Faccin-Galhardi LC, Lopes N, Rechenchoski DZ, de Almeida RR, Ricardo NM, et al. Antiviral activity of sulfated polysaccharide of *Adenantherapavonina* against poliovirus in HEp-2 cells. Evid Based Complement Altern Med 2014;2014:712634.

[73] Pu X, Wang H, Li Y, Fan W, Yu S. Antiviral activity of GuiQi polysaccharides against enterovirus 71 in vitro. Virol Sin 2013;28(6):352–9.

[74] Li J, Zhong Y, Li H, Zhang N, Ma W, Cheng G, et al. Enhancement of Astragalus polysaccharide on the immune responses in pigs inoculated with foot-and-mouth disease virus vaccine. Int J Biol Macromol 2011;49(3):362–8.

[75] Wang J, Hu Y, Wang D, Zhang F, Zhao X, Abula S, et al. *Lyciumbarbarum* polysaccharide inhibits the infectivity of Newcastle disease virus to chicken embryo fibroblast. Int J Biol Macromol 2010;46(2):212–16.

[76] Liu C, Chen J, Li E, Fan Q, Wang D, Zhang C, et al. Solomonseal polysaccharide and sulfated *Codonopsispilosula* polysaccharide synergistically resist Newcastle disease virus. PLoS One 2015;10(2): e0117916.

[77] Song X, Yin Z, Zhao X, Cheng A, Jia R, Yuan G, et al. Antiviral activity of sulfated *Chuanmingshenviolaceum* polysaccharide against Newcastle disease virus. J Gen Virol 2013;94(Pt 10):2164–74.

[78] Baek SH, Lee JG, Park SY, Bae ON, Kim DH, Park JH. Pectic polysaccharides from Panax ginseng as the antirotavirus principals in ginseng. Biomacromolecules. 2010;11(8):2044–52.

[79] Xue H, Gan F, Zhang Z, Hu J, Chen X, Huang K. Astragalus polysaccharides inhibits PCV2 replication by inhibiting oxidative stress and blocking NF-κB pathway. Int J Biol Macromol 2015;81:22–30.

[80] Feng H, Fan J, Yang S, Zhao X, Yi X. Antiviral activity of phosphorylated *Radix Cyathulae officinalis* polysaccharide against *Canine Parvovirus* in vitro. Int J Biol Macromol 2017;99:511–18.

[81] Lee JH, Park JS, Lee SW, Hwang SY, Young BE, Choi HJ. Porcine epidemic diarrhea virus infection: inhibition by polysaccharide from *Ginkgo biloba* exocarp and mode of its action. Virus Res 2015;195:148–52.

[82] Huang X, Wang D, Hu Y, Lu Y, Guo Z, Kong X, et al. Effect of sulfated astragalus polysaccharide on cellular infectivity of infectious bursal disease virus. Int J Biol Macromol 2008;42(2):166–71.

[83] Song X, Yin Z, Li L, Cheng A, Jia R, Xu J, et al. Antiviral activity of sulfated *Chuanminshenviolaceum* polysaccharide against duck enteritis virus in vitro. Antivir Res 2013;98(2):344–51.

Antiviral peptides against dengue virus

Michelle Felicia Lee[1], Mohd Ishtiaq Anasir[2] and Chit Laa Poh[1]

[1]Centre for Virus and Vaccine Research, School of Medical and Life Science, Sunway University, Bandar Sunway, Selangor, Malaysia [2]National Institutes of Health (NIH), Ministry of Health Malaysia, Shah Alam, Selangor, Malaysia

24.1 Introduction

Dengue is a mosquito-borne disease endemic in over 100 countries, especially in tropical and subtropical countries [1]. According to the World Health Organization (WHO), there are approximately 390 million dengue infections reported annually, with 500,000 hospitalization cases and 25,000 deaths [2]. Globally, the number of reported dengue cases recorded an eightfold increase within the last 20 years from approximately 505,430 cases in 2000 to 4.2 million cases in 2019 [3].

Various strategies have been undertaken to reduce DENV infections in the community. One such strategy is by controlling the breeding of the *Aedes* mosquito vectors (*Aedes aegypti* and *Aedes albopictus*) [4]. Methods such as fogging (spraying of pesticides to eliminate adult *Aedes* mosquitoes), eliminating the breeding habitats of *Aedes* mosquitoes, utilizing biological controls such as mosquito-eating fishes, utilizing larvicides to eliminate *Aedes* mosquitoes in the larval stage, releasing genetically modified *Aedes* mosquitoes to reduce the production of their progenies, and infection of female *Aedes* mosquitoes with Wolbachia have been undertaken but have unfortunately failed to reduce the *Aedes* mosquito population [5–7]. Another strategy that has been undertaken to reduce DENV infections in the community is the development of dengue vaccines. The first and only licensed dengue vaccine, Dengvaxia (also known as CYD-TDV, chimeric yellow fever virus-tetravalent dengue vaccine) was approved for clinical use in December 2015. It is a live-attenuated tetravalent vaccine which is madeup of the DENV prM and E genes inserted into the genome of the yellow fever 17D virus [8,9]. Although the phase 3 clinical trials of Dengvaxia revealed that it was able to significantly reduce dengue hospitalizations by

Viral Infections and Antiviral Therapies
DOI: https://doi.org/10.1016/B978-0-323-91814-5.00010-6

80%, its efficacies against DENV-1 and DENV-2 were relatively low at approximately 50% and 39%, respectively. On the other hand, its efficacies against DENV-3 and DENV-4 were higher at 75% and 77%, respectively [10,11]. Furthermore, previous phase 4 clinical trials with Dengvaxia in the Philippines had shown increased risks of hospitalization for children below nine years old [12]. Although Dengvaxia was recommended for use in countries with a high burden of dengue cases by the WHO, fatalities in children in the Philippines led to the withdrawal of Dengvaxia from the country [13,14].

Apart from vaccines, various small molecule compounds have been developed as antiviral agents against DENV. These antiviral agents are divided into two types, namely direct-acting antivirals (DAAs) and host-directed antivirals (HDAs). DAAs are compounds that directly interacted with viral proteins to exert their antiviral functions [15]. Several examples of DAAs include rolitetracycline and doxycycline which targeted the DENV envelope (E) protein, resulting in inhibition of viral-host membrane fusion [16]. Although a variety of DAAs have been identified using in vitro assays, only a few DAAs have been evaluated in vivo. To date, only one DAA, balapiravir, has been evaluated in clinical trials. However, there were no significant differences reported for fever clearance, viremia, and NS1 antigenemia between the control and balapiravir-treated groups [17]. On the other hand, HDAs are compounds that interfere with host mechanisms or cellular pathways that are required for productive virus replication [18]. A wide variety of antivirals targeting different stages of the DENV replication cycle have been identified. Examples of HDAs include duramycin and Met-RANTES which targeted the TIM1 receptor and CC-chemokine receptor CCR5, respectively, resulting in inhibition of DENV entry into target cells [19,20]. The most extensively studied target is α-glucosidase which functions to facilitate viral protein maturation and appropriate folding of the viral protein [15]. In addition, there are a few studies on inhibiting inosine monophosphate dehydrogenase which has vital functions in nucleotide biosynthesis [17]. The α-glucosidase inhibitor, UV-4B, is a HDA which has been evaluated in clinical trials. The data from the clinical trials of UV-4B exhibited good tolerability and there were no adverse side effects after administration of up to 1000 mg. However, a followup study to further determine its safety and pharmacokinetics in healthy volunteers had discontinued its evaluation as an effective antiviral agent [15].

The lack of efficient vector control strategies, unequal protective efficacies of Dengvaxia against all four DENV serotypes, and no approved antiviral drugs with potent inhibitory effects against all four DENV serotypes are driving the urgent need for the discovery of novel and effective dengue therapeutics. Besides small molecule drugs, various peptides have shown antiviral potentials to treat DENV infections. In contrast to small molecule drugs, peptides have good pharmacological profiles due to their high specificity and selective properties. Antiviral peptides targeting DENV can be divided into four categories, namely peptides derived from animals, peptides derived from plants, synthetic peptides which are artificially synthesized based on a specific target of interest, and recombinant peptides which are produced in an expression system such as *Escherichia coli*. Various methods can be utilized to design and develop antiviral peptides against DENV such as biopanning of phage display peptide libraries and structure-based design. Direct interactions between antiviral peptides and DENV host cell receptors or enzymes that have been discovered in several studies are highlighted as these studies can guide the development of antidengue therapeutics. Lastly, the advantages and limitations of using antiviral peptides are discussed as well as potential ways to overcome the limitations.

24.1.1 Dengue virus

Dengue virus (DENV) is an arbovirus belonging to the Flaviviridae family [21]. DENV is transmitted to humans via arthropod vectors such as *A. aegypti and A. albopictus* mosquitoes. It consists of four genetically and antigenically distinct serotypes, DENV-1−4 [22]. Infection with one serotype induces long-term immune protection specific to only that particular serotype and short-term cross-protections against the other three heterologous serotypes [23]. Each serotype generates similar disease manifestations during infection, ranging from a mild flu-like illness to severe complications such as dengue hemorrhagic fever (DHF) and dengue shock syndrome (DSS) in certain individuals [22,24]. The incubation period of dengue infection is from 3 to 7 days, and the clinical symptoms include fever, muscle pain, chills, retro-orbital pain, frontal headache, nausea, vomiting, and arthralgia [25]. On the third or fourth day of infection, a skin rash is often observed and its cutaneous feature may start appearing on the extremities or the trunk, followed by spreading to other parts of the body including the face. Severe infection can escalate to DHF and DSS which are characterized by plasma leakage and hemorrhage [26,27]. It can also lead to shock or death in patients who were previously infected by one DENV serotype and were subsequently infected by another heterologous serotype. This phenomenon is known as antibody-dependent enhancement (ADE) or original antigenic sin [28,29]. Other clinical manifestations of DENV infection include ocular complications, cardiovascular impairment, hepatic injury, and oral lesions [30−32].

The dengue virion is made up of a lipid bilayer, an outer shell with icosahedral symmetry, and a nucleocapsid core encapsulating the RNA genome. DENV is an enveloped, spherical virus consisting of a positive-sense single-stranded RNA genome which is approximately 11 kb in size with a type I cap at the 5′ end and lacking a poly(A) tail at the 3′ end [33]. It encodes a polyprotein made up of approximately 3400 amino acids. This polyprotein consists of three structural proteins known as the nucleocapsid (C), precursor membrane (prM) or membrane (M), and envelope (E) proteins, and seven nonstructural (NS) proteins known as the NS1, NS2A, NS2B, NS3, NS4A, NS4B, and NS5 proteins (Fig. 24.1) [34]. The structural proteins function to form the DENV viral particle whereby both prM/M and E proteins are located on the surface of the viral particle and the C protein is located inside the viral envelope [35]. On the other hand, the NS proteins are important components of various stages of the DENV life cycle which include viral replication, cleavage of the polyprotein, virion assembly, maturation of viral particle, and defense against the host immune system [36−39].

24.1.2 The life cycle of dengue virus

DENV infection in humans begins with a DENV-infected *Aedes* mosquito bite. DENV can replicate in a variety of cells present in various organs such as the liver, kidney, and spleen [40,41]. However, the primary targets of DENV include dendritic cells (DC), macrophages, and monocytes [42,43]. The DENV cycle is initiated by viral attachment via the interaction between

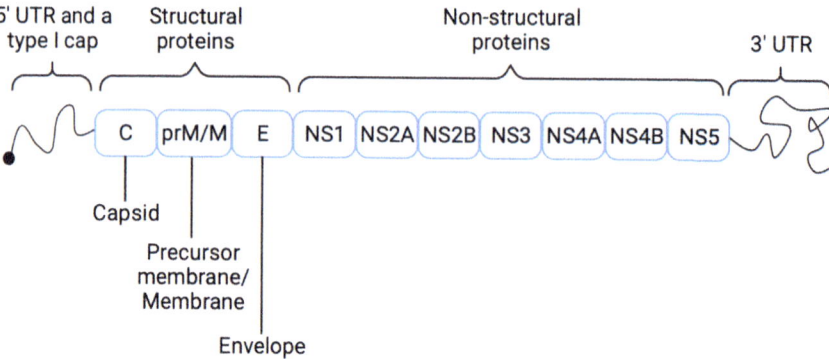

FIGURE 24.1 DENV RNA genome and its encoded structural and nonstructural proteins. The DENV genome is a positive-sense single-stranded RNA with a type I cap and the 5′ untranslated region at the 5′ end as well as the 3′ untranslated region at the 3′ end. The DENV RNA genome encodes a polyprotein that is cleaved into three structural proteins known as the nucleocapsid (C), precursor membrane (prM) or membrane (M), and envelope (E) proteins, and seven nonstructural (NS) proteins known as the NS1, NS2A, NS2B, NS3, NS4A, NS4B, and NS5 proteins. *The figure has been created with Biorender.com (ON, Canada).*

viral surface proteins and their respective receptors on the surface of the target cells. Studies have shown that receptor recognition is mediated by domain III of the DENV E protein which enables the virus to enter host cells via clathrin-dependent or receptor-mediated endocytosis [44,45]. Apart from that, viral entry could also occur via fusion of the virus and the host cell membrane [46–48]. After virus internalization into the host cell, the E homodimers dissociate due to the acidic pH in the endosome. Domain II of the DENV E protein will then project outwards followed by the insertion of the hydrophobic fusion loop of domain II into the endosomal membrane [49,50]. These events will cause the folding back of domain III, thus forcing the viral particle and the endosomal membrane to move towards each other and fuse together [51,52]. The fusion of the viral particle with the endosomal membrane will trigger the release of the nucleocapsid into the cytoplasm of the target cell, thus uncoating and releasing the RNA genome [53]. Next, the viral RNA will be translated into a polyprotein and processed by viral-derived and cellular proteases into the three structural proteins and seven NS proteins. After protein translation, the NS proteins will initiate the genome replication process at the intracellular membranes to produce more viral RNA [54]. The newly produced RNA will be packed by C proteins to form the nucleocapsid [55]. On the other hand, the prM and E proteins will form heterodimers that orient into the lumen of the endoplasmic reticulum (ER) and subsequently induce a curved surface lattice that serves as a guide for virion budding [56]. Therefore, viral assembly takes part in the ER, followed by budding from the ER and migrating to the trans-Golgi network (TGN) for further maturation. The slightly acidic environment of the TGN causes the prM/E heterodimers to dissociate to form 90 dimers, whereby prM caps the fusion peptide at domain II of the DENV E protein [57]. prM will then be cleaved at Arg-X-(Lys/Arg)-Arg (X is any amino acid) by furin (otherwise known as cellular endoprotease) to give rise to the M protein and "pr" peptide [58,59]. Apart from that, the E protein is also stabilized by both prM and "pr" peptide during the secretory pathway due to their roles in the prevention of premature membrane fusion. Upon the release of the virus progeny via exocytosis, the "pr" peptide will then dissociate [54].

24.1.3 Antiviral peptides as potential therapeutic agents against dengue virus

Scientific research in recent years has highlighted the ability of peptides to inhibit the growth of a broad range of viruses. These findings initiated the research on antiviral peptides as a new form of therapeutics [60]. Antiviral peptides function by directly inhibiting viral particles, competing with viruses for the receptor in host cells, or by suppressing viral gene expressions [61,62]. Peptides are biologically active short chains of amino acids connected by peptide bonds. They are usually small in size, but there are also some large peptides such as the naturally-occurring lactoferrin from milk which was reported to display antiviral activities against both DNA and RNA viruses [63]. Peptides also have good pharmacological profiles because they are highly selective and relatively safe to use [64]. Previously, peptides were not considered a source of potential drugs due to their instability and ease of degradation by proteases [65]. However, technological advancements in recent years have made it possible to overcome these limitations and thus, initiating an interest in the field of antiviral peptide drug development.

At present, research on DENV antivirals is focused on targeting both the DENV structural and NS proteins. The main structural protein target of DENV antivirals that are most studied is the E protein and it is due to its vital role in attachment and fusion to host cells [66]. Apart from that, two other NS proteins, NS3 and NS5, have been extensively investigated as targets for antivirals. The NS3 protein has a wide variety of functions and most studies on DENV antivirals are focused on inhibiting its NS2B/NS3 serine protease function. The NS5 protein is the most conserved and largest NS protein and has dual functions, namely methyltransferase (MTase) and RNA-dependent RNA polymerase (RdRP) activity. Although both functions are studied as potential DENV antiviral targets, the most extensively studied function of the NS5 protein is the inhibition of its RdRP activity [67,68].

24.2 Antiviral peptides targeting dengue virus

24.2.1 Peptides from animal origins

Antiviral peptides targeting DENV which were derived from animal origins are listed in Table 24.1. Rothan et al. [69] evaluated the anti-DENV activity of a cationic cyclic peptide, protegrin-1 (PG-1), which was originally isolated from porcine white blood cells against DENV-2. It was discovered that PG-1 significantly inhibited DENV NS2B/NS3 protease in a dose-dependent manner. The highest inhibition (95.7%) of DENV NS2B/NS3 protease was observed with 40 μM of PG-1. Subsequent in vitro studies of PG-1 against DENV-2 revealed that DENV-2 viral copy number decreased with increasing PG-1 concentrations ranging from 2.5 to 12.5 μM at 24, 48, and 72 hours. The highest percentage of inhibition of DENV-2 (100%) in MK2 cells was observed at the PG-1 concentration of 12.5 μM at 24 hours [69].

Xu et al. [70] screened potential DENV NS2B/NS3 protease inhibitors from a natural peptide library consisting of conotoxins (a mixture of peptide neurotoxins produced by the cone snails from *Conus* species) and evaluated their respective antiviral activities against DENV-2. The crude venom extract of *Conus marmoreus* exhibited potent inhibitory activity against DENV-2 NS2B/NS3 protease and the specific component within the extract with DENV-2 NS2B/NS3

TABLE 24.1 Antiviral peptides targeting DENV which were derived from animal origins.

Peptides	Sequences	Target	References
Protegrin-1 (PG-1)	RGGRLCYCRRRFCVCVGR	DENV NS2B/NS3 protease	[69]
Peptide 9	CAGKRKSG	DENV NS2B/NS protease	[70]
Latarcin (Ltc 1)	SMWSGMWRRKLKKLRNALKKKLKGE	DENV NS2B/NS3 protease	[71]
Bovine lactoferrin (bLF)	ND	Heparan sulfate (HS), dendritic cell-specific intercellular adhesion molecule 3-grabbing nonintegrin (DC-SIGN), and low-density lipoprotein (LDLR) receptors	[72]
HS-1	FLPLILPSIVTALSSFLKQG	DENV E protein	[73]
Smp76	GWINEKKMQQKIDEKIGKNIIGGMAKA VIHKMAKNEFQCVANVDTLGNCKKHC AKTTGEKGYCHGTKCKCGIELSY	ND	[74]

ND, Not determined.

protease inhibitory activity was identified to be conotoxin MrIA. Conotoxin MrIA was found to inhibit DENV-2 NS2B/NS3 protease with a K_i (inhibitory constant) of $9.0 \pm 0.4\,\mu M$ [70]. Several other peptides were further derived from MrIA such as the single loop peptide with one disulfide bond, MrIA-SL, and ten peptides with different point mutations (peptides 1, 2, 3, 4, 5, 6, 7, 8, 9, and 10). Peptide 9 was identified to be the strongest inhibitor of DENV-2 NS2B/NS3 protease with a K_i value of $2.2 \pm 0.2\,\mu M$. Additionally, peptide 9 exhibited great resistance to digestion from a variety of proteases such as trypsin and cathepsin S, which suggested that it might have a long half-life in physiological conditions [70].

Next, Rothan et al. [71] evaluated the anti-DENV activity of another peptide, latarcin (Ltc 1), which was produced in the venom gland of *Lachesana tarabaeve* (central Asian spider) against DENV-2. It was found that Ltc 1 significantly inhibited DENV-2 in a dose-dependent manner. Additionally, there was a dose-dependent reduction of DENV-2 viral copy numbers after treatment with Ltc 1 peptide at concentrations ranging from $2.5-80\,\mu M$ at 24, 48, and 72 hours. The Ltc 1 peptide inhibited DENV-2 replication at $EC_{50} = 8.3 \pm 1.2\,\mu M$ at 24 hours, $EC_{50} = 7.6 \pm 2.7\,\mu M$ at 48 hours, and at 72 hours with $EC_{50} = 6.8 \pm 2.5\,\mu M$. Ltc 1 peptide was shown to have significant inhibitory effects during pretreatment (reduction in viral loads to $4.5 \pm 0.6 \times 10^7$ pfu/mL versus control $6.9 \pm 0.5 \times 10^7$ pfu/mL), virucidal (reduction in viral loads to $0.7 \pm 0.3 \times 10^7$ pfu/mL versus control $7.2 \pm 0.5 \times 10^7$ pfu/mL), and posttreatment assays (reduction in viral loads to $1.8 \pm 0.7 \times 10^7$ pfu/mL versus control $6.8 \pm 0.6 \times 10^7$ pfu/mL) [71].

Another study was conducted by Chen et al. [72] to evaluate the anti-DENV activity of bovine lactoferrin (bLF) which is naturally present in milk against all four DENV serotypes. The addition of bLF to Vero cells infected with DENV-1 to DENV-4 resulted in a significant reduction of DENV infection rates, most notably for DENV-2 (76%) and DENV-3 (61%). Subsequent in vitro studies of bLF against DENV-2 revealed that treatment with $200\,\mu g/mL$ of bLF significantly reduced the number of DENV-2 plaque-forming units by 82% during the viral infection stage. This suggested that the mode of action of bLF was by inhibiting the binding of DENV to the host cell membrane during the early stages of DENV infection (notably during viral entry). Apart from that, Chen et al. [72] also discovered that heparan sulfate (HS), dendritic cell-specific intercellular adhesion molecule 3-grabbing nonintegrin (DC-SIGN), and low-density lipoprotein receptors (LDLR) receptors played vital roles in the anti-DENV-2 activity of bLF. Additionally, antiviral effects of bLF against DENV in vivo were demonstrated to reduce morbidity in a suckling mouse challenge model [72].

Monteiro et al. [73] evaluated the anti-DENV activity of HS-1 peptide which was derived from the skin of a Brazilian rainforest tree frog (anuran *Hypsiboas semilineatus*) against DENV-2 and DENV-3. HS-1 peptide was able to exhibit significant virucidal activities against DENV-2 (100%) and DENV-3 (95%) at peptide concentrations ranging from 15.625–125 μg/mL. Preinfection treatment of Vero cells with 125 μg/mL of HS-1 peptide was observed to cause a 50% reduction of plaques for DENV-2, but no reduction of plaques was observed for DENV-3. HS-1 peptide did not exert any antiviral effects against both serotypes in postinfection treatments. Thus, the HS-1 peptide was able to significantly prevent DENV-2 infections when applied in the early stages of infection. Subsequent in vivo studies in neonatal Balb/c mice demonstrated that HS-1 peptide could effectively confer 80% protection against DENV-2 when HS-1 peptide was administered at concentrations ranging from 15.625–125 μg/mL. However, a lower efficacy (40%) of protection against DENV-3 was observed even when the peptide concentration was raised to 125 μg/mL. Additionally, it was revealed by atomic force microscopy that HS-1 peptide could induce the appearance of invaginations on the surface of the DENV-3 viral envelope and a reduction in the size of the DENV-3 viral particle, thereby suggesting that the antiviral activity of the HS-1 peptide was due to destabilization of the viral envelope [73].

A study conducted by El-Bitar et al. [74] evaluated the antiviral activity of Smp76, a peptide derived from the venom of an Egyptian scorpion, *Scorpio maurus palmatus*, against DENV and hepatitis C virus (HCV). Among the various high-performance liquid chromatography (HPLC) fractions that were isolated from the crude venom of *S. m. palmatus*, the eluted fraction containing Smp76 identified at 36.4 minutes showed the most potent antiviral activity against DENV with an IC_{50} value of 0.01 μg/mL [74].

24.2.2 Peptides from plant origins

Apart from animals, plants are also good sources of antiviral peptides against DENV (Table 24.2). Gao et al. [75] designed four polypeptides (polypeptides 1, 2, 3, and 4) based on kalata B1 (plant-derived proteins or cyclotides) by adding several modifications such as inserting DENV protease substrate sequences and substituting its disulfide bond with a salt bridge. These polypeptides were then evaluated for their antiviral activities against DENV-2 NS2B/NS3 protease. Polypeptide 4 exhibited the best inhibition of the DENV-2 NS2B/NS3 protease and it was then subjected to cyclization to obtain cyclopeptide 1 which exhibited greater inhibition of the DENV-2 NS2B/NS3 protease at a low micromolar concentration as compared to polypeptide 4. Additionally, cyclopeptide 1 was converted into its oxidized forms, isomers 1B and 1C, which showed potent inhibition of DENV-2 NS2B/NS3 protease with K_i values of 1.39 ± 0.35 μM and 3.03 ± 0.75 μM, respectively [75].

TABLE 24.2   Antiviral peptides targeting DENV derived from plant origins.

Peptides	Plant sources	Sequences	Target	References
Cyclopeptide 1	Kalata B1	GLPVCGSEESRRGCNTPGCRRSWPVCTRR	DENV NS2B/NS3 protease	[75]
Peptides 2 and 4	*Acacia catechu*	Peptide 2: DHVTPDIAYNPRTYM Peptide 4: DHVTPDIAYNPWAYF	ND	[76]
Pep-RTYM	*Acacia catechu*	DHVTPDIAYNPRTMY	DENV-2 E protein	[77]
NICTABA and UDA	*Nicotiana tabacum* and *Urtica dioica*	ND	ND	[78]
As1	*Alstonia scholaris*	CRPYGYRCDGVINQCCDPYHCTPPLIGICL	DENV-2 NS3 protein	[79]

ND, Not determined.

Panya et al. [76] evaluated the potential antidengue activities of novel bioactive peptides identified from 33 Thai medicinal plants. Extracts from *Thunbergia laurifolia Lindl.* and *Acacia catechu* were found to significantly reduce DENV-2 viral titers in a dose-dependent manner by 44-fold and >100-fold, respectively. Subsequent in vitro studies with the extract from *A. catechu* demonstrated a significant reduction in DENV-2 viral titers (>1000-fold reduction) from pretreatment, thereby indicating that the extract from *A, catechu* inhibited DENV-2 during the early stages of viral infection. The extract from *A. catechu* was also shown to lower the levels of intracellular E proteins in all four DENV serotypes and it was most potent against DENV-3. Fraction 13 of the extract from *A. catechu* isolated by HPLC was found to significantly lower the DENV-2 E antigen level to 32.24% compared to the untreated control. A total of 4 peptides were identified from fraction 13 and each peptide was synthesized and their antiviral activities against DENV-2 were evaluated. Peptides 2 and 4 exhibited the most significant inhibitory effects against DENV-2 with >98% and 100% reduction in the formation of DENV-2 foci, respectively [76].

In another study, Panya et al. [77] evaluated the antidengue activities of Pep-RTYM, Pep-WAYF, and Pep-CORE which are synthetic bioactive peptides derived from the *A. catechu* extract against all four dengue serotypes. It was reported that Pep-RTYM had the best inhibitory activity against DENV-2 with an IC_{50} of 7.9 μM. Pretreatment of DENV-2 with Pep-RTYM before infecting Vero cells resulted in significant inhibition of DENV infection, thereby suggesting that Pep-RTYM potentially inhibited DENV infection during an early stage of the infection. Subsequently, it was found that 25 μM of Pep-RTYM was able to significantly reduce virus internalization by approximately 40%. It was also revealed that treatment with 50 μM of Pep-RTYM inhibited DENV-2 production by >1000-fold [77].

Gordts et al. [78] evaluated the antiviral activity of NICTABA, a tobacco agglutinin isolated from the *Nicotiana tabacum* plant and UDA, a chitin-binding protein isolated from the rhizome of the stinging nettle, *Urtica dioica* against DENV-2. Both NICTABA and UDA displayed inhibitory activity against DENV-2 with EC_{50} values of $526 \pm 2.3\,\mu$M and $1176 \pm 423\,\mu$M, respectively [78].

The antiviral activity of an alstotide (As1), a cystine knot α-amylase inhibitor discovered from the *Alstonia scholaris* plant of the Apocynaceae family was evaluated against DENV-2. Based on the densitometry analysis, it was revealed that As1 moderately reduced the production of DENV-2 NS3 protein in DENV-2-infected Vero cells with 57% inhibition at a concentration of 100 μM. The EC$_{50}$ value was estimated to be 90 μM [79].

24.2.3 Synthetic peptides

Apart from peptides derived from natural sources, there are also various synthetic peptides being reported to target DENV. Yin et al. [80] synthesized tetrapeptide inhibitors with various electrophilic warheads such as boronic acid, aldehyde, and trifluoromethyl ketone and evaluated their respective antiviral activities against the DENV-2 NS2B/NS3 protease. Tetrapeptide 21 with a boronic acid warhead and tetrapeptide 7 with an aldehyde warhead exhibited potent inhibitory activities against DENV-2 NS2B/NS3 protease with K_i values of 43 nM and 5.8 μM, respectively [81]. Subsequently, Yin et al. [80] evaluated the antiviral activities of tetrapeptide inhibitors with aldehyde electrophilic warheads based on the lead tetrapeptide 1 (K_i value of 5.8 μM) against the DENV-2 NS2B/NS3 protease. Tetrapeptide 21 and 22 exhibited potent inhibitory activities against DENV-2 NS2B/NS3 protease with K_i values of 1.5 and 12 μM, respectively [80].

Schuller et al. [82] synthesized 17 tripeptides with sequences X-KRR-H and X-KKR-H (X was a varying cap group) and evaluated their antiviral activities against DENV-2 and West Nile virus (WNV). It was discovered that the most potent tripeptide aldehyde inhibitor of the DENV-2 protease was tripeptide 2 (phenylacetyl-KRR-H) with an IC$_{50}$ value of 6.7 $\pm$ 1.1 μM [82]. Nitsche et al. [83] synthesized retro dipeptide and tripeptide hybrids made up of a retro peptide sequence and an alternative structural component for one of the two obligatory arginine residues and evaluated their antiviral activities against DENV-2 and WNV NS2B/NS3 proteases. Retro dipeptide 3 exhibited the most potent inhibitory activity against DENV-2 NS2B-NS3 protease with a K_i value of 15.4 $\pm$ 1.8 μM. On the other hand, retro tripeptide 11 exhibited the most potent inhibitory activity against DENV-2 NS2B/NS3 protease with a K_i value of 4.9 $\pm$ 0.3 μM [83].

Four anionic, antiviral septapeptides (DDHELQD, DETELQD, DEVMLQD, and DEVLMQD) were identified from viprolaxikine and their antiviral activities against DENV-2 were evaluated. At a concentration of 0.1 μg, the mixture of peptides provided 98.3% protection of Vero cells against DENV-2 whereas the crude filtrate of viprolaxikine only provided 8.3% protection [84].

Panya et al. [85] synthesized a DENV membrane (M) protein-mimicking peptide, MLH40, and evaluated its antiviral activity against all four DENV serotypes. At a concentration of 50 μM, the MLH40 peptide reduced significantly the DENV-2 titers and intracellular DENV E antigens to 28.75% and 49.73%, respectively. At a concentration of 100 μM, the inhibitory effects were more significant whereby the DENV-2 titers and intracellular DENV E antigens were further reduced to 14.29% and 28.67%, respectively. MLH40 peptide exhibited significant inhibitory effects during pretreatment as the amount of intracellular DENV E antigen was reduced to 44.75%, thereby indicating that MLH40 peptide played a significant role in inhibiting initial DENV entry. Since MLH40 peptide was found to be highly conserved among all four DENV serotypes, its inhibitory effects were also evaluated against all four DENV serotypes. The

MLH40 peptide exhibited similar inhibition rates against all four DENV serotypes with IC_{50} values of $30.35 \pm 1.25\,\mu M$, $31.41 \pm 1.09\,\mu M$, $27.95 \pm 1.41\,\mu M$, and $24.45 \pm 1.2\,\mu M$ for DENV-1, DENV-2, DENV-3, and DENV-4, respectively [85].

Lin et al. [86] synthesized 12 cyclic peptides targeting both the P and P′ sites of the DENV NS2B/NS3 protease. The K_i values of these cyclic peptides varied from $2.9\,\mu M$ (cyclic peptide 7) to $780.3\,\mu M$ (cyclic peptide 9), and thus cyclic peptide 7 was identified as the best inhibitor of the DENV NS2B/NS3 protease [86].

The antiviral activity of an endogenous dipeptide, carnosine, was evaluated against DENV-2 and Zika virus. Carnosine significantly inhibited the production of DENV-2 viral particles by >70% at the highest concentration tested ($200\,\mu M$) with an EC_{50} value of $52.3\,\mu M$. Additionally, carnosine also significantly reduced DENV-2 viral replication in DENV-2-infected Huh7 cells at concentrations of 100 and $200\,\mu M$. Carnosine exhibited significant virucidal effects whereby infection of Huh7 cells with DENV-2 treated with $200\,\mu M$ of carnosine showed a reduction of DENV-2 viral titer by 89.7%. Additionally, postinfection treatment of DENV-2-infected Huh7 cells with $200\,\mu M$ of carnosine also exhibited a significant reduction of DENV-2 viral titer by 89% [87] (Table 24.3).

TABLE 24.3   Synthetic peptides targeting DENV.

Peptides	Sequences	Target	References
Tetrapeptides 7 and 21	Tetrapeptide 7: Bz-Nle-Lys-Arg-Arg-H Tetrapeptide 21: Bz-Nle-Lys-Arg-Arg-B(OH)$_2$	DENV NS2B/NS3 protease	[81]
Tetrapeptides 21 and 22	Tetrapeptide 21: Bz-Lys-Arg-Arg-H Tetrapeptide 22: Bz-Arg-Arg-H	DENV NS2B/NS3 protease	[80]
Tripeptide 2	Phenylacetyl-KRR-H	DENV NS2B/NS3 protease	[82]
Retro dipeptide 3 and retro tripeptide 11	Retro dipeptide 3: R-Arg-Lys-NH$_2$ with an arylcyano-acrylamide group as the N-terminal cap Retro tripeptide 11: R-Arg-Lys-Nle-NH$_2$ with an arylcyano-acrylamide group as the N-terminal cap	DENV NS2B/NS3 protease	[83]
Anionic septapeptides	DDHELQD, DETELQD, DEVMLQD, and DEVLMQD	ND	[84]
MLH40	SVALVPHVGMGLETRTETWMSSEGAWKHVQRIETWILRHPG	DENV M protein	[85]
Cyclic peptide 7	ND	DENV NS2B/NS3 protease	[86]
Carnosine	β-alanine and L-histidine	DENV NS2B/NS3 protease	[87]

ND, Not determined.

24.2.4 Recombinant peptides

Several studies had utilized expression systems such as *E. coli* and HEK293T cells to produce recombinant peptides against DENV (Table 24.4). Retrocyclin-1 (RC-1), a disulfide cyclic peptide, was cloned and expressed as a soluble recombinant protein in *E. coli*. RC-1 exhibited 100% inhibition of DENV-2 NS2B/NS3 protease activity at $100\,\mu$M at $37°$C. The inhibition of the recombinant protease was temperature-dependent. The lowest IC_{50} value of RC-1 was reported at $40°$C ($14.1 \pm 1.2\,\mu$M) while the IC_{50} value of $21.4 \pm 1.6\,\mu$M was recorded at $37°$C. RC-1 exhibited high viral RNA reductions after simultaneous treatment (DENV-2 supernatant was treated with RC-1 before infecting the Vero cells with the RC-1-treated DENV-2 virus supernatant) at 48 hours ($70\% \pm 6.3$) and 72 hours ($85\% \pm 7.1$). A moderate reduction in viral replication was reported for pretreatment after 48 hours (40%) and posttreatment after 72 hours (45%) [88].

Plectasin (PLSN) was cloned and expressed as tandem recombinant peptides in *E. coli* and its antiviral activity against DENV-2 was evaluated. Recombinant PLSN was shown to inhibit the activity of DENV NS2B/NS3 protease at a K_i value of $5.03 \pm 0.98\,\mu$M. The recombinant PLSN exhibited the highest reduction of viral RNA replication at 24, 48, and 72 hours postinfection at a concentration of $20\,\mu$M [89].

Rothan et al. [71] fused protegrin-1 (PG-1) and plectasin (PLSN) with MAP30 protein to produce the recombinant peptide-fusion protein (PG-1-MAP30-PLSN) as inclusion bodies in *E. coli* and evaluated its antiviral activity against DENV-2. The recombinant peptide-fusion protein exhibited potent inhibitory activity against DENV-2 NS2B/NS3 protease with an IC_{50} value of $0.50 \pm 0.1\,\mu$M. Treatment of DENV-2-infected cells with the recombinant peptide-fusion protein resulted in high viral RNA reductions to $1.21 \pm 0.3 \times 10^7$ pfu/mL at 24 hours, $0.93 \pm 0.2 \times 10^7$ pfu/mL at 48 hours, and $0.82 \pm 0.2 \times 10^7$ pfu/mL at 72 hours. The results from the virus binding assays revealed that the recombinant peptide-fusion protein could potentially inhibit the binding of DENV-2 to Vero cells as evidenced by significant reductions of viral copy number to $0.74 \pm 0.2 \times 10^5$ at 24 hours, $0.65 \pm 0.1 \times 10^5$ at 48 hours, and $0.49 \pm 0.1 \times 10^5$ at 72 hours. The maximum inhibition activity of the recombinant peptide-fusion protein against DENV-2 was determined to be $18.6\% \pm 7.2$ at $0.75\,\mu$M with an EC_{50} of approximately $0.43\,\mu$M. Additionally, it was found that the recombinant peptide-fusion protein was able to protect the ICR mice from DENV-2 infection in a dose-dependent manner. ICR mice treated with 50 mg/kg of the recombinant peptide-fusion protein exhibited a 100% survival rate after 7 days post-DENV-2 infection [90].

The pr peptide was cloned and expressed as a recombinant protein in HEK293T cells and its antiviral activity against DENV-1 and DENV-2 was evaluated. It was found that the pr peptide strongly inhibited DENV-1 and DENV-2 fusion and infection in a dose-dependent manner with 45%−49% inhibition at $6\,\mu$M and 81%−85% inhibition at $30\,\mu$M [91].

24.3 Strategies to identify and develop antiviral peptides against dengue virus

24.3.1 Biopanning of phage display peptide libraries

The screening of phage display libraries utilizes phages to display random foreign peptides and it is widely used for the rapid identification of phages carrying specific targets [92].

TABLE 24.4 Recombinant peptides targeting DENV.

Peptides	Expression system	Sequences	Target	References
Retrocyclin-1 (RC-1)	*Escherichia coli*	GICRCICGRGICRCICGR	DENV NS2B/NS3 protease	[88]
Plectasin	*Escherichia coli*	MGFGCNGPWDEDDMQCHNHCKSIKGYKGGYCAKGGFVCKCY	DENV NS2B/NS3 protease	[89]
Protegrin-1-MAP30-plectasin	*Escherichia coli*	RGGRLCYCRRRFCVCVGRVPGVGVPGVGDVNFDLSTATAKTYT KFIEDFRATLPFSHKVYDIPLLYSTISDSRRFILLNLTSYAYETISV AIDVTNVVVAYRTRDVSYFFKESPPEAYNILFKGTRKITLPYTG NYENLQTAAHKIRENIDLGLPALSSAITTLFYYNAQSAPSALLVLI QTTAEAARFKYIERHVAKYVATNFKPNLAIISLENQWSALSKQIF LAQNQGGKFRNPVDLIKPTGERFQVTNVDSDVVKGNIKLLLNSR ASTADENFITTMTLLGESVVNVPGVGVPGVGGFGCNGPWDEDD MQCHNHCKSIKGYKGGYCAKGGFVCKCY	DENV-2 NS2B/NS3 protease	[90]
pr	HEK293T cells	ND	DENV-2 prM protein	[91]

ND, Not determined.

A selection of antiviral peptides targeting DENV was identified by biopanning of phage display peptide libraries (Table 24.5). Chew et al. [93] identified a novel antiviral peptide against DENV-2 from a phage display peptide library. The peptide gg-ww was discovered to have the highest frequency of occurrence (26.7%) during the four rounds of screening of the phage display peptide library, followed by peptide vd-yp (6.7%) and peptide eg-gp (6.7%). The phages that displayed peptide gg-ww exhibited the highest binding affinity for DENV-2 which was fivefold higher when compared to the wild-type phage. This indicated that peptide gg-ww could have a stronger interaction with DENV-2. The addition of peptide gg-ww to DENV-2-infected Vero cells resulted in a decrease in the cytopathic effects and plaque formation in a concentration-dependent manner at concentrations ranging from $60-250 \mu mol/I$. Treatment of DENV-2-infected Vero cells with $250 \mu mol/I$ of peptide gg-ww resulted in a decrease in plaque formation and intracellular viral RNA by 99% and 96%, respectively. Simultaneous treatment of DENV-2 with peptide gg-ww prior to infecting Vero cells exhibited high inhibitory effects ($>95\%$ inhibition at $250 \mu mol/I$), thereby indicating that peptide gg-ww inhibited DENV-2 during the viral entry stage of DENV infection [93].

A study conducted by de la Guardia et al. [94] identified three peptides with potent antiviral activities against DENV-2 from a phage display peptide library. At the concentration of $500 \mu M$, peptide ELLASPW exhibited $>90\%$ inhibition of DENV-2 infectivity while peptide STSFWIT inhibited $>70\%$ and peptide SYQSHYY inhibited $>50\%$. Further docking simulation studies revealed that these peptides were binding to domain III of the DENV-2 E protein [94].

Biopanning of a random library of phage-displayed peptides conducted by Songprakhon et al. [95] identified novel antiviral peptides against all four DENV serotypes. A total of 11 peptides that could bind effectively to DENV-2 NS1 protein were identified. It was revealed by in silico analysis that peptides 3, 4, 10, and 11 were able to bind spontaneously to DENV-2 NS1 protein due to their high negative values of binding free energy for peptide-receptor complex formation. At $10 \mu M$, all 4 peptides were able to reduce DENV-2 virion production significantly by $42\%-57\%$ at 4 hours postinfection. In addition, an increase in the concentration of the 4 peptides to $20 \mu M$ resulted in a more significant decrease in DENV-2 production by $69\%-79\%$ at 4 hours postinfection. Peptide 3 was observed to exhibit the highest efficacy in lowering DENV-2 production. However, treatment with peptide 3 at $20 \mu M$ also caused high cell death. Besides DENV-2, peptides

TABLE 24.5 Antiviral peptides targeting DENV which were identified by biopanning of phage display peptide libraries.

Peptides	Sequences	Target	References
Peptide gg-ww	GGARDAGKAEWW	ND	[93]
Peptides ELLASPW, STSFWIT, and SYQSHYY	ELLASPW, STSFWIT, and SYQSHYY	Domain III of the DENV E protein	[94]
Peptides 3 and 4	Peptide 3: QFGPVFTWLNHA Peptide 4: SFVNLWTPRYSL	DENV-1, DENV-2, and DENV-4 NS1 protein	[95]

ND, Not determined.

3 and 4 were observed to exhibit significant inhibitions of DENV-1 and DENV-4 production. High reductions in DENV production were observed against DENV-1 (62%) and DENV-2 (58%) after postinfection treatment with peptide 3. Significant reductions of DENV-2 (69%) and DENV-4 (64%) were also observed after postinfection treatment with peptide 4. However, no peptides were found to inhibit DENV-3 production [95].

24.3.2 Structure-based design of antiviral peptides

Another approach that can be utilized to design and develop antiviral peptides against DENV is the structure-based design approach. The structure-based design approach can be divided into three categories which are consisted of structure-based virtual screening (molecular docking), de novo ligand design, and structure-based rational design. Structure-based virtual screening involves the use of molecular docking to identify potential antiviral agents that can bind to the target binding site [96]. On the other hand, de novo ligand design involves the computational design of potential antiviral agents to fit into the target binding site [97]. The structure-based rational design approach involves the design of antiviral agents that target specific regions in DENV such as the E protein, viral enzymes, and capsid protein based on the available three-dimensional (3D) structures [98].

24.3.2.1 Molecular docking

The antiviral peptides targeting DENV which were identified by molecular docking are listed in Table 24.6. Tambunan and Alamudi [99] designed seven cyclopentapeptide inhibitors (CKRRC, CGRRC, CRGRC, CRTRC, CTRRC, CKRKC, and CRRKC) of DENV-2

TABLE 24.6 Antiviral peptides targeting DENV identified by molecular docking.

Peptides	Sequences	Target	References
Cyclopentapeptides	CKRRC, CGRRC, CRGRC, CRTRC, CTRRC, CKRKC, and CRRKC	DENV-2 NS2B/NS3 protease	[99]
Cyclopentapeptides	CTWYC and CYEFC	DENV NS5 methyltransferase	[100]
Porcine BNP (7−32)	DSGCFGRRLDRIGSLSGLGCNVLRRY	DENV E protein	[101]
Human [Tyr123] Prepro Endothelin (110−130) amide and human Urotensin II	Human [Tyr123] Prepro Endothelin (110−130) amide: CQCASQKDKKWSYCQAGKEI Human Urotensin II: ETPDCFWKYCV	DENV NS5 methyltransferase	[102]
Peptides 13, 16, 20, 25, fragment-attached hexapeptide, and CMK-attached hexapeptide	Peptide 13: Ser-Met'-Met'-Gly Peptide 16: Ser-Ile-Lys-Phe-Ala Peptide 20: Phe-Ile-Lys-Ala-Ser Peptide 25: Ala-Ile-Lys-Lys-Phe-Ser Fragment-attached hexapeptide: Ala-Ile[1-hexyl-4-phenylpiperazine]Lys-Lys-Phe-Ser CMK-attached hexapeptide: Ala-Ile-Lys [CH2Cl]-Lys-Phe-Ser	DENV NS2B/NS3 protease	[103]
Peptide EF	EF (Glu-Phe)	DENV E protein	[104]

NS2B/NS3 protease by analyzing the binding pocket of the catalytic site of DENV-2 NS2B/NS3 protease and the substrate specificity for the DENV-2 NS2B/NS3 protease. Docking studies revealed that cyclopentapeptide CKRKC was the best inhibitor of the DENV-2 NS2B/NS3 protease with an estimated binding free energy value of -8.39 kcal/mol and a K_i value of 0.707 μM [99].

Another docking study by Idrus et al. [100] identified two cyclopentapeptides, CTWYC and CYEFC, as potential NS5 MTase inhibitors. From the screening of 1,635 cyclopenta-peptides as potential inhibitors of the SAM site of DENV NS5 MTase, CTWYC exhibited the highest binding affinity with an estimated binding free energy value of -30.72 kcal/mol. On the other hand, from the screening of 736 cyclopentapeptides as potential inhibitors of the RNA-cap site of DENV NS5 MTase, CYEFC exhibited the highest binding affinity with an estimated binding free energy value of -22.89 kcal/mol [100].

Parikesit et al. [101] screened potential peptide inhibitors of DENV E protein from a total of 301 cyclic peptides via molecular docking. The three best peptides which exhibited the highest binding affinity for DENV E protein were porcine BNP (7−32), human Big endothelin-1 (1−38), and human Amylin with binding free energy values of -42.2298 kcal/mol, -40.8696 kcal/mol, and -39.7079 kcal/mol, respectively. Since porcine BNP (7−32) exhibited the least significant probability of side effects, it was selected for further analysis. Molecular dynamics simulation found that at 37°C, porcine BNP (7−32) interacted consistently with the fusion peptide segments of DENV E protein. It was revealed that porcine BNP (7−32) actively interacted with amino acid residues Glu44, Gly104, Asn103, Lys246, and Lys247 of DENV E protein during both molecular docking and molecular dynamics simulations. At 39°C, it was found that porcine BNP (7−32) interacted with amino acid residues Val151, Asp154, and Lys246 of DENV E protein from the initial stage until the cooling stage. The peptide-receptor interactions did not undergo any significant changes at temperatures of 37°C and 39°C. However, it was found that porcine BNP (7−32) was much more interactive with DENV E protein at 39°C than 37°C [101].

A study conducted by Tambunan et al. [102] screened potential NS5 MTase inhibitors from a total of 300 cyclic peptides. These 300 cyclic peptides were docked into two binding sites of the NS5 MTase, namely the S-Adenosyl methionine (SAM) site, and the RNA-cap site. Human [Tyr123] Prepro Endothelin (110−130) amide exhibited the highest binding affinity for the SAM site with a highly negative binding free energy value of -24.73 kcal/mol. On the other hand, human Urotensin II displayed the highest binding affinity for the RNA-cap site with a highly negative binding free energy value of -19.04 kcal/mol. Apart from that, molecular dynamics simulation was conducted to determine the stability of the peptide-NS5 MTase complex at temperatures of 37°C and 39°C. The NS5 MTase-human [Tyr123] Prepro Endothelin (110−130) amide complex did not undergo any significant conformational changes at both temperatures. However, the NS5 MTase-human Urotensin II complex was observed to maintain a more stable conformation at 39°C as compared to 37°C [102].

Next, Velmurugan et al. [103] designed several tripeptides, tetrapeptides, pentapeptides, and hexapeptides that could act as potential DENV NS2B/NS3 protease inhibitors by analyzing the binding pocket of the catalytic site of DENV NS2B/NS3 protease and the specificity of the substrate for DENV NS2B/NS3 protease. The docking study revealed that peptides 13, 16, 20, and 25 as well as fragment-attached hexapeptide Ala-Ile[1-hexyl-4-phenylpiperazine] Lys-Lys-Phe-Ser and CMK-attached hexapeptide Ala-Ile-Lys[CH2Cl]-Lys-Phe-Ser provided highly favorable interactions with DENV NS2B/NS3 protease as they were able to inter-act with amino acid residues Ser135, Asp75, and His51 from the catalytic triad of DENV NS2B/NS3 protease. These interactions were considered to be highly favorable as the activity of DENV NS2B/NS3 protease was mainly carried out by amino acid residues in the catalytic triad. Therefore, interactions with amino acid residues in the catalytic triad could result in potential inhibition of the catalytic activity of the DENV NS2B/NS3 protease [103].

Seven small peptides (KEN, WEN, EEG, WET, EQ, EH, and EF) that bind specifically to the hydrophobic pocket of the DENV E protein via molecular docking were reported by Panya et al. [104]. Three distinct hot spots (H1, H2, and H3) which were hypothesized to be vital for binding and played an important role in the inhibitory effects of several com-pounds (doxycycline, relitetracycline, compound A5, compound 36, compound 11) were identified at the hydrophobic pocket of the DENV E protein. The top seven peptides with high docking scores that could bind to at least one of the hot spot sites were selected for further in vitro studies to evaluate their antiviral activities against DENV-2. Peptides EF and KEN were found to significantly reduce the formation of DENV-2 foci by 90% and their respective IC$_{50}$ values were determined to be 96.50 µM and 331.9 µM, respectively. Subsequent in vitro studies of peptide EF revealed that treatment of Vero cells with 200 µM of peptide EF significantly reduced DENV-2 RNA and protein production by 83.47% and 84.15%, respectively. This indicated that peptide EF played a vital role in the inhibition of viral entry into cells as well as inhibition of RNA and protein production in DENV-2. At a concentration of 100 µM, peptide EF was discovered to significantly reduce the amount of E proteins of DENV-1, DENV-3, and DENV-4 by approximately 20%−40% whereas it reduced the amounts of E protein of DENV-2 by approximately 70% [104].

However, the peptide inhibitors that were identified from the docking studies by Tambunan and Alamudi [99], Idrus et al. [100], Parikesit et al. [101], Tambunan et al. [102], and Velmurugan et al. [103] have not been validated and thus, future in vitro studies are necessary to further evaluate the efficacy of these antiviral peptides.

24.3.2.2 De novo design of antiviral peptides

The antiviral peptides targeting DENV which were developed by de novo design are pre-sented in Table 24.7. Alhoot et al. [105] utilized the BioMoDroid algorithm to design several

TABLE 24.7 Antiviral peptides targeting DENV which were developed by de novo design.

Peptides	Sequences	Target	References
Peptides DET2 and DET4	DET2: PWLKPGDLDL DET4: AGVKDGKLDF	Domain III of the DENV E protein	[105]
Peptides DN57opt and DN81opt	DN57opt: RWMVWRHWFHRLRLPYNPGKNKQNQQWP DN81opt: RQMRAWGQDYQHGGMGYSC	Domain I/domain II hinge of the DENV E protein	[106]
Peptide 1OAN1	FWFTLIKTQAKQPARYRRFC	Extended beta sheet connection of domain I/domain II of the DENV E protein	[106,107]

antiviral peptides that targeted domain III of the DENV-2 E protein and selected four peptides (DET1, DET2, DET3, and DET4) that exhibited the best scores (representing the in silico bonding strength of the peptide-receptor complex) for further evaluation of their antiviral activities against DENV-2. Peptides DET2 and DET4 showed 40.6% $\pm$ 24.8 and 84.6% $\pm$ 5.6 reductions of plaque formation, respectively. Peptides DET2 and DET4 also significantly reduced viral RNA copy number by 0.29 fold $\pm$ 0.16 (28.6% $\pm$ 16.3) and 0.81 fold $\pm$ 0.07 (81.0% $\pm$ 7.0), respectively. Peptide DET2 exhibited maximum inhibitory activity of 41.5% $\pm$ 20.0 at 200 μM against DENV-2 with IC_{50} above 500 μM. Peptide DET4 exhibited maximum inhibitory activity of 84.6% $\pm$ 5.6 at 500 μM against DENV-2 with IC_{50} of 35 μM. In addition, it was shown by transmission electron microscopy that the surface of DENV-2 viral particles treated with peptides DET2 and DET4 became irregular and with rough edges, thus suggesting a possible rearrangement of the DENV-2 E protein on the surface [105].

Costin et al. [106] designed seven peptides (DN57opt, DN80opt, DN81opt, 1OAN1, 1OAN2, 1OAN3, and 1OAN4) using the structural data from the prefusion DENV-2 E protein and evaluated their antiviral activities against DENV-2. The respective residue-specific all-atom probability discriminatory function scores obtained via this approach were used to identify in situ amino acid sequences that had the highest probability of having high structural and binding stability. Peptides DN57opt, DN80opt, and DN81opt were optimized variants of peptides that were originally designed from DENV inhibitory peptide sequences located in domain II near the domain I/domain II hinge region by Hrobowski et al. [106,108]. Peptides 1OAN1, 1OAN2, 1OAN3, and 1OAN4 were novel peptides designed from an extended beta-sheet region comprising the first connection between domains I and II. Peptides DN57opt and DN81opt exhibited significant inhibition of DENV-2 and their IC_{50} values were determined to be 8 $\pm$ 1 μM and 36 $\pm$ 6 μM, respectively. Both DN57opt and DN81opt displayed maximum inhibition of 97% at 20 μM and 57% at 50 μM against DENV-2, respectively. Among the four novel peptides that were designed, only peptide 1OAN1 inhibited DENV-2 effectively with an IC_{50} of 7 $\pm$ 4 μM and maximum inhibition of 99% at 50 μM. The results from cryo-electron microscopy revealed that treatment of DENV-2 with peptide 1OAN1 caused the surface of the DENV virion to transition from smooth to rough, indicating that the peptide caused an alteration in the arrangement of the surface of the E protein [106]. Additionally, Nicholson et al. [107] evaluated the ability of peptide 1OAN1, which was originally designed by Costin et al. [106], in blocking DENV ADE in cell cultures using Fc receptor II (FcRII) of human K562 cells. Incubation of anti-DENV serum with peptide 1OAN1 was found to exhibit a dose-responsive inhibition of ADE with 50% inhibitory concentration at 3 μM [106,107].

24.3.2.3 *Rational design of antiviral peptides*

Three approaches can be used for rational design which are physicochemical, template-based design, and de novo methods. The physicochemical approach uses a known peptide sequence to produce peptide analogs with different physicochemical properties. Similarly, the template-based design also uses a known peptide sequence to produce peptide analogs with increased selectivity and activity. This is performed by adding new amino acid residues or changing the positions of the original amino acid residues. On the other hand, the de novo method utilizes amino acid patterns or frequencies to generate novel peptides [109].

Antiviral peptides targeting DENV developed by rational design that showed inhibition of DENV are presented in Table 24.8. Hrobowski et al. [108] utilized the Wimley-White interfacial hydrophobicity scale to design several peptides based on the E protein sequences of DENV (peptides DN80, DN57, DN81, and DN59) and WNV and the peptides that exhibited significant WWHIS scores were evaluated for their antiviral activities against both DENV-2 and WNV. Peptide DN59 exhibited maximum inhibitory activity of $100.0 \pm 0.5\%$ at $20 \, \mu M$ against DENV-2. DN59 exhibited >99% inhibition of DENV-2 plaque formation at concentrations of $<25 \, \mu M$ and cross-inhibition of WNV fusion/infectivity by >99% inhibition at $<25 \, \mu M$ [108]. A followup study which was conducted by Nicholson et al. [107] evaluated the ability of peptide DN59 in blocking DENV ADE in cell cultures using Fc receptor II (FcRII) of human K562 cells. Anti-DENV serum which was incubated with peptide DN59 exhibited a dose-responsive inhibition of ADE with 50% inhibitory concentration at $6 \, \mu M$ [107]. Lok et al. [110] further evaluated the ability of peptide DN59 to cause the release of DENV RNA genome from the virus particle. It was shown that peptide DN59 reduced the infectivity of all four dengue serotypes by 50% (IC_{50}) at concentrations of $2-5 \, \mu M$. The cryo-electron microscopy data revealed that treatment of DENV-2 viral particles with peptide DN59 disrupted the symmetry of the virus particle. Subsequent three-dimensional (3D) icosahedral re-construction of the viral particles revealed the absence of the RNA genome and suggested the formation of holes at the fivefold vertices of the virus particle. It was also discovered that the peptide concentration ($17 \, \mu M$) required to yield 50% degradation of the RNA genome was approximately fourfold higher than the peptide concentration ($4.8 \, \mu M$) which caused a 50% reduction in DENV-2 infectivity. It was confirmed that peptide DN59 acted directly on the virus particle to release the DENV RNA genome and not on any other viral or cellular target. Peptide DN59 was also found to induce the formation of holes in the viral membrane [110].

Schimidt et al. [112] designed several peptides based on the stem region of DENV-2 ($DV2^{396-408}$, $DV2^{396-412}$, $DV2^{396-423}$, $DV2^{396-429}$, $DV2^{416-427}$, $DV2^{420-429}$, $DV2^{420-435}$, $DV2^{413-435}$, $DV2^{413-440}$, $DV2^{413-447}$, $DV2^{419-447}$, $DV2^{419-440}$, and $DV2^{429-447}$) and evaluated their binding affinities with the prefusion DENV-2 soluble E protein (sE) dimer and the postfusion DENV-2 sE trimer. A representative stem peptide, $DV2^{419-447}$, was selected and its antiviral activity was evaluated against DENV-2. The addition of $DV2^{419-447}$ peptide at a concentration of $1 \, \mu M$ exhibited potent inhibition of membrane fusion. $DV2^{413-447}$ and $DV2^{419-447}$ peptides were determined to be the strongest inhibitors of DENV-2 infectivity. It was also established that the inhibition of DENV-2 infectivity by $DV2^{419-447}$ peptide was due to a direct interaction between the peptide and DENV-2 virion which blocked an early step of DENV infection [111]. In another study, peptides corresponding

TABLE 24.8 Antiviral peptides targeting DENV which were developed by rational design.

Peptides	Sequences	Target	References
Peptide DN59	MAILGDTAWDFGSLGGVFTSIGKALHQVFGAIY	DENV E protein	[107,108,110]
Peptide $DV2^{419-447}$	AWDFGSLGGVFTSIGKALHQVFGAIYGAA	DENV E protein	[111,112]
Peptides P4 and P7	P4: CKIPFEIMDLEKRHV P7: GVEPGQLKLNWFKK	ND	[113]

ND, Not determined.

to residues 419 to 447 derived from the stem region of each of the four DENV serotypes ($DV^{419-447}$ peptides) were evaluated for their respective antiviral activities against all four DENV serotypes. These peptides were conjugated with a C-terminal solubility tag, RGKGR, which was known to be compatible with inhibitory activity. It was found that the $DV2^{419-447}$ peptide was the strongest inhibitor of all four dengue serotypes, followed by $DV1^{419-447}$, $DV3^{419-447}$, and $DV4^{419-447}$ peptides. Although all the DENV serotypes could be inhibited to some extent, DENV-3 was particularly insensitive to inhibition by $DV1^{419-447}$ and $DV4^{419-447}$ peptides [112].

Six peptides (P2 to P7) based on the amino acid sequence of domain III of the DENV-2 E protein (EDIII) and two control peptides (P1 and P8) based on domain II of DENV-2 E protein (EDII) and the stem region of DENV-2 E protein were designed and their antiviral activities were evaluated against DENV-1 and DENV-2. Peptides P4 and P7 at 40 μM were shown to reduce the growth of DENV-2 in human umbilical vein endothelial cells (HUVECs) by $70 \pm 9\%$ and $65 \pm 6\%$, respectively. Pretreatment of HUVECs confirmed that peptides P4 and P7 acted on the cells and not on the DENV-2 viral particles. Subsequent in vitro studies to evaluate the ability of peptides P4 and P7 to inhibit the entry of DENV-1 and DENV-2 into HUVECs revealed that P7 significantly inhibited the entry of DENV-1 into HUVECs with an inhibition rate of $65 \pm 12\%$ at a concentration of 40 μM [113].

24.4 Direct interactions between antiviral peptides with host cell receptors and enzymes

24.4.1 Interactions between antiviral peptides and dengue virus host cell receptors

Several interactions between antiviral peptides and DENV host cell receptors such as HS receptor, dendritic cell-specific intercellular adhesion molecule 3-grabbing nonintegrin (DC-SIGN) receptor, LDLR, and β3 integrin receptor have been identified. The HS receptor of host cells was reported to play a vital role in the inhibition of DENV-2 infection by bLF. The infection rate of HS-expressing Chinese hamster ovary (CHO)-K1 cells was observed to be significantly greater than that of HS-deficient CHO-pgsA745 cells without the addition of bLF [72]. This indicated that HS played a vital role in DENV-2 infection. Furthermore, with the addition of bLF, the infection rates of HS-expressing CHO-K1 cells were reduced significantly in a dose-dependent manner at concentrations ranging from 0.1–200 μg/mL. On the other hand, it was discovered that the infection rates of HS-deficient CHO-pgsA745 cells were not affected by the addition of bLF at concentrations ranging from 0.1–200 μg/mL, thus suggesting that HS expression played a vital role in bLF inhibition of DENV-2 infection [72].

The DC-SIGN receptor also played an essential role in the anti-DENV-2 activity of bLF. It was discovered that the infection rate of DC-SIGN-expressing THP-1 cells was significantly greater than that of DC-SIGN-deficient THP-1 cells without the addition of bLF. This indicated that DC-SIGN played a vital role in DENV-2 infection. With the addition of bLF at a concentration of 200 μg/mL, it was observed that the infection rate of DC-SIGN-expressing THP-1 cells was significantly reduced. However, the infection rates of DC-

SIGN-deficient THP-1 cells with or without the addition of bLF were identical to one another. This indicated that bLF interacted with DC-SIGN to protect DC-SIGN-expressing THP-1 cells from DENV-2 infection [72].

Apart from that, Chen et al. [72] discovered that the LDLR played a vital role in the anti-DENV-2 activity of bLF. It was revealed that anti-LDLR antibodies inhibited DENV-2 infection by 36%, thus suggesting the involvement of LDLR in DENV-2 entry into host cells. The role of bLF in inhibiting LDLR-DENV-2 interactions was demonstrated when a significant reduction in the number of plaques was observed when DENV-2 infected Vero cells were treated with 200 μg/mL of bLF (62% inhibition). The addition of 200 ng/mL of recombinant LDLR (rLDLR) to Vero cells was found to reduce the inhibitory effects of bLF and the number of plaques rose significantly from 38% in the absence of rLDLR to 58% upon addition of 200 ng/mL of rLDLR. This indicated that bLF potentially interacted with LDLR and rLDLR competed with host LDLR for bLF to reduce the inhibitory effects of bLF against DENV-2 [72].

Cui et al. [113] reported the potential interaction between DENV and β3 integrin receptor by observing immunofluorescence which revealed that there were high levels of expression of β3 integrin in HUVECs infected with DENV-2 at 24 hours postinfection. The expression levels of β3 integrin were higher at 48 and 72 hours postinfection. Colocalization of DENV-2 antigens with β3 integrin was also observed, thus indicating that there was a possible interaction between β3 integrin and DENV-2 proteins. The yeast two-hybrid membrane protein system was utilzed to identify the DENV-2 protein which interacted with the β3 integrin. DENV-2 E and NS2B proteins were shown to interact with β3 integrin. The specific viral protein that interacted with β3 integrin was identified by transfecting plasmids carrying genes encoding for each of the DENV-2 proteins into HUVECs. The results revealed that the DENV-2 E protein had the most significant colocalization with β3 integrin, thus further confirming that the DENV-2 E protein might be a vital component for the interaction between DENV-2 and β3 integrin. The sites at which the DENV-2 E protein interacted with β3 integrin were confirmed by surface plasmon resonance studies. Domain III of DENV-2 E protein (EDIII) was shown to exhibit the highest binding affinity with β3 integrin, thus indicating that the binding domain to β3 integrin was located in EDIII [113].

24.4.2 Interactions between antiviral peptides and dengue virus proteases

Interactions between antiviral peptides and the DENV NS2B/NS3 protease have also been identified. Docking studies by Rothan et al. [87] discovered that carnosine had a high binding affinity for the DENV-2 NS2B/NS3 protease. Computational modeling studies predicted that carnosine could potentially interact with amino acid residues from the catalytic triad of the DENV NS2B/NS3 protease. It was further predicted that carnosine was able to form hydrogen bonds with His107, Pro187, Ile92, Gly206, Ser190, and Gly208 of the DENV-2 NS2B/NS3 protease. Further analysis verified that His107 and Ser190 were amino acid residues from the catalytic triad of DENV NS2B/NS3 protease, thereby confirming the interactions between carnosine and amino acid residues from the catalytic triad of the DENV NS2B/NS3 protease. The interaction between carnosine and DENV-2 NS2B/NS3 protease was validated by biochemical analysis using fluorogenic substrates whereby carnosine exhibited DENV-2 NS2B/NS3 protease inhibition with an IC_{50} of 63.7 μM [87].

Another docking study conducted by Rothan et al. [71] discovered that the Ltc 1 peptide interacted with the DENV NS2B/NS3 protease near its active site. The binding affinity of Ltc 1 peptide depended on the hydrophobic interactions between its four leucine residues and two tryptophan residues with the hydrophobic residues of the DENV NS2B/NS3 protease. The interaction between Ltc 1 peptide and DENV NS2B/NS3 protease was validated by the dengue NS2B/NS3 protease assay whereby Ltc 1 peptide exhibited significant inhibition of DENV NS2B/NS3 protease in a dose-dependent manner at concentrations ranging from 20−120 μM [71].

Velmurugan et al. [103] discovered that peptides 13, 16, 20, and 25 as well as the fragment-attached hexapeptide Ala-Ile[1-hexyl-4-phenylpiperazine]Lys-Lys-Phe-Ser and CMK-attached hexapeptide Ala-Ile-Lys[CH2Cl]-Lys-Phe-Ser could interact with amino acid residues Ser135, Asp75, and His51 of the catalytic triad in the DENV NS2B/NS3 protease [103].

Next, docking studies conducted by Tambunan and Alamudi [99] discovered that the cyclopentapeptide CKRKC was able to interact with the DENV NS2B/NS3 protease by forming hydrogen bonds with 9 amino acid residues (including amino acid residues His51, Asp75, and Ser135 of the catalytic triad) in the active site of the protease. It was also revealed that the shape of cyclopentapeptide CKRKC complemented the shape of the binding pocket of DENV NS2B/NS3 protease, which indicated the formation of a favorable stable enzyme-ligand complex [99].

Docking studies conducted by Xu et al. [70] with peptide 9 revealed that peptide 9 was anchored tightly in the catalytic pocket of the DENV NS2B/NS3 protease. The lysine and arginine side chains of peptide 9 were inserted into the negatively-charged cleft of the binding pocket of the DENV NS2B/NS3 protease, forming an extensive hydrogen-bonding network [70]. Molecular dynamics simulations conducted by Lin et al. [86] had shown that cyclic peptide 7 interacted with the active site of the DENV NS2B/NS3 protease [86].

However, the interactions between antiviral peptides and the DENV NS2B/NS3 protease that were identified from studies conducted by Tambunan and Alamudi [99] and Velmurugan et al. [103] have not been validated experimentally thus, future in vitro studies are needed to further confirm these interactions.

24.4.3 Interactions between antiviral peptides and dengue virus methyltransferases

Interactions between antiviral peptides and the DENV NS5 methyltransferase have been identified. Docking studies conducted by Tambunan et al. [102] discovered two commercial cyclic peptides, human [Tyr123] Prepro Endothelin (110−130) amide and human Urotensin II, which were able to interact with the SAM and RNA-cap sites of DENV NS5 methyltransferase, respectively. The human [Tyr123] Prepro Endothelin (110−130) amide was able to form hydrogen bonds with Glu111, Phe113, Glu149, Ser150, Val132, Cys179, Lys181, Leu183, Glu217, Lys29, Ly42, Arg57, Lys105, Arg163, Trp171, Arg212, and Thr215 from the SAM site of the DENV NS5 methyltransferase. On the other hand, human

Urotensin II formed hydrogen bonds with Val156, Glu157, Lys14, Lys29, Ser151, Asn153, Arg160, Asn184, Tyr186, and Trp13 from the RNA-cap site of the DENV NS5 methyltransferase [102].

Idrus et al. [100] reported two cyclopentapeptides, CTWYC and CYEFC, which exhibited high binding affinities for the SAM and RNA-cap sites of DENV NS5 methyltransferase, respectively. CTWYC formed three polar interactions with the SAM site, namely, polar basic, acidic, and uncharged interactions. Polar basic residues such as Lys61, Lys105, and Lys181 of the SAM site of the DENV NS5 methyltransferase interacted with CTWYC by forming hydrogen bonds with its cysteine (Lys61 and Lys181) and tyrosine (Lys105) residues. Polar acidic Glu217 of the SAM site of the DENV NS5 methyltransferase interacted with CTWYC by forming hydrogen bonds with its cysteine residue while the polar uncharged Ser150 of the SAM site of DENV NS5 methyltransferase interacted with CTWYC by forming hydrogen bonds with its threonine residue [100]. CYEFC also formed three polar interactions with the RNA-cap site, namely, polar basic, acidic, and uncharged interactions. The polar basic Lys29 of the RNA-cap site of DENV NS5 methyltransferase interacted with CYEFC by forming hydrogen bonds with the side chain of its glutamate residue. The polar acidic Glu149 of the RNA-cap site of the DENV NS5 methyltransferase interacted with CYEFC by forming hydrogen bonds with its cysteine residue. The polar uncharged Asn18, Ser150, and Ser214 of the RNA-cap site of DENV NS5 methyltransferase interacted with CYEFC by forming hydrogen bonds with its tyrosine, cysteine, and glutamic acid residues, respectively [100].

However, the interactions between antiviral peptides and the DENV NS5 methyltransferase that were identified from studies conducted by Tambunan et al. [102] and Idrus et al. [100] have not been validated and thus, future in vitro studies are needed to further confirm these interactions.

24.5 Advantages of peptides as antiviral agents

The advantages of peptides over small molecule compounds are low toxicity, good safety, high selectivity and potency, predictable metabolism, commercial scalability, and balance of conformational rigidity and flexibility [61,114–116]. These advantages have made peptides promising candidates for drug development, especially to target binding interfaces that were proven to be challenging for small molecule compounds.

Several antiviral peptides have been approved by the United States Food and Drug Administration (FDA) for clinical use in recent years. The first antiviral peptide drug that was approved by the FDA in 2003 was Enfuvirtide (Enf), a peptide fusion inhibitor of HIV-1. Enf is a 36-amino acid peptide that corresponds to amino acid residues 127–162 of the heptad-repeat 2 (HR2) domain of the gp41 HIV E protein. It functions by blocking the interactions between HR1 and HR2 domains by binding to the hydrophobic grooves of the HR1 trimer, thus blocking the formation of HIV fusion with host cells, leading to inhibition of HIV infection [117]. Boceprevir and Telaprevir are antiviral peptide drugs approved by the FDA in 2011 for clinical use against HCV. They function by acting on the NS3/4A protein of HCV and inhibit viral replications [118]. Additionally, Myrcludex B, Adaptavir, and Aviptadil are antiviral peptide drugs that are currently undergoing phase II of clinical trials. Myrcludex B functions by inhibiting the sodium taurocholate cotransporting polypeptide to treat hepatitis B virus (HBV) infections, Adaptavir functions as an antagonist of the C-C chemokine receptor type 5 (CCR5) to treat HBV infections, and Aviptadil functions by inhibiting intraleukin-6, tissue necrotic factor α, and N-methyl-D-aspartate-induced caspase 3 activations to treat severe acute respiratory syndrome coronavirus 2 (SARS-CoV-2) infections [119–121].

24.6 Limitations of peptides

However, despite having advantages, peptides have their disadvantages as well. The disadvantages of peptides when compared to small molecule compounds include having a short half-life, a high tendency for aggregation, chemical and physical instability, low cell membrane permeability, low bioavailability, and potential immunogenicity [61,114,115].

24.6.1 Chemical modifications to overcome peptide limitations

Chemical modifications can be utilized to increase the in vivo half-life of peptides by preventing proteolytic degradation. First, the N- or C-terminal sequences of the peptides can be modified to reduce proteolytic degradation and to improve the bioavailability of the peptides [116]. Apart from that, substituting certain critical amino acid residues within the peptides with synthetic enantiomer amino acids could also help in enhancing peptide resistance to proteases. Peptide cyclization could also be used in enhancing the protease resistance of peptides [116]. Posttranslational modifications such as amidation, acetylation, and lipidization could also help to improve peptide stability and increase the half-life of the peptides by enhancing protease resistance of the peptides [122–124]. Additionally, stapling of peptides could also lead to an improved plasma half-life and an enhanced protease resistance [125]. Stapled peptides are α-helical peptides with linking residues such as olefinic amino acids (UAA) on the backbone of the hydrocarbon chain at positions such as i, i + 3, i + 4, or i + 7 [126]. The addition of PEG (polyethylene glycol) to the peptides could also aid in attaining dual solubility (PEG is a molecule with both hydrophilic and hydrophobic properties), improving peptide stability, and enhancing protease resistance of the peptides (due to steric hindrance) [127].

There are also various methods to improve the cell membrane permeability of peptides. The intracellular uptake of peptides could be improved by modulating the hydrophobicity and electrostatic charges to improve the passive uptake of the peptides. Conjugation of the active peptide drug to a cell-penetrating peptide could also facilitate the active transport of the peptides into the respective target cell [116]. Posttranslational modifications such as the addition of fatty acid chains or cholesterol could also help in improving cell membrane permeability [128].

24.6.2 Delivery of peptides using nanomaterials

Due to the challenges associated with peptide drug delivery, the use of nanomaterials to enhance the oral delivery of peptides has been studied. Nanoparticles that are made of polymeric particles such as ester amides, lactic acid, and fatty acid esters could provide hydrophobic interactions, resulting in an increase in its retention time at the mucosal lining. On the other hand, inorganic particles such as selenium, silica, and gold have better drug release profiles than polymeric particles as they can withstand an acidic environment [129]. Nanocrystallization or nanosuspension has also been proposed as potential ways to improve drug penetration, stability, and retention time of biological activity [130,131]. Nanoparticles are also a potential option to overcome the challenges associated with easily degradable or thermolabile peptides [131]. Although nanoparticles could provide improved delivery of peptide drugs, they could also potentially affect the pharmacodynamics, pharmacokinetics, and therapeutic properties of the encapsulated peptide drugs [130].

24.7 Conclusion

Inhibition of DENVs by antiviral peptides derived from natural resources such as plants and animals has been widely studied in vitro, but there is a lack of validation in animal models. High-throughput screening of phage libraries has contributed to the discovery of antiviral peptides that showed potent inhibition against dengue in in vitro studies. The structure-based design approach has also contributed to the discovery of antiviral peptides against dengue. This approach can be divided into three categories which are structure-based virtual screening (molecular docking), de novo ligand design, and structure-based rational design. However, despite the vast discovery of antiviral peptides from these approaches, there is also a lack of validation in animal models. Therefore, future in vivo studies is needed to select potent antidengue therapeutics for progression into clinical trials.

Apart from that, predictions of interactions between peptides and potential target sites such as host cell receptors and DENV enzymes were highlighted for future in vitro validation studies on the design of antiviral peptides against these alternative targets. Peptides have various advantages over small molecule compounds due to their high specificity and selectivity properties. However, peptides face various limitations such as having a short half-life and high susceptibility to protease degradation which could be overcome by encapsulation in nanomaterials or chemical modifications to enhance their oral bioavailability and in vivo stability. In addition, the production cost of antiviral peptides is high due to solid-phase synthesis [132]. Cost reductions could be achieved by utilizing newer methods of chemical synthesis or through the production of recombinant peptides [89,133].

Acknowledgments

We acknowledged that this book chapter has been produced under the FRGS grant (FRGS/1/2020/SKK06/SYUC/03/1) and the Research Centre grant for Centre for Virus and Vaccine Research.

Disclosure of interest

The authors declare no conflicts of interest.

References

[1] Beatty ME, Stone A, Fitzsimons DW, Hanna JN, Lam SK, Vong S, et al. Best practices in dengue surveillance: a report from the Asia-Pacific and Americas dengue prevention boards. PLoS Negl Trop Dis 2010;4(11):e890.

[2] Bhatt S, Gething PW, Brady OJ, Messina JP, Farlow AW, Moyes CL, et al. The global distribution and burden of dengue. Nature. 2013;496(7446):504−7.

[3] World Health Organization. Dengue and severe dengue, <https://www.who.int/news-room/fact-sheets/detail/dengue-and-severe-dengue>; 2021 [accessed 05.08.21].

[4] Murray NE, Quam MB, Wilder-Smith A. Epidemiology of dengue: past, present and future prospects. Clin Epidemiol 2013;5:299−309.

[5] Bouri N, Sell TK, Franco C, Adalja AA, Henderson DA, Hynes NA. Return of epidemic dengue in the United States: implications for the public health practitioner. Public Health Rep 2012;127(3):259−66.

[6] Frentiu FD, Zakir T, Walker T, Popovici J, Pyke AT, van den Hurk A, et al. Limited dengue virus replication in field-collected *Aedes aegypti* mosquitoes infected with Wolbachia. PLoS Negl Trop Dis 2014;8(2):e2688.

[7] Phuc HK, Andreasen MH, Burton RS, Vass C, Epton MJ, Pape G, et al. Late-acting dominant lethal genetic systems and mosquito control. BMC Biol 2007;5:11.

[8] Guirakhoo F, Pugachev K, Zhang Z, Myers G, Levenbook I, Draper K, et al. Safety and efficacy of chimeric yellow fever-dengue virus tetravalent vaccine formulations in nonhuman primates. J Virol 2004;78(9):4761−75.

[9] Guirakhoo F, Weltzin R, Chambers TJ, Zhang ZX, Soike K, Ratterree M, et al. Recombinant chimeric yellow fever-dengue type 2 virus is immunogenic and protective in nonhuman primates. J Virol 2000;74(12):5477−85.

[10] Capeding MR, Tran NH, Hadinegoro SR, Ismail HI, Chotpitayasunondh T, Chua MN, et al. Clinical efficacy and safety of a novel tetravalent dengue vaccine in healthy children in Asia: a phase 3, randomised, observer-masked, placebo-controlled trial. Lancet. 2014;384(9951):1358−65.

[11] Villar L, Dayan GH, Arredondo-García JL, Rivera DM, Cunha R, Deseda C, et al. Efficacy of a tetravalent dengue vaccine in children in Latin America. N Engl J Med 2015;372(2):113−23.

[12] Hadinegoro SR, Arredondo-García JL, Capeding MR, Deseda C, Chotpitayasunondh T, Dietze R, et al. Efficacy and long-term safety of a dengue vaccine in regions of endemic disease. N Engl J Med 2015;373(13):1195−206.

[13] Fatima K, Syed NI. Dengvaxia controversy: impact on vaccine hesitancy. J Glob Health 2018;8(2):010312.

[14] World Health Organization. Weekly epidemiological record, <http://apps.who.int/iris/bitstream/handle/10665/274315/WER9336.pdf?ua = 1>; 2018 [accessed 05.08.21].

[15] Low JG, Gatsinga R, Vasudevan SG, Sampath A. Dengue antiviral development: a continuing journey. Adv Exp Med Biol 2018;1062:319−32.

[16] Yang JM, Chen YF, Tu YY, Yen KR, Yang YL. Combinatorial computational approaches to identify tetracycline derivatives as flavivirus inhibitors. PLoS One 2007;2(5):e428.

[17] Botta L, Rivara M, Zuliani V, Radi M. Drug repurposing approaches to fight dengue virus infection and related diseases. Front Biosci (Landmark Ed) 2018;23:997−1019.

[18] Kaufmann SHE, Dorhoi A, Hotchkiss RS, Bartenschlager R. Host-directed therapies for bacterial and viral infections. Nat Rev Drug Discov 2018;17(1):35−56.

[19] Marques RE, Guabiraba R, Del Sarto JL, Rocha RF, Queiroz AL, Cisalpino D, et al. Dengue virus requires the CC-chemokine receptor CCR5 for replication and infection development. Immunology. 2015;145(4):583−96.

[20] Richard AS, Zhang A, Park SJ, Farzan M, Zong M, Choe H. Virion-associated phosphatidylethanolamine promotes TIM1-mediated infection by ebola, dengue, and West Nile viruses. Proc Natl Acad Sci U S A 2015;112(47):14682−7.

[21] Guzman MG, Halstead SB, Artsob H, Buchy P, Farrar J, Gubler DJ, et al. Dengue: a continuing global threat. Nat Rev Microbiol 2010;8(12 Suppl.):S7−16.

[22] Sayce AC, Miller JL, Zitzmann N. Targeting a host process as an antiviral approach against dengue virus. Trends Microbiol 2010;18(7):323−30.

[23] Messina JP, Brady OJ, Scott TW, Zou C, Pigott DM, Duda KA, et al. Global spread of dengue virus types: mapping the 70 year history. Trends Microbiol 2014;22(3):138−46.

[24] Halstead SB. Dengue. Lancet. 2007;370(9599):1644−52.

[25] Simmons CP, Farrar JJ, Nguyen V, Wills B. Dengue. N Engl J Med 2012;366(15):1423−32.

[26] Rothman AL. Immunity to dengue virus: a tale of original antigenic sin and tropical cytokine storms. Nat Rev Immunol 2011;11(8):532–43.

[27] St John AL, Abraham SN, Gubler DJ. Barriers to preclinical investigations of anti-dengue immunity and dengue pathogenesis. Nat Rev Microbiol 2013;11(6):420–6.

[28] Flipse J, Diosa-Toro MA, Hoornweg TE, van de Pol DP, Urcuqui-Inchima S, Smit JM. Antibody-dependent enhancement of dengue virus infection in primary human macrophages; balancing higher fusion against antiviral responses. Sci Rep 2016;6:29201.

[29] Midgley CM, Bajwa-Joseph M, Vasanawathana S, Limpitikul W, Wills B, Flanagan A, et al. An in-depth analysis of original antigenic sin in dengue virus infection. J Virol 2011;85(1):410–21.

[30] Bich TD, Pham OK, Hai DH, Nguyen NM, Van HN, The TD, et al. A pregnant woman with acute cardiorespiratory failure: dengue myocarditis. Lancet. 2015;385(9974):1260.

[31] Carod-Artal FJ, Wichmann O, Farrar J, Gascón J. Neurological complications of dengue virus infection. Lancet Neurol 2013;12(9):906–19.

[32] Yacoub S, Wertheim H, Simmons CP, Screaton G, Wills B. Cardiovascular manifestations of the emerging dengue pandemic. Nat Rev Cardiol 2014;11(6):335–45.

[33] Chambers TJ, Hahn CS, Galler R, Rice CM. Flavivirus genome organization, expression, and replication. Annu Rev Microbiol 1990;44:649–88.

[34] Falgout B, Markoff L. Evidence that flavivirus NS1-NS2A cleavage is mediated by a membrane-bound host protease in the endoplasmic reticulum. J Virol 1995;69(11):7232–43.

[35] Byk LA, Gamarnik AV. Properties and functions of the dengue virus capsid protein. Annu Rev Virol 2016;3(1):263–81.

[36] El Sahili A, Lescar J. Dengue virus non-structural protein 5. Viruses. 2017;9(4).

[37] Gopala Reddy SB, Chin WX, Shivananju NS. Dengue virus NS2 and NS4: minor proteins, mammoth roles. Biochem Pharmacol 2018;154:54–63.

[38] Rosales Ramirez R, Ludert JE. The dengue virus nonstructural protein 1 (NS1) is secreted from mosquito cells in association with the intracellular cholesterol transporter chaperone caveolin complex. J Virol 2019;93(4).

[39] Silva EM, Conde JN, Allonso D, Ventura GT, Coelho DR, Carneiro PH, et al. Dengue virus nonstructural 3 protein interacts directly with human glyceraldehyde-3-phosphate dehydrogenase (GAPDH) and reduces its glycolytic activity. Sci Rep 2019;9(1):2651.

[40] Upanan S, Kuadkitkan A, Smith DR. Identification of dengue virus binding proteins using affinity chromatography. J Virol Methods 2008;151(2):325–8.

[41] Wahid SF, Sanusi S, Zawawi MM, Ali RA. A comparison of the pattern of liver involvement in dengue hemorrhagic fever with classic dengue fever. Southeast Asian J Trop Med Public Health 2000;31(2):259–63.

[42] Marovich M, Grouard-Vogel G, Louder M, Eller M, Sun W, Wu SJ, et al. Human dendritic cells as targets of dengue virus infection. J Investig Dermatol Symp Proc 2001;6(3):219–24.

[43] Tassaneetrithep B, Burgess TH, Granelli-Piperno A, Trumpfheller C, Finke J, Sun W, et al. DC-SIGN (CD209) mediates dengue virus infection of human dendritic cells. J Exp Med 2003;197(7):823–9.

[44] Huerta V, Chinea G, Fleitas N, Sarría M, Sánchez J, Toledo P, et al. Characterization of the interaction of domain III of the envelope protein of dengue virus with putative receptors from CHO cells. Virus Res 2008;137(2):225–34.

[45] Swaminathan S, Khanna N. Dengue: recent advances in biology and current status of translational research. Curr Mol Med 2009;9(2):152–73.

[46] Chu JJ, Leong PW, Ng ML. Analysis of the endocytic pathway mediating the infectious entry of mosquito-borne flavivirus West Nile into Aedes albopictus mosquito (C6/36) cells. Virology. 2006;349(2):463–75.

[47] Chu JJ, Ng ML. Infectious entry of West Nile virus occurs through a clathrin-mediated endocytic pathway. J Virol 2004;78(19):10543–55.

[48] Krishnan MN, Sukumaran B, Pal U, Agaisse H, Murray JL, Hodge TW, et al. Rab 5 is required for the cellular entry of dengue and West Nile viruses. J Virol 2007;81(9):4881–5.

[49] Heinz FX, Stiasny K, Allison SL. The entry machinery of flaviviruses. Arch Virol Suppl 2004;18:133–7.

[50] Stiasny K, Allison SL, Schalich J, Heinz FX. Membrane interactions of the tick-borne encephalitis virus fusion protein E at low pH. J Virol 2002;76(8):3784–90.

[51] Harrison SC. Viral membrane fusion. Nat Struct Mol Biol 2008;15(7):690–8.

[52] Kielian M, Rey FA. Virus membrane-fusion proteins: more than one way to make a hairpin. Nat Rev Microbiol 2006;4(1):67–76.

[53] Clyde K, Kyle JL, Harris E. Recent advances in deciphering viral and host determinants of dengue virus replication and pathogenesis. J Virol 2006;80(23):11418−31.

[54] Rodenhuis-Zybert IA, Wilschut J, Smit JM. Dengue virus life cycle: viral and host factors modulating infectivity. Cell Mol Life Sci 2010;67(16):2773−86.

[55] Samsa MM, Mondotte JA, Iglesias NG, Assunção-Miranda I, Barbosa-Lima G, Da Poian AT, et al. Dengue virus capsid protein usurps lipid droplets for viral particle formation. PLoS Pathog 2009;5(10):e1000632.

[56] Zhang Y, Zhang W, Ogata S, Clements D, Strauss JH, Baker TS, et al. Conformational changes of the flavivirus E glycoprotein. Structure. 2004;12(9):1607−18.

[57] Bressanelli S, Stiasny K, Allison SL, Stura EA, Duquerroy S, Lescar J, et al. Structure of a flavivirus envelope glycoprotein in its low-pH-induced membrane fusion conformation. EMBO J 2004;23(4):728−38.

[58] Stadler K, Allison SL, Schalich J, Heinz FX. Proteolytic activation of tick-borne encephalitis virus by furin. J Virol 1997;71(11):8475−81.

[59] Zybert IA, van der Ende-Metselaar H, Wilschut J, Smit JM. Functional importance of dengue virus maturation: infectious properties of immature virions. J Gen Virol 2008;89(Pt 12):3047−51.

[60] Vilas Boas LCP, Campos ML, Berlanda RLA, de Carvalho Neves N, Franco OL. Antiviral peptides as promising therapeutic drugs. Cell Mol Life Sci 2019;76(18):3525−42.

[61] Galdiero S, Falanga A, Tarallo R, Russo L, Galdiero E, Cantisani M, et al. Peptide inhibitors against herpes simplex virus infections. J Pept Sci 2013;19(3):148−58.

[62] Qureshi A, Thakur N, Tandon H, Kumar M. AVPdb: a database of experimentally validated antiviral peptides targeting medically important viruses. Nucleic Acids Res 2014;42(Database issue):D1147−53.

[63] van der Strate BW, Beljaars L, Molema G, Harmsen MC, Meijer DK. Antiviral activities of lactoferrin. Antivir Res 2001;52(3):225−39.

[64] Uhlig T, Kyprianou T, Martinelli FG, Oppici CA, Heiligers D, Hills D, et al. The emergence of peptides in the pharmaceutical business: from exploration to exploitation. EuPA Open Proteom 2014;4:58−69.

[65] López-Otín C, Matrisian LM. Emerging roles of proteases in tumour suppression. Nat Rev Cancer 2007;7(10):800−8.

[66] Lim SP. Dengue drug discovery: progress, challenges and outlook. Antivir Res 2019;163:156−78.

[67] Chan CY, Ooi EE. Dengue: an update on treatment options. Future Microbiol 2015;10(12):2017−31.

[68] Kaptein SJ, Neyts J. Towards antiviral therapies for treating dengue virus infections. Curr Opin Pharmacol 2016;30:1−7.

[69] Rothan HA, Abdulrahman AY, Sasikumer PG, Othman S, Rahman NA, Yusof R. Protegrin-1 inhibits dengue NS2B-NS3 serine protease and viral replication in MK2 cells. J Biomed Biotechnol 2012;2012:251482.

[70] Xu S, Li H, Shao X, Fan C, Ericksen B, Liu J, et al. Critical effect of peptide cyclization on the potency of peptide inhibitors against dengue virus NS2B-NS3 protease. J Med Chem 2012;55(15):6881−7.

[71] Rothan HA, Bahrani H, Rahman NA, Yusof R. Identification of natural antimicrobial agents to treat dengue infection: in vitro analysis of latarcin peptide activity against dengue virus. BMC Microbiol 2014;14:140.

[72] Chen JM, Fan YC, Lin JW, Chen YY, Hsu WL, Chiou SS. Bovine lactoferrin inhibits dengue virus infectivity by interacting with heparan sulfate, low-density lipoprotein receptor, and DC-SIGN. Int J Mol Sci 2017;18(9).

[73] Monteiro JMC, Oliveira MD, Dias RS, Nacif-Marçal L, Feio RN, Ferreira SO, et al. The antimicrobial peptide HS-1 inhibits dengue virus infection. Virology. 2018;514:79−87.

[74] El-Bitar AMH, Sarhan M, Abdel-Rahman MA, Quintero-Hernandez V, Aoki-Utsubo C, Moustafa MA, et al. Smp76, a scorpine-like peptide isolated from the venom of the scorpion. Int J Pept Res Ther 2020;26(2):811−21.

[75] Gao Y, Cui T, Lam Y. Synthesis and disulfide bond connectivity-activity studies of a kalata B1-inspired cyclopeptide against dengue NS2B-NS3 protease. Bioorg Med Chem 2010;18(3):1331−6.

[76] Panya A, Yongpitakwattana P, Budchart P, Sawasdee N, Krobthong S, Paemanee A, et al. Novel bioactive peptides demonstrating anti-dengue virus activity isolated from the asian medicinal plant acacia catechu. Chem Biol Drug Des 2019;93(2):100−9.

[77] Panya A, Sawasdee N, Songprakhon P, Tragoolpua Y, Rotarayanont S, Choowongkomon K, et al. A synthetic bioactive peptide derived from the Asian medicinal plant. Viruses. 2020;12(11).

[78] Gordts SC, Renders M, Férir G, Huskens D, Van Damme EJ, Peumans W, et al. NICTABA and UDA, two GlcNAc-binding lectins with unique antiviral activity profiles. J Antimicrob Chemother 2015;70(6):1674−85.

[79] Nguyen PQT, Ooi JSG, Nguyen NTK, Wang S, Huang M, Liu DX, et al. Antiviral cystine knot α-amylase inhibitors from Alstonia scholaris. J Biol Chem 2015;290(52):31138−50.

[80] Yin Z, Patel SJ, Wang WL, Chan WL, Ranga Rao KR, Wang G, et al. Peptide inhibitors of dengue virus NS3 protease. Part 2: SAR study of tetrapeptide aldehyde inhibitors. Bioorg Med Chem Lett 2006;16(1):40−3.

[81] Yin Z, Patel SJ, Wang WL, Wang G, Chan WL, Rao KR, et al. Peptide inhibitors of dengue virus NS3 protease. Part 1: warhead. Bioorg Med Chem Lett 2006;16(1):36−9.

[82] Schüller A, Yin Z, Brian Chia CS, Doan DN, Kim HK, Shang L, et al. Tripeptide inhibitors of dengue and West Nile virus NS2B-NS3 protease. Antivir Res 2011;92(1):96−101.

[83] Nitsche C, Behnam MA, Steuer C, Klein CD. Retro peptide-hybrids as selective inhibitors of the dengue virus NS2B-NS3 protease. Antivir Res 2012;94(1):72−9.

[84] Laosutthipong C, Kanthong N, Flegel TW. Novel, anionic, antiviral septapeptides from mosquito cells also protect monkey cells against dengue virus. Antivir Res 2013;98(3):449−56.

[85] Panya A, Sawasdee N, Junking M, Srisawat C, Choowongkomon K, Yenchitsomanus PT. A peptide inhibitor derived from the conserved ectodomain region of DENV membrane (M) protein with activity against dengue virus infection. Chem Biol Drug Des 2015;86(5):1093−104.

[86] Lin KH, Ali A, Rusere L, Soumana DI, Kurt Yilmaz N, Schiffer CA. Dengue virus NS2B/NS3 protease inhibitors exploiting the prime side. J Virol 2017;91(10).

[87] Rothan HA, Abdulrahman AY, Khazali AS, Nor Rashid N, Chong TT, Yusof R. Carnosine exhibits significant antiviral activity against dengue and zika virus. J Pept Sci 2019;25(8):e3196.

[88] Rothan HA, Han HC, Ramasamy TS, Othman S, Rahman NA, Yusof R. Inhibition of dengue NS2B-NS3 protease and viral replication in vero cells by recombinant retrocyclin-1. BMC Infect Dis 2012;12:314.

[89] Rothan HA, Mohamed Z, Suhaeb AM, Rahman NA, Yusof R. Antiviral cationic peptides as a strategy for innovation in global health therapeutics for dengue virus: high yield production of the biologically active recombinant plectasin peptide. OMICS. 2013;17(11):560−7.

[90] Rothan HA, Bahrani H, Mohamed Z, Abd Rahman N, Yusof R. Fusion of protegrin-1 and plectasin to MAP30 shows significant inhibition activity against dengue virus replication. PLoS One 2014;9(4):e94561.

[91] Zheng A, Umashankar M, Kielian M. In vitro and in vivo studies identify important features of dengue virus Pr-E protein interactions. PLoS Pathog 2010;6(10):e1001157.

[92] Matsubara T. Potential of peptides as inhibitors and mimotopes: selection of carbohydrate-mimetic peptides from phage display libraries. J Nucleic Acids 2012;2012:740982.

[93] Chew MF, Tham HW, Rajik M, Sharifah SH. Anti-dengue virus serotype 2 activity and mode of action of a novel peptide. J Appl Microbiol 2015;119(4):1170−80.

[94] de la Guardia C, Quijada M, Lleonart R. Phage-displayed peptides selected to bind envelope glycoprotein show antiviral activity against dengue virus serotype 2. Adv Virol 2017;2017:1827341.

[95] Songprakhon P, Thaingtamtanha T, Limjindaporn T, Puttikhunt C, Srisawat C, Luangaram P, et al. Peptides targeting dengue viral nonstructural protein 1 inhibit dengue virus production. Sci Rep 2020;10(1):12933.

[96] Lionta E, Spyrou G, Vassilatis DK, Cournia Z. Structure-based virtual screening for drug discovery: principles, applications and recent advances. Curr Top Med Chem 2014;14(16):1923−38.

[97] Suryanarayanan V, Panwar U, Chandra I, Singh SK. De novo design of ligands using computational methods. Methods Mol Biol 2018;1762:71−86.

[98] Selzer PM, Marhöfer RJ, Koch O. Protein structures and structure-based rational drug design. Applied bioinformatics. Springer; 2018. p. 73−89.

[99] Tambunan US, Alamudi S. Designing cyclic peptide inhibitor of dengue virus NS3-NS2B protease by using molecular docking approach. Bioinformation. 2010;5(6):250−4.

[100] Idrus S, Tambunan US, Zubaidi AA. Designing cyclopentapeptide inhibitor as potential antiviral drug for dengue virus ns5 methyltransferase. Bioinformation. 2012;8(8):348−52.

[101] Parikesit AA, Kinanty, Tambunan US. Screening of commercial cyclic peptides as inhibitor envelope protein dengue virus (DENV) through molecular docking and molecular dynamics. Pak J Biol Sci 2013;16(24):1836−48.

[102] Tambunan US, Zahroh H, Utomo BB, Parikesit AA. Screening of commercial cyclic peptide as inhibitor NS5 methyltransferase of dengue virus through molecular docking and molecular dynamics simulation. Bioinformation. 2014;10(1):23−7.

[103] Velmurugan D, Mythily U, Rao K. Design and docking studies of peptide inhibitors as potential antiviral drugs for dengue virus ns2b/ns3 protease. Protein Pept Lett 2014;21(8):815−27.

[104] Panya A, Bangphoomi K, Choowongkomon K, Yenchitsomanus PT. Peptide inhibitors against dengue virus infection. Chem Biol Drug Des 2014;84(2):148−57.

[105] Alhoot MA, Rathinam AK, Wang SM, Manikam R, Sekaran SD. Inhibition of dengue virus entry into target cells using synthetic antiviral peptides. Int J Med Sci 2013;10(6):719–29.

[106] Costin JM, Jenwitheesuk E, Lok SM, Hunsperger E, Conrads KA, Fontaine KA, et al. Structural optimization and de novo design of dengue virus entry inhibitory peptides. PLoS Negl Trop Dis 2010;4(6):e721.

[107] Nicholson CO, Costin JM, Rowe DK, Lin L, Jenwitheesuk E, Samudrala R, et al. Viral entry inhibitors block dengue antibody-dependent enhancement in vitro. Antivir Res 2011;89(1):71–4.

[108] Hrobowski YM, Garry RF, Michael SF. Peptide inhibitors of dengue virus and West Nile virus infectivity. Virol J 2005;2:49.

[109] Porto WF. Prediction and rational design of antimicrobial peptides. sine loco: IntechOpen; 2012.

[110] Lok SM, Costin JM, Hrobowski YM, Hoffmann AR, Rowe DK, Kukkaro P, et al. Release of dengue virus genome induced by a peptide inhibitor. PLoS One 2012;7(11):e50995.

[111] Schmidt AG, Yang PL, Harrison SC. Peptide inhibitors of dengue-virus entry target a late-stage fusion intermediate. PLoS Pathog 2010;6(4):e1000851.

[112] Schmidt AG, Yang PL, Harrison SC. Peptide inhibitors of flavivirus entry derived from the E protein stem. J Virol 2010;84(24):12549–54.

[113] Cui X, Wu Y, Fan D, Gao N, Ming Y, Wang P, et al. Peptides P4 and P7 derived from E protein inhibit entry of dengue virus serotype 2 via interacting with β3 integrin. Antivir Res 2018;155:20–7.

[114] Castel G, Chtéoui M, Heyd B, Tordo N. Phage display of combinatorial peptide libraries: application to antiviral research. Molecules. 2011;16(5):3499–518.

[115] Fosgerau K, Hoffmann T. Peptide therapeutics: current status and future directions. Drug Discov Today 2015;20(1):122–8.

[116] Lee AC, Harris JL, Khanna KK, Hong JH. A comprehensive review on current advances in peptide drug development and design. Int J Mol Sci 2019;20(10).

[117] Teissier E, Penin F, Pécheur EI. Targeting cell entry of enveloped viruses as an antiviral strategy. Molecules. 2010;16(1):221–50.

[118] Divyashree M, Mani MK, Reddy D, Kumavath R, Ghosh P, Azevedo V, et al. Clinical applications of antimicrobial peptides (AMPs): where do we stand now? Protein Pept Lett 2020;27(2):120–34.

[119] Safety and efficacy of ADAPTAVIR's ability to eliminate treatment-resistant infectious virus in peripheral blood mononuclear cells (PBMCs). 2009 [Accessed 11 August 2021] < https://clinicaltrials.gov/ct2/show/NCT00951743 >.

[120] Myrcludex B plus pegylated interferon-alpha-2a in patients with HBeAg negative HBV/HDV co-infection. 2018 [Accessed 11 August 2021] < https://clinicaltrials.gov/ct2/show/NCT02637999 >.

[121] Intravenous aviptadil for critical COVID-19 with respiratory failure (COVID-AIV). 2021 [Accessed 11 August 2021] < https://clinicaltrials.gov/ct2/show/NCT04311697 >.

[122] Gentilucci L, De Marco R, Cerisoli L. Chemical modifications designed to improve peptide stability: incorporation of non-natural amino acids, pseudo-peptide bonds, and cyclization. Curr Pharm Des 2010;16(28):3185–203.

[123] Nguyen LT, Chau JK, Perry NA, de Boer L, Zaat SA, Vogel HJ. Serum stabilities of short tryptophan- and arginine-rich antimicrobial peptide analogs. PLoS One 2010;5(9).

[124] Shartouny JR, Jacob J. Mining the tree of life: host defense peptides as antiviral therapeutics. Semin Cell Dev Biol 2019;88:147–55.

[125] Bird GH, Madani N, Perry AF, Princiotto AM, Supko JG, He X, et al. Hydrocarbon double-stapling remedies the proteolytic instability of a lengthy peptide therapeutic. Proc Natl Acad Sci U S A 2010;107(32):14093–8.

[126] Nyanguile O. Peptide Antiviral strategies as an alternative to treat lower respiratory viral infections. Front Immunol 2019;10:1366.

[127] Suk JS, Xu Q, Kim N, Hanes J, Ensign LM. PEGylation as a strategy for improving nanoparticle-based drug and gene delivery. Adv Drug Deliv Rev 2016;99(Pt A):28–51.

[128] Papo N, Oren Z, Pag U, Sahl HG, Shai Y. The consequence of sequence alteration of an amphipathic alpha-helical antimicrobial peptide and its diastereomers. J Biol Chem 2002;277(37):33913–21.

[129] Brown TD, Whitehead KA, Mitragotri S. Materials for oral delivery of proteins and peptides. Nat Rev Mater 2020;5(2):127–48.

[130] Cao SJ, Xu S, Wang HM, Ling Y, Dong J, Xia RD, et al. Nanoparticles: oral delivery for protein and peptide drugs. AAPS PharmSciTech 2019;20(5):190.

[131] Lembo D, Donalisio M, Civra A, Argenziano M, Cavalli R. Nanomedicine formulations for the delivery of antiviral drugs: a promising solution for the treatment of viral infections. Expert Opin Drug Deliv 2018;15 (1):93−114.

[132] Fields GB. Introduction to peptide synthesis. Current protocols in protein science. Hoboken, NJ: John Wiley & Sons; 2001. p. 18.1.1−18.1.9.

[133] Pattabiraman VR, Bode JW. Rethinking amide bond synthesis. Nature. 2011;480(7378):471−9.

mRNA vaccines for COVID-19

Anamika Sengupta

Department of Health Science Education & Pathology, College of Medicine, University of
Illinois, Peoria, IL, United States

25.1 Introduction

The severe acute respiratory syndrome coronavirus 2 (SARS-CoV-2) was first identified in December 2019, in Wuhan, China. By March 2020, the rapid spread of the virus posed a serious threat to the public health infrastructure of major countries across the globe. In spontaneity, the quick progression of the debilitating COVID-19 disease (characterized by respiratory symptoms such as fever, unproductive cough, myalgia, and fatigue) had likewise spun into a global epidemic, affecting huge populations across international boundaries. It was summarily declared a pandemic by the World Health Organization (March 2020). By December 2020, the devastating effects of COVID-19 had resulted in over 80 million infections and 1.8 million deaths [1], highlighting the urgency for the development of a safe and efficacious vaccine against the SARS-CoV-2 infection.

As countries across the globe imposed total or partial lockdowns (to control the rapid spread of the deadly virus and its ensuing death toll), unprecedented global efforts to generate successful vaccines for SARS-CoV-2 sprung up in many technologically advanced countries (United States, Germany, United Kingdom, and China) along with some developing nations, like India. By the end of 2020, 61 SARS-CoV-2 vaccine candidates were already under clinical evaluation as were 172 others in the preclinical development phase [2]. While some traditional vaccine approaches (inactivated or live-attenuated viral vaccines, recombinant subunit vaccines) looked promising [3], novel platforms like the nucleic acid-based messenger RNA (mRNA) vaccines were also under investigation [4,5]. The mRNA vaccines aimed to deliver the genetic information to produce viral antigens like the spike protein (S protein), the main protease, for the induction of an immune response in the recipient. There was initial resistance, however, from a "fear of the unknown" related to this novel mRNA approach, while vaccine safety was unquestioned with the more traditional vaccine platforms (due to their ubiquitous use over several decades).

This chapter is an attempt to summarize the history of the development and the crucial aspects related to the design and delivery strategies related to the first two COVID-19 mRNA vaccines, that received "emergency use authorization" (EUA) from the Food and Drug Administration (FDA) in the United States, in early 2021. The vaccines include:

- SARS-CoV-2 mRNA vaccine mRNA-1273, codeveloped by Moderna (Cambridge, MA, USA) and the National Institutes of Allergy and Infectious Diseases Vaccine Research Center
- BNT162b1 and BNT162b2, codeveloped by BioNTech (Mainz, Germany) and Pfizer (New York City, NY, USA).

Furthermore, the chapter has attempted not only to place emphasis on the immune responses induced by the above-mentioned mRNA vaccines, but to discuss the strengths and challenges associated with the usage of these novel vaccines, and highlight associated safety issues as well as summarize their mechanism of action with emphasis on effective strategies to improve vaccine safety and efficacy.

25.2 General advantages associated with messenger RNA vaccines

The use of mRNA-based vaccines has several beneficial features over the subunits, "killed" and "live" attenuated virus, as well as DNA-based vaccines [6]. They are summarized as follows:

- *Safety*: Since mRNA is a noninfectious, nonintegrating platform, there is no potential risk of infection or insertional mutagenesis. Furthermore, as they carry minimal genetic vectors, antivector immunity can be avoided, thus increasing the possibility of repeated administration of the vaccine.
- *Efficacy*: mRNA vaccines can be readily modified by switching the target immunogenic epitopes, the only requirement being the template DNA sequence of the antigen. Various structural modifications can also be introduced into the antigen mRNA to make them stable and translatable. Additionally, the SARS-CoV-2 vaccine constructs can be rapidly adjusted to target newly emerging viral strains [7]. The two FDA-approved SARS-CoV-2 mRNA vaccines (mRNA-1273 and BNT162b) reported a greater than 94% efficacy [8,9].
- *Delivery*: Efficient delivery is achievable by formulating mRNA into carrier molecules like ionized lipid nanoparticles, permitting rapid uptake and expression of the mRNA into the cytoplasm of the recipient cells [10]. Complex delivery methods involving electroporation (required by DNA vaccines) or the addition of an adjuvant (required with protein vaccines), can be wholly avoided in the delivery process of mRNA vaccines.
- *Contamination*: With the lack of viral growth requirement, the possibility of other contaminating viruses from cell lines can be eliminated.
- *Manufacturing*: As the only need for the production of an mRNA vaccine candidate is the DNA template of the desired antigen, the manufacturing process of such vaccines is more rapid [6]. mRNA vaccines have the potential for inexpensive and scalable

manufacturing, owing to the high yields of in vitro transcription reactions. This is in stark contrast to the protracted development of traditional vaccine platforms due to the inherently slow nature of cultivating cell lines and generating virus and/or producing clinical-grade protein subunits.

Clinical testing of the first mRNA vaccine candidate (mRNA-1273) began in March 2020 [Safety and Immunogenicity Study of 2019-nCoV Vaccine (mRNA-1273)], within 66 days of the release of the SARS-CoV-2 sequence in January 2020 [11]. The second vaccine candidate (BNT162b1 and BN162b2) entered phases I & II of the clinical trials a month later [A Trial Investigating the Safety and Effects of Four BNT162 Vaccines. 2019; Study to Describe the Safety, Tolerability, Immunogenicity, and Efficacy of RNA Vaccine Candidates 905].

25.3 General concerns associated with messenger RNA vaccines

Exogenous mRNA is immunostimulatory. It is recognized by a variety of cell surfaces as well as endosomal and cytosolic innate immune receptors [12]. Thus, the initial concerns with these vaccines were their undesirable side effects, likely induced by the reactogenicity and immunogenicity associated with exogenous mRNA administrations. Some of the concerns were:

- Single-stranded mRNA molecules themselves act as pathogen-associated molecular pattern sensors (PAMPs) upon exogenous delivery to cells [6].
- Single-stranded oligoribonucleotides and their degradative products are also detected by the endosomal sensors [Toll-like receptor 7 (TLR7) and TLR8 (REFS)] [13−15] resulting in type I interferon production [16].
- Activation of other molecular components like the retinoic acid-inducible gene I [6,17] could activate an inflammatory response.
- Innate immune sensing of mRNA (which is yet another concern) has been associated with the inhibition of antigen expression and negative effects on immune response [17,18]. However, the positive side is that the inherent immunogenicity and reactogenicity associated with exogenous mRNA delivery as vaccines can be readily downmodulated to further increase the safety profile [17,19]. The immunogenicity of mRNAs could also be potentially advantageous for vaccinations as this feature may help elicit robust T- and B-cell immune responses through dendritic cell (DC) maturation [6].

25.4 The target viral antigen selection for the COVID-19 messenger RNA vaccines

An important factor underlying antigen selection for an mRNA vaccine is the choice of an immunogenic target antigen that would also elicit protective immunity against the pathogen [4]. The SARS-CoV-2 spike (S) glycoprotein, a 180−200 kDa major surface protein with key roles in receptor recognition and fusion to the host cell membrane [20], was the antigen of choice for the COVID-19 mRNA vaccine development [21]. Listed below are the structural and functional features of the natural SARS-CoV-2 spike (S) glycoprotein:

- The spike (S) glycoprotein is a 1273 amino acid long protein consisting of an extracellular N-terminus with a signal peptide (amino acids 1−13), a transmembrane (TM) domain anchored in the viral membrane, and a short intracellular C-terminal segment [22].
- The Spike glycoprotein is characterized by the presence of subunits S1 (14−685 residues) and S2 (686−1273 residues) both of which play profound roles in viral binding to the host cell, viral entry, and viral fusion. The S1 and S2 subunits are joined by a furin cleavage site (for posttranslational cleavage of the subunits) unique to this coronavirus [23,24].
- The S1 subunit contains a receptor-binding domain (RBD) that recognizes and binds to the host cell angiotensin-converting enzyme 2 (ACE2) receptor [25], mediating viral attachment to the host cells in the form of a trimer [26]. Nine ACE2-contacting residues in CoV RBD are fully conserved while four remain partially so. The RBD region is a critical target for neutralizing antibodies (nAbs) produced against the virus by the host cell during natural infections with the virus.

Functional significance of the S1 subunit: Responsible for the binding of viral particles to the ACE2 receptors on the surface of the host cell which marks the initiation of virus infection. The receptor recognition event is thus an important determinant for viral entry and often a target for drug designing.

- The S2 subunit is composed of the fusion peptide (FP-788−806 residues), heptapeptide repeat sequence 1 (HR1-912-984 residues), heptapeptide repeat sequence 2 (HR2-1163−1213 residues), TM domain (1213−1237 residues), and the cytoplasmic domain (1237−1273) [27]. HR1 is located at the C-terminus of a hydrophobic FP, while HR2 is located at the N-terminus of the TM domain [28]. The TM domain anchors the S protein to the viral membrane. The S2 subunit ends in a CT tail.

Functional significance of the S2 subunit: The S2 subunit is responsible for viral fusion (fusion of the viral and host cell membranes, resulting in the release of the viral genome into the host cell), thus marking the entry of the virus into the host cell. HR1 and HR2 form the six-helical bundle (6-HB), essential for the viral fusion and entry function of the S2 subunit.

- The S1 and S2 subunits of the S protein stay associated until bound to the ACE2 receptor via the RBD, leading to irreversible conformational changes. Cleavage of the S1 and S2 subunits at the furin cleavage site by host proteases is the basis of the viral fusion. The subunits exist in a noncovalent form until viral fusion occurs [29].

25.5 Development of the COVID-19 messenger RNA vaccines

December 11 of year 2020 was marked by a global public health breakthrough. On this date, the first protective measure against the largest global pandemic to strike in over 100 years was approved. The US FDA issued the first EUA for the vaccine to prevent COVID-19, manufactured by Pfizer-BioNTech (BNT162b2; trade name: Comirnaty; generic name: tozinameran) [30]. A week later, the approval of a second COVID-19 vaccine followed, manufactured by Moderna [(mRNA-1273) (Moderna COVID-19 Vaccine) [31]. These two vaccines are also documented as being the first approved synthetic mRNA vaccines.

Like natural mRNAs that carry genetic instruction in the form of codons for protein production in every cell of our body, synthetic mRNAs can do the same but are designed to encode proteins with therapeutic effect [32]. Both vaccines are nonreplicating synthetic mRNAs that are capable of encoding the production of a single component of the SARS-CoV-2 genome. The mRNA naturally decomposes without integration into the host cell genome and thus is incapable of causing COVID-19 infection upon administration of the vaccine in the recipient. The two vaccines encode the full-length SARS-CoV-2 S protein with two mutations (K986P and V987P). The mutations ensure the antigen produced remains in an antigenically favorable prefusion conformation [20,33]. The structure and function of the SARS-CoV-2 S protein were summarized in the previous section of the chapter.

25.5.1 The characteristic features of the sequence of the Pfizer-BioNTech (BNT162b2) mRNA vaccine

It is a 4284-nucleotide linear mRNA sequence with several structural modifications essential for protecting the extremely labile and rapidly degradable mRNA, thus increasing its half-life. The structural modifications, as described by Vogel et al. [34], are listed below.

- A 5'-cap (m7(3'OMeG)(5')ppp(5')(2'OMeA)pG), commonly referred to as trinucleotide "cap 1" helps recruit the ribosome for efficient protein production and also protect the mRNA from degradation [35].
- A 5'-untranslated region (UTR) derived from the human α-globin mRNA with an optimized Kozak sequence that drives increased translation from the correct start codon [36].
- A codon-optimized coding sequence that specifies the production of the TM-anchored immunogenic SARS-CoV-2 spike glycoprotein.
- A 3'-UTR consists of two sequences derived from the amino-terminal enhancer of split mRNA and the mitochondrial encoded 12S rRNA. Both these sequences promote increased protein expression by stabilizing the mRNA [37].
- An unusual 3'-terminus consisting of two segmented poly(adenosine) tracts. The poly(adenosine) stretches play an important role in regulating mRNA translation as well as providing stability. The segmented structure helps reduce unwanted recombination [38].
- In the case of BNT162b2, every uridine residue in the synthetic mRNA strand is replaced with N1-methyl-pseudouridine (m1Ψ) [39]. This was based on the work done by Karikó et al. [17], the first to test a variety of naturally occurring nucleoside modifications in mRNA molecules (pseudo uridine 5-methylcytidine, N6-methyladenosine, 5-methyluridine, and 2-thiouridine). Of these variants, the incorporation of N1-methyl-pseudouridine (m1Ψ) in place of uridine led to a 10-fold increase in translation over unmodified mRNA. Another significant association with the

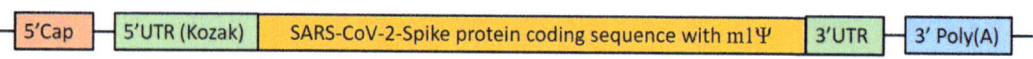

FIGURE 25.1   The design elements in the synthetic spike protein mRNA used in COVID-19 mRNA vaccines. UTR: Untranslated regions at 5′ and 3′ ends Kozak: Marks the position of the start codon for initiation of translation Poly (A): Segmented 3′ poly adenosine tail m1Ψ: N1-methyl-pseudouridine residues replacing uridine residues in the synthetic mRNA.

replacement of rare codons with frequently used synonymous codons was a reduced trigger of the PAMP mechanisms. Fig. 25.1 shows the design elements of the COVID-19 mRNA vaccines.

25.6 Lipid nanoparticles-mediated delivery of the COVID-19 messenger RNA vaccines

Biodegradable yet ionizable LNPs have been the preferred method for delivery of the mRNAs. The two authorized COVID-19 mRNA vaccines use LNPs to deliver the antigen mRNA [8,40] into the arms of the vaccine recipients. The choice of LNPs is based on the following:

- LNPs can efficiently deliver mRNA in vivo [13,41]. The ionizable lipid is positively charged and forms complexes with the negatively charged mRNA. This influences cellular uptake and endosomal escape [14,42] of the mRNA being delivered.
- LNPs protect mRNAs from degradation by nucleases, significantly improving the bioavailability of the mRNA (time of mRNA in the circulation) [42].
- The particle size and morphology of LNPs can easily be manipulated to influence the trafficking routes postadministration. This increases the efficiency of the mRNA delivery to cells [14,43].

25.6.1 Composition and functional roles of the components of the LNP delivery system

The LNP delivery system contains a combination of ionizable cationic lipids, cholesterol, phospholipid, and polyethylene glycol (PEGs) that self-assemble into ~100 nm nanoparticles encapsulating the antigen mRNA [44,45]. Listed below are the structural characteristics and functional roles of each component of the LNP system used for the delivery of the COVID-19 mRNA vaccines.

25.6.1.1 The cationic or ionizable lipids

These lipids have a positively charged head group that can protonate at low pH (rendering them with a positive charge) and yet remain neutral at physiological pH [46–48]. The pH sensitivity of the ionizable lipids is ideal for in vivo delivery of mRNA as it minimizes the interaction of the drug with the anionic membranes of blood cells, thus improving biocompatibility. SM-102 and ALC-0315 are the ionizable delivery components in the mRNA-1273 and BNT162b COVID-19 vaccines, respectively [49].

25.6.1.2 Phospholipids

Phosphatidylcholine and phosphatidylethanolamine are the preferred phospholipids used in LNPs. They improve article stability, delivery efficacy, tolerability, and biodistribution of the nanoparticle [48,49]. 1,2- distearoyl-*sn*glycero-3-phosphocholine used in the mRNA-1273 and BNT162b2 COVID-19 vaccines [49] are phosphatidylcholine molecules with saturated tails. They have a melting temperature of 54°C and a cylindrical geometry allowing the formation of a lamellar phase that stabilizes the structure of the LNPs [50].

25.6.1.3 Cholesterol

Contributes to enhanced particle stability by modulating membrane integrity and rigidity [48,49]. The molecular geometry of cholesterol derivatives (such as C-24 alkyl phytosterols) used in LNPs affect in vivo delivery efficacy and biodistribution of LNPs [51].

25.6.1.4 Polyethylene glycol lipids

PEGs have multiple significant effects on the properties of LNPs [49,52,53]. The quantity of PEG lipids used in the LNP formulation can affect particle size and zeta potential [49,52]. They augment particle stability by decreasing particle aggregation [48,49,53]. Some PEG modifications can reduce clearance mediated by the kidneys and the mononuclear phagocyte system [48,49,53,54], thus prolonging the circulation time of LNPs (bioavailability). Additionally, PEG lipids are often used to conjugate specific ligands to the particle for targeted delivery of the drug. An article by Hou et al. [10] provides an in-depth insight into the LNP delivery system employed by the novel COVID-19 specific mRNA vaccines.

25.7 Vaccine uptake at the injection site and translation at the cellular level

The mRNA vaccines (unlike DNA vaccines) do not require delivery into the cell nucleus via electroporation or other devices. That is because protein translation from an mRNA occurs in the cell cytoplasm and thus regular needle injection into the muscle is sufficient [42].

Upon intramuscular injection of the vaccine, all cells expressing low-density lipoprotein receptors have been linked to the uptake of LNPs. The process is also facilitated by ApoE lipoprotein which is present in serum and tissues [55,56]. This would suggest that most cells can be direct recipients of LNPs (with the synthetic mRNA vaccine), given that they have the appropriate endocytic mechanisms to internalize them [42].

The LNPs with the synthetic mRNA vaccine are internalized by muscle cells at the injection site and by the infiltrating immune cells (at both the injection site and in the draining lymph nodes). These cells (at the site of injection) translate the synthetic mRNA into the encoded antigen which is the viral S protein [57]. In addition, cells that have endocytosed mRNA in LNPs have been shown to secrete extracellular vesicles containing the mRNA which may be an alternative mechanism for delivering mRNA between cells [58] for ultimately inducing a widespread translation of the S protein in different cell types. Detectable protein production (for up to ten days) is found at the site of injection after subcutaneous, intramuscular, and intradermal methods of administration.

Once translated by the different cell types, a TM anchor causes the S protein to be displayed on the surface of these cells, thus allowing the specific cells to be recognized by the immune system. This promotes the initiation of strong adaptive immune responses [59], triggering the production of B-cell antibodies and the activation of T-cells. The strong immune response generated protects against natural infection with the virus and also prevents any serious disease progression.

25.8 Immune responses induced by COVID-19 messenger RNA vaccines

25.8.1 Humoral immunity and germinal center reactions

A characteristic feature of the humoral immune response is a time-dependent progressive increase in the affinity of antibodies produced [60,61]. This can be attributed to cytidine deaminase-driven random somatic hypermutations in the antigen-binding variable regions of the immunoglobulin genes [62,63]. Antibodies with higher affinity recognize antigens faster, bind strongly to the specific antigen, and remain bound under adverse conditions, providing a stronger immune response [61].

Antibody (Ab) diversification and affinity maturation occur at sites called the germinal centers (GCs) which are microanatomical structures [64,65] that emerge in multiple copies within secondary lymphoid organs upon exposure to antigens (either by infection or due to immunization). Formed at the center of B-cell follicles of secondary lymphoid organs, they are interspersed within a network of stromal cells called the follicular dendritic cells (FDCs) [66]. It is in the GCs that the B cells compete for an array of signals delivered in an affinity-dependent manner. Those B cells with higher affinity receptors progressively outcompete the lower-affinity B cells and are positively selected and rescued from apoptosis. This sophisticated Darwinian process contributes to the persistence of high-affinity B cells which thereafter proliferate and differentiate into Ab-secreting plasma cells and memory-B cells. Over time, further differentiation of both the plasma and memory-B cells from this evolving B-cell population drives the increase in the overall affinity of serum antibodies during the primary antigen response upon immunization or reinfection [62]. FDCs located within the GCs support this affinity-dependent selection process via long-term retention of the intact antigen within the complement-coated immune complexes thereby affecting the overall efficiency of the GC reactions. Experimental blocking of FDC activation has resulted in smaller GC reactions and lower Ab titers in response to immunizations [67].

Both the approved SARS-CoV-2 mRNA vaccines (Moderna and BioNTech) confer protection from COVID-19 infection by eliciting protective Ab responses that last over time [61,68]. Massive GC reactions produce affinity-matured, long-lived plasma cells and memory-B against the S protein translated from the synthetic mRNA delivered through these vaccines. The plasma cells thus produced are predicted to have the ability to survive for years and even several decades, continuously secreting affinity-matured, long-lasting, and nAbs against the SARS-CoV-2 virus without the need for further antigen stimulation [69]. The memory-B cells are expected to get activated in response to subsequent pathogen exposure and to produce high-affinity Ab-secreting cells [70]. Extensive studies are currently in progress to further confirm these expectations. Although the immune response to natural SARS-CoV-2 infection is not the focus of this chapter, it is worth mentioning that the humoral immunity induced by natural SARS-CoV-2 infection is reported to be short-lived with limited durability [71−73], and a timespan limited to 5−8 months postinfection [74,75]. This is attributed to the limited somatic hypermutations occurring in the less efficient GC reactions [73,76] that ensue in response to the natural infection with the virus. This is in sharp contrast to the SARS-CoV-2 mRNA vaccine that induces a robust GC response, spurring the production of high-affinity Ab-secreting cells and long-lasting serological memory.

25.8.2 Innate immune response induced by the messenger RNA vaccines

As robust T cell responses to SARS-CoV-2 natural infection were readily detected [77], the induction of innate immunity by mRNA vaccines against SARS-CoV-2 infection was widely explored in mice, humans, and nonhuman primates during the past year. The mRNA vaccines developed both by Moderna and BioNTech are reported to elicit strong CD4 + and CD8 + T responses with Th1-type biased responses in mice [7,78]. Humans immunized with the Moderna vaccine were reported to induce a strong Th1-biased CD4 + T cell response, with low detectable Th2 or CD8 + T cell response [79]. An extensive study by Painter et al. [80] reported rapid induction of SARS-CoV-2-specific CD4 + T-cells within two weeks of receiving the first mRNA vaccine dose (in SARS-CoV-2-naive individuals). The study also identified the antigen-specific CD4 + T-cells as possible contributors to the initial vaccine-induced protection against symptomatic SARS-CoV infection, since the neutralizing Ab titers would remain low in many such individuals during this initial infection period [8,9,81]. Anderson et al. [40] studied the dose-dependent effect of the mRNA vaccination in different age groups of elderly human volunteers (56, 70, and above 71 years). The study indicated that Th2 and CD8 + T cell responses varied minimally by age and were measurable only after two vaccine doses.

Multiple studies [82] and reviews have focused on the influence of the T follicular helper (Tfh) cells on vaccine-induced humoral immune responses against the SARS-CoV-2 virus. The Tfh cell is a specialized type of CD4 + T cell that, through the production of interleukin 21 (IL-21) [83], influences the formation of GCs, the selection of affinity-matured GC-B cells, and their differentiation into long-lasting plasma and memory cells [84,85]. Although the generation of Tfh cells post-SARS-CoV-2 mRNA vaccination was initially associated only with the RBD-encoding mRNA vaccination [82], Tfh cell responses in the draining lymph nodes, the spleen, and the blood, were later confirmed postimmunization with the S protein-encoding mRNA vaccine [7,86].

Several aspects of the immune functions induced by the COVID-19 mRNA vaccines remain unexplored or are under current investigation. Provided below is a rough overview of the stepwise induction of immune functions by the SARS-CoV-2 mRNA vaccines, modified from the comprehensive review published by Bettini and Locci [4]. Fig. 25.2 provide a diagrammatic representation of the overview.

1. The synthetic mRNA vaccine upon administration is taken up by antigen-presenting cells such as DCs, which traffic to the lymph nodes to prime CD4 + and CD8 + T lymphocytes [42].

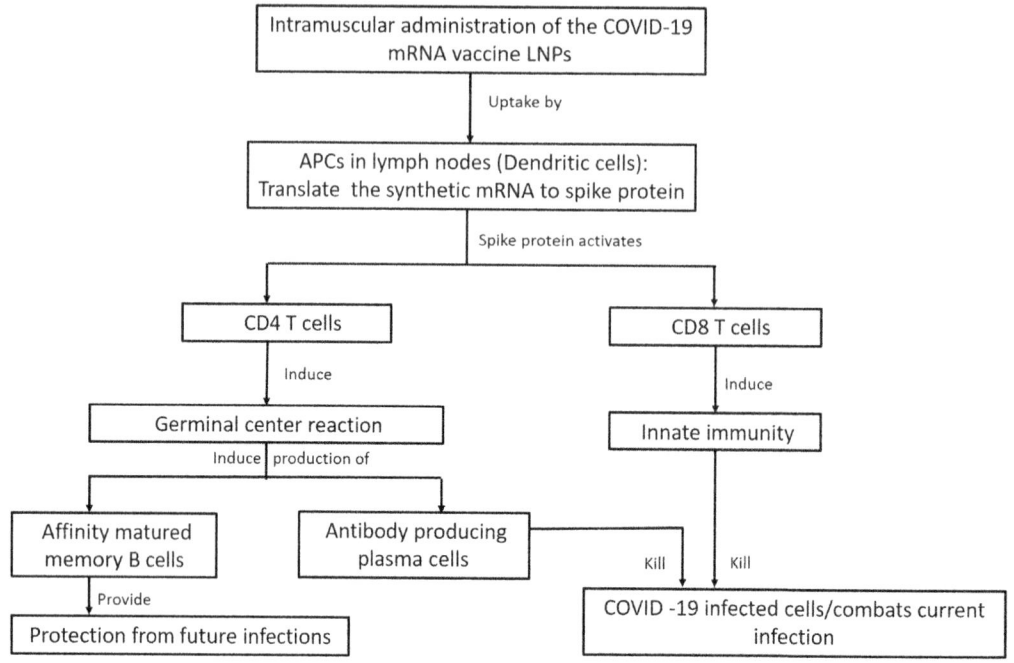

FIGURE 25.2   Overview of the stepwise induction of immune functions by COVID-19 mRNA vaccines. *APCs,* Antigen-presenting cells; *LNPs,* lipid nanoparticles.

2. The priming of CD8 T-cells induce innate immunity by the formation of cytotoxic T lymphocytes which are capable of the direct killing of infected cells.
3. Antigen-primed CD4 T-cells differentiate into Th1 cells or Tfh cells of which the latter helps to initiate a GC reaction.
4. GC reactions induce the formation of affinity-matured memory-B cells and the Ab-secreting long-lived plasma cells.
5. Tfh cells could affect the class switching of Abs produced by the plasma cells to either Th1- or Th2-associated antibodies.

25.9 Conclusion

The idea that mRNAs could be used both to activate and prevent protein production dates as far back as 1987. Back then, the use of synthetic RNA for blocking the expression of target genes was explored to treat future diseases. But the prevailing conventional wisdom that mRNA was fragile, prone to degradation, and thus would likely be associated with high production costs, presented too formidable a challenge to consider its use in therapeutics. The actual idea of mRNA vaccines surfaced in the 1990s and had more favorable reception in the field of Oncology, more so as a therapeutic agent rather than for disease prevention. In the 2000s, nearly every vaccine company that considered working with mRNA opted to invest its resources elsewhere.

With the rapid spread of the COVID-19 virus in late 2019, the two companies, Moderna (in collaboration with the US National Institute of Allergy and Infectious Diseases) and BioNTech (in collaboration with the New York-based drug company, Pfizer) moved at a record pace (within days of the availability of the viral genome sequence) to develop the two mRNA vaccines and conduct animal studies, immediately followed by phases I and II of the clinical trials. Thus, within a short span of 8−9 months from their initiation, both vaccines received EUA in the United States. The authorized vaccines used modified synthetic mRNA that encoded a form of the SARS-CoV-2 S protein capable of adopting a more amenable form (for inducing protective immunity against the virus) when injected into the arms of recipients. Both vaccines were formulated in lipid nanoparticles for rapid delivery.

This chapter has provided (within the prevailing guidelines) a synopsis of the mRNA vaccine therapy currently in use in the United States to contain the spread of the deadly COVID-19 virus. The chapter has focused specifically on the Moderna and BioNTech SARS-CoV-2 mRNA vaccines, with particular emphasis on the pros and cons associated with the first mRNA vaccine approach, the selected viral antigen and the mRNA sequence employed, the adopted delivery approach, and above all, the general immune response elicited by the two mRNA vaccines.

References

[1] Johns Hopkins Coronavirus Resource Center. (n.d.). COVID-19; United States cases by county. Johns Hopkins University & Medicine. Retrieved on August 31, 2021. Available from: https://coronavirus.jhu.edu/us-map.

[2] Covid-19 Vaccine tracker and Landscape (n.d.); July 2022; World Health Organization; R&D Blue Print; Retrieved on August 31, 2021. Available from: https://www.who.int/publications/m/item/draft-landscape-of-covid-19-candidate-vaccines.

[3] Krammer F. SARS-CoV-2 vaccines in development. Nature 2020;586:516−27.

[4] Bettini E, Locci M. SARS-CoV-2 mRNA vaccines: immunological mechanism and beyond. Vaccines (Basel) 2021;9(2):147. Available from: https://doi.org/10.3390/vaccines9020147 PMID: 33673048; PMCID: PMC7918810.

[5] Dong Y, Dai T, Wei Y, et al. A systematic review of SARS-CoV-2 vaccine candidates. Sig Transduct Target Ther 2020;5:237. Available from: https://doi.org/10.1038/s41392-020-00352-y.

[6] Pardi N, Hogan MJ, Porter FW, Weissman D. mRNA vaccines—a new era in vaccinology. Nat Rev Drug Discov 2018;17(4):261−79. Available from: https://doi.org/10.1038/nrd.2017.243.

[7] Corbett KS, Edwards DK, Leist SR, et al. SARS-CoV-2 mRNA vaccine design enabled by prototype pathogen preparedness. Nature. 2020;586:567−71. Available from: https://doi.org/10.1038/s41586-020-2622-0.

[8] Baden LR, El Sahly HM, Essink B, Kotloff K, Frey S, Novak R, et al. Efficacy and safety of the mRNA-1273 SARS-CoV-2 vaccine. N Engl J Med 2021;384:403−16.

[9] Polack FP, Thomas SJ, Kitchin N, Absalon J, Gurtman A, et al. Safety and efficacy of the BNT162b2 mRNA COVID-19 vaccine. N Engl J Med 2020;383:2603−15.

[10] Hou X, Zaks T, Langer R, et al. Lipid nanoparticles for mRNA delivery. Nat Rev Mater 2021;. Available from: https://doi.org/10.1038/s41578-021-00358-0.

[11] Wu Q, Dudley MZ, Chen X, et al. Evaluation of the safety profile of COVID-19 vaccines: a rapid review. BMC Med 2021;19:173. Available from: https://doi.org/10.1186/s12916-021-02059-5.

[12] Chen N, Xia P, Zhu J, et al. RNA sensors of the innate immune system and their detection of pathogens. IUBMB Life 2017;69:297−304.

[13] Pardi N, Hogan MJ, Naradikian MS, Parkhouse K, Cain DW, Jones L, et al. Nucleoside-Modified MRNA vaccines induce potent T follicular helper and germinal center B cell responses. J Exp Med 2018;215:1571−88.

[14] Reichmuth AM, Oberli MA, Jaklenec A, Langer R, Blankschtein D. mRNA vaccine delivery using lipid nanoparticles. Ther Deliv 2016;7:319−34.

[15] Tanji H, et al. Toll-like receptor 8 senses degradation products of single-stranded RNA. Nat Struct Mol Biol 2015;22:109−15.

[16] Isaacs A, Cox RA, Rotem Z. Foreign nucleic acids as the stimulus to make interferon. Lancet. 1963;2:113−16.

[17] Kariko K, et al. Incorporation of pseudouridine into mRNA yields superior nonimmunogenic vector with increased translational capacity and biological stability. Mol Ther 2008;16:1833−40.

[18] Kariko K, Muramatsu H, Ludwig J, Weissman D. Generating the optimal mRNA for therapy: HPLC purification eliminates immune activation and improves translation of nucleoside-modified, protein-encoding mRNA. Nucleic Acids Res 2011;39:e142.

[19] Thess A, et al. Sequence-engineered mRNA without chemical nucleoside modifications enables an effective protein therapy in large animals. Mol Ther 2015;23:1456−64.

[20] Huang Y, Yang C, Xu XF, Xu W, Liu SW. Structural and functional properties of SARS- CoV-2 spike protein: potential antivirus drug development for COVID-19. Acta Pharmacol Sin 2020;41(9):1141−9.

[21] Liu C, Zhou Q, Li Y, Garner LV, Watkins SP, Carter LJ, et al. Research and development on therapeutic agents and vaccines for COVID-19 and related human coronavirus diseases. ACS Cent Sci 2020;6:315−31.

[22] Bosch BJ, van der Zee R, de Haan CA, Rottier PJ. The coronavirus spike protein is a class I virus fusion protein: structural and functional characterization of the fusion core complex. J Virol 2003;77:8801−11.

[23] Kirchdoerfer RN, Cottrell CA, Wang N, Pallesen J, Yassine HM, Turner HL, et al. Pre-fusion structure of a human coronavirus spike protein. Nature. 2016;531:118−21.

[24] Walls AC, Park YJ, Tortorici MA, Wall A, McGuire AT, Veesler D. Structure, function, and antigenicity of the SARS-CoV-2 spike glycoprotein. Cell. 2020;181:281−92 e6.

[25] Hoffmann M, Kleine-Weber H, Schroeder S, Kruger N, Herrler T, Erichsen S, et al. SARS- CoV-2 cell entry depends on ACE2 and TMPRSS2 and is blocked by a clinically proven protease inhibitor. Cell. 2020;181:271−80 e8.

[26] Tang T, Bidon M, Jaimes JA, Whittaker GR, Daniel S. Coronavirus membrane fusion mechanism offers a potential target for antiviral development. Antivir Res 2020;178:104792.

[27] Xia S, Zhu Y, Liu M, Lan Q, Xu W, Wu Y, et al. Fusion mechanism of 2019-nCoV and fusion inhibitors targeting HR1 domain in spike protein. Cell Mol Immunol 2020;17:765−7.

[28] Robson B. Computers and viral diseases. Preliminary bioinformatics studies on the design of a synthetic vaccine and a preventative peptidomimetic antagonist against the SARS-CoV-2 (2019-nCoV, COVID-19) coronavirus. Comput Biol Med 2020;119:103670.

[29] Tortorici MA, Walls AC, Lang Y, Wang C, Li Z, Koerhuis D, et al. Structural basis for human coronavirus attachment to sialic acid receptors. Nat Struct Mol Biol 2019;26:481−9.

[30] U.S. Food and Drug Administration. Pfizer-BioNTech COVID-19 vaccine EUA letter of authorization. U.S. Food and Drug Administration; 2021.

[31] U.S. Food and Drug Administration. Moderna COVID-19 vaccine EUA letter of authorization. U.S. Food and Drug Administration; 2021.

[32] Sahin U, Karikó K, Türeci Ö. mRNA-Based therapeutics developing a new class of drugs. Nat Rev Drug Discov 2014;13(13):759−80.

[33] Hsieh CL, Goldsmith JA, Schaub JM, DiVenere AM, Kuo HC, Javanmardi K, et al. Structure-based design of prefusion-stabilized SARS-CoV-2 spikes. Science. 2020;369(6510):1501−5.

[34] Vogel AB, Kanevsky I, Che Y, Swanson KA, Muik A, Vormehr M, et al. BNT162b vaccines protect rhesus macaques from SARS-CoV-2. Nature 2021;592(7853):283−9. Available from: https://doi.org/10.1038/s41586-021-03275-y.

[35] Henderson JM, Ujita A, Hill E, Yousif-Rosales S, Smith C, Ko N, et al. Cap 1 Messenger RNA synthesis with co-transcriptional CleanCap® analog by in vitro transcription. Curr Protoc 2021;1(2):e39.

[36] Babendure JR, Babendure JL, Ding JH, Tsien RY. Control of mammalian translation by mRNA structure near caps. RNA. 2006;12(5):851−61.

[37] Orlandini von Niessen AG, Poleganov MA, Rechner C, Plaschke A, Kranz LM, Fesser S, et al. Improving mRNA-based therapeutic gene delivery by expression-augmenting 3′ UTRs identified by cellular library screening. Mol Ther 2019;27(4):824−36.

[38] Trepotec Z, Geiger J, Plank C, Aneja MK, Rudolph C. Segmented poly(A) tails significantly reduce recombination of plasmid DNA without affecting mRNA translation efficiency or half- life. RNA. 2019;25(4):507–18.

[39] Nance LD, Meier JL. Modifications in an emergency: the role of N1-methylpseudouridine in COVID-19 vaccines. ACS Cent Sci 2021;7(5):748–56. Available from: https://doi.org/10.1021/acscentsci.1c00197.

[40] Anderson EJ, Rouphael NG, Widge AT, Jackson LA, Roberts PC, Makhene M, et al. Safety and immunogenicity of SARS- CoV-2 mRNA-1273 vaccine in older adults. N Engl J Med 2020;383:2427–38.

[41] Pardi N, Tuyishime S, Muramatsu H, Kariko K, Mui BL, Tam YK, et al. Expression Kinetics of nucleoside-modified MRNA delivered in lipid nanoparticles to mice by various routes. J Control Release 2015;217:345–51.

[42] Cagigi A, Loré K. Immune responses induced by mRNA vaccination in mice, monkeys and humans. Nato Adv Sci Inst 2021;9:61.

[43] Eygeris Y, Patel S, Jozic A, Sahay G. Deconvoluting lipid nanoparticle structure for messenger RNA delivery. Nano Lett 2020;20:4543–9.

[44] Kauffman KJ, Webber MJ, Anderson DG. Materials for non-viral intracellular delivery of messenger RNA therapeutics. J Control Releaseease 2016;240:227–34.

[45] Cullis PR, Hope MJ. Lipid nanoparticle systems for enabling gene therapies. Mol Ther 2017;25:1467–75.

[46] Hajj KA, Whitehead KA. Tools for translation: non- viral materials for therapeutic mRNA delivery. Nat Rev Mater 2017;2:17056.

[47] Kowalski PS, Rudra A, Miao L, Anderson DG. Delivering the messenger: advances in technologies for therapeutic mRNA delivery. Mol Ther 2019;27:710–28.

[48] Meng C, Chen Z, Li G, Welte T, Shen H. Nanoplatforms for mRNA therapeutics. Adv Ther 2021;4:2000099.

[49] Kim J, Eygeris Y, Gupta M, Sahay G. Self-assembled mRNA vaccines. Adv Drug Deliv Rev 2021;170:83–112.

[50] Koltover I, Salditt T, Rädler JO, Safinya CR. An inverted hexagonal phase of cationic liposome-DNA complexes related to DNA release and delivery. Science. 1998;281:78–81.

[51] Patel S, et al. Naturally-occurring cholesterol analogues in lipid nanoparticles induce polymorphic shape and enhance intracellular delivery of mRNA. Nat Commun 2020;11:983.

[52] Ryals RC, et al. The effects of PEGylation on LNP based mRNA delivery to the eye. PLoS ONE 2020;15: e0241006.

[53] Knop K, Hoogenboom R, Fischer D, Schubert US. Poly(ethylene glycol) in drug delivery: pros and cons as well as potential alternatives. Angew Chem Int Ed 2010;49:6288–308.

[54] Zhu X, et al. Surface de-PEGylation controls nanoparticle- mediated siRNA delivery in vitro and in vivo. Theranostics. 2017;7:1990–2002.

[55] Akinc A, Querbes W, De S, Qin J, Frank-Kamenetsky M, Jayaprakash KN, et al. Targeted delivery of RNAi therapeutics with endogenous and exogenous ligand-based mechanisms. Mol Ther 2010;18:1357–64.

[56] van den Elzen P, Garg S, Leon L, Brigl M, Leadbetter EA, Gumperz JE, et al. Apolipoprotein-mediated pathways of lipid antigen presentation. Nature. 2005;437:906–10.

[57] Brito LA, Chan M, Shaw CA, Hekele A, Carsillo T, Schaefer M, et al. A cationic nanoemulsion for the delivery of next-generation RNA vaccines. Mol Ther 2014;22:2118–29.

[58] Maugeri M, Nawaz M, Papadimitriou A, Angerfors A, Camponeschi A, Na M, et al. Linkage between endosomal escape of LNP-mRNA and loading into EVs for transport to other cells. Nat Commun 2019;10:4333.

[59] Liang F, Lindgren G, Lin A, Thompson EA, Ols S, Röhss J, et al. Efficient targeting and activation of antigen-presenting cells in vivo after modified MRNA vaccine administration in *Rhesus macaques*. Mol Ther 2017;25:2635–47.

[60] Eisen HN. Affinity enhancement of antibodies: how low-affinity antibodies produced early in immune responses are followed by high-affinity antibodies later and in memory B-cell responses. Cancer Immunol Res 2014;2:381–92.

[61] Mesin L, Ersching J, Victora GD. Germinal center B cell dynamics. Immunity. 2016;45:471–82.

[62] Berek C, Milstein C. Mutation drift and repertoire shift in the maturation of the immune response. Immunol Rev 1987;96:23–41.

[63] Muramatsu M, Kinoshita K, Fagarasan S, Yamada S, Shinkai Y, Honjo T. Class switch recombination and hypermutation require activation- induced cytidine deaminase (AID), a potential RNA editing enzyme. Cell. 2000;102:553–63.

[64] Berek C, Berger A, Apel M. Maturation of the immune response in germinal centers. Cell. 1991;67:1121–9.

[65] Jacob J, Kelsoe G, Rajewsky K, Weiss U. Intraclonal generation of antibody mutants in germinal centres. Nature. 1991;354:389–92.

[66] Heesters BA, Myers RC, Carroll MC. Follicular dendritic cells: dynamic antigen libraries. Nat Rev Immunol 2014;14:495–504.

[67] Garin A, Meyer-Hermann M, Contie M, Figge MT, Buatois V, Gunzer M, et al. Toll-like receptor 4 signaling by follicular dendritic cells is pivotal for germinal center onset and affinity maturation. Immunity. 2010;33:84–95.

[68] Plotkin SA. Correlates of protection induced by vaccination. Clin Vaccine Immunol 2010;17:1055–65.

[69] Amanna IJ, Carlson NE, Slifka MK. Duration of humoral immunity to common viral and vaccine antigens. N Engl J Med 2007;357:1903–15.

[70] Sallusto F, Lanzavecchia A, Araki K, Ahmed R. From vaccines to memory and back. Immunity. 2010;33:451–63.

[71] Seow J, Graham C, Merrick B, Acors S, Pickering S, Steel KJA, et al. Longitudinal observation and decline of neutralizing antibody responses in the three months following SARS-CoV-2 infection in humans. Nat Microbiol 2020;5:1598–607.

[72] Beaudoin-Bussières G, Laumaea A, Anand SP, Prévost J, Gasser R, Goyette G, et al. Decline of humoral responses against SARS-CoV-2 spike in convalescent individuals. Mbio. 2020;11.

[73] Kaneko N, Kuo HH, Boucau J, Farmer JR, Allard-Chamard H, Mahaja VS, et al. Loss of Bcl-6-expressing t follicular helper cells and germinal centers in COVID-19. Cell 2020;183.

[74] Dan JM, Mateus J, Kato Y, Hastie KM, Yu ED, Faliti CE, et al. Immunological memory to SARS-CoV-2 assessed for up to eight months after infection. Biorxiv 2020;.

[75] Wajnberg A, Amanat F, Firpo A, Altman DR, Bailey MJ, Mansour M, et al. Robust Neutralizing antibodies to SARS-CoV-2 infection persist for months. Science. 2020;370:1227–30.

[76] Brouwer PJM, Caniels TG, van der Straten K, Snitselaar JL, Aldon Y, Bangaru S, et al. Potent neutralizing antibodies from COVID-19 patients define multiple targets of vulnerability. Science 2020;369:643–50.

[77] Sekine T, Perez-Potti A, Rivera-Ballesteros O, Stralin K, Gorin JB, Olsson A, et al. Robust T cell immunity in convalescent individuals with asymptomatic or mild COVID-19. Cell. 2020;183:158–68 e114.

[78] Sahin U, Muik A, Derhovanessian E, Vogler I, Kranz LM, Vormehr M, et al. COVID-19 vaccine BNT162b1 elicits human antibody and TH1 T cell responses. Nature. 2020;586:594–9.

[79] Corbett KS, Flynn B, Foulds KE, Francica JR, Boyoglu-Barnum S, Werner AP, et al. Evaluation of the mRNA-1273 vaccine against SARS-CoV-2 in nonhuman primates. N Engl J Med 2020;383:1544–55.

[80] Painter MM, Mathew D, Goel RR, et al. Immunity 2021;54:2133–42. Available from: https://doi.org/10.1016/j.immuni.2021.08.001.

[81] Goel R, Apostolidis A, Painter MM, et al. Distinct antibody and memory B cell responses in SARS-CoV-2 naïve and recovered individuals after mRNA vaccination. Sci Immunology 2021;6:58.

[82] Tai W, Zhang X, Drelich A, Shi J, Hsu JC, Luchsinger L, et al. Novel receptor-binding domain (RBD)-based MRNA vaccine against SARS-CoV-2. Cell Res 2020;30:932–5.

[83] Crotty S. A brief history of T cell help to B cells. Nat Rev Immunol 2015;15:185–9.

[84] Crotty ST. Follicular helper cell biology: a decade of discovery and diseases. Immunity 2019;50:132–1148.

[85] Vinuesa CG, Linterman MA, Yu D, MacLennan ICM. Follicular helper T cells. Annu Rev Immunol 2015;34:1–34.

[86] Vogel AB, Kanevsky I, Che Y, Swanson KA, Muik A, Vormehr M, et al. A Prefusion SARS-CoV-2 spike RNA vaccine is highly immunogenic and prevents lung infection in non-human primates. Biorxiv 2020;.

Immunotherapy as an emerging and promising tool against viral infections

Vahid Reza Askari[1,2,3], *Roghayeh Yahyazadeh*[4] *and Vafa Baradaran Rahimi*[3,5]

[1]Department of Pharmaceutical Sciences in Persian Medicine, School of Persian and Complementary Medicine, Mashhad University of Medical Sciences, Mashhad, Iran [2]Applied Biomedical Research Center, Mashhad University of Medical Sciences, Mashhad, Iran [3]Pharmacological Research Center of Medicinal Plants, Mashhad University of Medical Sciences, Mashhad, Iran [4]Departments of Pharmacodynamics and Toxicology, School of Pharmacy, Mashhad University of Medical Sciences, Mashhad, Iran [5]Department of Cardiovascular Diseases, Faculty of Medicine, Mashhad University of Medical Sciences, Mashhad, Iran

Abbreviation

ALV	Avian leukosis virus
AP-2	adapter protein-2
CAR-T cells	chimeric antigen receptor T cells
COVID-19	CORONAVIRUS disease
CRS	cytokine release syndrome
CTLA-4	cytotoxic T-lymphocyte-associated protein 4
FDA	Food and Drug Administration
gpB	glycoprotein B
HBD	human β-defensin
HBsAg	HBV surface antigen
HBV	hepatitis B virus
HIV	human immunodeficiency virus
HPV	human papillomavirus
HSV-1	herpes simplex virus 1
ICI	immune checkpoint inhibitor
IFN-γ	interferon-γ
IFNs	interferons

Viral Infections and Antiviral Therapies
DOI: https://doi.org/10.1016/B978-0-323-91814-5.00004-0

IVIG	intravenous immunoglobulin
mAbs	monoclonal antibodies,
MAS	macrophage activation syndrome
MCP-1	monocyte chemoattractant protein 1
MHC	antigen-major histocompatibility complex
MHC	main histocompatibility complex
MIP-1β	macrophage inflammatory protein-1β
NK	natural kille
PD-L1	programmed death-ligand-1
PD-1	programmed death-1
rPV	recombinant protein-based vaccine
RSV	respiratory syncytial virus
S-CARs	S domain receptor
scFv	single-chain variable fragment
TCL	T cell lines
TCR	T cell receptor
Th	T helper
TIM3	mucin domain-containing protein-3
TNF-α	tumor necrosis factor-α

26.1 Introduction

Immunotherapy manages the immune system to treat diseases or viral infections by deleting pathogens or unhealthy host cells. Also, it regulates the host's adaptive and innate immune responses to disease regression or progression. In other words, immunotherapy is a therapeutic attitude that manipulates and controls the immune system to overcome diseases or disorders [1]. Some studies have classified immunotherapy as active (allergen-specific and vaccine therapy) and passive immunotherapies (adoptive and antibody-based immune). It is necessary to mention that passive immunotherapies utilize components generated ex vivo (e.g., recombinant antibody produces or immune cells) for administrating to patients, whereas it cannot motivate the host immune response. However, host immunological responses were triggered by active immunotherapies (humoral response) duo to generate specific immune effectors such as antibodies and T lymphocytes [2].

As intracellular pathogens are entirely dependent on the host, the viruses provide human or veterinary problems and are not easily improved via pharmaceuticals. Hence, adjusting specific or nonspecific immune responses is vital for intervention in viral infections [3].

Vaccination is considered the primary approach in the management of infectious diseases. Vaccination enforced rapid and effective defense against a pathogen via simulating natural interaction through the human immune system [4]. Hence, the vaccine diminishes the risk of mortality, morbidity, and burden of diseases subsequently exposed to an infectious pathogen. Furthermore, induction and maintenance of immunity and the memories against noxious stimuli are performed during vaccination related to T and B-lymphocytes [5]. As an actual result of vaccination, immunization programs have caused the elimination of some infectious and more dangerous diseases such as smallpox and polio. Recently, scientific discoveries have affected the formation of new vaccine platforms by recombinant antibodies, improving adjuvant and nucleic acid-based vaccines [6].

Monoclonal antibodies (mAbs), as exchangeable antibodies, are produced in the laboratory to manipulate the immune system's attack by viral or cancer cells [7]. Firstly, it was detected by Dr. Edward Jenner from a smallpox pustule [8] and have been used generally for therapeutic aspect since 1986. As a result, the antibody development stage and hybridoma were begun [9]. The early mAbs were created via a murine protein that humans did not endure. Therefore, this caused the new technology stage (fully humanized antibodies) [8]. Structurally, mAbs consist of two light and two heavy polypeptide chains bound together by a disulfide bond. The consequent "Y" consists of the "Fab" region, which varies antigen bound to it, whereas the "Fc" region in antigen structure causes to link the antibody with cells which interferes with providing the immune response [10]. In another way, the binding of the antigen to the antibody and the "Fc" region can trigger complement-system cytotoxicity and inhibit the intracellular signaling pathway [11].

Immune checkpoint inhibitor (ICI)-based therapy, as a kind of immunotherapy, suppressed the immune inhibitory signaling pathways, such as the programmed death-ligand-1 (PD-L1) axis, programmed cell death protein-1 (PD-1), cytotoxic lymphocyte antigen proteins (CTLA-4), T-cell immunoglobulin, mucin domain-containing protein-3 (TIM3) and OX40 (CD134) pathway [12], which encourage effector T-cell activation. During the past decade, ICI therapy in 2011 was mainly presented for the management of many solid tumors and hematological malignancies (such as Merkel cell carcinoma, colorectal cancer, renal cell carcinoma, and Hodgkin lymphoma) by numerous medicines that were approved by the US Food and Drug Administration (FDA) [13,14]. However, although ICI, as a famous costimulatory and coinhibitory, regulates T lymphocyte activation, T cell injury is a typical signature of chronic viral diseases such as hepatitis B virus (HBV), human immunodeficiency virus (HIV) [15], tuberculosis, and malaria [16].

Cytokines are described as soluble glycoproteins or polypeptides, which improve the biological processes such as differentiation, growth, and stimulation of cells. Also, they exhibit immunosuppressive and pro- or antiinflammatory roles referred to in Fig. 26.1 [17]. Also, it has been used for viral and infectious disease therapies since 1986 [16]. The immune reaction against viruses is started with the cognition of viral compartments. Then signal system culminating is initiated by activating transcription factors that create an antiviral state and an inflammation process. Thus, the cytokine network for viruses initiates with cytokines secretion by virus-infected cells, such as interleukin (IL)-1, IL-18, IL-8, IL-6, IL-12, IL-2, and IL-23, granulocyte-macrophage colony-stimulating factor, tumor necrosis factor-α (TNF-α), and interferons (IFNs), inducing potent inflammatory factors, and activating phagocyte cells, natural killer (NK) cells, and mast cells, to the site of infection. Besides, these cytokines induce an immune activity T helper (Th1)/T cell lines (TCL) intending to eliminate the extracellular virus and infected cells. In contrast, cytokines such as IL-4, IL-10, IL-13, IL-37, and transforming growth factor adjust the immune response to a Th2 phenotype, producing antiinflammatory activity [18].

As an artificial production of T cell receptors (TCR), chimeric antigen receptor T cells (CAR-T cells) have been used for immunotherapy aims [19]. The United States FDA approved the CAR-T cell therapies in 2017 for two purposes: first for treating adults with advanced lymphomas and the second for children with acute lymphoblastic leukemia (refractory pre-B cell to diffuse large B cell lymphoma) [20]. In this therapy, engineered T cells can recognize an antigen to have an impressive effect on target cells and perish them [19]. CAR-T cells involve:

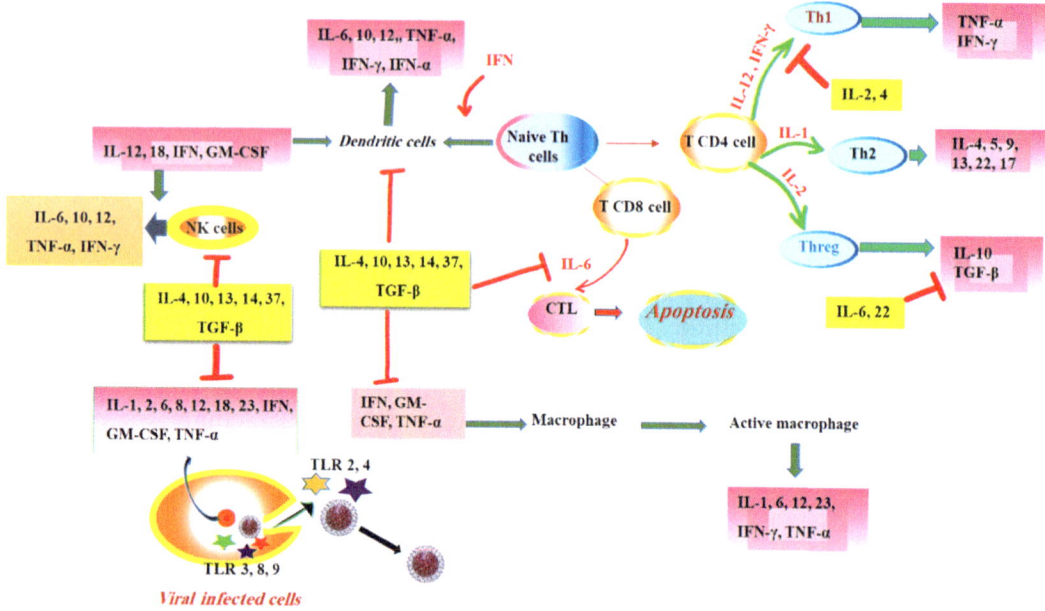

FIGURE 26.1   The schematic pattern of virus recognition and cytokine production.

1. Extracellular domains for cell assessment goal that illustrated a mAb by a single-chain variable fragment (scFv).
2. Intracellular domain for signaling motifs capable of T-cell activation [21].

It is necessary to mention that CAR-T cells are the main histocompatibility complex (MHC) for antigen recognition. CAR-T cell manufacturing for therapeutic aim is performed by T cell dissociation from peripheral blood of patients, followed by insertion of CAR genes (viral or nonviral) into the T cell genome. Finally, the rapidly expanded CAR-T cell production is induced and returned to the patient [22].

This chapter describes clinically related examples of adoptive therapies to highlight the various corresponding for viral infectious diseases. Significant successes in the treatment of viral infection have led to a growing scientific and clinical interest in using immunotherapy to treat other types of infections. Despite differences in etiology, there are common features in the pathogenesis of the diseases explained below that potentially permit them to be treated with various types of immunotherapy.

26.2 Vaccines

As a most cost-effective strategy, vaccines have been used for managing and inhibiting viral infections against viruses, causing acute, chronic, and self-limited infections. So, it has been adequate for the vaccine to emulate the nature of the virus [23]. Most vaccines are based on inactivated or live-attenuated technologies [24]. Nevertheless, some vaccines rely on viral vectors or nucleic acid, including mRNA or plasmid DNA (pDNA) [25].

As an intermediate translator, mRNA is categorized into self-amplifying RNA (saRNA) and nonreplicating mRNA [26]. The alpha-virus genome is the origin of the saRNA vaccine. The saRNA vaccine is originated from the alpha-virus genome. This vaccine contains a transgenetic encoding of the therapeutic antigen and the other gene responsible for the viral RNA replication [27]. Also, the changed structure of mRNA happens via replacing the mRNA sequence with scarce codons and presenting shifted nucleotides that influence the translational accuracy. Indeed, these changes further impress the type and severity of immune responses [28]. Because of the presence of mRNAs in the cytosol, these kinds of vaccines are being used in disease-related immunotherapy in which mRNAs are entered into host cells, mimicking virus-infection-like cellular immunity and humoral immunity [29].

Furthermore, the mRNA vaccine improves the host's antitumor and antivirus effects by enhancing T cells' antigen reactiveness. In fact, mRNA vaccines prompt adaptive and innate immunity and encourage humoral immune responses without antibody-dependent boosting activity [30]. However, due to the inconstancy in mRNA vaccines delivery and effectiveness, scientists have introduced some different carrier methods, such as virus-like replicon particle delivery, polymer-based delivery, lipid-based delivery, peptide-based, and cationic nanoemulsion delivery.

To date, pure mRNA vaccines are newly expanded to improve adaptive immunity that are directly injected into cells [31], such as HIV-1 vaccine for HIV [32] and electroporation-related mRNA vaccine for inflammatory diseases [31]. In addition, many studies have been performed to discover the impressive antiviral agents by testing compound libraries or screening factors, which are effective against viruses (Tables 26.1 and 26.2).

pDNA vaccines, as a potent genetic immunization, have been introduced to induce humoral and cellular immune responses. DNA vaccines are based on DNA encoding in eukaryotic cassettes for viral antigens [43]. However, because of the integration risk of vaccine DNA sequences into the host genome and inducing the risk of an anti-DNA immune response [44], currently, these are not permitted for human use. However, there are several continuing clinical trials considering safety cases [16].

Recombinant viral-based vector vaccines have been used in nonreplicative or live attenuated forms' to achieve a target antigen [45]. Furthermore, viral vectors present some superiorities over traditional vaccines, such as inducing prominent antibody reactions and cytotoxic T lymphocytes induction that is influential for managing cancer and intracellular pathogens that are not performed by protein-based vaccines, such as HIV [46] and hetero subtopics influenza vaccines [47].

A recombinant protein-based vaccine (rPV) contains immunogenic proteins and antibodies that assist the immunization against a pathogen. The immunogenic proteins are achieved from the serum of infected patients [48]. It may also be a better choice for rapid production in an epidemiological emergency. Also, rPV may include recombinant-virus subunits obtained from viral capsids [45]. Presentation of high numbers of antigen epitopes causes maintenance of the original conformation in viral-like particles. Thereby, viral immunogenicity is preserved via crosslinking of B-cell receptors and uptake into antigen-presenting cells. Chimeric protein expression is displayed when self-assembly is impossible [45]. It is important to emphasize that the monomeric forms of most rPV are poorly

TABLE 26.1   Evaluation of antiviral drugs in the in-vitro study.

Factor	Cells	Study design	IC $_{50/90/99}$	CC $_{50}$	Reference
Ribavirin	Vero	Yield reduction	IC$_{99}$ = 263 μM	>1929 μM	[33]
	Huh7		IC$_{99}$ = 83 μM		
	U2OS		IC$_{99}$ = 78 μM		
	Vero		IC$_{99}$ = 424 μM	>1929 μM	
	Huh7		IC$_{99}$ = 63 μM		
	U2OS		IC$_{99}$ = 73 μM		
	Vero		IC$_{99}$ = 176 μM[a]	>2000 μM	[34]
	Vero		IC50 = 15.1−35.7 μM	>128 μM	[35]
	Vero		IC50 = 40.1 μM[b]	>100 μM	[36]
	Vero		IC50 = 49.7 μM	>320 μM	[37]
Favipiravir	Vero	Yield reduction	IC50 = 6 μM, IC90 = 22 μM	>1000 μM	[38]
2′-Fluoro, 2-′deoxycytidine	Vero	Yield reduction	IC90 = 3.7 μM	>320 μM	[37]
Benidipine	Vero	Yield reduction	IC50 = 1.412 μM[b]	>96.92 μM	[39]
Nifedipine	Vero	Yield reduction	IC50 = 98 μM[b]	>250 μM	

[a]*Ribavirin and IFNs combination, virus titers were decreased from 3.2−3.6 log.*
[b]*Titers were estimated by reverse transcriptase-PCR of the virus genome.*

immunogenic unless adjuvants are used to boost their efficacy and effectiveness. In contrast, high organized oligomeric structures have dramatic successful results in immunogenicity [49,50].

26.3 Antibody-based therapies

As a protective protein, an antibody is manufactured via the immune system in response to recognizing foreign substances as antigen [51]. Antibodies are recommended for pharmaceutical purposes because of their high potency to target specific epitopes with their various domains. Virologists describe the exact mechanisms of binding and entering of viruses into hosts and their antiviral immunity by antiviral antibody research [52]. Primarily, antibodies are created by inserting an antigen into a mouse, rat, rabbit, goat, sheep, horse [53], and chicken's egg yolk [54,55]. The mAbs [7], polyclonal antibodies [55], antiserum, and intravenous immunoglobulin (IVIG) are obtained from mentioned animals, after some process on isolated blood, such that all of these enhanced binding eagerness through multiple binding sites on one molecule [56].

Antiviral antibodies are mechanistically categorized as neutralizing and nonneutralizing antibodies. Neutralizing antibodies neutralize the virus infection by binding to a specific

TABLE 26.2 Evaluation of antiviral drugs in the in vivo study.

Factors	Animals	Dose/route	Duration (Day)	Virus dose/route	Sur. (%)	Reference
Ribavirin	STAT2-/-hamster	75 mg/kg/day (twice) p.o.	1–11	50 PFU/s.c.	0	[40]
Favipiravir	IFNAR-/-C57BL/6	60 mg/kg/day (once) i.p.	1–5	10^6 $TCID_{50}$/s.c.	100	[38]
		300 mg/kg/day (once)			100	
		60 mg/kg/day (once) p.o.			100	
		300 mg/kg/day (once)			100	
		300 mg/kg/day (once) p.o.	1–6		100	
			2–7		100	
			3–8		100	
			4–9		83	
			5–10		50	
		120 mg/kg/day (twice) p.o.	0–4	10^6 $TCID_{50}$/s.c.	100	[41]
			1–5		100	
			2–6		100	
			3–7		100	
			4–8		67	
			5–9		0	
		200 mg/kg/day (twice) p.o.	0–4		100	
			1–5		100	
			2–6		100	
			3–7		100	
			4–8		100	
			5–9		20	
2′, FdC[c]	IFNAR-/-C57BL/6	50 mg/kg/day (twice) i.p.	0–8	3 PFU/s.c.	80	[37]

(Continued)

TABLE 26.2 (Continued)

Factors	Animals	Dose/route	Duration (Day)	Virus dose/route	Sur. (%)	Reference
Benidipine	C57BL/6	15 mg/kg/day (once) i.g.	0–7	10^6 TCID$_{50}$/i.p.	100[a]	[39]
	Humanized mouse		0–10		83.3[b]	
Nifedipine	C57BL/6	100 mg/kg/day (once) i.g.	0–7		100[a]	
	Humanized mouse		0–10		100[b]	
IFN-γ	3 days old ICR	0.5 μg/animal (once) i.p.	1	1.5×10^3 TCID$_{50}$ i.c.	25	[42]
		0.05 μg/animal (once)			25	
		0.5 μg/animal (once)	+1	1.5×10^3 TCID$_{50}$ i.c.	0	
		0.05 μg/animal (once)			0	

[a]Non-lethal model. The viral loads in the spleen and serum were significantly decreased.
[b]The fatality rate of the vehicle control group was 57.1%.
[c]2'-Fluoro-2'-deoxycytidine.
i.g., inguinal; i.p., intraperitoneal; p.o., oral; PFU, plaque-forming unit; s.c., subcutaneous; Sur., survival; TCID, tissue culture infective dose.

epitope on the viral envelope via several mechanisms. Nevertheless, nonneutralizing antibodies bind with the area on the virus's surface and mark it without neutralizing the infection or total viral titer. Eventually, the virus-antibody complexes are entered into the cell by endocytosis. Other nonneutralizing humoral factors are cell-mediated cytotoxicity or antibody-mediated phagocytosis, agglomeration, and even immune activation [57—59]. Two mechanisms of antibody-dependent enhancement are described in Fig. 26.2.

After secretion of pro-inflammatory cytokines, the immune cell was recruited and activated the cascade factors. This aggressive inflammation can obstruct the airway and can lead to acute respiratory distress syndrome in severe conditions. On the other hand, coronavirus disease (COVID-19) immunopathology researches suggest that human macrophage infection by SARS-CoV-2 is unproductive.

Antiserum is a blood serum with monoclonal and polyclonal antibodies that are consumed to stimulate passive immunity to many diseases via plasmapheresis [60]. In some cases, antiserum is produced from viral polypeptides antigen. Antiserum incorporation with passive immunization has shown significant protection in mammals [61,62] and birds [63]. The study has proved that between the viral proteins (nucleoproteins, nonstructural protein, hemagglutinin, neuraminidase, polymerase, and a mixture of all these), nucleoproteins produced antiserum and presented better protection in birds via challenging with

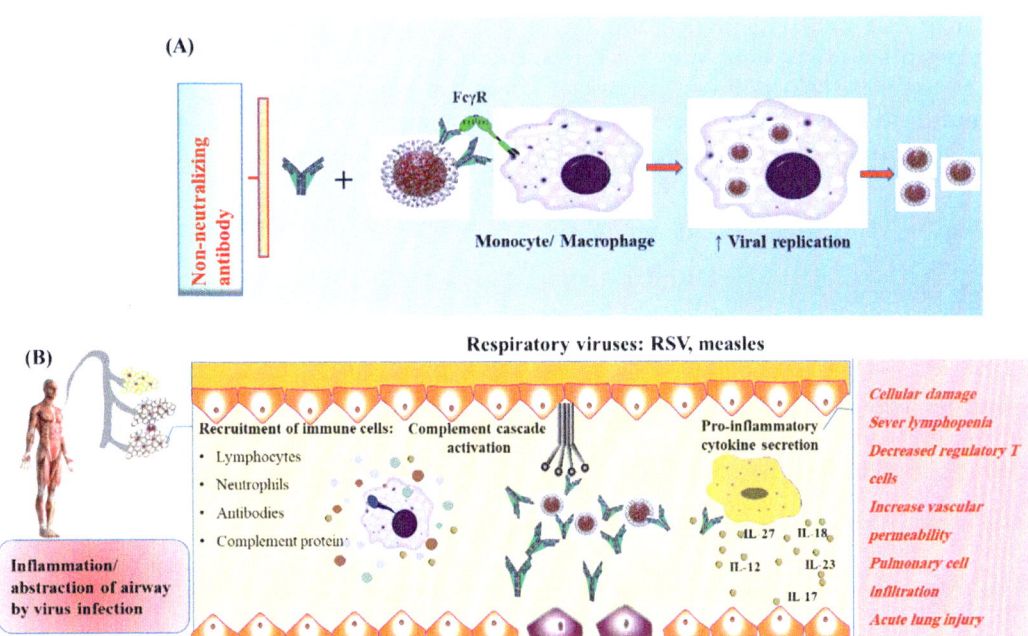

FIGURE 26.2   Two main antibody-dependent enhancement mechanisms in viral disease. (A). dengue virus and Feline infectious peritonitis virus (FIPV) as macrophage-tropic viruses, subneutralizing or nonneutralizing antibodies cause viral infection of macrophages or monocytes by FcγRIIa-mediated endocytosis. (B). Respiratory syncytial virus (RSV) and measles as nonmacrophage-tropic respiratory viruses, nonneutralizing antibodies can generate an immune response with viral antigens into airway tissues.

III. Antiviral agents and therapeutics

highly pathogenic avian influenza virus [62]. Another study indicated that azidothymidine, as a nucleoside reverse transcriptase inhibitor, lonely was insufficient for completely viral elimination and detoxification. Treatment of the avian leukosis virus (ALV) strains with antiserum and azidothymidine revealed that ALV replication was almost eliminated in the vaccine virus seeds [64].

Polyclonal antibodies therapy, as a prophylactic and therapeutic drug, are a group of neutralizing antibodies. It can purpose many epitopes on the viral envelope within the infection and cause strong neutralizing activity [65]. Therefore, the complex is considered as multiple targets for opsonization and activation of the complement system [66]. Because of its high stability and affinity, short production period, and low cost it is more used over other mAbs. In contrast, their use is limited due to a high degree of cross-reactivity and lack of specificity [52].

IVIG is a biological agent that controls various immunodeficiency states, such as autoimmune, inflammatory, and infectious diseases [67]. The source of IVIG is human immunoglobulins, which are mainly IgG-assembled and isolated from many healthy plasma donors that are convalesced after viral infections [68,69]. IVIG product concentrates on native IgG and its subclasses (IgG1, IgG2, IgG3, and IgG4) as the monomer whose half-life is 21 days and with some IgA and IgM. Noteworthy, 1%−10% of IgG levels are in dimer form [68,70]. Moreover, various idiotypes of the antibody with specific molecular variants vary in everybody, such as MHC molecules and blood group antigens from mother donors, and cause nonspecific reactions in recipients [70].

A mAb as a potential therapeutic agent, displays one of the most favorable classes of medicines due to their long-term track record of safety in humans, their exceptional specificity to the virus that reduces the risk of out-target effects, and their potency to manage the immune defense against many viral infections, such as SARS-CoV-2 pandemic [71,72] via the antigen-binding fragment (Fab). Moreover, mAb belongs to one type of immunoglobulin that has a single isotype [73]. Therefore, anti-viral mAbs features are included:

1. The capacity to target specific epitopes
2. High specify
3. High potency
4. Almost negligible side effects in clinical administration, which provides and persuades large scale production [74].

Tocilizumab is a recombinant humanized antihuman IL-6 receptor mAb used in combination with different groups of drugs or alone to attenuate the clinical features of COVID-19 in several clinical trials [75].

26.4 Chimeric antigen receptor T cells immunotherapy

CAR T-cell therapy has been defined as effective for patients who suffered from refractory or relapsed hematological malignancies [74]. As one of adoptive cell transfer, CAR T-cell operates against CD19 B-cell antigens. CAR-T cells are mostly separated from a patient's peripheral blood by multistep progression manufacturing for therapeutic application. It is highlighted that induction of CAR gens into the T cell genome occurred by viral or nonviral agents [22]. Also, CARs involved extracellular domains represented by scFv from a mAb capable of stimulating T cell activation by intracellular domains [21]. For the extracellular domain, CD28 and CD8α are utilized as spacers [76,77]. The intracellular signaling domain consists of CD3ζ from the TCR complex [78]. CD8$^+$ T cells can remove foreign cells by attractive CAR-Ts for improving infectious diseases. Several generations of CARs have been changed to earn better function in their intracellular signaling [79].

The first CAR T-cells were generated for HIV envelope protein (Env) to exchange the extracellular domain by CD4$^+$ (CD4-CAR) in the early 90th [19]. Nevertheless, second CARs consist of the intracellular domain (CD28) that stimulated higher cytokine manufacture and controlled HIV reproduction invitro whereas, they were sensitive to HIV infection. To reach this aim, viral fusion inhibitors or small hairpin RNAs were linked to CD4-CARs. So, it is capable of destroying viral RNA and CCR5 as HIV-1 coreceptor [80]. Mentioned strategies represent the reduction of viral infection by CD4-CARs treatment in animal models.

In addition, other improved techniques have been utilized to demolish CCR5. Second-generation CARs for HIV CD4$^+$ binding site were planned through single-chain variable fragments extracted from Env-specific bNAbs [81]. Whereas CAR-T cells indicated the destruction of HIV-infected cells, their antiviral properties were varied and strain-dependent. Antiviral efficacy was facilitated by mixing second-generation glycan CARs, targeting different glycan sites on the HIV cells with deleting of CCR5, capable of controlling the viral replication over the CAR alone. The third generation of CAR T-cells (gp120-specific CARs) had significant lysis over CD4$^+$ CARs and preserved uninfected cells over interaction with the safe cells. This procedure's chief problem is viral mutants that exhibit unqualified therapy [16,80] (Fig. 26.3). Therefore, bi- and tri-specific CARs targeting up to three functionally distinct HIV Env-binding domains were planned to define two particular CARs on the same T-cell via two targeting components to enhance the treatment efficacy [82] (Fig. 26.4).

In one study, Krebs et al. worked on CAR-T cells with S domain receptor (S-CARs) for HBV envelope proteins (S, M, and L), which caused an HBV surface antigen (HBsAg) on the surface of infected cells. The results showed that infusion of S-CARs in mice reduced hepatocytes with HBV core protein, circulating virions in the bloodstream. At the same time, after the S-CAR-Ts weakened, viral activities were enhanced in contrast to the human S-CAR domains [83].

Experimentally, the recognition ability of CARs has been evaluated on HBV$^+$ cells and HBs Ag particles. The results proved that CAR-T cells had potential efficacy on HBV-infected hepatocytes in human liver chimeric mice mode [84].

Hepatitis C virus (HCV) infection is a medical symptom for end-stage of liver transplantation in patients who have rejected the current therapies. In some cases of HCV, it has been seen that reinfection occurs after treatment. Therefore, there is a need for the development of alternative therapeutic approaches. In this regard, Sautto et al. developed CAR-T cells that recognize HCV E2 glycoprotein (HCV/E2), a major target of the host immune response and one of the most variable viral proteins exposed on the surface of infected cells. These anti-HCV/E2 CAR-T cells indicated that they have notable cytotoxic activity against HCV-infected cells [85].

Reactivation of human cytomegalovirus (HCMV) reduces survival and induces mortality after transplantations and haematopoietic stem cells [86,87]. To get advantages of HCMV adoptive transfer engineered T cells, the human leukocyte antigen-matched donor is required. Thus, this may provide memory cells against HCMV glycoprotein B (gB), but anti-gB CAR-T

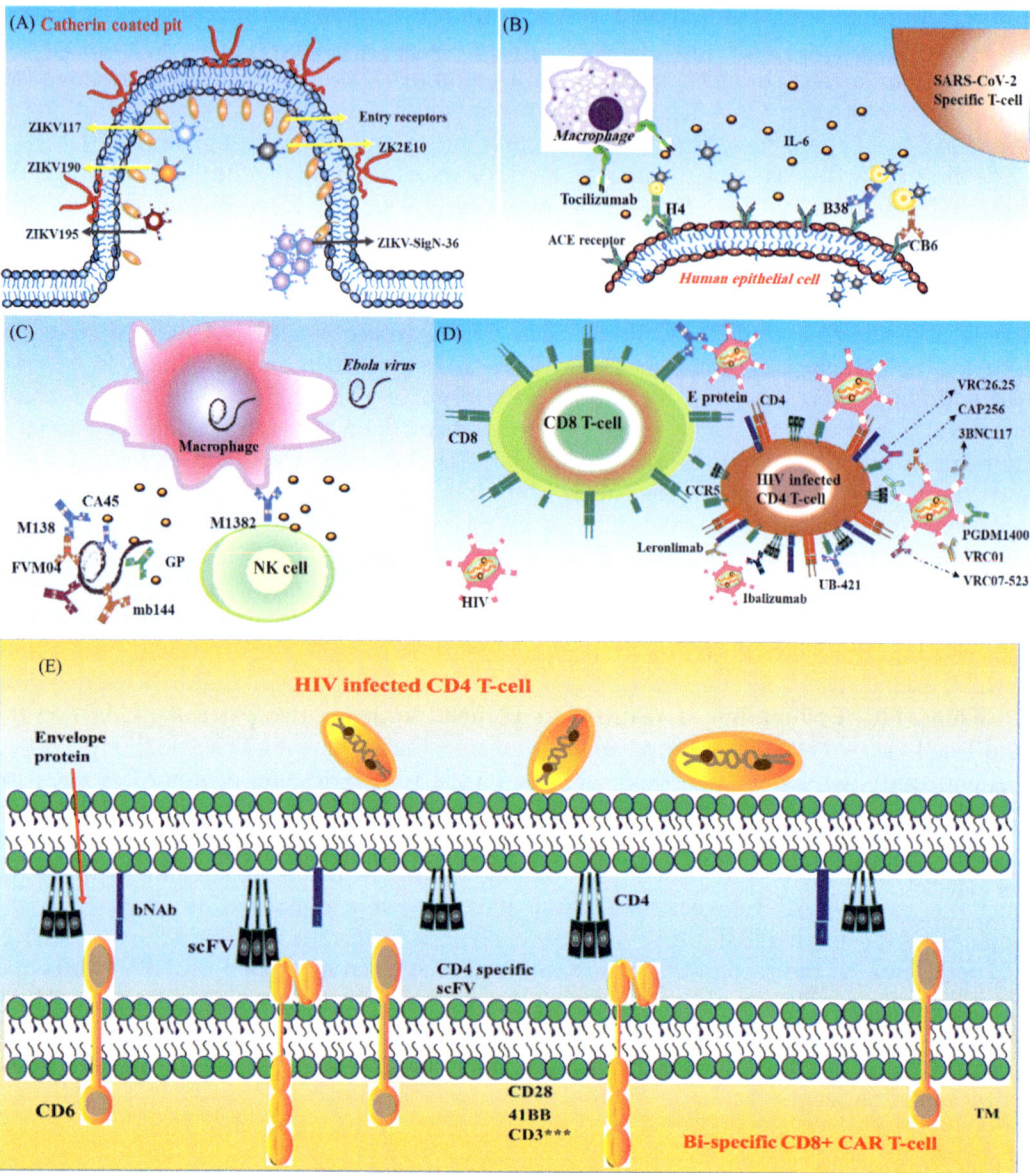

FIGURE 26.3 Immunotherapy against viral diseases. (A). Zika virus E trimers link to entry receptors formed in clathrin-coated holes of target cells in the primary stages of viral infection. Preventing viral binding and infection happens by binding mAbs (ZK2E10, ZIKV117, ZIKV195, and ZIKV190) with the entry receptors. ZIKV-195 crosslinks E protein which inhibits the construction of E trimers that are necessary for viral entry. Binding of ZIKV-SigN-36 to the E protein results in the constitution of bulks, which inhibits viral entry. (B) T-cells (activated by SARS-CoV-2) secrete IL-6 that contributes to cytokine storms during infection. The binding of tocilizumab to IL-6 inhibits the activation of the IL-6 receptor and reduces the inflammation from the cytokine storming. The binding of SARS-CoV-2 to the ACE2 receptor causes a viral entry in the cells. However, the mAbs (CB6, H4, and

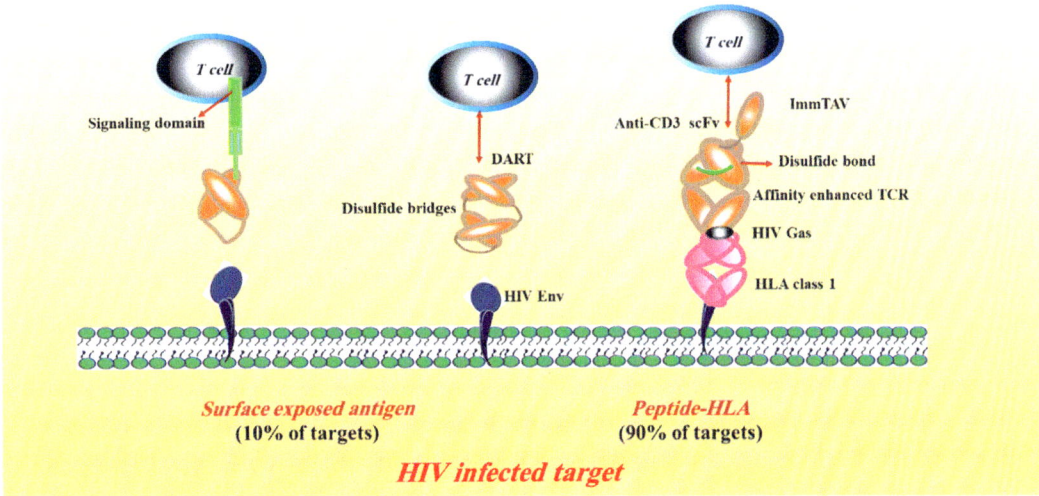

FIGURE 26.4 Schematic showing CAR T-cell. Immune-mobilizing monoclonal T cell receptor against viruses (ImmTAV) and dual affinity retargeting (DART) antigen recognition domains (T cell receptors or antibodies shown as blue ovals) and their related targets on HIV-infected cells. The CAR is fused to intracellular signaling domains. DARTs and ImmTAVs trigger signaling in T-cells via cell-surface CD3 via an anti-CD3 single-chain variable fragment (scFv) which is fused to the antibody/TCR by a transformative linker.

cells could not break unhealthy cells in in-vitro study [88]. Therefore, HCMV gB mainly represented on the infected cells in viral infection could be surveyed, as therapeutical means, in patients who suffer from drug-refractory HCMV reactivations using CAR-T cells. The HCMV envelope gB is the chief fusogenic protein in the HCMV-fusion multiplex, expressed early after HCMV infection on the cell membranes (72 to 96 hours after infection) [89,90]. Furthermore, the inhibition of HCMV replication by anti-gB CAR-T cells is independent of cytotoxic effector roles through the release of IFN-γ and TNF-α by activated CAR-Ts [91]. So examination of HCMV-specific CAR-T cells that include the CD28 or 4−1BB co-exciter end domains incorporated to an scFv obtained from the SM5−1 anti-gB antibody revealed that it

B38) against ACE2 inhibits the viral entry. (C) Sticking the Ebola virus to macrophages initiates the viral infection. mAbs (M138, CA45, mb144, and FVM04 specific) against Ebola glycoproteins inhibits virus binding interaction with macrophages and prevents disease. Binding of mAb (M1382) to death receptors expressed on the cell membranes of Ebola-infected macrophages degranulate and activate death signaling by recruitment of NK cells. (D) The binding of gp120 to the CD4 receptor initiates HIV infection. Leronlimab, Ibalizumab, and UB-421 are linked to the CD4 receptor on T-cells and inhibit binding the virus gp120 to its receptors. Afterwards, this binding prevents viral infection. The specific antibodies against gp120, including VRC07, VRC01, VRC26.25, PGDM1400, CAP256, and 3BNC117, are bound to gp120 on the virus's envelope and inhibit gp120 from binding to CD4, as a result, prevent to generate viral infection in target cells. The binding of CD8 and gp120 bi-specific antibodies to CD8 on one side and binding of gp120 with HIV on the other side gets HIV closer to cytotoxic T-cells, improving their ability to target and destroy the virus. (E). binding of CD4 and HIV E protein-specific CD8 CAR-T-cell to both CD4 and E proteins on T-cells (CD4 infected) cause to cell death of the infected T-cell. The C46 expression on the CAR-T-cell prevents the CAR T-cell itself from being HIV infected.

has potential affinity binding to a highly conserved, noncontiguous, and non-glycosylated domain of gB that is preserved for the duration of infection in both pre-and post-fusion conformations [90,92]. Also, it has been demonstrated that CD4$^+$ and CD8$^+$ T cells obtained from HCMV-seronegative adult blood or cord blood (CB) transferred by vectors effectively represented the gB-CARs. Therefore, gB-CAR-T cells can be used clinically to treat HCMV reactivations, especially when memory T cells are not available from the donors [93].

Accepting the immunogenicity of CAR-T cell therapy is vital to the prevention of infection and severe disease in a pandemic. In patients with cancer and haematopoietic cell transplant recipients, vaccination is required to prevent influenza and SARS-CoV-2 infections before and after CAR-T-cell therapy regardless of B-cell aplasia or hypogammaglobulinemia. Respiratory infections (especially with viruses) are the most common infectious complication after CAR-T-cell remedy, in which these patients are highly at a risk for influenza-caused death by 11%−33% of affected individuals [94,95]. Nevertheless, CAR T-cells in poor prognosis patients are potentially curative [74]. However, some studies indicated that autologous CAR-T cells therapy in coronavirus patients could have been used in urgent treatment requirements (less than 28 days). In contrast, allogeneic CAR-T cell treatment is suggested for patients wherein treatments could be delayed for 28 days. Also, the study revealed that cytokine release syndrome (CRS) is the adverse effect of CAR-T cell therapy in the treatment of SARS-CoV-2 and COVID-19 infection, which is stimulated by their target antigens [96,97]. Therefore, the study recommends CAR-T cell therapy should not be used for COVID-19 or hypoxemia patients [97]. As a reversible syndrome, CRS is triggered by the activation of epithelial cells produced in pulmonary edema (leading to acute respiratory failure in severe cases) [98]. Also, the study explored that many cytokines, such as ILs (IL-6, IL-8, and IL-10), macrophage inflammatory protein-1β (MIP-1β), IFN-γ, and monocyte chemoattractant protein 1 (MCP-1) participate in the progress of severe CRS [99,100]. Because of the viral reactivation of COVID-19 in CAR T-cell therapy, it was recommended that CAR T-cell therapy must be performed on patients who are unsymptomatic (symptom remission) for at least 14 days.

26.4.1 Checkpoint inhibition therapy

As an immune system regulator, inhibitory checkpoint molecules, such as CTLA-4 (ipilimumab), programmed cell death protein-1 (PD-1, i.e. pembrolizumab, nivolumab, cemiplimab), and PD-L1 (i.e. durvalumab, atezolizumab, avelumab), have been approved as first, second, or third-line treatments for various diseases [13,101,102]. Activation of T lymphocytes is adjusted by stimulators and inhibitors identified as immune checkpoints [103]. In otherwise, the association of TCR and antigen-MHC binding with the signaling of inhibitory receptors (i.e., CTLA-4) caused suppressing and exhaustion of the immune cells (mainly T cells). In turn, exhausted T cells lose their function, up-regulate the expression of immune checkpoint molecules, and poorly reminds responses and a distinct transcriptional from memory T cells [104]. In addition, it is highlighted that CTLA-4 is not generated in naïve T lymphocytes but is rapidly up-regulated during T cell activation. Significantly, CTLA-4 generally inhibits T cell autoreactive within the early priming phase in lymphoid organs to prevent autoimmunity [13,103]. Therefore, activation of T lymphocytes is related to the ratio between CD28 and CTLA-4 binding [103]. PD-1 causes the limitation of immune responses in peripheral tissues.

The binding of PD-1 with PD- L1/2 (its ligands) inhibits cytokine secretion (mediated by TCRs), survival and proliferation of CD8 cytotoxic T cell and promote the apoptosis of tumor-infiltrating lymphocytes and differentiation of CD4$^+$ T lymphocytes [13,14,96]. The activation of B lymphocytes, T lymphocytes, NK cells, and monocytes cause the PD-1 expression in several cells, such as tumor cells and some host cells (myeloid, lymphoid, and epithelial cells) [105,106]. T-cell inactivation pathways with immune checkpoints in viral infection are summarized in Table 26.3.

It has been indicated that ICI therapy for some chronic persistent viral infections, such as HIV, COVID-19, and HBV, stimulates inhibitory interactions between immune cells via immune checkpoint proteins to control the immune system. In this regard, some adverse events, such as autoimmune colitis, pneumonitis, hypophysitis, hepatitis, and thyroiditis, are reported in ICI therapy [118–120]. Therefore, managing checkpoint molecules could be helpful to viral infection clearance. Although combination therapy of PD-1 and CTLA4 is more effective in clinical trials but leads to induce adverse effects in patients [121]. In HIV-infected subjects receiving antiretroviral treatment, high expression of PD-1 is found on persisting infected T cells [121,122]. Indeed, the high amount of PD-1 and CTLA-4 expression has been related to a reduction in CD4$^+$ T lymphocytes during chronic and acute infection in untreated patients with HIV [15,121,123]. Some pre-clinical HIV persistence animal models demonstrated that T cell exhaustion is improved by PD-1 blockade in ICI therapy [106,121,124,125]. The cohort study revealed that HIV was controlled and stopped in 93% of patients and CD4$^+$ lymphocyte count increased during the ICI therapy. Also, the previous studies showed no immune reactivation inflammatory syndromes during the treatment [125]. Moreover, the clinical trial investigating proved in patients with chronic HIV infection, the tolerability and safety of CTLA-4 inhibition (ipilimumab) were reported in the lack of parallel malignancies [126]. Thus, much evidence suggests that ICI therapy could be improved immunity against HIV infection [127,128].

HBV, as a DNA virus, predominantly reproduces in the hepatocytes [129]. In chronic HBV infection, total and related CD8$^+$ T cells represent a high amount of CTLA4, PD-1, and TIM3 [122,123,129]. However, in acute HBV infection, intrahepatic CD8$^+$ T cells represent a high range of PD-1 [124]. The ligand for PD-1, PD-L1, has been exhibited to be induction on circulating CD19$^+$ B cells and CD14$^+$ monocytes in individuals with chronic viral infection in hepatic, therefore may participate to progress T cell exhaustion [125]. Antigen-specific CD8$^+$ T cells cause apoptosis via BIM (pro-apoptotic protein) expression [123]. HBV-specific CD4$^+$ T cells up-regulate the PD-1 levels without affection on CTLA4, TIM3, KLRG1, and CD244 levels [130]. Various ex vivo studies have illustrated that prevention of PD-1, CTLA4, TIM3, and 2B4 up-regulated the HBV-specific CD8$^+$ T cell function via blood collected from individuals with chronic HBV infection [118,123,126–128]. However, PD-1 blockade does not entirely enhance HBV-specific CD4$^+$ T cell operations via the production of IFN-γ, IL-2, and TNFα [130].

Therefore, ICI therapy alone or in combination with other medications may significantly improve the generation of HBV-specific CD8$^+$ T cells and antibodies against HBsAg. However, there are potential theoretical risks such as elevated reinvigorated T cells infiltration in the liver, which may stimulate liver inflammation [121], whereas this has not been investigated in recent clinical trials. In addition, the probable occurrence of immune-mediated hepatotoxicity in patients with virus hepatitis for cancer treatment is a critical adverse effect of ICI therapy. Furthermore, it is an essential diagnostic challenge in viral or autoimmune hepatitis

TABLE 26.3   Summarized immune checkpoints in viral infections and related pathways.

Immune checkpoint	Super family	Ligand	T-cell inactivation pathway	Presentation	References
CTLA-4	CD28	B7−1 (CD80), B7−2 (CD86)	CTLA-4 → PP2A → Akt → mTORc1 → regulates T-cell differentiation and proliferation	T cell-intrinsic	[107,108]
PD-1	CD28	PD-L1, PD-L2	a. PD-1(ITSMs) → SHP2 → PI3K/AKT →NFAT b. PD-1(ITSMs) → SHP2 → ZAP70/LCK → Ras-MEK-MAPK →AP1	T-cells, DCs and activated monocytes. T-cells, B-cells, natural killer.	[109,110]
LAG-3	IG	MHC-2	LAG- KIEELE motif/FXXL motif/C terminus EX repeat → unknown	T-cells Treg cells, B-cells, DCs and NK cells	[111]
TIM-3	TIM	Galcetin-9	a. TIM-3 → release Bat3 → LCK/FYN → ZAP70 → PLC-γ → NFAT b. TIM-3 → release Bat3 → LCK/FYN → ZAP70 → Ras-MEK-MAPK → AP1	Th1, Tc1, Treg cells	[112]
TIGIT	CD28	CD155 (on DCs)	TIGIT → SHIP1 → PI3K → AKT/PKC/IKK → NF-κB	T-cells and NK	[113–115]
BTA/CD160	CD28	HVEM	a. BTLA(ITIM/ITSM) → SHP-1/2 → LCK/FYN → ZAP70 → Ras-MEK-MAPK → AP1 b. BTLA(ITIM/ITSM) → SHP-1/2 → LCK/FYN → ZAP70 → PLC-γ → NFAT	T-cells, B-cells	[116,117]

AP1, Activator protein 1; *AP-2*, adapter protein-2; *BTLA*, B and T lymphocyte attenuator; *CTLA-4*, cytotoxic T-lymphocyte-associated protein 4; *HVEM*, herpesvirus entry mediator; *ITSM*, immune receptor tyrosine-based switch motif; *IG*, immunoglobulin; *IKK*, inhibitor of nuclear factor kappa-B kinase; *LAG-3*, lymphocyte activation gene-3; *LCK*, lymphocyte-specific protein tyrosine kinase; *NFAT*, nuclear factor of activated T-cells; *PLC-γ*, phospholipase C-γ; *PD-1*, programmed death-1; *PD-L1*, programmed death-ligand-1; *PKC*, protein kinase C; *SHP2*, SH2 domain-containing tyrosine phosphatase 2; *SHIP-1*, SH2-containing inositol phosphatase-1; *Th*, T helper; *TIGIT*, T-cell immunoglobulin and ITIM domain; *TCR*, T-cell receptor; *ZAP 70*, zeta chain of T-cell receptor-associated protein kinase 70.

patients that immune-mediated hepatitis may show [131−133]. Therefore, further studies are necessary to recognize whether ICI can be used to induce HBV remission.

Influenza A (H1N1, H3N2) and B (Victoria, Yamagata) viruses are the reason for seasonal epidemics, which affect the global population annually and cause more than 645,000 deaths worldwide [119,120,134]. Bersanelli et al. demonstrated that the ICI therapy had a more prolonged overall survival effect in patients vaccinated with the influenza vaccine and/or progressed influenza syndrome, which there were no reports of immune-related adverse events in these studies [134]. Also, Yu et al. revealed that influenza infection, to a great degree, elevates the amount of PD-1 positive innate lymphoid cells in the mice lungs. Therefore, the anti-PD-1 treatment causes a reduction of total lung innate lymphoid cells with approximately total depleting of PD-1 highly positive cells. So, they proposed that anti-PD-1 treatment or ICI therapy maybe have impressive effects for both treatment and prevention of disease [133].

the severe acute respiratory syndrome coronavirus 2 (SARS- COV-2) has been the reason for more than 200,000 deaths worldwide since March 2020 [106]. Diao et al. revealed that in patients with COVID-19, an increase of pro-inflammatory cytokines enhanced the decrease of T lymphocytes [135]. In turn, it caused a high amount of PD-1 and Tim-3 expression in progressed patients from prodromal to symptomatic stages (same as other viral diseases) [136,137]. So, inhibition of PD-1 on CD8$^+$ lymphocytes by ICI therapy [119] and controlling of TNF-α, IL-10, and other cytokines may have modulated T cell exhaustion through programmed cell death [137]. Also, it is worth mentioning that ICI therapy may affect just the initial and intermediate stages of this disease. Nevertheless, T lymphocyte exhaustion will be irreversible if the PD-1 expression level is high on CD8$^+$ T lymphocytes, and ICI cannot affect it [134]. Based on data, the possibility of autoinflammatory process in the advanced stage of COVID-19, ICI treatment could display the over activation of the immune system [106].

Alternatively, as a novel approach to ICI, administration of CD200- CD200R1 decreases the RNA virus (single-strand) sensor toll-like receptor seven (TLR-7) in myeloid cells. Also, it restores IFN-γ levels for elevating virus elimination [138,139]. Notably, ICI administration may effectively manage COVID-19 patients without cancer [135,137,140,141]. Furthermore, combination therapy with ICI and anti-IL-6 antibody is a practical approach to decrease the risks of irAEs and secretion of cytokine in severe COVID-19 cases [142].

26.5 Defensin therapy

Defensins play an essential role in the host immune response against viral infections by modulating innate and adaptive immune responses [143]. Defensins are a vital element in the immune system that appear before lymphocytes and immunoglobulin. Once secreted this peptide empowers its hosts' immunity against potential pathogens in mammals that are the principal antimicrobial and anti-viral activity from epithelial and neutrophil cells [144]. This peptide belongs to variably cationic and amphipathic proteins, which are in high amounts in the granules of mammalian phagocytes [145]. The large β-sheet structure is an expected future of all defensins. Defensins include three intramolecular cysteine disulfide bonds [144]. The base of structure and disulfide connectivity, vertebral defensins are subdivided into α (between 29 and 35 residues), β (up to 45 residues), and θ (most petite with 18 residues) defensins. Although in humans, β-defensins are expressed by epithelial cells [144,146], they are stimulated fewer by IL-1 [147]. Commonly, enveloped viruses are sensitive to α, β, and θ-defensins. In this way, interactions of the viral glycoproteins with cellular receptors are inhibited by defensins. For instance, inhibition of RSV is performed by human β-defensin (HBD) to duo the membrane disruption [148]. Nevertheless, in non-enveloped viruses, only α-defensins directly prevent uncoating and lead to removing the virus to the lysosome for degradation without any affection by β and θ-defensins [149]. Defensin activity has been studied on various enveloped and non-enveloped viruses, such as herpes simplex virus 1 (HSV-1) [150], HIV [132], Influenza A virus [151], human papillomavirus (HPV) [152].

The study of the G-protein signaling antagonist in the single-cycle virus system revealed that HBD inhibited HIV progression at an early post-entry stage with G-protein coupled receptor-mediated signaling. The result showed that monocyte-derived macrophages represent the shared chemokine-HBD receptors but at variable levels of donors. In addition, the signal of HBD is performed by the additional receptor. These findings propose that HBD treatment could suppress HIV in MDM [153].

The α-defensins, as innate immune effectors against nonnveloped viral pathogens, suppress the host-mediated proteolytic processing vital for HPV infection. In the presence of α-defensin, the HPV capsid was sensitized to antibodies and proteases via the internalization of this virus. Therefore, α-defensin inhibits the viral capsid decomposition from the genome. Finally, it decreased viral trafficking to the trans-Golgi network and catalyzed the degeneration of capsids in lysosomes. Interestingly, the mechanisms of inhibition of HPV are remarkably similar to mechanisms of inhibition of human adenovirus infection [149].

Measuring laboratory parameters in COVID-19 indicated an association between serum IL-1 receptor antagonist, IL-18, and α defensin levels and the clinical terms and prognosis of related viral infection. In patients with developed macrophage activation syndrome (MAS) and acute respiratory distress syndrome (ARDS), IL-1Ra, IL-18, and serum α-defensin levels were remarkably increased compared to the control group. In addition, the mentioned parameters levels were enormously elevated in COVID-19 patients with and without ARDS or MAS compared to the control group [154]. As an anti-inflammatory agent, α-defensin can be used to treat MAS, which is secreted by apoptotic and necrotic neutrophils at the inflammation site [143,155,156], with a chemotactic effect on T cells, monocytes, and dendritic cells. Also, α-defensin increases in influenza A infection, and it induces virus uptake with neutrophils [157]. In contrast, it minimizes oxidative damage during virus uptake [158]. In addition, the known antiviral activities of human defensins against enveloped and non-enveloped DNA viruses are summarized in Table 26.4 [158]. The study demonstrated that defensin causes inhibitory effects on cell virus entry and the transmission of viruses among cells. So, the anti-viral effect of defensins may be utilized as a treatment to control and prevent viral infections [143,155].

TABLE 26.4 Anti-viral activities of human defensins against various viruses [156].

Virus/defensin	α defensin				Enteric α defensin		β defensins			
	HNP1	HNP2	HNP3	HNP4	HD5	HD6	HBD1	HBD2	HBD3	HBD4
HSV2/ enveloped	Attachment inhibition via binding to gB, prevents VP16 transport and viral gene expression post-infection	Attachment inhibition via binding to gB, prevents VP16 transport and viral gene expression post-infection	Attachment inhibition via binding to gB, prevents VP16 transport and viral gene expression post-infection	Attachment inhibition via binding to heparan sulfate	Attachment inhibition via binding to gB, inhibits viral gene expression post-infection	Attachment inhibition via binding to heparan sulfate	Tested HDP had no effect	Tested HDP had no effect	Attachment inhibition via binding to gB and heparin	Yet to be tested
Adenovirus/ nonenveloped	–	–	–	–	–	–	–	–	–	–
A	Unknown mechanism	Yet to be tested	Yet to be tested	Yet to be tested	Inhibits endosomal lysis and viral uncoating via binding to the virus	Yet to be tested	Yet to be tested	Yet to be tested	Yet to be tested	Yet to be tested
B	Unknown mechanism	Yet to be tested	Yet to be tested	Yet to be tested	Inhibits endosomal lysis and viral uncoating via binding to the virus	Yet to be tested	No effect by HDP	No effect by HDP	Yet to be tested	Yet to be tested
C	Inhibits endosomal lysis	Yet to be tested	Yet to be tested	Yet to be tested	Inhibits endosomal lysis and viral uncoating via binding to the virus	Yet to be tested	Unknown mechanism	No effect by HDP	Yet to be tested	Yet to be tested
D	No effect by HDP	Yet to be tested	Yet to be tested	Yet to be tested	No effect by HDP	Yet to be tested	No effect by HDP	No effect by HDP	Yet to be tested	Yet to be tested
E	Unknown mechanism	Yet to be tested	Yet to be tested	Yet to be tested	Inhibits endosomal lysis and viral uncoating via binding to the virus	Yet to be tested	Yet to be tested	Yet to be tested	Yet to be tested	Yet to be tested

(Continued)

TABLE 26.4 (Continued)

Virus/ defensin	α defensin				Enteric α defensin		β defensins			
	HNP1	HNP2	HNP3	HNP4	HD5	HD6	HBD1	HBD2	HBD3	HBD4
F	No effect by HDP	Yet to be tested	Yet to be tested	Yet to be tested	No effect by HDP	Yet to be tested	Yet to be tested	Yet to be tested	Yet to be tested	Yet to be tested
HPV/ nonenveloped	Inhibits endosomal lysis and viral uncoating via binding to virus	Unknown mechanism	Unknown mechanism	Unknown mechanism	Inhibits endosomal lysis and viral uncoating via binding to virus	No effect by HDP	No effect by HDP	No effect by HDP	Yet to be tested	Yet to be tested
HCN	Inhibition of VE	Inhibition of VE	Inhibition of VE	Inhibition of VE	Yet to be tested	Yet to be tested	Inhibition of VE	Inhibition of VE	Inhibition of VE	Inhibition of VE
Influenza A virus	Prevents PKC activation and viral envelope fusion with endosome, increases neutrophil uptake of virus	Prevents PKC activation and viral envelope fusion with endosome, increases neutrophil uptake of virus	Increases and aggregates neutrophil uptake of virus	Unknown mechanism	Increases and aggregates neutrophil uptake of virus	Unknown mechanism	Unknown mechanism	Unknown mechanism	Prevents HA-mediated fusion and MPM	Yet to be tested
HIV-I	Decreases CD4 and CXCR4 via Binding to CD4 and gp120, upregulation of CCR5 ligands, inhibits PKC activation and VE	Decreases CD4 and CXCR4 via Binding to CD4 and gp120, upregulation of CCR5 ligands	Binds to gp120	Binds to CD4 and gp120	E	E	Unknown mechanism	Binds and competes to the virus for HSPGs, decreases CXCR4, induces APOBEC3G expression	Binds and competes to the virus for HSPGs, decreases CXCR4	Yet to be tested

APOBEC3G, apolipoprotein B mRNA editing enzyme, catalytic polypeptide-like 3G; *E*, tested HDP enhanced viral infection; *HBD*, human β-defensins; *HD*, human defensins; *HNP*, human neutrophil peptide; *HSPGs*, heparan sulfate proteoglycans; *MPM*, membrane protein mobility; *VE*, virus entry.

References

[1] Papaioannou NE, Beniata OV, Vitsos P, Tsitsilonis O, Samara P. Harnessing the immune system to improve cancer therapy. Ann Transl Med 2016;4(14):261.

[2] Tur MK, Barth S. Immunotherapy. In: Schwab M, editor. Encyclopedia of cancer. Berlin, Heidelberg: Springer; 2017. pp. 2237–2239.

[3] Hegde NR, Rao PP, Bayry J, Kaveri SV. Immunotherapy of viral infections. Immunotherapy 2009;1(4):691–711.

[4] Canouï E, Launay O. History and principles of vaccination. Rev Mal Respir 2019;36(1):74–81.

[5] Pollard AJ, Bijker EM. A guide to vaccinology: from basic principles to new developments. Nat Rev Immunol 2021;21(2):83–100.

[6] Rappuoli R, Pizza M, Del Giudice G, De Gregorio E. Vaccines, new opportunities for a new society. Proc Natl Acad Sci U S A 2014;111(34):12288–93.

[7] Shepard HM, Phillips GL, DT C, Feldmann M. Developments in therapy with monoclonal antibodies and related proteins. Clin Med (London, Engl) 2017;17(3):220–32.

[8] Kaunitz JD. Development of monoclonal antibodies: the dawn of mAb rule. Digestive Dis Sci 2017;62 (4):831–2.

[9] Hooks MA, Wade CS, Millikan Jr. WJ. Muromonab CD-3: a review of its pharmacology, pharmacokinetics, and clinical use in transplantation. Pharmacotherapy. 1991;11(1):26–37.

[10] Buss NAPS, Henderson SJ, McFarlane M, Shenton JM, de Haan L. Monoclonal antibody therapeutics: history and future. Curr Opin Pharmacol 2012;12(5):615–22.

[11] Basta M. Ambivalent effect of immunoglobulins on the complement system: activation vs inhibition. Mol Immunol 2008;45(16):4073–9.

[12] Pardoll DM. The blockade of immune checkpoints in cancer immunotherapy. Nat Rev Cancer 2012;12 (4):252–64.

[13] Darvin P, Toor SM, Sasidharan Nair V, Elkord E. Immune checkpoint inhibitors: recent progress and potential biomarkers. Exp Mol Med 2018;50(12):1–11.

[14] Granier C, De Guillebon E, Blanc C, Roussel H, Badoual C, Colin E, et al. Mechanisms of action and rationale for the use of checkpoint inhibitors in cancer. ESMO Open 2017;2(2):e000213.

[15] Dyck L, Mills KHG. Immune checkpoints and their inhibition in cancer and infectious diseases. Eur J Immunol 2017;47(5):765–79.

[16] Ramamurthy D, Nundalall T, Cingo S, Mungra N, Karaan M, Naran K, et al. Recent advances in immunotherapies against infectious diseases. Immunother Adv 2021;1(1):ltaa007.

[17] Ohno M, Natsume A, Wakabayashi T. Cytokine therapy. Adv Exp Med Biol 2012;746:86–94.

[18] Muñoz-Carrillo JL, Contreras-Cordero J, Gutierrez O, Villalobos-Gutiérrez P, Ramos-Gracia L, Hernández-Reyes V. Cytokine profiling plays a crucial role in activating immune system to clear infectious pathogens; 2018.

[19] Seif M, Einsele H, Löffler J. CAR T cells beyond cancer: hope for immunomodulatory therapy of infectious diseases. Front Immunol 2019;10:2711.

[20] Brudno JN, Kochenderfer JN. Recent advances in CAR T-cell toxicity: mechanisms, manifestations and management. Blood Rev 2019;34:45–55.

[21] Klebanoff CA, Rosenberg SA, Restifo NP. Prospects for gene-engineered T cell immunotherapy for solid cancers. Nat Med 2016;22(1):26–36.

[22] Rosenberg SA, Restifo NP. Adoptive cell transfer as personalised immunotherapy for human cancer. Science 2015;348(6230):62–8.

[23] Berzofsky JA, Ahlers JD, Janik J, Morris J, Oh S, Terabe M, et al. Progress on new vaccine strategies against chronic viral infections. J Clin Invest 2004;114(4):450–62.

[24] Francis MJ. Recent advances in vaccine technologies. Vet Clin North Am Small Anim Pract 2018;48 (2):231–41.

[25] Ulmer JB, Mason PW, Geall A, Mandl CW. RNA-based vaccines. Vaccine 2012;30(30):4414–18.

[26] Ballesteros-Briones MC, Silva-Pilipich N, Herrador-Cañete G, Vanrell L, Smerdou C. A new generation of vaccines based on alphavirus self-amplifying RNA. Curr Opvirol 2020;44:145–53.

[27] Lundstrom K. Self-amplifying RNA viruses as RNA vaccines. Int J Mol Sci 2020;21(14).

[28] Homma K, Noguchi T, Fukuchi S. Codon usage is less optimised in eukaryotic gene segments encoding intrinsically disordered regions than in those encoding structural domains. Nucleic Acids Res 2016;44 (21):10051–61.

[29] Cagigi A, Loré K. Immune responses induced by mRNA vaccination in mice, monkeys and humans. Vaccines 2021;9(1).

[30] Liehl P, Zuzarte-Luís V, Chan J, Zillinger T, Baptista F, Carapau D, et al. Host-cell sensors for Plasmodium activate innate immunity against liver-stage infection. Nat Med 2014;20(1):47−53.

[31] Wang Y, Zhang Z, Luo J, Han X, Wei Y, Wei X. mRNA vaccine: a potential therapeutic strategy. Mol Cancer 2021;20(1):33.

[32] Gay CL, DeBenedette MA, Tcherepanova IY, Gamble A, Lewis WE, Cope AB, et al. Immunogenicity of AGS-004 dendritic cell therapy in patients treated during acute HIV infection. AIDS Res Hum Retroviruses 2018;34(1):111−22.

[33] Shimojima M, Fukushi S, Tani H, Yoshikawa T, Fukuma A, Taniguchi S, et al. Effects of ribavirin on severe fever with thrombocytopenia syndrome virus in vitro. Jpn J Infect Dis 2014;67(6):423−7.

[34] Shimojima M, Fukushi S, Tani H, Taniguchi S, Fukuma A, Saijo M. Combination effects of ribavirin and interferons on severe fever with thrombocytopenia syndrome virus infection. Virol J 2015;12:181.

[35] Lee MJ, Kim KH, Yi J, Choi SJ, Choe PG, Park WB, et al. In vitro anti-viral activity of ribavirin against severe fever with thrombocytopenia syndrome virus. Korean J Intern Med 2017;32(4):731−7.

[36] Baba M, Toyama M, Sakakibara N, Okamoto M, Arima N, Saijo M. Establishment of an anti-viral assay system and identification of severe fever with thrombocytopenia syndrome virus inhibitors. Anti-viral Chem Chemother 2017;25(3):83−9.

[37] Smee DF, Jung KH, Westover J, Gowen BB. 2′-Fluoro-2′-deoxycytidine is a broad-spectrum inhibitor of bunyaviruses in vitro and in phleboviral disease mouse models. Anti-viral Res 2018;160:48−54.

[38] Tani H, Fukuma A, Fukushi S, Taniguchi S, Yoshikawa T, Iwata-Yoshikawa N, et al. Efficacy of T-705 (Favipiravir) in the treatment of infections with lethal severe fever with thrombocytopenia syndrome virus. mSphere 2016;1(1).

[39] Li H, Zhang LK, Li SF, Zhang SF, Wan WW, Zhang YL, et al. Calcium channel blockers reduce severe fever with thrombocytopenia syndrome virus (SFTSV) related fatality. Cell Res 2019;29(9):739−53.

[40] Gowen BB, Hickerson BT. Hemorrhagic fever of bunyavirus etiology: disease models and progress towards new therapies. J Microbiol (Seoul, Korea) 2017;55(3):183−95.

[41] Tani H, Komeno T, Fukuma A, Fukushi S, Taniguchi S, Shimojima M, et al. Therapeutic effects of favipiravir against severe fever with thrombocytopenia syndrome virus infection in a lethal mouse model: dose-efficacy studies upon oral administration. PLoS One 2018;13(10):e0206416.

[42] Ning YJ, Mo Q, Feng K, Min YQ, Li M, Hou D, et al. Interferon-γ-directed inhibition of a novel high-pathogenic phlebovirus and viral antagonism of the antiviral signaling by targeting STAT1. Front Immunol 2019;10:1182.

[43] Bolhassani A, Yazdi SR. DNA immunisation as an efficient strategy for vaccination. Avicenna J Med Biotechnol 2009;1(2):71−88.

[44] Lorenzen N, LaPatra SE. DNA vaccines for aquacultured fish. Rev Sci Tech 2005;24(1):201−13.

[45] Pitman MC, Lau JSY, McMahon JH, Lewin SR. Barriers and strategies to achieve a cure for HIV. Lancet HIV 2018;5(6):e317−28.

[46] Rollier CS, Reyes-Sandoval A, Cottingham MG, Ewer K, Hill AV. Viral vectors as vaccine platforms: deployment in sight. Curr Opimmunol 2011;23(3):377−82.

[47] Quiñones-Parra S, Loh L, Brown LE, Kedzierska K, Valkenburg SA. Universal immunity to influenza must outwit immune evasion. Front Microbiol 2014;5:285.

[48] Ohtake S, Arakawa T. Recombinant therapeutic protein vaccines. Protein Pept Lett 2013;20(12):1324−44.

[49] Roldão A, Mellado MC, Castilho LR, Carrondo MJ, Alves PM. Virus-like particles in vaccine development. Expert Rev Vaccines 2010;9(10):1149−76.

[50] Kushnir N, Streatfield SJ, Yusibov V. Virus-like particles as a highly efficient vaccine platform: diversity of targets and production systems and advances in clinical development. Vaccine 2012;31(1):58−83.

[51] Litman GW, Rast JP, Shamblott MJ, Haire RN, Hulst M, Roess W, et al. Phylogenetic diversification of immunoglobulin genes and the antibody repertoire. Mol Biol Evol 1993;10(1):60−72.

[52] Ali MG, Zhang Z, Gao Q, Pan M, Rowan EG, Zhang J. Recent advances in therapeutic applications of neutralising antibodies for virus infections: an overview. Immunol Res 2020;68(6):325−39.

[53] Huse K, Böhme H-J, Scholz GH. Purification of antibodies by affinity chromatography. J. Biochem. Biophys. Methods 2002;51(3):217−31.

[54] Erhard M, Von Quistorp I, Schranner I, Jüngling A, Kaspers B, Schmidt P, et al. Development of specific enzyme-linked immunosorbent antibody assay systems for the detection of chicken immunoglobulins G, M, and A using monoclonal antibodies. Poult Sci 1992;71(2):302–10.

[55] Wootla B, Denic A, Rodriguez M. Polyclonal and monoclonal antibodies in clinic. Methods Mol Biol (Clifton, NJ) 2014;1060:79–110.

[56] Zowail ME, El-Balshy RM, Asran AE-MA, Khidr FK, Zaki GGJTEJOEB. Effect of ginger against the toxicity of brodifacoum on albino male mice. J Chin Med Assoc 2019;15(2):253–62.

[57] Excler JL, Ake J, Robb ML, Kim JH, Plotkin SA. Nonneutralizing functional antibodies: a new "old" paradigm for HIV vaccines. Clin Vaccine Immunol: CVI 2014;21(8):1023–36.

[58] Burton DR, Mascola JR. Antibody responses to envelope glycoproteins in HIV-1 infection. Nat Immunol 2015;16(6):571–6.

[59] Mayr LM, Decoville T, Schmidt S, Laumond G, Klingler J, Ducloy C, et al. Non-neutralizing antibodies targeting the V1V2 domain of HIV exhibit strong antibody-dependent cell-mediated cytotoxic activity. Sci Rep 2017;7(1):12655.

[60] Mupapa K, Massamba M, Kibadi K, Kuvula K, Bwaka A, Kipasa M, et al. Treatment of ebola hemorrhagic fever with blood transfusions from convalescent patients. J Infect Dis 1999;179(Suppl. 1):S18–23.

[61] Foo DG, Alonso S, Chow VT, Poh CL. Passive protection against lethal enterovirus 71 infection in newborn mice by neutralising antibodies elicited by a synthetic peptide. Microbes Infect 2007;9(11):1299–306.

[62] Shahzad MI, Naeem K, Mukhtar M, Khanum A. Passive immunisation against highly pathogenic Avian Influenza Virus (AIV) strain H7N3 with antiserum generated from viral polypeptides protect poultry birds from lethal viral infection. Virol J 2008;5:144.

[63] Chambers TM, Kawaoka Y, Webster RG. Protection of chickens from lethal influenza infection by vaccinia-expressed hemagglutinin. Virology 1988;167(2):414–21.

[64] Cui S, Li Y, Wang Y, Cui Z, Chang S, Zhao P. Joint treatment with azidothymidine and antiserum for eradication of avian leukosis virus subgroup a contamination in vaccine virus seeds. Poult Sci 2019;98 (2):629–33.

[65] Marasco WA, Sui J. The growth and potential of human anti-viral monoclonal antibody therapeutics. Nat Biotechnol 2007;25(12):1421–34.

[66] Haurum JS. Recombinant polyclonal antibodies: the next generation of antibody therapeutics? Drug Discov Today 2006;11(13–14):655–60.

[67] Arumugham VB, Rayi A. Intravenous immunoglobulin (IVIG). Treasure Island, FL: StatPearls Publishing; 2021.

[68] Galeotti C, Kaveri SV, Bayry J. IVIG-mediated effector functions in autoimmune and inflammatory diseases. Int Immunol 2017;29(11):491–8.

[69] Ballow M. Clinical and investigational considerations for the use of IGIV therapy. Am J Health-Syst Pharm 2005;62(16 Suppl. 3):S12–18 quiz S9–21.

[70] Matucci A, Maggi E, Vultaggio A. Mechanisms of action of Ig preparations: immunomodulatory and anti-inflammatory effects. Front Immunol 2014;5:690.

[71] Halwe S, Kupke A, Vanshylla K, Liberta F, Gruell H, Zehner M, et al. Intranasal administration of a mono-clonal neutralising antibody protects mice against SARS-CoV-2 infection. bioRxiv 2021;.

[72] Cruz-Teran C, Tiruthani K, McSweeney M, Ma A, Pickles R, Lai SK. Challenges and opportunities for anti-viral monoclonal antibodies as COVID-19 therapy. Adv Drug Deliv Rev 2020;.

[73] Saylor C, Dadachova E, Casadevall A. Monoclonal antibody-based therapies for microbial diseases. Vaccine 2009;27(Suppl. 6):G38–46.

[74] Ortiz de Landazuri I, Egri N, Muñoz-Sánchez G, Ortiz-Maldonado V, Bolaño V, Guijarro C, et al. Manufacturing and management of CAR T-cell therapy in "COVID-19's Time": central vs point of care proposals. Front. Immunol., 11. 2020. p. 2496.

[75] Rothan HA, Byrareddy SN. The epidemiology and pathogenesis of coronavirus disease (COVID-19) out-break. J Autoimmun 2020;109:102433.

[76] Milone MC, Fish JD, Carpenito C, Carroll RG, Binder GK, Teachey D, et al. Chimeric receptors containing CD137 signal transduction domains mediate enhanced survival of T cells and increased antileukemic efficacy in vivo. Mol Therapy 2009;17(8):1453–64.

[77] Kochenderfer JN, Feldman SA, Zhao Y, Xu H, Black MA, Morgan RA, et al. Construction and preclinical evaluation of an anti-CD19 chimeric antigen receptor. J Immunother (Hagerstown, MD 2009;32(7):689–702 1997.

[78] Gong MC, Latouche JB, Krause A, Heston WD, Bander NH, Sadelain M. Cancer patient T cells genetically targeted to prostate-specific membrane antigen specifically lyse prostate cancer cells and release cytokines in response to prostate-specific membrane antigen. Neoplasia 1999;1(2):123–7.

[79] Subklewe M, von Bergwelt-Baildon M, Humpe A. Chimeric antigen receptor T cells: a race to revolutionize cancer therapy. Transfus Med Hemother 2019;46(1):15–24.

[80] Niessl J, Baxter AE, Mendoza P, Jankovic M, Cohen YZ, Butler AL, et al. Combination anti-HIV-1 antibody therapy is associated with increased virus-specific T cell immunity. Nat Med 2020;26(2):222–7.

[81] Liu L, Patel B, Ghanem MH, Bundoc V, Zheng Z, Morgan RA, et al. Novel CD4-based bispecific chimeric antigen receptor designed for enhanced anti-HIV potency and absence of HIV entry receptor activity. J Virol 2015;89(13):6685–94.

[82] Maldini CR, Claiborne DT, Okawa K, Chen T, Dopkin DL, Shan X, et al. Dual CD4-based CAR T cells with distinct costimulatory domains mitigate HIV pathogenesis in vivo. Nat Med 2020;26(11):1776–87.

[83] Festag MM, Festag J, Fräßle SP, Asen T, Sacherl J, Schreiber S, et al. Evaluation of a fully human, hepatitis B virus-specific chimeric antigen receptor in an immunocompetent mouse model. Mol Ther 2019;27 (5):947–59.

[84] Kruse RL, Shum T, Tashiro H, Barzi M, Yi Z, Whitten-Bauer C, et al. HBsAg-redirected T cells exhibit antiviral activity in HBV-infected human liver chimeric mice. Cytotherapy 2018;20(5):697–705.

[85] Sautto GA, Wisskirchen K, Clementi N, Castelli M, Diotti RA, Graf J, et al. Chimeric antigen receptor (CAR)-engineered T cells redirected against hepatitis C virus (HCV) E2 glycoprotein. Gut 2016;65 (3):512–23.

[86] Fuji S, Einsele H, Kapp M. Cytomegalovirus disease in hematopoietic stem cell transplant patients: current and future therapeutic options. Curr Opin Infect Dis 2017;30(4):372–6.

[87] Teira P, Battiwalla M, Ramanathan M, Barrett AJ, Ahn KW, Chen M, et al. Early cytomegalovirus reactivation remains associated with increased transplant-related mortality in the current era: a CIBMTR analysis. Blood 2016;127(20):2427–38.

[88] Proff J, Brey CU, Ensser A, Holter W, Lehner M. Turning the tables on cytomegalovirus: targeting viral Fc receptors by CARs containing mutated CH2-CH3 IgG spacer domains. J Transl Med 2018;16(1):26.

[89] Cooper RS, Heldwein EE. Herpesvirus gB: a finely tuned fusion machine. Viruses 2015;7(12):6552–69.

[90] Pötzsch S, Spindler N, Wiegers AK, Fisch T, Rücker P, Sticht H, et al. B cell repertoire analysis identifies new antigenic domains on glycoprotein B of human cytomegalovirus which are target of neutralising antibodies. PLoS Pathog 2011;7(8):e1002172.

[91] Kumaresan PR, Manuri PR, Albert ND, Maiti S, Singh H, Mi T, et al. Bioengineering T cells to target carbohydrate to treat opportunistic fungal infection. Proc Natl Acad Sci U S A 2014;111(29):10660–5.

[92] Spindler N, Rücker P, Pötzsch S, Diestel U, Sticht H, Martin-Parras L, et al. Characterisation of a discontinuous neutralising epitope on glycoprotein B of human cytomegalovirus. J Virol 2013;87(16):8927–39.

[93] Olbrich H, Theobald SJ, Slabik C, Gerasch L, Schneider A, Mach M, et al. Adult and cord blood-derived high-affinity gB-CAR-T cells effectively react against human cytomegalovirus infections. Hum Gene Ther 2020;31(7–8):423–39.

[94] Walti CS. Humoral immunogenicity of the seasonal influenza vaccine before and after CAR-T-cell therapy; 2021.

[95] Hill JA, Li D, Hay KA, Green ML, Cherian S, Chen X, et al. Infectious complications of CD19-targeted chimeric antigen receptor-modified T-cell immunotherapy. Blood 2018;131(1):121–30.

[96] Liu D, Zhao J. Cytokine release syndrome: grading, modeling, and new therapy. J Hematol Oncol 2018;11 (1):121.

[97] Hu Y, Tan Su Yin E, Yang Y, Wu H, Wei G, Su J, et al. CAR T-cell treatment during the COVID-19 pandemic: management strategies and challenges. Curr Res Transl Med 2020;68(3):111–18.

[98] Teachey DT, Lacey SF, Shaw PA, Melenhorst JJ, Maude SL, Frey N, et al. Identification of predictive biomarkers for cytokine release syndrome after chimeric antigen receptor T-cell therapy for acute lymphoblastic leukemia. Cancer Discov 2016;6(6):664–79.

[99] Zhang C, Wu Z, Li J-W, Zhao H, Wang G-Q. Cytokine release syndrome in severe COVID-19: interleukin-6 receptor antagonist tocilizumab may be the key to reduce mortality. Int J Antimicrob Agents 2020;55 (5):105954.

[100] Huang C, Wang Y, Li X, Ren L, Zhao J, Hu Y, et al. Clinical features of patients infected with 2019 novel coronavirus in Wuhan, China. Lancet 2020;395(10223):497–506.

[101] Zhang N, Tu J, Wang X, Chu Q. Programmed cell death-1/programmed cell death ligand-1 checkpoint inhibitors: differences in mechanism of action. Immunotherapy 2019;11(5):429–41.

[102] Vaddepally RK, Kharel P, Pandey R, Garje R, Chandra AB. Review of indications of FDA-approved immune checkpoint inhibitors per NCCN guidelines with the level of evidence. Cancers 2020;12(3).

[103] Qin S, Xu L, Yi M, Yu S, Wu K, Luo S. Novel immune checkpoint targets: moving beyond PD-1 and CTLA-4. Mol Cancer 2019;18(1):155.

[104] Blank CU, Haining WN, Held W, Hogan PG, Kallies A, Lugli E, et al. Defining 'T cell exhaustion'. Nat Rev Immunol 2019;19(11):665–74.

[105] Leung CS, Yang KY, Li X, Chan VW, Ku M, Waldmann H, et al. Single-cell transcriptomics reveal that PD-1 mediates immune tolerance by regulating proliferation of regulatory T cells. Genome Med 2018;10(1):71.

[106] Gambichler T, Reuther J, Scheel CH, Becker JC. On the use of immune checkpoint inhibitors in patients with viral infections including COVID-19. J Immunother Cancer 2020;8(2).

[107] Qureshi OS, Kaur S, Hou TZ, Jeffery LE, Poulter NS, Briggs Z, et al. Constitutive clathrin-mediated endocytosis of CTLA-4 persists during T cell activation. J Biol Chem 2012;287(12):9429–40.

[108] Chuang E, Fisher TS, Morgan RW, Robbins MD, Duerr JM, Vander Heiden MG, et al. The CD28 and CTLA-4 receptors associate with the serine/threonine phosphatase PP2A. Immunity 2000;13(3):313–22.

[109] Yokosuka T, Takamatsu M, Kobayashi-Imanishi W, Hashimoto-Tane A, Azuma M, Saito T. Programmed cell death 1 forms negative costimulatory microclusters that directly inhibit T cell receptor signaling by recruiting phosphatase SHP2. J Exp Med 2012;209(6):1201–17.

[110] Chemnitz JM, Parry RV, Nichols KE, June CH, Riley JL. SHP-1 and SHP-2 associate with immunoreceptor tyrosine-based switch motif of programmed death 1 upon primary human T cell stimulation, but only receptor ligation prevents T cell activation. J Immunol (Baltimore, Md: 1950) 2004;173(2):945–54.

[111] Maeda TK, Sugiura D, Okazaki I-M, Maruhashi T, Okazaki T. Atypical motifs in the cytoplasmic region of the inhibitory immune co-receptor LAG-3 inhibit T cell activation. J Biol Chem 2019;294(15):6017–26.

[112] Rangachari M, Zhu C, Sakuishi K, Xiao S, Karman J, Chen A, et al. Bat3 promotes T cell responses and autoimmunity by repressing Tim-3–mediated cell death and exhaustion. Nat Med 2012;18(9):1394–400.

[113] Tomkowicz B, Walsh E, Cotty A, Verona R, Sabins N, Kaplan F, et al. TIM-3 suppresses anti-CD3/CD28-induced TCR activation and IL-2 expression through the NFAT signaling pathway. PLoS One 2015;10(10):e0140694.

[114] Lee MJ, Woo M-Y, Chwae Y-J, Kwon M-H, Kim K, Park S. Down-regulation of interleukin-2 production by CD4 + T cells expressing TIM-3 through suppression of NFAT dephosphorylation and AP-1 transcription. Immunobiology 2012;217(10):986–95.

[115] Liu S, Zhang H, Li M, Hu D, Li C, Ge B, et al. Recruitment of Grb2 and SHIP1 by the ITT-like motif of TIGIT suppresses granule polarisation and cytotoxicity of NK cells. Cell Death Differ 2013;20(3):456–64.

[116] Cai G, Freeman GJ. The CD160, BTLA, LIGHT/HVEM pathway: a bidirectional switch regulating T-cell activation. Immunol Rev 2009;229(1):244–58.

[117] Gavrieli M, Murphy KM. Association of Grb-2 and PI3K p85 with phosphotyrosile peptides derived from BTLA. Biochem Biophys Res Commun 2006;345(4):1440–5.

[118] Liu J, Zhang E, Ma Z, Wu W, Kosinska A, Zhang X, et al. Enhancing virus-specific immunity in vivo by combining therapeutic vaccination and PD-L1 blockade in chronic hepadnaviral infection. PLoS Pathog 2014;10(1):e1003856.

[119] Bersanelli M, Giannarelli D, Castrignanò P, Fornarini G, Panni S, Mazzoni F, et al. Influenza vaccine indication during therapy with immune checkpoint inhibitors: a transversal challenge. The INVIDIa study. Immunotherapy 2018;10(14):1229–39.

[120] Bersanelli M, Giannarelli D, De Giorgi U, Pignata S, Di Maio M, Clemente A, et al. Influenza vaccine indication during therapy with immune checkpoint inhibitors: a multicenter prospective observational study (INVIDIa-2). J Immunother Cancer 2021;9(5).

[121] Wykes MN, Lewin SR. Immune checkpoint blockade in infectious diseases. Nat Rev Immunology 2018;18(2):91–104.

[122] Bengsch B, Martin B, Thimme R. Restoration of HBV-specific CD8$^+$ T cell function by PD-1 blockade in inactive carrier patients is linked to T cell differentiation. J Hepatol 2014;61(6):1212–19.

[123] Schurich A, Khanna P, Lopes AR, Han KJ, Peppa D, Micco L, et al. Role of the coinhibitory receptor cytotoxic T lymphocyte antigen-4 on apoptosis-Prone CD8 T cells in persistent hepatitis B virus infection. Hepatology (Baltimore, MD) 2011;53(5):1494–503.

[124] Zhang Z, Zhang JY, Wherry EJ, Jin B, Xu B, Zou ZS, et al. Dynamic programmed death 1 expression by virus-specific CD8 T cells correlates with the outcome of acute hepatitis B. Gastroenterology 2008;134(7):1938–49 49.e1–3.

[125] Huang ZY, Xu P, Li JH, Zeng CH, Song HF, Chen H, et al. Clinical significance of dynamics of programmed death ligand-1 expression on circulating CD14(+) monocytes and CD19(+) B cells with the progression of hepatitis B virus infection. Viral Immunol 2017;30(3):224–31.

[126] Boni C, Fisicaro P, Valdatta C, Amadei B, Di Vincenzo P, Giuberti T, et al. Characterisation of hepatitis B virus (HBV)-specific T-cell dysfunction in chronic HBV infection. J Virol 2007;81(8):4215–25.

[127] Fisicaro P, Valdatta C, Massari M, Loggi E, Biasini E, Sacchelli L, et al. Anti-viral intrahepatic T-cell responses can be restored by blocking programmed death-1 pathway in chronic hepatitis B. Gastroenterology 2010;138(2):682–93 93.e1–4.

[128] Raziorrouh B, Schraut W, Gerlach T, Nowack D, Grüner NH, Ulsenheimer A, et al. The immunoregulatory role of CD244 in chronic hepatitis B infection and its inhibitory potential on virus-specific CD8$^+$ T-cell function. Hepatology (Baltimore, Md) 2010;52(6):1934–47.

[129] Nebbia G, Peppa D, Schurich A, Khanna P, Singh HD, Cheng Y, et al. Upregulation of the Tim-3/galectin-9 pathway of T cell exhaustion in chronic hepatitis B virus infection. PLoS One 2012;7(10):e47648.

[130] Raziorrouh B, Heeg M, Kurktschiev P, Schraut W, Zachoval R, Wendtner C, et al. Inhibitory phenotype of HBV-specific CD4$^+$ T-cells is characterised by high PD-1 expression but absent coregulation of multiple inhibitory molecules. PLoS One 2014;9(8):e105703.

[131] Verderame F, Rizzo M, Guaitoli G, Fratino L, Accettura C, Mencoboni M, et al.

[132] Zhang L, Yu W, He T, Yu J, Caffrey RE, Dalmasso EA, et al. Contribution of human alpha-defensin 1, 2, and 3 to the anti-HIV-1 activity of CD8 anti-viral factor. Science 2002;298(5595):995–1000.

[133] Yu Y, Tsang JC, Wang C, Clare S, Wang J, Chen X, et al. Single-cell RNA-seq identifies a PD-1(hi) ILC progenitor and defines its development pathway. Nature 2016;539(7627):102–6.

[134] Bersanelli M, Buti S, De Giorgi U, Di Maio M, Giannarelli D, Pignata S, et al. State of the art about influenza vaccination for advanced cancer patients receiving immune checkpoint inhibitors: when common sense is not enough. Crit Rev Oncol/Hematol 2019;139:87–90.

[135] Diao B, Wang C, Tan Y, Chen X, Liu Y, Ning L, et al. Reduction and functional exhaustion of T cells in patients with coronavirus disease 2019 (COVID-19). Front Immunol 2020;11:827.

[136] Cao X. COVID-19: immunopathology and its implications for therapy. Nat Rev Immunol 2020;20(5):269–70.

[137] Chiappelli F, Khakshooy A, Greenberg G. CoViD-19 immunopathology and immunotherapy. Bioinformation 2020;16(3):219–22.

[138] Ceribelli A, Motta F, De Santis M, Ansari AA, Ridgway WM, Gershwin ME, et al. Recommendations for coronavirus infection in rheumatic diseases treated with biologic therapy. J Autoimmun 2020;109:102442.

[139] Karnam G, Rygiel TP, Raaben M, Grinwis GC, Coenjaerts FE, Ressing ME, et al. CD200 receptor controls sex-specific TLR7 responses to viral infection. PLoS Pathog 2012;8(5):e1002710.

[140] Kattan J, Kattan C, Assi T. Do checkpoint inhibitors compromise the cancer patients' immunity and increase the vulnerability to COVID-19 infection? Immunotherapy 2020;12(6):351–4.

[141] Bersanelli M. Controversies about COVID-19 and anticancer treatment with immune checkpoint inhibitors. Immunotherapy 2020;12(5):269–73.

[142] Bonam SR, Kaveri SV, Sakuntabhai A, Gilardin L, Bayry J. Adjunct immunotherapies for the management of severely Ill Covid-19 patients. Cell Rep Med 2020;1(2):100016.

[143] Holly MK, Diaz K, Smith JG. Defensins in viral infection and pathogenesis. Annu Rev virology 2017;4(1):369–91.

[144] Lehrer RI, Bevins CL, Ganz T. Defensins and other antimicrobial peptides and proteins. Mucosal Immunol 2005;95–110.

[145] Ganz T, Selsted ME, Lehrer RI. Defensins. Eur J Haematol 1990;44(1):1–8.

[146] Ouellette AJ, Satchell DP, Hsieh MM, Hagen SJ, Selsted ME. Characterisation of luminal paneth cell alpha-defensins in mouse small intestine. Attenuated antimicrobial activities of peptides with truncated amino termini. J Biol Chem 2000;275(43):33969–73.

[147] Liu L, Roberts AA, Ganz T. By IL-1 signaling, monocyte-derived cells dramatically enhance the epidermal antimicrobial response to lipopolysaccharide. J Immunol (Baltimore, Md: 1950) 2003;170(1):575–80.

[148] Kota S, Sabbah A, Chang TH, Harnack R, Xiang Y, Meng X, et al. Role of human beta-defensin-2 during tumor necrosis factor-alpha/NF-kappaB-mediated innate anti-viral response against human respiratory syncytial virus. J Biol Chem 2008;283(33):22417–29.

[149] Wiens ME, Smith JG. α-Defensin HD5 inhibits human papillomavirus 16 infection via capsid stabilization and redirection to the lysosome. mBio 2017;8(1).

[150] Ganz T, Selsted ME, Szklarek D, Harwig SS, Daher K, Bainton DF, et al. Defensins. Natural peptide antibiotics of human neutrophils. J Clin Invest 1985;76(4):1427−35.

[151] Mou Q, Jiang Y, Zhu L, Zhu Z, Ren T. EGCG induces β-defensin 3 against influenza A virus H1N1 by the MAPK signaling pathway. Exp Therap Med 2020;20(4):3017−24.

[152] Buck CB, Day PM, Thompson CD, Lubkowski J, Lu W, Lowy DR, et al. Human alpha-defensins block papillomavirus infection. Proc Natl Acad Sci U S A 2006;103(5):1516−21.

[153] Bharucha JP, Sun L, Lu W, Gartner S, Garzino-Demo A. Human beta-defensin 2 and 3 inhibit HIV-1 replication in macrophages. Front Cell Infect Microbiol 2021;11 535352.

[154] Kerget B, Kerget F, Aksakal A, Aşkın S, Sağlam L, Akgün M. Evaluation of alpha defensin, IL-1 receptor antagonist, and IL-18 levels in COVID-19 patients with macrophage activation syndrome and acute respiratory distress syndrome. J Med Virol 2021;93(4):2090−8.

[155] Wilson SS, Bromme BA, Holly MK, Wiens ME, Gounder AP, Sul Y, et al. Alpha-defensin-dependent enhancement of enteric viral infection. PLoS Pathog 2017;13(6):e1006446.

[156] Wilson SS, Wiens ME, Smith JG. Anti-viral mechanisms of human defensins. J Mol Biol 2013;425 (24):4965−80.

[157] Salvatore M, Garcia-Sastre A, Ruchala P, Lehrer RI, Chang T, Klotman ME. alpha-Defensin inhibits influenza virus replication by cell-mediated mechanism(s). J Infect Dis 2007;196(6):835−43.

[158] Tecle T, White MR, Gantz D, Crouch EC, Hartshorn KL. Human neutrophil defensins increase neutrophil uptake of influenza A virus and bacteria and modify virus-induced respiratory burst responses. J Immunol (Baltimore, Md: 1950) 2007;178(12):8046−52.

Role of nutraceuticals as immunomodulators to combat viruses

Benil P.B.[1], Vrenda Roy[2], Rajakrishnan Rajagopal[3] and Ahmed Alfarhan[3]

[1]Department of Agadatantra, Vaidyaratnam P.S. Varier Ayurveda College, Kottakkal, Kerala, India [2]Department of Indian System of Medicine, Government of Kerala, Kerala, India [3]Department of Botany and Microbiology, College of Science, King Saud University, Riyadh, Saudi Arabia

27.1 Introduction

A patent immune response in the host is essential to limit the insult of disease-causing agents. Among pathogenic agents, viruses produce a greater disease burden globally and are considered a major risk factor contributor to other major diseases (e.g., hepatitis B increases the risk of developing liver cancer and cirrhosis of liver) [1]. The natural defense mechanisms provided by the mucosal and skin barriers acting through the innate (nonspecific) and an acquired (specific) arm provides protection from these agents (Mucosal immunity) [2]. Immunomodulators are agents that can dampen or enhance the host's inherent potential to resist infections and or tumors by specific or nonspecific mechanisms. Immunomodulators can include those agents that stimulate the immune system (immunostimulatory) and those supresses the immune system (immunosuppressive agents) [3]. In the context of infectious diseases, immunostimulants holds more relevance. As the strategy of infectious agents like viruses has changed a lot through mutations and mechanisms like "drift" and "shift" to tie up several adversities in their path of evolution. These evolutionary advantages procured by the viruses has given them a tactical upper hand; like evading the immune system of the host and persist in the host longer and run a chronic coarse leading to several other morbidities and sequels [4]. Antiviral agents have advanced accordingly through the advancements in immunology. As we stand in the first quarter of the 21st century and after facing the havoc of a global pandemic of COVID-19, the armamentarium of antiviral agents falls short in achieving their target [5]. The limelight has been shifted to the body's own immunity to be enhanced to clear the viral infection. A natural way of boosting the immunity of the body to fight back and win over the viruses were sought out from the beneficial array of natural chemicals that reaches through food that we consume [6]. Earlier the role of phytochemicals derived from medicinal plants gained considerable momentum as antiviral agents. Sustenance of normal immune responses in the most sustainable and persistent manner shifted the attention from phytochemicals to the nutraceuticals, as they form a larger class of natural compounds that can contribute significantly in enhancing the immunity [7]. The term nutraceutical is a conglomeration of the terms "nutrition" and "pharmaceuticals," which ideally reflects on the dual pronged aspects of acting as food and medicine at the same time. Nutraceuticals includes all dietary articles which can provide nutrition and at the same time attributes health benefits also [8]. It is the adjunctive use of nutraceuticals along with the conventional treatments and preventive strategies gave them a significant boost in their understanding through researches. As viruses are continually getting altered due to mutations, combination therapies are the current treatment strategy. Additional approaches targeted as preventive and supplemental therapies are now being considered equally important as medicines. Adequate nutrition that ensures reduction in infection and potentiates immune response are at high demand to be used as alternative therapies to reduce the number and severity of infections [9]. As majority of these agents comes under the classification of dietary supplements as per FDA, they are sold without any prescriptions across the globe and includes such items in daily use like milk, vegetables fruits, etc. to more complex natural antioxidants, micronutrients, minerals, and synthetic compounds [10]. This chapter discusses the potential role of nutraceuticals as immunomodulators in combating viral infections.

27.2 Immunity and its classification

Immunity is derived from "immunis," a Latin word meaning "free from burden." In biological and medical sciences, immunity is the freeing from the burden of diseases caused by viruses, bacteria, protozoa, worms, and toxins and keeping them at bay from the physiological roles of variety of components that make up the immune system [11]. In humans, two types of immunity are identified [12];

1. Innate immunity
2. Acquired immunity

27.2.1 Innate immunity

Innate immunity is the body's primary defense against the invading microorganisms and are of four types:

1. Anatomic barriers
2. Physiologic barriers
3. Endocytic and phagocytic barriers and
4. Inflammatory barriers

Anatomic barriers include the defense system developed by the skin and the mucous membrane. This system enables itself in defending the body against the pathogenic microbes by identifying certain basic structural patterns on the microbes. These pattern recognition receptors allows the immune cells associated with this system to identify a large group of pathogens that shares common structures like the lipopolysaccharides in bacteria and double-stranded ribonucleic acid (RNA) in viruses during their replication.

Innate immunity rapidly attracts immune cells to the site of infection by secreting cytokines and chemokines. Tumor necrosis factor (TNF), interleukin 1 (IL-1), and interleukin 6 (IL-6) are the key cytokines secreted. They initiate cell recruitment and local inflammation which are essential for clearance of the pathogens.

Activation of the complement system is another way of identifying and opsonizing the pathogens. This process makes the pathogens susceptible for phagocytosis. Phagocytosis is considered to be one of the most useful process to remove dead cells, antibody complexes and foreign substances [13–15].

Cells associated with innate immune response are; phagocytes, dendritic cells, mast cells, basophils, eosinophils, natural killer (NK) cells, and lymphoid cells. Phagocytes are of two cell types; macrophages and neutrophils. Phagocytosis disintegrates bacteria using multiple biochemical pathways. Neutrophils are short-lived cells that perform phagocytosis along with enzymatic digestion with the help of the granules present in the cytoplasm. Macrophages on the other hand are lineage of cells with longer life span performing the dual role of phagocytosis and antigen presentation. Among other cell types of innate immune response, dendritic cells also exhibit phagocytosis and antigen presentation and hence they act as a messenger between the innate and acquired immune systems. Mast cells and basophils initiates acute inflammatory responses like allergy and asthma. Mast cells are present in the connective tissue surrounding the blood vessels and are considered "sentinel cells" since they produce cytokines during the early phase of infection. Eosinophils and granulocytes, apart from their usual role in allergic reactions, have a definitive role in destroying parasites that are too large to be phagocytosed. NK cells destroy virus infected cells by secreting porins and granzymes which induces apoptosis and are also a good source of essential cytokines like interferon-gamma (IFN-γ), which activates antigen-presenting cells (APCs) and thus provides antiviral immunity. Innate lymphoid cells are secretors of cytokines IL-4, IFN-γ, and IL-17 that directs the immune response to specific pathogens [13].

27.2.2 Adaptive immunity

Adaptive immunity is influenced by the actions of the innate immunity. The functions of adaptive immunity are:

1. To differentiate the "non-self" antigens from the "Self" antigens.
2. To induce pathogen-specific immunologic effector pathways to eliminate the pathogen, pathogen infected cells, and
3. To develop and maintain immunologic memory for subsequent infections from the pathogen [14].

27.2.2.1 The cells involved in adaptive immune responses are

27.2.2.1.1 T cells and antigen-presenting cells

T cells are progenies of bone marrow derived hematopoietic cells that matures in the thymus and expresses antigen-binding receptors on their surface; T-cell receptors (TCR). T cells proliferate rapidly on signals received from APCs. APCs expresses two types of proteins on their surface known as major histocompatibility complex (MHC) class I (also called Human leukocyte antigen, HLA) and MHC class II found only on certain cells like macrophages, dendritic cells and B cells. MHC class I molecules are present on all nucleated cell surface and presents endogenous peptides while MHC class II molecules are present only on APCs and presents exogenous peptides to T cells. MHC proteins displays the antigens from an infected cell that are phagocytosed [14,15]. TCRs bind to specific foreign peptides expressed on MHC and activates the T cells to secrete cytokines. Subsequent to antigen presentation, the T cell differentiates into either a cytotoxic T cell (CD8 + cells) or T-helper cells (CD4 + cells). CD8 + cells are involved in the destruction of infected cells by secreting cytokines that induce apoptosis of target cell. CD8 + cells achieve this after cloning into a large subset of cytotoxic cells and after the resolution of the infection, these cells die and are cleared off by phagocytosis. Some of these cells are retained as memory cells that can differentiate and clone into cytotoxic cells on subsequent exposure [14,15].

27.2.2.1.2 T helper cells

CD4 + T-helper cells (Th cells) are activated when TCRs are activated by the antigen bound to MHC class II molecules. On activation, an APC can induce the formation of several types of Th cells. Th1, Th2, and Th17 are the most frequent responses. The Th1 response is characterized by the formation of IFN-γ which activates the macrophages and enhances the antiviral immunity. They also help in the differentiation of B cells and promotes the opsonization. Th2 response releases IL-4, IL-5, and IL-13 involved in the immunoglobulin E (IgE) antibody production from B cells and recruitment of mast cells and eosinophils essential for antiparasitic activity. Regulatory T-cells (T reg) are subsets of CD4 + T cells and helps in the suppression of immune responses. It controls the responses of self-antigens and prevents the autoimmune disease manifestations [14,15].

27.2.2.1.3 B cells

B cells originate from the hematopoietic stem cells of bone marrow and upon maturation expresses a unique antigen-binding receptor on its surface. B cells can recognize antigens without the help of APCs. The main function of the B cells is to produce antibodies against the foreign antigens. When activated, B cells undergo proliferation and differentiate into antibody-producing plasma cells. Some of the B cells forms the memory B cells which survives for a longer duration and continue expressing antigen-binding receptors and can be handy in case of a re-exposure or infection. Plasma cells are short lived and they undergo apoptosis after the primary inciting antigen is eliminated [14,15].

27.2.2.1.4 Natural killer cells

NK cells are lymphocytes showing cytotoxic activity and can recognize virus-infected cells and tumor cells. They possess a number of multiple activating as well as inhibitory receptors that recognizes several ligands expressed by virus infected cells and tumor cells [16]. NK cells have killer cell immunoglobulin-like receptors (KIRs) which are inhibitory receptors and recognizes MHC Class I as a suppressing ligand. As MHC Class I molecules are expressed on the cell surface of normal cells and can present the viral peptides in case of an infection and subsequent proteolytic enzyme activity inside the cell. Generally, MHC Class I expression is downregulated in cells infected with virus to evade the immune response from the host cell. Thus, KIRs on NK cells fail to recognize MHC Class I, and NK cells react to these cells as "missing self" and damage them [16,17]. Though MHC class I downregulation is a mechanism adapted by viruses to evade the host immune responses like Cytotoxic T cell activation, they become more susceptible to pertinent damages from NK cells. Thus, downregulation of MHC Class I molecule expression on cell surface has both beneficial and deleterious effects on viruses. Certain groups of viruses have evolved specific mechanisms to evade the activity of NK cells by decreasing the surface expression of MHC Class I molecules and due to which NK cells can target these cells and to avoid damage from NK cells, the infected cells will suppress the expression of NK cell activating ligands or upregulate the expression of NK cell inhibitory ligands. NK cell activating and inhibitory ligands are species specific. Upregulation of the surface expression of NK inhibitory ligands are also exhibited by certain viruses.

27.2.2.2 Mediators in immune response

27.2.2.2.1 Cytokines

Cytokines are substances produced by the influence of immune responses and helps in cell-to-cell communication, flaring up of inflammatory response, and regulation of immune response. Cytokines are short lived, small polypeptide molecules exhibiting varied biological activities on a variety of cells. Cytokines are produced by CD4 + cells, CD8 + cells, and macrophages in response to exogenous, endogenous antigens, or bacterial products. Cytokines binds to its receptor by its specific chain and tyrosine kinases associated with the cytoplasmic region of the receptors termed Janus-activated kinases (JAKs) are phosphorylated. They in turn phosphorylates a family of cytoplasmic proteins called signal transducers and activators of transcription (STATs). After phosphorylation, STATs translocate to the nucleus where they act as DNA-binding proteins and rapidly transcribes the related genes. Cytokines having affinity to the same type of receptors also exhibit similar activities.

27.2.2.2.2 Chemokines

Chemokines are cytokines capable of inducing migration of cells and their intra-tissue movement and localization of immune cells [18]. Chemokines plays an incessant role in the manifestation of several diseases like inflammatory diseases, autoimmune diseases and infectious diseases. Chemokines act through the stimulation of the G protein coupled receptors which are made of seven transmembrane units that transmits signals on activation. Leukocytes expresses chemokine receptors on their surface and can invade inflamed and virus-infected cells. They can also regulate and modulate their expression based on various stimuli. So, chemokines play a dual role of inducing cell chemotaxis by leukocytes, and also activate leukocytes and promote release of mediators [19–21].

27.2.2.2.3 Interferons

Interferons are a family of proteins with prominent antiviral activity. They have anti-proliferative, antiviral and immunomodulatory effects. IFN-γ is the secreted by the CD4 + cells and are predominantly immunomodulatory mediators. IFN signals after receptors binding and phosphorylation of the JAK/STAT pathway and transcribes the IFN-responsive genes within 15 minutes. IFNs inhibit all phases of viral interactions within cells including viral entry, viral uncoating and synthesis of mRNA and proteins. specific mediators like 2'5'-OAS is a latent cellular nuclease that cleaves the double-stranded viral RNA. EIF-2, another mediator blocks the translation of the viral proteins and thus blocks viral replication [22–25].

27.2.2.2.4 Complement activation

Complement system forms a major constituent of the innate immune response comprising of extracellular and surface membrane proteins. More than 30 proteins form the component of the complement system. Complement system becomes activated on recognition of the PAMPs or abnormal or damaged cells. There are three pathways of complement activation; the classical, lectin, and alternative pathways. Each pathway has specific protease zymogens associated and respond to different antigens. All pathways, however, culminate in opsonizing pathogens, lyse pathogens and infected cells, regulate inflammatory response and enhance the clearance of immune complexes and cell debris. The complement activation can directly neutralize virus particles through opsonization, membrane attack complex (MAC) formation on the virion, MAC formation on virus infected cells, targeting intracellular viral components for proteasomal degradation. Regulation of inflammation or chemotaxis, induction of antiviral state and enhancement of adaptive immune responses specific to viral antigens are other modes of antiviral activity of complement activation [26–30].

27.2.2.2.5 Oxidative stress

A shift in the redox status takes place during the cellular processes like cell proliferation, differentiation, signaling, and metabolism. Accumulation of ROS and/or RNS and the depletion of the natural scavenging systems leads to oxidative stress, chronic immune system activation, and inflammatory responses. High reactivity of ROS with all biomolecules like proteins, lipids and nucleic acids leads to genome instability, dysfunction of organelles, and apoptosis when their levels escalate. The natural antioxidant that scavenges superoxide radicals are the superoxide dismutase (SOD) system present in the mitochondria which converts the superoxide to hydrogen peroxide, which later gets transformed to water and oxygen in the presence of other cellular enzymes like catalase, peroxiredoxins, or glutathione peroxidases. All viral infections, especially respiratory virus infections are associated with high elevation in the ROS production. Virus-induced activation of phagocytosis is also associated with high release of pro-oxidant cytokines such as TNF-α and IL-1 which promotes iron uptake by the reticuloendothelial system [31,32]. Lipid peroxidation resultant of ROS generation can cross the cell membranes and reaches circulation resulting in the dysfunction of vital cellular processes such as membrane transport and mitochondrial respiration [33].

27.3 Virus evasion of the host immune system

Generally, in viral infections, humoral immunity, and cell-mediated immunity are at stake. In host cells, the virus activates a complex mechanism to tie over the inevitable host immune responses. The mechanisms undertaken by the viruses to evade the host immunity are broadly classified into; passive and active mechanisms [34,35]. Passive evasion is carried out by the latent expression of viral antigens making the infected cell invisible to the body's own immune system [36]. The active evasion mechanisms involve complex interactions of the viral gene products (immunevasins) interacting with the specific immune functions like antigen presentation [37−39]. The evasion strategy adapted by the viruses in the host cells are not absolute, but it can be considered as blunting the effectiveness of the immune responses. Striking a perfect balance between the evasion responses and the host immune mechanisms allow the viruses to persist in living systems as carriers with very minimal pathogenic consequences [40,41]. The sum total of the immune responses in an organism following a viral attack can be summarized as the Naïve CD8 + T cells, which forms an integral part of humoral immunity, recognizes the MHC class I showcasing the antigens and enhances the cloning and differentiation of these naïve T cells into Cytotoxic T lymphocytes (CTLs) (Fig. 27.1) [42]. CTLs helps in elimination of the virus infected cells and thereafter they gets converted to the memory CD8 + cells and protects from probable reinfection from the same agent by maintaining a subpopulation of these cells in the host. while the cellular immunity exhibited by cells like NK cells lack specificity as there are no involvement of antigen specific receptors unlike T cells, but they express several other activating and inhibitory receptors that controls their functions [43,44]. NK cells are keen in detecting the viral infection induced cellular stress responses like the expression of ligands identifiable by NK cell activating receptors. On the contrary, NK cells express multiple inhibitory receptors that can recognize ligands like MHC class I molecules. NKT cells are another subset of immunologically active T cells capable of expressing marker molecules of some T cells and NK cells. The NKT cell TCR recognizes glycolipids and phospholipid presented by antigen-presenting cells through the CD1d molecules [45].

27.3.1 Mechanism of evasion of major histocompatibility complex class I and cytotoxic T lymphocytes

Viruses on gaining entry into the host cells gets degraded to peptides by the cytosolic proteasomes. These peptides are then transported to the endoplasmic reticulum (ER) with the help of transporter associated antigen processing protein complex. MHC class I molecules on the ER in association with Tapasin, ERp57, and calreticulin forms MHC class I-peptide-loading complex and translocate the viral peptide into the peptide-binding groove of MHC class I molecules. The MHC class I molecules are then transported to cell surface and presented with the help of Golgi apparatus. CTLs on recognizing the foreign antigens

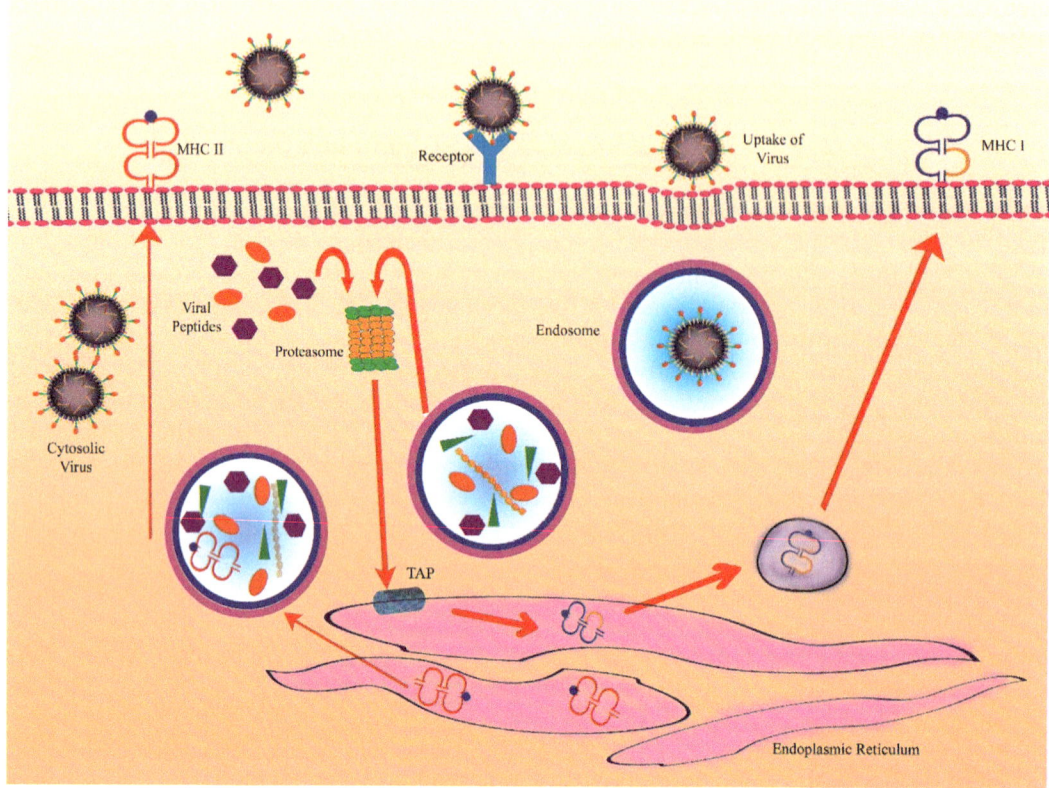

FIGURE 27.1 General immune response to viral infections. *MHC I*, Major histocompatibility complex I; *MHC II*, major histocompatibility complex II; *TAP*, transporter associated with antigen processing.

on the MHC class I releases substances like perforin, granzyme A and B and interferon-γ to induce apoptosis and kill the virus infected cells [46]. Thus, tampering the MHC class I antigen presentation in infected cells seems to be a crucial step in evading the host immune responses and establish pathogenicity by viruses [47,48].

Immunevasins produced by several viruses targeting the MHC class I antigen presentation pathways are well identified and functionally characterized. It is evident from the above descriptions that these viruses encode multiple immunevasins which act together at different steps of MHC class I antigen presentation pathways. But, the role of MHC Class II is less understood in this regard. A few researches on MHC Class II molecules and their role in viral immunology is described here.

MHC Class II expression is restricted only to the professional APCs and this limits the protective action of cytotoxic CD4 + T cells. Conventionally, cytotoxic CD8 + T cells are considered to be the most crucial cell component helping in clearing the infections by killing the infected host cells. The role of CD4 + T cells, however, is very little known and has been believed to be indirectly acting by ancillary support to the functions of the B cells and CD8 + T cells. Evidences are growing to support the role of CD4 + T cell's cytotoxic functions and their major antiviral roles. Several studies have proved that the CD4 + T cells exhibit antiviral properties independently and in association with CD8 + T cells. Recent evidences show that CD4 + T cells effectively brings in antiviral control by two independent mechanisms, IFN-γ production and cytotoxicity. CD4 + T cells controls replication of virus, prevent lethal infection and inhibit establishing latent infections. Several animal studies have shown that CD4 + T cells helps in the clearance of viruses from the animal bodies probably by inducing apoptosis. IFN-γ and other inflammatory cytokines inducted after viral infections upregulates MHC-II expression on cell surfaces of epithelial and endothelial cells, which targets viruses and enables them present their antigen to CD4 + T cells. Expression of MHC class II on cell surface occurs through several coordinated biosynthetic steps [49]. mRNA for MHC class II α and β chains are regulated by IFN-γ and other cytokines. IFN-γ enhances the expression of MHC class II antigens on APCs [50,51]. Several viruses reduce the secretion of IFN-γ and thus reduces the expression of mRNA related to MHC class II antigens [52,53].

27.3.2 Molecular mimicry and immune evasion

Viruses exhibit multiple mechanisms to evade host defense using cytokines. Molecular mimicry demonstrated by viral-IL10 inhibits cytokine synthesis by Th-1 cells at mRNA level and inhibits antigen-specific T-cell proliferation. IL-1β converting enzyme inhibitor encoding by cow pox virus and a homolog of a cytokine receptor expression to interfere with receptor-ligand interactions are also examples of molecular mimicry [54].

27.3.3 Complement evasion

Complement system functions by killing and containment of invading pathogens. Viruses from Herpesviridae and Coronaviridae families prevents complement binding by shedding the antibody-antigen complexes from the cell surface of the infected cell or by the expression of Fc receptors. Certain viruses encode and express proteins and complement regulators like RCA proteins and protect their lipid envelopes and the membranes of the cells they have infected. Certain viruses incorporate host complement control proteins in their viral envelope and upregulate the expression of these proteins on the infected cells [55].

27.4 Mechanism of action of nutraceuticals

27.4.1 Inhibiting NOX-2

A nutraceutical ingredient from spirulina, Phycocyanobilin (PCB), inhibits NOX2-dependent NADPH oxidase with subsequent TLR7-mediated stimulation of type 1 interferon and antiviral antibodies. Thus, they can reduce the production of hydrogen peroxide and can prevent the damage to Cys98 in TLR7. PCB also shares a structural similarity with bilirubin and thus mimics its NADPH oxidase inhibiting activity. They could also inhibit NOX2 system through unconjugated bilirubin maintained intracellularly by the activation of hemeoxygenase 1.

27.4.2 Enhancing MAVS

Several preclinical studies have proved the regulation of relatively higher concentration of UDP-N-acetylglucosamine on administering glucosamine-rich diet. UDP-N-acetylglucosamine is a substrate for O-GlcNAcylation due which the activation of MAVS is achieved.

27.4.3 Antioxidant potency

Viruses in general and Influenza viruses in particular alters the redox sensitive signaling pathways inside the cell upon infection. This event is realized through the depletion of intracellular glutathione or by enhancing the reactive oxygen species (ROS) production. Nutraceuticals with antiviral property suppresses the spread of the virus locally in lung tissue and downregulates the pro-inflammatory signaling pathways in the alveolar endothelial cells. Thus, the recruiting and influx of pro-inflammatory cells are tremendously reduced. The antioxidant property of nutraceuticals is by any or all of these factors [56]

1. ROS scavenging activity
2. Inhibition of superoxide formation
3. Inducing the production of peroxidase enzymes for restoring the cysteine site on TLR7 receptor, and
4. Inducing the formation of glutathione, which maintains the redox state of the cells, acts as a major factor for peroxidases and increases reconversion of oxidized cysteine to its original state on TLR7 receptor [57].

27.5 Nutraceuticals

27.5.1 Definition

The term nutraceutical was coined as an amalgamation of two terms nutrition and pharmaceutical, which was coined by DeFelice in 1989 and defined it as "any substance that may be considered a food or part of a food and provides medical or health benefits, including the prevention and treatment of disease" [58]. Nutraceuticals range from isolated nutrients, dietary supplements and genetically engineered "designer" foods, herbal products and processed foods such as soups, and beverages [59]. The terms "nutraceuticals" and "functional food" have been used interchangeably and are considered to be synonymous. The term "functional" in this context means food sources with established health benefits for those who consume it [60]. "Foods or dietary components that may provide a health benefit beyond basic nutrition" is the definition attributed by International Food Information Council for nutraceuticals [61]. A similar definition is assigned by the Life Sciences Institute of North America to nutraceuticals which defines it as "foods that by virtue of physiologically active food components provide health benefits beyond basic nutrition" [62]. A more comprehensive definition ascribed by Health Canada which states that functional foods are "similar in appearance to conventional food, consumed as part of the usual diet, with demonstrated physiological benefits, and/or to reduces the risk of chronic disease beyond basic nutritional functions" [63]. A more commercially driven definition of functional foods is given by the *Nutrition Business Journal* as "food fortified with added or concentrated ingredients to functional levels, which improves health or performance [64]." Functional food includes enriched food items like cereals, breads, sport drinks, bars, fortified snack foods, baby foods, prepared meals and the like [65,66].

27.5.2 Classification of nutraceuticals

The general classification of nutraceuticals can be many and can go beyond the scope of this chapter. Yet, for the sake of classification and academic instructional purposes, nutraceuticals can be broadly classified as;

a) Potential nutraceuticals
b) Established nutraceuticals

A potential nutraceutical is the class of items which have specific health benefits but they are not positively established by clinical data so that they can be marketed. Majority of nutraceuticals belongs to this group and they are briefly subclassified as [67];

1. Probiotic
2. Prebiotic
3. Dietary fiber (DF)
4. Omega 3 fatty acid
5. Antioxidant probiotics

Based on the origin; nutraceuticals are classified as;

1. Plant origin
2. Animal origin and
3. Microbial origin
4. Nonfood sources

The food products under nutraceuticals can be classified under the following major groups [66];

1. DF
2. Probiotics
3. Prebiotics
4. Polyunsaturated fatty acids (PUFAs)
5. Antioxidant vitamins
6. Polyphenols
7. Spices

The natural sources of nutraceuticals can be classified under;

1. Carbohydrates and fiber
2. Fat and essential fatty acids
3. Protein
4. Macro elements and trace elements

5. Vitamins
6. Water
7. Antioxidants
8. Photochemical
9. Intestinal bacterial flora
10. Recombinant nutraceuticals

27.5.2.1 Dietary fiber

The total fiber content represents the nondigestible carbohydrates, lignins and resistant starches naturally present in a plant. Among them, DF constitutes high molecular weight carbohydrates and lignins while the isolated nondigestible carbohydrates including resistant starches and oligosaccharides having specific physiological effects are called Functional fibers. Recommended Dietary Reference Intake for adult males is 38 grams/day that for adult female is 25 grams/day [68].

27.5.2.1.1 Probiotics

Mammalian gut harbors 10^{14} diverse microorganisms which has been acquired even before the birth of the individual and are termed gut microbiota. The complex ecosystem maintained by these resident microorganisms contains microorganisms that are beneficial to the human body and contributes significantly to the normal physiology and metabolism of the host. These beneficial microorganisms can be supplemented to the host by ingesting adequate quantity of live microbes and constitutes what is known as "Probiotics." Probiotics chiefly consists of Lactobacillus, Bifidobacterium, and other complex sugar fermenters which are part of the normal gut microbiota. The beneficial effect of probiotics includes infection prevention and enhancement of immune response in the host and there is considerable scientific evidence pointing towards the role of probiotics in enhancing defense against pathogenic microbes gaining entry from the external environment and also in maintaining balance of the intestinal immune system. The criteria for considering any bacterial strain to be a probiotic is that; (1) they are able to survive in the gastrointestinal tract and that they can multiply and form a considerable colony in the intestines; (2) their growth and multiplication inside the gut must benefit the host; (3) these microorganisms must be nonpathogenic and nontoxic; (4) they should provide ample protection to the host against pathogenic microorganisms utilizing multiple mechanisms; and (5) they should not develop or neither transfer antibiotic resistance. The probiotic strains of bacteria modulate the mucosal and systemic immune systems and are thus considered as immunobiotics. Probiotics protect the host by producing antimicrobial and antiadhesion substances that prevents the binding of the viruses to the host-receptors by preoccupying them. Probiotic-virus interactions, secretion of metabolites with viricidal properties, immune system stimulation are other pathways of antiviral activity of probiotics. They also suppress molecular signaling pathways and alters the state of the cell and stimulates the innate and adaptive immunity. Suppressing the inflammatory pathways with the help of Toll-like receptor (TLR) modulation and their associated signaling pathways is another way of regulating innate immunity. These regulatory activities escalate phagocytic activity, activity of polymorphonuclear and monocytic leukocytes, expression of receptors increasing phagocytosis and the promotion of microbicidal activity of neutrophils. Viral entry into the host cell stimulates the innate immune response by creating an inflammasome, which acts in biphasic manner by producing Type I interferons (IL-1b and IL-18) initially followed by the secretion of cytokines and other inflammatory agents that modulates the immune response and causes antiviral effect in the second phase. Another pathway of action of probiotics is by activating and maturing the mucosal immune axis by secreting pH reducing metabolites like organic acids and short chain fatty acids and by secreting cytotoxic agents like hydrogen peroxide, coagulation molecules and bacteriocins. Thus, prebiotics stimulates both innate and adaptive immunity by regulating NK cells, macrophages, granulocytes, dendritic cells, and epithelial cells which are part of the innate immunity and the Th1, Th2, Th17, Treg cells, and lymphocytes which are part of the adaptive immunity [69].

Health benefits of probiotics include irritable bowel syndrome, inflammatory bowel disease, infectious diarrhea, and antibiotic-related diarrhea. They are also beneficial in eczema and for urinary and vaginal health [70,71].

27.5.2.1.2 Prebiotics

Prebiotics are nondigestible micronutrients, like oligosaccharides, which selectively stimulate the growth and colonization of beneficial bacterial species of the intestinal flora leading to the reduction in the intestinal pH and thus making the environment inhospitable for the pathogenic bacteria. Within the definitional frame work of functional food, both pre and postbiotics stands in an intermediate position between foods and drugs. All prebiotics are fibers yet all fibers are not prebiotics. A stringent classification criterion excludes other fibers from prebiotics. They are; (1) These fibers are resistant to gastric acidity, (2) They resists hydrolysis by mammalian enzymes, (3) They resist absorption from the upper gastrointestinal tract. (4) They are fermented by the intestinal microflora, and (5) Selectively stimulates the proliferation of benevolent bacterial species associated with health and well-being [72].

Health benefits of prebiotics includes: (1) reduces antibiotic-associated diarrhea prevalence and disease severity, (2) in inflammatory bowel disease, the inflammatory response and flare up of symptoms are reduced, (3) prevention and protection against colon cancer, (4) improves absorption and enhances bioavailability of micronutrients and trace elements, (5) reduces cardiovascular disease risk factors, and (6) prevents obesity by promoting satiety and weight loss [68].

27.5.2.2 *Polyunsaturated fatty acids*

PUFAs are of two major categories—omega-3 and omega-6 PUFAs. Linoleic acid and α-linolenic acid are the two PUFAs considered to be essential as they cannot be synthesized in human body and are essential for the physiological integrity of the body [73,74].

27.5.2.3 *Antioxidants*

Free radical induced cellular damages are pivotal in the ageing process and disease progression. ROS generated during cellular processes damages the biomolecules initiating many disease pathologies. Antioxidants provides the first line of defense against oxidative stressors. They stabilize or deactivates free radicals and are thus essential for maintaining optimal wellbeing of an individual. Humans have evolutionarily acquired a complex antioxidant system of endogenous and exogenous entities that intricately and synergistically integrate their functions to neutralize the free radicals [75]. The antioxidant systems in humans are of the following components;

Nutrient derived antioxidants: ascorbic acid, tocopherols, tocotrienols, carotenoids, glutathione, and lipoic acid.

Antioxidant enzymes: Enzymes that catalyzes the quenching of free radicals generated in the body like SOD, glutathione peroxidase, and glutathione reductase.

Metal binding proteins: They are proteins that can sequester free ferrous and copper ions that can catalyze oxidative reactions like ferritin, lactoferrin, and ceruloplasmin.

Phytonutrient antioxidants: natural plant-based food like fruits, vegetables, whole grains, cereal, legumes, tea, coffee, wine, and cocoa are rich source of natural phytochemicals with antioxidant properties. The array of chemicals presents in natural foods include phenolic acids, flavonoids [76], tilbenes, lignans, and polymeric lignans. These secondary metabolites provide protection against UV radiation, redox reactions, and pathogens [77].

Polyphenols, one of the most abundant secondary metabolite, are of different classes based on the number of phenol rings and the side chain functional groups [78]. About one-third of the dietary sources contains phenolic acids represented by hydroxybenzoic derivatives (protocatechuic acid, gallic acid, and p-hydroxybenzoic acid) and hydroxycinnamic acid derivatives (caffeic acid, chlorogenic acid, coumaric acid, ferulic acid, sinapic acid). Berry fruits, kiwi, cherry, apple, pear, chicory, and coffee are phenolic acid-rich sources.

Flavonoids are classified into six groups—anthocyanins, flavonols, flavanols, flavanones, flavones, and isoflavones. Berries, red wine, red cabbage, cherry, black grape, and strawberry are rich sources of different anthocyanins like cyanidin, pelargonidin, delphinidim, and malvidin [76,77].

Apart from these, there are physiologically active group of endogenous antioxidants are also considered; like bilirubin, thiols (glutathione, lipoic acid, N-acetyl cysteine, NADPH, NADH, ubiquinone (coenzyme Q10), uric acid and enzymes like Cu/Zn SOD, Mn-SOD, catalase, and glutathione peroxidase).

Dietary antioxidants include vitamin C, vitamin E, β-carotene, carotenoids and oxycarotenoids, lycopene and lutein, polyphenols (flavonoids, flavones, and flavonols), and proanthocyanidins.

27.5.2.4 Egg as a functional food

Even when egg cholesterol content is directly associated with cardiovascular disease risk and with the recent research findings showing that the results are otherwise, the upper limit of recommended cholesterols levels have been modified. Yet, there is still a looming risk of cardiovascular diseases associated with dietary cholesterols especially those from eggs. The importance of egg as a nutrient source has now increased as the attention over it has shifted from cholesterol to more healthy ingredients like vitamin D, vitamin E, selenium, lutein, zeaxanthin, choline, and good source of high-quality protein that helps in providing satiety and benefits weight reduction [79]. The carotenoids in egg, lutein and zeaxanthin are made more bioavailable by the lipid matrix surrounding them that helps in faster solubilization into micelles and chylomicrons and helps absorption. Egg intake is also associated with much higher high-density lipoproteins which transports lutein and zeaxanthin and helps in the reverse transport of cholesterol. The carotenoids in eggs also possess high antioxidant property and is found helpful in preventing age related macular degeneration. Lutein and zeaxanthin are believed to modulate inflammatory responses by inhibiting the expression of TNF-α, IL-6, and IL-1β in macrophages. Human studies have also shown reduction in CRP, serum amyloid A, AST, and TNF-α. Choline is an essential ingredient for the synthesis of membrane phospholipids, including phosphatidylcholine and sphingomyelin. Deficient intake of choline in diet has been positively associated with cognitive impairment, neural tube defects, muscle damage, and fatty liver. Dietary supplementation of egg seems to prevent such changes as egg is the one of the richest forms of choline as phosphatidylcholine in its yolk. Choline is essential for neurotransmitter synthesis and is found beneficial in the treatment of dementia, Alzheimer's disease and also helps in the formation of new synapses in aged population. Newer evidences show that phosphatidyl choline present in eggs prevents the lymphatic absorption of cholesterol and helps the liver to secrete more of very-low-density lipoprotein and provides protection against hepatic steatosis. PC has also show cardio-protective effect by lowering plasma homocysteine levels. Even with these much of benefits, PC is associated with a major risk of a cardiovascular event probably from the gut micro-biome mediated metabolism of choline to trimethylamine (TMA). TMA is oxidized by flavin monooxygenases in the liver and convert them to trimethylamine N-Oxide (TMAO). Higher TMAO levels are associated with patients at risk of cardiovascular events [80]. Eggs are a store house of varied proteins distributed in the white and the yolk that executes varied functions ranging from promoting immunity, protection against microbes, and cancers to controlling hypertension. These proteins contain all the essential amino acids and are shown to be beneficial in building up of skeletal muscle mass and are hence considered to prevent sarcopenia in elderly. They reduce the glycemic index and prevent malnourishment among children as well [81]. Egg-derived proteins also showcase a variety of immunological functions; lysozyme, hydrolyzes the structural proteins, peptidoglycans, in the bacterial cell walls. They also exhibit anticancerous property. Bacterial infections derive iron from the host cell to establish infection; ovo-transferrin, inhibits this by scavenging the iron [82]. Avidin, another functional protein, prevents bacteria from accessing biotin by binding on to this vitamin. These are some of the proteins derived from the white of the egg. Egg yolk proteins contain immunoglobulin Y, which prevents the growth and colonization of species like *Escherichia coli* and *Staphylococcus aureus*. It also contains phosvitin, a phosphoprotein having antimicrobial property.

Hen eggs are considered the most pleiotropic and affordable functional and balanced food providing all essential nutrients required for the sustained growth and development. They store a wide variety of molecular species with beneficial biological functions like antioxidant, anticancer, antihypertensive, antimicrobial, and immunomodulatory activities. Eggs are equipped with physical and physiologically and biologically functional defense mechanisms to protect the embryo. The egg white contains around 60% of its weight occupied by water and proteins. The chief protein contents of egg white are ovalbumin, ovotransferrin (Fig. 27.2), ovomucoid, ovomucin, lysozyme, avidin, cystatin, ovoinhibitor, and ovostatin. The functional aspects of these proteins against viral infections are summarized in Table 27.1.

Egg yolk consists of lipids and proteins. α-lipovitellin, β-lipovitellin, phosvitin, and low-density lipoproteins. Egg yolk immunoglobulins (IgY) is homologous to mammalian immunoglobulin IgG and plays a key role in avian acquired immunity. IgY has been used to induce passive immunization to immunocompromised patients. This aspect has been considered in addressing the issue of antimicrobial and drug-resistant microbial infections. Passive

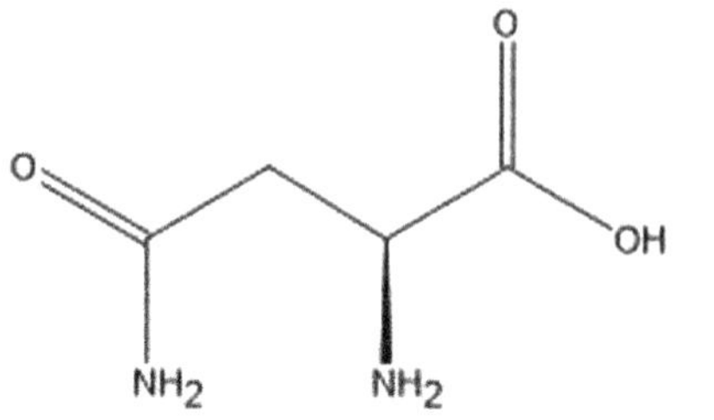

FIGURE 27.2   Chemical structure of ovotransferrin.

TABLE 27.1   Antiviral activity of egg proteins.

Protein	Class of protein	Mechanism of antiviral activity	Activity of specific viruses
Ovomucin	Glycoprotein	Hemagglutination inhibitory activity	Newcastle virus, Bovine rotavirus, human influenza virus
Ovotransferrin	Glycoprotein	Superoxide dismutase like antioxidant activity. Regulation of innate immune response by modulating macrophage and heterophil cell activity in infections and inflammations.	Avian Marek's disease virus
Lysozyme	Glycoside hydrolase	Restoration of cellular and humoral defense systems from immunosuppressive action of viruses.	Herpes simplex virus, chicken pox virus, norovirus, hepatitis A virus, HIV-1 virus
Cystatin	Globulin	Inhibits cysteine active sites in proteases Reduce virus loads by modulating intracellular proteolytic process	Poliovirus

immunization is achieved through vaccination and in many instances, they are unable to induce the necessary level of immune response in patients especially when their immunity is impaired. Use of IgY has found the following applications in humans; (1) Egg yolks are a cheap and abundant source for producing IgY, (2) the evolutionary link between chicken and mammals represented by the phylogenetic distance is closer to consider as a target in mammals, and (3) IgY spares the Fc receptors in mammals and does not produce any allergic and inflammatory manifestations. Mounting evidences are present on the antiviral property of IgY antibodies. Passive immunization with IgY antibodies has been studied against simian rotavirus SA 11, poliomyelitis type 2, and Coxsackie B2 viruses [83,84]. Dietary supplementation in children infected with RV-induced diarrhea was successfully controlled using IgY antibodies derived from hyperimmunized chickens. Oral administration of IgY has promising results in preventing rotavirus infections in low-birth weight infants and bestows protection against virus-induced gastroenteritis [84]. In influenza A virus infections H5N1 and H1N1, IgY antibodies showed strong neutralization activity. SARS coronavirus-specific IgY antibodies showed high stability and strong antiviral property. IgY isolated from immunized chicken could prove as a cost-effective, safe, and convenient alternative therapy for controlling infections like influenza [85]. Researches to optimize the antibody production with efficient protocols are going to broaden the scope of immunotherapy in viral infections.

Egg yolk also contains a large amount of sialic acids like N-acetylneuraminic acid, and glycoproteins, glycolipids and oligosaccharides. The inhibition of hemagluttination of erythrocytes exhibited by ovomucin is due to its high sialic acid content [86]. Egg yolk derived sialic acid glycoproteins inhibit rotavirus replication and can be effectively used in the treatment of viral gastroenteritis. Sialic acid-rich glycopeptides and phospholipids inhibited the simian (SA-11) and human (MO) strains of rotaviruses [87].

27.5.2.5 Nutraceuticals from microbes

Microbe interactions have played a key role in the development of food industry. Several micro-organisms like Agrobacterium and Rhizobium species produces low molecular weight exopolysaccharides like curdlan. Curdlan is a homopolysaccharide and is used in food industry to form stable gels that remains stable even after several freeze-thawing cycles, deep fat frying. They are with low calorific value and is a good texturizer and water holding agent in several food items like pasta, tofu, jellies, fish pastes, etc. Sulfated derivatives of curdlan show immunostimulatory, antitumor and antiviral properties [88,89]. Aquatic microorganisms like microalgae and cyanobacteria can biosynthesize antioxidants, antiviral agents, antibiotics and antiinflammatory agents as secondary metabolites. Filamentous fungi like *Sclerotium glucanicum* and *Sclerotium rolfsii* can produce large molecular weight extracellular glucose polysaccharide named scleroglucan. They are used as a food stabilizer in food industry in making ice-creams jellies, desserts, and sauces. Scleroglucan is also extensively used in pharmaceutical industry as edible films and tablet coatings that are stable, biocompatible, and biodegradable. These b-glucans are also found to have significant antitumor and antiviral activity [90].

Bacteriocins, produced by nonpathogenic bacteria has gained therapeutic interest recently owing to their anticancerous and antibiotic potentials. Antiviral studies of bacteriocins have been undertaken with promising results recently. The results of these studies have shown that there is ample evidence to use bacteriocins alone or as an accompanying therapy against viruses [91].

Spirulina, a blue-green-algae is one of the highly marketed nutraceutical agent in the world. Several natural nutrients present in spirulina are having immune-modulatory activities. Bioactive proteins, vitamin B12, b-carotenes and minerals like iron, phenolic acids, tocopherols and g-linolenic acid, polysaccharides, glycolipids, and sulfolipids are the main ingredients found in Spirulina [92]. PCB, phycocyanin, and allophycocyanin are the compounds exhibiting significant antiviral activity. Calcium-spirulan isolated from *Spirulina platensis* showed inhibitory activity against HIV-1, mumps virus, measles virus, herpes simplex virus type-1, human cytomegalovirus, and influenza virus. In vivo studies shown that there was increased macrophage mobility, accumulation of NK cells, and increased production of antibodies and secretion of cytokines were noted [93].

Beta-Glucans are polysaccharides derived from yeasts which could activate the innate and adaptive immune responses in humans. They evoke a foreign body response in the body by acting as an antigen and activates dectin-1 receptor, complement receptor 3 (CR3) and TLR present on monocytes, macrophages, neutrophils, eosinophils, dendritic cells, and NK cells, which are the chief immune cells. The resultant immune responses from the activation of these cells like phagocytosis, oxidative burst, cytokine, and chemokine production kills the pathogens [94].

27.5.2.6 Citrus fruits

Lemon juice and lemon oil were investigated for their antibacterial and antiviral properties. Among these lemon oil showed more bacteriostatic and antiviral property than lemon juice. High citral and linalool content have been attributed to this activity. Polyphenol from the citrus fruits have been proved as having very good antiinflammatory, anticancerogenic, neuroprotective, antiallergic, estrogenic, antithrombotic, hepatoprotective, antibiotic, antiviral, antiulcer, antilipidaemic, and vasorelaxing activities [95]. Polymethoxy flavones (PMFs) derived from citrus fruits also exhibited anticancer, antiviral, and antiinflammatory effects. These PMFs are different from other citrus flavonoids in that they are not glycosylated. These PMFs are abundant in cold pressed citrus fruit juices [96]. Flavonoids generally derived from citrus family exhibits physiological actions like antifungal and antiviral activities in the plants. Many of them are phytolexins and phytoanticipins and are formed as antimicrobial barriers against microbial invasion ion plants. These compounds are proved to resist the spore generation in plants and they show similar efficacy as an antimicrobial agent in man [97]. Coumarins are the other pharmacologically active phytoconstituents in citrus plants. They are having the basic nucleus of benzo-a-pyrene and are categorized as (1) simple coumarins, (2) furanocoumarins, and (3) pyrone-substituted coumarins. Even though their physiological role in plants are not yet established, they exhibit several pharmacological activities including antiviral property in humans [98]. Quercetin 7-rhamnoside and quercetin 3-rhamnoside and quercetin glycosides with reported antiviral activity from citrus family. These compounds have been shown effective against porcine epidemic diarrhea and influenza A virus replication [99]. Limonoids, are the group of compounds extensively studied in citrus fruits due to their pronounced anticancerous activities. Recent investigations have proved their antiviral activity as well [100].

Hesperidin belongs the family of flavanone glycoside and are widely distribute among the citrus fruits (Fig. 27.3). They have exhibited antiatherogenic, antihyperlipidemic, antidiabetic, cardioprotective, antioxidant, and antiinflammatory activities. They showed inhibition of influenza A virus by reducing virus replication by inhibiting the sialidase activity that helps in the entry and release of the virus. Hesperidin has strong affinity to bind to the ACE2 interface preventing it to bind with the SARS-CoV-2 virus. High antiinflammatory activity of hesperidin prevents the release of IFN-γ, IL-6, IL-1β, and TNF $-$ α and prevents cytokine storming which is a characteristic feature of COVID-19 infection.

27.5.2.7 Nutraceuticals from marine organisms

Marine carbohydrates synthesized by photosynthetic organisms are considered vital organic compounds and a good energy source for the heterotrophs. Carbohydrates exists as monosaccharides, disaccharides, and polysaccharides in marine organisms. Among them polysaccharides are considered rich bioactive molecules. Carrageenan, chitosan, fucoidan, chitin, and alginate are complex polysaccharides from marine organisms having pharmacological activities. These compounds also exhibit immune-stimulatory, antioxidative, anticancer and antiviral properties. Epoxypolysaccharides isolated from cyanobacteria

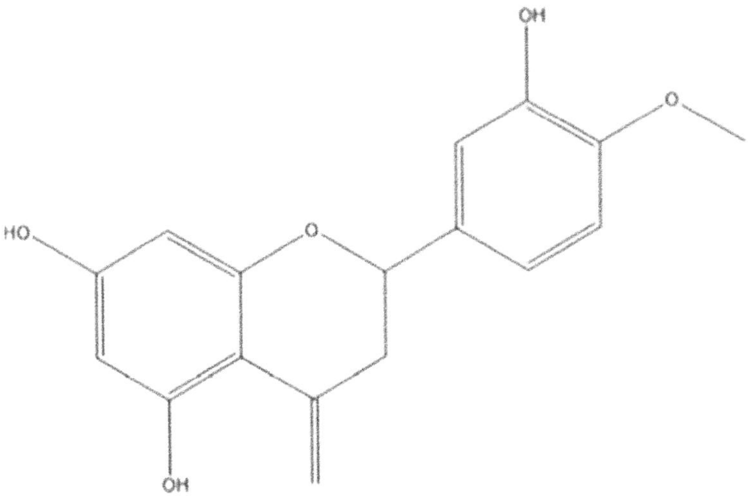

FIGURE 27.3 Chemical structure of hesperidin.

and marine carbohydrates such as algins are used in the food industry for stabilizing emulsions and as bioflocculants. Hydrocolloids, fucans/fucanoids, carrageenans, and glycosaminoglycans extracted from crustaceans, marine algae show antiproliferative, antiinflammatory, and antiviral activities [101].

27.5.2.8 Herbs and spices

27.5.2.8.1 Capsicum

Capsicum annuum L., or pepper fruit are rich source of flavonoids. Flavonoids in general are compounds with strong antioxidant potential depending upon the position and number of hydroxyl groups present. Along with antioxidant properties they also exhibit anticancerogenic and immune-stimulating and antiviral properties [102].

27.5.2.8.2 Resveratrol

Resveratrol or 3,4,5-trihydroxy-trans-stilbene are polyphenolic compounds widely distributed in plant kingdom (Fig. 27.4). The chief source of resveratrol is grapes and the health benefits of wine is attributed to its presence. It is a strong antioxidant, antiinflammatory, antiplatelet, cardioprotective, anticarcinogenic, and immunomodulatory agent. In immune system they induce the proliferation of lymphocytes, activate NK cells, and also regulate the process of apoptosis. Studies have shown their efficacy against several viruses including varicella zoster, herpes simplex, polyomavirus, influenza A, and HIV. Considering the emergent disease of COVID-19, resveratrol has shown strong interaction with the spike protein and ACE-2 receptor complex of SARS-CoV-2 virus [67].

27.5.2.8.3 Glycyrrhizin

Glycyrrhizin (Fig. 27.5) interferes with the release steps of viruses in hepatitis C virus (HCV) infections and they also inhibit the viral full-length particles and core gene expression. Glycyrrhizin is a strong immunostimulatory agent as it stimulates T lymphocytes

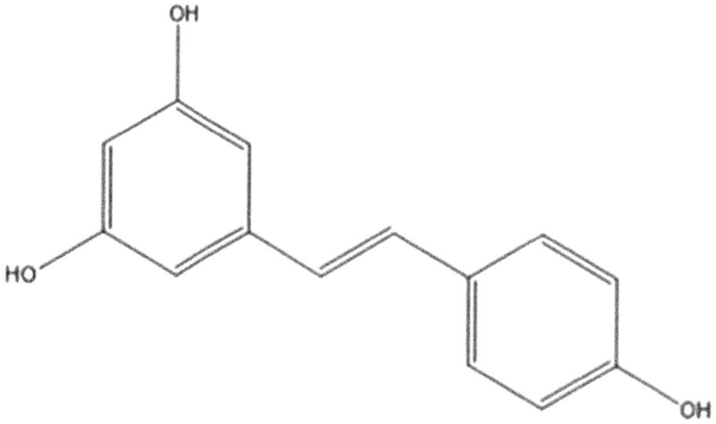

FIGURE   27.4   Chemical structure of resveratrol.

production and inhibits chemokine ligand 10 (CXCL10), IL-6, chemokine (C-Cmotif) ligand 5(CCL5), and reduce virus-induced apoptosis [103].

27.5.2.8.4 Black caraway

Nigella sativa or black caraway seeds is a rich source of thymoquinone (Fig. 27.6), which boosts humoral immune system and enhances the expression of cytokines and helps in the early clearance of virus from the host [104]. Studies have shown its efficacy in avian influenza, murine cytomegalovirus and HCV infections. A significant elevation in CD4 + T helper cell counts and macrophages were observed [105,106].

27.5.2.8.5 Garlic

Allium sativum has long been used as a therapeutic and prophylactic dietary ingredient in different cultures. Garlic bulbs are rich in organosulfur compounds that exhibit antiviral properties and is widely documented. Researches have shown that the organosulfur compounds present in garlic prevent cell attachment, transcription, and translation of viral genome and viral assembly in the host cells. Garlic's activity against influenza a virus, cytomegalovirus, rhinovirus, HIV, herpes simplex virus, viral pneumonia, and toravirus are well established [107,108].

27.5.2.8.6 Cinnamon

The bark of *Cinnamomum zeylanicum* contains cinnamaldehyde, cinnamic acid, cinnamyl alcohol, coumarin, and eugenol as their major ingredients (Fig. 27.7). In HIV-1 and HIV-2 infections, they have been proved to prevent viral replication by inhibiting HIV protease, integrase and reverse transcriptase. They have also shown activity against Dengue and chikungunya viruses by inhibiting the viral protease and neuraminidase expression. Coumarins from cinnamon has shown to be preventing the entry of the virus to the host cell [109,110].

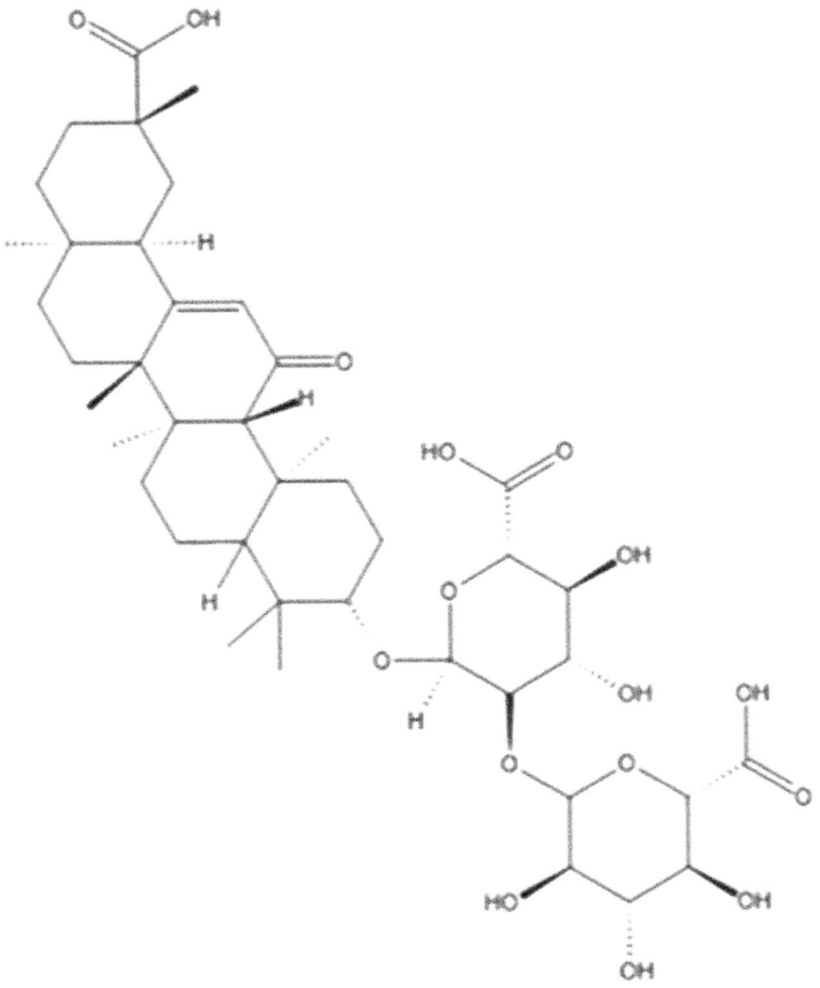

FIGURE 27.5   Chemical structure of glycyrrhizin.

27.5.2.8.7 Black pepper

Considered as the "king of spices," *Piper nigrum* is extensively used as a spice since ancient times. Being a store house of a variety of bioactive ingredients they have also found medicinal uses. Piperine (Fig. 27.8), the chief ingredient of black pepper, has been shown to regulate IL-10 and NF-κB with GABA and activated p38, JNK and MAPK pathway to increase the release of EPO and EPO-R expression. Strong antiviral activity is expressed by black pepper against coxsackie virus type B3, vesicular stomatitis Indiana virus and human parainfluenza virus [111,112].

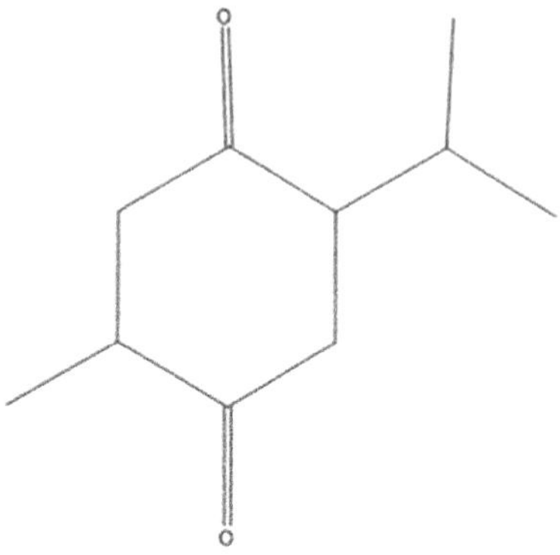

FIGURE 27.6   Chemical structure of thymoquinone.

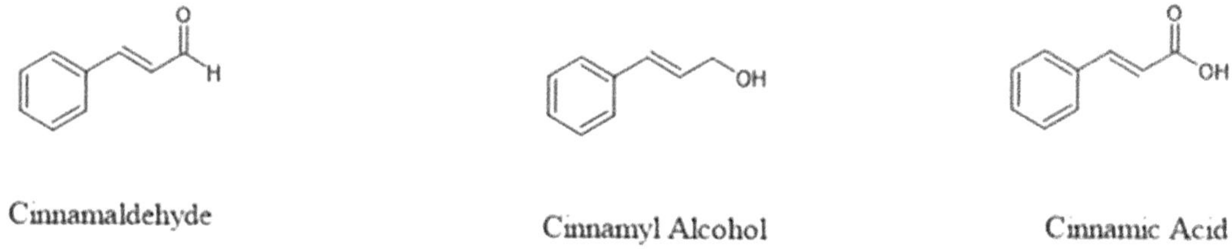

Cinnamaldehyde

Cinnamyl Alcohol

Cinnamic Acid

FIGURE 27.7   Chemical structures of cinnamaldehyde, cinnamyl alcohol, and cinnamic acid.

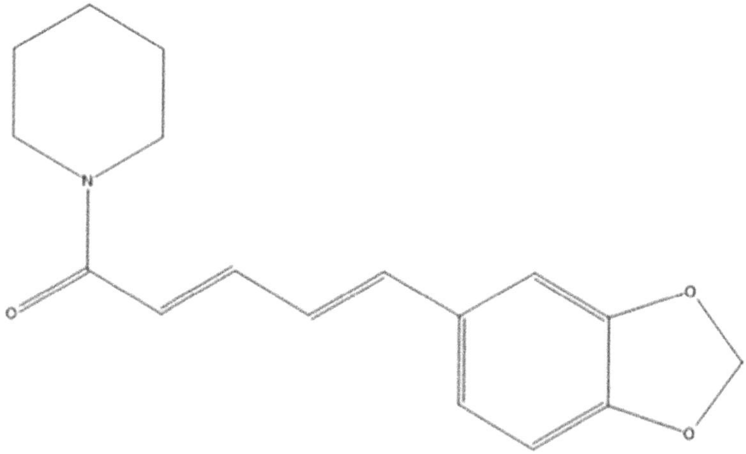

FIGURE 27.8 Chemical structure of piperine.

27.5.2.8.8 Moringa

Moringa oleifera is a nutrient-rich plant widely distributed in sub-Himalayan regions of India, Pakistan, Bangladesh, and Afghanistan. It contains zeatin, quercetin, beta-sitosterol, caffeoylquinic acid, and kaempferol. Thiocarbamate and niaziminin obtained from the leaves of *M. oleifera* inhibited Epstein Barr virus. Reported activities against HSV-1,

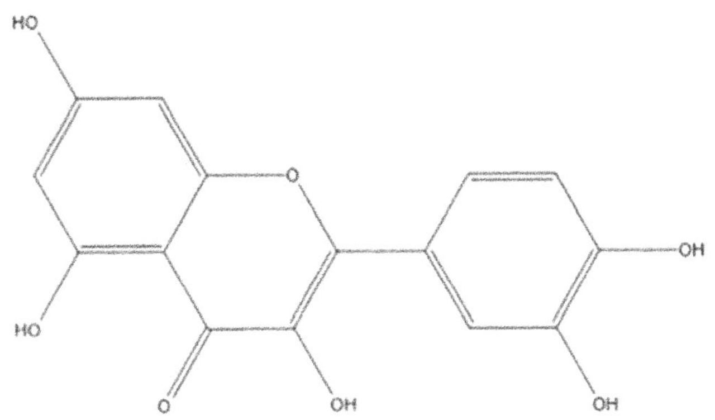

FIGURE 27.9 Chemical structure of quercetin.

hepatitis B virus, Newcastle disease virus, and infectious bursal disease virus are reported. The leaves of *M. oleifera* has shown potential to be used in antiviral regimens already followed against HIV infections as they immensely support the immunity of the patient [113–116].

27.5.2.8.9 Quercetin

Quercetin is a natural polyphenol coming under the flavonoid and is present richly in several fruits and vegetables (Fig. 27.9). The abundant sources of quercetin are onions, broccoli, buckwheat, capers, peppers, brassica vegetables, apples, grapes, berries, tea, wine, nuts, seeds, barks, flowers, leaves, and spices [117]. Quercetin is responsible for inhibiting hemagglutinin A which is essential for the initial phase of infection by the fusion of virus and host cell membranes. Blocking this step could prevent the entry of the virus into the cell. Evidences are strong in suggesting the role of quercetin in inhibiting several inflammatory mediators like TNF-α, IL-1β, IL-6, and MMP-9. The bulk of quercetin intake comes from fruits and vegetables and from tea. They get metabolized in intestinal and hepatic pathways and due to which they attain low plasma levels. Natural quercetin consists of a monosaccharide or disaccharide molecule linked to the 3, 7, and 4′ positions as an O-glycosidic form. Quercetin inhibits the NLRP3 inflammasome-mediated IL production like IL-1β. They are believed to act by exciting different pathways (Table 27.2) [117].

27.5.2.9 *Mushrooms*

Mushrooms in the group of basidiomycetes have been traditionally used as food and also used in traditional health care systems. They are rich source of polysaccharides, lipids, steroids, organic acids and tetracyclic triterpenes [144]. Many of the constituents of mushrooms can directly interact with the viral enzymes, nucleic acid synthesis, adsorption and intake of virus into host cells, and direct stimulation of immune activity. Ganoderma and Piptoporus species of mushrooms inhibited HIV and pox virus infection by preventing virus adsorption to cells and by blocking the replication of virus by inhibiting the nucleic acid synthesis [145].

TABLE 27.2   Antiviral activities of quercetin.

Experimental model	Target virus	Activity
Mice	Mengo virus	Enhanced graft-versus-host reaction but failed to excite humoral antibody response [118]
In vitro	HSV-1, HSV-2	Cytopathic effect inhibitory activity [119]
In vitro	Canine Distemper Virus	Inhibition of virus nucleoprotein gene expression and increase in cellular viability [120]
In vitro	Rhino virus, echovirus type 7, 11, 12 and 19, coxsackie virus A21 and B1 and poliovirus type 1 Sabin	RNA polymerase inhibition [121]
Mice	Rhino virus	Viral replication and attenuation of virus-induced airway cholinergic hyperresponsiveness [122]
In silico	Influenza A H1N1 and H7N9 virus	Neuraminidase inhibition against drug-resistant mutants [123]
In silico	HCV	NS3 helicase, NS5B polymerase and p7 proteins [124]
In vitro	HCV	NS3 helicase and heat-shock protein inhibition [125].
Phase I clinical trial	Hepatitis C virus	Serum AST and ALT levels were not elevated and the drug showed antiviral activity against hepatitis C [126].
Clinical trial	HSV oral herpes	Reduced mean number of outbreaks, mean duration of outbreaks per year with no adverse events [127].
In vivo	Influenza H3N2	Viral infection induced reduction in pulmonary catalase, reduced glutathione and superoxide dismutase enzyme levels was elevated [128]
In vivo	Influenza	Protection of lung morphology, reduced oxidative stress and decrease in the number of infiltrating cells [129]
Clinical trial (randomized placebo controlled double blinded trial)	Lower respiratory tract infections	Reduction in infection severity (36%), reduction in sick days (31%) >40 individuals with taking an herbal formula containing 1000 mg Que compared to placebo [130].
Clinical trial	Anti-inflammatory and immune modulating influences	Plasma levels of DHA and DPA increased, upregulation of interferon-induced antiviral mechanism [131].

(Continued)

TABLE 27.2 (Continued)

Experimental model	Target virus	Activity
In vitro	Varicella-zoster virus and Human cytomegalovirus.	Inhibited replication of varicella-zoster virus. Cytomegalovirus replication was not inhibited [132]
In vitro	HSV-1, HSV-2 and acyclovir-resistant HSV-1	Blocks viral binding and penetration to the host cell and suppresses NF-κB activation essential for gene expression [133].
In vitro	HSV, Newcastle disease virus, vesicular stomatitis virus	Induce secretion of type 1 interferon (IFN), reduces replication [134].
In vivo	Influenza A subtypes H1N1, H5N2, H7N3 and H9N2	Induce secretion of type 1 interferon (IFN), reduces replication [134].
In vitro	Cytomegalovirus inoculated HeLa cells	Replication inhibition and selectivity index [135].
In vitro	Dengue virus type 2 (DENV-2) infected vero cells	Replication inhibition and RNA reduction by blocking viral entry and inhibiting viral polymerases [136].
In vitro	DENV-2 and DENV-3	Inhibition in the absence or presence of enhancing antibody [137].
Clinical trial	Upper respiratory tract infections	Athletes supplemented with Que showed protection against stress-induced susceptibility to respiratory tract infections [138]
In vitro	Influenza virus	Inhibits replication by blocking endocytosis by phosphatidylinositol 3-kinase inhibition, blocking transcription of viral RNA and increases viral clearance [139]
In vitro	Influenza	Inhibiting cell fusion by inhibiting influenza hemagglutinin protein [140]
In vitro	Japanese encephalitis virus and human T-lymphotropic virus 1	Reduced viral adhesion to the host cell, produced direct extracellular virucidal property [141]
In vitro	DENV-2 and HCV	Suppressing nonstructural protein 3 protease activity [137]
In vitro	Porcine epidemic diarrhea virus (PEDV) and Influenza A virus	Direct antioxidant property of the quercetin [142]
In vitro	Norovirus and feline calicivirus (FCV)	FCV titer was reduced [143].

27.5.2.10 *Vitamins and minerals*

Vitamins belongs to the group of micronutrients which are essential for the proper structuring and functioning of proteins and helps in the physiological processes and signaling pathways in the body [146].

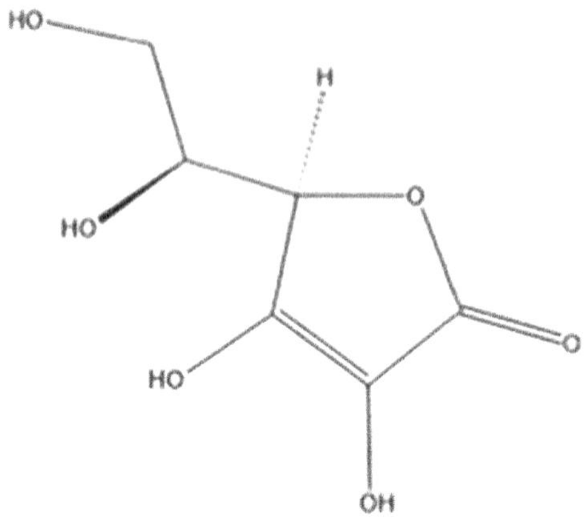

FIGURE 27.10   Chemical structure of ascorbic acid.

27.5.2.10.1 Vitamin C

Vitamin C or ascorbic acid is advocated to be administered in higher doses in viral infections. It is a water-soluble vitamin and have antioxidant, antiinflammatory, and immunomodulatory activities (Fig. 27.10). They act as a cofactor for numerous biosynthetic pathways including those of catecholamines. Historically vitamin C is linked to the cure of several viral infections like herpes zoster virus and influenza infection [147]. Recently, in the COVID pandemic their efficacy against the SARS CoV 2 virus has been studied. It is found that high dose of vitamin C increased the ventilator free days and decreased the 28-day mortality on intravenous administration [148]. They exert a dual role in such situations by inhibiting the release of pro-inflammatory mediators by the immune cells by acting as a pro-oxidant and enhances alveolar fluid clearance. The second role of vitamin C is to act as an antioxidant to enhance the functions of the lung epithelium [149].

27.5.2.10.2 Vitamin D

Vitamin D shows antiinflammatory and antioxidant properties. Vitamin D modulates the innate immunity and its supplementation in diet has reduced the risk of ARDS including upper respiratory tract infections (Fig. 27.11) [150]. They prevent respiratory infection by producing antibiotic peptides in lung tissue. Vitamin D also promotes ACE2 gene expression and regularizes the renin-angiotensin-aldosterone system [151].

27.5.2.10.3 Vitamin A

Vitamin A is present in abundance in organ meat, milk and cheese. It supports the integrity of gastrointestinal epithelial cell integrity against infectious agents. This vitamin is important for the regulation of NK cells, macrophages, and neutrophils. Vitamin A helps in the early differentiation of NK cells and downregulates IFN-g and upregulates IL-5. It helps in the differentiation of dendritic cell precursors and promotes secretion of pro-inflammatory cytokines IL-12 and IL-23. Vitamin A primes the helper T cells to enhance antibody production (Fig. 27.12).

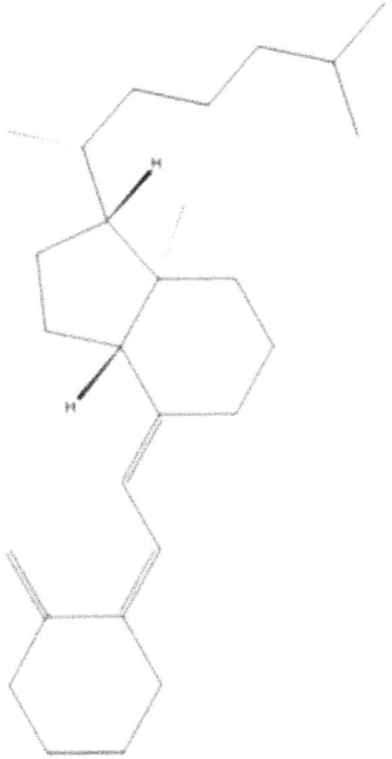

FIGURE 27.11 Chemical structure of vitamin D.

FIGURE 27.12 Chemical structure of retinol.

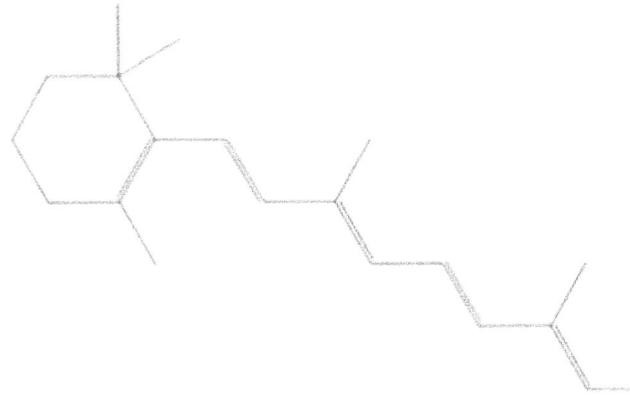

27.5.2.10.4 Zinc

Zinc is crucial in the proper immune activity due to its involvement in the development, differentiation and functioning of immune cells. Zinc upregulates the expression of TNF $-$ α, TNF-γ and is found effective against viruses affecting the upper respiratory tract.

Presence of zinc upregulates apoptosis as they link with metallothioneins. It can prevent adhesion of the viruses to the nasopharyngeal mucosa and prevents their replication [152].

27.5.3 Other nutraceutical sources with antiviral properties

27.5.3.1 Honey

Honey is a unique combination of sugars mainly glucose and fructose admixture with organic acids like gluconic acid and amino acid proline [153]. Manuka variety of honey from New Zealand exerts significant antiviral property against influenza virus. They act synergistically with other NA inhibitors and are also found to be strong inhibitors of NF-κB, IκBα, COX-2, and TNF $-$ α in murine models. The studies also revealed its strong anti-inflammatory activity and protective activity against UVB radiations by inhibiting IL-1β, IL-6, and TNF-α release from PAM212 keratinocytes and a decreased expression of COX-2 and prostaglandin E2. Methyl glyoxal present in honey has prominent inhibitory effect against respiratory syncytial virus and influenza virus [154].

27.5.3.2 Bee propolis

Propolis, also called bee putty or bee glue, is a resinous mixture collected by bees from various sources of trees and shrubs enriched with the salivary and enzymatic secretions. The main ingredients in the propolis are various groups of phenolic acid and flavonoids. Bee propolis is a mixture of several antiviral agents like chlorogenic acid, caffeic acid, arte-pillin C, baccharin, 3,4,5-tricaffeoylquinic acid, isosakuranetin, p-coumaric acid, drupanin, and ferulic acid. These compounds are capable of preventing cytokine storming by inhibiting several cytokines. A derivative of chlorogenic acid, caffeic acid phenethyl ester effectively inhibits TNF-α, IL-8, and IκB-α.

27.5.3.3 Seed storage proteins

SSP forms the major source of protein in human and livestock globally. These proteins are abundantly distributed in nature and are the preferred source of food based on the quality of proteins. Several essential amino acids are deficient among different sources of food; like methionine and cysteine are deficient in legumes and cereals lack tryptophan, lysine, and threonine. Hence, supplementing diets with SSPs provide sufficient reserve of essential amino acids. SSPs are stored in protein storage vacuoles called protein bodies and are synthesized in endosperm or the cotyledons and are made available as per the physiological requirements and are usually seen embedded with protein deposits. Seeds are a rich source of stored proteins, carbohydrates and lipids. These primary metabolites provide necessary protection and nutrition to the developing embryo and other tissues. Among these stored primary metabolites, proteins play a pivotal role in maintaining the metabolic functions of a cell. The bioactive peptides stored in the cells of a seed are capable of performing diverse physiological functions like antiviral, antitumor, and immune-modulatory activities. Several commercially viable bioactive peptides having potential antimicrobial and antiviral activities are developed, which can overcome the menace of developing resistance to antimicrobial and antiviral agents. The seed storage proteins include four major classes—albumins, globulins, glutelins, and prolamins. These SSPs releases bioactive peptides while undergoing GI digestion with the help of hydrolytic enzymes. Albumins derived from plant proteins can coagulate and are soluble in water. The most characterized albumins from plant seeds include leucosin, phaselin, legumelin and ricin derived from wheat, kidney bean, legumes, and castor bean, respectively. Albumins obtained from seeds easily disintegrate during germination to compounds providing sulfur and nitrogen for the developing embryo. Globulin storage proteins found in seeds are also of high medical importance. These globulins bind to human IgE antibodies and accelerates the hypersensitivity reactions. Globulins from leguminaceae like legumin, vicilin, and convicilin and some globulins from graminae are well studied.

27.5.3.3.1 Glutenins

Glutenins are widely distributed among the members of the poaceae family. The wheat gluten proteins are of two types gliadins and glutenins. The water soluble glutenins are rich in amino acids like phenylalanine, valine, proline, tyrosine, and leucine and it forms the chief energy-rich proteins store in the seed endosperm.

27.5.3.3.2 Prolamins

Prolamins forms a major class of SSP among plants. Based on their molecular size they are classified as four groups; Triticeae, oats, rice, and Panicoideae. Prolamins from Triticeae family are good sources of amino acids proline and glutamine. Prolamins are further classified into; sulfur-rich amino acids prolamines (S-rich), high molecular weight prolamines, very low amount sulfur amino acids prolamins (S-poor), and low molecular weight.

Several biological activities of SSPs have generated a wealth of evidences for their use in human diseases. Ricin and abrin are two albumins showed significant antiviral activity and also showed antitumor activity. Globulins like vicilin and concanavalin A showed diverse biological activities. Vicilin showed tobacco mosaic virus infection inhibition in plants. Concanavalin A in association with bivalent cations showed antiviral activity against simian virus 40 and polyoma virus by exhibiting strong agglutination activity. The agglutination activity is mediated through the presence of a-methyl-L-fucopyranoside (a-MG) carbohydrate which has strong binding affinity to Con-A protein. Upon binding with a-MG, the multiplication of the virus is inhibited. Cystatins, the smallest protein belonging to the glutenin family also exhibits antiviral activity through the inhibition of proteinases. Several related cystatin molecules have been isolated from rice, corn and wheat. Cystatin and oryzacystatin, isolated from wheat and rice structurally demonstrated their capability to inactivate poliovirus. World is now waiting for the revolution of new transgenic rive cultivars that could showcase antiviral activity. The role of cereals in health and wellness is yet to unfurl.

Ribosome-inactivating proteins (RIPs) are naturally occurring toxic proteins in several plant species. Cereals like corn expresses an RIP in an inactive form (proRIP1) which is activated by the proteolytic activity leading to the cleavage of N- and C-termini and internal sequences. RIP2 and RIP3 are two other forms of RIPs expressed in maize. Several studies have showed antifungal and antiviral activities of RIPs [129].

27.5.3.3.3 Mechanisms of antiviral activity of seed storage proteins

The mechanism of action of SSPs as antiviral agents are carried out by two major mechanisms; ribosome-inactivating proteins (RIPs) and agglutination and agglutinin activity. RIPs are seed proteins with capability to inhibit or reduce the translation of ribosomal subunits. RIPs generally belong to three classes type I, II, and III. The most potent among them is the RIP type II which are made up of two chains, chain A and B linked through disulfide bonds. The major group of RIP type II are shiga toxins and ricin, which inactivate the 60S subunit of ribosomes through N-glycosidic cleavage that knockouts a specific adenine moiety from the sugar-phosphate backbone. RIP type I is characterized by the presence of either an active A chain or B chain and several SSPs with antiviral properties like Pokeweed antiviral protein, gelonin (*Gelonium multiflorum*), β-luffin (*Luffa cylindrica*), Trichosanthin (*Trichosanthes kirilowii*), and saporin (*Saponaria officinalis*) are groups of type I RIPs. Trichosanthin has demonstrated activity against HIV-1 infected T cells and macrophages. This activity is attributed to the presence of a conserved glutamic and arginine residues side by side which is essential for the catalytic action of this biochemical process.

Agglutination occurs either by a direct or an indirect interaction of the specific antigen and antibody resulting the formation of cellular clumps. Antigens including bacterial polysaccharides or antigens of viruses and rickettsia are capable of producing agglutination. Agglutination is a major mechanism of antiviral action of seed proteins. Agglutinins (antibodies) are particles or substances that coagulate the antigens through antigen-binding interactions.

The antiviral activity of seed proteins effects through host cell adsorption, or by its action on the viral envelope. One of the most important strategy of antiviral activity of seed peptides are through the interactions with the host cell or virus receptors which prevents the entry of the virus. An equally important strategy is through the electrostatic interactions between the negatively charged mammalian cell surface and the virus.

27.5.3.4 *Yogurt and lactoferrin*

Yogurt is a product of coagulated milk with the help of *Lactobacillus bulgaricus* and *Streptococcus thermophilus* with the help of lactic acid. The milk derived proteins are good immunostimulants and are hence therapeutically utilized. In viral infections, yogurt proteins showed increased cytokine production, antibody production, phagocytosis, T cell, and NK cell activities. Consumption of lactobacilli and bifidobacterial has significantly reduced the frequency and severity of respiratory infections [155]. The activity of yogurt is always understood in the purview of probiotics. Lactoferrin is a glycoprotein present in breast milk and bovine milk. A high concentration of lactoferrin is found in colostrum. They have the capability of binding to unbound iron in body fluids and can thus prevent the generation of toxic oxygen radicals. Lactoferrin has antibacterial, antiviral, antioxidant, and immunomodulatory functions. They seem to inhibit the binding of the virus to human receptors. Against SARS-CoV-2 they exhibit their action in three ways; (1) directly binding on to virus and inactivating it, (2) binding to lactoferrin with heparan sulfate proteoglycans which are the adhesion molecules of the viruses to the cell surface, and (3) intracellular inhibition of viral replication.

27.6 Conclusion

The disease burden due to viral infections are rising. Understanding the viral pathology and specific evolutionary advantages gained by viruses in establishing infection and the mechanisms evolved to evade the host immune responses has given them an advantage of standing against the existing treatment options. The success against viral infections gained through the development of specific therapeutic agents that acted at specific locations and time of viral replications and release inside the host cells are now jeopardized by the continuous mutations in them. Mere medicaments alone are not alone in gaining success against viruses. Strategic activation of the host immune response against the invading virus gives a more natural and effective way of combating viral disease burden. Vast researches in the field of food science and nutraceuticals have given us a thorough understanding of the action of food and their constituents on various physiological systems of the body. This understanding has given us a leverage in advancing the fight against viruses through food by boosting the natural immunity of the host. All that we know about the immune responses is only the "tip of an iceberg." As the knowledge is advancing, more and more precise and targeted approach will be implemented against pathological processes and targets to achieve health. Always, nutraceuticals will stand in the frontline of this battle field in guarding human lives against diseases.

References

[1] Lopez AD, Mathers CD, Ezzati M, Jamison DT, Murray CJ. Global and regional burden of disease and risk factors, 2001: systematic analysis of population health data. Lancet 2006;367(9524):1747–57.

[2] Basset C, Holton J, O'Mahony R, Roitt I. Innate immunity and pathogen–host interaction. Vaccine 2003;21: S12–23.

[3] Masihi KN. Immunomodulators in infectious diseases: panoply of possibilites. Int J Immunopharmacol 2000;22(12):1083–91.

[4] Hilleman MR. Strategies and mechanisms for host and pathogen survival in acute and persistent viral infections. Proc Natl Acad Sci 2004;101(Suppl. 2):14560–6.

[5] Pradhan D, Biswasroy P, Naik PK, Ghosh G, Rath G. A review of current interventions for COVID-19 prevention. Arch Med Res 2020;51(5):363–74.

[6] Thirumdas R, Kothakota A, Pandiselvam R, Bahrami A, Barba FJ. Role of food nutrients and supplementation in fighting against viral infections and boosting immunity: a review. Trends Food Sci Technol 2021; Feb 4.

[7] Savant S, Srinivasan S, Kruthiventi AK. Potential Nutraceuticals for COVID-19. Nutr Diet Suppl 2021;13:25.

[8] Dutta S, Ali KM, Dash SK, Giri B. Role of nutraceuticals on health promotion and disease prevention: a review. J Drug Deliv Therap 2018;8(4):42–7.

[9] Ebrahimzadeh-Attari V, Panahi G, Hebert JR, Ostadrahimi A, Saghafi-Asl M, Lotfi-Yaghin N, et al. Nutritional approach for increasing public health during pandemic of COVID-19: a comprehensive review of antiviral nutrients and nutraceuticals. Health Promot Perspect 2021;11(2):119–36.

[10] Shelke K, Hewes C. Regulations for Nutraceuticals and Functional Foods in the Indian Subcontinent. Nutraceutical and functional food regulations in the United States and around the world. Academic Press; 2008. p. 323–40. Jan 1.

[11] Peakman M, Vergani D. Basic and clinical immunology E-book. Elsevier Health Sciences; 2009. Apr 24.

[12] Marshall JS, Warrington R, Watson W, Kim HL. An introduction to immunology and immunopathology. Allergy, Asthma Clin Immunol 2018;14(2):1 0.

[13] Turvey SE, Broide DH. Innate immunity. J Allergy Clin Immunol 2010;125(2):S24–32.

[14] Bonilla FA, Oettgen HC. Adaptive immunity. J Allergy Clin Immunol 2010;125(Suppl 2):S33–40.

[15] Murphy KM, Travers P, Walport M. Janeway's immunobiology. 7th ed. New York: Garland Science; 2007.

[16] Vivier E, Tomasello E, Baratin M, Walzer T, Ugolini S. Functions of natural killer cells. Nat Immunol 2008;9 (5):503−10.

[17] Shimasaki N, Jain A, Campana D. NK cells for cancer immunotherapy. Nat Rev Drug Discov 2020;19 (3):200−18.

[18] Griffith JW, Sokol CL, Luster AD. Chemokines and chemokine receptors: positioning cells for host defense and immunity. Annu Rev Immunol 2014;32:659−702.

[19] Thapa M, Carr DJ. Chemokines and chemokine receptors critical to host resistance following genital herpes simplex virus type 2 (HSV-2) infection. Open Immunol J 2008;1:33.

[20] Wuest TR, Carr DJ. The role of chemokines during herpes simplex virus-1 infection. Front Biosci 2008;13:4862.

[21] Melchjorsen J, Sørensen LN, Paludan SR. Expression and function of chemokines during viral infections: from molecular mechanisms to in vivo function. J Leukoc Biol 2003;74(3):331−43.

[22] Schindler C, Shuai K, Prezioso VR, Darnell Jr JE. Interferon-dependent tyrosine phosphorylation of a latent cytoplasmic transcription factor. Science. 1992;809−13.

[23] Soh J, Donnelly RJ, Kotenko S, Mariano TM, Cook JR, Wang N, et al. Identification and sequence of an accessory factor required for activation of the human interferon γ receptor. Cell. 1994;76(5):793−802.

[24] Uze G, Lutfalla G, Gresser C. Activation of transcription factors by interferon-alpha in cell-free system. Science 1992;257:7.

[25] Isaacs A, Lindenmann J. Virus interference. I. The interferon. cytokines. Proc R Soc Lond B Biol Sci 1957;147:258−67.

[26] Schiela B, Bernklau S, Malekshahi Z, Deutschmann D, Koske I, Banki Z, et al. Active human complement reduces the Zika virus load via formation of the membrane-attack complex. Front Immunol 2018;9:2177.

[27] Tam JC, Bidgood SR, McEwan WA, James LC. Intracellular sensing of complement C3 activates cell autonomous immunity. Science. 2014;345:6201.

[28] Wetsel RA, Kildsgaard J, Haviland DL. Complement anaphylatoxins (C3a, C4a, C5a) and their receptors (C3aR, C5aR/CD88) as therapeutic targets in inflammation. Therapeutic interventions in the complement system. Totowa, NJ: Humana Press; 2000. p. 113−53.

[29] Harris SL, Frank I, Vee A, Cohen GH, Eisenberg RJ, Friedman HM. Glycoprotein C of herpes simplex virus type 1 prevents complement-mediated cell lysis and virus neutralization. J Infect Dis 1990;162(2):331−7.

[30] Ji X, Olinger GG, Aris S, Chen Y, Gewurz H, Spear GT. Mannose-binding lectin binds to Ebola and Marburg envelope glycoproteins, resulting in blocking of virus interaction with DC-SIGN and complement-mediated virus neutralization. J Gen Virol 2005;86(9):2535−42.

[31] Gmünder H, Eck HP, Benninghoff B, Roth S, Dröge W. Macrophages regulate intracellular glutathione levels of lymphocytes. Evidence for an immunoregulatory role of cysteine. Cell Immunol 1990;129(1):32−46.

[32] Golenbock DT, Hampton RY, Qureshi N, Takayama K, Raetz CR. Lipid A-like molecules that antagonize the effects of endotoxins on human monocytes. J Biol Chem 1991;266(29):19490−8.

[33] Halliwell B. Oxidants and human disease: some new concepts 1. FASEB J 1987;1(5):358−64.

[34] Judson KA, Lubinski JM, Jiang M, Chang Y, Eisenberg RJ, Cohen GH, et al. Blocking immune evasion as a novel approach for prevention and treatment of herpes simplex virus infection. J Virol 2003;77(23):12639−45.

[35] Kotwal GJ. Microorganisms and their interaction with the immune system. J Leukoc Biol 1997;62(4):415−29.

[36] Johnson WE, Desrosiers RC. Viral persistence: HIV's strategies of immune system evasion. Annu Rev Med 2002;53(1):499−518.

[37] Ploegh HL. Viral strategies of immune evasion. Science. 1998;280(5361):248−53.

[38] Virgin HW, Wherry EJ, Ahmed R. Redefining chronic viral infection. Cell. 2009;138(1):30−50.

[39] Zuo J, Rowe M. Herpesviruses placating the unwilling host: manipulation of the MHC class II antigen presentation pathway. Viruses. 2012;4(8):1335−53.

[40] Kikkert M. Innate immune evasion by human respiratory RNA viruses. J Innate Immun 2020;12(1):4−20.

[41] Nelemans T, Kikkert M. Viral innate immune evasion and the pathogenesis of emerging RNA virus infections. Viruses. 2019;11(10):961.

[42] Andersen MH, Schrama D, Thor Straten P, Becker JC. Cytotoxic T cells. J Investig Dermatol 2006;126 (1):32−41.

[43] Finlay D, Cantrell DA. Metabolism, migration and memory in cytotoxic T cells. Nat Rev Immunol 2011;11 (2):109−17.

[44] Kos FJ, Engleman EG. Requirement for natural killer cells in the induction of cytotoxic T cells. J Immunol 1995;155(2):578–84.

[45] Koyanagi N, Kawaguchi Y. Evasion of the cell-mediated immune response by alphaherpesviruses. Viruses. 2020;12(12):1354.

[46] Neefjes J, Jongsma ML, Paul P, Bakke O. Towards a systems understanding of MHC class I and MHC class II antigen presentation. Nat Rev Immunol 2011;11(12):823–36.

[47] Hansen TH, Bouvier M. MHC class I antigen presentation: learning from viral evasion strategies. Nat Rev Immunol 2009;9(7):503–13.

[48] Horst D, Verweij MC, Davison AJ, Ressing ME, Wiertz EJ. Viral evasion of T cell immunity: ancient mechanisms offering new applications. Curr Opimmunol 2011;23(1):96–103.

[49] Neefjes JJ, Momburg F. Cell biology of antigen presentation. Curr Opimmunol 1993;5(1):27–34.

[50] Ting JP, Baldwin AS. Regulation of MHC gene expression. Curr Opimmunol 1993;5(1):8–16.

[51] Glimcher LH, Kara CJ. Sequences and factors: a guide to MHC class-II transcription. Annu Rev Immunol 1992;10(1):13–49.

[52] Scholz M, Hamann A, Blaheta RA, Auth MK, Encke A, Markus BH. Cytomegalovirus-and interferon-related effects on human endothelial cells: cytomegalovirus infection reduces upregulation of HLA class II antigen expression after treatment with interferon-γ. Hum Immunol 1992;35(4):230–8.

[53] Buchmeier NA, Cooper NR. Suppression of monocyte functions by human cytomegalovirus. Immunology. 1989;66(2):278.

[54] Hsu DH, Malefyt RD, Fiorentino DF, Dang MN, Vieira P. Expression of interleukin-10 activity by Epstein-Barr virus protein BCRF1. Science 1990;250(4982):830.

[55] Favoreel HW, Van de Walle GR, Nauwynck HJ, Pensaert MB. Virus complement evasion strategies. J Gen Virol 2003;84(1):1–5.

[56] Marin FR, Frutos MJ, Perez-Alvarez JA, Martinez-Sanchez F, Del Rio JA. Flavonoids as nutraceuticals: structural related antioxidant properties and their role on ascorbic acid preservation. Studies in natural products chemistry, Vol. 26. Elsevier; 2002. p. 741–78.

[57] Sgarbanti R, Amatore D, Celestino I, Elena Marcocci M, Fraternale A, Ciriolo MR, et al. Intracellular redox state as target for anti-influenza therapy: are antioxidants always effective? Curr Top Med Chem 2014;14(22):2529–41.

[58] Jain P, Pundir RK. Nutraceuticals: recent developments and future prospectives. Recent trends in biotechnology and therapeutic applications of medicinal plants 2013;213–24.

[59] Andlauer W, Fürst P. Nutraceuticals: a piece of history, present status and outlook. Food Res Int 2002;35(2–3):171–6.

[60] Singh AK, Chaturvedani AK, Singh NP, Baranawal A. Nutraceuticals: meaning and regulatory scenario. Pharm Innov J 2018;7:448 5.

[61] Kalra EK. Nutraceutical-definition and introduction. Aaps Pharmsci 2003;5(3):27–8.

[62] Blomhoff R. Dietary antioxidants and cardiovascular disease. Curr Opin Lipidol 2005;16(1):47–54.

[63] Roberfroid MB. Global view on functional foods: European perspectives. Br J Nutr 2002;88(S2):S133–8.

[64] Bogue J, Collins O, Troy AJ. Market analysis and concept development of functional foods. Developing new functional food and nutraceutical products. Academic Press; 2017. p. 29–45. Jan 1.

[65] Sun-Waterhouse D. The development of fruit-based functional foods targeting the health and wellness market: a review. Int J Food Sci Technol 2011;46(5):899–920.

[66] Kalia AN. Textbook of industrial pharmacognosy. New Delhi: CBS Publisher and Distributor; 2005. p. 204–8.

[67] Souyoul SA, Saussy KP, Lupo MP. Nutraceuticals: a review. Dermatol Ther 2018;8(1):5–16.

[68] Verma G, Mishra MK. A review on nutraceuticals: classification and its role in various diseases. Int J Pharm Therap 2016;7(4).

[69] Akatsu H, Iwabuchi N, Xiao JZ, Matsuyama Z, Kurihara R, Okuda K, et al. Clinical effects of probiotic *Bifidobacterium longum* BB536 on immune function and intestinal microbiota in elderly patients receiving enteral tube feeding. J Parenter Enter Nutr 2013;37(5):631–40.

[70] Sanders ME. Probiotics: definition, sources, selection, and uses. Clin Infect Dis 2008;46(Suppl. 2):S58–61 Feb 1.

[71] Gibson GR, Roberfroid MB. Dietary modulation of the human colonic microbiota: introducing the concept of prebiotics. J Nutr 1995;125(6):1401–12.

[72] Jacob RA. The integrated antioxidant system. Nutr Res 1995;15(5):755–66.

[73] Goodnight Jr SH, Harris WS, Connor WE, Illingworth DR. Polyunsaturated fatty acids, hyperlipidemia, and thrombosis. Arteriosclerosis 1982;2(2):87–113.

[74] James MJ, Gibson RA, Cleland LG. Dietary polyunsaturated fatty acids and inflammatory mediator production. Am J Clin Nutr 2000;71(1):343s–348ss.

[75] Peterhans E. Oxidants and antioxidants in viral diseases: disease mechanisms and metabolic regulation. J Nutr 1997;127(5):962S–965SS.

[76] Ndiaye M, Chataigneau T, Chataigneau M, Schini-Kerth VB. Red wine polyphenols induce EDHF-mediated relaxations in porcine coronary arteries through the redox-sensitive activation of the PI3-kinase/Akt pathway. Br J Pharmacol 2004;142(7):1131–6.

[77] Wang L, Zhu LH, Jiang H, Tang QZ, Yan L, Wang D, et al. Grape seed proanthocyanidins attenuate vascular smooth muscle cell proliferation via blocking phosphatidylinositol 3-kinase-dependent signaling pathways. J Cell Physiol 2010;223(3):713–26.

[78] Bravo L. Polyphenols: chemistry, dietary sources, metabolism, and nutritional significance. Nutr Rev 1998;56 (11):317–33.

[79] Watson R, DeMeester F, Fernandez M, Andersen C. Handbook of eggs in human function, human health handbooks. Wageningen Academic. Publication; 2015. Nov 28.

[80] Tang WW, Wang Z, Levison BS, Koeth RA, Britt EB, Fu X, et al. Intestinal microbial metabolism of phosphatidylcholine and cardiovascular risk. N Engl J Med 2013;368(17):1575–84.

[81] Rebello CJ, Liu AG, Greenway FL, Dhurandhar NV. Dietary strategies to increase satiety. Adv Food Nutr Res 2013;69:105–82.

[82] Wu J, Acero-Lopez A. Ovotransferrin: structure, bioactivities, and preparation. Food Res Int 2012;46 (2):480–7.

[83] Shirman G, Graevskaya N, Lavrova I, Ginevskaya V, Baeva L. Obtention d'immunoglobulines à partir du jaune d'oeuf de poulets immunisés avec des virus entériques. Vopr 1988;33(6):725–9.

[84] Hatta H, Ozeki M, Tsuda K. Egg yolk antibody IgY and its application. Hen Eggs. CRC Press; 2018. p. 151–78. May 4.

[85] Roberts JR. The nutritional and physiological functions of egg yolk components Yasumi Horimoto, University of Guelph, Canada and Hajime Hatta, Kyoto Women's University, Japan. Achieving sustainable production of eggs, Volume 1. Burleigh Dodds Science Publishing; 2017. p. 69–116. Oct 31.

[86] Yazawa S, Hosomi O, Takeya A. Isolation and characterization of anti-H antibody from egg yolk of immunized hens. Immunol Invest 1991;20(7):569–81.

[87] Wu J, editor. Eggs as functional foods and nutraceuticals for human health. Royal Society of Chemistry; 2019. May 1.

[88] Zhan XB, Lin CC, Zhang HT. Recent advances in curdlan biosynthesis, biotechnological production, and applications. Appl Microbiol Biotechnol 2012;93(2):525–31.

[89] Goodridge HS, Wolf AJ, Underhill DM. β-glucan recognition by the innate immune system. Immunol Rev 2009;230(1):38–50.

[90] Jong SC, Donovick R. Antitumor and antiviral substances from fungi. Adv Appl Microbiol 1989;34:183–262.

[91] Todorov SD, Wachsman MB, Knoetze H, Meincken M, Dicks LMT. An antibacterial and antiviral peptide produced by *Enterococcus mundtii* ST4V isolated from soy beans. Int J Antimicrob Agents 2005;25(6):508–13.

[92] Hayashi O, Katoh T, Okuwaki Y. Enhancement of antibody production in mice by dietary Spirulina platensis. J Nutr Sci Vitaminol 1994;40(5):431–41.

[93] Luescher-Mattli M. Algae, a possible source for new drugs in the treatment of HIV and other viral diseases. Curr Med Chem-Anti-Infect Agents 2003;2(3):219–25.

[94] Murphy EA, Davis JM, Brown AS, Carmichael MD, Carson JA, Van Rooijen N, et al. Benefits of oat β-glucan on respiratory infection following exercise stress: role of lung macrophages. Am J Physiol-Regul Integr Comp Physiol 2008;294(5):R1593–9.

[95] Heilmann J. New medical applications of plant secondary metabolites. In: Wink M, editor. Annual plant reviews, functions and biotechnology of plant secondary metabolites. New York: John Wiley & Sons; 2009. p. 348–66.

[96] Manthey JA, Guthrie N, Grohmann K. Biological properties of citrus flavonoids pertaining to cancer and inflammation. Curr Med Chem 2001;8(2):135–53.

[97] Harborne JB, Williams CA. Advances in flavonoid research since 1992. Phytochemistry. 2000;55(6):481–504.

[98] Rajabi F, Feiz A, Luque R. An efficient synthesis of coumarin derivatives using a SBA-15 supported cobalt (II) nanocatalyst. Catal Lett 2015;145:1621−5.

[99] Choi SY, Ko HC, Ko SY, Hwang JH, Park JG, Kang SH, et al. Correlation between flavonoid content and the NO production inhibitory activity of peel extracts from various citrus fruits. Biol Pharm Bull 2007;30(4):772−8.

[100] Rohr AC, Wilkins CK, Clausen PA, Hammer M, Nielsen GD, Wolkoff P, et al. Upper airway and pulmonary effects of oxidation products of (+)-α-pinene, d-limonene, and isoprene in BALB/c mice. Inhal Toxicol 2002;14(7):663−84.

[101] Ruocco N, Costantini S, Guariniello S, Costantini M. Polysaccharides from the marine environment with pharmacological, cosmeceutical and nutraceutical potential. Molecules. 2016;21(5):551.

[102] Howard LR, Wildman RE. Antioxidant vitamin and phytochemical content of fresh and processed pepper fruit (Capsicum annuum). Nutraceuticals 2007;.

[103] Fiore C, Eisenhut M, Krausse R, Ragazzi E, Pellati D, Armanini D, et al. Antiviral effects of Glycyrrhiza species. Phytother Res 2008;22(2):141−8.

[104] Umar S, Munir MT, Subhan S, Azam T, Nisa Q, Khan MI, et al. Protective and antiviral activities of Nigella sativa against avian influenza (H9N2) in turkeys. J Saudi Soc Agric Sci 2016;10.

[105] Salem ML, Hossain MS. Protective effect of black seed oil from Nigella sativa against murine cytomegalovirus infection. Int J Immunopharmacol 2000;22(9):729−40.

[106] Barakat EM, El Wakeel LM, Hagag RS. Effects of Nigella sativa on outcome of hepatitis C in Egypt. World J Gastroenterol: WJG 2013;19(16):2529.

[107] Tsai Y, Cole LL, Davis LE, Lockwood SJ, Simmons V, Wild GC. Antiviral properties of garlic: in vitro effects on influenza B, herpes simplex and coxsackie viruses. Planta Med 1985;51(05):460−1.

[108] Bayan L, Koulivand PH, Gorji A. Garlic: a review of potential therapeutic effects. Avicenna J Phytomed 2014;4(1):1.

[109] Usta J, Kreydiyyeh S, Barnabe P, Bou-Moughlabay Y, Nakkash-Chmaisse H. Comparative study on the effect of cinnamon and clove extracts and their main components on different types of ATPases. Hum Exp Toxicol 2003;22(7):355−62.

[110] Mishra S, Pandey A, Manvati S. Coumarin: an emerging antiviral agent. Heliyon. 2020;6(1):e03217.

[111] Priya NC, Kumari PS. Antiviral activities and cytotoxicity assay of seed extracts of Piper longum and Piper nigrum on human cell lines. Int J Pharm Sci Rev Res 2017;44(1):197−202.

[112] Lee YM, Choi JH, Min WK, Han JK, Oh JW. Induction of functional erythropoietin and erythropoietin receptor gene expression by gamma-aminobutyric acid and piperine in kidney epithelial cells. Life Sci 2018;215:207−15.

[113] Biswas D, Nandy S, Mukherjee A, Pandey DK, Dey A. Moringa oleifera Lam. and derived phytochemicals as promising antiviral agents: a review. South Afr J Bot 2020;129:272−82.

[114] Waiyaput W, Payungporn S, Issara-Amphorn J, Nattanan T, Panjaworayan T. Inhibitory effects of crude extracts from some edible Thai plants against replication of hepatitis B virus and human liver cancer cells. BMC Complement Altern Med 2012;12(1):1−7.

[115] Ahmad W, Ejaz S, Anwar K, Ashraf M. Exploration of the in vitro cytotoxic and antiviral activities of different medicinal plants against infectious bursal disease (IBD) virus. Open Life Sci 2014;9(5):531−42.

[116] Chollom SC, Olawuyi AK, Danjuma LD, Nanbol LD, Makinde IO, Hashimu GA, et al. Antiviral potential of aqueous extracts of some parts of Momordica balsamina plant against Newcastle disease virus. J Adv Pharm Educ Res 2012;2(3):82−92.

[117] Agrawal PK, Agrawal C, Blunden G. Quercetin: antiviral significance and possible COVID-19 integrative considerations. Nat Product Commun 2020;15(12) 1934578X20976293.

[118] Güttner J, Veckenstedt A, Heinecke H, Pusztai R. Effect of quercetin on the course of Mengo virus infection in immunodeficient and normal mice. A histologic study. Acta Virol 1982;26(3):148−55.

[119] Lyu SY, Rhim JY, Park WB. Antiherpetic activities of flavonoids against herpes simplex virus type 1 (HSV-1) and type 2 (HSV-2) in vitro. Arch Pharm Res 2005;28(11):1293−301.

[120] González-Búrquez MDJ, González-Díaz FR, García-Tovar CG, Carrillo-Miranda L, Soto-Zárate CI, Canales-Martínez MM, et al. Comparison between in vitro antiviral effect of Mexican propolis and three commercial flavonoids against canine distemper virus. Evid-Based Complement Altern Med 2018;2018.

[121] Ishitsuka H, Ohsawa C, Ohiwa T, Umeda I, Suhara Y. Antipicornavirus flavone RO 09−0179. Antimicrob Agents Chemother 1982;22(4):611−16.

[122] Ganesan S, Faris AN, Comstock AT, Wang Q, Nanua S, Hershenson MB, et al. Quercetin inhibits rhinovirus replication in vitro and in vivo. Antivir Res 2012;94(3):258–71.

[123] Liu Z, Zhao J, Li W, Wang X, Xu J, Xie J, et al. Molecular docking of potential inhibitors for influenza H7N9. Computat Math Methods Med 2015;2015 Mar 15.

[124] Mathew S, Fatima K, Fatmi MQ, Archunan G, Ilyas M, Begum N, et al. Computational docking study of p7 Ion channel from HCV genotype 3 and genotype 4 and its interaction with natural compounds. PLoS One 2015;10(6):e0126510.

[125] Bachmetov L, Gal-Tanamy M, Shapira A, Vorobeychik M, Giterman-Galam T, Sathiyamoorthy P, et al. Suppression of hepatitis C virus by the flavonoid quercetin is mediated by inhibition of NS3 protease activity. J Viral Hepat 2012;19(2):e81–8.

[126] Lu NT, Crespi CM, Liu NM, Vu JQ, Ahmdieh Y, Wu S, et al. A phase I dose escalation study demonstrates quercetin safety and explores potential for bioflavonoid antivirals in patients with chronic hepatitis C. Phytother Res 2016;30(1):160–8.

[127] Polansky H, Javaherian A, Itzkovitz E. Clinical trial of herbal treatment Gene-Eden-VIR/Novirin in oral herpes. J Evid-Based Integr Med 2018;23 :2515690X18806269.

[128] Kumar P, Khanna M, Srivastava V, Tyagi YK, Raj HG, Ravi K. Effect of quercetin supplementation on lung antioxidants after experimental influenza virus infection. Exp Lung Res 2005;31(5):449–59.

[129] Kumar P, Sharma S, Khanna M, Raj HG. Effect of Quercetin on lipid peroxidation and changes in lung morphology in experimental influenza virus infection. Int J Exp Pathol 2003;84(3):127–34.

[130] Heinz SA, Henson DA, Austin MD, Jin F, Nieman DC. Quercetin supplementation and upper respiratory tract infection: a randomized community clinical trial. Pharmacol Res 2010;62(3):237–42.

[131] Cialdella-Kam L, Nieman DC, Knab AM, Shanely RA, Meaney MP, Jin F, et al. A mixed flavonoid-fish oil supplement induces immune-enhancing and anti-inflammatory transcriptomic changes in adult obese and overweight women—a randomized controlled trial. Nutrients. 2016;8(5):277.

[132] Kim CH, Kim JE, Song YJ. Antiviral activities of quercetin and isoquercitrin against human herpesviruses. Molecules. 2020;25(10):2379.

[133] Hung PY, Ho BC, Lee SY, Chang SY, Kao CL, Lee SS, et al. *Houttuynia cordata* targets the beginning stage of herpes simplex virus infection. PLoS One 2015;10(2):e0115475.

[134] Cho WK, Weeratunga P, Lee BH, Park JS, Kim CJ, Ma JY, et al. Epimedium koreanum Nakai displays broad spectrum of antiviral activity in vitro and in vivo by inducing cellular antiviral state. Viruses. 2015;7(1):352–77.

[135] Yang Q, Gao L, Si J, Sun Y, Liu J, Cao L, et al. Inhibition of porcine reproductive and respiratory syndrome virus replication by flavaspidic acid AB. Antivir Res 2013;97(1):66–73.

[136] Zandi K, Teoh BT, Sam SS, Wong PF, Mustafa MR, AbuBakar S. Antiviral activity of four types of bioflavonoid against dengue virus type-2. Virol J 2011;8(1):1.

[137] Jasso-Miranda C, Herrera-Camacho I, Flores-Mendoza LK, Dominguez F, Vallejo-Ruiz V, Sanchez-Burgos GG, et al. Antiviral and immunomodulatory effects of polyphenols on macrophages infected with dengue virus serotypes 2 and 3 enhanced or not with antibodies. Infect Drug Resist 2019;12:1833.

[138] Boettger S, Puta C, Yeragani VK, Donath L, Mueller HJ, Gabriel HH, et al. Heart rate variability, QT variability, and electrodermal activity during exercise. Med Sci Sports Exerc 2010;42(3):443–8.

[139] Chaabi M. Antiviral effects of quercetin and related compounds. Naturopathic Curr 2020;2020:1–4.

[140] Wu W, Li R, Li X, He J, Jiang S, Liu S, et al. Quercetin as an antiviral agent inhibits influenza A virus (IAV) entry. Viruses. 2016;8(1):6.

[141] Johari J, Kianmehr A, Mustafa MR, Abubakar S, Zandi K. Antiviral activity of baicalein and quercetin against the Japanese encephalitis virus. Int J Mol Sci 2012;13(12):16785–95.

[142] Song JH, Shim JK, Choi HJ. Quercetin 7-rhamnoside reduces porcine epidemic diarrhea virus replication via independent pathway of viral induced reactive oxygen species. Virol J 2011;8(1):1–6.

[143] Seo DJ, Jeon SB, Oh H, Lee BH, Lee SY, Oh SH, et al. Comparison of the antiviral activity of flavonoids against murine norovirus and feline calicivirus. Food Control 2016;60:25–30.

[144] Teplyakova TV, Psurtseva NV, Kosogova TA, Mazurkova NA, Khanin VA, Vlasenko VA. Antiviral activity of polyporoid mushrooms (higher Basidiomycetes) from Altai Mountains (Russia). Int J Med Mushrooms 2012;14(1).

[145] Stamets P. Antiviral activity from medicinal mushrooms. Shelton, WA. US. Patent No 20050276815; 2005.

[146] Carr AC. Micronutrient status of COVID-19 patients: a critical consideration. Crit Care 2020;24(1):1−2.

[147] Colunga Biancatelli RM, Berrill M, Marik PE. The antiviral properties of vitamin C. Expert Rev Anti-infect Ther 2020;18(2):99−101.

[148] Liu F, Zhu Y, Zhang J, Li Y, Peng Z. Intravenous high-dose vitamin C for the treatment of severe COVID-19: study protocol for a multicentre randomised controlled trial. BMJ Open 2020;10(7):e039519.

[149] Kim H, Jang M, Kim Y, Choi J, Jeon J, Kim J, et al. Red ginseng and vitamin C increase immune cell activity and decrease lung inflammation induced by influenza A virus/H1N1 infection. J Pharm Pharmacol 2016;68 (3):406−20.

[150] Rizvi S, Raza ST, Faizal Ahmed AA, Abbas S, Mahdi F. The role of vitamin E in human health and some diseases. Sultan Qaboos Univ Med J 2014;14(2):e157.

[151] Cannell JJ, Vieth R, Umhau JC, Holick MF, Grant WB, Madronich S, et al. Epidemic influenza and vitamin D. Epidemiol Infect 2006;134(6):1129−40.

[152] Ibs KH, Rink L. Zinc-altered immune function. J Nutr 2003;133(5):1452S−1456SS.

[153] Rahman MM, Mosaddik A, Alam AK. Traditional foods with their constituent's antiviral and immune system modulating properties. Heliyon. 2021;7(1):e05957.

[154] Zareie PP. Honey as an antiviral agent against respiratory syncytial virus [Doctoral dissertation]. University of Waikato, 2011.

[155] Calder PC. Nutrition, immunity and COVID-19. BMJ Nutr Prev Health 2020;3(1):74.

SECTION IV

Others

In vitro and in vivo approaches for evaluating antiviral efficacy

Ram Gopal Nitharwal

Department of Biotechnology, Central University of Haryana, Mahendergarh, Haryana, India

28.1 Introduction

Several antiviral therapies ($\sim$118) have been developed and approved so far since the approval of the first antiviral drug idoxuridine (to treat herpes simplex infection) by the US Food and Drug Administration (FDA) in 1963 [1]. Most of the approved antiviral drugs target different parts of a virus life cycle, such as host-cell binding (betulinic acid), uncoating of capsid (pleconaril), replication[reverse transcriptase inhibitors (zidovudine), integrase inhibitors (raltegravir), RNA dependent RNA polymerase inhibitors (favipiravir), DNA or RNA polymerase inhibitors (acyclovir, cidofovir, ribavirin, etc.), proteases/polyprotein processing inhibitors (lopinavir)], assembly/maturation inhibitors (indinavir, ritonavir, and rimantadine, etc.), and virus release (oseltamivir). Despite the spectacular progress in antiviral therapies, viral diseases are the leading cause of mortality and morbidity globally. Therefore, the development of new antiviral therapy is a pressing need due to the non-availability of drugs against existing viruses (e.g., Chikungunya virus, Dengue virus, West Nile virus, Zika viruses, Rabies virus, Nipah virus, and Epstein-Barr virus), and due to emergence of new viruses like Ebola, corona viruses (SARS-CoV, SARS-CoV-2). The emergence of drug resistance especially against RNA viruses is of special concern. Therefore, there is a constant need to design and test new inhibitors, targeting different steps of viral life cycles. Since viruses are intimately connected to host cells, therefore, designing effective antivirals, that will target the viral enzymes or viralreplication, without affecting the host cell has been proved to be difficult. There are technological limitations of antiviral testing due to the need for biosafety containment facilities (BSL 3 and BSL 4). These facilities are expensive to build and maintain and require a high level of oversight [2]. Drug discovery and development is a lengthy and costly undertaking along with a high degree of uncertainty for a drug to succeed and reach the market. Typically, a rational drug discovery path

Viral Infections and Antiviral Therapies
DOI: https://doi.org/10.1016/B978-0-323-91814-5.00016-7

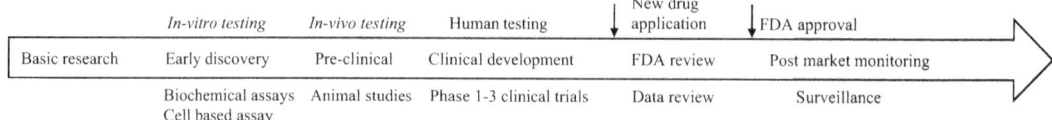

	In-vitro testing	In-vivo testing	Human testing	New drug application	FDA approval
Basic research	Early discovery	Pre-clinical	Clinical development	FDA review	Post market monitoring
	Biochemical assays Cell based assay	Animal studies	Phase 1-3 clinical trials	Data review	Surveillance

FIGURE 28.1 Sketch showing rational drug development and approval pipeline. In the beginning basic research and early discovery (in vitro testing) identify and validate the target, determine target structure and provide a rationale for drug discovery (mechanism of action). In vivo testing involves efficacy testing employing animal models, after which clinical trials start. Based on the clinical trials data, new drug application is filed with FDA for approval. After the drug is approved and launched in the market, performance of the drug is monitored (postmarket surveillance). Source: *Adapted from Phases of drug development process, Drug discovery process. Nebiolab. com. 2020. <https://www.nebiolab.com/drug-discovery-and-development-process/> [accessed 27.11.21].*

follows various steps ;(1) Discovery and development, (2) Preclinical research, (3) Clinical research, (4) Review and approval, and (5) Postmarket safety monitoring (Fig. 28.1) [3,4].

Basic research dominates in the "discovery and development" stage and mainly involve target identification and validation, specific assay development, high-throughput screening, etc. A drug target is a key molecule (protein/DNA/RNA) that is involved in a disease condition or pathology, or to the infectivity or survival of a pathogen. The structure and function of the target molecule are also explored in this step. If the structure of the target is known then in silico screening of various databases is performed using computer-based programs. Based on this, various lead compounds are identified. Once this is done then screening assays are developed. Assays are test systems that evaluate the effect of a new drug candidate at the molecular, biochemical, and cellular levels. Nonclinical and clinical virology study reports are essential for the FDA's review of antiviral drug investigational product application and marketing application. Before initiating the phase 1 of clinical trials, the data pertaining to investigational product, are required by FDA on the mechanism of action, the specific antiviral activity of the investigational product, effects of serum protein binding on antiviral activity, antagonism of other antiviral products, development of viral resistance, cross-resistance to approved antiviral products having the same target, cytotoxicity and therapeutic indexes. [5]. Assessing the antiviral activity of the investigational product is mainly done through in vitro and in vivo testing at the preclinical stage. In vitro testing involves testing the efficacy and toxicity of a new drug using the biochemical/cellular model systems. Before testing the new drug in humans, in vivo testing is done in animal models (in vivo testing) to determine the efficacy, toxicity, and pharmacokinetics. Here, insights into in vitro and in vivo (animal model) approaches that are generally employed to test the efficacy of new antiviral drugs are provided.

28.1.1 Antiviral activity

Before initiating clinical trials on humans, the antiviral activity of the compounds should be tested in vitro (biochemical/cellular) or in vivo (animal models). The antiviral activity should be tested at the drug concentrations that can be achieved in vivo. This will help in the selection of appropriate dose ranges in early clinical trials. Where cells are used for antiviral activity data, the use of primary human target cells is advantageous. The antiviral activity of the test drugs should be tested for multiple laboratories and clinical viral isolates. It is also recommended that the antiviral activity of the investigational product should be tested against mutant viruses that are resistant to the same target molecule as the investigational product as well as to the viruses resistant to other approved products for the same indication. The antiviral activity of drugs can be evaluated using in vitro (biochemical and cell-based) and in vivo (animal models) approaches. Efficacy of antiviral compound is measured as EC50 or IC50 value, which is the concentration of investigational product at which virus replication is inhibited by 50% (EC50 for cell-based assays; IC50 for biochemical assays). Nonclinical virology studies (in vitro and in vivo approaches) help in evaluating the safety and efficacy of an investigational product before it is tested in humans [5].

28.2 In vitro approaches

In vitro approaches aid in the evaluation of the safety and efficacy of an investigational product before it is tested in animals/humans. In vitro assays usually exploit the virus's ability to infect and replicate in specific cell lines (cell culture systems) or specific biochemical assay are used (where no cell culture systems are available). The cell culture systems provide rapid and less cumbersome methods for growing viruses and testing of antiviral compounds. Here, an overview of in vitro methods, including cell-based assays, that may be suitable for screening of antiviral compounds are discussed (Table 28.1 and 28.2).

28.2.1 Cell-based assays

These assays are used for viruses where there are cell culture systems in which the virus can undergo a complete life cycle. The cell culture system can be employed to measure specific, quantifiable antiviral activity [e.g., inhibition of virus replication, inhibition of cytopathic effect (CPE), etc.] in the presence of increasing concentrations of the test compound and can be compared to replication in the absence of the compound. These tests can also suggest if there is more than one target (if a test compound inhibits virus

TABLE 28.1 Commonly used in vitro approaches which are used for testing efficacy of antiviral drugs.

S. No	Virus	In vitro assay	Reference
1	HSV, (herpes)	Plaque Assay (Human foreskin fibroblast), Vero, MRC-5, BHK, HEp-2	[12,13]
2.	Influenza (Flu)	Plaque Assay (MDCK), Hemagglutination Inhibition assay	[14–16]
3.	Polio virus (Paralysis, Aseptic meningitis)	CPE assay (LLC-MK2 rhesus monkey kidney epithelial cells), Plaque Assay (L20B cells)	[17]
4.	HCV (Liver cirrhosis, carcinoma)	Huh-7 hepatoma cell line, RT-PCR	[18]
5.	HIV (AIDS)	CPE (C8166 cells), Reverse transcriptase assay	[19]

replication at concentrations lower than biochemical data) [5]. Following are the routinely used cell-based assays to evaluate the antiviral activity of test compounds (Table 28.1).

28.2.1.1 *Antiviral assay by cytopathic effect*

Observable changes in the host cell due to virus infection/replication are referred as CPE. CPE includes structural changes in host cell (e.g., swelling, rounding, fusion with neighboring cells to form syncytia, formation of nuclear/cytoplasmic inclusion bodies, detachment from the culture plate, etc.), cell death, cell lysis, apoptosis, etc. Since CPEs are easily observable using microscopy (with or without staining), CPE-based assay is a well-recognized assay that has been used in antiviral drug discovery against a number of viruses that induced rapid and observable CPE/apoptosis [6]. Due to its simplicity, this test is used for initial screening of antiviral compounds. The antiviral activity is assessed based on the ability of the compound to prevent virus from causing viral CPE in cell culture (Fig. 28.2). Eight dilutions of test compound are evaluated, and the effective antiviral concentration (EC50) can be calculated by regression analysis (Fig. 28.2). The toxicity of the test compound is determined in parallel. An abbreviated test with four dilutions of compound may be employed to screen large numbers of compounds quickly and at a

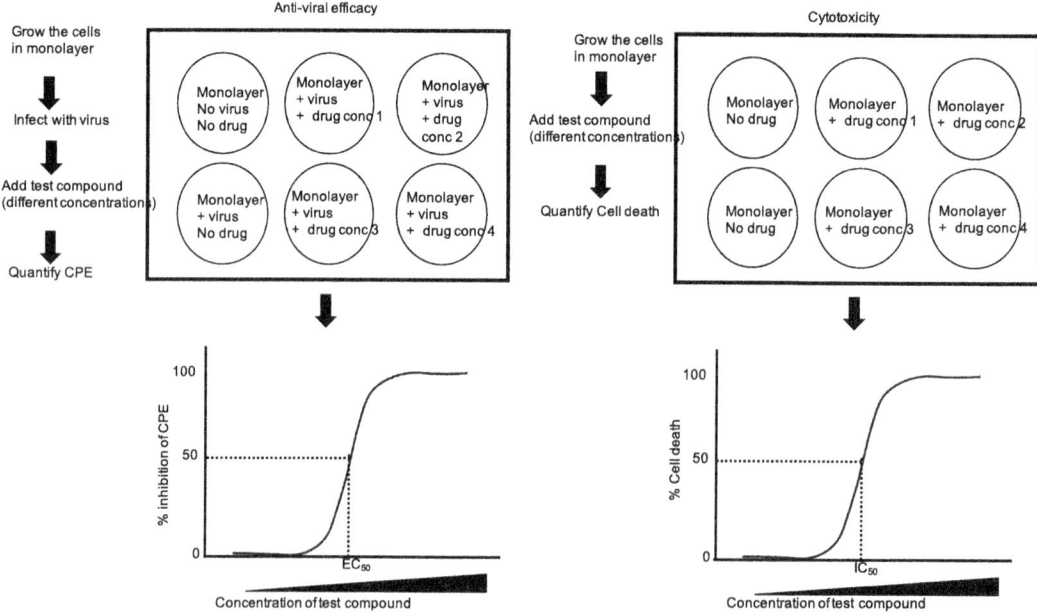

FIGURE 28.2   Sketch of CPE inhibition assay. Monolayer of suitable cells are infected with virus. CPE (observable changes in virus infected cell) are quantified using suitable method (for example, uptake of neutral red dye by live cells). Inhibition of CPE values (due to test drug) are plotted with the test drug concentrations. EC50 value can be calculated by regression analysis. Similarly the cytotoxicity is determined in parallel and CC50 value calculated. Selectivity index (SI) (defined by CC50 /EC50) of more than three is considered good. The higher the SI ratio, the more effective and safe a drug would be.

reduced cost. CPE is determined by microscopic observation of cell culture monolayers as well as uptake of neutral red dye (live cells take up this dye). When a potential antiviral compound has been identified using the CPE-based assay, it will need to be subjected to the ten-concentration dose-response assay to determine the range of antiviral efficacy and cytotoxicity. The antiviral efficacy, represented as the 50% inhibitory concentration (IC50) or the 50% effective concentration (EC50), is the concentration of a drug that inhibits virus-induced CPE halfway between the baseline and maximum. The cytotoxicity of the antivirals, i.e., the 50% cytotoxicity concentration (CC50), is the concentration of a drug inducing 50% of cytotoxicity between the baseline and maximum. The selective index (SI), denoted as 50% SI (SI50) is calculated from CC50/EC50, which determines the specificity of the antiviral against virus-induced CPE. The IC50 (or EC50), CC50, and SI50 values are critical measures to determine whether an antiviral compound is potent and selective for further drug development (Fig. 28.2).

28.2.1.2 Plaque reduction assay

Although molecular tests such as RT-PCR-detecting viral DNA/RNA can be used to quantify viral loads in clinical samples and virus titers of cell culture supernatants, they cannot quantify infectious virus. Plaque assay is used to quantitate the replication competent live virus titer. It is developed by Renato Dulbecco in 1952 for the animal viruses (Nobel Prize in 1975). Plaque reduction assay measures the plaque forming efficiency of a virus in the presence of different concentrations of a test compound (Fig. 28.3). In this assay, after virus adsorption, an immobilizing overlay is used to cover the infected monolayer. This isto restrict virus growth at the sites of initial infection. During incubation, zones of cell death develop due to viral infection and replication, leading to plaque formation. After incubation, cells are stained to enhance the contrast between plaques and the uninfected monolayer. Depending on the type of the overlay there are variations of the plaque assay. Here a basic plaque assay protocol (in brief) is described as follows [7,8]. Virus host cell are seeded in six well culture dishes (approximately 1×10^6 cells per mL)

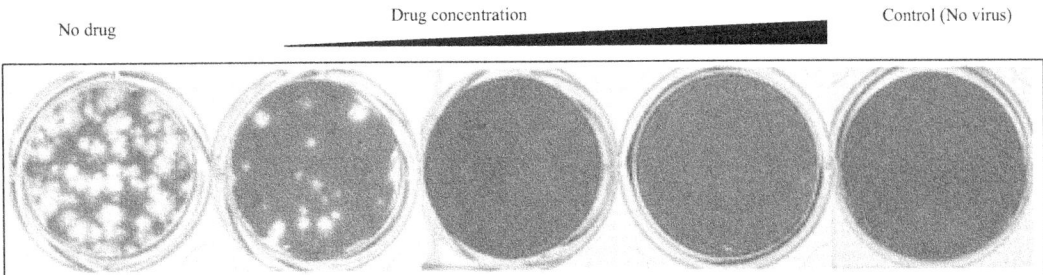

FIGURE 28.3 A typical plaque reduction assay. Monolayer of MDCK cells were infected with influenza virus (equal amount in each well). After viral adsorption, cell monolayer was covered with overlay medium containing drug compound (increasing concentrations) and further cultured. Overlay medium was removed, and the cell monolayer was fixed, stained with 1% crystal violet, and plaques were counted [9]. Source: *Adapted from Parhira S, Yang Z-F, Zhu G-Y, Chen Q-L, Zhou B-X, Wang Y-T, et al. In vitro anti- influenza virus activities of a new lignan glycoside from the latex of* Calotropis gigantea. *PLoS One 2014;9(8):e104544.*

and grown to atleast 80% confluency. Various dilutions of virus are made (10^{-1} to 10^{-8}) in the appropriate culture medium without FBS. The culture medium is aspirated out and the cell are washed twice with PBS. Then add virus dilution (500 μL) to each well. In the negative control do not add the virus. Incubate the infected monolayer for one hour. Prepare 3% agarose solution (dH_2O) and sterlize by autoclaving. Keep the melted agarose at 45°C incubator. Dilute the melted agarose ten times in the growth medium, mix it and maintain at 42°C in waterbath. Take out the virus infected monolayer from the incubator and aspirate out the medium. Add 2 mL of the agarose medium mixture gently over the monolayer. Allow the agarose to solidify for 15 minutes at room temperature and then move the plate to incubator. Plaque visualization is done after two days. Fix the cells by 3.65% formaldehyde at room temperature for one hour. Discard the formaldehyde and agarose. Wash the wells with water. Stain the monolayer with 5% crystal violet for 5 minutes. Wash out the stain with water. The plaques will be visible as white patches in blue background. Choose the well which have 5–100 plaques in one well. Note the virus dilution used for this well. The following formulae will be used to calculate plaque forming units (PFU)/mL (original stock) = no. of plaques/dilution factor X volume of viral dilution used (in mL). This method can be performed in the persence of various test antiviral coumpounds. The compund can be applied to the cells before the virus infection or after the virus infection. Using the above method IC50 value of particular test compound can be determined. The compound showing the better IC50 values can be used for further studies.

Plaque reduction neutralization test (PRNT), a variation of this assay, is used to determine the presence and concentration of neutralizing antibodies in a serum sample or antibody solution. In this assay, the antibody sample is incubated with the virus and then this mixture is added to cell culture monolayers . The cell monolayer is overlaid (with agar or carboxymethyl cellulose) to prevent spreading of the virus. The concentration of serum or antibody solution needed to reduce the number of plaques by 50% is calculated as the PRNT50. The number of plaques are determined by either visual observation or by staining the cell monolayer.

28.2.1.3 Hemagglutination inhibition assay

This assay can be used to find drugs that inhibit the hemagglutination of RBCs by viruses having HA proteins (e.g., influenza virus). This assay can also be used to determine the neutralizing antibody dilutions. When added to a round-bottom 96-well microplate, the RBCs will normally settle to the bottom of the plate forming a small "button" (Fig. 28.4). When an influenza virus sample is added to the RBCs, then it will hemagglutinate the RBCs (forms a lattice-type structure) that prevents the formation of "button". If a sample containing antiviral compound/antibodies (against the influenza virus) is placed in the micro-well prior to addition of virus andthe RBCs, the antibodies (in the serum sample) will bind the influenza virus and thus inhibit the hemagglutination process leading to the re-appearance of "button". Visual observation of the plate can determine an Hemagglutination inhibition assay (HAI) titer that indicates the presence of influenza virus neutralizing antibodies. While less accurate than a plaque assay, it is cheaper and quicker (taking just 30 minutes) (Fig. 28.4).

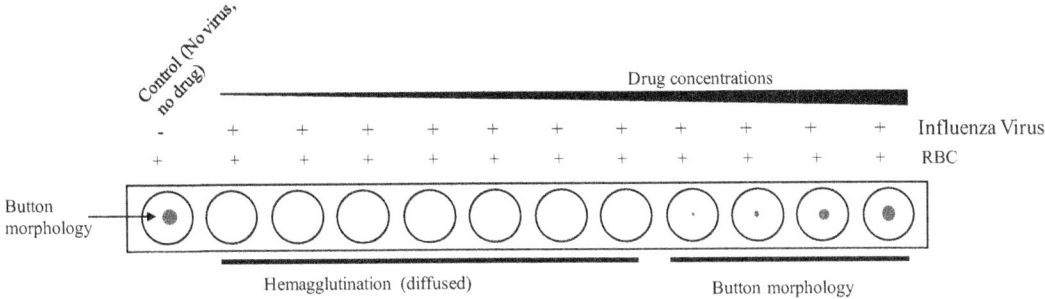

FIGURE 28.4 Sketch showing inhibition of hemagglutination of RBCs. Influenza virus inhibits the button morphology of RBC due to hemagglutination. Virus is incubated with different dilutions of drug in the wells (~30 min). Then the chicken RBCs are added. At a certain drug concentration, the virus loses its ability to agglutinate indicating HA protein interaction with the drug.

28.2.1.4 Cell-based immunodetection assay

Plaque reduction assay and CPE, are considered as the gold standard for antiviral screening. These assays are cell-based and use live viruses therefore require high-level biosafety containment. These assays also require experienced technicians, and high-turnaround times. Cell-based immnunodetection assay is a fast and accurate cell-based immunodetection assay (IA) allowing quantification of viral antigen by monoclonal antibody. For example, IA has been developed to detect Zika and Dengue antigen by a specific monoclonal antibody to E protein domain II [10]. Briefly, Human cell lines (A549, Huh7, and LN-18) are grown in a 96-well plate to obtain 90% confluence. After which, virus is adsorbed to target cells (for 1 hour at 37°C, in 5% CO_2 humidified chamber). After virus removal, serial dilutions (0.03−100 μM) test drug are added to the cell culture media and the plates are incubated at 37°C with 5% CO_2 for 48−96 hours. For the immunodetection of virus antigen, the supernatant was removed and cells were fixed (10% formaldehyde, 30 minutes), rinsed with 1% PBS, and permeabilized with 1% Triton X-100 (for 10 minutes). Cells are incubated (for 1 hour) with monoclonal mouse antibodies to "E protein domain II" (1:400 dilution in PBS containing 1% BSA and 0.1% Tween 20). After washing, cells are incubated (for 1 hour) with a polyclonal horseradish peroxidase-coupled antimouse IgG secondary antibody. Cells are washed and the 3,3′,5,5′-tetramethylbenzidine substrate is added to each well. After 15 minutes of incubation in the dark, the reaction is stopped with one volume of 0.5 M sulfuric acid. Absorbance is measured at 450 nm. The IC50 value can be calculated plotting the drug concentration and % virus replication. The well which is infected with the virus but does not have any drug is considered 100% virus replication. This assay can also be used to distinguish early and late antivital effects [10]. A related assay is cell-based immunofluorescence assay in which immunofluorescent antibodies are used to directly visualize (using immunofluorescence microscope) the virus inside the virus infected cell [11] in the presence of different concentration of antiviral drugs.

28.2.2 Biochemical assays

For many human viruses (e.g., hepatitis B and hepatitis C viruses), no satisfactory cell culture or animal models exist, and in such cases, inhibition of an essential viral function

TABLE 28.2   Some of biochemical in vitro assays used to test efficacy of antiviral drugs.

S. No	Biochemical test	References
1.	Polymerase inhibitor	[23,24]
2.	Methyltransferases inhibitor	[25]
3.	Helicase inhibitor	[26,27]
4.	Integrase inhibitor	[28,29]
5.	Protease inhibitor	[30–33]

or activity against related viruses can be used to indicate potential activity [5]. An alternative approach is to test the effect of test drug compounds in vitro on target viral proteins to determine if they can inhibit enzymes such as viral DNA/RNA polymerases, proteases or block key protein—protein interactions [5,20]. Recombinant viral proteins can safely be produced in larger amounts and used in a variety of assays using biophysical methods such as surface plasmon resonance. There are various in vitro biochemical assays depending on the particular virus (Table 28.2). These assays are reviewed by Michaela Rumlováa, 2017 [21].

28.2.3 Neuraminidase inhibition assay

Neuraminidase (NA) releases influenza virus by cleaving the sialic acid from the galactose on the cell surface. The NA inhibitors based influenza antivirals were rationally designed to bind tightly to the NA enzymatic active site, thereby preventing the release and spread of virus progeny. Nuraminidase enzyme (NA) cleaves the 2'-(4-Methylumbelliferyl)-α-D-N-acetylneuraminic acid (MUNANA) substrate to release the fluorescent product 4-methylumbelliferone (4-MU) [22]. Using this assay, the inhibitory effect of an NA inhibitor (on the influenza virus NA) is estimated. Based on this IC50 value of inhibitor concentrated can be calculated.

28.3 In vivo assays approaches

The use of animal models in drug discovery is a need for target validation, estimation of efficacy, margin of safety, toxicity, metabolism, and pharmacokinetics of a new drug. Regulatory authorities require data on these parameters in appropriate animal models for evaluating the new drug application (NDA) [5]. Various animal models can be used for different virus infections such as mouse, guinea pigs, ferrets, rabbit, primates, etc. The important parameter in in vivo testing is to establish a virus infection at the appropriate site in animal model. The primary consideration while choosing animal model of infection is that it should be able to recreate all/maximum aspects of the human disease/latency. This may be difficult to achieve for many of the infections. The other considerations include ease of handling (smaller animals are preferred), cost-effectiveness, and ethical issues. For infection, an appropriate viral dose should be used. Generally, it is expressed

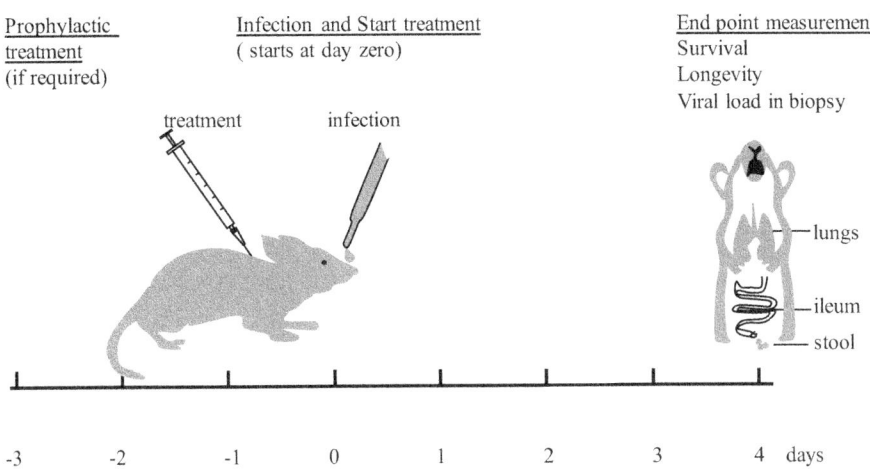

FIGURE 28.5   General outline of in vivo efficacy test. An appropriate animal model is infected at the appropriate site. Drug treatment can start before, same time or after the infection. End point measurement (survival, longevity, viral load, etc.) should show the significant difference between control and treated group of animals.

in LD50 (the amount which kills 50% of infected animals) or PFU. Depending on the virus, it can be given through various routes such as injection, ingestion, inhalation, intranasal, eye infection, topical application, etc. The infection should be confirmed by specific assay (e.g., RT-PCR) or by measuring viral titers. The drug intervention can be prophylactic (starting before infection), starting at the same time as infection, or can be postinfection. The drug dose may be determined emperically and it can be given in the form of gavage, injection, inhalation, drops, topical application, etc. For the evaluation of efficacy of the drug there should be a specific endpoint measurement. It should show a significant desired difference in the treated and control groups. The virus-infected organs are harvested after sacrificing the animal. The viruses are purified by either mincing the tissue, or treating with colagenase and freeze-thawing in media. Titers are measured by any of the in vitro tests such as immunofluorescence assay on tissue sections, plaque reduction assay and quantitative PCR. Indirect measure of inhibition of viral infection can also be done by monitoring the fever, activation of immune response, inflammatory cytokines and mortality post virus infection. The outline of general in vivo efficacy test is shown in Fig. 28.5. Specific in vivo efficacy tests for various viruses are summarised (Table 28.3) and are described below.

28.3.1 Coronavirus (SARS-CoV-2)

SARS-CoV-2 is a single-stranded RNA virus (+ssRNA) that causes severe illness in humans. The clinical symptoms associated with SARS-CoV-2 may include mild illness, pneumonia (mild to severe)and Acute Respiratory Distress Syndrome involving sepsis, septic shock, and multiorgan failure. It is very urgent to identify new drugs for the treatment of the SARS-CoV-2 outbreak. Although Rhesus and Cynomolgus macaques are relevant models for studying the early stages of COVID-19 in humans [34], Syrian hamsters (smaller animals) are preferred due to cost-effectiveness and ethical considerations [35,36].

TABLE 28.3   Animal models for testing efficacy of drugs against some common viral infections.

Sr. no.	Virus	Animal model	References
1.	SARS-Cov-2	Syrian hamsters (*Mesocricetus auratus*)	[37]
2.	Herpes simplex virus type 2	Mouse and guinea pig models	[42]
3.	Influenza	Mouse, cotton rats, guinea pigs, and ferrets	[42]
4.	Polio virus	Mice	[43,44]
5.	HCV	Chimpanzee, human liver chimeric mouse	[18]
6.	HIV	Transgenic mice	[40]

Syrian hamsters (*Mesocricetus auratus*), are highly permissive to SARS-CoV-2 and develop bronchopneumonia and strong inflammatory responses in the lungs with neutrophil infiltration and edema [33,37]. The animals are housed and manipulated in isolators in a biosafety level-3 facility, with ad libitum access to water and food. Before manipulation, animals should be acclimatized for one week. Briefly, index hamsters (6—10 week old) are infected intranasal by 50 μL of virus. Hamsters are euthanized at day 4 pi and the viral load in lung, ileum, and stool is determined using RT PCR. The infective load of the virus can be determined by virus titer assay. Using Syrian hamster model, it was discovered that favipiravir and ivermectin have potent anti-SARS-CoV-2 activity [35].

28.3.2 Herpes virus

Herpes virus is an enveloped virus having icosahedral capsid and a relatively large (184 kbs), linear dsDNA genome. Two kind of herpes simplex virus; HSV-1 and HSV-2 cause infection in humans of which HSV-1 is highly contagious. It causes oral herpes (cold sores) and transmitted by oral-to-oral contact. HSV-2 causes genital herpes which is characterized by genital blisters/sores. It also often includes fever, bodyaches, and swollen lymph nodes. It is transmitted through genital contacts during sex. According to WHO data, in 2016, an estimated 3.7 billion people under the age of 50, had HSV-1 infection (oral or genital). Sometimes, HSV-1 infection can also lead to encephalitis (brain infection) or keratitis (eye infection). Although the drugs acyclovir, famciclovir, and valacyclovir, are the most effective treatment for HSV infection but it cannot cure the infection. People who contract HSV will have the virus for the rest of their lives (in nerve cells). Development of better treatment , as well as increasing awareness about HSV infection and its symptoms are needed. Due to broad host range of HSV different animal models such as mice, rabbit, guinea pigs, etc. are available [38]. The ideal model should demonstrate to establish a localized initial HSV infection, neuronal spread and establishment of latency. The most extensively used mouse model is cost effective and easy to handle but suffers from the lack of efficient in vivo reactivation. The model for latency must show virus reactivation similar to humans. Both rabbit and the guinea pig are appropriate for this ideal situation with some limitations [38]. For testing the efficay of the drugs on the cutaneous herpes (skin), guinea pig model is used [38]. First of all, the possibility of skin irritation of the test compound has to be determined. Guinea pigs (male and female in equal numbers) of body weight around 200–250 g are used. Body hair are removed from the dorsal side and thenaked skin (6 × 7 cm) is washed (with warm water), dried, and abraded (with dermal Seven-Star needles). Then the test compound is applied to the area directly or in a cream form. After 24 hour, the creams are removed (with warm water) and the animals are examined for erythema and edema 1 hour later. The animals are observed for the next 72 hour. No sign of edema or erythema in any of the animals indicating that the test compound has no dermal toxicity, therefore the test compound can be tested for the efficacy. To study the efficacy of test compound on HSV-1 induced cutaneous lesion, the dorsal skin of guinea pigs can be prepared and abraded as above. The abraded area should be divided into four quadrants and each of the quadrants will be infected with 30 μL of tenfold diluted HSV-1. The animals will be observed for typical herpes lesion development (upto 10 days). On the basis of above results, the amount of stock virus can be determined (usually 150 μL of 10^8 PFU) to infect 42 cm^2 skin area. The cohorts of animals ($n = 15$) will be infected as above and treated with the (i) test drug or cream based on it, (ii) 3% (wt./wt.) acyclovir cream (positive control), or (iii) base cream (negative cream) to the infected area with sterile cotton swabs twice daily for a 6-days period. The extent of lesion will be scored daily as follows- (a) lesions on 1/4 of infected area score 1.0–1.6, (b) lesions on 1/2 of infected area score 1.7–2.4, (c) lesions on 3/4 of infected area score 2.5–3.2, (d) lesions on the entire infected area score 3.3–4.0. Generally, all the animals develop typical herpes lesions on day 4 postinfection and the lesion scores on day 4 are very similar in all groups. The % inhibition of lesion is calculated by the following formulae = [(lesion score of virus control group − lesion score of treatment group)/(lesion score of virus control group)] × 100 [34].

28.3.3 Influenza virus

Influenza viruses (types A, B, C and thogotovirus) are enveloped and belong to the family Orthomyxoviridae. The genome consists of segmented single-strand RNA (negative-sense) segments. The influenza types A and B (clinically relevant for humans) have eight genome segments that are loosely encapsidated by the nucleoprotein. The influenza viruses, especially influenza A, evolve rapidly which leads to its great variability. The disease (flu or influenza) spreads from person to person through the air (via coughs or sneezes) or through contact with infected surfaces. The drugs being used for treatment oseltamivir and zanamivir act by inhibiting the essential NA protein. For checking the efficacy of anti-influenza test drugs, the animal model, mice are intranasally (i.n.) inoculated with 0.1 mL of mouse-adapted influenza virus at 10 × MID50 (50%-mouse infectious dose; sublethal). Mice are treated by drugs (positive control, test drug, negative control) by gavage with 20 mg/kg at day 1 and 3 postinfection or 10 mg/kg two times daily after infection. The mice are sacrificed at day 2 and 4 pi, and BAL fluids are collected to determine the amount of secreted virus and pro-inflammatory cytokines. For determination of virus titre in lungs (at day 5 or 6), mice were euthanized and lungs are collected. For the survival study, mice were infected intranasally with 10 × LD50 (50%-lethal dose) and treated with 5 or 10 mg/kg verdinexor (test drug) at day 1 and 3 or 10 mg/kg oseltamivir (positive control) twice daily for 4 days starting at day 1 pi. Mice were evaluated for weight, clinical signs, and survival for 14 days [39].

28.3.4 Human immunodeficiency virus

The human immunodeficiency viruses (HIV-1 and HIV-2) are species of *Lentivirus* (a subgroup of retrovirus) that infect humans. HIV is roughly spherical (diameter ~ 120 nm) having two copies of $+$ssRNA that codes for the virus's nine genes. It is enclosed by a conical capsid composed of protein p24. The single-stranded RNA is tightly bound to proteins p7, reverse transcriptase, proteases, ribonuclease, and integrase. HIV is mainly a sexually transmitted infection but nonsexual transmission can also occur. HIV infects CD-4 T-cells of the immune system due to which the progressive failure of the immune system allows life-threatening opportunistic infections and cancers to thrive. In vivo testing (preclinical testing) of anti-HIV drugs can be done using simian immunodeficiency virus/HIV and rhesus macaque model. The use of monkies is severly limited by ethical concerns in using nonhuman primates, high cost, and the problem in animal handling. Since native mice and rats are nonpermissive for HIV therefore the in vivo testing is hampered due to lack of a small animal that is readily available, easy to handle. Sprague–Dawley rats that transgenically express the HIV-1 receptor complex on CD4 T cells are used as a model for the preclinical evaluation of inhibitors [40]. A human CD4/CCR5 transgenic rat model that express the transgenes in CD4 T lymphocytes, macrophages, and microglia has been extensively used to analyze different aspects of HIV infection and treatment. Infection is established through tail vein. A test drug is administered in treated group of animals and PBS is given to the control group animals. HIV load is determined using cDNA from splenocytes by quantitative PCR analysis. Prophylactic treatment can also be tested using the similar approach.

28.3.5 Hepatitis B virus

Hepatitis B which causes acute and chronic infection of the liver, remains a major public health concern. The currently approved drugs lamivudine, adefovir dipivoxil, entecavir, telbivudine, and tenofovir suppress Hepatitis B virus (HBV) replication. There is a significant need for new anti-HBV drugs. High expenses, animal ethics, and availability issues limit the use of Chimpanzees (in vivo model for HBV) for investigating virologic response and the progression of HBV-related liver diseases. Therefore, there is a motivation to develop mice model. Recently, in vivo anti-HBV activity of KCT-01 (an herbal formulation) was evaluated in mice [41].

Infection was established by injecting HBV genome hydrodynamically. KCT-01 was administered by oral gavage into the stomach. The endpoint measurement was by quantitation of genomeic DNA using qPCR in serum and in liver tissue samples [41] (Table 28.3).

28.4 Conclusion

The pausity of treatment of viral diseases is a major health concern of our times. Therefore, there is constant search for new antiviral drugs. Using the rational drug discovery pipeline, there is an inherent need to show data on the efficacy of new antiviral drugs. This is done using the appropriate in vitro and in vivo methods. In vitro testing is done using the cell-based and biochemical assay systems. In vivo testing is done using appropriate animal models. Plaque reduction assay and inhibition of CPE are the most commonly used in vitro approaches. Other assays include HAI, immunodetection assay, and biochemical assay, such as inhibition of polymerase, protease, and reverse transcriptase activity.

In addition to efficacy, in vivo assay also provides information on the toxicity, mechanism, drug dosing, and pharmacokinetics. Depending on the virus there are various model systems. Smaller animal is preferred due to cost effectiveness, ease of handling, and ethical concerns.

Acknowledgment

Author is recepient of the grants from Science and Engineering Research Board India (SERB-SRG-2020/001904) and University Grants Commission India (UGC-BSR No. F.30–545/2021).

References

[1] Tompa DR, Immanuel A, Srikanth S, Kadhirvel S. Trends and strategies to combat viral infections: a review on FDA approved antiviral drugs. Int J Biol Macromol 2021;172:524–41.

[2] Adamson CS, Chibale K, Goss RJM, Jaspars M, Newman DJ, Dorrington RA. Antiviral drug discovery: preparing for the next pandemic. Chem Soc Rev 2021;50(6):3647–55.

[3] Phases of drug development process, Drug discovery process. Nebiolab.com, <https://www.nebiolab.com/drug-discovery-and-development-process/>; 2020 [accessed 27.11.21].

[4] Office of the Commissioner. The drug development process. Fda.gov, <https://www.fda.gov/patients/learn-about-drug-and-device-approvals/drug-development-process>; 2020 [accessed 27.11.21].

[5] Center for Drug Evaluation and Research. Antiviral product development–conducting and submitting virology studies to the agency 2006. Fda.gov, <https://www.fda.gov/regulatory-information/search-fda-guidance-documents/antiviral-product-development-conducting-and-submitting-virology-studies-agency>; 2020 [accessed 27.11.21].

[6] Gu L, Schneller SW, Li Q. Assays for the identification of novel antivirals against bluetongue virus. J Vis Exp 2013;(80). Available from: https://doi.org/10.3791/50820.

[7] Mendoza EJ, Manguiat K, Wood H, Drebot M. Two detailed plaque assay protocols for the quantification of infectious SARS-CoV-2. Curr Protoc Microbiol. 2020;57(1) ecpmc105.

[8] Plaque Assay for Influenza Virus. Youtube, <https://www.youtube.com/watch?v = QThpXUHjcZM>; 2013 [accessed 26.11.21].

[9] Parhira S, Yang Z-F, Zhu G-Y, Chen Q-L, Zhou B-X, Wang Y-T, et al. In vitro anti- influenza virus activities of a new lignan glycoside from the latex of Calotropis gigantea. PLoS One 2014;9(8):e104544.

[10] Vicenti I, Dragoni F, Giannini A, Giammarino F, Spinicci M, Saladini F, et al. Development of a cell-based immunodetection assay for simultaneous screening of antiviral compounds inhibiting Zika and dengue virus replication. SLAS Discov 2020;25(5):506–14.

[11] Wang M, Cao R, Zhang L, Yang X, Liu J, Xu M, et al. Remdesivir and chloroquine effectively inhibit the recently emerged novel coronavirus (2019-nCoV) in vitro. Cell Res 2020;30(3):269–71.

[12] Snyder MB, Saravolatz LD, Markowitz N, Pohlod D, Taylor RC, Ward SG. The in- vitro and in-vivo efficacy of cisplatin and analogues in the treatment of herpes simplex virus- II infections. J Antimicrob Chemother 1987;19(6):815–22.

[13] Prichard MN, Kern ER, Hartline CB, Lanier ER, Quenelle DC. CMX001 potentiates the efficacy of acyclovir in herpes simplex virus infections. Antimicrob Agents Chemother 2011;55(10):4728–34.

[14] Hayden FG, Cote KM, Douglas ARG. Plaque inhibition assay for drug susceptibility testing of influenza viruses antimicrobial agents and chemotherapy. 1980;17:865–70.

[15] Muratore G, Goracci L, Mercorelli B, Foeglein Á, Digard P, Cruciani G, et al. Small molecule inhibitors of influenza A and B viruses that act by disrupting subunit interactions of the viral polymerase. Proc Natl Acad Sci U S A 2012;109(16):6247–52.

[16] Shi W-Z, Jiang L-Z, Song G-P, Wang S, Xiong P, Ke C-W. Study on the antiviral activities and hemagglutinin-based molecular mechanism of novel chlorogenin 3-O-β- chacotrioside derivatives against H5N1 subtype viruses. Viruses 2020;12(3):304.

[17] Oberste MS, Moore D, Anderson B, Pallansch MA, Pevear DC, Collett MS. In vitro antiviral activity of V-073 against polioviruses. Antimicrob Agents Chemother 2009;53(10):4501−3.

[18] Gerold G, Pietschmann T. The HCV life cycle: in vitro tissue culture systems and therapeutic targets. Dig Dis 2014;32(5):525−37.

[19] Patience C, Moore J, Boyd M. In vitro assessment of compounds for anti-HIV activity. Mol Biotechnol 1994;1 (1):49−58.

[20] Schinazi RF, Bassit L, Gavegnano C. HCV drug discovery aimed at viral eradication. J Viral Hepat 2010;17 (2):77−90.

[21] Rumlová M, Ruml T. In vitro methods for testing antiviral drugs. Biotechnol Adv 2018;36(3):557−76.

[22] Leang S-K, Hurt AC. Fluorescence-based neuraminidase inhibition assay to assess the susceptibility of influenza viruses to the neuraminidase inhibitor class of antivirals. J Vis Exp 2017;(122). Available from: https:// doi.org/10.3791/55570.

[23] Amraiz D, Zaidi N-U-SS, Fatima M. Development of robust in vitro RNA-dependent RNA polymerase assay as a possible platform for antiviral drug testing against dengue. Enzyme Microb Technol 2016;92:26−30.

[24] Campagnola G, Gong P, Peersen OB. High-throughput screening identification of poliovirus RNA-dependent RNA polymerase inhibitors. Antiviral Res 2011;91(3):241−51.

[25] Coutard B, Barral K, Lichière J, Selisko B, Martin B, Aouadi W, et al. Zika virus methyltransferase: structure and functions for drug design perspectives. J Virol 2017;(5):91. Available from: https://doi.org/10.1128/jvi.02202-16.

[26] Cao X, Li Y, Jin X, Li Y, Guo F, Jin T. Molecular mechanism of divalentmetal-induced activation of NS3 helicase and insights into Zika virus inhibitor design. Nucleic Acids Res 2016;44(21):10505−14.

[27] Tani H, Fujita O, Furuta A, Matsuda Y, Miyata R, Akimitsu N, et al. Real-time monitoring of RNA helicase activity using fluorescence resonance energy transfer in vitro. Biochem Biophys Res Commun 2010;393 (1):131−6.

[28] Han Y-S, Xiao W-L, Quashie PK, Mesplède T, Xu H, Deprez E, et al. Development of a fluorescence-based HIV-1 integrase DNA binding assay for identification of novel HIV-1 integrase inhibitors. Antiviral Res 2013;98(3):441−8.

[29] He HQ, Ma XH, Liu B, Zhang XY, Chen WZ, Wang CX, et al. Highthroughput real- time assay based on molecular beacons for HIV-1 integrase 3′-processing reaction. Acta Pharmacol Sin 2007;28(6):811−17.

[30] Konvalinka J, Kräusslich H-G, Müller B. Retroviral proteases and their roles in virion maturation. Virology 2015;479−480:403−17.

[31] Midde NM, Patters BJ, Rao P, Cory TJ, Kumar S. Investigational protease inhibitors as antiretroviral therapies. Expert Opin Investig Drugs 2016;25(10):1189−200.

[32] Foote BS, Spooner LM, Belliveau PP. Boceprevir: a protease inhibitor for the treatment of chronic hepatitis C. Ann Pharmacother 2011;45(9):1085−93.

[33] Pillaiyar T, Manickam M, Namasivayam V, Hayashi Y, Jung S-H. An overview of severe acute respiratory syndrome−Coronavirus (SARS-CoV) 3CL protease inhibitors: peptidomimetics and small molecule chemotherapy. J Med Chem 2016;59(14):6595−628.

[34] Maisonnasse P. Hydroxychloroquine in the treatment and prophylaxis of SARS-CoV- 2 infection in non-human primates. Res Sq 2020;10 21203(RS):27223 1.

[35] de Melo GD, Lazarini F, Larrous F, Feige L, Kergoat L, Marchio A, et al. Anti-COVID- 19 efficacy of ivermectin in the golden hamster. bioRxiv 2020. Available from: https://doi.org/10.1101/2020.11.21.392639.

[36] Kaptein SJF, Jacobs S, Langendries L, Seldeslachts L, Ter Horst S, Liesenborghs L, et al. Favipiravir at high doses has potent antiviral activity in SARS-CoV-2-infected hamsters, whereas hydroxychloroquine lacks activity. Proc Natl Acad Sci U S A 2020;117(43):26955.

[37] Boudewijns R, Thibaut HJ, Kaptein SJF, Li R, Vergote V, Seldeslachts L, et al. STAT2 signaling restricts viral dissemination but drives severe pneumonia in SARS-CoV-2 infected hamsters. Nat Commun 2020;11(1):5838.

[38] Chattopadhyay D, Sarkar MC-, Chatterjee T, Sharma Dey R, Bag P, Chakraborti S, et al. Recent advancements for the evaluation of anti-viral activities of natural products. N Biotechnol 2009;25(5):347−68.

[39] Perwitasari O, Johnson S, Yan X, Register E, Crabtree J, Gabbard J, et al. Antiviral efficacy of verdinexor in vivo in two animal models of influenza A virus infection. PLoS One 2016;11(11):e0167221.

[40] Goffinet C, Allespach I, Keppler OT. HIV-susceptible transgenic rats allow rapid preclinical testing of antiviral compounds targeting virus entry or reverse transcription. Proc Natl Acad Sci U S A 2007;104(3):1015–20.

[41] Kim H, Jang E, Kim S-Y, Choi J-Y, Lee N-R, Kim D-S, et al. Preclinical evaluation of in vitro and in vivo antiviral activities of KCT-01, a new herbal formula against hepatitis B virus. Evid Based Complement Alternat Med 2018;2018:1073509.

[42] Edwin Yunhao, G., editor. Antiviral methods and protocols. Meth. Mol. Biol. 2013;1030. DOI 10.1007/978-1-62703-484-5_24.

[43] McKinlay MA, Miralles JV, Brisson CJ, Pancic F. Prevention of human poliovirus-induced paralysis and death in mice by the novel antiviral agent arildone. Antimicrob Agents Chemother 1982;22(6):1022–52.

[44] Gong EY, editor. Antiviral methods and protocols. Totowa, NJ: Humana Press; 2013.

Clinical Trials and Regulatory considerations of Antiviral agents

Samir Bhargava[1], Bhavna[1], Neeraj Sethiya[1], Amal Kumar Dhara[2], Jagannath Sahoo[1], H. Chitme[1], Mayuri Gupta[1], Navraj Upreti[1] and Yusra Ahmad[3]

[1]Faculty of Pharmacy, DIT University, Dehradun, Uttarakhand, India [2]Department of Pharmacy, Contai Polytechnic, Purba Medinipur, West Bengal, India [3]Faculty of Pharmacy, Uttarakhand Technical University, Dehradun, Uttarakhand, India

Abbreviations

ARDS	Acute Respiratory Distress Syndrome
ART	Anti Retro Viral
CI	Confidence Interval
CMV	Cytomegalovirus
DAV	Division of Antiviral
DAVP	Division of Antiviral Products
DNA	deoxyribonucleic acid
DOR	Doravirine
DRM	Division of Risk Management
EBV	Epstein−Barr Virus
EFV	Efavirenz
FTC	Emtricitabine
HBV	Hepatitis B virus
HCV	Hepatitis C virus
HDL-c	High density lipoprotein cholesterol
HDV	Hepatitis D virus
HIV	Human Immunodeficiency Virus
HSCT	Hematopoietic stem-cell transplantation
HSV-1	Herpes Simplex Virus-1
HSV-2	Herpes Simplex Virus-2
LDL-c	Low density lipoprotein cholesterol
NAIs	neuraminidase inhibitors

Viral Infections and Antiviral Therapies
DOI: https://doi.org/10.1016/B978-0-323-91814-5.00021-0

NDA	New Drug Application
NRTIs	Nucleoside reverse transcriptase inhibitors
NNRTIs	Nonnucleoside reverse transcriptase inhibitors
OALWH	Older Adults living with HIV
PCR	Polymerase chain reaction
PTLD	Posttransplant lymphoproliferative disease
PrEP	Pre-Exposure Prophylaxis
RAMs	Resistance-associated mutations
RAS	Resistance-associated substitutions
REMS	Risk Evaluation and Mitigation Strategy
TDF	Tenofovir disoproxil fumarate
3TC	Lamivudine

29.1 Introduction

Antiviral drugs are the molecules, inhibiting the regulation of virus in the human cellular system. In past decade many plant metabolites have also been researched for potential source of antiviral agents, based on usage depicted in traditional systems [1−3]. It is important to know that virus undergoes rapid mutation in their nucleic acids. This problem needs to be rectified through proper dosage forms. Antiviral agents, which are in initial phases for IND Application, are required to pass through stringent regulatory requirements for better patient compliance, improved pharmacokinetics and low toxicity. The chapter provides a collected evidence of the undergoing clinical trials in recent years on different types of antiviral agents or their combinations.

29.2 Classification of antiviral agents

The major category of antiviral agents has been subdivided according to their chemical nature or mechanisms of drug actions against specific viral proteins. The classification has been done based on recent several scientific usage of antiviral drugs and effect obtained through their synergistic blends [4]. Based on our research, antiviral agents can be classified [5] as

1. **Category 1—Chemical analog of nucleoside moiety within viral DNA or RNA. These chemical molecules are either synthetic analogs or are converted by an enzyme to a metabolic derivative, which incorporates in virus and inhibits its replication.**
   - Anti-Herpes Virus Drugs—Idoxuridine; Trifluridine; Acyclovir; Valacyclovir; Famciclovir
   - Anticytomegalovirus (CMV) Drugs—Ganciclovir; Valganciclovir; Cidofovir
   - Anti-Hepatitis Viral Drugs—Entecavir; Adefovir dipivoxil; Tenofovir disoproxil fumarate
   - Anti-Hepatitis C Drug—Ribavarin
   - Anti-HIV Drug—Abacavir (NRTIs)
   - Anti-Influenza Virus Drugs—Molnupiravir
2. **Category 2—Enzymatic blockers of viral replication supporters. These molecules bind with clefts within viral enzymes, by inhibiting the potential viral replication and assembly.**
   - Anti-CMV Drug—Foscarnet
   - Anti-Influenza Virus Drugs—Oseltamivir; Zanamivir; Peramivir; Baloxavir marboxil

- Anti-Hepatitis Drugs—Sofosbuvir (NS5B polymerase inhibitor); Simeprevir; Glecaprevir; Voxilaprevir (NS3/4A Protease inhibitor); Daclatasvir, Ledipasvir, Velpatasvir (NS5A inhibitors); Morphothiadin
- Anti-HIV Virus—Zidovudine, Didanosine, Stavudine, Lamivudine, Emtricitabine (NRTIs); Nevirapine, Efavirenz, Rilpivirine, Doravirine Delavirdine, Etravirine; Islatravir (NNRTIs); Atazanavir, Indinavir, Nelfinavir, Saquinavir, Ritonavir, Lopinavir, Fosamprenavir, Darunavir (Retroviral protease inhibitors); Raltegravir, Dolutegravir Elvitegravir (Integrase inhibitors); Enfuvirtide [Entry (fusion) inhibitor]; Cabotegravir
- Others—Interferon α; Small interfering RNAs

29.3 Clinical trials and Food and Drug Administration in the development of antiviral agents

Clinical trials lead to in-depth examination of new or modified molecules. These trials help to acquire knowledge about new research-based treatment strategies, for everyone's access. This becomes useful, when standard therapies fail over a period of time [6]. Progressive and complex clinical trials have initiated new development strategies for a new product. But many times, clinical trials fails to report sufficient quality, safety, and efficacy issues, due to which concern of FDA came in existence [7]. FDA's guidance documents emphasize on quality issues in clinical trials and some risk-based methods which may help sponsors to avoid futuristic failures. These guidance documents (Table 29.1) are based on ICH Efficacy guidelines, bringing synchronization during study developments. Application submitted by manufacturers to the FDA requires some synergistic advisories and regular reviews. This enables the manufacturer to manufacture the drug for market, followed by its follow-up studies and surveillance. FDA ensures the quality, safety, and efficacy assessment of drug from clinical phases till drug reaches to patient.

29.4 The US regulator (Food and Drug Administration)

The FDA, an agency under US Government, is entitled to enforced rules, regulations, and laws for promotion of public health and encounter rapidly developing revolutions. The FDA encounters and manages its activities through its various Offices. The Office of—"Centre for Drug Evaluation and Research" is responsible for promotion of tasks to explore challenges for overcoming the burden of diseases. The office clarifies the drug developers about different components of application for its submission till its complete assessment. The center issues reports on the activities related to applications received and their status. The center includes 12 Offices, including the Office for New Drug [8]. Office of New Drugs is an intermediary advisor between pharmaceutical Industry and drug user. The office, contacts the Pharmaceutical industry for decisively determining whether risks of approved drugs outweigh uses. The Office includes Office of Infectious Diseases (OID) as one of its review offices. OID is an important office under FDA, responsible for reviewing activities of Division of

TABLE 29.1   Clinical data guidance documents endorsed by Food and Drug Administration (FDA).

Guideline	FDA endorsement
ICH E5	Document enables processing of drug registration for assessing the effect of ethnic factors (Factors effecting people with similar cultural characters) on drug's effect. The guidance enables effect of dosage form and prevents duplication of clinical data by accelerating the approval process.
ICH E3	For Industry Guidance—Structure & Content of Clinical Study Reports This document defines the content of the clinical report, to be acceptable by FDA. This includes the explanations, tables, analytical representations, forms, statistics and related publications. The document assists all stakeholders for ready acceptability.
ICH E6	The guidance document in FDA is applicable for proper designing, conduct and reporting of trials. The documents assures the subjects of their rights and provides unified standard for acceptability in European Union, USA, and Japan
ICH E7	Guidance defining for studies to be conducted in special populations—Geriatrics
ICH E10	Document defines the consideration of ethics in selection of control groups. The selection of control groups effects trial design and demonstration of efficacy. The document minimizes bias, defines recruitments, and kind of endpoints on which credibility is defined.
ICH Q9	The document is intended to provide principles and tools of quality risk management enabling effective and consistent risk-based decisions, but not effecting product stability.
VICH GL9	The guidance document is aimed for conduct of clinical trials related to veterinary medicinal products. The document is focused for all involved in clinical studies, and ensures that studies are as per Good Clinical Practice (GCP).

Antivirals (DAV) [9]. The Table 29.2 includes the list of Antiviral applications (includes drugs only) approved by FDA from 2015 to 2022, to be used in clinical practice.

29.5 Applications submitted to division of antiviral products (US FDA)

The DAV is entitled to review applications related to drugs and therapeutics for treatment of possible and present viral infections. Hence, a communication-based consultation program on Antivirals by DAV for IND Applications is the Pre-IND consultation program on Antivirals. Historically, the program was started in 1988 by OID. The program is an important initiative to support sponsors for developing a scheme toward successful development of Antiviral [30]. Pre IND application helps the sponsor for obtaining an early opinion from FDA and saving unnecessary prospective costs involvements. The sponsor can communicate with division for collecting required data (related to potential required drug candidate) toward successful submission of IND application. However, DAV conducts pre-IND meetings as teleconference meetings, where strong supporting documents are to be provided from sponsor to familiarize FDA [31]. Pre IND application for antiviral drugs must include the following [30]

- Sponsor is required to make complete available information regarding descriptions of chemistry and manufacturing methods using the drug for the finished dosage form.

TABLE 29.2 New Drug Application approved by Food and Drug Administration for different antivirals (2015−22).

Molecule	Viral disease	Dosage form	Applicant application number	Recommended population	Risks involved	References
Remdesivir	COVID-19	Lyophilized formulation (Inj)—100 mg Solution formulation (Inj)—5 mg/mL	Gilead Sciences, Inc. 214787	For adults and pediatric ≥ 12 Years (Weight—AtF least 40 kg) Administered in hospital giving acute care	Hepatoxicity; hypersensitivity	[10]
Baloxavir Marboxil	Influenza	Tablet—20 mg/40 mg	GENENTECH INC 210854 214410	For patients with acute uncomplicated influenza ≥ 12 years of age	Risk of bacterial infection; hypersensitivity	[11,12]
Maribavir	CMV Infection	Tablet—200 mg	Takeda Pharmaceuticals USA, Inc. 215596	Treatment of adults and pediatric patients ≥ 12 Years Weight—At least 35 kg (If CMV Infection is unmanageable by Ganciclovir, Valganciclovir, Cidofovir, or Foscarnet).	Virologic failure; may casue accumulation of immune suppresents	[13]
Letermovir	CMV Virus	Tablets—480 mg (Once daily) IV Infusion—480 mg over 1 hour through 100 days posttransplant	Merck Sharp & Dohme Corp. 209939 209940	For treatment of CMV Infection in adult CMV—seropositive recipients of an allogeneic hematopoietic stem cell transplant patient.	Reduced adverse reactions due to drug interactions	[14]
Combination of Atoltivimab; Maftivimab; Odesivimab-ebgn	Ebola Virus	(Recommended dose—50 mg/drug/kg) Injection IV Use	Regeneron Pharmaceuticals, Inc 761169	Infected with Ebola in adults and oediatrics	Hypersensitivity	[15]
Cabotegravir + Rilpivirine Combination In 2 Kits	HIV-1	Available as Tablet—30 mg Extended-release injectable suspension	VIIV HEALTH CARE Company 212887 212888	For HIV-1 infection in adults with no resistance against cabotegravir or Rilpivirine.	Hypersensitivity; hepatoxicity; Depressive orders	[16,17]
Cabotegravir	For PrEP against HIV-1	Extended-release injectable suspension	VIIV HEALTH CARE Company 215499	For individuals at-risk adults and adolescents Weight ≥ 35 kg for PrEP weighing at least 35 kg for PrEP Individuals must have a negative HIV-1 test	Hypersensitivity; hepatoxicity; depressive orders; development of resistance	[18]

(Continued)

TABLE 29.2 (Continued)

Molecule	Viral disease	Dosage form	Applicant application number	Recommended population	Risks involved	References
Fostemsavir	HIV-1	Extended Release tablet—600 mg	VIIV HEALTH CARE Company 212950	For treatment of HIV-1 infection in adults with multidrug-resistant HIV-infection	Immune reconstitution syndrome; QTc prolongation; Enhancement in hepatic transaminases in HBC/HBB Infected patients	[19]
Combination of Bictegravir; Emtricitabine; Tenofovir alafenamide	HIV-1	Tablets—25 mg/50 mg/200 mg	Gilead Sciences, Inc. 210251	For treatment of HIV-1 infection in adults with no known substitutions associated with resistance to the individual components of this drug.	Immune reconstitution syndrome; renal Impairment; lactic acidosis	[20]
(Ibalizumab-uiyk)	HIV-1	Injection for IV Use 200 mg/1.33 mL (150 mg/mL) in a single-dose vial	TaiMed Biologics USA Corp. 761065	For treatment in Individuals with multidrug-resistant HIV-1 infection failing with their present ARV treatment	Immune reconstitution syndrome	[21]
Doravirine	HIV-1	Tablets—100 mg	Merck Sharp & Dohme Corp. 210806	For treatment of HIV-1 infection (Adults)	Immune reconstitution syndrome	[22]
Combination of Doravirine; Lamivudine; Tenofovir disoproxil fumarate	HIV-1	Tablets	Merck Sharp & Dohme Corp. 210807	For treatment of HIV-1 infection (Adults)	Renal impairment; Immune reconstitution syndrome; bone loss and mineralization defects	[22]
Dolutegravir; Lamivudine	HIV-1	Tablets	VIIV HEALTH CARE Company 211994	For treatment of HIV-1 infection (Adults)	Hypersensitivity; hepatotoxicity; Embryo-fetal toxicity; immune reconstitution syndrome; lactic acidosis	[23]

Drug	Virus	Form	Company/Number	Indication	Adverse Effects	Reference
Combination of Elvitegravir; Cobicistat; Emtricitabine; Tenofovir alafenamide	HIV-1	Tablets	GILEAD SCIENCES INC 207561	For treatment in Individuals (adults and pediatric patients ≥ 12 Years)	Immune reconstitution syndrome; renal impairment; bone loss and mineralization defects; not recommended with other ARVs	[24]
Tenofovir alafenamide	Heaptitis B Virus	Tablets—25 mg	GILEAD SCIENCES INC 208464	For treatment of hepatitis B infection—adults with liver disease	HBV and HIV-1 coinfection; renal impairment; lactic acidosis	[25]
Daclatasvir (discontinued)	Hepatitis C Virus	Tablets—60 mg in combination with Sofosbuvir	BRISTOL-MYERS SQUIBB 206843	For treatment of HCV Genotype 3 infection	Bradycardia; hepatitis B virus reactivation	[26]
Elbasvir and Grazoprevir	Hepatitis C Virus	Tablets (To be administered with Ribavirin)	Merck Sharp & Dohme Corp. 208261	For treatment of HCV Genotype 1–4 in adults and pediatrics	Increase in alanine aminotransferase; hepatitis B virus reactivation	[27]
Sofosbuvir and Velpatasvir	Hepatitis C Virus	Tablets	Gilead Sciences, Inc. 208341	For treatment of HCV Genotype 1–6 in adults and pediatrics	Bradycardia; hepatitis B virus reactivation	[28]
Glecaprevir and Pibrentasvir	Hepatitis C Virus	Tablets	AbbVie Inc. 209394	For treatment of HCV Genotype 1 in adults and pediatrics	Hepatitis B virus reactivation	[29]

- Since pharmacology and toxicological studies of antiviral drugs will formulate an essential component of IND application. Hence clinical studies planned and discoursing about good practices to be included are significant requirements for IND application.
- Research strategy describing the microbiological studies as a preclinical plan. The study must address mechanism, possibility of resistance development and activities in cell-based models. Hence, depending on drug molecule the study must be included as per FDA guidelines. The data to be included in Pre IND applications must be included with strong understanding data.
- Animal studies in such applications form an important evidence for product evaluation and formation of safety sheets.
- Plans for proposed clinical developments impacting study and designs of clinical studies. Sponsors receive guidance in such issues.

It may also be noted that in case of antivirals, request for an emergency IND application can be made by physician based on following [32]

- Urgent requirement of drug by patient due to a life threatening state physician may request for emergency Investigational New Drug Application (INDA).
- If no acceptable substitute therapy is available.
- Patient fails to receive drugs through prevailing trials or expanded protocols.

29.5.1 Investigational new drug application

The application is filed before FDA for assessing the activity of potential drug candidate in humans. By filing an INDA, investigator possess legal rights to transport drug molecule to states, where clinical studies have to be carried out [33]. However once the filed application comes in effect then only the new drug can be administered to human subjects. The INDA application must include

- Preclinical data of animal studies
- Controls for manufacturing the required drug substance
- Protocols; guidance and investigator information

29.5.2 New drug application

The application is a step ahead then IND, as it includes data on animal studies. Sponsors use New Drug Application (NDA) application for requesting FDA to approve a drug for sale and marketing in USA. The information included in application must provide complete details to prove safety and effectiveness of drug in its native state, information to be included on labels and whether information about drug can be revealed from data of manufacturing processes. In addition, FDA also provides few guidance documents as an aid to sponsor and its staff [34]. The sponsors use these directional documents as guidelines for developing content of their applications, and also for regulation-based manufacturing, processing of products.

29.5.3 Abbreviated new drug application

These applications are not comprised of preclinical (animal) and clinical (human) data to establish safety and effectiveness. The application is for potential approval of generic product as a cost alternative to reference drug. Moreover, these applications are required to possess similar bioequivalence data, for further approval by FDA [35]. The guidance documents such as Product-Specific Guidance's for Generic Drug Development are required to be followed by manufacturers before submission of Abbreviated New Drug Application (ANDA) application [36]. The 21 CFR regulations—21 CFR Part 314 and 21 CFR Part 320 are applicable for ANDA process. The standardized practices to be followed by CDER staff for generic drugs are in Chapter 5200 of Manual of Policies and Procedures.

29.6 Clinical trials and Food and Drug Administration recommendations for antiherpes viral drugs

Herpes virus is a common infection, which may exist in a patient's body throughout his/her life. It is a DNA-based virus, including varicella-zoster virus; Epstein—Barr virus; CMV; Kaposi's sarcoma herpesvirus; HSV-1, and HSV-2.

29.6.1 Drugs against cytomegalo viral infections

CMV caused infection is a communal source of infection and death. The existing drugs have a range of adverse effects, limiting the usage of these drugs. In contrast to existing drugs, Maribavir does not possess side effects like nephrotoxicity and myelosuppression. A Phase 2 open-label trial was conducted and involved randomization of 161 participants, with 4 Blocks. Hence, 79% of patients from 62% of patients receiving Maribavir showed a response with overall risk ratio of 1.2 (95% CI). Thus, 22% of patients receiving Maribavir showed relapse of viral infection. However, the dosage of Maribavir at 400 mg/day—12 weeks resulted in elimination of Virus. Moreover, the drug has been cleared for emergency usage, as it resulted in viral elimination from immunocompromised users (Hematopoietic cell transplant receivers) when administered for 6 weeks. The drug can be associated to gastrointestinal disturbances and dysgeusia as its side effects [37]. In another study, Maribavir was administered to 120 randomized patients, in a 3 Block design. The dosage of Maribarvir was assigned as 400, 800 and 1200 mg/day (bid), respectively. Patients selected for study were mostly immunocompromised with Age ≥ 12 years, RR (refractory or resistant) CMV infections and having plasma CMV DNA ≥ 1000 copies/mL. Hence, finally 67% (95% CI, 57%−75%) patients reached undetectable CMV DNA i.e. ≤ 200 copies/mL till 6th Week. The study was effective, after including a varied group of transplant receivers infected severely with CMV. The authors reported that Maribavir dosage ≥ 400 mg bid proved to be efficacious, while comparing it with different anti-CMV dosage protocols in the study [38].

Maribavir is approved by FDA [13], as an alternative to treatment against available regimens. Viral resistance, posttransplantation to available regimens have emerged. Although, Antimicrobial Advisory Committee (October 7, 2021) discussed Maribavir due to development of resistance virus against it. But due to nature of outweighing benefits, drug has been approved for posttransplant CMV infections. The potential drawbacks found within clinical trial programmes were existence of open-label nature of studies and lack of diversity. [39]. Similarly, another drug letermovir approved by FDA in 2017, for CMV positives of a patient with HSCT. Patients receiving the drug have received HSCT for blood related cancers. It was noted that letermovir interacts with organic anion-transporting polypeptide transporters and CYP3A substrates. Hence, coadministration of drug with substrates increases risk. Thus approval-based recommendations can be amended due to existence of unmanageable risks by healthcare professionals recommending letermovir [14].

29.6.2 Drugs against herpes simplex virus infections

Herpes simplex virus-2 (HSV-2) is usually transmitted from a seropositive individual during sexual interaction. The viral entry in body is either through mucosal skin or through skin abrasions, with further proliferation in epidermal or dermal cells. Any presence of HSV-2 is enhanced to multifold in enhances risk of acquiring HIV to many folds [40,41].

An open-label randomized trial on AIDS Clinical Trial Group was conducted, with different antiretroviral treatment (ART) involving combination with TDF. The study was conducted with an objective to identify TDF, as a potential candidate for reducing risk of HSV-2 in patients [41]. Thus, reduction of HSV-2 will prevent acquiring of HIV in such patients [42]. Hence in study by Celum et al., out of eligible participants who were HIV seropositive but HSV-2 seronegative were assessed during the study. The $CD4^+$ counting in TDF assessment group was lower than non-TDF assessment group. The results revealed that low oral doses of TDF prevent HSV-2 infections, which may fail for patients who have acquired HSV-2-based infection. Thus, TDF-based regimen reduced HSV-2 incidence by 56% in patients with $CD4^+ \leq 200$ cells/μL [41].

Another single-blinded, randomized controlled clinical trial was completed to evaluate effectiveness of photodynamic therapy along with acyclovir. The patients included in study were adolescent age groups, infected with herpes labialis (HSV-1 virus). The therapy was given, by breaking the lesioned sores and applying laser at 660 nm. The spotted sore was exposed to laser for 30 seconds through a tip, placed perpendicularly over the exposed site. Patients receiving combination therapy, displayed enormous reduction in HSV-1 concentration and pain score as assessed at 4 week, with follow-up continued till 6 months. It was also found pro-inflammatory cytokine levels of IL-6 reduced from 152.6 ± 89.0 (baseline) to 40.4 ± 12.5 (after 6 months), Similarly levels of TNF-α reduced from 173.2 ± 89.2 (Baseline) to 42.6 ± 25.6 (after 6 months) [43]. In similar study, a low-power laser beam was applied for 120 seconds and 1 cm before the effected site. The parameters like reduction in pain, tingling, and edema were reported for conclusion. It was revealed that application of photodynamic therapy reduced pain, tingling, and edema only on Day 1, in comparison to acyclovir [44]. It should be noted that for management of acute herpetic keratitis (due to HSV-1 and HSV-2), most of FDA-approved ophthalmic antiviral ointments (other than acyclovir-based ophthalmic ointments) have been discontinued [45]. However, the acyclovir-based ointment has been discontinued for further marketing, due to safety concerns [46,47]. Based on a public petition in Feb 2021, FDA has been notified about ANDA-based applications for acyclovir ophthalmic ointments. The applications will be implemented along with all legal and regulatory necessities [47].

29.6.3 Drugs against Kaposis sarcoma infections

Kaposis sarcoma, a cancer commonly found in HIV Infected patients. The virus behind the cause of Kaposis sarcoma is herpes virus 8. The virus can effect multiple cells like monocytes, endothelial cells, B cells, and dendritic cells, by possessing multiple endocytic pathways to enter host cells [48]. A 3 + 3 dose escalation design of proteasome inhibitor, having many activities in several malignancies. In the trial, bortezomib was administered intravenously in patients with relapsed/refractory (r/r) AIDS—Kaposi sarcoma taking antiretroviral therapy. The patients selected for trial were having at least some quantifiable sarcoma lesions. During study, dose limiting toxicity (DLT) was kept at 2 cycles (8 weeks). Among the patients who were able to abide with dosage during 2 cycles, without any DLT were allowed to remain at therapy for further 8 cycles. The greatest response to therapy happened as partial response in 60% (9 out of 15) patients; while partial response for maximum dose was seen in 83% (5 out of 6) patients. The remainder were said to have stable disease. During the study, it was assessed that there was no sign of adverse effect related to viral related cellular lysis after administration of bortezomib [49]. Local treatment approaches are always favored, as they are more patient compliant. A single-blinded, noncontrolled clinical trial was conducted over patients with classic Kaposi sarcoma using a topical immunomodulatory agent Imiquimod. Thus, 12 out of 13 patients with 50 lesions were involved in study, during which 24 were treated with cryotherapy and imiquimod was used for remaining 26 lesions. Hence, thorough response was observed in 42.3% lesions receiving Imiquimod; and 50% receiving cryotherapy. Hence, total lesion score reduced by further 50% in 73.1% lesions receiving imiquimod and in 87.5% receiving crytotherapy. Statistically, the results between the 2 results were not significantly related. Both the therapies were performed 3 times per week, and evaluated at end of 12 Weeks in HIV-negative patients. Hence Imiquimod proved to be a successful, useful, and safe therapy for classical lesions [50].

On basis of ongoing trials, FDA (14 May 2020) has expanded the indication for including pomalidomide for treatment of Kaposi sarcoma. FDA, based on the severity of disease, has changed designation for drug as priority review and breakthrough therapy [51]. The decisions were made on basis of ongoing clinical trials (ID-NCT01495598; 120047; 12-C-0047) of pomalidomide [52]. Hence, based on ANDA applications—204026 and 210249 the generic drugs were approved on basis of postmarketing requirements, for administration in combination with dexamethasone. Pomalidomide is now part of FDA Cancer Accelerated Approval program, where results of ongoing clinical trials are estimated for marketing requirements and labeling purposes.

29.6.4 Drugs against Epstein–Barr viral infections

Epstein–Barr virus is another herpes virus known as Human Herpes Virus 4. The effect of virus results in several painful complications. Infection with EBV can be a route cause of many diseases like infectious mononucleosis and many kinds of malignancies. However, no specific FDA-approved therapeutic is available against Epstein–Barr virus. It has been found through molecular modeling studies, that targeting membrane bound protein (gH) may assist in reducing complications and development of possible therapeutics [53]. Hence, several clinical trials are underway to find a possible therapeutics against EBV.

A randomized, double-blind, placebo-controlled study was conducted on 24 men with oropharyngeal HHV 8 were enrolled to evaluate in vitro effect of valganciclovir on EBV. The 62% infected persons were receiving at least 3 ART drugs and were having average CD_4 T-cell count as 434 cells/μL. Valganciclovir administration resulted in reduction of EBV blood level compared to placebo by 72%. Thus, an oral administration of valganciclovir per day resulted as an important prophylactic agent [54]. Epstein–Barr virus initiates B cell proliferation, resulting in an immune dysregulation and usage of immune suppressors. This initiates the mode for PTLD. Hence, the EBV-related lymphoproliferative disorder is multifaceted, causing uncontrolled proliferation of B cells [55]. A potential Phase 2 clinical trial was piloted with 4 weekly infusions of rituximab (375 mg/m^2), for prolonged management in patients with PTLD. The assessment of diagnosis after the therapy was based on local pathological evaluation and disease-specific survival. The clinical survival of patients was assessed after a follow-up of 13 years, while patients in real-world were evaluated after a followup of 6.5 years. It was found for patients under clinical trial, who have attained complete diminution after receiving therapy showed disease-specific survival of 94.4% and 88.1%, after 5 and 10 years, respectively. The results proved the significance of Rituximab [56].

29.7 Clinical trials and Food and Drug Administration recommendations for anti-HIV drugs

The emerging spread of HIV, states the ability of virus to efficiently neutralize immune system of body. With a modest genome size, HIV also leads in betraying the biological immune components. Thus, constantly developing HIV diversity poses an immense challenge for developing therapeutics against the virus [57]. Based on the data submission through trials, FDA approves different products for HIV based on its guidance documents and policies. The role of FDA cannot be underestimated in controlling HIV by determining the molecules, in which mighty benefits always outweigh associated risks. In January 2020, FDA launched a mobile friendly database including approved/tentatively approved lifesaving drugs for HIV. It was for the first time FDA reviewed the accession of product labeling, along with updates of manufacturing sites [58].

Several clinical trials are underway to find a possible and permanent solution for HIV-1. A Phase 3 trial was conducted to compare antiviral efficiency of cabotegravir and rilpivirine dosage every 8 weeks (as in Phase 2 of their trial) compared to every per week dosing. The trial aimed to reduce dosing frequency of ARTs, from daily to weekly. The secondary efficiency end point included plasma HIV-1 RNA copies < 50 copies/mL at week 48; confirmed virological failure (RNA > 200 copies/mL) at week 48; and resistance levels after virological failure. The combination acted as a long-lasting regimen dosed every 8 weeks and was noninferior to dosing done at every 4 weeks. This has been proved with no-treatment associated death [59]. In January 2021, FDA approved Vocabria and Cabenuva for short-term treatment of HIV-1 for adults [17]. In addition, during risk-based evaluation it was recommended that Cabotegravir and rilpivirine combination must be given to appropriate individuals as missing doses may lead to treatment failures. It was recommended that a full ARV regimen must be initiated within 1 month, after last dose of cabotegravir and rilpivirine combination to avoid further resistance development. However, due to several prominent benefits in spite of several noted quality issues during manufacturing, the combination has been approved without REMS requirements [60]. Similarly, in December 2021 FDA approved a cabotegravir-based i.m. injection as PrEP [18]. The APRETUDE also includes boxed caution for being required to administer to HIV negative patients only. This marks an important step for those patients who missed oral pills [61]. In February 2020, fostemsavir tromethamine was approved by FDA as a tablet dosage form for HIV-1. The reviewers recommended to improve the labels, by including adverse effects and their management. The reviews signified the need to modify Section 17 of Patient Counseling Information to inform patients [62].

29.7.1 Doravirine

A novel, NNRTI (doravirine) has been investigated for its tolerable dose in HIV patients. Double-blinded studies were conducted in multiple parts to explore dose range for the drug. In Part I identified single doses of drug ranging from 6−450 mg, were administered to subjects after fasting (except for dose of 50 mg). However, Part II explored multiple doses of drug as 30, 60, 240 mg, respectively. The second study involved study on higher tolerable doses of Doravirine which is 600 or 1200 mg. The pharmacokinetics revealed that drug was rapidly absorbed, with median T_{max} value as 1−5 hours after the dose. The blood level reduced in single exponential phase with terminal $t_{1/2}$ of 12−21 hours. The results concluded that doravirine was well tolerated at single doses upto 1200 mg, while in multiple doses upto 750 mg only. Also, based on week 24 data for once daily dosing (from 25, 50, 100, and 200 mg dosing) in HIV infected patients, 100 mg dose was further selected for improvement [63]. Another Phase 3 multicenter, double-blinded trial was conducted in HIV Infected subjects using ART combination including doravirine. Participants were divided for a fixed-dose tablet/day in combination with doravirine (100 mg); lamivudine (300 mg); TDF (300 mg) (DOR/3TC/TDF) vs efavirenz (600 mg), emtricitabine (200 mg); TDF (300 mg) (EFV/FTC/TDF). The results at week 96 revealed that percentage of partcipants with HIV-1 RNA | < 50 copies/mL was 77.5% within DOR/3TC/TDF group and 73.6% in the EFV/FTC/TDF group. Hence, when comparing the lipid profiles in DOR/3TC/TDF group vs EFV/FTC/TDF group from week 48 with week 96, it was revealed that DOR/3TC/TDF had promising lipid profile change (LDL-c; non-HDL-c) with respect to baseline as (−0.6 and −2.1 mg/dL, respectively) [64]. Thus combination of ART involving dorivarine resulted in better efficacy and low adverse events [64−69]. Based on NDA application 210807, FDA recommended to increase usage of DOR/3TC/TDF combination for pediatric patients (Weight ≥ 35 kg). The supplemental approval by FDA is part of postmarketing requirements [70]. Hence, as per risk-based recommendations the possibilities of vulnerabilities from combination, are required to be included in Warnings and Precautions section of label.

29.7.2 Dolutegravir

A long-term administration of established ART, specifically in OALWH resulted in comorbidity events which may be life threatening. Such additional morbidities also resulted in administration of multiple drugs with subsequent interactions or adverse effects. Hence data from 6 Phase III/IIIb clinical trials were evaluated for efficacy and safety of dolutegravir in old adults. There were different sets of comparative drug regimens in different trial groups. It was found that administration of ART-based regimens in OALWH, resulted in an entire intake of ≥ 5 concomitant medicines. The response rate for dolutegravir administration was near to similar in all age groups [71]. In another significant study, dolutegravir with lamivudine was given to suppress virological response in patients with recorded lamivudine resistance and continuing history of marginal lamivudine RAMs. The results were recorded at week 96 of this open-label, single-arm pilot trial. At week 96, it was concluded that dolutegravir plus lamivudine combination could be safe and effective for preserving viral suppression even in patients with historical lamivudine RAMs as detected in proviral DNA by population sequencing [72]. Similar trials are underway in recent years due to viral suppression by ARTs, indicating usage of dolutegravir and lamivudine combinations as first line treatment. The combination will refine metabolism when compared to other combinations [73–75].

In one of the recent developments, a dolutegravir-based tablet (Tivicay—approved by FDA in 2013) has been further recommended by FDA for changes in efficacy label for including clinical data. In December 2020, Tivicay was approved by FDA for pediatric patients (Age ≥ 4 Weeks; Weight ≥ 3 kg) [76]. The minimum recommended dose for pediatric patients was 5 mg. In November 2021, changes in label of Dovato and Triumeq (dolutegravir-based combination) were recommended for including dosing in patients with creatinine clearance between 39–40 mL/min [77]. In 2022, another dolutegravir-based combination has been amended for change in efficacy label. The clinical data was required to be updated to 96 weeks (based on TANGO study) and 144 weeks (based on GEMINI-1 and GEMINI-2 studies) [78,79]. In addition, combination products (Prezcobix and Evotaz) of enzyme inhibitors against HIV-1 were recommended by FDA in April 2020 for pediatric patients, with weight at least 40 kg and at least 35 kg, respectively [80].

29.8 Clinical trials and Food and Drug Administration recommendations for Antiinfluenza viral drugs

Influenza viruses are significant human respiratory pathogens, effecting human respiratory epithelia. [81,82]. Multiple mutations within viral genome supports it to escape existing immunity. Mainly, three types of influenza viruses (IAV, IBV, and ICV) that infect the humans [81]. Influenza virus (if untreated) may have the potential to cause ARDS, multiorgan dysfunctions, with a possibility leading to death [83]. It is mandatory for healthcare professionals to be aware about the complications of symptoms from influenza. There are 4 FDA-approved drugs for socializing influenza viruses. However, developing circulating virus strains are resistant to amantadine and rimantadine, inhibiting their usage for viral management.

29.8.1 Baloxavir marboxil

The existing antiviral agents belonging to—adamantane and NAIs (neuraminidase inhibitors) are acceptable for influenza treatment, but both are confined by certain limitations. Influenza strains have acquired resistance against adamantanes, while NAIs are effective if administered within 48 hours of onset of symptoms. NAIs fail to maintain viral shedding for a prolonged length of time [84]. A Phase I study was designed to assess safety; tolerability and pharmacokinetics of a cap-dependent endonuclease inhibitor (Baloxavir marboxil), including effect of food during administration of drug. The study 1 was a randomized, double-blinded and single center study, while study 2 was randomized, three-dosing arrangement, and with crossover food effects study. The single dose of Baloxavir marboxil was accepted by participants, as an oral dose (6—80 mg), or as an tablet (20 mg). Single dose administration resulted in better therapeutic benefits and quicker removal of virus [84]. Similarly, in another study Phase 2 and Phase 3 double-blinded trials were conducted with results indicating lessening of influenza symptoms in Baloxavir group within shorter duration. Patients with positive test result was criteria for their entry in study in Phase 2 but not in Phase 3. The Phase 3 trial was oseltamivir-controlled. In Phase 2 trials, patients were administered with 3 types of single doses—10, 20, and 40 mg leading to reduction in viral titers on 2—3 day of dosage administration. In Phase 3 trial, patients were administered with single oral dose of 40 and 80 mg depending on body weight of patients, while oseltamivir was given twice daily at dose of 75 mg. The primary outcome of trial was defined as condition when patient showed mild absence of symptoms or existence of symptoms for at least 21.5 hours, while secondary outcome was defined as time to return toward normal health. In trials, it was concluded that Baloxavir showed greater effectiveness in comparison to oseltamivir and placebo. The treatment is not suitable for patients infected with influenza virus strains unaffected to neuraminidase inhibitors [85]. Another metaanalysis study revealed the support for usage of Baloxavir for stopping the replication of influenza virus. According to primary efficacy interpretation, when compared to placebo, Baloxavir took significantly fewer time to dismiss symptoms like fever. In this metaanalysis, 3771 patients from three RCTs were included for significant findings. The findings recommended that Baloxavir has potential to overcome existing influenza treatment strategies proven through clinically and virologically reactions [86]. Several other trials have very well supported in recent 2 years the use of Baloxavir marboxil for prophylaxis against influenza [87—90].

29.8.2 Oseltamivir

In many cases, antiviral therapy alone may be unsuccessful in correcting the clinical course. Single therapy may lead to development of resistance against virus. As a result, medicines with antiinflammatory and/or immunomodulatory properties have been proposed as supplementary therapy. Among several trials, a double-blinded trial was conducted over patients with influenza like viral illness and positive antigen test. The trial used an antipyretic agent 500 mg Paracetamol (4 times/day), along with oseltamivir 75 mg tablet (bid) with diet for 5 days. Primary result of trial was PCR influenza log10 viral concentration till 5th day, while secondary outcomes were measured at extreme temperature recorded/day, viral indicator scores/day, time taken for treatment and visual analog measurement. The study provided a wide angled view that consistent dosage of paracetamol did not affect viral load and clinical influenza symptoms, even when the antipyretic is administered with oseltamivir [91]. In another open-label trial, participants from hospital with laboratory confirmed Influenza such as infected with A/H3N2 were included. Macrolides with antiinflammatory characteristics were used in trial, for their synergistic effect in treatment of influenza. Study treatments were I—oseltamivir—75 mg bid and azithromycin—500 mg for 5 days (orally for 5 days); II—oseltamivir—75 mg bid (orally for 5 days). The outcomes were changes induced due to plasma cytokine/chemokine and proinflammatory mediator changes, along with resolving of symptoms due to virus were compared in both patient groups. Faster resolution of symptoms was observed in patient group offered with azithromycin, with regular analysis of blood levels of cytokines. Proinflammatory cytokines such as IL-6 reduced faster from baseline to 59.5%; IL-8 reduced to 58%; IL-17 to 34% along with others. The results of study were very significant requiring further progress [83].

Recently, FDA recommended changes in labeling of oseltamivir (TAMIFLU; NDA-021246 & 021247) for including clinical data, proving safety in immunocompromised individuals. The subjects included have solid organ or hematopoietic cell transplantations [92]. FDA has approved several ANDA-based applications for oseltamivir phosphate as drug can be used for treatment of influenza A and B in patients of age 1 year or younger.

29.8.3 ARMS-1 (A patented formulation)

In another double-blinded, placebo controlled trial of ARMS-I (patented formulation) to prevent influenza infection. ARMS-I is an oral steam-based spray, for mucosal surfaces. The ARMS-I was administered to one of the groups intra-orally (thrice daily), while another group was given placebo. Severity of symptoms was recorded as an outcome. It was found that upper respiratory tract infections were 28% more in placebo group then ARMS-I. The cough period, aching throat, or runny nose had lower duration in ARMS-I group then placebo group ($P \leq .019$). The results showed that ARMS-I proved safe and tolerated formulation, and reduced influenza symptoms. [93].

29.8.4 Cocoa-based plant extracts

Plant-based medicinally active components have strong antiinfluenza properties. This has been subsequent after huge amount of active research by scientific community [94]. A research focusing on several regulatory functional foods containing active components. Cocoa extract, a reservoir of polyphenols showed an antiviral effect in human trials after vaccination. Patients with noninfection history due to A(H1N1) pdm2009, was condition for involvement in trial. The average age of participants in trial was 39.2 ± 8.6 years (cocoa intake group). The cocoa group was administered with cocoa for 3 weeks, earlier and later to vaccine administration. The results concluded the supplementary defensive effect of cocoa in human involvement study [95].

29.9 Clinical trials and Food and Drug Administration recommendations for Antihepatitis viral drugs

Hepatitis is an inflammation of liver tissue, where chronic effect can cause deadly diseases like fibrosis, cirrhosis, and hepatocellular carcinoma. With technological advancements, the effect on liver can be diagnosed molecularly. The hepatitis causing majority of death globally is due to 5 types of viruses (A, B, C, D, E). [96]. The pathogenesis, clinical features, management, and inhibition of different types of viral hepatitis has been reviewed by many authors in recent years [96−104].

The US-based regulator, FDA has provided many guidance documents for industry against hepatitis. Getting vaccination appropriately and counseling from healthcare providers is the suitable option in fight against this deadly disease. Many clinical trials are underway for a successful molecule, but still final approval through regulators makes a long way for development of successful antiviral drug.

29.9.1 Drugs against hepatitis B infections

Hepatitis B infections caused by the hepatitis B virus (HBV) can become either acute or chronic diseases. Chronic HBV (CHB) carriers are at risk for developing HBV-related liver complications, such as chronic hepatitis, cirrhosis, and primary hepatocellular carcinoma, during their lifetimes [105]. Patients with hepatitis B can be classified as compensated or decompensated. This depends on severity of liver complications experienced by them. Zhao et al. have defined recompensation of decompensated hepatitis B cirrhosis as a state, obtained after vigorous management with no problem related to cirrhosis decompensation. The liver disease of patient does not show any significant progress during this period [106].

Bentysrepinine Y101 is a derivative of repensine (a compound isolated from Dichondra repens Forst), which has the potential of anti-HBV activity. Y101 was able to reduce the amount of DNA-HBV and cccDNA, which means that it may has the ability to clear HBV infection by eradicating cccDNA from the liver [107]. Two randomized, double-blind, placebo-controlled trials was conducted involving 94 subjects using bentysrepinine was conducted. A single oral dose (50−900 mg, Study-01), multiple doses (300 mg and 600 mg, Study-02), and a randomized, open, crossover food-effect study (600 mg, Study-03) of bentysrepinine was evaluated. It was evaluated that bentysrepinine was rapidly absorbed and metabolized with a mean time to reach maximum concentration (Tmax) between 1−2 hours and a mean terminal elimination half-life (t1/2) of approximately 1−3 hours. Bentysrepinine exhibited acceptable safety and tolerability in healthy subjects in the dose range of 50−900 mg in both single and multiple-dose studies. There was no accumulation of drug after the administration of multiple 300 and 600 mg doses [108].

It is also known that achievement of cure from hepatitis B-based liver infection is detectable absence of hepatitis B Antigens, which makes treatment lifelong. In a Phase III double-blinded trial, results of tenofovir alafenamide (new prodrug of tenofovir) was compared with tenofovir disoproxil fumarate based on 96 week outcome. The study included patients with positive HBeAg and negative HBeAg. The dose of Tenofovir alafenamide (25 mg), was 90% lower than that of tenofovir disoproxil (300 mg) during the trial. The outcomes for this 96 week long study included ALT analysis with parameters such as HBV DNA < 29 IU/mL, antigen loss and formation of viral antibodies. The resistance analysis was also included during the study. It was also found that rate of ALT normalization was higher in patients receiving tenofovir alafenamide. Tenofovir alafenamide caused lesser systemic exposure than tenofovir, which may be due to its sustained release. Hence there was lesser renal and bone toxicity then tenofovir disoproxil fumarate [109].

The regulatory administrator, FDA has issued a guidance documents to assist sponsors for treatment of Hepatitis B Infections. In one of the notification from Drug Safety Commissions, FDA induced boxed warnings for direct acting antivirals for treatment of hepatitis C. The warning included the risk of HBV reactivation leading to death. FDA urged the healthcare providers and patients for reporting the history of HBV [110]. In addition to existing drugs for adults, FDA approved HEPSERA in 2007 for pediatrics ≥ 12 years.

29.9.2 Drugs against hepatitis C infections

It has been determined that there are seven major genotypes of HCV, with genotype 1, 3 constituting the majority of all HCV infected infections [111]. The US FDA defined direct acting antivirals as class of drugs (prescription based) for treatment of adults against hepatitis C infections [110]. The usage of directly acting antivirals for a short term has led to resistance-associated substitutions (RAS) in infected individuals. In RAS the peptide chain associated with antiviral treatment undergoes exchange/swapping of amino acids leading to resistance [112].

A single-center, controlled trial has been conducted among patients infected with hepatitis C but with absence of liver stiffness. The trial aimed to compare 2 treatment regimens—Glecaprevir/Pibrentasvir (GLE/PIB) treatment with and without ribavirin for patients with chronic hepatitis C. Dosing of drugs was given with 2 weeks interval, while ribavirin was given as 15 mg/kg (bid) [112]. Ribavirin (1-β-D-ribofuranosyl-1,2,4-thiazole-3-carboxamide) is a synthetic analog of guanosine used to treat patients with chronic HCV infection since the 1990s [113]. Blood for samples were collected at 4 and 12 weeks. At week 12 sustained virological response was measured in both treatment groups (With HCV RNA > 15 IU/mL), and offered retreatment with sofosbuvir-based regimens for further 12 weeks. Hence, ribavirin-based regimens lead to greater probability of sustained virological response (SVR12) but values of hemoglobin declined by 2.24 g/dL (mean reduction from start to end of treatment) [112].

In infections with HCV, achievement of SVR12 through a short duration of antiviral-based regimens can be targeted by using combination of drugs with different mechanism-approaches. The restriction to access these drugs must be resolved along with regular treatment observance. In another Phase 1b trial, the simultaneous administration of CDI-

31244 (400 mg/day for 2 weeks) is studied when given with combination of sofosbuvir 400 mg/velpatasvir 100 mg (as combination daily for 6 weeks). The patient group involved in study were infected with chronic genotype 1 HCV (HCV RNA > 1000 IU/mL) without advanced fibrosis or cirrhosis. Primary end point was SVR12 (HCV RNA < lower limit of quantification (LLOQ)—15 IU/mL), while secondary outcome was at SVR24. It was found that usage of CDI-31244 helped in rapidly reducing the HCV RNA within initial 2 days, and maintained HCV RNA below LLOQ during the study period [114]. It was suggested that usage of CDI-31244 as an adjunct therapy for some more longer duration will prevent development of resistant strains against sofosbuvir/velpatasvir.

The elbasvir and grazoprevir combination has been approved by FDA in 2016. Due to notable liver-based toxicity concerns, physicians should obtain laboratory assessments at week 8 and week 12 from patients who have been treated on 16-week regimen. Dosing with ribavirin has been recommended for certain populations. In addition to including the required warnings the combination drug has been approved by FDA in 2021 for pediatrics ≥ 12 years of age (Weight ≥ 30 kg) [27]. Based on reviewer's recommendation, FDA approved sofosbuvir and velpatasvir-based regimen for pediatrics ≥ 3 years on basis of clinical trials conducted over 214 subjects. The required dosage was of sofosbuvir and velpatasvir was recommended on basis of body weight. In pediatric subjects with severe cirrhosis ribavirin was required to be included to improve virological responses at end of 12 weeks [28].

29.9.3 Drugs against hepatitis D infections

Hepatitis D infection is caused to people effected by hepatitis B infection. HDV has a 1.7 kb single-stranded circular genomic RNA, with lipid envelope [115]. HDV requires HBV for its transmission and continued replication process to happen in cell. Such patients are at greater risk to progressive cirrhosis and hepatocellular carcinoma. With such co-infection, risk of therapy failure always exists in case of unidentified cases. An open-label study was piloted to assess effect of prolonged peginterferon therapy in patients infected with HDV. The presence of HDV antigens in liver as identified through immunoperoxidase staining and pretreatment for 6 months with peginterferon was opted as one of the inclusion norms. Pegylated interferon α-2a was dosed at 180 μg (sc route)/week with maximum dose till 360 μg, but for a period of 24 weeks. The length of treatment was allowed to prolonged till 5 years. It was found that extending the treatment provided a sustained effect, with continued loss of hepatitis antigen. The treatment effected patients, individually due to which dose adjustments is requirement to maintain tolerability for such long-term treatments. [116].

29.10 Clinical trials of herbal molecules as antiviral agents

The use of herbal medicine since ancient time includes a varied range of plant and plant-derived products (botanical materials and bioactive) for medicinal purposes. At present, there is still lack of any effective antiviral therapy against many significant emerging viruses. The potential source of antiviral compounds could be considered naturally from vegetables, fruits, flowers, herbal plants, marine organisms, and microorganisms as from past evidence [117]. Many researches have shown that various compounds obtained from natural product exhibit antiviral activities, with good tolerability and minimal side effects [118]. With an abundance of natural products to screen for new antiviral compounds, it is highly optimistic that natural products will continue to play an important role in contributing to antiviral drug development and in reducing the global infection burden of many viruses [119]. In this connection many natural products and their derived compound have been screened against several kinds of DNA and RNA-based viruses. Several of such medicinal plant-based formulations and bioactive constituents are enlisted in Table 29.3.

Some of useful medicinal plants and bioactive compounds such as inulin-type fructans [136], roots of *Isatis indigotica* [137], EGYVIR (a potent herbal extract against SARS-CoV-2) [138], lupin, salvia, garlic, and EVOO [139], saponin [140], acteoside [141], *Rheum tanguticum* [142], and liquorice [143] find useful applications against many virus including SARS-CoV-2. There are several essential oils tested against many virus and are found suitable for prevention. These medicinal plants and bioactive compound further need to evaluated for safety in near future. This may lead to develop several value added products including nutraceutical for prevention and treatment of many emerging viral infectious diseases.

29.11 Conclusions and future prospects

In the current of emerging new viral infections, it seems that resistance against specific molecular entities is developed due to several factors such as patient noncompliance, inadequate usage, insufficient data on ethnicity, and virus suppressions. Patient compliance arises due to nonappropriate administration of dosage and multiple dosages/day. Hence, the designing of single available dosage to counteract multiple dosage administration may help patients escape dosage loss. Drug resistance (In any case) is costly to healthcare sector, to patient who tried to achieve benefits and for community where resistance develops [144]. It is the needed that healthcare providers should provide appropriate counseling to patients on antiviral dosage. The guidance documents by regulators must be followed for primary usage of antivirals. The Industry partners must enhance the usage of technical capabilities in initial phases of clinical trials as per GCP guidelines. However, FDA has prepared a guidance document on use of digital technologies in obtaining data during clinical trial investigations [145]. The results in initial phases of trails must be critically analyzed statistically, to reveal data-based decision for a positive outcome [146]. The regulators must also encourage trials for antivirals, in combination with other possible agents [147]. The emergence of mutations in virus has initiated the need for progressive approaches and usage of analytical tools for early decisive processes.

TABLE 29.3 Antiviral activity of bio actives alone or in combination as dosage form.

Disease	Patients	Medication (dose)	Effects	References
Oral Herpes	68 Human volunteers (36 male; 32 female)	Gene-Eden-VIR—patented herbal treatment (1–4 capsule/day)-with 325.1 mg composition including 5 Ingredients	Antiviral outbreaks reduced further, with better efficacy and greater safety	[120]
Hepatitis B Cirrhosis;	85 Patients HQD granules group—28 male; 14 female control group—26 male; 17 female	Huangqi Decoction (HQD) Granules HQD—composition of Radix et Rhizoma Glycyrrhizae and Radix Astragali (6:1 w/w).	CDC42—A GTPase protein of Rho family, CDC42 regulates hepatitis C infectivity. GLI1—up-regulated transcription factor (Zinc Family 1, GLI1) HQD reduces CDC42 and GLI1 levels; while effecting quantities of other factors	[121]
HCoV-229E Induced effect	40 Bal/c mice, SPF grade weight—14 ± 1 g (20 males; 20 females)	Shufeng Jiedu capsules—Patented drug comprising of eight medicinal plants 6.24 g of dry extract = 32.4 g of crude drug- For Human Dose High dose—11.8 g/kg crude drug Dry Herbal extract Low Dose—1.14 g/kg crude dry herbal extract	Produced antiviral; antiinflammatory effect and increased CD4; CD8 when compared to model groups.	[122]
Liver disorders from viral infection	67 volunteers (54 male; 13 female)	Chunggan extract (CGX) – Oral administration of 1 g/2 g/Placebo per day for 6 months. Preclinical studies revealed CGX is safe for Humans (60 kg) upto 3243 mg/day.	Liver stiffness is measured, which reduced significantly. Hence CGX can be used synergistically with antiviral agents	[123]
Hepatitis B virus (HBV) infection	395 Patients (253 males; 142 female)	Chinese herbal formula—comprising of 12 herbs (Sachet of sachet of (32.67 g) granules is same as 190 g raw herbs) once a day in 2 doses for 12 Weeks.	High rate of viral clearance has been obtained with insistent Plasma aminotransferase levels.	[124]
Common Cold due to Viral Infection (With Wind—Heat Syndrome)	240 Patients	Lian-Ju-Gan-Mao capsules (LJGMC) comprising of 3 Herbs. High dose group dose—1.12 g medium dose group dose—0.56 gm LJGMC + 0.56 gm Placebo low dose group dose—0.28 g LJGMC + 0.84 gm placebo placebo group—1.12 gm Placebo. Dose—3 times/day + 5 day follow-up	Dosage found safe, recommended for next level of clinical trial groups.	[125]
	240 Patients	Binafuxi granules mainly comprising from 5 herbs	Dosage recommended for next level of clinical trial groups.	[126]

(Continued)

TABLE 29.3 (Continued)

Disease	Patients	Medication (dose)	Effects	References
Common Cold with Heat Syndrome		minimum effective dose—2.75 gm (1 sachet) High dose group—2 Sachets bid Low dose group—[1 sachet + 1 placebo sachet (2.75 gm)] bid placebo group—2 sachets placebo bid		
common cold with wind-heat syndrome	72 Patients (in 2 groups)	Reduqing granules comprising of 14 Herbs. 1 Bag Reduqing granules = 20 g 1 Bag Lianhuaqingwen capsule = 0.35 gm	Percentage of symptom score reduction was parameter for assessment. Reduqing granules did not showed any adverse effect. Hence the proposed granules are safe for treatment.	[127]
COVID-19	131 Patients rehabilitated 66 BFHX group 65 placebo group	Bufei Huoxue capsule (BFHX)—Chinese Patented Medicine with 3 herbs 1 capsule = 0.35 gm of Herb patient receive = 1.4 gm (4 doses) for 90 days	Patients in BFHX group exhibits significant reduction in lung lesions, followed by improved in lung pneumonia. Hence the drug supported patient recovery.	[128]
Hepatitis C	82 Patients (3 groups) low dose—27; medium dose—28; high dose—27	Viron (1 gm tablet composed of 1 g film-coated tablet contains dried powdered 6 herbs) Highest dose—3 tablets bid; medium dose—2 tablets bid; lowest dose—1 tablet bid	60 eligible patients responded, with a significant linear relation between virological reaction and high dose. Also Health related life quality enhanced, as predicted through Chronic Liver Disease Questionnaire (CLDQ Score). The herbal formula is safe.	[129]
COVID-19	140 Participants EXPECTED 70—MEDICINE GROUP 70—PLACEBO GROUP	Maxingshigan-Weijing decoction made of 14 Herbs—200 mL (2 times) per day for continuous 14 days.	An open-label study, with results expected to come out with futuristic possibilities.	[130]
COVID-19	72 out of 75 patients were enrolled 37—Medicine Group 35—Placebo Group	Intervention Arm—given 2 capsules investigational product 1—400 mg—2 times a day till First 15 day investigational product 2—450 mg containing water and CO_2 extract of 8 Herbs with Zinc.—2 times a day continuously till 30 days. Placebo Arm: Edible starch ~ 450 mg.	A randomized controlled trial to be further explored for opportunities.	[131]

Disease	Patients/Groups	Treatment	Outcome	Ref.
COVID-19	120 out of 125 Patients 40—Placebo 40—Nilavembu Kudineer 40—Kaba Sura Kudineer	Nilavembu Kudineer Decoction—9 Drugs Kaba Sura Kudineer Decoction—15 Drugs 60 mL/day—bid for 10 days, followed by a follow up for 30 Days (Phone). Plcaebo—Decaffienated Tea	RT PCR; LFT; ECG and electrolytes was done at 0, 3, 6, 10 days. Kaba Sura Kudineer showed a significant effect. Patients administered with Kaba showed reduced hospital stay time as compared to patients administered Nilavembu and Placebo. However, both Siddha drugs are safe and will provide a possible synergistic combination with allopathy.	[132]
COVID-19	195 Patients Included 99—Treatment Group 96—Control Group	Hua Shi Bai Du granule (Q-14) granules along with standard care—10 g granules—bid for 14 days. Hua Shi Bai Du granule (Q-14) granules— composed of 14 Herbs.	The use of granules could be a supportive therapy in COVID-19 management. There was no significant adverse effect, except for diarrohea reported in 8 Patients.	[133]
COVID-19	150 Patients included 75—treatment group 75—control group	Kampo medicine—Kakkonto (2.5 gm) with Shosaikotokakikyosekko (2.5 gm)—3 times a day for 14 days in addition to conventional treatment.	While estimating the symptoms the median survival time in treatment group was 1.5 more than control group. The treatment improved patient's symptoms to a significant extent.	[134]
COVID-19	284 Patients 142—treatment group 142—control group	Lianhuaqingwen capsule—composed of 13 drugs—14 days dosage—4 capsules thrice daily for 14 days	Treatment with capsules enhanced recovery of symptoms significantly promptly (Including the chest irregularities).	[135]

References

[1] Sagaya Jansi R, Khusro A, Agastian P, Alfarhan A, Al-Dhabi NA, Arasu MV, et al. Emerging paradigms of viral diseases and paramount role of natural resources as antiviral agents. Sci Total Env 2021;759. Available from: https://doi.org/10.1016/j.scitotenv.2020.143539.

[2] Lowe H, Steele B, Bryant J, Fouad E, Toyang N, Ngwa W. Antiviral activity of Jamaican medicinal plants and isolated bioactiv compounds. Molecules 2021;26. Available from: https://doi.org/10.3390/molecules26030607.

[3] Bhattacharya R, Dev K, Sourirajan A. Antiviral activity of bioactive phytocompounds against coronavirus: an update. J Virol Methods 2021;290. Available from: https://doi.org/10.1016/j.jviromet.2021.114070.

[4] Li G, Jing X, Zhang P, De Clercq E. Antiviral classification. Encycl Virol 2021;121−30. Available from: https://doi.org/10.1016/b978-0-12-814515-9.00126-0.

[5] Tripathi K. Essentials of Medical Pharmacology. 8th ed. Delhi: JayPee;; 2019. Available from: https://doi.org/10.5005/jp/books/12256.

[6] Novitzke JM. The significance of clinical trials. J Vasc Interv Neurol 2008;1:31.

[7] Chaitanya MVNL, Ali HS, Usamo FB. Regulatory considerations of herbal biomolecules. Herb Biomol Healthc Appl 2022;669−76. Available from: https://doi.org/10.1016/b978-0-323-85852-6.00009-3.

[8] CDER Offices and Divisions. <https://www.fda.gov/about-fda/center-drug-evaluation-and-research-cder/cder-offices-and-divisions>; n.d. [accessed 8.2.22].

[9] Office of Infectious Diseases (OID). <https://www.fda.gov/about-fda/center-drug-evaluation-and-research-cder/office-infectious-diseases-oid>; n.d. [accessed 8.2.22].

[10] NDA 214787. Cross Discipline Team Leader, Division Director and ODE Director Summary review. <https://www.accessdata.fda.gov/drugsatfda_docs/nda/2020/214787Orig1s000Sumr.pdf>; n.d. [accessed 7.2.22].

[11] New Drug Application (NDA): 210854. <https://www.accessdata.fda.gov/drugsatfda_docs/label/2018/210854s000lbl.pdf>; n.d. [accessed 8.2.22].

[12] New Drug Application (NDA): 214410. <https://www.accessdata.fda.gov/drugsatfda_docs/label/2020/214410s000,210854s004s010lbl.pdf>; n.d. [accessed 8.2.22].

[13] New Drug Application (NDA): 215596. <https://www.accessdata.fda.gov/drugsatfda_docs/label/2021/215596lbl.pdf>; n.d. [accessed 8.2.22].

[14] New Drug Application (NDA): 209939. <https://www.accessdata.fda.gov/drugsatfda_docs/label/2020/209939s008,209940s006lbl.pdf>; n.d. [accessed 8.2.22].

[15] Biologic License Application (BLA): 761169. <https://www.accessdata.fda.gov/drugsatfda_docs/label/2020/761169s000lbl.pdf>; n.d. [accessed 8.2.22].

[16] New Drug Application (NDA): 212888. <https://www.accessdata.fda.gov/drugsatfda_docs/label/2022/212888s002lbl.pdf>; n.d. [accessed 8.2.22].

[17] New Drug Application (NDA): 212888. <https://www.accessdata.fda.gov/scripts/cder/daf/index.cfm?event = overview.process&varApplNo = 212888>; n.d. [accessed 8.2.22].

[18] Drugs@FDA. FDA-approved drugs new drug application (NDA). 215499. <https://www.accessdata.fda.gov/drugsatfda_docs/label/2021/215499s000lbl.pdf>; n.d. [accessed 8.2.22].

[19] New Drug Application (NDA): 212950. <https://www.accessdata.fda.gov/drugsatfda_docs/label/2020/212950s000lbl.pdf>; n.d. [accessed 8.2.22].

[20] New Drug Application (NDA): 210251. <https://www.accessdata.fda.gov/drugsatfda_docs/label/2021/210251s014lbl.pdf>; n.d. [accessed 8.2.22].

[21] Biologic License Application (BLA): 761065. <https://www.accessdata.fda.gov/drugsatfda_docs/label/2021/761065s011lbl.pdf>; n.d. [accessed 8.2.22].

[22] Drug Approval Package. PIFELTRO (doravirine). <https://www.accessdata.fda.gov/drugsatfda_docs/nda/2018/210806Orig1s000,210807Orig1s000TOC.cfm>; n.d. [accessed 8.2.22].

[23] New Drug Application (NDA): 211994. <https://www.accessdata.fda.gov/drugsatfda_docs/label/2022/211994s012s013lbl.pdf>; n.d. [accessed 8.2.22].

[24] New Drug Application (NDA): 207561. <https://www.accessdata.fda.gov/drugsatfda_docs/label/2022/207561s029lbl.pdf>; n.d. [accessed 8.2.22].

[25] New Drug Application (NDA): 208464. <https://www.accessdata.fda.gov/drugsatfda_docs/label/2021/208464s013lbl.pdf>; n.d. [accessed 8.2.22].

[26] New Drug Application (NDA): 206843. <https://www.accessdata.fda.gov/drugsatfda_docs/label/2019/206843s008lbl.pdf>; n.d. [accessed 8.2.22].

[27] New Drug Application (NDA): 208261. <https://www.accessdata.fda.gov/drugsatfda_docs/label/2021/208261s007lbl.pdf>; n.d. [accessed 8.2.22].

[28] New Drug Application (NDA): 208341. <https://www.accessdata.fda.gov/drugsatfda_docs/label/2021/208341s017lbl.pdf>; n.d. [accessed 8.2.22].

[29] New Drug Application (NDA): 209394. <https://www.accessdata.fda.gov/drugsatfda_docs/label/2021/209394s014,215110s001lbl.pdf>; n.d. [accessed 8.2.22].

[30] Division of Anti-Viral (DAV). Pre-IND letter of instruction. <https://www.fda.gov/drugs/pre-ind-consultation-program/division-anti-viral-dav-pre-ind-letter-instruction>; n.d. [accessed 8.2.22].

[31] Grignolo A, Choe S. Meetings with the FDA. FDA Regul. Aff., 2014; 105−24. <https://doi.org/10.1201/b16471-8>.

[32] Emergency Investigational New Drug (EIND). Applications for antiviral products. <https://www.fda.gov/drugs/investigational-new-drug-ind-application/emergency-investigational-new-drug-eind-applications-antiviral-products>; n.d. [accessed 8.2.22].

[33] Investigational New Drug (IND) Application. <https://www.fda.gov/drugs/types-applications/investigational-new-drug-ind-application>; n.d. [accessed 8.2.22].

[34] New Drug Application (NDA). <https://www.fda.gov/drugs/types-applications/new-drug-application-nda>; n.d. [accessed 8.2.22].

[35] Abbreviated new drug application (ANDA). <https://www.fda.gov/drugs/types-applications/abbreviated-new-drug-application-anda>; n.d. [accessed 8.2.22].

[36] Product-specific guidances for generic drug development. <https://www.fda.gov/drugs/guidances-drugs/product-specific-guidances-generic-drug-development>; n.d. [accessed 8.2.22].

[37] Maertens J, Cordonnier C, Jaksch P, Poiré X, Uknis M, Wu J, et al. Maribavir for preemptive treatment of cytomegalovirus reactivation. N Engl J Med 2019;381:1136−47. Available from: https://doi.org/10.1056/nejmoa1714656.

[38] Papanicolaou GA, Silveira FP, Langston AA, Pereira MR, Avery RK, Uknis M, et al. Maribavir for refractory or resistant cytomegalovirus infections in hematopoietic-cell or solid-organ transplant recipients: a randomized, dose-ranging, double-blind, phase 2 study. Clin Infect Dis 2019;68:1255−64. Available from: https://doi.org/10.1093/cid/ciy706.

[39] 215596Orig1s000 risk assessment and risk mitigation review(s). <https://www.accessdata.fda.gov/drugsatfda_docs/nda/2021/215596Orig1s000RiskR.pdf>; n.d. [accessed 8.2.22].

[40] Barnabas RV, Celum C. Infectious co-factors in HIV-1 transmission herpes simplex virus type-2 and HIV-1: new insights and interventions. Curr HIV Res 2012;10:228−37. Available from: https://doi.org/10.2174/157016212800618156.

[41] Celum C, Hong T, Cent A, Donnell D, Morrow R, Baeten JM, et al. Herpes simplex virus type 2 acquisition among HIV-1-infected adults treated with Tenofovir disoproxil fumarate as part of combination antiretroviral therapy: results from the ACTG A5175 PEARLS study. J Infect Dis 2017;215:907−10. Available from: https://doi.org/10.1093/infdis/jix029.

[42] Reynolds SJ, Risbud AR, Shepherd ME, Zenilman JM, Brookmeyer RS, Paranjape RS, et al. Recent herpes simplex virus type 2 infection and the risk of human immunodeficiency virus type 1 acquisition in India. J Infect Dis 2003;187:1513−21. Available from: https://doi.org/10.1086/368357.

[43] Ajmal M. Effectiveness of photodynamic therapy as an adjunct to topical antiviral therapy in the treatment of herpes labialis: a randomized controlled clinical trial. Photodiag Photodyn Ther 2021;34. Available from: https://doi.org/10.1016/j.pdpdt.2021.102302.

[44] Ramalho KM, Cunha SR, Gonçalves F, Escudeiro GS, Steiner-Oliveria C, Horliana ACRT, et al. Photodynamic therapy and Acyclovir in the treatment of recurrent herpes labialis: a controlled randomized clinical trial. Photodiag Photodyn Ther 2021;33. Available from: https://doi.org/10.1016/j.pdpdt.2020.102093.

[45] New drug therapy approvals. Advancing health through innovation, <https://www.fda.gov/drugs/new-drugs-fda-cders-new-molecular-entities-and-new-therapeutic-biological-products/new-drug-therapy-approvals-2019>; 2019 [accessed 8.2.22].

[46] New drug application (NDA): 202408 <https://www.accessdata.fda.gov/drugsatfda_docs/label/2021/202408s011lbl.pdf>; n.d. [accessed 8.2.22].

[47] Determination that AVACLYR (acyclovir ophthalmic ointment), 3 percent, was not withdrawn from sale for reasons of safety or effectiveness, <https://www.federalregister.gov/documents/2021/05/20/2021-10593/determination-that-avaclyr-acyclovir-ophthalmic-ointment-3-percent-was-not-withdrawn-from-sale-for>; n.d. [accessed 8.2.22].

[48] Dittmer DP, Damania B. Kaposi sarcoma-associated herpesvirus: immunobiology, oncogenesis, and therapy. J Clin Invest 2016;126:3165−75. Available from: https://doi.org/10.1172/JCI84418.

[49] Reid EG, Suazo A, Lensing SY, Dittmer DP, Ambinder RF, Maldarelli F, et al. Pilot trial AMC-063: safety and efficacy of bortezomib in AIDS-associated Kaposis Sarcoma. Clin Cancer Res 2020;26:558−65. Available from: https://doi.org/10.1158/1078-0432.CCR-19-1044.

[50] Odyakmaz Demirsoy E, Bayramgürler D, Çağlayan Ç, Bilen N, Şikar Aktürk A, Kıran R. Imiquimod 5% cream vs cryotherapy in classic Kaposi sarcoma. J Cutan Med Surg 2019;23:488−95. Available from: https://doi.org/10.1177/1203475419847954.

[51] FDA grants accelerated approval to pomalidomide for Kaposi sarcoma. <https://www.fda.gov/drugs/resources-information-approved-drugs/fda-grants-accelerated-approval-pomalidomide-kaposi-sarcoma>; n.d. [accessed 8.2.22].

[52] ClinicalTrials.gov Identifier. NCT01495598 Pomalidomide for Kaposi sarcoma in people with or without HIV. <https://www.clinicaltrials.gov/ct2/show/NCT01495598?term = Pomalidomide&cond = Kaposis + Sarcoma&draw = 2&rank = 3>; n.d. [accessed 8.2.22].

[53] Jakhmola S, Hazarika Z, Jha AN, Jha HC. In silico analysis of antiviral phytochemicals efficacy against Epstein−Barr virus glycoprotein H. J Biomol Struct Dyn 2021;13:1−14. Available from: https://doi.org/10.1080/07391102.2020.1871074.

[54] Yager JE, Magaret AS, Kuntz SR, Selke S, Huang ML, Corey L, et al. Valganciclovir for the suppression of Epstein-Barr virus replication. J Infect Dis 2017;216:198−202. Available from: https://doi.org/10.1093/infdis/jix263.

[55] Gupta D, Mendonca S, Chakraborty S, Chatterjee T. Post transplant lymphoproliferative disorder. Indian J Hematol Blood Transfus 2020;36:229−37. Available from: https://doi.org/10.1007/s12288-019-01182-x.

[56] González-Barca E, Capote FJ, Gómez-Codina J, Panizo C, Salar A, Sancho JM, et al. Long-term follow-up of a prospective phase 2 clinical trial of extended treatment with rituximab in patients with B cell post-transplant lymphoproliferative disease and validation in real world patients. Ann Hematol 2021;100:1023−9. Available from: https://doi.org/10.1007/s00277-020-04056-9.

[57] Simon V, Ho DD, Abdool Karim Q. HIV/AIDS epidemiology, pathogenesis, prevention, and treatment. Lancet 2006;368:489−504. Available from: https://doi.org/10.1016/S0140-6736(06)69157-5.

[58] President's emergency plan for AIDS relief (PEPFAR). <https://www.fda.gov/international-programs/presidents-emergency-plan-aids-relief-pepfar>; n.d. [accessed 8.2.22].

[59] Overton ET, Richmond G, Rizzardini G, Jaeger H, Orrell C, Nagimova F, et al. Long-acting cabotegravir and rilpivirine dosed every 2 months in adults with HIV-1 infection (ATLAS-2M), 48-week results: a randomised, multicentre, open-label, phase 3b, non-inferiority study. Lancet 2020;396:1994−2005. Available from: https://doi.org/10.1016/S0140-6736(20)32666-0.

[60] 212888Orig1s000 risk assessment and risk mitigation review(s). <https://www.accessdata.fda.gov/drugsatfda_docs/nda/2021/212887Orig1s000,212888Orig1s000RiskR.pdf>; n.d. [accessed 8.2.22].

[61] FDA approves first injectable treatment for HIV pre-exposure prevention. <https://www.fda.gov/news-events/press-announcements/fda-approves-first-injectable-treatment-hiv-pre-exposure-prevention>; n.d. [accessed 8.2.22].

[62] 212950Orig1s000 risk assessment and risk mitigation review(s). <https://www.accessdata.fda.gov/drugsatfda_docs/nda/2020/212950Orig1s000RiskR.pdf>; n.d. [accessed 8.2.22].

[63] Anderson MS, Gilmartin J, Cilissen C, De Lepeleire I, Van Bortel L, Dockendorf MF, et al. Safety, tolerability and pharmacokinetics of doravirine, a novel HIV non-nucleoside reverse transcriptase inhibitor, after single and multiple doses in healthy subjects. Antivir Ther 2015;20:397−405. Available from: https://doi.org/10.3851/IMP2920.

[64] Orkin C, Squires KE, Molina JM, Sax PE, Sussmann O, Lin G, et al. Doravirine/lamivudine/tenofovir disoproxil fumarate (TDF) vs Efavirenz/Emtricitabine/TDF in treatment-naive adults with human immunodeficiency virus type 1 infection: week 96 results of the randomized, double-blind, phase 3 DRIVE-AHEAD noninferiority. Clin Infect Dis 2021;73:33−42. Available from: https://doi.org/10.1093/cid/ciaa822.

[65] Yee KL, Cabalu TD, Kuo Y, Fillgrove KL, Liu Y, Triantafyllou I, et al. Physiologically based pharmacokinetic modeling of doravirine and its major metabolite to support dose adjustment with rifabutin. J Clin Pharmacol 2021;61:394–405. Available from: https://doi.org/10.1002/jcph.1747.

[66] Molina JM, Yazdanpanah Y, Afani Saud A, Bettacchi C, Chahin Anania C, DeJesus E, et al. Islatravir in combination with doravirine for treatment-naive adults with HIV-1 infection receiving initial treatment with islatravir, doravirine, and lamivudine: a phase 2b, randomised, double-blind, dose-ranging trial. Lancet HIV 2021;8:e324–33. Available from: https://doi.org/10.1016/S2352-3018(21)00021-7.

[67] Matthews RP, Jackson Rudd D, Fillgrove KL, Zhang S, Tomek C, Stoch SA, et al. A Phase 1 study to evaluate the drug interaction between Islatravir (MK-8591) and doravirine in adults without HIV. Clin Drug Investig 2021;41:629–38. Available from: https://doi.org/10.1007/s40261-021-01046-1.

[68] Nelson M, Winston A, Hill A, Mngqibisa R, Bassa A, Orkin C, et al. Efficacy, safety and central nervous system effects after switch from efavirenz/tenofovir/emtricitabine to doravirine/tenofovir/lamivudine. AIDS 2021;35:759–67. Available from: https://doi.org/10.1097/QAD.0000000000002804.

[69] Kumar P, Johnson M, Molina JM, Rizzardini G, Cahn P, Bickel M, et al. Brief report: switching to DOR/3TC/TDF maintains HIV-1 virologic suppression through week 144 in the DRIVE-SHIFT trial. J Acquir Immune Defic Syndr 2021;87:801–5. Available from: https://doi.org/10.1097/QAI.0000000000002642.

[70] NDA 210807/S-008 supplement approval. <https://www.accessdata.fda.gov/drugsatfda_docs/appletter/2022/210807Orig1s008ltr.pdf>; n.d. [accessed 8.2.22].

[71] Spinelli F, Prakash M, Slater J, van der Kolk M, Bassani N, Grove R, et al. Dolutegravir-based regimens in treatment-naive and treatment-experienced aging populations: analyses of 6 phase III clinical trials. HIV Res Clin Pract 2021;22:46–54. Available from: https://doi.org/10.1080/25787489.2021.1941672.

[72] Rial-Crestelo D, De Miguel R, Montejano R, Dominguez-Dominguez L, Aranguren-Rivas P, Esteban-Cantos A, et al. Long-term efficacy of dolutegravir plus lamivudine for maintenance of HIV viral suppression in adults with and without historical resistance to lamivudine: week 96 results of ART-PRO pilot study. J Antimicrob Chemother 2021;76:738–42. Available from: https://doi.org/10.1093/jac/dkaa479.

[73] Rojas J, de Lazzari E, Negredo E, Domingo P, Tiraboschi J, Ribera E, et al. Efficacy and safety of switching to dolutegravir plus lamivudine vs continuing triple antiretroviral therapy in virologically suppressed adults with HIV at 48 weeks (DOLAM): a randomised non-inferiority trial. Lancet HIV 2021;8:e463–73. Available from: https://doi.org/10.1016/S2352-3018(21)00100-4.

[74] Rolle CP, Berhe M, Singh T, Ortiz R, Wurapa A, Ramgopal M, et al. Dolutegravir/lamivudine as a first-line regimen in a test-and-treat setting for newly diagnosed people living with HIV. AIDS 2021;35:1957–65. Available from: https://doi.org/10.1097/QAD.0000000000002979.

[75] van Wyk J, Ait-Khaled M, Santos J, Scholten S, Wohlfeiler M, Ajana F, et al. Brief report: improvement in metabolic health parameters at week 48 after switching from a Tenofovir Alafenamide-based 3- or 4-drug regimen to the 2-drug regimen of dolutegravir/lamivudine: the TANGO study. J Acquir Immune Defic Syndr 2021;87:794–800. Available from: https://doi.org/10.1097/QAI.0000000000002655.

[76] FDA approves drug to treat infants and children with HIV. <https://www.fda.gov/news-events/press-announcements/fda-approves-drug-treat-infants-and-children-hiv>; n.d. [accessed 8.2.22].

[77] The Food and Drug Administration approved revisions to the TRIUMEQ (abacavir/dolutegravir/lamivudine) and DOVATO (dolutegravir/lamivudine) labels. <https://www.fda.gov/drugs/human-immunodeficiency-virus-hiv/food-and-drug-administration-approved-revisions-triumeq-abacavirdolutegravirlamivudine-and-dovato>; n.d. [accessed 8.2.22].

[78] NDA 211994/S-012 NDA 211994/S-013 supplement approval. <https://www.accessdata.fda.gov/drugsatfda_docs/appletter/2022/211994Orig1s012.s013ltr.pdfo>; n.d. [accessed 8.2.22].

[79] FDA approved changes to the DOVATO (dolutegravir/lamivudine) product labeling. <https://www.fda.gov/drugs/human-immunodeficiency-virus-hiv/fda-approved-changes-dovato-dolutegravirlamivudine-product-labeling>; n.d. [accessed 8.2.22].

[80] FDA approved the product labeling for PREZCOBIX and EVOTAZ for use in pediatric patients. <https://www.fda.gov/drugs/human-immunodeficiency-virus-hiv/fda-approved-product-labeling-prezcobix-and-evotaz-use-pediatric-patients>; n.d. [accessed 8.2.22].

[81] Hutchinson EC. Influenza virus. Trends Microbiol 2018;26:809–10. Available from: https://doi.org/10.1016/j.tim.2018.05.013.

[82] Taubenberger JK, Morens DM. The pathology of influenza virus infections. Annu Rev Pathol Mech Dis 2007;0. Available from: https://doi.org/10.1146/annurev.pathol.3.121806.154316 071015171337001.

[83] Lee N, Wong CK, Chan MCW, Yeung ESL, Tam WWS, Tsang OTY, et al. Anti-inflammatory effects of adjunctive macrolide treatment in adults hospitalized with influenza: a randomized controlled trial. Antivir Res 2017;144:48–56. Available from: https://doi.org/10.1016/j.antiviral.2017.05.008.

[84] Koshimichi H, Ishibashi T, Kawaguchi N, Sato C, Kawasaki A, Wajima T. Safety, tolerability, and pharmacokinetics of the novel anti-influenza agent baloxavir marboxil in healthy adults: phase I study findings. Clin Drug Investig 2018;38:1189–96. Available from: https://doi.org/10.1007/s40261-018-0710-9.

[85] Hayden FG, Sugaya N, Hirotsu N, Lee N, de Jong MD, Hurt AC, et al. Baloxavir marboxil for uncomplicated influenza in adults and adolescents. N Engl J Med 2018;379:913–23. Available from: https://doi.org/10.1056/nejmoa1716197.

[86] Kuo YC, Lai CC, Wang YH, Chen CH, Wang CY. Clinical efficacy and safety of baloxavir marboxil in the treatment of influenza: a systematic review and *meta*-analysis of randomized controlled trials. J Microbiol Immunol Infect 2021;54:865–75. Available from: https://doi.org/10.1016/j.jmii.2021.04.002.

[87] Ikematsu H, Hayden FG, Kawaguchi K, Kinoshita M, de Jong MD, Lee N, et al. Baloxavir marboxil for prophylaxis against influenza in household contacts. N Engl J Med 2020;383:309–20. Available from: https://doi.org/10.1056/nejmoa1915341.

[88] Ison MG, Portsmouth S, Yoshida Y, Shishido T, Mitchener M, Tsuchiya K, et al. Early treatment with baloxavir marboxil in high-risk adolescent and adult outpatients with uncomplicated influenza (CAPSTONE-2): a randomised, placebo-controlled, phase 3 trial. Lancet Infect Dis 2020;20:1204–14. Available from: https://doi.org/10.1016/S1473-3099(20)30004-9.

[89] Yang T. Baloxavir Marboxil: the first cap-dependent endonuclease inhibitor for the treatment of influenza. Ann Pharmacother 2019;53:754–9. Available from: https://doi.org/10.1177/1060028019826565.

[90] Baker J, Block SL, Matharu B, Burleigh Macutkiewicz L, Wildum S, Dimonaco S, et al. Baloxavir marboxil single-dose treatment in influenza-infected children: a randomized, double-blind, active controlled phase 3 safety and efficacy trial (miniSTONE-2). Pediatr Infect Dis J 2020;39:700–5. Available from: https://doi.org/10.1097/INF.0000000000002747.

[91] Jefferies S, Braithwaite I, Walker S, Weatherall M, Jennings L, Luck M, et al. Randomized controlled trial of the effect of regular paracetamol on influenza infection. Respirology 2016;21:370–7. Available from: https://doi.org/10.1111/resp.12685.

[92] New Drug Application (NDA): 021246. <https://www.accessdata.fda.gov/drugsatfda_docs/label/2019/021087s071,021246s054lbl.pdf>; n.d. [accessed 8.2.22].

[93] Mukherjee PK, Esper F, Buchheit K, Arters K, Adkins I, Ghannoum MA, et al. Randomized, double-blind, placebo-controlled clinical trial to assess the safety and effectiveness of a novel dual-action oral topical formulation against upper respiratory infections. BMC Infect Dis 2017;17. Available from: https://doi.org/10.1186/s12879-016-2177-8.

[94] Shoji M, Woo SY, Masuda A, Win NN, Ngwe H, Takahashi E, et al. Anti-influenza virus activity of extracts from the stems of Jatropha multifida Linn.collected in Myanmar. BMC Complement Altern Med 2017;17. Available from: https://doi.org/10.1186/s12906-017-1612-8.

[95] Kamei M, Nishimura H, Takahashi T, Takahashi N, Inokuchi K, Mato T, et al. Anti-influenza virus effects of cocoa. J Sci Food Agric 2016;96:1150–8. Available from: https://doi.org/10.1002/jsfa.7197.

[96] Razavi H. Global epidemiology of viral hepatitis. Gastroenterol Clin North Am 2020;49:179–89. Available from: https://doi.org/10.1016/j.gtc.2020.01.001.

[97] Ganem D, Prince AM, Hepatitis B. Virus infection—natural history and clinical consequences. N Engl J Med 2004;350:1118–29. Available from: https://doi.org/10.1056/nejmra031087.

[98] McMahon BJ. Chronic hepatitis B virus infection. Med Clin North Am 2014;98:39–54. Available from: https://doi.org/10.1016/j.mcna.2013.08.004.

[99] Lanini S, Ustianowski A, Pisapia R, Zumla A, Ippolito G. Viral hepatitis: etiology, epidemiology, transmission, diagnostics, treatment, and prevention. Infect Dis Clin North Am 2019;33:1045–62. Available from: https://doi.org/10.1016/j.idc.2019.08.004.

[100] Thuener J. Hepatitis A and B infections. Prim Care—Clin Pract 2017;44:621–9. Available from: https://doi.org/10.1016/j.pop.2017.07.005.

[101] Mysore KR, Leung DH. Hepat B C Clin Liver Dis 2018;22:703−22. Available from: https://doi.org/10.1016/j.ejogrb.2020.11.052.

[102] Kamar N, Izopet J, Pavio N, Aggarwal R, Labrique A, Wedemeyer H, et al. Hepatitis E virus infection. Nat Rev Dis Prim 2017;3. Available from: https://doi.org/10.1038/nrdp.2017.86.

[103] Chilaka VN, Konje JC. Viral hepatitis in pregnancy. Eur J Obstet Gynecol Reprod Biol 2021;256:287−96. Available from: https://doi.org/10.1016/j.ejogrb.2020.11.052.

[104] Seto MTY, Cheung KW, Hung IFN. Management of viral hepatitis A, C, D and E in pregnancy. Best Pract Res Clin Obstet Gynaecol 2020;68:44−53. Available from: https://doi.org/10.1016/j.bpobgyn.2020.03.009.

[105] Seto WK, Lo YR, Pawlotsky JM, Yuen MF. Chronic hepatitis B virus infection. Lancet 2018;392:2313−24. Available from: https://doi.org/10.1016/S0140-6736(18)31865-8.

[106] Zhao H, Wang Q, Luo C, Liu L, Xie W. Recompensation of decompensated hepatitis B cirrhosis: current status and challenges. Biomed Res Int 2020;2020. Available from: https://doi.org/10.1155/2020/9609731.

[107] Stice S, Liu G, Matulis S, Boise LH, Cai Y. Determination of multiple human arsenic metabolites employing high performance liquid chromatography inductively coupled plasma mass spectrometry. J Chromatogr B Anal Technol Biomed Life Sci 2016;1009−1010:55−65. Available from: https://doi.org/10.1016/j.jchromb.2015.12.008.

[108] Liu X, Xue L, Zhang H, Xu Q, Zhang S, Ma S, et al. Phase I, first-in-human, single and multiple ascending dose- and food-effect studies to assess the safety, tolerability and pharmacokinetics of a novel anti-hepatitis B virus drug, Bentysrepinine (Y101), in healthy Chinese subjects. Clin Drug Investig 2020;40:555−66. Available from: https://doi.org/10.1007/s40261-020-00909-3.

[109] Agarwal K, Brunetto M, Seto WK, Lim YS, Fung S, Marcellin P, et al. 96 weeks treatment of tenofovir alafenamide vs. tenofovir disoproxil fumarate for hepatitis B virus infection. J Hepatol 2018;68:672−81. Available from: https://doi.org/10.1016/j.jhep.2017.11.039.

[110] FDA Drug Safety Communication. FDA warns about the risk of hepatitis B reactivating in some patients treated with direct-acting antivirals for hepatitis C <https://www.fda.gov/media/100702/download>; n. d. [accessed 8.2.22].

[111] Cornberg M, Razavi HA, Alberti A, Bernasconi E, Buti M, Cooper C, et al. A systematic review of hepatitis C virus epidemiology in Europe, Canada and Israel. Liver Int 2011;31:30−60. Available from: https://doi.org/10.1111/j.1478-3231.2011.02539.x.

[112] Madsen LW, Christensen PB, Fahnøe U, Pedersen MS, Bukh J, Øvrehus A. Inferior cure rate in pilot study of 4-week glecaprevir/pibrentasvir treatment with or without ribavirin of chronic hepatitis C. Liver Int 2021;41:2601−10. Available from: https://doi.org/10.1111/liv.14991.

[113] Mejer N, Galli A, Ramirez S, Fahnøe U, Benfield T, Bukh J. Ribavirin inhibition of cell-culture infectious hepatitis C genotype 1-3 viruses is strain-dependent. Virology 2020;540:132−40. Available from: https://doi.org/10.1016/j.virol.2019.09.014.

[114] Chua JV, Ntem-Mensah A, Abutaleb A, Husson J, Mutumbi L, Lam KW, et al. Short-duration treatment with the novel non-nucleoside inhibitor CDI-31244 plus sofosbuvir/velpatasvir for chronic hepatitis C: an open-label study. J Med Virol 2021;93:3752−60. Available from: https://doi.org/10.1002/jmv.26652.

[115] Koh C, Canini L, Dahari H, Zhao X, Uprichard SL, Haynes-Williams V, et al. Oral prenylation inhibition with lonafarnib in chronic hepatitis D infection: a proof-of-concept randomised, double-blind, placebo-controlled phase 2A trial. Lancet Infect Dis 2015;15:1167−74. Available from: https://doi.org/10.1016/S1473-3099(15)00074-2.

[116] Hercun J, Kim GE, Da BL, Rotman Y, Kleiner DE, Chang R, et al. Durable virological response and functional cure of chronic hepatitis D after long-term peginterferon therapy. Aliment Pharmacol Ther 2021;54:176−82. Available from: https://doi.org/10.1111/apt.16408.

[117] El-Tantawy WH, Temraz A. Natural products for the management of the hepatitis C virus: a biochemical review. Arch Physiol Biochem 2020;126:116−28. Available from: https://doi.org/10.1080/13813455.2018.1498902.

[118] Langeder J, Grienke U, Chen Y, Kirchmair J, Schmidtke M, Rollinger JM. Natural products against acute respiratory infections: strategies and lessons learned. J Ethnopharmacol 2020;248. Available from: https://doi.org/10.1016/j.jep.2019.112298.

[119] Goh VSL, Mok CK, Chu JJH. Antiviral natural products for arbovirus infections. Molecules 2020;25. Available from: https://doi.org/10.3390/molecules25122796.

[120] Polansky H, Javaherian A, Itzkovitz E. Clinical trial of herbal treatment gene-eden-VIR/Novirin in oral herpes. J Evid-Based Integr Med 2018;23. Available from: https://doi.org/10.1177/2515690X18806269.

[121] Cheng Y, Liu P, Hou TL, Maimaitisidike M, Ababaikeli R, Abudureyimu A. Mechanisms of Huangqi decoction granules (黄芪汤颗粒剂) on Hepatitis B cirrhosis patients based on RNA-sequencing. Chin J Integr Med 2019;25:507–14. Available from: https://doi.org/10.1007/s11655-018-3013-3.

[122] Xia L, Shi Y, Su J, Friedemann T, Tao Z, Lu Y, et al. A promising herbal therapy for moderate COVID-19: antiviral and anti-inflammatory properties, pathways of bioactive compounds, and a clinical real-world pragmatic study. Phytomedicine 2021;85. Available from: https://doi.org/10.1016/j.phymed.2020.153390.

[123] Joung JY, Kim HG, Lee JS, Cho JH, Ahn YC, Lee DS, et al. Anti-hepatofibrotic effects of CGX, a standardized herbal formula: a multicenter randomized clinical trial. Biomed Pharmacother 2020;126. Available from: https://doi.org/10.1016/j.biopha.2020.110105.

[124] Xing YF, Wei CS, Zhou TR, Huang DP, Zhong WC, Chen B, et al. Efficacy of a Chinese herbal formula on hepatitis B e antigen-positive chronic hepatitis B patients. World J Gastroenterol 2020;26:4501–22. Available from: https://doi.org/10.3748/WJG.V26.I30.4501.

[125] Wang S, Jiang H, Yu Q, She B, Mao B. Efficacy and safety of Lian-Ju-Gan-Mao capsules for treating the common cold with wind-heat syndrome: study protocol for a randomized controlled trial. Trials 2017;18. Available from: https://doi.org/10.1186/s13063-016-1747-9.

[126] Min J, She B, Zhang X, Mao B, Chen Y. Binafuxi Granules in the treatment of common cold with heat syndrome based on traditional Uighur medicine: study protocol for a multicenter randomized controlled trial, <https://doi.org/10.21203/rs.1.5/v2>; 2019.

[127] Ma Y, Zhang Z, Wei L, He S, Deng X, Ji A, et al. Efficacy and safety of reduqing granules in the treatment of common cold with wind-heat syndrome: a randomized, double-blind, double-dummy, positive-controlled trial. J Tradit Chin Med 2017;37:185–92. Available from: https://doi.org/10.1016/s0254-6272(17)30043-2.

[128] Chen Y, Liu C, Wang T, Qi J, Jia X, Zeng X, et al. Efficacy and safety of Bufei Huoxue capsules in the management of convalescent patients with COVID-19 infection: a multicentre, double-blind, and randomised controlled trial. J Ethnopharmacol 2022;284. Available from: https://doi.org/10.1016/j.jep.2021.114830.

[129] Shawkat H, Yakoot M, Shawkat T, Helmy S. Efficacy and safety of a herbal mixture (Viron® tablets) in the treatment of patients with chronic hepatitis C virus infection: a prospective, randomized, open-label, proof-of-concept study. Drug Des Devel Ther 2015;9:799–804. Available from: https://doi.org/10.2147/DDDT.S77168.

[130] Zeng C, Yuan Z, Pan X, Zhang J, Zhu J, Zhou F, et al. Efficacy of Traditional Chinese Medicine, Maxingshigan-Weijing in the management of COVID-19 patients with severe acute respiratory syndrome: a structured summary of a study protocol for a randomized controlled trial. Trials 2020;21. Available from: https://doi.org/10.1186/s13063-020-04970-3.

[131] Rangnekar H, Patankar S, Suryawanshi K, Soni P. Safety and efficacy of herbal extracts to restore respiratory health and improve innate immunity in COVID-19 positive patients with mild to moderate severity: a structured summary of a study protocol for a randomised controlled trial. Trials 2020;21. Available from: https://doi.org/10.1186/s13063-020-04906-x.

[132] Srivastava A, Rengaraju M, Srivastava S, Narayanan V, Gupta V, Upadhayay R, et al. Efficacy of two siddha polyherbal decoctions, Nilavembu Kudineer and Kaba Sura Kudineer, along with standard allopathy treatment in the management of mild to moderate symptomatic COVID-19 patients—a double-blind, placebo-controlled, clinical trial. Trials 2021;22. Available from: https://doi.org/10.1186/s13063-021-05478-0.

[133] Liu J, Yang W, Liu Y, Lu C, Ruan L, Zhao C, et al. Combination of Hua Shi Bai Du granule (Q-14) and standard care in the treatment of patients with coronavirus disease 2019 (COVID-19): a single-center, open-label, randomized controlled trial. Phytomedicine 2021;91. Available from: https://doi.org/10.1016/j.phymed.2021.153671.

[134] Takayama S, Namiki T, Ito T, Arita R, Nakae H, Kobayashi S, et al. A multi-center, randomized controlled trial by the Integrative Management in Japan for Epidemic Disease (IMJEDI study-RCT) on the use of Kampo medicine, kakkonto with shosaikotokakikyosekko, in mild-to-moderate COVID-19 patients for symptomatic relief and. Trials 2020;21. Available from: https://doi.org/10.1186/s13063-020-04746-9.

[135] Hu K, Guan W, Jie BY, Zhang W, Li L, Zhang B, et al. Efficacy and safety of Lianhua Qingwen capsules, a repurposed Chinese herb, in patients with Coronavirus disease 2019: a multicenter, prospective, randomized controlled trial [Phytomedicine 85 (2021) 153242]. Phytomedicine 2022;94:153242. Available from: https://doi.org/10.1016/j.phymed.2021.153800.

[136] Dobrange E, Peshev D, Loedolff B, Van Den Ende W. Fructans as immunomodulatory and antiviral agents: the case of Echinacea. Biomolecules 2019;9. Available from: https://doi.org/10.3390/biom9100615.

[137] Zhang ZJ, Morris-Natschke SL, Cheng YY, Lee KH, Li RT. Development of anti-influenza agents from natural products. Med Res Rev 2020;40:2290–338. Available from: https://doi.org/10.1002/med.21707.

[138] Roshdy WH, Rashed HA, Kandeil A, Mostafa A, Moatasim Y, Kutkat O, et al. EGYVIR: an immunomodulatory herbal extract with potent antiviral activity against SARS-CoV-2. PLoS One 2020;15:e0241739. Available from: https://doi.org/10.1371/journal.pone.0241739.

[139] Rizzo A, Sciorsci RL, Magrone T, Jirillo E. Exploitation of Some Natural Products for the Prevention and/or Nutritional Treatment of SARS-CoV2 Infection. Endocr Metab Immune Disord—Drug Targets 2020;21:1171–82. Available from: https://doi.org/10.2174/1871530320999200831231029.

[140] Sharma P, Tyagi A, Bhansali P, Pareek S, Singh V, Ilyas A, et al. Saponins: extraction, bio-medicinal properties and way forward to anti-viral representatives. Food Chem Toxicol 2021;150. Available from: https://doi.org/10.1016/j.fct.2021.112075.

[141] Song X, He J, Xu H, Hu XP, Wu XL, Wu HQ, et al. The antiviral effects of acteoside and the underlying IFN-γ-inducing action. Food Funct 2016;7:3017–30. Available from: https://doi.org/10.1039/c6fo00335d.

[142] Shen MX, Ma N, Li MK, Liu YY, Chen T, Wei F, et al. Antiviral properties of *R. Tanguticum* nanoparticles on herpes simplex virus type I in vitro and in vivo. Front Pharmacol 2019;10. Available from: https://doi.org/10.3389/fphar.2019.00959.

[143] Bisht D, Rashid M, Arya RKK, Kumar D, Chaudhary SK, Rana VSSN. Revisiting liquorice (*Glycyrrhiza glabra* L.) as anti-inflammatory, antivirals and immunomodulators: potential pharmacological applications with mechanistic insight. Phytomed Plus 2022;2:100206.

[144] Pillay D, Zambon M. Antiviral drug resistance. Br Med J 1998;317:660–2. Available from: https://doi.org/10.1136/bmj.317.7159.660.

[145] Digital Health Technologies for Remote Data Acquisition in Clinical Investigations. <https://www.fda.gov/regulatory-information/search-fda-guidance-documents/digital-health-technologies-remote-data-acquisition-clinical-investigations>; n.d. [accessed 14.02.22].

[146] Homer V, Yap C, Bond S, Holmes J, Stocken D, Walker K, et al. Early phase clinical trials extension to guidelines for the content of statistical analysis plans. BMJ 2022;e068177. Available from: https://doi.org/10.1136/bmj-2021-068177.

[147] Kow CS, Ramachandram DS, Hasan SS. Future of antivirals in COVID-19: the case of favipiravir. Int Immunopharmacol 2022;103. Available from: https://doi.org/10.1016/j.intimp.2021.108455.

CHAPTER

30

Future perspectives of antiviral therapy

Debesh Chandra Bhattacharya

Department of Microbiology, Vidyasagar University, Midnapore, West Bengal, India

30.1 Introduction

From time immemorial, infectious diseases were known to human beings in all the civilizations to date. The organisms responsible for these diseases were varied such as bacteria, fungi, viruses, etc [1]. Unlike other microorganisms, viruses were most difficult to be understood and visualized till the development of the electron microscope by Knoll & Ruska. The life cycle of viruses was also another common impediment to study the etiology and control of viral infections in most cases. Among the viruses, some have DNA and others have RNA as the genetic material which is again confined within a coat of proteins commonly known as a capsid. Capsids are made up of individual protein subunits known as capsomeres. In some viruses, known as enveloped viruses, the capsid is confined by another membrane cover made up of lipid bilayers, whereas others, which do not contain these lipid bilayers, are known as naked viruses. Lipid bilayers (envelope) contain thorn-like glycoprotein projections commonly known as spikes. These spikes facilitate the molecular attachment to the host cell receptors for viral entry. The life cycle of the viruses depends on the availability of nutrients from the host cell and other molecular host cell factors to complete the replication of viral genetic materials. To survive the host cell defense mechanisms, viruses are bestowed with a remarkable natural quality to evolve rapidly. Hence, developing effective methods to kill viruses and control the viral infections within the host completely are quite challenging.

Antiviral drugs are a common terminology extensively used to cover all therapeutic compounds intended to be used to treat specific viral diseases (e.g., under various phases of trial); have already been used and approved by the regulatory authorities such as FDA, United States, etc. (Bogner, Holzenberg) [2]. Primarily all of these chemotherapeutic

Viral Infections and Antiviral Therapies
DOI: https://doi.org/10.1016/B978-0-323-91814-5.00022-2

compounds have to fulfill some basic principles to be classified as antiviral drugs. These principles are:

1. The compound must functionally be targeted only to the infected organ,
2. The compound must be stable enough to withstand metabolic changes in vivo,
3. This must not have nontarget effects and
4. It should be readily absorbed by the host metabolic system without any contra-indications like carcinogenicity, teratogenicity or mutagenicity, etc.

Keeping in pace with the advancement of viral infections and complexity, antiviral development approaches can be broadly categorized into three different levels: physiological, mechanistic, and genetic approaches. The attempts to identify the antiviral targets come under the physiological approach. The mechanistic approach constitutes the identification of the antiviral targets within various metabolic pathways inside the host/model organisms and their spatio-temporal variability under different complex host responses. The genetic approach involves the identification of potential antiviral targets through the comparison of differential gene expressions & protein expressions under the healthy and infected physiological status of the host [2,3]. A compact history of the development of antiviral compounds is given in the table below (Table 30.1).

TABLE 30.1 History of development of antiviral compounds.

Year	Antiviral compound
1951	Synthesis of Thiosemicarbazone
1957	Discovery of Interferon
1959	Synthesis of Idoxuridine
1961	Synthesis of Hydroxybenzyl benzimidazole, Guanidine
1962	Confirmation of clinical effectiveness of Idoxuridine
1964	Synthesis of Viderabine
1965	Synthesis of Amantadine
1969	Synthesis of Rimantadine
1973	Development of Phosphonoacaetic acid
1975	Development of Trifluridine, Ribavirin
1978	Development of DHPA
1979	Development of Aciclovir, Foscarnet
1982	Development of Ganciclovir
1986	Development of Zidovudine
1987	Development of Didanosine, Stavudine, Penciclovir, Zalcitabine
1988	Development of Cidofovir
1989	Development of Famciclovir

(*Continued*)

TABLE 30.1   (Continued)

Year	Antiviral compound
1990	Development of Nevirapine
1991	Development of Lamivudine
1992	Development of Saquinavir
1993	Development of Valaciclovir, Zanamivir, Delaviridine
1994	Development of Indinavir, Docosanol, Cyclosporine
1995	Development of Ritonavir, Amprenavir, Nelfinavir, Efavirenz
1996	Development of Enfuvirtide
1997	Development of Oseltamivir, Abacavir, Adefovir, Tenofovir,
1998	Development of Palivizumab, Tipranavir
1999	Development of Resveratrol, Valganciclovir, Emtricitabine
2000	Developmeent of Entecavir, Atazanavir
2002	Developmeent of Etravirine
2004	Developmeent of Telbivudine, Darunavir
2005	Developmeent of Maraviroc
2006	Developmeent of Elvitegravir
2007	Developmeent of Raltegravir

Adapted from De Clercq E, Li G. Approved antiviral drugs over the past 50 years. Clin Microbiol Rev 2016;29: 695−747; Saxena SK, Mishra N, Saxena R. Advances in anti-viral drug discovery and development, part I: advancements in antiviral drug discovery. Future Virol 2009; 4 (2):101−07; Chaudhuri S, Symons JA, Deval J. Innovation and trends in the development and approval of antiviral medicines: 1987−2017 and beyond. Antivir Res (2018), https://doi.org/10.1016/j.antiviral.2018.05.005.

30.2 General classification of antiviral drugs

The general classification of antiviral drugs can again be discussed under the following separate subgroups:

1. Antiviral drugs: Direct-acting and host-directed.
2. Antiviral drugs: Small molecule (mostly synthetic) and other large molecules (proteins, etc.).
3. Antiviral drugs: Mono (single compound) and combination drugs (more than one compound).
4. Antiviral drugs: Different groups based upon the molecular mechanism of action such as cell fusion inhibitors, viral DNA polymerase inhibitors (DPIs), viral protease inhibitors (PIs), viral reverse transcriptase inhibitors (RTIs), viral integrase inhibitors, etc.

30.2.1 Direct-acting antiviral compounds

Most of the approved antiviral drugs ($\sim 70\%$) come under this group as the compound directly targets either the viral polymerase, viral integrase, reverse transcriptase (RT), protease or nonstructural protein 5A (NS5A), etc. or any other viral enzymes (e.g., Oseltamivir—targets viral neuraminidase).

30.2.2 Host acting antiviral compounds

Only a small percentage of approved antiviral drugs come under this category, including different immunomodulators (e.g., interferon alphacon, alpha-2b, alpha-n3; peginterferon Alfa-2A, Alfa-2B) host-targeting small molecules (e.g., Imiquimod; Docosanol, Ribavirin, etc.).

30.2.3 Small molecules and large molecules

The majority of the approved drugs contain small molecules whereas 8 drugs were included under large molecules. Large molecules were mostly synthesized after 1997 and they include unique interferons, oligonucleotides, and monoclonal antibody (Mab). The first large molecule to be approved was interferon alpha-2b. Later developed interferons were interferon-alpha n-3, alfacon-1, pegylated interferons such as peginterferon Alpha-2B, while one approved antiviral oligonucleotide was fomivirsen [4,5].

30.2.4 Mono and combination drug therapy

With the advent of emerging drug resistance, the need for combination therapy was felt by the clinical researchers especially in dealing with HIV-1 infections. Before these drug resistance problems, monotherapy was routinely administered to HIV-1 infected patients. In 1997 Lamivudine/zidovudine was first approved as a fixed-dose combination therapy against HIV-1 infections. In case of HCV-infections, most of the approved therapies were combination therapy, while Sofosbuvir was the one exception that got approved in 2013 as monotherapy. Some examples of successful combination therapy are summarized in table (Table 30.2) [4].

30.2.5 Polymerase inhibitors

Viral polymerase inhibitors are popular choices in designing antiviral drugs. The polymerase inhibitors are further divided into two groups—nucleoside analogs and nonnucleoside inhibitors. The first nucleoside analog to be approved for HIV-1 treatment was Zidovudine, while other nucleoside analogs like acyclovir was also approved against HCV infections.

TABLE 30.2 Successful combination therapies against HCV.

Antiviral drug combination	Viral infection	% Efficacy	Mode of action
Sofosbuvir + velpatasvir	HCV genotype 1–6	97.4	Inhibits NS5A, NS5B
Daclatasvir + asunaprevir	HCV genotype 1	86.4	Inhibits NS5A, NS5B
Daclatasvir + asunaprevir + beclabuvir	HCV genotype 1	91.5	Inhibits NS5A, NS3/4A, protease &NS5B

These nucleoside analogs are prodrugs which again get recognized by viral polymerase and subsequently phosphorylated for the synthesis of viral nucleic acid. Long use of the first-generation anti-HIV nucleoside analogs like zalcitabine and didanosine has been restricted due to the proven toxicity resulting from the insufficient selectivity in between viral and human DNA polymerases [6]. Lamivudine and emitricitabine are second-generation anti-HIV nucleoside analogs that showed improved selectivity and tolerability in patients. Nevirapine was the first anti-HIV nonnucleoside polymerase inhibitor that successfully got FDA approval in the year 1996 [4,5].

30.2.6 Reverse transcriptase inhibitors

RTIs act on viral RT either through binding to the active site or blocking the elongation step of growing DNA. They are further classified into three subclasses:

1. Nucleoside-analog RTIs—for example, Stavudine, Zalcitabine, Didanosine, Lamivudine, Zidovudine.
2. Nonnucleoside analog RTIs—for example, Nevirapine, Delavirdine, Efavirenz, Etravirine.
3. Nucleotide analog RTIs—for example, Adefovir, Tenofovir

30.2.7 Protease inhibitors

PIs are agents that successfully stop the vital maturation step of polyprotein processing in the viral life cycle. In 1995, the first anti-HIV PI saquinavir was developed and thereafter another nine PIs have been approved for commercial use. Some of them are ritonavir, nelfinavir, amprenavir, lopinavir, atazanavir, tipranavir, and darunavir [4,5].

30.2.8 Integrase inhibitors

Integrase inhibitors stop the integration of the HIV genome into the human chromosomal DNA by inhibiting the viral integrase. Three anti-HIV integrase inhibitors are approved, they are raltegravir, elvitegravir and dolutegravir. Among these dolutegravir is the only second-generation integrase inhibitor that can show a higher degree of genetic barrier to resistance and retains the effective activity against several (but not all) raltegravir-elvitegravir resistant strains [7].

30.2.9 Nonstructural protein 5A inhibitors

Unlike other enzymes, NS5A is a dimeric membrane protein that cannot show any enzymatic activity. It can interact with viral RNA and other nonstructural proteins for the replication of HCV viral RNA. Thus NS5A inhibitors can be an effective choice for a viral drug. Daclatasvir and ledipasvir were the first NS5A inhibitors to be approved in 2015/2014 [4,5].

A new class of inhibitors known as Portmanteau inhibitors that can inhibit both RT and viral integrase are being extensively tested by researchers for human trials. Table 30.3 has summarized the detailed classification of antiviral drugs based on the mode of action of the virus.

TABLE 30.3 Antiviral groups based on mode of action.

Year	Trade name	Generic name (abbreviation)	Virus	MOA	Type	Mono/combo
1987	Retrovir	Zidovudine (AZT)	HIV-1	Pol	Small molecule	Mono
1988	Intron A	Interferon Alfa-2B (INT2B)	HPV	Host	Protein	Mono
1989	Cytovene	Ganciclovir sodium (GAN)	CMV	Pol	Small Molecule	Mono
1989	Alferon N inj	Interferon Alfa N3	HPV	Host	Protein	Mono
1991	Intron A	Interferon Alfa-2B (INT2B)	HCV	Host	Protein	Mono
1991	Foscavir	Foscarnet sodium (FOS)	CMV	Pol	Small molecule	Mono
1991	Videx#	Didanosine (ddI)	HIV-1	Pol	Small molecule	Mono
1992	Hivid#	Zalcitabine (ddC)	HIV-1	Pol	Small molecule	Mono
1992	Intron A	Interferon Alfa-2B (INT2B)	HBV	Host	Protein	Mono
1993	Flumadine	Rimantadine (RIM)	Influenza	O	Small molecule	Mono
1994	Zerit	Stavudine (d4T)	HIV-1	Pol	Small molecule	Mono
1994	Famvir#	Famciclovir (FAM)	HSV	Pol	Small molecule	Mono
1995	Valtrex	Valacyclovir hydrochloride (VAL)	HSV	Pol	Small molecule	Mono
1995	Epivir	Lamivudine (3TC)	HIV-1	Pol	Small molecule	Mono
1995	Invirase	Saquinavir mesylate (SQV)	HIV-1	Pr	Small molecule	Mono
1996	Norvir	Ritonavir (RTV)	HIV-1	Pr	Small molecule	Mono
1996	Crixivan	Indinavir sulfate (IDV)	HIV-1	Pr	Small molecule	Mono
1996	Viramune	Nevirapine (NVP)	HIV-1	Pol	Small molecule	Mono
1996	Vistide#	Cidofovir (CDV)	CMV	Pol	Small molecule	Mono
1996	Denavir	Penciclovir (PEN)	HSV	Pol	Small molecule	Mono
1997	Aldara	Imiquimod (IMI)	HPV	Host	Small molecule	Mono
1997	Viracept#	Nelfinavir Mesylate (NFV)	HIV-1	Pr	Small molecule	Mono
1997	Rescriptor	Delavirdine Mesylate (DLV)	HIV-1	Pol	Small molecule	Mono
1997	Combivir	Lamivudine (3TC)/Zidovudine (AZT)	HIV-1	Pol/Pol	Small molecule	Combo

Year	Brand	Generic	Virus	Target	Type	
1997	Infergen	Interferon Alfacon-1 (INTA1)	HCV	Host	Protein	Mono
1998	Synagis	Palivizumab (PAV)	RSV	O	Protein	Mono
1998	Rebetol	Ribavirin (RIB)	HCV	Host	Small molecule	Mono
1998	Vitravene#	Fomivirsen Sodium (FOM)	CMV	O	Oligonucleotide	Mono
1998	Sustiva	Efavirenz (EFV)	HIV-1	Pol	Small molecule	Mono
1998	Epivir-HBV	Lamivudine (3TC)	HBV	Pol	Small molecule	Mono
1998	Ziagen	Abacavir Sulfate (ABC)	HIV-1	Pol	Small molecule	Mono
1999	Agenerase#	Amprenavir (APV)	HIV-1	Pr	Small molecule	Mono
1999	Relenza	Zanamivir (ZAN)	Influenza	O	Small molecule	Mono
1999	Tamiflu	Oseltamivir (OSE)	Influenza	O	Small molecule	Mono
2000	Abreva	Docosanol (DOC)	HSV	O	Small molecule	Mono
2000	Kaletra	Lopinavir (LPV)/Ritonavir (RTV)	HIV-1	Pr/Pr	Small molecule	Combo
2000	Trizivir	Abacavir Sulfate (ABC)/Lamivudine (3TC)/Zidovudine (AZT)	HIV-1	Pol/Pol/Pol	Small molecule	Combo
2001	Pegintron/Sylatron	Peginterferon Alpha-2B (PEG2B)	HCV	Host	Protein	Mono
2001	Valcyte	Valganciclovir Hydrochloride (VALG)	CMV	Pol	Small molecule	Mono
2001	Viread	Tenofovir Disoproxil Fumarate (TDF)	HIV-1	Pol	Small molecule	Mono
2002	Hepsera	Adefovir Dipivoxil (ADE)	HBV	Pol	Small molecule	Mono
2002	Pegasys	Peginterferon Alfa-2A (PEG2A)	HCV	Host	Protein	Mono
2003	Fuzeon	Enfuvirtide (T20)	HIV-1	O	Peptide	Mono
2003	Reyataz	Atazanavir Sulfate (ATV)	HIV-1	Pr	Small molecule	Mono
2003	Emtriva	Emtricitabine (FTC)	HIV-1	Pol	Small molecule	Mono
2003	Lexiva	Fosamprenavir Calcium (FPV)	HIV-1	Pr	Small molecule	Mono
2004	Epizicom	Abacavir Sulfate (ABC)/Lamivudine (3TC)	HIV-1	Pol/Pol	Small molecule	Combo
2004	Truvada	Emtricitabine (FTC)/Tenofovir Disoproxil Fumarate (TDF)	HIV-1	Pol/Pol	Small molecule	Combo

(Continued)

TABLE 30.3 (Continued)

Year	Trade name	Generic name (abbreviation)	Virus	MOA	Type	Mono/combo
2005	Baraclude	Entecavir (ENT)	HBV	Pol	Small molecule	Mono
2005	Pegasys	Peginterferon Alfa-2A (PEG2A)	HBV	Host	Protein	Mono
2005	Aptivus	Tipranavir (TPV)	HIV-1	Pr	Small molecule	Mono
2006	Prezista	Darunavir Ethanolate (DRV)	HIV-1	Pr	Small molecule	Mono
2006	Atripla	Efavirenz (EFV)/Emtricitabine (FTC)/Tenofovir Disoproxil Fumarate (TDF)	HIV-1	Pol/Pol/Pol	Small molecule	Combo
2006	Tyzeka#	Telbivudine (TEL)	HBV	Pol	Small molecule	Mono
2006	Veregen	Sinecatechins (SIN)	HPV	Host	Small molecule	Mono
2007	Selzentry	Maraviroc (MVC)	HIV-1	Host	Small molecule	Mono
2007	Isentress	Raltegravir Potassium (RAL)	HIV-1	Int	Small molecule	Mono
2008	Intelence	Etravirine (ETR)	HIV-1	Pol	Small molecule	Mono
2008	Viread	Tenofovir Disoproxil Fumarate (TDF)	HBV	Pol	Small molecule	Mono
2009	Xerese	Acyclovir Hydrocortisone (ACY)	HSV	Pol	Small molecule	Combo
2011	Victrelis	Boceprevir (BOC)	HCV	Pr	Small molecule	Mono
2011	Edurant	Rilpivirine Hydrochloride (RPV)	HIV-1	Pol	Small molecule	Mono
2011	Incivek#	Telaprevir (TELA)	HCV	Pr	Small molecule	Mono
2011	Complera	Emtricitabine (FTC)/Rilpivirine Hydrochloride (RPV)/Tenofovir Disoproxil Fumarate (TDF)	HIV-1	Pol/Pol/Pol	Small molecule	Combo
2012	Stribild	Cobicistat (COBI)	HIV-1	O/Int/Pol	Small molecule	Combo
2013	Tivicay	Dolutegravir Sodium (DTG)	HIV-1	Int	Small molecule	Mono
2013	Olysio	Simprevir Sodium (SIM)	HCV	Pr	Small molecule	Mono
2013	Sovaldi	Sofosbuvir (SOF)	HCV	Pol	Small molecule	Mono
2014	Triumeq	Abacavir Sulfate (ABC)/Dolutegravir Sodium (DTG)/Lamivudine (3TC)	HIV-1	Pol/Int/Pol	Small molecule	Combo
2014	Vitekta	Elvitegravir (EVG)	HIV-1	Int	Small molecule	Mono

Year	Name	Composition	Virus	Target	Type	Therapy
2014	Harvoni	Ledipasvir (LED)/Sofosbuvir (SOF)	HCV	NS5A/Pol	Small molecule	Combo
2014	Rapivab	Peramivir (PER)	Influenza	O	Small molecule	Mono
2014	Viekira Pak	Dasabuvir Sodium (DAS)/Ombitasvir (OMB)/Paritaprevir (PAR)/Ritonavir (RTV)	HCV	Pr/Pr/NS5A/Pol	Small molecule	Combo
2015	Prezcobix	Cobicstat (COBI)/Darunavir Ethanolate (DRV)	HIV-1	Pr	Small molecule	Combo
2015	Evotaz	Atazanavir Sulfate (ATV)/Cobicstat (COBI)	HIV-1	Pr	Small molecule	Combo
2015	Dutrebis#	Lamivudine (3TC)/Raltegravir (RAL)	HIV-1	Pol/Int	Small molecule	Combo
2015	Daklinza	Daclatasvir Dihydrochloride (DAC)	HCV	NS5A	Small molecule	Mono
2015	Teknivie	Cobicstat (COBI)/Ombitasvir (OMB)/Paritaprevir (PAR)/Ritonavir (RTV)	HCV	Pr/Pr/NS5A	Small molecule	Combo
2015	Genvoya	Cobicstat (COBI)/Elvitegravir (EVG)/Emtricitabine (FTC)/Tenofovir Alafenamide Fumarate (TAF)	HIV-1	O/Int/Pol/Pol	Small molecule	Combo
2016	Zepatier	Elbasvir (ELB)/Grazoprevir (GRA)	HCV	Pr/NS5A	Small molecule	Combo
2016	Odefsey	Emtricitabine (FTC)/Rilpivirine Hydrochloride (RPV)/Tenofovir Alafenamide Fumarate (TAF)	HIV-1	Pol/Pol/Pol	Small molecule	Combo
2016	Descovy	Emtricitabine (FTC)/Tenofovir Alafenamide Fumarate (TAF)	HIV-1	Pol	Small molecule	Combo
2016	Epclusa	Sofosbuvir (SOF)/Velpatasvir (VEL)	HCV	Pol/NS5A	Small molecule	Combo
2016	Vemlidy	Tenofovir Alafenamide Fumarate (TAF)	HBV	Pol	Small molecule	Mono
2017	Vosevi	Sofosbuvir (SOF)/Velpatasvir (VEL)/Voxilaprevir (VOX)	HCV	Pr/Pol/NS5A	Small molecule	Combo
2017	Mavyret	Glecaprevir (GLE)/Pibrentasvir (PIB)	HCV	Pr/NS5A	Small molecule	Combo
2017	Previmis	Letermovir (LET)	CMV	O	Small molecule	Mono
2017	Zuluka	Dolutegravir (DTG)/Rilpivirine (RPV)	HIV-1	Pol/Int	Small molecule	Combo

Viruses: CMV, cytomegalovirus; HBV, Hepatitis B virus; HCV, Hepatitis C virus; HIV-1, Human Immunodeficiency virus-1; HPV, human papilloma virus; HSV, Herpes simplex virus.

Targets: Int, Integrase; NS5A, Non structural protein 5A; O, Other; Pol, Polymerase; Pr, Protease.

Combo, combination therapy; MOA, mode of action; Mono, monotherapy; #, discontinued or withdrawn.

Adapted from Chaudhuri S, Symons JA, Deval J. Innovation and trends in the development and approval of antiviral medicines: 1987–2017 and beyond. Antivir Res (2018). doi 10.1016/j.antiviral.2018.05.005.

30.3 Problems and limitations in antiviral drugs

Despite of the development of more than a hundred approved antiviral drugs and their combinations, there remains a considerable number of challenges to be solved in the coming years.

30.3.1 Resistance shown after long-term use of antivirals

Development of resistance is almost universal in all the viruses against their respective antiviral drugs. HIV developed resistance against lamivudine, abacavir, tenofovir, and emtricitabine whereas acyclovir—resistance has been detected in herpes virus infections. Mainly the cellular pool of resistance strains shows high variability in their degree of resistance towards a certain antiviral drug. This variation occurs due to the differential rate of mutation in the resistance strains that paves the way for possible alterations in viral enzymes, viral structural proteins, etc. At this juncture, the high selective pressure replaces the wild-type strains with more and more resistant strains. The second most vital reason is the observation of cross-resistance (CR) among antiviral drugs. Nelfinavir/saquinavir has already shown CR towards indinavir/ritonavir [8] and delaviridine showed CR towards nevirapine [9].

30.3.2 Toxicity and immunosupression

Almost all the antiviral drugs have the drawback of lacking absolute specificity against the viruses, hence toxicity is the major associated problem with the use of antivirals. Many anti-HIV drugs show high toxicity to WBCs and lead to megaloblastic anemia, drug-induced immunosuppresson, mitochondrial toxicity, nerve injury, etc. Thus immunosuppressive properties of antivirals must be verified to minimize the damage caused to humans.

30.3.3 Viral latency

Viral latency is another common problem associated with the prolonged use of antivirals. As soon as the medication stops, the latent viruses start proliferating again within the host, for example, the use of acyclovir does not guarantee the prevention of future HCV infections.

30.3.4 Time-consuming, tedious, and associated with risks

A single antiviral development takes around 10–15 years with an expenditure of around 1 billion USD and thereafter associated with risks of being disapproved in various stages of trials. Cases of contraindications and other safety issues may hinder the final commercial approval.

30.4 Modern perspectives in the development approaches of antivirals

In the omics era, the design and development approaches of antivirals have already taken a major shift from the earlier ones during the 1980s or 1990s. It has further been accelerated through the availability of in silico approaches, computational biology/chemistry-based software, and high throughput screening (HTS) facilities. Studies on the chemico-biological interfaces were further advanced through QSAR-based applications [10,11].

30.4.1 The antisense approach

The use of antisense oligonucleotides involves the targeting of RNA by RNA interference (RNAi), Here, the viral RNA is targeted by inhibiting the gene expression either at the transcriptional or translational level. siRNA and miRNA are the key components of the RNAi process in the formation of RISC (RNAi induced silencing complex). Many applications of RNAi have been reported [12,13].

30.4.2 The aptameric approach

Specific oligonucleic acids or peptide molecules that are capable to bind a specific target molecule to interfere with the interaction of proteins inside the cell are known as aptamers. They contain distinct variable peptide loops attached at both ends to a protein scaffold. Inhibition of HCV through targeting of NS-3 protease and helicase has already been reported [14,15].

30.4.3 The ribozyme approach

RNAs working as enzymes are commonly termed ribozymes. They can catalyze the hydrolysis of one phosphodiester bond on their own or in any other RNAs. Hairpin ribozymes have been reported to reduce HBV replication [16] and have been reported to protect against HIV and HCV [17,18].

30.4.4 The CRISPR/Cas9 approach

The ability of the CRISPR/cas9 system to directly target the viral DNA or RNA is a unique advantage that can be exploited in the fight against viral infections. CRISPR/Cas 9 system has shown promising results in different stages of viral infections. Engineered *F. novicida* Cas9 (Fncas9) can successfully target the positive sense single-strand HCV in eukaryotic host cells. CRISPR/cas9 editing tool is also helpful in providing alternative way to get rid of HBV latent infections in humans [19,20].

TABLE 30.4   Modern technology-driven approaches for antiviral development.

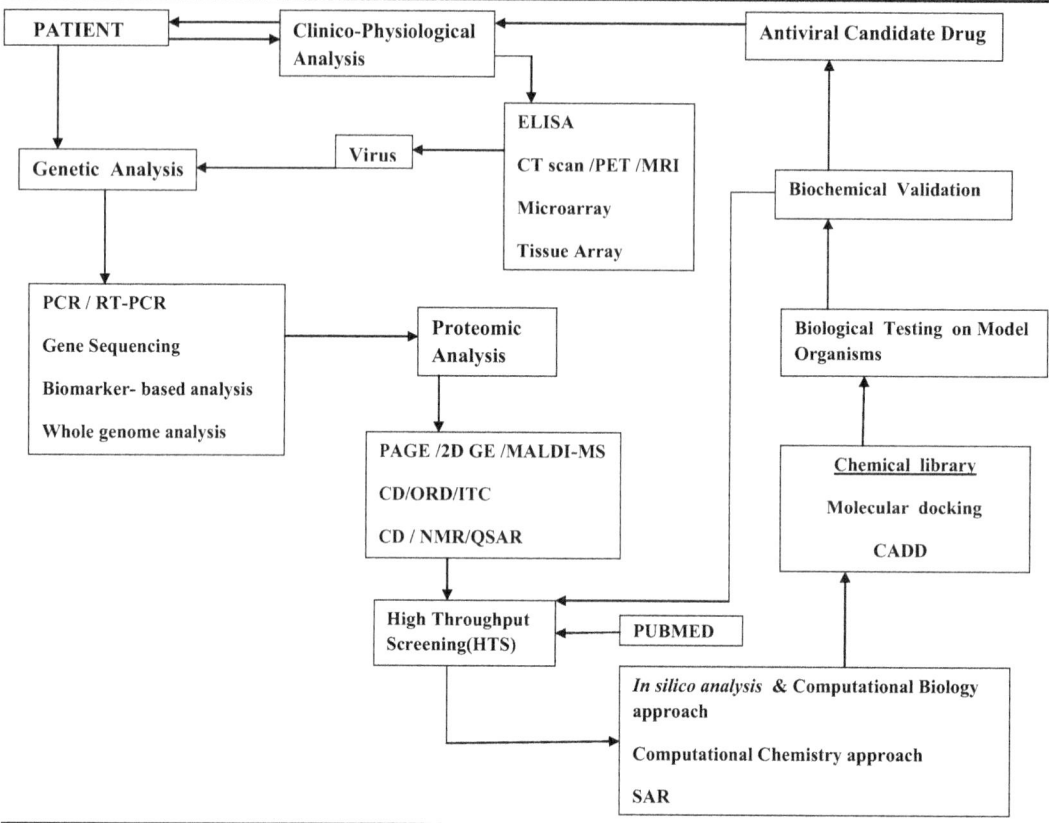

30.4.5 The technological shift in the omics era

The technological ease of identification of viral targets and the development of antiviral candidate drugs have become somewhat less lengthy with the advent of the HTS approach augmented by in silico/computational biology/chemistry approach. The completely summarized scheme has been given below in the table (Table 30.4).

30.5 Conclusion

The continual onslaught and re-emergence of various viral infections have made it a technological imperative to search for better antivirals in the coming years. This has also been perfect for economically, socially, and scientifically driven societies that are experiencing newer mutant strains, higher number of resistance patterns, and of course death of patients. Hence, the rate of discovery of antivirals must have that much encouragement which makes the scientific efforts worthy in terms of a holistic fight against the enemy. Technology is always supportive, with the above-discussed developments, tools and softwares, etc., one can hope for the development of better antivirals for a better future.

References

[1] Nene YL. A glimpse at viral diseases in ancient period. Asian Agrihist 2007;11(01):35—46.

[2] Bogner E, Holzenberg A, editors. New concepts of anti-viral therapy. Springer; 2006.

[3] Saxena SK, Mishra N, Saxena R. Advances in Anti-viral drug discovery and development, Part I: advancements in antiviral drug discovery. Future Virol 2009;4(2):101—7.

[4] De Clercq E, Li G. Approved antiviral drugs over the past 50 years. Clin Microbiol Rev 2016;29:695—747.

[5] Chaudhuri S, Symons JA, Deval J. Innovation and trends in the development and approval of antiviral medicines: 1987—2017 and beyond. Antivir Res 2018;2018. Available from: https://doi.org/10.1016/j.antiviral.2018.05.005.

[6] Johnson AA, Ray AS, Hanes J, et al. Toxicity of antiviral nucleoside analogs and the human mitochondrial DNA polymerase. J Biol Chem 2001;276:40847—57.

[7] Anstett K, Brenner B, Mesplede T, Wainberg MA. HIV drug resistance against strand transfer integrase inhibitors. Retrovirology 2017;14:36.

[8] Delaugerre C, Wirden M, Simon A et al. Resistance profile & cross-resistance to HIV-1 among 104 patients failing a non-nucleoside reverse transcriptase inhibitor-containing regimen. Proceedings of the eighth conference on retroviruses and opportunistic infections, Chicago, IL, USA, 8th August 2001.

[9] Jeffrey S, Baker D, Tritch R. et al. A resistance and cross-resistance profile for Sustiva (efavirenz, DMP 266): roceedings of the fifth conference on retroviruses & opportunistic infections. Chicago, IL, USA, 5th August 1998.

[10] Kapetanovic IM. Computer aided drug discovery and development (CADDD): in silico-chemico biological approach. Chem Biol Interact 2008;171:165—76.

[11] Yuan H, Parrill AL. QSAR development to describe HIV - 1 integrase inhibition. Theochem 2000;529:273—82.

[12] Kumar A. RNA-interference: a multifaceted innate antiviral defense. Retrovirology 2008;5:17.

[13] Leonard JN, Schaffer DV. Antiviral RNAi therapy: emerging approaches for hitting a moving target. Gene Ther 2006;13:532—40.

[14] Crawford M, Woodman R, Ko Ferringo P. Peptide aptamers: tools for biology and drug discovery. Brief Funct Genom Proteom 2003;2(1):72—9.

[15] Hwang B, Cho JS, Yeo AJ, et al. Isolation of specific and high-affinity RNA aptamers against NS3 helicase domain of HCV. RNA 2004;10:1277—90.

[16] Zu Putliz J, Yu Q, Burke MJ, et al. Combinatorial screening and intracellular antiviral activity of hairpin ribozymes directed against hepatitis B virus. J Virol 1999;75:5381—7.

[17] Nazari R, Ma XZ, Joshi S. Inhibition of human immunodeficiency virus-1 entry using vectors expressing a hammerhead ribozyme targeting the CCR5 mRNA. J Gen Virol 2008;89:2252—61.

[18] Welch PJ, Yei S, Barber JR. Ribozyme gene therapy for hepatitis C virus infections. Clin Diagn Virol 1998;10:163—71.

[19] Price AA, Sampson TR, Ratner HK, Grakoui A, Weiss DB. Cas-9 mediated targeting of viral RNA in eukaryotic cells. Proc Natl Acad Sci USA 2015;112(9):6164—9.

[20] Ramanan V, Shlomai A, Cox DB, Schwartz RE, Michailidis E, Bhatta A, et al. Crispr/cas-9 cleavage of viral DNA efficiently suppresses hepatitis B virus. Sci Rep 2015;5:10833. Available from: https://doi.org/10.1038/srep10833.

.

Index

CPI Antony Rowe
Eastbourne, UK
November 12, 2022